KU-400-506

Volume I
Gas Dynamics

Maurice J. Zucrow
Joe D. Hoffman

School of Mechanical Engineering Purdue University

JOHN WILEY & SONS

A NOTE TO THE READER
This book has been electronically reproduced from
digital information stored at John Wiley & Sons, Inc.
We are pleased that the use of this new technology
will enable us to keep works of enduring scholarly
value in print as long as there is a reasonable demand
for them. The content of this book is identical to
previous printings.

All rights reserved. Published simultaneously in Canada.

No part of this publication may be reproduced, stored in a retrieval system or transmitted
in any form or by any means, electronic, mechanical, photocopying, recording, scanning
or otherwise, except as permitted under Sections 107 or 108 of the 1976 United States
Copyright Act, without either the prior written permission of the Publisher, or
authorization through payment of the appropriate per-copy fee to the Copyright
Clearance Center, 222 Rosewood Drive, Danvers, MA 01923, (978) 750-8400, fax
(978) 750-4470. Requests to the Publisher for permission should be addressed to the
Permissions Department, John Wiley & Sons, Inc., 111 River Street, Hoboken, NJ 07030,
(201) 748-6011, fax (201) 748-6008, E-Mail: PERMREQ@WILEY.COM.

To order books or for customer service please, call 1(800)-CALL-WILEY (225-5945).

Library of Congress Cataloging in Publication Data:

Zucrow, Maurice Joseph, 1899–1975.
    Gas dynamics.

    Includes bibliographical references and index.
    1. Gas dynamics.   I. Hoffman, Joe D., 1934–
joint author.   II. Title.
QC168.Z8 1976   533'.2   76-6855
ISBN 0-471-98440-X (v. 1)

preface

This book is an introduction to the science of gas dynamics, designed for a first course at the senior or graduate level. It is also written for the practicing engineer and scientist.

The presentation of the subject matter is sufficiently clear so that the student can understand the material without the assistance of an instructor. Therefore, a review of fundamental principles is included, the derivations of equations are presented in detail, and many illustrative examples are worked out completely. Exercise problems are presented at the end of each chapter. Tables of the functions required for solving the illustrative examples and the exercise problems are included in the appendixes.

A background in fluid mechanics and thermodynamics is desirable for the study of gas dynamics. Introductory undergraduate courses in those disciplines are sufficient.

The material in this book is too lengthy to cover completely in a one-semester course; at Purdue University, we discuss in detail the basic concepts presented in each chapter, and the application of those concepts to the flow of a perfect gas. As time permits, real gas effects are discussed, and applications such as nozzles, diffusers, and the like, are presented.

The book is completely self-contained, beginning with a review of fundamental principles (Chapter 1), followed by a detailed derivation of the governing equations of fluid flow (Chapter 2), and an extensive treatment of the classical topics of steady one-dimensional gas dynamics (Chapters 3 to 9). The general features of the steady multidimensional adiabatic flow of an inviscid fluid are discussed (Chapter 10). An introduction to the concept of linearized flow (Chapter 11), the method of characteristics and steady two-dimensional supersonic flow (Chapter 12), and unsteady one-dimensional flow (Chapter 13) are presented. An appendix gives a brief introduction to the concepts of numerical analysis that are employed in the book. Extensive tables are included to aid in working problems.

The material in this book closely resembles that in the several books already written about gas dynamics, for example, Zucrow's *Aircraft and Missile Propulsion, Volume 1, Thermodynamics of Fluid Flow and Application to Propulsion Engines*, and Shapiro's *Dynamics and Thermodynamics of Compressible Fluid Flow*. The major difference between this book and the existing books is the emphasis on the application of numerical methods for solving real gas dynamic problems.

We believe that the student learns by doing. Consequently, numerous illustrative examples, worked out in complete detail, are presented in each chapter to illustrate the application of the theoretical analysis to real problems.

Because of the inevitable conversion to the international system of units (i.e., the SI system), the illustrative examples and exercise problems employ that system of units.

The tables that are included to aid in working problems include tabulations of the conventional gas dynamic flow functions, the physical properties of the standard atmosphere, and the thermodynamic properties of air. Computer programs are presented for the generation of each of the tables, so the user can construct additional tables if he desires.

v

The concepts of numerical analysis are introduced briefly. The material includes all of the numerical methods employed in the book: approximation and interpolation, the solution of systems of linear algebraic equations, the solution of nonlinear functions, integration, and the solution of ordinary differential equations. Numerical examples are presented to illustrate these methods.

Volume 1 of this book is based on a combination of Maurice Zucrow's book *Aircraft and Missile Propulsion, Volume I, Thermodynamics of Fluid Flow and Application to Propulsion Engines* and class notes employed in a gas dynamics course taught at Purdue University over the past 12 years by Joe Hoffman. Many of the figures and several of the illustrative examples in this book are taken from Zucrow's book. The comments of the students at Purdue University who have used Zucrow's book and the class notes were most helpful in the preparation of this book. Two colleagues at Purdue University deserve special acknowledgments for their generous assistance. Professor H. Doyle Thompson contributed ideas, comments, and criticisms during the writing of the book. Professor Peter Liley provided invaluable assistance in obtaining thermodynamic property data. Their generous assistance is greatly appreciated. The excellent work performed by our typist, Miss Cynthia Hoffman, deserves special note.

This book was the joint effort of a teacher and one of his students. For Professor Zucrow, it was a labor of love for his profession, performed after his retirement from Purdue University. For Joe Hoffman, it was an opportunity to work with his teacher, counselor, and friend, Doc Zucrow, in a warm personal relationship that only a few people ever achieve. Doc Zucrow passed away during the final review of the manuscript. He will be missed by his colleagues, students, and friends. This book, on which he worked diligently even on the last day of his life, is a fitting final contribution to his illustrious career.

Joe D. Hoffman

West Lafayette, Indiana, 1976

contents

Volume 1
Gas Dynamics

1
review of fundamental principles

1–1 PRINCIPAL NOTATION FOR CHAPTER 1

a acoustic speed.

$\mathbf{a}$ acceleration.

A area.

c_p specific heat at constant pressure.

$\bar{c}_p$ $= \bar{m}c_p$, molar specific heat at constant pressure.

c_v specific heat at constant volume.

$\bar{c}_v$ $= \bar{m}c_v$, molar specific heat at constant volume.

C_i $= m_i/m$, mass fraction of species i.

e specific stored energy.

E stored energy.

F   force.

F the dimension *force*.

g local acceleration due to gravity.

g_c factor of proportionality defined by Newton's second law.

g_o standard acceleration due to gravity (at sea level, 45 deg latitude, in a vacuum) $= 9.80665 \text{ m/s}^2$ (32.1740 ft/sec^2).

h specific enthalpy.

k Boltzmann's constant, $1.38054 \cdot 10^{-23} \text{ J/K}$.

K $= -v(dp/dv)$, bulk modulus.

Kn $= \lambda/L$, Knudsen number.

L characteristic length, or the dimension *length*.

m mass.

$\bar{m}$ molecular weight.

M $= V/a$, Mach number, or the dimension *mass*.

M   momentum.

n   unit normal vector.

N moles.

N_A Avogadro's number, $6.02252 \cdot 10^{23}$ molecules/mol.

N_i moles of species i.

p static pressure.

Q heat.

R $= \bar{R}/\bar{m}$, gas constant for a specific gas.

$\bar{R}$ universal gas constant, 8314.3 J/kmol-K.

Re $= LV\rho/\mu$, Reynolds number.

s specific entropy.

S entropy.

t absolute temperature, or time.

t   unit tangential vector.

T the dimension *time*.

u specific internal energy.

U internal energy.

v specific volume.

$\mathscr{V}$ volume.

V velocity magnitude.

V   velocity.

W work.

X_i $= N_i/N$, mole fraction of species i.

Greek Letters

γ $= c_p/c_v$, specific heat ratio.

Θ the dimension *temperature*.

κ thermal conductivity.

λ mean free path of gas molecules.

μ absolute or dynamic viscosity.

v $= \mu/\rho$, kinematic viscosity.

ρ density.

σ stress.

τ shear stress.

ϕ $= \int_{t_o}^{t} c_p \dfrac{dt}{t}$.

Subscripts

i denotes species i.

Other

$^{-}$ denotes molar basis.

1–2 INTRODUCTION

Gas dynamics is the science concerned with the causes and effects arising from the motion of compressible fluids, particularly gases, and is a branch of the more general science of fluid dynamics. Gas dynamics brings together concepts and principles from several branches of science, including mechanics, thermodynamics, aerodynamics, and chemical kinetics. Nuclear, electrical, and magnetic effects generally are not considered part of the science of gas dynamics.

This chapter reviews the fundamental concepts and principles underlying the theoretical basis for the analytical methods of gas dynamics. As in the case of any problem in fluid dynamics, the analysis of a gas dynamic problem is based on the interrelations between the following four fundamental physical laws.

1. The law of the conservation of mass.
2. Newton's second law of motion.
3. The first law of thermodynamics.
4. The second law of thermodynamics.

These laws are independent of either the properties of the flowing fluid or the particular flow process under consideration. The law of the conservation of mass is self-explanatory. The underlying concepts of the other three laws, as applied to a system of fixed mass, are reviewed briefly in this chapter.

In applying the above fundamental laws to a flowing fluid, the properties of the fluid must be considered. Consequently, brief reviews of the continuum concept and of the elastic and thermodynamic properties of perfect substances are presented in this chapter. Furthermore, because of its major importance to gas dynamics, the properties of the perfect gas are discussed in some detail.

1–3 DIMENSIONS AND UNITS[1–4]

Dimensions are the names employed for characterizing physical quantities. The physical quantities of interest in gas dynamics are *force F*, *mass M*, *length L*, *time T*, and *temperature* Θ. *Units* are the names given to certain magnitudes of a dimension chosen as a standard of measurement. For example, the dimension length may be measured in units of meters. The basic units of measurement are arbitrary but, once chosen, they must be employed in a consistent manner. The present section is concerned with several of the inherent characteristics of dimensions and units.

1–3(a) Dimensions

Two kinds of quantities enter quite generally into engineering measurements: *dimensional quantities* and *dimensionless quantities*. A *dimensional quantity* has its magnitude expressed in terms of one or more basic units of measurement; for example, a length of so many meters, a velocity of so many meters per second, or an acceleration of so many meters per second per second, etc. On the other hand, a *dimensionless quantity* has no dimensional category whatsoever and is, therefore, a pure number.

A dimensionless quantity may be a coefficient, such as the discharge coefficient for an orifice plate, the ratio of two similar dimensional quantities, or the product of several dimensional quantities arranged to give a dimensionless result. The numerical magnitude of a dimensionless quantity is independent of the size of the fundamental units employed for evaluating it, provided that a *consistent* set of units is employed.

The choice of either the *principal dimensions* or the size of the *basic units* for expressing the magnitude of a physical measurement is arbitrary. It is based entirely on convenience. In general, the principal dimensions may be any mutually independent set that is convenient to use. Experience has demonstrated, however, that in the field of fluid mechanics, four *principal dimensions* suffice. They are *length L*, *time T*, either *mass M* or *force F*, and *temperature Θ*. The magnitude of any physical measurement in the field of fluid mechanics is expressible in terms of units having the foregoing principal dimensions.

Mass and force are related by Newton's second law of motion; that is, *force $\propto$ mass $\times$ acceleration*. In dimensional form, Newton's second law of motion is $F = ML/T^2$. If *mass M* is taken as the *primary* dimension, then the dimensions of *force F* are ML/T^2. If *force F* is chosen as the *primary* dimension, then the dimensions of *mass M* are FT^2/L. Thus, in any system of dimensions, either *mass M* or *force F* may be chosen as the *primary* dimension; the other quantity is then a *secondary* or *derived* dimension.

1–3(b) Systems of Units

Like the selection of the primary dimensions, the selection of the basic units of measurement, hereafter termed the *basic units* for brevity, is arbitrary, being based solely on convenience. Evidence of this is the large number of different units employed in engineering and physics for the four dimensions *M, L, T,* and *F*. There is, however, a restriction imposed by Newton's second law of motion, termed the *condition for consistency*, that must be satisfied by a system of basic units. Consistency requires that the basic units must satisfy *numerically* the following relationship: 1 *unit force* = 1 *unit mass* $\times$ 1 *unit acceleration*.

Prior to 1971, several systems of units were widely used in the United States. They were (1) the *English Engineering* (EE) system, (2) the *English Absolute* (EA) system, (3) the *English Gravitational* (EG) system, and (4) the *Metric* system. In 1971 the U.S. Bureau of Standards recommended that the United States adopt and convert to a modified Metric system called the *Standard International* system (the SI system). Table 1.1 presents the basic units of the aforementioned five systems of units, and the recommended symbol for each unit.

Table 1.1 Basic Units in Several Systems of Units

Quantity	System of Units				
	English Engineering (EE)	English Absolute (EA)	English Gravitational (EG)	Metric System (Metric)	International System (SI)
Length	foot (ft)	foot (ft)	foot (ft)	centimeter (cm)	meter (m)
Mass	pound (lbm)	pound (lbm)	slug (slug)	gram (g)	kilogram (kg)
Time	second (sec)	second (sec)	second (sec)	second (s)	second (s)
Force	pound (lbf)	poundal (pdl)	pound (lbf)	dyne (dyne)	newton (N)
Temperature	rankine (R)	rankine (R)	rankine (R)	kelvin (K)	kelvin (K)

A brief history of measurement systems, including the new SI system, is given in Reference 4. The standard for each unit of measurement in the SI system was adopted by the General Conference of Weights and Measures in 1960. The standard for the *unit of mass* is the kilogram (kg), a cylinder of platinum-iridium alloy kept at the International Bureau of Weights and Measures in Paris. Of all the standard units, the mass unit is the only basic unit still defined by an artifact. The standard for the *unit of length* is the meter (m), which is defined as 1,650,763.73 wavelengths, in vacuum, of the radiation corresponding to the transition between the levels $2p_{10}$ and $5d_5$ of the krypton-86 atom. The standard for the *unit of time*, the second (s), is defined as the time interval required for 9,192,631,770 cycles of radiation corresponding to the transition between the two hyperfine levels of the ground state of the cesium-133 atom. The degree kelvin (K), the standard for the *unit of temperature*, is defined as 1/273.16 of the thermodynamic temperature of the triple point of water. The newton (N) is the standard for the *unit of force* in the SI system. Its magnitude *is defined* as the force required to accelerate a mass of 1 kg at the rate of 1 m/s². The mole (mol) is the standard for the *amount of substance of a system*. The mole contains as many elementary entities as there are atoms in 0.012 kilogram of carbon-12. When the mole is used, the elementary entities must be specified and may be atoms, molecules, ions, electrons, other particles, or specified groups of such particles. The standard for the *unit of measure of a plane angle* is the radian (rad), which is defined as the angle with its vertex at the center of a circle and subtended by an arc equal in length to the radius.

During the period that the SI system is being adopted in the United States, there will be at least five different consistent systems of basic units being employed. The relationships between them (and other systems of units) are readily determinable by applying the consistency condition to them.

1. *The EE system.*
     1 *pound force* (1 *lbf*) $=$ 1 *pound mass* (1 *lbm*) $\times$ 32.1740 *ft/sec²*
2. *The EA system.*
     1 *poundal force* (1 *pdl*) $=$ 1 *pound mass* (1 *lbm*) $\times$ 1 *ft/sec²*
3. *The EG system.*
     1 *pound force* (1 *lbf*) $=$ 1 *slug mass* (1 *slug*) $\times$ 1 *ft/sec²*
4. *The Metric system.*
     1 *dyne force* (1 *dyne*) $=$ 1 *gram mass* (1 *g*) $\times$ 1 *cm/s²*
5. *The SI system.*
     1 *newton force* (1 *N*) $=$ 1 *kilogram mass* (1 *kg*) $\times$ 1 *m/s²*

Table C.1* presents the dimensional formulas for several of the physical quantities of interest in gas dynamics. Also included in Table C.1 are the units of the physical quantities in the English Engineering (EE), English Absolute (EA), English Gravitational (EG), and Standard International (SI) systems of units.

Table C.2 presents the values of some of the *universal physical constants* that are pertinent to fluid mechanics.

In Section 1–3(a), it is pointed out that Newton's second law of motion states that

$$\mathbf{F} \propto m\mathbf{a}$$

To convert the above proportionality into an equation, a *factor of proportionality*, denoted by $1/g_c$, is introduced into the above relationship. Thus

$$\mathbf{F} = \frac{1}{g_c} m\mathbf{a} \tag{1.1}$$

* Tables labeled C.1, C.2, etc., are in Appendix C.

From equation 1.1, $g_c = ma/F$. The magnitude and units of g_c depend on the units employed for m, $\mathbf{a}$, and $\mathbf{F}$. Table 1.2 presents the values of g_c for several systems of units that are commonly employed in gas dynamics.

Table 1.2   The Factor of Proportionality g_c for Several Systems of Units

System of Units	Length	Time	Mass	Force	Factor of Proportionality g_c
EE	ft	sec	lbm	lbf	32.1740 lbm-ft/lbf-sec^2
EA	ft	sec	lbm	pdl	1.0 lbm-ft/pdl-sec^2
EG	ft	sec	slug	lbf	1.0 slug-ft/lbf-sec^2
SI	m	s	kg	N	1.0 kg-m/N-s^2
Metric	cm	s	g	dyne	1.0 g-cm/dyne-s^2

The factor g_c is not a physical quantity but merely a conversion factor similar to 1 m = 100 cm, 1 N-m = 1 J, 1 kg = 1000 g, and so forth. For a consistent set of units, the numerical value of $g_c = 1$ and may, therefore, be omitted from equation 1.1, to yield

$$\mathbf{F} = m\mathbf{a} \tag{1.2}$$

Since g_c is not a physical quantity, it will be omitted from the equations of this book. Remember, however, that when making numerical calculations, the appropriate values of g_c and other factors of proportionality must be determined and employed.

The conversion factors for the units of many of the physical parameters pertinent to gas dynamics are presented on the inside back cover of this book.

Example 1.1.   A body of mass 5 kg is given an acceleration of 10 m/s^2. Calculate the external force acting on the body in these units: (a) newtons, (b) dynes, (c) pounds force, and (d) poundals.

Solution

(a) $\quad F = ma = (5 \text{ kg})\left(10 \dfrac{m}{s^2}\right)\left(\dfrac{N\text{-}s^2}{m\text{-}kg}\right) = 50 \text{ N}$

(b) $\quad F = ma = (5 \text{ kg})\left(10 \dfrac{m}{s^2}\right)\left(\dfrac{N\text{-}s^2}{m\text{-}kg}\right)\left(\dfrac{10^5 \text{ dyne}}{N}\right) = 5 \cdot 10^6 \text{ dyne}$

(c) $\quad F = ma = (5 \text{ kg})\left(10 \dfrac{m}{s^2}\right)\left(\dfrac{N\text{-}s^2}{m\text{-}kg}\right)\left(0.224809 \dfrac{lbf}{N}\right) = 1.124 \text{ lbf}$

(d) $\quad F = ma = (5 \text{ kg})\left(10 \dfrac{m}{s^2}\right)\left(\dfrac{N\text{-}s^2}{m\text{-}kg}\right)\left(0.224809 \dfrac{lbf}{N}\right)\left(32.174 \dfrac{pdl}{lbf}\right) = 36.165 \text{ pdl}$

In this example, the units of each physical quantity are written in the equation together with its numerical value. By so doing, the units of the resulting quantity appear explicitly, and it is immediately obvious whether or not a conversion factor such as g_c is required. The technique of including the units in the equations when making numerical computations is highly recommended. The required conversion factors, if any, then suggest themselves automatically in terms of the units of the original physical quantities and the desired units of the calculated physical quantity.

Despite the foregoing strong recommendation, that practice, because of space limitations, is not followed in the illustrative examples presented in this book. All of the examples were, however, originally worked in the recommended manner to assure that the desired units were obtained in the answers.

1-3(c) Principle of Dimensional Homogeneity

An important principle pertaining to the mathematical relationships between physical quantities is that of *dimensional homogeneity*, which states that all of the terms of an equation expressing an actual physical relationship between physical variables must have the same dimensions. Thus, the units employed in evaluating each term in an equation may be quite different as long as the dimensions of each term are the same. However, when the terms of a dimensionally homogeneous equation are to be combined, each term must be expressed in the same units. At that time, the conversion factors are introduced into the numerical computations.

1-4 THE CONTINUUM POSTULATE[5, 6]

From a microscopic viewpoint a material substance is composed of discrete particles; that is, molecules or atoms. In dealing with fluids, liquid or gaseous, it is assumed quite generally that the detailed molecular structure of the fluid may be replaced by a *continuum* that makes it possible to deal with a fluid on a macroscopic scale. The *continuum postulate* assumes that every differential element of a body of fluid contains a tremendous number of molecules and that the average statistical properties of the molecules contained in an elementary volume represent the macroscopic properties of the fluid in the region of that elementary volume. Consequently, the continuum model is a satisfactory one only for those situations where the characteristic dimensions of the body of fluid under consideration, or of a material body within the fluid, are very large when compared with the average molecular distance between the molecules comprising the fluid. In other words, the continuum postulate is satisfied if an infinitesimal change in the volume of the fluid influences an exceedingly large number of molecules.

Before defining some of the properties of a fluid based on the continuum postulate, it is helpful to review briefly the forces that may act on a body of fluid. Such forces deform or *strain* it. If the forces are substantially invariant with time, they are termed *static* forces; otherwise they are termed *dynamic* forces.

1-4(a) Classification of the Forces Acting on a Body of Fluid

The external forces that act on a body of fluid may be segregated into two groups: (1) *surface forces*, and (2) *body forces*. A *surface force* may have any orientation with respect to the surface of the body of the fluid and may be decomposed into a *normal force* acting perpendicular to the surface and a *tangential*, or *shearing*, *force* acting parallel to the surface. A *body force* is one that is distributed over the entire volume of a body of material; for example, the forces caused by gravitational attraction, magnetic and electrostatic fields, and the like.

The force per unit area acting on a body (dimensions F/L^2) is termed a *stress*. If the force is normal to the area, the stress is called a *normal stress*. If a normal stress acts in the direction for bringing the particles comprising the body in closer contact with each other, it is termed a *compressive stress*. If it acts in the direction that tends to cause separation of the particles, it is known as a *tensile stress*.

1-4(b) Density at a Point in a Continuum

The smallest elementary volume $\delta\mathscr{V}$ comprising an elementary mass of fluid that contains a sufficiently large number of molecules for satisfying the continuum postulate is termed a *fluid particle*; let $\delta\mathscr{V}'$ denote that *limiting volume*.

Figure 1.1 illustrates schematically a large mass of fluid enclosed by a volume $\mathscr{V}$. Let $P(x,y,z)$ denote an arbitrary point inside of $\mathscr{V}$. Assume that the point P is surrounded by an elementary volume $\delta\mathscr{V}$ that contains the elementary mass of fluid δm. Let $\bar{\rho}$ denote the average density of the mass enclosed by $\delta\mathscr{V}$. Then

$$\bar{\rho} \equiv \frac{\delta m}{\delta\mathscr{V}} \tag{1.3}$$

Now let $\delta\mathscr{V}$ shrink about the point $P(x,y,z)$, and plot the ratio $\delta m/\delta\mathscr{V}$, as is done qualitatively in Fig. 1.2. It is seen that as $\delta\mathscr{V}$ shrinks, the curve presenting $\delta m/\delta\mathscr{V}$ as a function of $\delta\mathscr{V}$ exhibits the following characteristics.

1. $\delta m/\delta\mathscr{V}$ approaches an asymptotic value; that is, the material becomes more homogeneous.

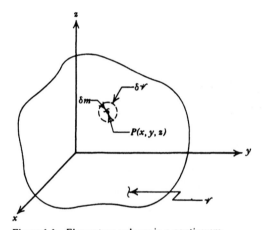

Figure 1.1   Elementary volume in a continuum.

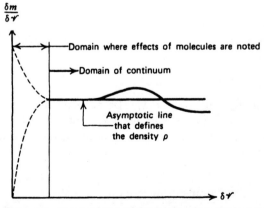

Figure 1.2   Determination of the density at a point in a continuum.

2. After $\delta\mathscr{V}$ becomes smaller than some minimum value where it contains only a few molecules, the value of $\delta m/\delta\mathscr{V}$ fluctuates widely as one or more molecules either enter or leave $\delta\mathscr{V}$. Consequently, the magnitude of $\delta m/\delta\mathscr{V}$ becomes indeterminate.

Introducing the *limiting volume* $\delta\mathscr{V}'$, then the density of the fluid at the point P, denoted by ρ, is given by

$$\rho \equiv \underset{\delta\gamma \to \delta\gamma''}{\text{Limit}} \frac{\delta m}{\delta\mathscr{V}} \tag{1.4}$$

1–4(c) Velocity at a Point in a Body of Fluid

Let **V** denote the average value of the velocity of the fluid instantaneously contained in the limiting volume $\delta\mathscr{V}'$ surrounding an arbitrary point $P(x,y,z)$. In general,

$$\mathbf{V} = \mathbf{V}(x,y,z,t) \tag{1.5}$$

Refer to Fig. 1.3. Let **r** denote the *radius vector* to the point P at any instant of time t. Then

$$\mathbf{V} = \mathbf{V}(\mathbf{r},t) \tag{1.6}$$

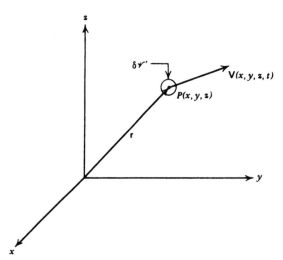

Figure 1.3   Velocity at a point in a flowing fluid.

The flow is called *unsteady* if **V** varies with time t, and it is termed *steady* if **V** is invariant with t. In the latter case,

$$\mathbf{V} = \mathbf{V}(\mathbf{r}) = \mathbf{V}(x,y,z) \tag{1.7}$$

It should be noted that the velocity at the point $P(x,y,z)$ is independent of the instantaneous velocity of the molecule of the fluid particle closest to point P. **V** is the velocity of the *center of mass* of the fluid particle, enclosed by $\delta\mathscr{V}'$, at the instant that it coincides with the point P. By definition,

$$\mathbf{V} \equiv \frac{\sum\limits_{i=1}^{N} m_i \mathbf{V}_i}{\sum\limits_{i=1}^{N} m_i} = \frac{\text{total momentum contained in } \delta\mathscr{V}'}{\text{total mass inside } \delta\mathscr{V}'} \tag{1.8}$$

where N denotes the number of particles inside $\delta\mathscr{V}'$, and $\mathbf{V}_i$ denotes the velocity of the ith particle, m_i denotes its mass, and $m_i\mathbf{V}_i$ denotes its momentum.

1-4(d) Stress at a Point

Figure 1.4a illustrates a solid body, assumed to be a continuum, in equilibrium under the action of several external forces, $\mathbf{F}_1$, $\mathbf{F}_2$, $\mathbf{F}_3$, $\mathbf{F}_4$, $\mathbf{F}_5$, and $\mathbf{F}_6$. Because of the actions of those external forces, *internal forces* are transmitted through the material composing the body; the expression internal force as employed here does not refer to the individual forces between molecules, but to their combined effect.

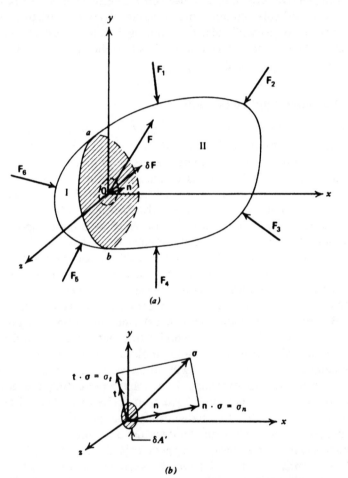

Figure 1.4 Stress at a point in a continuum. (a) Forces acting on a body. (b) Stress at a point.

Assume that the material of the body is divided into two parts, labeled I and II, by an imaginary plane ab that passes through point O, and let point O be the origin of the Cartesian coordinate axes x, y, z. Part I of the body is assumed to be in equilibrium under the combined actions of the external forces $\mathbf{F}_5$ and $\mathbf{F}_6$ and the internal forces distributed over the area A of the plane ab, caused by the action of the material of Part II on that of Part I. It is convenient to express the magnitude of an internal force in terms of the *stress* (force per unit area) it produces on the area over which it acts. Let the stress be denoted by σ. In the general case the stress is not distributed uniformly over the area A of plane ab (see Fig. 1.4a). Let $\mathbf{F}$ denote the *resultant* force acting over the area A. Then the stress σ at the arbitrary point O in plane ab is defined by

$$\sigma = \underset{\delta A \to \delta A'}{\text{Limit}} \frac{\delta \mathbf{F}}{\delta A} \qquad (1.9)$$

where $\delta A'$ is an area comparable in size to the volume $\delta \mathscr{V}'$.

The stress σ, illustrated in Fig. 1.4b, is a vector having the same direction as the resultant force $\delta \mathbf{F}$, and it is ordinarily inclined with respect to the *unit normal vector* $\mathbf{n}$, which indicates the direction of the surface δA. By convention, the outwardly drawn normal points in the positive direction, and vice versa. The stress σ can be resolved into two mutually perpendicular components, a *normal stress* parallel to $\mathbf{n}$, and a *tangential* or *shearing stress* perpendicular to $\mathbf{n}$; that is, it is in the direction of the *unit* tangential vector $\mathbf{t}$. Hence, the stress σ is defined by the following vector equation.

$$\sigma = \mathbf{n}\sigma_n + \mathbf{t}\sigma_t \qquad (1.10)$$

1–4(e) The Continuum Postulate Applied to a Gas

In the case where a body of gas is at a low pressure (e.g., atmospheric air at high altitudes), the gas density may be so low that the applicability of the continuum postulate may be open to question. It is, therefore, essential that analytical criteria be available for determining the limitations to the application of the continuum postulate. To do that some of the results obtained from the microscopic viewpoint of the structure of a gas are considered briefly (see also Section 1–14).

According to the kinetic theory, a gas may be conceived as being composed of a very large number of molecules (e.g., Avogadro's number $N_A = 6.02252 \cdot 10^{23}$ molecules/mol). The gas molecules move with extremely rapid random motions, but the motion of each molecule is greatly impeded by collisions with the tremendous number of neighboring molecules and the walls of the containing vessel. The collision process is assumed to be elastic and to take a very short time; it also conserves energy and momentum. Because of the collisions, two extreme conditions are possible. Some molecules may transfer their kinetic energies to other molecules and attain velocities that are practically zero, while other molecules may attain very high velocities. The remainder of the molecules will attain velocities (kinetic energies) between the aforementioned two extremes. Functions representing the equilibrium velocity distribution for the molecules have been derived by Maxwell and by Boltzmann, by applying the methods of statistical mechanics.[7–10]

Under normal conditions of pressure and temperature a gas molecule moves only a short linear distance, called the *molecular free path*, before it collides with another gas molecule. Consequently, the time interval between consecutive collisions is likewise very short. The average value of the free path for an assemblage of molecules is termed the *mean free path*, denoted by λ, and several of the transport properties of a gas are related to λ; for example, its viscosity, thermal conductivity, and diffusion coefficient. In general, for a given number of molecules per unit volume, λ will be larger for small molecules than for large molecules; the probability of a collision decreases with a reduction in the size (diameter) of the molecule.

For the elastic collisions of molecules having a Maxwellian velocity distribution,[10] the *mean free path* λ is given by

$$\lambda = \frac{1}{\sqrt{2}\,(\pi N\, d^2)} \text{ (meters)} \qquad (1.11)$$

where N is the number of molecules per cubic meter and d is the effective molecular diameter in meters. If m denotes the mass of a molecule and ρ the density of the gaseous continuum, then

$$\lambda = \frac{m}{\sqrt{2}\,(\rho\pi\,d^2)} \tag{1.12}$$

In general, all of the quantities for calculating the mean free path are readily obtainable except the molecular diameter d. The latter is generally calculated from the application of the kinetic theory to the calculation of the transport properties of the gas, such as its viscosity, heat conductivity, and self-diffusion.

Table 1.3 presents typical approximate values of the molecular diameters for several gases at standard temperature and pressure (STP); that is, at $p = 1$ atm and $t = 298.15$ K. It is evident from Table 1.3 that λ at STP is an extremely small distance.

Table 1.3   Molecular Diameter d and Mean Free Path λ (at STP) for Several Gases

Gas	d, m $\cdot$ 10^{-10}	λ, m $\cdot$ 10^{-8}	Gas	d, m $\cdot$ 10^{-10}	λ, m $\cdot$ 10^{-8}
Argon (A)	2.90	10.9	Oxygen (O_2)	2.95	10.5
Helium (He)	2.00	22.9	Carbon dioxide (CO_2)	3.30	8.39
Nitrogen (N_2)	3.50	7.46	Ammonia (NH_3)	3.00	10.2

1-4(f) The Knudsen Number

For a gas to satisfy the continuum postulate, the molecular mean free path must be small compared to a significant characteristic linear dimension L pertinent to the flow field. By definition, the ratio λ/L is termed the *Knudsen number*, and is denoted by Kn. Thus,[11]

$$\text{Kn} \equiv \lambda/L \tag{1.13}$$

The Knudsen number Kn can be related to the *Reynolds number* $\text{Re} = LV\rho/\mu$ and the *Mach number* $M = V/a$, where a is the *acoustic speed* in the gas. Employing the *specific heat ratio* for the gas, $\gamma = c_p/c_v$, it can be shown that[11]

$$\text{Kn} = 1.26\,\sqrt{\gamma}\,(M/\text{Re}) \tag{1.14}$$

For small values of the Reynolds number Re, the characteristic length L may ordinarily be assumed to be either a dimension of a body immersed in the flowing gas or a dimension of the flow passage. For large values of Re (i.e., $\text{Re} \gg 1$), a more significant characteristic length is the thickness δ of the layer of fluid adjacent to the solid walls where viscous effects are important; this layer of fluid is termed the *boundary layer* (see Section 5-10). The variables L, Re, and δ are related generally, as indicated below.[11]

$$\delta/L \cong \sqrt{1/\text{Re}} \qquad \text{Re} \gg 1 \tag{1.15}$$

Equation 1.15 is not applicable to *hypersonic* flows ($M > 7$, approximately). For large values of Re, Kn is, therefore, given by

$$\text{Kn} \cong M/\sqrt{\text{Re}} \tag{1.16}$$

The continuum postulate is applicable to those flows for which the appropriate Knudsen number is less than approximately 0.01. Accordingly, a gas may be assumed to be a continuum if

$$Kn \cong M/Re \quad < 0.01 \quad \text{and} \quad Re \cong 1 \qquad (1.17a)$$

$$Kn \cong M/\sqrt{Re} < 0.01 \quad \text{and} \quad Re \gg 1 \qquad (1.17b)$$

When $Kn > 0.01$, the gas should be treated as an assemblage of discrete particles. Table 1.4 presents a classification of the flow regimes of a gas based on the value of the Knudsen number.

Table 1.4 Flow Regimes Based on the Knudsen Number

$Kn = \lambda/L$	Flow Regime
$Kn < 0.01$	Continuum
$0.01 < Kn < 0.1$	Slip flow
$0.1 < Kn < 3.0$	Transition regime
$3.0 < Kn$	Free molecule flow

Based on Reference 11.

Example 1.2. An electric-arc heated wind tunnel is constructed as illustrated schematically in Fig. 1.5. A model 0.0254 m high and 0.1524 m long is placed in the tunnel and tested at a Mach number of 7. At that Mach number, $V = 1743$ m/s, $\rho = 0.0182$ kg/m^3, $t = 154$ K, and $p = 833$ N/m^2. A pitot probe made of hypodermic tubing having an outside diameter of $7.62 \cdot 10^{-5}$ m, for measuring the free-stream stagnation pressure, is placed at the front of the model. Determine whether or not rarefied flow effects should be expected.

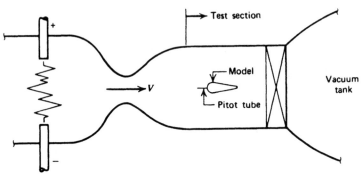

Figure 1.5 Sketch for Example 1.2.

Solution

For the model itself, the characteristic dimension of concern is the boundary layer thickness δ on the model. The Reynolds number based on the model length is

$$Re = \frac{LV\rho}{\mu} = \frac{(0.1524)(1743)(0.0182)}{(12.2 \cdot 10^{-6})} = 3.963 \cdot 10^5$$

where $\mu = 12.2 \cdot 10^{-6}$ kg/m-s at $t = 154$ K. Thus, the appropriate value of Kn for the model is

$$\text{Kn} = \frac{M}{\sqrt{\text{Re}}} = \frac{7}{(3.963 \cdot 10^5)^{1/2}} = 0.0111$$

For this value of Kn, rarefied flow effects would probably not be important.

For the probe the appropriate characteristic dimension is the probe diameter, because boundary layer effects are not present at the probe inlet. Thus

$$\text{Re} = \frac{DV\rho}{\mu} = \frac{(7.62 \cdot 10^{-5})(1743)(0.0182)}{(12.2 \cdot 10^{-6})} = 198.14$$

The appropriate value of Kn for the probe is

$$\text{Kn} = \frac{M}{\text{Re}} = \frac{7}{198.14} = 0.03533$$

Rarefied flow effects should be investigated for the probe.

1–5 PERFECT SUBSTANCES

In general, engineering problems are concerned with material substances that are subjected to certain physical and chemical processes. Exact mathematical descriptions of the substance and of the processes involved either lead to unsolvable equations or to ones that cannot be dealt with conveniently. To make the problem amenable to mathematical analysis, the actual substance is generally replaced by a *perfect substance* that behaves in accordance with certain simple laws, and the actual processes are replaced by ones that have a more or less simple mathematical description.

A perfect substance is, of course, a fictitious one, but its consideration leads to important information giving a first approximation, at least, to the behavior of a real substance. The equations derived for the behavior of the perfect substance can be made applicable to a real substance by modifying them with experimentally determined correction coefficients.

A perfect substance is one that is *homogeneous* in its composition, and *isotropic*. By virtue of its homogeneity, its chemical composition and physical condition are uniform throughout its entire mass. Because of its isotropicity, its elastic properties are identical at all points and in all directions; the stresses developed in the perfect substance depend only on the magnitude of the strains. A perfect substance is frequently termed a *simple system*, and possesses the following two fundamental characteristics.

1. Its elastic properties are completely specified by two elastic moduli: (a) the shear or rigidity modulus N, and (b) the bulk modulus K.
2. Its thermodynamic state is completely defined by its macroscopic properties, also called *thermodynamic coordinates*, only two of which are independent.

It will be assumed in all subsequent discussions, unless the contrary is stated explicitly, that any infinitesimal change in the dimensions of a perfect substance, such as a volumetric change $d\mathscr{V}$, involves such a large number of molecules that the substance may be considered to be a continuum.

Figure 1.6 illustrates a unit cube of perfect substance subjected to a shearing stress τ. *By definition*, the *shear modulus N* is given by

$$N \equiv \frac{\text{shearing stress}}{\text{shearing strain}} = \frac{\tau}{\Delta L/L} = \frac{\tau}{\tan \alpha} \cong \frac{\tau}{\alpha} \tag{1.18}$$

since for the small values of the angle α involved, $\tan \alpha \cong \alpha$.

The definition of the *bulk modulus K* is obtained by considering a cube of material subjected to a uniform hydrostatic pressure p, as illustrated in Fig. 1.7. If the hydrostatic pressure is increased from p to $p + \Delta p$, the volume of the cube diminishes

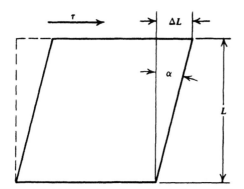

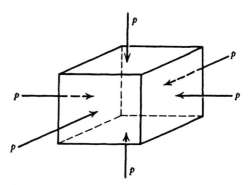

Figure 1.6 Deformation of a unit cube because of a tangential stress.

Figure 1.7 Unit cube of a material subjected to a uniform hydrostatic pressure.

from $\mathscr{V}$ to $\mathscr{V} - \Delta\mathscr{V}$. The unit volumetric strain is accordingly $-\Delta\mathscr{V}/\mathscr{V}$, the negative sign indicating that the volume decreases as the pressure increases. *By definition,* noting that $\Delta\mathscr{V}/\mathscr{V} = \Delta v/v$

$$K \equiv \lim_{(\Delta p \to 0)} \left(-\frac{\Delta p}{\Delta v/v} \right) = -v \frac{dp}{dv} \tag{1.19}$$

The reciprocal of the bulk modulus is termed the *compressibility*, and is denoted by κ. Thus,

$$\kappa = \frac{1}{K} = -\frac{1}{v}\left(\frac{dv}{dp}\right) \tag{1.20}$$

When the compression is conducted under isothermal conditions, then $K = K_t =$ the *isothermal bulk modulus* and, when the compression is a reversible adiabatic process, then $K = K_s =$ the *isentropic bulk modulus*.

1–5(a) The Perfect Solid

A perfect substance possessing both bulk elasticity ($K \neq 0$) and rigidity ($N \neq 0$) is by definition a *perfect solid*. When a perfect solid is stressed, the external forces acting on it are balanced by the internal forces arising from the elastic strains produced. Since a perfect solid can resist shear (tangential) strains, the resultant stress on an arbitrary elementary plane either within or on the solid can have any direction with respect to that plane. Furthermore, a perfect solid obeys Hooke's law: *the stress is proportional to the strain.*

1–5(b) The Perfect Fluid

From an elasticity viewpoint a perfect substance possessing bulk elasticity ($K \neq 0$) but no rigidity ($N = 0$) is defined as a *perfect fluid*. Since the perfect fluid cannot resist a tangential stress no matter how small it may be, the only forces it can resist are *normal* forces (i.e., pressure forces). Consequently, the direction of the stress acting on an elementary plane drawn in the perfect fluid is always normal to that plane. Hence, the magnitude of the pressure at an arbitrary point in a perfect fluid is independent of the orientation of a plane drawn through that point.

Because $N = 0$, a perfect fluid possesses the following three characteristics.

1. It can transmit only pressure forces.
2. It is frictionless, because it cannot transmit tangential (shearing) forces.
3. The pressure at a point is the same in all directions.

Perfect fluids may be subdivided into *perfect liquids* and *perfect gases*.

1–5(c) The Perfect Liquid

An *incompressible* perfect fluid is called a *perfect liquid*. For such a substance,

$$K = -v\left(\frac{\partial p}{\partial v}\right) = \infty \tag{1.21}$$

It follows from equation 1.21 that for a perfect liquid $\kappa = 1/K = 0$. No real liquid is truly incompressible, but the numerical values of K for most real liquids are sufficiently large to justify assuming that they are incompressible, especially when the pressure changes involved are small. Moreover, for compressions over a small range of temperature, the value of K may be assumed to remain constant at the value prior to the compression.

1–5(d) The Perfect Gas

From the viewpoint of its elastic moduli, a *perfect gas* may be defined as a perfect fluid ($N = 0$) having an *isothermal bulk modulus* K_t that is equal to its static pressure p. Hence, for a perfect gas,

$$K_t = -v\left(\frac{\partial p}{\partial v}\right)_t = p \tag{1.22}$$

The molecules of a perfect gas exert no intermolecular forces; their only interactions arise from instantaneous elastic collisions (see Section 1–14). From thermodynamics, it is shown that for an isentropic change of state, p and v for a perfect gas are related by $pv^\gamma =$ constant (see equation 1.127). Hence, for a perfect gas, the *isentropic bulk modulus* K_s is given by

$$K_s = -v\left(\frac{\partial p}{\partial v}\right)_s = \gamma p \tag{1.23}$$

In equation 1.23, the *specfic heat ratio* $\gamma = c_p/c_v$ is assumed to be a constant.

For a perfect gas, the ratio K_s/K_t is given by

$$K_s/K_t = \gamma p/p = \gamma = c_p/c_v = \text{constant} \tag{1.24}$$

Hence, γ is the ratio of the slopes of the isentropic and isothermal processes, for the perfect gas, drawn on the pv plane.

Summarizing, a perfect gas is a homogeneous compressible medium having the following properties: $K_t = p$, $K_s = \gamma p$, $\gamma = $ constant, and $N = 0$. It is shown in Section 1–15, where the thermodynamic properties of gases are discussed, that the relation between p, v, and t for a perfect gas is given by equation 1.99, which is known as the *thermal equation of state*.

1–6 REAL FLUIDS

The perfect fluid discussed in Section 1–5(b) is an idealization, because all real fluids possess a finite shear modulus. The general characteristics of real fluids are discussed in this section.

1–6(a) General Comments on Perfect and Real Fluids

Real fluids differ from the fictitious perfect fluid in that they are not frictionless. Shear stresses arise whenever there is relative motion between the fluid particles, and the shear stresses persist as long as such motion prevails. Since the shear stresses arise only when there is relative motion, a real fluid at rest behaves exactly as the perfect fluid, that is, when at rest, a real fluid cannot sustain a tangential stress.

Because the perfect fluid is frictionless, a body immersed therein cannot create a resistance or *drag*, a force component acting in the same direction as the fluid velocity, even when there is relative motion between the body and the fluid. The only forces that can occur in a perfect liquid, for example, in steady flow with the velocity V, are (1) inertia forces that depend on $\rho V^2/2$, and (2) the static pressure forces that depend on the pressure intensity p. It is shown in any text on fluid mechanics that, when a perfect liquid having a density ρ is in steady flow, the aforementioned two types of forces are related by the Bernoulli equation

$$dp + \rho\, d\left(\frac{V^2}{2}\right) = 0 \tag{1.25}$$

Equation 1.25 shows that a change in the inertia force along a streamline is balanced by a change in the static pressure intensity.

Consider the case where a body, such as the infinitely long cylinder illustrated in Fig. 1.8, is immersed in a perfect liquid. The streamline pattern formed around the cylinder by the fluid flowing past it is symmetrical around the vertical plane $O–O$. Because of the symmetry of the streamline pattern, the forces acting on the left-hand half of the cylinder are exactly balanced by those acting on the right-hand half. Consequently, as stated above, there is no drag force or hydrodynamic resistance created by the cylinder.

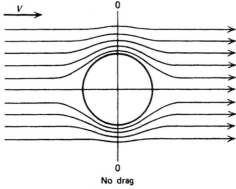

Figure 1.8 Symmetrical streamline pattern around an infinitely long cylinder immersed in a flowing perfect fluid.

In a real liquid, however, because of its viscosity [see Section 1-6(b)], viscous forces are present in addition to the inertia and static pressure forces. The viscous forces manifest themselves as shearing stresses between the adjacent streamlines. Because of those shearing stresses, a body such as a cylinder immersed in a real liquid experiences a resistance or drag force. In general, the total drag force may be subdivided into two different types of drags.

1. The *form* or *pressure drag* caused by the pressure force acting on the front (upstream side) of the body being larger than that acting on its rear.
2. The *skin friction drag* caused by shear stresses in the fluid adjacent to the surface of the immersed body.

Figure 1.9 illustrates schematically the conditions prevailing when a real liquid flows past a cylinder immersed within it. The fluid flows toward the cylinder with the uniform velocity V, but some of the streamlines are brought to *stagnation* by the front portion of the cylinder, thereby producing the pressure force F_p, tending to move the cylinder to the right. Some of the streamlines will be unable to follow the contour of the cylinder, so that the flow separates from it at some point, such as point s in Fig. 1.9. Behind the cylinder a low velocity or separation region is formed,

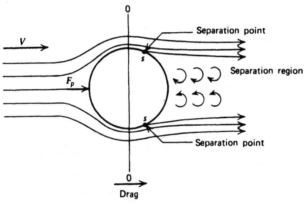

Figure 1.9 Streamline pattern and eddy formation for an infinitely long cylinder immersed in a flowing real fluid.

and in this region the static pressure intensity does not differ markedly from that at the *separation point s*. If the speed of the fluid is increased, the separation point s moves upstream, and the pressure intensity at the separation point decreases, as does the pressure intensity in the separation region. Consequently, there arises a difference in the pressure forces acting on the front and back of the cylinder, and the resultant of those pressure forces constitutes the *form drag*. The separation point is usually defined as the point, on the (solid) surface, where a streamline that is very close to the surface leaves the surface.

The shear stresses giving rise to the skin friction are confined to the *boundary layer* adjacent to the cylinder, being caused by the steep velocity gradient dV/dn in that thin layer. Outside of the boundary layer the velocity gradient is comparatively negligible, so that the flow outside of the boundary layer may be assumed to be nonviscous, that is, comparable to the flow of a perfect fluid. It is the behavior of the boundary layer that causes all flow separations with the attendant increase in drag. Moreover, the boundary layer plays a predominant role in the transfer of heat by forced convection where relative motion between a fluid and a solid surface is involved.

1-6(b) Viscosity

The property of a real fluid known as its *dynamic* or *absolute viscosity*, hereafter called *viscosity* for brevity, is the source of all fluid friction. A real fluid, by virtue of its viscosity, can sustain a shearing stress when in motion, whereas the perfect fluid cannot. Consequently, in formulating the equations of motion for a real fluid, its viscosity must be taken into account. Unfortunately the equations of motion for a viscous fluid are amenable to mathematical analysis in only a few special cases.

The definition of viscosity is based on the ability of a real fluid to distort or deform when it is subjected to a tangential (shearing) force and thereby resist the shearing stresses imposed on it. Thus, consider the steady parallel flow of a real fluid past a solid surface, as illustrated in Fig. 1.10. At different normal distances from the surface, the velocities of the fluid will be different, giving rise to a velocity profile such as that illustrated schematically in Fig. 1.10. It appears that the fluid layer in contact with the solid surface adheres to it and consequently assumes the same velocity as the surface. The fluid velocity increases to the main stream velocity as one proceeds along the line $O-O'$ normal to the surface.

Consider two parallel surfaces Y and Z a distance n apart, and assume that the space between them is occupied by a homogeneous viscous fluid as illustrated in Fig. 1.11. Let the tangential force F_t be applied to surface Y and cause it to move with

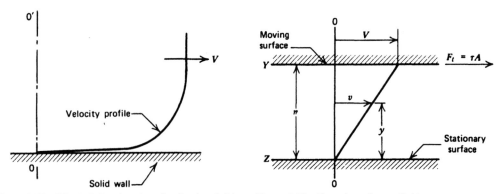

Figure 1.10   Effect of the viscosity of a flowing fluid on its velocity profile.

Figure 1.11   Shear in a viscous fluid.

the constant velocity V, while surface Z is held stationary. The fluid in contact with surface Y, owing to its viscosity, will assume the velocity V while that in contact with surface Z will have zero velocity with respect to surface Z. The layers of fluid between surfaces Z and Y, as first pointed out by Newton, will assume velocities that are proportional to their respective perpendicular distances from surface Z. The *rate of shear* V/n is, therefore, constant throughout the homogeneous fluid. Accordingly, for a *Newtonian fluid* the relative velocity v of a fluid layer located at a distance y from surface Z is

$$v = y \frac{V}{n} \tag{1.26}$$

If A is the area of surface Y, then the shear stress on that surface is given by

$$\tau = \frac{F_t}{A} \qquad (1.27)$$

To maintain the surface Y at the constant relative velocity V, the shear stress τ must be opposed by an equal and opposite stress arising from the internal resistance of the fluid opposing the shear stress distorting it. The viscosity of the fluid is defined as the ratio of the shear stress to the rate of shear and is denoted by μ. Hence, *by definition,*

$$\mu \equiv \frac{\text{shear stress}}{\text{rate of shear}} = \frac{\tau}{V/n} \qquad (1.28a)$$

or, in differential form,

$$\tau = \mu \frac{dv}{dy} \qquad (1.28b)$$

A fluid that obeys equation 1.28b is termed a *Newtonian fluid.*

The dimensions and units of viscosity in the MLT system are M/LT (e.g., kg/m-s), and in the FLT system they are FT/L^2 (e.g., N-s/m²). Much of the physical data pertaining to the viscosity of fluids are stated in centipoises, where 100 centipoises = 1 poise = 1 g/cm-s = 0.1 N-s/m².

In general, the viscosities of fluids depend primarily on the temperature and only slightly on the pressure. For liquids the viscosity decreases with increasing temperature, while the reverse is the case for gases. The independence of the viscosity of a gas from its pressure holds only in the range where the gas is far removed from its *critical pressure.*

It follows from equation 1.28b that the shear stress in a real fluid depends on the viscosity μ and the *normal velocity gradient dv/dy.* In the case of fluids of small viscosity, such as water and air, large shearing stresses can occur only in those regions of the fluid where the normal velocity gradient is considerable. Experience has shown that large velocity gradients occur close to a solid surface past which the fluid flows. At the wall the fluid velocity relative to it is zero, whereas in a small distance normal to the wall the velocity increases to practically the main stream velocity as illustrated schematically in Fig. 1.10. L. Prandtl (1904) applied the term *boundary layer* to the thin layer wherein the velocity gradient dv/dy is large. As mentioned earlier, outside of the boundary layer the normal velocity gradients are so small that the effects of viscosity may be neglected; that is, the fluid flowing outside of the boundary layer may be assumed to be a perfect fluid.

The viscosity μ is a measure of the shearing stress τ in a flowing fluid. In most fluid flow problems it is the *ratio* of the *shearing force* to the *inertia force* that is of prime importance. The inertia force is proportional to the density of the fluid ρ and the velocity squared. The ratio μ/ρ for a fluid is called its *kinematic viscosity* and is denoted by v. Thus

$$v = \text{kinematic viscosity} = \mu/\rho \qquad (L^2/T) \qquad (1.29)$$

In metric units, the unit of v is the *stoke,* 1 cm²/s, but it is more customary to use the *centistoke,* 0.01 stoke, which is 10^{-6} m²/s.

Example 1.3. The 5.08 cm diameter shaft illustrated in Fig. 1.12 is rotated centrally at an angular velocity $n = 4000$ rev/min in a bearing having a clearance $C = 0.00508$ cm. The viscosity of the lubricating oil is 0.186 poise, and its specific gravity is 0.92 under the operating conditions. The bearing is 7.62 cm long. Neglecting end effects and assuming that the bearing is lightly loaded so that the only resistance to

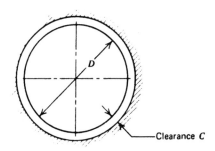

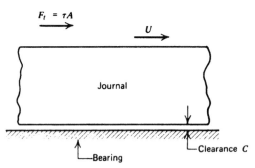

Figure 1.12   Sketch for Example 1.3.

motion is due to the viscosity of the lubricant, calculate the friction torque and the power consumed in friction. Refer to Fig. 1.12.

Solution

The viscosity of the oil μ is

$$\mu = 0.186(0.1) = 0.0186 \text{ N-s/m}^2$$

The peripheral velocity of the journal, $U = \pi Dn$, equals the maximum velocity of the oil in the clearance. The tangential stress in the oil τ is given by

$$\tau = \mu \frac{U}{C} = \frac{\pi \mu Dn}{C}$$

The tangential force acting on the journal F_t is

$$F_t = \tau A = \frac{\pi^2 L \mu n D^2}{C} = \frac{\pi^2 (7.62/100)(0.0186)(4000/60)(5.08/100)^2}{(0.00508/100)} = 43.37 \text{ N}$$

The friction torque acting on the journal $\mathscr{T}_f$ is

$$\mathscr{T}_f = \frac{F_t D}{2} = \frac{47.37(5.08/100)}{2} = 1.203 \text{ m-N}$$

The friction power $\mathscr{P}_f$ is

$$\mathscr{P}_f = 2\pi n \mathscr{T}_f = 2\pi (4000/60)(1.203) = 503.9 \text{ watts} \ (0.675 \text{ hp})$$

1-6(c)   Viscosity Data for Gases

The older published experimental data on the viscosity of gases as well as more recent experimental work have been reviewed critically by Watson.[12] A correlation equation was derived that gives a satisfactory representation of the reassessed experimental data over the temperature range 270 to 2200 K. That correlating equation is as follows.

$$\mu = \frac{\sqrt{t}}{A_0 + \dfrac{A_1}{t} + \dfrac{A_2}{t^2} + \dfrac{A_3}{t^3} + \dfrac{A_4}{t^4}} \quad \text{N-s/m}^2 \cdot 10^{-6} \tag{1.30}$$

where the viscosity μ is in N-s/m$^2 \cdot 10^{-6}$, the temperature t is in K, and the coefficients A_0, A_1, etc., have the values presented in Table 1.5 for the different gases listed therein.

Table 1.5 Values of the Coefficients in the Viscosity Equation for Several Gases

$$\mu = \frac{\sqrt{t}}{A_0 + \dfrac{A_1}{t} + \dfrac{A_2}{t^2} + \dfrac{A_3}{t^3} + \dfrac{A_4}{t^4}} \quad \text{N-s/m}^2 \, 10^{-6} \, (t \text{ in K})$$

Gas	A_0	$A_1 \cdot 10^{-2}$	$A_2 \cdot 10^{-4}$	$A_3 \cdot 10^{-6}$	$A_4 \cdot 10^{-8}$
Air	0.552795	2.810892	−13.508340	39.353086	−41.419387
O_2	0.476823	2.785625	−15.410077	51.909942	−63.049470
N_2	0.579561	2.847486	−13.232490	37.106107	−37.549675
CO_2	0.624736	1.402570	0.553161	0.0	0.0
H_2	1.074552	6.569139	−19.639020	22.851072	0.0
CH_4	0.867788	3.783913	−14.473411	42.547274	−44.474225

Reproduced from Reference 12.

According to Reference 12, the values of μ for air calculated by equation 1.30, in conjunction with Table 1.5, have the following accuracies: from $t = 270$ to 600 K within 0.25 percent; from $t = 610$ to 1600 K within 1 percent; and from $t = 1610$ to 2200 K within 2 percent.

Although equation 1.30 and Table 1.5 are based on measurements of viscosity μ at or below atmospheric pressure, the calculated results for air are accurate to within 1 percent at pressures up to $1.2 \cdot 10^6$ N/m^2 at 300 K and $2.2 \cdot 10^6$ N/m^2 at 500 K. Further increases in temperature reduce the effect of the air pressure on its viscosity.

Figure 1.13 presents the viscosities of some common gases as functions of the gas temperature; μ in N-s/m$^2 \cdot 10^{-6}$ is plotted as a function of t.

1-7 THE SIMPLE THERMODYNAMIC SYSTEM

The *simple thermodynamic system* is one for which the equilibrium state is determined by specifying the values of any two independent thermodynamic properties. The inherent characteristics of the simple thermodynamic system are discussed in this section.

1-7(a) Properties of the Simple Thermodynamic System

In thermodynamics attention is focused on either a restricted region of space or a finite portion of matter, which is assumed to be separated from all other space or matter by an imaginary bounding surface called a *control surface*. No consideration is given to the molecular structure of the matter so enclosed. The region or matter

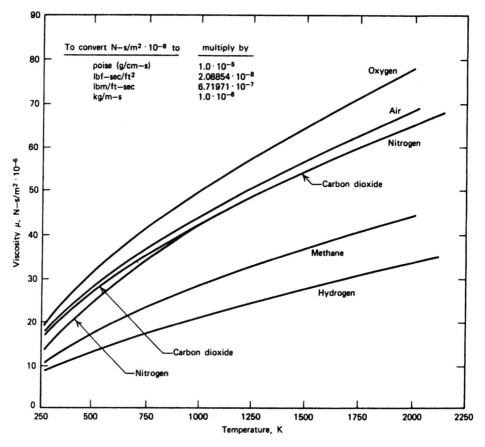

Figure 1.13 Viscosities of some common gases as functions of temperature (based on Reference 12).

surrounded by the control surface is termed the *system*, and everything external to it is called either the *surroundings* or the *environment*. In general, the behavior of the system and its interactions with the environment are described in terms of readily observable or measurable characteristics called either the *properties* or the *coordinates* of the system. A property is, in general, a quantity having a value that depends only on the state of a system. Since thermodynamics is concerned only with the macroscopic behavior of a system when it passes from one equilibrium state to another, the properties selected are those giving a suitable macroscopic description of the changes produced in the characteristics of the system after it has passed from one equilibrium state to another.

The *equilibrium* state of a system may be described by specifying the magnitudes of its corresponding *independent properties* or their equivalents. Hence, the state of a system is expressed in the form of functional relationships between each of its dependent properties and the pertinent independent properties. In the case of a *simple system*, one that is free of the effects of relativity, nuclear, capillarity, electric, magnetic, or gravitational effects and shearing stresses, the state is fully specified by any two independent properties. In general, a thermodynamic system is assumed to be a simple system (see Section 1–15). Thus, if z is a dependent property of a simple system defined by the independent properties x and y, then

$$z = f(x, y) \tag{1.31}$$

The functional relationship between properties of the type symbolized by equation 1.31 is termed an *equation of state*, and its actual form is determined by experiment. An equation of state presents the equilibrium conditions connecting the dependent property z and the independent properties x and y. In general, the thermodynamic properties x and y are scalars, but they may be either *intensive*, independent of the mass involved, or *extensive*, dependent on the mass involved.

If the values of the properties x and y are changed by the small amounts dx and dy, then the dependent property z will change by the small amount dz, where the operator d denotes an *exact differential*. An expression for dz is obtained by the partial differentation of equation 1.31. Thus

$$dz = \left(\frac{\partial f}{\partial x}\right)_y dx + \left(\frac{\partial f}{\partial y}\right)_x dy = M\,dx + N\,dy \qquad (1.32)$$

If equation 1.32 has a solution, it must be possible to derive the total differential dz by differentiating some function of the independent variables (properties) x and y. In those cases where the aforementioned procedure is possible, the quantity dz is called an *exact* or *perfect differential*. In general, if the functions M and N are selected arbitrarily, dz will not be an exact differential. Hereafter, when a differential, such as dz, is *inexact*, it will be written as δz, signifying that the differential is not the result of differentiating an actual function of the independent variables.

If a functional relationship $z = f(x,y)$ exists for a simple system, and if the function and its partial derivatives ($M = \partial f/\partial x$ and $N = \partial f/\partial y$) are continuous throughout the region of the xy plane under consideration, then, for the linear differential expression $M\,dx + N\,dy$ to be exact, it must satisfy the *reciprocal relationship*

$$\frac{\partial M}{\partial y} = \frac{\partial N}{\partial x} \qquad (1.33)$$

It is shown in the calculus that, if equation 1.33 is satisfied, then

$$\int_C (M\,dx + N\,dy) = \int_{a,b}^{x,y} (M\,dx + N\,dy) = f(x,y) - f(a,b) \qquad (1.34)$$

where (a,b) and (x,y) are the initial and final points for the integration along curve C. Equation 1.34 shows that for an exact differential expression the value of the *line integral* $\int_C (M\,dx + N\,dy)$ depends only on the initial and final coordinates of the path C. Hence, *the line integral of an exact differential expression is independent of the paths (curves) connecting the end points* of the integration.

If the line integral of a differential expression has the aforementioned property, then the function from which the differential expression is formed is called either a *state* or a *point function*. A corollary to the foregoing is that *for a point function the line integral has the same value for all curves (paths) connecting the same end points.*

Consider now the special case where the initial and final points of the path for the line integral of an exact differential expression have identical coordinates. The path is then a *closed curve*, such as a complete thermodynamic cycle. In that case,

$$\oint_C (M\,dx + N\,dy) = 0 \qquad (1.35)$$

Hence, *the line integral of an exact differential expression taken over a closed curve (complete cycle) is zero*; the only limitation is that there are no singular points, such as sources or sinks, within the region bounded by the closed curve and that the partial derivatives $\partial M/\partial y$ and $\partial N/\partial x$ are continuous.

The properties of the exact linear differential expression may be summarized as follows.

1. Its line integral between two end points is independent of the curve (path) connecting them.
2. Its line integral around a closed curve (complete cycle) vanishes.
3. Its second partial derivatives satisfy the reciprocal relationship $\partial M/\partial y = \partial N/\partial x$.
4. If the line integral of a linear differential expression taken over an arbitrarily selected closed curve vanishes, then the expression is an exact differential.
5. If a linear differential expression satisfies any one of the conditions 1, 2, 3, or 4 above, it satisfies all of them.

In thermodynamics, a function that satisfies the above requirements is said to be a *property* of the system.

1-7(b) Change of State for a Simple Thermodynamic System

The equilibrium state of a simple thermodynamic system is completely defined by any two independent thermodynamic properties. Hence, each property may be taken to be a function of any two independent properties. Thus a thermodynamic property satisfies the requirements for a point function, as discussed in Section 1–7(a).

Let the properties selected for describing the state of a simple thermodynamic system be its absolute pressure p, its specific volume v, and its absolute temperature t; actually any three properties may be employed for defining its equilibrium state. Accordingly, its equation of state, expressed in terms of p, v and t, is given by any one of the following functional relationships. Thus,

$$p = p(v,t), \qquad v = v(p,t), \qquad t = t(p,v) \tag{1.36}$$

If there is a change from one equilibrium state to another, then

$$dp = \left(\frac{\partial p}{\partial v}\right)_t dv + \left(\frac{\partial p}{\partial t}\right)_v dt \tag{1.37a}$$

$$dv = \left(\frac{\partial v}{\partial p}\right)_t dp + \left(\frac{\partial v}{\partial t}\right)_p dt \tag{1.37b}$$

$$dt = \left(\frac{\partial t}{\partial p}\right)_v dp + \left(\frac{\partial t}{\partial v}\right)_p dv \tag{1.37c}$$

For the total differentials dp, dv, and dt to be exact, the reciprocal relationship (equation 1.33) must be satisfied, and the corresponding line integral for the pertinent differential expression taken around a closed curve must vanish.

As an example, consider the equation of state

$$v = v(p,t) = \text{constant} \tag{1.38}$$

In that case, $dv = 0$, and equation 1.37(b) becomes

$$dv = \left(\frac{\partial v}{\partial p}\right)_t dp_v + \left(\frac{\partial v}{\partial t}\right)_p dt_v = 0 \tag{1.39}$$

where the subscripts v attached to dp and dt denote that v remains constant during the indicated changes. Rearrangement of equation 1.39 yields

$$\left(\frac{\partial p}{\partial t}\right)_v = -\frac{(\partial v/\partial t)_p}{(\partial v/\partial p)_t} \tag{1.40}$$

or

$$\left(\frac{\partial p}{\partial t}\right)_v \left(\frac{\partial v}{\partial p}\right)_t \left(\frac{\partial t}{\partial v}\right)_p = -1 \qquad (1.41)$$

The partial derivatives $(\partial p/\partial t)_v$ and $(\partial v/\partial t)_p$ are related to the *thermometric coefficients* of the system. Thus,

$$\beta_v = \frac{1}{p}\left(\frac{\partial p}{\partial t}\right)_v = \text{coefficient of pressure change at constant volume} \qquad (1.42)$$

and

$$\beta_p = \frac{1}{v}\left(\frac{\partial v}{\partial t}\right)_p = \begin{array}{l}\text{coefficient of volume change at constant} \\ \text{pressure } (\textit{dilatation coefficient})\end{array} \qquad (1.43)$$

In general, the coefficients β_v and β_p are functions of the temperature. If the changes in temperature are small, it may be assumed that β_v and β_p are constants.

1-8 REVERSIBLE AND IRREVERSIBLE PROCESSES

A *reversible process* is defined as one that is in a state of equilibrium at all points in its path so that any infinitesimal change in the *influence* producing the change will cause the process to proceed in the direction of that change. When heat is added to a system, the influence causing the system to change its state is the temperature difference applied to the equilibrium state. Consequently, if an arbitrarily small temperature difference produces either an expansion or a compression, depending on its sign, and the corresponding absolute values of the mechanical effects produced for the same quantities of heat transferred from or to the system are equal to each other, then the process is *reversible*.

It is apparent from this that a process involving friction cannot be reversible because a finite amount of work equal to the work required for overcoming the friction must always be supplied before the direction of the process can be reversed. It is generally assumed that the work expended in overcoming friction is instantly converted into heat. Since friction invariably accompanies the flow of real fluids, *real flow processes are always irreversible*. As a matter of fact, all natural or spontaneous processes are irreversible. They tend to proceed in a given direction toward equilibrium and never reverse themselves unless work is done, at the expense of some other system, to reverse them.

Consider the following examples of natural processes. First, consider a closed volume containing an inert gas that has different flow velocities in different parts of the volume. Because of the irreversible process of *viscous momentum flux*, the variations in the velocity are eventually reduced to zero over the entire volume. Second, consider a closed region containing a material having variations in temperature; these variations are finally eliminated by the irreversible process of *heat conduction*. Third, consider a closed volume containing two different species of inert gases nonuniformly distributed so that there are variations in the concentrations of the species; these variations are eliminated in time by the irreversible process of *mass diffusion*. Irreversible processes, such as the three discussed here, are characterized by involving the flux of some property and thus require a finite time interval to establish the state of equilibrium. Reversible processes are characterized by the absence of fluxes and thus require infinite time for their completion; that is, they are infinitely slow processes.

It will be recalled from elementary thermodynamics that the definition of temperature is based on the concept of thermal equilibrium. Thus, *bodies are at the same temperature when they are in thermal equilibrium with each other*. Since the temperature t is one of the coordinates of the simple system, the above definition of temperature restricts the thermodynamic study of the system to its equilibrium states. Moreover, the processes causing the system to change its state must take place at an infinitely slow rate so that in all of its intermediate states the system may be assumed to be in equilibrium. Actual processes do not satisfy the above requirement, even if they could be frictionless, because they complete themselves in a finite time. In general, the more rapid an actual process, the greater is its degree of irreversibility and the poorer is its efficiency from a thermodynamic standpoint.

The reversible processes of thermodynamics are, therefore, imaginary frictionless processes that take infinite time to complete themselves. Such processes are, however, the most efficient for accomplishing changes of state. Although they depart considerably from real processes, they are of great value because they furnish a basis for judging the efficiency of actual processes. Furthermore, the results obtained by assuming reversible processes, which processes can be dealt with mathematically, can be made applicable to real processes by introducing modifications based on experience.

1–9 WORK

In thermodynamics, *work* is defined as *energy in transit across the boundaries of a system where the sole effect external to the system could have been the raising or lowering of a weight.*

Classical thermodynamics is based primarily on the concept of a *closed system*, which signifies that mass can neither enter nor leave the system. The latter is an identifiable collection of matter that can change both its shape and volume. Figure 1.14 illustrates a simple closed system enclosed in an impermeable control surface A. Assume that there is no friction, and let p denote the uniform hydrostatic pressure exerted by the system on A. Let it be further assumed that p is larger than p_0, the

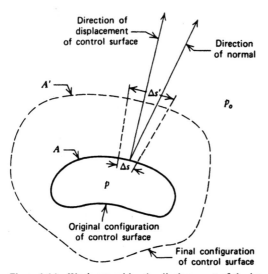

Figure 1.14   Work caused by the displacement of the boundaries of a system.

hydrostatic pressure exerted by the environment on A, by the differential amount dp. Because of the pressure difference dp, the control surface A will be displaced and distorted to some new configuration A'. The work performed by the system as a whole in displacing from A to A' is called *moving boundary work*. Because there is no friction, the displacement of the system boundaries is a reversible process; that is, the work performed is reversible.

For the reversible displacement of the control surface A, assumed here, the static pressure p of the system may be plotted as a function of its volume $\mathscr{V}$, in the manner illustrated in Fig. 1.15. The result is the curve 12, which relates p and $\mathscr{V}$ for the change from the equilibrium state 1 to the equilibrium state 2.

Let W_R denote the work done by a system that undergoes a reversible change of state, from state 1 to state 2. Then

$$W_R = \int_1^2 p\, d\mathscr{V} \tag{1.44}$$

The subscript R attached to W denotes that the change of state is a reversible process. Hereafter, the subscript R will be omitted, and it should be understood that $W = \int p\, d\mathscr{V}$ applies only to a reversible change of state.

Equation 1.44 is the expression for the moving boundary work done by a closed system against its environment. The general convention is to regard *work done by the system* on its surroundings as *positive work*, and vice versa. The representation of work on the $p\mathscr{V}$ plane, as in Fig. 1.15, leads to the following rule for distinguishing positive work from negative work. *If the area $\int p\, d\mathscr{V}$ lies to the right of an observer moving in the direction of the process, the work is positive*, and vice versa.

In the case where the path is a closed curve, such as that illustrated in Fig. 1.16, the *net work* is the difference between the positive work and the negative work. It is given by the area enclosed by the curves (paths) forming the closed curve (cycle). If the expression $p\, d\mathscr{V}$ were an exact differential expression, its line integral around the closed curve of Fig. 1.16 would vanish instead of being equal to the enclosed area [see Section 1–7(a)]. Hence, the differential $\delta W = p\, d\mathscr{V}$ is inexact. The work depends on the process or path, that is, on the functional relationship between p and $\mathscr{V}$.

Equation 1.44 is applicable to either a system or its working substance, provided there is no friction. In the case where the friction is confined to the components of a machine, but the working fluid is frictionless, equation 1.44 may be applied only to the fluid.

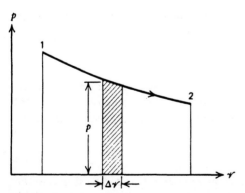

Figure 1.15 Determination of work for a process plotted on the $p\mathscr{V}$ plane.

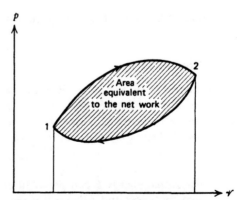

Figure 1.16 Work for a cyclic process.

1–10 HEAT

Experience has shown that energy may be transferred across the boundaries of a system in the absence of macroscopic changes in the system. For example, when a system at a given temperature is brought into contact with its surroundings at a different temperature, energy flows from the higher temperature region to the lower temperature region, unaccompanied by any macroscopic changes in the system. This energy transfer process is termed *heat*. Hence, *heat* is defined as *energy in transit across the boundaries of a system caused by a temperature difference between the system and its surroundings*.

By convention, the heat transferred to a system is defined as *positive heat*, and that transferred from a system is defined as *negative heat*. A process that permits heat transfer is termed a *diabatic process*, while one that does not permit heat transfer is termed an *adiabatic process*.

Heat, like work, depends on the process or path. Consequently, heat is not a property. Its differential is written as δQ to indicate that the differential is inexact.

1–11 THE FIRST LAW OF THERMODYNAMICS

The first law of thermodynamics is based entirely on experience, and its validity is based on the fact that no phenomenon has been uncovered that contradicts it. Its simplest statement is that heat and work are mutually interconvertible and that the ratio of conversion is a fixed quantity termed the mechanical equivalent of heat, denoted by J (note that the symbol J is also used to denote the unit of energy the joule). The first law gives no information regarding the amount of heat transferred to a system that can be converted into work (or vice versa). It merely states that when a system converts heat into work, the ratio of the amount of heat so converted to the amount of work produced (and vice versa) is a constant denoted by J.

1–11(a) Mathematical Statement of the First Law of Thermodynamics

Consider a closed system of mass m that is subjected to a series of processes involving the transfer of heat and the performance of work. Assume that the series of processes form a complete cycle. Let δQ denote the heat transferred to or from the system and δW the corresponding amount of work performed during the cycle. Then, experience shows that

$$\oint (\delta Q - \delta W) = 0 \qquad (1.45)$$

Equation 1.45 is the basic mathematical statement of the first law of thermodynamics.

1–11(b) Stored Energy

It follows from Section 1–7(a) that because the line integral of the differential expression in equation 1.45 vanishes for a closed curve, the quantity $(\delta Q - \delta W)$ defines a property of a closed system. Let E, termed the *stored energy*, denote the property defined by equation 1.45. Hence,

$$dE = \delta Q - \delta W \qquad (1.46)$$

The stored energy E comprises all of the forms of energy that can be stored in the system. Both δQ and δW are not storable; they are energies in transit. In general, E may include the following kinds of stored energy: thermal energy (internal energy

U); kinetic energy ($mV^2/2$); potential energy caused by the effect of gravity (mgz); electrostatic, chemical, nuclear, and other forms of energy. For most of the situations encountered in fluid mechanics, one need consider only the first three of the aforementioned forms of energy. Hence, let

$$E = U + m\frac{V^2}{2} + mgz \qquad (1.47)$$

On a unit mass basis, the stored energy E is denoted by e, where

$$e = \frac{E}{m} = u + \frac{V^2}{2} + gz$$

where u is the internal energy per unit mass. On a unit mass basis, equation 1.46 becomes

$$de = \delta Q - \delta W \qquad (1.48)$$

where δQ and δW are also determined on a unit mass basis.

Equations 1.46 and 1.48 apply to both reversible and irreversible processes. In an irreversible process the energy δE_f is expended in overcoming dissipative effects such as friction, turbulence, etc. Such forms of energy are completely converted into the random molecular motions of the molecules. Moreover, the mechanical work δW will be smaller than $p\,d\mathcal{V}$ (see equation 1.44). Hence, when energy dissipating effects are involved, equations 1.46 and 1.48 become

$$dE = \delta Q - \delta E_f - p\,d\mathcal{V} \qquad \text{and} \qquad de = \delta Q - \delta e_f - p\,dv \qquad (1.49)$$

1-11(c) Internal Energy

Energy can be stored in the atoms and molecules of a system because of their motions. That form of energy is termed *internal energy* U (it is also called *thermal energy*), and comprises the kinetic energies of translation, vibration, and rotation of the particles, and the electronic energy caused by the state of the particle electron cloud. Consider a closed system of mass m having no kinetic and potential energies. Let δQ denote the heat transferred to the system and δW the corresponding work performed during a change of state. Then, from equation 1.49, the resulting change in the internal energy of the system is given by

$$dU = \delta Q - \delta W = \delta Q - \delta U_f - p\,d\mathcal{V} \qquad \text{and} \qquad du = \delta Q - \delta u_f - p\,dv \quad (1.50)$$

where δU_f denotes the increase in the internal energy caused by the dissipative effects of friction.

1-11(d) Flow Work

Figure 1.17 illustrates a system of fixed mass m enclosed by a fixed boundary A. Adjacent to A is a small mass δm. Assume that the mass δm is forced across A and joins the system mass m. Let the pressure of the surroundings, denoted by p, force δm into the system. The work performed by the surroundings on the combined system, comprising m and δm, by moving δm the distance dx into the original system, is called the flow work δW_{flow}. It is given by

$$\delta W_{\text{flow}} = F\,dx = (pdA)\,dx = p\,d\mathcal{V} = pv\,\delta m \qquad (1.51)$$

where $v = d\mathcal{V}/\delta m$. Physically, $p\,d\mathcal{V}$ is the work required to compress the volume $d\mathcal{V}$ into a region where the pressure is p. The value of $p\,d\mathcal{V}$ is negative when mass is forced

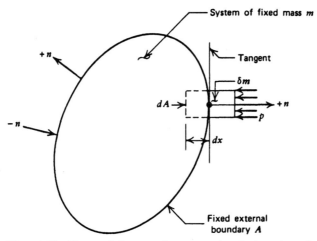

Figure 1.17   Flow work because of mass crossing the boundary of a system.

into the system (see Fig. 1.17), and positive when mass is forced out of the system. On a unit mass basis, equation 1.51 becomes

$$\delta W_{flow} = pv \tag{1.52}$$

1–11(e) Enthalpy

By definition, the *enthaply* per unit of mass of a substance, denoted by h, is

$$h \equiv u + pv \tag{1.53}$$

Like the specific internal energy u, both h and pv are thermodynamic properties per unit mass. Differentiating equation 1.53 yields

$$dh = du + (p\,dv + v\,dp) \tag{1.54}$$

From equation 1.50, a change in the specific internal energy is given by

$$du = \delta Q - \delta u_f - p\,dv \tag{1.50}$$

Combining equations 1.50 and 1.54 yields

$$dh = \delta Q - \delta u_f + v\,dp \tag{1.55}$$

For a *reversible process*, $\delta u_f = 0$, and equation 1.55 reduces to

$$\delta Q_R = dh - v\,dp \tag{1.56}$$

where the subscript R denotes that Q is transferred in a reversible manner. If the process is adiabatic and reversible, that is, *isentropic*, then

$$dh - v\,dp = dh - \frac{1}{\rho}\,dp = 0 \tag{1.57}$$

1–11(f) Specific Heats

In general, the *specific heat* of a substance, denoted by c, is defined by

$$c \equiv \frac{\delta Q}{dt} \tag{1.58}$$

Since δQ is an inexact differential, the value of c depends on how the heat transfer process is conducted. Of particular interest are the specific heats c_v and c_p, corresponding to constant-volume and constant-pressure heating, respectively. It can be readily shown that[13]

$$c_v = \left(\frac{\partial u}{\partial t}\right)_v \tag{1.59}$$

$$c_p = \left(\frac{\partial h}{\partial t}\right)_p \tag{1.60}$$

The name specific heat implies a heating process to attain the change in the thermodynamic state. The specific heats c_v and c_p defined by equations 1.59 and 1.60 are partial dervatives of thermodynamic properties, and thus are themselves thermodynamic properties, which are independent of the actual process taking place. The term specific heat is thus a misnomer when applied to such partial derivatives. Because general usuage has firmly established the name specific heat for c_v and c_p, that terminology will be adhered to in this book.

1–12 THE SECOND LAW OF THERMODYNAMICS AND ENTROPY

Even in a reversible process it is not possible to transform all of the heat added to a system into useful work. Experience has demonstrated that the transformation of heat into work is inevitably accompanied by a degradation of a portion of the energy supplied into a form that is less useful. For example, the thermal energy associated with a gas cannot be converted entirely into mechanical work by permitting it to expand; this is because the thermal energy associated with the random movements of the gas molecules is not subject to complete control. As already mentioned, the First Law is not concerned with the quantity of heat that can be converted into work and gives no information regarding that matter. The First Law is concerned only with the portion of the heat that is converted into work. It is the Second Law that is concerned with the quantity of heat that can be converted into useful work. Like the First Law, the Second Law is based entirely on experience.

According to the Second Law, every natural system, if left undisturbed, will change spontaneously and approach a state of equilibrium or rest, and the process is irreversible. As the change or process moves the system toward the equilibrium state, the system loses capability for spontaneous change. The property associated with the capability of a system for spontaneous change, is called its *entropy s*.

For a closed system that undergoes a cyclic change where the heat per unit mass δQ_R is received reversibly at the absolute temperature t, we may write[13]

$$\oint \frac{\delta Q_R}{t} = 0 \tag{1.61}$$

Equation 1.61 is of the same form as equation 1.35, which defines a property of a simple system. Hence, $\delta Q_R/t$ defines a thermodynamic property of the system, termed the *entropy* by Clausius (1850). If s denotes the entropy per unit mass for a system then, *by definition*,

$$ds \equiv \frac{\delta Q_R}{t} \tag{1.62}$$

Equation 1.61 shows that the change in entropy between two given states is the same for all processes connecting them. Hence, the entropy change for an irreversible process can be measured by the value of $\int \delta Q/t$ for any arbitrarily selected reversible processes connecting the same initial and final states.

For a closed system that experiences a reversible change ($\delta u_f = 0$) from one equilibrium state to another, equation 1.50 reduces to

$$\delta Q_R = du + p\,dv \tag{1.63}$$

Substituting for δQ_R from equation 1.62 into equation 1.63, we obtain

$$t\,ds = du + p\,dv \tag{1.64}$$

Substituting for du from equation 1.54 into equation 1.64 gives

$$t\,ds = dh - v\,dp = dh - \frac{1}{\rho}\,dp \tag{1.65}$$

Although equations 1.64 and 1.65 are derived for reversible processes, they are also valid for irreversible processes because they involve only thermodynamic properties; that is, they involve only exact differentials having integrated values that are independent of the process.

Combining equation 1.50, expressed on a unit mass basis, with equation 1.64 gives the following equation for both reversible and irreversible processes.

$$ds = \frac{du + p\,dv}{t} = \frac{\delta Q - \delta u_f}{t} \tag{1.66}$$

Thus, the entropy change during a process change from state 1 to state 2 is given by

$$s_2 - s_1 = \int_1^2 \frac{du + p\,dv}{t} \tag{1.67}$$

where the *integration* between state 1 and state 2 is performed along any arbitrarily selected reversible process or processes connecting the states 1 and 2.

It follows from equation 1.66 that for a reversible process, $\delta u_f = 0$ and $\delta Q = t\,ds$, whereas for an irreversible process, $\delta u_f > 0$ and $\delta Q < t\,ds$. In the case where $\delta Q = 0$, the process is said to be *adiabatic*. When $\delta u_f = 0$, the process is *reversible*. In the special case where $\delta Q = \delta u_f = 0$, the process is said to be *isentropic*, a *reversible adiabatic* process, and $ds = 0$. Consequently,

$$ds \geqq \frac{\delta Q}{t} \tag{1.68}$$

where the equality applies to reversible processes and the inequality applies to irreversible (natural) processes.

Example 1.4. Air is compressed from the initial state where $t_1 = 288.9$ K and $p_1 = 1.38 \cdot 10^5$ N/m^2 to a new state where $t_2 = 566.7$ K and $p_2 = 4.14 \cdot 10^5$ N/m^2. Calculate the change in the entropy of the gas for the following cases: (a) a reversible adiabatic compression from p_1 to p_2 followed by a reversible isobaric heating to the final temperature t_2, and (b) an irreversible process between the same states where the work expended in overcoming friction is equivalent to 232,600 J/kg and the heat caused by friction remains in the gas. These two processes are illustrated in Fig. 1.18. Assume $\gamma = 1.40$.

Solution

(a) If the air is assumed to be a perfect gas (see Section 1–15) then, for an isentropic compression (see Fig. 1.18),

$$t_3' = t_1 \left(\frac{p_3}{p_1}\right)^{(\gamma-1)/\gamma} = t_1 \left(\frac{p_2}{p_1}\right)^{(\gamma-1)/\gamma} = 288.9 \left(\frac{4.14}{1.38}\right)^{0.28571} = 395.4 \text{ K}$$

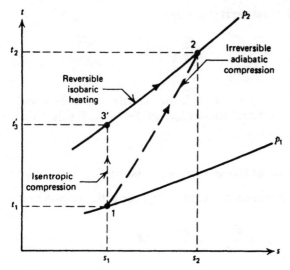

Figure 1.18   Sketch for Example 1.4.

From equation 1.65, for an isobaric process ($dp = 0$)

$$ds = \frac{dh}{t} \tag{a}$$

From Section 1–15(d), for a perfect gas, $dh = c_p \, dt$, where $c_p = 1004$ J/kg-K for air. Thus,

$$s_2 - s_1 = s_2 - s_3' = \int_{3'}^{2} \frac{c_p \, dt}{t} = c_p \ln \left(\frac{t_2}{t_3'} \right)$$

$$s_2 - s_1 = 1004 \ln \left(\frac{566.7}{395.4} \right) = 361.4 \text{ J/kg-K} \ (0.0864 \text{ Btu/lbm-R})$$

(b)   For the irreversible process the entropy change ds is not given by $\delta Q/t$. The reversible process in part a, however, connects the same state points as the irreversible process. Since entropy is a property, the reversible process and the irreversible process have the same entropy change.

1–13 NEWTON'S LAWS OF MOTION

Before presenting Newton's three laws of motion, the reference frame to which they are to be referred will be specified. The laws have their simplest and customary form when they are referred to either a *primary inertial coordinate system*, which comprises a set of rigid axes fixed relative to the average position of fixed stars, or to a *secondary inertial coordinate system*, which comprises a set of axes that moves with a constant velocity and no rotation relative to the primary inertial system (see Reference 14).

If the motion of a system of particles is referred to either inertial system, Newton's laws can be written as follows.

1. *Every particle persists in its state of rest or uniform motion in a straight line except insofar as it may be compelled by force to change that state.*
2. *The rate of change of momentum (quantity of motion) is proportional to the force applied to the body, and it takes place in the direction of the line of action of the force.*
3. *To every action there is an equal and opposite reaction.*

According to Einstein, the mass m is given by

$$m = \frac{m_o}{\left[1 - \left(\frac{V}{c}\right)^2\right]^{1/2}} \tag{1.69}$$

where m_o is the mass of the body at rest, V is the velocity of the body, and c is the speed of light. For all fluid mechanic applications in this book, the fluid mass m will be assumed to be equal to its rest mass m_o.

1-13(a) Mathematical Expressions of Newton's Laws of Motion

Let $\mathbf{M} = m\mathbf{V}$ denote the momentum of a particle or a body. By Newton's second law of motion,

$$\mathbf{F} = \frac{d(m\mathbf{V})}{dt} = \frac{d\mathbf{M}}{dt} = \dot{\mathbf{M}} \tag{1.70}$$

Equation 1.70 shows that the external force vector $\mathbf{F}$ is equal to the derivative with respect to time of the momentum vector $\mathbf{M}$. From equation 1.70, it follows that the component of $\mathbf{F}$ acting in an arbitrary direction, for example, the x direction, denoted by X, is

$$X = \frac{dM_x}{dt} = m\frac{dV_x}{dt} = ma_x \tag{1.71}$$

where a_x denotes the acceleration of the mass m in the x direction.

Newton's third law, customarily termed the *reaction principle*, implies that if body A exerts the force $\mathbf{F}_{AB}$ on body B, then body B exerts the equal and opposite force $\mathbf{F}_{BA}$ on body A. Hence,

$$\mathbf{F}_{AB} + \mathbf{F}_{BA} = 0 \tag{1.72}$$

1-13(b) Momentum of a Particle

Consider a particle of mass m, and let x, y, and z be its Cartesian coordinates. At a given time t let $\mathbf{V}$ denote its velocity vector. The components of $\mathbf{V}$ referred to the three coordinate axes will be denoted by u, v, and w, respectively. If X, Y, and Z denote the components of the external force $\mathbf{F}$ acting on the particle, then

$$X = \frac{d}{dt}(mu) = \frac{d}{dt}(m\dot{x}) \tag{1.73}$$

$$Y = \frac{d}{dt}(mv) = \frac{d}{dt}(m\dot{y}) \tag{1.74}$$

$$Z = \frac{d}{dt}(mw) = \frac{d}{dt}(m\dot{z}) \tag{1.75}$$

where $\dot{x} = dx/dt$, $\dot{y} = dy/dt$, and $\dot{z} = dz/dt$. If the mass of the particle is constant, invariant with time, then

$$X = m\dot{u} = m\ddot{x} \tag{1.76a}$$

$$Y = m\dot{v} = m\ddot{y} \tag{1.76b}$$

$$Z = m\dot{w} = m\ddot{z} \tag{1.76c}$$

where $\ddot{x} = \dot{u} = d^2x/dt^2$, etc.

Let a denote the corresponding *acceleration vector*. Then

$$\mathbf{a} = \mathbf{i}\ddot{x} + \mathbf{j}\ddot{y} + \mathbf{k}\ddot{z} \tag{1.77}$$

Hence,

$$\mathbf{F} = m\mathbf{a} \tag{1.78}$$

Equation 1.78 shows that if the mass m is constant, $\mathbf{F} = m\mathbf{a}$.

1-13(c) Momentum of a System of Particles

Equation 1.70 is based on the dynamics of a single particle. In most engineering problems the interest is in a system comprising such a large number of particles that it may be assumed to be a continuum. In that case, in addition to the external forces acting on the particles, there are internal forces because of the actions and reactions between the particles. By Newton's Third Law, each pair of particles satisfies equation 1.72, so that the net effect of the internal forces in a body is zero. For an arbitrary system of particles having the masses $m_1, m_2, \ldots, m_n$ with the corresponding velocities $\mathbf{V}_1, \mathbf{V}_2, \ldots, \mathbf{V}_n$, the external force acting on the jth particle is $\mathbf{F}_j$ and its corresponding change in momentum is $m_j(d\mathbf{V}_j/dt)$. Hence, for the system of particles, the net external force, denoted by $\mathbf{F}$, acting on the assemblage of particles is

$$\mathbf{F} = \sum_{j=1}^{n} \mathbf{F}_j = \sum_{j=1}^{n} \frac{d}{dt}(m_j \mathbf{V}_j) \tag{1.79}$$

or

$$\mathbf{F} = \frac{d\mathbf{M}}{dt} = \frac{d}{dt}(m\mathbf{V}) = \dot{\mathbf{M}} \tag{1.80}$$

Equation 1.80 shows that if a mass of fluid composed of a large number of particles is contained within a control surface A at the instant of time t, the rate of change in the momentum for that mass of fluid is $d\mathbf{M}/dt$, which is equal to the net external force $\mathbf{F}$ acting on that mass of fluid. As noted earlier, the velocity and acceleration terms in equation 1.80 must be measured in an *inertial frame of reference*.

1-14 KINETIC THEORY AND THE PERFECT GAS

The macroscopic viewpoint of classical thermodynamics as applied to gases, when harmonized with the microscopic viewpoint of the kinetic theory of gases, gives an insight into some of the important characteristics of the perfect gas that cannot be obtained from thermodynamic considerations alone.

1-14(a) Basic Considerations of the Kinetic Theory

One of the aims of the kinetic theory of gases is the interpretation of the thermal equation of state for the *perfect gas* (see equation 1.99). According to that theory, which is based on the molecular structure of matter, a gas may be conceived as being composed of a very large number of molecules that move with extremely rapid random motions. For example, at standard conditions the number of molecules in a cubic centimeter of air is approximately $2.5 \cdot 10^{19}$. Although the molecular velocities are extremely large (e.g., the average velocity of the hydrogen molecule under standard conditions is 1840 m/s approximately), the motion of each molecule is greatly impeded by collisions with the tremendous number of neighboring molecules. Consequently, as pointed out in Section 1-4(e), the *molecular mean free path* λ is very small.

According to equation 1.11, λ is given by

$$\lambda = \frac{1}{\sqrt{2}\,(\pi N\,d^2)} \text{ (meters)} \tag{1.81}$$

where N is the number of molecules per cubic meter and d is the effective diameter of the molecule in meters.

Because of the large number of collisions with the neighboring molecules, each gas molecule follows an extremely torturous path. For that reason, the diffusion of a gas, the movement of gas molecules through neighboring gas molecules, is inherently a rather slow process.

A gas molecule will have different independent motions, called *degrees of freedom*, depending on its structure. In general, it may have translational motions and rotational motions, and the atoms of a polyatomic molecule may vibrate with respect to each other. A monatomic molecule has 3 degrees of translational freedom only; a rigid diatomic molecule has 5 degrees of freedom, 3 of translation and 2 of rotation about two axes at right angles to the long axis of the molecule. If the atoms of a diatomic molecule can vibrate with respect to each other, the degrees of freedom are increased to 7 because of the vibrational degrees of freedom. A nonlinear triatomic molecule will have at least 6 degrees of freedom, 3 for translation and 3 for rotation about three mutually perpendicular axes and, if the atoms vibrate, 3 for vibration, or 9 degrees of freedom in all.[7, 8]

For the perfect gas it is assumed that the only degrees of freedom are the 3 for translation (same as for a monatomic gas). It is further assumed that the molecules are perfectly elastic and that the only interactions between the molecules are the collisions; that is, there are no attractive forces.

The kinetic theory is based on applying the laws of Newtonian mechanics to the molecules, but because of the large number of molecules involved it becomes a mathematical necessity to apply statistical methods to the system with the consequence that the results obtained are the averaged values of the properties of the gas.

The effective diameter of a molecule d exerts a predominating influence on the *transport properties* of the gas. Physically speaking, if two molecules approach each other and the distance between their centers is smaller than d, then the molecules must collide. The collision is accompanied by changes in the original directions and speeds of the colliding molecules. If the molecules are deformable, then the effective diameter d will be smaller for molecules experiencing high-speed collisions than it is for those experiencing low-speed collisions. Consequently, the effective diameter of a molecule will be a function of the speed of the collision.

The *rms* velocity $\sqrt{\overline{C^2}}$ is different from the mean velocity $\bar{c}$. They are related by

$$\bar{c} = \left(\frac{8\overline{C^2}}{3\pi}\right)^{1/2} = 0.921\sqrt{\overline{C^2}} \tag{1.82}$$

In this section, results derived from the kinetic theory for such thermodynamic properties of a gas as temperature, pressure, internal energy, etc., are presented. In addition, the results for the transport properties viscosity, thermal conductivity, and diffusion of a perfect gas are discussed.

1–14(b) Pressure Intensity

Consider a mass of gas enclosed in a container having the volume $\mathscr{V}$. If N is the number of molecules in the volume $\mathscr{V}$, and m is the mass of each molecule, then the density of the gas ρ is given by

$$\rho = \frac{Nm}{\mathscr{V}} \tag{1.83}$$

According to the kinetic theory, the pressure intensity p acting on the walls of the container is due to the impacts of the gas molecules on those walls. Let $\overline{C^2}$ denote the mean of the square of the velocities of the particles. It can be shown that[7]

$$p\mathscr{V} = \frac{2}{3}\left(\frac{1}{2}Nm\overline{C^2}\right) = \frac{2}{3}(KE)_{tr} \tag{1.84}$$

Hence, in view of equation 1.83, the pressure intensity p is given by

$$p = \frac{2}{3}\left(\frac{1}{2}\rho\overline{C^2}\right) \tag{1.85}$$

Equation 1.85 shows that the pressure intensity p is directly proportional to the kinetic energy of translation per unit volume of the gas.

1-14(c) Temperature

According to the kinetic theory, the absolute temperature t of a perfect gas is proportional to the translational kinetic energy $(KE)_{tr}$ associated with the random motions of the gas particles, which is

$$(KE)_{tr} = \frac{1}{2}Nm\overline{C^2} \propto t \tag{1.86}$$

It follows from equation 1.86 that, for a given mass of gas, the velocity term $\sqrt{\overline{C^2}}$ increases with the temperature of the gas. It should be noted that $(KE)_{tr}$ constitutes the *internal energy* of the gas. Hence, if the temperature is fixed, then the internal energy of the gas is fixed.

It is shown in Section 1-15(a) that (see equation 1.99)

$$p\mathscr{V} = mRt \tag{1.99}$$

Combining equations 1.84 and 1.99 yields

$$mRt = \frac{2}{3}(KE)_{tr} \tag{1.87}$$

1-14(d) Avogadro's Number and Boltzmann's Constant

According to Avogadro, equal volumes of all gases, when held at the same temperature t and pressure p, contain the same number of molecules. The *mole*[3] is defined as the amount of a substance that contains as many elementary entities as there are carbon atoms in 0.012 kg of carbon 12. That number, denoted by N_A, is called Avogadro's number. It has been shown experimentally that $N_A = 6.02252 \cdot 10^{23}$ particles/mol.

The universal gas constant $\bar{R}$, defined in Section 1-15(a), and Avogadro's number N_A are both universal constants. Their ratio is likewise a universal constant, termed *Boltzmann's constant*, which is denoted by k. Hence

$$k \equiv \frac{\bar{R}}{N_A} = 1.38054 \cdot 10^{-23} \text{ J/K} = \text{Boltzmann's constant} \tag{1.88}$$

Boltzmann's constant k is the gas constant for a single molecule of a gas. If z denotes the *molecular density*, the number of molecules per unit volume, then equation 1.99 becomes

$$p = zkt \tag{1.89}$$

1–14(e) Specific Heats

According to the kinetic theory, the heat δQ required to raise the temperature of a gas by an amount dt is stored in the gas in the form of the kinetic energy of its molecules. For a *monatomic gas* the kinetic energy is obtained from equation 1.87; thus, on a molar basis,

$$(\overline{KE})_{tr} = \frac{3}{2} \bar{R}t \tag{1.90}$$

Let $\bar{c}_v = \bar{m}c_v = $ the *molar heat capacity* of the gas at constant volume. Since $\bar{c}_v$ is the increase in the internal kinetic energy per mole of gas per unit change in temperature, it follows from equations 1.58 and 1.90 that

$$\bar{m} \, \delta Q = \bar{m}c_v \, dt = c_v \, dt = \frac{3}{2} \bar{R} \, dt \tag{1.91}$$

and

$$\bar{c}_v = \frac{3}{2} \bar{R} = \frac{3}{2}(8314.3) = 12{,}471 \text{ J/kmol-K} \tag{1.92}$$

Equation 1.92 indicates that for a perfect gas the specific heat is independent of p and t, and gives values of $\bar{c}_v$ in good agreement with the measurements for monatomic gases. Polyatomic gases have values of $\bar{c}_v$ larger than those given by equation 1.92. The difference is attributed to the neglect of the energy required for supporting the rotational and vibrational motions of the polyatomic molecule.

According to Maxwell's *equipartition law*,[9, 10] the kinetic energy of a gas molecule is equally divided among all of its degrees of freedom f. Hence, since the monatomic molecule has 3 degrees of freedom (translation in 3 directions), the energy per mole of gas is $\bar{c}_v t = 3/2 \ \bar{R}t$. Consequently, for a rigid diatomic molecule with $f = 5$ (3 translational and 2 rotational degrees of freedom), $\bar{c}_v = 5/2\bar{R} = 20{,}785$ J/kmol-K, a value in much better agreement with measurements. For a rigid polyatomic molecule with $f = 6$ (3 translational and 3 rotational degrees of freedom), $\bar{c}_v = 6/2\bar{R}$. For non-rigid molecules, there are additional degrees of freedom because of the vibration of the atoms; the latter phenomenon becomes more and more important as the gas temperature increases.

1–14(f) Viscosity

It is pointed out in Section 1–4(e) that the kinetic theory relates the mean free path of the gas molecules to such molecular transport properties of the gas as its viscosity, heat conductivity and diffusion coefficient.

The viscosity of a gas arises from the transport of the momentum of the gas molecules. It can be shown that for a perfect gas the viscosity μ is given by[9]

$$\mu = \frac{1}{2} \rho \lambda \bar{c} \tag{1.93}$$

If the gas density ρ decreases, the mean free path λ increases in such a manner that the product $\rho\lambda$ tends to remain constant. The viscosity μ is, therefore, proportional to

$\bar{c}$, which depends entirely on the gas temperature t. Consequently, for a perfect gas, the viscosity μ depends only on t and increases with t.

1-14(g) Thermal Conductivity

The thermal conductivity in a gas arises from spatial differences in its temperature. The latter cause the molecules to transport the mean energy of their random and internal motions. According to the kinetic theory, the thermal conductivity of a gas, denoted by κ, is given by[9]

$$\kappa = \frac{5}{4} \rho \bar{c} \lambda c_v \cong \frac{5}{2} \mu c_v \qquad (1.94)$$

A parameter that is important in the study of viscous heat transfer problems is the *Prandtl number* Pr, defined by

$$\text{Pr} \equiv c_p \mu / \kappa \qquad (1.95)$$

Combining equations 1.93, 1.94, and 1.95 gives

$$\text{Pr} = \frac{2}{5} \frac{c_p}{c_v} = \frac{2}{5} \gamma \qquad (1.96)$$

Equation 1.96 applies strictly to a monatomic gas ($\gamma = 5/3$). A better approximation is the *Eucken relationship*, which is given by[9]

$$\text{Pr} = \frac{4\gamma}{9\gamma - 5} \qquad (1.97)$$

Equation 1.97 gives the same result as equation 1.96 for a monatomic gas and is a good approximation for other gases at ordinary temperatures.

1-14(h) Diffusion

Diffusion in a gas mixture arises from gradients in the concentrations of the individual species comprising the mixture. According to the kinetic theory, the coefficient of self-diffusion of a gas, denoted by D, is given by

$$D = \beta_D \bar{c} \lambda \qquad (1.98)$$

where β_D is a parameter that depends on the intermolecular model. Equation 1.98 may also be employed for the diffusion of gases that are nearly identical. For nonsimilar gases, a much more complicated relationship applies.

1-15 THERMODYNAMIC PROPERTIES OF THE PERFECT GAS

The thermodynamic state of a perfect gas is described in terms of the following properties; p, v, t, u, h, and s.

1-15(a) Thermal Equation of State

It has been demonstrated experimentally that for a homogeneous system, composed of a single chemical species of gas of molecular weight $\bar{m}$, the pressure p, specific volume v, and temperature t are related by[13]

$$R \equiv \underset{p \to 0}{\text{Limit}} \left(\frac{pv}{t} \right) = \textit{the gas constant}$$

The gas constant R depends only on the kind of gas involved. For real gases, R varies with the gas pressure and temperature because of the volume occupied by the gas molecules and the presence of intermolecular forces. A gas for which R is a constant, or very nearly so, is termed a *thermally perfect gas*. For such a gas

$$pv = Rt, \qquad p = \rho Rt, \qquad p\mathscr{V} = mRt \qquad (1.99)$$

Equation 1.99 is known as the *thermal equation of state* for a perfect gas.

The dimensionless ratio pv/Rt is termed the *compressibility factor*, which is denoted by Z. Hence,

$$Z \equiv \frac{pv}{Rt} = \frac{p}{\rho Rt}$$

The compressibility factor Z is, in general, a function of two independent properties of a simple thermodynamic system. Usually Z is plotted as a function of p/p_{cr} for different constant values of t/t_{cr}, where p_{cr} denotes the *critical pressure* and t_{cr} is the *critical temperature*.* For values of $t/t_{cr} > 2$ or $p/p_{cr} < 0.05$ approximately, the curve of Z versus p/p_{cr} is practically a horizontal line having the value $Z = 1$. Consequently, for pressures that are low relative to p_{cr} and temperatures that are high compared with t_{cr}, it may be assumed that $Z \approx 1$; that is, the gas behaves in accordance with equation 1.99. Because a gas is thermally perfect does not imply that its specific heats c_p and c_v are constant [see Section 1–15(b)].

The mass unit known as the *mole*[3], denoted by mol, is defined as the amount of a substance that contains as many elementary entities as there are atoms in 0.012 kg of carbon-12. That number of elementary entities is Avogadro's number N_A, discussed in Section 1–14(d). The *molecular weight* $\bar{m}$ is the measure of the mass of a mole of a substance. Thus, the units of molecular weight are (mass per mole), for example, kg/kmol, g/mol, etc. If the number of moles of a substance in a volume $\mathscr{V}$ is denoted by N, and the corresponding mass is denoted by m, then

$$m = \bar{m}N$$

It has been shown empirically that the gas constant R is related to a universal gas constant $\bar{R}$ and the molecular weight $\bar{m}$ of the gas as follows.

$$R = \frac{\bar{R}}{\bar{m}}$$

Thus, equation 1.99 may be transformed to read

$$pv = \left(\frac{\bar{R}}{\bar{m}}\right)t, \qquad p = \rho\left(\frac{\bar{R}}{\bar{m}}\right)t, \qquad p\mathscr{V} = m\left(\frac{\bar{R}}{\bar{m}}\right)t \qquad (1.100)$$

The values of $\bar{R}$ for different systems of units of measurement are presented on the inside cover of this book. The thermodynamic properties for several common gases are presented in Table C.3.

Substituting $m = \bar{m}N$ into equation 1.100 yields

$$p\mathscr{V} = N\bar{R}t \qquad (1.101)$$

Equation 1.101 expresses the thermal equation of state for a thermally perfect gas in terms of the number of moles N in the volume $\mathscr{V}$. Equation 1.101 is generally employed by chemists and physicists in preference to equation 1.99.

---

* t_{cr} is the temperature above which a gas cannot be liquefied no matter how large a pressure is applied; p_{cr} is the minimum pressure for liquefying a gas at the temperature t_{cr}.

1–15(b) Caloric Equation of State

The following functional relationship, known as the *caloric equation of state*, may be written for the internal energy of any gas. Thus

$$u = u(t,v)$$

Differentiating the above expression yields

$$du = \left(\frac{\partial u}{\partial t}\right)_v dt + \left(\frac{\partial u}{\partial v}\right)_t dv \qquad (1.102)$$

It can be shown that,[13] in the case of a thermally perfect gas,

$$\left(\frac{\partial u}{\partial v}\right)_t = 0$$

Hence, it follows from the above equations that the internal energy u of a thermally perfect gas depends only on its temperature, a conclusion in harmony with the kinetic theory of gases [see Section 1–14(c)]. Accordingly, for a thermally perfect gas,

$$u = u(t)$$

Combining equations 1.59 and 1.102, gives for a thermally perfect gas,

$$du = c_v \, dt \qquad (1.103)$$

Integrating equation 1.103, we obtain

$$u - u_o = \int_{t_o}^{t} c_v(t) \, dt \qquad (1.104)$$

where t_o is an arbitrary reference temperature where $u = u_o$. If c_v is constant, equation 1.104 becomes

$$u - u_o = c_v(t - t_o) \qquad (1.105)$$

If t_o and u_o are chosen as zero, equation 1.105 becomes

$$u = c_v t \qquad (1.106)$$

A gas having a constant value of c_v is termed a *calorically perfect gas*, and equation 1.106 is one form of the *caloric equation of state* for such a gas. In the remainder of this book, only a gas that is both *thermally* and *calorically perfect* will be termed a *perfect gas*, without qualification. Otherwise the gas will be called an *imperfect* gas.

1–15(c) Specific Heat Relationships

For a thermally perfect gas, the First Law yields for a reversible process,

$$\delta Q = c_v \, dt + p \, dv \qquad (1.107)$$

Differentiation of the thermal equation of state $pv = Rt$ gives

$$p \, dv + v \, dp = R \, dt \qquad (1.108)$$

Combining equation 1.107 and equation 1.108 yields

$$\frac{\delta Q}{dt} = c_v + R - v\frac{dp}{dt} \qquad (1.109)$$

For the case where δQ is added isobarically ($dp = 0$), we obtain

$$\left(\frac{\delta Q}{\delta t}\right)_p = c_p = c_v + R \tag{1.110}$$

Let $\gamma = c_p/c_v$ denote the *specific heat ratio* for the gas; then it follows from equation 1.110 that

$$c_v = R/(\gamma - 1) \tag{1.111}$$

and

$$c_p = \gamma R/(\gamma - 1) \tag{1.112}$$

Substituting $\gamma = c_p/c_v$ into equation 1.110 gives

$$\gamma = 1 + \frac{R}{c_v} = 1 + \frac{\bar{R}}{\bar{c}_v} = 1 + \frac{2}{f} = \frac{f + 2}{f} \tag{1.113}$$

where f is the number of degrees of freedom in the molecules, as discussed in Section 1–14(e). For a monatomic gas, $f = 3$, for a rigid diatomic gas, $f = 5$, and for a rigid polyatomic gas, $f = 6$. Thus, according to the kinetic theory,

1. For monatomic gases $\gamma = 5/3 = 1.67$.
2. For diatomic gases $\gamma = 7/5 = 1.40$.
3. For polyatomic gases $\gamma = 8/6 = 1.33$.

The above results are accurate for monatomic gases, less accurate for diatomic gases, and least accurate for polyatomic gases.

1–15(d) Enthalpy Change

The thermodynamic property *enthalpy h* is defined in Section 1–11(e). A differential change in enthalpy dh is given by equation 1.54. For a thermally perfect gas, since $du = c_v \, dt$ and $c_p = c_v + R$, equation 1.54 transforms to

$$dh = c_v \, dt + d(pv) = (c_v + R) \, dt = c_p \, dt \tag{1.114}$$

Equation 1.114, when integrated, yields

$$h - h_o = \int_{t_o}^{t} c_p(t) \, dt \tag{1.115}$$

For a calorically perfect gas, c_v is constant [see Section 1–15(b)]. Consequently, from equation 1.110, c_p is also constant, and equation 1.115 becomes

$$h - h_o = c_p(t - t_o) \tag{1.116}$$

where t_o is an arbitrary reference temperature where $h = h_o$. If t_o and h_o are chosen as zero, equation 1.116 becomes

$$h = c_p t \tag{1.117}$$

Equation 1.117 is a useful form of the *caloric equation of state* for a *perfect gas*.

Example 1.5.   The enthalpy of air, assumed to be a perfect gas, is measured above 300 K as the datum. If its value at 400 K is $h = 100.83$ kJ/kg-K, calculate the average value of c_p.

Solution

$$h_2 - h_1 = c_p(t_2 - t_1)$$

$$c_p = \frac{h_2 - h_1}{t_2 - t_1} = \frac{100.83}{400 - 300} = 1.0083 \text{ kJ/kg-K} \ (0.2408 \text{ Btu/lbm-R})$$

1–15(e) Entropy Change

The entropy change for a reversible process is given by equations 1.62, 1.64, and 1.65. For a thermally perfect gas, $pv = Rt$, $du = c_v\, dt$, and $dh \doteq c_p\, dt$. Substituting for du and p into equation 1.64, we obtain

$$ds = c_v \frac{dt}{t} + R \frac{dv}{v} \qquad (1.118)$$

Substituting for dh and v in equation 1.65 yields

$$ds = c_p \frac{dt}{t} - R \frac{dp}{p} \qquad (1.119)$$

Integration of equation 1.119 yields

$$s = \int c_p \frac{dt}{t} - R \ln p + \text{constant} \qquad (1.120)$$

For a perfect gas, $c_p = \text{constant}$, so that equation 1.120 integrates to

$$s = c_p \ln t - R \ln p + \text{constant} \qquad (1.121)$$

Equation 1.121 may be transformed to read

$$s = c_p \ln \left[\frac{t}{p^{R/c_p}} \right] + \text{constant} = c_p \ln \left[\frac{t}{p^{(\gamma - 1)/\gamma}} \right] + \text{constant} \qquad (1.122)$$

For an isentropic change of state, $ds = 0$ and, from equations 1.112 and 1.119, it follows that

$$c_p \frac{dt}{t} = \left(\frac{\gamma R}{\gamma - 1} \right) \frac{dt}{t} = R \frac{dp}{p} \qquad (1.123)$$

from which

$$\frac{dt}{t} = \frac{\gamma - 1}{\gamma} \left(\frac{dp}{p} \right) \qquad (1.124)$$

For a perfect gas, γ is constant, so that integration of equation 1.124 yields

$$\ln t = \left(\frac{\gamma - 1}{\gamma} \right) \ln p + \text{constant} \qquad (1.125)$$

Hence

$$t p^{-(\gamma - 1)/\gamma} = \text{constant} \qquad (1.126)$$

Furthermore, for a perfect gas, $pv = Rt$, so that equation 1.126 may be transformed to read

$$pv^\gamma = p\rho^{-\gamma} = \text{constant} \qquad (1.127)$$

Hence, if a perfect gas undergoes an isentropic change of state, then its pressure and temperature are related by equation 1.126, and its pressure and volume are related by equation 1.127. To indicate that a state is arrived at by an isentropic change of state, the superscript prime (') will be hereafter attached to the property attained by that change of state. Hence, for an isentropic change of state from state 1 to state 2′

$$\frac{t_2'}{t_1} = \left(\frac{p_2'}{p_1} \right)^{(\gamma - 1)/\gamma} \qquad (1.128)$$

1-15(f) Process Equations

Processes by which a gas can pass from one state to another are, in the abstract at least, infinite in number. In general, the following specific processes are of the greatest significance in engineering.

1. The isothermal change ($dt = 0$).
2. The adiabatic change ($\delta Q = 0$).
3. The isobaric change ($dp = 0$).
4. The isovolumic change ($dv = 0$).
5. The isentropic change ($ds = 0$).

It will be recalled from elementary thermodynamics that all of the above changes of state may be regarded as special cases of the general so-called *polytropic change* $pv^n =$ constant. Thus,

$$n = 1 \text{ corresponds to } dt = 0$$
$$n = 0 \text{ corresponds to } dp = 0$$
$$n = \infty \text{ corresponds to } dv = 0$$
$$n = \gamma \text{ corresponds to ds} = 0$$

Flow processes that take place rapidly, such as the flow of gases through the exhaust nozzle of a rocket motor or turbojet engine, the nozzles of a turbine, the blade passages of an axial compressor or turbine, air inlet ducts, and diffusers, are substantially adiabatic ($\delta Q = 0$) because of the extremely short time interval available for the transfer of heat. Such processes, however, are *never* isentropic ($ds = 0$) because they are invariably irreversible, since they are accompanied by internal friction, turbulence, wall friction, and eddies, all of which cause energy losses that increase the entropy of the fluid stream. Consequently, in discussing the flow of gases through a nozzle, turbine, or compressor, it is tacitly assumed as a first approximation that the changes of state are reversible adiabatics, that is, isentropics, and the resulting equations are then modified to take into account the effects of irreversibility.

1-15(g) Entropy-Enthalpy Diagram

Most gas dynamic processes may be illustrated graphically by plotting the process on an entropy-enthalpy diagram, generally called a *Mollier diagram*. For a perfect gas, however, h, u and t are all directly proportional. Hence, any of the three may be employed in constructing the *entropy-enthalpy diagram*. The general characteristics of the entropy-enthalpy diagram for a perfect gas are developed in the present section.

Integrating equation 1.118 between a general state and a *reference state* denoted by the subscript o yields

$$s(v,t) = c_v \ln (t/t_o) + R \ln (v/v_o) \tag{1.129}$$

where the constant of integration is chosen to be zero. Lines of constant specific volume v on the ts plane may be determined by means of equation 1.129 by employing constant values of v as a parameter. Then

$$t = t_o(v_o/v)^{\gamma - 1}e^{s/c_v} = \text{constant } v^{-(\gamma - 1)}e^{s/c_v} \tag{1.130}$$

Equation 1.130 shows that curves of $v = 1/\rho =$ constant, termed *v-lines* or *ρ-lines*, are exponential curves on the ts plane.

Integrating equation 1.119, yields

$$s(p,t) = c_p \ln (t/t_o) - R \ln (p/p_o) \tag{1.131}$$

where the constant of integration is chosen to be zero. Curves of constant pressure, called either *p-lines* or *isobars*, are thus given by

$$t = t_o(p/p_o)^{(\gamma-1)/\gamma}e^{s/c_p} = \text{constant } p^{(\gamma-1)/\gamma}e^{s/c_p} \qquad (1.132)$$

Figure 1.19 presents curves, calculated from equations 1.130 and 1.132, for air, assumed to be a perfect gas having $\gamma = 1.40$ and $R = 287.04$ J/kg-K. The ordinate is the air temperature t and the abcissa is the dimensionless entropy s/R. The figure

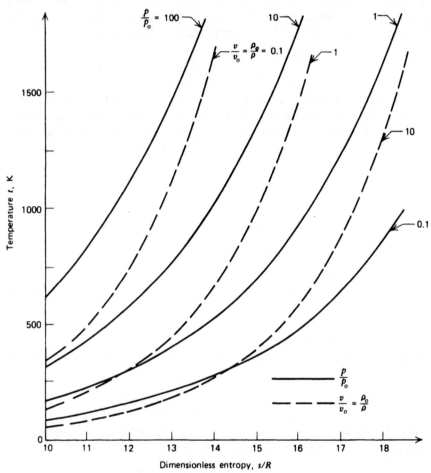

Figure 1.19 Characteristics of the isovolumic (v = constant) and the isobaric (p = constant) lines in the ts plane (also hs and us planes for a perfect gas).

presents three curves of constant volume ratio $v/v_o = \rho_o/\rho$, and four curves of constant pressure ratio p/p_o. It follows from equations 1.130 and 1.132 that, as stated earlier, both sets of curves are exponential. The curves of constant v/v_o are steeper than those for constant p/p_o; this is because $c_p > c_v$, so that exp $(s/c_v) >$ exp (s/c_p). At constant values of s, the values of p increase with t, while the values of $v = 1/\rho$ decrease.

The reference conditions for the curves of Fig. 1.19 are $t_o = 294.4$ K and $p_o = 1$ atm. Since the reference value for s/R is arbitrary, it was chosen as $s_o/R = 12$ so that a convenient range of values is obtained for s/R.

1-15(h) Mixtures of Thermally Perfect Gases

Consider a mixture composed of several perfect gases that do not interact chemically, as for example atmospheric air. Assume that the constituent gases, hereafter termed the *species*, are in thermal equilibrium with each other. Accordingly, each of the species is at the mixture temperature t. The thermodynamic properties of such a gas mixture are determined in this section.

The conditions specified above satisfy the requirements for *Dalton's law* to be applicable; that is, each species behaves as if all the others were absent, so that the static pressure of the gas mixture, denoted by p, is equal to the sum of the *partial pressures* p_i of each of the species. The partial pressure p_i is defined as the pressure that each species would exert if it occupied the entire mixture volume $\mathscr{V}$ by itself at the mixture temperature t. For n constituents, it follows that

$$p = \sum_{i=1}^{n} p_i \tag{1.133}$$

A mixture of thermally perfect gases also satisfies *Amagat's law*; that is, the volume of the gas mixture, denoted by $\mathscr{V}$, is equal to the sum of the *partial volumes* $\mathscr{V}_i$ of each of the species. The partial volume $\mathscr{V}_i$ is defined as the volume that each species would occupy by itself at the pressure p and temperature t of the mixture. Thus

$$\mathscr{V} = \sum_{i=1}^{n} \mathscr{V}_i$$

Let m denote the mass of the gas mixture. Then

$$m = \sum_{i=1}^{n} m_i = m_1 + m_2 + \cdots + m_n$$

Let v denote the specific volume of the gas mixture, and let ρ denote its density. Then

$$v = \frac{1}{\rho} = \frac{\mathscr{V}}{m}$$

By definition, the *partial specific volume* v_i and the *partial density* ρ_i of species i, are given by

$$v_i = \frac{1}{\rho_i} \equiv \frac{\mathscr{V}}{m_i}$$

For the ith species of the mixture, having the molecular weight $\bar{m}_i$, the partial pressure p_i is obtained from equation 1.99. Thus,

$$p_i = \rho_i R_i t = \left(\frac{m_i}{\mathscr{V}}\right)\left(\frac{\bar{R}}{\bar{m}_i}\right) t = \frac{\bar{R}t}{\mathscr{V}}\left(\frac{m_i}{\bar{m}_i}\right) \tag{1.134}$$

By definition, the *mass fraction* C_i is given by

$$C_i \equiv \frac{m_i}{m} = \frac{\rho_i}{\rho} \tag{1.135}$$

Substituting for m_i from equation 1.135 into equation 1.134 gives

$$p_i = \frac{\bar{R}t}{\mathscr{V}}\left(\frac{C_i m}{\bar{m}_i}\right) = \frac{mt}{\mathscr{V}}\left(\frac{C_i \bar{R}}{\bar{m}_i}\right) = \frac{mt}{\mathscr{V}}(C_i R_i) \tag{1.136}$$

where $R_i = \bar{R}/\bar{m}_i$.

The pressure p of the gas mixture, for brevity termed the *mixture pressure*, is obtained by substituting equation 1.136 into equation 1.133. Thus,

$$p = \frac{mt}{\mathscr{V}} \sum_{i=1}^{n} C_i R_i = \rho t \sum_{i=1}^{n} C_i R_i \qquad (1.137)$$

where ρ is the density of the gas mixture. Equation 1.137 may be put in the form of the thermal equation of state for a perfect gas. Thus,

$$p = \rho R t$$

where R is the *effective gas constant* for the gas mixture and, in view of equation 1.137, is given by

$$R = \sum_{i=1}^{n} C_i R_i = \frac{\bar{R}}{\bar{m}} \qquad (1.138)$$

where $\bar{m}$ is the *effective molecular weight* of the gas mixture. Hence, a mixture of *inert thermally perfect gases* behaves as a thermally perfect gas having the effective molecular weight $\bar{m}$ defined by equation 1.138.

For a mixture of *chemically reacting thermally perfect gases*, the definitions and results presented above are still valid. The effective gas constant R in that case, however, is not a constant, since each of the mass fractions C_i changes.

The thermal equation of state for a perfect gas, on a molar basis, is given by equation 1.101, which is repeated below.

$$p\mathscr{V} = N\bar{R}t \qquad (1.101)$$

Let N_i denote the number of moles of the ith species present in the mixture. Then, for species i, equation 1.101 becomes

$$p_i\mathscr{V} = N_i\bar{R}t$$

Combining the above equation and equation 1.101 gives

$$\frac{p_i}{p} = \frac{N_i}{N} = X_i \qquad (1.139)$$

where X_i is the *mole fraction* of the ith species, and

$$N = \sum_{i=1}^{n} N_i$$

In terms of the partial volume $\mathscr{V}_i$, equation 1.101 may be written as

$$p\mathscr{V}_i = N_i\bar{R}t$$

Combining the above equation with equation 1.101 yields

$$X_i = \frac{\mathscr{V}_i}{\mathscr{V}} = \frac{N_i}{N} \qquad (1.140)$$

Equation 1.139 relates the partial pressure p_i and the *mixture pressure* p to the mole fraction X_i, and equation 1.140 relates the partial volume $\mathscr{V}_i$ and the mixture volume $\mathscr{V}$ to the mole fraction X_i.

The effective molecular weight of the mixture $\bar{m} = m/N$ may be related to the mole fractions X_i as follows.

$$m = \sum_{i=1}^{n} m_i = \sum_{i=1}^{n} \bar{m}_i N_i = \sum_{i=1}^{n} \bar{m}_i(X_i N)$$

Solving the above expression for $\bar{m}$, we obtain

$$\bar{m} = \frac{m}{N} = \sum_{i=1}^{n} X_i \bar{m}_i \qquad (1.141a)$$

The mole fraction X_i may be related to the mass fraction C_i as follows.

$$m_i = C_i m = \bar{m}_i N_i = \bar{m}_i(X_i N) = \bar{m}_i \left(\frac{X_i m}{\bar{m}} \right)$$

Solving the above expression for X_i yields

$$X_i = \frac{C_i \bar{m}}{\bar{m}_i} \qquad (1.141b)$$

The internal energy and the enthalpy of the gas mixture are determined by summing over all of the species. On a mass basis, for each species i,

$$u_i = \int_{t_o}^{t} c_{vi} \, dt + u_{io} \qquad \text{and} \qquad h_i = \int_{t_o}^{t} c_{pi} \, dt + h_{io} \qquad (1.142a,b)$$

Summing over the n species gives

$$u = \sum_{i=1}^{n} C_i u_i \qquad \text{and} \qquad h = \sum_{i=1}^{n} C_i h_i \qquad (1.143a,b)$$

Alternately, the internal energy and the enthalpy of the gas mixture may be determined from the specific heats c_v and c_p for the mixture. Thus,

$$c_v = \sum_{i=1}^{n} C_i c_{vi} \qquad \text{and} \qquad c_p = \sum_{i=1}^{n} \cdot C_i c_{pi} \qquad (1.144a,b)$$

The internal energy and the enthalpy of the gas mixture are obtained by integrating equations 1.144. Thus,

$$u = \int_{t_o}^{t} c_v \, dt + u_o \qquad \text{and} \qquad h = \int_{t_o}^{t} c_p \, dt + h_o \qquad (1.145a,b)$$

If all of the species in the mixture are perfect gases, then every c_{vi} and c_{pi} is a constant, and if $u_o = h_o = 0$ when $t = t_o$, then equations 1.145 become

$$u = c_v t \qquad \text{and} \qquad h = c_p t \qquad (1.146a,b)$$

Equations 1.146 are identical to equations 1.106 and 1.117 for a pure perfect gas. The specific heat ratio $\gamma = c_p/c_v$ for a mixture of perfect gases may be determined from the values of c_p and c_v obtained from equations 1.144.

On a molar basis, equations 1.142 through 1.146 become:

$$\bar{u}_i = \int_{t_o}^{t} \bar{c}_{vi} \, dt + \bar{u}_{io} \qquad \text{and} \qquad \bar{h}_i = \int_{t_o}^{t} \bar{c}_{pi} \, dt + \bar{h}_{io} \qquad (1.142c,d)$$

$$\bar{u} = \sum_{i=1}^{n} X_i \bar{u}_i \qquad \text{and} \qquad \bar{h} = \sum_{i=1}^{n} X_i \bar{h}_i \qquad (1.143c,d)$$

$$\bar{c}_v = \sum_{i=1}^{n} X_i \bar{c}_{vi} \qquad \text{and} \qquad \bar{c}_p = \sum_{i=1}^{n} X_i \bar{c}_{pi} \qquad (1.144c,d)$$

$$\bar{u} = \int_{t_o}^{t} \bar{c}_v \, dt + \bar{u}_o \qquad \text{and} \qquad \bar{h} = \int_{t_o}^{t} \bar{c}_p \, dt + \bar{h}_o \qquad (1.145c,d)$$

$$\bar{u} = \bar{c}_v t \qquad \text{and} \qquad \bar{h} = \bar{c}_p t \qquad (1.146c,d)$$

The entropy of a mixture of perfect gases may be determined in two steps. First, each species is subjected to a change of state from a reference pressure and temperature of p_o and t_o, respectively, to the pressure p and temperature t of the mixture, but each species occupies a volume equal to its partial volume $\mathscr{V}_i$ in the mixture. Second, each species is expanded adiabatically to fill the entire volume $\mathscr{V}$ of the mixture, at a pressure equal to its pressure p_i in the mixture. For a perfect gas, that expansion process is isothermal, since

$$p\mathscr{V}_i = p(X_i\mathscr{V}) = (X_i p)\mathscr{V} = p_i\mathscr{V} = mRt \tag{1.147}$$

The entropy change of a pure substance is given by equation 1.119. Thus,

$$ds = c_p \frac{dt}{t} - R \frac{dp}{p} \tag{1.148}$$

Integrating equation 1.148 for the first step discussed above, we obtain

$$s = \int_{t_o}^{t} c_p \frac{dt}{t} - R \ln \frac{p}{p_o} = \phi - R \ln p \tag{1.149}$$

where $s_o = 0$, $p_o = 1$ atm, and ϕ is defined as

$$\phi \equiv \int_{t_o}^{t} c_p \frac{dt}{t} \tag{1.150}$$

On a molar basis, $\bar{s} = \bar{m}s$. Thus,

$$\bar{s} = \bar{\phi} - \bar{R} \ln p \tag{1.151}$$

For a mixture of n species,

$$S = \sum_{i=1}^{n} N_i \bar{s}_i = \sum_{i=1}^{n} N_i \bar{\phi}_i - N\bar{R} \ln p \tag{1.152}$$

Equation 1.152 determines the entropy of the mixture when the individual species are at the pressure p and temperature t of the mixture.

For the second step, $t = $ constant, and equation 1.148 gives

$$ds = -R \frac{dp}{p} \tag{1.153}$$

Integrating equation 1.153 between the limits $p = p$ and $p = p_i$, we obtain for species i,

$$\Delta s_{i,\text{mixing}} = -R \ln \frac{p_i}{p} = -R \ln X_i \tag{1.154}$$

For a mixture of n species, on a molar basis,

$$\Delta S_{\text{mixing}} = \sum_{i=1}^{n} N_i \Delta \bar{s}_{i,\text{mixing}} = -N\bar{R} \sum_{i=1}^{n} X_i \ln X_i \tag{1.155}$$

The entropy of the mixture may be obtained by adding equations 1.152 and 1.155. Thus,

$$S = \sum_{i=1}^{n} N_i \bar{\phi}_i - N\bar{R} \ln p - N\bar{R} \sum_{i=1}^{n} X_i \ln X_i \tag{1.156}$$

On a mass basis, equation 1.156 becomes

$$s = \sum_{i=1}^{n} C_i \phi_i - R \ln p - R \sum_{i=1}^{n} X_i \ln X_i \qquad (1.157a)$$

On a molar basis, we obtain,

$$\bar{s} = \sum_{i=1}^{n} X_i \bar{\phi}_i - \bar{R} \ln p - \bar{R} \sum_{i=1}^{n} X_i \ln X_i \qquad (1.157b)$$

Example 1.6. The composition of dry air, by volume, is given in Table 1.10. The properties of air are described adequately by considering it to consist of only N_2, O_2, and A. Thus, the volume percentages in Table 1.10 may be adjusted to neglect the other species. For that idealized model of the composition of air, compute the following properties at a pressure of 1 atm ($1.0133 \cdot 10^5$ N/m^2) and a temperature of 298.15 K: (a) the partial pressures and densities of the three gases; (b) the mixture pressure and density; (c) the mass fractions of the individual species; (d) the gas constant and molecular weight of the mixture; and (e) its specific heats and specific heat ratio. Use the thermodynamic properties presented in Table C.3.

Solution

The composition of air by volume, as obtained from Table 1.10 and corrected to consider only N_2, O_2, and A, is presented in the following table. The molecular weights, specific heats, and specific heat ratios obtained from Table C.3 are also tabulated.

Species	Percent by Volume, Table 1.10	Percent by Volume, Corrected	$\bar{m}$	c_p, J/kg-K	γ	$c_v = c_p/\gamma$, J/kg-K	$R = \bar{R}/\bar{m}_i$, J/kg-K
N_2	78.084	78.11	28.013	1038.3	1.400	741.64	296.80
O_2	20.9476	20.96	32.000	916.90	1.395	657.28	259.82
A	0.934	0.93	39.944	524.61	1.658	316.41	208.15

(a) From equations 1.139 and 1.140,

$$p_i = X_i p = p \left(\frac{\mathscr{V}_i}{\mathscr{V}} \right)$$

$$p_{N_2} = 0.7811(1.0133 \cdot 10^5) = 79{,}149 \text{ N/m}^2$$

$$p_{O_2} = 0.2096(1.0133 \cdot 10^5) = 21{,}238 \text{ N/m}^2$$

$$p_A = 0.0093(1.0133 \cdot 10^5) = 942 \text{ N/m}^2$$

From equation 1.134

$$\rho_i = \frac{p_i}{R_i t} = \frac{p_i \bar{m}_i}{\bar{R} t}$$

$$\rho_{N_2} = \frac{(79{,}149)(28.013)}{(8314.3)(298.15)} = 0.89442 \text{ kg/m}^3$$

$$\rho_{O_2} = \frac{(21,238)(32.000)}{(8314.3)(298.15)} = 0.27417 \text{ kg/m}^3$$

$$\rho_A = \frac{(942)(39.944)}{(8314.3)(298.15)} = 0.01518 \text{ kg/m}^3$$

(b) $p = \sum p_i = 79,149 + 21,238 + 942 = 101,329 \text{ N/m}^2$

$\rho = \sum \rho_i = 0.89442 + 0.27417 + 0.01518 = 1.18377 \text{ kg/m}^3$

(c) $C_i = \dfrac{m_i}{m} = \dfrac{\rho_i}{\rho}$

$$C_{N_2} = \frac{0.89442}{1.18377} = 0.7556$$

$$C_{O_2} = \frac{0.27417}{1.18377} = 0.2316$$

$$C_A = \frac{0.01518}{1.18377} = 0.0128$$

(d)   From equation 1.138,

$R = \sum C_i R_i = 0.7556(296.80) + 0.2316(259.82) + 0.0128(208.15)$
$= 287.09 \text{ J/kg-K}$

$\bar{m} = \dfrac{\bar{R}}{R} = \dfrac{8314.3}{287.09} = 28.96$

(e)   From equation 1.144,

$c_p = \sum C_i c_{pi} = 0.7556(1038.3) + 0.2316(916.90) + 0.0128(524.61)$
$= 1003.6 \text{ J/kg-K}$

$c_v = \sum C_i c_{vi} = 0.7556(741.64) + 0.2316(657.28) + 0.0128(316.41)$
$= 716.6 \text{ J/kg-K}$

$\gamma = \dfrac{c_p}{c_v} = \dfrac{1003.6}{716.6} = 1.4005$

1–16 IMPERFECT GASES AND GAS TABLES

For a thermally perfect gas the ratio $pv/Rt = 1$ for all pressures and temperatures [see Section 1–15(a)]. For most real gases the compressibility factor $Z = pv/Rt \neq 1$. When Z is plotted as a function of p, with the temperature as a parameter, it is seen that, except at low pressures, there is considerable deviation from the value unity.[13] Several equations such as the van der Waals, Beattie-Bridgman, and polynomial equations have been proposed for the equation of state of an imperfect gas.[15] In general, those equations are either too inaccurate or too complicated for general use. Moreover, as pointed out in Section 1–15(a), the equation $pv = Rt$ gives accurate results for all gases when the pressure is low, and it gives satisfactory accuracy when the pressure is a large fraction of the critical pressure, provided that the temperature greatly exceeds the critical temperature. Consequently, for calculating p, v, and t relationships for many imperfect gases of practical interest, the thermal equation of state $pv = Rt$ is satisfactory for most engineering purposes.

For imperfect gases, however, the specific heats c_p and c_v vary significantly with the gas temperature and, to a much smaller extent, with the pressure.

1–16(a) Specific Heat Data

Air is an important fluid in many practical engineering applications. It is important, therefore, to have accurate data on its thermodynamic properties. Measurements demonstrate that the effect of temperature on c_p for air is appreciable, but the influence of pressure is of much less significance.[16, 17]

Figure 1.20 presents the specific heat c_p for air as a function of the temperature for several different constant pressures ranging from $p = 0$ to $p = 1034 \cdot 10^5$ N/m². The curves show that, except at very low temperatures, the effect of increasing the pressure from $p = 0$ to more than $68.95 \cdot 10^5$ N/m² is less than 1 percent at $t = 1000$ K. At the higher temperatures, above 1500 K, increasing the pressure from $p = 0$ to $p > 689.5 \cdot 10^5$ N/m² has little influence on the values of c_p.

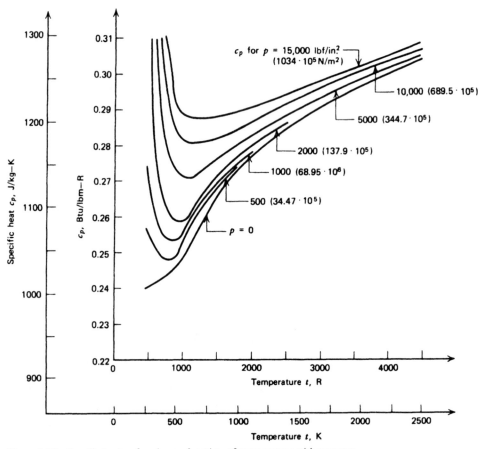

Figure 1.20 Specific heat c_p for air as a function of temperature with pressure as a parameter (based on Reference 16).

Figure 1.21 presents the specific heat ratio γ for air as a function of t, with the pressure p as a parameter. The curves show, as one would expect, that the effect of pressure on γ is significant only at low temperatures.

Because of the relatively small influence of pressure on c_p for air, it is generally assumed that the instantaneous values of c_p are defined by the curve for zero pressure ($p = 0$). Moreover, that conclusion is reasonably correct for most gases. Conse-

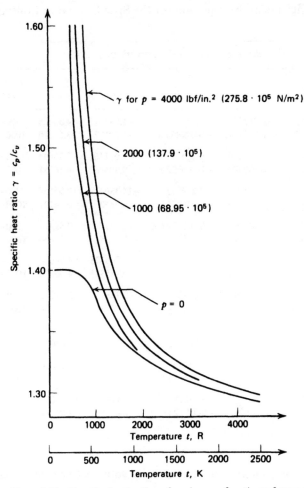

Figure 1.21 Specific heat ratio γ for air as a function of temperature with pressure as a parameter (based on Reference 16).

quently, the equations for correlating the experimental data on c_p and γ for gases assume quite generally that c_p is a function of the gas temperature alone.[18–20]

Several equations have been proposed for correlating the empirical c_p data for different gases as functions of the gas temperature t. Equations having polynomial, and also exponential, forms are available. The correlating equations presented in this book are based on polynomials of the following form:

$$c_p = (a + bt + ct^2 + dt^3 + et^4)R \tag{1.158}$$

where c_p has the same units as R. Table 1.6 presents values of the coefficients $a, b, \ldots$ for several pure gases.[21] The specific heat data in Reference 21 were obtained by a least squares curve fit of the data in Reference 20. The curve fits were obtained for two adjoining temperature intervals, the low temperature interval extending from 300 to 1000 K, and the high temperature interval extending from 1000 to 5000 K. The data are constrained to be equal at 1000 K. Curve fits for 483 substances are presented in Reference 21. Also presented in Table 1.6 are values of the properties h_o and ϕ_o, which are defined by equations 1.160 and 1.166, respectively.

Table 1.6 Values of the Coefficients in the Equation for the Specific Heat at Constant Pressure c_p for Several Gases

	$c_p = (a + bt + ct^2 + dt^3 + et^4)R$, t is in K, c_p has the units of R							
Gas	a	$b \cdot 10^3$	$c \cdot 10^6$	$d \cdot 10^9$	$e \cdot 10^{12}$	$h_o \cdot 10^{-3}$	ϕ_o	Temperature Range, K
O_2	3.62560	−1.87822	7.05545	−6.76351	2.15560	−1.04752	4.30528	300–1000
	3.62195	0.736183	−0.196522	0.0362016	−0.00289456	−1.20198	3.61510	1000–5000
N_2	3.67483	−1.20815	2.32401	−0.632176	−0.225773	−1.06116	2.35804	300–1000
	2.89632	1.51549	−0.572353	0.0998074	−0.00652236	−0.905862	6.16151	1000–5000
CO	3.71009	−1.61910	3.69236	−2.03197	0.239533	−14.3563	2.95554	300–1000
	2.98407	1.48914	−0.578997	0.103646	−0.00693536	−14.2452	6.34792	1000–5000
CO_2	2.40078	8.73510	−6.60709	2.00219	0.000632740	−48.3775	9.69515	300–1000
	4.46080	3.09817	−1.23926	0.227413	−0.0155260	−48.9614	−0.986360	1000–5000
A	2.50000					−0.745375	4.36600	300–1000
	2.50000					−0.745375	4.36600	1000–5000
H_2	3.05745	2.67652	−5.80992	5.52104	−1.81227	−0.988905	−2.29971	300–1000
	3.10019	0.511195	0.0526442	−0.0349100	0.00369453	−0.877380	−1.96294	1000–5000
H_2O	4.07013	−1.10845	4.15212	−2.96374	0.807021	−30.2797	−0.322700	300–1000
	2.71676	2.94514	−0.802243	0.102267	−0.00484721	−29.9058	6.63057	1000–5000
CH_4	3.82619	−3.97946	24.5583	−22.7329	6.96270	−10.1450	0.866901	300–1000
	1.50271	10.4168	−3.91815	0.677779	−0.0442837	−9.97871	10.7071	1000–5000
C_2H_4	1.42568	11.3831	7.98900	−16.2537	6.74913	5.33708	14.6218	300–1000
	3.45522	11.4918	−4.36518	0.761551	−0.0501232	4.47731	2.69879	1000–5000

Example 1.7. Consider the composition of dry air, by volume, to be that tabulated below.

Species	Chemical Formula	Molecular Weight	Composition, Percent by Volume
Nitrogen	N_2	28.016	78.11
Oxygen	O_2	32.000	20.96
Argon	A	39.94	0.93

Employ the specific heat data presented in Table 1.6 and determine the specific heat at constant pressure of air as a function of temperature.

Solution

For air, equation 1.144 on a molar basis, gives

$$\bar{c}_p = \sum_{i=1}^{3} X_i \bar{c}_{pi} = X_{N_2}\bar{c}_{p,N_2} + X_{O_2}\bar{c}_{p,O_2} + X_A\bar{c}_{p,A} \tag{a}$$

Introducing the values presented in Table 1.6 yields

Gas	$X_i a_i$	$X_i b_i \cdot 10^3$	$X_i c_i \cdot 10^6$	$X_i d_i \cdot 10^9$	$X_i e_i \cdot 10^{12}$	Temperature Range, K
N_2	2.87041	−0.943686	1.81528	−0.493793	−0.176351	300–1000
O_2	0.759926	−0.393675	1.47882	−1.41763	0.451813	300–1000
A	0.023250					300–1000
Air	3.65359	−1.33736	3.29421	−1.91142	0.275462	300–1000
N_2	2.26232	1.18375	−0.447065	0.0779596	−0.00509462	1000–5000
O_2	0.759161	0.154304	−0.041191	0.00758786	−0.00060670	1000–5000
A	0.023250					1000–5000
Air	3.04473	1.33805	−0.488256	0.0855475	−0.00570132	1000–5000

From the above table, in the range 300 to 1000 K

$$\bar{c}_p = \left(3.65359 - \frac{1.33736t}{10^3} + \frac{3.29421t^2}{10^6} - \frac{1.91142t^3}{10^9} + \frac{0.275462t^4}{10^{12}} \right) \bar{R} \quad \text{(b)}$$

and in the range 1000 to 5000 K

$$\bar{c}_p = \left(3.04473 + \frac{1.33805t}{10^3} - \frac{0.488256t^2}{10^6} + \frac{0.0855475t^3}{10^9} - \frac{0.00570132t^4}{10^{12}} \right) \bar{R} \quad \text{(c)}$$

where, in both equations, t is in K. The units of $\bar{R}$ determine the units of $\bar{c}_p$. On a unit mass basis, $c_p = \bar{c}_p/\bar{m}$. Values of $\bar{R}$ in different units are presented on the inside cover of this book.

1–16(b) Gas Tables

For making the thermodynamic analyses pertinent to high-temperature gas dynamic processes, the equations for the specific heats of gases presented in Table 1.6 are too cumbersome for general use. Considerable labor is saved by employing tabulated values of the pertinent thermodynamic properties of the gases, as is done for air in Table C.4.

It is shown in Sections 1–15(b) and 1–15(d) that for a thermally perfect gas $du = c_v \, dt$ and $dh = c_p \, dt$. Consequently, the values of u and h for a real gas may be tabulated as functions of temperature if the instantaneous values of c_v and c_p are employed in calculating u and h. Substituting equation 1.158 into equation 1.115 gives

$$h = h_o R + \int_{t_o}^{t} (a + bt + ct^2 + dt^3 + et^4)R \, dt \quad (1.159)$$

where $h = h_o R$ at $t = t_o$. Values of h_o, including the energy of formation [see Section 14–3(a)], are presented in Table 1.6. Integrating equation 1.159 yields

$$h = \left(h_o + at + \frac{bt^2}{2} + \frac{ct^3}{3} + \frac{dt^4}{4} + \frac{et^5}{5} \right) R \quad (1.160)$$

From equations 1.53 and 1.99, we obtain

$$u = h - pv = h - Rt \tag{1.161}$$

When the values of h_o, a, b, etc., presented in Table 1.6 are substituted into equations 1.160 and 1.161, h and u are determinable. Table C.4 presents the values of u and h, accounting for variable c_v and c_p, for air at one atmosphere pressure as a function of temperature. Tables for several other gases are presented in Reference 22.

The entropy change ds for a perfect gas is given by equation 1.119 and is repeated here for convenience. Thus,

$$ds = c_p \frac{dt}{t} - R \frac{dp}{p} \tag{1.162}$$

Equation 1.162 may be applied to a real gas by expressing c_p as a function of temperature. It is apparent from equation 1.162 that the entropy of a gas depends on both its temperature and its pressure. Consequently, the compilation of a table of entropy values requires calculating them at different temperatures for several values of constant pressure. If t_o and p_o are the base temperature and pressure, respectively, for the entropy table, where $s = 0$, then

$$s = \int_{t_o}^{t} c_p \frac{dt}{t} - R \ln \frac{p}{p_o} = \phi - R \ln \frac{p}{p_o} \tag{1.163}$$

where $s_o = 0$ and the parameter ϕ is defined by

$$\phi \equiv \int_{t_o}^{t} c_p \frac{dt}{t} \tag{1.164}$$

and is a function only of the gas temperature. Substituting equation 1.158 into equation 1.164 gives

$$\phi = \phi_o R + \int_{t_o}^{t} \left(\frac{a}{t} + b + ct + dt^2 + et^3 \right) R \, dt \tag{1.165}$$

Integrating equation 1.165 yields

$$\phi = \left(\phi_o + a \ln t + bt + \frac{ct^2}{2} + \frac{dt^3}{3} + \frac{et^4}{4} \right) R \tag{1.166}$$

where $\phi = \phi_o R$ at $t = t_o$. Values of ϕ_o are presented in Table 1.6. Table C.4 presents values of ϕ for air.

A change in entropy is calculated from the tabulated values of ϕ by means of equation 1.163. Thus,

$$s_2 - s_1 = \phi_2 - \phi_1 - R \ln \left(\frac{p_2}{p_1} \right) \tag{1.167}$$

For an isentropic process ($ds = 0$), equation 1.162 reduces to

$$\frac{dp}{p} = \frac{c_p}{R} \frac{dt}{t} \tag{1.168}$$

Employing the subscript o to denote the base condition for the table, then, for an isentropic process,

$$\ln \left(\frac{p}{p_o} \right) = \ln p_r = \frac{1}{R} \int_{t_o}^{t} c_p \frac{dt}{t} = \frac{\phi}{R} \tag{1.169}$$

The pressure ratio $p_r = p/p_o$ is termed the *relative pressure* and is seen to be a single-valued function of the temperature t. Solving equation 1.169 for p_r gives

$$p_r = e^{\phi/R} \qquad (1.170)$$

For an isentropic change of state between the temperatures t_1 and t_2, the pressure ratio p_2/p_1 is given by

$$\frac{p_2}{p_1} = \frac{(p_2/p_o)}{(p_1/p_o)} = \frac{p_{r2}}{p_{r1}} \qquad (1.171)$$

Thus, only the ratios of the relative pressures are of significance. Hence, equation 1.170 may be written as

$$p_r = \frac{e^{\phi/R}}{(p_r)_{ref}} = e^{(\phi - \phi_{ref})/R} \qquad (1.172)$$

where $(p_r)_{ref}$ is chosen so that a convenient range of p_r is obtained in the table. Values of p_r for air are presented in Table C.4, where $(p_r)_{ref}$ is defined at $t = 298.15$ K.

The *relative volume* v_r is defined as

$$v_r = \frac{v}{v_o} = \frac{p_o}{p} \frac{t}{t_o} = \frac{1}{p_r} \frac{t}{t_o} \qquad (1.173)$$

The relative volume v_r is a single-valued function of the temperature t. For an isentropic change of state,

$$\frac{v_2}{v_1} = \frac{(v_2/v_o)}{(v_1/v_o)} = \frac{v_{r2}}{v_{r1}} \qquad (1.174)$$

Thus, only the ratios of the relative volumes are of significance. Hence, equation 1.173 may be rewritten as

$$v_r = \frac{Rt}{p_r} \qquad (1.175)$$

Equation 1.175 was employed by Keenan and Kaye.[22] The above definition of v_r satisfies equation 1.173 except for a constant, which is acceptable for use in equation 1.174. Coincidentally, it has the feature that v_r is the specific volume when p_r is the pressure.

Table C.4 presents the thermodynamic properties of air in SI units and also in EE units.

From equation 1.110,

$$c_v = c_p - R \qquad (1.176)$$

The specific heat ratio γ is defined as

$$\gamma = \frac{c_p}{c_v} \qquad (1.177)$$

For a perfect gas, the speed of sound is given by equation 1.181. Thus,

$$a = (\gamma R t)^{1/2} \qquad (1.178)$$

Values of c_p, c_v, γ, and a for air are tabulated as functions of temperature in Table C.4.

A FORTRAN computer program is presented below for determining the thermodynamic properties of a thermally perfect gas with variable specific heats, based on the equations developed in this section. The program has the capability of considering a mixture of such gases consisting of up to nine species. The input variables are presented in Table 1.7. Values of h_o and ϕ_o from Reference 21 are presented in Table

1.6. Table 1.8 presents the input data deck that was employed in the calculations for Table C.4. The first card in the data deck is employed for identifying information and may contain up to 54 characters. The table may have up to 15 different temperature increments, as specified by the variables M(J) and DT(J), J = 1, . . . , 15, where M(J) specifies the number of entries in each section of the table and DT(J) specifies the corresponding temperature increment.

Table 1.7 Input Variables for the Thermodynamic Properties Program

NX	Number of species, 1 to 9.
X(J)	Mole fractions of up to nine species.
W(J)	Molecular weights of up to nine species.
IUNITS	= 1 specifies EE units, = 2 specifies modified SI units employing calories instead of Joules, and = 3 specifies SI units.
MASS	= 1 specifies mass basis, = 2 specifies molar basis.
M(J)	Number of temperature entries for each temperature increment, up to 15.
DT(J)	Temperature increments, K or R, up to 15.
TI	Initial temperature, K or R.
TM	Temperature at which the two $\bar{c}_p$ curve fits meet, K. TREF The reference temperature where h is defined to equal HBASE.
A(I,J,K)	Coefficients in equation 1.158. $I = 1, . . . , 5$ denotes the coefficients a, b, etc., $J = 1, . . . , 9$ denotes up to nine species, and $K = 1,2$ denotes the low and high temperature range, respectively. Values of A(I,J,K) must be compatible with t in K.
HI(J,K)	Value of h_0 in equation 1.160. J denotes the species, and K denotes the temperature range.
SI(J,K)	Value of ϕ_0 in equation 1.166. J denotes the species, and K denotes the temperature range.
HBASE	The reference value of h corresponding to t_{ref}.

Table 1.8 Input Data Deck for Table C.4

```
AIR (K, KJ, KG, M, S)
$DATA NX=4, IUNITS=3, MASS=1, TI=50.0, TM=1000.0, TREF=277.77778,
HBASE=279.0158,
M(1)=191, 100, 20, 0, DT(1)=5.0, 10.0, 50.0,
A(1,1,1)=3.67483E+00, -1.20815E-03,  2.32401E-06, -6.32176E-10, -2.25773E-13,
A(1,1,2)=2.89632E+00,  1.51549E-03, -5.72353E-07,  9.98074E-11, -6.52236E-15,
A(1,2,1)=3.62560E+00, -1.87822E-03,  7.05545E-06, -6.76351E-09,  2.15560E-12,
A(1,2,2)=3.62195E+00,  7.36183E-04, -1.96522E-07,  3.62016E-11, -2.89456E-15,
A(1,3,1)=2.50000E+00,  0.00000E+00,  0.00000E+00,  0.00000E+00,  0.00000E+00,
A(1,3,2)=2.50000E+00,  0.00000E+00,  0.00000E+00,  0.00000E+00,  0.00000E+00,
A(1,4,1)=2.40078E+00,  8.73510E-03, -6.60709E-06,  2.00219E-09,  6.32740E-16,
A(1,4,2)=4.46080E+00,  3.09817E-03, -1.23926E-06,  2.27413E-10, -1.55260E-14,
HI(1,1)=-1.06116E+03,         HI(1,2)=-9.05862E+02,
HI(2,1)=-1.04752E+03,         HI(2,2)=-1.20198E+03,
HI(3,1)=-7.45375E+02,         HI(3,2)=-7.45375E+02,
HI(4,1)=-4.83775E+04,         HI(4,2)=-4.89614E+04,
SI(1,1)= 2.35804E+00,         SI(1,2)= 6.16151E+00, X(1)=0.780863, W(1)=28.016,
SI(2,1)= 4.30528E+00,         SI(2,2)= 3.61510E+00, X(2)=0.209482, W(2)=32.000,
SI(3,1)= 4.36600E+00,         SI(3,2)= 4.36600E+00, X(3)=0.009340, W(3)=39.950,
SI(4,1)= 9.69515E+00,         SI(4,2)=-9.86360E-01, X(4)=0.000314, W(4)=44.010 $
```

A modified version of subroutine THERMO for the calculation of $\bar{c}_p$, $\bar{h}$, and $\bar{\phi}$, on a molar basis, in both SI and EE units, is presented in Section 14–3. Table D.1, Volume 2, presents the values of $\bar{c}_p$, $\bar{h}$, and $\bar{\phi}$ for 13 substances, calculated by the modified version of subroutine THERMO.

Example 1.8. Air is compressed from 1 atm pressure and 300 K to 3 atm pressure. Calculate the compression work (a) if the compression is isentropic, and (b) if the isentropic compression efficiency $\eta_c = 0.80$. (c) Calculate the entropy change corresponding to part b. Assume $R = 287.04$ J/kg-K.

```
        SUBROUTINE THERMO

C       THERMODYNAMIC PROPERTY DATA FOR A THERMALLY PERFECT GAS

        DIMENSION A(5,9,2),HI(9,2),SI(9,2),X(9),W(9),GAS(9),M(15),DT(15)
        NAMELIST /DATA/NX,X,W,IUNITS,MASS,TI,TM,TREF,M,DT,A,HI,SI,HBASE
        CPF(J,K)=X(J)*(A(1,J,K)+A(2,J,K)*T+A(3,J,K)*T**2+A(4,J,K)*T**3+
       1A(5,J,K)*T**4)*R
        HF(J,K)=X(J)*(HI(J,K)+A(1,J,K)*T+A(2,J,K)*T**2/2.0+A(3,J,K)*T**3/
       13.0+A(4,J,K)*T**4/4.0+A(5,J,K)*T**5/5.0)*R*GT
        SF(J,K)=X(J)*(SI(J,K)+A(1,J,K)*ALOG(T)+A(2,J,K)*T+A(3,J,K)*T**2/
       12.0+A(4,J,K)*T**3/3.0+A(5,J,K)*T**4/4.0)*R

C       READ INPUT DATA AND INITIALIZE PARAMETERS

    10  READ (5,1000) GAS $ WRITE (6,2000) GAS $ READ (5,DATA) $ WW=0.0
        CPREF=0.0 $ HREF=0.0 $ SREF=0.0 $ DO 20 J=1,NX $ WW=WW+X(J)*W(J)
    20  CONTINUE $ IF (MASS.EQ.1) GM=WW $ IF (MASS.EQ.2) GM=1.0
        TI=TI-DT(1) $K=1 $L=0 $N=1 $GD=1.0 $GT=1.8 $GO TO(30,40,40),IUNITS
    30  R=1.98586/GM $ GC=32.174 $ GJ=778.16 $ GL=144.0 $ GO TO 60
    40  R=8.31434/GM $ GC=1.0 $ GJ=1000.0 $ GL=1.0 $ GT=1.0 $ GD=1.0
        IF (IUNITS.EQ.3) GO TO 60 $ R=R/4.184 $ GJ=GJ*4.184
    60  DO 70 J=1,NX $ T=298.15 $ CPREF=CPREF+CPF(J,K) $ SREF=SREF+SF(J,K)
        HI(J,1)=HI(J,1)+(0.0-HF(J,1))/(X(J)*R*GT) $ T=1000.0
        HI(J,2)=HI(J,2)+(HF(J,1)-HF(J,2))/(X(J)*R*GT)
        T=TREF/GT $ HREF=HREF+HF(J,K)
    70  CONTINUE $ DELH=(HREF-HBASE)/GD

C       CALCULATE THERMODYNAMIC PROPERTIES

    80  NT=M(N) $ DO 100 I=1,NT $ T=(TI+FLOAT(I)*DT(N))/GT $ CP=0.0 $H=0.0
        S=0.0 $ IF (T.GT.TM) K=2 $ DO 90 J=1,NX $ CP=CP+CPF(J,K)
        H=H+HF(J,K)/GD $ IF (T.NE.0.0) S=S+SF(J,K)
    90  CONTINUE $ T=T*GT $ H=H-DELH $ U=H-R*T/GD $ P=EXP((S-SREF)/R)
        V=GJ*R*T/P/GL $ CV=CP-R $G=CP/CV $ C=SQRT(G*GC*R*T*GJ*GM/WW) $L=L+1
        WRITE (6,2010) T,H,U,S,P,V,CP,CV,G,C $ IF (L.LT.55) GO TO 100
        WRITE (6,2000) GAS $ L=0
   100  CONTINUE $ N=N+1 $ TI=T $ IF((N.EQ.15).OR.(M(N).EQ.0)) GO TO 10
        GO TO 80

  1000  FORMAT (9A6)
  2000  FORMAT (1H1,23X,28HTHERMODYNAMIC PROPERTIES OF ,9A6//8X,1HT,7X,
       111HH,9X,1HU,8X,1HS,9X,2HPR,9X,2HVR,8X,2HCP,6X,2HCV,7X,1HG,9X,
       21HA/1H )
  2010  FORMAT (F10.0,2F10.3,F9.4,E15.5,E13.5,F7.4,F8.4,F9.4,F10.2)
        END
```

Solution

(a)   From Table C.4, for $t_1 = 300$ K, $h_1 = 301.329$ kJ/kg, $p_{r1} = 1.0219$, and $\phi_1 = 6.7035$ kJ/kg-K. Thus,

$$p_{r2} = 3p_{r1} = 3(1.0219) = 3.0657$$

Entering Table C.4 at $p_{r2} = 3.0657$, we obtain

$$t_2' = 410.12 \text{ K} \qquad h_2' = 412.418 \text{ kJ/kg}$$
$$\Delta h_c' = h_2' - h_1 = 412.418 - 301.329 = 111.089 \text{ kJ/kg (47.760 Btu/lbm)}$$

(b)   For an isentropic compression efficiency $\eta_c = 0.8$

$$\Delta h_c = \frac{\Delta h_c'}{\eta_c} = \frac{111.089}{0.8} = 138.861 \text{ kJ/kg (59.700 Btu/lbm)}$$

Hence,

$$h_2 = h_1 + \Delta h_c = 301.329 + 138.861 = 440.190 \text{ kJ/kg}$$

Entering Table C.4 at $h_2 = 440.190$ kJ/kg, gives $t_2 = 437.43$ K.

(c)   From equation 1.167,

$$\Delta s = \phi_2 - \phi_1 - R \ln \frac{p_2}{p_1}$$

From Table C.4, for $t_2 = 437.43$ K, $\phi_2 = 7.0844$ kJ/kg-K. Hence,

$$\Delta s = (7.0844 - 6.7035) - 0.28704 \ln (3.0) = 0.0655 \text{ kJ/kg-K} (0.01564 \text{ Btu/lbm-R})$$

The entropy change may also be computed from the values of ϕ_2 and ϕ_2', the latter being the value of ϕ for an isentropic change of state. From part c, $t_2 = 437.43$ K and $\phi_2 = 7.0844$ kJ/kg-K. From part a, for the isentropic compression, $t_2' = 410.12$ K, and from Table C.4, $\phi_2' = 7.0189$ kJ/kg-K. Hence,

$$\Delta s = \phi_2 - \phi_2' = 7.0844 - 7.0189 = 0.0655 \text{ kJ/kg-K} (0.01564 \text{ Btu/lbm-R})$$

Example 1.9.   Carbon dioxide is compressed isentropically from an initial pressure and temperature of 1 atm and 300 K, respectively, to a final pressure of 50 atm. Calculate the final temperature and the compression work (a) assuming that the specific heats of carbon dioxide are constant, and (b) accounting for the variation of the specific heats with temperature.

Solution

(a)   From Table C.3, $\bar{m} = 44.010$, $c_p = 845.73$ J/kg-K, and $\gamma = 1.288$. From equation 1.128,

$$t_2' = t_1 \left(\frac{p_2'}{p_1}\right)^{(\gamma-1)/\gamma} = 300 \left(\frac{50}{1}\right)^{(1.288-1.0)/1.288} = 719.5 \text{ K}$$

$$\Delta h_c' = h_2' - h_1 = c_p(t_2' - t_1) = 845.73(719.5 - 300) = 354.8 \text{ kJ/kg}$$

(b)   From Table D.1B, Volume 2, for $t_1 = 300$ K, $\bar{h}_1 = 69$ kJ/kmol, and $\bar{\phi}_1 = 213.927$ kJ/kmol-K. From equation 1.167, for $s_2 - s_1 = 0$, we obtain

$$\bar{\phi}_2 = \bar{\phi}_1 + \bar{R} \ln \left(\frac{p_2}{p_1}\right) = 213.927 + 8.3143 \ln \left(\frac{50}{1}\right) = 246.453 \text{ kJ/kmol-K}$$

From Table D.1B, Volume 2, for $\bar{\phi}_2 = 246.453$ kJ/kmol-K,

$$t_2' = 643.7 \text{ K} \qquad \text{and} \qquad \bar{h}_2' = 15{,}029 \text{ kJ/kmol}$$

$$\Delta h_c' = \frac{\bar{h}_2' - h_1}{\bar{m}} = \frac{15{,}029 - 69}{44.010} = 339.9 \text{ kJ/kg}$$

Example 1.10.   A 15-m^3 tank contains air at $p_1 = 5.0 \cdot 10^5$ N/m^2 and $t_1 = 500$ K. The air is discharged into the atmosphere through a nozzle until the mass of air contained in the tank is reduced to one half of its original value. Assuming that the process is adiabatic and frictionless, calculate the pressure and temperature of the air remaining in the tank. Take into account the variation of the specific heat of air with temperature.

Solution

The process is assumed to be isentropic. From Table C.4, for $t_1 = 500$ K, $v_{r1} = 23{,}124$ and $p_{r1} = 6.2062$. Since the final mass of the gas in the tank is one half of the original mass, the final specific volume $v_2 = 2v_1$. Hence,

$$v_{r2} = 2v_{r1} = 2(23{,}124) = 46{,}248$$

From Table C.4, for $v_{r2} = 46,248$, $p_{r2} = 2.3655$ and $t_2 = 381.05$ K (685.9 R). Hence, the final pressure p_2 is:

$$p_2 = p_1 \left(\frac{p_{r2}}{p_{r1}} \right) = 5.0 \cdot 10^5 \left(\frac{2.3655}{6.2062} \right) = 1.9058 \cdot 10^5 \ \text{N/m}^2 (27.64 \ \text{lbf/in.}^2)$$

1–17 THE ACOUSTIC SPEED AND THE MACH NUMBER

For a compressible flow, the speed of propagation of small disturbances, called the *acoustic speed*, and the ratio of the flow velocity to the acoustic speed, called the *Mach number*, are important properties of the flow process. Those properties are defined and discussed in this section.

1–17(a) The Acoustic Speed

The *acoustic speed*, also called the *sonic* speed, is the speed with which a sound wave, or a small pressure disturbance, is propagated in a fluid medium. When the fluid may be considered to be a continuum, it can be shown (see any text on elementary physics) that the acoustic speed, denoted by a, is given by

$$a = \left(\frac{K_s}{\rho} \right)^{1/2} \tag{1.179}$$

where the isentropic bulk modulus $K_s = -v(dp/dv)_s$, the density $\rho = 1/v$, and $d\rho/\rho = -dv/v$. Hence, equation 1.179 may be written in the form

$$a^2 = \left(\frac{\partial p}{\partial \rho} \right)_s \tag{1.180}$$

It is pointed out in Section 1–5(d) that, if a fluid is compressible, the magnitude of K will depend on the manner in which the compression is executed. If the attention is focused on the main body of compressible fluid, the fluid external to the boundary layer, and the fluid is assumed to be a perfect gas, and if the compression process is isentropic, then $K_s = \gamma p$ where $\gamma = c_p/c_v$. Introducing $K_s = \gamma p$ into equation 1.179 gives the following expression for the acoustic speed in a perfect gas. Thus

$$a = \left(\frac{\gamma p}{\rho} \right)^{1/2} = (\gamma p v)^{1/2} = (\gamma R t)^{1/2} \tag{1.181}$$

The speed at which a sound wave or small pressure disturbance travels in a fluid depends on its wave length. The pressure wave creates a disturbance in the fluid that imparts directional energy to the fluid particles in the immediate vicinity of the disturbance. By colliding with the contiguous particles, the energized particles impart directional energy to them and, by that process, the disturbance is propagated throughout the body of fluid. Briefly, the collisions between the fluid particles constitute the mechanism by which the pressure wave propagates itself and gives direction to the motion of the fluid particles. Since the speed of propagation of a small pressure disturbance (sound wave) is of the same order of magnitude as the mean translational speed of the fluid particles, the transmission of a small pressure disturbance is influenced by the molecular density of the fluid.

If the molecular density is large, so that the mean free path of the molecules is very small, a small pressure disturbance is propagated in the medium with small energy loss. The foregoing would apply to the propagation of a disturbance in air at standard sea level, for example, where 1 m³ of standard sea-level air contains $2.5 \cdot 10^{24}$ molecules and the corresponding value for the molecular mean free path is $7.37 \cdot 10^{-8}$ m. At increasing altitudes, the molecular density of atmospheric air decreases and, at extreme altitudes, as for example in the ionosphere, it is vanishingly small.

At the high altitudes where the molecular density is very small, the molecular mean free path may be so large that it is equal to or longer than the wave length of the sound wave. If that is the case, a large number of air molecules can move, in a distance equal to 1 wavelength, from the high-pressure region of the sound wave to its low-pressure region without colliding with other particles. Because the temperatures and pressures in the high- and low-pressure regions of a wave are different, the transport of molecules from a high-pressure region of the wave to a low-pressure region without collisions tends to equalize the pressures and temperatures in the wave. The net effect is a dissipation of the energy of the wave, causing it to become damped. Since the length of the mean free path λ governs the propagation of sound waves, it is the criterion that distinguishes the realm of fluid mechanics (continuum) from that of *free particle flow*.

The importance of the acoustic speed in the flow of compressible fluids, jet propulsion, and supersonic flight arises from the fact that the phenomena occurring when the relative velocity between a fluid and a body is large can be related to the acoustic speed. Section 3–8(b) discusses briefly the phenomena associated with the speed of propagation of small pressure disturbances in a compressible fluid. Section 3–8(a) presents a rigorous derivation of equation 1.180.

1–17(b) The Mach Number

Where there is a large relative speed between a body and the compressible fluid surrounding it, the *compressibility* of the fluid, the variation of its density with speed, influences the properties of the flow field. The ratio of the local speed of the fluid V, to its acoustic speed a, called the local *Mach number* $M = V/a$, is a dimensionless criterion of the flow phenomena. For a perfect gas $a = (\gamma Rt)^{1/2}$, so that

$$M = \frac{V}{a} = \frac{V}{(\gamma Rt)^{1/2}} \tag{1.182}$$

or

$$M^2 = \frac{V^2}{a^2} = \frac{V^2}{\gamma Rt} = \frac{\text{directed kinetic energy}}{\text{random kinetic energy}} \tag{1.183}$$

The physical significance of the Mach number can be readily grasped by considering equation 1.183. The velocity V measures the directed motion of the gas particles, and V^2 measures the kinetic energy of the directed flow. The acoustic velocity a for a given gas is proportional to $\sqrt{t}$, which is proportional to the random velocity of the gas particles (see equation 1.86). Hence, a^2 is a measure of the kinetic energy associated with the random motions of the gas molecules. Consequently, $M^2 = V^2/a^2$, for a given set of conditions, may be regarded as a measure of the ratio of the kinetic energy of directed fluid flow to the kinetic energy of random molecular motion.

Example 1.11.   A gas flows through a passage with a speed of 800 m/s. Its local static temperature is 1800 K, its specific heat ratio $\gamma = 1.25$, and the gas constant $R = 322.8$ J/kg-K. Calculate the local sonic velocity and the Mach number.

Solution

From equation 1.181,

$$a = (\gamma R t)^{1/2} = [1.25(322.8)(1800)]^{1/2} = 852.2 \text{ m/s (2796.0 ft/sec)}$$

From equation 1.182,

$$M = \frac{V}{a} = \frac{800}{852.2} = 0.9387$$

1–18 PROPERTIES OF THE ATMOSPHERE

Atmospheric air is one of the important gas mixtures pertinent to the science of gas dynamics. It is of value, therefore, to discuss some of its properties.

The atmospheric density ρ is a function of both the pressure and temperature of the atmospheric air, while the viscosity depends only on the atmospheric temperature. Since the pressure and temperature of atmospheric air are, however, functions of the altitude above the surface of Earth, it has become the common practice to express the physical properties of atmospheric air as functions of the altitude rather than of its pressure and temperature. Since the pressure and temperature of the atmosphere at a given altitude change somewhat with the latitude, season, time of day, and weather, it has become necessary in the interests of standardization and convenience to adopt the so-called International Standard Atmosphere as a standard of reference. Based on that standard, which has been adopted by practically all major countries, the National Aeronautics and Space Administration (NASA) has developed a series of empirical equations for defining the temperature, pressure, and density of atmospheric air as functions of the altitude.[23] Table 1.9 presents the basic characteristics of the standard atmosphere, and Table 1.10 presents its chemical composition. Table C.5 presents the properties of the atmosphere as a function of geometric altitude. It should be noted that the reliability of the tabulated properties of the atmosphere for altitudes exceeding 20,000 m has not been completely established.

Table 1.9   Basic Characteristics of the Sea Level U.S. Standard Atmosphere[23]

		SI Units	*EE* Units
Temperature	t_o	288.16 K	59 F
Pressure	p_o	$1.01325 \cdot 10^5$ N/m^2	2116.22 lbf/ft^2
Density	ρ_o	1.225 kg/m^3	0.076474 lbm/ft^3
Gas constant	R	287.04 J/kg-K	53.35 ft-lbf/lbm-R
Specific heat ratio	$\gamma = c_p/c_v$	1.40	1.40
Acoustic speed	a_o	340.29 m/s	1116.4 ft/sec
Viscosity	μ_o	$1.7894 \cdot 10^{-5}$ kg/m-s	$1.2024 \cdot 10^{-5}$ lbm/ft-sec
Kinematic viscosity	$v_o = \mu_o/\rho_o$	$1.4607 \cdot 10^{-5}$ m^2/s	$1.5723 \cdot 10^{-4}$ ft^2/sec
Mean free path of air molecule	λ_o	$6.6328 \cdot 10^{-8}$ m	$2.1761 \cdot 10^{-7}$ ft
Molecular weight	$\bar{m}_o$	28.9644	28.9644

Table 1.10 Composition of
Clean Dry Atmospheric Air Near
Sea Level[23]

Constituent	Percent by Volume
Nitrogen	78.084
Oxygen	20.9476
Argon	0.934
Carbon dioxide	0.0314
Hydrogen	0.00005
Neon	0.001818
Krypton	0.000114
Xenon	0.0000087
Helium	0.000524
Methane	0.0002
Nitrous oxide	0.00005

For convenience the atmosphere surrounding Earth may be divided into four gaseous layers having markedly different characteristics. Each layer approximates a spherical shell of different thickness, but there is no sharply dividing bounding surface separating one layer from the next. Instead, the characteristics of one layer gradually merge into those of the next layer. The characteristics of each of the four layers are, however, distinctly different. In the order of their distances from the surface of Earth, the layers are called the *troposphere*, the *stratosphere*, the *ionosphere*, and the *exosphere*. All of the layers have the common characteristic that, as one moves farther and farther from the surface of Earth, the density of the gas in each layer continues to diminish and, at extreme altitudes, it approaches zero.

The characteristics of atmospheric air are subject to wide variations from sea level to extreme altitudes. At the lower altitudes, the frequency of molecular collisions is very large, and the air behaves as a continuum (see Section 1–4). As the altitude is increased, the mean free path increases, and the frequency of molecular collisions diminishes; and, at the extreme altitudes, the air is no longer a continuum. Consequently, a body moving through the lower regions of the atmosphere collides with a large number of air molecules, some of which rebound and collide with other molecules, and some of which are deflected into the path of the moving body to repeat the process. At altitudes above 150,000 m, it may be assumed that the atmosphere is practically free space, signifying that in such regions there can be no aerodynamic resistance to flight, nor can there be any lift.

Analyses of the problems of high-speed flight indicate that, where the atmospheric density is substantial, the drag becomes large at very high speeds, and surface heating arises owing to frictional effects in the boundary layer. These limitations are removed if the flight is conducted at altitudes several kilometers above the surface of Earth. The propulsion engines for accomplishing such flights cannot, however, be air-consuming engines. Currently, rocket jet propulsion is the only conceivable means for conducting such flights.

REFERENCES

1. P. W. Bridgman, *Dimensional Analysis*, Yale University Press, New Haven, 1922.
2. E. Buckingham, "Model Experiments and Forms of Empirical Equations," *Transactions of the American Society of Mechanical Engineers*, Vol. 37, 1915.

3. E. A. Mechtly, "The International System of Units," NASA SP-7012, Second Edition, National Aeronautics and Space Administration, 1973.

4. "Brief History of Measurement Systems," NBS SP 304A, National Bureau of Standards, October 1972.

5. H. Rouse, *Advanced Mechanics of Fluids*, Chap. 2, Wiley, New York, 1959.

6. V. L. Streeter, *Handbook of Fluid Mechanics*, Section 2 by A. H. Shapiro, McGraw-Hill, New York, 1961.

7. L. B. Loeb, *The Nature of a Gas*, Wiley, New York, 1931.

8. R. L. Sproull, *Modern Physics*, Wiley, New York, 1956.

9. W. G. Vincenti and C. H. Kruger, *Introduction to Physical Gas Dynamics*, Chap. 2, Wiley, New York, 1965.

10. J. F. Clarke and M. McChesney, *The Dynamics of Real Gases*, Chapter 2, Butterworths, London, 1964.

11. S. A. Schaaf and P. L. Chambre, *Fundamentals of Gas Dynamics*, edited by H. Emmons, pp. 687–692, Princeton University Press, Princeton, N.J., 1958.

12. J. T. R. Watson, "Viscosity of Gases in Metric Units," National Engineering Laboratory, Edinburgh, Her Majesty's Stationary Office, 1972.

13. G. N. Hatsopoulos and J. H. Keenan, *Principles of General Thermodynamics*, Wiley, New York, 1964.

14. L. Page, *Introduction to Theoretical Physics*, D. Van Nostrand, Princeton, N.J., 1959.

15. K. K. Kelly, U.S. Bureau of Mines, Bulletins 476, 1949; 477, 1950; and 584, 1960.

16. F. O. Ellenwood, N. Kulik, and N. R. Gay, "The Specific Heats of Certain Gases over Wide Ranges of Pressures and Temperatures," Bulletin 30, Cornell University Engineering Experiment Station, October 1942.

17. A. A. Vasserman, Y. Z. Kazavchinskii, and V. A. Robinovich, "Thermophysical Properties of Air and Air Components," TT70–50095, National Technical Information Service, Springfield, Va., 1970.

18. T. P. Thinh, J. L. Duran, R. S. Ramalho, and S. Kaliaguine, "Equations Improve C_p^* Predictions," *Hydrocarbon Processing*, pp. 98–103, January 1971.

19. P. E. Liley and W. R. Gambill, "Physical and Chemical Data," *Chemical Engineers Handbook*, Fifth Edition, edited by R. H. Perry and C. H. Chilton, Section 3, McGraw Hill, New York, 1973.

20. D. R. Stull, et al., *JANAF Thermochemical Tables*, Second Edition, NSRDS-NBS 37, National Standard Reference Data Series, National Bureau of Standards, June 1971.

21. S. Gordon and B. J. McBride, "Computer Program for Calculation of Complex Chemical Equilibrium Compositions, Rocket Performance, Incident and Reflected Shocks, and Chapman-Jouguet Detonations," NASA SP-273, National Aeronautics and Space Administration, 1971.

22. J. H. Keenan and J. Kaye, *Gas Tables*, Wiley, New York, 1945.

23. "U.S. Standard Atmosphere, 1962," Superintendent of Documents, U.S. Government Printing Office, Washington, D.C., December 1962.

PROBLEMS

1. The mass of a body at sea level, at the equator, is 10,000 kg. The acceleration because of gravity at sea level is given by

$$g_o = 9.8066 - 0.0259 \cos 2\theta \text{ m/s}^2$$

where θ is the latitude in degrees. Calculate the weight of the body at sea level (a) at 45° latitude, (b) at the North Pole.

2. Derive an equation relating the weight of a body at any altitude z to its weight W_o at sea level. Plot a curve of g/g_o, where g is the local gravitational acceleration and g_o is its sea-level value, as a function of the altitude z. At the equator the diameter of Earth ($2R$) is 12,756 km.

3. Calculate the *mean free path* λ for the gases considered in Table 1.3 (a) at 10 atm and 298 K, (b) at 0.1 atm and 298 K, and (c) at 1 atm and 596 K.

4. Most solid propellants for rocket motor applications contain finely powdered aluminum as a fuel additive, which results in Al_2O_3 appearing in the combustion products. At the temperatures occurring in typical solid propellant rocket motors, Al_2O_3 exists in the liquid phase as very small spherical particles. Those particles are accelerated by the drag force exerted on them by the accelerating gas. Determine whether or not rarefied gas-flow effects should be expected during the expansion process in the propulsive nozzle. Assume that the mean particle diameter $D_p = 1$ micron (10^{-6} m), and that the gas properties at a point near the exit of the propulsive nozzle are $V = 3000$ m/s, $p = 0.35 \cdot 10^5$ N/m², $t = 980$ K, $\gamma = 1.30$, $R = 430$ J/kg-K, and $\mu = 3.0 \cdot 10^{-5}$ kg-m/s. The particle velocity $V_p = 2000$ m/s.

5. A cubic meter of water originally at 293 K and a static pressure of $1.013 \cdot 10^5$ N/m² is subjected to a pressure of $4826 \cdot 10^5$ N/m². Assuming that the walls of the container are not deformed and that there is no change in the temperature of the water, calculate the final volume of the water.

6. The viscosity μ, in N-s/m², of air as a function of the temperature t in K is given by equation 1.30. Calculate the viscosity μ of air, in N-s/m², and the kinematic viscosity v of air, in m²/s, as a function of the altitude z for the range $z = 0$ to 30,000 km. Plot the results.

7. A variable force F, in N, acts on a particle moving along the curve $y = 2x$, in m. The x and y components of F are given by

$$X = x^2 + 2xy \quad \text{and} \quad Y = 10x^2 - 2y$$

Calculate (a) the work required to move the particle from the point where $x = 1$ m to that where $x = 2$ m, (b) the magnitude of F at the point where $x = 2$ m.

8. The circulation Γ is the line integral of the tangential velocity along a given path. Find the circulation about the rectangle enclosed by the lines $x = \pm 2$, $y = \pm 1$ for a two-dimensional flow having the velocity components

$$u = x - y \quad \text{and} \quad v = x^2 - y$$

where u and v are the components, in m/s, in the x and y directions, respectively, in m.

9. Calculate the circulation around a circular path of radius R about the center of a free vortex ($VR = $ constant).

10. Air is allowed to expand from an intial state A (where $p_A = 2.068 \cdot 10^5$ N/m² and $t_A = 333$ K) to state B (where $p_B = 1.034 \cdot 10^5$ N/m² and $t_B = 305$ K). Calculate the change in the specific entropy of the air, and show that the change in entropy is the same for (a) an isobaric process from A to some intermediate state C followed by an isovolumic change from C to B, and (b) an isothermal change from A to some intermediate state D followed by an isentropic change from D to B.

11. The specific heat at constant pressure of a body is determined by heating it internally with an electric coil connected to a 12 V source of electricity. When the current flowing through the coil is 2 A, the temperature of the mass of the body plus coil increases by 30 K in 2 min. If the mass of the system is 0.25 kg (includes mass of the coil) and the specific heat of the coil and the body are identical, calculate the mean specific heat of the body.

12. The specific heat of air at constant pressure, in the range 300 to 1000 K, is given by equation (b) in Example 1.7. How many joules are required to heat 5 kg of air at constant pressure from 300 K to 350 K? (Check the answer by means of Table C.4).

13. During a process, 9000 J of heat leaves the system and enters the surroundings, which are at a temperature of 300 K. (a) If the change in the entropy of the system is -30.0 J/kg-K, is the process reversible, irreversible, or impossible? (b) If the change in the entropy of the system is -20 J/kg-K is the process reversible, irreversible, or impossible? (c) If the change in the entropy of the system is -40.0 J/kg-K, is the process reversible, irreversible, or impossible?

14. A body with a mass of 500 kg moves so that its component velocities in the directions of the x, y, z coordinate axes are $V_x = 150$ m/s, $V_y = 200$ m/s, and $V_z = 250$ m/s. Calculate the components of its momentum in the directions of the coordinate axes and its total momentum, in kg-m/s.

15. An object with a mass of 50 kg has a velocity of 20 m/s in the positive direction of the y axis. It collides with a body having a mass of 75 kg and moving with a velocity of 30 m/s in the negative direction of the y axis. What are the velocities of the bodies after impact, assuming a perfectly elastic collision?

16. Air is compressed isentropically in a centrifugal compressor from a pressure of $1.0 \cdot 10^5$ N/m^2 to a pressure of $6.0 \cdot 10^5$ N/m^2. The initial temperature is 290.0 K. Using Table C.4, calculate (a) the change in temperature, (b) the change in internal energy, (c) the work imparted to the air, neglecting the velocity change, and (d) the average value of the specific heat c_p for the compression process.

17. Air is expanded in an insulated cylinder equipped with a frictionless piston. The initial temperature of the air is 1400 K. The original volume is 1/10 the final volume. Using Table C.4, calculate (a) the change in temperature, (b) the work removed from the gas, and (c) the pressure ratio.

18. The isentropic efficiency of the compressor of a turbojet engine is 0.86. The pressure ratio is 5.2 and the unit is being operated at 10,000-m altitude. (a) Calculate the work per kg for compressing the air (neglect the velocity change). (b) What is the exit temperature?

19. The isentropic efficiency of the turbine of a turbojet engine is 0.90. The pressure ratio of the turbine is 3.0. The inlet temperature to the turbine is 1000 K, and the unit is operated at 10,000-m altitude. (a) Will this turbine drive the compressor and auxiliaries of the turbojet of Problem 18? (Neglect velocity change.) (b) What is the temperature of the gas leaving the turbine?

20. A gas-turbine cycle consists of an isentropic compression from 295 K and 1 atm pressure to 5 atm pressure, an isobaric heating from the compressor outlet conditions to 1000 K, and an isentropic expansion to atmospheric pressure. Use Table C.4, and calculate (a) the compressor work, in J/kg, (b) the heat added in isobaric heating, in J/kg, (c) the turbine work, in J/kg, (d) the net work (turbine work − compressor work), (e) the thermal efficiency (net work/heat added). (f) Compare the answer of part e with the thermal efficiency calculated assuming $c_p = 1.0043$ J/kg-K and $\gamma = 1.4$ throughout the cycle.

21. Calculate the enthalpy change for a compressor with an isentropic efficiency of 0.83, a pressure ratio of 5.2, and an inlet temperature of 265 K (a) by using Table C.4, and (b) by using $c_p = 1.0036$ kJ/kg-K and $\gamma = 1.40$.

22. Calculate the enthalpy change for a turbine operating on heated air with an isentropic efficiency of 0.88, a pressure ratio of 2.3, and an inlet temperature of 1090 K (a) by using Table C.4, (b) by using c_p and γ at 1090 K from Table C.4, and (c) by using an average value of c_p and γ from Table C.4.

23. An airplane is flying at an altitude of 10,000 m at a velocity of 270 km/hr. Calculate the Mach number of the airplane relative to the air.

24. Example A.1 in Appendix A illustrates the determination of a second-order interpolating polynomial for c_p versus t for N_2 in the temperature range 1000 to 5000 K. Determine the corresponding equations for (a) N_2 from 300 to 1000 K, (b) O_2 from 300 to 1000 K, and (c) O_2 from 1000 to 5000 K.

25. In illustrative Example 1.7, two fourth-order interpolating polynomials for c_p versus t for air for the temperature ranges 300 to 1000 K and 1000 to 5000 K are determined. Determine two second-order interpolating polynomials for air for the aforementioned temperature ranges, employing the results obtained in Problem 24.

2
governing equations for compressible fluid flow

2–1 PRINCIPAL NOTATION FOR CHAPTER 2

$\mathbf{B}$ vector body force per unit mass.
$d\mathbf{A}$ vector differential area.
e specific stored energy.
E total stored energy of a system.

F   vector force.

g   acceleration caused by gravity.

h   specific enthalpy.

i,j,k unit vectors in *x*, *y*, and *z* directions, respectively.

m   mass of a system.

ṁ   mass rate of flow.

M   vector momentum of a system.

n   general intensive property.

n$_i$ unit normal vector in *i* direction.

N   general extensive property.

p   absolute static pressure.

Q   heat.

s   specific entropy.

S   total entropy of a system.

u   specific internal energy, or *x*-component of velocity.

v   specific volume, or *y*-component of velocity.

V $= (u^2 + v^2 + w^2)^{1/2}$, magnitude of velocity.

V   velocity vector.

$\mathscr{V}$ control volume.

w   *z*-component of velocity.

W   work.

Greek letters

α angle between **V** and *d***A**.

ρ density.

Other

δ inexact differential.

∇ vector del or nabla operator.

$\dfrac{\partial}{\partial(\ )}$ partial derivative with respect to ( ).

$\dfrac{D}{Dt}$ substantial derivative; that is, differentiation following the motion of a fluid particle.

Vector Operators in Cartesian Coordinates

$\mathbf{A} = \mathbf{i}A_x + \mathbf{j}A_y + \mathbf{k}A_z$

$\mathbf{A} \cdot \mathbf{B} = A_xB_x + A_yB_y + A_zB_z$

$\mathbf{A} \times \mathbf{B} = \mathbf{i}(A_yB_z - A_zB_y) + \mathbf{j}(A_zB_x - A_xB_z) + \mathbf{k}(A_xB_y - A_yB_x)$

$\nabla = \mathbf{i}\dfrac{\partial}{\partial x} + \mathbf{j}\dfrac{\partial}{\partial y} + \mathbf{k}\dfrac{\partial}{\partial z}$

$\nabla\phi = \text{grad } \phi = \mathbf{i}\dfrac{\partial\phi}{\partial x} + \mathbf{j}\dfrac{\partial\phi}{\partial y} + \mathbf{k}\dfrac{\partial\phi}{\partial z}$

$\nabla \cdot \mathbf{A} = \text{div } \mathbf{A} = \dfrac{\partial A_x}{\partial x} + \dfrac{\partial A_y}{\partial y} + \dfrac{\partial A_z}{\partial z}$

$\nabla \times \mathbf{A} = \text{curl } \mathbf{A} = \mathbf{i}\left(\dfrac{\partial A_z}{\partial y} - \dfrac{\partial A_y}{\partial z}\right) + \mathbf{j}\left(\dfrac{\partial A_x}{\partial z} - \dfrac{\partial A_z}{\partial x}\right) + \mathbf{k}\left(\dfrac{\partial A_y}{\partial x} - \dfrac{\partial A_x}{\partial y}\right)$

$\dfrac{D(\ )}{Dt} = \dfrac{\partial(\ )}{\partial t} + u\dfrac{\partial(\ )}{\partial x} + v\dfrac{\partial(\ )}{\partial y} + w\dfrac{\partial(\ )}{\partial z}$

$\mathbf{i} \cdot \mathbf{j} = \mathbf{i} \cdot \mathbf{k} = \mathbf{j} \cdot \mathbf{k} = 0$

$\mathbf{i} \cdot \mathbf{i} = \mathbf{j} \cdot \mathbf{j} = \mathbf{k} \cdot \mathbf{k} = 1$

$$\mathbf{i} \times \mathbf{i} = \mathbf{j} \times \mathbf{j} = \mathbf{k} \times \mathbf{k} = 0$$
$$\mathbf{i} \times \mathbf{j} = \mathbf{k}, \mathbf{j} \times \mathbf{k} = \mathbf{i}, \mathbf{k} \times \mathbf{i} = \mathbf{j}$$

Vector Identities

$$\nabla(ab) = a \nabla b + b \nabla a$$
$$\nabla \cdot (\phi \mathbf{A}) = \phi \nabla \cdot \mathbf{A} + \nabla \phi \cdot \mathbf{A}$$
$$\nabla \times (\phi \mathbf{A}) = \phi \nabla \times \mathbf{A} + \nabla \phi \times \mathbf{A}$$
$$\nabla \cdot (\mathbf{A} \times \mathbf{B}) = (\nabla \times \mathbf{A}) \cdot \mathbf{B} - (\nabla \times \mathbf{B}) \cdot \mathbf{A}$$
$$\nabla \times (\mathbf{A} \times \mathbf{B}) = (\mathbf{B} \cdot \nabla)\mathbf{A} - \mathbf{B}(\nabla \cdot \mathbf{A}) + \mathbf{A}(\nabla \cdot \mathbf{B}) - (\mathbf{A} \cdot \nabla)\mathbf{B}$$
$$(\mathbf{A} \cdot \nabla)\mathbf{A} = \nabla\left(\frac{A^2}{2}\right) - \mathbf{A} \times (\nabla \times \mathbf{A})$$
$$\nabla \cdot (\nabla \times \mathbf{A}) = \operatorname{div} \operatorname{curl} \mathbf{A} = 0$$
$$\nabla \times (\nabla \phi) = \operatorname{curl} \operatorname{grad} \phi = 0$$
$$\nabla \times (\nabla \times \mathbf{A}) = \nabla(\nabla \cdot \mathbf{A}) - \nabla^2 \mathbf{A}$$
$$\frac{D(\ )}{Dt} = \frac{\partial(\ )}{\partial t} + (\mathbf{V} \cdot \nabla)(\ )$$

Vector Calculus

Stokes' theorem

$$\oint_C B \cdot d\mathbf{l} = \int_A (\nabla \times B) \cdot dA$$

Divergence theorem

$$\int_A \mathbf{B} \cdot d\mathbf{A} = \int_{\mathscr{V}} (\nabla \cdot \mathbf{B}) \, d\mathscr{V}$$

2-2 INTRODUCTION

The analysis of a physical situation involving the flow of a fluid is based on determining, for the specific situation at hand, the forms taken by the equations expressing the following physical laws.

1. The law of the conservation of mass (*the continuity equation*).
2. Newton's second law of motion (*the momentum equation*).
3. The first law of thermodynamics, which expresses the principle of the conservation of energy (*the energy equation*).
4. The second law of thermodynamics (*the entropy equation*).
5. The thermodynamic properties of the fluid, from tables of its properties, empirical equations, or an idealized model, such as the equation $pv = Rt$ for a thermally perfect gas.

The mathematical expressions corresponding to principles 1 to 5 are then combined so that relationships are obtained between the variables that are pertinent to the specific problem being analyzed. In many cases, it is unnecessary to derive the equations representing all of the principles listed above.

This chapter is concerned primarily with deriving the general forms of the equations for each of the first four principles for the case where a fluid flows through a rigid, nonaccelerating (i.e., inertial) control volume. In this book it will be assumed, unless the contrary is stated explicitly, that the fluid being studied satisfies the continuum postulate discussed in Section 1-4. The aforementioned general equations, which are hereafter termed the *governing equations*, are derived first in their *integral forms*, which are applicable to any type of flow situation, and then in their corresponding *differential forms* for the case of the adiabatic flow of an inviscid compressible fluid.

The equations derived in Chapter 2 are fundamental to the analytical techniques developed in the remainder of this book.[1–9, 17, 18]

2–3 MATHEMATICAL DESCRIPTION OF A CONTINUUM

Before deriving the *governing equations* for the flow of a fluid, it is useful to formalize in mathematical terms concepts such as *system, control volume, control surface, extensive* and *intensive properties, property field, Lagrangian* and *Eulerian* descriptions of flow, and the *substantial derivative*; some of those concepts are discussed to a limited extent in Chapter 1.

2–3(a) System, Control Volume, and Control Surface

The basic laws governing flow processes are related to a fixed identifiable assemblage of matter that is termed a *system*. Mass neither enters nor leaves a system. Everything external to the system is called either the *surroundings* or the *environment*. Figure 2.1 illustrates a *system* moving through space.

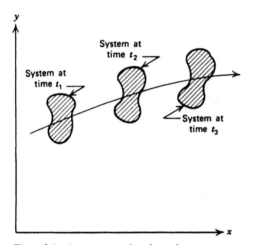

Figure 2.1 A system moving through space.

A *control volume* is defined as an imaginary fixed volume through which a fluid may flow. In general, the control volume may change its shape and its position in space. In this book, however, only rigid control volumes are considered, and if the control volume is moving, it is assumed to move at a constant velocity (i.e., with respect to an *inertial frame of reference*, see Section 1–13); no accelerating, or non-inertial, control volumes are considered.

A *control surface* is the imaginary permeable surface completely enclosing a control volume. Figure 2.2a illustrates the concepts of *system, control volume,* and *control surface* when the control volume is isolated in a body of flowing fluid. Mass can flow across the control surface A except where A is parallel to the fluid velocity vector $\mathbf{V}$. Figure 2.2b illustrates the case where the control surface is in contact with one or more solid boundaries; no fluid can be transported across the latter boundaries. Fluid may flow, of course, across portions of the control surface A that are either not in contact with solid boundaries or parallel to the fluid velocity vector $\mathbf{V}$.

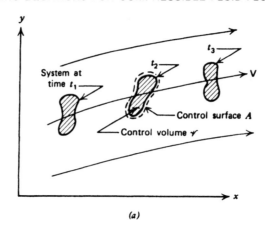

(a)

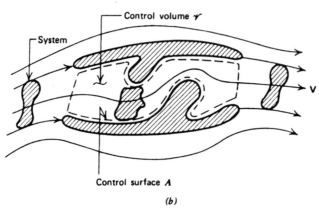

(b)

Figure 2.2 The relationship between a system, control surface, and control volume. (a) System, control surface, and control volume in a flowing fluid with no solid boundaries. (b) System, control surface, and control volume in the presence of solid boundaries.

2–3(b) Extensive and Intensive Properties

An *extensive property* is one that depends on the quantity of mass under consideration. For example, the volume of a system, the mass of a system, and the momentum of a system. In this book, as far as is possible, extensive properties are denoted by capital letters; for example, U for internal energy, S for entropy, etc. An exception is mass, which is denoted by m even though it is an extensive property. A *general extensive property* is denoted by the symbol N.

An *intensive property* is one having a value that is independent of the amount of mass under consideration. There are two types of intensive properties. First, those like pressure and temperature, which are not explicitly dependent on the amount of mass involved, but have magnitudes that are representative of the overall state of the system. Second, there are intensive properties that are the specific values of extensive properties; for example, *specific internal energy u* (internal energy per unit mass), *specific entropy s* (entropy per unit mass), and *specific enthalpy h* (enthalpy per unit mass). The second type of intensive property is generally denoted by lower case letters. A *general intensive property* is denoted by the symbol n.

For a material substance that satisfies the continuum postulate (see Section 1–4), n is defined by

$$n \equiv \lim_{\Delta m \to 0} \frac{\Delta N}{\Delta m} = \frac{dN}{dm} \tag{2.1}$$

Hence, the value of the general extensive property N for a system is given by

$$N = \int_{\text{System}} n \, dm \tag{2.2}$$

Since density is defined by $\rho = dm/d\mathscr{V}$ [see Section 1–4(b)], equation 2.2 transforms to

$$N = \int_{\mathscr{V}} n\rho \, d\mathscr{V} \tag{2.3}$$

where the integration is performed over the volume $\mathscr{V}$ occupied by the system.

The corresponding values of N and n are presented below for the following properties associated with a flowing fluid; mass, momentum, stored energy, and entropy.

$$\text{mass} = m = \int_{\mathscr{V}} \rho \, d\mathscr{V} \qquad (n = 1) \tag{2.4}$$

$$\text{momentum} = \mathbf{M} = \int_{\mathscr{V}} \mathbf{V}\rho \, d\mathscr{V} \qquad (n = \mathbf{V}) \tag{2.5}$$

$$\text{stored energy} = E = \int_{\mathscr{V}} e\rho \, d\mathscr{V} \qquad \left(n = e = u + \frac{V^2}{2} + gz\right) \tag{2.6}$$

$$\text{entropy} = S = \int_{\mathscr{V}} s\rho \, d\mathscr{V} \qquad (n = s) \tag{2.7}$$

The above relationships are utilized later in deriving the governing equations for the flow of a fluid through a control volume.

2–3(c) The Property Field

To simplify the specification of the fluid properties, the concept of a *property field* is employed. In rigid body dynamics, the properties of each particle or body are specified as a function of time. Such an approach is impractical in fluid mechanics because of the tremendously large number of individual particles moving relative to one another in a mass of flowing fluid. Consequently, the properties of a flowing fluid, hereafter termed the *flow properties*, are described in terms of a position in space, thereby establishing a *property field*. The properties of the space are assumed to be those of the fluid flowing through that space. Obviously, a position in space itself has no properties because it is merely a geometrical point. Nevertheless, the artificial concept of a property field is a helpful one, which enables us to develop the mathematical techniques for describing the flow of a fluid. Consequently, instead of employing equations such as $\mathbf{V}_i = \mathbf{V}_i(t)$ for the velocity of an arbitrary fluid particle i, the concept of a property field makes it possible to assign a value of $\mathbf{V} = \mathbf{V}(x,y,z,t)$ to every spatial location in the flow region under consideration. Thus, $\mathbf{V}$ is considered to be a function of (x,y,z,t) rather than of the fluid particles.

2–3(d) The System or Lagrangian Approach

One approach to studying the flow of a fluid is to consider one fluid particle at a time and follow its path through the flow field, as is illustrated schematically for a single particle in Fig. 2.3. In that approach, a fluid particle is an aggregation of discrete particles of fixed identity that satisfies the continuum postulate (see Section 1–4). The approach based on following the paths of the individual particles of the fluid is termed either the *system* or the *Lagrangian* approach. In most cases, however, the Lagrangian approach is not very useful because of the difficulties entailed in keeping track of the tremendously large number of particles comprising a body of fluid. Because the basic laws governing the flow of a fluid are valid, however, only for a system having a fixed mass, the Lagrangian approach must be employed for obtaining the forms of the governing equations that are valid for fixed positions in space (i.e., for control volumes).

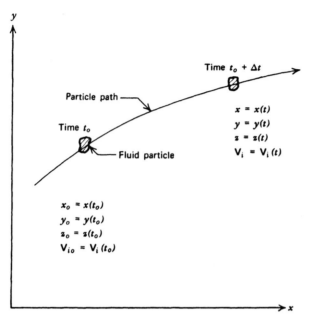

Figure 2.3   The system or Lagrangian approach to the study of fluid motion.

2–3(e) The Control Volume or Eulerian Approach

The difficult problems associated with the *Lagrangian* approach for analyzing the flow of a fluid have been indicated. An alternate approach, termed the Eulerian approach, is based on concentrating the attention on a fixed volume in space and determining the properties of the fluid instantaneously occupying that volume, as illustrated in Fig. 2.4. The Eulerian approach eliminates the need for identifying any particular fluid particle.

The properties of the flow field are completely described in terms of the field description. For example, $V(x,y,z,t)$ defines the velocity at the point $P(x,y,z)$ of the fluid particle that is at that location at time t. Since we do not follow the individual particles as a function of time, the properties of the individual particles are not obtained explicitly. Once the velocity field $V(x,y,z,t)$ is known, however, the trajectories of the individual particles can be traced through the known velocity field, thereby making the properties of the individual particles determinable. In most flow situations, however, detailed information on the individual particles is not required, so that the *Eulerian* description of the flow is entirely adequate; the *Eulerian* approach is the one employed in this book.

Because the basic laws are expressed, however, in a frame of reference for a system, a relationship is needed between the *Eulerian* and *Lagrangian* approaches that will enable expressing the basic laws in terms of the variables applicable to a control volume (see Section 2–4).

2–3(f) The Substantial Derivative

Since the governing equations for the flow of a fluid, hereafter for brevity called *fluid flow*, are valid for a system, the time derivative of the properties of a fluid particle must be expressed in terms of the field description of the flow properties. The time derivative of a flow property is known as its *substantial derivative*; it is also called either the *material* or the *particle derivative*, and is denoted by $D(\ )/Dt$, where $(\ )$ denotes the property under consideration. The differentiation process itself is commonly termed either *differentiation by pursuit* or *differentiation along a particle path*.

Figure 2.5 is an illustration of the process in a Cartesian Lagrangian space. Let N denote an arbitrary *extensive property*; in a field representation N is specified by

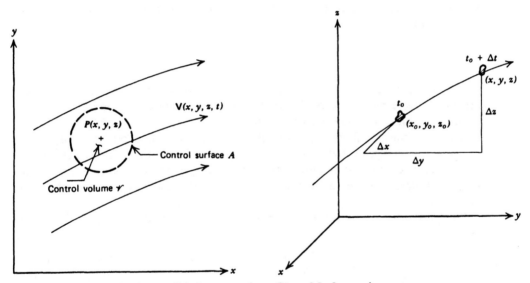

Figure 2.4 The control volume or Eulerian approach to the study of fluid motion.   **Figure 2.5** Lagrangian space.

$N(x,y,z,t)$. In Lagrangian space, however, the coordinates of a fluid particle (x,y,z) are also functions of time, so that

$$x = x(t) \qquad y = y(t) \qquad z = z(t) \qquad (2.8)$$

The change in N during the increment of time Δt may be approximated by the first-order terms of a Taylor series. Thus,

$$\Delta N_{\text{System}} = N(t_o + \Delta t) - N(t_o)$$

$$= \left(\frac{\partial N}{\partial x}\right)_{t_o} \Delta x + \left(\frac{\partial N}{\partial y}\right)_{t_o} \Delta y + \left(\frac{\partial N}{\partial z}\right)_{t_o} \Delta z + \left(\frac{\partial N}{\partial t}\right)_{t_o} \Delta t \qquad (2.9)$$

The instantaneous time rate of change of N for the system at time t_o may be obtained by dividing equation 2.9 by Δt and taking the limit as Δt approaches zero. Thus,

$$\left(\frac{dN}{dt}\right)_{\text{System at } t_o} = \lim_{\Delta t \to 0} \frac{N(t_o + \Delta t) - N(t_o)}{\Delta t} \tag{2.10}$$

In Lagrangian space, since $x = x(t)$ for the system,

$$\lim_{\Delta t \to 0} \left(\frac{\Delta x}{\Delta t}\right)_{\text{System}} = \left(\frac{dx}{dt}\right)_{\text{System}} = u(t_o) \tag{2.11}$$

where $u(t_o)$ is the particle velocity in the x direction at the time t_o. By applying similar considerations to the second and third terms of equation 2.9, we obtain $v(t_o)$ and $w(t_o)$, where $v(t_o)$ and $w(t_o)$ denote the velocities parallel to the y and z axes, respectively. Thus, equation 2.10 transforms to

$$\left(\frac{dN}{dt}\right)_{\text{System at } t_o} = \left[\left(\frac{\partial N}{\partial t}\right) + u\left(\frac{\partial N}{\partial x}\right) + v\left(\frac{\partial N}{\partial y}\right) + w\left(\frac{\partial N}{\partial z}\right)\right] \tag{2.12}$$

The derivative expressed by equation 2.12 is the *substantial derivative* of the arbitrary extensive property N and, as pointed out earlier, is denoted by the special symbol $D(\ )/Dt$. Thus,

$$\frac{DN}{Dt} \equiv \left(\frac{dN}{dt}\right)_{\text{System at } t_o} \tag{2.13}$$

The first term in equation 2.12 arises from any unsteady flow effects that may be present and is termed the *local change*. The last three terms each arise from a change in the position of the fluid particle in the flow field and are called the *convective change*. The substantial derivative of a flow property, therefore, relates the instantaneous rate of change of that flow property to derivatives of the property field. Since the property field may be employed directly in an Eulerian description of a flow, the substantial derivative, therefore, relates the *Lagrangian* and *Eulerian* descriptions of fluid flow.

The substantial derivative may be considered to be an *operator* of the following form.

$$\frac{D(\ )}{Dt} = \frac{\partial(\ )}{\partial t} + u\frac{\partial(\ )}{\partial x} + v\frac{\partial(\ )}{\partial y} + w\frac{\partial(\ )}{\partial z} \tag{2.14}$$

In vector notation, the substantial derivative is

$$\frac{D(\ )}{Dt} = \frac{\partial(\ )}{\partial t} + (\mathbf{V} \cdot \nabla)(\ ) \tag{2.15}$$

where the order of $\mathbf{V}$ and ∇ in the expansion of the above *dot* or *scalar product* must be preserved. The operator ∇, which is called either *del* or *nabla*, is defined in Cartesian coordinates by

$$\nabla \equiv \mathbf{i}\frac{\partial}{\partial x} + \mathbf{j}\frac{\partial}{\partial y} + \mathbf{k}\frac{\partial}{\partial z}$$

Example 2.1. An Eulerian description of an unsteady two-dimensional velocity field in Cartesian coordinates is given by

$$\mathbf{V}(x,y,t) = \mathbf{i}e^{xt} + \mathbf{j}e^{yt}$$

Determine the acceleration of the particle at location (1,2) at time $t = 2$.

Solution

The acceleration of a particle is the time rate of change of the velocity of the particle, which is the substantial derivative of the velocity. Thus,

$$\mathbf{a}_p = \frac{D\mathbf{V}}{Dt} = \frac{\partial \mathbf{V}}{\partial t} + (\mathbf{V} \cdot \nabla)\mathbf{V} \tag{a}$$

For a two-dimensional flow in a Cartesian coordinate system $\mathbf{V} = \mathbf{i}u + \mathbf{j}v$. Then,

$$\frac{\partial \mathbf{V}}{\partial t} = \frac{\partial}{\partial t}(\mathbf{i}u + \mathbf{j}v) = \mathbf{i}\frac{\partial u}{\partial t} + \mathbf{j}\frac{\partial v}{\partial t} \tag{b}$$

$$(\mathbf{V} \cdot \nabla)\mathbf{V} = u\frac{\partial \mathbf{V}}{\partial x} + v\frac{\partial \mathbf{V}}{\partial y} \tag{c}$$

$$(\mathbf{V} \cdot \nabla)\mathbf{V} = u\frac{\partial}{\partial x}(\mathbf{i}u + \mathbf{j}v) + v\frac{\partial}{\partial y}(\mathbf{i}u + \mathbf{j}v) \tag{d}$$

$$(\mathbf{V} \cdot \nabla)\mathbf{V} = \mathbf{i}u\frac{\partial u}{\partial x} + \mathbf{j}u\frac{\partial v}{\partial x} + \mathbf{i}v\frac{\partial u}{\partial y} + \mathbf{j}v\frac{\partial v}{\partial y} \tag{e}$$

Adding equations (b) and (e) and rearranging gives

$$\mathbf{a}_p = \mathbf{i}\left(u\frac{\partial u}{\partial x} + v\frac{\partial u}{\partial y} + \frac{\partial u}{\partial t}\right) + \mathbf{j}\left(u\frac{\partial v}{\partial x} + v\frac{\partial v}{\partial y} + \frac{\partial v}{\partial t}\right) \tag{f}$$

Substituting $u = e^{xt}$ and $v = e^{yt}$ into equation (f) yields

$$\mathbf{a}_p = \mathbf{i}[e^{xt}te^{xt} + e^{yt}(0) + xe^{xt}] + \mathbf{j}[e^{xt}(0) + e^{yt}te^{yt} + ye^{yt}]$$

$$\mathbf{a}_p = \mathbf{i}e^{xt}(te^{xt} + x) + \mathbf{j}e^{yt}(te^{yt} + y) \tag{g}$$

Substituting the values $x = 1$, $y = 2$, and $t = 2$ into equation (g) gives

$$\mathbf{a}_p = \mathbf{i}e^{(1)(2)}[2e^{(1)(2)} + 1] + \mathbf{j}e^{(2)(2)}[2e^{(2)(2)} + 2]$$

$$\mathbf{a}_p = \mathbf{i}e^2(2e^2 + 1) + \mathbf{j}2e^4(e^4 + 1)$$

2–4 RELATIONSHIP BETWEEN THE SYSTEM AND THE CONTROL VOLUME APPROACHES

The discussion in the present section is concerned with determining how the time rate of change of an arbitrary extensive fluid property N for a system may be expressed as the variation of that property for a control volume. Figure 2.6 illustrates diagrammatically an arbitrary velocity field $\mathbf{V}(x,y,z,t)$ measured with respect to the x, y, z *coordinate system*. The control volume is denoted by $\mathscr{V}$ and the corresponding control surface by A. A system of finite size is illustrated at the times t and $t + \Delta t$; the streamline pattern is that applicable to the instantaneous time t. The control volume is assumed to be fixed in x, y, z space. The system and the control volume occupy the identical space at the time t, the two regions denoted by I and II. At the later time $t + \Delta t$, the system occupies the regions II and III.

The time rate of change of the extensive property N for the system is determined by applying the following limiting process.

$$\left(\frac{dN}{dt}\right)_{System} = \lim_{\Delta t \to 0}\left(\frac{\Delta N}{\Delta t}\right)_{System} = \left\{\begin{array}{l}\text{Time rate of} \\ \text{change of } N \text{ for} \\ \text{the System at } t\end{array}\right\} = \left\{\begin{array}{l}\text{Time rate of change of} \\ N \text{ for the mass in the} \\ \text{control volume at } t\end{array}\right\} \tag{2.16}$$

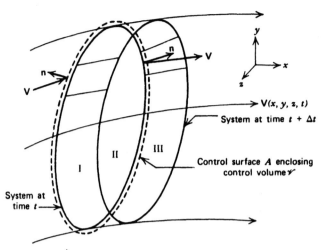

Figure 2.6 The relationship between a system and a control volume.

As pointed out in Section 2–3(f), a derivative of the type presented in equation 2.16 is a total derivative following a fluid particle and is consequently the substantial derivative of N. Hence, by introducing the operator $D(\ )/Dt$ (see equation 2.13), we may write

$$\frac{DN}{Dt} = \left(\frac{dN}{dt}\right)_{\text{System}} \tag{2.17}$$

Application of equation 2.16 yields

$$\frac{DN}{Dt} = \lim_{\Delta t \to 0} \left\{ \frac{N_{t+\Delta t} - N_t}{\Delta t} \right\}_{\text{System}} \tag{2.18}$$

The terms in equation 2.18 are given by the following expressions, where $dN = n\rho\, d\mathcal{V}$. Thus,

$$(N_{t+\Delta t})_{\text{System}} = (N_{\text{II}} + N_{\text{III}})_{t+\Delta t} = \left(\int_{\text{II}} n\rho\, d\mathcal{V} + \int_{\text{III}} n\rho\, d\mathcal{V} \right)_{t+\Delta t} \tag{2.19}$$

$$(N_t)_{\text{System}} = (N_{\text{I}} + N_{\text{II}})_t = \left(\int_{\text{I}} n\rho\, d\mathcal{V} + \int_{\text{II}} n\rho\, d\mathcal{V} \right)_t \tag{2.20}$$

Combining equations 2.18 to 2.20, and noting that the limit of the sum is equal to the sum of the limits, the following expression is obtained.

$$\frac{DN}{Dt} = \lim_{\Delta t \to 0} \frac{\left[\left(\int_{\text{II}} n\rho\, d\mathcal{V} \right)_{t+\Delta t} - \left(\int_{\text{II}} n\rho\, d\mathcal{V} \right)_t \right]}{\Delta t}$$
$$+ \lim_{\Delta t \to 0} \frac{\left(\int_{\text{III}} n\rho\, d\mathcal{V} \right)_{t+\Delta t}}{\Delta t} - \lim_{\Delta t \to 0} \frac{\left(\int_{\text{I}} n\rho\, d\mathcal{V} \right)_t}{\Delta t} \tag{2.21}$$

In the limit, as Δt approaches zero, region II becomes identical to the control volume $\mathcal{V}$, and the first term on the right-hand side of equation 2.21 becomes

$$\frac{\partial}{\partial t} \int_{\mathcal{V}} n\rho\, d\mathcal{V} \tag{2.22}$$

The value of the second term on the right-hand side of equation 2.21 is determined by noting that the integral is the amount of the extensive property N that entered region III during the time Δt. It is, therefore, the amount of N that left the control volume during the time Δt. Thus,

$$\left(\int_{\text{III}} n\rho \, d\mathcal{V}\right)_{t+\Delta t} = \int (dN_{\text{III}})_{t+\Delta t} \tag{2.23}$$

Dividing equation 2.23 by Δt and taking the limit, as Δt approaches zero, yields the instantaneous rate at which N leaves the control volume.

$$\lim_{\Delta t \to 0} \frac{\left(\int_{\text{III}} n\rho \, d\mathcal{V}\right)_{t+\Delta t}}{\Delta t} = \int \left(\frac{dN}{dt}\right)_{\text{out}} = \text{Rate of efflux of } N \tag{2.24}$$

Since $dN = n \, dm$, the rate of efflux of N may be expressed as the product of the mass flow rate out $d\dot{m}_{\text{out}}$, and n, the specific value of the general extensive property N. Hence,

$$\left(\frac{dN}{dt}\right)_{\text{out}} = n \, d\dot{m}_{\text{out}} \tag{2.25}$$

The mass rate of flow out of the control volume $\mathcal{V}$ may be expressed in terms of ρ, $\mathbf{V}$, and the differential area dA. Consider a differential portion of the control surface dA, as illustrated in Fig. 2.7. The flow of mass through the differential area dA is given by

$$d\dot{m} = \rho V \cos \alpha \, dA \tag{2.26}$$

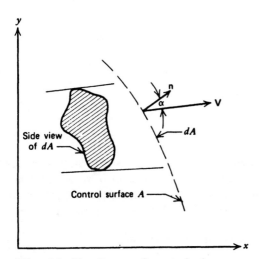

Figure 2.7 Mass flow out of a control volume.

Expressing dA as a vector in the direction of the normal to the surface (using the convention that the outward normal points in the positive direction), equation 2.26 may be rewritten as

$$d\dot{m} = \rho \mathbf{V} \cdot d\mathbf{A} \tag{2.27}$$

where $d\mathbf{A} = \mathbf{n} \, dA$, and $\mathbf{n}$ denotes the unit normal in the direction of the outward normal to the control surface. Combining equations 2.25 and 2.27 yields

$$\left(\frac{dN}{dt}\right)_{\text{out}} = n\rho \mathbf{V} \cdot d\mathbf{A}_{\text{out}} \tag{2.28}$$

Substituting equation 2.28 into equation 2.24, the second term on the right-hand side of equation 2.21 becomes

$$\lim_{\Delta t \to 0} \frac{\left(\int_{III} n\rho \, d\mathcal{V} \right)_{t+\Delta t}}{\Delta t} = \int_{A_{out}} \left(\frac{dN}{dt} \right)_{out} = \int_{A_{out}} n\rho \mathbf{V} \cdot d\mathbf{A}_{out} \qquad (2.29)$$

Equation 2.29 converts the original *volume integral* over region III into a *surface integral* over that portion of the control surface where mass flows out of the control volume.

The value of the third term on the right-hand side of equation 2.21 is obtained in an analogous manner; that is, by noting that the integrand is the amount of N that entered the control volume during the time Δt. Figure 2.8 illustrates a differential

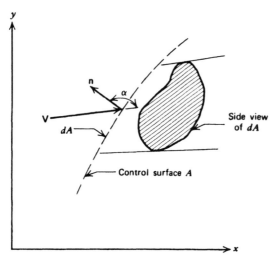

Figure 2.8 Mass flow into a control volume.

portion of the control surface through which mass, and consequently N, is flowing into the control volume. Note that for fluid *entering* the control volume, the angle α will always be greater than 90 deg and less than 270 deg; consequently, $\cos \alpha$ is always negative. Hence, the mass flow rate crossing the differential area dA is given by

$$d\dot{m} = -\rho V \cos \alpha \, dA = -\rho \mathbf{V} \cdot d\mathbf{A} \qquad (2.30)$$

where, as before, $d\mathbf{A} = \mathbf{n} \, dA$. The negative sign appears because, as pointed out above, $\cos \alpha$ is always negative; the mass flow rate, of course, is always a positive quantity.

By employing the procedure utilized for determining the value of the second term in equation 2.21, we obtain for the third term of that equation

$$-\int_{A_{in}} \left(\frac{dN}{dt} \right)_{in} = \int_{A_{in}} n\rho \mathbf{V} \cdot d\mathbf{A}_{in} \qquad (2.31)$$

Note that the negative sign preceding the third term in equation 2.21 combines with the negative sign of equation 2.30. As a result, a positive expression is obtained in equation 2.31. Furthermore, it should be noted that the original volume integral over region II has been transformed into a surface integral that applies to that portion of the control surface across which mass flows into the control volume.

Combining equations 2.29 and 2.31 gives

$$\int_{A_{out}} \left(\frac{dN}{dt}\right)_{out} - \int_{A_{in}} \left(\frac{dN}{dt}\right)_{in} = \int_A n\rho \mathbf{V} \cdot d\mathbf{A} \tag{2.32}$$

In equation 2.32, the subscript A attached to the integral sign denotes that the integral is to be extended over the entire control surface A. For mass efflux, the *scalar* product of $\mathbf{V}$ and $d\mathbf{A}$ yields a positive contribution to DN/Dt, while for mass influx it yields a negative contribution, which is in agreement with the signs in equation 2.21. The single term on the right-hand side of equation 2.32 accounts, therefore, for both the second and third terms on the right-hand side of equation 2.21.

The foregoing shows that the results of the limiting process indicated in equation 2.21 are given by the terms in equation 2.22 and 2.32. Hence,

$$\frac{DN}{Dt} = \frac{\partial}{\partial t} \int_{\mathscr{V}} n\rho \, d\mathscr{V} + \int_A n\rho \mathbf{V} \cdot d\mathbf{A} \tag{2.33}$$

Hence, the instantaneous time rate of change for any extensive property N for a system, at time t, has been expressed in terms of two effects: one relative to the control volume, and the other relative to its control surface. The first term expresses the time rate of change of N inside the control volume at the time t, and the second term expresses the net rate of flow of N across the control surface A at the same instant of time. Equation 2.33 is the desired relationship between the system and control volume approaches to fluid flow problems. For an inertial control volume of fixed shape, the time derivative may be taken inside the integral, yielding

$$\frac{DN}{Dt} = \int_{\mathscr{V}} \frac{\partial(n\rho)}{\partial t} \, d\mathscr{V} + \int_A n\rho \mathbf{V} \cdot d\mathbf{A} \tag{2.34}$$

A useful alternate form of equation 2.34 may be obtained in terms of volume integrals entirely, by employing the *divergence theorem*. From vector analysis (see Section 2–1), for any vector $\mathbf{B}$

$$\int_A \mathbf{B} \cdot d\mathbf{A} = \int_{\mathscr{V}} \nabla \cdot \mathbf{B} \, d\mathscr{V} \tag{2.35}$$

If $\mathbf{B}$ is chosen to be $(n\rho \mathbf{V})$, then

$$\int_A (n\rho \mathbf{V}) \cdot d\mathbf{A} = \int_{\mathscr{V}} \nabla \cdot (n\rho \mathbf{V}) \, d\mathscr{V} \tag{2.36}$$

Combining equations 2.34 and 2.36 yields

$$\frac{DN}{Dt} = \int_{\mathscr{V}} \left[\frac{\partial(n\rho)}{\partial t} + \nabla \cdot (n\rho \mathbf{V}) \right] d\mathscr{V} \tag{2.37}$$

Once the extensive property N has been specified (i.e., mass, momentum, energy, or entropy), its specific value n can be determined. The integrand of the second term in equation 2.37 may be expanded further by employing the following vector identity. Let ϕ be any scalar and $\mathbf{A}$ any vector; then, from Section 2–1,

$$\nabla \cdot (\phi \mathbf{A}) = \phi(\nabla \cdot \mathbf{A}) + \mathbf{A} \cdot (\nabla \phi) \tag{2.38}$$

Another useful concept is that of the instantaneous time rate of change of N for a system of differential size. If the control volume $\mathscr{V}$ is shrunk to differential size $d\mathscr{V}$, then the limiting value of the integral in equation 2.37 is the integrand itself. Dividing that result by $d\mathscr{V}$ yields an expression for the time rate of change of N per unit volume. Thus,

$$\frac{DN/Dt}{d\mathscr{V}} = \frac{\partial(n\rho)}{\partial t} + \nabla \cdot (n\rho \mathbf{V}) = (n\rho)_t + \nabla \cdot (n\rho \mathbf{V}) \tag{2.39}$$

where $(n\rho)_t = \partial(n\rho)/\partial t^*$. The concept embodied in equation 2.39 is often employed for obtaining a differential equation from a corresponding integral equation, such as equation 2.37.

In summary, equation 2.34 expresses the time rate of change of any extensive property N of a system in terms of the properties of a control volume. Equation 2.34 will be employed for expressing the basic equations of mechanics and thermodynamics for systems as equations that are valid for control volumes.

2-5 CONSERVATION OF MASS

The law of the conservation of mass for a system states that, in the absence of nuclear and relativity effects, the mass of a system is a constant. Thus,

$$(\text{mass})_{\text{System}} = \text{constant} \tag{2.40}$$

Let the extensive property N denote the mass of a system; that is, $N = \text{mass}$. Accordingly, equation 2.40 gives

$$\frac{DN}{Dt} = \frac{D(\text{mass})}{Dt} = 0 \tag{2.41}$$

The mass contained inside the control volume $\mathscr{V}$ is given by $N = \text{mass} = \int_{\mathscr{V}} \rho \, d\mathscr{V}$; the corresponding intensive property is $n = 1$. From equation 2.34, it follows that $DN/Dt = 0$ is given by

$$\int_{\mathscr{V}} \frac{\partial \rho}{\partial t} d\mathscr{V} + \int_{A} \rho \mathbf{V} \cdot dA = 0 \tag{2.42}$$

Equation 2.42 is the integral form of the *law of the conservation of mass*. It is valid for a control volume, and is frequently called the *continuity equation*. The only restrictions on that equation are that the continuum postulate is valid, and that nuclear and relatively effects are absent. For *steady flows*, the first term is zero.

In words, the basic principle expressed by equation 2.42 is that the rate at which (fluid) mass *accumulates* inside an inertial control volume $\mathscr{V}$ of fixed size is equal to the difference between the rate at which mass enters and leaves $\mathscr{V}$ by crossing the control surface A, which encloses $\mathscr{V}$.

A differential form for the continuity equation, valid for control volumes, is required in the study of multidimensional flows. Two procedures are available for obtaining the pertinent differential equations from the integral equations. Both approaches are illustrated in deriving the equation for the *law of the conservation of mass*. The remaining basic laws are obtained in their differential form by applying the first of the two approaches.

The *first approach* employs the *divergence theorem* (see equation 2.35). The latter is employed for converting a surface integral into a volume integral. The limiting value of the volume integral is then obtained as $\mathscr{V}$ approaches zero; the technique is employed in deriving equation 2.39. For $N = \text{mass}$, $n = 1$, and $DN/Dt = 0$. Equation 2.39 becomes

$$\rho_t + \nabla \cdot (\rho \mathbf{V}) = 0 \tag{2.43}$$

where $\rho_t \equiv (\partial \rho / \partial t)$.

---

* The convention of denoting partial derivatives by subscripts will be employed when there can be no confusion with other notation.

Equation 2.43 is the desired differential form of the continuity equation that is valid for a control volume. The corresponding equation for any coordinate system may be obtained by applying the *operator* ∇ for that system. For steady flows, $\rho_t = 0$.

· The *second* approach, which will be employed here only for the purpose of illustrating the technique, is based on applying the integral equation to a control volume of differential size. Figure 2.9 illustrates a differential control volume in a Cartesian

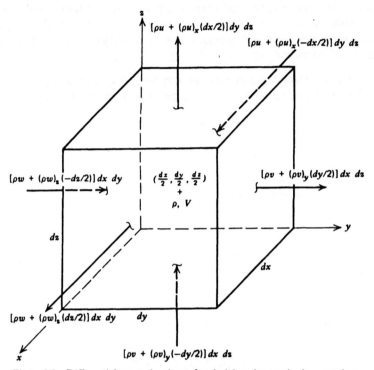

Figure 2.9 Differential control volume for deriving the continuity equation.

coordinate system. Let ρ and V denote the properties at the *center* of the differential control volume that has the coordinates $(dx/2, dy/2, dz/2)$. For the differential control volume, it may be assumed (to a first-order approximation) that the properties of the fluid are constant over each of its six faces. On its back face, that facing the negative x direction, the density is $\rho + \rho_x(-dx/2)$, and on the front face it is $\rho + \rho_x(dx/2)$. Similar expressions may be written for the products (ρu), (ρv), and (ρw), where u, v, and w are the Cartesian components of the velocity V. Hence, on the back face, the value of (ρu) is $(\rho u) + (\rho u)_x(-dx/2)$, and on the front face it is $(\rho u) + (\rho u)_x(dx/2)$. The mass flow entering the back face is, from equation 2.42,

$$\int_A \rho V \cdot dA = \left[\rho u - (\rho u)_x \left(\frac{dx}{2} \right) \right] dy \, dz \qquad (2.44)$$

Similar expressions may be written for the other five faces. The appropriate expressions are indicated in Fig. 2.9. The unsteady term in equation 2.42 is simply $\rho_t(dx \, dy \, dz)$. Combining those results and dividing by $(dx \, dy \, dz)$ yields

$$\rho_t + (\rho u)_x + (\rho v)_y + (\rho w)_z = 0 \qquad (2.45)$$

Equation 2.45 is the *differential form of the continuity equation* for a Cartesian coordinate system. Results for other coordinate systems may be obtained in a similar manner.

In vector notation, equation 2.45 becomes

$$\rho_t + \nabla \cdot (\rho \mathbf{V}) = 0 \qquad (2.46)$$

which is identical to equation 2.43. The alternate approach, as to be expected, yields the same result as that obtained by the first approach. Because it is more involved than the latter, it will not be employed hereafter. It should be pointed out, however, that in those cases where transport phenomena (i.e., viscosity, conductivity, and diffusion) must be considered, the alternate approach may prove to be the more straightforward (see Section 5–9).

Example 2.2.   Air flows out of a tank through a rectangular outlet duct, as illustrated in Fig. 2.10. The combined volume of the tank and the outlet duct is denoted by $\mathscr{V}$.

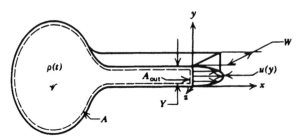

Figure 2.10   Sketch for Example 2.2.

If the height of the outlet duct Y is small compared to the width W, then the velocity profile at the outlet section is given by

$$u(y) = (4U/Y^2)(yY - y^2)$$

where U is the maximum velocity. Assume that the flow rate is so small that the density of the air throughout the tank and outlet duct may be considered to be uniform [i.e., $\rho \neq \rho(x,y,z)$]. (a) Develop expressions for the instantaneous mass flow rate at the outlet and for the time rate of change of the density of the air in the tank and outlet duct. (b) A specific tank and outlet duct has a combined volume of 0.283 m^3, an outlet duct height $Y = 0.00254$ m, and an outlet duct width $W = 0.3048$ m. Calculate $\dot{m}_{out}$ and $\partial \rho / \partial t$ at an instant when the pressure $p = 1.379 \cdot 10^5$ N/m^2, the temperature $t = 277.8$ K, and the outlet maximum velocity $U = 30.48$ m/s. The gas constant for air is 287.04 J/kg-K.

Solution

(a)   Define the inside walls of the tank and outlet duct and the outlet area itself as the control surface A. The control volume $\mathscr{V}$ comprises the volume of the tank plus the volume of the outlet duct. The continuity equation for this control volume is obtained from equation 2.42. Thus,

$$\int_{\mathscr{V}} \frac{\partial \rho}{\partial t} d\mathscr{V} + \int_{A_{out}} \rho \mathbf{V} \cdot d\mathbf{A} = 0 \qquad (a)$$

where A_{out} denotes that portion of the total control surface A across which mass flows.

The instantaneous mass flow rate out of the control volume $\mathscr{V}$ is obtained from the second term in equation (a).

$$\dot{m}_{out} = \int_{A_{out}} \rho \mathbf{V} \cdot d\mathbf{A} = \int_{A_{out}} d\dot{m}_{out} \tag{b}$$

The time rate of change of the density of the air in the tank is obtained from the first term in equation (a). If ρ is uniform over the control volume, then $\partial\rho/\partial t$ is not a function of x, y, z. The first term in equation (a) becomes

$$\int_{\mathscr{V}} \frac{\partial\rho}{\partial t} d\mathscr{V} = \frac{\partial\rho}{\partial t} \mathscr{V} \tag{c}$$

Combining equations (a), (b), and (c), gives

$$\frac{\partial\rho}{\partial t} \mathscr{V} + \dot{m}_{out} = 0 \tag{d}$$

To determine $\dot{m}_{out}$, note that

$$\mathbf{V} = \mathbf{i} \left(\frac{4U}{Y^2}\right)(yY - y^2) \tag{e}$$

$$d\mathbf{A} = \mathbf{i}(dz\,dy)$$

so that equation (b) becomes

$$\dot{m}_{out} = \rho \int_0^Y \int_0^W \left(\frac{4U}{Y^2}\right)(yY - y^2)\,dz\,dy \tag{g}$$

$$\dot{m}_{out} = \rho W \left(\frac{4U}{Y^2}\right)(Yy^2/2 - y^3/3)\bigg|_0^Y = \frac{2\rho UWY}{3} \tag{h}$$

Substituting equation (h) into equation (d) gives

$$\frac{\partial\rho}{\partial t} = -\frac{2\rho UWY}{3\mathscr{V}} \tag{i}$$

(b) The instantaneous density of the air in the control volume is calculated by applying the thermal equation of state for a perfect gas [equation 1.99, Section 1–15(a)].

$$\rho = \frac{p}{Rt} = \frac{1.379 \cdot 10^5}{287.04(277.8)} = 1.729 \text{ kg/m}^3$$

From equation (h),

$$\dot{m}_{out} = \frac{2\rho UWY}{3} = \frac{2(1.729)(30.48)(0.3048)(0.00254)}{3} = 0.0272 \text{ kg/s (0.0600 lbm/sec)}$$

From equation (d),

$$\frac{\partial\rho}{\partial t} = -\frac{\dot{m}_{out}}{\mathscr{V}} = -\frac{0.0272}{0.283} = -0.0961 \text{ kg/m}^3\text{-s } (-0.00600 \text{ lbm/ft}^3\text{-sec})$$

2-6 NEWTON'S SECOND LAW OF MOTION

The equation obtained by applying Newton's second law of motion to the fluid flowing through a control volume is called the *momentum equation*. From Section

1–13(c), equation 1.80, Newton's second law of motion for a system of particles is

$$\mathbf{F}_{external} = \frac{d\mathbf{M}}{dt} \tag{2.47}$$

The derivative in equation 2.47 is a derivative following a system of fluid particles. Consequently, it is the substantial derivative, and equation 2.47 may be rewritten as follows.

$$\mathbf{F}_{external} = \frac{D\mathbf{M}}{Dt} \tag{2.48}$$

In terms of the general properties N and n, $N = \mathbf{M} = \int_{\mathscr{V}} \mathbf{V} \rho \, d\mathscr{V}$ and $n = \mathbf{V}$. Substituting these results into equation 2.34 yields

$$\mathbf{F}_{external} = \int_{\mathscr{V}} (\rho \mathbf{V})_t \, d\mathscr{V} + \int_A \mathbf{V}(\rho \mathbf{V} \cdot d\mathbf{A}) \tag{2.49}$$

where the subscript t denotes partial differentiation with respect to time t. To complete the transformation from the system approach to the control volume approach, the external forces acting on the system must be expressed in terms of the forces acting upon the control volume.

2–6(a) The Net External Force Acting on a Body of Fluid

It is pointed out in Section 1–4(a) that the external forces acting on a body of fluid may be divided into body forces $\mathbf{F}_B$ and surface forces $\mathbf{F}_S$.

$$\mathbf{F}_{external} = \mathbf{F}_B + \mathbf{F}_S \tag{2.50}$$

Body forces are those that act on the entire mass of the fluid; they are frequently called *forces at a distance*. The most common body force is gravitational attraction; other important examples are electrical and magnetic forces on a charged fluid. In general, $\mathbf{F}_B$ may be written as

$$\mathbf{F}_B = \int_{\mathscr{V}} \mathbf{B}\rho \, d\mathscr{V} \tag{2.51}$$

where $\mathbf{B}$ denotes the body force per unit mass.

Surface forces are those forces that act on the boundaries of the system by virtue of their contact with the surroundings. Surface forces may be subdivided into *normal forces* $\mathbf{F}_n$ and *tangential forces* $\mathbf{F}_t$, where $\mathbf{F}_S = \mathbf{F}_n + \mathbf{F}_t$. In the case of the flow of an *inviscid fluid*, the normal forces are caused by pressure. Thus,

$$\mathbf{F}_n = -\int_A p \, d\mathbf{A} \tag{2.52}$$

The *negative sign* is introduced in equation 2.52 because pressure is a compressive stress acting in the inward direction on a surface and, for the surface element $d\mathbf{A}$ the positive normal acts in the outward direction. For an inviscid flow $\mathbf{F}_t = 0$ because there are no shear forces. For a viscous flow the normal force has components caused by the viscous shear stresses. Detailed expressions for the normal and tangential forces acting on a viscous fluid are developed in Section 5–9. The interested reader is also referred to the many excellent works in the literature dealing with the flow of viscous fluids.[10–13] Accordingly, $\mathbf{F}_t$ will be denoted by $\mathbf{F}_{shear}$ to denote its origin and will be retained in that general form when analyzing one-dimensional flows.

2-6(b) The Momentum Equation for a Control Volume

Combining all of the forces discussed above yields the following expression for the net external force $\mathbf{F}_{\text{external}}$ acting on a system. Thus,

$$\mathbf{F}_{\text{external}} = \int_{\mathscr{V}} \mathbf{B}\rho \, d\mathscr{V} - \int_{A} p \, d\mathbf{A} + \mathbf{F}_{\text{shear}} \tag{2.53}$$

Substituting equation 2.53 into equation 2.49 yields *the momentum equation*. Thus,

$$\int_{\mathscr{V}} \mathbf{B}\rho \, d\mathscr{V} - \int_{A} p \, d\mathbf{A} + \mathbf{F}_{\text{shear}} = \int_{\mathscr{V}} (\rho \mathbf{V})_{t} \, d\mathscr{V} + \int_{A} \mathbf{V}(\rho \mathbf{V} \cdot d\mathbf{A}) \tag{2.54}$$

Equation 2.54 is a vector relationship having three components; each component is an independent relationship corresponding to one of the directions of the three mutually orthogonal coordinate axes.

To obtain a general form for the component equations, consider the x_i direction denoted by the unit vector $\mathbf{n}_i$ (e.g., in Cartesian coordinates, $x_i = x$, y, or z and $\mathbf{n}_i = \mathbf{i}, \mathbf{j}$, or $\mathbf{k}$). Denote the corresponding velocity components by u_i. To obtain the component of equation 2.54 in the x_i direction, the x_i direction component of each term is determined by forming the scalar product of $\mathbf{n}_i$ and equation 2.54. Thus,

$$\int_{\mathscr{V}} B_i \rho \, d\mathscr{V} - \int_{A} p\mathbf{n}_i \cdot d\mathbf{A} + \mathbf{n}_i \cdot \mathbf{F}_{\text{shear}} = \int_{\mathscr{V}} (\rho u_i)_{t} \, d\mathscr{V} + \int_{A} u_i(\rho \mathbf{V} \cdot d\mathbf{A}) \tag{2.55}$$

For *steady flow*, the first integral on the right-hand side of equation 2.55 is zero.

In many applications it is neither possible nor necessary to determine in detail the external surface forces due to pressure and shear. The only force of interest may be the *net resultant surface force*. Examples are the thrust developed by a jet propulsion engine (rocket, turbojet, or ramjet), the total force exerted on a body immersed in a fluid stream, and the net force required for holding in place a duct in which a fluid is flowing. In such cases, it is more convenient to retain the surface forces in the general form $\mathbf{F}_{\text{surface}}$, which is the sum of all external forces acting on the surface of the control volume. Hence, we obtain the following expression for the *momentum equation*.

$$\mathbf{F}_{\text{surface}} + \int_{\mathscr{V}} \mathbf{B}\rho \, d\mathscr{V} = \int_{\mathscr{V}} (\rho \mathbf{V})_{t} \, d\mathscr{V} + \int_{A} \mathbf{V}(\rho \mathbf{V} \cdot d\mathbf{A}) \tag{2.56}$$

Equation 2.56 may be employed for calculating the net resultant surface force acting on the surface of a control volume, provided that the momentum change inside the control volume and the momentum flux crossing the control surface can be determined. No detailed knowledge is required of the distributions of either the pressure or the shear stress.

2-6(c) Differential Form of the Momentum Equation

Before the differential form of the momentum equation can be derived, a decision must be made concerning the shear stresses. If they are to be retained, it is essential that they be related to the properties of the flow field. The resulting differential equations are known as the *Navier-Stokes equations* (see References 10–13, and Section 5–9). For an inviscid fluid, $\mathbf{F}_{\text{shear}} = 0$.

The surface integrals in equation 2.55 are converted into volume integrals by applying the divergence theorem (see equation 2.35). The pressure term becomes

$$-\int_{A} p\mathbf{n}_i \cdot d\mathbf{A} = -\int_{\mathscr{V}} \frac{\partial p}{\partial x_i} \, d\mathscr{V} \tag{2.57}$$

The momentum flux term becomes

$$\int_A u_i(\rho \mathbf{V} \cdot d\mathbf{A}) = \int_{\mathcal{V}} \nabla \cdot (\rho u_i \mathbf{V}) \, d\mathcal{V} \tag{2.58}$$

Substituting the above results into equation 2.55 and considering a control volume of differential size yields the following result.

$$\rho B_i - \frac{\partial p}{\partial x_i} = (\rho u_i)_t + \nabla \cdot (\rho u_i \mathbf{V}) \tag{2.59}$$

The right-hand side (RHS) of equation 2.59 is simplified by expanding the terms as follows.

$$(\text{RHS}) = \rho(u_i)_t + u_i[\rho_t + \nabla \cdot (\rho \mathbf{V})] + \rho(\mathbf{V} \cdot \nabla)u_i \tag{2.60}$$

The expression inside the square brackets is identically equal to zero, since it expresses the law of the conservation of mass, equation 2.46. Hence, equation 2.60 reduces to

$$(\text{RHS}) = \rho[(u_i)_t + (\mathbf{V} \cdot \nabla)u_i] = \rho \frac{Du_i}{Dt} \tag{2.61}$$

Substituting equation 2.61 into 2.59 yields

$$\rho \frac{Du_i}{Dt} + \frac{\partial p}{\partial x_i} - \rho B_i = 0 \tag{2.62}$$

Equation 2.62 is the component form of the differential form of the momentum equation.

By summing the component equations vectorially, the following vector equation is obtained.

$$\rho \frac{D\mathbf{V}}{Dt} + \nabla p - \rho \mathbf{B} = 0 \tag{2.63}$$

Equation 2.63 is known as Euler's equation of motion. It is valid for inviscid flow. For *steady flow*, $(u_i)_t = 0$, and the first term becomes $\rho(D\mathbf{V}/Dt) = \rho(\mathbf{V} \cdot \nabla)\mathbf{V}$.

Example 2.3. A small model rocket is illustrated in Fig. 2.11. The rocket is propelled by a jet of water that is forced out of the nozzle by a volume of compressed air inside the rocket. The velocity of the water surface inside the rocket has been measured, and is given by $V_c = V_o - kt$. The inside cross-sectional area of the water chamber is A_c, and the exit area of the converging nozzle is $A_e = A_c/2$. The initial mass of the rocket is M_o, and the density ρ of the water is constant. Determine an expression for the restraining force R required to hold the rocket stationary. Note that the rocket thrust is equal in magnitude to R and acts in the opposite direction.

Solution

Several simplifying assumptions must be made before the problem can be solved. It is assumed that the momentum of the compressed air and the frictionless piston are negligible, that the velocities V_c and V_e are uniform over their respective cross sections, and that the pressure of the jet p_e at the nozzle exit cross section is uniform and equal to the atmospheric pressure. First, it is necessary to determine $V_e(t)$. For that purpose, define control volume $\mathcal{V}_1$, as illustrated in Fig. 2.11. Applying the continuity equation, equation 2.42, to $\mathcal{V}_1$ yields

$$\int_{\mathcal{V}_1} \left(\frac{\partial \rho}{\partial t}\right)^0 d\mathcal{V} + \int_{A_1} \rho \mathbf{V} \cdot d\mathbf{A} = 0 \tag{a}$$

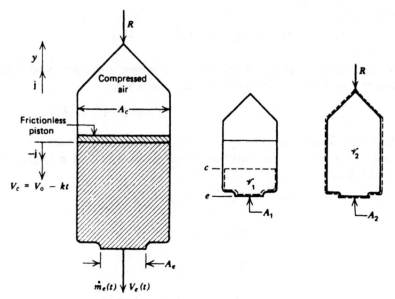

Figure 2.11 Sketch for Example 2.3.

$$V_e = -\mathbf{j}V_e \qquad \text{and} \qquad d\mathbf{A}_e = -\mathbf{j}\,dA_e \tag{b}$$

$$V_c = -\mathbf{j}V_c \qquad \text{and} \qquad d\mathbf{A}_c = \mathbf{j}\,dA_c \tag{c}$$

Substituting these values into equation (a) and simplifying yields

$$\rho A_e V_e - \rho A_c V_c = 0 \tag{d}$$

$$V_e = \frac{A_c V_c}{A_e} = 2(V_o - kt) \tag{e}$$

The instantaneous mass of the rocket may be determined from control volume $\mathscr{V}_2$, illustrated in Fig. 2.11, for which equation 2.42 becomes

$$\int_{\mathscr{V}_2} \frac{\partial \rho}{\partial t}\,d\mathscr{V} + \int_{A_2} \rho \mathbf{V} \cdot d\mathbf{A} = 0 \tag{f}$$

The first term in equation (f) may be written as

$$\int_{\mathscr{V}_2} \frac{\partial \rho}{\partial t}\,d\mathscr{V} = \frac{\partial}{\partial t}(\rho \mathscr{V}) = \frac{\partial M(t)}{\partial t} \tag{g}$$

where $M(t)$ is the total mass of water, air, and solid rocket hardware in control volume $\mathscr{V}_2$. The second term in equation (f) is simply

$$\int_{A_2} \rho \mathbf{V} \cdot d\mathbf{A} = \int_{A_e} \rho(-\mathbf{j}V_e) \cdot (-\mathbf{j}\,dA_e) = \rho V_e A_e = 2\rho A_e(V_o - kt) \tag{h}$$

Substituting equations (g) and (h) into (f) yields

$$\partial M/\partial t + 2\rho A_e(V_o - kt) = 0 \tag{i}$$

Equation (i) may be integrated to yield

$$\int_{M_o}^{M} dM = -\int_0^t 2\rho A_e(V_o - kt)\,dt \tag{j}$$

from which

$$M(t) = M_o - 2\rho A_e \left(V_o t - \frac{kt^2}{2} \right) \tag{k}$$

The reaction force on the rocket may now be determined by applying the momentum equation, equation 2.56, to control volume $\mathscr{V}_2$. Thus,

$$\mathbf{F}_{\text{surface}} + \int_{\mathscr{V}_2} \mathbf{B}\rho \, d\mathscr{V} = \int_{\mathscr{V}_2} (\rho \mathbf{V})_t \, d\mathscr{V} + \int_{A_2} \mathbf{V}(\rho \mathbf{V} \cdot d\mathbf{A}) \tag{l}$$

The only surface force acting on A_2 is $(-R)$. The body force caused by gravity is $-Mg$, and the velocity of the water inside the control volume is $(-jV_c)$. Hence, equation (1) becomes

$$-R - Mg = \frac{\partial}{\partial t}(-V_c M) - \rho A_e V_e^2 \tag{m}$$

The third term in equation (m) may be written as

$$\frac{\partial}{\partial t}(-V_c M) = -M\frac{\partial V_c}{\partial t} - V_c \frac{\partial M}{\partial t} \tag{n}$$

Substituting equation (i) and $V_c = (V_o - kt)$ into equation (n) yields

$$\frac{\partial}{\partial t}(-V_c M) = kM + 2\rho A_e (V_o - kt)^2 \tag{o}$$

Substituting equations (k) and (o) into equation (m) and solving for R yields

$$R = -(g + k)\left[M_o - 2\rho A_e \left(V_o t - \frac{kt^2}{2} \right) \right] + 2\rho A_e (V_o - kt)^2 \tag{p}$$

The above example is a good illustration of the application of the momentum equation to an unsteady flow problem. It demonstrates vividly that to obtain a solution, careful attention must be given to the assumptions and approximations required.

2-7 THE FIRST LAW OF THERMODYNAMICS

It is shown in Section 1–11(b) that for a unit mass of a continuum, the first law of thermodynamics may be written in the form (see equation 1.48)

$$de = \delta Q - \delta W \tag{2.64}$$

where de comprises the change in all of the forms of stored energy associated with the fluid. For a flowing fluid, the stored energy per unit mass will be assumed to be limited to thermal, kinetic, and potential energies. Consequently, the stored energy per unit mass of fluid is

$$e = u + \frac{V^2}{2} + gz \tag{2.65}$$

For a system having the total mass m, the total stored energy is $E = me$, and equation 2.64 becomes

$$dE = \delta Q - \delta W \tag{2.66}$$

where δQ and δW apply to the entire system. To be mathematically unambiguous, it might be preferable to employ other symbols for Q and W in equation 2.64 to denote that they are intensive properties. Nevertheless, the convention adopted here is to employ Q and W in the energy equation for representing the heat and work terms, respectively, regardless of whether they are based on a unit mass or the total mass of the system. When Q and W are to be determined on a rate basis (i.e., per unit time), the symbols $\dot{Q}$ and $\dot{W}$ will be employed. Any disadvantages arising from the lack of precision in the meaning of the symbols Q and W is more than offset by not having to define different symbols for the many different conditions under which Q and W may have to be determined. In a specific application, the basis for the terms Q and W is clearly defined by the symbol for the stored energy term. To illustrate, when e is employed, Q and W refer to a unit mass basis; when E is employed, Q and W apply to the entire mass of the system; when de/dt is employed, $\dot{Q}$ and $\dot{W}$ are rates per unit mass; and when dE/dt is employed, $\dot{Q}$ and $\dot{W}$ are rates for the entire mass of the system.

Figure 2.12 illustrates the physical model for the control volume employed for developing the expression for the first law of thermodynamics for a control volume. The figure indicates schematically shaft work, shear work, heat transfer, and the flux

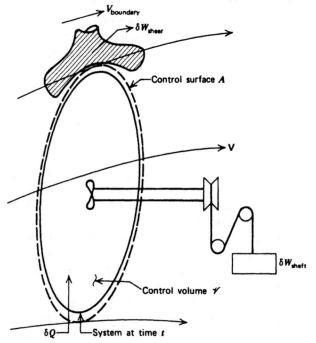

Figure 2.12   Control volume for deriving the first law of thermodynamics for a flowing fluid.

of mass across the boundaries of the control volume. Equation 2.66 is valid for a fluid particle, but it may be transformed into a rate equation for a system by dividing by dt. Thus,

$$\frac{dE}{dt} = \delta\dot{Q} - \delta\dot{W} \tag{2.67}$$

Equation 2.67 expresses the time rate of change of E following a fluid particle and is, therefore, a substantial derivative. Hence,

$$\frac{DE}{Dt} = \delta\dot{Q} - \delta\dot{W} \tag{2.68}$$

In the present case, $N = E = \int_{\mathscr{V}} (u + V^2/2 + gz)\rho \, d\mathscr{V}$ and $n = e = u + V^2/2 + gz$. Substituting for N and n into equation 2.34, yields

$$\frac{DE}{Dt} = \int_{\mathscr{V}} \frac{\partial}{\partial t} \left[\rho \left(u + \frac{V^2}{2} + gz \right) \right] d\mathscr{V} + \int_{A} \left[\rho \left(u + \frac{V^2}{2} + gz \right) \right] (\rho \mathbf{V} \cdot d\mathbf{A}) \quad (2.69)$$

To complete the transformation of the first law of thermodynamics for the system to a control volume, the heat and work terms must be expressed in terms of the flow properties pertinent to the control volume.

2–7(a) Heat Transfer to a Control Volume

The amount of heat transferred to a control volume, denoted by δQ, is the amount of heat transferred to the mass instantaneously occupying the control volume. Heat transferred to a system is considered positive. In general, the heat transfer may be caused by conduction, convection, and/or radiation. Thus,

$$Q = \text{conduction} + \text{convection} + \text{radiation} \quad (2.70)$$

To express δQ explicitly in terms of the flow properties of the control volume, the different heat transfer processes and the flow properties must be related to each other, as is done in References 10 and 14 for conduction and convection and in Reference 15 for heat transfer by radiation. In this book the details of the heat transfer processes are not considered; the effect of heat transfer is, however, retained in the general form δQ. By so doing, the general influence of heat transfer is included in the governing equations and may be considered in solving one-dimensional problems.

2–7(b) Work Done by a Control Volume

The work done by a control volume on the surroundings is defined as positive work and is of two main types. First, there is *shaft work*, denoted by W_{shaft}; that is, the work done by a rotating shaft crossing the boundaries of the system. Examples of shaft work are the works done to operate compressors, hoists, and other machines. Second, there is the work done by the surface forces where a fluid crosses the boundaries of a control volume, or where the boundary of a control volume is moving; the former is called *flow work* [see Section 1–11(d)]. The surface forces, as shown earlier, can be resolved into normal forces due to pressure stresses and tangential forces due to shear stresses. Expressions for the work done by shear stresses will not be developed further herein; for a detailed discussion of shear work, see Reference 10. Instead, the work done by shear stresses will be denoted by W_{shear} and retained in that general form in the appropriate equations.

The work done by the pressure stresses may be either the work done in pushing mass out of the control volume (*positive work*), or the work received when the surroundings push mass into the control volume (*negative work*).

Figure 2.13 illustrates the determination of the flow work because of normal stresses. The normal force acting on the infinitesimal vector area $d\mathbf{A}$ is given by

$$d\mathbf{F}_n = p \, d\mathbf{A} \quad (2.71)$$

The differential force $d\mathbf{F}_n$ given by equation 2.71 is the force exerted by the fluid inside the control volume on the surroundings and is, therefore, a positive term. The rate of doing work is obtained by multiplying the component of $d\mathbf{F}_n$ in the direction of the velocity $\mathbf{V}$ by the magnitude of the latter. Thus,

$$\delta \dot{W}_n = d F_n V \cos \alpha = d\mathbf{F}_n \cdot \mathbf{V} \quad (2.72)$$

$$\delta \dot{W}_n = p \mathbf{V} \cdot d\mathbf{A} = pv(\rho \mathbf{V} \cdot d\mathbf{A}) \quad (2.73)$$

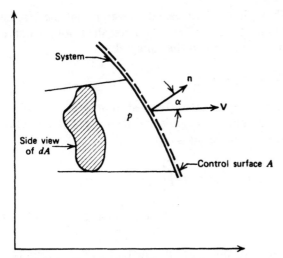

Figure 2.13 Flow work done by normal forces when mass crosses a control surface.

Since ($\rho\mathbf{V} \cdot d\mathbf{A}$) is positive for mass leaving the control volume and negative for mass entering, the work δW_n done by pressure forces is likewise positive for mass flowing out of the control volume and negative for mass flowing into it, which is in accordance with the convention that the work done by the system is positive, and vice versa. In summary, the system does work when it pushes mass out of the control volume, across the control surface, and vice versa.

The total work done by the normal forces is obtained by integrating $\delta\dot{W}_n$ over the entire control surface A. Thus,

$$\dot{W}_n = \int_A \delta\dot{W}_n = \int_A pv(\rho\mathbf{V} \cdot d\mathbf{A}) \tag{2.74}$$

Equation 2.74 includes both positive and negative work. The sign of the work is determined by that of the scalar product $\mathbf{V} \cdot d\mathbf{A}$.

The total rate at which work is done by the fluid that is instantaneously occupying the control volume is given by

$$\dot{W} = \dot{W}_{\text{shaft}} + \dot{W}_{\text{shear}} + \int_A pv(\rho\mathbf{V} \cdot d\mathbf{A}) \tag{2.75}$$

2-7(c) The Energy Equation for a Control Volume

Equations 2.68, 2.69, 2.70, and 2.75, when combined, yield the equation representing the *first law of thermodynamics* for a control volume; it is called the *energy equation*. Thus,

$$\dot{W}_{\text{shaft}} + \dot{W}_{\text{shear}} - \dot{Q} + \int_v \frac{\partial}{\partial t}\left[\rho\left(u + \frac{V^2}{2} + gz\right)\right]dv$$
$$+ \int_A \left(h + \frac{V^2}{2} + gz\right)(\rho\mathbf{V} \cdot d\mathbf{A}) = 0 \quad (2.76)$$

Note that the flow work caused by pressure stresses, $\int_A pv(\rho\mathbf{V} \cdot d\mathbf{A})$, is combined with the specific internal energy u in the stored energy term in equation 2.69 to yield the expression $(u + pv)$. By definition [see Section 1-11(e)] the latter is the *specific enthalpy h*. Since the internal energy u and the flow work pv always occur in combination in the surface integral, it is convenient to employ the specific enthalpy h of the fluid, as in equation 2.76. It should be recalled, however, that the first integral specifies the unsteady stored energy term. Since the latter does not contain a flow work term, the internal energy u appears in that integral.

2-7(d) Differential Form of the Energy Equation

The differential form of the energy equation is obtained from equation 2.76 by converting the surface integrals into volume integrals and then applying the result to a control volume of differential size. The result is

$$\delta \dot{W}_{shaft} + \delta \dot{W}_{shear} - \delta \dot{Q} + \frac{\partial}{\partial t}\left[\rho\left(u + \frac{V^2}{2} + gz\right)\right]$$
$$+ \nabla \cdot \left[\left(h + \frac{V^2}{2} + gz\right)\rho \mathbf{V}\right] = 0 \quad (2.77)$$

By adding $\pm[\rho(pv)]_t$ to the unsteady flow term in equation 2.77, that term becomes

$$\frac{\partial}{\partial t}\left[\rho\left(h + \frac{V^2}{2} + gz\right)\right] - p_t \quad (2.78)$$

where $p_t = \partial p/\partial t$. Expanding both the unsteady flow term and the energy flux term in equation 2.77, as in the derivation of the differential momentum equation (see equation 2.60), an expression comprising the product of the specific stored energy terms $(h + V^2/2 + gz)$ and the continuity equation is obtained, and that product may be discarded. The final form for the energy equation (equation 2.77) is

$$\delta \dot{W}_{shaft} + \delta \dot{W}_{shear} - \delta \dot{Q} + \rho \frac{D}{Dt}\left(h + \frac{V^2}{2} + gz\right) - p_t = 0 \quad (2.79)$$

For steady flow, $D/Dt = (\mathbf{V} \cdot \nabla)$, and $p_t = 0$. The shaft work, shear work, and heat transfer terms in equation 2.79 only indicate that those processes may be present. Their calculation, however, requires deriving the appropriate equations for expressing the work and heat transfer processes in terms of the flow properties and their gradients.

Example 2.4. Air enters the compressor of a stationary gas turbine power plant with a static temperature of 300 K and an average velocity of 150 m/s. It is discharged from the compressor with an average velocity of 50 m/s. The compressor receives 100,000 J/kg of work from the turbine, and the heat loss to the surroundings is 10,000 J/kg. Assuming steady flow, calculate the enthalpy and temperature of the air leaving the compressor, taking into account the variation of the specific heat of the air with temperature.

Solution

From Table C.4, for $t_1 = 300$ K, $h_1 = 301,329$ J/kg. For steady flow in the absence of shear work and gravitational effects, equation 2.76 reduces to

$$\dot{W}_{shaft} - \dot{Q} + \int_A \left(h + \frac{V^2}{2}\right)\rho \mathbf{V} \cdot d\mathbf{A} = 0 \quad (a)$$

Assuming uniform flow at the entrance 1 and the exit 2, equation (a) becomes

$$\dot{W}_{shaft} - \dot{Q} + \dot{m}\left(h_2 + \frac{V_2^2}{2} - h_1 - \frac{V_1^2}{2}\right) = 0 \quad (b)$$

Dividing equation (b) by $\dot{m}$ and solving for h_2 gives

$$h_2 = Q - W_{\text{shaft}} + h_1 + \frac{V_1{}^2 - V_2{}^2}{2} \tag{c}$$

$$h_2 = (-10{,}000) - (-100{,}000) + 301{,}329 + \frac{(150)^2 - (50)^2}{2}$$

$$h_2 = 401{,}329 \text{ J/kg (172.54 Btu/lbm)} \tag{c}$$

From Table C.4, for $h_2 = 401{,}329$ J/kg, $t_2 = 399.18$ K (718.53 R).

Example 2.5. Consider the tank and outlet duct described in Example 2.2. Assume that the mass discharge rate is small enough so that all of the properties inside of the tank and outlet duct may be considered to be uniform (i.e., not functions of x, y, and z), and that heat is transferred to the tank at the rate of 1055 J/s. Determine the instantaneous rate of change of the internal energy of the air in the tank (i.e., $\partial u/\partial t$) for the conditions specified in Example 2.2.

Solution

Define the inside of the tank and outlet duct as the control volume $\mathscr{V}$. There is no work, and the effects of body forces are negligible. For the low speeds involved, the kinetic energy may also be neglected. Thus, equation 2.76 becomes

$$-\dot{Q} + \int_{\mathscr{V}} \frac{\partial}{\partial t}(u\rho)\, d\mathscr{V} + \int_{A_{\text{out}}} h(\rho \mathbf{V} \cdot d\mathbf{A}) = 0 \tag{a}$$

The unsteady term in equation (a) may be written as

$$\frac{\partial E}{\partial t} = \int_{\mathscr{V}} \frac{\partial}{\partial t}(u\rho)\, d\mathscr{V} = \frac{\partial}{\partial t}\int_{\mathscr{V}} u\rho\, d\mathscr{V} = \frac{\partial}{\partial t}(u\rho\mathscr{V}) \tag{b}$$

$$\frac{\partial E}{\partial t} = u\mathscr{V}\frac{\partial \rho}{\partial t} + \rho\mathscr{V}\frac{\partial u}{\partial t} \tag{c}$$

where $\partial\rho/\partial t = -\dot{m}_{\text{out}}/\mathscr{V}$ from equation (d) of Example 2.2. Hence,

$$\frac{\partial E}{\partial t} = -u\dot{m}_{\text{out}} + \rho\mathscr{V}\frac{\partial u}{\partial t} \tag{d}$$

The energy flux term in equation (a) may be written as

$$\dot{E}_{\text{out}} = \int_{A_{\text{out}}} h(\rho\mathbf{V} \cdot d\mathbf{A}) = h\int_{A_{\text{out}}} d\dot{m}_{\text{out}} = h\dot{m}_{\text{out}} \tag{e}$$

Combining equations (a), (d), and (e), gives

$$-\dot{Q} - u\dot{m}_{\text{out}} + \rho\mathscr{V}\frac{\partial u}{\partial t} + h\dot{m}_{\text{out}} = 0 \tag{f}$$

For a perfect gas, $u = c_v T$ and $h = c_p T$, where T denotes the static temperature in *this problem* to distinguish it from the time t. Solving equation (f) for $\partial u/\partial t$ gives

$$\rho\mathscr{V}\frac{\partial u}{\partial t} = \dot{Q} - \dot{m}_{\text{out}}(h - u) = \dot{Q} - \dot{m}_{\text{out}}T(c_p - c_v) \tag{g}$$

$$\frac{\partial u}{\partial t} = \frac{\dot{Q} - \dot{m}_{\text{out}}RT}{\rho\mathscr{V}} \tag{h}$$

where $(c_p - c_v) = R$ for a perfect gas.

Equation (h) is the desired relationship between $\partial u/\partial t$ and the other flow properties. From Example 2.2, $\mathscr{V} = 0.283 \text{ m}^3$, $\dot{m}_{out} = 0.0272 \text{ kg/s}$, $\rho = 1.729 \text{ kg/m}^3$, and $T = 277.8 \text{ K}$. Substituting those values into equation (h), we obtain

$$\frac{\partial u}{\partial t} = \frac{1055 - (0.0272)(287.04)(277.8)}{(1.729)(0.283)} = -2277 \text{ J/kg} \ (-0.979 \text{ Btu/lbm})$$

2–8 THE SECOND LAW OF THERMODYNAMICS

In Section 1–12 the second law of thermodynamics (i.e., the *entropy equation*) is developed for a system (equation 1.68). Thus,

$$ds \geqq \frac{\delta Q}{t} \tag{1.68}$$

where the equality applies to reversible processes and the inequality to irreversible (natural) processes.

As in the development of the energy equation in Section 2–7, the symbol Q will be employed for heat transfer on a general basis, regardless of the exact mechanism of the heat transfer. Obviously, the basis for determining Q must correspond to that for determining the entropy change.

For a system of mass m, the total entropy of the system is $S = ms$, and equation 1.68 becomes

$$dS \geqq \frac{\delta Q}{t} \tag{2.80}$$

Since equation 2.80 is valid for a fluid particle, the time rate of change in the entropy S of the system is given by the substantial derivative of S. Hence,

$$\frac{DS}{Dt} \geqq \frac{\delta \dot{Q}}{t} \tag{2.81}$$

In terms of the properties N and n, $N = S = \int_{\mathscr{V}} s\rho \, d\mathscr{V}$, and $n = s$. Substituting $n = s$ into equation 2.34, yields

$$\int_{\mathscr{V}} (s\rho)_t \, d\mathscr{V} + \int_A s(\rho \mathbf{V} \cdot d\mathbf{A}) \geqq \frac{\dot{Q}}{t} \tag{2.82}$$

where $(s\rho)_t = \partial(s\rho)/\partial t$, and $\dot{Q} = \int_{\mathscr{V}} \delta \dot{Q}$. Equation 2.82, called the *entropy equation*, is the integral form of the second law of thermodynamics for a control volume.

The differential form of the entropy equation is obtained by converting the surface integral to a volume integral, by means of the divergence theorem (equation 2.35), and applying the result to a control volume of differential size. Thus,

$$(s\rho)_t + \nabla \cdot (s\rho \mathbf{V}) \geqq \frac{\delta \dot{Q}}{t} \tag{2.83}$$

Expanding equation 2.83 and discarding the term containing the continuity equation yields

$$\rho \frac{Ds}{Dt} \geqq \frac{\delta \dot{Q}}{t} \tag{2.84}$$

For steady flow, equation 2.84 becomes

$$\rho(\mathbf{V} \cdot \nabla)s \geqq \frac{\delta \dot{Q}}{t} \tag{2.85}$$

As pointed out in deriving the energy equation, the heat transfer Q in equation 2.84 must be expressed in terms of the flow properties. Equation 2.84 is suitable, however, for determining the overall effect of heat transfer.

2-9 SUMMARY

In this chapter the equations expressing the laws governing the flow of a fluid are derived for a control volume in *integral* and in *differential* forms. For convenience of reference, those equations are presented in Tables 2.1 and 2.2, respectively.

The *integral forms* of the *governing equations* are employed in the studies of *one-dimensional flows*. They are presented in Table 2.1.

Table 2.1 Integral forms of the governing equations for fluid flow

Continuity equation

$$\int_{\mathcal{V}} \rho_t \, d\mathcal{V} + \int_A \rho \mathbf{V} \cdot d\mathbf{A} = 0 \tag{2.86}$$

Momentum equation

$$\int_{\mathcal{V}} \mathbf{B}\rho \, d\mathcal{V} - \int_A p \, d\mathbf{A} + \mathbf{F}_{\text{shear}} = \int_{\mathcal{V}} (\rho\mathbf{V})_t \, d\mathcal{V} + \int_A \mathbf{V}(\rho\mathbf{V} \cdot d\mathbf{A}) \tag{2.87}$$

Energy equation

$$W_{\text{shaft}} + W_{\text{shear}} - Q + \int_{\mathcal{V}} \frac{\partial}{\partial t}\left[\rho\left(u + \frac{V^2}{2} + gz\right)\right] d\mathcal{V}$$
$$+ \int_A \left(h + \frac{V^2}{2} + gz\right)(\rho\mathbf{V} \cdot d\mathbf{A}) = 0 \tag{2.88}$$

Entropy equation

$$\int_{\mathcal{V}} (s\rho)_t \, d\mathcal{V} + \int_A s(\rho\mathbf{V} \cdot d\mathbf{A}) \geq \frac{Q}{t} \tag{2.89}$$

The *differential forms* of the *governing equations* are derived with $\mathbf{F}_{\text{shear}}$, W_{shear}, W_{shaft}, and Q retained in functional form. Consequently, the influence of the latter processes may be considered in a qualitative manner only. Those processes must either be omitted or expressed in terms of the fluid flow properties and their gradients before solving the pertinent differential equations. The *differential forms* of the *governing equations* are employed in the study of *multidimensional flows*. They are presented in Table 2.2.

Table 2.2 Differential forms of the governing equations for fluid flow

Continuity equation

$$\rho_t + \nabla \cdot (\rho\mathbf{V}) = 0 \tag{2.90}$$

Momentum equation

$$\rho \frac{D\mathbf{V}}{Dt} + \nabla p - \rho\mathbf{B} - d\mathbf{F}_{\text{shear}} = 0 \tag{2.91}$$

Energy equation

$$\delta\dot{W}_{\text{shaft}} + \delta\dot{W}_{\text{shear}} - \delta\dot{Q} + \rho\frac{D}{Dt}\left(h + \frac{V^2}{2} + gz\right) - p_t = 0 \tag{2.92}$$

Entropy equation

$$\rho\frac{Ds}{Dt} \geqq \frac{\delta\dot{Q}}{t} \tag{2.93}$$

A useful collection of the *differential forms* of the governing equations (including viscous stresses) is presented in Reference 16. The basic differential equations are presented there in the following coordinate systems: Cartesian tensor, vector, orthogonal curvilinear, Cartesian (rectangular), cylindrical, and spherical.

REFERENCES

1. M. J. Zucrow, *Aircraft and Missile Propulsion*, Vol. 1, Wiley, New York, 1958.
2. A. H. Shapiro, *The Dynamics and Thermodynamics of Compressible Fluid Flow*, Vol. 1, Ronald Press, New York, 1953.
3. K. Oswatitsch, *Gas Dynamics*, Academic Press, New York, 1956.
4. R. von Mises, *Mathematical Theory of Compressible Fluid Flow*, Academic Press, New York, 1958.
5. I. H. Shames, *Mechanics of Fluids*, McGraw-Hill, New York, 1962.
6. R. W. Fox and A. T. McDonald, *Introduction to Fluid Mechanics*, Wiley, New York, 1973.
7. H. W. Liepmann and A. Roshko, *Elements of Gas Dynamics*, Wiley, New York, 1957.
8. J. A. Owczarek, *Fundamentals of Gas Dynamics*, International Textbook Co., Scranton, Pa., 1964.
9. R. Aris, *Vector, Tensors, and the Basic Equations of Fluid Mechanics*, Prentice-Hall, Englewood Cliffs, N.J., 1962.
10. H. Schlichting, *Boundary Layer Theory*, Sixth Edition, McGraw-Hill, New York, 1968.
11. S. I. Pai, *Viscous Flow Theory, I. Laminar Flow*, Van Nostrand, Princeton, N.J., 1956.
12. L. Prandtl and O. J. Tietjens, *Fundamentals of Hydro- and Aeromechanics*, McGraw-Hill, New York, 1934.
13. J. W. Daily and D. R. F. Harleman, *Fluid Dynamics*, Addison-Wesley, Reading, Mass., 1966.
14. E. R. G. Eckert and R. M. Drake, *Heat and Mass Transfer*, McGraw-Hill, New York, 1959.
15. W. G. Vincenti and C. H. Kruger, *Introduction to Physical Gas Dynamics*, McGraw-Hill, New York, 1965.
16. W. F. Hughes and E. W. Gaylord, *Basic Equations of Engineering Science*, Schaum Publishing Co., New York, 1964.
17. L. Milne-Thompson, *Theoretical Hydrodynamics*, Macmillan, New York, 1950.
18. *Handbook of Supersonic Aerodynamics*, NAVORD Report 1488, Vol. 1, U.S. Government Printing Office, April 1950.

PROBLEMS

1. A fluid particle in steady flow has the vector velocity **V**. It is found that **V** depends only on the position of the particle in space, and its magnitude is given by the functional relationship

$$V = 2x^2y + 3yz - 4xz^2$$

where x, y, and z are distances measured along the coordinate axes. Calculate (a) the value of V at point 1, 2, 3, and (b) the values of $\partial V/\partial x$, $\partial V/\partial y$, and $\partial V/\partial z$ at the same point.

2. In a two-dimensional flow the scalar component of the velocity vector **V** in the direction of the x axis is denoted by u, and the component along the y axis by v. The components u and v are defined by $u = y^2 - x^2$, and $v = 2xy$. Calculate at the point $(2,2)$ (a) the magnitude V of the velocity vector, (b) the angle θ made by **V** with the x axis, and (c) the acceleration of the fluid particles as they pass that point.

3. By definition a streamline is a curve such that its tangent at each point has the same direction as the velocity vector **V**. If u and v are defined by the relationships given in Problem 2, derive the differential equation for the streamlines. Solve the differential equation, plot the streamlines in the region bounded by $x = -10$ to 10, $y = -10$ to 10, and discuss the results.

4. Figure 2.14 illustrates a tank of volume $\mathscr{V}$ containing a liquid having an initial density denoted by ρ_i. A second liquid having a density denoted by ρ enters the tank steadily with the mass flow rate $\dot{m}$, and mixes thoroughly with the fluid in the tank. The liquid level in the tank is maintained constant by allowing flow out of the side of the tank. Assume that $\rho_i > \rho$. Derive

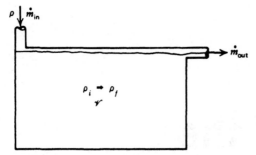

Figure 2.14 Sketch for Problem 2.4.

an expression for (a) the time rate of change of the density of the liquid in the tank, and (b) the time required for the density in the tank to reach the value $\rho_f < \rho_i$.

5. An incompressible inviscid fluid flows steadily through a converging conical flow passage having an inlet diameter D_o and a wall semiangle α, where the distance x is measured from the inlet. The mass flow rate is $\dot{m}$ and the density is ρ. Assume that body forces are negligible, and that the flow is uniform at each cross section. Derive an expression for the acceleration of the fluid particles as a function of x.

6. Figure 2.15 illustrates a cylindrical tank from which water flows through a well-rounded circular orifice at the bottom of the tank. Determine an expression for the height h of the water column in terms of A_1, A_2, h_o, g, and the time t. Assume that $A_1 \gg A_2$.

7. A flat plate oriented perpendicularly to a horizontal jet of water is moved toward the jet at a velocity $V = 6.0$ m/s. The water jet strikes the plate and flows vertically from the surface of the plate. The mass flow rate of the jet from a stationary nozzle is 50 kg/s, and the velocity of the jet is 15.0 m/s. Determine the force that must be applied to the plate to maintain it at a constant velocity.

8. Figure 2.16 illustrates a reducing elbow. Water at the pressure $p_1 = 14.80 \cdot 10^5$ N/m^2 enters the elbow with a uniform velocity $V_1 = 6.0$ m/s normal to the inlet area. The direction of the discharge is rotated 90 deg from that of the inlet. The inlet diameter $D_1 = 0.30$ m and that of the exit $D_2 = 0.15$ m. The pressure of the water at the exit section $p_2 = 12.00 \cdot 10^5$ N/m^2, and the atmospheric pressure $p_o = 1.0135 \cdot 10^5$ N/m^2. Assume that the flow is uniform at the exit section. Calculate (a) the force exerted by the pipe on the water, (b) the force exerted by the water on the pipe, and (c) the force required to hold the pipe stationary.

9. A rocket motor is fired statically on a thrust stand. The mass flow rate of propellants $\dot{m} = 9.0$ kg/s, the average jet velocity at the nozzle exit plane $V_e = 2500$ m/s, and the static pressure acting on the exit plane is equal to the pressure of the surrounding atmosphere. Determine the force transmitted to the thrust stand by the rocket motor.

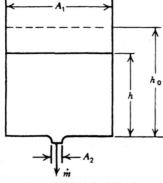

Figure 2.15 Sketch for Problem 2.6.

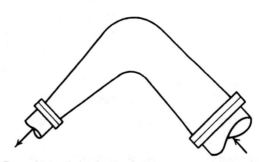

Figure 2.16 Sketch for Problem 2.8.

3

general features of the steady one-dimensional flow of a compressible fluid

3-1 PRINCIPAL NOTATION FOR CHAPTER 3

a	speed of sound.
a_o	stagnation speed of sound.
$a*$	critical (i.e., sonic) speed of sound.
A	flow cross-sectional area.
c_p	specific heat at constant pressure.
c_v	specific heat at constant volume.
D	drag force.
$\mathscr{D}$	hydraulic diameter.
f	friction factor in the Fanning equation
$\mathbf{F}$	vector force.
$\mathscr{F}$	$= pA + \dot{m}V$, stream thrust, and impulse function.
g	acceleration due to gravity.
G	$= \dot{m}/A$, mass flux.
h	static specific enthalpy.
H	stagnation specific enthalpy.
m	mass.
$\dot{m}$	mass flow rate.
$\bar{m}$	molecular weight.
M	Mach number.

M^* $= V/a^*$, dimensionless velocity.

p absolute static pressure.

P stagnation pressure.

q $= \rho V^2/2$, dynamic pressure.

Q heat.

R gas constant.

$\bar{R}$ universal gas constant.

s specific entropy.

t absolute static temperature.

T stagnation temperature.

u static specific internal energy.

V velocity magnitude.

$\mathbf{V}$ vector velocity.

V'_{max} maximum isentropic speed.

$\mathscr{V}$ volume of a control volume.

W work.

x displacement along flow path.

z position in a gravitational field.

Greek Letters

α $= \sin^{-1}(1/M)$, Mach angle.

γ $= c_p/c_v$, specific heat ratio.

ρ static density.

ρ_0 stagnation density.

Superscripts

$'$ denotes that the state is reached by an isentropic process.

$*$ critical condition, where $M = 1$.

3–2 INTRODUCTION

In Chapter 2, the governing equations for the flow of a fluid are developed in both integral and differential form for a control volume (see Tables 2.1 and 2.2). Those equations are a highly nonlinear set of equations in three space variables and time, and their general solutions are practically nonexistent. Consequently, they must be simplified before they may be employed for obtaining useful results.

Thus, as a first approximation, it is assumed that (1) the flow is *steady*, (2) the fluid properties are *uniform* at each location where mass crosses the control surface, and (3) the effects of *body forces are negligible*.

Assumption (1) signifies that, at each cross section of a flow passage, the magnitudes of the flow properties are invariant with time. In reality, steady flow exists only if the fluid particles move along streamlines. Because of the turbulence and eddies that accompany most of the practical cases of fluid flow, a practical flow is not steady in a strict sense, even though no variations in the rate of flow are detectable with the most refined measuring instruments. If the measured rate of flow of a fluid is constant, it may be imagined that fixed streamlines can be drawn in the fluid in such a manner that they are tangent to the average direction of the flow. The actual fluid motion may thereby be conceived to be composed of a steady motion along the aforementioned streamlines with irregular random disturbances of velocity superimposed on them, the net effect of the disturbances on the mass rate of flow of the main fluid being zero.

Assumption (2) signifies that the area of the cross section of a flow passage is either constant or changes so gradually that at each cross section the fluid properties change appreciably only in the flow direction under consideration. In all other directions the changes in the flow properties take place so slowly that they may be ignored. Such flows are called *one-dimensional flows.*

Assumption (3) is justified for the flow of gases where the effects of body forces are usually insignificant.

The results obtained under the aforementioned three assumptions are quite accurate for many *internal flows* (i.e., flows inside of solid boundaries), and the results are useful qualitatively for understanding *external flows.*

In this chapter, the governing equations are developed, subject to the aforementioned three assumptions, in a form applicable to the flow of any fluid.[1–6] In general, the solution of those equations must be obtained by either graphical or numerical techniques. A fourth assumption is introduced that the working fluid behaves thermodynamically in accordance with both the thermal and caloric equations of state for a perfect gas (see Section 1–15). The results obtained from the governing equations derived by applying the aforementioned four assumptions may be modified, if necessary, so that any deviations from the simplifying assumptions may be taken into account.

The developments of this chapter show that the local *static* properties of a flowing fluid p, ρ, and t, when expressed as ratios of their corresponding local *stagnation* values P, ρ_o, and T, depend only on the local Mach number M. Consequently, for a compressible fluid, each flow variable may be considered to be a function of the Mach number M. Consequently, if the Mach number M is either known, or calculable at some point in a flow field, then the ratios p/P, ρ/ρ_o, and t/T at that point are known, and vice versa.

3–3 THE ONE-DIMENSIONAL FLOW CONCEPT

The governing equations for analyzing fluid flow involve four independent variables: three space dimensions and time. Furthermore, because of the mathematical difficulties encountered in deriving general solutions of those equations, simplifying assumptions are introduced, more often than not, to obtain a physical model that is more amenable to mathematical analysis. The most important and most commonly employed approximation is that the flow is *one-dimensional*; that is, that all of the fluid properties are *uniform over every cross section* of the flow passage.

Figure 3.1a illustrates diagrammatically three different velocity profiles for a fluid flowing in a circular duct of diameter $D = 2R$. They correspond to a *one-dimensional* flow, which has a uniform velocity profile, a fully developed *turbulent* flow (with the Reynolds number Re $= DV\rho/\mu$, ranging from 10^4 to 10^5), and a fully developed *laminar* flow.

The equation for the laminar velocity profile illustrated in Fig. 3.1a is an exact solution of the Navier-Stokes equations (see equation 5.172), and is

$$u = u_{\max}\left(1 - \frac{y^2}{R^2}\right) \tag{3.1}$$

The velocity profile for the fully developed turbulent flow is based on experiment. From the vast amount of experimental data that has been accumulated on the

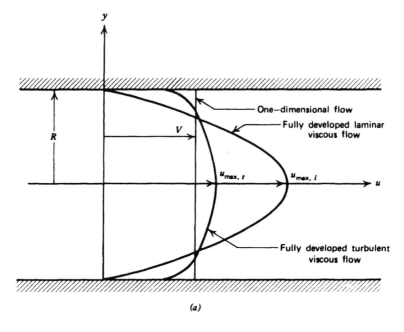

(a)

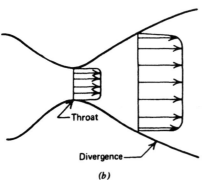

(b)

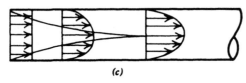

(c)

Figure 3.1.   Velocity profiles in several typical flow passages. (a) A comparison of one-dimensional, laminar, and turbulent velocity profiles in a round pipe. (b) Flow in a converging-diverging nozzle. (c) Developing flow in the entrance region of a pipe.

velocity profiles for the turbulent flow of fluids in pipes, the following empirical equation has been derived.[7, 8]

$$u = u_{max}\left(1 - \frac{y}{R}\right)^{1/n} \tag{3.2}$$

Equation 3.2 is known as the *power law equation* (see Section 5–10). In that equation the value of n is determined by experiment. The following table presents n as a function of the Reynolds number Re.

Exponent n for Power Law Velocity Profiles

$Re = \dfrac{DV\rho}{\mu}$	$4 \cdot 10^3$	$2.3 \cdot 10^4$	$1.1 \cdot 10^5$	$1.1 \cdot 10^6$	$2 \cdot 10^6$	$3.2 \cdot 10^6$
n	$6 \cdot 0$	6.6	7.0	8.8	10	10

Figure 3.1b illustrates schematically the velocity profiles in the throat and divergence of a converging-diverging nozzle. The bulk of the flow is inviscid, the viscous effects being confined to the thin *boundary layer* near the wall. Figure 3.1c illustrates the development of the velocity profile in the entrance region of a pipe.

From the foregoing few examples, it is apparent that different configurations are possible for the velocity profile of a flowing medium. Whether or not the assumption of *one-dimensional flow* is justified in a particular case depends on how significantly the actual velocity profile differs from the uniform velocity profile of one-dimensional flow (see Fig. 3.1a).

In a one-dimensional flow, because of its uniform velocity profile (see Fig. 3.1a), the flow properties are calculated so that they represent, as closely as possible, the actual overall values of the mass flow rate, momentum, and kinetic energy corresponding to the actual flow. If the velocity profile of the actual flow deviates appreciably from the uniform velocity profile of the corresponding one-dimensional flow, the calculated values of the flow properties will be in error. For most practical turbulent flows, however, the error is ordinarily within the desired accuracy tolerance.

For a fully developed turbulent flow in a pipe, having a velocity profile similar to that illustrated in Fig. 3.1a, the one-dimensional approximation is quite accurate. It is also quite accurate for the flows in the throats of converging and converging-diverging nozzles, and in the divergence of the latter nozzles (see Fig. 3.1b). The approximation is also satisfactory for the flow in diffusers where the bulk of the flow remains outside of the boundary layer adjacent to the walls. The one-dimensional assumption may lead to significant errors in laminar flows, flows inside passages having complicated shapes, and in ducts with developing flows, such as that illustrated in Fig. 3.1c.

The one-dimensional flow concept as employed in gas dynamics is an approximation only as far as the flow model is concerned, but not insofar as the governing flow equations are concerned. Once the approximation of uniform flow properties at each cross section is made, the integral forms of the governing equations (see Table 2.1) can be applied to the simplified flow model. The exact differential forms of the governing equations (see Table 2.2) are derived from the integral equations without any restrictions being placed on the flow model. Consequently, the aforementioned differential equations are invalid for the subject *one-dimensional* (i.e., a uniform flow) model. A new set of differential equations are, therefore, derived in this chapter for the *steady one-dimensional* flow model by employing the integral forms of the governing equations presented in Table 2.1.

The one-dimensional flow approximation is exact for the flow through an infinitesimal stream tube. Thus, many of the general features that characterize large-scale one-dimensional flows are also present along the streamlines of a multidimensional flow. In general, the one-dimensional approximation is reasonable if the rate of change of the so-called *flow driving potential* is small in the direction of flow. The following flow driving potentials are considered in this book: *area change, wall friction, heat transfer,* and *mass addition.* Furthermore, the radius of curvature of the flow passage should be large, and the profiles of the flow properties should remain similar at each flow cross section. Accordingly, it is to be expected that the one-dimensional flow model will be poor for flow passages such as pipe elbows, and in a duct wherein a transition from laminar to turbulent flow occurs. It should be noted that the one-dimensional flow model considers changes only in the *average* or *bulk* values of the flow properties in the direction of flow; it disregards completely the variations in the flow properties in the direction *normal* to the streamlines.

Example 3.1. Consider the incompressible flow of a fluid in a pipe for three different cases, all having the same mass flow rate $\dot{m}$: (1) one-dimensional flow, (2) fully developed laminar flow, and (3) fully developed turbulent flow. Refer to Fig. 3.1a. The laminar velocity profile is given by equation 3.1. Thus,

$$u = u_{max}\left(1 - \frac{y^2}{R^2}\right) \tag{a}$$

where R is the inside radius of the pipe, and u_{max} denotes the centerline velocity. For Reynolds numbers in the neighborhood of 10^5, the turbulent velocity profile is well approximated by equation 3.2. Thus,

$$u = u_{max}\left(1 - \frac{y}{R}\right)^{1/7} \tag{b}$$

Assume that all three velocity profiles yield the same mass flow rate $\dot{m}$. Determine u_{max} for both the laminar and turbulent flows in terms of the one-dimensional velocity V, and calculate the ratios of the actual to the one-dimensional flow momentum and kinetic energy for each velocity profile.

Solution

For the one-dimensional flow case (subscript 1-D):

$$\dot{m}_{1-D} = \int_A \rho V \cdot dA = \int_0^R \rho u(2\pi y\, dy) = 2\pi\rho \int_0^R uy\, dy = \pi\rho R^2 V \tag{c}$$

$$\dot{M}_{1-D} = \text{Momentum} = \int_A V(\rho V \cdot dA) = 2\pi\rho \int_0^R u^2 y\, dy = \pi\rho R^2 V^2 \tag{d}$$

$$\dot{E}_{1-D} = \text{Kinetic energy} = \int_A \frac{V^2}{2}(\rho V \cdot dA) = 2\pi\rho \int_0^R \frac{u^3}{2} y\, dy = \frac{\pi}{2}\rho R^2 V^3 \tag{e}$$

(a) For the laminar flow case (subscript l):

$$\dot{m}_l = 2\pi\rho \int_0^R u_{max}\left(1 - \frac{y^2}{R^2}\right) y\, dy = \frac{\pi}{2}\rho R^2 u_{max} \tag{f}$$

Combining equations (c) and (f) yields

$$u_{max} = 2V \tag{g}$$

The calculation of the momentum and kinetic energy yields

$$\dot{M}_l = 2\pi\rho \int_0^R u_{max}^2\left(1 - \frac{y^2}{R^2}\right)^2 y\, dy = \frac{\pi}{3}\rho R^2 u_{max}^2 \tag{h}$$

$$\dot{E}_l = 2\pi\rho \int_0^R \frac{u_{max}^3}{2}\left(1 - \frac{y^2}{R^2}\right)^3 y\, dy = \frac{\pi}{8}\rho R^2 u_{max}^3 \tag{i}$$

Combining equations (d) and (h) yields

$$\frac{\dot{M}_l}{\dot{M}_{1-D}} = \frac{4}{3} \tag{j}$$

Combining equations (e) and (i) yields

$$\frac{\dot{E}_l}{\dot{E}_{1-D}} = 2 \tag{k}$$

(b) For the turbulent flow case (subscript t):

$$\dot{m}_t = 2\pi\rho \int_0^R u_{max} \left(1 - \frac{y}{R}\right)^{1/7} y\,dy = \frac{49\pi}{60}\rho R^2 u_{max} \tag{l}$$

Combining equations (c) and (l) yields

$$u_{max} = \left(\frac{60}{49}\right) V \tag{m}$$

For the momentum and the kinetic energy of the fluid in the case of a turbulent flow:

$$\dot{M}_t = 2\pi\rho \int_0^R u_{max}^2 \left(1 - \frac{y}{R}\right)^{2/7} y\,dy = \frac{49\pi}{72}\rho R^2 u_{max}^2 \tag{n}$$

$$\dot{E}_t = 2\pi\rho \int_0^R \frac{u_{max}^3}{2} \left(1 - \frac{y}{R}\right)^{3/7} y\,dy = \frac{49\pi}{170}\rho R^2 u_{max}^3 \tag{o}$$

From equations (d) and (n) the ratio $\dot{M}_t/\dot{M}_{1-D}$ is

$$\frac{\dot{M}_t}{\dot{M}_{1-D}} = 1.020 \tag{p}$$

Combining equations (e) and (o), the ratio $\dot{E}_t/\dot{E}_{1-D}$ is

$$\frac{\dot{E}_t}{\dot{E}_{1-D}} = 1.058 \tag{q}$$

Equations (p) and (q) indicate that the one-dimensional flow model may be appropriate for fully developed turbulent pipe flow, but equations (j) and (k) indicate that it may lead to considerable error for a fully developed laminar flow.

3-4 CONSERVATION OF MASS FOR STEADY ONE-DIMENSIONAL FLOW

In this section, the equation governing the conservation of mass is derived for a steady one-dimensional flow. The resultant equation is called the *continuity equation*. From Table 2.1, the integral form of the law of the conservation of mass is given by equation 2.86. Thus,

$$\int_V \rho_t \, d\mathcal{V} + \int_A \rho \mathbf{V} \cdot d\mathbf{A} = 0 \tag{2.86}$$

For steady flows, the first term in equation 2.86 is zero.

Refer to Fig. 3.2, which illustrates the flow through a stream tube. Let A_1 and A_2 denote the inlet and exit areas, V_1 and V_2 the mean velocities over the areas A_1 and A_2, and ρ_1 and ρ_2 the corresponding values of the density of the fluid. Applying equation 2.86 to A_1 and A_2 yields

$$\int_{A_2} \rho \mathbf{V} \cdot d\mathbf{A} + \int_{A_1} \rho \mathbf{V} \cdot d\mathbf{A} = 0 \tag{3.3}$$

Integrating equation 3.3 gives the following equation for the mass rate of flow through the stream tube.

$$\dot{m} = \rho_1 A_1 V_1 = \rho_2 A_2 V_2 = \rho A V = \text{constant} \tag{3.4}$$

Equation 3.4 applies to the steady one-dimensional flow of any fluid.

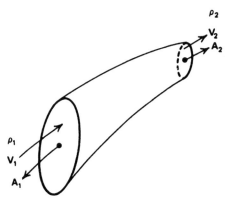

Figure 3.2. Steady flow through a stream tube.

3-5 DYNAMICS OF STEADY ONE-DIMENSIONAL FLOW

The momentum equation for a steady one-dimensional flow is derived below. Inviscid flows are considered first. The results are then extended to account for the effects of friction. Finally, a method is presented for determining the external forces acting on a flowing fluid.

3-5(a) Frictionless Flow

The integral form of the momentum equation for the x_i direction (i.e., equation 2.55) is repeated and renumbered below.

$$\int_{\mathscr{V}} B_i \rho \, d\mathscr{V} - \int_A p \mathbf{n}_i \cdot d\mathbf{A} + \mathbf{n}_i \cdot \mathbf{F}_{\text{shear}} = \int_{\mathscr{V}} (\rho u_i)_t \, d\mathscr{V} + \int_A u_i (\rho \mathbf{V} \cdot d\mathbf{A}) \quad (3.5)$$

Figure 3.3 illustrates a portion of a frictionless one-dimensional flow passage through which fluid flows in the x direction. For a steady frictionless flow, $(\rho u_i)_t = \mathbf{F}_{\text{shear}} = 0$, and equation 3.5 reduces to

$$\int_{\mathscr{V}} B_i \rho \, d\mathscr{V} - \int_A p \mathbf{n}_i \cdot d\mathbf{A} = \int_A u_i (\rho \mathbf{V} \cdot d\mathbf{A}) \quad (3.6)$$

Hereafter, a flow in which there are no shear forces will be termed an *inviscid flow*.

For the flow model illustrated in Fig. 3.3, $\mathbf{n}_i = \pm \mathbf{i}$, $u_i = V$, and $\mathbf{V} = \mathbf{i}V$. On the face where mass leaves the control volume, $d\mathbf{A} = \mathbf{i} \, dA$, and on the face where mass enters, $d\mathbf{A} = -\mathbf{i} \, dA$. The only body force considered is that caused by gravitational attraction, so that $\mathbf{B} = \mathbf{g}$, which, by convention, is assumed to act in the negative z direction, $-\mathbf{k}$, where the z axis makes the angle β with the x axis. Thus, the component of the body force in the x direction is $-\mathbf{i}\rho A g \, dz$; note that in Fig. 3.3 the z axis is not orthogonal to the x axis. For the assumed one-dimensional flow, each of the variables p, ρ, and V is uniform over any arbitrary cross-sectional area. For the element of differential volume of length dx, the surfaces over which equation 3.6 must be integrated are the inlet area A, the exit area $A + dA$, and the stream tube boundary area $(dA/\sin \alpha)$, where α is the angle made by the passage boundary with respect to the x axis.

A few comments on the conventions employed in analyzing control volumes of differential size are in order. By convention, all of the flow properties on the surfaces across which mass enters will be assigned the nominal values p, ρ, V, $V^2/2$, pA, etc., and it will be assumed that positive changes in these properties occur in the direction of flow. Thus, at the exit area, the properties are $p + dp$, $\rho + d\rho$, $V + dV$, $V^2/2 + d(V^2/2)$, $pA + d(pA)$, etc. In addition, positive changes in the driving potentials are

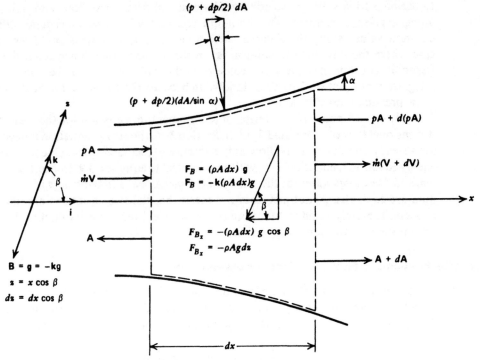

Figure 3.3. Forces acting on a differential element of a stream tube in a frictionless flow.

assumed. Hence, the inlet area is A and the exit area is $A + dA$. In a specific flow situation, any or all of the property changes may turn out to be negative. If, however, a consistent convention is employed in developing the governing equations, the results obtained will be consistent. The convention discussed above will be followed throughout this book.

Returning to equation 3.6, assign the properties pA and $\dot{m}V$ to the inlet area A, and $pA + d(pA)$ and $\dot{m}(V + dV)$ to the exit area $A + dA$. On the stream tube boundary the average static pressure intensity is $p + dp/2$, which acts on the area $dA/\sin \alpha$. However, only the component of that force acting in the direction of flow is desired. Consequently, the total force acting on the boundary surface of the stream tube is given by $(p + dp/2)\, dA$. Hence, equation 3.6 becomes

$$-\rho Ag\, dz + pA + \left(p + \frac{dp}{2}\right) dA - pA - d(pA) = \dot{m}(V + dV) - \dot{m}V \quad (3.7)$$

Equation 3.7 is simplified by canceling like terms and neglecting the higher-order terms. The result is

$$-\rho Ag\, dz - A\, dp = \dot{m}\, dV \tag{3.8}$$

Substituting $\dot{m} = \rho AV$, from equation 3.4, into equation 3.8 gives

$$dp + \rho V\, dV + \rho g\, dz = 0 \tag{3.9}$$

Equation 3.9 is the well-known *Bernoulli equation*.

The integrated form of equation 3.9 is

$$\int \frac{dp}{\rho} + \frac{V^2}{2} + gz = \text{constant} \qquad (\text{i.e., } \textit{the Bernoulli constant}) \tag{3.10}$$

Equation 3.10 is valid for steady one-dimensional frictionless flow, and holds only along a given streamline. The Bernoulli constant will have, in general, different constant values along the different streamlines in a multidimensional flow. In the case where the flow field is uniform, as in the assumed one-dimensional flow, the Bernoulli constant has the same value over the entire flow field. To complete the integration of equation 3.10, the relationship between the density of the fluid ρ and its static pressure p must be known.

Equations 3.9 and 3.10 are forms of the momentum equation for the steady one-dimensional flow of an inviscid fluid. In Section 3–6, the energy equation is developed for a steady one-dimensional flow in the absence of heat and work. One form of that energy equation is equation 3.40, which is identical to equation 3.9. Hence, the *momentum* and *energy equations* for the steady one-dimensional adiabatic flow of an inviscid fluid are equivalent. That fact does not mean that the two equations are not independent. It merely signifies that both equations yield the same result, and that they are satisfied by the same expression.

3–5(b) Frictionless Flow of an Incompressible Fluid

If the steady inviscid flow of an incompressible liquid is considered, or gases under conditions where the changes in density are negligible (e.g., low-speed subsonic gas flows) then, because ρ = constant, the integration of equation 3.10 yields

$$\frac{p}{\rho} + \frac{V^2}{2} + gz = \text{constant} \tag{3.11}$$

Equation 3.11 is the integrated form of Bernoulli's equation for the steady frictionless flow of an incompressible fluid.

3–5(c) Adiabatic Frictionless Flow of a Compressible Fluid

When the flowing medium is a compressible fluid, it is desirable, as stated in Section 3–2, to express the properties of the flow in terms of its Mach number $M = V/a$. Rewriting equation 3.9, we obtain

$$\frac{dp}{d\rho}\frac{d\rho}{\rho} + V\,dV + g\,dz = 0 \tag{a}$$

For the special case of an adiabatic frictionless flow, the entropy remains constant (see Section 1–12). Consequently, from equation 1.180,

$$a^2 = \left(\frac{\partial p}{\partial \rho}\right)_s = \frac{dp}{d\rho} \tag{b}$$

For most flow processes involving gases, the body forces are negligible, so that $g\,dz \approx 0$. Combining equations (a) and (b) and introducing $g\,dz = 0$, we obtain

$$\frac{d\rho}{\rho} + M^2\frac{dV}{V} = 0 \tag{3.12}$$

Equation 3.12 is an alternate form of Bernoulli's equation for the steady one-dimensional adiabatic frictionless (i.e., isentropic) flow of a compressible fluid.

It may be assumed, according to Prandtl, that all of the frictional effects are confined to the *boundary layer* where the velocity gradient is large. Hence, the equations derived in Sections 3–5(a), 3–5(b), and 3–5(c) apply to the region of a flow field that is located outside of the boundary layer; in that particular region the fluid flow may be assumed to be an inviscid flow.

3-5(d) Flow with Friction

Consider the steady one-dimensional flow of a fluid through a passage having solid boundaries such as that illustrated in Fig. 3.4. Let δF_f denote the x component of the *wall friction force* acting on the fluid element, and let δD denote the x component of the *internal drag forces* caused by obstructions, such as screens and struts, submerged in the fluid. The forces δF_f and δD oppose the motion of the fluid and thus

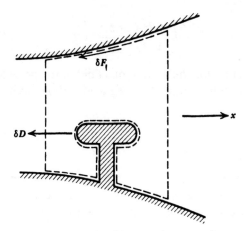

Figure 3.4. Additional forces acting on an element of a stream tube due to friction and internal obstructions.

act in the same direction as the pressure force $A\,dp$ [see Section 3-5(a)]. Hence, for the flow with friction illustrated in Fig. 3.4, the net external force acting in the x direction is given by

$$-(\rho g A\,dz + A\,dp + \delta F_f + \delta D) \tag{3.13}$$

Substituting from equation 3.13 into equation 3.8 and introducing $\dot{m} = \rho A V$, we obtain

$$A\,dp + \rho A V\,dV + \rho g A\,dz + \delta F_f + \delta D = 0 \tag{3.14}$$

The *wall friction force* δF_f may be expressed in terms of the hydraulic characteristics of the flow passage and an experimental *friction coefficient* f, which is defined later by equation 3.16. Let (WP) denote the average value of the *wetted perimeter* for the flow passage. The length of the fluid element is dx, so that the surface wetted by the fluid element is $(WP)\,dx$. If m denotes the *hydraulic radius* for the flow passage and $\mathscr{D}$ denotes the corresponding *hydraulic diameter* then, by definition,

$$m = \frac{\mathscr{D}}{4} = \frac{\text{flow area}}{\text{wetted perimeter}} = \frac{A}{(WP)} \tag{3.15}$$

The definition of $\mathscr{D}$ is chosen so that it is equal to the diameter for a flow passage having a circular cross section.

The friction coefficient f is defined by the *Fanning equation*. Thus,

$$f \equiv \frac{\text{tangential force}}{\frac{1}{2}\rho V^2\,(\text{wetted area})} = \frac{\delta F_f}{\frac{1}{2}\rho V^2 (WP)\,dx} = \frac{\tau}{\frac{1}{2}\rho V^2} \tag{3.16}$$

where τ is the shear stress at the wall. Hence,

$$\delta F_f = f\,\frac{\rho V^2}{2}\,(WP)\,dx = \frac{\rho V^2}{2}\left(\frac{4f\,dx}{\mathscr{D}}\right) A \tag{3.17}$$

The friction coefficient $\mathfrak{f}$ defined by equation 3.16 is the *Fanning* friction coefficient. Darcy defined a *friction factor* $\mathfrak{f}'$ by the equation

$$\mathfrak{f}' \equiv \frac{4\tau}{\frac{1}{2}\rho V^2} = 4\mathfrak{f} \tag{3.18}$$

Hence, the *Darcy friction factor* $\mathfrak{f}'$ is four times the *Fanning friction coefficient* $\mathfrak{f}$.

Substituting for $\delta F_{\mathfrak{f}}$ from equation 3.17 into equation 3.14 and dividing by A gives the following equation.

$$dp + \rho V \, dV + \rho g \, dz + \frac{\rho V^2}{2}\left(\frac{4\mathfrak{f}\,dx}{\mathscr{D}}\right) + \frac{\delta D}{A} = 0 \tag{3.19}$$

Equation 3.19 is the differential equation for the motion of a fluid in the presence of wall friction and obstructions. The equation cannot be integrated without additional knowledge relating p, ρ, V, $\mathfrak{f}$, and $\mathscr{D}$.

3–5(e) Incompressible Flow with Friction

In the case of the flow of an incompressible liquid, or gas at low speeds where density changes are negligible, the density $\rho = $ constant. Moreover, if there are no obstructions in the flow passage, then the drag term $\delta D = 0$. For those conditions, the integration of equation 3.19 gives

$$\frac{p_2}{\rho} + \frac{V_2^2}{2} + g z_2 + \int_1^2 \frac{V^2}{2}\left(\frac{4\mathfrak{f}\,dx}{\mathscr{D}}\right) = \frac{p_1}{\rho} + \frac{V_1^2}{2} + g z_1 = \text{constant} \tag{3.20}$$

The integral term in equation 3.20 represents a loss because of the effects of wall friction. All of the parameters in the integrand of that term are positive, so that the value of the loss because of such frictional effects increases as the length x of the flow passage increases. To integrate the aforementioned term, the relationship between V, $\mathfrak{f}$, $\mathscr{D}$, and x must be known. In general, the integral term cannot be determined without employing empirical data. The determination of the aforementioned frictional term is discussed in Section 5–4, which deals with flow in constant area ducts.

3–5(f) Compressible Flow of a Perfect Gas in the Presence of Friction

In the special case where the flowing fluid is a perfect gas, equation 3.19 can be expressed in terms of the flow Mach number M. Multiplying throughout by $1/p$, noting that $\gamma p/\rho = a^2$, and neglecting body forces, we obtain

$$\frac{dp}{p} + \frac{\rho\gamma}{p\gamma}\frac{V^2}{V^2}d\left(\frac{V^2}{2}\right) + \frac{\rho\gamma}{p\gamma}\frac{V^2}{2}\left(\frac{4\mathfrak{f}\,dx}{\mathscr{D}}\right) + \frac{\delta D}{Ap} = 0 \tag{3.21}$$

Hence,

$$\frac{dp}{p} + \frac{\gamma M^2}{2}\frac{dV^2}{V^2} + \frac{\gamma M^2}{2}\left(\frac{4\mathfrak{f}\,dx}{\mathscr{D}}\right) + \frac{\delta D}{Ap} = 0 \tag{3.22}$$

Since $V^2 = M^2 a^2$ and $da^2/a^2 = dt/t$, the second term in equation 3.22 becomes

$$\frac{\gamma M^2}{2}\frac{dV^2}{V^2} = \frac{\gamma}{2}dM^2 + \frac{\gamma M^2}{2}\frac{dt}{t} \tag{3.23}$$

Substituting equation 3.23 into equation 3.22 yields

$$\frac{dp}{p} + \frac{\gamma}{2} dM^2 + \frac{\gamma M^2}{2} \frac{dt}{t} + \frac{\gamma M^2}{2} \left(\frac{4f\, dx}{\mathcal{D}}\right) + \frac{\delta D}{Ap} = 0 \qquad (3.24)$$

Equation 3.24 is a form of the momentum equation for the flow of a perfect gas in the presence of friction; its integration is presented in Section 5.5.

3-5(g) External Forces Acting on a Flowing Fluid

Consider a fluid flowing through the passage illustrated in Fig. 3.5a. If there are no obstructions within the fluid, and the effects of body forces are neglected, then the only forces acting on the fluid are: (1) the pressure forces acting over the inlet and

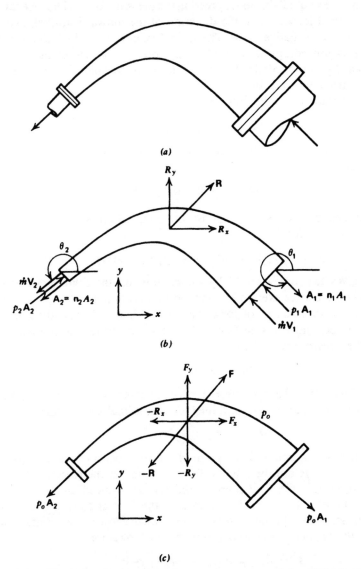

(a)

(b)

(c)

Figure 3.5. Determination of the forces acting on a conduit and the fluid inside of it. (a) Flow of a fluid through a conduit. (b) Free-body diagram of the fluid inside the conduit. (c) Free-body diagram for the conduit.

exit areas A_1 and A_2, and (2) the reaction force $\mathbf{R}$ exerted by the inner surface of the conduit on the fluid.

Figure 3.5b is a free-body diagram of the fluid within the conduit. Note that a coordinate system has been specified, and a direction of action for the resultant force $\mathbf{R}$ has been chosen. The components R_x and R_y are arbitrarily chosen in the positive directions. The choice of the assumed direction is immaterial because, if the direction is chosen incorrectly, the numerical results for R_x and R_y will then be negative.

The momentum equation for the flow situation illustrated in Fig. 3.5a is given by equation 2.56, which becomes

$$\mathbf{F}_{\text{Surface}} = \int_A \mathbf{V}(\rho\mathbf{V} \cdot d\mathbf{A}) \tag{3.25}$$

There are two choices available for expressing the pressure forces. The first assumes that all of the normal forces act in the direction of the outward normal, and then considers pressure as a negative stress. The second makes use of the knowledge that all pressure forces are compressive; that is, the pressure forces in the free-body diagram act inwardly, as indicated in Fig. 3.5b. The second technique is recommended, and is followed throughout this book.

Thus, the external forces on the fluid are:

$$-p_1(\mathbf{n}_1 A_1) - p_2(\mathbf{n}_2 A_2) + \mathbf{R} \tag{a}$$

where $\mathbf{n}_1$ and $\mathbf{n}_2$ are the respective unit normals to A_1 and A_2. The change in momentum of the fluid is

$$\dot{m}(\mathbf{n}_2 V_2) - \dot{m}(-\mathbf{n}_1 V_1) \tag{b}$$

Note that the direction of $\mathbf{V}_1$ is $(-\mathbf{n}_1)$. Substituting equations (a) and (b) into equation 3.25 and solving for $\mathbf{R}$ yields

$$\mathbf{R} = \mathbf{n}_2(p_2 A_2 + \dot{m}V_2) + \mathbf{n}_1(p_1 A_1 + \dot{m}V_1) \tag{3.26}$$

Equation 3.26 gives the force exerted by the conduit on the fluid. Reversing the direction of $\mathbf{R}$ gives the force exerted by the fluid on the conduit.

Since $\dot{m}V$ has the dimensions of force (i.e., ML/T^2), it is frequently called the *stream effective force*, and the sum $(pA + \dot{m}V)$ is often termed the *stream thrust*. Let $\mathscr{F}$ denote the stream thrust; then

$$\mathscr{F} \equiv pA + \dot{m}V \tag{3.27}$$

Equation 3.26 shows that the net external force $\mathbf{R}$ acting on a fluid is the vector sum of the stream thrusts at the exit and inlet cross-sectional areas. Thus,

$$\mathbf{R} = \mathbf{n}_2\mathscr{F}_2 + \mathbf{n}_1\mathscr{F}_1 \tag{3.28}$$

Figure 3.5c illustrates a free-body diagram of the solid walls of the flow passage. The forces on the passage are the internal force $\mathbf{R}$ acting in the reversed direction, the external restraining force $\mathbf{F}$, if present, and the unbalanced ambient pressure forces acting on the back of the projected areas of the inlet and exit cross sections. The latter forces act along the same lines as the pressure forces within the fluid in Fig. 3.5b, but in the opposite directions. Summing the forces, we obtain

$$\mathbf{F} - \mathbf{R} + \mathbf{n}_2 p_o A_2 + \mathbf{n}_1 p_o A_1 = 0 \tag{c}$$

Substituting for $\mathbf{R}$ from equation 3.28 gives

$$\mathbf{F} = \mathbf{n}_2(\mathscr{F}_2 - p_o A_2) + \mathbf{n}_1(\mathscr{F}_1 - p_o A_1) \tag{3.29}$$

If $\mathscr{F}'$ is defined by

$$\mathscr{F}' \equiv \mathscr{F} - p_o A = (p - p_o)A + \dot{m}V \qquad (3.30)$$

then equation 3.29 becomes

$$\mathbf{F} = \mathbf{n}_2 \mathscr{F}'_2 + \mathbf{n}_1 \mathscr{F}'_1 \qquad (3.31)$$

In other words, when a conduit is immersed in a uniform ambient pressure field, the net force required to hold the conduit stationary is calculated in the same manner as the net internal force of the conduit acting on the fluid, except that *absolute pressures* are replaced by *gage pressures*.

Example 3.2. Water at a static pressure of $14.80 \cdot 10^5$ N/m^2 enters a reducing section of pipe with an average velocity of 6.096 m/s normal to the entrance section (see Fig. 3.6). The entrance and exit diameters of the pipe are 0.3048 m and 0.1524 m, respectively. The direction of discharge is 90 deg to that at the entrance, and the static pressure of the water in the exit section is $12.05 \cdot 10^5$ N/m^2. The atmospheric pressure surrounding the pipe is $1.013 \cdot 10^5$ N/m^2, and the water density is 999.55 kg/m^3. Calculate the force exerted by the water on the pipe section.

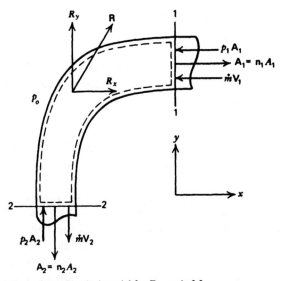

Figure 3.6.   Physical model for Example 3.2.

Solution

$$A_1 = \frac{\pi D_1{}^2}{4} = \frac{\pi(0.3048)^2}{4} = 0.07297 \text{ m}^2$$

$$A_2 = \frac{\pi D_2{}^2}{4} = \frac{\pi(0.1524)^2}{4} = 0.01824 \text{ m}^2$$

For steady one-dimensional flow:

$$\dot{m} = \rho_1 A_1 V_1 = \rho_2 A_2 V_2$$

$$V_2 = \frac{V_1 A_1}{A_2} = \frac{6.096(0.07297)}{(0.01824)} = 24.385 \text{ m/s}$$

$$\dot{m} = \rho_1 A_1 V_1 = 999.55(0.07297)(6.096) = 444.62 \text{ kg/s (980.2 lbm/sec)}$$

The units normals n_1 and n_2 are given by

$$n_1 = (+1)i + (0)j = i \quad \text{and} \quad n_2 = (0)i + (-1)j = -j$$

From equation 3.27,

$$\mathscr{F}_1 = p_1 A_1 + \dot{m} V_1 = (14.80 \cdot 10^5)(0.07297) + 444.62(6.096) = 110,706 \text{ N}$$

$$\mathscr{F}_2 = p_2 A_2 + \dot{m} V_2 = (12.05 \cdot 10^5)(0.01824) + 444.62(24.385) = 32,821 \text{ N}$$

From equation 3.28, the components R_x and R_y of the force exerted on the fluid by the pipe are obtained as follows.

$$R = iR_x + jR_y = n_2\mathscr{F}_2 + n_1\mathscr{F}_1 = -j\mathscr{F}_2 + i\mathscr{F}_1$$

$$R_x = \mathscr{F}_1 = 110,706 \text{ N (24,890 lbf)}$$

$$R_y = -\mathscr{F}_2 = -32,821 \text{ N } (-7,378 \text{ lbf})$$

The component forces exerted by the fluid on the pipe have the same magnitudes as R_x and R_y, but they act in the opposite direction. The corresponding forces required to hold the elbow stationary are calculated in the same manner as R_x and R_y, with the exception that $\mathscr{F}_1$ and $\mathscr{F}_2$ are based on gage pressures. Thus,

$$F_x = (\mathscr{F}_1 - p_o A_1) = [110,706 - 1.013 \cdot 10^5 (0.07297)] = 103,314 \text{ N (23,226 lbf)}$$

$$F_y = -(\mathscr{F}_2 - p_o A_2) = -[32,821 - 1.013 \cdot 10^5 (0.01824)] = -30,973 \text{ N } (-6960 \text{ lbf})$$

Example 3.3.   Derive an equation for the net static thrust F produced by the rocket engine of a rocket-propelled vehicle when the system is operated under static conditions in the atmosphere.

Solution

The selection of the control volume is arbitrary. It is selected so that it will make the solution of the problem convenient. Accordingly, consider the control volume illustrated in Fig. 3.7a.

The restraining force F' applied by the thrust stand (not shown) is equal in magnitude to the net thrust F, and acts in the opposite direction. Thus, $F' = -F$. The force R exerted by all of the fluids inside of the rocket engine on the control surface is the internally generated thrust. It is transmitted to the walls of the rocket engine through the pressure and shear forces acting on those walls. A simple force balance yields the result $F' = ip_o A - R$, and thus $F = R - ip_o A$. The calculation of the internal pressure and shear forces, however, is extremely difficult.

An alternate method for determining the internally generated thrust R is to replace all internal forces by the net resultant force exerted by the walls of the engine upon the fluid, $-R$, and then apply the momentum equation to the fluid occupying the control volume shown in Fig. 3.7b. Thus, equation 3.25 yields

$$-R + ipA = \rho(-iV) \cdot (-iA)(-iV) = -i\rho A V^2 = -i\dot{m}V \tag{a}$$

$$R = i(\dot{m}V + pA) \tag{b}$$

The net thrust F on the vehicle is then given by

$$F = R - ip_o A = i[\dot{m}V + (p - p_o)A] \tag{c}$$

Consider the same problem when an annular surface of area A_b surrounds the exit section of the exhaust nozzle at the base of the missile. The area A_b is called the *base area*, and the external pressure p_b acting on it is called the *base pressure*. That situation is illustrated in Fig. 3.7c. In this case, the net restraining force F' is found directly

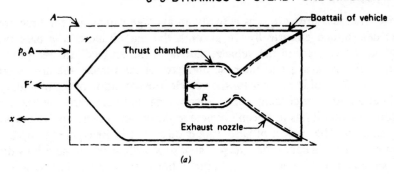

(a)

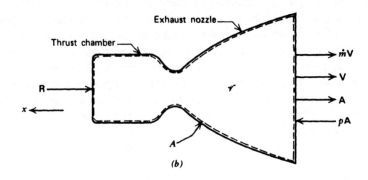

(b)

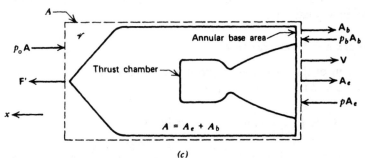

(c)

Figure 3.7. Control volumes for Example 3.3. (a) Control volume for relating the net thrust and the internally generated thrust. (b) Control volume for determining the internally generated thrust. (c) Control volume for determining the net thrust, including base thrust.

from equation 3.25. Thus,

$$\mathbf{F}' - p_o\mathbf{A} + p\mathbf{A}_e + p_b\mathbf{A}_b = -\mathbf{i}\dot{m}V \tag{d}$$

$$\mathbf{F} = -\mathbf{F}' = \mathbf{i}[\dot{m}V + (p - p_o)A_e + (p_b - p_o)A_b] \tag{e}$$

This approach eliminates the intermediate step required by the first approach to find **R**.

When a base region is present, an additional term appears in the thrust equation, the *base thrust* given by $(p_b - p_o)A_b$. Frequently the *base thrust* is called the *base drag*, since p_b is generally much lower than p_o, and the term $(p_b - p_o)A_b$ yields a negative contribution to the thrust acting on the propelled vehicle. The base region is a very complicated flow region because the flow from the nozzle separates from the wall at the nozzle lip. A highly viscous recirculation region develops over the base area. The base pressure p_b is actually an average value over the base area and must be determined empirically. In general, the magnitude of the base pressure depends on the velocity of the vehicle relative to the ambient air, the ratio of the diameter of the exhaust jet to the base diameter of the vehicle, the nozzle pressure ratio p_e/p_o, the Reynolds number for the vehicle, the condition of the boundary layer on the boattail, and other factors. The phenomena that determine the *base pressure* are so complex that it has not been possible to derive any purely analytic expression for evaluating it. The subject of base drag is discussed in References 9 to 12.

Under static testing conditions the pressure p_b is ordinarily not materially different from p_o, so the term $(p_b - p_o)A_b \approx 0$. In flight, however, because of the relative speed of the vehicle with respect to the ambient atmospheric air, p_b may be much smaller than p_o. The base drag may be reduced by shaping the *boattail* of the vehicle; that is, the cylindrical aft portion of the vehicle that is reduced in diameter as it approaches the base (rear end) of the vehicle. When the rocket engine is operating and the vehicle is in flight, the presence of the jet ejected through the exit section A_e of the nozzle is effective in reducing the total base area so that it is limited to the annular base area A_b. However, when the rocket engine completes its operating period, called the *powered flight*, there is no exhaust jet and the base pressure acts on the entire base area of the vehicle, which includes the area inside of the rocket motor. Accordingly, careful attention must be given to the design of the nozzle-base-boattail combination so that the total thrust acting on the aft end of the vehicle is as large as possible.

3–6 THERMODYNAMICS OF STEADY ONE-DIMENSIONAL FLOW

The concepts of steady one-dimensional flow discussed in Section 3–3 will now be applied to the energy equation.

3–6(a) Flow with Work and Heat Transfer

The integral form of the energy equation is presented in Table 2.1, equation 2.88, which is repeated and renumbered below.

$$\dot{W}_{shaft} + \dot{W}_{shear} - \dot{Q} + \int_{\mathscr{V}} \frac{\partial}{\partial t}\left[\rho\left(u + \frac{V^2}{2} + gz\right)\right] d\mathscr{V}$$
$$+ \int_A \left(h + \frac{V^2}{2} + gz\right)(\rho \mathbf{V} \cdot d\mathbf{A}) = 0 \quad (3.32)$$

Figure 3.8 illustrates the steady one-dimensional flow model to which equation 3.32 will be applied. The shear and shaft works are combined into a single term and simply called work W. As explained in Section 2–7, the basis for determining the work and heat terms is assumed to agree with that for determining the remaining energy terms. Accordingly, the single set of symbols W and Q will be employed for representing work and heat, respectively.

Applying equation 3.32 to the flow model presented in Fig. 3.8 yields

$$\delta W - \delta Q + \left[h + dh + \frac{V^2}{2} + d\left(\frac{V^2}{2}\right) + gz + g\, dz\right]\dot{m} - \left[h + \frac{V^2}{2} + gz\right]\dot{m} = 0$$

Cancelling terms and dividing by $\dot{m}$, we obtain

$$\delta W - \delta Q + dh + d\left(\frac{V^2}{2}\right) + g\, dz = 0 \qquad (3.33)$$

Equation 3.33 is the differential form of the energy equation for a steady one-dimensional flow.

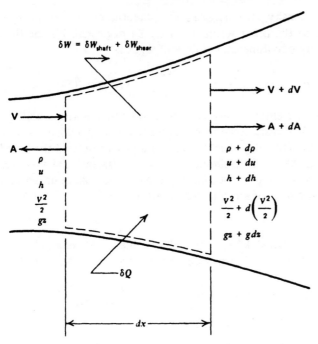

Figure 3.8.   Energy balance for a fluid element of a stream tube.

Equation 3.32 may also be applied to a control volume of finite size such as that illustrated in Fig. 3.9. In that case, equation 3.32 becomes

$$\dot{W} - \dot{Q} + \int_{A_2}\left(h + \frac{V^2}{2} + gz\right)\rho V\, dA - \int_{A_1}\left(h + \frac{V^2}{2} + gz\right)\rho V\, dA = 0 \quad (3.34)$$

where Q denotes the heat added to the fluid flowing between stations 1 and 2, and W corresponds to the work performed. Let the subscripts 1 and 2 denote the values of the variables h, V, and z, at A_1 and A_2 respectively. The steady one-dimensional energy equation is obtained from equation 3.34 by evaluating the integrals and dividing throughout by ρAV. The result is

$$Q + h_1 + \frac{V_1^2}{2} + gz_1 = W + h_2 + \frac{V_2^2}{2} + gz_2 \qquad (\textit{for the flow of any fluid}) \quad (3.35)$$

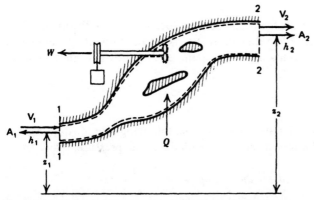

Figure 3.9.   Energy balance for a control volume of finite size.

For fluids of small density, such as gases, the potential energy term $g(z_2 - z_1)$ is negligible compared to the other terms and may be neglected. Hence, the energy equation for the steady one-dimensional flow of a gas is

$$Q + h_1 + \frac{V_1^2}{2} = W + h_2 + \frac{V_2^2}{2} \quad \text{(for gas flow)} \quad (3.36)$$

Example 3.4. Heated air enters a turbine with a static temperature of 1000 K and a velocity of 60.0 m/s. It is discharged from the turbine with a velocity of 120.0 m/s. The turbine delivers 175,000 J/kg of shaft work to a generator, and the heat loss to the surroundings is 25,000 J/kg. Assuming steady one-dimensional flow, calculate the enthalpy and temperature of the air leaving the turbine, taking into account the variation of the specific heat of air with temperature.

Solution

From Table C.4, $h_1 = 1047.248$ kJ/kg. From the problem statement, $Q = -25.000$ kJ/kg and $W = 175.000$ kJ/kg. Substituting those values into equation 3.36 yields

$$h_2 = -25.000 - 175.000 + 1047.248 - \frac{(120.0)^2 - (60.0)^2}{2(1000)}$$

$$h_2 = 841.848 \text{ kJ/kg (361.94 Btu/lbm)}$$

From Table C.4, for $h_2 = 841.848$ kJ/kg, $t_2 = 817.03$ K (1470.6 R).

3–6(b) Adiabatic Flow with No External Work

In many flow passages such as diffusers, ducts, and nozzles, the working fluid flows under substantially adiabatic conditions and does no external work. For such flow passages $\delta Q = \delta W = 0$, and equation 3.33 reduces to

$$dh + d\left(\frac{V^2}{2}\right) + g\,dz = 0 \quad (3.37)$$

Since equation 3.33 applies to *steady flow with or without friction*, equation 3.37 also applies to steady adiabatic flow (with no external work) with or without friction. Unless stated to the contrary, the phrase *steady adiabatic flow* will henceforth refer to adiabatic flow with no external work. Integrating equation 3.37 between states 1 and 2 gives the integrated form of the energy equation for steady adiabatic flow. Thus,

$$h_1 + \frac{V_1^2}{2} + gz_1 = h_2 + \frac{V_2^2}{2} + gz_2 = h + \frac{V^2}{2} + gz = \text{constant} \quad (3.38)$$

Since $(h + V^2/2 + gz)$ is the total energy associated with a flowing fluid, it follows from equation 3.38 that for a fluid in steady adiabatic flow, the total energy of the fluid is constant at all points in its flow path; such a flow is termed an *isoenergetic* flow.

3–6(c) Isentropic Flow

For isentropic changes of state, we obtain the following relationship from equation 1.65, which relates changes in entropy, enthalpy, and pressure. Thus,

$$t\,ds = dh - v\,dp = dh - \frac{dp}{\rho} = 0 \quad (3.39)$$

Substituting for dh in equation 3.37 from equation 3.39 gives

$$\frac{dp}{\rho} + d\left(\frac{V^2}{2}\right) + g\,dz = 0 \tag{3.40}$$

Equation 3.40 is a form of the energy equation, per unit mass of fluid, for a steady one-dimensional isentropic flow; the equivalence of equation 3.40 and equation 3.9 is discussed in Section 3–5(a).

In equation 3.40, the term dp/ρ represents the change in pressure energy, $d(V^2/2)$ is the change in kinetic energy, and $g\,dz$ is the change in potential energy. The equation shows that the total energy per unit mass of the flowing fluid remains constant, and that the only changes possible are those between the pressure and the kinetic and potential energies. To complete the integration of equation 3.40, the relationship between the density ρ and the pressure p must be known.

For an adiabatic flow ($\delta Q = \delta W = 0$), the flow field outside of the boundary layer may be assumed to be isentropic ($ds = 0$). The assumption of *isentropicity* is valid for both subsonic and supersonic flows in the inviscid portions of the flow field for *adiabatic flow expansions*. It should be noted that for an *adiabatic flow compression*, the flow may be assumed to be isentropic provided that the flow is *subsonic* throughout. If, however, a *supersonic flow* is being compressed, the flow must be examined closely to ascertain that the assumption of isentropicity is permissible, because of the possibility of the occurrence of *compression shock wave phenomena*. Such phenomena increase the entropy of the flowing fluid, thereby invalidating the assumption that $ds = 0$. Shock wave phenomena are discussed in Chapter 7.

3-6(d) Adiabatic Flow of a Perfect Gas

If the flowing fluid is a perfect gas, $dh = c_p\,dt$, and the effects of body forces may be neglected; that is, $g\,dz = 0$. Then equation 3.37 becomes

$$c_p\,dt + d\left(\frac{V^2}{2}\right) = 0 \tag{3.41}$$

Assume that c_p is not a function of t. Integrating equation 3.41 between states 1 and 2 then yields

$$c_p t_1 + \frac{V_1^2}{2} = c_p t_2 + \frac{V_2^2}{2} = c_p t + \frac{V^2}{2} = \text{constant} \qquad (c_p = \text{constant}) \tag{3.42}$$

In equation 3.42 the kinetic energy terms $V_1^2/2$ and $V_2^2/2$ represent the energy associated with the directed motion of the gas molecules, while the terms $c_p t_1$ and $c_p t_2$ measure the energy associated with their random motions (see Section 1–17). Consequently, the energy equation for steady adiabatic flow may be regarded as a statement relating the interchangeability of the energy associated with the fluid molecules from random into directed motion.

Since the Mach number M is a measure of the aforementioned energy ratio, it is convenient to express equation 3.42 in terms of the Mach number. For a perfect gas, $c_p = \gamma R/(\gamma - 1)$ and $a^2 = \gamma R t$. Substituting for c_p and a^2 into equation 3.42 yields

$$\frac{V_1^2}{2} + \frac{a_1^2}{\gamma - 1} = \frac{V_2^2}{2} + \frac{a_2^2}{\gamma - 1} = \frac{V^2}{2} + \frac{a^2}{\gamma - 1} = \text{constant} \tag{3.43}$$

Introducing $M = V/a$ into equation 3.43 and rearranging gives

$$a_1{}^2\left(1 + \frac{\gamma - 1}{2}M_1{}^2\right) = a_2{}^2\left(1 + \frac{\gamma - 1}{2}M_2{}^2\right)$$

$$= a^2\left(1 + \frac{\gamma - 1}{2}M^2\right) = \text{constant} \qquad (3.44)$$

Recalling that $a^2 = \gamma Rt$, equation 3.44 becomes

$$t_1\left(1 + \frac{\gamma - 1}{2}M_1{}^2\right) = t_2\left(1 + \frac{\gamma - 1}{2}M_2{}^2\right)$$

$$= t\left(1 + \frac{\gamma - 1}{2}M^2\right) = \text{constant} \qquad (3.45)$$

Equations 3.41 to 3.45 are different forms of the energy equation for the adiabatic flow of a perfect gas; they are directly applicable, of course, only when it may be assumed that c_p is a constant independent of the gas temperature.

3–6(e) Isentropic Discharge Speed for a Perfect Gas

Consider the case where a perfect gas is expanded in a frictionless nozzle under adiabatic conditions, and it is desired to determine the speed attained by the gas. Since the aforementioned process is adiabatic, equation 3.42 may be applied directly. Let point 1 denote the conditions in the supply reservoir where $V_1 \approx 0$, $t_1 = T$, and $p_1 = P$. Let point 2 denote any other point in the isentropic flow where $V_2 = V'$, $t_2 = t'$, and $p_2 = p'$. Solving for V', the *isentropic discharge speed*, from equation 3.42, gives

$$V' = \left[2c_pT\left(1 - \frac{t'}{T}\right)\right]^{1/2} \qquad (3.46)$$

For a perfect gas, $c_p = \gamma R/(\gamma - 1)$. For an isentropic change of state of a perfect gas, equation 1.128 gives

$$\frac{t'}{T} = \left(\frac{p'}{P}\right)^{(\gamma - 1)/\gamma} \qquad (3.47)$$

Substituting for c_p and t'/T into equation 3.46 yields

$$V' = \left\{\frac{2\gamma RT}{\gamma - 1}\left[1 - \left(\frac{p'}{P}\right)^{(\gamma - 1)/\gamma}\right]\right\}^{1/2} \qquad (3.48)$$

Equation 3.48 was derived independently by both St. Venant and Wantzel (1839) and is known as the *St. Venant-Wantzel equation*. It gives the maximum speed attainable by a perfect gas that expands isentropically from an infinite reservoir where $V = 0$, $p = P$, and $t = T$, to the static pressure $p = p'$. In a real nozzle, because of friction, the speed obtained by expanding a real gas is always smaller than V'.

Since $R = \bar{R}/\bar{m}$, where $\bar{R}$ is the universal gas constant and $\bar{m}$ is the molecular weight of the gas, equation 3.48 may be expressed in terms of $\bar{m}$. Thus,

$$V' = \left\{\frac{2\gamma}{\gamma - 1}\frac{\bar{R}T}{\bar{m}}\left[1 - \left(\frac{p'}{P}\right)^{(\gamma - 1)/\gamma}\right]\right\}^{1/2} \qquad (3.49)$$

Equation 3.49 shows that, for given values of the pressure ratio p'/P and the specific heat ratio γ, the *isentropic discharge speed* depends directly on $\sqrt{T/\bar{m}}$.

Equations 3.48 and 3.49 apply for expansions from a stagnation pressure P where $V = 0$ to a static pressure p' when body forces are negligible.

3-7 THE SECOND LAW OF THERMODYNAMICS FOR STEADY ONE-DIMENSIONAL FLOW

From Table 2.1, equation 2.89, the second law of thermodynamics, or the entropy equation, for a control volume is

$$\int_{\mathscr{V}} (s\rho)_t \, d\mathscr{V} + \int_A s(\rho \mathbf{V} \cdot d\mathbf{A}) \geqq \dot{Q}/t \tag{3.50}$$

The steady one-dimensional flow model illustrated in Fig. 3.8 may be employed for obtaining the applicable form of equation 3.50. Proceding in a manner analogous to that employed in deriving the steady one-dimensional forms for the continuity, momentum, and energy equations, the following result is obtained.

$$ds \geqq \delta Q/t \tag{3.51}$$

To apply equation 3.51 to a control volume of finite size, an explicit expression must be derived for δQ in terms of the thermodynamic properties of the flowing fluid for the particular heat transfer process that is pertinent to the situation (i.e., conduction, convection, or radiation).

In the absence of heat transfer, equation 3.51 reduces to the important statement

$$ds \geqq 0 \tag{3.52}$$

Equation 3.52 states that, in the absence of heat transfer, the entropy of a fluid flowing through a control volume must either increase or remain constant.

Generally, equations 3.51 and 3.52 are applied only qualitatively: for determining either the direction of a process or whether a given process is physically possible. The actual magnitude of the entropy change is quite often of no particular interest.

3-8 SOME GENERAL EFFECTS OF COMPRESSIBILITY ON FLUID FLOW

In the preceding sections of this chapter, the governing equations are derived for the steady one-dimensional flow of a fluid. In this section, several general effects caused by the compressibility of a flowing fluid are considered.

3-8(a) Speed of Propagation of Small Disturbances

In Section 1-17(a) it is stated without proof that a small disturbance is propagated with the speed of sound a, which is given by equation 1.180. Thus,

$$a^2 = \left(\frac{\partial p}{\partial \rho}\right)_s \tag{3.53}$$

The equations derived in the preceding sections of Chapter 3 will now be employed for demonstrating that small pressure disturbances are propagated with the speed a.

Figure 3.10a illustrates schematically the one-dimensional propagation of a plane-fronted small pressure disturbance. Let dp denote the infinitesimal amplitude of the disturbance, which is, of course, accompanied by infinitesimal changes in the other flow properties of the fluid, denoted by $d\rho$, dt, etc. As a consequence of the

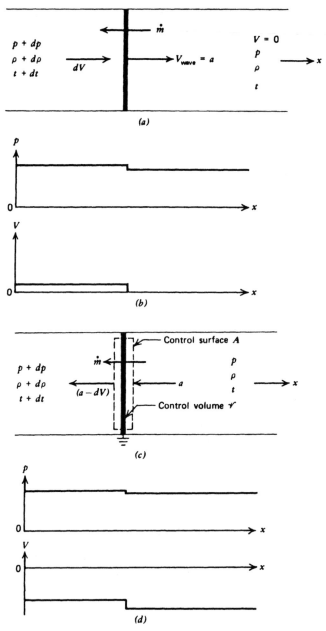

Figure 3.10. Propagation of a small pressure disturbance. (a) Moving pressure wave of small amplitude. (b) Distributions of static pressure and fluid velocity. (c) Stationary pressure wave. (d) Distributions of static pressure and fluid velocity.

passage of the pressure wave through the fluid, an infinitesimal velocity change dV is imparted to the fluid in the direction of propagation for the wave. Figure 3.10b illustrates the distributions of the pressure and velocity in the fluid, for the process illustrated in Fig. 3.10a. The latter is, in effect, an *unsteady* one-dimensional flow that cannot be analyzed by the steady flow equations developed in this chapter. To obviate the effects caused by the flow being *unsteady* to a *stationary* observer, the coordinate system must be transformed by assuming that the observer moves at the same speed as the pressure wave. Hence, from the viewpoint of the *moving* observer, the process appears as a steady flow of fluid moving past the moving observer; that is, the situation illustrated schematically in Fig. 3.10c. Mathematically, the transformation is effected by subtracting the wave velocity, denoted by a, from all of the velocities in Fig. 3.10a. It is a *dynamic transformation* because only the dynamical properties of the fluid are affected (the fluid velocity, the stagnation pressure, the stagnation temperature, etc.). The static thermodynamic properties of the fluid are unaffected by the transformation. Furthermore, note that the positive direction for the coordinate system remains unchanged. If desired, a new coordinate system may be defined with its positive direction pointing toward the left, but the results are unaffected by such a change. Figure 3.10d illustrates, for the transformed wave, the distributions of the properties of the fluid. Note that the direction of the velocities a and $(a - dV)$ are to the left in Fig. 3.10c.

The latter figure also illustrates a stationary control volume $\mathscr{V}$ surrounding the wave; the actual width of the control volume is immaterial, but it must include the wave. From Table 2.1, for $\mathbf{B} = \mathbf{F}_{\text{shear}} = \partial \mathbf{V}/\partial t = 0$, the momentum equation for the control volume $\mathscr{V}$ becomes

$$-\int_A p\, d\mathbf{A} = \int_A \mathbf{V}(\rho\mathbf{V} \cdot d\mathbf{A}) \tag{3.54}$$

Figure 3.11 illustrates in detail the pressure forces and the momentum fluxes pertinent to the control volume $\mathscr{V}$. Applying equation 3.54 to that control volume, we obtain

$$-pA + (p + dp)A = \dot{m}[-(a - dV)] - \dot{m}(-a) \tag{3.55}$$

An important point arises in applying equation 3.54 to obtain equation 3.55. The momentum flux term in equation 3.54 is given by $\int_A \mathbf{V}(\rho\mathbf{V} \cdot d\mathbf{A})$. As discussed

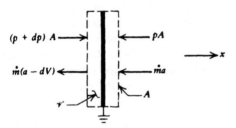

Figure 3.11. Control volume for determining the speed of sound.

in Section 2-4, the term $(\rho\mathbf{V} \cdot d\mathbf{A})$ is inherently positive for the mass leaving a control volume and negative for the mass entering it. In addition to that sign convention, the velocity $\mathbf{V}$ itself must agree with the sign convention established by the assumed coordinate system. In the subject problem, this results in the negative signs being associated with the velocities in equation 3.55.

Simplifying equation 3.55 yields

$$dp = \dot{m}\frac{dV}{A} = (\rho A a)\frac{dV}{A} = \rho a\, dV \tag{3.56}$$

From the continuity equation,

$$\rho A a = (\rho + d\rho)A(a - dV) \tag{3.57}$$

Neglecting higher-order terms, simplifying, and solving for dV, we obtain

$$dV = a \frac{d\rho}{\rho} \qquad (3.58)$$

Combining equations 3.56 and 3.58 and solving for a^2 yields

$$a^2 = \frac{dp}{d\rho} = \left(\frac{\partial p}{\partial \rho} \right)_s \qquad (3.59)$$

The subscript s, denoting an isentropic process, is attached because the model postulated by Fig. 3.11 applies to an adiabatic inviscid flow with no external work or body forces; that is, an isentropic process. Equation 3.59 is, therefore, the speed with which small disturbances are propagated in a compressible fluid.

3–8(b) Pressure Disturbances in a Compressible Fluid

The basic difference in the phenomena associated with subsonic and supersonic speeds may be demonstrated qualitatively in the following manner. Consider the propagation of a spherical sound wave emanating from the stationary point source 0, as illustrated in Fig. 3.12. Assume that the fluid surrounding the point source 0 is at rest before the creation of the small pressure disturbance, so that point 0 is the *source* from which sound waves are propagated in all directions with the constant sonic or *acoustic speed a*. Each sound wave, therefore, moves with the speed a and, in the same time interval Δt, each wave advances an equal radial distance $r = a \Delta t$.

Now consider the case where a body moves with the uniform subsonic speed V in an isothermal homogeneous fluid, as illustrated in Fig. 3.13. Assume that the pressure disturbances created by the moving body are small, so that they may be treated as sound waves. As each point of the body contacts previously undisturbed fluid, it becomes a source from which pressure disturbances are propagated through the fluid with the local sonic speed a. Since the pressure disturbances have been assumed to be small, it may be assumed that the sonic speed a remains constant. Let the instant when point P on the body is at A (see Fig. 3.13) be the starting point of the discussion. The pressure disturbance created by point P will traverse the distance $a \Delta t$ from the starting point A in the time interval Δt. During the same time interval Δt,

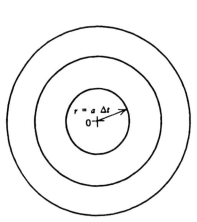

Figure 3.12. Propagation of sound waves from a stationary point source.

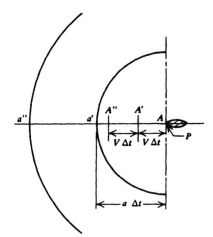

Figure 3.13. Pressure disturbances produced by a body moving with subsonic speed.

the point P moves the distance $AA' = V\,\Delta t$. In a second time interval Δt, point P moves to A''; that is, the distance $A'A'' = V\,\Delta t$. The initial disturbance created by point P at the starting point A, in the second time interval Δt, moves the distance $a'a'' = a\,\Delta t$. Since the acoustic speed a is always larger than V, the speed of the body, the wave front produced by the moving body is always ahead of the latter. The net result is that the body always moves into a fluid that has already undergone changes because of the motion of the body. Consequently, if a body moves in a compressible fluid with subsonic speed, the fluid ahead of the body may be said to become aware of the presence of the body because the latter propagates disturbance signals ahead of itself. Moreover, when a body moves with subsonic speeds, the disturbances it creates are said to "clear away" from it.

Consider next the case where a body moves with the sonic speed a; that is, $V = a$, as illustrated in Fig. 3.14. During any time interval Δt, the front of the pressure disturbance moves the same distance $a\,\Delta t$ as the point $V\,\Delta t$ that created the disturbance. Consequently, all of the disturbances created by the body coalesce at the point P on the body, and they are all contained in the half-plane to the right of the body. This coalescense of waves creates some extremely complicated flow patterns in the sonic and transonic flow regimes.

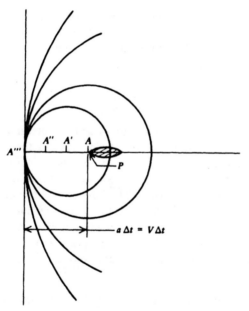

Figure 3.14. Pressure disturbances produced by a body moving at the sonic speed.

Now consider the case where the body moves with the uniform speed V greater than the sonic speed a, as illustrated in Fig. 3.15. In the time interval Δt the front of the pressure disturbance created by a point P on the body is a spherical surface of radius $a\,\Delta t$. The point P on the body, in the interval Δt, moves from A to the position A'''', where $AA'''' = V\,\Delta t$. Since $V > a$, the wave front of the disturbance created by the body lags behind the point on the body that created the disturbance, and the disturbance wave front cannot overtake the moving body. Consequently, in all of its successive positions A', A'', etc., the moving body is ahead of the disturbance wave fronts it has produced. In fact, the body, in moving along its flight path, must

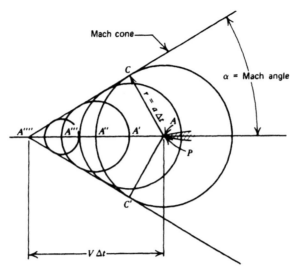

Figure 3.15.   Pressure disturbances produced by a body moving with supersonic speed.

pass through every wave front emanating from the successive positions it occupied in moving along the path. The different disturbance wave fronts are enveloped by a conical surface, called a Mach cone, and the half-angle of the cone, denoted by α, is called the Mach angle. From the geometry it is seen that

$$\sin \alpha = \frac{AC}{AA''''} = \frac{a\,\Delta t}{V\,\Delta t} = \frac{a}{V} = \frac{1}{M} \tag{3.60}$$

The foregoing discussion shows that, when a body moves with supersonic speed, all of the disturbances in the flow are confined to the Mach cone. In the regions external to the Mach cone, the fluid medium is unaffected by the moving body. The conical separating surface, therefore, forms a wave front, termed a *Mach wave*, which is an extremely weak compression shock wave. The foregoing considerations are based on small disturbances that are transmitted with the speed of sound waves. For more intense disturbances the phenomena are analogous but more complicated.

The Mach cone is actually a three-dimensional conoid in space, as is illustrated schematically in Fig. 3.16. Only in the case where a disturbance moves into a uniform region is the Mach conoid a right circular cone.

Example 3.5.   A supersonic aircraft flies horizontally at 1500 m altitude with a constant velocity of 750 m/s. The aircraft passes directly over a stationary ground observer. How much time elapses after it has passed over the observer before the latter hears the aircraft? Assume that the average speed of sound is 335 m/s and that the airplane creates a small disturbance that may be treated as a sound wave.

Solution

The physical situation is illustrated in Fig. 3.17. The observer at point D will first hear the sound wave generated by the airplane when it was at point C. During the time interval required for the sound wave to travel from point C to point D, the airplane travels from point C to point A. The Mach angle α is

$$\alpha = \sin^{-1}\left(\frac{1}{M}\right) = \sin^{-1}\left(\frac{a}{V}\right) = \sin^{-1}\left(\frac{335}{750}\right) = 26.53 \text{ deg}$$

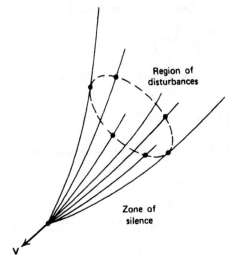

Figure 3.16. The Mach conoid created by a non-uniform disturbance moving into a nonuniform region.

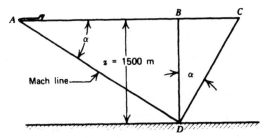

Figure 3.17. Sketch for Example 3.5.

The distance from point B to point A is

$$BA = \frac{BD}{\tan \alpha} = \frac{1500}{\tan (26.53)} = 3004.6 \text{ m}$$

The time interval required for the airplane to travel from point B to point A is

$$\Delta t = \frac{BA}{V} = \frac{3004.6}{750} = 4.006 \text{ s}$$

3-8(c) Compressibility Factor

The compressibility of a fluid influences many of its thermodynamic and dynamic properties, especially its stagnation pressure P (i.e., the pressure attained by an isentropic deceleration to $V = 0$). The pressure increase caused by the deceleration is called the *dynamic pressure q* [see Section 3-13(a)]. From equation 3.11, for the incompressible flow of an inviscid gas (neglecting body forces), the dynamic pressure q_{inc} is given by

$$q_{inc} = P - p = \tfrac{1}{2}\rho V^2 \tag{3.61}$$

When a compressible fluid is brought to rest isentropically, the stagnation pressure P is higher than in the equivalent incompressible case. The equation for the stagnation pressure of a perfect gas is derived in Section 3.9 (see equation 3.76). Thus,

$$\frac{P}{p} = \left(1 + \frac{\gamma - 1}{2} M^2\right)^{\gamma/(\gamma - 1)}$$ (3.62)

where M is the local Mach number. Hence, the dynamic pressure q for a compressible fluid is given by

$$q_{comp} = P - p = p\left[\left(1 + \frac{\gamma - 1}{2} M^2\right)^{\gamma/(\gamma - 1)} - 1\right]$$ (3.63)

For small values of the Mach number, the quantity in parentheses may be expanded by the binomial theorem. Thus,

$$q_{comp} = p\left(1 + \frac{\gamma M^2}{2} + \frac{\gamma M^4}{8} + \cdots - 1\right)$$ (3.64)

Taking $\gamma M^2/2$ outside of the brackets, and noting that $\gamma p M^2/2 = \rho V^2/2$, equation 3.64 becomes (for $\gamma = 1.40$)

$$q_{comp} = P - p = \frac{\rho V^2}{2}\left(1 + \frac{M^2}{4} + \frac{M^4}{40} + \frac{M^6}{1600} + \cdots\right)$$ (3.65)

By definition, the *compressibility factor*, denoted by $\mathscr{F}_c$, is given by

$$\mathscr{F}_c \equiv \frac{P - p}{\frac{1}{2}\rho V^2}$$ (3.66)

For incompressible flows, $\mathscr{F}_c = 1.0$. Substituting from equation 3.65 into equation 3.66 gives, for compressible flow,

$$\mathscr{F}_c = \left(1 + \frac{M^2}{4} + \frac{M^4}{40} + \frac{M^6}{1600} + \cdots\right)$$ (3.67)

Equation 3.67 applies where M is small. The assumption of incompressibility introduces significant errors when the Mach number exceeds approximately 0.3 to 0.4.

3-9 STAGNATION (OR TOTAL) CONDITIONS

The concept of *stagnation conditions* (sometimes called *total conditions*) is extremely useful in that it defines a convenient reference state for a flowing fluid. The stagnation state is defined as that state where the flow velocity is zero. Accordingly, a static fluid is in its stagnation state. The latter state for a flowing fluid is defined as the state attained by the fluid when it is decelerated to zero speed. Alternately, it can be defined as the *static state* from which a fluid must be accelerated from an infinite reservoir in order to attain the actual state for a given flow. These two equivalent definitions are illustrated schematically in Fig. 3.18. Obviously, the nature of the flow process during the deceleration (or acceleration) of the fluid determines its final state. To obtain a unique final state, restrictions must be placed on the deceleration process.

By definition, the stagnation energy state is the state attained by the fluid when it is decelerated to zero speed by an adiabatic (with no external work) deceleration. With the aforementioned restriction, the stagnation internal energy and enthalpy of the fluid are uniquely determined. By further restricting the deceleration process to an isentropic process (i.e., one that is both adiabatic and frictionless), a unique stagnation temperature, pressure, and density are determined. Figure 3.19 illustrates the stagnation conditions on the *Mollier diagram*; that is, the *hs* plane.

The concept of the stagnation state as a *reference state* is independent of the type of fluid being considered. It is also independent of the processes taking place in the

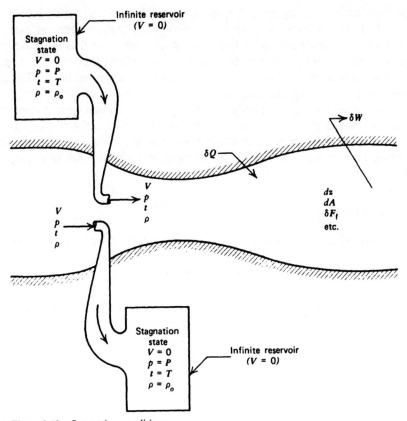

Figure 3.18. Stagnation conditions.

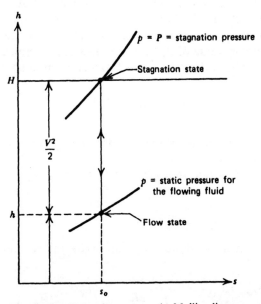

Figure 3.19. Stagnation state on the Mollier diagram.

actual flow under investigation. Thus, the actual flow process may involve work, heat, body forces, friction, etc., along the flow path. Nevertheless, at every point along the actual flow path, there is a local stagnation state that is defined by the imaginary deceleration process discussed above. Consequently, the stagnation conditions are point functions that may be defined at every point in any flow process for any fluid. Obviously the stagnation conditions may vary from one point to another in an actual flow. Usually the changes in the stagnation properties of a given flow process may be related to the *flow driving potential* of the actual flow process; that is, work, heat, area change, friction, etc. Consequently, the concept of the stagnation condition is extremely useful in analyzing the flow of a compressible fluid.

3–9(a) Stagnation Enthalpy

For a steady adiabatic flow ($\delta Q = \delta W = g \, dz = 0$), equation 3.38 reduces to

$$h_1 + \frac{V_1^2}{2} = h_2 + \frac{V_2^2}{2} \tag{3.38}$$

Equation 3.38 applies to both isentropic and irreversible adiabatic processes. Consider the case where the fluid is decelerated adiabatically to zero speed from the initial speed V_1. In that case $V_2 = 0$; the corresponding value of h_2 is called the stagnation, or total, enthalpy. Let H denote the *stagnation enthalpy*; then

$$H \equiv (h_2)_{V_2=0} = h + \frac{V^2}{2} \tag{3.68}$$

Hence, the stagnation enthalpy at any point in any flow field is the sum of the static enthalpy and the kinetic energy at that point, as is illustrated in Fig. 3.19.

3–9(b) Stagnation Temperature

The stagnation temperature is the temperature where the speed of the fluid is zero. When the working fluid is a perfect gas, the energy equation for adiabatic flow can be written in the form of equation 3.45. Consider now the special case where a perfect gas is decelerated isentropically so that its final speed, and hence its Mach number, is zero. The corresponding value of the final temperature is termed either the *stagnation* or the *total temperature* and is denoted by T. Hence, from equation 3.45,

$$T = (t_2)_{V_2=0} = t \left(1 + \frac{\gamma - 1}{2} M^2 \right) \tag{3.69}$$

Values of the ratio t/T as a function of the Mach number M for several different values of γ are presented in Table C.6.

For a perfect gas, $h = c_p t$. Substituting that expression into equation 3.68 and solving for T yields

$$T = \frac{H}{c_p} = t + \frac{V^2}{2c_p} \tag{3.70}$$

The ratio $V^2/2c_p$ is frequently termed the *impact temperature rise* and is denoted by Δt_{imp}. Hence,

$$\Delta t_{imp} \equiv \frac{V^2}{2c_p} = \frac{\gamma - 1}{\gamma} \frac{V^2}{2R} \tag{3.71}$$

In terms of Δt_{imp}, the stagnation temperature T is given by

$$T = t + \Delta t_{imp} \tag{3.72}$$

The preceding discussion shows that the stagnation temperature T corresponds to the static temperature that the gas will attain when it is decelerated isentropically to zero speed. As indicated earlier in Section 3–9, it may also be regarded as the static temperature at which the gas must be held in an infinite reservoir from which it is accelerated isentropically (with $\delta W = 0$) to the values of V and t for an actual flow. For any given values of V and t, therefore, a corresponding value of T can always be computed for the flow.

Note that the definition of the stagnation temperature specified an isentropic deceleration. For a perfect gas, $T = H/c_p$, so that any deceleration process that attains a unique value for H will also attain a unique value for T. As discussed in Section 3–9(a), the only restriction required for defining a unique value for H is that the deceleration process be adiabatic. Thus, for a perfect gas, the stagnation temperature is actually attained by any adiabatic deceleration process. However, for other fluids for which $t = f(h,s)$, the restriction is required that the deceleration process be isentropic so that a unique stagnation temperature will be defined.

If the actual flow process is adiabatic, then equation 3.45, which is repeated and renumbered here for convenience, is applicable. Thus,

$$t_1\left(1 + \frac{\gamma - 1}{2} M_1{}^2\right) = t_2\left(1 + \frac{\gamma - 1}{2} M_2{}^2\right) = \text{constant} = T \tag{3.73}$$

Equation 3.73 indicates that for an adiabatic flow process involving a perfect gas, the stagnation temperature T remains constant.

An expression for the relative change in stagnation temperature dT/T may be obtained by logarithmic differentiation of equation 3.69. Thus,

$$\frac{dT}{T} = \frac{dt}{t} + \frac{\dfrac{\gamma - 1}{2} M^2}{1 + \dfrac{\gamma - 1}{2} M^2} \frac{dM^2}{M^2} \tag{3.74}$$

Example 3.6. An airplane flies at a constant speed of 900 km/hr at 10,000 m altitude. A pressure traverse shows that the air is brought to rest at a particular location on the fuselage. Calculate the temperature of the air in the stagnation region and the temperature rise caused by impact. Assume air is a perfect gas and $\gamma = 1.4$.

Solution

From Table C.5, at 10,000 m altitude, $t = 223.25$ K and $a = 299.53$ m/s. Therefore,

$$M = \frac{(900)(1000)}{(3600)(299.53)} = 0.8346$$

In the stagnation region, the stagnation temperature is T. From equation 3.69

$$T = 223.25\left[1 + \left(\frac{1.4 - 1}{2}\right)(0.8346)^2\right] = 254.35 \text{ K} \qquad (457.8 \text{ R})$$

The impact temperature rise is found from equation 3.72. Thus,

$$\Delta t_{imp} = 254.35 - 223.25 = 31.1 \text{ K} \qquad (56.0 \text{ R})$$

3-9(c) Stagnation Pressure

It is pointed out in Section 3-9(b) that the stagnation temperature T corresponds to the static temperature in an infinite reservoir from which the gas is accelerated isentropically to its actual speed V. The static pressure corresponding to the stagnation temperature T is called either the *stagnation* or the *total pressure* and is denoted by P. Since the flow process is isentropic, for a perfect gas the stagnation pressure P and the stagnation temperature T are related by equation 1.128. Thus,

$$\frac{P}{p} = \left(\frac{T}{t}\right)^{\gamma/(\gamma-1)} \tag{3.75}$$

Substituting for T/t from equation 3.69 into equation 3.75 gives the following relationship between p, P, and M.

$$\frac{P}{p} = \left(1 + \frac{\gamma-1}{2}M^2\right)^{\gamma/(\gamma-1)} \tag{3.76}$$

Because equation 3.75 is employed in deriving equation 3.76, the latter equation likewise applies only to isentropic flow processes, and to a point in a field of flow. Values of the *pressure* ratio p/P as a function of the Mach number M, for different values of γ, are presented in Table C.6.

The *static pressure ratio* corresponding to the isentropic flow of a perfect gas from the initial Mach number M_1 to the final Mach number M_2 is given by

$$\frac{p_2}{p_1} = \frac{(p_2/P)_{M_2}}{(p_1/P)_{M_1}} \tag{3.77}$$

In equation 3.77, the numerator is calculated for the condition $M = M_2$ and the denominator for the condition $M = M_1$

As indicated in the foregoing, for any flow, its stagnation pressure at any point in the flow field can *always* be calculated by assuming that the flow is decelerated isentropically from the values of p and M, at the point in question, to zero velocity where $p = P$.

An expression for the relative change in stagnation pressure dP/P may be obtained from equation 3.76 by logarithmic differentiation. Thus,

$$\frac{dP}{P} = \frac{dp}{p} + \frac{\gamma M^2/2}{1 + \frac{\gamma-1}{2}M^2}\frac{dM^2}{M^2} \tag{3.78}$$

Example 3.7. Air at atmospheric pressure $1.0133 \cdot 10^5$ N/m^2 and a temperature of 300 K flows into an ideal diffuser. The entrance speed is 180.0 m/s. Calculate the maximum pressure (ram pressure) that can be achieved if the air is diffused to zero speed. Assume $\gamma = 1.40$ and $R = 287.04$ J/kg-K for air.

Solution

$$a = (\gamma R t)^{1/2} = [1.40(287.04)(300)]^{1/2} = 347.21 \text{ m/s}$$

$$M = \frac{V}{a} = \frac{180.0}{347.21} = 0.5184$$

The maximum pressure that can be achieved is the isentropic stagnation pressure.

Thus, from equation 3.76,

$$\frac{P}{p} = \left(1 + \frac{\gamma - 1}{2} M^2\right)^{\gamma/(\gamma - 1)} = [1 + 0.20(0.5184)^2]^{3.50} = (1.05375)^{3.50} = 1.20111$$

$$P = (1.0133)10^5(1.20111) = 1.2171 \cdot 10^5 \text{ N/m}^2 \ (17.65 \text{ lbf/in}^2.)$$

Example 3.8. Air flows through an adiabatic frictionless passage. At station 1, the Mach number $M_1 = 0.9$, and the static pressure is $p_1 = 4.15 \cdot 10^5$ N/m^2. At station 2, the Mach number $M_2 = 0.2$. Calculate the change in the static pressure between stations 1 and 2.

Solution

The flow is isentropic, so that Table C.6 applies. Thus, at $M_1 = 0.9$, $p_1/P = 0.59126$, and at $M_2 = 0.2$, $p_2/P = 0.97250$. From equation 3.77

$$p_2/p_1 = \frac{(p_2/P)}{(p_1/P)} = \frac{0.97250}{0.59126} = 1.6448$$

$$p_2 = (4.15 \cdot 10^5)(1.6448) = 6.826 \cdot 10^5 \text{ N/m}^2$$

$$p_2 - p_1 = (6.826 - 4.15)10^5 = 2.676 \cdot 10^5 \text{ N/m}^2 \ (38.81 \text{ lbf/in.}^2)$$

3-9(d) Stagnation Density

The *stagnation* or *total density*, denoted by ρ_o, is the density corresponding to the stagnation temperature and pressure. From the thermal equation of state for a perfect gas,

$$\rho_o = P/RT \tag{3.79}$$

Substituting equations 3.69 and 3.76 into equation 3.79 yields

$$\frac{\rho_o}{\rho} = \left(1 + \frac{\gamma - 1}{2} M^2\right)^{1/(\gamma - 1)} \tag{3.80}$$

Values of the ratio ρ/ρ_o as a function of the Mach number M, for different values of γ, are presented in Table C.6.

3-9(e) Stagnation Acoustic Speed

It is possible to define a *stagnation* or *total acoustic speed* a_o corresponding to the stagnation state for the gas. Thus,

$$a_o^2 = \gamma RT \tag{3.81}$$

The *stagnation acoustic speed* a_o is a characteristic of the state of the gas and remains constant for an adiabatic flow. The local acoustic speed a is related to the stagnation acoustic speed a_o by the relationship

$$\left(\frac{a}{a_o}\right)^2 = \frac{t}{T} = \left(1 + \frac{\gamma - 1}{2} M^2\right)^{-1} \tag{3.82}$$

The energy equation for adiabatic flow, expressed in terms of the stagnation temperature T, is given by equation 3.42. Thus,

$$\frac{V^2}{2} + c_p t = c_p T = \text{constant} \tag{3.83}$$

Substituting for t and T in terms of a and a_o into equation 3.83 gives the energy equation for adiabatic flow in terms of a, a_o, and V. Thus,

$$\frac{V^2}{2} + \frac{a^2}{\gamma - 1} = \frac{a_o^2}{\gamma - 1} \tag{3.84}$$

3-9(f) Entropy Change in Terms of Stagnation Properties

The entropy change for a perfect gas that changes its state under static conditions may be obtained by differentiating equation 1.122. Thus,

$$ds = c_p \, d \left\{ \ln \left[\frac{t}{p^{(\gamma - 1)/\gamma}} \right] \right\} \tag{1.122}$$

By definition, the entropy of a fluid in a static state is equal to the stagnation entropy corresponding to that state. The static entropy change between two states is equal, therefore, to the stagnation entropy change between those two states, as illustrated in Fig. 3.20. Equation 1.122 may, therefore, be applied between any two stagnation

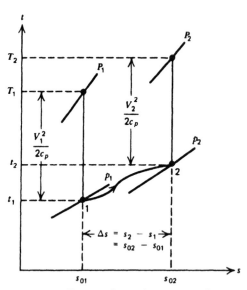

Figure 3.20. Entropy change for a process in terms of stagnation properties.

states. Hence,

$$ds = ds_o = c_p \, d \left\{ \ln \left[\frac{T}{P^{(\gamma - 1)/\gamma}} \right] \right\} \tag{3.85}$$

Rewriting equation 3.85 gives

$$\frac{ds}{c_p} = \frac{dT}{T} - \frac{\gamma - 1}{\gamma} \frac{dP}{P} \tag{3.86}$$

For a calorically perfect gas (c_p = constant), equation 3.86 can be integrated between states 1 and 2.

$$\Delta s = s_2 - s_1 = c_p \ln \left(\frac{T_2}{T_1} \right) - R \ln \left(\frac{P_2}{P_1} \right) \tag{3.87}$$

For a steady adiabatic flow, the stagnation temperature T remains constant. In that case, equation 3.87 relates the entropy change to the change in stagnation pressure.

$$\Delta s = s_2 - s_1 = -R \ln \left(\frac{P_2}{P_1} \right) \tag{3.88}$$

For a frictionless adiabatic process the entropy of a flowing fluid remains constant. From equation 3.88, it follows that for an isentropic change of state for a perfect gas, the stagnation pressure P likewise remains constant. If friction is involved, then according to the second law of thermodynamics, as expressed by equation 3.52, the entropy of the gas must increase. Since $\Delta s > 0$, equation 3.88 requires that $P_2 < P_1$. Hence, the *stagnation pressure decreases in the presence of friction*. The decrease in stagnation pressure of the gas may be regarded, therefore, as a measure of the increase in its specific entropy; that is, of the irreversibility of a steady adiabatic flow process. The relationship between the stagnation pressure ratio and the entropy change for a steady adiabatic flow is obtained directly from equation 3.88. The result is

$$\frac{P_2}{P_1} = \exp \left(-\frac{\Delta s}{R} \right) \tag{3.89}$$

It will be shown later that, if heat is transferred to a flowing gas, or if external work is involved, there cannot be a conservation of stagnation pressure. Heat transfer and work cause the stagnation pressure to change even when the process is reversible. If the heat transfer or work is irreversible, then the stagnation pressure change is greater than it is for a corresponding reversible process.

3–10 CHARACTERISTIC SPEEDS OF GAS DYNAMICS

In the analysis of gas dynamic processes, there are several different characteristic speeds that are representative of the overall properties of the flow. In this section, those characteristic speeds are derived, and their applications to the analysis of compressible flows are discussed.

3–10(a) Maximum Isentropic Speed

The isentropic speed V' attained by a gas in expanding isentropically is given by equation 3.48. The speed V' attains its maximum value when the final pressure p is zero; that is, when the gas expands into a vacuum. Let V'_{max} denote the isentropic speed in that special case. Then

$$V'_{max} = \left(\frac{2\gamma RT}{\gamma - 1} \right)^{1/2} = a_o \left(\frac{2}{\gamma - 1} \right)^{1/2} \tag{3.90}$$

The speed V'_{max}, termed the *maximum isentropic speed*, represents the speed corresponding to the complete transformation of the kinetic energy associated with the random motions of the gas molecules into directed kinetic energy. The static temperature corresponding to that condition is zero, and a real gas would, of course, liquefy before that speed is attained.

It should be realized that the speed V'_{max} applies to the flow of a gas out of an infinite reservoir. It does not apply to the motion of a body in a gas, such as a missile in the air of the atmosphere. In the latter case, the limitation on the maximum attainable speed is the thrust available for propelling the vehicle. On the other hand, if a body is at rest and the air moves toward it with the speed V, the air cannot accelerate locally around the body to a speed exceeding V'_{max}. The maximum isentropic speed is a fictitious speed unattainable even by an isentropic flow. It is of significance, however, because it has a definite value for a given set of flow conditions. The speed V'_{max} may, therefore, be regarded as a characteristic of the flow.

A useful relationship between V'_{max}, a, and V is obtained by combining equations 3.84 and 3.90. Thus,

$$\frac{V^2}{2} + \frac{a^2}{\gamma - 1} = \frac{V'^2_{max}}{2} \tag{3.91}$$

3–10(b) Critical Speed of Sound

Equation 3.44, derived in Section 3–6(d), for a perfect gas, is repeated below and renumbered for convenience. Thus,

$$a_1^2 \left(1 + \frac{\gamma - 1}{2} M_1^2\right) = a_2^2 \left(1 + \frac{\gamma - 1}{2} M_2^2\right) = \text{constant} \tag{3.92}$$

Assume that a gas initially in subsonic flow ($M_1 < 1$) is expanded adiabatically as it flows in a frictionless passage (stream tube), as illustrated in Fig. 3.21. As the gas expands isentropically, its speed increases and its static temperature falls. At some location the speed V_2 attains the same magnitude as the local acoustic speed a_2. The corresponding value for the local Mach number will accordingly be $M_2 = 1$.

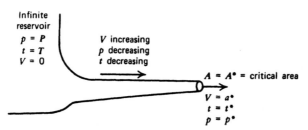

Infinite
reservoir
$p = P$
$t = T$
$V = 0$

V increasing
p decreasing
t decreasing

$A = A^* = $ critical area

$V = a^*$
$t = t^*$
$p = p^*$

Figure 3.21.   Isentropic expansion of a gas to the critical condition.

By definition, let $a_2 = a^*$ when $V_2 = a_2$. The acoustic speed a^* is called the *critical speed of sound* and is a characteristic speed for the flow. The value of a^* may be obtained from equation 3.92 by substituting into it the values $a_2 = a^*$ and $M_2 = 1$. Thus, dropping the subscript 1 from the left-hand side, equation 3.92 becomes

$$a^2 \left(1 + \frac{\gamma - 1}{2} M^2\right) = \frac{\gamma + 1}{2} a^{*2} \tag{3.93}$$

Substituting for $M^2 = V^2/a^2$ into equation 3.93, the energy equation in terms of V, a, and a^* is

$$\frac{V^2}{2} + \frac{a^2}{\gamma - 1} = \frac{(\gamma + 1)a^{*2}}{2(\gamma - 1)} \tag{3.94}$$

From equations 3.84 and 3.94, it follows that the stagnation acoustic speed a_o is related to the critical speed of sound a^* by the following equation. Thus,

$$a_o^2 = \frac{\gamma + 1}{2} a^{*2} \tag{3.95}$$

3-10(c) Thermodynamic Properties Corresponding to a^*

The critical temperature t^* may be determined from equation 3.69 by setting $t = t^*$ and $M = 1$. Thus, t^* is given by

$$t^* = \frac{2}{\gamma + 1} T \tag{3.96}$$

The critical pressure p^* and the critical density ρ^* are readily obtained by applying the thermal equation of state for a perfect gas, $p = \rho R t$, and the relationships for an isentropic change of state to the critical and the stagnation conditions for the gas. Thus,

$$\frac{p^*}{P} = \left(\frac{t^*}{T}\right)^{\gamma/(\gamma-1)} = \left(\frac{2}{\gamma+1}\right)^{\gamma/(\gamma-1)} \tag{3.97}$$

$$\frac{\rho^*}{\rho_o} = \left(\frac{p^*}{P}\right)^{1/\gamma} = \left(\frac{2}{\gamma+1}\right)^{1/(\gamma-1)} \tag{3.98}$$

Figure 3.22 presents the ratios t^*/T, p^*/P, and ρ^*/ρ_o as functions of γ.

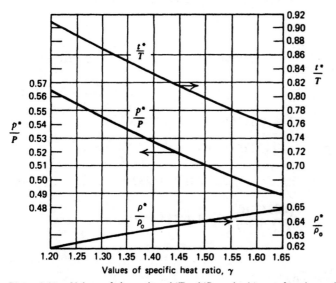

Figure 3.22. Values of the ratios t^*/T, p^*/P, and ρ^*/ρ_o as functions of the specific heat ratio γ.

3-10(d) Relationships between the Characteristic Speeds

The critical speed of sound a^*, like the stagnation acoustic speed a_o, is a characteristic speed for the flow. The maximum isentropic speed V'_{max} is also related to a^* and a_o. From equation 3.90,

$$V'^2_{max} = \frac{2}{\gamma - 1} a_o^2 \tag{3.99}$$

Substituting for a_o from equation 3.95 gives

$$V'^2_{max} = \frac{\gamma + 1}{\gamma - 1} a^{*2} \tag{3.100}$$

It follows from equations 3.95, 3.99, and 3.100 that, for gases having $\gamma = 1.4$, $a^* = 0.91287a_o$, $a_o = 1.0954a^*$, and $V'_{max} = 2.2361a_o = 2.4495a^*$.

Example 3.9.   Air, assumed to be a perfect gas with $\gamma = 1.4$, flows through a frictionless passage. The speed of the air increases in the direction of flow. At station 1, the static temperature is 333.3 K, the static pressure is $2.0684 \cdot 10^5$ N/m^2, and the mean speed is 152.4 m/s. At station 2 the local Mach number is unity. Calculate the values of the following at station 2: (a) the static temperature, (b) the mean speed of the air, (c) the static pressure, and (d) the density of the air.

Solution

From equation 1.181, $a_1 = (\gamma R t_1)^{1/2}$. For air $R = 287.04$ J/kg-K. Hence,

$$a_1 = [1.4(287.04)(333.3)]^{1/2} = 366.0 \text{ m/s}$$

$$M_1 = \frac{V_1}{a_1} = \frac{152.4}{366.0} = 0.4164$$

From equation 3.69,

$$T = 333.3[1 + 0.2(0.4164)^2] = 344.9 \text{ K}$$

From equation 3.75,

$$P = p_1 \left(\frac{T}{t_1}\right)^{\gamma/(\gamma-1)} = (2.0684)10^5 \left(\frac{344.9}{333.3}\right)^{3.5} = 2.3305 \cdot 10^5 \text{ N/m}^2$$

At the location where $M_2 = 1$, $t_2 = t^*$, $V_2 = a^*$, $p_2 = p^*$, and $\rho_2 = \rho^*$. Thus,

(a) $t_2 = t^* = \left(\dfrac{2}{\gamma + 1}\right) T = 0.83333(344.9) = 287.4 \text{ K}$

(b) $V_2 = a^* = (\gamma R t^*)^{1/2} = [1.4(287.04)(287.4)]^{1/2} = 339.9 \text{ m/s}$

(c) $p_2 = p^* = P \left(\dfrac{2}{\gamma + 1}\right)^{\gamma/(\gamma-1)} = (2.3305)10^5(0.52828) = 1.2312 \cdot 10^5 \text{ N/m}^2$

(d) $\rho_2 = \rho^* = \dfrac{p^*}{R t^*} = \dfrac{(1.2312)10^5}{(287.04)(287.4)} = 1.4924 \text{ kg/m}^3$

3-10(e) Dimensionless Velocity M^*

The Mach number $M = V/a$, when employed as a flow characteristic, suffers from the disadvantage that it is not directly proportional to the velocity of the fluid because a is a variable. For that reason the so-called *dimensionless velocity M^** is found to be a useful parameter. It is *defined* by

$$M^* \equiv \frac{V}{a^*} \tag{3.101}$$

The dimensionless velocity M^* is related to the local Mach number M by

$$M^{*2} = \left(\frac{V}{a^*}\right)^2 \left(\frac{a}{a}\right)^2 = M^2 \left(\frac{a}{a^*}\right)^2 \tag{3.102}$$

Writing the energy equation in the form of equation 3.94 and dividing throughout by $a^{*2}/2$ gives

$$\left(\frac{V}{a^*}\right)^2 + \frac{2}{\gamma - 1}\left(\frac{a}{a^*}\right)^2 = \frac{\gamma + 1}{\gamma - 1} \tag{3.103}$$

Introducing the definitions for M and M^* into equation 3.103 yields

$$\frac{\gamma - 1}{2} M^{*2} + \left(\frac{M^*}{M}\right)^2 = \frac{\gamma + 1}{2} \tag{3.104}$$

Solving equation 3.104 for. M^{*2} gives

$$M^{*2} = \frac{(\gamma + 1)M^2}{2 + (\gamma - 1)M^2} \tag{3.105}$$

Solving equation 3.104 for M^2 yields

$$M^2 = \frac{\dfrac{2}{\gamma + 1} M^{*2}}{1 - \dfrac{\gamma - 1}{\gamma + 1} M^{*2}} \tag{3.106}$$

Values of M^* as a function of M, for different values of γ, are presented in Table C.6.

It is apparent from equation 3.105 that, if $M = 0$, then $M^* = 0$, and that, when $M = 1$, $M^* = 1$. As M approaches infinity, however, M^* approaches the finite value

$$\lim_{M \to \infty} M^* = \left(\frac{\gamma + 1}{\gamma - 1}\right)^{1/2} \tag{3.107}$$

Another useful dimensionless velocity is given by the ratio V/a_o. From the definition $M^* = V/a^*$ and equation 3.95,

$$\frac{V}{a_o} = \frac{V}{a^*} \frac{a^*}{a_o} = M^* \frac{a^*}{a_o} = \left(\frac{2}{\gamma + 1}\right)^{1/2} M^* \tag{3.108}$$

Substituting equation 3.105 into equation 3.108 gives

$$\left(\frac{V}{a_o}\right)^2 = \frac{M^2}{1 + \dfrac{\gamma - 1}{2} M^2} \tag{3.109}$$

Useful alternate forms of the stagnation property ratios in terms of the dimensionless velocity M^* may be obtained from equation 3.84. Thus,

$$\frac{t}{T} = \left(\frac{a}{a_o}\right)^2 = 1 - \frac{\gamma - 1}{2} \left(\frac{V}{a_o}\right)^2 \tag{3.110}$$

Substituting equation 3.108 into equation 3.110 gives

$$\frac{t}{T} = \left(1 - \frac{\gamma - 1}{\gamma + 1} M^{*2}\right) \tag{3.111}$$

The ratios p/P and ρ/ρ_o in terms of M^* may be found by combining equations 3.111, 3.75, and 3.79. Thus,

$$\frac{p}{P} = \left(1 - \frac{\gamma - 1}{\gamma + 1} M^{*2}\right)^{\gamma/(\gamma - 1)} \tag{3.112}$$

$$\frac{\rho}{\rho_o} = \left(1 - \frac{\gamma - 1}{\gamma + 1} M^{*2}\right)^{1/(\gamma - 1)} \tag{3.113}$$

The stagnation property ratios may also be expressed in terms of the ratio V/a_o by combining equations 3.110, 3.75, and 3.79.

3–11 THE CONTINUITY RELATIONSHIPS FOR A PERFECT GAS

In Section 3–4, the *continuity equation* for steady one-dimensional flow is derived (see equation 3.4); it is repreated here for convenience.

$$\dot{m} = \rho A V = \text{constant} \tag{3.4}$$

In this section, several alternate forms of the continuity equation for the flow of a perfect gas are derived.

3–11(a) The Continuity Equation for a Perfect Gas

For a perfect gas, $p = \rho R t$, and the continuity equation may be transformed to read

$$\dot{m} = \rho A V = p A V / R t = \text{constant} \tag{3.114}$$

Substituting $a = \sqrt{\gamma R t}$ and $M = V/a$ into equation 3.114 yields

$$\dot{m} = \frac{A p V}{R t} = \frac{A p V}{(\gamma R t)^{1/2}} \left(\frac{\gamma}{R t}\right)^{1/2} = A p M \left(\frac{\gamma}{R t}\right)^{1/2} = \text{constant} \tag{3.115}$$

Equation 3.115 is the continuity equation for a perfect gas expressed in terms of the Mach number M for a flow crossing the area A.

The actual flow at cross section A may be conceived to have originated in an infinite reservoir, where $p = P$ and $t = T$, from which the gas expanded isentropically to the actual values V, p, and t. From equation 3.69, it follows that

$$t = T \left(1 + \frac{\gamma - 1}{2} M^2\right)^{-1} \tag{3.116}$$

Hence, in terms of the stagnation temperature T, equation 3.115 becomes

$$\dot{m} = A p M \left[\frac{\gamma}{R T} \left(1 + \frac{\gamma - 1}{2} M^2\right)\right]^{1/2} = \text{constant} \tag{3.117}$$

From equation 3.76, it follows that

$$p = P \left(1 + \frac{\gamma - 1}{2} M^2\right)^{-\gamma/(\gamma - 1)} \tag{3.118}$$

Hence, the continuity equation expressed in terms of the stagnation pressure P for the actual flow becomes

$$\dot{m} = A M P \left(\frac{\gamma}{R T}\right)^{1/2} \left(1 + \frac{\gamma - 1}{2} M^2\right)^{-(\gamma + 1)/2(\gamma - 1)} = \text{constant} \tag{3.119}$$

Equation 3.119 applies to each section of a flow passage, provided there is no change in the mass flow rate. In the special case where it can be assumed that γ and R remain constant at all flow cross-sectional areas, then for any cross-sectional area A, equation 3.119 can be written in the form

$$\frac{A M P / \sqrt{T}}{\left(1 + \dfrac{\gamma - 1}{2} M^2\right)^{(\gamma + 1)/2(\gamma - 1)}} = \text{constant} \tag{3.120}$$

3-11(b) Critical Flow Area

Consider now the problem of establishing the magnitude of the flow area of the passage illustrated in Fig. 3.21, where the velocity is equal to the critical speed of sound a^*. The flow area where $V = a^*$ is denoted by A^* and is called the *critical flow area*. Applying the continuity equation between the two areas A and A^* gives

$$\rho A V = \rho^* A^* a^* = \text{constant} \tag{3.121}$$

The area ratio A/A^* is, accordingly,

$$\frac{A}{A^*} = \frac{\rho^* a^*}{\rho V} = \frac{\rho^* \sqrt{\gamma R t^*}}{\rho V} \tag{3.122}$$

Substituting for t^*, ρ^*, ρ, and V from equations 3.96, 3.98, 3.80, and 3.48, respectively, into equation 3.122, gives the following equation for isentropic flow.

$$\frac{A}{A^*} = \left\{ \left(\frac{\gamma - 1}{2} \right) \frac{\left(\frac{2}{\gamma + 1} \right)^{(\gamma + 1)/(\gamma - 1)}}{\left(\frac{p}{P} \right)^{2/\gamma} \left[1 - \left(\frac{p}{P} \right)^{(\gamma - 1)/\gamma} \right]} \right\}^{1/2} \quad (\textit{isentropic flow}) \tag{3.123}$$

Equation 3.123 presents the *critical area ratio* A/A^* that causes an isentropic flow to attain a speed equal to the critical speed of sound a^*. The critical area ratio depends on the pressure ratio p/P and the specific heat ratio γ. Figure 3.23 presents A/A^* as a function of the pressure ratio p/P for gases having $\gamma = 1.4$.

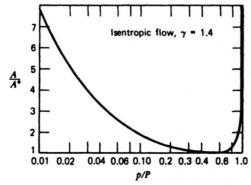

Figure 3.23. Critical area ratio A/A^* as a function of pressure ratio p/P for $\gamma = 1.40$.

The critical area ratio may be expressed as a function of the flow Mach number M by eliminating the pressure ratio p/P from equation 3.123 by means of equation 3.76. The result is

$$\frac{A}{A^*} = \frac{1}{M} \left[\left(\frac{2}{\gamma + 1} \right) \left(1 + \frac{\gamma - 1}{2} M^2 \right) \right]^{(\gamma + 1)/2(\gamma - 1)} \tag{3.124}$$

Values of A/A^* as a function of the Mach number M, for different values of γ, are presented in Table C.6.

It is demonstrated in Section 3-11(d) that, when the gas speed is equal to a^*, the passage discharges the maximum mass rate of flow, $\dot{m}_{max} = \dot{m}^*$, called the *critical* *mass flow.* When the mass flow rate is $\dot{m}^*$, the minimum flow area of the passage is the critical area A^*.

It follows from equations 3.119 and 3.124 that equation 3.119 may be transformed to read

$$\dot{m} = PA \left(\frac{\gamma}{RT}\right)^{1/2} \left(\frac{2}{\gamma + 1}\right)^{(\gamma+1)/2(\gamma-1)} \left(\frac{1}{A/A^*}\right) \tag{3.125}$$

The advantage of equation 3.125 is that Table C.6, which lists the values of A/A^* for steady one-dimensional isentropic flow, may be employed for determining the flow Mach number M directly.

Applying equation 3.119 at the critical condition, where $M = 1$, yields

$$\dot{m}^* = PA^* \left(\frac{\gamma}{RT}\right)^{1/2} \left(\frac{2}{\gamma + 1}\right)^{(\gamma+1)/2(\gamma-1)} \tag{3.126}$$

Equation 3.126 may be rewritten in the form

$$\dot{m}^* = \frac{PA^*}{\sqrt{\gamma RT}} \gamma \left(\frac{2}{\gamma + 1}\right)^{(\gamma+1)/2(\gamma-1)} = \frac{\Gamma PA^*}{a_o} \tag{3.127}$$

where a_o is the stagnation speed of sound and Γ is defined as

$$\Gamma \equiv \gamma \left(\frac{2}{\gamma + 1}\right)^{(\gamma+1)/2(\gamma-1)} \tag{3.128}$$

Values of Γ are presented in Table C.15. Equation 3.127 demonstrates the functional dependence of the critical mass flow rate $\dot{m}^*$ on the other flow parameters. Thus, in functional form,

$$\dot{m}^* = f(P, A^*, 1/\sqrt{R}, 1/\sqrt{T}, \Gamma) \tag{3.129}$$

Example 3.10. At a certain station in a converging flow passage the mass rate of flow of air is 11.34 kg/s, the stagnation temperature is 500 K, the stagnation pressure is $2.068 \cdot 10^5$ N/m^2, the cross-sectional area is 0.03716 m^2, and the flow is subsonic. Determine the Mach number at that station.

Solution

For air, $\gamma = 1.4$ and $R = 287.04$ J/kg-K. Equation 3.125 reduces to

$$\dot{m} = \frac{PA}{24.743\sqrt{T}} \frac{1}{(A/A^*)}$$

$$\frac{A}{A^*} = \frac{(2.068 \cdot 10^5)(0.03716)}{24.743\sqrt{500}\,(11.34)} = 1.225$$

From Table C.6, for $A/A^* = 1.225$, the Mach number is 0.571.

Example 3.11. A perfect gas flows through a conduit under isothermal conditions. At section 1, the area is A_1, the Mach number is $M_1 = 0.4$, and the static pressure is $2.0684 \cdot 10^5$ N/m^2. Further downstream where the area is $0.8A_1$, the static pressure is $1.3790 \cdot 10^5$ N/m^2. Calculate the Mach number at section 2.

Solution

From equation 3.115,

$$A_1 p_1 M_1 \left(\frac{\gamma}{Rt_1}\right)^{1/2} = A_2 p_2 M_2 \left(\frac{\gamma}{Rt_2}\right)^{1/2}$$

Since $t_1 = t_2$ and $A_2 = 0.8A_1$, the above equation reduces to

$$p_1 M_1 = 0.8 p_2 M_2$$

Solving for M_2 gives

$$M_2 = \frac{p_1 M_1}{0.8 p_2} = \frac{2.0684(0.4)}{(0.8)(1.3790)} = 0.750$$

Example 3.12. An ideal air diffuser is to reduce the Mach number of the air entering it from $M_1 = 0.8$ to $M_2 = 0.2$. Determine its area ratio.

Solution

From Table C.6, at $M_1 = 0.8$, $A_1/A^* = 1.0382$, and at $M_2 = 0.2$, $A_2/A^* = 2.9635$. Hence, the area ratio A_2/A_1 is found to be

$$\frac{A_2}{A_1} = \frac{(A_2/A^*)}{(A_1/A^*)} = \frac{2.9635}{1.0382} = 2.854$$

Example 3.13. A perfect gas flows through a duct with a velocity of 182.88 m/s. The mass flow rate is 9.072 kg/s. The area of the duct is 0.0516 m^2, and the flow Mach number is $M = 0.5$. Calculate the static pressure of the air. Assume $\gamma = 1.4$ and $R = 287.04$ J/kg-K.

Solution

First, a relationship between T, M, and V must be developed. Thus,

$$\frac{V}{a_o} = \frac{V}{(\gamma RT)^{1/2}} = \frac{V}{(\gamma Rt)^{1/2}} \left(\frac{t}{T}\right)^{1/2} = \frac{V}{a}\left(\frac{t}{T}\right)^{1/2} = M\left(\frac{t}{T}\right)^{1/2} = \frac{M}{\left(1 + \dfrac{\gamma - 1}{2} M^2\right)^{1/2}}$$

Solving for T gives

$$T = \frac{V^2\left(1 + \dfrac{\gamma - 1}{2} M^2\right)}{\gamma R M^2} = \frac{(182.88)^2[1 + 0.2(0.5)^2]}{1.4(287.04)(0.5)^2} = 349.6 \text{ K}$$

From equation 3.117,

$$p = \frac{\dot{m}\sqrt{RT}}{AM\left[\gamma\left(1 + \dfrac{\gamma - 1}{2} M^2\right)\right]^{1/2}} = \frac{9.072[287.04(349.6)]^{1/2}}{0.0516(0.5)\{1.4[1 + 0.2(0.5)^2]\}^{1/2}}$$

$$p = 0.9184 \quad 10^5 \text{ N/m}^2 \text{ (13.32 lbf/in.}^2\text{)}$$

3-11(c) Effect of Compressibility on the Mass Flow Rate

The fact that the density of a compressible fluid changes with its flow speed introduces considerations not present in the case of incompressible fluids. It is of interest to investigate the effect of the compressibility of the flowing fluid on its mass rate of flow $\dot{m}$ through a converging-diverging passage, such as that illustrated in Fig. 3.24. To do this, it will be assumed that the flow is isentropic, and use will be made of the continuity equation (equation 3.4), rewritten in the following form,

$$G = \dot{m}/A = \rho V \tag{3.130}$$

where G is the *mass flux* (mass flow rate per unit area). The magnitudes of ρ and V depend on the pressure ratio governing the flow process. There are two extreme

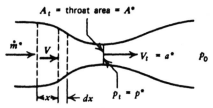

Figure 3.24. Isentropic flow through a converging-diverging passage through which the maximum mass flow rate $\dot{m}^*$ is flowing.

operating conditions: (1) when the pressure ratio for the passage is $p/P = 1$, and (2) when $p/P = 0$. When $p/P = 1$, the fluid is at rest (*stagnation*), so that $\rho = \rho_0$ and $V = 0$, so that $\rho V = 0$. On the other hand, when $p/P = 0$, the back pressure into which the gas expands is $p = 0$. In that case the specific volume of the gas after expansion is infinite, so that $\rho = 1/v = 0$, and the fluid speed is the maximum isentropic speed V'_{max} (see equation 3.90), and again $\rho V = \rho V'_{max} = 0$. Hence, for both limiting cases, the product $\rho V = 0$.

It follows from the foregoing that the mass flux G is zero when the back pressure p is equal to the initial pressure P, and also when $p = 0$. Between those two values of the pressure ratio the curve representing G as a function of p/P must have at least one maximum point. Investigations show that there is only one maximum point between the values $p/P = 1$ and $p/P = 0$. The maximum value of G is attained at a particular value of p/P, called the *critical pressure ratio*.

3–11(d) Condition for Maximum Isentropic Mass Rate of Flow

It is of interest to determine the speed that is attained by a gas when the *isentropic mass rate of flow* reaches its maximum value $\dot{m}'_{max}$. For that purpose, consider the one-dimensional converging-diverging flow passage illustrated in Fig. 3.24 and assume that the flow is isentropic. Let A denote the cross-sectional area normal to the flow, A_t its smallest value (*throat*), and x the distance from the entrance cross section A_1 to any cross section A. The criterion for determining the condition when $\dot{m}'_{max}$ is reached is the maximum value of the mass flux G'. Thus, when $\dot{m}' = \dot{m}'_{max}$, $G' = G'_{max}$. Obviously, G' attains its largest value at the throat section of the converging-diverging passage; the curve of flow area A as a function of the length x has a minimum point at the throat section, signifying that at the *throat section $dA/dx = 0$*.

Assume now that the converging-diverging passage passes the flow $\dot{m}'_{max}$. Then, by continuity,

$$\dot{m}'_{max} = A\rho'V' = \text{constant} \tag{3.131}$$

Logarithmic differentiation of equation 3.131 and division by dx gives

$$\frac{1}{A}\frac{dA}{dx} + \frac{1}{V'}\frac{dV'}{dx} + \frac{1}{\rho'}\frac{d\rho'}{dx} = 0 \tag{3.132}$$

For the throat, since $dA/dx = 0$, equation 3.132 reduces to

$$\frac{dV'_t}{dx} = -\frac{V'_t}{\rho'_t}\frac{d\rho'_t}{dx} \tag{3.133}$$

where the subscript t denotes that the properties are evaluated at the throat. Since the flow has been assumed to be isentropic, the dynamic equation for frictionless flow (equation 3.9 with $g \, dz = 0$) is applicable. Let $V = V'_t$ and $dp = dp'_t$; then rewriting equation 3.9, dividing throughout by dx, and rearranging the result gives

$$\frac{dV'_t}{dx} = -\frac{1}{\rho'_t V'_t} \frac{dp'_t}{dx} \tag{3.134}$$

Equating equations 3.133 and 3.134 and solving for V'_t, yields

$$(V'_t)^2 = \left(\frac{dp'}{d\rho'}\right)_t = a^2 = a^{*2} \tag{3.135}$$

Since the flow is isentropic, $(dp'/d\rho')_t$ is identical to the isentropic derivative $(\partial p/\partial \rho)_s = a^2$.

Equation 3.135 shows that, when a converging-diverging passage passes the isentropic flow $\dot{m}'_{max}$, the speed of the fluid in the throat section is equal to the local speed of sound; that is, $V'_t = a^*$. Hence, for the isentropic flow in a converging-diverging passage, it follows that, when $\dot{m} = \dot{m}'_{max}$, $V'_t = a^*$, so that $M^*_t = V'_t/a^* = 1$. The throat area A_t is then equal to the critical area A^*, and the properties of the gas are p^*, t^*, and ρ^*; the latter parameters are determined by the appropriate equations presented in Section 3–10(c). The mass flow rate at the critical flow condition is given by equation 3.127. Thus,

$$\dot{m}'_{max} = \dot{m}^* = \frac{\Gamma P A^*}{\sqrt{\gamma R T}} = \frac{\Gamma P A^*}{a_0} \tag{3.136}$$

Example 3.14.   A gas flows one-dimensionally and adiabatically in a frictionless variable area duct. At station 1, $M_1 = 2.5$, $t_1 = 311.11$ K, $p_1 = 0.5068 \cdot 10^5$ N/m², and $A = 0.09293$ m². At the locations where $M_2 = 2.5$, 1.2, and 1.0, find (a) the static pressure, (b) the static temperature, (c) the gas density, (d) the gas velocity, and (e) the area. Assume $\gamma = 1.4$ and $R = 287.04$ J/kg-K.

Solution

From Table C.6.

M	p/P	t/T	ρ/ρ_0	A/A*	M* = V/a*
2.5	0.05853	0.44444	0.13169	2.6367	1.82574
1.2	0.41238	0.77640	0.53114	1.0304	1.15828
1.0	0.52828	0.83333	0.63394	1.0000	1.00000

The reference conditions for this isentropic flow may be determined from the initial conditions and the property ratios at $M = 2.5$. Thus,

$$P = \frac{p_1}{0.05853} = \frac{(0.5068)10^5}{0.05853} = 8.658 \cdot 10^5 \text{ N/m}^2$$

$$T = \frac{t_1}{0.44444} = \frac{311.11}{0.44444} = 700.0 \text{ K}$$

$$\rho_o = \frac{P}{RT} = \frac{(8.658)10^5}{(287.04)(700.00)} = 4.3091 \text{ kg/m}^3$$

$$A^* = \frac{A_1}{(A_1/A^*)} = \frac{0.09293}{2.6367} = 0.0352448 \text{ m}^3$$

$$a_1 = [1.4(287.04)(311.11)]^{1/2} = 353.58 \text{ m/s}$$

$$V_1 = M_1 a_1 = 2.5(353.58) = 883.96 \text{ m/s}$$

$$a^* = \frac{V_1}{M_1^*} = \frac{833.96}{1.82574} = 484.17 \text{ m/s}$$

The properties at the other two stations may be calculated from the reference conditions and the property ratios. The results are presented in the following table.

M	p, N/m$^2 \cdot 10^5$	t, K	ρ, kg/m^3	V, m/s	A, m^2
2.5	0.5068	311.11	0.56753	883.92	0.09293
1.2	3.5706	543.48	2.28872	560.80	0.03632
1.0	4.5741	583.33	2.73179	484.17	0.03524

3-12 STREAM THRUST AND THE IMPULSE FUNCTION

It is pointed out in Section 3-5(g) that the sum $(pA + \dot{m}V)$ for a flowing fluid is termed the *stream thrust*. For problems involving the calculation of the reaction force caused by the one-dimensional flow of a fluid in a duct, it is convenient to regard that sum as a single function, termed the *impulse function*, which is denoted by $\mathscr{F}$. Thus,

$$\mathscr{F} \equiv pA + \dot{m}V = pA + \rho A V^2 = pA\left(1 + \frac{\rho V^2}{p}\right) \tag{3.137}$$

where p is the *absolute* static pressure acting over the cross section A.

Consider the internal flow between the inlet (section 1) and the exit (section 2) of a passage such as a nozzle or a diffuser. For a one-dimensional flow in the positive x direction, $\mathbf{n}_1 = -\mathbf{i}$ and $\mathbf{n}_2 = \mathbf{i}$. From equation 3.28, the net force F_{fluid} applied to the fluid by the internal walls of the passage in the x direction is

$$F_{\text{fluid}} = \Delta\mathscr{F} = \mathscr{F}_2 - \mathscr{F}_1 \tag{3.138a}$$

The reaction force exerted by the fluid on the internal walls of the passage is $F_{\text{int}} = -F_{\text{fluid}}$. From equation 3.29, the external restraining force F_{ext} acting on the duct is

$$F_{\text{ext}} = (\mathscr{F}_2 - p_o A_2) - (\mathscr{F}_1 - p_o A_1) \tag{3.138b}$$

For a perfect gas it is convenient to express $\mathscr{F}$ in terms of the flow Mach number M. Noting that $\gamma p/\rho = a^2$, equation 3.137 transforms to

$$\mathscr{F} = pA(1 + \gamma M^2) \tag{3.139}$$

The value of $\mathscr{F}$ when $M = 1$ is denoted by $\mathscr{F}^*$. The ratio $\mathscr{F}/\mathscr{F}^*$ is given by

$$\frac{\mathscr{F}}{\mathscr{F}^*} = \frac{p}{P} \frac{P}{p^*} \frac{A}{A^*}\left(\frac{1 + \gamma M^2}{\gamma + 1}\right) \tag{3.140}$$

To simplify equation 3.140 further, the ratios p/P, P/p^*, and A/A^* must be related to the Mach number M for the flow. Those relationships depend on the driving potentials (area change, friction, heat transfer, etc.) acting on the flow. For example, for isentropic flow with area change, the aforementioned ratios are given by equations 3.76, 3.97, and 3.124. Substituting those equations into equation 3.140, we obtain

$$\frac{\mathcal{F}}{\mathcal{F}^*} = \frac{1 + \gamma M^2}{M\left[2(\gamma + 1)\left(1 + \frac{\gamma - 1}{2}M^2\right)\right]^{1/2}} \tag{3.141}$$

Equation 3.141 is applicable to steady one-dimensional isentropic flow. The ratio $\mathcal{F}/\mathcal{F}^*$ for isentropic flow is tabulated as a function of the Mach number M for several values of γ in Table C.6.

3-13 DYNAMIC PRESSURE AND THE PRESSURE COEFFICIENT

In problems involving the interaction of a fluid with a solid body, such as an airfoil, it is useful to present the results in terms of nondimensional coefficients. Three of the more important nondimensional coefficients in common use are the *pressure coefficient* C_p, the *lift coefficient* C_L, and the *drag coefficient* C_D. Those parameters are defined in the present section.

3-13(a) Dynamic Pressure

Consider first the flow of an inviscid incompressible fluid when the *force* caused by gravitational attraction is negligible. Then, equation 3.11 reduces to

$$p + \frac{1}{2}\rho V^2 = \text{constant} \tag{3.142}$$

Equation 3.142 shows that the fluid motion depends on the relative magnitudes of the pressure force, which is proportional to p, and the inertia force, which is proportional to $\rho V^2/2$, acting on the fluid particle. The inertia force is a *dynamic* force, and the term $\rho V^2/2$ is termed the *dynamic pressure* q_{inc}. Thus, by *definition*,

$$q_{\text{inc}} \equiv \frac{1}{2}\rho V^2 = \text{the dynamic pressure for incompressible flow} \tag{3.143}$$

and equation 3.142 may be rewritten as

$$p + q_{\text{inc}} = \text{constant} \tag{3.144}$$

If a fluid flowing at a pressure p and a velocity V is brought to rest, then, as stated earlier, the pressure at the rest state is called the *stagnation pressure P*. Thus, equation 3.144 gives

$$p + q_{\text{inc}} = P = \text{constant} \tag{3.145}$$

Equation 3.145 shows that the dynamic pressure q_{inc} is equal to the increase in static pressure that occurs when an inviscid incompressible fluid is brought to rest.

For an inviscid compressible flow in the absence of body forces, equation 3.10 becomes

$$\int \frac{dp}{\rho} + \frac{V^2}{2} = \text{constant} \tag{3.146}$$

To complete the integration of equation 3.146, the relationship between pressure p and density ρ must be specified. That integration is performed in Section 3–9(c) for isentropic flow (see equation 3.76). Equation 3.146 may be put in the same form as equation 3.145, but the dynamic pressure q_{comp} for a compressible fluid is larger than the dynamic pressure q_{inc} for an incompressible fluid. In fact, the ratio q_{comp}/q_{inc} is identical to the compressibility factor $\mathscr{F}_c$ defined by equation 3.66.

3–13(b) The Pressure Coefficient

If ρ_∞ is the density of an undisturbed flowing medium, and V_∞ is the corresponding free-stream velocity, then the pressure coefficient C_p is *defined by*

$$C_p \equiv \frac{p - p_\infty}{q_\infty} \tag{3.147}$$

where q_∞ is the *free-stream dynamic pressure*. Equation 3.147 is simply a convenient means of nondimensionalizing the pressure p in terms of the properties of the free stream. In that form, the results are independent of the actual magnitudes of p, ρ, and V, and depend only on the relative magnitudes of the pressure and inertia forces.

For an incompressible flow, equation 3.147 becomes

$$C_p = \frac{p - p_\infty}{q_{inc}} = \frac{p - p_\infty}{\frac{1}{2}\rho_\infty V_\infty^2} \tag{3.148}$$

while, for a compressible flow,

$$C_p = \frac{p - p_\infty}{q_{comp}} = \frac{p - p_\infty}{P_\infty - p_\infty} \tag{3.149}$$

It is conventional practice to employ the pressure coefficient defined by equation 3.148 even for compressible flows. Little or no use is made of the definition presented in equation 3.149. Since $q_{inc} < q_{comp}$, the value of C_p based on equation 3.148 may exceed unity for a compressible flow.

In a like manner, the *lift* and *drag coefficients* C_L and C_D, respectively, for a body, such as an airfoil, immersed in a flowing medium, are *defined by*

$$C_L \equiv \frac{L/A}{\frac{1}{2}\rho_\infty V_\infty^2} \tag{3.150}$$

and

$$C_D \equiv \frac{D/A}{\frac{1}{2}\rho_\infty V_\infty^2} \tag{3.151}$$

where L is the lift force, D is the drag force, and A is some characteristic area associated with the body. Both C_L and C_D are based on the incompressible dynamic pressure q_{inc}, although analogous expressions could have been obtained in terms of q_{comp}. Equations 3.150 and 3.151 are the definitions in common usage.

For a perfect gas, the pressure coefficient C_p may be expressed in terms of the free-stream Mach number M_∞ and the velocity ratio (V/V_∞) as follows. From equation 3.147,

$$C_p = \frac{p - p_\infty}{\frac{1}{2}\rho_\infty V_\infty^2} = \frac{2(p - p_\infty)}{\gamma p_\infty M_\infty^2} = \frac{2}{\gamma M_\infty^2}\left(\frac{p}{p_\infty} - 1\right) \tag{3.152}$$

where $a_\infty^2 = \gamma p_\infty/\rho_\infty$ and $M_\infty = V_\infty/a_\infty$. For p/p_∞ we may write

$$\frac{p}{p_\infty} = \frac{p}{P}\frac{P}{p_\infty} \tag{3.153}$$

For an isentropic flow the ratios p/P and p_∞/P may be determined by means of equation 3.76. Hence,

$$\frac{p}{p_\infty} = \left[\frac{2 + (\gamma - 1)M_\infty^2}{2 + (\gamma - 1)M^2}\right]^{\gamma/(\gamma - 1)} \tag{3.154}$$

where $M = V/a$ and $M_\infty = V_\infty/a_\infty$. Substituting equation 3.154 into equation 3.152 yields

$$C_p = \frac{2}{\gamma M_\infty^2}\left\{\left[\frac{2 + (\gamma - 1)M_\infty^2}{2 + (\gamma - 1)M^2}\right]^{\gamma/(\gamma - 1)} - 1\right\} \tag{3.155}$$

It is desirable to express C_p in terms of M_∞^2 and $(V/V_\infty)^2$. To do this use is made of equation 3.43, which is a form of the energy equation. Thus,

$$\frac{V^2}{2} + \frac{a^2}{\gamma - 1} = \frac{V_\infty^2}{2} + \frac{a_\infty^2}{\gamma - 1} \tag{3.156}$$

Combining equation 3.156 with equation 3.155, we obtain

$$C_p = \frac{2}{\gamma M_\infty^2}\left\{\left[1 + \frac{\gamma - 1}{2}M_\infty^2\left(1 - \frac{V^2}{V_\infty^2}\right)\right]^{\gamma/(\gamma - 1)} - 1\right\} \tag{3.157}$$

Equation 3.157 shows that the local value of C_p at a given point in a flowing fluid depends strongly on the free-stream Mach number M_∞.

3-14 SUMMARY

In this chapter the governing equations are derived for the steady one-dimensional (or uniform) flow of any type of fluid. The forms taken by those equations for special flow conditions, such as flow with and without friction, flow with and without work and heat transfer, and flow with and without body forces are derived herein. The principal forms of all of the aforementioned equations are assembled in Table 3.1, for convenience of reference.

Some of the general effects of the propagation of a small pressure disturbance in a compressible fluid are discussed in Section 3-8. It is shown that a small pressure disturbance is propagated in a stationary body of compressible fluid with the acoustic speed a. The disturbance waves "clear away" from a disturbance source if it moves with a subsonic speed ($M < 1$). If a small disturbance source moves, however, with a supersonic speed ($M > 1$) into a stationary compressible fluid, the disturbance source must cross each disturbance wave propagated by it. A Mach cone becomes attached to the source and the half-angle of the cone α is given by $\sin \alpha = 1/M$.

It is shown in Section 3-9 that the stagnation state for a fluid (the state where $V = 0$) is a convenient reference state, which is independent of the type of fluid and the actual flow being studied. Consequently, a local stagnation state may be defined at each point in the actual flow field. In the special case where the flowing fluid is a perfect gas, the ratios t/T, p/P, and ρ/ρ_o for a given perfect gas are functions only of the local Mach number M. Table 3.2 presents, for ready reference, the governing equations for the steady one-dimensional flow of a perfect gas. Table 3.3 presents the equations for the stagnation properties H, T, P, and ρ_o.

It is shown in Section 3-10 that when the flowing fluid is a perfect gas, there are certain characteristic speeds associated with a steady one-dimensional flow (subject to the restrictions of absence of work, heat transfer, and body forces). Such flows are quite generally termed *adiabatic flows*. If there is the additional restriction that the

Table 3.1 Governing Equations for the Steady One-Dimensional Flow of a Fluid

---

Continuity equation

$$\dot{m} = \rho A V \tag{3.158}$$

Momentum equation

(1) *Frictionless flow*

$$\frac{dp}{\rho} + V\,dV + g\,dz = 0 \tag{3.159}$$

(2) *Flow with friction*

$$dp + \rho V\,dV + \rho g\,dz + \frac{\rho V^2}{2}\left(\frac{4f\,dx}{\mathscr{D}}\right) + \frac{\delta D}{A} = 0 \tag{3.160}$$

Energy equation

(1) *Flow with work and heat transfer*

$$\delta W - \delta Q + dh + d\left(\frac{V^2}{2}\right) + g\,dz = 0 \tag{3.161}$$

$$Q + h_1 + \frac{V_1^2}{2} + gz_1 = W + h_2 + \frac{V_2^2}{2} + gz_2 \tag{3.162}$$

(2) *Flow without work and heat transfer (adiabatic flow)*

$$dh + d\left(\frac{V^2}{2}\right) + g\,dz = 0 \tag{3.163}$$

$$h + \frac{V^2}{2} + gz = \text{constant} \tag{3.164}$$

Entropy equation

(1) *Flow with heat transfer*

$$ds \geqq \frac{\delta Q}{t} \tag{3.165}$$

(2) *Adiabatic flow*

$$ds \geqq 0 \tag{3.166}$$

---

flow is frictionless, then the flow is termed an *isentropic flow*, since $ds = 0$. Isentropic flow is discussed in detail in Chapter 4. Table 3.4 presents the equations for the aforementioned characteristic speeds and the relationships between them.

Section 3–11 discusses the effect of the compressibility of the flowing fluid on the continuity equation, the flow area, and the mass flow rate. In addition, it discusses the concepts of the critical speed of sound a^*, the critical flow area A^* (where $V = a^*$), and the critical mass flow rate $\dot{m}^*$. It is shown that the ratio A/A^*, for a given gas, is a function only of the local Mach number M. The equations for $\dot{m}^*$ and A/A^* are presented in Table 3.2, for a perfect gas. The critical properties for a perfect gas are presented in Table 3.5.

Section 3–11(c) discusses the effect of the compressibility of a flowing fluid on its mass flow rate and demonstrates that for an isentropic flow of a perfect gas, $A^* = A_t$, where A_t is the area of the throat (smallest flow area) of the flow passage. Moreover, when the velocity crossing A_t is $V = a^*$, the mass flow rate through the flow passage is a maximum and is equal to $\dot{m}^*$.

Table 3.2 Governing Equations for the Steady One-Dimensional Flow of a Perfect Gas

---

Continuity equation

(1) *Mass flow rate*

$$\dot{m} = \rho A V = \frac{ApV}{Rt} = ApM \left(\frac{\gamma}{Rt}\right)^{1/2} = ApM \left[\frac{\gamma}{RT}\left(1 + \frac{\gamma - 1}{2} M^2\right)\right]^{1/2}$$

$$\dot{m} = AMP \left(\frac{\gamma}{RT}\right)^{1/2} \left(1 + \frac{\gamma - 1}{2} M^2\right)^{-(\gamma+1)/2(\gamma-1)} \tag{3.167}$$

(2) *Critical mass flow rate*

$$\dot{m}^* = \frac{\Gamma PA^*}{(\gamma RT)^{1/2}} = \frac{\Gamma PA^*}{a_o} \qquad \text{where} \qquad \Gamma = \gamma \left(\frac{2}{\gamma + 1}\right)^{(\gamma+1)/2(\gamma-1)} \tag{3.168}$$

(3) *Critical flow area ratio*

$$\frac{A}{A^*} = \frac{1}{M}\left[\frac{2}{\gamma + 1}\left(1 + \frac{\gamma - 1}{2} M^2\right)\right]^{(\gamma+1)/2(\gamma-1)} \tag{3.169}$$

Momentum (or dynamic) equation

(1) *Flow with friction and submerged bodies*

$$\frac{dp}{p} + \frac{\gamma}{2} dM^2 + \frac{\gamma M^2}{2}\frac{dt}{t} + \frac{\gamma M^2}{2}\left(\frac{4\mathfrak{f}\, dx}{\mathscr{D}}\right) + \frac{\delta D}{Ap} = 0 \tag{3.170}$$

(2) *Frictionless flow (no body forces)*

$$\frac{d\rho}{\rho} + M^2 \frac{dV}{V} = 0 \tag{3.171}$$

Energy equation

(1) *Flow with work and heat transfer*

$$\delta W - \delta Q + c_p\, dt + d\left(\frac{V^2}{2}\right) + g\, dz = 0 \tag{3.172}$$

$$W - Q + c_p(t_2 - t_1) + \frac{V_2^2 - V_1^2}{2} + g(z_2 - z_1) = 0 \tag{3.173}$$

(2) *No external work, heat transfer, or body forces*

$$c_p\, dt + d\left(\frac{V^2}{2}\right) = c_p\, dT = dH = 0 \tag{3.174}$$

$$c_p t + \frac{V^2}{2} = c_p T = H \tag{3.175}$$

Entropy equation

(1) *In terms of static properties*

$$\frac{ds}{c_p} = d\left\{\ln\left[\frac{t}{p^{(\gamma-1)/\gamma}}\right]\right\} = \frac{dt}{t} - \frac{\gamma - 1}{\gamma}\frac{dp}{p} \tag{3.176}$$

$$\Delta s = s_2 - s_1 = c_p \ln\left(\frac{t_2}{t_1}\right) - R \ln\left(\frac{p_2}{p_1}\right) = c_p \ln\left[\frac{t_2/t_1}{(p_2/p_1)^{(\gamma-1)/\gamma}}\right] \tag{3.177}$$

(2) *In terms of stagnation properties*

$$\frac{ds}{c_p} = d\left\{\ln\left[\frac{T}{P^{(\gamma-1)/\gamma}}\right]\right\} = \frac{dT}{T} - \frac{\gamma - 1}{\gamma}\frac{dP}{P} \tag{3.178}$$

$$\Delta s = s_2 - s_1 = c_p \ln\left(\frac{T_2}{T_1}\right) - R \ln\left(\frac{P_2}{P_1}\right) = c_p \ln\left[\frac{T_2/T_1}{(P_2/P_1)^{(\gamma-1)/\gamma}}\right] \tag{3.179}$$

Table 3.3 Stagnation Properties for a Perfect Gas

(1) *Stagnation enthalpy*

$$H = h + \frac{V^2}{2}$$

(3.180)

(2) *Stagnation temperature*

$$T = \frac{H}{c_p} = t + \frac{V^2}{2c_p} = t\left(1 + \frac{\gamma - 1}{2} M^2\right)$$

(3.181)

(3) *Stagnation pressure*

$$P = p\left(\frac{T}{t}\right)^{\gamma/(\gamma-1)} = p\left(1 + \frac{\gamma - 1}{2} M^2\right)^{\gamma/(\gamma-1)}$$

(3.182)

(4) *Stagnation density*

$$\rho_o = \frac{P}{RT} = \rho\left(1 + \frac{\gamma - 1}{2} M^2\right)^{1/(\gamma-1)}$$

(3.183)

Table 3.4 Characteristic Speeds for the Steady One-Dimensional Flow of a Perfect Gas

Isentropic speeds

(1) *Isentropic discharge speed* (St. Venant-Wantzel equation)

$$V' = \left\{\frac{2\gamma RT}{\gamma - 1}\left[1 - \left(\frac{p'}{P}\right)^{(\gamma-1)/\gamma}\right]\right\}^{1/2}$$

(3.184)

(2) *Maximum isentropic speed*

$$(V'_{\text{max}})^2 = \frac{2\gamma RT}{\gamma - 1} = \frac{2}{\gamma - 1} a_o^2 = V^2 + \frac{2}{\gamma - 1} a^2$$

(3.185)

Acoustic speeds

(1) *Stagnation acoustic speed*

$$a_o^2 = \gamma RT = a^2\left(\frac{T}{t}\right) = a^2\left(1 + \frac{\gamma - 1}{2} M^2\right) = a^2 + \frac{\gamma - 1}{2} V^2$$

(3.186)

(2) *Critical speed of sound*

$$a^{*2} = \gamma Rt^* = \frac{2\gamma RT}{\gamma + 1} = \frac{2}{\gamma + 1} a_o^2 = \frac{2}{\gamma + 1} a^2\left(1 + \frac{\gamma - 1}{2} M^2\right)$$

(3.187)

(3) *Relationships between the characteristic speeds*

$$(V'_{\text{max}})^2 = \frac{2}{\gamma - 1} a_o^2 = \frac{\gamma + 1}{\gamma - 1} a^{*2}$$

(3.188)

Dimensionless velocities

(1) *Dimensionless velocity* ($M^* = V/a^*$)

$$M^{*2} = \frac{(\gamma + 1)M^2}{2 + (\gamma - 1)M^2} \quad \text{or} \quad M^2 = \frac{\dfrac{2}{\gamma + 1} M^{*2}}{1 - \dfrac{\gamma - 1}{\gamma + 1} M^{*2}}$$

(3.189)

(2) *Dimensionless velocity* (V/a_o)

$$\frac{V}{a_o} = \left(\frac{2}{\gamma + 1}\right)^{1/2} M^* = \left(\frac{M^2}{1 + \dfrac{\gamma - 1}{2} M^2}\right)^{1/2}$$

(3.190)

Table 3.5   Critical Properties for a Perfect Gas

(1) *Critical temperature*

$$t^* = \left(\frac{2}{\gamma + 1}\right) T \qquad (3.191)$$

(2) *Critical pressure*

$$p^* = \left(\frac{2}{\gamma + 1}\right)^{\gamma/(\gamma - 1)} P \qquad (3.192)$$

(3) *Critical density*

$$\rho^* = \left(\frac{2}{\gamma + 1}\right)^{1/(\gamma - 1)} \rho_o \qquad (3.193)$$

In Section 3–12, the impulse function $\mathscr{F}$ is defined, and it is shown that, for the steady one-dimensional flow of a given perfect gas, the ratio $\mathscr{F}/\mathscr{F}^*$ is a function only of the local Mach number.

Table C.6 presents the ratios t/T, p/P, ρ/ρ_o, $\mathscr{F}/\mathscr{F}^*$, and A/A^* as functions of M, for different values of the specific heat ratio γ.

REFERENCES

1. M. J. Zucrow, *Aircraft and Missile Propulsion*, Vol. 1, Wiley, New York, 1958.
2. A. H. Shapiro, *The Dynamics and Thermodynamics of Compressible Fluid Flow*, Vol. 1, Ronald Press, New York, 1953.
3. K. Oswatitsch, *Gas Dynamics*, Academic Press, New York, 1956.
4. R. von Mises, *Mathematical Theory of Compressible Fluid Flow*, Academic Press, New York, 1958.
5. H. W. Liepmann and A. Roshko, *Elements of Gas Dynamics*, Wiley, New York, 1973.
6. J. A. Owczarek, *Fundamentals of Gas Dynamics*, International Textbook Co., Scranton, Pa., 1964.
7. H. Schlichting, *Boundary Layer Theory*, Sixth Edition, McGraw-Hill, New York, 1968.
8. J. W. Daily and D. R. F. Harleman, *Fluid Dynamics*, Addison-Wesley, Reading, Mass., 1966.
9. E. A. Bonney, M. J. Zucrow, and C. W. Besserer, *Aerodynamics, Propulsion, Structures, and Design Practice*, Principles of Guided Missile Design, edited by G. Merrill, p. 123, D. Van Nostrand, New York, 1956.
10. H. Korst, R. H. Page, and M. E. Childs, "A Theory of Base Pressures", in *Transonic and Supersonic Flow*, ME TN 392–2, University of Illinois, May 1962.
11. T. J. Mueller, "Determination of the Turbulent Base Pressure in Supersonic Axisymmetric Flow," *Journal of Spacecraft and Rockets*, Vol. 5, No. 1, pp. 101–107, January 1968.
12. W. P. Sule and T. J. Mueller, "Truncated Plug Nozzle Flowfield and Base Pressure Characteristics", *Journal of Spacecraft and Rockets*, Vol. 10, No. 11, pp. 689–695, November 1973.

PROBLEMS

1. The applicability of the one-dimensional flow concept is investigated in illustrative Example 3.1, where the ratios of the actual to the one-dimensional values of the flow momentum and kinetic energy are determined for the fully developed turbulent flow in a pipe when the power law exponent $n = 7$. Calculate the aforementioned ratios for the values of n presented in the table in Section 3–3. Plot the values of those ratios as a function of Reynolds number.

2. A liquid having the density ρ flows through a conduit. At section 1 the area is A_1, and at section 2 it is $A_2 \neq A_1$. The pressure at A_1 is p_1 and at A_2 it is p_2. The mass rate of flow of liquid is $\dot{m}$. Show that

$$\dot{m} = A_1 A_2 \left[\frac{2\rho(p_1 - p_2)}{A_1^2 - A_2^2}\right]^{1/2}$$

3. Consider the vane illustrated in Fig. 3.25. Assume that the fluid impinging on the vane is incompressible and frictionless, and that the vane is stationary. Determine an expression for the components F_x and F_y of the reaction force $\mathbf{F}$. Calculate $\mathbf{F}$ when $\dot{m} = 22.68$ kg/s, $V_1 = 9.144$ m/s, and $\theta = 150$ deg.

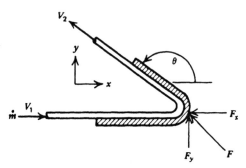

Figure 3.25. Sketch for Problem 3.3.

4. Water enters the elbow illustrated in Fig. 3.26 at a static pressure of $27.58 \cdot 10^5$ N/m^2 with a volume flow rate of 5.663 m^3/s. The inlet and outlet areas are 0.1858 m^2 and 0.0929 m^2, respectively. The atmospheric pressure $p_o = 1$ atm. Assume that friction and body forces are negligible. Calculate (a) V_1, V_2, and p_2, (b) the force exerted by the elbow on the fluid, (c) the force exerted by the fluid on the elbow, and (d) the force required to hold the elbow stationary.

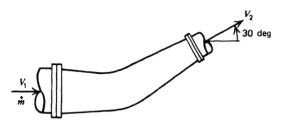

Figure 3.26. Sketch for Problem 3.4.

5. Water flows through a frictionless converging conduit. The cross section in the conduit where the pressure is $5.171 \cdot 10^5$ N/m^2 and the velocity is 15.24 m/s is 15.24 m below the entrance to the conduit. At a cross section located 137.2 m below the entrance of the conduit, the static pressure of the water is $14.824 \cdot 10^5$ N/m^2. Calculate the velocity at the cross section where the pressure is $14.824 \cdot 10^5$ N/m^2 by (a) the dynamic equation, and (b) the energy equation. Assume that the density of the water is 1000 kg/m^3 and that there is no heat transfer.

6. A cylindrical water supply tank is 10.668 m high and 1.829 m in diameter, and the top is open to the atmosphere. The water level in the tank is at 7.62 m. The horizontal discharge pipe is 0.3048 m in diameter and 21.336 m long, and the center line of the pipe is 1.524 m above the bottom of the tank. At the end of the pipe is a converging nozzle with an outlet area of 0.03716 m^2. How long will it take to lower the water level to 3.048 m? Assume no friction or losses. Plot the discharge rate as a function of time for the period under discussion.

7. Water at a static pressure of $14.82 \cdot 10^5$ N/m^2 enters a 90 deg reducing elbow with a velocity of 6.096 m/s. The diameters of the entrance and the exit sections are 0.3048 m and 0.1524 m, respectively. Calculate the x and y components of the reaction force acting on the elbow, and make a sketch of the elbow indicating therein the momentum and force vectors.

8. A Pitot tube is inserted in a pipe through which air is flowing. The stagnation gage pressure reading is 0.100 m of water, and the static gage pressure reading is 0.020 m of water. The

atmospheric pressure is 1 atm, and the static temperature is 300 K. Calculate (a) the dynamic pressure, and (b) the air velocity.

9. Air at 300 K and 1 atm is compressed adiabatically to a pressure and temperature of 5 atm and 490 K, respectively. Employing Table C.4, determine (a) the change in entropy of the gas, and (b) the efficiency of the compressor. The velocity of the air entering and leaving the compressor is negligible.

10. A diffuser of a turbojet engine operating at 10,000 m altitude passes a mass flow rate of 25 kg/s. The inlet velocity is 200 m/s, the inlet static pressure is $0.35 \cdot 10^5$ N/m², and the inlet static temperature is 230 K. The exit area of the diffuser is 0.5 m². Assuming frictionless flow, calculate the reaction force acting on the diffuser.

11. Show that for the adiabatic flow of a perfect gas

$$\frac{t}{T} = 1 - \frac{\gamma - 1}{2}\left(\frac{V}{a_o}\right)^2$$

$$\frac{t}{T} = 1 - \frac{\gamma - 1}{\gamma + 1}\left(\frac{V}{a^*}\right)^2$$

12. An airplane has a velocity of 1931.2 km/hr at sea level. Calculate the stagnation temperature of the air (a) assuming that c_p and γ are constant at their initial values, (b) based on an average value of c_p and γ, and (c) accounting for the variation of c_p with temperature.

13. Superheated steam at $82.74 \cdot 10^5$ N/m² and 811.1 K flows in a 0.1524 m diameter horizontal pipe with a speed of 91.44 m/s. Determine its stagnation temperature and stagnation pressure.

14. A stream of air has a velocity of 250 m/s, a static temperature $t = 300$ K, and a static pressure $p = 1$ atm. Calculate the dynamic pressure (a) assuming that the air is incompressible, and (b) accounting for the effects of the compressibility of the air.

15. Air flows in a duct at a static temperature $t = 300$ K and a static pressure $p = 3.50 \cdot 10^5$ N/m². Calculate the stagnation temperature, pressure, and density assuming that (a) c_p is constant at its initial value, (b) an appropriate average value of c_p is employed, and (c) c_p is a function of temperature (employ Table C.4).

16. An airplane flies at an altitude of 15,000 m with a velocity of 800 km/hr. Calculate (a) the maximum possible temperature on the airplane skin, (b) the maximum possible pressure intensity on the airplane body, (c) the critical velocity of the air relative to the airplane, and (d) the maximum possible velocity of the air relative to the airplane.

17. A perfect gas having $c_p = 1017.4$ J/kg-K and $\bar{m} = 28.97$ flows adiabatically in a converging passage with a mass flow rate $\dot{m} = 29.188$ kg/s. At a particular cross section, $M = 0.6$, $T = 550$ K, and $P = 2.0 \cdot 10^5$ N/m². Calculate the area of the cross-section of the passage at that point.

18. Air having the stagnation conditions $T = 300$ K and $P = 35.0 \cdot 10^5$ N/m² is available in a large storage tank. A converging nozzle is attached to the outlet of the tank to deliver a mass flow rate of 5.0 kg/s. What is the throat area of the nozzle?

19. A stream of air flowing in a duct has $p = 1.40 \cdot 10^5$ N/m², $\dot{m} = 0.25$ kg/s, and $M = 0.6$, at a section where the flow area $A = 6.50$ cm². Calculate the stagnation temperature of the air (a) assuming that c_p is constant and (b) accounting for the variation of c_p with temperature.

4

steady one-dimensional isentropic flow with area change

4–1 PRINCIPAL NOTATION FOR CHAPTER 4

The notation presented in Section 3–1 applies to the discussions of this chapter. The additional notation required for this chapter are listed below.

C_D = $\dot{m}/\dot{m}_{1-D,\,\text{isentropic}}$, discharge coefficient.
C_F = F/PA_t, thrust coefficient.
F thrust of a propulsive nozzle.
I_{sp} = $F/\dot{m}$, specific impulse.
p_o ambient (back) pressure.

Greek Letters

$$\Gamma \quad = \psi^* = \gamma\left(\frac{2}{\gamma+1}\right)^{(\gamma+1)/2(\gamma-1)}$$

ε = A_e/A_t, nozzle area ratio.
ψ flow factor, defined by equation 4.75.
ψ^* = Γ, critical value of ψ.

Subscripts

e nozzle exit.
p planar flow.
s source flow.
t nozzle throat.

Superscripts

$'$ denotes state attained by an isentropic process.
$*$ denotes critical (sonic) condition.

4–2 INTRODUCTION

In Chapter 3, the basic equations governing steady one-dimensional flow are developed, and several important concepts such as the stagnation conditions, the continuity relationships, etc., are introduced. Although the equations presented in

Chapter 3 are simpler than the original equations developed in Chapter 2, nevertheless, they are too complicated to permit obtaining general closed form solutions, because they involve several independent *driving potentials*; for example, area change, heat transfer, friction, and body forces. In many cases of practical interest, one of the driving potentials may predominate, so that the others may be neglected entirely, thereby permitting even further simplifications of the governing equations.

This chapter considers steady one-dimensional flows wherein friction, heat transfer, and body forces are negligible, that is, adiabatic and frictionless flows. Such flows may be assumed to be isentropic, and the sole driving potential responsible for the variations in the flow properties is *area change*, that is, the changes in the area of the flow passage. The equations and concepts developed in this chapter are of great practical value in analyzing many real flow situations. Moreover, the accuracy obtainable by applying the steady one-dimensional isentropic analysis is quite satisfactory for many engineering problems, especially when the results are modified by means of appropriate experimental correction coefficients.

4-3 STEADY ONE-DIMENSIONAL ISENTROPIC FLOW WITH AREA CHANGE

In this section, the equations governing the steady one-dimensional isentropic flow of a fluid in a variable area passage are derived. Figure 4.1 illustrates schematically the flow model for isentropic flow with area change. The equations governing this flow model are derived from the general one-dimensional flow equations derived in Chapter 3, by simplifying them where it is appropriate.

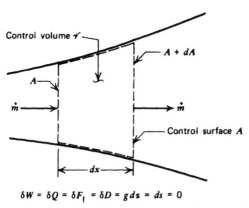

$$\delta W = \delta Q = \delta F_f = \delta D = g \, dz = ds = 0$$

Figure 4.1. Flow model for steady one-dimensional isentropic flow with area change.

4-3(a) Governing Equations

From Table 3.1, the following governing equations are obtained for a steady one-dimensional isentropic flow, in the absence of body forces; that is, a flow for which $\delta F_f = \delta D = \delta Q = \delta W = g \, dz = 0$.

Continuity equation

$$\dot{m} = \rho A V = \text{constant} \tag{4.1}$$

Momentum or dynamic equation

$$dp + \rho V \, dV = 0 \tag{4.2}$$

Energy equation

$$h + \frac{V^2}{2} = \text{constant} = H \tag{4.3}$$

Entropy equation

$$s = \text{constant} \tag{4.4}$$

Equations 4.1 to 4.4 are the basic equations governing the isentropic flow of a compressible fluid in a variable area passage. There are no restrictions on the thermodynamic properties of the working fluid. All that is required is that its equations of state be known in some form: graphical, tabular, or algebraic. When the equations of state are known in either graphical or tabular form, numerical methods are required for solving the flow equations; that is, no general closed form solutions can be obtained (see Section 4–5). When the equations of state are specified in algebraic form, however, general closed form solutions may be possible. Such solutions are developed in Section 4–4 for a perfect gas. Similar results have been obtained for gases having other algebraic equations of state.[1-3]

4-3(b) The Isentropic Process

In general, for a given flow process, the initial flow properties and the mass flow rate are known. Those data establish a point, such as point 1, on the *hs* plane, called the *Mollier diagram*, which is illustrated schematically in Fig. 4.2. The locus of all other points along the isentropic flow path, termed the *isentropic locus*, is specified by the condition that the entropy remains constant: that is, $s = s_1 = \text{constant}$. It is seen in Fig. 4.2 that the isentropic locus for all of the possible thermodynamic states for the flowing fluid is a vertical line passing through point 1. From equation 4.3, it follows that for the flowing fluid, the sum of the static enthalpy h and the kinetic

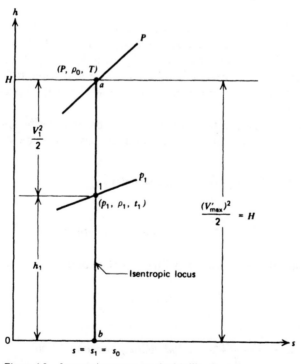

Figure 4.2. Isentropic process on the Mollier diagram.

energy $V^2/2$ is a constant, which is equal to the stagnation enthalpy H. Consequently, the upper limit for the enthalpy h for the fluid experiencing an isentropic flow process occurs at point a (see Fig. 4.2) where the flow velocity is zero, so that $h = H$. The lower enthalpy limit of the isentropic process occurs at point b, where the enthalpy $h = 0$, and the flow velocity is, accordingly, the maximum isentropic speed V'_{max} [see Section 3–10(a)].

Because the isentropic flow process is restricted to a vertical line on the Mollier diagram, the unique stagnation state, denoted by point a in Fig. 4.2, is attained by decelerating the flow from any point on the isentropic locus. Consequently, for a given isentropic flow, the stagnation properties H, P, ρ_0, T, and s_0 are constants. Figure 4.2 illustrates the aforementioned intrinsic features of an isentropic flow. The driving potential causing the changes in state, along the isentropic locus, is *area change*; that is, a variation in the cross-sectional area of the flow passage.

The isentropic process illustrated in Fig. 4.2 is an approximation to a real flow process, which always involves friction. From Table 3.1, the differential form of the energy equation (equation 3.163) is valid for both frictionless and frictional flows. The equation does not indicate explicitly, however, the effect of friction on the flow process. The effect of friction appears in the integration of equation 3.163. Consider a given initial condition denoted by 1. If the flow is adiabatic and frictionless, then the flow process is isentropic, and the upper limit of integration is denoted by 2'. Conversely, if the process involves friction, then the upper limit of integration is denoted by 2. Hence, for an isentropic flow, involving no body forces, we obtain

$$\int_{h_1}^{h'_2} dh = -\int_{V_1}^{V'_2} V \, dV \qquad \text{(isentropic flow)} \qquad (4.5)$$

The corresponding relationship for flow with friction is

$$\int_{h_1}^{h_2} dh = -\int_{V_1}^{V_2} V \, dV \qquad \text{(adiabatic flow with friction)} \qquad (4.6)$$

For identical values of the initial and final static pressures p_1 and p_2, respectively, the magnitudes of h'_2 and h_2 are different, and so are those for V'_2 and V_2. In all cases, either flow expansion or flow compression, $h_2 > h'_2$, and $V_2 < V'_2$. Figure 4.3 compares adiabatic flows with no work or body forces, with and without friction. Figure 4.3a applies to a *flow expansion* (*nozzle action*), and Fig. 4.3b applies to a *flow compression* (*diffuser action*).

For an adiabatic flow expansion in the presence of friction, the specific enthalpy change is denoted by Δh_e and, without friction (isentropic flow), it is denoted by $\Delta h'_e$. Hence,

$$\Delta h'_e = h_1 - h'_2 = \frac{(V'_2)^2 - V_1^2}{2} \qquad \text{(isentropic)} \qquad (4.7)$$

and

$$\Delta h_e = h_1 - h_2 = \frac{V_2^2 - V_1^2}{2} \qquad \text{(adiabatic flow with friction)} \qquad (4.8)$$

Similarly, for an adiabatic flow compression with no work or body forces, we obtain

$$\Delta h'_c = h'_2 - h_1 = \frac{V_1^2 - (V'_2)^2}{2} \qquad \text{(isentropic)} \qquad (4.9)$$

and

$$\Delta h_c = h_2 - h_1 = \frac{V_1^2 - V_2^2}{2} \qquad \text{(adiabatic flow with friction)} \qquad (4.10)$$

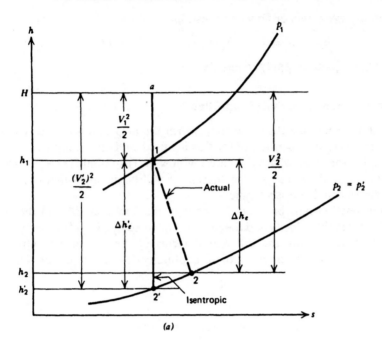

(a)

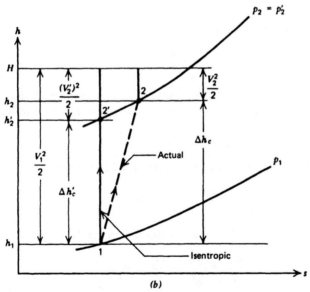

(b)

Figure 4.3. Comparison of expansion and compression processes on the Mollier diagram. (a) Expansion process ($V'_2 > V_2$, $h_2 > h'_2$). (b) Compression process ($V'_2 > V_2$, $h_2 > h'_2$).

The difference between the specific enthalpy changes for an adiabatic flow with friction and the corresponding isentropic flow must be equal to the heat equivalent δQ_f of the energy expended in overcoming the friction. Hence, for *nozzle action* or flow expansion,

$$h_1 - h_2 = h_1 - h'_2 - {}_1(E_f)_2 = \Delta h'_e - \delta Q_f \qquad (4.11)$$

Similarly, for *diffuser action* or flow compression,

$$h_2 - h_1 = h_2' - h_1 + {}_1(E_f)_2 = \Delta h_c' + \delta Q_f \tag{4.12}$$

The preceding equations apply to any fluid.

4–3(c) Effect of Area Change on Flow Properties

It is desirable to determine the manner in which the area of a flow passage must change to achieve (1) an expansion of the flowing fluid, and (2) a compression of the flowing fluid, for the cases where the entering flow is (1) subsonic ($M < 1$), and (2) supersonic ($M > 1$). As a first approximation, it will be assumed that the flow is steady one-dimensional and isentropic.

The desired relationships between a change in the flow area dA and the corresponding changes in the velocity and the pressure, denoted by dV and dp, respectively, are obtained by combining the differential forms of the appropriate governing equations. Logarithmic differentiation of the continuity equation, equation 4.1, yields

$$\frac{d\rho}{\rho} + \frac{dA}{A} + \frac{dV}{V} = 0 \tag{4.13}$$

The momentum equation, equation 4.2, may be rewritten as

$$\frac{dp}{\rho} + V^2 \frac{dV}{V} = 0 \tag{4.14}$$

The speed of sound is given by equation 1.180. Thus,

$$a^2 = \left(\frac{\partial p}{\partial \rho}\right)_s \tag{1.180}$$

Since the entire flow process under consideration is isentropic, the partial derivatives at constant entropy in equation 1.180 may be written as total derivatives, and equation 1.180 may be rewritten as

$$dp = a^2 \, d\rho \tag{4.15}$$

Substituting for $d\rho$ from equation 4.15 into equation 4.13 gives

$$\frac{dp}{\rho a^2} + \frac{dA}{A} + \frac{dV}{V} = 0 \tag{4.16}$$

Substituting for dp from equation 4.14 into equation 4.16 yields

$$-\frac{V^2}{a^2}\frac{dV}{V} + \frac{dA}{A} + \frac{dV}{V} = 0 \tag{4.17}$$

Introducing $M = V/a$ into equation 4.17 and rearranging gives

$$\frac{dA}{A} = (M^2 - 1)\frac{dV}{V} \tag{4.18}$$

Equation 4.18 is the desired relationship between dA and dV.

The corresponding relationship involving dA and dp is obtained by substituting for dV/V, from equation 4.14, into equation 4.18. Thus,

$$\frac{dA}{A} = -(M^2 - 1)\frac{dp}{\rho V^2} \tag{4.19}$$

Rearranging equation 4.19 gives

$$\frac{dA}{A} = (1 - M^2)\left(\frac{p}{\rho V^2}\right)\frac{dp}{p} \tag{4.20}$$

By means of equations 4.18 and 4.20, the manner in which the flow area must change to accomplish either *nozzle action* or *diffuser action* may be readily determined. Equations 4.18 and 4.20 apply to the *isentropic* flow of any fluid. It should be noted, however, that the qualitative results obtained in the following discussion apply also to *irreversible adiabatic* flow.

Consider the area change required when a flow passage is to be employed for expanding (accelerating) a flowing compressible fluid; that is, the passage is to be a *nozzle*. For such a passage the velocity must increase in the direction of flow (dV positive) and simultaneously the pressure must fall (dp negative). Hence, for nozzle action, dV/V must be positive and dp/p negative in the direction of flow. If a compressible fluid enters a passage with a *subsonic velocity* ($M < 1$) then, for dV/V to be positive and dp/p to be negative, equations 4.18 and 4.20 show that the area term dA/A must be negative; that is, the passage must converge. Conversely, if the compressible fluid enters a passage with a *supersonic velocity* ($M > 1$) then, to produce nozzle action, the term dA/A must have a positive value; that is, the passage must diverge in the direction of flow. These conclusions are illustrated schematically in Fig. 4.4*a*.

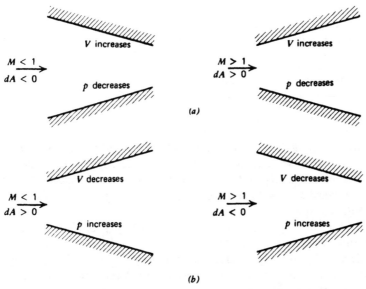

Figure 4.4. Effect of area change on steady one-dimensional isentropic compressible flow. (a) Nozzle flow. (b) Diffuser flow.

Consider now the case where a flowing compressible fluid is to be compressed (decelerated); that is, the flow passage is to be a *diffuser*. In this case the pressure must increase (dp positive) and the velocity decrease (dV negative) in the direction of flow. If the velocity is subsonic ($M < 1$), it follows from equations 4.18 and 4.20 that dA/A must be positive. Flow compression (diffusion) of a subsonic compressible fluid, therefore, requires a diverging passage. Conversely, if the flow is supersonic ($M > 1$), to obtain *diffusion*, the area term dA/A must be negative; that is, the passage must converge. Figure 4.4b illustrates the foregoing conclusions.

4-3(d) Limiting Conditions (Choking)

In Section 4–3(c) the area changes required to achieve nozzle and diffuser actions are determined. No limitation is placed on the magnitude of the effective area change. In this section, the limiting conditions on the area change for achieving nozzle and diffuser actions are determined.

First, it is necessary to obtain a relationship between the changes in velocity V and changes in the Mach number M. From Fig. 4.2, it is seen that for an isentropic process, as V increases, h (and, therefore, t) decreases. For all compressible fluids, the speed of sound a decreases as t decreases, and vice versa. Therefore, for an isentropic flow process, as V increases, a decreases, and the ratio $M = V/a$ increases, and vice versa. Hence, V and M change in the same direction.

The information presented in Fig. 4.4 is expressed in quantitative form in Table 4.1. The limiting conditions for isentropic flow with area change may be determined from the information presented in Table 4.1.

Table 4.1 Relationship between dA and dM for Steady One-Dimensional Isentropic Flow

dA	M	
	Less Than 1.0	Greater Than 1.0
$dA < 0$	$dM > 0$	$dM < 0$
$dA > 0$	$dM < 0$	$dM > 0$

Consider the diverging flow passage illustrated in Fig. 4.5a. If M_1 is subsonic and $dA > 0$, then M decreases and $M_2 < M_1$. Conversely, if M_1 is supersonic and $dA > 0$, M increases and $M_2 > M_1$. A diverging passage, therefore, decelerates a subsonic flow toward zero velocity and accelerates a supersonic flow toward the maximum isentropic speed.

Now refer to the converging passage illustrated in Fig. 4.5b. If M_1 is less than 1.0 and $dA < 0$, then M increases, and $M_2 > M_1$. Conversely, if M_1 is supersonic and $dA < 0$, M decreases, and $M_2 < M_1$. Consequently, a converging passage accelerates a subsonic flow toward the sonic speed, and decelerates a supersonic flow toward the sonic speed.

Suppose now that the Mach number M reaches 1.0, either by accelerating a subsonic flow in a converging passage or by decelerating a supersonic flow in a converging passage, as illustrated in Fig. 4.5b. Can such a sonic flow continue in a converging duct, as illustrated in Fig. 4.6a, and still satisfy the governing equations for steady one-dimensional isentropic flow? If we assume that M decreases so that M_2 becomes less than 1.0, Table 4.1 shows that $dM > 0$, which contradicts the assumption that M decreases. Conversely, if we assume that M increases so that M_2 becomes larger than 1.0, Table 4.1 shows that $dM < 0$, which contradicts the assumption that M increases. Consequently, a sonic flow cannot enter a converging

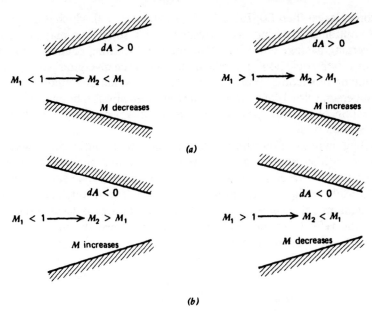

Figure 4.5.   Effect of area change on Mach number. (a) Diverging passage.
(b) Converging passage.

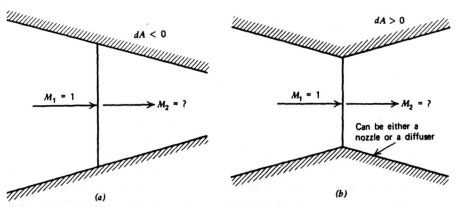

Figure 4.6.   Choking conditions in steady one-dimensional isentropic compressible flow. (a) Sonic flow entering a converging passage. (b) Sonic flow entering a diverging passage.

passage and still satisfy the governing equations for steady one-dimensional isentropic flow. The aforementioned contradictions arise from a phenomenon known as *choking*. We conclude from the foregoing analysis that for a flow in a converging passage, the Mach number continuously changes toward the value $M = 1$ for either a subsonic or a supersonic flow; that is, there is a limiting Mach number for a flow in a converging passage, and its value is $M = 1$. When the Mach number $M = 1$, no further decreases in area are possible without violating the governing equations for steady one-dimensional isentropic flow, and the flow is said to be *choked*.

The other possibility to be examined is illustrated in Fig. 4.6b. Can a sonic flow enter a diverging passage and remain isentropic? If we assume that M decreases so that M_2 is less than 1.0, Table 4.1 shows that $dM < 0$, which is in agreement with the assumption that M decreases. Conversely, if we assume that M increases so that M_2 becomes larger than 1.0, Table 4.1 shows that $dM > 0$, which is in agreement with the assumption that M increases. Accordingly, a sonic flow may enter a continuously diverging passage and keep satisfying the governing equations for steady one-dimensional isentropic flow. The Mach number in the diverging passage may be either subsonic or supersonic. Which of those two possibilities actually occurs (i.e., subsonic or supersonic flow) depends on the downstream physical boundary condition at the exit section of the flow passage. The downstream boundary conditions are discussed in Section 4–3(e).

The preceding analysis shows that a flow passage that continuously accelerates a gas from an initially subsonic flow to a supersonic flow must comprise a converging section followed by a diverging section; that is, it must have the general form illustrated schematically in Fig. 4.7. When the mass rate of flow through such a passage attains its maximum value ($\dot{m} = \dot{m}'_{max} = \dot{m}^*$), the throat area is termed the *critical area A^**, and the velocity of the fluid in the throat, termed the *throat velocity*, is $V'_t = a^*$ [see Section 3–11(b) and Fig. 3.24].

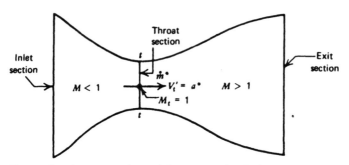

Figure 4.7.   Flow passage for attaining supersonic velocity.

If a diffusion process is to start from a supersonic velocity and end with a subsonic velocity, the flow passage must first converge and then diverge. If a diffusion begins and ends with supersonic velocities, the flow passage must converge continuously.

Summarizing, a converging-diverging passage, known as a De Laval *nozzle*, can produce the following flow phenomena. If the entering velocity is subsonic, the velocity continues to increase and the pressure continues to decrease up to the throat section. If the velocity in the throat is the critical speed of sound a^*, either a flow expansion with supersonic velocities or a diffusion with subsonic velocities is possible in its diverging portion. Which of these two flow phenomena will occur depends on the static pressure at the exit section of the converging-diverging passage.

4–3(e) The Downstream Physical Boundary Conditions

Up to this point, no account has been taken of the conditions at the exit cross section of the flow passage. It is mentioned in the preceding section that the downstream conditions determine whether a sonic flow accelerates or decelerates in a diverging passage. The application of those boundary conditions is discussed here.

Figure 4.8 illustrates the conditions that may occur at the exit from a flow passage when the design Mach number M_e is subsonic (i.e., $M_e \leq 1.0$), and Fig. 4.9 illustrates the conditions that may exist at the exit from the flow passage when the design Mach number M_e is supersonic (i.e., $M_e \geq 1.0$). Figure 4.8a illustrates the internal flow conditions immediately upstream to the exit section of the passage; the Mach number

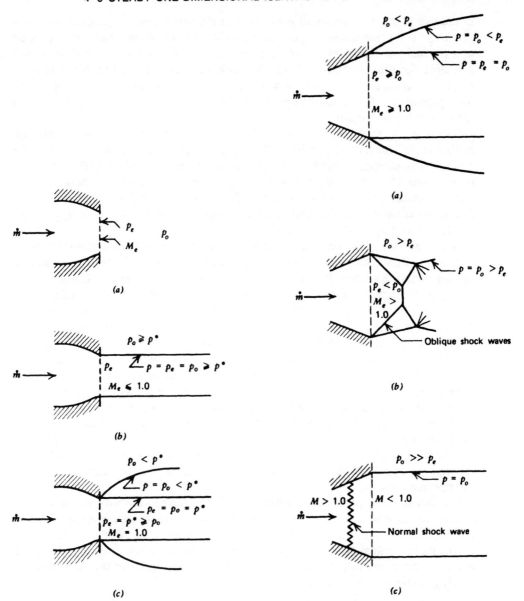

Figure 4.8. Conditions at the exit of a flow passage when $M_e \leq 1.0$. (a) Flow model. (b) $M_e \leq 1.0$ and $p_e = p_o$. (c) $M_e = 1.0$ and $p_e \geq p_o$.

Figure 4.9. Conditions at the exit of a flow passage when $M_e \geq 1.0$. (a) $M_e \geq 1.0$ and $p_e \geq p_o$. (b) $M_e > 1$ and $p_o > p_e$. (c) $M_e > 1$ and $p_o \gg p_e$.

is M_e and the corresponding pressure is p_e. The pressure of the ambient medium into which the fluid flows, termed the *back pressure* or the *ambient pressure*, is denoted by p_o. Obviously, to maintain mechanical equilibrium between the discharging fluid and the surrounding medium, the pressure in the fluid stream must equalize with the ambient pressure. The requirement that the pressures must equalize determines the downstream physical boundary condition that must be satisfied by the internal flow.

Figure 4.8*b* illustrates the situation where the flow in the exit plane is *subsonic*. In that case, the presence of the ambient pressure p_o is communicated to the fluid in the exit plane, because small pressure disturbances travel at the sonic speed (see Section 1–17). Consequently, p_e equals p_o, and the fluid flows into the ambient region as a parallel jet. That flow pattern is characteristic of all subsonic flows, up to and including the sonic condition. Once the sonic condition is reached, the maximum mass flow rate is obtained, and the flow in the exit plane becomes independent of the back pressure p_o. The foregoing applies only if $p_o > p^*$ or $p_o = p^*$, where p^* denotes the critical (sonic) pressure.

If $p^* > p_o$, the expansion from $p_e = p^*$ to p_o takes place outside of the flow passage, as illustrated schematically in Fig. 4.8c. Figures 4.8b and 4.8c demonstrate that the downstream physical boundary condition for an initially subsonic flow is either $p_e = p_o$ and $M_e < 1$, or $M_e = 1$ and $p_e = p^* > p_o$. Equation 3.192 shows that the critical pressure p^* is a function of the specific heat ratio γ and the stagnation pressure P. Consequently, it is readily determinable whether $p_e = p_o$ and $M_e < 1$ (i.e., $p_o > p^*$) or $M_e = 1$ and $p_e = p^*$ (i.e., $p_o \leqq p^*$).

When the flow in the passage is supersonic, the exit boundary conditions change, because the flow velocity exceeds the propagation speed of small pressure waves and the fluid arriving at the exit plane is unaware of the ambient pressure. Consequently, if $p_e \geqq p_o$, the expansion from p_e to p_o occurs outside of the flow passage, as illustrated in Fig. 4.9a, for a diverging passage. However, if $p_o > p_e$, the sudden compression of the fluid flowing out of the passage results in a discontinuous pressure increase known as a shock wave (see Chapter 7). If p_o is only slightly larger than p_e, as illustrated in Fig. 4.9b, the shock waves are attached to the divergence of the nozzle in the vicinity of its exit plane. If p_o is much larger than p_e, as illustrated in Fig. 4.9c, then the shock wave system moves up into the flow passage, and may eventually become a *normal shock wave*. In all cases where the flow is supersonic anywhere in the flow passage, the mass flow rate equals that for a choked flow; the latter condition makes it possible to determine the flow properties at the minimum area, or *throat*, of the flow passage. Should p_o increase to a sufficiently large value, then the normal shock wave will move through the minimum area, unchoking the flow, and the flow in the entire passage will become subsonic.

The downstream physical boundary conditions discussed above are developed here for steady one-dimensional isentropic flow in a variable area passage. Identical conditions are applicable, however, to other types of flow (e.g., frictional flows, and flows with heat transfer). Accordingly, the effects of the boundary conditions discussed above are quite general, and they are applied where applicable in the discussions of steady one-dimensional nonisentropic flows.

4–4 STEADY ONE-DIMENSIONAL ISENTROPIC FLOW OF A PERFECT GAS

The general features of the steady one-dimensional isentropic flow of a compressible fluid are presented in Section 4–3. The present section is concerned with the steady one-dimensional isentropic flow of a perfect gas.

4–4(a) Governing Equations for the Isentropic Flow of a Perfect Gas

For a perfect gas, the equations of state are given by equations 1.99 and 1.117. Thus,

$$p = \rho R t \tag{4.21}$$

and

$$h = c_p t \tag{4.22}$$

where h is measured above $t = 0$. When a perfect gas undergoes an isentropic change of state [see Section 1–15(e)], then

$$tp^{-(\gamma-1)/\gamma} = \text{constant} \tag{4.23}$$

and

$$p\rho^{-\gamma} = \text{constant} \tag{4.24}$$

Combining equations 4.1 to 4.4 with equations 4.21 to 4.24 yields a set of closed form algebraic equations that completely specifies the steady one-dimensional isentropic flow for a perfect gas. The results are developed in Chapter 3 (Sections 3–9 through 3–12) and are summarized here.

In Section 3–9, the stagnation conditions T, P, and ρ_o are defined as the final state of the perfect gas after an isentropic deceleration to zero velocity from the local flow conditions. Thus, T, P, and ρ_o are local properties that are calculable at each point in a flow field and may, in general, vary along the flow path. It is shown in Section 4–3(b) that for an isentropic flow, the stagnation properties remain constant throughout the flow process. Accordingly, the equations that relate T, P, and ρ_o to the local Mach number M are valid throughout the flow process, with T, P, and ρ_o having constant values. From Table 3.3, we obtain the following:

$$T = t\left(1 + \frac{\gamma-1}{2}M^2\right) = \text{constant} \tag{4.25}$$

$$P = p\left(1 + \frac{\gamma-1}{2}M^2\right)^{\gamma/(\gamma-1)} = \text{constant} \tag{4.26}$$

$$\rho_o = \rho\left(1 + \frac{\gamma-1}{2}M^2\right)^{1/(\gamma-1)} = \text{constant} \tag{4.27}$$

By means of equations 4.25 to 4.27, we may determine the values of the static properties t, p, and ρ corresponding to the flow Mach number M, for a given value of γ. Table C.6 presents the ratios t/T, p/P, and ρ/ρ_o as functions of the Mach number M, for different values of γ.

From Table 3.4, the isentropic discharge speed V' attained by expanding a perfect gas isentropically from the stagnation state is

$$V' = \left\{\frac{2\gamma RT}{\gamma-1}\left[1 - \left(\frac{p'}{P}\right)^{(\gamma-1)/\gamma}\right]\right\}^{1/2} \tag{4.28}$$

The speed of sound in a perfect gas is given by $a^2 = \gamma Rt$, and the Mach number by $M = V/a$.

From Table 3.2, the critical flow area ratio A/A^* is given by

$$\frac{A}{A^*} = \frac{1}{M}\left[\frac{2}{\gamma+1}\left(1 + \frac{\gamma-1}{2}M^2\right)\right]^{(\gamma+1)/2(\gamma-1)} \tag{4.29}$$

Values of A/A^* as a function of M, for different values of γ, are tabulated in Table C.6.

Several forms of the continuity equation are developed in Section 3–11(a). All of the forms presented there are, of course, valid for isentropic flow. Because T and P remain constant for such a flow, the following equation for the mass flow rate $\dot{m}$ is especially convenient in dealing with a steady one-dimensional isentropic flow of a perfect gas (see equation 3.167, Table 3.2).

$$\dot{m} = AMP\left(\frac{\gamma}{RT}\right)^{1/2}\left(1 + \frac{\gamma-1}{2}M^2\right)^{-(\gamma+1)/2(\gamma-1)} \tag{4.30}$$

Two additional flow functions of interest in studying the steady one-dimensional isentropic flow of a perfect gas are the dimensionless velocity M^*, and the impulse function ratio $\mathscr{F}/\mathscr{F}^*$. From Table 3.4 and equation 3.141, respectively,

$$M^* = \left[\frac{(\gamma + 1)M^2}{2 + (\gamma - 1)M^2}\right]^{1/2} \tag{4.31}$$

and

$$\frac{\mathscr{F}}{\mathscr{F}^*} = \frac{1 + \gamma M^2}{M\left[2(\gamma + 1)\left(1 + \frac{\gamma - 1}{2}M^2\right)\right]^{1/2}} \tag{4.32}$$

Values of M^* and $\mathscr{F}/\mathscr{F}^*$ are tabulated as functions of M, for different values of γ, in Table C.6.

Equations 4.25 to 4.32 determine the eight properties p, ρ, t, V, A, $\dot{m}$, M^*, and $\mathscr{F}$ as functions of the Mach number M.

An alternative form of the energy equation that is valid for the flow of a perfect gas is obtained by combining equations 4.3 and 4.22. Thus,

$$c_p t + \frac{V^2}{2} = \text{constant} = c_p T \tag{4.33}$$

For a *choked flow*, the Mach number $M = 1$, and the values of t, p, and ρ are their respective critical values, which are denoted by t^*, p^*, and ρ^*. From Table 3.5,

$$t^* = \left(\frac{2}{\gamma + 1}\right)T \tag{4.34}$$

$$p^* = \left(\frac{2}{\gamma + 1}\right)^{\gamma/(\gamma - 1)}P \tag{4.35}$$

$$\rho^* = \left(\frac{2}{\gamma + 1}\right)^{1/(\gamma - 1)}\rho_o \tag{4.36}$$

The *critical velocity* $V^* = a^*$ is given by

$$V^* = a^* = (\gamma R t^*)^{1/2} = \left(\frac{2\gamma R T}{\gamma + 1}\right)^{1/2} = a_o\left(\frac{2}{\gamma + 1}\right)^{1/2} \tag{4.37}$$

where $a_o = (\gamma R T)^{1/2}$ is the stagnation speed of sound. The critical mass flow rate $\dot{m}^*$ is obtained from equation 4.30. Thus,

$$\dot{m}^* = \frac{\Gamma A^* P}{(\gamma R T)^{1/2}} = \frac{\Gamma A^* P}{a_o} \tag{4.38}$$

where

$$\Gamma = \gamma\left(\frac{2}{\gamma + 1}\right)^{(\gamma + 1)/2(\gamma - 1)} \tag{4.39}$$

Values of Γ as a function of the specific heat ratio γ are presented in Table C.15.

4-4(b) Isentropic Flow Tables for a Perfect Gas

Equations are presented in Section 4-4(a) for the flow property ratios in a steady one-dimensional isentropic flow as a function of the local flow Mach number M for a given value of the specific heat ratio γ. Those equations are summarized in Table 4.2.

Table 4.2 Property Ratios for the Steady One-Dimensional Isentropic Flow of a Perfect Gas

$$\frac{T}{t} = \left(1 + \frac{\gamma - 1}{2} M^2\right) \tag{4.25}$$

$$\frac{P}{p} = \left(1 + \frac{\gamma - 1}{2} M^2\right)^{\gamma/(\gamma - 1)} \tag{4.26}$$

$$\frac{\rho_o}{\rho} = \left(1 + \frac{\gamma - 1}{2} M^2\right)^{1/(\gamma - 1)} \tag{4.27}$$

$$\frac{A}{A^*} = \frac{1}{M}\left[\left(\frac{2}{\gamma + 1}\right)\left(1 + \frac{\gamma - 1}{2} M^2\right)\right]^{(\gamma + 1)/2(\gamma - 1)} \tag{4.29}$$

$$M^* = \left[\frac{(\gamma + 1)M^2}{2 + (\gamma - 1)M^2}\right]^{1/2} \tag{4.31}$$

$$\frac{\mathscr{F}}{\mathscr{F}^*} = \frac{1 + \gamma M^2}{M\left[2(\gamma + 1)\left(1 + \frac{\gamma - 1}{2} M^2\right)\right]^{1/2}} \tag{4.32}$$

Values of the isentropic flow property ratios are tabulated as functions of the Mach number M in Table C.6, for different values of the specific heat ratio γ. It is recommended that Table C.6 be employed, whenever possible, in solving steady one-dimensional isentropic flow problems. Table 4.3 summarizes the general form of Table C.6 and also presents the limiting values of the property ratios.

Table 4.3 Limiting Values of the Property Ratios for the Steady One-Dimensional Isentropic Flow of a Perfect Gas

M	M^*	t/T	p/P	ρ/ρ_o	$\mathscr{F}/\mathscr{F}^*$	A/A^*
0	0	1	1	1	∞	∞
1	1	$\left(\frac{2}{\gamma + 1}\right)$	$\left(\frac{2}{\gamma + 1}\right)^{\gamma/(\gamma - 1)}$	$\left(\frac{2}{\gamma + 1}\right)^{1/(\gamma - 1)}$	1	1
∞	$\left(\frac{\gamma + 1}{\gamma - 1}\right)^{1/2}$	0	0	0	$\dfrac{\gamma}{(\gamma^2 - 1)^{1/2}}$	∞

In most cases the initial conditions at the inlet cross section of the flow passage are specified. Employing those conditions, the corresponding values of t/T, p/P, etc., may be read directly from Table C.6. The reference conditions T, P, ρ_o, A^*, and $\mathscr{F}^*$ may then be calculated.

One condition at the exit cross section of the flow passage must be known, so that Table C.6 may be employed for determining the property ratios, and hence the flow properties, at the exit cross section. Many problems of practical interest are not that straight-forward, but an approach of the same general nature is applicable.

A FORTRAN computer program, subroutine ISEN, for calculating the flow property ratios for steady one-dimensional isentropic flow is presented below. The input

```
      SUBROUTINE ISEN

C     ISENTROPIC FLOW PROPERTY RATIOS FOR A PERFECT GAS

      REAL M,MS $ DIMENSION J(9),DM(9) $ DATA INF/6H INFIN/
      NAMELIST /DATA/ G,J,DM $ READ (5,DATA)
      G1=(G-1.0)/2.0 $ G2=2.0/(G+1.0) $ G3=G/(G-1.0) $ G4=2.0*(G+1.0)
      G5=(G+1.0)/(2.0*(G-1.0)) $ G6=G2**G5
      M=0.0 $ MS=0.0 $ T=1.0 $ P=1.0 $ R=1.0 $ I=1 $ L=0
      WRITE (6,2000) G $ WRITE (6,2010) M,MS,T,P,R,INF,INF,INF

C     CALCULATE PROPERTY RATIOS

   40 N=J(I) $ DO 50 K=1,N $ M=M+DM(I) $ C=1.0+G1*M**2 $ MS=M/SQRT(G2*C)
      T=1.0/C $ P=T**G3 $ R=P/T $ F=(1.0+G*M**2)/(M*SQRT(G4*C))
      A=G6*C**G5/M $ PA=P*A $ WRITE (6,2020) M,MS,T,P,R,F,PA,A $ L=L+1
      IF (L.LT.50) GO TO 50 $ WRITE (6,2000) G $ L=0
   50 CONTINUE $ I=I+1 $ IF((I.EQ.10).OR.(J(I).EQ.0)) GO TO 60 $GO TO 40
   60 MS=SQRT((G+1.0)/(G-1.0)) $T=0.0 $P=0.0 $R=0.0 $ F=G/SQRT(G**2-1.0)
      PA=0.0 $ WRITE (6,2030) INF,MS,T,P,R,F,PA,INF $ RETURN

 2000 FORMAT (1H1,28X,42HISENTROPIC FLOW WITH AREA CHANGE,  GAMMA =,F5.2
     1//14X,1HM,5X,2HM*,9X,4HT/TO,9X,4HP/PO,9X,4HR/RO,9X,4HF/F*,6X,
     212H(P/PO)(A/A*),4X,4HHA/A*/)
 2010 FORMAT (1H ,F15.2,F9.5,2X,3E13.5,1X,A6,2(7X,A6))
 2020 FORMAT (1H ,F15.2,F9.5,2X,6E13.5)
 2030 FORMAT (10X,A6,F9.5,2X,5E13.5,1X,A6)
      END
```

to the program comprises the specific heat ratio γ, denoted by G, the number of entries in each section of the table, denoted by J(1), J(2), etc., and the Mach number increment DM(1), DM(2), etc., in each section of the table corresponding to the values of J(1), J(2), etc. The results presented in Table C.6 for $\gamma = 1.4$ are obtained by the following specifications.

$G = 1.4$, $J(1) = 200$, $J(2), = 50$, $J(3) = 30$, $J(4) = 8$, $J(5) = 10$, $J(6) = 0$,
$DM(1) = 0.01$, $DM(2) = 0.02$, $DM(3) = 0.1$, $DM(4) = 0.5$, $DM(5) = 1.0$

Example 4.1. Air at a stagnation temperature and pressure of 288.9 K and $6.895 \cdot 10^5$ N/m², respectively, flows from a reservoir through a converging nozzle that exhausts into the atmosphere where the pressure is $1.014 \cdot 10^5$ N/m². Calculate (a) the pressure in the nozzle exit plane, (b) the minimum stagnation pressure for which the flow remains choked, and (c) the exit plane pressure if the stagnation pressure is reduced to $1.724 \cdot 10^5$ N/m².

Solution

(a) From Table C.6, the critical pressure ratio for sonic conditions at the nozzle exit is $p^*/P = 0.52828$. If the nozzle is choked, then

$$p_2 = p^* = (0.52828)(6.895 \cdot 10^5) = 3.642 \cdot 10^5 \text{ N/m}^2 \ (52.83 \text{ lbf/in.}^2)$$

Since $p^* > p_{atm}$, the flow is choked, so that the pressure in the exit plane is $p_2 = p^*$.
(b) The minimum stagnation pressure for choked flow occurs when $p_2 = p^* = p_{atm}$.
Thus,

$$P_{min} = \frac{p^*}{0.52828} = \frac{p_{atm}}{0.52828} = \frac{1.014 \cdot 10^5}{0.52828} = 1.919 \cdot 10^5 \text{ N/m}^2 \ (27.83 \text{ lbf/in.}^2)$$

(c) For a stagnation pressure of $1.724 \cdot 10^5$ N/m^2,

$$p^* = (0.52828)(1.724)10^5 = 0.9106 \cdot 10^5 \text{ N/m}^2 \ (13.21 \text{ lbf/in.}^2)$$

Since $p^* < p_{atm}$, the flow will not be choked, and $p_2 = p_{atm} = 1.014 \cdot 10^5$ N/m^2
(14.7 lbf/in.^{2}).

Example 4.2. A spaceship cabin may be considered to be a rigid pressurized vessel, which contains the atmosphere required to support the life of its occupants. A puncture in the cabin when it is in vacuum space must be detected quickly, so that the cabin occupants can seal the puncture or switch to their individual life support systems. The time interval that elapses before a dangerously low pressure is reached is a critical quantity for designing the warning devices and countermeasure systems. Derive an equation giving an estimate for the pressure P after puncture as a function of the time t, the initial pressure P_i, the initial temperature T_i, the area A of the puncture, and the volume $\mathscr{V}$ of the cabin. Consider the atmosphere of the cabin to be a perfect gas (see Reference 4). If the cabin atmosphere is air at an initial temperature of 300 K, determine the ratio P/P_i as a function of time with the ratio $(A/\mathscr{V})$ as a parameter. Assume that for air, $\gamma = 1.4$ and $R = 287.04$ J/kg-K.

Solution

Assume that the puncture can be treated as a converging flow passage with a minimum area A, that the depressurization of the cabin is isentropic, that the flow through the puncture is isentropic, and that the flow is quasi-steady (i.e., it consists of a series of steady states, each originating from the local stagnation conditions in the cabin). Since the ambient pressure is a vacuum, the flow through the puncture will be choked, and the area A of the puncture may be considered to be the critical flow area. These assumptions will yield the maximum mass flow rate, and thus the minimum time after puncture to reach a given pressure.

For a choked isentropic flow, equation 4.38 yields

$$\dot{m}^* = \frac{dm}{dt} = \frac{\Gamma AP}{(\gamma RT)^{1/2}} \tag{a}$$

where P and T are the stagnation pressure and temperature, respectively, in the cabin at the time t after it is punctured. For an isentropic depressurization of the cabin, equation 4.23 yields

$$\frac{T}{T_i} = \left(\frac{P}{P_i}\right)^{(\gamma - 1)/\gamma} \tag{b}$$

Substituting equation (b) into equation (a) gives

$$\frac{dm}{dt} = \frac{\Gamma AP}{(\gamma RT_i)^{1/2}} \left(\frac{P_i}{P}\right)^{(\gamma - 1)/2\gamma} \tag{c}$$

From equation 4.21, the mass m_c of gas in the cabin at any time is

$$m_c = \frac{P\mathscr{V}}{RT} \tag{d}$$

since P and T are also the static pressure and temperature, respectively, in the cabin. Substituting equation (b) into equation (d) gives

$$m_c = \frac{\mathscr{V} P_i^{(\gamma - 1)/\gamma}}{R T_i} P^{1/\gamma} \qquad \text{(e)}$$

Differentiating equation (e) with respect to time t, we obtain

$$\frac{dm_c}{dt} = \frac{\mathscr{V} P_i^{(\gamma - 1)/\gamma}}{\gamma R T_i} P^{(1 - \gamma)/\gamma} \frac{dP}{dt} \qquad \text{(f)}$$

The rate of decrease of mass in the cabin is equal to the rate of mass flow out of the puncture. Thus,

$$-\frac{dm_c}{dt} = \frac{dm}{dt} \qquad \text{(g)}$$

Substituting equations (c) and (f) into equation (g) yields

$$-\frac{\mathscr{V} P_i^{(\gamma - 1)/\gamma}}{\gamma R T_i} P^{(1 - \gamma)/\gamma} \frac{dP}{dt} = \frac{\Gamma A P}{(\gamma R T_i)^{1/2}} \left(\frac{P_i}{P}\right)^{(\gamma - 1)/2\gamma} \qquad \text{(h)}$$

Equation (h) may be simplified to

$$-P^{(1 - 3\gamma)/2\gamma} dP = \Gamma (\gamma R T_i)^{1/2} \left(\frac{A}{\mathscr{V}}\right) P_i^{(1 - \gamma)/2\gamma} dt \qquad \text{(i)}$$

Integrate equation (i) between the limits $P = P_i$ when time $t = 0$ and $P = P$ when $t = t$. The result is

$$-\frac{2\gamma}{(\gamma - 1)} P^{(\gamma - 1)/2\gamma} \Big|_{P_i}^{P} = \Gamma (\gamma R T_i)^{1/2} \left(\frac{A}{\mathscr{V}}\right) P_i^{(1 - \gamma)/2\gamma} \, t \Big|_0^t \qquad \text{(j)}$$

Simplifying equation (j), and solving for P/P_i, gives

$$\frac{P}{P_i} = \frac{1}{\left[1 + \left(\dfrac{\gamma - 1}{2\gamma}\right) (\gamma R T_i)^{1/2} \Gamma (A/\mathscr{V}) t\right]^{2\gamma/(\gamma - 1)}} \qquad \text{(k)}$$

For air at 300 K,

$$(\gamma R T_i)^{1/2} = [1.4(287.04)(300)]^{1/2} = 347.21 \text{ m/s}$$

$$\Gamma = \gamma \left(\frac{2}{\gamma + 1}\right)^{(\gamma + 1)/2(\gamma - 1)} = 1.4\left(\frac{2}{1.4 + 1}\right)^{(1.4 + 1)/2(1.4 - 1)} = 1.4(0.83333)^{3.0} = 0.81018$$

Equation (k) becomes

$$\frac{P}{P_i} = \frac{1}{[1 + (0.14286)(347.21)(0.81018)(A/\mathscr{V}) t]^{2(1.4)/(1.4 - 1)}}$$

$$\frac{P}{P_i} = \frac{1}{[1 + 40.19(A/\mathscr{V}) t]^{7.0}} \qquad \text{(l)}$$

4–4(c) Effect of Area Change on the Flow Properties of a Perfect Gas

The influence of area change dA/A on the corresponding changes in the dependent variables dM/M, dp/p, dt/t, etc., may be determined by expressing all of the governing equations in differential form, and then solving for the change in each dependent variable as a function of dA/A. Differentiating equation 4.1 logarithmically yields

$$\frac{d\rho}{\rho} + \frac{dA}{A} + \frac{dV}{V} = 0 \tag{4.40}$$

Rearranging equation 4.2 gives

$$\frac{dp}{p} + \frac{\rho V}{p} dV = \frac{dp}{p} + \gamma M^2 \frac{dV}{V} = 0 \tag{4.41}$$

Differentiating equation 4.33 and rearranging the result yields

$$\frac{dt}{t} + \frac{V \, dV}{c_p t} = \frac{dt}{t} + (\gamma - 1)M^2 \frac{dV}{V} = 0 \tag{4.42}$$

Differentiating equation 4.21 logarithmically gives

$$\frac{dp}{p} - \frac{d\rho}{\rho} - \frac{dt}{t} = 0 \tag{4.43}$$

Logarithmic differentiation of $a^2 = \gamma Rt$, the equation for the speed of sound in a perfect gas, yields

$$\frac{da}{a} - \frac{1}{2}\frac{dt}{t} = 0 \tag{4.44}$$

From the definition of the Mach number, $M = V/a$,

$$\frac{dM}{M} - \frac{dV}{V} + \frac{da}{a} = 0 \tag{4.45}$$

Differentiating equation 3.139, which defines the impulse function $\mathscr{F}$, yields

$$\frac{d\mathscr{F}}{\mathscr{F}} - \frac{dp}{p} - \frac{dA}{A} - \frac{2\gamma M^2}{1 + \gamma M^2}\frac{dM}{M} = 0 \tag{4.46}$$

Equations 4.40 to 4.46 comprise a set of seven first-order ordinary differential equations relating the seven dependent variables dp/p, dt/t, $d\rho/\rho$, dV/V, da/a, dM/M, and $d\mathscr{F}/\mathscr{F}$, and the independent variable dA/A. Those seven equations are linear in the derivatives of the flow properties. Hence, if any one of the eight property changes is specified, the remaining seven are determinable in terms of the one specified change by solving the seven equations simultaneously. Choosing the area change dA/A as the independent variable, equations 4.40 to 4.46 may be put into matrix form as follows.

$$\begin{bmatrix} 0 & 0 & 1 & 1 & 0 & 0 & 0 \\ 1 & 0 & 0 & \gamma M^2 & 0 & 0 & 0 \\ 0 & 1 & 0 & (\gamma - 1)M^2 & 0 & 0 & 0 \\ 1 & -1 & -1 & 0 & 0 & 0 & 0 \\ 0 & -\frac{1}{2} & 0 & 0 & 1 & 0 & 0 \\ 0 & 0 & 0 & -1 & 1 & 1 & 0 \\ -1 & 0 & 0 & 0 & 0 & -\frac{2\gamma M^2}{1 + \gamma M^2} & 1 \end{bmatrix} \begin{bmatrix} dp/p \\ dt/t \\ d\rho/\rho \\ dV/V \\ da/a \\ dM/M \\ d\mathscr{F}/\mathscr{F} \end{bmatrix} = \begin{bmatrix} -\dfrac{dA}{A} \\ 0 \\ 0 \\ 0 \\ 0 \\ 0 \\ \dfrac{dA}{A} \end{bmatrix} \tag{4.47}$$

Equation 4.47, when expressed in matrix notation, becomes

$$Ax = b$$

where A is the 7×7 coefficient matrix, x is the column matrix containing the property changes, and b is the column matrix containing the nonhomogeneous terms.

The seven property changes may be expressed in terms of the area change dA/A by applying Cramer's rule (see Appendix A–3) to solve the seven linear nonhomogeneous equations. Cramer's rule states that

$$x_j = \frac{\det(A^j)}{\det(A)} \qquad (j = 1, 2, \ldots, 7)$$

where $\det(A)$ is the value of the determinant of the coefficient matrix in equation 4.47 and $\det(A^j)$ is the value of the determinant of the matrix we get on replacing the jth column of A by the column matrix b. The value of $\det(A)$ is

$$\det(A) = 1 - M^2 \tag{4.48}$$

To illustrate the procedure, solve for dM/M. Thus,

$$(1 - M^2)\frac{dM}{M} = \begin{vmatrix} 0 & 0 & 1 & 1 & 0 & -\dfrac{dA}{A} & 0 \\ 1 & 0 & 0 & \gamma M^2 & 0 & 0 & 0 \\ 0 & 1 & 0 & (\gamma - 1)M^2 & 0 & 0 & 0 \\ 1 & -1 & -1 & 0 & 0 & 0 & 0 \\ 0 & -\frac{1}{2} & 0 & 0 & 1 & 0 & 0 \\ 0 & 0 & 0 & -1 & 1 & 0 & 0 \\ -1 & 0 & 0 & 0 & 0 & \dfrac{dA}{A} & 1 \end{vmatrix} \tag{4.49}$$

Solving equation 4.49 for dM/M yields

$$\frac{dM}{M} = -\frac{1 + \dfrac{\gamma - 1}{2}M^2}{1 - M^2}\frac{dA}{A}$$

The six remaining property changes are obtained in a similar manner. The results are presented in Table 4.4. The coefficients of dA/A are known as *influence coefficients*, since each coefficient of dA/A is the partial derivative of a particular flow property with respect to the driving potential dA/A.

Equation 4.54 is the same as equation 4.18 and, on substituting $a^2 = \gamma p/\rho$, equation 4.51 becomes the same as equation 4.20. Equation 4.50 expresses in equation form the relationships between dA and dM that are exhibited in Table 4.1.

Equations 4.50 to 4.56 illustrate two intrinsic features of isentropic flows: (1) the effects caused by increasing area ($dA > 0$) are opposite to the effects caused by decreasing area ($dA < 0$), and (2) except for the change in the impulse function $d\mathscr{F}$, the changes for all of the flow properties are of opposite sign for a subsonic flow ($M < 1$) and a supersonic flow ($M > 1$). The second result is due to the term $(1 - M^2)$ in the denominator of equations 4.50 to 4.55, which is positive for subsonic flows and negative for supersonic flows.

Table 4.5 summarizes the effects of area change on the flow properties for the steady one-dimensional isentropic flow of a *perfect gas*. It should be noted that the general study of the influence of area change on an isentropic flow is not restricted to a specific compressible fluid, but is based on the general features of the Mollier diagram for an isentropic process pertinent to any fluid [see Section 4–3(c)].

Table 4.4 Influence Coefficients for Steady One-Dimensional Isentropic Flow with Simple Area Change (Perfect Gas)

$$\frac{dM}{M} = -\frac{1 + \dfrac{\gamma - 1}{2} M^2}{1 - M^2} \frac{dA}{A} \tag{4.50}$$

$$\frac{dp}{p} = \frac{\gamma M^2}{1 - M^2} \frac{dA}{A} \tag{4.51}$$

$$\frac{d\rho}{\rho} = \frac{M^2}{1 - M^2} \frac{dA}{A} \tag{4.52}$$

$$\frac{dt}{t} = \frac{(\gamma - 1)M^2}{1 - M^2} \frac{dA}{A} \tag{4.53}$$

$$\frac{dV}{V} = \frac{-1}{1 - M^2} \frac{dA}{A} \tag{4.54}$$

$$\frac{da}{a} = \frac{(\gamma - 1)M^2}{2(1 - M^2)} \frac{dA}{A} \tag{4.55}$$

$$\frac{d\mathscr{F}}{\mathscr{F}} = \frac{1}{1 + \gamma M^2} \frac{dA}{A} \tag{4.56}$$

Table 4.5 Effect of Simple Area Change on the Flow Properties for a Perfect Gas (Steady One-Dimensional Isentropic Flow)

Property Ratio	$dA < 0$		$dA > 0$	
	$M < 1$	$M > 1$	$M < 1$	$M > 1$
dM/M	+	−	−	+
dp/p	−	+	+	−
dt/t	−	+	+	−
$d\rho/\rho$	−	+	+	−
dV/V	+	−	−	+
da/a	−	+	+	−
$d\mathscr{F}/\mathscr{F}$	−	−	+	+

4-5 STEADY ONE-DIMENSIONAL ISENTROPIC FLOW OF IMPERFECT GASES

The essential features of the steady one-dimensional isentropic flow of a *perfect gas* are discussed in Section 4-4. The present section is concerned with the steady one-dimensional isentropic flow of *imperfect gases*.

The term *imperfect gas* denotes a compressible fluid having a behavior that deviates from that for a *perfect gas*. The deviations are termed *imperfect gas effects*, and arise from the following.

1. Variations in the specific heats because of the activation of vibrational and electronic modes of energy stored in the gas molecules.
2. Variations in the gas constant of a single species because of the intermolecular forces and the effects of the molecular volume.
3. Variations in the molecular weight of a gas mixture because of chemical reactions.

The basic equations of Section 4–3(a), equations 4.1 to 4.4, are independent of the type of fluid being considered. As mentioned in Section 4–3(a), the equations of state for an imperfect gas may be in algebraic, tabular, or graphical form. In the following discussion, a generalized model of an imperfect gas is considered; that is, its equations of state are specified in general functional form only. Thus,

$$t = t(p,s) \tag{4.57}$$

$$\rho = \rho(p,s) \tag{4.58}$$

$$h = h(p,s) \tag{4.59}$$

$$a = a(p,s) \tag{4.60}$$

For an isentropic process,

$$s = \text{constant} \tag{4.61}$$

If the equations of state are known, the analysis is independent of the source causing the imperfect gas effects.

In Section 3–5(a), it is shown that for a steady one-dimensional isentropic flow, the *momentum equation* and the *energy equation* are equivalent. Thus, only one of those equations may be employed. Because the energy equation is an algebraic equation, whereas the momentum equation is a differential equation, the energy equation is employed in the subsequent analysis.

The continuity and energy equations for a steady one-dimensional isentropic flow are equations 4.1 and 4.3, respectively. They are repeated here for convenience.

$$\dot{m} = \rho A V \tag{4.62}$$

$$h + \frac{V^2}{2} = H \tag{4.63}$$

Equations 4.57 to 4.63 comprise a system of seven equations involving the variables t, p, s, ρ, h, a, and V, and the flow area A. When the flow properties are given at area A in an isentropic flow passage (i.e., H, $\dot{m}$, A, t_1, p_1, $s = s_1 = s_2$, ρ_1, h_1, a_1, and V_1), the aforementioned system of equations may be solved for the corresponding flow properties t_2, p_2, ρ_2, h_2, a_2, and V_2 at specified values of the flow area A_2. In general, those equations must be solved numerically.

Various numerical methods may be employed for solving systems of nonlinear equations. An extremely rapid method is one based on assuming an *initial predicted value* of the solution, and then substituting that predicted solution into the nonlinear equations, thereby obtaining the *first calculated solution*. The latter solution value is then chosen as the *second predicted solution*, which is then substituted into the non-linear equations to give a *second calculated solution*. Those two sets of predicted and calculated solutions are then employed in conjunction with the *secant method* [see Appendix A–4(b)] to determine a *third predicted solution*. The latter solution is then substituted into the nonlinear equations to furnish a *third calculated solution*. The secant method may be applied repetitively until the predicted and calculated solutions agree to within an acceptable tolerance.

The form of the numerical algorithm employed in the analysis depends on the data that are provided and the desired results. Two typical problems are discussed here.

Problem a. Consider the problem of determining the critical properties p^*, t^*, etc., and the critical mass flux G^* corresponding to a given set of stagnation conditions. The following sequence of steps leads to the solution.

1. Specify the equations of state, and the stagnation properties T, P, etc.
2. Assume a value for the critical pressure p^*. To obtain the first predicted value of

p^*, the results obtained by assuming that the gas is a perfect gas may be employed. The second predicted value of p^* is obtained from the results of the first calculation, and subsequent values of p^* are obtained by the secant method.

3. Determine the values of t^*, ρ^*, h^*, and a^* from the equations of state.
4. Determine V^* from equation 4.63, and calculate $M^* = V^*/a^*$.
5. Check to see if M is within the desired tolerance for $M = 1.0$.
6. If not, steps 2, 3, and 4 must be repeated to attain convergence.
7. When convergence is attained, G^* is determined from equation 4.62. The calculation of G^* completes the solution.

Problem b. Consider the problem where the stagnation conditions and mass flow rate are known, and one condition (e.g., M_2, A_2, p_2, etc.) is specified in the flow passage. The remaining flow properties are to be determined. The solution is obtained by utilizing the following sequence of steps.

1. Specify the equations of state, the stagnation properties T, P, etc., and the one final condition.
2. Assume a value for the pressure p. For the first predicted value of p, employ the results of a perfect gas analysis. The second predicted value of p is obtained from the results of the first calculation, and subsequent values of p are obtained by applying the secant method.
3. Determine the values of t, ρ, h, and a from the equations of state.
4. Determine V from equation 4.63, and the flow area A from equation 4.62. Calculate $M = V/a$.
5. Check to see if the calculated value of the one specified property is within the desired tolerance for the specified value of that property.
6. If not, steps 2, 3, and 4 must be repeated to achieve convergence.
7. When convergence is achieved, the solution is complete.

The numerical algorithms discussed above are simple to apply, and converge rapidly. By employing such methods, *imperfect gas* effects in steady isentropic flow may be taken into account.

Example 4.3.   A blow-down wind tunnel employs a pebble bed heater for raising the stagnation temperature of air to 2000 K before it enters the wind tunnel. The stagnation pressure is $35 \cdot 10^5$ N/m². Determine the critical mass flux G^* (i.e., $\dot{m}^*/A^* = \rho^* V^*$) corresponding to these stagnation conditions. Assume that the air behaves as a thermally perfect gas with $R = 287.04$ J/kg-K, but take into account the variation of the specific heats of air with temperature.

Solution

If dissociation is neglected, the only *imperfect gas effect* of significance is the variation of the specific heats of the air with temperature. Table C.4 may, therefore, be employed for the equations of state relating t, h, and p. For a thermally perfect gas,

$$p = \rho Rt \qquad (a)$$

and

$$a = (\gamma Rt)^{1/2} \qquad (b)$$

where the specific heat ratio γ is a function of temperature. For air, values of γ as a function of temperature are presented in Table C.4. The numerical method discussed under Problem a in Section 4.5 will be employed.

1. *Initial conditions.* The equations of state are specified by equations (a) and (b), and Table C.4. The stagnation temperature $T = 2000$ K, and the stagnation pressure

$P = 35 \cdot 10^5$ N/m². From Table C.4, for $T = 2000$ K, $H = 2252.414$ kJ/kg and $P_r = 1520.8$.

2. *Assume a value for p*.* As a first trial, calculate p^* for a perfect gas. From Table C.4, at $T = 2000$ K, $\gamma = 1.2975$. Thus, assume $\gamma = 1.30$. From Table C.6, for $M = 1.0$, $p^*/P = 0.54573$. Thus,

$$p^* = (0.54573)(35)10^5 = 19.101 \cdot 10^5 \text{ N/m}^2.$$

The first predicted value for $p^* = 19.101 \cdot 10^5$ N/m². The remaining critical properties are determined for comparison with the *imperfect gas solution*. From Table C.6, $t^*/T = 0.86957$. Thus,

$$t^* = (0.86957)(2000) = 1739.1 \text{ K}$$

From equation (a),

$$\rho^* = \frac{p^*}{Rt^*} = \frac{(19.101)10^5}{(287.04)(1739.1)} = 3.8263 \text{ kg/m}^3$$

From equation (b),

$$a^* = V^* = (\gamma R t^*)^{1/2} = [1.3(287.04)(1739.1)]^{1/2} = 805.57 \text{ m/s}$$

The critical mass flux is thus

$$G^* = \frac{\dot{m}^*}{A^*} = \rho^* V^* = (3.8263)(805.57) = 3082.4 \text{ kg/s-m}^2$$

3. *Calculation of t*, h*, and a*.*

$$p_r^* = \frac{p^*}{P} P_r = \frac{(19.101)(1520.8)}{35} = 829.97$$

From Table C.4, for $p_r^* = 829.97$, $t^* = 1738.9$ K, $h^* = 1927.921$ kJ/kg, and $\gamma = 1.3036$. From equation (b),

$$a^* = (\gamma R t^*)^{1/2} = [1.3036(287.04)(1738.9)]^{1/2} = 806.64 \text{ m/s}$$

4. *Determine V* and M*.* From equation 4.63,

$$V^* = [2(H - h^*)]^{1/2} = [2(2252.414 - 1927.921)(1000)]^{1/2} = 805.60 \text{ m/s}$$

$$M^* = \frac{V^*}{a^*} = \frac{805.60}{806.64} = 0.99871$$

5. *Convergence check.* The calculated value of M^* differs from 1.0 by -0.13 percent. For most engineering purposes, that agreement is satisfactory. Nevertheless, a second trial is made for the purpose of illustrating the procedure.

6. *Repetition of steps 2, 3, and 4.* A second predicted value p_2^* may be estimated from the first predicted value of p_1^* and the pressure ratio for a perfect gas, obtained from Table C.6, for attaining $M_2 = 1.0$ from $M_1 = 0.99871$. Thus, for $\gamma = 1.30$,

$$\frac{p_2^*}{p_1^*} = \frac{(p/P) \text{ for } M_2 = 1.0}{(p/P) \text{ for } M_1 = 0.99871} = \frac{0.54573}{0.54653} = 0.99854$$

$$p_2^* = 0.99854(19.101)10^5 = 19.073 \cdot 10^5 \text{ N/m}^2$$

$$p_r^* = \frac{(19.073)(1520.8)}{35} = 828.75$$

From Table C.4, $t^* = 1738.3$ K, $h^* = 1927.181$ kJ/kg, and $\gamma = 1.3036$.

$$a^* = [1.3036(287.04)(1738.3)]^{1/2} = 806.50 \text{ m/s}$$
$$V^* = [2(2252.414 - 1927.181)(1000)]^{1/2} = 806.52 \text{ m/s}$$

$$M^* = \frac{806.52}{806.50} = 1.00002$$

The last value of M^* is close enough to 1.0.

7. *Calculate G^*.* From equation (a),

$$\rho^* = \frac{p^*}{Rt^*} = \frac{(19.073)10^5}{(287.04)(1738.3)} = 3.8225 \text{ kg/m}^3$$

From equation 4.62,

$$G^* = \frac{m^*}{A^*} = \rho^* V^* = (3.8225)(806.52) = 3082.9 \text{ kg/s-m}^2$$

The results for the *imperfect gas* are compared with those for the *perfect gas* in the following table.

Property	Perfect Gas	Imperfect Gas	Error, Percent
p^*, N/m$^2 \cdot 10^5$	19.101	19.073	0.15
t^*, K	1739.1	1738.3	0.05
ρ^*, kg/m^3	3.8263	3.8225	0.10
V^*, m/s	805.57	806.52	−0.12
G^*, kg/s-m^2	3082.4	3082.9	−0.02

It is evident from the table that the errors caused by neglecting the imperfect gas effects in this problem are insignificant for most engineering work. Hence, for Example 4.3, the solution for the perfect gas is adequate.

Example 4.4. The area ratio A/A^* of the expansion section of the blow-down wind tunnel described in Example 4.3 is to be chosen so that air can be provided to a test section with a Mach number of 6.0. Determine the required area ratio, assuming air to be a thermally perfect gas, but take into account the variation of the specific heats of air with temperature.

Solution

The problem is of the type discussed under Problem b in Section 4–5, with the Mach number specified. The numerical method discussed under Problem b is employed.

1. *Initial conditions.* The equations of state are specified by equations (a) and (b) in Example 4.3 and Table C.4. The stagnation temperature $T = 2000$ K and the stagnation pressure $P = 35 \cdot 10^5$ N/m^2. From Table C.4, $H = 2252.414$ kJ/kg and $P_r = 1520.8$.

2. *Assume a value for p.* To obtain the first predicted value for p, calculate p at a Mach number $M = 6.0$, for a perfect gas. Assume $\gamma = 1.40$. From Table C.6, $p/P = 0.00063336$. Thus,

$$p = (0.00063336)(35)10^5 = 2216.8 \text{ N/m}^2$$

The first predicted value for $p = 2216.8$ N/m^2. The remaining properties at the exit cross section of the nozzle are calculated for comparison with the imperfect gas solution. From Table C.6, $t/T = 0.12195$. Thus,

$$t = (0.12195)(2000) = 243.90 \text{ K}$$

From equation (a), Example 4.3,

$$\rho = \frac{p}{Rt} = \frac{2216.8}{(287.04)(243.90)} = 0.031664 \text{ kg/m}^3$$

From equation (b), Example 4.3,

$$a = (\gamma Rt)^{1/2} = [1.4(287.04)(243.90)]^{1/2} = 313.07 \text{ m/s}$$

$$V = Ma = 6.0(313.07) = 1878.4 \text{ m/s}$$

$$G = \frac{\dot{m}}{A} = \rho V = (0.031664)(1878.4) = 59.478 \text{ kg/s-m}^2$$

From Table C.6, for $M = 6$ and $\gamma = 1.40$, $A/A^* = 53.180$.

3. *Calculate t, h, and a.*

$$p_r = \frac{p}{P} P_r = \frac{(0.022168)(1520.8)}{35} = 0.96323$$

From Table C.4, for $p_r = 0.96323$, $t = 294.97$ K, $h = 296.281$ kJ/kg and $\gamma = 1.4002$. From equation (b), Example 4.3,

$$a = (\gamma Rt)^{1/2} = [1.4002(287.04)(294.97)]^{1/2} = 344.31 \text{ m/s}$$

4. *Determine V and M.* From equation 4.63,

$$V = [2(H - h)]^{1/2} = [2(2252.414 - 296.281)(1000)]^{1/2} = 1977.9 \text{ m/s}$$

$$M = \frac{V}{a} = \frac{1977.9}{344.31} = 5.7445$$

5. *Convergence check.* The calculated value of M differs from the specified value of $M = 6.0$ by -4.25 percent. Thus, a second trial is warranted.

6. *Repetition of steps 2, 3, and 4.* A second predicted value for p_2 can be estimated from the first predicted value p_1 and the perfect gas pressure ratio (obtained from Table C.6) required to attain $M_2 = 6.0$ from $M_1 = 5.7445$. Thus, for $\gamma = 1.40$,

$$\frac{p_2}{p_1} = \frac{(p/P) \text{ for } M_2 = 6.0}{(p/P) \text{ for } M_1 = 5.7445} = \frac{0.00063336}{0.00082766} = 0.76524$$

$$p_2 = (0.76524)(2216.8) = 1696.4 \text{ N/m}^2$$

$$p_r = \frac{(0.016964)(1520.8)}{35} = 0.73711$$

From Table C.4, $t = 273.23$ K, $h = 274.445$ kJ/kg, and $\gamma = 1.4005$.

$$a = [1.4005(287.04)(273.23)]^{1/2} = 331.42 \text{ K}$$

$$V = [2(2252.414 - 274.445)(1000)]^{1/2} = 1989.0 \text{ m/s}$$

$$M = \frac{1989.0}{331.42} = 6.0014$$

The calculated value of M differs from the specified value of $M = 6.0$ by 0.02 percent, which is close enough.

7. *Calculate A/A^*.* From equation (a), Example 4.3,

$$\rho = \frac{p}{Rt} = \frac{1696.4}{(287.04)(273.23)} = 0.021630 \text{ kg/m}^3$$

$$G = \frac{\dot{m}}{A} = \rho V = (0.021630)(1989.0) = 43.022 \text{ kg/s-m}^2$$

From equation 4.62,

$$\frac{A}{A^*} = \frac{\dot{m}/\rho V}{\dot{m}^*/\rho^* V^*} = \frac{\rho^* V^*}{\rho V} = \frac{G^*}{G}$$

where $\dot{m} = \dot{m}^*$, since the flow is choked. From Example 4.3, $G^* = 3082.9$ kg/s-m^2. Thus,

$$\frac{A}{A^*} = \frac{3082.9}{43.022} = 71.659$$

The following table compares the results for the *imperfect gas* with those for the *perfect gas*.

Property	Perfect Gas	Imperfect Gas	Error, Percent
p, N/m^2	2216.8	1696.4	30.68
t, K	243.90	273.23	-10.73
ρ, kg/m^3	0.031664	0.021630	46.39
V, m/s	1878.4	1989.0	-5.56
G, kg/s-m^2	59.478	43.022	38.25
$\dfrac{A}{A^*}$	53.180	71.659	-25.79

It is obvious that the large errors incurred by assuming that the flowing fluid is a perfect gas are unacceptable for the design of the wind tunnel. Example 4.4 demonstrates vividly that *imperfect gas effects* must be taken into account if the flow is a high-energy, high-Mach number flow.

4-6 THE CONVERGING NOZZLE

Figure 4.10 illustrates schematically the case where a compressible fluid, after flowing through a conduit, is discharged through a *converging nozzle* into a region having the ambient static pressure p_o, called the *back pressure*. The cross-sectional areas immediately upstream from the nozzle and in the exit plane, or *throat*, of the nozzle, are denoted by A_1 and A_t, respectively.

4-6(a) Isentropic Throat Speed

It will be assumed unless stated to the contrary that the flow is a steady one-dimensional isentropic flow. Hence, the velocity of the fluid crossing the plane of the throat A_t, called the *isentropic throat speed* V_t', is given by

$$V_t' = [2(H - h_t')]^{1/2} \tag{4.64}$$

where $H = h_1 + V_1^2/2$ is the stagnation enthalpy.

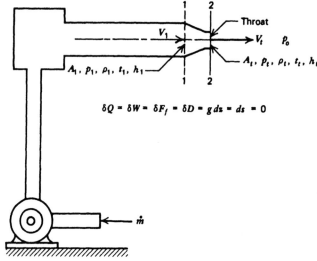

Figure 4.10. Steady isentropic flow through a converging nozzle.

In the special case where the fluid is a perfect gas, the isentropic throat speed V'_t is (see equation 4.28)

$$V'_t = \left\{\frac{2\gamma RT}{\gamma - 1}\left[1 - \left(\frac{p'_t}{P}\right)^{(\gamma - 1)/\gamma}\right]\right\}^{1/2} = a_o\left\{\frac{2}{\gamma - 1}\left[1 - \left(\frac{p'_t}{P}\right)^{(\gamma - 1)/\gamma}\right]\right\}^{1/2} \quad (4.65)$$

where a_o is the stagnation acoustic speed. The Mach number for the flow in the throat section A_t is

$$M'_t = \frac{V'_t}{a'_t} = \frac{a_o}{a'_t}\left\{\frac{2}{\gamma - 1}\left[1 - \left(\frac{p'_t}{P}\right)^{(\gamma - 1)/\gamma}\right]\right\}^{1/2} \quad (4.66)$$

For the isentropic flow of a perfect gas,

$$\frac{a_o}{a'_t} = \left(\frac{T}{t'_t}\right)^{1/2} = \left(\frac{P}{p'_t}\right)^{(\gamma - 1)/2\gamma} \quad (4.67)$$

Substituting for a_o/a'_t into equation 4.66 gives

$$M'_t = \left\{\frac{2}{\gamma - 1}\left[\left(\frac{P}{p'_t}\right)^{(\gamma - 1)/\gamma} - 1\right]\right\}^{1/2} \quad (4.68)$$

If the back pressure p_o into which the converging nozzle discharges is decreased while the stagnation pressure P at the entrance to the nozzle is held constant, the static pressure in the throat of the nozzle p'_t remains equal to p_o until $p'_t = p_o = p^*$, the *critical pressure*. Thereafter, further reduction of p_o has no influence on the static pressure in the throat; the latter remains constant at the value $p'_t = p^*$. Moreover, when the critical pressure p^* prevails in the throat, the isentropic throat speed V'_t is equal to a^*, the critical speed of sound, and the mass flow rate is the maximum or *critical mass flow rate* $\dot{m}^*$.

Reducing the back pressure p_o below p^* causes a rarefaction wave to be propagated from the outside toward the nozzle to stimulate an increase in the rate of discharge. The wave is propagated into the jet with the absolute local sonic speed, which is a^*, so that the speed of the wave relative to the jet is $(a^* - a^*) = 0$. Consequently, the static pressure in the throat of the converging nozzle is unaffected by the reduction in the back pressure. The critical pressure ratio $r = p^*/P$ for the nozzle does not decrease to the actual value of p_o/P but remains constant at the value p^*/P, *no matter how much the back pressure p_o is reduced, or the inlet pressure P is increased*; if P is increased, p^* also increases, but the ratio p^*/P remains fixed. The Mach number in the throat section of the converging nozzle remains constant, therefore, at the value unity, and the mass flow rate remains at the critical value $\dot{m}^*$ corresponding to the stagnation pressure P and the stagnation temperature T. The preceding discussion applies not only to converging nozzles, but also to the throat section of a converging-diverging nozzle.

The maximum discharge speed attainable by a gas expanding isentropically in a converging nozzle is the throat speed corresponding to the critical pressure ratio p^*/P and the value of the stagnation temperature T. Let $(V'_t)_{max}$ denote the *maximum isentropic throat speed*. When $(V'_t)_{max}$ occurs, the pressure ratio p'_t/P is given by equation 3.192, which is repeated here for convenience. Thus,

$$\frac{p'_t}{P} = \frac{p^*}{P} = \left(\frac{2}{\gamma + 1}\right)^{\gamma/(\gamma - 1)} \tag{4.69}$$

where p^*/P is the *critical pressure ratio*. Substituting equation 4.69 into equation 4.65 and simplifying gives

$$(V'_t)_{max} = a^* = \left(\frac{2\gamma RT}{\gamma + 1}\right)^{1/2} = a_o \left(\frac{2}{\gamma + 1}\right)^{1/2} \tag{4.70}$$

When any form of nozzle, converging or converging-diverging, operates with the critical speed of sound a^* in its throat, the nozzle is said to operate with *complete nozzling*, and the flow is said to be *choked*. The corresponding critical mass rate of flow $\dot{m}^*$ is also termed the *choking mass flow rate*.

4-6(b) Isentropic Mass Rate of Flow

The mass rate of flow for either a converging nozzle or a converging-diverging nozzle is obtained by applying the continuity equation to the throat area A_t. Assuming steady one-dimensional isentropic flow, and letting $\dot{m}'$ denote the corresponding *isentropic mass flow rate* then, from equation 4.1,

$$\dot{m}' = A_t \rho'_t V'_t \tag{4.71}$$

Substituting for V'_t from equation 4.64 gives

$$\dot{m}' = A_t \rho'_t [2(H - h'_t)]^{1/2} \tag{4.72}$$

In the special case where the fluid is a perfect gas, the density ρ'_t is obtained from equations 4.21 and 4.24. Thus,

$$\rho'_t = \rho_o \left(\frac{p'_t}{P}\right)^{1/\gamma} = \left(\frac{P}{RT}\right) \left(\frac{p'_t}{P}\right)^{1/\gamma} \tag{4.73}$$

Substituting equations 4.65 and 4.73 into equation 4.71 and rearranging yields

$$\dot{m}' = \frac{PA_t}{\sqrt{\gamma RT}} \left\{ \frac{2\gamma^2}{\gamma - 1} \left(\frac{p'_t}{P}\right)^{2/\gamma} \left[1 - \left(\frac{p'_t}{P}\right)^{(\gamma - 1)/\gamma}\right] \right\}^{1/2} \tag{4.74}$$

Examination of equation 4.74 shows that for fixed entrance conditions the mass flow rate $\dot{m}'$ depends only on the expression enclosed by the brackets. That expression is termed the *flow factor* and is denoted by ψ. Hence,

$$\psi \equiv \left\{ \frac{2\gamma^2}{\gamma - 1} \left(\frac{p_t'}{P} \right)^{2/\gamma} \left[1 - \left(\frac{p_t'}{P} \right)^{(\gamma - 1)/\gamma} \right] \right\}^{1/2} \tag{4.75}$$

Figure 4.11 presents values of ψ as a function of p_t'/P for different values of γ. Although values of ψ can be calculated for values of $p_t'/P < p^*/P$, those values of ψ have no physical significance because the mass flow rate remains constant at the critical value $\dot{m}^*$ for all back pressures less than p^*.

In terms of the flow factor ψ, equation 4.74 becomes

$$\dot{m}' = \frac{\psi P A_t}{\sqrt{\gamma R T}} = \frac{\psi P A_t}{a_o} \tag{4.76}$$

It is evident from equation 4.76 that for a fixed value of the entrance stagnation pressure P, and a fixed value of the pressure ratio $r = p_t'/P$ (i.e., $\psi = $ constant), the isentropic mass rate of flow $\dot{m}'$ for a specific gas depends only on the stagnation temperature T. Raising T decreases the value of $\dot{m}'$, and vice versa.

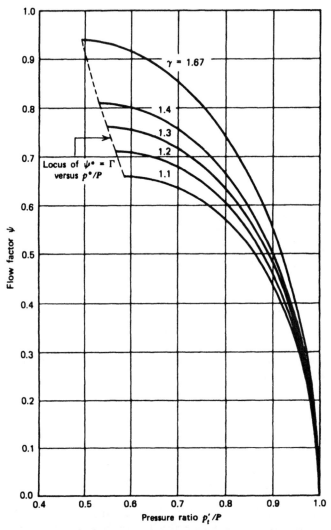

Figure 4.11. Flow factor ψ as a function of the pressure ratio p_t'/P.

The maximum value of $\dot{m}'$ for the assumed isentropic conditions occurs when the throat speed $V'_t = a^*$. When that condition prevails, the flow factor $\psi = \psi^*$; the equation for ψ^* is obtained by substituting for p^*/P from equation 4.69 into equation 4.75. Hence,

$$\psi^* = \left[\frac{2\gamma^2}{\gamma + 1} \left(\frac{2}{\gamma + 1} \right)^{2/(\gamma - 1)} \right]^{1/2} = \gamma \left(\frac{2}{\gamma + 1} \right)^{(\gamma + 1)/2(\gamma - 1)} = \Gamma \qquad (4.77)$$

where Γ is defined by equation 4.39. For air, $\gamma = 1.4$, and $\psi^* = \Gamma$ is

$$\psi^* = \Gamma = 1.4 \left(\frac{2}{2.4} \right)^{3.0} = 0.81019 \qquad (4.78)$$

Values of $\psi^* = \Gamma$, as a function of γ, are tabulated in Table C.15. The critical mass flow rate $\dot{m}^*$ is accordingly given by

$$\dot{m}^* = \frac{\psi^* P A_t}{\sqrt{\gamma R T}} = \frac{\psi^* P A_t}{a_o} = \frac{\Gamma P A_t}{a_o} \qquad (4.79)$$

For air, $R = 287.04$ J/kg-K, and equation 4.79 may be written as

$$\frac{\dot{m}^* \sqrt{T}}{P A_t} = 0.040416 \ (0.53175 \text{ in EE units}) \qquad (4.80)$$

The conditions at the throat of a converging-diverging nozzle are governed by the same relationships as those for the throat of a converging nozzle. Consequently, the equations derived in the present section for the isentropic mass flow rate apply not only to the flow in a converging nozzle but also to the flow in a converging-diverging nozzle.

4–6(c) Effect of Decreasing the Back Pressure with Constant Inlet Pressure

The foregoing demonstrates that the value of $\dot{m}'$ obtained from equation 4.76 and the plot of ψ versus p'_t/P in Fig. 4.11, for the range $p'_t/P = 1$ to zero, is correct until the pressure ratio p'_e/P is reduced to the critical pressure ratio. Thereafter, because disturbances beyond the throat can no longer be propagated upstream, the flow behaves as if no further reduction in the back pressure had occurred; the flow has no way of becoming aware of the reduction in pressure. Consequently, when the back pressure is equal to or less than the critical pressure, the nozzle discharges the same mass flow rate, the *critical mass flow rate* $\dot{m}^* = \dot{m}'_{max}$. This fact was known to St. Venant and Wantzel (1839) and was explained both theoretically and experimentally by Grashof (1875) and Zeuner (1900).

Experiments show that the shape of the jet discharging from a converging nozzle is different, depending on whether the back pressure is above or below the critical pressure. When the back pressure is equal to or above the critical pressure ($p_o \geqq p^*$), the jet issues as a cylindrical parallel stream, its surface being gradually retarded by the surrounding gas, so that a mixing zone is produced in which the velocity of the jet finally drops to that of the surroundings, as illustrated in Fig. 4.12.

When the back pressure is less than the critical pressure ($p_o < p^*$), the jet expands as it discharges from the nozzle, as indicated in Fig. 4.13. In that case, the pressure of the gas in the jet leaving the nozzle is the critical pressure p^*, which is larger than the back pressure. The sudden reduction in pressure causes the gaseous jet to expand in an explosive fashion. The gas particles are accelerated radially and, owing to their inertia, they are displaced from their equilibrium positions, thereby creating a pressure reduction in the core of the jet that causes the particles to reverse the

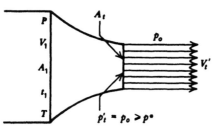

Figure 4.12. Shape of the jet discharged from a converging nozzle with subcritical operation ($p_t' = p_o \geqq p^*$).

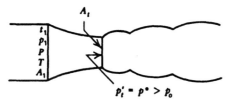

Figure 4.13. Shape of the jet discharged from a converging nozzle with supercritical operation ($p_t' = p^* > p_o$).

directions of their motions. The above phenomenon is periodic. As a result, the jet becomes thinner in some sections and thicker in others. Standing waves are observed that are associated with loud noise and with a decrease in the available energy of the gas.

4–6(d) Effect of Increasing the Inlet Pressure with Constant Back Pressure

Consider the case where the flow through a nozzle is isentropic and the back pressure p_o is held constant while the inlet stagnation pressure P is varied. As P is increased above p_o, the flow is initially subsonic and $p_t' = p_o$. The mass flow rate increases in accordance with equation 4.76. At some value of P the throat pressure is such that $p_t'/P = p^*/P$. Thereafter, increasing P has no influence on that ratio; the latter remains constant for all values of P larger than

$$P = \left(\frac{\gamma + 1}{2}\right)^{\gamma/(\gamma - 1)} p_o \qquad (4.81)$$

For air, with $\gamma = 1.4$, the *critical pressure ratio* is 0.52828, so that the throat pressure is given by 0.52828 P.

The speed of the gas flowing in the throat section of a choked nozzle is given by equation 4.70; thus,

$$V_t' = a^* = \left(\frac{2\gamma RT}{\gamma + 1}\right)^{1/2} = a_o\left(\frac{2}{\gamma + 1}\right)^{1/2} \qquad (4.82)$$

It is seen from equation 4.82 that for a given gas, V_t' depends only on T; it is independent of P. Consequently, after the critical condition is attained in the throat, increasing P has no effect on $V_t' = a^*$. Since the static pressure in the throat section of a choked nozzle is given by equation 4.69, then

$$p_t' = p^* = \left(\frac{2}{\gamma + 1}\right)^{\gamma/(\gamma - 1)} P \qquad (4.83)$$

Equation 4.83 shows that, for a given gas, $p_t' = p^*$ increases linearly with P. Furthermore, if T is held constant, the density of the gas in the throat is $\rho_t' = \rho^* = p^*/Rt^*$, which increases linearly with P because $t^* = 2T/(\gamma + 1)$ remains constant. Because of the increased value of $\rho_t' = \rho^*$ with increasing P, the mass flow rate $\dot{m}^*$ likewise increases linearly with P.

When the critical conditions prevail in the nozzle, the volumetric flow rate is $\dot{Q}^*$, which is given by

$$\dot{Q}^* = \frac{\dot{m}^*}{\rho^*} = A_t a^* = A_t \left(\frac{2}{\gamma + 1}\right)^{1/2} (\gamma RT)^{1/2} \tag{4.84}$$

All of the terms except $\sqrt{T}$, for a given gas, remain constant, indicating that $\dot{Q}^*$ is independent of the initial pressure. If T is held constant, $\dot{Q}^*$ remains constant. Increasing P so that P/p_o exceeds the critical value, assuming that the back pressure p_o is constant, has no effect on $\dot{Q}^*$, but $\dot{m}^*$ increases linearly with P; the increase in $\dot{m}^*$ is due to the increased density ρ^* for the fluid crossing the throat.

4-6(e) Nondimensional Mass Flow Rate

For many engineering purposes, particularly for a gas turbine engine, it is common to express the mass rate of flow in a nondimensional form so that the mass flow rate may be referred to arbitrarily selected standard, or reference, conditions.

The continuity equation for a perfect gas in terms of the local Mach number and stagnation values of pressure and temperature is given by equation 4.76.

$$\frac{\dot{m}\sqrt{\gamma RT}}{PA_t} = \psi \tag{4.85}$$

where ψ is the flow factor defined by equation 4.75. Equation 4.85 is applicable to both converging and converging-diverging nozzles, and the parameter $\dot{m}\sqrt{\gamma RT}/PA_t$ is termed the *nondimensional mass flow rate*. In the case where the operation of the nozzle is supercritical (i.e., the flow is choked), $\dot{m} = \dot{m}^*$, $A_t = A^*$, and $M_t = 1$. Under choking conditions, equation 4.85 becomes

$$\frac{\dot{m}^*\sqrt{\gamma RT}}{PA_t} = \psi^* \tag{4.86}$$

where ψ^* is defined by equation 4.78.

Let T and P denote the actual stagnation temperature and pressure, respectively, at the entrance to the nozzle, T_{std} the standard (or reference) temperature, and P_{std} the standard (or reference) pressure. *By definition,*

$$\theta \equiv \frac{T}{T_{std}} \quad \text{and} \quad \delta \equiv \frac{P}{P_{std}} \tag{4.87}$$

In terms of the parameters θ and δ, the standard (or reference) mass flow rate is given by

$$\dot{m}_{std} = \frac{\dot{m}\sqrt{\theta}}{\delta} \tag{4.88}$$

The *standard* (or *reference*) *critical mass flow rate* is accordingly given by

$$\dot{m}_{std}^* = \frac{\dot{m}^*\sqrt{\theta}}{\delta} \tag{4.89}$$

Example 4.5.   Air enters a converging nozzle with a static pressure of $2.00 \cdot 10^5$ N/m², a static temperature of 300 K, and a Mach number 0.2. It discharges into the atmosphere where the pressure is $1.0135 \cdot 10^5$ N/m². The nozzle throat area A_t is 0.0040 m². Calculate (a) the mass rate of flow under the actual operating conditions, and (b) the standard mass rate of flow. Assume $T_{std} = 298.15$ K and $P_{std} = 1.0133 \cdot 10^5$ N/m². Assume that the flow is isentropic, and $\gamma = 1.40$.

Solution

(a) Determine the stagnation conditions and nozzle pressure ratio. From equations 4.24 and 4.25,

$$T = t_1 \left(1 + \frac{\gamma - 1}{2} M_1{}^2 \right) = [1 + 0.2(0.2)^2](300.0) = 302.4 \text{ K}$$

$$P = p_1 \left(\frac{T}{t_1} \right)^{\gamma/(\gamma - 1)} = (2.00)10^5 \left(\frac{302.4}{300} \right)^{3.5} = 2.0566 \cdot 10^5 \text{ N/m}^2$$

$$\frac{P_o}{P} = \frac{1.0133}{2.0566} = 0.4927 < 0.5283 \text{ (} \textit{flow is supercritical} \text{)}$$

From equation 4.80,

$$\frac{\dot{m}^* \sqrt{T}}{PA^*} = \frac{\dot{m}^* \sqrt{302.4}}{(2.0566 \cdot 10^5)(0.0040)} = 0.040416$$

Hence,

$$\dot{m}^* = \frac{(0.040416)(2.0566 \cdot 10^5)(0.0040)}{\sqrt{302.4}} = 1.912 \text{ kg/s (4.215 lbm/sec)}$$

(b) From equations 4.87

$$\sqrt{\theta} = \left(\frac{T}{T_{std}} \right)^{1/2} = \left(\frac{302.4}{298.15} \right)^{1/2} = 1.0071 \quad \text{and} \quad \delta = \frac{P}{P_{std}} = \frac{2.0566}{1.0133} = 2.0296$$

From equation 4.89,

$$\dot{m}^*_{std} = \frac{\dot{m}^* \sqrt{\theta}}{\delta} = \frac{(1.912)(1.0071)}{2.0296} = 0.9488 \text{ kg/s (2.0916 lbm/sec)}$$

4–6(f) Multidimensional Flow Effects

The preceding discussion of the converging nozzle, being based on the assumption of one-dimensional flow, assumes that the flow properties are uniform across every plane perpendicular to the mean flow direction. In an actual converging nozzle of cylindrical cross section, the assumption of one-dimensional flow may not be valid, especially if the wall of the nozzle makes a large angle with the axis (or meridional plane for a rectangular cross section) of the nozzle.

A practical example of such a nozzle is the converging propulsive nozzle of a subsonic turbojet engine [see Section 4–9(a)]. In the interest of reducing its length and, accordingly, its weight, the wall angle of the nozzle may be quite large. Consequently, the flow through the nozzle deviates from the assumed one-dimensional flow; that is, the surfaces of constant flow properties are not planes perpendicular to the nozzle axis.

Furthermore, the inward radial momentum of the fluid causes the formation of a *vena contracta* downstream from the exit plane of the nozzle; the vena contracta is often called the aerodynamic throat of the nozzle. Moreover, the area of the vena contracta may be considerably smaller than the nozzle exit area. The combination of the nonuniform flow and the vena contracta reduces the mass flow rate to a value smaller than that for a one-dimensional flow. The shape of the vena contracta, because it is located external to the nozzle, is influenced by the conditions in the surrounding atmosphere even at pressure ratios considerably larger than the one-dimensional choking pressure ratio P/p^*.

Figure 4.14, based on theoretical calculations by Thornock and Brown[5], presents y/y_t, the normalized radial coordinate, as a function of x/y_t, the normalized axial coordinate, with the Mach number as a parameter, for a converging conical nozzle having a cone semiangle $\alpha = 40$ deg. The Mach number curves are for the case where the nozzle operates with a pressure ratio $P/p_o = 4.0$. It is customary in the gas turbine industry to employ the pressure ratio P/p instead of the pressure ratio

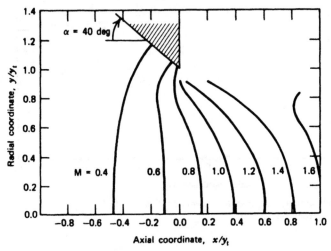

Figure 4.14. Lines of constant Mach number for a 40 degree converging nozzle at a pressure ratio of 4.0 (taken from Reference 5).

p/P. Thus, the *choking pressure ratio* for air, from equation 4.81, is 1.893. Figure 4.15 presents y/y_t as a function of x/y_t, and illustrates a plot of experimental values of the sonic lines for a conical nozzle, having a 25 degree semiangle, at various pressure ratios. Figure 4.15 indicates that the measured values of the location of the sonic line are not independent of the back pressure until a pressure ratio of 4.0 is attained. The deviation in the location of the sonic line from that corresponding to the one-dimensional location is large. The nozzle does not operate truly choked until the pressure ratio equals 4.0.

The effect of flow nonuniformity on the nozzle mass flow rate is accounted for by a discharge coefficient C_D, defined by

$$C_D \equiv \frac{\dot{m}_{actual}}{\dot{m}_{one\text{-}dimensional,\ isentropic}} \tag{4.90}$$

Figure 4.16 presents the measured values of the discharge coefficient C_D, defined by equation 4.90, as a function of nozzle pressure ratio P/p_o for three conical converging nozzles, having cone semiangles of 15, 25, and 40 deg. In all cases, C_D increases with the pressure ratio up to the pressure ratio for a choked mass flow rate. Even with

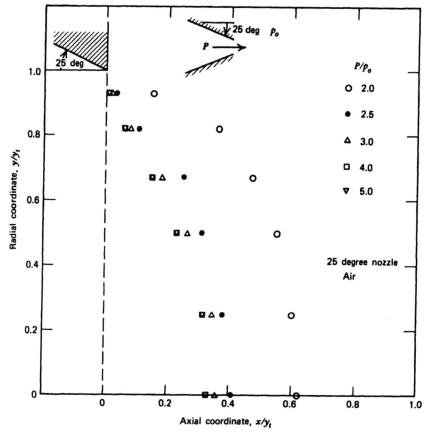

Figure 4.15. Experimental sonic lines at various pressure ratios for a 25 degree converging nozzle (taken from Reference 5).

choked flow, C_D may be significantly smaller than unity due to the adverse effects of multidimensional flow. An approximate locus of the choking condition is indicated in the figure by the broken line. It is seen from the experimental data plotted in Fig. 4.16, that the adverse effects of multidimensional flow in a conical converging nozzle are more pronounced for the nozzles having the larger cone angles, which is to be expected.

It is apparent from the foregoing that the assumption of one-dimensional flow for a conical converging nozzle having a large cone semiangle α may lead to serious errors. Nevertheless, the assumption of one-dimensional flow for such nozzles gives the correct trends qualitatively and, when combined with experimentally determined correction factors, such as the discharge coefficient C_D, makes it possible to obtain accurate quantitative results.

4-7 THE CONVERGING-DIVERGING OR DE LAVAL NOZZLE

It is shown in Section 4-6 that, if a nozzle ends at the throat (converging nozzle), the maximum isentropic speed attainable is $V_t' = a^*$. In the present section, it is shown that velocities exceeding a^* may be obtained in the divergence of a *converging-diverging* or *De Laval nozzle*.

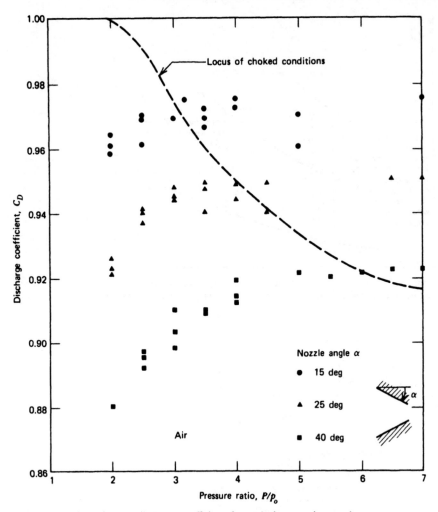

Figure 4.16.   Experimental discharge coefficients for conical converging nozzles (taken from Reference 5).

4–7(a)   Isentropic Discharge Properties

Consider the isentropic expansion illustrated schematically on the hs plane in Fig. 4.17. With a converging nozzle the maximum amount of enthalpy that can be transformed into jet kinetic energy is limited to the enthalpy change between points 1 and * illustrated in the figure.

It was shown by the Swedish engineer De Laval that the enthalpy change corresponding to an isentropic expansion from p^* to p_o could be converted into exit kinetic energy if a correctly designed diverging section is added to the converging nozzle, as illustrated in Fig. 4.18.

The phenomena that take place in a De Laval type of nozzle are explainable by the equations derived in Sections 4–3 and 4–4. The speed of the compressible fluid V'_e at the exit plane, called the *isentropic discharge speed*, is given by equation 4.3.

$$V'_e = [2(H - h'_e)]^{1/2} \tag{4.91}$$

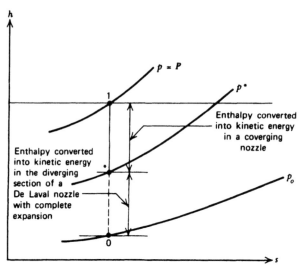

Figure 4.17. Enthalpy that is converted into kinetic energy by the isentropic expansion from the critical pressure p^* in the throat to the back pressure p_o.

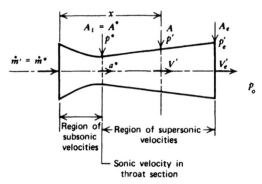

Figure 4.18. De Laval nozzle passing the maximum isentropic mass flow rate $\dot{m}^*$.

In the special case where the fluid is a perfect gas, the isentropic discharge speed is given by equation 4.28.

$$V_e' = \left\{ \frac{2\gamma RT}{\gamma - 1} \left[1 - \left(\frac{p_e'}{P} \right)^{(\gamma - 1)/\gamma} \right] \right\}^{1/2} \tag{4.92}$$

$$V_e' = a_o \left\{ \frac{2}{\gamma - 1} \left[1 - \left(\frac{p_e'}{P} \right)^{(\gamma - 1)/\gamma} \right] \right\}^{1/2} = V_{max}' \left[1 - \left(\frac{p_e'}{P} \right)^{(\gamma - 1)/\gamma} \right]^{1/2} \tag{4.93}$$

where a_o is the stagnation acoustic speed and V_{max}' is the maximum isentropic speed. The exit Mach number M_e' is

$$M_e' = \frac{V_e'}{a_e'} = \frac{a_o}{a_e'} \left\{ \frac{2}{\gamma - 1} \left[1 - \left(\frac{p_e'}{P} \right)^{(\gamma - 1)/\gamma} \right] \right\}^{1/2} \tag{4.94}$$

For the isentropic flow of a perfect gas,

$$\frac{a_o}{a_e'} = \left(\frac{T}{t_e'} \right)^{1/2} = \left(\frac{P}{p_e'} \right)^{(\gamma - 1)/2\gamma} \tag{4.95}$$

Substituting equation 4.95 into equation 4.94 yields

$$M'_e = \left\{ \frac{2}{\gamma - 1} \left[\left(\frac{P}{p'_e} \right)^{(\gamma - 1)/\gamma} - 1 \right] \right\}^{1/2} \tag{4.96}$$

As pointed out in Section 4-6(b), if the mass flow rate $\dot{m}'$ in a De Laval nozzle is choked, then the mass flow rate is the critical mass flow rate $\dot{m}^*$. From equation 4.1,

$$\dot{m}' = \dot{m}^* = \rho^* A_t a^* = \rho' A V' = \rho'_e A_e V'_e \tag{4.97}$$

Figure 4.19 presents the ratios V'_e/V'_{max}, ρ'_e/ρ_o, t'_e/T, and $(\dot{m}^*/A_e)/G^* = G_e/G^* = \rho'_e V'_e/\rho^* a^*$ as functions of the pressure ratio $r = p'_e/P$, for gases having a specific heat ratio $\gamma = 1.4$. For values of p'_e/P larger than p^*/P, the results are identical to those for a converging nozzle, except for $(\dot{m}^*/A_e)/G^*$, since $\dot{m}'$ does not equal $\dot{m}^*$ until $p'_t = p^*$.

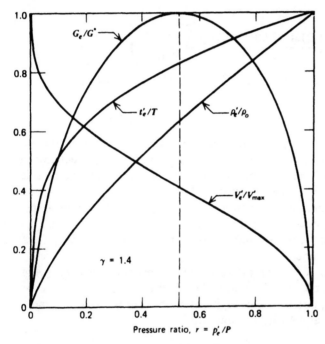

Figure 4.19. Ratios V'_e/V'_{max}, ρ'_e/ρ_o, t'_e/T, and G'_e/G^* as functions of the pressure ratio $r = p'_e/P$ for gases having $\gamma = 1.4$.

The continuity equation for a De Laval nozzle, equation 4.97, may be expressed in terms of the pressure ratio $r = p'/P$ For a perfect gas, combining equations 4.21 and 4.24 gives

$$\rho' = \rho_o \left(\frac{p'}{P} \right)^{1/\gamma} = \left(\frac{P}{RT} \right) \left(\frac{p'}{P} \right)^{1/\gamma} \tag{4.98}$$

Substituting equations 4.92 and 4.98 into equation 4.97 and simplifying yields

$$\dot{m}' = \frac{\psi P A}{\sqrt{\gamma R T}} \tag{4.99}$$

where

$$\psi = \left\{ \frac{2\gamma^2}{\gamma - 1} \left(\frac{p'}{P} \right)^{2/\gamma} \left[1 - \left(\frac{p'}{P} \right)^{(\gamma - 1)/\gamma} \right] \right\}^{1/2} \tag{4.100}$$

Applying equation 4.99 at the throat A_t gives

$$\dot{m}^* = \frac{\psi^* P A_t}{\sqrt{\gamma R T}} \qquad (4.101)$$

where ψ^* is defined by equation 4.77. Equations 4.99 to 4.101 are identical to equations 4.76, 4.75, and 4.79, respectively, which apply to the converging nozzle [see Section 4–6(b)].

Example 4.6. Air at a stagnation pressure of $10.0 \cdot 10^5$ N/m^2 and a stagnation temperature of 300 R is discharged through an isentropic De Laval nozzle. The air is expanded completely to the surrounding atmosphere that has a pressure of $1.0133 \cdot 10^5$ N/m^2. Assuming that $\gamma = 1.4$ and $R = 287.04$ J/kg-K for the air, calculate the isentropic discharge speed.

Solution

From equation 4.92

$$V'_e = \left\{ \frac{2(1.4)(287.04)(300)}{0.4} \left[1 - \left(\frac{1.0133}{10.0} \right)^{0.28571} \right] \right\}^{1/2}$$

$$V'_e = [602,784(0.48009)]^{1/2} = 537.95 \text{ m/s } (1764.9 \text{ ft/sec})$$

4–7(b) Area Ratio for Complete Expansion (Optimum Area Ratio)

For a De Laval nozzle, the pressure ratio $r = p'_e/P$ depends on the *nozzle area ratio* $\varepsilon = A_e/A_t$. Hence, for a given gas, the exit pressure p'_e is determined by the stagnation pressure P and the nozzle geometry. When the flowing fluid leaves the nozzle, its pressure must make an adjustment to the ambient pressure p_o, as discussed in Section 4–3(e). A nozzle for which $p'_e = p_o$, so that no adjustment to the ambient pressure is required, is said to operate with either *optimum*, or *complete expansion*. The corresponding area ratio ε is termed either the *area ratio for complete expansion* or the *optimum area ratio*.

Refer to Fig. 4.18. It is desirable to relate the *optimum area ratio* A/A^* to the pressure ratio p'/P and the Mach number M. When the throat of an isentropic converging-diverging nozzle is choked, it passes the critical mass flow rate $\dot{m}' = \dot{m}^*$. From equations 4.99 and 4.101, it follows that

$$\frac{A}{A^*} = \frac{\psi^*}{\psi} = \frac{\rho^* a^*}{\rho' V'}$$

Substituting for ψ and ψ^* from equations 4.100 and 4.77, respectively, into the above equation yields

$$\frac{A}{A_t} = \frac{A}{A^*} = \frac{\left(\dfrac{2}{\gamma + 1} \right)^{1/(\gamma - 1)} \left(\dfrac{\gamma - 1}{\gamma + 1} \right)^{1/2}}{\left(\dfrac{p'}{P} \right)^{1/\gamma} \left[1 - \left(\dfrac{p'}{P} \right)^{(\gamma - 1)/\gamma} \right]^{1/2}} \qquad (4.102)$$

Equation 4.102 specifies the area ratio $A/A_t = A/A^*$ for expanding the gas isentropically, from the stagnation conditions P and T to the static pressure p', when the nozzle mass flow rate is $\dot{m}' = \dot{m}^*$. Figure 4.20 presents the area ratio $A/A^* = A/A_t$ as a function of P/p' for different values of γ; the curves are for the condition $\dot{m}' = \dot{m}^*$.

For some purposes it is more convenient to express the area ratio at any section in terms of the local flow Mach number M. This may be done by applying the results

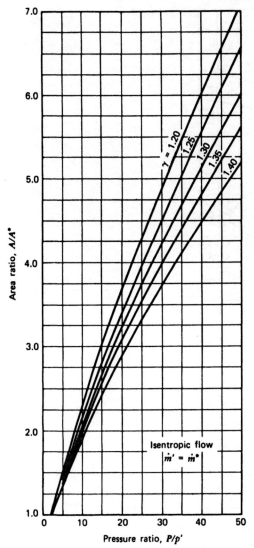

Figure 4.20. Nozzle area ratio for complete expansion for various values of the specific heat ratio.

of Section 4–4(a). Thus, equation 4.29 gives

$$\frac{A}{A_t} = \frac{A}{A^*} = \frac{1}{M}\left[\left(\frac{2}{\gamma+1}\right)\left(1 + \frac{\gamma-1}{2}M^2\right)\right]^{(\gamma+1)/2(\gamma-1)} \tag{4.29}$$

The velocity ratio V'/a^*, denoted by M^*, is obtained by dividing equation 4.92 by $a^* = [2\gamma RT/(\gamma+1)]^{1/2}$. The result is

$$M^* = \frac{V'}{a^*} = \left\{\frac{\gamma+1}{\gamma-1}\left[1 - \left(\frac{p'}{P}\right)^{(\gamma-1)/\gamma}\right]\right\}^{1/2} \tag{4.103}$$

Figure 4.21 presents V'/a^* as a function of P/p' for different values of γ.

When $\dot{m}' < \dot{m}^*$, the throat area $A_t > A^*$, the throat Mach number M_t is less than unity, and the flow is subsonic throughout the converging-diverging nozzle. The static pressure of the fluid decreases from the entrance to the throat, where it reaches its smallest value, and then increases in the diverging section.

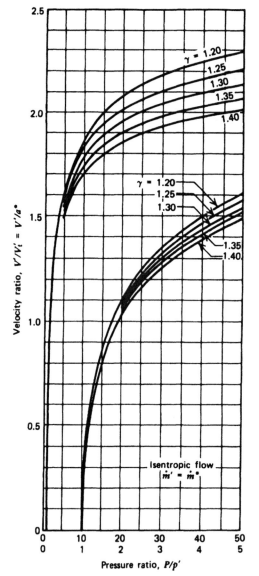

Figure 4.21. Ratio of exit velocity to throat velocity for different values of the specific heat ratio.

In the particular case where the area A coincides with the exit section of the nozzle ($A = A_e$), the ratio A_e/A^* is determined from equations 4.102 and 4.29 by setting $p' = p'_e$ and $M = M_e$. Similarily, the velocity ratio V'_e/a^* is obtained from equation 4.103 by setting $p' = p'_e$.

The maximum value for the isentropic discharge speed V'_e is obtained when $p'_e = p_o$, a condition that is obtained by designing the nozzle so that A_e/A^* gives $p'_e = p_o$. Such a nozzle, as stated earlier, has the *optimum area ratio* or the *area ratio for complete expansion*, which is obtained from equation 4.102 by setting $p' = p_o$. Figure 4.20 presents the area ratio A/A^* for a given value of P/p'; by setting $p' = p_o$, the optimum area ratio A_e/A^* may be read directly from the curve for the pertinent value of γ.

When the exit area A_e is selected so that $p'_e = p_o$, all of the available enthalpy of the flowing fluid is transformed into jet kinetic energy. It follows, therefore, that for the isentropic flow in a converging-diverging nozzle having the optimum area ratio, the exit velocity V'_e is given by

$$V'_e = [2(H - h'_e)]^{1/2} \quad (\textit{for optimum } A_e/A_t)$$

The condition when the back pressure p_o is reduced to zero is of academic interest. In that case, V'_e is the maximum isentropic speed which, from equation 3.188, Table 3.4, is given by

$$(V'_e)_{max} = V'_{max} = a^* \left(\frac{\gamma + 1}{\gamma - 1}\right)^{1/2} = a_o \left(\frac{2}{\gamma - 1}\right)^{1/2} \text{ (for } p_o = 0)$$

Example 4.7. A converging-diverging nozzle discharges air into a receiver where the static pressure is $1.034 \cdot 10^5$ N/m^2. The nozzle passes the maximum mass flow rate, and the static pressure and temperature in the constant-area duct to which the nozzle is attached are $p_1 = 6.895 \cdot 10^5$ N/m^2 and $t_1 = 444.4$ K; the duct area is 0.0929 m^2. The Mach number for the air flowing through the duct is $M_1 = 0.3$. Assuming isentropic flow and $\gamma = 1.4$, calculate (a) the stagnation properties of the fluid (T, P, ρ_o, and a_o), (b) the mass rate of flow through the nozzle, (c) the values of t, p, ρ, and V for the throat section, (d) the area of the throat section, and (e) the area of the exit section A_e that gives complete expansion of the air, and the corresponding values of p, ρ, V, M, and t. For air, $\gamma = 1.4$ and $R = 287.04$ J/kg-K.

Solution

(a) $\quad a_1 = (\gamma R t_1)^{1/2} = [1.4(287.04)(444.4)]^{1/2} = 422.6$ m/s

$\quad V_1 = M_1 a_1 = (0.3)(422.6) = 126.8$ m/s

$\quad \rho_1 = \dfrac{p_1}{R t_1} = \dfrac{(6.895)10^5}{287.04(444.4)} = 5.405$ kg/m^3

$\quad T = t_1 \left(1 + \dfrac{\gamma - 1}{2} M_1{}^2\right) = 444.4(1.018) = 452.4$ K (814.3 R)

$\quad P = p_1 \left(1 + \dfrac{\gamma - 1}{2} M_1{}^2\right)^{\gamma/(\gamma - 1)} = 6.895 \cdot 10^5(1.0644)$

$\quad\quad\quad = 7.339 \cdot 10^5$ N/m^2 (106.45 lbf/in.^{2})

$\quad \rho_o = \dfrac{P}{RT} = \dfrac{(7.339)10^5}{287.04(452.4)} = 5.652$ kg/m^3 (0.3529 lbm/ft^3)

$\quad a_o = (\gamma R T)^{1/2} = [1.4(287.04)(452.4)]^{1/2} = 426.4$ m/s (1398.9 ft/sec)

(b) By continuity,

$\quad \dot{m}' = \rho_1 A_1 V_1 = \dot{m}^* = 5.405(0.0929)(126.8) = 63.67$ kg/s (140.4 lbm/sec)

(c) Since $\dot{m}' = \dot{m}^*$, critical conditions prevail in the throat. From equation 4.34,

$$t'_t = t^* = \left(\frac{2}{\gamma + 1}\right) T = (0.83333)(452.4) = 377.0 \text{ K (678.6 R)}$$

From equation 4.35,

$$p'_t = p^* = \left(\frac{2}{\gamma + 1}\right)^{\gamma/(\gamma - 1)} P = (0.52828)(7.339 \cdot 10^5)$$

$$= 3.877 \cdot 10^5 \text{ N/m}^2 \text{ (56.23 lbf/in.}^2)$$

From equation 4.36,

$$\rho_t' = \rho^* = \left(\frac{2}{\gamma + 1}\right)^{1/(\gamma - 1)} \rho_o = (0.63394)(5.652) = 3.583 \text{ kg/m}^3 \ (0.2237 \text{ lbm/ft}^3)$$

From equation 4.37,

$$V_t' = a^* = \left(\frac{2}{\gamma + 1}\right)^{1/2} a_o = (0.91287)(426.4) = 389.2 \text{ m/s} \ (1277.0 \text{ ft/sec})$$

(d) From equation 4.79, since $\dot{m}' = \dot{m}^*$ and $A_t = A^*$,

$$A^* = \frac{\dot{m}^* a_o}{\psi^* P} = \frac{(63.67)(426.4)}{(0.81019)(7.339)10^5} = 0.04566 \text{ m}^2 \ (70.77 \text{ in.}^2)$$

(e) For complete expansion, $p_e' = p_o = 1.034 \cdot 10^5 \text{ N/m}^2$. Then,

$$\left(\frac{p_o}{P}\right)^{(\gamma - 1)/\gamma} = \left(\frac{1.034}{7.339}\right)^{0.2857} = 0.5713$$

From equation 4.103,

$$M_e^* = \frac{V_e'}{a^*} = \left\{\left(\frac{2.4}{0.4}\right)[1 - (0.5713)]\right\}^{1/2} = 1.6038$$

Hence,

$$V_e' = M_e^* a^* = 1.6038(389.2) = 624.2 \text{ m/s} \ (2047.9 \text{ ft/sec})$$

$$t_e' = t^* \left(\frac{p_e'}{p^*}\right)^{(\gamma - 1)/\gamma} = 377.0 \left(\frac{1.034}{3.877}\right)^{0.2857} = 258.4 \text{ K} \ (465.2 \text{ R})$$

$$a_e' = (\gamma R t_e')^{1/2} = [1.4(287.04)(258.4)]^{1/2} = 322.2 \text{ m/s} \ (1057.2 \text{ ft/sec})$$

$$M_e' = \frac{V_e'}{a_e'} = \frac{624.2}{322.2} = 1.937$$

$$\rho_e' = \frac{p_e'}{R t_e'} = \frac{(1.034)10^5}{287.04(258.4)} = 1.394 \text{ kg/m}^3 \ (0.08705 \text{ lbm/ft}^3)$$

From equation 4.29,

$$\frac{A_e}{A_t} = \frac{A_e}{A^*} = \frac{1}{1.937}\left\{\left(\frac{2}{2.4}\right)[1 + 0.2(1.937)^2]\right\}^{3.0} = 1.6023$$

Hence,

$$A_e = 1.6023(0.04566) = 0.07316 \text{ m}^2 \ (113.4 \text{ in.}^2)$$

4–7(c) Effect of Varying the Back Pressure on a Converging-Diverging Nozzle

Figure 4.22 presents the pressure ratio p_e'/P and the exit Mach number M_e' as functions of the area ratio A_e/A^* for a converging-diverging nozzle that passes the maximum isentropic mass flow rate $\dot{m}^*$. It is apparent from the figure that to each value of A_e/A^* there correspond two values of M_e', one being subsonic and the other supersonic. Which of the two values of M_e' occurs depends on the back pressure p_o [see Section 4–3(e)].

When the isentropic mass flow rate $\dot{m}'$ for a converging-diverging nozzle is such that $\dot{m}' < \dot{m}^*$, the flow is subsonic throughout the nozzle. Only a small range of exit pressures is attainable with a subsonic isentropic flow, and that range of exit

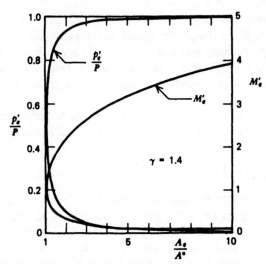

Figure 4.22. Pressure ratio p'_e/P and exit Mach number M'_e as functions of A_e/A^* for a converging-diverging nozzle passing the maximum isentropic mass flow rate $\dot{m}^*$.

pressures is indicated schematically in Fig. 4.23. The nozzle, in that case, operates in much the same manner as it would if the fluid flowing through it were incompressible, the divergent section of the nozzle acting as a *diffuser*, since $M < 1$ (see equation 4.19).

Consider now the case where the nozzle operates under steady isentropic flow conditions but passes the mass flow rate $\dot{m}^*$. Figure 4.23 shows that only two different values of exit pressure p'_e are possible if the flow is to be isentropic. The gas expands

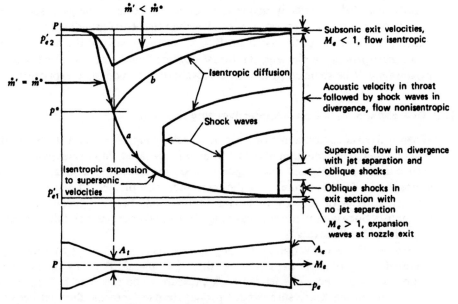

Figure 4.23. Pressure distribution in a converging-diverging nozzle under different operating conditions.

isentropically to p^* in the throat and may either continue its isentropic expansion along curve a to the exit pressure p'_{e1} or it may be diffused isentropically along curve b to the exit pressure p'_{e2}. If the gas expands along curve a, the Mach number of the gas crossing the exit area A_e is supersonic whereas, if the gas is compressed along curve b, then $M_e < 1$. None of the values of exit pressure between p'_{e1} and p'_{e2} (in Fig. 4.23) are attainable if the flow is isentropic.

A given nozzle has a fixed value for its area ratio A_e/A^*; the latter is given by equation 4.29, which is repeated here for convenience. Thus, for an isentropic flow,

$$\frac{A_e}{A^*} = \frac{1}{M_e} \left[\left(\frac{2}{\gamma + 1} \right) \left(1 + \frac{\gamma - 1}{2} M_e^2 \right) \right]^{(\gamma + 1)/2(\gamma - 1)} \tag{4.29}$$

If the nozzle is operated with the inlet stagnation pressure P held constant, the exit area A_e will give complete expansion of the gas only when $p'_e = p_o$; that is, for only one value of the back pressure p_o. As described in Section 4-3, when $p'_e > p_o$, the expansion of the fluid from p'_e to p_o takes place beyond the exit section of the nozzle in the free jet. A fixed-area nozzle operating with $p'_e > p_o$ is said to operate with *underexpansion*. Conversely, if the nozzle operates with $p'_e < p_o$, the nozzle is said to operate with *overexpansion*. The phenomena of overexpansion and under-expansion occur in propulsive nozzles when the back pressure is either higher or lower, respectively, than the value assumed in designing the nozzle.

4-7(d) Underexpansion in Converging-Diverging Nozzles

As stated earlier, when a nozzle passes the maximum mass flow rate $\dot{m}^*$ and expands the gas completely to the back pressure ($p'_e = p_o$), the velocities in the divergence are supersonic throughout. Consequently, reducing p_o (so that $p_o < p'_e$) has no effect on the flow conditions inside the nozzle, because a small pressure change is propagated only with the sonic velocity. Beyond the exit section of the nozzle, the supersonic jet expands in much the same manner as does a jet emerging from a converging nozzle operating with a supercritical pressure ratio (see Fig. 4.13). Because the gas is not expanded completely inside the nozzle, the kinetic energy associated with the jet crossing the exit section A_e is smaller than that corresponding to the complete isentropic expansion of the gas from P to p_o. Consequently, if the nozzle is utilized as a thrust-producing element, the thrust obtained when it is operated with *under-expansion* will be reduced from that corresponding to complete expansion.

4-7(e) Overexpansion in Converging-Diverging Nozzles

When a fixed-area converging-diverging nozzle is employed as a thrust-producing element, there is a reduction of thrust when it is operated with overexpansion. Under such conditions the gases may expand inside the nozzle to a pressure below the atmospheric back pressure, and they are then compressed to the back pressure through a series of shock waves. This problem has been discussed by several investigators, but a complete understanding of the flow process under different operating conditions is still lacking.[6, 7]

Assume that a converging-diverging nozzle having a fixed area ratio A_e/A_t is operated with complete expansion so that $p'_e = p_o$ and the flow in the diverging section is supersonic. Now let the back pressure p_o be raised so that $p_o > p'_e$. The increase in the back pressure cannot propagate itself into the fluid jet because the latter emerges with a supersonic speed. It can, however, propagate itself upstream through the boundary layer surrounding the jet since, within the boundary layer, the fluid velocities vary from a supersonic value at its interface with the main core of fluid to zero at the wall of the nozzle (see Section 5–10). Consequently, inside the boundary layer there are regions where the gas velocities are subsonic. A pressure wave can, therefore, start at the exit edge of the nozzle and be transmitted upstream through the regions of the boundary layer where the flow is subsonic. If p_o is only slightly larger than p'_e, oblique shock waves are formed at the corner of the exit section, and the pressure at the wall of the nozzle increases sharply from the exit pressure p'_e to p_o. The range of back pressures giving operation with oblique shocks formed in the exit section is indicated qualitatively in Fig. 4.23. If the back pressure p_o is raised further, the jet detaches itself from the divergent section in the manner illustrated schematically in Fig. 4.24. The overexpanded gas is compressed to the back pressure p_o by flowing through the oblique shock waves. The separation of the jet takes place in a more or less regular manner, as illustrated in the latter figure. The static pressure of the gas at the point where separation occurs is called the *separation pressure* and is denoted by p_s. Proper regulation of the separation of the jet is desirable in rocket motors, which must operate over an extremely large range of altitudes.

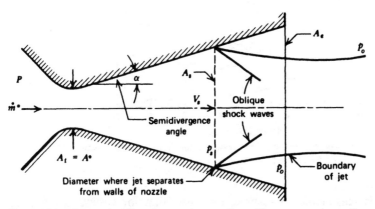

Figure 4.24. Converging-diverging nozzle operating with overexpansion and jet separation.

Figure 4.25 presents the results obtained by A. Stodola from experiments on the effect of back pressure on the static pressure along the axis of a converging-diverging nozzle.[8] Figure 4.25 indicates that when the back pressure is correct, curve M, the static pressure along the nozzle axis was the lowest obtained in the experiments, and curve M is the limiting condition for all of the measurements. As the back pressure was slightly increased, the pressure along the axis of the nozzle showed a sharp increase or shock wave close to the nozzle exit. Further increases in the back pressure intensified the shock wave and caused it to move further inside the nozzle; see curves L to E. Curves such as C, B, and A are typical for subsonic flow throughout the nozzle.

More recent experiments with overexpanded rocket motor nozzles indicate that the separation pressure p_s is approximately 0.40 p_o. As pointed out above, the flow separation phenomenon results from the action of the back pressure on the boundary layer, which flows in the direction of a positive pressure gradient. When momentum is removed from the boundary layer faster than it receives momentum from the main stream, the phenomenon of jet separation with its attendant shock waves

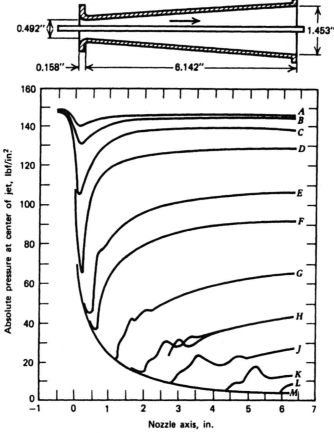

Figure 4.25. Effect of back pressure on the pressure along the axis of a De Laval nozzle (reproduced from Reference 8).

occurs. It is reasonable to assume, therefore, that for conical nozzles, the exact value of the separation pressure will be influenced by the character of the boundary layer and the semiangle α of the conical divergence. The reported experiments indicate that, for a given value of α, the *separation pressure ratio* p_s/p_o is a function of the nozzle pressure ratio P/p_o. Figure 4.26 presents a correlation of the results obtained by different experimenters with rocket motor nozzles having $\alpha = 15$ deg, by plotting $(p_o - p_s)/P$ as a function of P/p_o. The figure shows that the results obtained by the different experimenters are in substantial agreement.[7]

Figure 4.27 illustrates diagrammatically the flow phenomena occuring in a converging-diverging nozzle when it is operated under different conditions of back pressure p_o.

4–7(f) Multidimensional Flow Effects

The analysis of the flow field for a converging-diverging nozzle presented in the preceding discussion assumes that the flow is one-dimensional. It is shown in Section 4–6(f) that the latter assumption may lead to significant errors in the case of converging nozzles. For converging-diverging nozzles, however, the one-dimensional flow model is generally quite accurate. Deviations from one-dimensional flow do occur, however, and they may be significant in some applications, as discussed in Chapters 12, 15, 16, and 17.

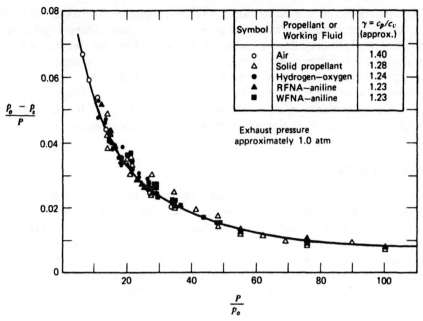

Figure 4.26. Correlation of data on flow separation in nozzles (taken from Reference 7).

p_e' denotes calculated exit pressure for isentropic flow

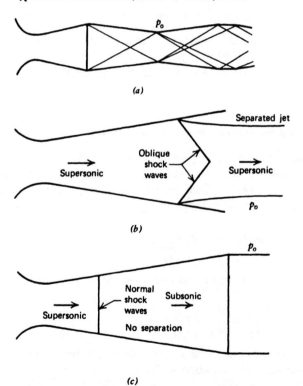

Figure 4.27. Effect of the back pressure p_o on the jet discharged from a De Laval nozzle. (a) p_o slightly greater than p_e'. (b) p_o sufficiently larger than p_e' to cause jet separation and oblique shock waves. (c) p_o sufficiently larger than p_e' to cause strong normal shock wave in the nozzle divergence.

In summary, the one-dimensional analysis generally predicts the flow field in a converging-diverging nozzle with good accuracy. Multidimensional flow effects typically produce deviations of only a few percent from the one-dimensional flow results; this is a degree of error that is acceptable in many engineering applications. For those engineering applications that require a higher degree of accuracy, the one-dimensional approach may be inadequate for the final design, but it is quite useful for the preliminary design phase and for studying the trends resulting from making changes in the design variables and operating conditions.

4–8 LOSSES IN NOZZLES[9, 10, 11]

The function of a nozzle in either a steam turbine or a gas turbine, or as the propulsive element of a jet propulsion engine, is to convert thermal energy into the maximum amount of jet kinetic energy corresponding to the nozzle inlet and exit operating conditions. Although it is reasonable to assume that the expansion of the fluid in a nozzle takes place adiabatically, the expansion process is never isentropic. Because of friction and turbulence, even a well-designed nozzle operates with energy losses. As a result, the jet kinetic energy developed by an actual nozzle is always smaller than that computed for an isentropic expansion between the same inlet pressure and back pressure that prevail in the operation of the actual nozzle.

The effect of friction on the operation of a nozzle may be taken into account by either of the following parameters.

1. The stagnation pressure ratio for the nozzle P_2/P_1.
2. The kinetic energy efficiency η_n, or velocity coefficient ϕ, of the nozzle.

In general, the conditions at the inlet to the nozzle are either known or readily calculable and, in most cases, the problem is to determine the throat area required for passing a specified mass flow rate of fluid, and the flow properties in the nozzle exit plane.

4–8(a) Stagnation Pressure Ratio for Nonisentropic Flow in Nozzles

Because of friction, the isentropic area ratio, given by equation 4.29, must be modified if the nozzle is to pass the desired mass flow rate. The modification is made in the following manner. Between any two stations in the flow passage, denoted by the subscripts 1 and 2, the continuity equation for perfect gases, equation 3.167 (Table 3.2) gives the following:

$$\frac{p_1 A_1 M_1}{\sqrt{t_1}} = \frac{p_2 A_2 M_2}{\sqrt{t_2}}$$

Hence, the actual area ratio for changing the flow from $M = M_1$ to $M = M_2$ is given by

$$\frac{A_2}{A_1} = \frac{M_1}{M_2} \frac{p_1}{p_2} \left(\frac{t_2}{t_1}\right)^{1/2} \tag{a}$$

Since the flow through the nozzle may be assumed to be adiabatic ($\delta Q = 0$, but $ds \neq 0$), $T_1 = T_2 = T$. Hence, from equation 4.25,

$$\left(\frac{t_2}{t_1}\right)^{1/2} = \left(\frac{t_2}{T}\right)^{1/2} \left(\frac{T}{t_1}\right)^{1/2} = \left(\frac{1 + \dfrac{\gamma - 1}{2} M_1^2}{1 + \dfrac{\gamma - 1}{2} M_2^2}\right)^{1/2} \tag{b}$$

The static pressure ratio p_1/p_2 may be written in the form

$$\frac{p_1}{p_2} = \frac{p_1}{P_1}\frac{P_2}{p_2}\frac{P_1}{P_2} = \left(\frac{1 + \dfrac{\gamma - 1}{2}M_2{}^2}{1 + \dfrac{\gamma - 1}{2}M_1{}^2}\right)^{\gamma/(\gamma - 1)}\left(\frac{P_1}{P_2}\right) \tag{c}$$

where p_1/P_1 and p_2/P_2 are obtained from equation 4.26. Substituting for $(t_2/t_1)^{1/2}$ and p_1/p_2 from equations (b) and (c) into equation (a) yields

$$\frac{A_2}{A_1} = \frac{M_1}{M_2}\left(\frac{1 + \dfrac{\gamma - 1}{2}M_2{}^2}{1 + \dfrac{\gamma - 1}{2}M_1{}^2}\right)^{(\gamma + 1)/2(\gamma - 1)}\left(\frac{P_1}{P_2}\right) \tag{d}$$

From equation 4.29,

$$\left(\frac{A'}{A^*}\right)_M = \frac{1}{M}\left[\left(\frac{2}{\gamma + 1}\right)\left(1 + \frac{\gamma - 1}{2}M^2\right)\right]^{(\gamma + 1)/2(\gamma - 1)} \tag{4.29}$$

where $(A'/A^*)_M$ denotes the area ratio for an isentropic flow ($s = $ constant) from the Mach number M to $M = 1$. Hence, equation (d) may be rewritten in the form

$$\frac{A_2}{A_1} = \frac{P_1}{P_2}\left[\frac{(A'/A^*)_{M_2}}{(A'/A^*)_{M_1}}\right] = \frac{P_1}{P_2}\left(\frac{A'_2}{A'_1}\right) \tag{4.104}$$

where $(A'/A^*)_{M_1}$ and $(A'/A^*)_{M_2}$ are the area ratios corresponding to isentropic flows from M_1 and M_2 to $M = 1$, respectively, and P_1/P_2 is the actual ratio of the stagnation pressures at A_1 and A_2, respectively.

Equation 4.104 shows that, because of the decrease in the stagnation pressure (i.e., $P_2 < P_1$), the area ratio of an actual nozzle must be larger than that for the corresponding isentropic nozzle.

The results derived above apply to any flow channel wherein the flow may be assumed adiabatic and one-dimensional.

In many cases, the flow may be considered isentropic up to the throat and non-isentropic thereafter. A case in point is when there is a shock wave system in the diverging portion of a De Laval nozzle. In that case, let $A_1 = A_t$; then $M_1 = M_t = 1$, and equation 4.104 becomes

$$\frac{A_2}{A_t} = \frac{P_1}{P_2}\left(\frac{A'}{A^*}\right)_{M_2} \tag{e}$$

To determine the effects of nonisentropicity in the flow for any actual nozzle, some experimental measurement of the flow properties in the exit plane is necessary. The easiest measurement to make is that of the static pressure p_2. Hence, it may be assumed that the ratio p_2/P_1 is generally known. The static pressure p_2 is related to the stagnation pressure P_2 by equation 4.26. Thus,

$$P_2 = p_2\left(1 + \frac{\gamma - 1}{2}M_2{}^2\right)^{\gamma/(\gamma - 1)} \tag{f}$$

Combining equations (e) and (f) yields

$$\left(\frac{A_2}{A_t}\right)\left(\frac{p_2}{P_1}\right) = \left(\frac{p_2}{P_2}\right)\left(\frac{A'}{A^*}\right)_{M_2} \tag{4.105}$$

When A_2/A_t and p_2/P_1 are known, the term $[(p_2/P_2)(A'/A^*)_{M_2}]$ may be determined from equation 4.105. Since (p_2/P_2) and $(A'/A^*)_{M_2}$ are both functions of M_2 alone,

M_2 may be determined. To simplify such calculations, the function $[(p/P)(A/A^*)]$ for isentropic flow is tabulated in Table C.6, as a function of M, for several values of the specific heat ratio γ.

Example 4.8. Air flows through a frictionless converging-diverging nozzle. The area of the exit section A_e is three times the area of the throat section A_t, and the ratio of the stagnation pressure at the entrance to the static pressure in the exit section $P_1/p_2 = 2.5$. Assume that air is a perfect gas with a constant specific heat ratio $\gamma = 1.4$, and that the only loss is that due to a normal shock wave formed in the diverging portion of the nozzle. Calculate (a) the Mach number at the nozzle exit, and (b) the entropy increase.

Solution

(a)   From equation 4.105,

$$\left(\frac{p_2}{P_2}\right)\left(\frac{A'}{A^*}\right)_{M_2} = \left(\frac{A_2}{A_t}\right)\left(\frac{p_2}{P_1}\right) = \frac{3.0}{2.5} = 1.20$$

From Table C.6, for $[(p/P)(A/A^*)] = 1.20$, $M_2 = 0.472$.
(b)   From equation 3.179, Table 3.2, since $T_1 = T_2$

$$\Delta s = -R \ln\left(\frac{P_2}{P_1}\right)$$

From Table C.6, for $M_2 = 0.472$, $p_2/P_2 = 0.85849$. Thus,

$$\frac{P_2}{P_1} = \frac{P_2}{p_2}\frac{p_2}{P_1} = \frac{1}{(0.85849)(2.5)} = 0.4659$$

Hence,

$$\Delta s = -287.04 \ln(0.4659) = 219.2 \text{ J/kg-K } (0.05236 \text{ Btu/lbm-R})$$

4–8(b) Velocity Coefficient and Kinetic Energy Efficiency

Let V_e denote the actual velocity of the gases crossing the exit area A_e of a converging-diverging nozzle, and V'_e the corresponding isentropic velocity. *By definition*, the *velocity coefficient* for the nozzle, denoted by ϕ, is given by

$$\phi \equiv \frac{V_e}{V'_e} = \left(\frac{H - h_e}{H - h'_e}\right)^{1/2} \tag{4.106}$$

The *kinetic energy efficiency* for the nozzle, denoted by η_n, is defined by

$$\eta_n \equiv \frac{V_e^2/2}{(V'_e)^2/2} = \frac{H - h_e}{H - h'_e} = \phi^2 \tag{4.107}$$

The characteristics of a nonisentropic nozzle flow process are developed in the following discussion in terms of the kinetic energy efficiency η_n.

Figure 4.28 illustrates schematically the thermodynamic model for the non-isentropic expansion process on the Mollier diagram (*hs* plane). A nonisentropic and a reference isentropic process are plotted in the figure. The expansions occur between the stagnation pressure P and the static pressure p_e. For an adiabatic process, which is assumed here, the stagnation enthalpy H remains constant. The energy equations for the adiabatic and the isentropic processes are

$$H = h'_e + \frac{(V'_e)^2}{2} = h_e + \frac{V_e^2}{2} = \text{constant} \tag{a}$$

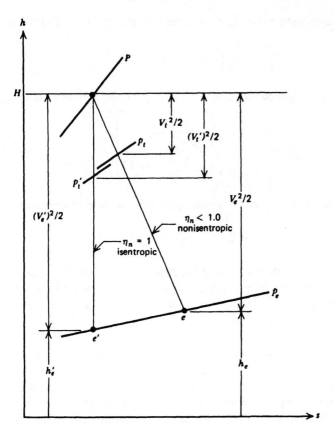

Figure 4.28. Mollier diagram for a nonisentropic expansion process.

The nonisentropic flow velocity V_e is related to the isentropic flow velocity V'_e by the kinetic energy efficiency η_n (defined by equation 4.107). The effects of the nonisentropicity on the flow are accounted for by η_n. It reduces the kinetic energy attainable by a compressible fluid that is expanded between given pressure levels. In practice, the value for η_n is usually estimated from experience with similar flow passages; its value usually ranges from 0.930 to 0.996. The kinetic energy efficiency may be related to the pressure ratio $r = p_e/P$ and the Mach number M_e for a nonisentropic flow. Thus, from equation 4.107,

$$\eta_n = \frac{V_e^2/2}{(H - h'_e)} = \frac{V_e^2/2}{c_p T(1 - t'_e/T)} = \frac{\left(\dfrac{\gamma - 1}{2}\right) V_e^2}{\gamma R T(1 - t'_e/T)} \tag{b}$$

where, for a perfect gas, $h'_e = c_p t'_e$, $H = c_p T$, and $c_p = \gamma R/(\gamma - 1)$. From equation 3.186, we obtain

$$\gamma R T = a_o^2 = a_e^2 \left(1 + \frac{\gamma - 1}{2} M_e^2\right) \tag{c}$$

Substituting equation (c) into equation (b) yields

$$\eta_n = \frac{\left(\dfrac{\gamma - 1}{2}\right) M_e^2}{\left(1 + \dfrac{\gamma - 1}{2} M_e^2\right)(1 - t'_e/T)} \tag{d}$$

Expressing the temperature ratio t'_e/T in terms of the pressure ratio $r = p_e/P$, we obtain

$$\frac{t'_e}{T} = \left(\frac{p_e}{P}\right)^{(\gamma-1)/\gamma} = r^{(\gamma-1)/\gamma} \tag{e}$$

Substituting equation (e) into equation (d) and solving for the pressure ratio r yields

$$r = \frac{p_e}{P} = \left[1 - \frac{\left(\frac{\gamma-1}{2}\right)M_e^2}{\eta_n\left(1 + \frac{\gamma-1}{2}M_e^2\right)}\right]^{\gamma/(\gamma-1)} \tag{4.108}$$

Equation 4.25, when solved for the nonisentropic temperature ratio, yields

$$\frac{T}{t_e} = 1 + \frac{\gamma-1}{2}M_e^2 \tag{4.109}$$

The Mach number at the throat M_t and the mass flow rate for the nozzle $\dot{m}$ may be determined from the continuity equation (note that the symbol t denotes the *minimum area*, which is not the sonic condition for a nonisentropic flow). From equations 4.1, 4.21, and 4.25, we obtain

$$\dot{m} = \rho AV = \frac{pAV}{Rt} = PAM\left(\frac{\gamma}{RT}\right)^{1/2}\left(1 + \frac{\gamma-1}{2}M^2\right)^{1/2}\left(\frac{p}{P}\right) \tag{f}$$

The flow in the nozzle is assumed to be choked. Hence, the throat conditions are determined by differentiating $\dot{m}/A$ with respect to M and setting the result equal to zero. We obtain the following expression for M_t, the Mach number in the minimum area.

$$M_t^4(\eta_n - 1) + M_t^2\left[3\eta_n - \left(\frac{3\gamma-1}{\gamma-1}\right)\right](\gamma-1)^{-1} + 2\eta_n(\gamma-1)^{-2} = 0 \tag{4.110}$$

Equation 4.110 is a simple quadratic equation for M_t^2, which may be solved numerically. Figure 4.29 presents curves of M_t as a function of η_n, for several constant values of γ. The curves show that M_t decreases as η_n decreases and is always less than unity for $\eta_n < 1$. M_t decreases with increasing values of γ.

The effect of the nonisentropicity in the flow process on the mass flow rate may be expressed in terms of the *mass flow efficiency factor* C_η, which is *defined* by

$$C_\eta \equiv \frac{\dot{m}_{(\eta_n < 1)}}{\dot{m}_{(\eta_n = 1)}} \tag{g}$$

The isentropic mass flow rate $\dot{m}_{(\eta_n = 1)} = \dot{m}^*$ is given by equation 4.38.

$$\dot{m}_{(\eta_n = 1)} = \dot{m}^* = \frac{\Gamma A_t P}{(\gamma RT)^{1/2}} \tag{4.38}$$

At the minimum area A_t, the mass flow rate $\dot{m}_{(\eta_n)}$ may be determined from equation (f). Thus,

$$\dot{m}_{(\eta_n)} = \frac{\gamma PA_tM_tr_t}{(\gamma RT)^{1/2}}\left(1 + \frac{\gamma-1}{2}M_t^2\right) \tag{h}$$

Substituting equations 4.38 and (h) into equation (g) gives

$$C_\eta = \gamma\Gamma^{-1}M_tr_t\left(1 + \frac{\gamma-1}{2}M_t^2\right)^{1/2} \tag{4.111}$$

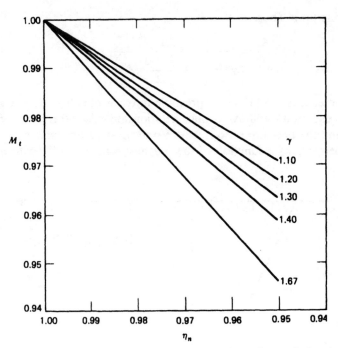

Figure 4.29. Mach number at the nozzle throat in nonisentropic flow (taken from Reference 9).

All of the terms in equation 4.111 are functions of M_t, which may be calculated from equation 4.110. Hence, C_η is a function only of γ and η_n. Figure 4.30 presents C_η as a function of η_n for several values of γ. The curves show that C_η decreases with decreasing η_n and increasing γ, and is equal to unity only when $\eta_n = 1$.

The *geometrical area ratio* $\varepsilon = A_e/A_t$ for a De Laval nozzle may be related to the flow properties by applying the continuity equation [equation (f)] at the areas A_t

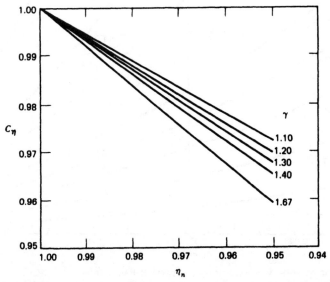

Figure 4.30. Mass flow efficiency factor c_η for nonisentropic flow (taken from Reference 9).

and A_e, and solving for ε. Thus,

$$\varepsilon = \frac{A_e}{A_t} = \frac{C_\eta \Gamma}{\gamma M r \left(1 + \frac{\gamma - 1}{2} M^2\right)^{1/2}} \tag{4.112}$$

All of the terms in equation 4.112 are expressible in terms of either M or r, and M is related to r by equation 4.108. An explicit relationship exists, therefore, between the area ratio ε and the pressure ratio r, with γ and η_n as parameters. Figure 4.31 presents $\Delta\varepsilon$, in percent, as a function of the *pressure ratio* $r = p/P$, with η_n as a

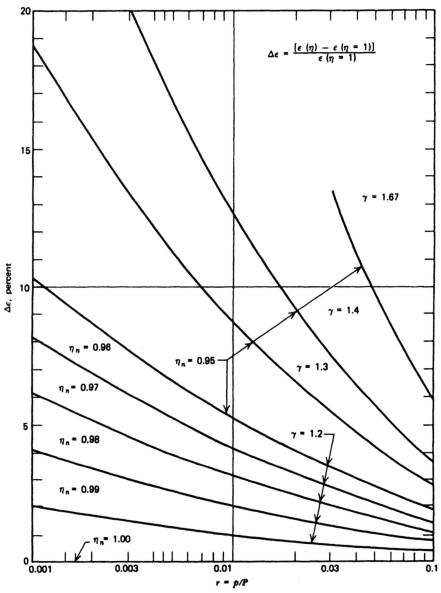

Figure 4.31. Area ratio increase for nonisentropic expansion (taken from Reference 9).

parameter, for different values of γ. The ordinate $\Delta\varepsilon$ is the required percentage increase in the area ratio from the corresponding isentropic area ratio ε, for expanding a nonisentropic flow to a given pressure ratio. The curves show that $\Delta\varepsilon$ increases with increasing γ and decreasing η_n, and is most significant at the small pressure ratios.

Example 4.9. Air flows in a converging-diverging nozzle having a kinetic energy efficiency $\eta_n = 0.95$. At the nozzle inlet, the stagnation pressure $P = 7.0 \cdot 10^5$ N/m^2 and the stagnation temperature $T = 500$ K. Calculate (a) the throat Mach number, (b) the throat pressure, (c) the throat temperature, (d) the throat density, (e) the throat velocity, (f) the nozzle mass flow rate per unit area, and (g) the corresponding isentropic flow values for comparison.

Solution

(a) From equation 4.110,

$$M_t^4(0.95 - 1) + M_t^2\left\{3(0.95) - \frac{[3(1.4) - 1]}{(1.4 - 1)}\right\}(1.4 - 1)^{-1} + 2(0.95)(1.4 - 1)^{-2} = 0$$

Solving the above expression for M_t yields $M_t = 0.95867$.

(b) From equation 4.108,

$$\frac{p_t}{P} = \left\{1 - \frac{(0.2)(0.95867)^2}{(0.95)[1 + 0.2(0.95867)^2]}\right\}^{3.5} = 0.5355$$

$$p_t = (0.5355)(7.0)10^5 = 3.749 \cdot 10^5 \text{ N/m}^2 \text{ (54.37 lbf/in.}^2)$$

(c) From equation 4.109,

$$\frac{t_t}{T} = \frac{1}{[1.0 + (0.2)(0.95867)^2]} = 0.8447$$

$$t_t = (0.8447)(500) = 422.4 \text{ K (760.2 R)}$$

(d) $\rho_t = \dfrac{p_t}{Rt_t} = \dfrac{(3.749)10^5}{(287.04)(422.4)} = 3.092$ kg/m^3 (0.1931 lbm/ft^3)

(e) $a_t = (\gamma Rt_t)^{1/2} = [1.4(287.04)(422.4)]^{1/2} = 412.0$ m/s

$V_t = M_t a_t = (0.9586)(412.0) = 394.9$ m/s (1295.7 ft/sec)

(f) $\dfrac{\dot{m}}{A_t} = \rho_t V_t = (3.092)(394.9) = 1221$ kg/s-m^2 (250.2 lbm/sec-ft^2)

(g) For isentropic flow, $M_t = 1.0$. From Table C.6, $t^*/T = 0.83333$ and $p^*/P = 0.52828$. Thus,

$$p^* = (0.52828)(7.0)10^5 = 3.698 \cdot 10^5 \text{ N/m}^2 \text{ (53.63 lbf/in.}^2)$$

$$t^* = (0.83333)(500) = 416.7 \text{ K (750.0 R)}$$

$$\rho^* = \frac{p^*}{Rt^*} = \frac{(3.698)10^5}{(287.04)(416.7)} = 3.092 \text{ kg/m}^3 \text{ (0.1931 lbm/ft}^3)$$

$$V_t = a^* = (\gamma Rt^*)^{1/2} = [1.4(287.04)(416.7)]^{1/2} = 409.2 \text{ m/s (1342.5 ft/sec)}$$

$$\frac{\dot{m}^*}{A_t} = \rho^* a^* = (3.092)(409.2) = 1265 \text{ kg/s-m}^2 \text{ (259.3 lbm/sec-ft}^2)$$

The following table presents a comparison of the flow properties for nonisentropic flow with those for the corresponding isentropic flow.

Property	Isentropic	Nonisentropic	Change, Percent
M_t	1.0000	0.9586	-4.32
p_t, 10^5 N/m^2	3.698	3.749	1.36
t_t, K	416.7	422.4	1.35
ρ_t, kg/m^3	3.092	3.092	0.00
V_t, m/s	409.2	394.9	-3.62
$\dfrac{\dot{m}}{A_t}$, kg/s-m^2	1265	1221	-3.60

As a check, from Fig. 4.30, $C_\eta = 0.965$. Thus,

$$\frac{\dot{m}}{A_t} = C_\eta \left(\frac{\dot{m}^*}{A_t}\right)_{\text{isen}} = (0.965)(1265) = 1221 \text{ kg/s-m}^2$$

4–8(c) Nozzle Discharge Coefficient

The nozzle discharge coefficient C_D is *defined* by equation 4.90, repeated below.

$$\dot{m} \equiv C_D \dot{m}'_{1-D} \tag{4.90}$$

The following are the most important factors that reduce the actual mass flow rate below the isentropic value.

1. Nonisentropic flow effects.
2. Three-dimensional flow effects.
3. The blockage of the flow area caused by the boundary layer displacement thickness (see Section 5–10).

Item 1 is accounted for by C_η, the mass flow efficiency factor defined in Section 4–8(b). Item 2 is accounted for by a *geometrical contraction factor* C_c, defined by

$$\dot{m}'_{3-D} = C_c \dot{m}'_{1-D} \tag{4.139}$$

The factor C_c is determined by calculating the isentropic three-dimensional mass flow rate through the nozzle throat A_t by an appropriate method (see Section 15–5), and then dividing that result by the corresponding isentropic one-dimensional mass flow rate. Consequently, C_c is a measure solely of the deviations from a one-dimensional flow.

By definition, the boundary layer displacement thickness δ^* is the distance normal to the nozzle wall through which, in an inviscid calculation, there is effectively no mass flow rate of fluid because of the slow moving viscous fluid in the boundary layer adjacent to the wall (see Section 5–10). The net effect of the viscous boundary layer is to "block off" an annulus of the geometrical flow area of the throat. The amount of the blockage may be determined by analyzing the boundary layer. The "blockage effect" may be expressed by

$$C_{b\ell} = \frac{A_t - (A \text{ blocked by } \delta^*)}{A_t}$$

Multiplying the three aforementioned factors together yields the nozzle *discharge coefficient* C_D. Thus,

$$C_D = C_\eta C_c C_{bl}$$

In general, the magnitude of the discharge coefficient C_D is determined experimentally. The procedure discussed above, however, permits a detailed study of the role of the individual effects for an actual flow process.

4–9 SOURCE AND SINK FLOWS

The concept of one-dimensional flow is a useful approximation to many real flows. All real flows, however, generally violate the one-dimensional flow assumption unless the streamlines are straight and parallel. In the special case where the streamlines are straight but either diverge from or converge toward a unique source or sink point, the flow may be considered as behaving like a one-dimensional flow.

Figure 4.32 illustrates schematically the source and sink flow models in a converging-diverging nozzle. Point A denotes the *source* (point) from which all of the streamlines in the diverging portion of the nozzle are assumed to originate, and point B denotes the *sink* (point) into which all of the streamlines in the converging portion of the nozzle are assumed to terminate. The flow properties are assumed to be constant on each spherical surface having either a source or sink point as its center of curvature. Accordingly, the one-dimensional approximation may be applied to the flow surfaces of the source and sink.

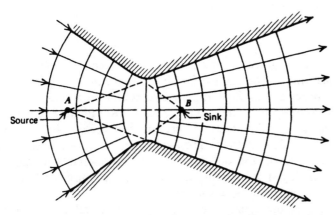

Figure 4.32. Source and sink flows in a converging-diverging conical nozzle.

The flow model described above has the obvious advantage that the streamlines are parallel to the solid boundaries; that is, at the passage walls. However, the model is inadequate in the throat region, and has no physical meaning as either the source or the sink point is approached. Consequently, the source or sink flow model must be restricted to regions of a flow passage where the streamlines are fairly straight and either diverge, or converge, uniformly toward a point. Those conditions are met to some extent in the convergence and divergence of a conical De Laval nozzle. When either the inlet contraction ratio, or the exit expansion ratio, is large, the streamlines have sufficient time to adjust themselves to the geometry of the sink and source flow. When the area ratios are small, however, the source and sink flow models may be no better, and possibly even less accurate, than the planar one-dimensional flow model.

An extension of the source and sink flow concept discussed above is the concept of a distributed source or sink flow; that is, a flow in which the source and sink points move in such a manner that the velocity remains tangent to the wall, as the wall angle varies. Figure 4.33 illustrates the concept of a distributed sink model applied to a converging nozzle; the model reduces to the planar one-dimensional model when the wall has zero slope. The outer streamline remains parallel to the wall at all points. In many flow passages, especially when estimates of the multidimensional flow effects are required, the concept of the distributed source and sink flow is a useful one. Care must always be exercised, however, when interpreting the results obtained for a quasi-multidimensional flow solution to a truly multidimensional flow field.

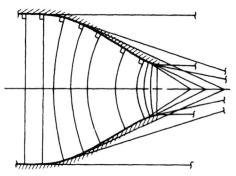

Figure 4.33.   Distributed sink flow in a converging nozzle.

The mass flow rate in a source or sink flow must be specified before the one-dimensional flow equations can be applied for obtaining the flow properties on the flow surfaces for the source or sink flow. If the nozzle is choked, it may be assumed that it passes the critical mass flow rate; the latter mass flow rate may be calculated by either the one-dimensional flow methods presented in Section 4–7, or by the multidimensional transonic flow methods discussed in Section 15–5. Once the mass flow rate has been established, the one-dimensional flow methods developed in this chapter may be employed for relating the areas of the source and sink flow surfaces to the flow properties on those surfaces.

The spherical flow area A_s of a source or sink flow will now be related to the planar flow area A_p and the wall angle α at each point in the flow passage. Figure 4.34 presents the pertinent geometric parameters; α denotes the angle of the wall, R the distance to the source, A_p the planar flow area, and A_s the corresponding spherical flow area.

The area of the spherical surface A_s is given by

$$A_s = \int dA_s = \int_0^\alpha 2\pi y R \, d\theta = 2\pi R^2 \int_0^\alpha \sin \theta \, d\theta \tag{a}$$

where $y = R \sin \theta$. Integrating the above equation gives

$$A_s = -2\pi R^2 \cos \theta \Big|_0^\alpha = 2\pi R^2 (1 - \cos \alpha) \tag{b}$$

The area A_p of the planar surface is

$$A_p = \pi y_w^2 = \pi R^2 \sin^2 \alpha \tag{c}$$

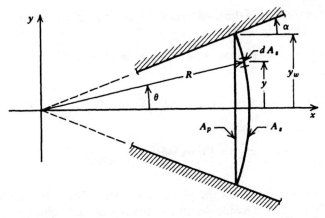

Figure 4.34. Relationship between planar flow area A_p, wall angle α, and spherical flow area A_s.

The area ratio A_p/A_s is

$$\frac{A_p}{A_s} = \frac{\pi R^2 \sin^2 \alpha}{2\pi R^2 (1 - \cos \alpha)} = \frac{1 - \cos^2 \alpha}{2(1 - \cos \alpha)} = \frac{1 + \cos \alpha}{2} \qquad (4.113)$$

Equation 4.113 relates A_p, α, and A_s. If the wall shape $y = y(x)$ is specified, the spherical flow area A_s may be determined from equation 4.113.

Example 4.10. A rocket engine is equipped with a conical propulsive nozzle that has a 15 deg half-angle and a throat radius $y_t = 0.025$ m. The throat has a circular arc radius of curvature $\rho_t = 0.0125$ m. The nozzle length $x_e = 0.25$ m. Figure 4.35 illustrates the geometry. The working fluid is a perfect gas for which $\gamma = 1.20$ and $R = 320.0$ J/kg-K. The stagnation temperature $T = 3000$ K and the stagnation pressure $P = 70.0 \cdot 10^5$ N/m². Calculate the flow properties at the exit section of the nozzle, based on (a) planar one-dimensional flow, and (b) source flow.

Solution

Assume that in both cases the nozzle passes the critical mass flow rate $\dot{m}^*$ corresponding to the planar throat area A_t. Hence, in both cases, $A^* = A_t = \pi y_t^2$. The

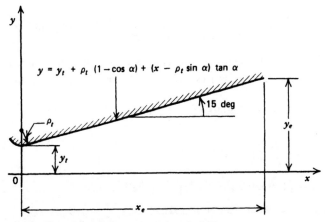

Figure 4.35. Nozzle contour for Example 4.10.

equation specifying the contour of the conical portion of the nozzle wall as a function of the nozzle length x (see Fig. 4.35) is

$$y = y_t + \rho_t(1 - \cos \alpha) + (x - \rho_t \sin \alpha) \tan \alpha \tag{a}$$

Thus, when $x_e = 0.25$ m,

$$y_e = 0.025 + (0.0125)[1 - \cos(15)] + [0.25 - (0.0125)\sin(15)]\tan(15) = 0.091546 \text{ m}$$

(a) The planar area ratio ε_p is given by

$$\varepsilon_p = \frac{A_p}{A_t} = \frac{\pi y_e^2}{\pi y_t^2} = \frac{(0.091546)^2}{(0.025)^2} = 13.40907$$

The exit Mach number M_{ep} based on the planar area ratio ε_p may be determined from tables such as Table C.6, if available. In the absence of such tables, M_{ep} may be determined by solving equation 4.29 by an iterative method, such as the *Newton-Raphson method* [see Appendix A-4(a) and Example A.3]. The result for $\gamma = 1.20$ and $\varepsilon_p = 13.40907$ is $M_{ep} = 3.48216$. From equation 4.25,

$$\frac{T}{t_{ep}} = 1 + \frac{\gamma - 1}{2} M_{ep}^2 = 1 + (0.1)(3.48216)^2 = 2.21254$$

$$t_{ep} = \frac{3000.0}{2.21254} = 1355.91 \text{ K}$$

$$a_{ep} = (\gamma R t_{ep})^{1/2} = [1.2(320.0)(1355.91)]^{1/2} = 721.57 \text{ m/s}$$

$$V_{ep} = M_{ep} a_{ep} = (3.48216)(721.57) = 2512.62 \text{ m/s}$$

From equation 4.26,

$$\frac{P}{p_{ep}} = \left(1 + \frac{\gamma - 1}{2}M_{ep}^2\right)^{\gamma/(\gamma - 1)} = \left(\frac{T}{t_{ep}}\right)^{\gamma/(\gamma - 1)} = (2.21254)^{6.0} = 117.313$$

$$p_{ep} = \frac{70.0 \cdot 10^5}{117.313} = 59,669 \text{ N/m}^2$$

$$\rho_{ep} = \frac{p_{ep}}{R t_{ep}} = \frac{59,669}{320.0(1355.91)} = 0.137521 \text{ kg/m}^3$$

(b) The spherical area ratio ε_s is given by equation 4.113.

$$\varepsilon_s = \frac{A_s}{A_t} = \frac{A_s}{A_p}\frac{A_p}{A_t} = \frac{2\varepsilon_p}{(1 + \cos \alpha)} = \frac{2(13.40907)}{[1 + \cos(15)]} = 13.64148$$

M_{es} may also be determined by solving equation 4.29 by the Newton-Raphson method. For $\gamma = 1.20$ and $\varepsilon_s = 13.64148$, $M_{es} = 3.49406$. From equation 4.25,

$$\frac{T}{t_{es}} = 1 + 0.1(3.49406)^2 = 2.22085$$

$$t_{es} = \frac{3000.0}{2.22085} = 1350.83 \text{ K}$$

$$a_{es} = [1.2(320.0)(1350.83)]^{1/2} = 720.22 \text{ m/s}$$

$$V_{es} = 3.49406(720.22) = 2516.49 \text{ m/s}$$

$$\frac{P}{p_{es}} = (2.22085)^{6.0} = 119.982$$

$$p_{es} = \frac{70.0 \cdot 10^5}{119.982} = 58,342 \text{ N/m}^2$$

$$\rho_{es} = \frac{58,342}{320.0(1350.83)} = 0.134968 \text{ kg/m}^3$$

4–10 JET PROPULSION ENGINES[12–14]

Experience has shown that all methods for propelling a body in a fluid medium are based on the *reaction principle* first formulated by Sir Isaac Newton (1642–1727). The reaction principle (see Section 1–13) states that to *every action there is an equal and opposite reaction*. Examples of the reaction principle are quite familiar.[12] Thus, the screw propelling a ship, the propeller for moving an airplane, and jet propulsion are applications of the same basic principle. In these examples, the application of the reaction principle involves increasing the *momentum* of a mass of fluid in such a manner that the reaction to the *time rate of increase in the momentum of the fluid produces a force*, called the *thrust*, acting in the direction of motion for the body [see Section 2–6(b)]. The thrust is the result of increasing the fluid momentum in the direction opposite to that desired for the body.

The available means for propelling a body in a fluid medium differ only in the methods and mechanisms utilized for increasing the flow of momentum (*momentum flux*) associated with the working fluid. For example, the conventional airplane propeller increases the momentum flux of the atmospheric air passing through the propeller disk circle, and the thrust it develops is equal to the increase in the momentum flux of the air moved by the propeller.

Jet propulsion differs from propeller propulsion in that it achieves the increase in the momentum flux of the working fluid by ejecting it as a high-speed jet from within the propelled body; in all of the other methods the working fluid passes around the propelled body. There is no restriction, at least in the abstract, on the character of the fluid jet ejected from within the jet-propelled body. It can be water, steam, heated air, the gases produced by chemical reactions, charged particles, and their combinations. The selection of the most appropriate fluid and the means for creating the high-speed jet is dictated by the requirements of the propulsion problem.

Although the jet propulsion principle may be applied with any type of fluid jet, there are limitations imposed on the choice of a suitable fluid when it is to be applied to the propulsion of bodies in the atmosphere. Experience indicates that two types of fluids are particularly suitable for that purpose.

1. There is the jet consisting of heated compressed atmospheric air, admixed with the products of the combustion produced by burning a chemical fuel in that air, the thermochemical energy of the fuel being utilized for increasing the temperature of the air to the desired value. Jet propulsion engines utilizing atmospheric air will be called *air-breathing engines*.

2. Another class of jet propulsion engines employs a jet of gases produced by the chemical reaction of a fuel and oxidizer, each of which is carried in the propelled body. No atmospheric air is used in forming the jet. The equipment wherein the chemical reaction takes place, including the exhaust nozzle, is called a *rocket*

motor. The complete apparatus, including the means for forcing the chemicals into the rocket motor, is called a *rocket engine*.

It is seen that chemical jet propulsion engines, to which the discussions are limited, may, therefore, be grouped into two broad classes: (1) air-breathing engines, which utilize or consume air from the atmosphere, and (2) rocket engines, which do not utilize atmospheric air.

From a general point of view all chemical jet propulsion engines comprise two main subassemblies: (1) a means or apparatus for generating high-pressure, high-temperature gas, called the *hot gas generator*, and (2) an exhaust nozzle, which is the *propulsive element*, for expanding the gas furnished it by the hot gas generator and for discharging the expanded gas to the surrounding atmosphere. The *propulsive nozzle* may be either of the converging or converging-diverging type, depending on the mission for the engine.

4–10(a) Air-Breathing Engines

Examples of air-breathing engines are the turbojet, turbofan, ramjet, and pulsejet engines. Zucrow[12, 13] discusses the general characteristics of each of those engines. In their operation, all of them induct air from the free stream through an inlet, burn fuel in it, thus increasing the stagnation enthalpy of the working fluid, and expel the heated air through a propulsive nozzle. The turbojet and turbofan engines also increase the stagnation pressure of the air by passing it through a compressor, which is driven by a gas turbine. The internal reaction force between the fluid and the solid boundaries of the engine is called the *net thrust*, and it is that force that propels the engine and the vehicle to which it is attached.

Every air-breathing engine has its own characteristic features. Engines operating at supersonic velocities have shock waves in the inlet. Turbofan engines bypass a portion of the inducted air around the combustion chamber. Turboprop engines deliver the bulk of their power to a propeller, and only a small part of their thrust is developed by jet propulsion. The model illustrated in Fig. 4.36 is intended to account for only the basic common features of air-breathing engines. In propulsion practice, more accurate flow models are employed for representing the specific engines.

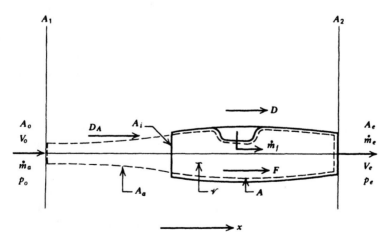

Figure 4.36. Determination of the thrust developed by an air-breathing jet propulsion engine.

For determining the net thrust developed by the air-breathing engine illustrated in Fig. 4.36, it is convenient to employ a *relative system of coordinates*, which assumes that the propulsion system is at rest, and the fluid flows toward it with a velocity V_o relative to the propulsion system; that is, V_o is equal to the velocity of the propulsion system in the actual case.

Since the interest is in the thrust F developed in the direction of flight, the longitudinal axis of the engine is made coincident with the x direction. Only the x component of the rate of change in the momentum for the fluid momentarily within the control surface is pertinent to the problem at hand. Since the positioning of the control surface is arbitrary, it is convenient, therefore, to arrange it so that only the x components of the momentum fluxes need be considered. The latter objective is accomplished by placing the propulsion system between the two planes A_1 and A_2, each of which is perpendicular to the streamlines and extends to infinity. All of the momentum changes in the x direction then occur across and perpendicular to the planes A_1 and A_2.

For convenience, the plane A_1 is located far enough upstream from the propulsion system so that the static pressure p_o acting on A_1 has the same value as it had before the propulsion system was introduced between A_1 and A_2. Plane A_2 is located at the nozzle exit plane.

To develop an equation for the *net thrust F* created by the fluid flowing through the propulsion system, called the *internal flow*, it is convenient to choose the control volume $\mathscr{V}$ so that it includes only the fluid that passes through the engine, assumed here to be a single stream of fluid. Changes in the pressure and momentum of the *external flow* have no effect on the internal flow, but produce the external drag D.

When the free-stream *capture area* A_o differs from the *inlet area* A_i, the free-stream streamlines curve as illustrated in the figure. The pressure forces acting along the curved portion A_a of the control surface create an additional force component in the axial direction; that component is termed the additive drag and denoted by D_a.

$$D_a = -\int_{A_a} p\mathbf{i} \cdot d\mathbf{A} \tag{a}$$

The total force exerted by the internal surfaces of the engine on the fluid flowing through the engine is denoted by F. The reaction to that force is the force exerted by the fluid on the engine; that is, the *net thrust*.

Equation 2.56 may be employed for determining the surface forces acting on a control volume. Assuming steady flow and neglecting body forces, equation 2.56 becomes

$$\mathbf{F}_{\text{surface}} = \int_A \mathbf{V}(\rho\mathbf{V} \cdot d\mathbf{A}) \tag{b}$$

Equation (b), when applied to the configuration illustrated in Fig. 4.36, yields

$$F + p_o A_o + D_a - p_e A_e = \dot{m}_e V_e - \dot{m}_a V_o \tag{c}$$

Solving equation (c) for the internal reaction force F gives

$$F = \dot{m}_e V_e - \dot{m}_a V_o - D_a + p_e A_e - p_o A_o \tag{4.114}$$

The *net thrust* is equal to $-F$. It is conventional, however, to neglect the negative sign, so that positive values are obtained for the net thrust even though it always acts in the direction opposite to the flow of the propulsive fluid through the engine.

Equation 4.114 is the desired equation for the *net thrust* of an air-breathing jet propulsion engine. It is common practice to assume that the engine operates in an ambient environment having the pressure p_o. All pressures in equation 4.114 are, therefore, measured above p_o; the drag forces D and D_a are evaluated in an appropriate manner. In that case, equation 4.114 becomes

$$F = \dot{m}_e V_e - \dot{m}_a V_o - D'_a + (p_e - p_o)A_e \qquad (4.115)$$

In equation 4.115

$$\dot{m}_e V_e = \text{the jet thrust (}\textit{also called the gross thrust}\text{)} \qquad (4.116a)$$

$$\dot{m}_a V_o = \text{the ram drag} \qquad (4.116b)$$

$$(p_e - p_o)A_e = \text{the pressure thrust} \qquad (4.116c)$$

$$D'_a = \text{the additive drag} = -\int_{A_a} (p - p_o)\mathbf{i} \cdot d\mathbf{A} \qquad (4.116d)$$

The sum of the jet thrust and the pressure thrust is the total thrust developed by the propulsive nozzle. Consequently, equation 4.115 may be written as

$$F = F_{\text{nozzle}} - \dot{m}_a V_o - D'_a \qquad (4.117)$$

where

$$F_{\text{nozzle}} = \dot{m}_e V_e + (p_e - p_o)A_e \qquad (4.118)$$

Air-breathing engines induct air at the rate $\dot{m}_a$ and burn fuel at the rate $\dot{m}_f$. Thus,

$$\dot{m}_e = \dot{m}_a + \dot{m}_f \qquad (4.119)$$

For a simple turbojet burning a hydrocarbon fuel, $\dot{m}_f$ is generally between 2 to 5 percent of $\dot{m}_a$.

The thrust developed by a unit mass flow rate of air through an air-breathing propulsion system is termed either the *air specific impulse*, denoted by I_a, or the *specific thrust*. Hence,

$$I_a \equiv \frac{F}{\dot{m}_a} \qquad (4.120)$$

Another performance parameter in general use is the thrust specific fuel consumption, denoted by $TSFC$, and defined as

$$TSFC \equiv \frac{\dot{m}_f}{F} \qquad (4.121)$$

To determine F, I_a, and $TSFC$, V_e and p_e must be known. Those properties of the exhaust jet may be determined by applying the isentropic flow analysis developed in this chapter to the flow through the nozzle.

A propulsive nozzle develops its maximum thrust when the area ratio $\varepsilon = A_e/A_t$ is designed to yield *optimum*, or *complete*, *expansion* (i.e., $p_e = p_o$). Hence, at that design point, the pressure thrust is zero. Even at off-design conditions, the pressure thrust is only a small fraction of the net thrust. In many applications, the additive drag D'_a is also a small fraction of the net thrust. Also $\dot{m}_a \approx \dot{m}_e$, since $\dot{m}_f$ is only a few percent of $\dot{m}_a$. Consequently, equation 4.115 may be expressed as

$$F \approx \dot{m}_a(V_e - V_o) \qquad (4.122)$$

Equation 4.122 shows that the net thrust F is the difference between the jet thrust (which is the nozzle thrust) and the ram drag, both of which may be several times larger than the net thrust itself. Consequently, a change in the jet thrust is magnified several times in its effect on the net thrust, showing that great care must be exercised in the design of the propulsive nozzle.

Example 4.11. An airplane is propelled at a constant speed of 268.2 m/s at 10,668 m altitude by a turbojet engine. The mass flow rate of air through the engine is 45.36 kg/s, and the fuel flow rate is 0.907 kg/s. The exhaust jet has a velocity of 609.6 m/s, and the static pressure in the nozzle exit plane is equal to that of the surrounding atmosphere. Assume that the additive drag is negligible. Calculate (a) the ram drag, (b) the jet thrust, (c) the net thrust, (d) the air specific impulse, and (e) the thrust specific fuel consumption.

Solution

(a)   From equation 4.116b,

$$\dot{m}_a V_o = 45.36(268.2) = 12,166 \text{ N}$$

(b)   From equation 4.116a,

$$\dot{m}_e V_e = (45.36 + 0.91)(609.6) = 28,206 \text{ N}$$

(c)   From equation 4.115,

$$F = \dot{m}_e V_e - \dot{m}_a V_o = 28,206 - 12,166 = 16,040 \text{ N}$$

(d)   From equation 4.120,

$$I_a = \frac{F}{\dot{m}_a} = \frac{16,040}{45.36} = 353.62 \text{ N-s/kg}$$

(e)   From equation 4.121,

$$TSFC = \frac{\dot{m}_f}{F} = \frac{(0.907)(3600)}{16,040} = 0.2036 \text{ kg/hr-N}$$

4–10(b) Rocket Engines[12–14]

Rocket engines differ from air-breathing engines in that no air is inducted into the engine, and all of the mass ejected through the propulsive nozzle is carried within the rocket propelled vehicle. The basic features of a rocket engine are illustrated in Fig. 4.37. Rocket engines are broadly classified by the type of propellants burned. The major classes are liquid propellant engines, solid propellant engines, and hybrid rocket engines. Also included are nuclear rocket engines and electric arc jet engines. In a broader sense, electric propulsion engines are also rocket engines (i.e., the plasma jet and the ion engine).

To determine the thrust developed by the rocket engine illustrated in Fig. 4.37, assume that the engine is restrained from moving by the external force F. The thrust acting on the rocket engine is then equal to $-F$. Choose a control surface that

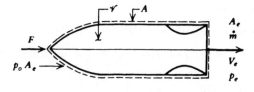

Figure 4.37. Determination of the thrust developed by a rocket jet propulsion engine.

completely encloses the rocket engine. Equation 2.56, when applied to the control volume illustrated in Fig. 4.37, yields

$$F + p_o A_e - p_e A_e = \dot{m} V_e \tag{a}$$

where $\dot{m}$ denotes the rate at which propellants are consumed. Solving equation (a) for the external restraining force F gives

$$F = \dot{m} V_e + (p_e - p_o) A_e \tag{4.122}$$

The rocket engine thrust is then $-F$. As mentioned earlier, it is conventional to neglect the negative sign so that the thrust of the rocket engine always has a positive value, which is given by equation 4.122. In equation 4.122, $\dot{m} V_e$ is the jet thrust and $(p_e - p_o) A_e$ is the pressure thrust. The thrust developed by consuming a unit mass of the propellants in unit time is termed the specific impulse, denoted by I_{sp}. Thus,

$$I_{sp} \equiv \frac{F}{\dot{m}} \tag{4.123}$$

For a maximum thrust nozzle [see Section 4–10(e)], the pressure thrust is equal to zero. In that case, equation 4.122 becomes

$$F = \dot{m} V_e \tag{4.124}$$

To determine F and I_{sp}, the terms V_e and p_e must be known.

Example 4.12. A liquid propellant rocket engine consumes liquid hydrogen and liquid oxygen as its propellants. The nozzle throat area is 0.09290 m^2, and the nozzle area ratio is 20. The propellant mass flow rate is 362.9 kg/s, the pressure in the nozzle exit plane is $p_e = 0.4827 \cdot 10^5$ N/m^2, and the jet exit velocity is 3048 m/s. Calculate (a) the thrust developed by the rocket engine at sea level, (b) the thrust developed at an altitude of 30,000 m, and (c) the specific impulse at both altitudes.

Solution

The nozzle exit area is

$$A_e = \varepsilon A_t = 20(0.09290) = 1.8580 \text{ m}^2$$

(a)   At sea level, $p_o = 1.0133 \cdot 10^5$ N/m^2. From equation 4.122,

$$F_{\text{sea level}} = \dot{m} V_e + (p_e - p_o) A_e$$
$$= 362.9(3048) + (0.4827 - 1.0133)(1.8580)10^5 = 1,007,500 \text{ N}$$

At sea level, the pressure thrust is negative due to overexpansion of the gases. Since $p_e/p_o = 0.4827/1.0133 = 0.476$, the flow would probably not separate from the nozzle wall [see Section 4–7(e)].

(b)   From Table C.5, at 30,000 m altitude, $p_o = 1970$ N/m^2. Since the flow inside of the nozzle is unaffected by a change in altitude, the jet thrust is the same as in part (a); only the pressure thrust changes. Thus,

$$F_{30,000} = 362.9(3048) + (0.4827 - 0.0197)(1.8580)10^5 = 1,192,100 \text{ N}$$

Note that the pressure thrust is positive in this case, since the flow is underexpanded.

(c)   From equation 4.123,

$$I_{sp, \text{sea level}} = \frac{F_{\text{sea level}}}{\dot{m}} = \frac{1,007,500}{362.9} = 2776 \text{ N-s/kg } (283.1 \text{ lbf-sec/lbm})$$

$$I_{sp, 30,000} = \frac{1,192,100}{362.9} = 3285 \text{ N-s/kg } (335.4 \text{ lbf-sec/lbm})$$

4–10(c) Performance of Propulsive Nozzles

It is shown in Sections 4–10(a) and 4–10(b) that the thrust-producing component of a jet propulsion engine is the exhaust nozzle. The net thrust developed by a propulsive nozzle depends on the mass flow rate of fluid through the nozzle, the exit jet velocity, the exit static pressure, the exit area, and the ambient pressure. Figure 4.38 illustrates schematically the essential features of a propulsive nozzle. Figure 4.38a applies to a converging nozzle, which is employed in low-pressure ratio applications typical of many air-breathing engines, and Fig. 4.38b applies to a converging-diverging nozzle, which is employed in high-pressure ratio applications that are typical of rocket engines. The characteristic behaviors of converging and converging-diverging nozzles are discussed in Sections 4–6 and 4–7, respectively.

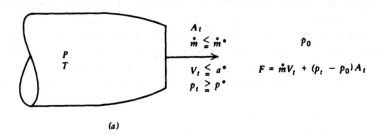

(a)

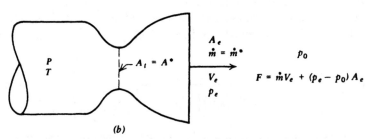

(b)

Figure 4.38.   Essential features of propulsive nozzles. (a) Converging propulsive nozzle. (b) Converging-diverging propulsive nozzle.

Two performance criteria are commonly employed for correlating the performance of propulsive nozzles, for both air-breathing and rocket engines. They are the specific impulse (see equation 4.123)

$$I_{sp} = \frac{F}{\dot{m}} = V_e + \frac{(p_e - p_o)A_e}{\dot{m}} \tag{4.125}$$

and the thrust coefficient, which is *defined* by

$$C_F \equiv \frac{F}{PA_t} = \frac{\dot{m}V_e}{PA_t} + \left(\frac{p_e}{P} - \frac{p_o}{P}\right)\frac{A_e}{A_t} \tag{4.126a}$$

It is desirable to express C_F for a converging-diverging nozzle in terms of V_e' and $\dot{m}^*$, because that type of nozzle is designed for operation with choked flow. Substituting from equations 4.101 and 4.93, respectively, into equation 4.126a, yields

$$C_F = \frac{\psi^* PA_t}{a_o PA_t} a_o \left\{\frac{2}{\gamma - 1}\left[1 - r_e^{(\gamma - 1)/\gamma}\right]\right\}^{1/2} + (r_e - r_o)\varepsilon_e \tag{4.126b}$$

where $r_e = p_e/P$, $r_o = p_o/P$, and $\varepsilon_e = A_e/A_t$. Substituting for ψ^* from equation 4.77 into equation 4.126b and simplifying yields

$$C_F = \gamma \left\{ \frac{2}{\gamma - 1} \left(\frac{2}{\gamma + 1} \right)^{(\gamma+1)/(\gamma-1)} \left[1 - r_e^{(\gamma-1)/\gamma} \right] \right\}^{1/2} + (r_e - r_o)\varepsilon_e \quad (4.126c)$$

For a given gas, the pressure ratio r_e depends only on the area ratio ε_e. Hence, the nozzle thrust coefficient depends only on the specific heat ratio γ, the pressure ratio r_o, and the area ratio ε_e.

Figure 4.39 presents C_F as a function of ε_e for gases with $\gamma = 1.2$. It is evident from equation 4.126c that the maximum thrust for a given area ratio occurs when $p_o = 0$. It is also apparent that, for a given value of r_o, there is a unique area ratio ε_e that yields the maximum thrust. The locus of the values of ε_e for maximum thrust is indicated in Fig. 4.39. In Section 4–10(e), it is shown that the maximum thrust is achieved when ε_e is chosen so that $p_e = p_o$ (i.e., when the pressure thrust is zero). Figure 4.39 also shows that the thrust curves are quite flat in the vicinity of the thrust maxima, so that some variation in the area ratio is possible without decreasing the thrust significantly. By selecting the area ratio so that it is slightly smaller than the optimum area ratio, the size and weight of the nozzle may be reduced with no substantial reduction in the thrust. If the area ratio exceeds the optimum area ratio, the exit pressure p_e becomes lower than the back pressure p_o, so that flow separation eventually occurs in the divergence, and equation 4.126c is no longer valid. In Fig. 4.39, an approximate region of separation is illustrated schematically, between the limits where $p_e = 0.4 \, p_o$ and $p_e = 0.2 \, p_o$. The exact separation point, as well as the thrust developed after separation occurs, can be determined only by experiment.

The thrust coefficient C_F may be expressed in terms of the exit Mach number M_e, by substituting for r_e and ε_e from equations 4.26 and 4.29, respectively, into equation 4.126c. Thus

$$C_F = (\gamma + 1) \left(\frac{2}{\gamma + 1} \right)^{\gamma/(\gamma-1)} \left\{ \frac{(1 + \gamma M_e^2)}{M_e \left[2(\gamma + 1) \left(1 + \frac{\gamma - 1}{2} M_e^2 \right) \right]^{1/2}} \right\} - r_o \varepsilon_e \quad (4.126d)$$

Comparing equation 4.126d with equation 4.32, which defines the impulse function ratio $\mathscr{F}/\mathscr{F}^*$, it is seen that

$$C_F = (\gamma + 1) \left(\frac{2}{\gamma + 1} \right)^{\gamma/(\gamma-1)} \left(\frac{\mathscr{F}}{\mathscr{F}^*} \right) - r_o \varepsilon_e \quad (4.126e)$$

Values of $\mathscr{F}/\mathscr{F}^*$ are presented in Table C.6, and values of $(\gamma + 1)[2/(\gamma + 1)]^{\gamma/(\gamma-1)}$ are tabulated in Table C.15.

Several correction factors are employed for correcting the calculated values of the performance parameters for a converging-diverging propulsive nozzle, based on assuming one-dimensional isentropic flow, for the effects of multidimensional flow, nonisentropic flow, and deviations of the flowing fluid from perfect gas behavior. The principal correction factors in current use are:

$$C_D = \frac{\dot{m}}{\dot{m}_{1-D, \text{isentropic}}} \quad (4.127a)$$

$$\eta_I = \frac{I_{sp}}{I_{sp, 1-D, \text{isentropic}}} \quad (4.127b)$$

$$\eta_F = \frac{F}{F_{1-D, \text{isentropic}}} \quad (4.127c)$$

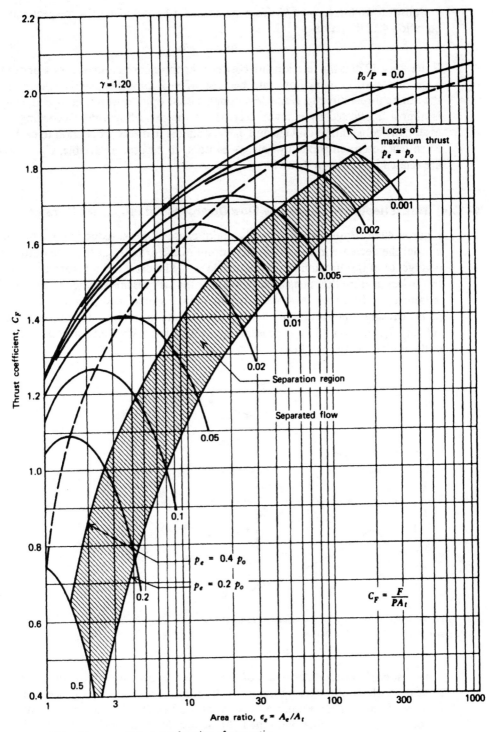

Figure 4.39. Thrust coefficient as a function of area ratio.

Substituting for $I_{sp} = F/\dot{m}$ from equations 4.127c and 4.127a, respectively, into equation 4.127b, yields

$$\eta_F = C_D \eta_I \qquad (4.127d)$$

Equation 4.127d is the relationship between the three correction factors. For example, a decrease in the mass flow rate because of *multidimensional flow effects* reduces the nozzle thrust directly. If the flow remains isentropic, however, the specific impulse is almost unaffected by a slight decrease in the mass flow rate. Accordingly, η_I is essentially constant, so that C_D and η_F vary almost in direct proportion to each other. Even when nonisentropic flow effects cause η_I to become a variable, C_D, η_I, and η_F are related by equation 4.127d.

4–10(d) Thrust Reduction Caused by Flow Divergence in Propulsive Nozzles

Figure 4.40 illustrates schematically a major cause of thrust reduction in a propulsive nozzle: the decrease in the axial momentum of the exhaust jet because of the radial divergence of the streamlines at the nozzle exit. The decrease in thrust, often termed *thrust loss*, is a multidimensional flow effect that is present whether or not the flow is isentropic. In an actual nozzle, the magnitudes of V_e and p_e vary over the exit plane, causing other multidimensional flow effects in addition to divergence of the velocity V_e. Section 16–4(a) presents a method for taking into account the effect of the multidimensional flow.

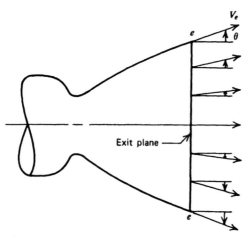

Figure 4.40. Divergence of the velocity vector at the exit plane of a propulsive nozzle.

A reasonably accurate estimate of the thrust loss because of flow divergence in a conical nozzle is obtained by assuming that the flow in the divergence is a source flow (see Section 4–9). Figure 4.41 illustrates schematically the source flow model for determining the aforementioned thrust loss for a conical nozzle. Section 4–9 discusses the method for determining the mass flow rate, exit surface velocity V_s, and exit surface pressure p_s for a source flow.

The axial thrust developed on the differential area dA_s is

$$dF = d\dot{m}\, V_s \cos\theta + (p_s - p_o)\, dA_s \cos\theta \qquad (a)$$

Substituting $d\dot{m} = \rho_s V_s\, dA_s$ and $dA_s = 2\pi R^2 \sin\theta\, d\theta$ [see equation (a), Section 4–9]

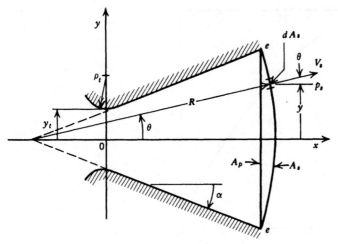

Figure 4.41. Model for the determination of the divergence loss in a conical nozzle.

into equation (a) and integrating yields

$$F = 2\pi R^2 \int_0^\alpha [\rho_s V_s^2 + (p_s - p_o)] \cos \theta \sin \theta \, d\theta \tag{b}$$

Since ρ_s, V_s, and p_s are constant on the spherical surface A_s, equation (b) may be integrated; the result is

$$F = 2\pi R^2 [\rho_s V_s^2 + (p_s - p_o)] \frac{1}{2} \sin^2 \theta \Big|_0^\alpha = \pi R^2 \sin^2 \alpha [\rho_s V_s^2 + (p_s - p_o)] \tag{4.128a}$$

The area of the spherical surface is given by equation (b) in Section 4–9. Substituting that expression into equation 4.128a and simplifying yields

$$F = \left(\frac{1 + \cos \alpha}{2}\right) A_s [\rho_s V_s^2 + (p_s - p_o)] \tag{4.128b}$$

Since $\dot{m} = \rho_s A_s V_s$, the term in brackets may be written as

$$F_s = \dot{m} V_s + (p_s - p_o) A_s \tag{4.129}$$

where F_s is the one-dimensional thrust based on the source flow model. The *divergence factor*, denoted by λ, is *defined* by

$$\lambda \equiv \frac{1 + \cos \alpha}{2} \tag{4.130}$$

In terms of λ and F_s, equation 4.128b becomes

$$F = \lambda F_s \tag{4.131}$$

Hence, the divergence factor λ corrects the one-dimensional source flow thrust F_s to account for the loss caused by the radial divergence of the streamlines. Figure 4.42 presents the divergence factor λ as a function of the cone angle α. It is apparent that the divergence factor decreases rapidly, and hence the so-called *thrust divergence loss* increases rapidly, as the cone angle exceeds 15 deg.

Example 4.13. Calculate the thrust F, thrust efficiency η_F, and specific impulse efficiency η_I of the conical nozzle described in Example 4.10, assuming (a) planar flow and (b) source flow. The ambient pressure is $p_o = 0.0$.

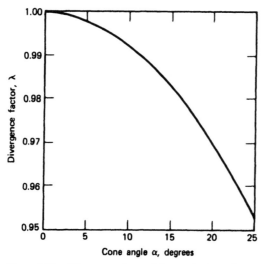

Figure 4.42. Divergence factor λ for conical nozzles as a function of nozzle cone angle α.

Solution

The mass flow rate in the nozzle is the critical mass flow rate $\dot{m}^*$ corresponding to the planar throat area A_t. From Table C.6, for $M = 1$, $t^*/T = 0.90909$ and $p^*/P = 0.56447$. Thus,

$$t^* = (0.90909)(3000.0) = 2727.27 \text{ K}$$

$$p^* = (0.56447)(70.0)10^5 = 39.5129 \cdot 10^5 \text{ N/m}^2$$

$$\rho^* = \frac{p^*}{Rt^*} = \frac{(39.5129)10^5}{320.0(2727.27)} = 4.52752 \text{ kg/m}^3$$

$$a^* = (\gamma Rt^*)^{1/2} = [1.2(320.0)(2727.27)]^{1/2} = 1023.36 \text{ m/s}$$

The area A_t of the planar throat of the nozzle is

$$A_t = \pi y_t^2 = \pi(0.025)^2 = 0.00196350 \text{ m}^2$$

$$\dot{m}^* = \rho^* A_t a^* = (4.52752)(0.00196350)(1023.36) = 9.09745 \text{ kg/s}$$

From Example 4.10,

$$\varepsilon_p = 13.40907 \qquad \varepsilon_s = 13.64148$$
$$V_{ep} = 2512.62 \text{ m/s} \qquad V_{es} = 2516.49 \text{ m/s}$$
$$p_{ep} = 59,669 \text{ N/m}^2 \qquad p_{es} = 58,342 \text{ N/m}^2$$

$$A_p = \varepsilon_p A_t = 13.40907(0.00196350) = 0.026329 \text{ m}^2$$

$$A_s = \varepsilon_s A_t = 13.64148(0.00196350) = 0.026785 \text{ m}^2$$

(a)   From equation 4.122, for planar flow,

$$F = \dot{m}V_{ep} + (p_{ep} - p_o)A_p = (9.09745)(2512.62) + (59,669)(0.026329) = 24,429 \text{ N}$$

From equation 4.123,

$$I_{sp} = \frac{F}{\dot{m}} = \frac{24,429}{9.09745} = 2685.3 \text{ N-s/kg}$$

By definition, $\eta_F = \eta_I = 1.0$ for planar isentropic flow.

(b)   From equation 4.129, for source flow,

$$F_s = \dot{m}V_{es} + (p_{es} - p_o)A_s = (9.09745)(2516.49) + (58,342)(0.026785) = 24,456 \text{ N}$$

From equation 4.130,

$$\lambda = \frac{1 + \cos\alpha}{2} = \frac{1 + \cos(15)}{2} = 0.98296$$

From equation 4.131,

$$F = \lambda F_s = (0.98296)(24,456) = 24,039 \text{ N}$$

From equation 4.123,

$$I_{sp} = \frac{F}{\dot{m}} = \frac{24,039}{9.09745} = 2642.4 \text{ N-s/kg}$$

From equations 4.127b and 4.127c we obtain

$$\eta_F = \frac{F_{\text{source}}}{F_{\text{planar}}} = \frac{24,039}{24,429} = 0.98404$$

$$\eta_I = \frac{I_{sp,\text{source}}}{I_{sp,\text{planar}}} = \frac{2642.4}{2685.3} = 0.98404$$

Since $C_D = 1.0$ by assumption, equation 4.127d shows that $\eta_F = \eta_I$, which is the case here.

Example 16.6 presents C_D, F, η_I, and η_F based on a two-dimensional analysis. In that case, $\eta_I = 0.9828$ and $\eta_F = 0.9773$. The value of $\eta_I = 0.9840$ determined by the source flow analysis presented here is in error by only 0.12 percent, while the value of $\eta_I = 1.0000$ determined by the planar flow analysis is in error by 1.72 percent.

4-10(e) Design of Propulsive Nozzles for Maximum Thrust

The thrust developed by a propulsive nozzle is given by equation 4.118, which is repeated below.

$$F = \dot{m}V_e + (p_e - p_o)A_e \tag{4.118}$$

It is of interest to determine the nozzle exit area that yields maximum thrust. Differentiating equation 4.118 yields

$$dF = \dot{m}\,dV_e + A_e\,dp_e + p_e\,dA_e - p_o\,dA_e \tag{a}$$

Substituting $\dot{m} = \rho AV$ into equation (a) and rearranging gives

$$dF = (dp_e + \rho_e V_e\,dV_e)A_e + (p_e - p_o)\,dA_e \tag{b}$$

Substituting equation 4.2, Bernoulli's equation, into equation (b) and solving for dF/dA_e gives

$$\frac{dF}{dA_e} = p_e - p_o \tag{c}$$

The maximum thrust is obtained when $dF/dA_e = 0$. Thus,

$$p_e = p_o \quad (\textit{for maximum thrust}) \tag{4.132}$$

Accordingly, the nozzle area ratio must be chosen so that the pressure ratio $p_e/P = p_o/P$.

The design criterion presented in equation 4.132 is based on planar flow. Scofield and Hoffman[15] derived the design equation for maximizing the thrust of a conical nozzle, including the divergence factor λ. Their result shows that some underexpansion is desirable. Hoffman[9] derived the design condition for nonisentropic flow in conical nozzles, and Byington and Hoffman[16] derived the design equation for isentropic flow in conical nozzles when an annular base region exists between the nozzle exit lip and the vehicle airframe lip. In the cases mentioned, the possible thrust gains were in the range from 0 to 2 percent above the thrust obtained from a nozzle designed to make $p_e = p_o$.

Example 4.14. Calculate the altitude for which the nozzle presented in Example 4.12 becomes a *maximum thrust nozzle*.

Solution

A maximum thrust nozzle, according to equation 4.132, is one designed so that $p_e = p_o$. Thus, the nozzle is a maximum thrust nozzle when $p_o = 0.4827 \cdot 10^5$ N/m². From Table C.5, at $p_o = 0.4827 \cdot 10^5$ N/m², the design altitude is found to be 5850 m (19,190 ft).

4–11 SUMMARY

In this chapter the primary subject discussed is the steady one-dimensional isentropic flow of a compressible fluid, in the absence of body forces ($g\,dz = 0$). The *driving potential* for such a flow is the *area change* of the flow passage. On the *Mollier diagram* (i.e., the *hs* plane), an isentropic flow is represented by a vertical line ($s = $ constant), termed the *isentropic locus* for the flow (see Fig. 4.2). The possible states for the flowing fluid are restricted to the isentropic locus and, in all of its states, the stagnation properties of the flowing fluid are constant. The differential form of the energy equation for an isentropic flow (equation 4.3) has the same form as that for a non-isentropic adiabatic flow ($\delta Q = 0$, but $ds \neq 0$). Their integral forms are different, however, because of the heat equivalent of the energy expended in overcoming frictional effects (δQ_f) in the case of nonisentropic adiabatic flow.

The effects produced on the flow Mach number by area change depend on whether the flow is initially subsonic ($M < 1$) or supersonic ($M > 1$), and whether the area of the flow passage decreases ($dA < 0$) or increases ($dA > 0$). If the initial flow is subsonic and $dA < 0$, the velocity V of the fluid increases in the direction of flow while its static pressure p decreases. Conversely, if the initial flow is supersonic, then the velocity V decreases and the static pressure p increases in the flow direction. On the other hand, if the flow passage area increases ($dA > 0$) in the flow direction, then with an initially subsonic flow ($M < 1$), V decreases and p increases as the fluid flows along the passage; the reverse effects are produced if the initial flow is supersonic ($M > 1$). The aforementioned effects are illustrated schematically in Figs. 4.4 and 4.5. A flow passage that causes the fluid velocity to increase and the static pressure to decrease is termed a *nozzle*, and one that causes the fluid velocity to decrease and the static pressure to increase is called a *diffuser*. Whether a converging passage is a nozzle or a diffuser depends on whether the initial flow is subsonic ($M_1 < 1$) or supersonic ($M_1 > 1$). The same remarks are applicable to diverging flow passages.

It is shown in Sections 4–3(c) and 4–3(d) that a converging passage ($dA < 0$) accelerates an initially subsonic flow to the sonic speed in the smallest area, called the *throat*, and decelerates an initially supersonic flow to the sonic speed in the throat. When $M = 1$ in the throat, the flow is said to be *choked*. A sonic flow cannot enter a converging passage and satisfy the conditions for a steady isentropic flow. A steady isentropic sonic flow may, however, enter a diverging flow passage and the flow remain steady and isentropic. The flow in the diverging passage may become either subsonic or supersonic, depending on the static pressure p_e in the exit section of the diverging passage. Actually, there are only two values of the exit pressure that satisfy the conditions for an isentropic flow in the diverging passage; one corresponds to a flow expansion with a supersonic exit velocity; the other corresponds to a flow compression (diffusion) with a subsonic exit velocity.

A converging-diverging flow passage is known as a De Laval nozzle. When an isentropic flow in a converging-diverging flow passage passes its maximum mass flow rate $\dot{m} = \dot{m}'_{max} = \dot{m}^*$, the throat is termed the *critical area* A^* and the fluid velocity in the throat, termed the *throat velocity*, is $V'_t = a^*$.

The effect of the back pressure p_o of the surroundings into which either a converging or a converging-diverging nozzle discharges is discussed in Section 4–3(e). For a converging nozzle the exit velocity may vary from a subsonic value up to and including the sonic velocity in the exit plane. As long as the flow crossing the exit plane is subsonic, the exit pressure $p_e = p_o$, and the fluid discharges into the ambient region as a parallel jet until $p_e = p^*$. When $p_e = p^* > p_o$, there is an expansion of the fluid comprising the jet from p^* to p_o beyond the exit plane of the nozzle.

For an isentropic flow in a choked converging-diverging nozzle, the exit velocity V'_e exceeds the sonic speed in the nozzle exit plane (i.e., $M_e > 1$). Consequently, if $p'_e \geqq p_o$, the expansion from p'_e to p_o occurs downstream from the exit plane of the nozzle. If $p'_e < p_o$ by only a small amount, that is, the nozzle operates with slight overexpansion, oblique shock waves become attached to the inside of the divergence of the nozzle in the vicinity of its exit plane. Should p'_e be significantly less than p_o, the oblique shock wave system may move inside the nozzle and a normal shock wave may form. If p_o becomes large enough, the normal shock wave will move into the throat, unchoking the flow, so that the flow in the divergence becomes subsonic throughout. The effects of the back pressure p_o discussed above for nozzles are also applicable qualitatively to nonisentropic one-dimensional flows.

In the special case where the flowing fluid is a perfect gas, quantitative relationships are readily derived between the flow variables pertinent to steady one-dimensional isentropic flow with area change. Those relationships, expressed as property ratios that depend on the Mach number M and the specific heat ratio γ, are presented in Table 4.2. Those property ratios are tabulated in Table C.6 as functions of M for different values of γ.

The differential equations relating the area change dA/A to such changes in the flow variables as dt/t, dp/p, etc., for the steady one-dimensional isentropic flow of a perfect gas, are derived in Section 4–4(c), equations 4.50 to 4.56. The effect of area change on the flow properties is summarized in Table 4.5.

The equations for steady one-dimensional isentropic flow may be applied to an imperfect gas by employing a suitable numerical procedure. For a large number of flow situations, the perfect gas isentropic flow equations yield results that are sufficiently accurate without modification. For high-energy, high-Mach-number flows, however, imperfect gas effects must be accounted for if serious errors are to be avoided (see Section 4–5).

Table 4.6 presents the principal equations for the steady one-dimensional isentropic flow of a perfect gas in converging and converging-diverging nozzles. To apply those equations to real nozzles, they must be modified by applying appropriate experimentally determined correction factors; the latter are discussed in Section 4–8.

Table 4.6 Steady One-Dimensional Isentropic Flow of a Perfect Gas in Nozzles

Converging nozzle

Isentropic throat velocity

$$V_t' = \left\{ \frac{2\gamma RT}{\gamma - 1} \left[1 - \left(\frac{p_t'}{P} \right)^{(\gamma - 1)/\gamma} \right] \right\}^{1/2} \tag{4.133}$$

$$(V_t')_{max} = a^* = a_o \left(\frac{2}{\gamma + 1} \right)^{1/2} \tag{4.134}$$

Isentropic mass flow rate

$$\dot{m}' = A_t \rho_t' V_t' = \frac{\psi P A_t}{\sqrt{\gamma RT}} = \frac{\psi P A_t}{a_o} \tag{4.135}$$

where

$$\psi = \left\{ \frac{2\gamma^2}{\gamma - 1} \left(\frac{p_t'}{P} \right)^{2/\gamma} \left[1 - \left(\frac{p_t'}{P} \right)^{(\gamma - 1)/\gamma} \right] \right\}^{1/2} \tag{4.136}$$

Maximum isentropic mass flow rate

$$\dot{m}^* = \frac{\psi^* P A_t}{\sqrt{\gamma RT}} = \frac{\Gamma P A_t}{\sqrt{\gamma RT}} = \frac{\Gamma P A_t}{a_o} \tag{4.137}$$

where

$$\psi^* = \Gamma = \gamma \left(\frac{2}{\gamma + 1} \right)^{(\gamma + 1)/2(\gamma - 1)} \tag{4.138}$$

Critical conditions in the throat

$$p_t' = p^* = \left(\frac{2}{\gamma + 1} \right)^{\gamma/(\gamma - 1)} P \tag{4.139}$$

$$t_t' = t^* = \left(\frac{2}{\gamma + 1} \right) T \tag{4.140}$$

$$\rho_t' = \rho^* = \left(\frac{2}{\gamma + 1} \right)^{1/(\gamma - 1)} \rho_o \tag{4.141}$$

Converging-diverging (or De Laval) nozzle

Isentropic throat velocity   (see equations 4.133 and 4.134)
Maximum isentropic mass flow rate   (see equations 4.137 and 4.138)
Properties in the exit plane (A_e)

$$V_e' = \left\{ \frac{2\gamma RT}{\gamma - 1} \left[1 - \left(\frac{p_e'}{P} \right)^{(\gamma - 1)/\gamma} \right] \right\}^{1/2} \tag{4.142}$$

$$M_e' = \left\{ \frac{2}{\gamma - 1} \left[\left(\frac{P}{p_e'} \right)^{(\gamma - 1)/\gamma} - 1 \right] \right\}^{1/2} \tag{4.143}$$

$$\frac{a_e}{a_o} = \left(\frac{t_e'}{T} \right)^{1/2} = \left(\frac{\rho_e'}{\rho_o} \right)^{(\gamma - 1)/2} = \left(\frac{p_e'}{P} \right)^{(\gamma - 1)/2\gamma} \tag{4.144}$$

Area ratio for complete expansion

$$\frac{A}{A_t} = \frac{A}{A^*} = \frac{\psi^*}{\psi} = \frac{\left(\dfrac{2}{\gamma + 1} \right)^{1/(\gamma - 1)} \left(\dfrac{\gamma - 1}{\gamma + 1} \right)^{1/2}}{\left(\dfrac{p'}{P} \right)^{1/\gamma} \left[1 - \left(\dfrac{p'}{P} \right)^{(\gamma - 1)/\gamma} \right]^{1/2}} \tag{4.145}$$

$$\frac{A}{A_t} = \frac{A}{A^*} = \frac{1}{M} \left[\left(\frac{2}{\gamma + 1} \right) \left(1 + \frac{\gamma - 1}{2} M^2 \right) \right]^{(\gamma + 1)/2(\gamma - 1)} \tag{4.146}$$

Nozzles are the propulsive components of all jet propulsion engines. Consequently, their characteristics as thrust-producing devices are of great practical interest. Accordingly, equations are derived for calculating the thrust F, the air-specific impulse I_a, and the thrust specific fuel consumption $TSFC$ for air-breathing engines. For rocket engines, the equations are derived for calculating the thrust F, the specific impulse I_{sp}, and the thrust coefficient C_F. For convenience of reference, the aforementioned equations are assembled in Table 4.7.

Table 4.7   Performance Equations for Jet Propulsion Engines

Air-breathing engines

Thrust

$$F = \dot{m}_e V_e - \dot{m}_a V_o - D'_a + (p_e - p_o)A_e \approx \dot{m}_a(V_e - V_o) \qquad (4.147)$$

Air specific impulse

$$I_a = \frac{F}{\dot{m}_a} \qquad (4.148)$$

Thrust specific fuel consumption

$$TSFC = \frac{\dot{m}_f}{F} \qquad (4.149)$$

Rocket engines

Thrust

$$F = \dot{m}V_e + (p_e - p_o)A_e \qquad (4.150)$$

Specific impulse

$$I_{SP} = \frac{F}{\dot{m}} \qquad (4.151)$$

REFERENCES

1. A. J. Eggers, "One-Dimensional Flows of an Imperfect Diatomic Gas," NACA Report 959, 1950.
2. H. S. Tsien, "One-Dimensional Flows of a Gas Characterized by van der Waal's Equation of State," *Journal of Mathematics and Physics*, Vol. XXV, No. 6, pp. 301–324, January 1947.
3. C. Donaldson, "Note on the Importance of Imperfect-Gas Effects and Variation of Heat Capacities on the Isentropic Flow of Gases," NACA RM No. L8J14, 1948.
4. S. T. Demetriades, "On the Decompression of a Punctured Pressurized Cabin in Vacuum Flight," *Jet Propulsion*, Vol. 24, No. 1, pp. 35–36, January–February 1954.
5. R. L. Thornock and E. F. Brown, "An Experimental Study of Compressible Flow Through Convergent-Conical Nozzles, Including a Comparison with Theoretical Results," *Journal of Basic Engineering*, Transactions ASME, Vol. 94, Series D, No. 4, pp. 926–932, December 1972.
6. M. Summerfield, C. R. Foster, and W. G. Swan, "Flow Separation in Overexpanded Supersonic Exhaust Nozzles," Institute of Fluid Mechanics and Heat Transfer, Los Angeles, Calif., June 22, 1948. Reproduced in *Jet Propulsion*, p. 319, September–October 1954.
7. L. Green, "Flow Separation in Rocket Nozzles," *Journal of the American Rocket Society*, p. 34–35, January–February 1953.
8. A. Stodola, *Steam and Gas Turbines*, Vol. 1, McGraw-Hill, New York, 1927, p. 95.
9. J. D. Hoffman, "Approximate Analysis of Nonisentropic Flow in Conical Nozzles," *Journal of Spacecraft and Rockets*, Vol. 6, No. 11, pp. 1329–1334, November 1969.

10. R. P. Frazer, P. N. Rowe, and M. O. Coulter, "Efficiency of Supersonic Nozzles for Rockets and Some Unusual Designs," Inst. Mech. Engrs., London, reprint, 1957.

11. M. J. Zucrow, *Aircraft and Missile Propulsion*, Vol. 1, Chap. 4, Section 5, Wiley, New York, 1958.

12. M. J. Zucrow, *Aircraft and Missile Propulsion*, Vol. 1, Chap. 2, Wiley, New York, 1958.

13. M. J. Zucrow, *Elements of Aircraft and Missile Propulsion*, Engineering Design Handbook, U.S. Army Material Command, AMC 706–285, July 1969.

14. M. J. Zucrow, "Space Propulsion Engines: Their Characteristics and Problems," *Science in Progress*, Yale University Press, New Haven, pp. 240–278, 1964.

15. M. P. Scofield and J. D. Hoffman "Optimization of Conical Thrust Nozzles," *Journal of Spacecraft and Rockets*, Vol. 4, No. 11, pp. 1547–1549, November 1967.

16. C. M. Byington and J. D. Hoffman, "Effects of Base Pressure on Conical Thrust Nozzle Optimization," *Journal of Spacecraft and Rockets*, Vol. 7, No. 3, pp. 380–382, March 1970.

PROBLEMS

1. Atmospheric air at standard sea level conditions is expanded until its velocity equals the critical speed of sound. Calculate the velocity of the air and its static temperature.

2. Air assumed to be a perfect gas flows into a frictionless passage. The speed of the air increases in the direction of the flow. At station 1 the static temperature is 450 K, the static pressure is $2.00 \cdot 10^5$ N/m², and the mean velocity is 200 m/s. At station 2 the velocity is equal to the speed of sound. Calculate (a) the static temperature, (b) the velocity, (c) the static pressure, and (d) the density of the air at station 2.

3. At a point in a stream tube where the area is A, the velocity of the air is 200 m/s, the temperature is 300 K, and the static pressure is $1.00 \cdot 10^5$ N/m². Further downstream the area is $A_2 = 0.8\,A$. Calculate (a) for A_2: (1) the stagnation pressure, (2) the static temperature, (3) the velocity, (4) the Mach number, and (5) M^*; (b) the minimum possible area for the stream tube, and (c) the quantities listed in (a) for that minimum area.

4. Air flows through an adiabatic frictionless passage. At station 1, $M_1 = 0.9$ and $p_1 = 4.00 \cdot 10^5$ N/m². At station 2 the velocity of the air corresponds to $M_2 = 0.2$. Calculate the change in static pressure between stations 1 and 2.

5. Atmospheric air at a temperature of 300 K, moving with a Mach number of 3, is brought to rest isentropically. Calculate the values of t/T, p/P, and ρ/ρ_0 (a) assuming that c_p remains constant, and (b) assuming that c_p is a function of temperature.

6. A converging-diverging passage is to be designed to give the critical mass flow rate (i.e., the maximum mass flow rate). Assuming the flow process to be isentropic, and a contraction ratio for the passage of $A_t/A_1 = 0.7$, what value of the inlet Mach number M_1 will give the critical mass flow rate? Assume that the fluid is a perfect gas having $\gamma = 1.4$.

7. Air flows through a small insulated tube in which two orifice plates are located in series, and in each the sonic velocity is maintained at its vena contracta. The static pressures measured in the vena contracta of each orifice plate are found to be as follows: static pressure downstream to first orifice plate $= 3.50 \cdot 10^5$ N/m², static pressure downstream to second orifice plate $= 2.75 \cdot 10^5$ N/m². The stagnation temperature of the air is 450 K. Assuming $\gamma = 1.4$, find the effective area ratio of the two orifice plates.

8. Air flows through a duct with a velocity of 200 m/s. The mass flow rate is 9.0 kg/s. The area of the duct is 0.050 m², and the initial Mach number is $M = 0.50$. Calculate the static and stagnation pressures of the air.

9. Air, assumed to be a perfect gas with $\gamma = 1.4$, enters a converging frictionless channel with $M = 0.2$. The mass flow rate is 0.90 kg/s, and the static pressure is $1.40 \cdot 10^5$ N/m². The exit area of the channel is 0.090 m² and the static pressure in the surroundings is $1.00 \cdot 10^5$ N/m². It is desired to increase the mass flow rate to 3.60 kg/s, maintaining the exit pressure at $1.00 \cdot 10^5$ N/m², without changing the temperature at the inlet section. How can this be accomplished?

10. A stream of air flowing in a duct is at a pressure of $1.40 \cdot 10^5$ N/m², its Mach number $M = 0.6$, and its mass flow rate is 0.20 kg/s. The cross-sectional area of the duct is 6.0 cm². Assuming no

friction and no heat transfer, (a) calculate the stagnation temperature of the air. (b) If the area of the duct is reduced so as to form a convergent nozzle, find the maximum percentage reduction of area which may be introduced without reducing the mass flow rate of the stream. (c) For the area reduction in part b, find the velocity and pressure at the minimum area.

11. Show that for the isentropic flow of a perfect gas:

(a) $\dfrac{p}{P} = \left[1 - \dfrac{\gamma - 1}{2} \left(\dfrac{V}{a_o} \right)^2 \right]^{\gamma/(\gamma - 1)}$

(b) $\dfrac{\rho}{\rho_o} = \left[1 - \dfrac{\gamma - 1}{2} \left(\dfrac{V}{a_o} \right)^2 \right]^{1/(\gamma - 1)}$

(c) $\dfrac{p}{P} = \left[1 - \dfrac{\gamma - 1}{\gamma + 1} \left(\dfrac{V}{a^*} \right)^2 \right]^{\gamma/(\gamma - 1)}$

(d) $\dfrac{\rho}{\rho_o} = \left[1 - \dfrac{\gamma - 1}{\gamma + 1} \left(\dfrac{V}{a^*} \right)^2 \right]^{1/(\gamma - 1)}$

12. Air at a pressure and temperature of $1.40 \cdot 10^5$ N/m² and 300 K, respectively, contained in a large vessel is to discharge through an ideal nozzle into a space at a pressure of $0.95 \cdot 10^5$ N/m². Find the mass flow rate if the nozzle is (a) convergent, (b) convergent-divergent with an area ratio of 1.25, and (c) convergent-divergent with the optimum expansion ratio giving the maximum mass flow rate. In each case assume that the minimum cross-sectional area of the nozzle is 0.00065 m².

13. Air at a stagnation pressure of $7.0 \cdot 10^5$ N/m² and a stagnation temperature of 300 K expands through a frictionless convergent-divergent nozzle to the exhaust pressure of $5.0 \cdot 10^5$ N/m². The area ratio of the nozzle is 2.0. Calculate (a) the exit velocity, and (b) the exit Mach number.

14. Air flows through a frictionless De Laval nozzle under adiabatic conditions. The static pressure at the inlet to the nozzle is $5.50 \cdot 10^5$ N/m², and the static pressure of the environment into which the air is discharged is $1.033 \cdot 10^5$ N/m². The static temperature of the air at the inlet section of the nozzle is 310 K. Assuming complete expansion of the air leaving the nozzle, calculate the local acoustic velocity (a) for the inlet section of the nozzle, (b) for the throat section of the nozzle, (c) for the exit section of the nozzle.

15. Air having $\gamma = 1.4$ and $R = 287.04$ J/kg-K flows through a straight duct having an inside diameter of 0.30 m. The gas is discharged from the duct through a frictionless nozzle having an exit Mach number of 2.0. At the inlet section of the nozzle the static pressure and temperature of the air are $4.0 \cdot 10^5$ N/m² and 350 K, respectively, and its Mach number is 0.2. Calculate (a) the area of the throat of the nozzle, (b) the pressure, density, temperature, and velocity of the gas in the throat, (c) the pressure, density, and velocity in the exit section, (d) the mass flow rate of air.

16. The thrust nozzle of a turbojet engine is convergent and has an exit area of 0.070 m². The nozzle inlet stagnation temperature is $T_1 = 1000$ K, and the nozzle inlet stagnation pressure is $0.90 \cdot 10^5$ N/m². Find the following quantities, assuming that the nozzle is operating at an altitude of 12,000 m, the gas is air with $\gamma = 1.4$ and $c_p = 1004.8$ J/kg-K, and the flow process is reversible and adiabatic: (a) the nozzle exit static temperature, (b) the nozzle exit velocity, (c) the nozzle exit static pressure, (d) the mass flow rate through the nozzle.

17. For the nozzle of Problem 16 assume that the gases are completely expanded to atmospheric pressure by a reversible adiabatic process. Find the following quantities: (a) the nozzle exit Mach number, (b) the area of the nozzle exit, (c) the static temperature at the nozzle exit, (d) the velocity at the nozzle exit, (e) the mass flow rate through the nozzle.

18. Consider the conical propulsive nozzle described in Example 4.10 and illustrated schematically in Fig. 4.35. For the same conditions specified in Example 4.10, calculate the flow properties for the following nozzle lengths: (a) 0.10 m, (b) 0.15 m, (c) 0.20 m, (d) 0.25 m, and (e) 0.30 m. Plot the results as a function of nozzle length.

19. A turbojet engine is mounted for a static test on a thrust stand. Air is brought into the engine through a duct mounted perpendicular to the axis of the engine. The average velocity of the air inducted by the engine is 250 m/s, the average velocity of the gas discharged by the propulsive nozzle is 590 m/s, the mass flow rate is 22.0 kg/s, and the pressure in the exit plane of the propulsive nozzle is atmospheric. Calculate the axial force transmitted from the engine to the thrust stand.

20. A rocket motor operates with a combustion pressure $P = 70.0 \cdot 10^5$ N/m^2, a propellant consumption rate of $\dot{m} = 45.0$ kg/s, and a nozzle exit velocity $V_e = 2450$ m/s. The specific heat ratio of the combustion gases $\gamma = 1.30$. The area ratio of the propulsive nozzle is 10, and the throat area $A_t = 0.020$ m^2. Calculate the specific impulse I_{sp} at (a) sea level, and (b) 15,000 m. (c) For what altitude does the nozzle yield optimum expansion? (d) Calculate the I_{sp} at the altitude determined in part c.

5
steady one-dimensional flow with friction

5-1 PRINCIPAL NOTATION FOR CHAPTER 5

The notation employed in Chapter 3 (Section 3-1) applies to the present chapter. The additional notation pertinent to the discussion of flow with friction are listed below.

D	diameter of a circular duct (pipe).
$\mathscr{D}$	hydraulic diameter of a flow passage.
$\mathfrak{f}$	Fanning friction coefficient.
$\mathfrak{f}'$	$= 4\mathfrak{f}$, Darcy friction coefficient.
$\bar{\mathfrak{f}}$	average value of the Fanning friction coefficient.
L	duct length.
L^*	duct length required to choke a flow by frictional effects.
m	hydraulic radius.
Re	$= \mathscr{D}V\rho/\mu$, Reynolds number.
$4\bar{\mathfrak{f}}L/\mathscr{D}$	friction parameter.

Greek Letters

ε	pipe surface roughness.

Subscripts

L	limiting state (choking).
1	state before the action of friction.
2	state after the action of friction.

5-2 INTRODUCTION

The isentropic flow of a compressible fluid with area change is investigated in Chapter 4, under the conditions $\delta Q = \delta W = ds = g\,dz = \delta D = \delta F_{\mathfrak{f}} = 0$, the driving potential affecting the thermodynamic states of the fluid being the area change dA. In all real flow passages, however, wall friction is present. But, in many cases, the effects of wall friction are so small compared to those caused by area change that the assumption of frictionless flow is quite satisfactory. Nevertheless, there are many practical flow processes in which the flow area is constant and the major driving potential is wall friction. For such flows, $dA = \delta Q = \delta W = g\,dz = \delta D = 0$, but $\delta F_{\mathfrak{f}} \neq 0$. Such flow conditions exist, for example, in ducts that serve as fluid transmission lines. Flows that satisfy the above restrictions are termed *simple frictional flows*.

The analysis of a simple frictional flow is based on the following assumptions:

1. The area of the duct is constant ($dA = 0$).
2. The flow is steady and one-dimensional.
3. There is no work or heat transfer ($\delta W = \delta Q = 0$), body forces are negligible ($g\,dz = 0$), and there are no obstructions within the flow ($\delta D = 0$).
4. Wall friction is the sole driving potential for the flow.

Sections 5-3 to 5-6 are concerned with the analysis of adiabatic simple frictional flows. Section 5-7 presents an analysis of the special problem of isothermal simple

w. In analyzing simple frictional flows, the wall frictional force is related ...k flow *properties* through an empirically determined *friction coefficient*. frictal flows are more complicated than the aforementioned simple frictional flow. ...e analysis of multidimensional frictional flow requires including the effect of friction in the momentum equations; the resulting momentum equations are known as the *Navier-Stokes* equations. The latter equations are introduced in Section 5–9. In many practical flows, it may be assumed that the effects of friction are confined to a thin layer near the solid boundaries, termed the *boundary layer*, and in the boundary layer the Navier-Stokes equations may be somewhat simplified. The boundary layer concept is discussed in Section 5–10.

5–3 THERMODYNAMICS OF STEADY ONE-DIMENSIONAL ADIABATIC FLOW WITH FRICTION IN A CONSTANT-AREA DUCT

Adiabatic flow in a constant-area duct may be studied conveniently by considering first the thermodynamics of the flowing compressible fluid. From that study, the locus of all of the possible thermodynamic states attainable by the fluid may be determined, and also the limiting conditions at choking. No information is, however, obtained regarding the effects of the driving potential, which in this case is wall friction. The dynamic effect of wall friction is determined by means of the momentum equation for the flow. This section deals with the thermodynamics of adiabatic flow with wall friction in a constant-area duct. The dynamics of such a flow are considered in Section 5–4.

5–3(a) Governing Equations

Consider the steady one-dimensional flow of a fluid with $dA = \delta Q = \delta W = \delta D = g\,dz = 0$. Figure 5.1 illustrates the control surface surrounding an elementary control volume ($A\,dx$) of the flowing fluid. The corresponding energy, continuity, and momentum or dynamic equations take the following forms (see Table 3.1).

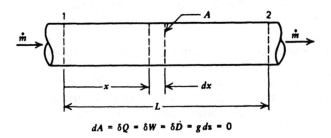

$$dA = \delta Q = \delta W = \delta \dot{D} = g\,dz = 0$$

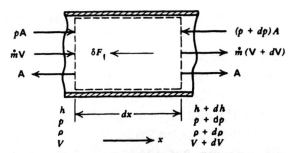

Figure 5.1. Steady one-dimensional adiabatic flow in a constant-area duct in the presence of wall friction (simple frictional flow).

(1) *Energy equation*

$$dh + d\left(\frac{V^2}{2}\right) = 0$$

(2) *Continuity equation*

$$G = \frac{\dot{m}}{A} = \rho V = \frac{V}{v} = \text{constant} \tag{5.1}$$

(3) *Momentum equation*

$$dp + \rho V \, dV + \frac{\rho V^2}{2}\left(\frac{4\mathfrak{f} \, dx}{\mathscr{D}}\right) = 0 \tag{5..}$$

Because the interest here is in determining the thermodynamic states of the fluid, only the energy and continuity equations need be considered. Rewriting the energy equation in its integrated form gives

$$h + \frac{V^2}{2} = H = \text{constant} \tag{5.4}$$

Substituting for V from equation 5.2 into equation 5.4 yields

$$h + \frac{G^2 v^2}{2} = H = \text{constant} \tag{5.5}$$

Applying equation 5.5 between stations 1 and 2 of the constant-area duct yields

$$h_1 + \frac{G^2 v_1^2}{2} = h_2 + \frac{G^2 v_2^2}{2} = H = \text{constant} \tag{5.6}$$

Equation 5.6 relates the enthalpy h and the specific volume v of the fluid for specified values of the mass flux G. It gives no explicit information concerning the law of friction between the stations being investigated. The effect of friction, although not stated explicitly, manifests itself by the influence it exerts on the thermodynamic properties of the fluid. A curve of h as a function of s for a constant value of G, for adiabatic flow with wall friction, is termed a *Fanno line*, and equation 5.5 is known as the *Fanno line equation*.[1]

The equations derived in this section are valid for any fluid, because no restrictions have been placed on the character of the equations of state for the fluid.

Example 5.1. Superheated steam at 1000 lbf/in.^{2} (68.95 · 10^5 N/m^2) and 900 F (755.6 K) enters a 10 in. (0.254 m) ID pipe with a velocity of 500 ft/sec (152.4 m/s). Determine the locus of the possible thermodynamic states that satisfy the continuity and energy equations (i.e., the Fanno line).

Solution

The thermodynamic properties of steam are taken from *Thermodynamic Properties of Steam*, Keenan and Keyes, Wiley, 1958. The initial steam properties are:

$$t_1 = 900 \text{ F } (755.6 \text{ K}),$$
$$p_1 = 1000 \text{ lbf/in.}^2 (68.95 \cdot 10^5 \text{ N/m}^2),$$
$$V_1 = 500 \text{ ft/sec } (152.4 \text{ m/s})$$
$$v_1 = 0.7604 \text{ ft}^3/\text{lbm } (0.04747 \text{ m}^3/\text{kg}),$$
$$h_1 = 1448.2 \text{ Btu/lbm } (3,368,500 \text{ J/kg}),$$
$$s_1 = 1.6121 \text{ Btu/lbm-R } (6749.5 \text{ J/kg-K})$$

The mass flux G is

$$G = \frac{V_1}{v_1} = \frac{500}{0.7604} = 657.5 \text{ lbm/sec-ft}^2 \quad (3209.3 \text{ kg/s-m}^2)$$

From equation 5.6,

$$H = h_1 + \frac{G^2 v_1^{\,2}}{2} = 1448.2 + \frac{(657.5)^2(0.7604)^2}{2(32.174)(778.16)}$$

$$= 1453.2 \text{ Btu/lbm} \quad (3{,}380{,}000 \text{ J/kg})$$

Hence, the equation for the Fanno line is given by

$$h = H - \frac{G^2 v^2}{2} = 1453.2 - 8.6335\, v^2 \quad (3{,}380{,}000 - 5{,}149{,}800\, v^2)$$

Assume selected values of v, as shown in Table 5.1, and by means of the Fanno line equation, calculate the value of h. Then, from the steam tables, determine the corresponding values of s, t, and p. The results are presented in Table 5.1. Figure 5.2 is a plot of the corresponding Fanno line on the hs plane.

Table 5.1 Steam Properties Based on the Fanno Line Equation (Example 5.1)

v, ft^3/lbm	h, Btu/lbm	s, Btu/lbm-R	t, F	p, lbf/in.^{2}
0.7604	1448.2	1.6121	900.0	1000.0
1.0	1444.6	1.6382	877.9	757.8
1.5	1433.8	1.6743	840.1	497.9
2.0	1418.7	1.6960	801.5	364.8
2.5	1399.2	1.7075	757.4	282.6
2.8	1385.5	1.7108	727.6	246.4
3.0	1375.5	1.7116	706.2	226.0
3.2	1364.8	1.7114	683.7	207.8
3.5	1347.4	1.7091	647.5	183.9
4.0	1315.1	1.7002	579.4	150.9
4.5	1278.4	1.6846	503.1	124.0
5.0	1237.4	1.6618	418.9	101.2
5.5	1192.0	1.6301	328.4	81.6

ft^3/lbm $\times$ 0.06248 = m^3/kg $\quad$ $(t + 459.6)/1.8$ = K
Btu/lbm $\times$ 2325.97 = J/kg $\quad$ lbf/in.^{2} $\times$ 6894.76 = N/m^2
Btu/lbm-R $\times$ 4186.8 = J/kg-K

5–3(b) The Fanno Line

In general, the initial values of h, v, and G are known and the stagnation enthalpy H remains constant. To plot a Fanno line, a specific value is assigned to the mass flux G, and a series of values are assigned to the specific volume v. The values of enthalpy h are then computed by means of equation 5.5. From the aforementioned values of h and v, the corresponding values of s are either taken from tabulated values of the thermodynamic properties of the fluid or calculated. The Fanno line is the locus of the possible thermodynamic states that are attainable by the fluid for the selected value of G.

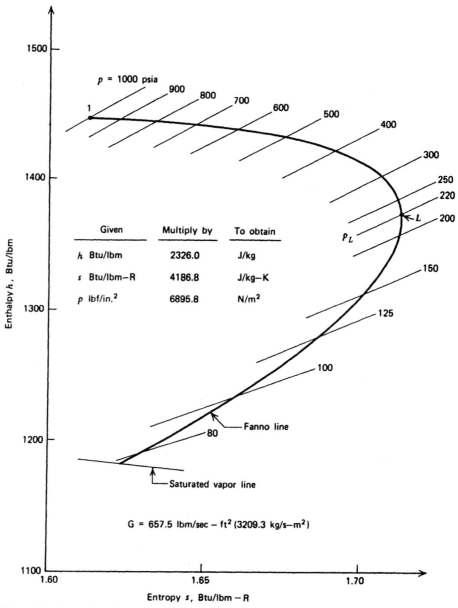

Figure 5.2. Fanno line for steam plotted on the hs plane (Example 5.1).

Example 5.1 illustrates the method of computation for steam as the flowing fluid. Referring to Fig. 5.2, it is seen that the Fanno line gives no information regarding the distance the fluid must traverse to attain any of the thermodynamic states lying on the Fanno line. The length of constant-area passage, with no obstructions, required for attaining a state on the Fanno line depends on the entropy gradient ds/dx, which depends on the wall friction. The general shape of the Fanno line presented in Fig. 5.2 for steam is typical of the Fanno line for all compressible fluids.

Reference to Fig. 5.2 shows that, with G held constant and starting at state 1, increasing the length of the constant-area duct causes the pressure and the enthalpy to decrease, but the flow velocity and the entropy to increase. At a particular value of the static pressure, denoted by $p = p_L$, the entropy reaches a maximum value.

From the shape of the curve it might be inferred that when the back pressure at the duct exit is reduced below p_L, the pressure and the entropy would continue to decrease. In view of the second law of thermodynamics, however, such a process is impossible. The foregoing indicates that thermodynamic states lying on the lower portion of the Fanno line, to the left of the maximum entropy state, cannot be reached by a flow expansion starting from state 1.

It is further concluded from the foregoing that, for an adiabatic flow in a constant-area duct in the presence of wall friction, with increasing flow velocity, the exit pressure for the duct cannot be reduced below the limiting static pressure p_L, no matter how small the back pressure is into which the duct discharges. When that condition is reached, the flow is said to be "*choked*." Moreover, since p_L is a limiting value for the static pressure, there is also a limiting value to the maximum velocity attainable by the fluid in the exit section of the duct by reducing the back pressure at the duct exit. Let V_L denote that *limiting velocity*, also called the *choking velocity*.

The lower portion of a Fanno line (see Fig. 5.2) lying to the left of the maximum entropy state, does, however, have physical significance. It represents the thermodynamic states that are attainable by a flow process starting at lower values of enthalpy and static pressure than the maximum entropy state. Such a process proceeds in the direction of higher static pressures; that is, it is a *flow compression*.

Each value assigned to G gives rise to a separate Fanno line but, if the initial values of h and s are held constant at the values $h = h_1$ and $s = s_1$, then all of the Fanno lines intersect at the initial point (h_1,s_1). Figure 5.3 illustrates, for air as the flowing fluid, three different Fanno lines corresponding to three different values of G and the same initial conditions (h_1,s_1).

A flow that originates at a fixed stagnation state may enter a constant-area duct at different static conditions, for example, after expansion through an isentropic nozzle, because the mass flux G and the initial static properties (h_1,s_1) are different for each nozzle exit condition. Figure 5.4 illustrates six such Fanno lines for air. Note that all six Fanno lines originate from an isentropic expansion from the same stagnation state.

5-3(c) Limiting Velocity for a Fanno Line

Consider an infinitesimal change of state that takes place in the neighborhood of the maximum entropy state; for example, state L in Fig. 5.5. Let the subscript L denote the state where the entropy is a maximum. For such an infinitesimal change of state, the corresponding change in enthalpy may be assumed to take place at constant entropy. Hence, for an infinitesimal change of state on a Fanno line in the neighborhood of the maximum entropy point, we may write equation 1.65 in the form

$$t_L \, ds_L = dh_L - v_L \, dp_L = 0 \tag{5.7}$$

At state L, equation 5.5, which defines the state of the fluid at any point on a Fanno line, becomes

$$h_L + \frac{G^2 v_L{}^2}{2} = H = \text{constant} \tag{5.8}$$

Hence,

$$dh_L + G^2 v_L \, dv_L = 0 \tag{5.9}$$

Solving equation 5.7 for dh_L and substituting the result into equation 5.9 gives

$$v_L \, dp_L + G^2 v_L \, dv_L = 0 \tag{5.10}$$

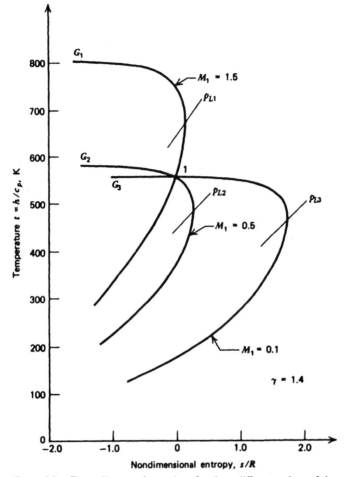

Figure 5.3. Fanno lines on the ts plane for three different values of the mass flux G and the same initial static conditions.

Noting that $v_L = 1/\rho_L$, $dv_L = -d\rho_L/\rho_L^2$, and $G^2 = V_L^2 \rho_L^2$, equation 5.10 may be rewritten as

$$\frac{dp_L}{\rho_L} - V_L^2 \frac{d\rho_L}{\rho_L} = 0 \tag{5.11}$$

Solving equation 5.11 for V_L^2, and remembering that it is assumed that $s = $ constant gives

$$V_L^2 = \frac{dp_L}{d\rho_L} = \left(\frac{\partial p}{\partial \rho}\right)_s = a^2 = \text{(local acoustic speed)}^2 = a^{*2} \tag{5.12}$$

Equation 5.12 shows that, in the case of the adiabatic flow of a compressible fluid in a constant-area duct in the presence of wall friction alone, the entropy of the fluid is at a maximum when its speed is equal to the local acoustic speed, and the Mach number $M_L = 1.0$. Thus, for the upper portion of the Fanno line, $V < a^*$, and $M < 1$. Conversely, for the lower portion of the Fanno line, $V > a^*$, and $M > 1$. It follows, therefore, that if the initial velocity is subsonic, the effect of friction is to accelerate the flow to the local acoustic velocity as a limit. Conversely, if the initial velocity is supersonic, the effect of friction is to decelerate the flow to the local acoustic velocity as a limit. The Fanno line shows that for a subsonic flow, the pressure decreases and the velocity increases, while for a supersonic flow, the reverse occurs.

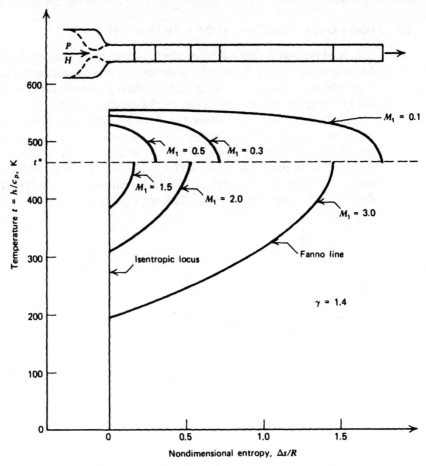

Figure 5.4. Fanno lines on the *ts* plane for different values of the mass flux *G* and the same initial stagnation conditions.

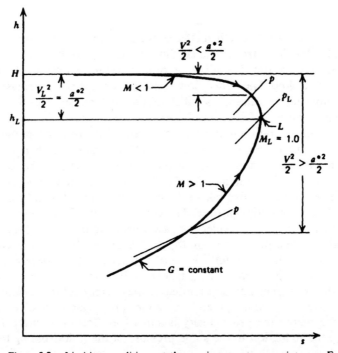

Figure 5.5. Limiting conditions at the maximum entropy point on a Fanno line.

251

It is of interest to consider the effects caused by increasing the resistance to flow because of increasing the length of the duct. Let it be assumed that in all cases the length of the constant-area duct and the mass flux are such that the velocity of the gas crossing the exit section is the local acoustic velocity.

Consider first the case where the duct operates with constant inlet conditions (p_1, V_1, h_1, s_1) and with the limiting value of exit pressure $p = p_L$, corresponding to the limiting velocity $V_L = a^*$. In the exit section the thermodynamic properties of the fluid are denoted by s_L, h_L, and v_L. The entropy increase for the fluid between states 1 and L is, accordingly, $s_L - s_1$ and remains constant if G remains constant; that is, if V_1 remains constant. The corresponding Fanno line is curve 1-L1, in Fig. 5.6.

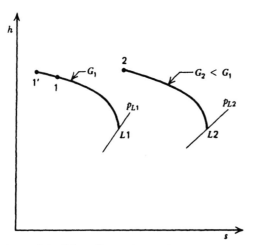

Figure 5.6. Effect of increasing the flow resistance on flow along a Fanno line.

Increasing the duct length increases the flow resistance, which is reflected in a larger value for the entropy change. Since p_1 and p_L are fixed, however, the entropy change $s_L - s_1$ is also fixed. The preceding conditions cannot be satisfied by the original Fanno line. The added flow resistance causes G to decrease with the result that the flow process lies on a different Fanno line having a smaller value of G, illustrated by curve 2-L2 in Fig. 5.6. If it is desired to maintain the original value of G, the initial conditions must be changed. The upstream pressure must be increased until the entropy at the new initial state differs from that for the original initial state by an amount equal to the increase in entropy caused by the added flow resistance. As illustrated in Fig. 5.6, the new initial state point, point 1', therefore, lies on the same Fanno line but to the left of the original initial state; that is, at the decreased value of initial entropy. The corresponding Fanno line is curve 1'-L1.

Refer to Fig. 5.7 and consider the case where the initial conditions are held constant and the back pressure p_o on the duct is gradually reduced. The flow velocity in the exit section increases until $V_L = a^*$ is reached, the corresponding exit pressure being $p_L = p_o = p^*$. Reducing p_o below p_L has no influence on the value of G, because the expansion waves generated by the decrease of the pressure p_o travel only with the acoustic speed and, therefore, cannot propagate themselves upstream to the exit section where the exit velocity is sonic [see Section 4-6(c)]. If p_o is reduced below p_L, the reduction in the static pressure of the fluid from p_L to p_o takes place

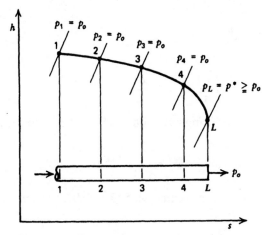

Figure 5.7. Effect of decreasing the back pressure on flow along a Fanno line.

beyond the exit section where the limitation imposed by the constant duct area is no longer applicable.

The preceding discussions show that, for a flow having the characteristics $dA = \delta Q = \delta W = \delta D = g\,dz = 0$ and $G =$ constant, there is a definite length of duct that causes the flow in the exit section to become *choked*; that length of duct is called the *critical length of duct* and is denoted by L^*. Hereafter, the values of the flow variables when the exit section of the duct operates with $M = 1$ will be denoted by a^*, p^*, ρ^*, s^*, t^*, etc. The reader should be careful not to confuse the critical conditions for the Fanno line with those for isentropic flow.

5–4 DYNAMICS OF STEADY ONE-DIMENSIONAL ADIABATIC FLOW WITH FRICTION IN A CONSTANT-AREA DUCT

The thermodynamics of steady adiabatic flow with wall friction in a constant-area duct is investigated in Section 5–3. In this section, the dynamic equation for flow with wall friction is derived, and data for the wall *friction coefficient* f are presented.

5–4(a) The Dynamic or Momentum Equation

Table 3.1 presents the general dynamic or momentum equation for the steady one-dimensional flow of a fluid (equation 3.160). In the special case where wall friction is the sole driving potential for the flow, $\delta D/A = g\,dz = 0$, and equation 3.160 reduces to

$$dp + \rho V\,dV + \frac{\rho V^2}{2}\left(\frac{4\mathfrak{f}\,dx}{\mathscr{D}}\right) = 0 \tag{5.13}$$

It is convenient to express equation 5.13 in terms of the *mass flux* $G = \dot{m}/A$ by eliminating V and dV from that equation. Expressions for V and dV in terms of G are readily derived from the continuity equation for steady flow. Thus, from equation 5.2,

$$V = \frac{\dot{m}}{A}v = Gv \tag{5.14}$$

where

$$G = \frac{\dot{m}}{A} \tag{5.15}$$

and

$$dG = -\dot{m}\frac{dA}{A^2} = -G\frac{dA}{A} \tag{5.16}$$

Differentiating equation 5.14 and introducing equations 5.15 and 5.16 into the result yields

$$dV = G\,dv + v\,dG = G\left(dv - v\frac{dA}{A}\right) \tag{5.17}$$

Substituting from equations 5.14 and 5.17 into equation 5.13 and dividing throughout by the specific volume v gives

$$\frac{dp}{v} + G^2\left(\frac{dv}{v} - \frac{dA}{A}\right) + \frac{G^2}{2}\left(\frac{4\mathfrak{f}\,dx}{\mathscr{D}}\right) = 0 \tag{5.18}$$

Equation 5.18 is the desired general dynamic equation. To integrate equation 5.18, the relationships between p, v, A, G, and $\mathfrak{f}$ are needed.

For a duct having a constant cross-sectional area, which is assumed here, $dA = 0$, and equation 5.18 reduces to

$$\frac{dp}{v} + G^2\left(\frac{dv}{v} + \frac{1}{2}\frac{4\mathfrak{f}\,dx}{\mathscr{D}}\right) = 0 \tag{5.19}$$

If either a constant value of $\mathfrak{f}$, or the mean value of $\mathfrak{f}$, denoted by $\bar{\mathfrak{f}}$, is employed, then the integrated form of equation 5.19 for a constant-area duct may be written as

$$\int_1^2 \frac{dp}{v} + G^2\left(\ln\frac{v_2}{v_1} + \frac{1}{2}\frac{4\bar{\mathfrak{f}}L}{\mathscr{D}}\right) = 0 \tag{5.20}$$

where

$$L = \int_1^2 dx = x_2 - x_1 \tag{5.21}$$

and

$$\bar{\mathfrak{f}} = \frac{1}{L}\int_{x_1}^{x_2} \mathfrak{f}\,dx \tag{5.22}$$

To complete the integration of equation 5.20, a relationship between p and v is needed. The dynamic equation is integrated in Section 5–5(b) for a perfect gas, and a method for applying the equation to the flow of an imperfect gas is presented in Section 5–6.

It is pointed out in Section 3–5(d) that the *friction coefficient* $\mathfrak{f}$ is that employed in the Fanning equation for the decrease in the pressure of the fluid caused by the effect of friction in a duct of hydraulic diameter $\mathscr{D}$. If dx is an infinitesimal length of the duct, and $dp_\mathfrak{f}$ denotes the corresponding *pressure drop*, then

$$dp_\mathfrak{f} = \frac{\rho V^2}{2}\left(\frac{4\mathfrak{f}\,dx}{\mathscr{D}}\right) = \frac{G^2 v}{2}\left(\frac{4\mathfrak{f}\,dx}{\mathscr{D}}\right) \tag{5.23}$$

Hydraulic data pertaining to the decrease in the pressure of a fluid flowing in a pipe are frequently correlated in terms of the D'Arcy friction factor $\mathfrak{f}' = 4\mathfrak{f}$. The D'Arcy equation is

$$dp_\mathfrak{f} = \frac{\rho V^2}{2}\left(\frac{\mathfrak{f}'\,dx}{\mathscr{D}}\right) \tag{5.24}$$

5-4(b) The Friction Coefficient f

It may be shown by dimensional analysis (see Section 5-8) that the friction coefficient f for a compressible fluid flowing in a duct is a function of the Reynolds number Re and the Mach number M. Experiments with steam and also with air indicate, however, that in the regime of well-developed turbulence, f is practically independent of the Mach number. Hence, it may be assumed that f depends only on the Reynolds number Re for both compressible and incompressible fluids. It also depends on the roughness of the pipe surface as demonstrated by Nikuradse, who conducted experiments with water on the flow through pipes having surfaces with measurable roughness produced by sand grains of diameter ε; the ratio ε/D is termed the *relative roughness*.[2]

For small values of the Reynolds number (Re less than 2000, approximately), the flow in a pipe is entirely laminar and the friction coefficient is a function of Re alone (see equation 5.175).

If the flow is turbulent, Re > 2000 approximately, but ε/D is very small, all of the experimental data for f = f(Re) lie on a single curve. Such a wall surface is said to be of *ultimate smoothness*, and the flow is termed *smooth pipe flow*. For smooth pipe flow, von Karman gives the following equation.[3]

$$\frac{1}{\sqrt{f'}} = 1.74 - 2 \log\left(\frac{18.6}{\mathrm{Re}\sqrt{f'}}\right) \tag{5.25}$$

If the surface of the pipe is very rough, that is, large values for ε/D, and if Re also has large values, experiments show that f' is independent of Re. Such a surface is said to be *wholly rough*. For the flow in a rough pipe, Prandtl gives the following equation.[4]

$$\frac{1}{\sqrt{f'}} = 1.74 + 2 \log (\varepsilon/D) \tag{5.26}$$

Colebrook conducted experiments with clean commercial pipes in the turbulent flow regime between ultimate smooth and wholly rough pipe surfaces; in that regime f' depends on both Re and ε/D. He derived the following empirical equation for that regime.[5]

$$\frac{1}{\sqrt{f'}} = -2 \log\left(\frac{\varepsilon/D}{3.7} + \frac{2.51}{\mathrm{Re}\sqrt{f'}}\right) \tag{5.27}$$

Values of ε for various surfaces are presented in Table 5.2.

Table 5.2 Values of the Roughness ε for Clean Commercial Pipes

Type of Pipe	Equivalent Sand Grain Diameter	
	ε, ft	ε, m
Asphalted cast iron	0.0004	0.000122
Cast iron	0.00085	0.000259
Commercial steel	0.00015	0.0000457
Concrete	0.001 to 0.010	0.000305 to 0.00305
Drawn tubing	0.000005	0.00000152
Galvanized iron	0.0005	0.000152
Riveted steel	0.003 to 0.030	0.000914 to 0.00914

Glass, drawn brass, aluminum, lead: smooth.

(Based on Reference 6).

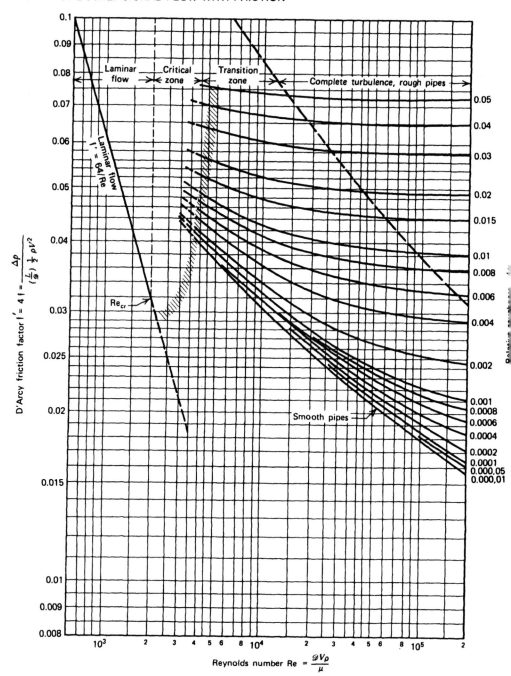

Figure 5.8 is the so-called *Moody chart*, which presents $\mathfrak{f}'$ as a function of Re for several different constant values of ε/D, the relative roughness.[6]

Genereaux proposed the following relationship, in the range of Re $= 4 \cdot 10^3$ to $20 \cdot 10^6$, as being sufficiently accurate for most engineering work.[7] Thus,

$$\mathfrak{f} = 0.04 \, Re^{-0.16} \tag{5.28}$$

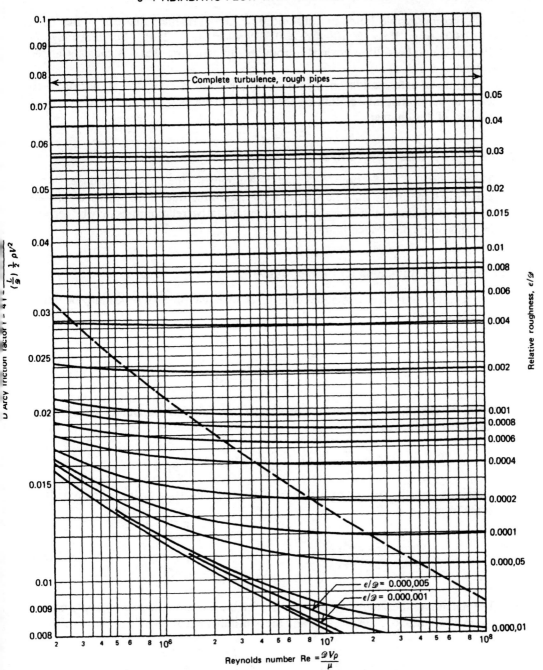

Figure 5.8. The D'Arcy friction coefficient for pipe flow as a function of the Reynolds number for constant values of the relative roughness $\varepsilon/\mathscr{D}$ (taken from Reference 6).

For compressible fluids in subsonic flow ($M < 1$), the value of $\mathfrak{f}$ varies from $\mathfrak{f} = 0.005$ when $\mathrm{Re} = 5 \cdot 10^4$ to $\mathfrak{f} = 0.0039$ when $\mathrm{Re} = 2 \cdot 10^5$. For that range of Reynolds numbers, little error is introduced by assuming that $\mathfrak{f}$ is constant at the value $\mathfrak{f} = 0.005$.

Additional information on pipe flow with friction is presented in Section 5–10 and, for additional information on the friction coefficient, refer to the references at the end of this chapter.

5–5 THE FANNO LINE FOR PERFECT GASES

The discussions in Sections 5–3 and 5–4 apply to the steady one-dimensional adiabatic flow of any fluid flowing in a constant-area duct in the presence of wall friction. If the flowing fluid is a perfect gas, quantitative relationships may be derived in terms of the flow Mach number.

5–5(a) Fanno Line Equations for Perfect Gases

For a perfect gas the energy equation is given by equation 3.45, which may be transformed to read

$$\frac{t_2}{t_1} = \frac{1 + \dfrac{\gamma - 1}{2} M_1{}^2}{1 + \dfrac{\gamma - 1}{2} M_2{}^2} \tag{5.29}$$

From the continuity equation for a perfect gas (see Table 3.2), we obtain the following for a duct of constant area.

$$G = \frac{\dot{m}}{A} = p_1 M_1 \left(\frac{\gamma}{R t_1}\right)^{1/2} = p_2 M_2 \left(\frac{\gamma}{R t_2}\right)^{1/2} \tag{5.30}$$

Assuming that the specific heat ratio γ is constant, equation 5.30 may be rewritten in the form

$$\frac{p_2}{p_1} = \frac{M_1}{M_2}\left(\frac{t_2}{t_1}\right)^{1/2} = \frac{M_1}{M_2}\left(\frac{1 + \dfrac{\gamma - 1}{2} M_1{}^2}{1 + \dfrac{\gamma - 1}{2} M_2{}^2}\right)^{1/2} \tag{5.31}$$

The remaining thermodynamic properties for the gas are obtained by applying the thermal equation of state $p = \rho R t$ and the continuity equation $\rho V = \text{constant}$. Thus,

$$\frac{\rho_2}{\rho_1} = \frac{v_1}{v_2} = \frac{p_2}{p_1}\frac{t_1}{t_2} = \frac{V_1}{V_2} = \frac{M_1}{M_2}\left(\frac{1 + \dfrac{\gamma - 1}{2} M_2{}^2}{1 + \dfrac{\gamma - 1}{2} M_1{}^2}\right)^{1/2} \tag{5.32}$$

The ratio of the stagnation pressures for two states on the same Fanno line is given by (see Table 3.3)

$$\frac{P_2}{P_1} = \frac{p_2}{p_1}\left(\frac{1 + \dfrac{\gamma - 1}{2} M_2{}^2}{1 + \dfrac{\gamma - 1}{2} M_1{}^2}\right)^{\gamma/(\gamma - 1)} = \frac{M_1}{M_2}\left(\frac{1 + \dfrac{\gamma - 1}{2} M_2{}^2}{1 + \dfrac{\gamma - 1}{2} M_1{}^2}\right)^{(\gamma + 1)/2(\gamma - 1)} \tag{5.33}$$

The entropy change as a function of the Mach number follows immediately from equations 3.88 and 5.33. Thus,

$$\frac{s_2 - s_1}{R} = \ln \frac{P_1}{P_2} = \ln \left[\frac{M_2}{M_1} \left(\frac{1 + \dfrac{\gamma - 1}{2} M_1{}^2}{1 + \dfrac{\gamma - 1}{2} M_2{}^2} \right)^{(\gamma + 1)/2(\gamma - 1)} \right] \tag{5.34}$$

The impulse function given by equation 3.139, after noting that $A_1 = A_2$, and substituting for p_2/p_1 from equation 5.31, becomes

$$\frac{\mathscr{F}_2}{\mathscr{F}_1} = \frac{p_2}{p_1} \frac{A_2}{A_1} \frac{(1 + \gamma M_2{}^2)}{(1 + \gamma M_1{}^2)} = \frac{M_1(1 + \gamma M_2{}^2)}{M_2(1 + \gamma M_1{}^2)} \left(\frac{1 + \dfrac{\gamma - 1}{2} M_1{}^2}{1 + \dfrac{\gamma - 1}{2} M_2{}^2} \right)^{1/2} \tag{5.35}$$

5–5(b) The Friction Parameter for a Constant-Area Duct

From equation 3.170, Table 3.2, the momentum equation for the steady one-dimensional flow of a perfect gas (in the absence of obstructions or body forces) may be transformed to read as follows:

$$\frac{dp}{p} + \gamma M^2 \frac{dV}{V} + \frac{\gamma M^2}{2} \left(\frac{4 \mathfrak{f} \, dx}{\mathscr{D}} \right) = 0 \tag{5.36}$$

since (see equation 5.43)

$$\frac{\gamma}{2} dM^2 + \frac{\gamma M^2}{2} \frac{dt}{t} = \gamma M^2 \frac{dV}{V}$$

It is desirable to relate the *friction parameter* $4\mathfrak{f} \, dx/\mathscr{D}$ to the flow Mach number M so that the effect of friction on the flow Mach number may be studied conveniently. To accomplish that, dp/p and dV/V must be eliminated from equation 5.36. It is assumed that the area of the duct is constant ($dA = 0$).

To eliminate dp/p, the continuity equation for perfect gases, $\dot{m} = VAp/Rt =$ constant, is differentiated logarithmically. Thus, for a constant-area duct,

$$\frac{dp}{p} = \frac{dt}{t} - \frac{dV}{V} \tag{5.37}$$

Combining equations 5.36 and 5.37 and rearranging gives

$$\frac{\gamma M^2}{2} \left(\frac{4 \mathfrak{f} \, dx}{\mathscr{D}} \right) = (1 - \gamma M^2) \frac{dV}{V} - \frac{dt}{t} \tag{5.38}$$

For the adiabatic flow assumed here, the energy equation (see equation 3.41) may be employed for eliminating dt/t. Thus, for a perfect gas,

$$c_p \, dt + V \, dV = \frac{\gamma R}{\gamma - 1} dt + V \, dV = 0 \tag{5.39}$$

Hence,

$$\gamma R \, dt + (\gamma - 1)V^2 \frac{dV}{V} = 0 \tag{5.40}$$

Dividing equation 5.40 by $a^2 = \gamma R t$, introducing $M^2 = V^2/a^2$, and rearranging gives

$$\frac{dt}{t} = (1 - \gamma)M^2 \frac{dV}{V} \tag{5.41}$$

The term dV/V may be expressed in terms of dM/M and dt/t. Thus,

$$V^2 = M^2 a^2 = M^2(\gamma R t) \tag{5.42}$$

Hence,

$$\frac{dV}{V} = \frac{dM}{M} + \frac{1}{2}\frac{dt}{t} \tag{5.43}$$

Substituting for dt/t from equation 5.41 into equation 5.43 and rearranging gives

$$\frac{dV}{V} = \frac{dM}{M\left(1 + \dfrac{\gamma - 1}{2} M^2\right)} \tag{5.44}$$

Substituting for dV/V from equation 5.44 into equation 5.41 gives

$$\frac{dt}{t} = \frac{(1 - \gamma)M^2}{1 + \dfrac{\gamma - 1}{2} M^2} \frac{dM}{M} \tag{5.45}$$

Substituting equations 5.44 and 5.45 into equation 5.38 and dividing by M^2 gives the desired relationship. Thus,

$$\frac{4\mathfrak{f}\,dx}{\mathscr{D}} = \frac{2(1 - M^2)\,dM}{\gamma M^3 \left(1 + \dfrac{\gamma - 1}{2} M^2\right)} \tag{5.46}$$

Equation 5.46 relates the *friction parameter* $4\mathfrak{f}\,dx/\mathscr{D}$ to the flow Mach number M for adiabatic flow in a constant-area duct in the presence of wall friction alone. The changes in the state of the gas are due entirely to overcoming the shearing stresses in the boundary layer.

It is convenient for the purpose of calculation to integrate the right-hand side of equation 5.46 between the limits $M = M$ and $M = 1$, the corresponding limits for the left-hand side being $x = 0$ and $x = L^*$. Thus,

$$\int_0^{L^*} \frac{4\mathfrak{f}\,dx}{\mathscr{D}} = \left[-\frac{1}{\gamma M^2} - \frac{\gamma + 1}{2\gamma} \ln\left(\frac{M^2}{1 + \dfrac{\gamma - 1}{2} M^2} \right) \right]_{M=M}^{M=1} \tag{5.47}$$

The result is,

$$\frac{4\bar{\mathfrak{f}}L^*}{\mathscr{D}} = \frac{1 - M^2}{\gamma M^2} + \frac{\gamma + 1}{2\gamma} \ln\left(\frac{\dfrac{\gamma + 1}{2} M^2}{1 + \dfrac{\gamma - 1}{2} M^2} \right) \tag{5.48}$$

where $\bar{\mathfrak{f}}$ is the mean value of $\mathfrak{f}$ for the length L^* as defined by equation 5.22, and L^* is the length of duct required to change the Mach number of the flow from $M = M$ to $M = 1$. The friction parameter $4\bar{\mathfrak{f}}L^*/\mathscr{D}$ is tabulated in Table C.7 as a function of the Mach number M for several values of the specific heat ratio γ.

5-5(c) Relationship between Duct Length and Mach Number in a Constant-Area Duct

It is desirable to relate the length of duct L to the Mach numbers at the inlet and exit cross sections of the duct. Assume that the fluid is a perfect gas and that the change in the static pressure of the fluid is due entirely to wall friction, so that $dA = \delta Q = \delta W = g\, dz = \delta D = 0$, and $G = \text{constant}$.

Let M_1 and M_2 denote the values of M at two arbitrary cross sections of a constant-area duct. From equation 5.48, it follows that the value of $4\bar{f}L/\mathscr{D}$ for accomplishing the Mach number change is given by

$$\frac{4\bar{f}L}{\mathscr{D}} = \left(\frac{4\bar{f}L^*}{\mathscr{D}}\right)_{M_1} - \left(\frac{4\bar{f}L^*}{\mathscr{D}}\right)_{M_2} \tag{5.49}$$

where the subscripts denote the value of M to be substituted into the right-hand side of equation 5.48. Hence,

$$\frac{4\bar{f}L}{\mathscr{D}} = \frac{1}{\gamma M_1{}^2}\left(\frac{M_2{}^2 - M_1{}^2}{M_2{}^2}\right) + \frac{\gamma+1}{2\gamma}\ln\left[\frac{M_1{}^2\left(1 + \dfrac{\gamma-1}{2}M_2{}^2\right)}{M_2{}^2\left(1 + \dfrac{\gamma-1}{2}M_1{}^2\right)}\right] \tag{5.50}$$

By means of equation 5.50, the length L of constant-area duct for accomplishing the Mach number change from M_1 to M_2 is readily calculated.

For some problems it is more convenient to relate $4\bar{f}L/\mathscr{D}$ to M_1 and V_1/V_2 instead of to M_2. Thus, combining equations 5.32 and 5.50, we obtain

$$\frac{4\bar{f}L}{\mathscr{D}} = \frac{1}{\gamma M_1{}^2}\left(1 + \frac{\gamma-1}{2}M_1{}^2\right)\left[1 - \left(\frac{V_1}{V_2}\right)^2\right] + \frac{\gamma+1}{2\gamma}\ln\left[\left(\frac{V_1}{V_2}\right)^2\right] \tag{5.51}$$

For a perfect gas, the velocity ratio V_1/V_2 and the static-pressure ratio p_1/p_2 may be related to M_1, the initial Mach number for the flow. Thus,

$$(h_2 - h_1) = c_p(t_2 - t_1) = \frac{\gamma}{\gamma-1}(Rt_2 - Rt_1) = \frac{\gamma}{\gamma-1}(p_2v_2 - p_1v_1) \tag{5.52}$$

Substituting equation 5.52 into the Fanno line equation (see equation 5.6) gives

$$\frac{\gamma}{\gamma-1}p_2v_2 + \frac{G^2v_2{}^2}{2} = \frac{\gamma}{\gamma-1}p_1v_1 + \frac{G^2v_1{}^2}{2} \tag{5.53}$$

Solving for p_2/p_1 yields

$$\frac{p_2}{p_1} = \frac{v_1}{v_2}\left[1 + \left(\frac{\gamma-1}{2\gamma p_1 v_1}\right)G^2(v_1{}^2 - v_2{}^2)\right] \tag{5.54}$$

But, for the Fanno line, $G^2 = (V_1/v_1)^2 = (V_2/v_2)^2$. Substituting $G^2 = (V_1/v_1)^2$ into equation 5.54 and noting that $V_1{}^2/\gamma p_1 v_1 = M_1{}^2$ yields

$$\frac{p_2}{p_1} = \frac{v_1}{v_2}\left\{1 + \frac{\gamma-1}{2}M_1{}^2\left[1 - \left(\frac{v_2}{v_1}\right)^2\right]\right\} \tag{5.55}$$

or

$$\frac{p_2}{p_1} = \frac{V_1}{V_2}\left\{1 + \frac{\gamma-1}{2}M_1{}^2\left[1 - \left(\frac{V_2}{V_1}\right)^2\right]\right\} \tag{5.56}$$

Equation 5.56 relates p_2/p_1, V_1/V_2, and M_1.

Example 5.2. Air flows in a 0.100 m ID duct under adiabatic conditions. Calculate the length of duct required to raise the Mach number of the air from $M_1 = 0.5$ to $M_2 = 0.9$ if the average value of the friction coefficient is $\bar{f} = 0.005$. Here $\mathcal{D} = D$.

Solution

From equation 5.48, to increase M from 0.5 to 1.0,

$$\frac{4\bar{f}L_1^*}{\mathcal{D}} = \frac{1 - (0.5)^2}{1.4(0.5)^2} + \frac{1.4 + 1}{2(1.4)} \ln \left[\frac{\dfrac{1.4 + 1}{2}(0.5)^2}{1 + \dfrac{1.4 - 1}{2}(0.5)^2} \right]$$

$$\frac{4\bar{f}L_1^*}{\mathcal{D}} = 2.1429 + (0.85714) \ln (0.28571) = 1.0691$$

To increase from $M = 0.9$ to $M = 1$,

$$\frac{4\bar{f}L_2^*}{\mathcal{D}} = \frac{1 - (0.9)^2}{1.4(0.9)^2} + (0.85714) \ln \left[\frac{\dfrac{1.4 + 1}{2}(0.9)^2}{1 + \dfrac{1.4 - 1}{2}(0.9)^2} \right]$$

$$\frac{4\bar{f}L_2^*}{\mathcal{D}} = 0.1675 + (0.85714) \ln (0.83649) = 0.0145$$

Hence, to accomplish the Mach number change from $M_1 = 0.5$ to $M_2 = 0.9$,

$$\Delta L = L_1^* - L_2^* = \frac{\mathcal{D}}{4\bar{f}} (1.0691 - 0.0145) = 1.0546 \frac{\mathcal{D}}{4\bar{f}}$$

$$\Delta L = \frac{1.0546(0.100)}{4(0.005)} = 5.273 \text{ m } (17.30 \text{ ft})$$

5–5(d) Entropy Change Caused by Friction in a Constant-Area Duct

The energy expended in overcoming friction by a gas flowing through a duct will, of course, increase its entropy as the gas flows downstream. The increase in entropy is given by equation 5.34 in terms of the Mach numbers M_1 and M_2 at the inlet and the outlet of the duct, respectively. If the duct inlet Mach number $M_1 = M$, and the duct length is L^*, so that choking occurs and $M_2 = 1$, then $s_1 = s$ and $s_2 = s^*$, and equation 5.34 becomes

$$\frac{s^* - s}{R} = \ln \left\{ \frac{1}{M} \left[\frac{2}{\gamma + 1} \left(1 + \frac{\gamma - 1}{2} M^2 \right) \right]^{(\gamma + 1)/2(\gamma - 1)} \right\} \tag{5.57}$$

Equation 5.57 gives the increase in entropy caused by wall friction for an adiabatic flow that experiences a change in flow Mach number from $M = M$ to $M = 1$.

The rate of change in the entropy of the flowing gas with Mach number is given by the derivative ds/dM. Hence, from equation 5.57,

$$\frac{ds}{dM} = \frac{R(1 - M^2)}{M \left(1 + \dfrac{\gamma - 1}{2} M^2 \right)} \tag{5.58}$$

Equations 5.57 and 5.58 may be employed for investigating the effect of changes in the flow Mach number caused by wall friction on the entropy of a perfect gas flowing under adiabatic conditions in a constant-area duct. From equation 5.58, the entropy is a maximum when $ds/dM = 0$; that is, when $M = 1$. Since all natural processes, according to the second law of thermodynamics, tend to approach the state of maximum entropy, the foregoing indicates that, in the case of steady adiabatic flow with wall friction alone, the speed of the gas tends to approach its local acoustic speed. The latter conclusion applies when the initial velocity is either subsonic or supersonic. Hence, the effect of wall friction is to raise the Mach number for an initially subsonic flow to $M = 1$ as a limit, and to decrease that for an initially supersonic flow to $M = 1$ as a limit.

5–5(e) Effect of Friction on the Flow Properties in a Constant-Area Duct

By methods similar to those presented in Section 4–4(c), relationships may be derived for the flow variables dM/M, dp/p, $d\rho/\rho$, dt/t, dV/V, dP/P, $d\mathscr{F}/\mathscr{F}$, and ds/R, with the friction parameter $4\mathfrak{f}\, dx/\mathscr{D}$ as the independent variable. The results are presented in Table 5.3. Table 5.4 summarizes the effect of wall friction on the flow properties for a perfect gas.

Table 5.3 Influence Coefficients for the Simple Frictional Flow of a Perfect Gas in a Constant-Area Duct

$$\frac{dM}{M} = \frac{\gamma M^2 \left(1 + \dfrac{\gamma - 1}{2} M^2\right)}{2(1 - M^2)}\left(\frac{4\mathfrak{f}\, dx}{\mathscr{D}}\right) \tag{5.59}$$

$$\frac{dp}{p} = -\frac{\gamma M^2[1 + (\gamma - 1)M^2]}{2(1 - M^2)}\left(\frac{4\mathfrak{f}\, dx}{\mathscr{D}}\right) \tag{5.60}$$

$$\frac{d\rho}{\rho} = -\frac{\gamma M^2}{2(1 - M^2)}\left(\frac{4\mathfrak{f}\, dx}{\mathscr{D}}\right) \tag{5.61}$$

$$\frac{dt}{t} = -\frac{\gamma(\gamma - 1)M^4}{2(1 - M^2)}\left(\frac{4\mathfrak{f}\, dx}{\mathscr{D}}\right) \tag{5.62}$$

$$\frac{dV}{V} = \frac{\gamma M^2}{2(1 - M^2)}\left(\frac{4\mathfrak{f}\, dx}{\mathscr{D}}\right) \tag{5.63}$$

$$\frac{dP}{P} = -\frac{\gamma M^2}{2}\left(\frac{4\mathfrak{f}\, dx}{\mathscr{D}}\right) \tag{5.64}$$

$$\frac{d\mathscr{F}}{\mathscr{F}} = -\frac{\gamma M^2}{2(1 + \gamma M^2)}\left(\frac{4\mathfrak{f}\, dx}{\mathscr{D}}\right) \tag{5.65}$$

$$\frac{ds}{R} = \frac{\gamma M^2}{2}\left(\frac{4\mathfrak{f}\, dx}{\mathscr{D}}\right) \tag{5.66}$$

5–5(f) Fanno Line Tables for a Perfect Gas

For calculation purposes it is convenient to employ the limiting condition, where $M = 1$, as a reference state. The corresponding values of p/p^*, t/t^*, P/P^*, etc., are readily derived from the appropriate equations of Section 5–5(a) by inserting the value $M_1 = 1$. The resulting equations are presented in Table 5.5.

Table 5.4 Effect of Friction on
the Flow Properties for a Perfect
Gas in a Constant-Area Duct

Property	$M < 1$	$M > 1$
dp/p	$-$	$+$
dt/t	$-$	$+$
$d\rho/\rho$	$-$	$+$
dP/P	$-$	$-$
dV/V	$+$	$-$
dM/M	$+$	$-$
$d\mathscr{F}/\mathscr{F}$	$-$	$-$
ds/c_p	$+$	$+$

Table 5.5 Property Ratios for the Simple Frictional Flow of a Perfect Gas

$$\frac{p}{p^*} = \frac{1}{M}\left[\left(\frac{2}{\gamma+1}\right)\left(1 + \frac{\gamma-1}{2}M^2\right)\right]^{-1/2} \tag{5.67}$$

$$\frac{\rho}{\rho^*} = \frac{V^*}{V} = \frac{v^*}{v} = \frac{1}{M}\left[\left(\frac{2}{\gamma+1}\right)\left(1 + \frac{\gamma-1}{2}M^2\right)\right]^{1/2} \tag{5.68}$$

$$\frac{t}{t^*} = \frac{a^2}{a^{*2}} = \left[\left(\frac{2}{\gamma+1}\right)\left(1 + \frac{\gamma-1}{2}M^2\right)\right]^{-1} \tag{5.69}$$

$$\frac{P}{P^*} = \frac{1}{M}\left[\left(\frac{2}{\gamma+1}\right)\left(1 + \frac{\gamma-1}{2}M^2\right)\right]^{(\gamma+1)/2(\gamma-1)} \tag{5.70}$$

$$\frac{\mathscr{F}}{\mathscr{F}^*} = \frac{1+\gamma M^2}{M\left[2(\gamma+1)\left(1 + \frac{\gamma-1}{2}M^2\right)\right]^{1/2}} \tag{5.71}$$

$$\frac{s-s^*}{R} = -\ln\left\{\frac{1}{M}\left[\left(\frac{2}{\gamma+1}\right)\left(1 + \frac{\gamma-1}{2}M^2\right)\right]^{(\gamma+1)/2(\gamma-1)}\right\} \tag{5.72}$$

$$\frac{4\bar{f}L^*}{\mathscr{D}} = \frac{1-M^2}{\gamma M^2} + \left(\frac{\gamma+1}{2\gamma}\right)\ln\left\{M^2\left[\left(\frac{2}{\gamma+1}\right)\left(1 + \frac{\gamma-1}{2}M^2\right)\right]^{-1}\right\} \tag{5.73}$$

Values of the above property ratios, with the exception of $(s - s^*)/R$, are presented as a function of the Mach number M in Table C.7 for several values of γ. The general arrangement of Table C.7 and the limiting values of the property ratios are presented in Table 5.6. For more details on the application of tables such as Table C.7, the discussion in Section 4–4(b) should be consulted. The discussion in Section 4–3(e), concerning the downstream physical boundary conditions at the exit of a nozzle, is also applicable to the exit of a constant-area duct.

A FORTRAN program, subroutine FANNO, is presented below for calculating the property ratios for a simple frictional flow. The input to this program is identical to that discussed in Section 4–4(b) for the program for the property ratios for isentropic flow. The results presented in Table C.7 for $\gamma = 1.4$ are obtained by the following specification.

Table 5.6 Limiting Values of the Property Ratios for the Simple Frictional Flow of a Perfect Gas

M	$M^* = \dfrac{V}{V^*} = \dfrac{\rho^*}{\rho}$	$\dfrac{p}{p^*}$	$\dfrac{P}{P^*}$	$\dfrac{t}{t^*}$	$\dfrac{\mathscr{F}}{\mathscr{F}^*}$	$\dfrac{4\bar{f}L^*}{\mathscr{D}}$
0	0	∞	∞	$\dfrac{\gamma+1}{2}$	∞	∞
1	1	1	1	1	1	0
∞	$\left(\dfrac{\gamma+1}{\gamma-1}\right)^{1/2}$	0	∞	0	$\dfrac{\gamma}{(\gamma^2-1)^{1/2}}\left[-\dfrac{1}{\gamma}+\left(\dfrac{\gamma+1}{2\gamma}\right)\ln\left(\dfrac{\gamma+1}{\gamma-1}\right)\right]$	

G = 1.4, J(1) = 50, J(2) = 50, J(3) = 10, J(4) = 20, J(5) = 12, J(6) = 5, J(7) = 0, DM(1) = 0.01, DM(2) = 0.02, DM(3) = 0.05, DM(4) = 0.1, DM(5) = 0.5, DM(6) = 1.0

```
      SUBROUTINE FANNO

C     FANNO LINE PROPERTY RATIOS FOR A PERFECT GAS

      REAL M,MS $ DIMENSION J(9),DM(9) $ DATA INF/6H INFIN/
      NAMELIST /DATA/ G,J,DM $ READ (5,DATA)
      G1=(G-1.0)/2.0 $ G2=2.0/(G+1.0) $ G3=1.0/G2
      G4=0.5*(G+1.0)/(G-1.0) $ G5=G2**G4 $ G6=2.0*(G+1.0) $ G7=G3/G
      M=0.0 $ MS=0.0 $ T=G3 $ I=1 $ L=0
      WRITE (6,2000) G $ WRITE (6,2010) M,MS,T,INF,INF,INF,INF

C     CALCULATE PROPERTY RATIOS

   40 N=J(I) $ DO 50 K=1,N $ M=M+DM(I) $ C=1.0+G1*M**2 $ MS=M/SQRT(G2*C)
      T=G3/C $ P=MS/M**2 $ PO=G5*C**G4/M $ F=(1.0+G*M**2)/(M*SQRT(G6*C))
      X=(1.0-M**2)/(G*M**2)+G7*ALOG(M**2*T)
      IF (ABS(X).LT.1.0E-12) X=0.0 $ WRITE (6,2020) M,MS,T,P,PO,F,X
      L=L+1 $ IF (L.LT.50) GO TO 50 $ WRITE (6,2000) G $ L=0
   50 CONTINUE $ I=I+1 $ IF((I.EQ.10).OR.(J(I).EQ.0)) GO TO 60 $GO TO 40
   60 MS=SQRT((G+1.0)/(G-1.0)) $ T=0.0 $ P=0.0 $ F=G/SQRT(G**2-1.0)
      X=G7*ALOG((G+1.0)/(G-1.0))-1.0/G
      WRITE (6,2030) INF,MS,T,P,INF,F,X $ RETURN

 2000 FORMAT (1H1,19X,20HFANNO LINE,  GAMMA =,F5.2//4X,1HM,4X,7HM*,R*/R,
     14X,4HT/T*,9X,4HP/P*,9X,6HPO/PO*,7X,4HF/F*,9X,6H4FL*/D/)
 2010 FORMAT (1H ,F5.2,F10.5,E13.5,3X,A6,3(7X,A6))
 2020 FORMAT (1H ,F5.2,F10.5,5E13.5)
 2030 FORMAT (A6,F10.5,2E13.5,3X,A6,4X,2E13.5)
      END
```

Example 5.3. A gas for which $\gamma = 1.4$ flows through a perfectly insulated constant-area duct under steady conditions. Inside the duct there is an obstruction for which the drag is given by $D = \frac{1}{2}\rho_1 V_1^2 C_D A$, where A is the cross-sectional area of the duct. The Mach number of the gas in the plane immediately upstream to the obstruction is $M = 0.5$, and the drag coefficient for the obstruction is $C_D = 1.0$. The static pressure immediately upstream to the obstruction is $2.0 \cdot 10^5$ N/m^2. (a) Derive the general equation for the pressure change caused by the obstruction. (b) Calculate the static pressure drop caused by the obstruction.

Solution

(a) To obtain the general equation for the pressure change caused by the obstruction, the dynamic equation and the continuity equation for the steady flow of a perfect gas in a constant-area duct are employed. *Dynamic equation* (see equation 3.170, Table 3.2)

$$dp + p\frac{\gamma}{2}dM^2 + p\frac{\gamma M^2}{2}\frac{dt}{t} + p\frac{\gamma M^2}{2}\left(\frac{4\mathfrak{f}\,dx}{\mathscr{D}}\right) + \frac{\delta D}{A} = 0 \qquad (3.170)$$

For the case being considered,

$$\frac{\gamma M^2}{2}\left(\frac{4\mathfrak{f}\,dx}{\mathscr{D}}\right) = 0$$

Hence, the dynamic equation for the obstruction is

$$dp + p\frac{\gamma}{2}dM^2 + p\frac{\gamma M^2}{2}\frac{dt}{t} + \frac{\delta D}{A} = 0 \qquad (a)$$

Logarithmic differentiation of equation 3.167, Table 3.2, the continuity equation for perfect gases, yields

$$\frac{1}{2}\frac{dt}{t} = \frac{dp}{p} + \frac{dM}{M} \qquad (b)$$

Combining equations (a) and (b) gives

$$dp + \gamma\,d(pM^2) + \frac{\delta D}{A} = 0 \qquad (c)$$

Integrating, substituting for $D = \frac{1}{2}\rho_1 V_1{}^2 C_D A = (\gamma/2)p_1 M_1{}^2 C_D A$, and solving for p_2/p_1 gives

$$\frac{p_2}{p_1} = \frac{1 + \gamma\left(1 - \dfrac{C_D}{2}\right)M_1{}^2}{1 + \gamma M_2{}^2} \quad (\textit{for the obstruction}) \qquad (d)$$

For $\gamma = 1.4$, $C_D = 1$, and $M_1 = 0.5$, equation (d) for the pressure ratio created by the obstruction becomes

$$\frac{p_2}{p_1} = \frac{1 + 1.4(1 - 0.5)(0.5)^2}{1 + 1.4M_2{}^2} = \frac{1.175}{1 + 1.4M_2{}^2} \qquad (e)$$

(b)   The pressure ratio p_2/p_1 must also satisfy the relationship given by equation 5.31 which, for the case in question, becomes

$$\frac{p_2}{p_1} = \frac{0.5\sqrt{1.05}}{M_2(1 + 0.2M_2{}^2)^{1/2}} = \frac{0.513}{M_2(1 + 0.2M_2{}^2)^{1/2}} \qquad (f)$$

Plot equations (e) and (f) for assumed values of M_2. The point of intersection of the two curves establishes the value of p_2/p_1. The point of intersection gives

$$M_2 = 0.685 \quad \text{and} \quad \frac{p_2}{p_1} = 0.713$$

$$p_2 = (0.713)(2.0)10^5 = 1.426 \cdot 10^5 \text{ N/m}^2$$

$$\Delta p = p_2 - p_1 = (1.426 - 2.000)10^5 = -0.574 \cdot 10^5 \text{ N/m}^2 \text{ (8.325 lbf/in.}^2)$$

Example 5.4.   In some solid propellant rocket motors, the combustion chamber may be located some distance from the thrust nozzle. The two are connected by a constant diameter duct, called a blast tube, which is illustrated in Fig. 5.9. Consider a case where the nozzle throat diameter is designed to be 0.075 m without a blast tube, assuming isentropic flow in the nozzle inlet. Due to design considerations,

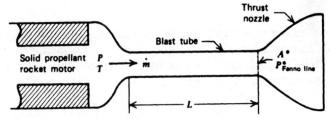

Figure 5.9. Rocket motor with a blast tube (Example 5.4).

a blast tube 3.0 m long is inserted between the motor and the thrust nozzle. Determine the mass flow rate reduction that occurs because of the presence of the blast tube. Assume $P = 68.95 \cdot 10^5$ N/m², $T = 3333.3$ K, $\gamma = 1.2$, $R = 322.82$ J/kg-K, and $\bar{f} = 0.005$.

Solution

From Table 3.2, for a choked flow,

$$\dot{m}^* = \frac{\Gamma P^* A^*}{a_o} \qquad (3.168)$$

Assume that the flow in the blast tube is adiabatic, so that T, and thus a_o, is the same for both the isentropic flow and the Fanno line flow. However, because of the presence of wall friction in the blast tube, P^* for the case with the blast tube will be less than P for the isentropic case. The gas constant R and specific heat ratio γ are constants. Thus, equation 3.168 gives

$$\frac{\dot{m}_{\text{blast tube}}}{\dot{m}_{\text{isentropic}}} = \frac{P^*_{\text{Fanno line}}}{P_{\text{isentropic}}} \qquad (a)$$

Because of the large stagnation pressure in the rocket motor, the flow is choked at the blast tube exit plane. Hence, $M = 1$ at that point, and

$$\frac{4\bar{f}L^*}{\mathcal{D}} = \frac{4(0.005)(3.0)}{(0.075)} = 0.80$$

From Table C.7, for $(4\bar{f}L^*/\mathcal{D}) = 0.80$

$$M_1 = 0.56329 \quad \text{and} \quad \frac{P_1}{P^*} = 1.2481$$

where $P_1 = P_{\text{isentropic}}$. From equation (a),

$$\frac{\dot{m}_{\text{blast tube}}}{\dot{m}_{\text{isentropic}}} = \frac{P^*_{\text{Fanno line}}}{P_{\text{isentropic}}} = \frac{1}{1.2481} = 0.8012$$

The presence of the blast tube reduces the mass flow rate by approximately 20 percent.

Example 5.5. A circular duct having an ID of 0.030 m, a surface roughness of 0.000030 m, and a length of 1.500 m is connected to the exit of a converging nozzle. Air at a stagnation pressure $P = 7.0 \cdot 10^5$ N/m² and a stagnation temperature $T = 300$ K is supplied to the nozzle. The ambient pressure at the exit of the duct is $1.0135 \cdot 10^5$ N/m². Determine the maximum mass flow rate of air through the nozzle and duct. Assume that the air behaves as a perfect gas with $\gamma = 1.40$ and $R = 287.04$ J/kg-K.

Solution

Assume that the flow in the converging nozzle is isentropic, and that the flow in the duct is a simple frictional flow (i.e., a Fanno line flow). Furthermore, assume that the frictional effects may be accounted for by an average value of the friction coefficient between the duct inlet and exit sections

$$\bar{f} = \frac{1}{2}(f_{inlet} + f_{exit}) \tag{a}$$

where f is determined from the Moody diagram (Fig. 5.8). The pipe roughness ratio is $\varepsilon/D = (0.000030/0.030) = 0.001$. The viscosity of air may be obtained from equation 1.30, repeated below,

$$\mu = \frac{\sqrt{t}}{0.552795 + \dfrac{2.810892 \cdot 10^2}{t} - \dfrac{13.508340 \cdot 10^4}{t^2} + \dfrac{39.353086 \cdot 10^6}{t^3} - \dfrac{41.419387 \cdot 10^8}{t^4}} \tag{b}$$

where t is in K and μ is in $(N\text{-}s/m^2)10^{-6}$. Because of the low back pressure, assume that the flow chokes at the exit section of the duct.

Because the friction coefficient depends on the Reynolds number which, in turn depends on the mass flow rate, this problem must be solved iteratively. The following procedure is one of several possibilities.

1. Determine a trial value for the average friction coefficient $\bar{f}$. For the first trial, assume that isentropic flow exists in both the nozzle and the duct. The flow properties throughout the duct are, therefore, the critical properties for isentropic flow. Calculate the corresponding Re and obtain $\bar{f}$ from the Moody chart.
2. Calculate $4\bar{f}L^*/\mathcal{D}$ and determine the duct inlet Mach number and the property ratios from Table C.7 (Fanno line).
3. From Table C.6 (isentropic flow), determine the property ratios at the nozzle exit (which is the duct inlet) corresponding to the Mach number obtained in step 2.
4. Calculate the properties at the duct inlet section (denoted by 1) and exit section (denoted by 2) from the property ratios determined in steps 3 and 4.
5. Determine the value of $\bar{f}$ corresponding to the properties obtained in step 4. Compare that value of $\bar{f}$ with the trial value of $\bar{f}$ determined in step 1. Repeat steps 2 through 5 until the value of $\bar{f}$ converges.
6. Calculate the corresponding mass flow rate.

The above procedure is now applied to the present problem.
(1) From Table C.6, for isentropic flow throughout,

$$\frac{t^*}{T} = 0.83333 \qquad \frac{\rho^*}{\rho_o} = 0.63394$$

For a perfect gas,

$$\rho_o = \frac{P}{RT} = \frac{(7.0)10^5}{(287.04)(300)} = 8.1289 \text{ kg/m}^3$$

$$t^* = (0.83333)(300) = 250.0 \text{ K}$$

$$V^* = a^* = (\gamma R t^*)^{1/2} = [1.4(287.04)(250.0)]^{1/2} = 316.96 \text{ m/s}$$

$$\rho^* = (0.63394)(8.1289) = 5.1532 \text{ kg/m}^3$$

Note that for the adiabatic flow of a perfect gas, t^* and V^* have the same values for both isentropic flow and simple frictional flow. From equation (b), for $t^* = 250.0$ K,

$$\mu^* = \frac{(250.0)^{1/2}}{0.552795 + \dfrac{2.810892 \cdot 10^2}{250.0} - \dfrac{13.508340 \cdot 10^4}{(250.0)^2} + \dfrac{39.353086 \cdot 10^6}{(250.0)^3} - \dfrac{41.419387 \cdot 10^8}{(250.0)^4}}$$

$$\mu^* = 16.232 \cdot 10^{-6} \text{ N-s/m}^2$$

$$\text{Re}^* = \frac{DV^*\rho^*}{\mu^*} = \frac{(0.030)(316.96)(5.1532)}{16.232 \cdot 10^{-6}} = 3.019 \cdot 10^6$$

From Fig. 5.8, for $\varepsilon/\mathscr{D} = 0.001$ and $\text{Re} = 3.019 \cdot 10^6$, $\bar{f}' = 0.0198$ and $\bar{f} = \bar{f}'/4 = 0.00495$.

(2) $$\frac{4\bar{f}L^*}{\mathscr{D}} = \frac{4(0.00495)(1.500)}{(0.030)} = 0.990$$

From Table C.7, for $4\bar{f}L^*/\mathscr{D} = 0.990$,

$$M_1 = 0.510 \qquad \frac{\rho^*}{\rho_1} = 0.54469$$

(3)   From Table C.6, for $M_1 = 0.510$,

$$M_1^* = 0.54469 \qquad \frac{t_1}{T} = 0.95055 \qquad \frac{\rho_1}{\rho_o} = 0.88093$$

(4) $t_1 = (0.95055)(300) = 285.17$ K

$V_1 = M_1^* a^* = (0.54469)(316.96) = 172.64$ m/s

$\rho_1 = (0.88093)(8.1289) = 7.1610$ kg/m^3

$\rho^* = (0.54469)(7.1610) = 3.9005$ kg/m^3

(5)   From equation (b),

$$\mu_1 = 17.813 \cdot 10^{-6} \text{ N-s/m}^2$$

$$\text{Re}_1 = \frac{DV_1\rho_1}{\mu_1} = \frac{(0.030)(172.64)(7.1610)}{17.813 \cdot 10^{-6}} = 2.082 \cdot 10^6$$

$$\text{Re}^* = \frac{DV^*\rho^*}{\mu^*} = \frac{(0.030)(316.96)(3.9005)}{16.232 \cdot 10^{-6}} = 2.285 \cdot 10^6$$

From Fig. 5.8, it is seen that, for $\varepsilon/\mathscr{D} = 0.001$, both the inlet and exit sections of the pipe are in the wholly rough region where f is independent of Re. Thus, for the entire pipe, $\bar{f} = f_1 = f^* = 0.00495$, and no further iterations are required to determine $\bar{f}$. However, if either or both values of Re had been out of the wholly rough region, values of f_1 and f^* would be determined from Fig. 5.8, and $\bar{f}$ would be their average. Steps 2 to 5 would be repeated with that value of $\bar{f}$.

(6) $\dot{m}^* = \rho^* A V^* = (3.9005)\pi(0.015)^2(316.96) = 0.8739$ kg/s (1.927 lbm/sec)

   Check to insure that the duct exit section is choked.

$$p^* = \rho^* R t^* = (3.9005)(287.04)(250.0) = 2.799 \cdot 10^5 \text{ N/m}^2$$

Thus, $p^* > p_o$, and the assumption of choked flow is valid. For comparison, the maximum isentropic mass flow rate achievable for $\mathfrak{f} = 0$ is

$$\dot{m}^*_{\text{isen}} = \rho^* A V^* = (5.1532)\pi(0.015)^2(316.96) = 1.1546 \text{ kg/s}$$

The effect of friction in the constant-area duct, therefore, reduces the mass flow rate by 24.3 percent.

Note that the symbols t^*, p^*, ρ^*, V^*, and $\dot{m}^*$ are employed to denote the critical conditions for both isentropic flow and simple frictional flow. However, only t^* and V^* have the same numerical values for both flows. Care must be taken to insure that the proper critical values are employed in each of the two flow processes.

5-6 THE FANNO LINE FOR IMPERFECT GASES

In Section 5-5, the properties of the Fanno line are developed for *perfect gases*. In this section, the influence of *imperfect gas effects* on simple frictional flow, in a constant-area duct, is considered. Several sources of imperfect gas effects are presented in Section 4-5.

A completely general model of an imperfect gas is considered. The equations of state for the imperfect gas are given by

$$v = v(p,t) \tag{5.74}$$

$$h = h(v,t) \tag{5.75}$$

$$s = s(v,t) \tag{5.76}$$

Equations 5.74, 5.75, and 5.76 may be in the form of equations, charts, tables, or a computer code. As long as the equations of state are known, the analysis does not depend on the source of the imperfect gas effects.

The continuity, momentum, and energy equations for the simple frictional flow of any fluid are given by equations 5.2, 5.6, and 5.20. Those equations may be rearranged to give

$$V_2 = \frac{V_1 v_2}{v_1} \tag{5.77}$$

$$h_2 = H - \frac{G^2 v_2^2}{2} \tag{5.78}$$

$$\frac{4\mathfrak{f}L}{\mathscr{D}} = 2\ln\frac{v_1}{v_2} - \frac{2}{G^2}\int_1^2 \frac{dp}{v} \tag{5.79}$$

Equations 5.74 to 5.79 comprise a set of six equations in terms of the flow properties p, v, t, h, s, and V, at point 1 upstream to the friction region and point 2 downstream to the friction region, and the friction parameter $4\mathfrak{f}L/\mathscr{D}$. When the upstream properties are given, those six equations may be solved for the six properties p_2, v_2, t_2, h_2, s_2, and V_2 for specified values of the friction parameter $4\mathfrak{f}L/\mathscr{D}$.

The term in equation 5.79 involving the integral of dp/v cannot be evaluated in closed form for an imperfect gas. A numerical integration of that term is, therefore, required for relating the friction parameter to the flow properties. In most cases, the trapezoidal rule presented in Appendix A-5(a) is satisfactory for performing that integration. When more accuracy is required, Simpson's rule, presented in Appendix A-5(b), may be employed.

Example 5.6. Consider the flow of superheated steam discussed in Example 5.1. Determine the length of duct required to choke the flow, and the velocity at the choked section. Assume $\mathfrak{f} = 0.005$.

Solution

Table 5.1 presents the solution of equations 5.74, 5.75, 5.76, and 5.78 for the present problem. From the data in Table 5.1, the state properties of the steam where the choking occurs are approximately $v = 3.0$ ft^3/lbm (0.1873 m^3/kg) and $p = 226$ lbf/in.^{2} (15.58 · 10^5 N/m^2). From equation 5.77,

$$V_2 = \frac{(500)(3.0)}{0.7604} = 1972.6 \text{ ft/sec (601.26 m/s)}$$

The length of duct required to decrease the pressure to 226 lbf/in.^{2} (15.58 · 10^5 N/m^2) is found from equation 5.79. The integral term is evaluated graphically by plotting $1/v$ versus p over the range of interest, and then measuring the area under the curve. The graphical integration is presented in Fig. 5.10. The area under the curve is

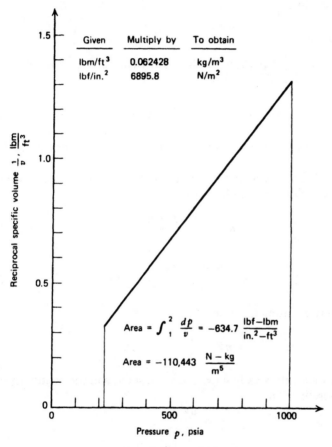

Figure 5.10. Determination of $\int_1^2 \frac{dp}{v}$ for Example 5.6.

-634.7 lbf-lbm/in.^{2}-ft^3 (-110, 443 N-kg/m^5). Hence, equation 5.79 gives

$$\frac{1}{2}\frac{4\bar{f}L}{\mathscr{D}} = \ln\left(\frac{0.7604}{2.8}\right) - \frac{(-634.7)(32.174)(144)}{(657.5)^2} = -1.3035 + 6.8022 = 5.4987$$

$$L = \frac{2(5.4987)(10)}{4(0.005)(12)} = 458.2 \text{ ft (139.7 m)}$$

5-7 STEADY ONE-DIMENSIONAL ISOTHERMAL FLOW WITH FRICTION IN A CONSTANT-AREA DUCT

There are some practical cases of gas flow in ducts where the flow may be assumed to be substantially isothermal. Natural gas pipelines extending for miles buried in the earth are examples of isothermal frictional flow.

The Reynolds number may be written in the form

$$\text{Re} = \frac{\rho V D}{\mu} = \frac{GD}{\mu} \tag{5.80}$$

For a constant (isothermal) temperature the viscosity of the flowing fluid μ is a constant. Since G is constant, the Reynolds number is constant at all points in the flow. Because the friction coefficient $\mathfrak{f}$ for a given pipe surface is a function of the Reynolds number alone [see Section 5–4(b)], it follows that in isothermal flow $\mathfrak{f}$ is invariant along the pipe.

5-7(a) Governing Equations

From Table 3.2, the energy equation for the isothermal ($dt = 0$) flow of a perfect gas in the absence of work and body forces is

$$\delta Q = d\left(\frac{V^2}{2}\right) = dH = c_p \, dT \tag{5.81}$$

Equation 5.81 shows that for an isothermal flow, the stagnation temperature T does not remain constant.

In the absence of body forces and the drag caused by submerged bodies, the dynamic equation, equation 3.160, reduces to

$$v \, dp + V \, dV + \frac{V^2}{2}\left(\frac{4\mathfrak{f} \, dx}{\mathcal{D}}\right) = 0 \tag{5.82}$$

Hence,

$$2\left(\frac{v}{V^2}\right) dp + \frac{2 \, dV}{V} + \left(\frac{4\mathfrak{f} \, dx}{\mathcal{D}}\right) = 0 \tag{5.83}$$

From the continuity equation (equation 5.2),

$$G = \frac{\dot{m}}{A} = \rho V = \frac{V}{v} = \text{constant} \tag{5.84}$$

For a constant-area duct with steady flow, G is a constant, and the continuity equation may be written in the form

$$\frac{dv}{v} = \frac{dV}{V} \tag{5.85}$$

For an isothermal process the equation of state for a perfect gas is $pv = \text{constant}$. Hence, in view of equation 5.85,

$$\frac{dp}{p} = -\frac{dv}{v} = -\frac{dV}{V} \tag{5.86}$$

To obtain a relationship between the friction term $4\mathfrak{f} \, dx/\mathcal{D}$ and the static pressure for the flow, the term dV/V is eliminated from equation 5.83 by means of equation 5.86. The result is

$$\left(\frac{2v}{pV^2}\right) p \, dp - \frac{2 \, dp}{p} + \frac{4\mathfrak{f} \, dx}{\mathcal{D}} = 0 \tag{5.87}$$

The initial conditions p_1, v_1, and V_1 may be assumed to be known constants and, for the reasons explained earlier, $\mathfrak{f}$ is a constant. The term $(2v/pV^2)$ may be rewritten as $(2v^2/V^2pv)$, which is equivalent to $(2/G^2Rt)$, which is a constant for isothermal flow. Consequently, $(2v/pV^2)$ is a constant and may be determined at the initial condition. Hence, equation 5.87 may be integrated to give

$$\frac{2v_1}{V_1{}^2p_1}\int_1^2 p\,dp - 2\int_1^2 \frac{dp}{p} + \frac{4\mathfrak{f}}{\mathscr{D}}\int_0^L dx = 0 \tag{5.88}$$

Integrating equation 5.88 and rearranging gives

$$\frac{4\mathfrak{f}L}{\mathscr{D}} = \frac{v_1}{V_1{}^2p_1}(p_1{}^2 - p_2{}^2) + \ln\left(\frac{p_2}{p_1}\right)^2 \tag{5.89}$$

But $\gamma p_1 v_1 = a_1{}^2$ and $a_1{}^2/V_1{}^2 = 1/M_1{}^2$. Hence,

$$\frac{4\mathfrak{f}L}{\mathscr{D}} = \frac{1}{\gamma M_1{}^2}\left[1 - \left(\frac{p_2}{p_1}\right)^2\right] + \ln\left(\frac{p_2}{p_1}\right)^2 \tag{5.90}$$

The stagnation temperature ratio between any two points in the flow is given by (recall that $t = $ constant)

$$\frac{T_2}{T_1} = \frac{1 + \dfrac{\gamma - 1}{2} M_2{}^2}{1 + \dfrac{\gamma - 1}{2} M_1{}^2} \tag{5.91}$$

The stagnation pressure ratio is given by

$$\frac{P_2}{P_1} = \frac{p_2}{p_1}\left(\frac{1 + \dfrac{\gamma - 1}{2} M_2{}^2}{1 + \dfrac{\gamma - 1}{2} M_1{}^2}\right)^{\gamma/(\gamma - 1)} \tag{5.92}$$

From the definition of the impulse function, equation 3.139

$$\frac{\mathscr{F}_2}{\mathscr{F}_1} = \frac{p_2\,(1 + \gamma M_2{}^2)}{p_1\,(1 + \gamma M_1{}^2)} \tag{5.93}$$

5-7(b) Limiting Conditions

As the gas flows along the pipe, its static pressure decreases. Eventually a limiting condition is reached where the pipe cannot be increased in length without altering the upstream conditions; that is, the flow has become *choked*. At that point, $dx/dp = 0$. Referring to equation 5.87, it is seen that at that limiting condition,

$$\frac{dx}{dp} = \frac{2\mathscr{D}}{4\mathfrak{f}}\left(\frac{v}{V^2} - \frac{1}{p}\right) = 0 \tag{5.94}$$

Denoting the properties at the limiting condition by the subscript L, equation 5.94 gives

$$V_L{}^2 = p_L v_L \tag{5.95}$$

Hence, the limiting or *choking velocity* for the isothermal flow of a perfect gas in a constant-area duct in the presence of wall friction alone is

$$V_L = \sqrt{p_L v_L} = \sqrt{Rt_L} = \sqrt{Rt_1} = \frac{a_1}{\sqrt{\gamma}} \tag{5.96}$$

It follows from equation 5.96 that when the gas velocity attains the value $a_1/\sqrt{\gamma}$, the flow becomes *choked*. The choking velocity V_L is seen to be less than the sonic velocity by the factor $1/\sqrt{\gamma}$. When $M < 1/\sqrt{\gamma}$, the fluid is receiving heat, and when $M > 1/\sqrt{\gamma}$, it rejects heat.

The pressure ratio at choking p_L/p_1 is readily determined from the equation of state, the continuity equation, and equation 5.96. Thus,

$$\frac{p_L}{p_1} = \frac{v_1}{v_L} = \frac{V_1}{V_L} = \frac{V_1\sqrt{\gamma}}{a_1} = M_1\sqrt{\gamma} \tag{5.97}$$

Substituting from equation 5.97 into equation 5.90 gives the equation relating the friction term $4\mathfrak{f}L_{max}/\mathscr{D}$ to the pressure ratio at choking flow; the length L_{max} is that corresponding to V_L and p_L. Thus,

$$\frac{4\mathfrak{f}L_{max}}{\mathscr{D}} = \left[\left(\frac{p_1}{p_L}\right)^2 - 1\right] - \ln\left(\frac{p_1}{p_L}\right)^2 \tag{5.98}$$

Equation 5.98 is applicable to compressible fluids which satisfy the relationship $pv = Rt$. Substituting equation 5.97 into equation 5.98 results in the following relationship between the maximum duct length L_{max} and the inlet Mach number M_1.

$$\frac{4\mathfrak{f}L_{max}}{\mathscr{D}} = \frac{1 - \gamma M_1{}^2}{\gamma M_1{}^2} + \ln(\gamma M_1{}^2) \tag{5.99}$$

If the gas flows isothermally under conditions where compressibility effects are negligible and only wall friction is encountered, then $dM^2 = 0$, $\delta D/Ap = 0$, and $dt/t = 0$, and the momentum equation (see equation 3.170, Table 3.2) reduces to

$$\frac{dp}{p} = -\frac{\gamma M_1{}^2}{2}\left(\frac{4\mathfrak{f}\,dx}{\mathscr{D}}\right) \tag{5.100}$$

Integrating equation 5.100 yields

$$\ln\left(\frac{p_2}{p_1}\right) = -\frac{\gamma M_1{}^2}{2}\left(\frac{4\mathfrak{f}L}{\mathscr{D}}\right) \tag{5.101}$$

For the flow to be considered incompressible, the pressure ratio p_2/p_1 cannot differ significantly from unity. Expanding $\ln(p_2/p_1)$ into a series gives

$$\ln\left(\frac{p_2}{p_1}\right) = \frac{p_2}{p_1} - 1 - \frac{1}{2}\left(\frac{p_2}{p_1} - 1\right)^2 + \frac{1}{3}\left(\frac{p_2}{p_1} - 1\right)^3 + \cdots \tag{5.102}$$

Hence, equation 5.101 may be written in the form

$$\frac{p_2}{p_1} \approx 1 - \frac{\gamma M_1{}^2}{2}\left(\frac{4\mathfrak{f}L}{\mathscr{D}}\right) \tag{5.103}$$

5-7(c) Tables for the Isothermal Frictional Flow of a Perfect Gas in a Constant-Area Duct

For calculation purposes, it is convenient to construct a table of flow property ratios for isothermal frictional flow, employing the limiting condition where $M = 1/\sqrt{\gamma}$ as a reference state. Substituting that value for M_2 into the equations derived in Sections 5-7(a) and 5-7(b), the results presented in Table 5.7 are obtained.

Table 5.7 Property Ratios for the Isothermal Frictional Flow of a Perfect Gas in a Constant-Area Duct

$$\frac{p_L}{p} = \frac{v}{v_L} = \frac{\rho_L}{\rho} = \frac{V}{V_L} = \sqrt{\gamma} M \tag{5.104}$$

$$\frac{P}{P_L} = \frac{1}{\sqrt{\gamma} M} \left[\left(\frac{2\gamma}{3\gamma - 1} \right) \left(1 + \frac{\gamma - 1}{2} M^2 \right) \right]^{\gamma/(\gamma - 1)} \tag{5.105}$$

$$\frac{T}{T_L} = \left(\frac{2\gamma}{3\gamma - 1} \right) \left(1 + \frac{\gamma - 1}{2} M^2 \right) \tag{5.106}$$

$$\frac{\mathscr{F}}{\mathscr{F}_L} = \frac{(1 + \gamma M^2)}{2\sqrt{\gamma} M} \tag{5.107}$$

$$\frac{4 f L_{max}}{\mathscr{D}} = \frac{(1 - \gamma M^2)}{\gamma M^2} + \ln (\gamma M^2) \tag{5.108}$$

Table 5.8 Limiting Values of the Property Ratios for the Isothermal Frictional Flow of a Perfect Gas in a Constant-Area Duct

$M = M^*$	$\dfrac{p}{p_L} = \dfrac{v}{v_L} =$ $\dfrac{\rho_L}{\rho} = \dfrac{V}{V_L}$	$\dfrac{T}{T_L}$	$\dfrac{P}{P_L}$	$\dfrac{\mathscr{F}}{\mathscr{F}_L}$	$\dfrac{4 f L_{max}}{\mathscr{D}}$
0	0	$\left(\dfrac{2\gamma}{3\gamma - 1} \right)$	∞	∞	∞
$\dfrac{1}{\sqrt{\gamma}}$	1	1	1	1	0
1	$\sqrt{\gamma}$	$\dfrac{\gamma(\gamma + 1)}{(3\gamma - 1)}$	$\dfrac{1}{\sqrt{\gamma}} \left[\dfrac{\gamma(\gamma + 1)}{(3\gamma - 1)} \right]^{\gamma/(\gamma - 1)}$	$\dfrac{(\gamma + 1)}{2\sqrt{\gamma}}$	$\left[\dfrac{(1 - \gamma)}{\gamma} + \ln (\gamma) \right]$
∞	∞	∞	∞	∞	∞

Values of the property ratios as a function of the Mach number M are presented in Table C.8 for $\gamma = 1.4$. Table 5.8 illustrates the general arrangement of Table C.8 and also presents the limiting values of the property ratios.

A FORTRAN program, subroutine TCFRIC, is presented below for calculating the property ratios presented in Table 5.7. The program input is the same as that discussed in Section 4–4(b). The results presented in Table C.8 for $\gamma = 1.4$ are obtained by the following specification.

```
G = 1.4, J(1) = 25, J(2) = 30, J(3) = 10, J(4) = 11, J(5) = 0,
DM(1) = 0.02, DM(2) = 0.05, DM(3) = 0.2, DM(4) = 1.0
```

Example 5.7. Air flows in a long pipe under isothermal conditions. At the pipe inlet, the static temperature and pressure are 300 K and $3.500 \cdot 10^5 \, \text{N/m}^2$, respectively, and the velocity is 175 m/s. The pipe ID is 0.150 m and the friction coefficient

```
      SUBROUTINE TCFRIC

C     ISOTHERMAL FRICTIONAL FLOW PROPERTY RATIOS FOR A PERFECT GAS

      REAL M,MS,MT $ DIMENSION J(9),DM(9) $ DATA INF/6H INFIN/
      NAMELIST /DATA/ G,J,DM $ READ (5,DATA)
      G1=(G-1.0)/2.0 $ G2=SQRT(G) $ G3=2.0*G/(3.0*G-1.0) $ G4=G/(G-1.0)
      M=0.0 $ V=0.0 $ TO=G3 $ MT=1.0/G2
      WRITE (6,2000) G $ WRITE (6,2010) M,V,TO,INF,INF,INF $ I=1 $ L=0

C     CALCULATE PROPERTY RATIOS

   40 N=J(I) $ DO 50 K=1,N $ M=M+DM(I) $ C=1.0+G1*M**2 $ V=G2*M
      TO=G3*C $ PO=TO**G4/V $ F=(1.0+G*M**2)/(2.0*V)
      X=(1.0-G*M**2)/V**2+ALOG(V**2) $ WRITE (6,2020) M,V,TO,PO,F,X
      L=L+1 $ IF (L.LT.50) GO TO 45 $ WRITE (6,2000) G $ L=0 $ GO TO 50
   45 IF ((MT-M).GT.DM(I)) GO TO 50 $ V=1.0 $ TO=1.0 $ PO=1.0 $ F=1.0
      X=0.0 $ WRITE (6,2040) MT,V,TO,PO,F,X $ MT=1.0E+30
   50 CONTINUE $ I=I+1 $ IF((I.EQ.10).OR.(J(I).EQ.0)) GO TO 60 $GO TO 40
   60 V=0.0 $ TO=0.0 $ WRITE (6,2030) INF,INF,INF,INF,INF,INF $ RETURN

 2000 FORMAT (1H1,17X,36HISOTHERMAL FRICTIONAL FLOW,  GAMMA =,F5.2//4X,
     11HM,7X,4HV/VL,7X,6HTO/TOL,7X,6HPO/POL,7X,4HF/FL,9X,6H4FL*/D/10X,
     29HPL/P,RL/R/)
 2010 FORMAT (1H ,F5.2,2X,2E13.5,3X,A6,2(7X,A6))
 2020 FORMAT (1H ,F5.2,2X,5E13.5)
 2030 FORMAT (A6,5X,A6,4(7X,A6))
 2040 FORMAT (1H ,F7.4,5E13.5)
      END
```

$f = 0.005$. Calculate (a) the length of pipe required to choke the flow, (b) the limiting velocity, (c) the limiting pressure, and (d) the length of pipe at the station where the Mach number is 0.60. Here $\mathscr{D} = D$.

Solution

The inlet speed of sound and Mach number are

$$a_1 = a = a_L = (\gamma R t_1)^{1/2} = [1.4(287.04)(300)]^{1/2} = 347.21 \text{ m/s}$$

$$M_1 = \frac{V_1}{a_1} = \frac{175}{347.21} = 0.50$$

From equation 5.108,

$$\frac{4 f L_{max}}{\mathscr{D}} = \frac{1 - \gamma M_1^2}{\gamma M_1^2} + \ln (\gamma M_1^2) = \frac{1 - (1.4)(0.5)^2}{(1.4)(0.5)^2} + \ln [(1.4)(0.5)^2]$$

$$\frac{4 f L_{max}}{\mathscr{D}} = 0.80732$$

(a) $$L_{max} = \frac{(0.80732)\mathscr{D}}{4 f} = \frac{(0.80732)(0.150)}{(4)(0.005)} = 6.055 \text{ m (19.865 ft)}$$

(b) For isothermal flow, $M_L = 1/\sqrt{\gamma} = 0.84515$. Therefore,

$$V_L = M_L a_L = (0.84515)(347.21) = 293.44 \text{ m/s (962.73 ft/sec)}$$

(c) From equation 5.104,

$$\frac{p_L}{p_1} = \sqrt{\gamma} M_1 = \sqrt{1.4}(0.5) = 0.59161$$

$$p_L = (0.59161)(3.50)10^5 = 2.0706 \cdot 10^5 \text{ N/m}^2 \text{ (30.03 lbf/in.}^2\text{)}$$

(d)   For a Mach number of 0.6, equation 5.108 yields

$$\frac{4fL_{max}}{\mathcal{D}} = \frac{1 - (1.4)(0.6)^2}{(1.4)(0.6)^2} + \ln\left[(1.4)(0.6)^2\right] = 0.29895$$

$$L_{max} = \frac{(0.29895)(0.150)}{(4)(0.005)} = 2.242 \text{ m}$$

$$\Delta L = L_{max}(M = 0.5) - L_{max}(M = 0.6) = 6.055 - 2.242 = 3.813 \text{ m (12.51 ft)}$$

5-8 DIMENSIONAL ANALYSIS

Many dynamic problems in engineering are solved by comparing the behaviors of two similar systems of widely different size, such as the prediction of the behavior of a prototype from experiments conducted with a model. For the results to be satisfactory, the model and the prototype must be not only *geometrically similar* but, in addition, any fluid flows that are involved must be *dynamically similar*; that is, the ratios of all of the forces acting on the fluid must have identical values for the model and the prototype.

Dimensional analysis[9-12] is a mathematical procedure for establishing the conditions for achieving dynamic similarity in geometrically similar systems. By applying dimensional analysis, we may determine how to arrange the variables that enter into a physical problem into dimensionless groups, which make it convenient for conducting an experimental investigation of the problem at hand without destroying the generality of the relationships between the pertinent variables. Dimensional analysis is particularly useful when a large number of variables is involved.

When a given number of physical variables enter into a physical relationship, the homogeneity principle [see Section 1-3(c)] limits the number of possible relationships in a special manner. It can be shown that, if a functional relationship exists between the variables $Q_1, Q_2, \ldots, Q_m$, then

$$F(Q_1, Q_2, \ldots, Q_m) = 0 \tag{5.109}$$

where $F(\ )$ denotes a functional relationship between the variables enclosed in the parentheses. If equation 5.109 represents a complete equation between the m different physical variables $Q_1, Q_2, \ldots, Q_m$, and if there are n principal dimensions [see Section 1-3(a)], the above functional relationship may be transformed to the form[9-12]

$$F(\pi_1, \pi_2, \ldots, \pi_{m-n}) = 0 \tag{5.110}$$

The relationship expressed by equation 5.110 is said to be a *complete equation* if it remains valid regardless of the consistent system of units employed for evaluating it. In the functional relationship given by equation 5.110, the π's represent $(m - n)$ independent *dimensionless groups* formed from the original variables $Q_1, Q_2, \ldots, Q_m$. Each π group is composed of $n + 1$ physical variables, and in each successive π group only one variable is changed.

If more than one kind of the same physical variable enters into the functional relationship, such as several lengths, each of the same kind of a variable may be represented by one of them and the ratios of the others to it. If there are l such ratios, then the functional relationship expressed by equation 5.110 transforms to

$$F(\pi_1, \pi_2, \ldots, \pi_{m-n-l}, r_1, r_2, \ldots, r_l) = 0 \tag{5.111}$$

where $r_1, r_2, \ldots, r_l$ are the pertinent ratios.

Each π group represents a dimensionless product of the form

$$\pi = Q_1{}^{e_1}Q_2{}^{e_2}\cdots Q_m{}^{e_m} \tag{5.112}$$

where $e_1, e_2, \ldots, e_m$ are exponents that have to be determined.

To illustrate the application of the method of dimensional analysis, consider the problem of determining the aerodynamic resistance, termed the *drag*, of an airfoil moving horizontally in the atmospheric air surrounding Earth.

The drag will, of course, depend on those variables that influence the resistance to motion of any body completely submerged to a considerable depth in a large body of fluid. Because of the depth of submergence of the airfoil in the air of Earth's atmosphere, no gravity waves will be formed on the free surface of the atmosphere. Hence, forces caused by the gravitational attraction of Earth on the airfoil will have no influence on the drag of the airfoil.

Assume that the airfoil moves with the uniform speed V relative to the atmospheric air, so that no accelerating forces act on it. Physical intuition would indicate that there is a functional relationship between the drag of the airfoil, denoted by D, and the following variables: V, the relative velocity of the airfoil with respect to the atmospheric air, termed the *relative wind velocity*; ρ, the density of the air; μ, the dynamic (or absolute) viscosity of the air; a, the local speed of sound or *acoustic speed* in the neighborhood of the airfoil; and the shape and size of the airfoil. For a family of geometrically similar airfoils, a single linear dimension, denoted by ℓ, will define each of them. The linear dimension ℓ is termed the *characteristic length* of the airfoil.

The following form of functional relationship is, therefore, indicated.

$$F(D, \ell, V, \rho, \mu, a) = 0 \tag{5.113}$$

The drag characteristics for the airfoil are determined from experiments conducted in a wind tunnel; the airfoil is held stationary and air is caused to flow past it with the relative wind velocity V. The most suitable method for conducting the experiments is determined by performing a dimensional analysis.

Referring to equation 5.113, it is seen that the number of physical variables that may influence the drag D is $m = 6$. The number of principal dimensions for a dynamical system is $n = 3$ (either F, L, and T or M, L, and T). Hence, there are $(m - n) = 3$ dimensionless groups; that is, there are 3 π groups. Therefore, we may write

$$F(\pi_1, \pi_2, \pi_3) = 0 \tag{5.114}$$

The form of each π group is determined by selecting three of the physical variables in equation 5.113, raising each variable to the power of an unknown exponent, and then combining them with a fourth physical variable raised to the exponent unity. Thus, selecting ℓ, V, and ρ as the three physical variables to be raised to unknown exponents, we may write the following equations:

$$\pi_1 = D\ell^{x_1}V^{y_1}\rho^{z_1} \tag{5.115}$$

$$\pi_2 = \mu\ell^{x_2}V^{y_2}\rho^{z_2} \tag{5.116}$$

$$\pi_3 = a\ell^{x_3}V^{y_3}\rho^{z_3} \tag{5.117}$$

To determine the values of the exponents x_1, y_1, and z_1 of π_1, each of the physical variables is expressed in terms of its dimensions raised to the exponent that is attached to the variable. For example,

$$(\pi_1) = (F^0L^0T^0) = (F^1)(L^{x_1})(L^{y_1}T^{-y_1})(F^{z_1}T^{2z_1}L^{-4z_1}) \tag{5.118}$$

Writing an equation for the exponent of each dimension, we obtain

$$F^0: 1 + z_1 = 0$$
$$L^0: x_1 + y_1 - 4z_1 = 0 \tag{5.119}$$
$$T^0: -y_1 + 2z_1 = 0$$

The simultaneous solution of equations 5.119 yields $x_1 = -2$, $y_1 = -2$, and $z_1 = 1$. Hence,

$$\pi_1 = \frac{D}{\ell^2 V^2 \rho} \tag{5.120}$$

Similarly, for π_2,

$$(\pi_2) = (F^1 T^1 L^{-2})(L^{x_2})(L^{y_2} T^{-y_2})(F^{z_2} T^{2z_2} L^{-4z_2}) \tag{5.121}$$

Solving for the exponents, we obtain $x_2 = -1$, $y_2 = 1$, and $z_2 = -1$. Hence,

$$\pi_2 = \frac{\mu}{\ell \rho V} \tag{5.122}$$

For π_3, we write

$$(\pi_3) = (L^1 T^{-1})(L^{x_3})(L^{y_3} T^{-y_3})(F^{z_3} T^{2z_3} L^{-4z_3}) \tag{5.123}$$

The numerical values of the exponents are $x_3 = 0$, $y_3 = -1$, and $z_3 = 0$. Hence,

$$\pi_3 = \frac{a}{V} \tag{5.124}$$

Substituting for π_1, π_2, and π_3 in the functional relationship expressed by equation 5.114 yields

$$F\left(\frac{D}{\ell^2 \rho V^2}, \frac{\mu}{\ell \rho V}, \frac{a}{V}\right) = 0 \tag{5.125}$$

Equation 5.125 indicates that the drag D for an airfoil moving through air with the uniform relative wind speed V is given by

$$\frac{D}{\ell^2 \rho V^2} = f\left(\frac{\ell \rho V}{\mu}, \frac{V}{a}\right) \tag{5.126}$$

The projected area of the airfoil, denoted by A, has the same dimensions as does ℓ^2 and may, therefore, replace ℓ^2 in equation 5.126. In addition, since equation 5.126 is a functional relationship, the general form of the relationship is unchanged by the introduction of a pure constant into any of the π groups. Only the specific form of the function changes when constants are introduced. Thus, the π_1 group defined by equation 5.120 may be rewritten as

$$\pi_1 = \frac{D}{\frac{1}{2}\rho V^2 A} = C_D \tag{5.127}$$

which is termed the *drag coefficient* and denoted by the symbol C_D.

Equation 5.126 may thus be written as

$$C_D = f(\text{Re}, M) \tag{5.128}$$

where

$$\text{Re} = \frac{\ell \rho V}{\mu}, \quad \text{the } Reynolds\ number \tag{5.129}$$

$$M = \frac{V}{a}, \quad \text{the } Mach\ number \tag{5.130}$$

Experiments are conducted, in a wind tunnel, for obtaining the characteristic curves of the drag coefficient C_D as a function of the Reynolds number Re, for several constant values of the Mach number M. Likewise, experiments are conducted for obtaining curves of C_D as a function of the Mach number M for several constant values of the Reynolds number Re.

At relative wind speeds less than $V = 150$ m/s approximately, it may be assumed that atmospheric air behaves as an *incompressible fluid*. For those conditions the compressibility of the air may be neglected and the Mach number does not enter into the functional relationship. Hence, for incompressible flow,

$$C_D = f(\text{Re}) \tag{5.131}$$

Equation 5.131 states that under incompressible flow conditions, the drag of an airfoil (or airplane) depends on the relative wind velocity V, the density ρ of the air, the viscosity μ of the air, the projected area A of the airfoil (or airplane wings), and an experimentally determined drag coefficient C_D.

For a more complete treatment of dimensional analysis, model testing, and dynamic similarity, refer to References 9 to 13 at the end of the chapter.

5-9 THE NAVIER-STOKES EQUATIONS

The preceding sections considered *simple frictional flows*; that is, flows that are steady, one-dimensional, and involve only wall friction. The effects of friction are taken into account by an empirically determined *friction coefficient* f. Most real flows are more complicated than the simple frictional flow. They may be unsteady multidimensional flows and involve fluids having significant viscosity. For such flows the momentum and energy equations must account for the effects of the fluid friction. In many flow situations, the effects of fluid shear in the energy equation (manifested by the shear work term in equation 2.88) is small and may be neglected. But the effects of fluid shear in the momentum equation (represented by the term $\mathbf{F}_{shear}$ in equation 2.87) may be significant. In the present section, the differential form of the momentum equation, including the effects of shearing stresses in the fluid, is presented. The resulting equations are known as the *Navier-Stokes equations*. One of the few exact solutions to those equations is presented herein; it is that for the steady fully developed laminar flow of an incompressible fluid in a circular duct, termed *simple laminar frictional flow*.

5-9(a) The Momentum Equation

Newton's second law of motion for an inertial rigid control volume is derived in Section 2-6; it is equation 2.49, which is repeated and renumbered below.

$$\mathbf{F}_{external} = \int_{\mathcal{V}} (\rho \mathbf{V})_t \, d\mathcal{V} + \int_A \mathbf{V}(\rho \mathbf{V} \cdot d\mathbf{A}) \tag{5.132}$$

The differential form of the momentum equation, including viscous stresses in the fluid, is obtained by applying equation 5.132 to an inertial rigid control volume of differential size, as illustrated in Fig. 5.11 for a Cartesian coordinate system. For that differential control volume, the right-hand side of equation 5.132 is determined by transforming the surface area term from a surface integral to a volume integral by means of the divergence theorem (see Section 2-1). The resulting volume integral is then evaluated in the limit as the control volume $\mathcal{V}$ approaches zero. Those steps are illustrated in Section 2-6, where the momentum equation for frictionless flow

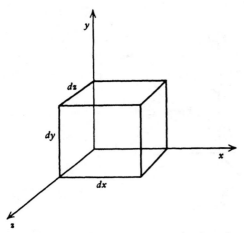

Figure 5.11. A differential size control volume in a Cartesian coordinate system.

(i.e., Euler's equation) is derived. For a differential control volume $d\mathcal{V}$, equation 5.132 becomes

$$\mathbf{F}_{\text{external}} = (\rho \, d\mathcal{V}) \frac{D\mathbf{V}}{Dt} \tag{5.133}$$

where $\mathbf{F}_{\text{external}}$ is the net external force acting on the differential control volume $d\mathcal{V}$.

The external forces acting on the differential control volume may be divided into two types of forces, body forces and surface forces. Hence,

$$\mathbf{F}_{\text{external}} = \mathbf{F}_B + \mathbf{F}_S \tag{5.134}$$

The surface forces may be further subdivided into normal forces and tangential forces.

$$\mathbf{F}_S = \mathbf{F}_n + \mathbf{F}_t \tag{5.135}$$

In the developments of Section 2–6(c), the tangential forces are neglected, and the resulting equation is Euler's equation, equation 2.63, which is repeated below.

$$\rho \frac{D\mathbf{V}}{Dt} + \nabla p - \rho \mathbf{B} = 0 \tag{2.63}$$

In the present section, the tangential forces are included in equations 5.133. The resulting equations are the *Navier-Stokes equations* mentioned above. They are the counterpart to Euler's equation for the case of flow with viscous shearing stresses. For a frictionless flow, the Navier-Stokes equations reduce to Euler's equation.

5–9(b) Stresses Acting on a Differential Control Volume

Figure 5.12 illustrates the forces acting on a differential control volume in a Cartesian coordinate system. The body force $\mathbf{F}_B$ is given by

$$\mathbf{F}_B = \mathbf{B}\rho \, d\mathcal{V} = (\mathbf{i}B_x + \mathbf{j}B_y + \mathbf{k}B_z)(\rho \, d\mathcal{V}) \tag{5.136}$$

where $\mathbf{B}$ is the body force per unit mass. The surface forces are decomposed into the forces acting on each of the six faces of the control surface, and denoted by $\mathbf{F}_{Sx}$, etc., where the subscripts x, y, and z denote a force acting on a surface *perpendicular* to the x, y, and z axis, respectively. The surface forces may be subdivided into normal

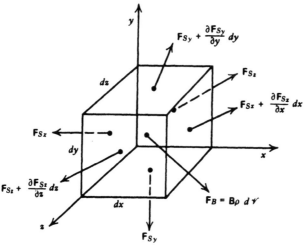

Figure 5.12. External forces acting on a rigid inertial control volume of differential size.

and tangential forces, and the tangential forces may be further subdivided into two orthogonal components parallel to the coordinate axes. Thus,

$$\mathbf{F}_S = \mathbf{F}_n + \mathbf{F}_t \tag{5.137}$$

Figure 5.13 illustrates the forces acting on the surfaces perpendicular to the x axis.

The following notation is employed for specifying the surface forces. For the normal force component, the subscript indicates the direction of the force; for example, F_{nx} is the normal force in the x direction acting on a surface perpendicular to the x axis. For the tangential forces, the first subscript denotes the direction of

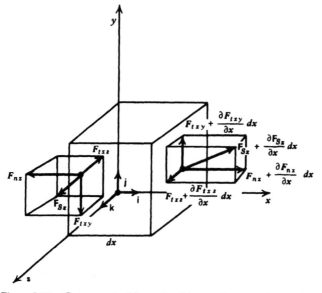

Figure 5.13. Components of the surface forces acting on surfaces perpendicular to the x axis.

the normal to the surface on which the force acts, and the second subscript indicates the direction in which the force acts. For example, F_{txy} is the tangential force on a surface perpendicular to the x axis but acting in the y direction. The surfaces themselves are called *faces*; they are termed *positive faces* when the outward normal points in a positive coordinate direction, and *negative faces* when the outward normal points in a negative coordinate direction. By convention, all forces are assumed to act in a positive direction on positive faces and in a negative direction on negative faces. Thus, on the negative x face in Fig. 5.13,

$$\mathbf{F}_{Sx} = -(\mathbf{i}F_{nx} + \mathbf{j}F_{txy} + \mathbf{k}F_{txz}) \tag{5.138}$$

Similar expressions may be written for the five remaining faces.

For a differential control volume, the surface forces may be expressed in terms of the surface force per unit area, or the surface stresses. Thus,

$$\sigma_i = \lim_{\Delta A_i \to 0} \frac{\Delta F_{ni}}{\Delta A_i} \qquad (i = x,y,z) \tag{5.139}$$

$$\tau_{ij} = \lim_{\Delta A_i \to 0} \frac{\Delta F_{tij}}{\Delta A_i} \qquad (i,j = x,y,z) \tag{5.140}$$

where σ_i denotes a normal stress in the i direction (i.e., F_{ni}) and τ_{ij} denotes a tangential, or shear, stress on the i face acting in the j direction (i.e., F_{tij}). In terms of σ_i and τ_{ij}, equation 5.138 becomes

$$\mathbf{F}_{Sx} = -(\mathbf{i}\sigma_x + \mathbf{j}\tau_{xy} + \mathbf{k}\tau_{xz})\, dy\, dz \tag{5.141a}$$

In a similar manner, we obtain

$$\mathbf{F}_{Sy} = -(\mathbf{i}\tau_{yx} + \mathbf{j}\sigma_y + \mathbf{k}\tau_{yz})\, dx\, dz \tag{5.141b}$$

$$\mathbf{F}_{Sz} = -(\mathbf{i}\tau_{zx} + \mathbf{j}\tau_{zy} + \mathbf{k}\sigma_z)\, dx\, dy \tag{5.141c}$$

Figure 5.14 illustrates the complete surface stress system acting on a differential control volume. The stresses on the positive faces are indicated by the prime superscript, where

$$\sigma_x' = \sigma_x + \frac{\partial \sigma_x}{\partial x}\, dx \tag{5.142}$$

$$\tau_{xy}' = \tau_{xy} + \frac{\partial \tau_{xy}}{\partial x}\, dx, \text{ etc.} \tag{5.143}$$

The general stress system illustrated in Fig. 5.14 requires nine scalar components (one normal stress and two shear stresses on each face). However, by summing moments and requiring that the angular acceleration of the fluid remain finite, it can be shown that[14]

$$\tau_{xy} = \tau_{yx}, \qquad \tau_{xz} = \tau_{zx}, \qquad \text{and} \qquad \tau_{yz} = \tau_{zy} \tag{5.144}$$

Consequently, only six independent stresses are obtained.

The momentum equation for the differential control volume is obtained by substituting equations 5.136 and 5.141 into equation equation 5.133. For the x direction, the substitution yields

$$B_x(\rho\, d\mathcal{V}) + \left(\sigma_x + \frac{\partial \sigma_x}{\partial x}\, dx\right) dy\, dz - \sigma_x\, dy\, dz + \left(\tau_{xy} + \frac{\partial \tau_{xy}}{\partial y}\, dy\right) dx\, dz$$

$$-\tau_{xy}\, dx\, dz + \left(\tau_{xz} + \frac{\partial \tau_{xz}}{\partial z}\, dz\right) dx\, dy - \tau_{xz}\, dx\, dy = (\rho\, d\mathcal{V})\frac{Du}{Dt} \tag{5.145}$$

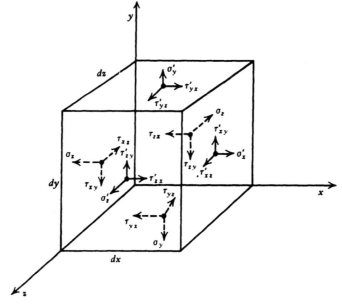

Figure 5.14. Surface stresses on a differential control volume.

Simplifying equation 5.145 yields

$$\rho \frac{Du}{Dt} = \rho B_x + \frac{\partial \sigma_x}{\partial x} + \frac{\partial \tau_{xy}}{\partial y} + \frac{\partial \tau_{xz}}{\partial z} \tag{5.146a}$$

Similarly, for the y and z directions,

$$\rho \frac{Dv}{Dt} = \rho B_y + \frac{\partial \tau_{xy}}{\partial x} + \frac{\partial \sigma_y}{\partial y} + \frac{\partial \tau_{yz}}{\partial z} \tag{5.146b}$$

$$\rho \frac{Dw}{Dt} = \rho B_z + \frac{\partial \tau_{xz}}{\partial x} + \frac{\partial \tau_{yz}}{\partial y} + \frac{\partial \sigma_z}{\partial z} \tag{5.146c}$$

5–9(c) The Navier-Stokes Equations

Equation 5.146 express Newton's second law of motion for a differential control volume in terms of the body forces and the normal and shear stresses in the fluid. It remains to relate those stresses to the physical properties of the fluid and its state of motion.

The static thermodynamic pressure is defined as the average value of the normal stresses. Thus,

$$-p = \bar{\sigma} = \frac{1}{3}(\sigma_x + \sigma_y + \sigma_z) \tag{5.147}$$

The shear stresses are related to the velocity gradients by employing the *Stokes' law of fluid friction*. The latter is a phenomenological hypothesis that states that shear stresses in a fluid are directly proportional to the velocity gradients in the fluid. Consider the flow field illustrated in Fig. 5.15, where a viscous fluid flows between two infinite flat plates such that $u = u(y)$. Stokes' law states that

$$\tau_{yx} \propto \frac{du}{dy} \tag{5.148}$$

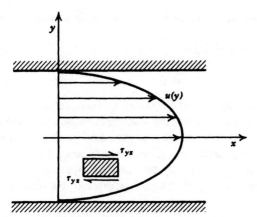

Figure 5.15. Stokes' law of friction.

For a *Newtonian fluid* [see Section 1–6(b)], equation 5.148 becomes

$$\tau_{yx} = \mu \frac{du}{dy} \tag{5.149}$$

where μ is the dynamic viscosity. For the general case where $u = u(x,y,z)$, the surface force $\mathbf{F}_S$ depends on the velocity gradients in all three directions (i.e., $\partial u/\partial x$, $\partial u/\partial y$, $\partial u/\partial z$, $\partial v/\partial x$, etc.). Stokes' law is analogous to Hooke's law for solids, which states that the stresses in a solid are directly proportional to strain. Thus, Stokes' law states that stresses in a fluid are proportional to the *time rate of strain*.

The relationship between the shear stresses and the velocity gradients in a fluid may be derived by first developing the relationship between shear stresses and strains in an elastic solid. An analogy is then drawn between the shear modulus of the solid and the viscosity of the fluid, and between the shear strain in the solid and the time rate of shear strain in the fluid.

Those relationships between shear stress and strain in a solid, and the analogous results for a fluid, are derived in many standard books on viscous flow.[14–17] The derivations, which are rather lengthy, are not repeated here. The results, however, are presented below.

$$\sigma_x = -p + 2\mu \frac{\partial u}{\partial x} - \frac{2}{3}\mu(\nabla \cdot \mathbf{V}) \tag{5.150a}$$

$$\sigma_y = -p + 2\mu \frac{\partial v}{\partial y} - \frac{2}{3}\mu(\nabla \cdot \mathbf{V}) \tag{5.150b}$$

$$\sigma_z = -p + 2\mu \frac{\partial w}{\partial z} - \frac{2}{3}\mu(\nabla \cdot \mathbf{V}) \tag{5.150c}$$

$$\tau_{xy} = \tau_{yx} = \mu\left(\frac{\partial v}{\partial x} + \frac{\partial u}{\partial y}\right) \tag{5.151a}$$

$$\tau_{xz} = \tau_{zx} = \mu\left(\frac{\partial u}{\partial z} + \frac{\partial w}{\partial x}\right) \tag{5.151b}$$

$$\tau_{yz} = \tau_{zy} = \mu\left(\frac{\partial w}{\partial y} + \frac{\partial v}{\partial z}\right) \tag{5.151c}$$

For a hypothetical inviscid fluid, or a viscous fluid at rest or in a state of uniform motion, we may write

$$-p = \bar{\sigma} = \sigma_x = \sigma_y = \sigma_z \tag{5.152}$$

and

$$\tau_{xy} = \tau_{xz} = \tau_{yz} = 0 \tag{5.153}$$

Accordingly, the normal stress is the same in all directions and is equal to the negative of the static thermodynamic pressure p. In a flowing viscous fluid, however, the shear stresses are nonzero, and the normal stresses are not the same in all directions. In that case, the static thermodynamic pressure p is given by equation 5.147.

The Navier-Stokes equations are obtained by substituting equations 5.150 and 5.151 into equation 5.146. The results are given below in Cartesian coordinates.

$$\rho \frac{Du}{Dt} = \rho B_x - \frac{\partial p}{\partial x} + \frac{\partial}{\partial x}\left[2\mu\frac{\partial u}{\partial x} - \frac{2}{3}\mu(\nabla \cdot \mathbf{V})\right] + \frac{\partial}{\partial y}\left[\mu\left(\frac{\partial u}{\partial y} + \frac{\partial v}{\partial x}\right)\right]$$
$$+ \frac{\partial}{\partial z}\left[\mu\left(\frac{\partial w}{\partial x} + \frac{\partial u}{\partial z}\right)\right] \tag{5.154a}$$

$$\rho \frac{Dv}{Dt} = \rho B_y - \frac{\partial p}{\partial y} + \frac{\partial}{\partial y}\left[2\mu\frac{\partial v}{\partial y} - \frac{2}{3}\mu(\nabla \cdot \mathbf{V})\right] + \frac{\partial}{\partial z}\left[\mu\left(\frac{\partial v}{\partial z} + \frac{\partial w}{\partial y}\right)\right]$$
$$+ \frac{\partial}{\partial x}\left[\mu\left(\frac{\partial u}{\partial y} + \frac{\partial v}{\partial x}\right)\right] \tag{5.154b}$$

$$\rho \frac{Dw}{Dt} = \rho B_z - \frac{\partial p}{\partial z} + \frac{\partial}{\partial z}\left[2\mu\frac{\partial w}{\partial z} - \frac{2}{3}\mu(\nabla \cdot \mathbf{V})\right] + \frac{\partial}{\partial x}\left[\mu\left(\frac{\partial w}{\partial x} + \frac{\partial u}{\partial z}\right)\right]$$
$$+ \frac{\partial}{\partial y}\left[\mu\left(\frac{\partial v}{\partial z} + \frac{\partial w}{\partial y}\right)\right] \tag{5.154c}$$

In vector form, equations 5.154 become

$$\rho \frac{D\mathbf{V}}{Dt} = \rho \mathbf{B} - \nabla p + \nabla\left[\frac{4}{3}\mu(\nabla \cdot \mathbf{V})\right] - \nabla \times [\mu(\nabla \times \mathbf{V})] \tag{5.155}$$

Several special forms of the Navier-Stokes equations are of interest. For flows with small temperature gradients, the viscosity μ, which is a function of temperature, may be approximated by a constant average value $\bar{\mu}$. In that case, equation 5.155 becomes

$$\rho \frac{D\mathbf{V}}{Dt} = \rho \mathbf{B} - \nabla p + \frac{4}{3}\bar{\mu}\nabla(\nabla \cdot \mathbf{V}) - \bar{\mu}\nabla \times (\nabla \times \mathbf{V}) \tag{5.156}$$

Another simplification is obtained for an incompressible fluid, or for the low-speed flow of a compressible fluid where compressibility effects are small. In both of those cases a constant average value $\bar{\rho}$ may be employed, and the continuity equation (see Table 2.2) is

$$\rho_t + \nabla \cdot (\rho\mathbf{V}) = \rho_t + \mathbf{V} \cdot \nabla\rho + \rho\nabla \cdot \mathbf{V} = 0 \tag{5.157}$$

For $\bar{\rho}$ = constant, $\bar{\rho}_t = \nabla\bar{\rho} = 0$, and equation 5.157 becomes

$$\nabla \cdot \mathbf{V} = 0 \ (incompressible \ flow) \tag{5.158}$$

Substituting equation 5.158 into equation 5.155 yields the Navier-Stokes equations for incompressible flow. Thus,

$$\bar{\rho} \frac{D\mathbf{V}}{Dt} = \bar{\rho}\mathbf{B} - \nabla p - \nabla \times [\mu(\nabla \times \mathbf{V})] \tag{5.159}$$

Many real flow fields may be approximated adequately by assuming ρ and μ both to be constants. In that case, equation 5.155 reduces to

$$\bar{\rho}\frac{D\mathbf{V}}{Dt} = \bar{\rho}\mathbf{B} - \nabla p - \bar{\mu}\nabla \times (\nabla \times \mathbf{V}) \tag{5.160}$$

Equation 5.160 may be written in an alternate form by employing the following vector identity (see Section 2–1).

$$\nabla \times (\nabla \times \mathbf{V}) = \nabla(\nabla \cdot \mathbf{V}) - \nabla^2\mathbf{V} \tag{5.161}$$

Substituting equation 5.161 into equation 5.160 and noting that $\nabla \cdot \mathbf{V} = 0$ for incompressible flow yields the following alternate form of the Navier-Stokes equations for the incompressible flow of a fluid having a constant viscosity.

$$\bar{\rho}\frac{D\mathbf{V}}{Dt} = \bar{\rho}\mathbf{B} - \nabla p + \bar{\mu}\nabla^2\mathbf{V} \tag{5.162}$$

The Navier-Stokes equations for viscous flow are similar to the Euler equations (see equation 2.63) for inviscid flow, except for the addition of the shear stress terms. The additional terms greatly complicate the mathematical form of the equations, so that it is difficult to obtain exact solutions to the Navier-Stokes equations. Exact solutions have been obtained for a few specialized cases. One such simple case is presented in the next section.

5–9(d) Hagen-Poiseuille Flow

Several of the exact solutions to the Navier-Stokes equations are presented in Reference 14. One exact solution is that for the fully developed steady flow of a viscous incompressible fluid having a constant viscosity, in a straight circular tube of constant diameter. Such a flow is called a *Hagen-Poiseuille flow*. The appropriate form of the Navier-Stokes equations is equation 5.162. The flow model is illustrated in Fig. 5.16. For a fully developed viscous flow, it may be assumed that: (1) the streamlines are parallel (which results in a constant pressure at each cross section); (2) that the velocity components in the radial and circumferential directions are zero; and (3) that the flow is steady.

Since the flow has axial symmetry, equation 5.162 is employed in cylindrical coordinates [see Section 10–3(b)]. For the assumptions listed above, $v_r = v_\theta = \partial/\partial r = \partial/\partial\theta = \partial/\partial t = 0$. Hence, the only equation that remains is that for the axial direction; that is, for the z direction. Hence, for the present case, equation 5.162

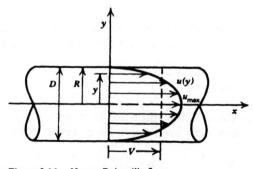

Figure 5.16. Hagen-Poiseuille flow.

reduces to

$$0 = -\frac{\partial p}{\partial z} + \mu\left(\frac{\partial^2 v_z}{\partial r^2} + \frac{1}{r}\frac{\partial v_z}{\partial r}\right) \tag{5.163}$$

Apply the following change of notation to equation 5.163; let $x = z$, $y = r$, and $u = v_z$. Equation 5.163 then transforms to

$$0 = -\frac{\partial p}{\partial x} + \mu\left(\frac{\partial^2 u}{\partial y^2} + \frac{1}{y}\frac{\partial u}{\partial y}\right) \tag{5.164}$$

with the following boundary conditions:

$$u(x,R) = 0 \qquad (no\ slip\ at\ the\ wall) \tag{5.165}$$

$$\frac{\partial u}{\partial y}(x,0) = 0 \qquad (symmetry\ about\ the\ axis) \tag{5.166}$$

For fully developed pipe flow, $\partial p/\partial x$ is known from experiments to be a constant. Since $u(x,y) = u(y)$ for a fully developed viscous flow, the partial derivatives in equation 5.164 may be replaced by total derivatives, and equation 5.164 may be rewritten to give

$$\frac{1}{y}\frac{d}{dy}\left(y\frac{du}{dy}\right) = \frac{1}{\mu}\frac{dp}{dx} \tag{5.167}$$

Integrating equation 5.167, we obtain

$$y\frac{du}{dy} = \frac{1}{\mu}\frac{dp}{dx}\frac{y^2}{2} + \text{constant} \tag{5.168}$$

Substituting equation 5.166 into equation 5.168 shows that the constant of integration is zero. Integrating equation 5.168 gives

$$u = \frac{1}{\mu}\frac{dp}{dx}\frac{y^2}{4} + \text{constant} \tag{5.169}$$

Evaluating the constant of integration, from equation 5.165, equation 5.169 yields

$$u = -\frac{R^2}{4\mu}\frac{dp}{dx}\left(1 - \frac{y^2}{R^2}\right) \tag{5.170}$$

Equation 5.170 is the velocity profile equation for the *Hagen-Poiseuille flow.*

Several additional features of Hagen-Poiseuille flow are of general interest. On the centerline, $u(x,0) = u_{max}$. From equation 5.170, we obtain

$$u_{max} = -\frac{R^2}{4\mu}\frac{dp}{dx} \tag{5.171}$$

Note that the pressure decreases in the flow direction, so that dp/dx is a negative number and u_{max} is a positive number. Hence, equation 5.170 may be written in the form

$$u = u_{max}\left(1 - \frac{y^2}{R^2}\right) \tag{5.172}$$

In Example 3.1, it is shown that the one-dimensional bulk flow velocity V that passes the same mass flow rate as the velocity profile given by equation 5.172 is

$$V = \frac{1}{2}u_{max} \tag{5.173}$$

Substituting equation 5.171 into equation 5.173 yields

$$V = -\frac{R^2}{8\mu}\frac{dp}{dx} = -\frac{D^2}{32\mu}\frac{dp}{dx} \tag{5.174}$$

In Section 5–4(b), equation 5.23, the Fanning friction coefficient $\mathfrak{f}$ is defined as

$$4\mathfrak{f} \equiv \frac{(dp/dx)D}{\frac{1}{2}\rho V^2} \tag{5.23}$$

Noting that dp/dx in equation 5.23 is actually $\overline{dp/dx}$, so that $\mathfrak{f}$ will be a positive number, solving for $\overline{dp/dx}$ from equation 5.174 and substituting the result into equation 5.23 yields

$$\mathfrak{f} = \frac{16\mu}{DV\rho} = \frac{16}{Re} \tag{5.175}$$

Equation 5.175 is quite accurate for laminar flow in a circular duct. It is in serious error, however, for turbulent flows.[14] A plot of equation 5.175 is presented in Fig. 5.8. For turbulent flows, the Stokes' law for simple friction is inadequate, and empirical data have to be employed for both the velocity distribution (see Section 3–3) and the friction coefficient, as discussed in Section 5–4(b).

5–10 THE BOUNDARY LAYER

A brief discussion of the physics of the boundary layer is presented in Section 1–6(a). The principal conclusion from that discussion is that the flow field for a fluid having a small viscosity may be subdivided into two contiguous regions. One is the bulk, or core, flow region wherein the fluid may be assumed to behave as a perfect fluid (i.e., $\mu = 0$). The other is a viscous boundary layer region that is adjacent to the solid boundaries wetted by the fluid. In the viscous boundary layer the flow is laminar and the shear stress τ on a fluid layer at the distance y from the wall is given by equation 1.28b which, for a two-dimensional flow in the xy plane, becomes

$$\tau = \mu\frac{\partial u}{\partial y}$$

where u is the velocity of the fluid in the x direction. At the wall itself, the velocity of the fluid in the boundary layer is zero and the shear stress at the wall is denoted by τ_o.

Even if the flowing fluid has only a small viscosity, such as is the case for air (see Fig. 1.13), the shear stress τ may be quite significant if the velocity gradient $\partial u/\partial y$ is sufficiently large.

The above concept of the boundary layer, due to Prandtl,[19] constitutes a major contribution to the analytical developments in the field of fluid mechanics. It is responsible for a major simplification of the Navier-Stokes equation that forms the basis of *boundary layer theory*.[4, 14, 15] The boundary layer equations are derived from the Navier-Stokes equations by neglecting terms that are demonstrably insignificant in the case of the fluid that flows inside the viscous boundary layer. The derivation of those equations is presented in detail in Reference 14.

Since the velocity of the fluid flowing within the viscous boundary layer must increase from zero at the wall to the free-stream velocity U_∞ in a distance equal to the boundary layer thickness, the magnitude of that thickness is an important parameter.

Figure 5.17 illustrates schematically a uniform flow having the free-stream velocity U_∞ over a flat plate that is at zero *angle of attack*, or zero *incidence*, with respect to U_∞. The origin of the coordinate system is coincident with the upper leading edge of the plate. Curve OA indicates schematically how the thickness of the boundary layer, denoted by $\delta(x)$, varies with the distance along the plate in the direction of the planar flow; that is, with increasing values of x. It is apparent that $\delta(x)$ increases or the boundary layer thickens as x increases. The thickness of $\delta(x)$, in Fig. 5.17, is greatly exaggerated compared to its actual dimensions. In the same figure the shapes of typical velocity profiles at different stations along the flat plate are illustrated diagrammatically. It is reemphasized here that the viscous boundary is the region of the overall flow field that is retarded in its motion by the viscous effects propagated from the wall wetted by the flowing fluid.

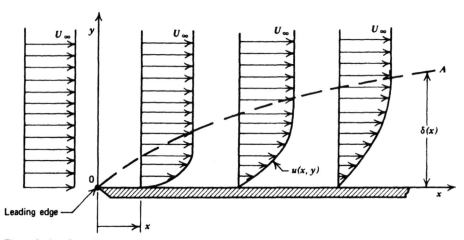

Figure 5.17.   Growth of a boundary layer on a flat plate in a uniform free stream.

By definition, the boundary layer thickness $\delta(x)$ is that normal distance from the wall (see Fig. 5.17) to the filament of fluid where the velocity $u(x,y)$ has the value $u = 0.99U_\infty$. Blasius showed that for the flow illustrated in Fig. 5.17, and his theoretical results were confirmed experimentally, that[20]

$$\frac{\delta}{x} = \frac{5}{\sqrt{\mathrm{Re}_x}} \tag{5.176}$$

where Re_x is the Reynolds number at the distance x from the leading edge of the flat plate. Thus,

$$\mathrm{Re}_x = \frac{U_\infty x \rho}{\mu} = \frac{U_\infty x}{\nu} \tag{5.177}$$

where $\nu = \mu/\rho$ is the kinematic viscosity of the fluid [see Section 1–6(a)].

Because the change in the velocity u in the boundary layer from zero at the wall to the value U_∞ takes place asymptotically, there is some indefiniteness in determining exactly where $u = 0.99U_\infty$.

A more precise measurement of the boundary layer thickness is the so-called *displacement thickness* δ^*. The latter is the distance that an equivalent inviscid flow region is displaced from the solid boundary as a consequence of the slow moving fluid in the boundary layer. The aforementioned displacement has the effect of causing a portion of the flow area of thickness $\delta*$ to become unavailable to the perfect fluid flow region. In other words, it is the *blockage* of the flow area referred to in Section 4–8(c).

For the flat plate illustrated in Fig. 5.17, the displacement thickness $\delta*$ is given by[20]

$$\frac{\delta*}{x} = \frac{1.73}{\sqrt{Re_x}} \approx \frac{\delta}{3} \tag{5.178}$$

Figure 5.18 illustrates diagrammatically the definition of $\delta*$ for the flow over a flat plate. It is the height $y = \delta*$ above the plate where the cross-hatched areas of the velocity profile curve, above and below $y = \delta*$ are equal.

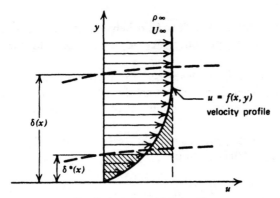

Figure 5.18. The displacement thickness $\delta*$.

Another measurement of the boundary layer thickness is the *momentum thickness*, denoted by Θ. Its definition is based on the reduction in the momentum of the flowing fluid caused by the shearing stresses in the boundary layer. For flow over a flat plate, the Blasius solution for the momentum thickness is[20]

$$\frac{\Theta}{x} = \frac{0.664}{\sqrt{Re_x}} \tag{5.179}$$

Hence, $\delta*/\Theta = 2.59$.

The viscous fluid in the boundary layer must continuously overcome the frictional resistance it encounters if it is to continue flowing downstream. If the pressure gradient is either favorable ($dp/dx < 0$) or zero ($dp/dx = 0$), there is always sufficient energy in the boundary layer to overcome the aforementioned resistance. On the other hand, if the pressure gradient is unfavorable ($dp/dx > 0$), the layers of fluid near the wall may actually come to rest and then reverse their flow direction, and *flow separation* occurs. Once flow separation occurs, there may be a radical alteration in the character of the flow field as, for example, the separated flow region downstream of the cylinder illustrated in Fig. 1.9. Separation phenomena in a converging-diverging nozzle are discussed in Section 4–7(e).

Figure 5.19 illustrates schematically the behavior of the boundary layer on the lower flat wall of a two-dimensional converging-diverging flow passage, where each of the three types of pressure gradients mentioned above are present. A flow separation point and a reverse flow region, in the viscous boundary layer, are also shown.

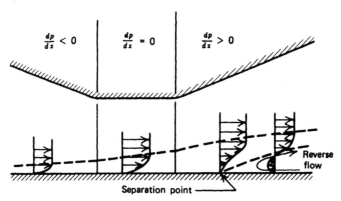

Figure 5.19. Flow separation in a boundary layer.

As long as there is no flow separation, the flow field behaves as that for a perfect fluid flowing between the thin viscous boundary layers adjacent to the solid boundaries.

Several analytical methods have been developed for predicting the *separation point* for the laminar boundary layer flow. Unfortunately, all of them are of limited usefulness.[14] Consequently, the separation point is ordinarily determined by experiment.

It can be shown that for a flat plate, such as that illustrated in Fig. 5.17, the shear stress at the wall τ_o (i.e., where $y = 0$) is given by[14]

$$\frac{\tau_o}{\rho U_\infty^2} = 0.332 \sqrt{\frac{v}{xU_\infty}} \tag{5.180}$$

Equation 5.180 shows that τ_o decreases as x increases and asymptotically approaches zero at $x = \infty$. Consequently, there will be no separation on a flat plate of finite length provided the pressure gradient for the flow is favorable.

Figure 5.20 shows that as the flow proceeds downstream along a flat plate, there is a transition from laminar to turbulent flow in the boundary layer. A *turbulent boundary layer* is generally thicker than a laminar one. Also, the velocity profile through most of the turbulent boundary layer is flatter, which results in a larger velocity gradient at the wall and a larger wall shear stress τ_o; that is, increased surface friction, also called *skin friction*. On the other hand, the turbulent boundary layer makes larger pressure recoveries possible. The behavior of turbulent boundary

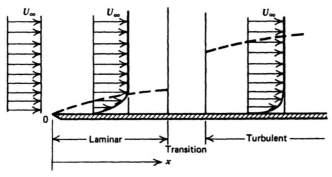

Figure 5.20. Transition from laminar to turbulent flow in a boundary layer.

layers is of great practical interest because most of the flows of importance in engineering are turbulent flows. Unfortunately, there is at this time no satisfactory analytic theory for investigating turbulent boundary layers that does not require experimental information.

In the cases for which the desired experimental information may be obtained by model testing (as, for example, determining the drag of an airfoil), the principles of dimensional analysis, discussed in Section 5–8, are employed for planning and conducting the experimental investigation.

A considerable amount of experimental data has been accumulated and correlated on the turbulent flow in pipes. In the case of fully developed turbulent flow in a horizontal pipe, the momentum equation (see equation 3.14) for the conditions $dV = \delta D = g\,dz = 0$, where V denotes the mean flow velocity in the pipe cross section, becomes

$$A\,dp + \delta F_f = 0$$

For a pipe of length $L = x_2 - x_1$ having the radius R,

$$A \int_1^2 dp + 2\pi R \tau_o \int_{x_1}^{x_2} dx = 0$$

where $\delta F_f = \tau_o(WP)\,dx = 2\pi R \tau_o\,dx$, and $A = \pi R^2$. Solving for the shear stress at the wall, we obtain

$$\tau_o = \frac{p_1 - p_2}{2}\frac{R}{L} \tag{5.181}$$

The shear stress τ at any radius $r < R$ is, accordingly,

$$\tau = \frac{p_1 - p_2}{2}\frac{r}{L} \tag{5.182}$$

Let $y = R - r$, the distance from the pipe wall to a point in the flow; then

$$\frac{\tau}{\tau_o} = \frac{r}{R} = 1 - \frac{y}{R} \tag{5.183}$$

Equation 5.182 indicates that for a fully developed turbulent pipe flow, important features of the flow can be determined by measuring the pressure drop $(p_1 - p_2)$.

It has been found convenient to divide the turbulent flow in a pipe into three regions that have different characteristics: (1) a thin *laminar sublayer* very close to the wall, where there cannot be any turbulent flow; (2) further inward from the wall, a so-called *turbulent shear region*; and (3) the region near the centerline of the pipe. The three regions are characterized by the following relationships.[14]

1. *Laminar sublayer*. It is very thin and it is assumed, according to Prandtl, that the shear stress τ, at any radius y where the velocity is u, is given by

$$\tau = \tau_o = \mu\frac{du}{dy} \approx \mu\frac{u}{y} \tag{5.184}$$

where τ_o is the shear stress at the wall. Introduce the friction velocity v^*, where

$$v^* = \sqrt{\frac{\tau_o}{\rho}} \tag{5.185}$$

The thickness of the laminar sublayer, denoted by δ_{SL}, expressed in terms of the friction velocity v^*, is given by

$$\delta_{SL} = \frac{5}{v^*} \tag{5.186}$$

Surface roughness has no appreciable effect on the flow, provided the roughness ε [see Section 5-4(b)] is smaller than δ_{SL}.

2. *Turbulent shear region.* The velocity distribution in the turbulent shear region is given by[14, 15]

$$\frac{u}{v^*} = a \ln y \frac{v^*}{v} + b \tag{5.187}$$

Equation 5.187 is known as the *universal velocity distribution law*. Measurements show that it is valid for the boundary layer on a flat plate, at zero incidence, and also for the turbulent flow in a pipe; for fully developed pipe flow the constants a and b have the values $a = 2.5$ and $b = 5.5$.

3. *Centerline region.* If U denotes the maximum velocity of the fluid, which occurs at the centerline of the pipe then, for the region in the vicinity of the centerline,

$$\frac{U - u}{v^*} = -2.5 \ln \frac{y}{R} \tag{5.188}$$

where R is the radius of the pipe, and y is the distance from the pipe wall.

For a fully developed turbulent pipe flow, the velocity profile is more nearly uniform than it is for laminar flow, for which it is parabolic, as shown in Section 5-9(d). Furthermore, the uniformity of the velocity profile increases with increasing Reynolds number. It has been shown that the turbulent velocity profile may be represented by the following *power law* (but not near the wall).[15]

$$\frac{u}{U} = \left(\frac{y}{R}\right)^{1/n} \tag{5.189}$$

where n varies from approximately $n = 6$ to $n = 10$ for $Re = 4 \cdot 10^3$ to $3 \cdot 10^6$. For $Re \approx 1 \cdot 10^5$, the velocity profile may be approximated with $n = 1/7$. If it is assumed that $Re \approx 1 \cdot 10^5$ and $\bar{u}/U = 0.8$, we obtain[14]

$$\frac{U}{v^*} = 8.74 \left(\frac{v^* Re}{v}\right)^{1/7} \tag{5.190}$$

It is shown in Section 5-4(b) that the friction coefficient $\mathfrak{f}$, for the surface of the pipe wall, is a function of Re and the relative roughness ε/D. The velocity profile equations, equations 5.185 to 5.190, do not appear to show that dependence. That dependence enters, however, through the magnitude of the friction velocity $v^* = \sqrt{\tau_o/\rho}$, because $\tau_o = f(\mu, \rho, D, \bar{u}, \varepsilon)$.

The experimental data pertaining to the turbulent flow over plates, at zero angle of attack, are rather meager compared to the data for fully developed pipe flow. Several methods of analysis have, therefore, been developed that are based on utilizing the data obtained from measurements of the flow in pipes as a basis for predicting the behavior of two-dimensional flows over plates. In general, those methods yield approximate results at best, and they have been unsuccessful in predicting separation phenomena.[14]

In summary, at this time there are no reliable purely theoretical methods for predicting the behavior of turbulent boundary layers in all their aspects. For additional information on that subject, see the references at the end of this chapter.

5–11 SUMMARY

A *simple frictional flow* is defined as steady one-dimensional flow in a constant-area duct in the absence of work, heat transfer, body forces, and obstructions in the flow, changes in the flow properties being caused by *wall friction* alone. Simple frictional flow for a compressible fluid is also called *Fanno line flow*. For such a flow, it is shown that if the flow in the constant-area duct is initially subsonic, the velocity and Mach number increase, while the static pressure, density, enthalpy, and temperature decrease. For a flow that is initially supersonic, the opposite trends are observed. For both subsonic and supersonic flows, the stagnation pressure decreases and the entropy increases.

The flow becomes *choked* in a definite length of the constant-area duct called the *limiting length*. In that cross section, the Mach number is unity, and all of the flow properties have their critical (i.e., sonic) values. If the duct length is to be increased beyond the limiting length, the upstream conditions must change. Equations are derived and tables are presented (see Table C.7) for determining the property ratios for the simple frictional flow of a perfect gas. A method for taking imperfect gas effects into account is presented in Section 5–6.

The wall frictional force is related to the bulk flow properties through an empirically determined friction coefficient, which depends on the Reynolds number, the Mach number, and the pipe surface roughness ratio.

The momentum equation for multidimensional viscous flow is presented (i.e., the Navier-Stokes equations). In addition, some of the effects of viscous boundary layer flow for fluids of small viscosity are discussed in Sections 5–9 and 5–10.

REFERENCES

1. A. Stodola, *Steam and Gas Turbines*, Vol. 1, translated by L. C. Loewenstein, McGraw-Hill, New York, 1927.

2. J. Nikuradse, "Strömungsgesetze in rauhen Rohren," VDI-Forschungsheft, No. 361, 1933. Translated as NACA TM 1292.

3. B. A. Bahkmeteff, *The Mechanics of Turbulent Flow*, Princeton University Press, Princeton, N.J., 1936.

4. L. Prandtl, *Strömungslehre*, Fried. Vieweg & Sohn, Braunschweig, 1949.

5. C. F. Colebrook, "Turbulent Flow in Pipes with Particular Reference to the Transition between Smooth and Rough Pipes," *Journal of the Institute of Civil Engineers*, London, February 1939.

6. L. F. Moody, "Friction Factors for Pipe Flow," *Transactions ASME*, Vol. 68, No. 8, pp. 671–684, November 1944.

7. R. P. Genereaux, "Fluid Flow Design Methods," *Industrial Engineering Chemistry*, Vol. 29, No. 4, pp. 385–388.

8. W. Frössel, "Strömung in glatten, geraden Rohren mit Uberund Untergeschallgeschwindigkeit," *Forsh, Gebiete Ingenieurw*, Vol. 7, pp. 75–84, 1936. Translated as NACA TM 844, 1938.

9. E. Buckingham, "On Physically Similar Systems; Illustrations of the Use of Dimensional Equations," *Physical Review*, Vol. 4, No. 4, pp. 345–378, 1914.

10. E. Buckingham, "Model Experiments and Forms of Empirical Equations," *Transactions ASME*, Vol. 37, 1915.

11. P. W. Bridgman, *Dimensional Analysis*, Yale University Press, New Haven, Conn., 1922.

12. L. Brand, "The Pi Theorem of Dimensional Analysis," *Archives Rational Mechanics Analysis*, Vol. 1, No. 1, pp. 35–45, 1957.

13. S. J. Kline, *Similitude and Approximation Theory*, McGraw-Hill, New York, 1965.

14. H. Schlichting, *Boundary Layer Theory*, Sixth Edition, McGraw-Hill, New York, 1968.

15. H. Schlichting, "Boundary-Layer Theory," translation by V. L. Streeter, *Handbook of Fluid Dynamics*, Section 9, McGraw-Hill, New York, 1961.
16. L. Prandtl, "The Mechanics of Viscous Fluids," *Aerodynamic Theory*, Vol. III, Section G, edited by W. F. Durand, California Institute of Technology, 1943.
17. J. A. Owczarek, *Fundamentals of Gas Dynamics*, International Textbook Co., Scranton, Pa., 1964.
18. S. I. Pai, *Viscous Flow Theory*, Vol. I, *Laminar Flow*, Van Nostrand, Princeton, N.J., 1956.
19. L. Prandtl, "Über Flüssigkeitsbewegung bei sehr kleiner Reibung," Vorhandlungen des III Internationalem Mathematiker-Kongresses, Heidlberg, pp. 484–491, 1904. Translated as NACA TM 452, 1928.
20. H. Blasius, "Grenzschichten in Flüssigkeiten mit kleiner Reibung," *Z. Math. u. Phys.*, Vol. 56, No. 1, 1908. Translated as NACA TM 1256.

PROBLEMS

1. Air flows adiabatically through a circular duct having a diameter $D = 0.30$ m. At the duct entrance, $t = 280$ K, $p = 2.0 \cdot 10^5$ N/m^2, and $M = 0.2$. Calculate (a) the length of duct required to choke the flow, and (b) the static and stagnation temperature and pressure, and the velocity at that condition. Assume $\bar{f} = 0.005$, and that c_p is constant.

2. Work Problem 1, taking into account the variation of the specific heat of air with the temperature.

3. A pipe 0.30 m in diameter and 1000 m long is attached to a large reservoir of air having a static pressure $p = 13.0 \cdot 10^5$ N/m^2 and a static temperature $t = 300$ K. The pipe exit is open to the atmosphere, and the average friction coefficient $\bar{f} = 0.005$. Assume that the flow is adiabatic and that c_p is constant. What is the mass flow rate through the pipe?

4. Air flows between a large reservoir and the atmosphere in an insulated tube 20.0 m long and 0.075 m in diameter. The average friction coefficient $\bar{f} = 0.005$. The stagnation pressure and temperature in the upstream reservoir are $P = 4.8 \cdot 10^5$ N/m^2 and $T = 325$ K, respectively. The static pressure in the surrounding atmosphere is 1 atm. Calculate (a) the duct inlet and exit Mach numbers, (b) the inlet and exit values of t, p, ρ, and V, and (c) the mass flow rate. Assume $\gamma = 1.40$ and $c_p = 1004.8$ J/kg-K.

5. Repeat Problem 4, accounting for the variation of c_p with t (employ Table C. 4).

6. Repeat Problem 4 for an upstream stagnation pressure of $2.75 \cdot 10^5$ N/m^2.

7. A pipe 0.050 m in diameter is attached to a large air tank where the pressure $p = 13.8 \cdot 10^5$ N/m^2 and the temperature $t = 310$ K. The exit of the pipe is open to the atmosphere. Assume adiabatic flow with an average friction coefficient $\bar{f} = 0.005$. What length of pipe is required to obtain a mass flow rate $\dot{m} = 2.25$ kg/s?

8. Plot a Fanno line on the ts plane for air entering a duct with a Mach number of 0.20, a static pressure of $7.0 \cdot 10^5$ N/m^2, and a static temperature of 550 K. Assume a duct diameter of 0.30 m and an average friction coefficient $\bar{f} = 0.005$.

9. Air enters a circular duct with a Mach number $M_1 = 3$. The average friction coefficient for the duct $\bar{f} = 0.0025$. How many diameters of pipe are required to reduce the Mach number to 2.0?

10. Atmospheric air at a pressure of $1.0135 \cdot 10^5$ N/m^2 and a temperature of 300 K is drawn through a frictionless bell-mouth entrance into a 3.0 m long tube having a 0.050 m diameter. The average friction coefficient $\bar{f} = 0.005$. The system is perfectly insulated. (a) Find the maximum mass flow rate and the range of back pressures that will produce this flow. (b) What exit pressure is required to produce 90 percent of the maximum mass flow rate, and what will be the stagnation pressure and the velocity at the exit for that mass flow rate?

11. Air enters an insulated tube of 0.025 m diameter through a De Laval nozzle with a throat diameter of 0.0125 m. The stagnation pressure and temperature are $7.0 \cdot 10^5$ N/m^2 and 300 K, respectively. The flow through the nozzle may be assumed to be frictionless, and the average friction coefficient in the duct $\bar{f} = 0.0025$. Calculate (a) the maximum length of the tube. For that condition, determine M, V, p, t, and P at (b) the inlet of the duct, (c) the outlet of the duct,

(d) a point 0.25 m from the inlet, and (e) a point where the stagnation pressure is $2.0 \cdot 10^5$ N/m^2. Plot those properties as a function of position in the tube.

12. A pipe having a roughness $\varepsilon = 0.00005$ m is to be designed for supplying compressed air to a building located 450 m from the air-storage tanks. The air-storage temperature is 300 K and the storage pressure is $9.0 \cdot 10^5$ N/m^2. It is essential that the pressure of the air at the building should not fall below $1.0 \cdot 10^5$ N/m^2 when the volume flow rate is 11.30 m^3/min. Determine the minimum internal diameter for the pipe that will satisfy the requirements.

13. A converging-diverging frictionless nozzle is connected to a large air reservoir by means of 2.85 m of 0.025 m ID pipe. The inside dimensions of the nozzle are as follows: inlet diameter 0.025 m, throat diameter 0.0225 m, outlet diameter 0.030 m. If the air is expanded to atmospheric pressure, and the average friction coefficient for the pipe $\bar{f} = 0.005$, find the stagnation pressure that must be maintained in the reservoir. Assume $\gamma = 1.4$.

14. Air at a pressure of $3.50 \cdot 10^5$ N/m^2 and a temperature of 300 K is to be transported at the rate of 0.090 kg/s over a distance of 600 m through a pipe. The final pressure is to be at least $1.40 \cdot 10^5$ N/m^2. Assuming isothermal flow and $\bar{f} = 0.004$, determine the minimum pipe diameter.

15. Air at $2.0 \cdot 10^5$ N/m^2 enters a 0.150 m diameter pipe at $M = 0.30$. Determine the maximum possible pipe length and the corresponding exit pressure, assuming $\bar{f} = 0.005$ for (a) isothermal flow, and (b) adiabatic flow.

16. A compressible fluid flows isothermally in a conduit of constant cross-sectional area in the presence of wall friction. The Mach number of the fluid at the entrance is $M_1 = 0.2$, and at station 2 further downstream it is $M_2 = 0.8$. If the molecular weight of the gas is 28, its constant-volume specific heat is 728.5 J/kg-K. $t_1 = 290$ K, and $p_1 = 1.0135 \cdot 10^5$ N/m^2, calculate (a) the heat added to the gas, (b) the change in the stagnation temperature of the gas, (c) the change in the stagnation pressure of the gas, and (d) the increase in its entropy.

17. In an oil refinery, hydrocarbon vapors are produced at 810 K. The vapors are conducted through a 0.10 m inside-diameter pipe to a condensing tower. The length of the pipe is 30.0 m. The vapors, which may be assumed to behave as a perfect gas, enter the pipe with a static pressure of $3.50 \cdot 10^5$ N/m^2, a density of 1.90 kg/m^3, and a viscosity of $6.30 \cdot 10^{-6}$ N-s/m^2. The mass flow rate is 7200 kg/hr. Sufficient heat is added to the pipe to maintain the static temperature of the vapors at 810 K. Calculate the pressure of the vapor entering the condensing tower. Assume that the inside surface of the pipe is smooth.

18. Derive the following relationships for the isothermal flow of a perfect gas in a constant-area duct in the presence of wall friction.

(a) $$\frac{dT}{T} = \frac{\gamma(\gamma - 1)M^4}{2(1 - \gamma M^2)\left(1 + \frac{\gamma - 1}{2}M^2\right)} \frac{4\bar{f}\,dx}{\mathcal{D}}$$

(b) $$\frac{dp}{p} = -\frac{\gamma M^2}{2(1 - \gamma M^2)}\frac{4\bar{f}\,dx}{\mathcal{D}}$$

(c) $$\frac{dP}{P} = \frac{\gamma M^2\left(1 - \frac{\gamma + 1}{2}M^2\right)}{2(\gamma M^2 - 1)\left(1 + \frac{\gamma - 1}{2}M^2\right)} \frac{4\bar{f}\,dx}{\mathcal{D}}$$

19. An incompressible viscous fluid flows steadily through the clearance of a plane thrust bearing as illustrated schematically in Fig. 5.21. The velocity profile for fully developed laminar flow is

$$u(y) = u_{max}\left[1 - \left(\frac{2y}{h}\right)^2\right]$$

Neglect entrance effects and determine an expression for the mass flow rate $\dot{m}$ in terms of p_i, p_o, μ, and the geometric parameters.

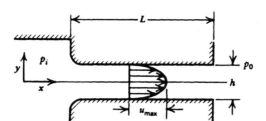

Figure 5.21. Sketch for Problem 5.19.

20. The pressure drop Δp in a constant-area duct because of the effect of wall friction alone is known to depend on the following variables: $\mathscr{D}$, the hydraulic diameter of the duct; ε, the pipe surface roughness; L, the length of the pipe; V, the one-dimensional flow velocity; a, the acoustic speed; ρ, the gas density; and μ, the dynamic viscosity of the fluid. Employing dimensional analysis, determine the functional dependence of the pressure drop on the remaining flow variables.

6

steady one-dimensional flow with heat transfer

6–1 PRINCIPAL NOTATION FOR CHAPTER 6

The notation presented in Section 3–1 is applicable to the discussions of this chapter. The following additional notation is employed.

$_1Q_2$ heat transferred between stations 1 and 2.

Greek Letters

α angle of tangent to a Rayleigh line.
δ angle of tangent to an *h-line*.
ϕ angle of tangent to an *s-line*.

Subscripts

R reservoir (stagnation) condition for an isothermal flow.
t minimum area (throat) for an isothermal flow.
1 state before heat transfer.
2 state after heat transfer.

6–2 INTRODUCTION

It is shown in Chapter 5 that a Fanno line, for brevity termed an *F-line*, depicts the thermodynamic states of a fluid in steady flow under the conditions $dA = \delta W = \delta Q = g\, dz = \delta D = 0$. In that case the only driving potential causing changes in the thermodynamic state of the fluid is the wall friction δF_f. There are many types of industrial equipment wherein there is a steady flow of a compressible fluid subject to the conditions $dA = \delta W = \delta F_f = dz = \delta D = 0$, but δQ is not zero. The heat δQ may be transferred either to or from the fluid and, for that type of flow, the stagnation enthalpy of the fluid does not remain constant. Such flow conditions occur in combustion chambers, heat exchangers, superheaters, and the like. Flows for which $\delta Q \neq 0$ are termed *diabatic flows*.[1–4] All practical cases of diabatic flow are invariably accompanied by friction. Moreover, if the heat addition is due to the flowing fluid reacting chemically with a fuel, the resulting combustion process causes not only a change in the mass of the flowing fluid but also produces a change in its chemical composition.

A first approximation to the effect of heat transfer on the flow of a compressible fluid is obtained by considering the flow to be a *simple diabatic flow*; that is, one that approximates certain practical diabatic flows. The analysis of simple diabatic flow is based on the following assumptions.

1. The area of the flow passage or duct is constant.
2. The flow is steady and one-dimensional.
3. There is no work, body forces are negligible, and the effects of friction are negligible.
4. Heat transfer is the only driving potential.

Because of the foregoing assumptions the changes in the stagnation enthalpy of the fluid are due entirely to the heat transfer. Assumption 3 may be assumed to represent either the case where the effects caused by heat transfer are considerably larger than the effects caused by friction or where the heat transfer is completed in such a short length of duct that the influence of the duct friction is negligible. In some cases there is such an abrupt change in the thermodynamic state, as in a detonation, for example, that from a mathematical standpoint the flow process may be conceived to involve a discontinuity. The analysis of *simple diabatic flows* is the object of the present chapter.

6-3 DYNAMICS OF STEADY ONE-DIMENSIONAL FLOW WITH HEAT TRANSFER IN A CONSTANT-AREA DUCT

To facilitate the study of diabatic flows in constant-area ducts, the effects because of the dynamics of the flow will be considered separately from the thermodynamic changes that occur in the flowing fluid. The locus of all of the attainable thermodynamic states for the fluid, as well as the limiting conditions at choking, may be determined solely from dynamic considerations. The determination of the actual states attained, however, requires the application of thermodynamic considerations; the latter are developed in Section 6-4. As mentioned earlier, the sole driving potential for the diabatic flows to be studied is the heat transfer Q.

6-3(a) Governing Equations

In view of the assumptions presented in the introduction to this chapter, the flow conditions for the diabatic flow of a fluid in a constant-area duct may be represented diagrammatically by Fig. 6.1. For that flow model the energy, momentum, and continuity equations take the following forms (see Table 3.1).

(1) *Energy equation*

$$\delta Q = dh + d\left(\frac{V^2}{2}\right) = dH \tag{6.1}$$

(2) *Momentum (or dynamic) equation for frictionless flow*

$$dp + \rho V \, dV = 0 \tag{6.2}$$

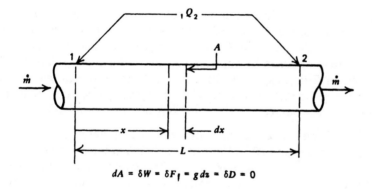

$$dA = \delta W = \delta F_f = g \, dz = \delta D = 0$$

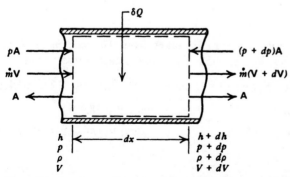

Figure 6.1 Steady one-dimensional frictionless flow in a constant-area duct with heat transfer, no external work, and no body forces (simple diabatic flow).

(3) *Continuity equation*

$$G = \frac{\dot{m}}{A} = \rho V = \frac{V}{v} = \text{mass flux} = \text{constant} \qquad (6.3)$$

Since the present section is concerned solely with the dynamics of the flow, only the continuity and momentum equations need to be considered. Combining equations 6.2 and 6.3 yields

$$dp + G\, dV = 0 \qquad (6.4)$$

Integrating equation 6.4 between states 1 and 2 gives

$$p_2 + GV_2 = p_1 + GV_1 = p + GV = \text{constant} \qquad (6.5)$$

Reference to Section 3–12 shows that the expression $pA + \rho A V^2$ is the *stream thrust*, denoted by $\mathscr{F}$. It follows from equation 6.5 that a basic characteristic of simple diabatic flow is that the *stream thrust per unit area remains constant*. Hence, equation 6.5 may be rewritten in the form

$$\frac{\mathscr{F}}{A} = p + GV = \text{constant} \qquad (6.6)$$

An expression for the pressure change $(p_2 - p_1)$ may be obtained by solving equation 6.5 for $(p_2 - p_1)$. Thus,

$$p_2 - p_1 = G(V_1 - V_2) \qquad (6.7)$$

Introducing the relationship $V = Gv$ into equation 6.5 yields

$$p_1 + G^2 v_1 = p_2 + G^2 v_2 = \text{constant} \qquad (6.8)$$

In most cases, p_1, v_1, and G are known initial conditions, so that equation 6.8 may be employed for plotting the thermodynamic states for the fluid on either the pv or hs plane. Each value of the mass flux G, also called the *flow density*, gives rise to a separate curve. A curve obtained by applying equation 6.8 is called a *Rayleigh line*, for brevity termed an *R-line*. Equation 6.8 is known as the *Rayleigh line* equation.

The equations derived in this section are valid for any fluid, because no restrictions have been placed on the character of the equations of state for the fluid.

Example 6.1. Superheated steam at an initial pressure of 400 lbf/in.^{2} ($27.58 \cdot 10^5$ N/m^2) and a specific volume of 1.800 ft^3/lbm (0.11237 m^3/kg) enters a 10 in. (0.244 m) ID pipe with a velocity of 1183.5 ft/sec (360.7 m/s). Determine the locus of the possible thermodynamic states that satisfies the continuity and momentum equations (i.e., the Rayleigh line).

Solution

The thermodynamic properties of steam are taken from *Thermodynamic Properties of Steam*, Keenan and Keyes, Wiley, 1958. The initial steam properties are:

$p_1 = 400$ lbf/in.^{2} ($27.58 \cdot 10^5$ N/m^2) $\qquad v_1 = 1.800$ ft^3/lbm (0.11237 m^3/kg)

$V_1 = 1183.5$ ft/sec (360.7 m/s) $\qquad t_1 = 790.1$ R (438.9 K)

$h_1 = 1411.17$ Btu/lbm ($3.282 \cdot 10^6$ J/kg) $\qquad s_1 = 1.6800$ Btu/lbm-R (7033.8 J/kg-K)

The mass flux G is

$$G = \frac{V}{v} = \frac{1183.5}{1.800} = 657.5 \text{ lbm/sec-ft}^2 \ (3209.3 \text{ kg/s-m}^2)$$

The above value of G is the same as that employed in Example 5.1 to determine a Fanno line for the flow of steam.

From equation 6.8,

$$p_1 + G^2 v_1 = 400 + \frac{(657.5)^2(1.800)}{(32.174)(144)} = 567.96 \text{ lbf/in.}^2$$

Hence, the equation for the Rayleigh line is given by

$$v = \frac{567.96 - p}{G^2} = \frac{(567.96 - p)(32.174)(144)}{(657.5)^2}$$

$$v = 6.0869 - 0.010717\, p \text{ ft}^3/\text{lbm} \ (0.38021 - 9.7091 \cdot 10^{-8}\, p \text{ m}^3/\text{kg})$$

Assume selected values of p, as shown in Table 6.1, and by means of the Rayleigh line equation, calculate the corresponding values of v. Then, from the steam tables, determine the corresponding values of t, h, and s. The results are presented in Table 6.1. Figure 6.2 is a plot of the corresponding Rayleigh line on the hs plane.

Table 6.1   Steam Properties Based on the Rayleigh Line Equation (Example 6.1)

p, lbf/in.^{2}	v, ft^3/lbm	t, R	h, Btu/lbm	s, Btu/lbm-R
425	1.5321	689.5	1355.53	1.6272
400	1.8000	790.1	1411.17	1.6800
375	2.0680	874.1	1456.63	1.7221
350	2.3359	938.8	1491.53	1.7551
325	2.6038	983.2	1515.68	1.7801
300	2.8718	1006.2	1528.48	1.7976
285	3.0326	1009.5	1530.64	1.8047
275	3.1397	1007.5	1529.84	1.8080
260	3.3005	998.1	1525.30	1.8111
250	3.4076	987.2	1519.94	1.8116
240	3.5148	973.0	1512.79	1.8111
230	3.6220	955.3	1503.85	1.8094
220	3.7292	934.0	1493.10	1.8066
200	3.9435	881.0	1466.40	1.7975
180	4.1578	864.0	1432.76	1.7832
150	4.4793	687.3	1369.89	1.7510
130	4.6937	571.1	1312.58	1.7155
100	5.0152	411.7	1233.73	1.6589
90	5.1224	348.9	1201.58	1.6317

lbf/in.^{2} × 6894.76 = N/m^2 Btu/lbm × 2325.97 = J/kg
ft^3/lbm × 0.062428 = m^3/kg   Btu/lbm-R × 4186.8 = J/kg-K
R/1.8 = K

6-3(b) The Rayleigh Line

Equation 6.8 for an *R-line* may be transformed to read

$$\frac{p_2 - p_1}{v_2 - v_1} = -G^2 \tag{6.9}$$

Equation 6.9 is the equation of a straight line plotted on the pv plane. If an *R-line*

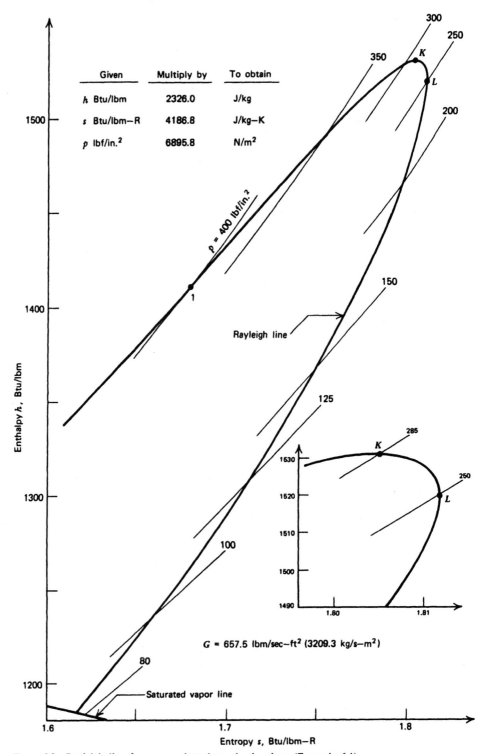

Figure 6.2 Rayleigh line for steam plotted on the *hs* plane (Example 6.1).

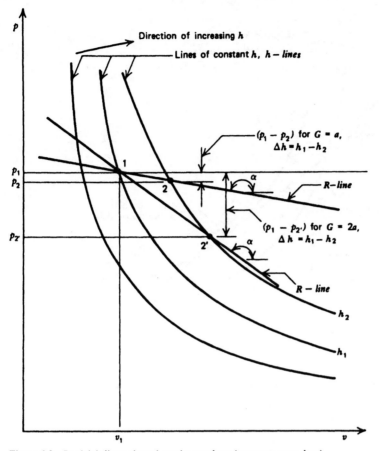

Figure 6.3 Rayleigh lines plotted on the pv plane (constant-area duct).

is plotted on the pv plane, as is done in Fig. 6.3, the resulting straight line has the slope $-G^2$. Consequently, if α is the angle made by the R-line and the v axis, then $\tan \alpha = -G^2$. Each value of the mass flux G gives rise to a different R-line, and the larger the value of G the steeper is the R-line. Since $G = V/v = $ constant for a given R-line, it follows from equation 6.9 that the fluid velocity corresponding to any state on an R-line plotted in the pv plane is given by

$$V = vG = v\left(\frac{p_2 - p_1}{v_1 - v_2}\right)^{1/2} = v(-\tan \alpha)^{1/2} \qquad (6.10)$$

Figure 6.3 illustrates diagrammatically three lines of constant specific enthalpy, hereafter referred to as h-lines, and two R-lines plotted on the pv plane. It is apparent from Fig. 6.3 that, if G is held constant, raising the enthalpy of the fluid from h_1 to h_2 by heat addition causes the static pressure of the fluid to fall and its specific volume to increase. Conversely, cooling the fluid, so that its specific enthalpy decreases from h_2 to h_1, is accompanied by an increase in its static pressure and a decrease in its specific volume. It is seen that, for the same increase in specific enthalpy ($\Delta h = h_2 - h_1$), increasing the mass flux from $G = a$ to $G = 2a$ quadruples the slope of the R-line and increases the pressure drop ($p_1 - p_{2'}$).

Figure 6.4a illustrates schematically a plot of two lines of constant enthalpy (h-lines) on the pv plane, and Fig. 6.4b illustrates two lines of constant entropy (s-lines)

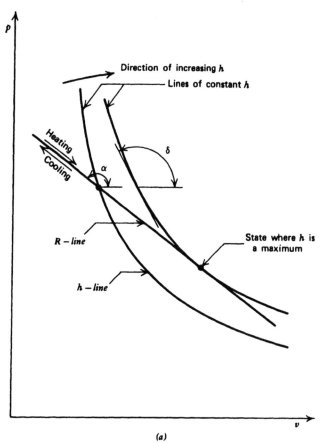

(a)

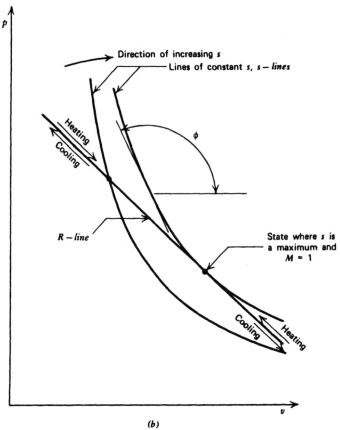

(b)

306

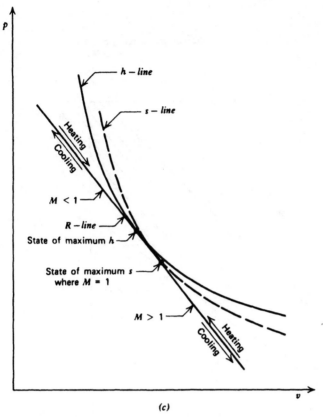

(c)

Figure 6.4 Conditions for maximum enthalpy and for maximum entropy for Rayleigh lines on the *pv* plane (constant-area duct). (*a*) Condition for maximum enthalpy. (*b*) Condition for maximum entropy. (*c*) States of maximum enthalpy and maximum entropy on the same Rayleigh line.

plotted on the same plane. It is seen that both the *h-lines* and the *s-lines* become more and more curved and concave upward for large values of *v* and small values of *p*. Consequently, if an *R-line* is extended in the direction of increasing values of *v*, it must become tangent to an *h-line*, as illustrated in Fig. 6.4*a*. Similarly, the *R-line* must also become tangent to an *s-line*, as illustrated in Fig. 6.4*b*. It is found that the specific volume of the fluid at the point where a given *R-line* is tangent to an *s-line* is larger than it is at the point where the *R-line* is tangent to an *h-line*, as illustrated in Fig. 6.4*c*. The foregoing shows that for a given *R-line* there is one state that corresponds to the condition for maximum enthalpy and another to the condition for maximum entropy.

Figure 6.5 illustrates diagrammatically the form of an *R-line* when plotted on the *hs* plane. The states of maximum enthalpy and of maximum entropy are indicated in the figure.

6–3(c) Condition for Maximum Enthalpy

When the fluid is in the maximum enthalpy state (see Fig. 6.5), any further addition of heat causes the kinetic energy of the fluid to increase at a rate equal to the rate of heat addition. Consequently, the enthalpy of the fluid is unaffected. If the velocity

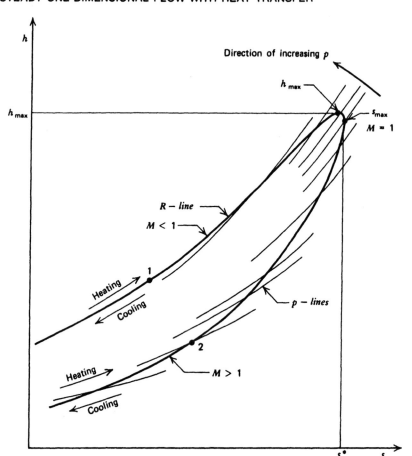

Figure 6.5 Rayleigh line plotted on the *hs* plane (constant-area duct).

of the fluid is increased at a rate faster than the heat addition rate (by reducing the back pressure, for example), the energy required for the corresponding increase in its kinetic energy must come from the energy stored in the fluid. As a result, the enthalpy of the fluid decreases to satisfy the demand for an increase in kinetic energy, and the state point for the fluid moves to the right, as illustrated in Fig. 6.5.

The slope of an *R-line* is given by equation 6.9. Thus,

$$\tan \alpha = \left(\frac{\partial p}{\partial v} \right)_R = -G^2 = -\frac{V^2}{v^2} \tag{6.11}$$

The slope of an *h-line* is denoted by tan δ, where

$$\tan \delta = \left(\frac{\partial p}{\partial v} \right)_h \tag{6.12}$$

At the state where the slopes of the *h-line* and the *R-line* are equal,

$$\tan \delta = \tan \alpha = -\frac{V^2}{v^2} \tag{6.13}$$

6–3(d) Condition for Maximum Entropy

If $\tan \phi$ denotes the slope of an *s-line*, then, in view of equation 1.180, it follows that

$$\tan \phi = \left(\frac{\partial p}{\partial v}\right)_s = -\frac{1}{v^2}\left(\frac{\partial p}{\partial \rho}\right)_s = -\frac{a^2}{v^2} \tag{6.14}$$

At the state where the entropy is a maximum, $\tan \phi = \tan \alpha$. Consequently,

$$\frac{V^2}{v^2} = \frac{a^2}{v^2}, \qquad \text{or} \qquad V = a \tag{6.15}$$

Equation 6.15 shows that when the entropy is a maximum the fluid velocity is equal to the local acoustic speed a; that is, $V = a^*$. The foregoing indicates that for a given set of initial conditions, there is a limiting state where the entropy is a maximum, and at that state $V = a^*$. When that condition is attained, the flow is said to be operating with *thermal choking*. Further addition of heat does not cause the fluid to attain states beyond the choking condition but produces a change in the upstream conditions.

The two branches of the *R-line* plotted in the *hs* plane (see Fig. 6.5) correspond to the following physical situations. The upper branch is the locus of the thermodynamic states that can be attained by heating a compressible fluid in subsonic flow ($M < 1$) from an initial state, such as state 1 in Fig. 6.5, which is located to the left of the state where thermal choking occurs. The lower branch of the *R-line* is the locus of the thermodynamic states that are attainable by a supersonic flow ($M > 1$) by heat addition from an initial state, such as state 2 in Fig. 6.5. It is seen that the effect of heat addition is to accelerate a subsonic flow to $M = 1$ as a limit, and to decelerate a supersonic flow to $M = 1$ as a limit.

6–4 THERMODYNAMICS OF STEADY ONE-DIMENSIONAL FLOW WITH HEAT TRANSFER

The dynamics of simple diabatic flow are investigated in Section 6–3. In this section, the energy equation for diabatic flow is developed, and some general observations are made concerning the intersection of a Fanno line and a Rayleigh line.

6–4(a) The Energy Equation

Table 3.1 shows that if $\delta W = g\,dz = 0$, for a simple diabatic flow, the energy equation is then given by

$$\delta Q = dh + d\left(\frac{V^2}{2}\right) \tag{6.16}$$

Referring to Fig. 6.1, if $_1Q_2$ is the heat transferred between stations 1 and 2, equation 6.16 may be integrated to yield

$$h_2 + \frac{V_2^2}{2} = h_1 + \frac{V_1^2}{2} + {}_1Q_2 \tag{6.17}$$

Combining the continuity equation, $G = V/v = $ constant, with equation 6.17 yields

$$h_2 + \frac{G^2 v_2^2}{2} = h_1 + \frac{G^2 v_1^2}{2} + {}_1Q_2 \tag{6.18}$$

From the definition of the stagnation enthalpy, $H = h + V^2/2$, it is apparent that equation 6.16 reduces to

$$dH = \delta Q \tag{6.19}$$

Equation 6.17, therefore, becomes

$$H_2 = H_1 + {}_1Q_2 \tag{6.20}$$

Equation 6.20 provides the connection between the driving potential (heat transfer) and the other flow properties. The actual mechanism of the heat transfer is unimportant in applying equation 6.20 to the calculation of the effects of a given ${}_1Q_2$ on the flow properties. In any real flow process, however, the detailed heat transfer mechanism must be specified (i.e., conduction, convection, radiation, or chemical reaction) so that the numerical value of ${}_1Q_2$ may be determined.

6–4(b) Intersection of a Fanno Line and a Rayleigh Line

If an *F-line* and an *R-line* for the same value of G are plotted on the *hs* plane, they will intersect at two states, such as the states 1 and 2 illustrated in Fig. 6.6. Since all of

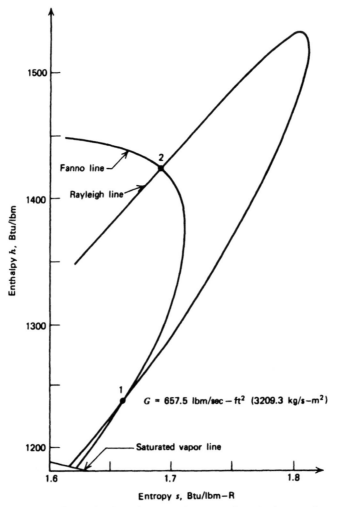

Figure 6.6 Determination of a normal compression shock wave in steam.

the states on the same *F-line* have the same stagnation enthalpy H, and all of the states on the same *R-line* have the same value of stream thrust $\mathscr{F}/A$, it follows that states 1 and 2 have identical values of $G, H,$ and $\mathscr{F}/A$. It will be seen later (see Chapter 7) that the flow through a shock wave satisfies the energy and continuity equations for the *F-line* and the momentum and continuity equations for the *R-line*. Consequently, the spontaneous change from state 1 to state 2 can be accomplished by a shock wave. State 2 is at a higher entropy than state 1, indicating that a spontaneous change from state 1 to state 2 does not violate the second law of thermodynamics. Since state 2 is at a higher static pressure than state 1, the spontaneous change of state is termed a *compression shock wave*. A change of state from state 2 to state 1 is impossible in view of the second law of thermodynamics: in other words, a spontaneous change from a subsonic to a supersonic flow is impossible; only the reverse can occur. The shock process is a sudden compression that increases the pressure and entropy of the fluid but decreases its velocity from a supersonic to a subsonic value.

Example 6.2. A Fanno line and a Rayleigh line for steam having a mass flux $G = 657.5$ lbm/sec-ft^2 (3209.3 kg/s-m^2) are determined in Examples 5.1 and 6.1, respectively. Plot those two processes on the *hs* plane and determine the properties ahead of and behind the normal shock wave corresponding to the intersection of the *F-line* and the *R-line*.

Solution

The results are presented in Fig. 6.6. The properties ahead of the shock wave (point 1 in Fig. 6.6) are:

$$h_1 = 1245.5 \text{ Btu/lbm } (2.897 \cdot 10^6 \text{ J/kg}) \qquad s_1 = 1.667 \text{ Btu/lbm-R } (6979.4 \text{ J/kg-K})$$

By interpolating from the steam tables,

$$p_1 = 107.46 \text{ lbf/in.}^2 (7.409 \cdot 10^5 \text{ N/m}^2), \qquad t_1 = 440.3 \text{ R } (244.6 \text{ K}),$$
$$v_1 = 4.8317 \text{ ft}^3/\text{lbm } (0.30163 \text{ m}^3/\text{kg})$$

The initial steam velocity is

$$V_1 = Gv_1 = (657.5)(4.8317) = 3176.84 \text{ ft/sec } (968.3 \text{ m/s})$$

The properties after the shock wave (point 2 in Fig. 6.6) are

$$h_2 = 1423 \text{ Btu/lbm } (3.310 \cdot 10^6 \text{ J/kg}) \qquad s_2 = 1.691 \text{ Btu/lbm-R } (7079.9 \text{ J/kg-K})$$

By interpolating from the steam tables,

$$p_2 = 394.11 \text{ lbf/in.}^2 (27.177 \cdot 10^5 \text{ N/m}^2), \qquad t_2 = 811.8 \text{ R } (451.0 \text{ K}),$$
$$v_2 = 1.8720 \text{ ft}^3/\text{lbm } (0.11687 \text{ m}^3/\text{kg})$$

The final steam velocity is

$$V_2 = Gv_2 = (657.5)(1.8720) = 1230.84 \text{ ft/sec } (375.16 \text{ m/s})$$

6-5 THE RAYLEIGH LINE FOR PERFECT GASES

In Sections 6-3 and 6-4, the general features of *simple diabatic flow* are discussed without reference to a specific fluid. In this section, the results obtained in the aforementioned two sections are applied to the flow of a perfect gas.

6–5(a) Rayleigh Line Equations for a Perfect Gas

For an *R-line* the stream thrust per unit area $\mathscr{F}/A$ is constant. Hence, if the fluid is a perfect gas, it follows from equation 6.5 that

$$\frac{\mathscr{F}}{A} = p + \rho V^2 = p\left(1 + \frac{\rho V^2}{p}\right) = p(1 + \gamma M^2)$$

$$= p_2(1 + \gamma M_2{}^2) = p_1(1 + \gamma M_1{}^2) = \text{constant} \tag{6.21}$$

From equation 6.21, it follows that

$$\frac{p_2}{p_1} = \frac{1 + \gamma M_1{}^2}{1 + \gamma M_2{}^2} \tag{6.22}$$

From Table 3.2, the continuity equation for the flow of a perfect gas in a constant-area duct becomes

$$\frac{p_2 M_2}{\sqrt{t_2}} = \frac{p_1 M_1}{\sqrt{t_1}} \tag{6.23}$$

From equations 6.22 and 6.23,

$$\frac{t_2}{t_1} = \frac{p_2{}^2 M_2{}^2}{p_1{}^2 M_1{}^2} = \frac{M_2{}^2}{M_1{}^2}\left(\frac{1 + \gamma M_1{}^2}{1 + \gamma M_2{}^2}\right)^2 \tag{6.24}$$

For a perfect gas, the equation of state yields

$$\frac{p_2 v_2}{t_2} = \frac{p_1 v_1}{t_1} \tag{6.25}$$

From equations 6.22, 6.24, 6.25, and the continuity equation $G = \rho V = \text{constant}$, we obtain

$$\frac{\rho_2}{\rho_1} = \frac{v_1}{v_2} = \frac{V_1}{V_2} = \frac{p_2 t_1}{p_1 t_2} = \frac{M_1{}^2}{M_2{}^2}\left(\frac{1 + \gamma M_2{}^2}{1 + \gamma M_1{}^2}\right) \tag{6.26}$$

From equation 3.177, we obtain the following equation for the entropy change for a perfect gas.

$$\frac{\Delta s}{R} = \ln\left[\left(\frac{t_2}{t_1}\right)^{\gamma/(\gamma-1)}\left(\frac{p_1}{p_2}\right)\right] \tag{6.27}$$

Substituting for p_1/p_2 and t_2/t_1 from equations 6.22 and 6.24 yields

$$\frac{\Delta s}{R} = \ln\left[\left(\frac{M_2}{M_1}\right)^{2\gamma/(\gamma-1)}\left(\frac{1 + \gamma M_1{}^2}{1 + \gamma M_2{}^2}\right)^{(\gamma+1)/(\gamma-1)}\right] \tag{6.28}$$

Combining equations 3.182 and 6.22, we obtain the following equation for the *stagnation pressure ratio* P_2/P_1. Thus,

$$\frac{P_2}{P_1} = \left(\frac{1 + \gamma M_1{}^2}{1 + \gamma M_2{}^2}\right)\left(\frac{1 + \dfrac{\gamma-1}{2}M_2{}^2}{1 + \dfrac{\gamma-1}{2}M_1{}^2}\right)^{\gamma/(\gamma-1)} \tag{6.29}$$

The slopes of the *R-line*, the *h-line*, and the *s-line* on the *pv* plane are given by equations 6.11, 6.12, and 6.14. When the fluid is a perfect gas, those equations transform as follows:

(a) *Slope of the R-line*

$$\tan \alpha = -\frac{V^2}{v^2} = \frac{\gamma p}{\gamma p} \frac{V^2}{v^2} = -\gamma \frac{p}{v} M^2 \tag{6.30}$$

(b) *Slope of the h-line (isotherm)*

$$\tan \delta = \left(\frac{\partial p}{\partial v}\right)_t = \left[\frac{\partial (Rt/v)}{\partial v}\right]_t = -\frac{Rt}{v^2} \tag{6.31}$$

(c) *Slope of the s-line (isentrope)*

$$\tan \phi = \left(\frac{\partial p}{\partial v}\right)_s = -\rho^2 \left(\frac{\partial p}{\partial \rho}\right)_s = -\frac{a^2}{v^2} \tag{6.32}$$

It follows from equations 6.30 and 6.32 that when the slopes of an *R-line* and an *s-line* are equal, then

$$\frac{\tan \alpha}{\tan \phi} = M^2 = 1 \tag{6.33}$$

Similarly, when the slopes of the *R-line* and an *h-line* are equal,

$$\frac{\tan \alpha}{\tan \delta} = \gamma M^2 = 1 \tag{6.34}$$

Hence, as shown in the preceding, at the state where the entropy is a maximum ($\tan \alpha = \tan \phi$), the flow Mach number is $M = 1$, and at the state where the enthalpy reaches its maximum value ($\tan \alpha = \tan \delta$), the flow Mach number $M = 1/\sqrt{\gamma}$.

6–5(b) Relationship between Heat Transfer, Stagnation Temperature, and Mach Number

For a perfect gas, the energy equation may be written in the form (see equation 3.173, Table 3.2)

$$c_p t_2 + \frac{V_2^{\,2}}{2} = c_p t_1 + \frac{V_1^{\,2}}{2} + {}_1 Q_2 \tag{6.35}$$

which may be transformed to read

$$t_2 \left(1 + \frac{\gamma - 1}{2} M_2^{\,2}\right) = t_1 \left(1 + \frac{\gamma - 1}{2} M_1^{\,2}\right) + \frac{{}_1 Q_2}{c_p} \tag{6.36}$$

For a perfect gas, $H = c_p T$. Consequently, equation 6.19 may be written as

$$c_p \, dT = \delta Q \tag{6.37}$$

In that case, equation 6.20 becomes

$$T_2 = T_1 + \frac{{}_1 Q_2}{c_p} \tag{6.38}$$

Equation 6.38 provides the connection between the heat transfer ${}_1 Q_2$ and the flow properties in the case of a perfect gas.

From the definition of the stagnation temperature [see Section 3–9(b)],

$$T = t \left(1 + \frac{\gamma - 1}{2} M^2\right) \tag{6.39}$$

The stagnation temperature ratio T_2/T_1 may be expressed in terms of the Mach numbers at stations 1 and 2. Thus,

$$\frac{T_2}{T_1} = \frac{t_2}{t_1} \frac{\left(1 + \dfrac{\gamma - 1}{2} M_2^2\right)}{\left(1 + \dfrac{\gamma - 1}{2} M_1^2\right)} \tag{6.40}$$

Substituting equation 6.24 into equation 6.40 yields

$$\frac{T_2}{T_1} = \frac{M_2^2 (1 + \gamma M_1^2)^2}{M_1^2 (1 + \gamma M_2^2)^2} \frac{\left(1 + \dfrac{\gamma - 1}{2} M_2^2\right)}{\left(1 + \dfrac{\gamma - 1}{2} M_1^2\right)} \tag{6.41}$$

6-5(c) Entropy Change Caused by Heat Transfer

For a perfect gas, the entropy change in terms of the stagnation properties of the flowing gas is given by equation 3.179, Table 3.2. Thus,

$$\frac{\Delta s}{c_p} = \ln\left[\frac{T_2/T_1}{(P_2/P_1)^{(\gamma - 1)/\gamma}}\right] \tag{6.42}$$

The entropy change may, therefore, be obtained by combining equations 6.29, 6.41, and 6.42.

Example 6.3. It is desired to achieve a stagnation temperature ratio of 4.0 in a ramjet combustor by the addition of heat, and not exceed a Mach number of 0.9 at its exit. The initial stagnation temperature is $T = 350$ K. Determine (a) the initial Mach number, and (b) the amount of heat which can be added. Assume $\gamma = 1.3$ and $R = 287.04$ J/kg-K.

Solution

(a)   Solve equation 6.41 for M_1. Thus,

$$M_1^2\left(1 + \frac{\gamma - 1}{2} M_1^2\right) = \frac{T_1}{T_2} \frac{M_2^2\left(1 + \dfrac{\gamma - 1}{2} M_2^2\right)}{(1 + \gamma M_2^2)^2} (1 + \gamma M_1^2)^2$$

$$M_1^4\left(\gamma^2 K - \frac{\gamma - 1}{2}\right) + M_1^2(2\gamma K - 1) + K = 0$$

where

$$K = \frac{T_1}{T_2} \frac{M_2^2\left(1 + \dfrac{\gamma - 1}{2} M_2^2\right)}{(1 + \gamma M_2^2)^2} = \frac{(0.9)^2[1 + 0.15(0.9)^2]}{4[1 + 1.3(0.9)^2]^2} = 0.0538823$$

$$M_1^4(-0.0589389) + M_1^2(-0.859906) + 0.0538823 = 0$$

$$M_1^2 = 0.0623939 \qquad M_1 = 0.24979$$

(b) $c_p = \dfrac{\gamma R}{\gamma - 1} = \dfrac{1.3(287.04)}{(0.3)} = 1243.8$ J/kg-K.

From equation 6.38,

$$Q = c_p(T_2 - T_1) = c_p T_1 \left(\frac{T_2}{T_1} - 1 \right)$$

$$Q = (1243.8)(350)(4 - 1) = 1306.0 \text{ kJ/kg } (561.49 \text{ Btu/lbm})$$

6–5(d) Effect of Heat Transfer on the Flow Properties

The effect of heat transfer on the flow properties may be determined by employing the method that is employed in Section 4–4(c). Relationships may be derived between the variables dM/M, dp/p, $d\rho/\rho$, dt/t, dV/V, dP/P, ds/c_p, and the heat transfer δQ expressed in terms of dT/T (see equation 6.37). The results are presented in Table 6.2.

Table 6.2 Influence Coefficients for the Simple Diabatic Flow of a Perfect Gas in a Constant-Area Duct

$$\frac{dM}{M} = \frac{(1 + \gamma M^2)\left(1 + \dfrac{\gamma - 1}{2} M^2\right)}{2(1 - M^2)}\left(\frac{dT}{T}\right) \tag{6.43}$$

$$\frac{dp}{p} = -\frac{\gamma M^2 \left(1 + \dfrac{\gamma - 1}{2} M^2\right)}{1 - M^2}\left(\frac{dT}{T}\right) \tag{6.44}$$

$$\frac{d\rho}{\rho} = -\frac{\left(1 + \dfrac{\gamma - 1}{2} M^2\right)}{1 - M^2}\left(\frac{dT}{T}\right) \tag{6.45}$$

$$\frac{dt}{t} = \frac{(1 - \gamma M^2)\left(1 + \dfrac{\gamma - 1}{2} M^2\right)}{1 - M^2}\left(\frac{dT}{T}\right) \tag{6.46}$$

$$\frac{dV}{V} = \frac{\left(1 + \dfrac{\gamma - 1}{2} M^2\right)}{1 - M^2}\left(\frac{dT}{T}\right) \tag{6.47}$$

$$\frac{dP}{P} = -\frac{\gamma M^2}{2}\left(\frac{dT}{T}\right) \tag{6.48}$$

$$\frac{ds}{c_p} = \left(1 + \dfrac{\gamma - 1}{2} M^2\right)\left(\frac{dT}{T}\right) \tag{6.49}$$

From those results, it is apparent that heat addition ($dT > 0$) has the opposite effect on the flow properties than does heat removal ($dT < 0$). The effect is quite similar to that observed for isentropic flow in a variable area duct where opposite effects are obtained for $dA > 0$ and $dA < 0$. Mathematically, it appears that a transition from subsonic to supersonic flow caused solely by heat transfer should be possible. Near the sonic condition, however, the heat transfer rate becomes extremely large, and the entire one-dimensional analysis becomes questionable. To date, no such transition has been reported.

Table 6.3 summarizes the effect of heat transfer on the flow properties for heat addition ($dT > 0$) to a perfect gas. For heat removal ($dT < 0$), all of the effects are reversed.

Table 6.3 Effect of Heat Addition ($dT > 0$) on the Flow Properties for a Perfect Gas in a Constant-Area Duct

Variable	$M < 1$	$M > 1$
dM/M	$+$	$-$
dp/p	$-$	$+$
$d\rho/\rho$	$-$	$+$
dt/t	$+$ for $M < \dfrac{1}{\sqrt{\gamma}}$ $-$ for $M > \dfrac{1}{\sqrt{\gamma}}$	$+$
dV/V	$+$	$-$
dP/P	$-$	$-$
ds/c_p	$+$	$+$

6–5(e) Rayleigh Line Tables for a Perfect Gas

As in the case of the Fanno line (*F-line*) for perfect gases, it is convenient to refer the state coordinates on an *R-line* to their values at the thermal choking limit; that is, where $M = 1$, $V = a^*$, $p = p^*$, etc. Substituting $M_1 = 1$ into the appropriate equations of Sections 6–3 and 6–4 gives the equations presented in Table 6.4.

Table 6.4 Property Ratios for the Simple Diabatic Flow of a Perfect Gas in a Constant-Area Duct

$$\frac{p}{p^*} = \frac{\gamma + 1}{1 + \gamma M^2} \tag{6.50}$$

$$\frac{\rho}{\rho^*} = \frac{v^*}{v} = \frac{V^*}{V} = \frac{1 + \gamma M^2}{(\gamma + 1)M^2} \tag{6.51}$$

$$\frac{t}{t^*} = \frac{(\gamma + 1)^2 M^2}{(1 + \gamma M^2)^2} \tag{6.52}$$

$$\frac{P}{P^*} = \frac{\gamma + 1}{1 + \gamma M^2}\left[\left(\frac{2}{\gamma + 1}\right)\left(1 + \frac{\gamma - 1}{2}M^2\right)\right]^{\gamma/(\gamma - 1)} \tag{6.53}$$

$$\frac{T}{T^*} = \frac{2(\gamma + 1)M^2}{(1 + \gamma M^2)^2}\left(1 + \frac{\gamma - 1}{2}M^2\right) \tag{6.54}$$

The ratios p/p^*, t/t^*, ρ/ρ^*, P/P^*, and T/T^* are presented as functions of the Mach number M in Table C.9 for different values of γ. The general arrangement of Table C.9 and the limiting values of the property ratios are presented in Table 6.5.

Table 6.5 Limiting Values of the Property Ratios for the Simple Diabatic Flow of a Perfect Gas in a Constant-Area Duct

M	$M^* = \dfrac{V}{V*} = \dfrac{\rho^*}{\rho}$	$\dfrac{t}{t*}$	$\dfrac{p}{p*}$	$\dfrac{P}{P*}$	$\dfrac{T}{T*}$
0	0	0	$(\gamma+1)$	$(\gamma+1)\left(\dfrac{2}{\gamma+1}\right)^{\gamma/(\gamma-1)}$	0
1	1	1	1	1	1
∞	$\dfrac{\gamma+1}{\gamma}$	0	0	∞	$\dfrac{\gamma^2-1}{\gamma^2}$

A FORTRAN program, subroutine RAY, for calculating the property ratios for the simple diabatic flow of a perfect gas is presented below. The input to this program is the same as that discussed in Section 4-4(b). The results presented in Table C.9 for $\gamma = 1.4$ are obtained by the following specification.

G = 1.4, J(1) = 50, J(2) = 50, J(3) = 10, J(4) = 20, J(5) = 12, J(6) = 5, J(7) = 0, DM(1) = 0.01, DM(2) = 0.02, DM(3) = 0.05, DM(4) = 0.1, DM(5) = 0.5, DM(6) = 1.0

```
      SUBROUTINE RAY

C     RAYLEIGH LINE PROPERTY RATIOS FOR A PERFECT GAS

      REAL M,MS $ DIMENSION J(9),DM(9) $ DATA INF/6H INFIN/
      NAMELIST /DATA/ G,J,DM $ READ (5,DATA)
      G1=(G-1.0)/2.0 $ G2=G+1.0 $ G3=G/(G-1.0) $ G4=2.0/(G+1.0)
      G5=G4**G3
      M=0.0 $ MS=0.0 $ T=0.0 $ P=G2 $ PO=G2*G5 $ TO=0.0
      WRITE (6,2000) G $ WRITE (6,2020) M,MS,T,P,PO,TO $ I=1 $ L=0

C     CALCULATE PROPERTY RATIOS

   40 N=J(I) $ DO 50 K=1,N $ M=M+DM(I) $ MS=G2*M**2/(1.0+G*M**2)
      T=(MS/M)**2 $ P=MS/M**2 $ C=1.0+G1*M**2 $ TO=G4*T*C $PO=P*G5*C**G3
      WRITE (6,2020) M,MS,T,P,PO,TO $ L=L+1 $ IF (L.LT.50) GO TO 50
      WRITE (6,2000) G $ L=0
   50 CONTINUE $ I=I+1 $ IF((I.EQ.10).OR.(J(I).EQ.0)) GO TO 60 $GO TO 40
   60 MS=G2/G $ T=0.0 $ P=0.0 $ TO=(G**2-1.0)/G**2
      WRITE (6,2030) INF,MS,T,P,INF,TO $ RETURN

 2000 FORMAT (1H1,21X,23HRAYLEIGH LINE,  GAMMA =,F5.2//4X,1HM,4X,7HM*,R*
     1/R,4X,4HT/T*,9X,4HP/P*,9X,6HPO/PO*,7X,6HTO/TO*/)
 2020 FORMAT (1H ,F5.2,F10.5,4E13.5)
 2030 FORMAT (A6,F10.5,2E13.5,3X,A6,E17.5)
      END
```

Example 6.4. Air leaves the subsonic diffuser of a ramjet engine with a static pressure of $0.5516 \cdot 10^5$ N/m^2. Between the diffuser exit and the combustion chamber inlet sections, a liquid fuel is sprayed into the air, the fuel-air ratio being $f = 1/29$. The fuel is vaporized and thoroughly mixed with the air and enters the constant-area combustion chamber with a static temperature of 333.3 K and an average velocity of 73.15 m/s. The fuel has a calorific value of $\Delta H_c = 41{,}867$ kJ/kg. Assume that the working fluid has the same thermodynamic properties as air before and after combustion, and that friction is negligible. Calculate (a) the stagnation temperature after combustion, (b) the Mach number after combustion, (c) the final static temperature, (d) the loss in stagnation pressure due to heat addition, (e) the entropy change, and (f) the final velocity of the combustion products. Assume $\gamma = 1.4$, $R = 287.04$ J/kg-K, and $c_p = 1.0048$ kJ/kg-K.

Solution

$$a_1 = (\gamma R t_1)^{1/2} = [1.4(287.04)(333.3)]^{1/2} = 366.0 \text{ m/s}$$

$$M_1 = \frac{V_1}{a_1} = \frac{73.15}{366.0} = 0.2000$$

From Table C.9 (Rayleigh line), for $M_1 = 0.2$,

$$\frac{T_1}{T^*} = 0.17355, \qquad \frac{t_1}{t^*} = 0.20661,$$

$$\frac{p_1}{p^*} = 2.2727, \qquad \frac{P_1}{P^*} = 1.2346, \qquad \frac{V_1}{V^*} = 0.09091$$

$$\frac{T_1}{t_1} = 1 + \frac{\gamma - 1}{2} M_1^2 = 1 + 0.2(0.2)^2 = 1.0080$$

$$T_1 = (1.0080)(333.3) = 336.0 \text{ K}$$

$$\frac{P_1}{p_1} = \left(\frac{T_1}{t_1}\right)^{\gamma/(\gamma - 1)} = (1.0080)^{3.5} = 1.02824$$

$$P_1 = (1.02824)(0.5516)10^5 = 0.5672 \cdot 10^5 \text{ N/m}^2$$

The energy of combustion ΔH for each unit mass of the air-fuel mixture is

$$\Delta H = \frac{\dot{m}_f \Delta H_c}{\dot{m}_f + \dot{m}_a} = \frac{\Delta H_c}{1 + \frac{\dot{m}_a}{\dot{m}_f}} = \frac{\Delta H_c}{1 + \frac{1}{f}} = \frac{41,867}{1 + 29} = 1,395.5 \text{ kJ/kg}$$

(a) Denote the stagnation temperature after combustion by T_2. Hence,

$$\Delta H = c_p(T_2 - T_1)$$

$$T_2 = T_1 + \frac{\Delta H}{c_p} = 336.0 + \frac{1,395.5}{1.0048} = 1724.8 \text{ K}$$

(b) $\dfrac{T_2}{T_1} = \dfrac{1724.8}{336.0} = 5.1333$ and $\dfrac{T_2}{T^*} = \dfrac{T_1}{T^*} \dfrac{T_2}{T_1} = (0.17355)(5.1333) = 0.89088$

From Table C.9, for $T_2/T^* = 0.89088$, $M_2 = 0.680$, and

$$\frac{t_2}{t^*} = 0.98144, \qquad \frac{p_2}{p^*} = 1.4569, \qquad \frac{P_2}{P^*} = 1.0489, \qquad \frac{V_2}{V^*} = 0.67367$$

(c) $t_2 = \dfrac{t_2/t^*}{t_1/t^*} t_1 = \dfrac{(0.98144)(333.3)}{(0.20661)} = 1583.2 \text{ K } (2849.8R)$

(d) $P_2 = \dfrac{P_2/P^*}{P_1/P^*} P_1 = \dfrac{1.0489(0.5672)10^5}{1.2346} = 0.4819 \cdot 10^5 \text{ N/m}^2 (6.989 \text{ lbf/in.}^2)$

$$\Delta P = P_2 - P_1 = (0.4819 - 0.5672)10^5 = -8530 \text{ N/m}^2 (1.24 \text{ lbf/in.}^2)$$

(e) $s_2 - s_1 = c_p \ln\left(\dfrac{T_2}{T_1}\right) - R \ln\left(\dfrac{P_2}{P_1}\right) = (1004.8)\ln(5.1333) - (287.04)\ln\left(\dfrac{0.4819}{0.5672}\right)$

$$s_2 - s_1 = 1690.4 \text{ J/kg-K } (0.4037 \text{ Btu/lbm-R})$$

(f) $V_2 = \dfrac{V_2/V^*}{V_1/V^*} V_1 = \dfrac{(0.67367)(73.15)}{(0.09091)} = 542.1 \text{ m/s } (1778.4 \text{ ft/sec})$

6-6 THE RAYLEIGH LINE FOR IMPERFECT GASES

The properties of the Rayleigh line are developed in Section 6-5 for *perfect gases*. In this section, the influence of *imperfect gas effects* on simple diabatic flow is considered. Several sources of imperfect gas effects are presented in Section 4-5.

A completely general *imperfect gas* model is considered. Assume that the equations of state for the flowing fluid are given by

$$\rho = \rho(p,t) \quad \text{and} \quad h = h(p,t) \tag{6.55}$$

Equations 6.55 may be in the form of equations, charts, tables, or a computer code. The source, or type, of the imperfect gas effect is immaterial as far as the analysis is concerned, provided that equations 6.55 are available.

The continuity, momentum, and energy equations for simple diabatic flow are given by equations 6.3, 6.5, and 6.17, respectively. Those equations may be rearranged to give

$$V_2 = \frac{\rho_1 V_1}{\rho_2} \tag{6.56}$$

$$p_2 = p_1 + \rho_1 V_1{}^2\left(1 - \frac{\rho_1}{\rho_2}\right) \tag{6.57}$$

$$h_2 = h_1 + \frac{V_1{}^2}{2}\left[1 - \left(\frac{\rho_1}{\rho_2}\right)^2\right] + {}_1Q_2 \tag{6.58}$$

Equations 6.55 to 6.58 comprise a set of five equations in terms of the flow properties p, ρ, h, t, and V at point 1 upstream to the heat transfer region and at point 2 downstream from it. When the upstream properties of the gas and the heat transfer ${}_1Q_2$ are specified, the aforementioned equations yield five relationships between the five properties p_2, ρ_2, h_2, t_2, and V_2. In general, those equations must be solved numerically.

Example 6.5. Consider the flow of steam discussed in Example 6.1. Determine the amount of heat addition required to choke the flow, and the properties of the steam at that section.

Solution

From Fig. 6.2, the choking condition is approximately

$p^* = 250 \text{ lbf/in.}^2 (17.24 \cdot 10^5 \text{ N/m}^2)$ and $h^* = 1520 \text{ Btu/lbm } (3.536 \cdot 10^6 \text{ J/kg})$

From the steam tables,

$$t^* = 987.4 \text{ R } (548.6 \text{ K}) \qquad v^* = 3.4079 \text{ ft}^3/\text{lbm } (0.21275 \text{ m}^3/\text{kg})$$

For flow in a constant-area duct,

$$V^* = Gv^* = (657.5)(3.4079) = 2240.7 \text{ ft/sec } (682.96 \text{ m/s})$$

From equation 6.17,

$$Q = h^* - h_1 + \frac{(V^*)^2 - V_1{}^2}{2}$$

$$Q = 1520 - 1411.17 + \frac{(2240.7)^2 - (1183.5)^2}{2(32.174)(778.16)}$$

$$Q = 108.83 + 72.30 = 181.13 \text{ Btu/lbm } (0.4213 \cdot 10^6 \text{ J/kg})$$

6-7 STEADY ONE-DIMENSIONAL ISOTHERMAL FLOW IN A VARIABLE-AREA PASSAGE

In Section 5–7, the steady isothermal flow of a perfect gas in a constant-area duct is investigated. In this section, the steady isothermal frictionless flow of a perfect gas in a variable-area passage is studied. At present, no practical examples of such a flow are available. Some devices have been considered, however, wherein the flow is maintained isothermal by either heat transfer or combustion. Because of the possible application of such a process in the future, the major points of interest are developed herein.

6-7(a) Governing Equations

For a steady isothermal flow of a perfect gas, in a variable-area duct, the energy, momentum, and continuity equations take the following form (see Table 3.2).

(1) *Energy equation*

$$\delta Q = d\left(\frac{V^2}{2}\right) = dH = c_p \, dT \tag{6.59}$$

which may be integrated to yield

$$c_p(T_2 - T_1) = {}_1Q_2 = \frac{V_2{}^2 - V_1{}^2}{2} \tag{6.60}$$

(2) *Momentum equation*

$$dp + \rho V \, dV = 0 \tag{6.61}$$

(3) *Continuity equation for a perfect gas*

$$\dot{m} = \rho A V = A p M \sqrt{\frac{\gamma}{Rt}} = \text{constant} \tag{6.62}$$

From equation 6.62, it follows that

$$p_2 A_2 M_2 = p_1 A_1 M_1 \tag{6.63}$$

For a perfect gas, the thermal equation of state is

$$pv = \frac{p}{\rho} = Rt = \text{constant} \tag{6.64}$$

Hence,

$$\frac{p_2}{p_1} = \frac{\rho_2}{\rho_1} = \frac{v_1}{v_2} \tag{6.65}$$

Combining equations 6.61 and 6.64 yields

$$dp + \frac{p}{Rt} d\left(\frac{V^2}{2}\right) = dp + \gamma p \, d\left(\frac{M^2}{2}\right) = 0 \tag{6.66}$$

Integration of equation 6.66 yields

$$\frac{p_2}{p_1} = \frac{\rho_2}{\rho_1} = \frac{v_1}{v_2} = e^{\gamma(M_1{}^2 - M_2{}^2)/2} \tag{6.67}$$

Substituting equation 6.67 into equation 6.63, we obtain

$$\frac{A_2}{A_1} = \frac{M_1}{M_2} e^{\gamma(M_2{}^2 - M_1{}^2)/2} \tag{6.68}$$

From the definition of the stagnation temperature, for $t_1 = t_2$, the stagnation temperature ratio T_2/T_1 is

$$\frac{T_2}{T_1} = \frac{1 + \dfrac{\gamma - 1}{2} M_2{}^2}{1 + \dfrac{\gamma - 1}{2} M_1{}^2} \tag{6.69}$$

From the definition of the stagnation pressures corresponding to p_1 and p_2, the stagnation pressure ratio P_2/P_1 is

$$\frac{P_2}{P_1} = \left(\frac{1 + \dfrac{\gamma - 1}{2} M_2{}^2}{1 + \dfrac{\gamma - 1}{2} M_1{}^2} \right)^{\gamma/(\gamma - 1)} e^{[\gamma(M_1{}^2 - M_2{}^2)/2]} \tag{6.70}$$

From the definition of the impulse function,

$$\frac{\mathscr{F}_2}{\mathscr{F}_1} = \frac{p_2 A_2 (1 + \gamma M_2{}^2)}{p_1 A_1 (1 + \gamma M_1{}^2)} = \frac{M_1 (1 + \gamma M_2{}^2)}{M_2 (1 + \gamma M_1{}^2)} \tag{6.71}$$

where p_2/p_1 and A_2/A_1 are obtained from equations 6.67 and 6.68, respectively.

6-7(b) Limiting Conditions

Let station 1 be some initial point in the flow where A_1 and M_1 are fixed, and let station 2 be any other station where $A_2 = A$ and $M_2 = M$. Then, equation 6.68 may be written in the form

$$A = \frac{1}{M} e^{\gamma M_2{}^2/2} (A_1 M_1 e^{-\gamma M_1{}^2/2}) \tag{6.72}$$

For fixed initial conditions, the term inside the parentheses is constant. At the minimum area (i.e., at the throat A_t), $dA_t = 0$. Differentiating equation 6.72 logarithmically, we obtain the following equation.

$$\frac{dA}{A} = -\frac{dM}{M} + \gamma M \, dM = (\gamma M^2 - 1)\frac{dM}{M} \tag{6.73}$$

Setting $dA_t = 0$ in equation 6.73, and solving for M_t, we obtain

$$M_t = \frac{1}{\sqrt{\gamma}} \tag{6.75}$$

By definition, $M_t = V_t/a_t$. For an isothermal flow the static temperature t is constant throughout the flow and so is $a = \sqrt{\gamma R t}$. At the throat, $M_t = 1/\sqrt{\gamma}$, and the velocity V_t is given by

$$V_t = M_t a_t = \frac{\sqrt{\gamma R t}}{\sqrt{\gamma}} = \frac{a}{\sqrt{\gamma}} = \sqrt{R t} \tag{6.75}$$

Consequently, the limiting velocity at the throat is less than the sonic velocity.

If equation 6.67 is applied between the reservoir pressure $p = P_R$ where $M = 0$, and the throat where $p = p_t$ and $M_t = 1/\sqrt{\gamma}$, we obtain

$$\frac{p_t}{P_R} = e^{\gamma(0 - 1/\gamma)/2} = e^{-1/2} = 0.60653 \qquad (6.76)$$

Hence, the choking pressure ratio is 0.60653 and is independent of the thermodynamic properties of the flowing fluid (i.e., the specific heat ratio γ).

6-7(c) Tables for the Isothermal Flow of a Perfect Gas in a Variable-Area Passage

The equations developed in Sections 6-7(a) and 6-7(b) may be applied between the sonic condition, denoted by *, and a general point in the flow to yield the property ratios presented in Table 6.6.

Table 6.6   Property Ratios for the Isothermal Flow of a Perfect Gas in a Variable-Area Passage

$$\frac{p}{p^*} = \frac{\rho}{\rho^*} = \frac{v^*}{v} = e^{\gamma(1 - M^2)/2} \qquad (6.77)$$

$$\frac{P}{P^*} = e^{\gamma(1 - M^2)/2}\left[\left(\frac{2}{\gamma + 1}\right)\left(1 + \frac{\gamma - 1}{2}M^2\right)\right]^{\gamma/(\gamma - 1)} \qquad (6.78)$$

$$\frac{T}{T^*} = \left(\frac{2}{\gamma + 1}\right)\left(1 + \frac{\gamma - 1}{2}M^2\right) \qquad (6.79)$$

$$\frac{\mathscr{F}}{\mathscr{F}^*} = \frac{1 + \gamma M^2}{(\gamma + 1)M} \qquad (6.80)$$

$$\frac{A}{A^*} = \frac{1}{M}e^{\gamma(M^2 - 1)/2} \qquad (6.81)$$

Values of the property ratios are presented as a function of the Mach number M for $\gamma = 1.4$ in Table C.10. The general arrangement of Table C.10 and the limiting values of the property ratios are presented in Table 6.7.

Table 6.7   Limiting Values of the Property Ratios for the Isothermal Flow of a Perfect Gas in a Variable-Area Passage

$M = M^*$	$\dfrac{p}{p^*}$	$\dfrac{P}{P^*}$	$\dfrac{A}{A^*}$	$\dfrac{\mathscr{F}}{\mathscr{F}^*}$	$\dfrac{T}{T^*}$
0	$e^{\gamma/2}$	$e^{\gamma/2}\left(\dfrac{2}{\gamma + 1}\right)^{\gamma/(\gamma - 1)}$	∞	∞	$\left(\dfrac{2}{\gamma + 1}\right)$
$\dfrac{1}{\sqrt{\gamma}}$	$e^{(\gamma - 1)/2}$	$e^{(\gamma - 1)/2}\left[\dfrac{(3\gamma - 1)}{\gamma(\gamma + 1)}\right]^{\gamma/(\gamma - 1)}$	$\sqrt{\gamma}\,e^{(1 - \gamma)/2}$	$\dfrac{2\sqrt{\gamma}}{(\gamma + 1)}$	$\dfrac{(3\gamma - 1)}{(\gamma + 1)}$
1	1	1	1	1	1
∞	0	∞	∞	∞	0

A FORTRAN program, subroutine TCAREA, for calculating the property ratios presented in Table 6.6 is presented below. The input to this program is the same as that discussed in Section 4–4(b). The results presented in Table C.10 for $\gamma = 1.4$ are obtained by the following specification.

G = 1.4, J(1) = 25, J(2) = 30, J(3) = 10, J(4) = 11, J(5) = 0,
DM(1) = 0.02, DM(2) = 0.05, DM(3) = 0.2, DM(4) = 1.0

```
      SUBROUTINE TCAREA

C     ISOTHERMAL VARIABLE AREA FLOW PROPERTY RATIOS FOR A PERFECT GAS

      REAL M,MS,MT,MI $ DIMENSION J(9),DM(9) $ DATA INF/6H INFIN/
      NAMELIST /DATA/ G,J,DM $ READ (5,DATA)
      G1=(G-1.0)/2.0 $ G2=G/2.0 $ G3=2.0/(G+1.0) $ G4=G/(G-1.0)
      G5=G3**G4 $ G6=G+1.0
      M=0.0 $ P=EXP(G2) $ PO=P*G5 $ TO=G3 $ MT=1.0/SQRT(G)
      WRITE (6,2000) G $ WRITE (6,2010) M,P,PO,INF,INF,TO $ I=1 $ L=0

C     CALCULATE PROPERTY RATIOS

   40 N=J(I) $ DO 50 K=1,N $ M=M+DM(I)
   42 C=1.0+G1*M**2 $ MS=M/SQRT(G2*C) $ P=EXP(G2*(1.0-M**2)) $ TO=G3*C
      PO=P*TO**G4 $ A=1.0/(P*M) $ F=(1.0+G*M**2)/(G6*M)
      IF (M.NE.MT) GO TO 44 $ WRITE (6,2040) MT,P,PO,A,F,TO
      M=MI $ MT=1.0E+30 $ GO TO 50
   44 WRITE (6,2020) M,P,PO,A,F,TO $ L=L+1 $ IF (L.LT.50) GO TO 45
      WRITE (6,2000) G $ L=0
   45 IF ((MT-M).GT.DM(I)) GO TO 50 $ MI=M $ M=MT $ GO TO 42
   50 CONTINUE $ I=I+1 $ IF((I.EQ.10).OR.(J(I).EQ.0)) GO TO 60 $GO TO 40
   60 P=0.0 $ PO=0.0 $ WRITE (6,2030) INF,P,PO,INF,INF,INF $ RETURN

 2000 FORMAT (1H1,14X,38HISOTHERMAL VARIABLE AREA FLOW, GAMMA =,F5.2//
     14X,1HM,7X,4HP/P*,9X,6HPO/PO*,7X,4HA/A*,9X,4HF/F*,9X,6HTO/TO*/)
 2010 FORMAT (1H ,F5.2,2X,2E13.5,3X,A6,7X,A6,E17.5)
 2020 FORMAT (1H ,F5.2,2X,5E13.5)
 2030 FORMAT (A6,2X,2E13.5,3X,A6,2(7X,A6))
 2040 FORMAT (1H ,F7.4,5E13.5)
      END
```

Example 6.6.   Air flows isothermally in a converging-diverging De Laval nozzle. The inlet static temperature, pressure, and Mach number are 300 K, $35 \cdot 10^5$ N/m², and 0.5, respectively. Calculate (a) the throat static temperature, (b) the throat velocity, (c) the inlet to throat area ratio, (d) the stagnation temperature at the throat, and (e) the heat transfer.

Solution

(a)   For isothermal flow, $t_t = t = 300$ K (540 R).

(b)   From equation 6.75,

$$V_t = (Rt)^{1/2} = [(287.04)(300)]^{1/2} = 293.4 \text{ m/s (962.8 ft/sec)}$$

(c)   From equation 6.74,

$$M_t = \frac{1}{\sqrt{\gamma}} = \frac{1}{\sqrt{1.4}} = 0.8452$$

From Table C.10, for $M_1 = 0.5$ and $M_t = 0.8452$, $A_1/A^* = 1.1831$ and $A_t/A^* = 0.96874$. Hence,

$$\frac{A_1}{A_t} = \frac{A_1}{A^*} \frac{A^*}{A_t} = \frac{1.1831}{0.96874} = 1.2213$$

(d)   From equation 3.181, the stagnation temperature T_t is given by

$$T_t = t_t\left(1 + \frac{\gamma - 1}{2}M_t^2\right) = 300[1 + 0.2(0.8452)^2] = 342.9 \text{ K } (617.1 \text{ R})$$

(e)   The inlet stagnation temperature T_1 is

$$T_1 = t_1\left(1 + \frac{\gamma - 1}{2}M_1^2\right) = 300[1 + 0.2(0.5)^2] = 315.0 \text{ K } (567.0 \text{ R})$$

The heat transfer is given by equation 6.60. Thus,

$$_1Q_t = c_p(T_t - T_1) = 1004.8(342.9 - 315.0) = 28.033 \text{ kJ/kg } (12.05 \text{ Btu/lbm})$$

6-8 SUMMARY

A steady one-dimensional compressible flow having heat transfer as its sole driving potential is termed a *simple diabatic flow*. Such a flow has different characteristic features, depending on whether the initial flow, before the transfer of heat, is subsonic or supersonic. The subsonic flow case will be considered first.

If the *initial flow is subsonic* ($M < 1$), then the effects of heat transfer to the flow are as follows.

1. Heating causes the flow Mach number M to increase and the corresponding static pressure p to decrease. Conversely, cooling a subsonic flow causes M to decrease and p to increase.
2. The enthalpy h of the fluid increases in the direction of the flow, attains a maximum, then decreases as the flow Mach number approaches $M = 1$ even though the fluid is receiving heat.
3. If the initial flow conditions (h_1, p_1, v_1, G_1) remain unaltered, the flow becomes *thermally choked* when $M = 1$.

If the *initial flow is supersonic* ($M > 1$), the effects of heat transfer are as follows.

1. Heat addition reduces the flow Mach number M and increases the static pressure p of the fluid. Cooling decreases p and increases M.
2. When the flow Mach number is reduced to $M = 1$, the flow is thermally choked, as in the subsonic case.

If an *R-line* and an *F-line* (see Chapter 5), for a given compressible fluid, are plotted on the *hs* plane for the same value of G, they will intersect at two points (see Fig. 6.6). One point corresponds to supersonic flow, the other to subsonic flow, indicating that a spontaneous change in state is possible. To satisfy the second law of thermodynamics, the spontaneous change, if it occurs, must be from a supersonic flow and a lower pressure to a subsonic flow and a higher pressure.

All of the comments presented up to this point in the Summary apply to any compressible fluid.

In the special case where the flowing fluid is a perfect gas, quantitative relationships are readily derived between the variables dp/p, dt/t, dM/M, etc. and the heat transfer δQ expressed in terms of dT/T [see Section 6–5(d)]. Furthermore, equations are derived for the property ratios for simple diabatic flow. Those equations are listed in Table 6.4, and the property ratios are tabulated as a function of M, for different values of γ, in Table C.9. The state where h has its maximum value corresponds to the Mach number $M = 1/\sqrt{\gamma}$.

If the flowing fluid is an imperfect gas, the deviations of the flow properties from those based on the perfect gas must be taken into account, as is done in Example 6.5.

If a perfect gas flows under isothermal conditions in a variable-area duct, there is a limiting Mach number at the throat; it is $M_t = 1/\sqrt{\gamma}$. Correspondingly, the limiting velocity is $V_t = a_t/\sqrt{\gamma}$, which is smaller than the sonic velocity for the throat. Table C.10 presents the flow property ratios as functions of M for $\gamma = 1.4$.

REFERENCES

1. M. J. Zucrow, *Aircraft and Missile Propulsion*, Vol. 1, Wiley, New York, 1958.
2. A. H. Shapiro and W. R. Hawthorne, "The Mechanics and Thermodynamics of Steady One-Dimensional Gas Flow," *Journal of Applied Mechanics*, Vol. 14, No. 4, pp. A-317 to A-336, December 1947.
3. B. L. Hicks and R. H. Wasserman, "The One-Dimensional Theory of Steady Compressible Fluid Flow in Ducts with Friction and Heat Addition," *Journal of Applied Physics*, pp. 891–902, October 1947.
4. J. V. Foa and G. Rudinger, "On the Addition of Heat to a Gas Flowing in a Pipe at Subsonic Speed," *Journal of the Institute of Aeronautical Sciences*, February 1949.

PROBLEMS

1. Air flows frictionlessly through a circular duct having an internal diameter $D = 0.30$ m. At the entrance to the duct, $t = 300$ K, $p = 2.0 \cdot 10^5$ N/m^2, and $M = 0.2$. Calculate (a) the quantity of heat transfer required to choke the flow, and (b) the static and stagnation temperature and pressure, and the velocity at that condition. Assume that c_p is constant.

2. Work Problem 1, taking into account the variation of the specific heat of air with the temperature.

3. A pipe 0.025 m in diameter and 3.0 m long is attached to a large air reservoir having a static pressure of $14.0 \cdot 10^5$ N/m^2 and a static temperature of 310 K. The pipe exit is open to the atmosphere. The pipe is coiled inside of a furnace where the temperature is 2000 K. Assuming that the flow is frictionless, calculate (a) the maximum amount of heat that can be transferred to the gas, and (b) the corresponding mass flow rate. Assume that c_p is constant.

4. Work Problem 3, taking into account the variation of the specific heat of air with the temperature.

5. Air flows from a large reservoir where the static pressure is $4.8 \cdot 10^5$ N/m^2 and the static temperature is 320 K through a 0.075 m internal diameter pipe and exhausts into the atmosphere (pressure = 1 atm). Heat in the amount 335 kJ/kg is transferred to the gas. Calculate (a) the duct inlet and exit Mach numbers, (b) the inlet and exit values of t, p, ρ, and V, and (c) the mass flow rate. Assume $\gamma = 1.40$ and $c_p = 1004.8$ J/kg-K.

6. Repeat Problem 5, taking into account the variation of c_p with the temperature.

7. Repeat Problem 5 for an upstream stagnation pressure of $2.75 \cdot 10^5$ N/m^2.

8. Plot a Rayleigh line for air on the ts plane for the following inlet conditions: pressure = $3.5 \cdot 10^5$ N/m^2, temperature = 550 K, area = 0.010 m^2, and mass flow rate = 30 kg/s. Plot this line for both the subsonic and supersonic region. Assume c_p is constant = 1004.8 J/kg-K, and that $s = 0$ at 220 K. From the Rayleigh-line plot, determine the amount of fuel that must be burned in the air to raise its temperature to the maximum value. Assume complete combustion, no dissociation, and that the heating value of the fuel is 3600 kJ/kg. Neglect the effect of the fuel addition on the mass flow rate and its specific heat.

9. What fuel-air ratio will produce "choking" $(M = 1)$? Assume the same conditions as in Problem 8.

10. How much heat must be added to or removed from air with $V = 1200$ m/s $(M = 3.0)$ to decrease the velocity to $V = 900$ m/s $(M = 1.275)$? Use the Rayleigh line plotted in Problem 8.

11. Plot a Rayleigh line on a pv diagram and employ the equation of the line to solve the following problem. Air flows in a tube of constant diameter. Across a section of the tube the air is heated by a process that does not involve friction. The pressures and temperatures on either side of the heating section are $1.0135 \cdot 10^5$ N/m², 300 K, $0.35 \cdot 10^5$ Nm², and 300 K, respectively. Is this a possible steady flow situation?

12. Air enters a frictionless duct at 290 K and a velocity of 700 m/s. Find the maximum amount of heat that may be transferred per kilogram of air passing through the duct without reducing the inlet velocity.

13. Air is flowing in a constant-area duct with an initial Mach number $M = 0.5$, stagnation temperature $T = 300$ K, and stagnation pressure $P = 1.75 \cdot 10^5$ N/m². The ambient pressure is 1 atm, and the duct may be considered frictionless. Calculate the maximum amount of heat that can be added to that flow (a) if c_p is constant, and (b) taking into account the variation of c_p with t.

14. Air flows frictionlessly through a duct having an internal diameter $D = 0.60$ m and a length $L = 10$ m. At the inlet, $M = 0.2$, $T = 390$ K, and $P = 2.75 \cdot 10^5$ N/m². Heat transfer occurs along the duct and, at the duct exit, $T = 2200$ K. Calculate (a) M_2, (b) P_2, and (c) the amount of heat transfer.

15. Air enters the constant-area combustor of a supersonic combustion ramjet (SCRAMJET) at a Mach number $M = 3.5$. The free-stream Mach number $M_o = 5.0$ and the free-stream temperature $t_o = 250$ K. The Mach number at the exit of the combustor must be maintained at 1.20 or larger. Calculate the maximum amount of thermal energy that can be added to the flowing air by combustion (a) if c_p is constant, and (b) taking into account the variation of c_p with t.

7

shock waves

7-1 PRINCIPAL NOTATION FOR CHAPTER 7

The notation presented in Section 3-1 is applicable to the present chapter. The additional notation listed below is peculiar to this chapter.

M' = V_N/a, component of the Mach number normal to an oblique shock wave.
u^* = u/a^*, nondimensional velocity component.
v^* = v/a^*, nondimensional velocity component.
V_N velocity component normal to an oblique shock wave.
V_T velocity component tangential to an oblique shock wave.

Greek Letters

β = $(p_2 - p_1)/p_1$, strength of a normal shock wave; or $\beta = (\varepsilon - \delta)$ for an oblique shock wave.
δ the flow deflection angle for an oblique shock wave.
δ_c semiangle of a cone.
δ_m maximum possible flow deflection angle.
δ_w semiangle of a wedge.
ε wave angle of an oblique shock wave.
ε_m wave angle corresponding to the maximum turning angle.
ε_s wave angle corresponding to sonic downstream flow.

Subscripts

1   state in front of a shock wave.
2   state in back of a shock wave.

7-2 INTRODUCTION

It has been observed for many years that a compressible fluid under certain conditions can experience an abrupt change of state. Familiar examples are the phenomena associated with detonation waves, explosions, and the wave system formed at the nose of a projectile moving with a supersonic speed. In all of those cases the wave front is very steep and there is a large pressure rise in traversing the wave, which is termed a *shock wave*. Because of the large pressure gradient in the shock wave, the gas experiences a large increase in its density with a corresponding change in its refractive index. Consequently, by employing suitable apparatus the shock wave may be photographed.[1, 2]

Since the shock wave is a more or less instantaneous compression of the gas, it cannot be a reversible process. The energy for compressing the gas flowing through the shock wave is derived from the kinetic energy it possessed upstream to the shock wave. Because of the irreversibility of the shock process, the kinetic energy of the gas leaving the shock wave is smaller than that for an isentropic flow compression between the same pressure limits. The reduction in the kinetic energy because of the shock wave appears as a heating of the gas to a static temperature above that corresponding to the isentropic compression value. Consequently, in flowing through the shock wave, the gas experiences a decrease in its available energy and, accordingly, an increase in its entropy.

There are several different types of shock waves, each having particular characteristics. In some cases, as illustrated in Fig. 7.1, the shock wave is stationary with respect to the body on which it is formed, signifying that the speed of propagation of the shock wave is equal to that of the body itself; otherwise the shock wave would not be attached to the body. When the shock wave is perpendicular to the direction of the flow, it is termed a *normal shock*.

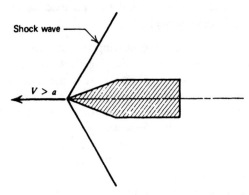

Figure 7.1   Shock wave attached to a moving body.

The theory relating to the effects produced by a shock wave is fairly well developed.[1-5] Equations have been derived relating the velocities and the thermodynamic properties of the gas immediately in front of and in back of the shock wave. However, the theory does not give complete information regarding the causes leading to the formation of the compression shock wave. Undoubtedly, the more or less spontaneous manner in which the shock process occurs makes it difficult to derive a complete theoretical explanation for the shock wave phenomena. In recent years considerable experimental work has been devoted to the accumulation of data regarding shock waves. Those data are being correlated with the results of theoretical studies to evolve a working theory for the shock wave phenomena.

In the present chapter, the basic equations describing the properties of shock waves are derived for both *normal* and *oblique shock waves*, and they are applied to the flow of both *perfect* and *imperfect* gases. The supersonic flow of a gas over both *wedges* and *cones* is analyzed. Several general rules pertinent to the *interaction* and *reflection* of oblique shock waves are presented.

7-3 NORMAL SHOCK WAVES

The present section is concerned with the general characteristics of the *normal shock wave*. For the perfect gas, explicit relationships are developed in Section 7-4 for the property changes across the normal shock wave in terms of the initial Mach number M_1. For imperfect gases, a numerical technique is presented in Section 7-5 for determining the changes in the flow properties caused by the normal shock wave; the numerical technique is applicable to any gas, irrespective of the form of its equation of state.

The flow through a normal shock wave may be analyzed by considering the one-dimensional flow through a stream tube. In the analysis, the following assumptions are made.

1. The boundary surface forming the stream tube is far removed from the boundary layers adjacent to any solid surface. Since all friction forces may be assumed to be confined to the shearing stresses in the boundary layer (see Section 5–10), the configuration under discussion is a *frictionless duct*.
2. The shock process takes place at *constant area*; that is, the streamlines forming the boundary of the stream tube are parallel.
3. The shock wave is perpendicular to the streamlines.
4. The flow process, including the shock wave, is adiabatic, no external work is performed, and the effects of body forces are negligible.

Figure 7.2a illustrates the model of a shock wave moving to the right with the velocity V_{shock} with respect to a stationary observer, and Fig. 7.2b illustrates the corresponding distributions of the flow properties V and p. It is demonstrated in the investigation of the speed of propagation of small disturbances in Section 3–8(a) that a moving wave problem may be transformed into a stationary wave problem by superimposing the wave velocity on all of the flow velocities, as is illustrated schematically in Fig. 7.2c. In effect, the observer in Fig. 7.2c moves at the same speed as the shock wave. Figure 7.2d illustrates the corresponding distributions of the flow properties V and p for the stationary normal shock wave.

The control volume enclosing the stationary normal shock wave is illustrated in Fig. 7.3. The analysis for the normal shock wave is similar to that for an acoustic wave, as discussed in Section 3–8(a). However, in the analysis of the acoustic wave, the differential forms of the governing equations are employed but, in the analysis of the normal shock wave, the integral forms of those equations are employed.

The analysis of the normal shock wave involves, as for any other flow problem, determining the pressure, density, and speed of the fluid at the location under consideration. In general, there are the four unknown properties p_2, ρ_2, h_2, and V_2, and for their determination there are available the continuity equation, the momentum or dynamic equation, the energy equation, and the equation of state for the fluid.

(1) *Continuity equation.* From Table 3.1,

$$\dot{m} = \rho_1 V_1 A_1 = \rho_2 V_2 A_2 = \text{constant} \qquad (7.1)$$

Since $A_1 = A_2 = A$, the mass rate of flow per unit area or *mass flux* $G = \dot{m}/A$ is given by

$$G = \frac{\dot{m}}{A} = \rho_1 V_1 = \rho_2 V_2 = \text{constant} \qquad (7.2)$$

(2) *Momentum equation.* The momentum equation for the fluid bounded by sections 1 and 2 and the stream tube wall (with $\mathbf{B} = \partial/\partial t = \mathbf{F}_{shear} = 0$), is given by

$$\sum (\text{pressure forces}) = \frac{D\,(\text{Momentum})}{Dt} \qquad (7.3)$$

For the control volume illustrated in Fig. 7.3, we obtain from equation 7.3,

$$-p_1 A_1 + p_2 A_2 = \dot{m}(-V_2) - \dot{m}(-V_1) = -\dot{m}V_2 + \dot{m}V_1 \qquad (7.4)$$

Introducing equation 7.1 into equation 7.4 gives

$$p_1 + \rho_1 V_1^2 = p_2 + \rho_2 V_2^2 = \text{constant} \qquad (7.5)$$

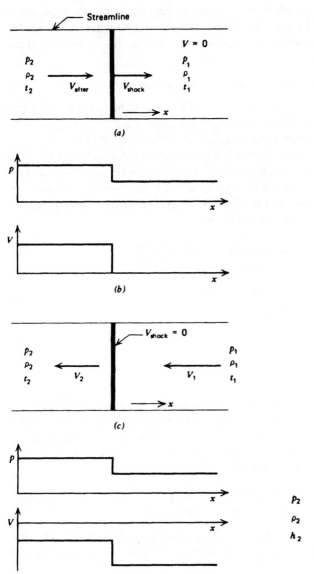

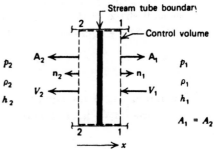

Figure 7.2 Model of a normal shock wave. (*a*) Moving normal shock wave. (*b*) Pressure and velocity distributions. (*c*) Stationary normal shock wave. (*d*) Pressure and velocity distributions.

Figure 7.3 Control volume for a normal shock wave.

(3) *Energy equation* (see Table 3.1, for $g\,dz = 0$)

$$h_1 + \frac{V_1^2}{2} = h_2 + \frac{V_2^2}{2} = H = \text{constant} \tag{7.6}$$

The enthalpy h is defined by

$$h = u + pv \tag{7.7}$$

where u is the specific internal energy. Equation 7.6 may be written as

$$u_1 + p_1v_1 + \frac{V_1^2}{2} = u_2 + p_2v_2 + \frac{V_2^2}{2} \tag{7.8}$$

(4) *Equation of state.* To complete the required system of equations, an equation of state for the fluid must be specified. For a simple thermodynamic system (see Section 1–7), each of the properties p, ρ, t, u, h, and s is uniquely related to any two of the other properties. Thus,

$$u = u(p,\rho) \qquad \text{and} \qquad h = h(p,\rho) \tag{7.9}$$

Equations 7.2, 7.5, 7.8, and 7.9 comprise a set of four equations involving the four flow properties p, ρ, u, and V. Those equations are quite general. They govern the behavior of a *normal shock wave* for a simple thermodynamic system.

The general features of a normal shock wave may be demonstrated by the *Hugoniot curve*, which is the locus, in the pv plane, of all of the allowable states that the fluid may attain after passing through a normal shock wave. The equation for the Hugoniot curve is obtained by combining equations 7.2, 7.5, 7.8, and 7.9 in the following manner. Combine equations 7.2 and 7.5, and solve for $V_1{}^2$.

$$V_1{}^2 = \frac{(p_2 - p_1)\rho_2}{(\rho_2 - \rho_1)\rho_1} = \frac{v_1{}^2(p_2 - p_1)}{(v_1 - v_2)} \tag{7.10a}$$

In a similar manner,

$$V_2{}^2 = \frac{(p_2 - p_1)\rho_1}{(\rho_2 - \rho_1)\rho_2} = \frac{v_2{}^2(p_2 - p_1)}{(v_1 - v_2)} \tag{7.10b}$$

After substituting equations 7.10a and 7.10b into equation 7.8 and simplifying, we obtain

$$u_2 - u_1 = \frac{1}{2}(p_1 + p_2)(v_1 - v_2) \tag{7.11}$$

Equation 7.11, in conjunction with equation 7.9, specifies, for a given initial state 1, the locus of the allowable states in the pv plane that satisfy the governing equations. That locus is known as the *Hugoniot curve*. Figure 7.4 illustrates schematically the general shape of the Hugoniot curve, called an *H-curve*, for all known fluids.

Combining equations 7.2 and 7.5 yields

$$G^2 = -\frac{(p_2 - p_1)}{(v_2 - v_1)} = -\left(\frac{\partial p}{\partial v}\right)_{R\text{-}line} \tag{7.12}$$

Equation 7.12 is the equation of a *Rayleigh line*, called an *R-line*, on the pv plane [see Section 6–3(b)]. Figure 7.5 illustrates schematically several *R-lines* on the

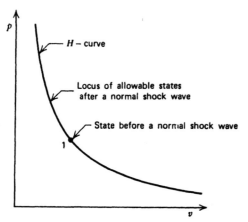

Figure 7.4   Sketch of a Hugoniot curve.

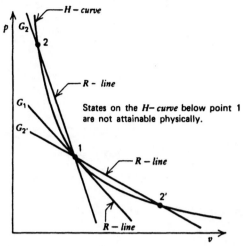

Figure 7.5 Intersection of several *R-lines* with the *H-curve*.

pv plane. G_1 is chosen so that the corresponding *R-line* is tangent to the *H-curve* at point 1, G_2' is less than G_1, and G_2 is greater than G_1. The state of the flow in back of a normal shock wave must satisfy the equations for both the *H-curve* and the *R-line*. Hence, for a given value of G, point 2 is uniquely determined, as indicated in Fig. 7.5.

The upper portion of the *H-curve*, above point 1, is the locus for compression shock waves, for which $p_2 > p_1$ and $\rho_2 > \rho_1$. The lower portion of that curve, below point 1, is the locus for expansion shock waves, for which, $p_2 < p_1$ and $\rho_2 < \rho_1$. Expansion shock waves, however, violate the second law of thermodynamics for an isolated system; that law requires that $\Delta s \geqq 0$. Consequently, the states lying on the portion of the *H-curve* below state 1 are not attainable physically.

The following argument demonstrates that an expansion shock wave violates the second law of thermodynamics. Differentiate the equation of the *H-curve*, equation 7.11, considering point 1 fixed and point 2 as a general point where $u_2 = u$, $p_2 = p$, etc. Thus,

$$du_H = -\frac{1}{2}(p_1 + p)\,dv_H + \frac{1}{2}(v_1 - v)\,dp_H \tag{7.13}$$

where the subscript H denotes that the derivatives apply along the *H-curve*. For a simple system, equation 1.64 applies. Thus,

$$du = t\,ds - p\,dv \tag{7.14}$$

Combining equations 7.13 and 7.14, we obtain

$$2t\left(\frac{ds}{dp}\right)_H = (v_1 - v)\left[1 - \frac{(p - p_1)}{(v - v_1)}\left(\frac{dv}{dp}\right)_H\right] \tag{7.15}$$

Substituting equation 7.12 into equation 7.15 yields

$$2t\left(\frac{ds}{dp}\right)_H = (v_1 - v)\left[1 + \frac{G^2}{(dp/dv)_H}\right] \tag{7.16}$$

Figure 7.6 illustrates the relationship between G and $(dp/dv)_H$ on the *H-curve*. Denote the angle between the negative abscissa and the *H-curve* by δ, and the angle

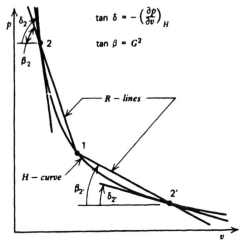

Figure 7.6 Relationship between G and $(dp/dv)_H$ on the *H-curve*.

between the negative abscissa and the *R-line* by β. At point $2'$, $\beta_{2'} > \delta_{2'}$. Hence,

$$\frac{G_{2'}^{2}}{\left(\dfrac{dp}{dv}\right)_{H,2'}} < -1$$

and the term in brackets in equation 7.16 is negative. Since $(v_1 - v_{2'}) < 0$, then $(ds/dp)_H > 0$ at each point on the lower portion of the *H-curve*. At point 2, $\delta_2 > \beta_2$, so that

$$0 \leqq \frac{G_2^{2}}{\left(\dfrac{dp}{dv}\right)_{H,2}} \leqq 1$$

and the term in brackets in equation 7.16 is positive. Since $(v - v_2) > 0$, then $(ds/dp)_H > 0$ on the upper portion of the *H-curve*.

The foregoing arguments demonstrate that on the *H-curve*,

$$\left(\frac{ds}{dp}\right)_H > 0 \tag{7.17}$$

for both branches. On the upper branch, the compression branch, $dp > 0$, requiring that $ds > 0$, which is in accordance with the second law of thermodynamics. However, on the lower branch, the expansion branch, $dp < 0$, requiring that $ds < 0$, which violates the second law of thermodynamics. Consequently, expansion shock waves are impossible.

The limiting point between the upper and the lower branches occurs at point 1 when the *R-line* is tangent to the *H-curve*. At that point,

$$\left(\frac{dp}{dv}\right)_{H,1} = -\rho_1^{2}\left(\frac{dp}{d\rho}\right)_{H,1} = -G_1^{2} = -\rho_1^{2}V_1^{2} \tag{7.18}$$

$$V_1^{2} = \left(\frac{dp}{d\rho}\right)_{H,1} \tag{7.19}$$

As point 2 approaches point 1, $(v_1 - v_2) \to 0$ and $(dp/dv)_H \to -G^2$. From equation 7.16, therefore, $ds \to 0$, and the property changes along the *H-curve* in the neighborhood close to point 1 are isentropic. Equation 7.19 may, therefore, be written as

$$V_1^2 = \left(\frac{\partial p}{\partial \rho}\right)_{s,1} = a_1^2 \tag{7.20}$$

and $M_1 = 1.0$. At all of the other points on the upper branch of the *H-curve*,

$$-G^2 = -\rho_1^2 V_1^2 > \left(\frac{dp}{dv}\right)_{H,1} = -\rho_1^2 a_1^2 \tag{7.21}$$

$$V_1^2 > a_1^2 \tag{7.22}$$

and $M_1 > 1.0$. Because the compression branch of the Hugoniot curve is the only one that is the locus for physically possible states, the initial Mach number for those states must be either larger or equal to unity. Shock waves occur, therefore, only in flows that are initially supersonic.

The preceding analysis assumes that there is a discontinuity in the flow properties, which actually cannot occur in a real fluid. The existence of the compression shock wave is due to the viscosity and thermal conductivity of the fluid, properties that make the shock type of disturbance possible. A detailed explanation of the processes taking place in the shock wave is very complicated and is beyond the scope of this book. Theoretical studies and experimental measurements indicate that a shock wave is extremely thin. For example, Linzer and Hornig[6] present a comparison of calculated and measured values of the shock wave thickness t in argon and nitrogen.

The experimental results obtained by Linzer and Hornig are presented in the reduced form (λ/t), where t is the shock wave thickness and λ is the Maxwellian mean free path ahead of the shock wave, defined by

$$\lambda \equiv \frac{16}{5} \frac{\mu}{n(2\pi mkt)^{1/2}}$$

where μ is the viscosity, n is the number density, m is the mass per molecule, k is the Boltzmann constant, and t is the absolute temperature. Figure 7.7a presents the results for argon, and Fig. 7.7b presents the corresponding results for nitrogen. For argon the reciprocal shock thickness (λ/t) rises smoothly and rapidly from a value of 0.0 at $M = 1$, peaks at a value of approximately 0.3 at $M = 3.5$, and decreases slightly thereafter. For nitrogen, a similar behavior is observed, with (λ/t) peaking at 0.4 when $M = 4.0$. Hence, for Mach numbers exceeding approximately 3, the shock wave thickness t is between 2.5 and 3 mean free paths, and the thickness increases as the Mach number decreases. For example, at $M = 2$, t/λ is approximately 5.

In the present section, several of the intrinsic features of normal shock waves are discussed. No restrictions, however, are placed on the equation of state for the gas other than that the latter must be a simple thermodynamic system. In Section 7-4, explicit equations are derived for the flow of a perfect gas through a normal shock wave and, in Section 7-5, a numerical technique is presented for taking into account imperfect gas effects.

7-4 NORMAL SHOCK WAVES IN PERFECT GASES

The present section is concerned with the application of the equations developed in Section 7-3 to normal shock waves in a *perfect gas*.

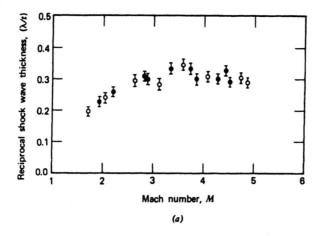

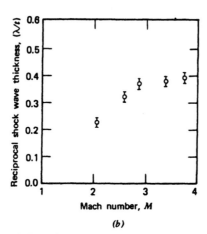

Figure 7.7 Variation of the reciprocal shock wave thickness (λ/t) with the Mach number M. (a) Argon. (b) Nitrogen. (Taken from Reference 6).

7-4(a) Governing Equations

For a perfect gas, the equations of state are

$$p = \rho Rt \tag{7.23}$$

and

$$h = c_p t \tag{7.24}$$

where $h = 0$ at $t = 0$. Hence, equation 7.6 becomes

$$t_1 + \frac{V_1^2}{2c_p} = t_2 + \frac{V_2^2}{2c_p} = T = \text{constant} \tag{7.25}$$

Equation 7.25 shows that the stagnation temperature T remains constant. From the definition of the stagnation temperature (see Table 3.3), for $T = \text{constant}$,

$$\frac{t_2}{t_1} = \frac{1 + \frac{\gamma - 1}{2} M_1^2}{1 + \frac{\gamma - 1}{2} M_2^2} \tag{7.26}$$

Substituting equation 7.23 into equation 7.2, we obtain

$$\frac{V_2}{V_1} = \frac{\rho_1}{\rho_2} = \frac{p_1 t_2}{p_2 t_1} \tag{7.27}$$

For a perfect gas, combining $V = Ma = M\sqrt{\gamma R t}$ and equation 7.27 gives

$$\frac{p_2}{p_1} = \frac{M_1}{M_2} \left(\frac{1 + \dfrac{\gamma - 1}{2} M_1^2}{1 + \dfrac{\gamma - 1}{2} M_2^2} \right)^{1/2} \tag{7.28}$$

Equation 7.5 may be rewritten as

$$p_1 \left(1 + \frac{\rho_1 V_1^2}{p_1} \right) = p_2 \left(1 + \frac{\rho_2 V_2^2}{p_2} \right) \tag{7.29}$$

Noting that $a^2 = \gamma p / \rho$, equation 7.29 yields

$$\frac{p_2}{p_1} = \frac{1 + \gamma M_1^2}{1 + \gamma M_2^2} \tag{7.30}$$

Equating the right-hand sides of equations 7.28 and 7.30, we obtain

$$\frac{M_1 \left(1 + \dfrac{\gamma - 1}{2} M_1^2 \right)^{1/2}}{1 + \gamma M_1^2} = \frac{M_2 \left(1 + \dfrac{\gamma - 1}{2} M_2^2 \right)^{1/2}}{1 + \gamma M_2^2} \tag{7.31}$$

Squaring both sides of equation 7.31 and solving for M_2 yields

$$M_2 = M_1 \tag{7.32}$$

$$M_2 = \left(\frac{M_1^2 + \dfrac{2}{\gamma - 1}}{\dfrac{2\gamma}{\gamma - 1} M_1^2 - 1} \right)^{1/2} \tag{7.33}$$

Equation 7.32 expresses the trivial result that no change occurs across the normal shock wave; that is, the latter has infinitesimal strength. Equation 7.33 gives the solution for a normal shock wave of finite strength. Figure 7.8a presents the Mach number after a normal shock wave M_2 as a function of the initial Mach number M_1, for $\gamma = 1.4$. Table C.11 presents M_2 as a function of M_1 for different values of γ.

For a normal shock wave to occur, the Mach number M_1 must be larger than unity, as pointed out in Section 7-3. Examination of equation 7.33, or Fig. 7.8a, shows that as M_1 in front of the normal shock wave is increased indefinitely, the Mach number M_2 in back of the normal shock wave continually decreases but approaches the limiting value $\sqrt{(\gamma - 1)/2\gamma}$. The limiting value for M_2 depends only on the specific heat ratio for the gas. Figure 7.8b presents the limiting value of M_2 as a function of γ.

The property ratios across a normal shock wave may be expressed in terms of the Mach number ahead of the shock wave M_1 by substituting equation 7.33 into equations 7.26, 7.27, and 7.30, to eliminate M_2. Thus,

$$\frac{t_2}{t_1} = \left(\frac{2\gamma}{\gamma + 1} M_1^2 - \frac{\gamma - 1}{\gamma + 1} \right) \left[\frac{\gamma - 1}{\gamma + 1} + \frac{2}{(\gamma + 1)M_1^2} \right] \tag{7.34}$$

$$\frac{\rho_2}{\rho_1} = \frac{v_1}{v_2} = \frac{V_1}{V_2} = \frac{(\gamma + 1)M_1^2}{2 + (\gamma - 1)M_1^2} \tag{7.35}$$

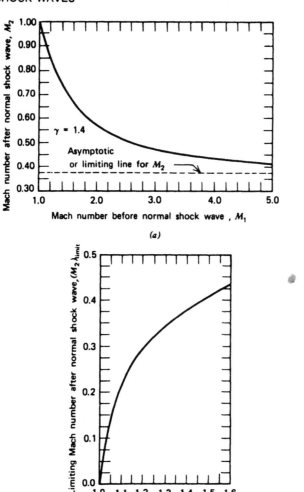

Figure 7.8 Mach number after a normal shock wave. (a) Mach number after a normal shock wave as a function of the Mach number before the shock wave ($\gamma = 1.4$). (b) Limiting values of the Mach number in back of a normal shock wave as a function of γ.

$$\frac{p_2}{p_1} = \frac{2\gamma}{\gamma + 1} M_1{}^2 - \frac{\gamma - 1}{\gamma + 1} \tag{7.36}$$

Values of the above property ratios are presented in Table C.11 as a function of M_1 for different values of γ.

It follows from equation 7.35 that as the initial Mach number M_1 approaches infinity, the density ratio ρ_2/ρ_1 approaches $(\gamma + 1)/(\gamma - 1)$ as a limiting value. On the other hand, the pressure ratio p_2/p_1 (see equation 7.36) increases continuously with M_1.

The flow conditions immediately in front of and in back of the normal shock wave are isentropic, in view of the assumptions presented at the beginning of Section 7–3. From Table 3.3,

$$\frac{P_2}{p_2} = \left(1 + \frac{\gamma - 1}{2} M_2{}^2\right)^{\gamma/(\gamma - 1)} \tag{7.37}$$

and

$$\frac{P_1}{p_1} = \left(1 + \frac{\gamma - 1}{2} M_1{}^2\right)^{\gamma/(\gamma - 1)} \tag{7.38}$$

The stagnation pressure ratio P_2/P_1 is

$$\frac{P_2}{P_1} = \frac{P_2}{p_2} \frac{p_1}{P_1} \frac{p_2}{p_1} \tag{7.39}$$

Substituting for P_2/p_2, P_1/p_1, and p_2/p_1 from equations 7.37, 7.38, and 7.36, respectively, into equation 7.39, and rearranging the result, gives

$$\frac{P_2}{P_1} = \left\{\left[\frac{\gamma - 1}{\gamma + 1} + \frac{2}{(\gamma + 1)M_1{}^2}\right]^\gamma \left(\frac{2\gamma}{\gamma + 1} M_1{}^2 - \frac{\gamma - 1}{\gamma + 1}\right)\right\}^{-1/(\gamma - 1)} \tag{7.40}$$

Values of the stagnation pressure ratio P_2/P_1 as a function of M_1 are presented in Table C.11 for different values of γ. The table shows that the decrease in stagnation pressure depends on the shock intensity; the more intense the shock (large values of M_1), the larger is the decrease in stagnation pressure.

The ratio P_2/p_1, the ratio of the stagnation pressure P_2 in back of the shock wave to the free-stream static pressure p_1, is what is indicated by a Pitot tube immersed in a supersonic stream. As illustrated schematically in Fig. 7.9a, when a fluid flows

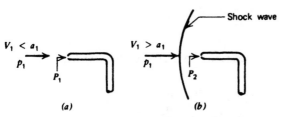

Figure 7.9   Pitot tube measurement for subsonic and supersonic flow. (a) Subsonic flow. (b) Supersonic flow.

toward the Pitot tube with a subsonic speed, the fluid may be assumed to decelerate isentropically to stagnation at the entrance to the instrument. When $M_1 > 1$, however, a shock wave is formed ahead of the mouth of the Pitot tube, as illustrated in Fig. 7.9b. Because the radius of curvature of the shock wave is large in the region directly in front of the entrance to the pitot tube, the portion of the shock wave in that region may be assumed to be a normal shock wave. After flowing through the latter normal shock wave, the stream of compressible fluid decelerates isentropically to the stagnation pressure P_2 at the entrance to the Pitot tube. The ratio P_2/p_1 is given by

$$\frac{P_2}{p_1} = \frac{P_2}{p_2} \frac{p_2}{p_1} \tag{7.41}$$

Substituting for P_2/p_2 and p_2/p_1 in terms of M_1 gives

$$\frac{P_2}{p_1} = \left(\frac{\gamma + 1}{2}\right) M_1{}^2 \left(\frac{\dfrac{\gamma + 1}{2} M_1{}^2}{\dfrac{2\gamma}{\gamma + 1} M_1{}^2 - \dfrac{\gamma - 1}{\gamma + 1}}\right)^{1/(\gamma - 1)} \tag{7.42}$$

The relationship given by equation 7.42 is known as the *Rayleigh Pitot tube equation*. Figure 7.10 presents the ratio p_1/P_2 as a function of M_1 for $\gamma = 1.4$. Table C.11 presents values of the ratio P_2/p_1 for different values of γ.

Since the flow process for the normal shock wave is adiabatic and the fluid is assumed to be a perfect gas, its entropy increase due to the normal shock wave is, from Table 3.2, given by

$$\Delta s = s_2 - s_1 = R \ln \left(\frac{P_1}{P_2} \right) \tag{7.43}$$

The stagnation pressure ratio P_1/P_2, in equation 7.43, is obtained from equation 7.40.

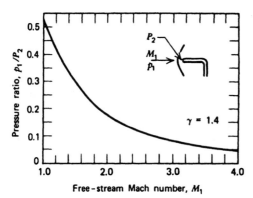

Figure 7.10   Rayleigh Pitot tube equation ($\gamma = 1.4$).

Shock waves are frequently described in terms of the *condensation* δ and the *shock strength* β, where

$$\delta \equiv \frac{\rho_2 - \rho_1}{\rho_1} = \frac{\rho_2}{\rho_1} - 1 \tag{7.44}$$

and

$$\beta \equiv \frac{p_2 - p_1}{p_1} = \frac{p_2}{p_1} - 1 \tag{7.45}$$

The term condensation has nothing whatever to do with a change of phase in the fluid. It denotes a compression of the fluid. Substituting equation 7.45 into equation 7.36, we obtain

$$\beta = \left(\frac{2\gamma}{\gamma + 1} \right) (M_1{}^2 - 1) \tag{7.46}$$

and

$$M_1{}^2 = 1 + \left(\frac{\gamma + 1}{2\gamma} \right) \beta \tag{7.47}$$

All of the property ratios presented in the foregoing analysis may be expressed in terms of the shock strength β by substituting equation 7.47 into the appropriate equations.

All of the flow properties for a *stationary* normal shock wave may be calculated by means of the equations derived above. There are many practical situations, however, such as blast waves and explosions, where the shock wave is not stationary as is assumed in the preceding derivations. In those cases the speed with which the shock wave travels through the still gas is V_1, the velocity with which the gas moves into a stationary shock wave. It is desired to relate the velocity V_1 to the strength of the shock wave.

Figures 7.11a and 7.11b illustrate, respectively, the conditions for a shock wave moving into a stationary gas, and the case where a gas flows into a stationary shock wave. Referring to Fig. 7.11a, the shock wave moves to the right with the velocity V_1. The gas that is stationary before being traversed by the shock wave is given a velocity $(V_1 - V_2)$ to the right, and its pressure and density are increased by the passage of the shock wave. On the other hand, the stationary gas toward which the shock wave is moving, to the right of the shock wave in Fig. 7.17a, is at a lower pressure and density than the gas to the left of the shock wave.

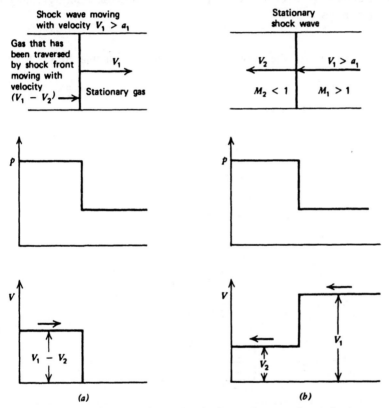

Figure 7.11 Transformation of a moving shock wave into a stationary frame of reference. (a) Shock wave moving into a stationary gas. (b) Gas flowing into a stationary shock wave.

The situation illustrated in Fig. 7.11a may be transformed to the case where the shock wave is stationary, as in Fig. 7.11b, by imagining that the observer travels with the speed of the shock wave V_1; that is, the observer moves to the right with the velocity V_1. To the moving observer, the shock wave appears to be stationary, while the gas appears to be flowing to the left with the velocity V_1. Moreover, as the gas flows through the shock wave, its velocity is reduced from V_1 to V_2 with respect to the moving observer. Simultaneously, the pressure and density of the gas increase to p_2 and ρ_2, respectively. Although the static properties are unchanged by the change in the frame of reference, the velocities and, consequently, the stagnation properties, do depend on the frame of reference.

The most straightforward manner for solving moving shock wave problems is to transform the problem to an equivalent stationary shock wave problem, solve for the changes in the flow properties across the stationary shock wave, and then transform the dynamic properties back into the reference frame for the moving shock wave.

7-4(b) Normal Shock Wave Tables

Table 7.1 presents the equations for the flow property ratios for a normal shock wave in a perfect gas. Table C.11 presents the values of the pertinent flow property ratios as functions of the Mach number M_1 for different values of γ. Table 7.2 presents the general arrangement of Table C.11, together with the limiting values of the flow property ratios.

Table 7.1 Equations for the Flow Property Ratios for Normal Shock Waves in a Perfect Gas

$$M_2 = \left(\frac{M_1^2 + \dfrac{2}{\gamma - 1}}{\dfrac{2\gamma}{\gamma - 1} M_1^2 - 1} \right)^{1/2} \tag{7.48}$$

$$\frac{p_2}{p_1} = \frac{2\gamma}{\gamma + 1} M_1^2 - \frac{\gamma - 1}{\gamma + 1} \tag{7.49}$$

$$\frac{\rho_2}{\rho_1} = \frac{v_1}{v_2} = \frac{V_1}{V_2} = \frac{(\gamma + 1)M_1^2}{2 + (\gamma - 1)M_1^2} \tag{7.50}$$

$$\frac{t_2}{t_1} = \left(\frac{2\gamma}{\gamma + 1} M_1^2 - \frac{\gamma - 1}{\gamma + 1} \right) \left[\frac{\gamma - 1}{\gamma + 1} + \frac{2}{(\gamma + 1)M_1^2} \right] \tag{7.51}$$

$$\frac{P_2}{P_1} = \left[\left(\frac{\gamma - 1}{\gamma + 1} + \frac{2}{(\gamma + 1)M_1^2} \right)^{\gamma} \left(\frac{2\gamma}{\gamma + 1} M_1^2 - \frac{\gamma - 1}{\gamma + 1} \right) \right]^{-1/(\gamma - 1)} \tag{7.52}$$

$$\frac{P_2}{p_1} = \left(\frac{\gamma + 1}{2} \right) M_1^2 \left(\frac{\dfrac{\gamma + 1}{2} M_1^2}{\dfrac{2\gamma}{\gamma + 1} M_1^2 - \dfrac{\gamma - 1}{\gamma + 1}} \right)^{1/(\gamma - 1)} \tag{7.53}$$

Table 7.2 Limiting Values of the Flow Property Ratios for Normal Shock Waves in a Perfect Gas

M_1	M_2	ρ_2/ρ_1 V_1/V_2	t_2/t_1	p_2/p_1	P_2/P_1	P_2/p_1
1	1	1	1	1	1	1
∞	$\sqrt{\dfrac{\gamma - 1}{2\gamma}}$	$\dfrac{\gamma + 1}{\gamma - 1}$	∞	∞	0	∞

A FORTRAN program, subroutine NSHOCK, is presented below for calculating the flow property ratios across a normal shock wave in a perfect gas. The input to the program is described in Section 4–4(b). The results presented in Table C.11 for $\gamma = 1.4$ are obtained by the following specification.

G = 1.4, J(1) = 100, J(2) = 75, J(3) = 25, J(4) = 8, J(5) = 15, J(6) = 0, DM(1) = 0.01, DM(2) = 0.02, DM(3) = 0.1, DM(4) = 0.5, DM(5) = 1.0

```
      SUBROUTINE NSHOCK
C     NORMAL SHOCK PROPERTY RATIOS FOR A PERFECT GAS
      REAL M1,M2 $ DIMENSION J(9),DM(9) $ DATA INF/6H INFIN/
      NAMELIST /DATA/ G,J,DM $ READ (5,DATA)
      G1=(G-1.0)/2.0 $ G2=(G+1.0)/2.0 $ G3=G/(G-1.0) $ G4=G/G2 $G5=G1/G2
      M1=1.0 $ M2=1.0 $ V=1.0 $ T=1.0 $ P=1.0 $ PO=1.0 $ POP=G2**G3
      WRITE (6,2000) G $ WRITE (6,2020) M1,M2,V,T,P,PO,POP $ I=1 $ L=0
C     CALCULATE PROPERTY RATIOS
   40 N=J(I) $ DO 50 K=1,N $ M1=M1+DM(I) $ C1=1.0+G1*M1**2
      V=G2*M1**2/C1 $ P=G4*M1**2-G5 $ T=P/V $ M2=M1/SQRT(P*V)
      C2=1.0+G1*M2**2 $ PO=P*(C2/C1)**G3 $ POP=P*C2**G3
      WRITE (6,2020) M1,M2,V,T,P,PO,POP $ L=L+1 $ IF (L.LT.50) GO TO 50
      WRITE (6,2000) G $ L=0
   50 CONTINUE $ I=I+1 $ IF((I.EQ.10).OR.(J(I).EQ.0)) GO TO 60 $GO TO 40
   60 M2=SQRT(G1/G) $ V=(G+1.0)/(G-1.0) $ PO=0.0
      WRITE (6,2030) INF,M2,V,INF,INF,PO,INF $ RETURN
 2000 FORMAT (1H1,19X,27HNORMAL SHOCK WAVE,  GAMMA =,F5.2//4X,2HM1,5X,
     12HM2,7X,5HV1/V2,8X,5HT2/T1,8X,5HP2/P1,8X,7HP02/P01,6X,6HP02/P1/
     220X,5HR2/R1,34X,7HA1*/A2*/)
 2020 FORMAT (1H ,F5.2,F10.5,5E13.5)
 2030 FORMAT (A6,F10.5,E13.5,3X,A6,7X,A6,E17.5,3X,A6)
      END
```

Example 7.1. A normal shock wave is propagated into still air with a velocity of 722.4 m/s. The still air pressure is 1 atm ($1.0133 \cdot 10^5$ N/m) and its temperature is 294.4 K. Calculate the Mach number, pressure, temperature, and velocity in back of the normal shock wave, relative to a stationary observer.

Solution

Consider the equivalent stationary normal shock wave problem where $V_1 = 722.4$ m/s.

$$a_1 = (\gamma R t_1)^{1/2} = [1.4(287.04)(294.4)]^{1/2} = 343.9 \text{ m/s}$$

$$M_1 = \frac{V_1}{a_1} = \frac{722.4}{343.9} = 2.10$$

From Table C.11, for $M_1 = 2.10$,

$$M_2 = 0.56128, \qquad \frac{p_2}{p_1} = 4.9783, \qquad \frac{t_2}{t_1} = 1.7704, \qquad \frac{V_1}{V_2} = 2.8119$$

$$p_2 = 4.9783(1.0133)10^5 = 5.045 \cdot 10^5 \text{ N/m}^2 \text{ (73.17 lbf/in.}^2)$$

$$t_2 = 1.7704(294.4) = 521.3 \text{ K (938.3 R)}$$

$$V_2 = \frac{722.4}{2.8119} = 256.9 \text{ m/s (842.9 ft/sec)}$$

Relative to a stationary observer,

$$V_{2R} = V_1 - V_2 = 722.4 - 256.9 = 465.6 \text{ m/s } (1527.2 \text{ ft/sec})$$

where the subscript R denotes a property relative to a stationary observer.

$$a_2 = (\gamma R t_2)^{1/2} = [1.4(287.04)(521.3)]^{1/2} = 457.7 \text{ m/s}$$

$$M_{2R} = \frac{V_{2R}}{a_2} = \frac{465.5}{457.7} = 1.017$$

The above result demonstrates that the Mach number in back of a moving shock wave may be supersonic relative to a stationary observer.

In this example, a moving shock wave is analyzed by employing the model of a stationary shock wave. The flow properties for the gas in front of and in back of a moving shock wave are determined by transforming the results for a stationary shock wave into the frame of reference for the moving shock wave. The aforementioned procedure for analyzing moving shock waves is recommended, rather than making a formal transformation of the equations governing the shock wave.

Example 7.2. Air flows through a frictionless adiabatic converging-diverging nozzle. The air stagnation pressure and temperature are $7.0 \cdot 10^5 \text{ N/m}^2$ and 500 K, respectively. The diverging portion of the nozzle has an area ratio of $A_e/A_t = 11.91$. A normal shock wave stands in the divergence where the Mach number is 3.0, as illustrated in Fig. 7.12. Calculate the Mach number and the static temperature and pressure in the nozzle exit plane. Assume $\gamma = 1.40$.

Solution

From Table C.11, for $M_1 = 3.00$, $M_2 = 0.4752$ and $P_2/P_1 = 0.3283$. Hence,

$$P_2 = (0.3283)(7.0)10^5 = 2.2981 \cdot 10^5 \text{ N/m}^2$$

Since the flow is adiabatic, $T_2 = T_1 = 500$ K. Since the flow is isentropic between Station 2 and the exit section, Station e, $T_e = T_2$ and $P_e = P_2$. From Table C.6, for $M_1 = 3.0$ and $M_2 = 0.4752$, $(A_1/A_1^*) = (A_1/A_t) = 4.2346$ and $A_2/A_2^* = 1.390$. Thus,

$$\left(\frac{A}{A^*}\right)_e = \frac{A_e}{A_2^*} = \left(\frac{A_e}{A_t}\right)\left(\frac{A_t}{A_1}\right)\left(\frac{A_2}{A_2^*}\right) = \frac{11.91(1.390)}{4.2346} = 3.9094$$

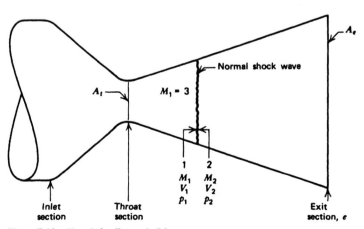

Figure 7.12   Sketch for Example 7.2.

From Table C.6, for $(A/A^*)_e = 3.9094$, for subsonic flow, $M_e = 0.15$, $t_e/T_e = 0.9955$, and $p_e/P_e = 0.9844$. Hence,

$$t_e = (0.9955)(500) = 497.8 \text{ K} (896.0 \text{ R})$$

$$p_e = (0.9844)(2.2981)10^5 = 2.2623 \cdot 10^5 \text{ N/m}^2 (32.81 \text{ lbf/in.}^2)$$

Example 7.3. Air at a static pressure of $1.50 \cdot 10^5$ N/m^2 and a static temperature of 300 K flows at 150 m/s in a pipe. The valve at the end of the pipe is suddenly closed, propagating a normal shock wave back into the pipe. Calculate the velocity of the shock wave relative to the pipe. Assume $\gamma = 1.4$.

Solution

The velocity in back of the normal shock wave is zero relative to the pipe, and the velocity of that wave relative to the pipe is V_{Wave}. The velocity of the air entering the normal shock wave relative to the shock wave is thus $V_1 = (V_{\text{Wave}} + 150)$. The velocity behind the normal shock wave, relative to the shock wave, is denoted by V_2. For the air in back of the normal shock wave to be stationary, V_2 must equal V_{Wave}. Thus,

$$M_1 = \frac{V_1}{a_1} = \frac{V_{\text{Wave}} + 150}{[1.4(287.04)(300)]^{1/2}} = \frac{V_{\text{Wave}} + 150}{347.21}$$

and

$$V_2 = V_{\text{Wave}}$$

An iterative solution may be employed for calculating V_{Wave}. Values of V_{Wave} are assumed, and V_1 and M_1 are calculated. V_1/V_2 is found from Table C.11, and V_2 is computed and compared with V_{Wave}. When $V_2 = V_{\text{Wave}}$, the solution is obtained. The results of the iteration are presented in the following table. The first assumed value of V_{Wave} is chosen arbitrarily, the second assumed value of V_{Wave} is chosen as the value of V_2 from the first trial, and the third assumed value of V_{Wave} is obtained by the secant method [see Appendix A–4(b)].

Trial	1	2	3
V_{Wave}, m/s (assumed)	400	274.3	298.7
V_1, m/s	550	424.3	448.7
M_1	1.584	1.222	1.292
V_1/V_2	2.0048	1.3799	1.5018
V_2, m/s	274.3	307.5	298.8

The shock wave travels upstream into the pipe at a velocity of $V_{\text{Wave}} = 298.8$ m/s relative to the pipe. Knowing the shock wave velocity, the flow properties behind the normal shock wave may be calculated. For example, from Table C.11, for $M_1 = 1.292$, $p_2/p_1 = 1.7808$. Hence,

$$p_2 = 1.7808(1.5)10^5 = 2.671 \cdot 10^5 \text{ N/m}^2 (38.74 \text{ lbf/in.}^2)$$

Example 7.4. A blast wave created by a nuclear explosion moves with a velocity of 15,240 m/s into still atmospheric air having a pressure of $1.0133 \cdot 10^5$ N/m^2 and a temperature of 294.4 K. Calculate (a) the Mach number of the shock wave relative to the stationary air, and (b) the stagnation values of the pressure and the temperature behind the shock wave relative to a stationary observer.

Solution

(a) The speed of sound in the still air is

$$a_1 = (\gamma R t_1)^{1/2} = [1.4(287.04)(294.4)]^{1/2} = 344.0 \text{ m/s}$$

The moving shock wave is transformed to a stationary shock wave by considering the still air to be moving into the stationary shock wave with a velocity of 15,240 m/s. Thus,

$$M_1 = \frac{V_1}{a_1} = \frac{15,240}{344.0} = 44.31$$

(b) Table C.11 for normal shock waves does not extend to such a large Mach number. Consequently, the equations for the property changes across a normal shock wave will be employed. From equation 7.33, the Mach number M_2 behind the normal shock wave relative to the shock wave is

$$M_2 = \left(\frac{M_1^2 + \dfrac{2}{\gamma - 1}}{\dfrac{2\gamma}{\gamma - 1} M_1^2 - 1} \right)^{1/2} = \left[\frac{(44.31)^2 + 5}{7(44.31)^2 - 1} \right]^{1/2} = 0.3785$$

From equation 7.49,

$$\frac{p_2}{p_1} = \frac{2\gamma}{\gamma + 1} M_1^2 - \frac{\gamma - 1}{\gamma + 1} = 1.1667(44.31)^2 - 0.1667 = 2290$$

$$p_2 = 2290(1.0133)10^5 = 2320 \cdot 10^5 \text{ N/m}^2 \ (33{,}648 \text{ lbf/in.}^2)$$

From equation 7.51,

$$\frac{t_2}{t_1} = \left(\frac{2\gamma}{\gamma + 1} M_1^2 - \frac{\gamma - 1}{\gamma + 1} \right) \left[\frac{\gamma - 1}{\gamma + 1} + \frac{2}{(\gamma + 1)M_1^2} \right]$$

$$\frac{t_2}{t_1} = 2290 \left[0.1667 + \frac{2}{2.4(44.31)^2} \right] = 382.64$$

$$t_2 = 382.64(294.4) = 112{,}650 \text{ K } (202{,}770 \text{ R})$$

$$a_2 = (\gamma R t_2)^{1/2} = [1.4(287.04)(112{,}650)]^{1/2} = 6728 \text{ m/s}$$

$$V_2 = M_2 a_2 = (0.3785)(6728) = 2547 \text{ m/s } (8355 \text{ ft/sec})$$

The velocity of the air behind the shock wave relative to a stationary observer, denoted by V_{2R}, is given by

$$V_{2R} = V_{\text{shock}} - V_2 = 15{,}240 - 2547 = 12{,}693 \text{ m/s } (41{,}643 \text{ ft/sec})$$

Hence, the Mach number M_{2R} relative to a stationary observer is

$$M_{2R} = \frac{V_{2R}}{a_2} = \frac{12{,}693}{6728} = 1.887$$

From Table C.6, for $M_{2R} = 1.887$, $t_2/T_{2R} = 0.58406$ and $p_2/P_{2R} = 0.15227$. Hence, the stagnation temperature and pressure behind the shock wave, relative to a stationary observer, are

$$T_{2R} = \frac{112{,}650}{0.58406} = 192{,}870 \text{ K} \qquad (347{,}170 \text{ R})$$

$$P_{2R} = \frac{(2.320)10^8}{0.15227} = 1.524 \cdot 10^9 \text{ N/m}^2 \qquad (220{,}980 \text{ lbf/in.}^2)$$

The above results are not quantitatively correct, since the air would cease to behave as a perfect gas at the high temperatures involved. The results are, however, qualitatively correct. They predict that extremely large values of pressure and temperature occur in back of such strong shock waves.

Example 7.5.   A blunt nose pitot tube is employed in a wind tunnel to estimate the flow Mach number. The stagnation pressure P_2 at the entrance to the pitot tube is $2.0 \cdot 10^5$ N/m². The free-stream static pressure p_1 ahead of the shock wave is measured by a static pressure tap in the wall of the tunnel, and is $0.15 \cdot 10^5$ N/m². Estimate the Mach number in the wind tunnel. Assume $\gamma = 1.4$.

Solution

$$\frac{P_2}{p_1} = \frac{2.0}{0.15} = 13.333$$

From Table C.11, for $P_2/p_1 = 13.333$,

$$M_1 = 3.16$$

Example 7.6.   Air enters a circular duct having a constant diameter of 0.610 m. The Mach number of the air at the duct entrance is 3.0, its temperature is 310 K, and its static pressure is $0.70 \cdot 10^5$ N/m². At a station in the duct where the Mach number for the air is 2.5, a normal shock wave occurs. The duct discharges into a receiver, and the Mach number of the air at the duct exit section is 0.8. Assume that the friction coefficient for the duct is constant at $f = 0.005$, and that the air is a perfect gas with a constant specific heat ratio of $\gamma = 1.4$. Calculate (a) the distance from the entrance section of the duct to the location of the normal shock wave, (b) the total length of the duct, (c) the stagnation pressure at the duct exit section, (d) the entropy change for the air between the entrance and exit sections of the duct, and (e) the static pressure at the duct exit section. Figure 7.13 illustrates the physical situation.

Solution

(a)   Let M_x and M_y denote the Mach numbers before and after the normal shock wave, respectively. From either Table C.11 or equation 7.33, for $M_x = 2.5$, $M_y = 0.51299$. The thermodynamic states of the fluid before and after the normal shock wave lie on the same Fanno line. Thus, from Table C.7,

M	3.0	2.5	0.51299	0.80
$\dfrac{4fL^*}{\mathscr{D}}$	0.52216	0.43197	0.96859	0.07229

The length of the duct from the entrance to the normal shock wave is that required to change the flow Mach number from $M_1 = 3.0$ to $M_x = 2.5$. Thus,

$$L_1 = L_1^* - L_x^* = \frac{\mathscr{D}}{4f}\left(\frac{4fL_1^*}{\mathscr{D}} - \frac{4fL_x^*}{\mathscr{D}}\right)$$

$$L_1 = \frac{(0.610)(0.55216 - 0.43197)}{4(0.005)} = 2.751 \text{ m (9.026 ft)}$$

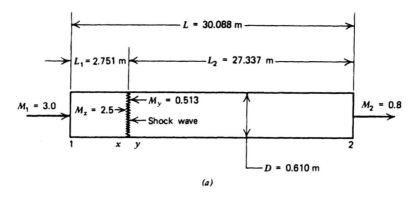

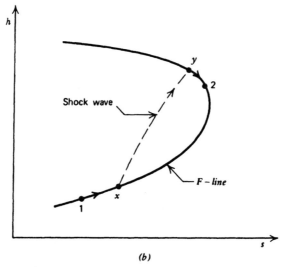

Figure 7.13 Sketches for Example 7.6. (*a*) Physical conditions. (*b*) Plot (schematic) of the flow process on the *hs* plane.

(b) The length of duct between the normal shock wave and the duct exit section corresponds to a change in the flow Mach number from $M_y = 0.51299$ to $M_2 = 0.80$ on the same Fanno line. Hence,

$$L_2 = L_y^* - L_2^* = \frac{\mathscr{D}}{4\mathfrak{f}}\left(\frac{4\mathfrak{f}L_y^*}{\mathscr{D}} - \frac{4\mathfrak{f}L_2^*}{\mathscr{D}}\right)$$

$$L_2 = \frac{(0.610)(0.96859 - 0.07229)}{4(0.005)} = 27.337 \text{ m } (89.688 \text{ ft})$$

Hence,

$$L = L_1 + L_2 = 2.751 + 27.337 = 30.088 \text{ m } (98.714 \text{ ft})$$

(c) Since states 1 and 2 are on the same Fanno line, $P_1^* = P_x^* = P_y^* = P_2^* = P^*$. The stagnation pressure ratio P_2/P_1 is given by

$$\frac{P_2}{P_1} = \frac{P_2}{P^*}\frac{P^*}{P_1}$$

From Table C.7 (Fanno line), for $M_1 = 3.0$ and $M_2 = 0.80$, we obtain

$$\frac{P_2}{P_1} = \frac{1.0382}{4.2346} = 0.24514$$

From Table C.6 (isentropic flow), for $M_1 = 3.0$, $p_1/P_1 = 0.027224$. Consequently,

$$P_2 = \frac{P_2}{P_1}\frac{P_1}{p_1} p_1 = \frac{(0.24514)(0.70)10^5}{(0.027224)} = 6.303 \cdot 10^5 \text{ N/m}^2 \ (91.42 \text{ lbf/in.}^2)$$

(d) From equation 7.43,

$$\Delta s = R \ln\left(\frac{P_1}{P_2}\right) = 287.04 \ln\left(\frac{1}{0.24514}\right) = 403.3 \text{ J/kg-K} \ (0.09633 \text{ Btu/lbm-R})$$

(e) From Table C.6 (isentropic flow), for $M_2 = 0.8$, $p_2/P_2 = 0.65602$. Hence,

$$p_2 = (0.65602)(6.303)10^5 = 4.135 \cdot 10^5 \text{ N/m}^2 \ (59.97 \text{ lbf/in.}^2)$$

7–4(c) The Rankine-Hugoniot Equation for a Normal Shock Wave

By combining equations 7.49 and 7.50, an expression is obtained relating the gas densities ρ_1 and ρ_2, in front of and in back of the normal shock wave, to the corresponding static pressures p_1 and p_2. Solving equation 7.49 for $M_1{}^2$ gives

$$M_1{}^2 = \frac{(\gamma + 1)p_2 + (\gamma - 1)p_1}{2\gamma p_1} \tag{7.54}$$

Substituting equation 7.54 into equation 7.50, after some rearrangement, yields

$$\frac{\rho_2}{\rho_1} = \frac{(\gamma + 1)p_2 + (\gamma - 1)p_1}{(\gamma + 1)p_1 + (\gamma - 1)p_2} \tag{7.55}$$

Equation 7.55 is known as the *Rankine-Hugoniot* equation. It relates the densities and static pressures for the gas in front of and in back of the normal shock wave. Equation 7.55 may be solved for the pressure ratio p_2/p_1. Thus,

$$\frac{p_2}{p_1} = \frac{(\gamma + 1)\rho_2 - (\gamma - 1)\rho_1}{(\gamma + 1)\rho_1 - (\gamma - 1)\rho_2} \tag{7.56}$$

Equations 7.55 and 7.56 for a perfect gas are equivalent to the Hugoniot equation (see Section 7–3, equation 7.11) for a general fluid.

The Rankine-Hugoniot equation does not specify whether the abrupt change of state is an expansion or a compression. It is shown in Section 7–3 from thermodynamic considerations, however, that the shock process is a compression.

The equation for an isentropic process ($p\rho^{-\gamma} = $ constant) shows that as $p_2 \to \infty$, then $\rho_2 \to \infty$. In the case of a normal shock wave, however, the ratio ρ_2/ρ_1 approaches a finite value. Thus,

$$\lim_{p_2 \to \infty}\left(\frac{\rho_2}{\rho_1}\right) = \frac{\gamma + 1}{\gamma - 1} \tag{7.57}$$

7–4(d) Prandtl's Velocity Equation for a Normal Shock Wave

A relationship may be obtained between the velocities V_1 and V_2 in front of and behind a normal shock wave, respectively. For a perfect gas, the relationship is obtained by substituting for M_1 and M_2, in equation 7.48, the corresponding values of the dimensionless velocity $M^* = V/a^*$ from equation 3.189, Table 3.4. Making the aforementioned substitutions and simplifying the results, we obtain *Prandtl's equation* for a normal shock wave in a perfect gas. Thus,

$$V_1 V_2 = \gamma R t^* = a^{*2} \tag{7.58}$$

Equation 7.58 shows that the product $V_1 V_2$ for a normal shock wave is equal to the square of the critical speed of sound a^{*2}; the latter is a constant for an adiabatic flow. From equation 7.58, it is seen that a normal shock wave approaches a sound wave as its strength becomes vanishingly small. Furthermore, the speed of the gas relative to the normal shock wave is supersonic upstream to the wave and subsonic downstream from it.

7–4(e) Entropy Change and Shock Strength for a Normal Shock Wave

For the steady one-dimensional flow of a perfect gas, the increase in the entropy of the gas because of the normal shock wave can be related to the shock strength $\beta = (p_2 - p_1)/p_1$. From Table 3.2, equation 3.177, the entropy change for a perfect gas, expressed in terms of the static pressure ratio p_2/p_1, is

$$\Delta s = c_p \ln \left(\frac{t_2}{t_1}\right) - R \ln \left(\frac{p_2}{p_1}\right) = c_v \ln \left[\left(\frac{t_2}{t_1}\right)^\gamma \left(\frac{p_2}{p_1}\right)^{-(\gamma-1)}\right] \tag{7.59}$$

Introducing $p = \rho R t$ into equation 7.59, we obtain

$$\frac{\Delta s}{c_v} = \ln \left[\frac{p_2}{p_1}\left(\frac{\rho_1}{\rho_2}\right)^\gamma\right] \tag{7.60}$$

Substituting for ρ_1/ρ_2 from equation 7.55 into equation 7.60 gives

$$\frac{\Delta s}{c_v} = \ln \left\{\left[\frac{(\gamma + 1) + (\gamma - 1)\dfrac{p_2}{p_1}}{(\gamma - 1) + (\gamma + 1)\dfrac{p_2}{p_1}}\right]^\gamma \left(\frac{p_2}{p_1}\right)\right\} \tag{7.61}$$

Introducing the normal shock wave strength β into equation 7.61 gives

$$\frac{\Delta s}{c_v} = \ln \left\{(1 + \beta)\left[\frac{2\gamma + (\gamma - 1)\beta}{2\gamma + (\gamma + 1)\beta}\right]^\gamma\right\}$$

Rearranging the last equation yields

$$\frac{\Delta s}{c_v} = \ln \left[(1 + \beta)\left(1 - \frac{\beta}{\gamma + \dfrac{\gamma + 1}{2}\beta}\right)^\gamma\right] \tag{7.62}$$

For a *weak normal shock wave* β is a small fraction, and equation 7.62 may be expanded into a Maclaurin series to give

$$\frac{\Delta s}{R} = \frac{\gamma + 1}{12\gamma^2}\beta^3 - \frac{\gamma + 1}{8\gamma^2}\beta^4 + \cdots \approx \frac{\gamma + 1}{12\gamma^2}\beta^3 \tag{7.63}$$

Equation 7.63 shows that for a *weak normal shock wave* (i.e., $\beta \ll 1$), the increase in the entropy of the gas is negligible. Consequently, a weak normal shock wave produces an efficient compression of the gas.

If β is expressed in terms of M_1, the Mach number in front of the normal shock wave, by means of equation 7.46, then equation 7.63 is transformed to read

$$\frac{\Delta s}{R} = \frac{2}{3}\frac{\gamma}{(\gamma + 1)^2}(M_1{}^2 - 1)^3 - \frac{2\gamma}{(\gamma + 1)^3}(M_1{}^2 - 1)^4 + \cdots \qquad (7.64)$$

Neglecting all powers of $(M_1{}^2 - 1)$ higher than the third yields

$$\frac{\Delta s}{R} \approx \frac{2}{3}\frac{\gamma}{(\gamma + 1)^2}(M_1{}^2 - 1)^3 \qquad (7.65)$$

which applies to weak normal shocks waves. Equation 7.65 shows that an increase in entropy results only when M_1 is significantly larger than unity. Otherwise, the increase in entropy is very small, and the shock process may be assumed to be isentropic.

Equation 7.63 shows that the increase in the entropy of a perfect gas because of a weak normal shock wave may be assumed to be proportional to the third power of the shock strength β, while equation 7.65 shows that it is proportional to the third power of the parameter $(M_1{}^2 - 1)$. Consequently, the magnitude of either β or $(M_1{}^2 - 1)$ may be employed as a criterion of the irreversibility of a weak normal shock wave. For a weak normal shock wave, $M_1 < 1.3$ approximately, the irreversibility of the shock wave may be ignored in calculating the states for the gas in front of and in back of the normal shock wave. The aforementioned pseudo-isentropic compression produced by a weak normal shock wave may be employed advantageously in designing the air-intake-diffuser systems for an air-breathing jet propulsion engine for propelling aircraft at supersonic speeds.

7–4(f) Shock or Wave Drag

Consider now the case where a shock wave is generated by a body moving through a frictionless fluid, as illustrated in Fig. 7.14. For convenience, a relative coordinate system is employed. Two control surfaces A_1 and A_2, extending to infinity, may be drawn perpendicular to the assumed one-dimensional flow. One plane is located upstream to the body where the pressure is p_1, and the other is located downstream from the body where the gas has expanded isentropically (after the shock wave) to the upstream static pressure p_1.

Since there is no unbalanced pressure force, the resultant pressure force acting on the control surface is zero. There is, however, a rate of change in momentum in the direction of flow because $V_2 < V_1$. Consequently, there is a resultant force acting in

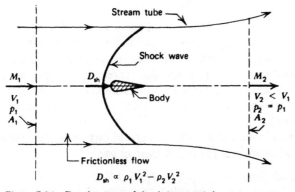

Figure 7.14 Development of shock (or wave) drag.

the direction opposite to the inflow velocity V_1. From the reaction principle it follows that the body producing the shock wave must be acted on by a force of equal magnitude acting in the opposite direction; that is, *a force acting in the same direction as the velocity* V_1 (see Fig. 7.14). In the relative coordinate system such a force is a *drag force*; it is a force that must be overcome by the thrust developed by the propulsion system. It is seen, therefore, that if a body moves in a perfect (frictionless) fluid and shock waves are generated, a drag force called either the *shock drag* or the *wave drag* is produced. The shock drag is independent of the skin friction drag and the form drag, which are always present when a body moves in a real fluid (see Section 1–6).

7–5 NORMAL SHOCK WAVES IN IMPERFECT GASES

The properties of normal shock waves are developed in Section 7–4 for the *perfect gas*. In this section, the influence of *imperfect gas effects* on the flow properties pertinent to a normal shock wave is considered. The term *imperfect gas* is employed here in a general sense to denote any gas that deviates from *perfect gas* behavior. Several sources of imperfect gas effects are discussed in detail in Section 4–5.

In the present section, a general *imperfect gas* model is considered for which there are available equations of state having the following forms.

$$\rho = \rho(p,t) \tag{7.66}$$

$$h = h(p,t) \tag{7.67}$$

The equations of state may be in the form of charts, tables, or a computer code. The source of the *imperfect gas effect* is unimportant. All that is required is that the aforementioned equations of state be known.

Substituting the continuity equation, equation 7.2, into the momentum equation, equation 7.5, and the energy equation, equation 7.6, we obtain

$$p_2 = p_1 + \rho_1 V_1^2 \left(1 - \frac{\rho_1}{\rho_2}\right) \tag{7.68}$$

$$h_2 = h_1 + \frac{V_1^2}{2}\left[1 - \left(\frac{\rho_1}{\rho_2}\right)^2\right] \tag{7.69}$$

Rearranging equation 7.2 yields

$$V_2 = \frac{\rho_1 V_1}{\rho_2} \tag{7.70}$$

Equations 7.66 to 7.70 comprise a set of five equations in terms of p, ρ, h, t, and V that is applicable at points 1 and 2 upstream to and downstream from the normal shock wave, respectively. When the upstream properties are specified, those equations yield five relationships between them and the five downstream properties p_2, ρ_2, h_2, t_2, and V_2; in general, those equations must be solved numerically. An extremely rapid numerical method for solving nonlinear equations is the secant method discussed in Appendix A–4(d).

The following steps are employed for obtaining the solutions for the five nonlinear equations governing normal shock waves in an imperfect gas.

1. Determine the initial conditions (i.e., p_1, ρ_1, h_1, t_1, and V_1).
2. *Assume* a value for ρ_2. For the first trial value, the value of ρ_2 based on the im-

perfect gas behaving as a perfect gas may be employed. The second trial value is the result of the first calculation, and subsequent trial values are obtained by the secant method.

3. Calculate the values of p_2 and h_2 by means of equations 7.68 and 7.69.
4. Determine t_2 and ρ_2 from the equations of state for the imperfect gas (equations 7.66 and 7.67). Those equations may be in the form of algebraic equations, graphs, or tables.
5. If the calculated value of ρ_2 from step 4 agrees to within the desired tolerance with the assumed value for ρ_2 assumed in step 2, the solution is obtained. If the agreement is unsatisfactory, steps 2 to 5 must be repeated until convergence of the assumed and calculated values is obtained.
6. When convergence is obtained, V_2 may be determined from equation 7.70. The calculation of V_2 completes the solution.

The procedure is simple to apply, and the assumed and calculated values converge rapidly.

Example 7.7. A reentry vehicle (RV) is at an altitude of 15,000 m and has a velocity of 1850 m/s. A bow shock wave envelopes the RV, as illustrated in Fig. 7.15a. Neglecting dissociation, determine the properties at a point just behind the shock wave on the RV centerline, where the shock wave may be treated as a normal shock wave. Assume $R = 287.04$ J/kg-K, and take into account the variation of the specific heat of air with temperature.

Solution

If dissociation is neglected, the only imperfect gas effect is that due to the variation of the specific heat of the atmospheric air with temperature. Consequently, Table C.4 may be employed as the caloric equation of state for the air. The numerical method discussed in Section 7–5 is employed for obtaining the solution.

(1) *Initial conditions.* From Table C.5, at an altitude of 15,000 m, $t_1 = 216.65$ K, $p_1 = 12,112$ N/m², and $\rho_1 = 0.19475$ kg/m³. From Table C.4, at $t = 216.65$ K, $\gamma = 1.4000$. Thus,

$$a_1 = (\gamma R t_1)^{1/2} = [1.4(287.04)(216.65)]^{1/2} = 295.1 \text{ m/s}$$

$$M_1 = \frac{V_1}{a_1} = \frac{1850}{295.1} = 6.2691$$

From Table C.4, at $t_1 = 216.65$ K, $h_1 = 217.660$ kJ/kg.

(2) *Assume a value for ρ_2.* As a first trial value, let ρ_2 be that based on assuming that the atmospheric air is a perfect gas. From Table C.11 for a normal shock wave, for $M_1 = 6.2691$, $(\rho_2/\rho_1) = (V_1/V_2) = 5.3227$, $t_2/t_1 = 8.5833$, and $p_2/p_1 = 45.687$. Hence,

$$\rho_2 = 5.3227(0.19475) = 1.0366 \text{ kg/m}^3$$

The above result yields the first trial value $\rho_2 = 1.0366$ kg/m³. The remaining properties are computed from the ratios determined above.

$$t_2 = 8.5833(216.65) = 1859.6 \text{ K}$$

$$p_2 = 45.687(12,112) = 5.534 \cdot 10^5 \text{ N/m}^2$$

$$V_2 = \frac{1850}{5.3227} = 347.57 \text{ m/s}$$

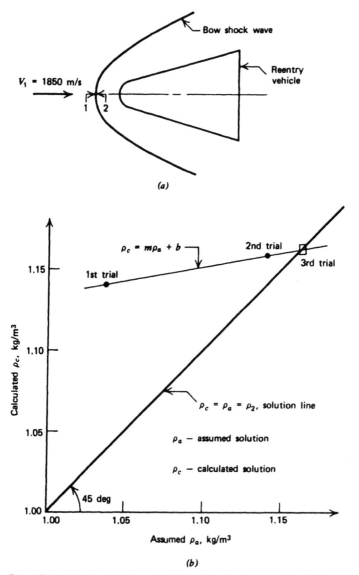

Figure 7.15 Sketches for Example 7.7. (*a*) Bow shock wave ahead of reentry vehicle. (*b*) Determination of ρ_2.

(3) *Calculation of p_2 and h_2.* From equation 7.68,

$$p_2 = 12{,}112 + (0.19475)(1850)^2 \left(1 - \frac{0.19475}{\rho_2} \right)$$

$$p_2 = 12{,}112 + 6.6653 \cdot 10^5 \left(1 - \frac{0.19475}{\rho_2} \right) \text{N/m}^2 \tag{a}$$

where ρ_2 is in kilograms per cubic meter. Equation 7.69, for h_2, yields

$$h_2 = 217.660 + \frac{(1850)^2}{2(1000)} \left[1 - \left(\frac{0.19475}{\rho_2} \right)^2 \right]$$

$$h_2 = 217.660 + 1711.3 \left[1 - \left(\frac{0.19475}{\rho_2} \right)^2 \right] \tag{b}$$

Substituting $\rho_2 = 1.0366$ kg/m^3 into equations (a) and (b) yields

$$p_2 = 12{,}112 + 6.6653 \cdot 10^5 \left(1 - \frac{0.19475}{1.0366} \right) = 5.534 \cdot 10^5 \text{ N/m}^2$$

$$h_2 = 217.660 + 1711.3 \left[1 - \left(\frac{0.19475}{1.0366} \right)^2 \right] = 1868.6 \text{ kJ/kg}$$

(4) *Calculation of t_2 and ρ_2.* From Table C.4, for $h_2 = 1868.6$ kJ/kg, $t_2 = 1690.6$ K. From the equation of state for a perfect gas $\rho = p/Rt$,

$$\rho_2 = \frac{p_2}{Rt_2} = \frac{(5.534)10^5}{(287.04)(1690.6)} = 1.1404 \text{ kg/m}^3$$

Comparison of the values for ρ_2 from steps 2 and 4 indicates that additional trials are needed.

(5) *Repetition of steps 3 and 4 with the value of ρ_2 from step 4.*

$$p_2 = 12{,}112 + 6.6653 \cdot 10^5 \left(1 - \frac{0.19475}{1.1404} \right) = 5.648 \cdot 10^5 \text{ N/m}^2$$

$$h_2 = 217.660 + 1711.3 \left[1 - \left(\frac{0.19475}{1.1404} \right)^2 \right] = 1879.1 \text{ kJ/kg}$$

From Table C.4, for $h_2 = 1879.1$ kJ/kg, $t_2 = 1699.2$ K. From the equation of state for a perfect gas,

$$\rho_2 = \frac{(5.648)10^5}{(287.04)(1699.2)} = 1.1580 \text{ kg/m}^3$$

Comparing the value of ρ_2 from step 5 with that from step 4 shows that a third trial for establishing the value of ρ_2 is warranted.

For the third trial, the *secant method* discussed in Appendix A–4(b) is employed. Denote the assumed value of ρ_2 for a trial by ρ_a, and the resulting calculated value by ρ_c. The results of the first two trials are plotted in the graph of ρ_c versus ρ_a, presented in Fig. 7.15b. The desired solution lies on the 45 deg line $\rho_c = \rho_a = \rho_2$. The intersection of the two lines gives the next trial value of ρ_2. Because of the nonlinear nature of the problem, the third calculated value of ρ_2 may not fall exactly on the 45 deg line, in which case the secant method may be applied again if required to obtain a fourth trial value of ρ_2. Convergence of the secant method is extremely rapid, and ordinarily only one or two iterations are required for achieving the desired accuracy.

For the present case, the equation of the solution curve is

$$\frac{\rho_c - 1.1580}{\rho_a - 1.1404} = \frac{1.1404 - 1.1580}{1.0366 - 1.1404} = 0.16956$$

$$\rho_c = 1.1580 + 0.16956(\rho_a - 1.1404) = 0.96464 + 0.16956\rho_a$$

Substituting $\rho_c = \rho_a = \rho_2$ into the above equation gives the third trial value $\rho_2 = 1.1616$ kg/m^3.

Repeat steps 3 and 4 with $\rho_2 = 1.1616$ kg/m^3.

$$p_2 = 12{,}112 + (6.6653)10^5 \left(1 - \frac{0.19475}{1.1616} \right) = 5.669 \cdot 10^5 \text{ N/m}^2$$

$$h_2 = 217.660 + 1711.3 \left[1 - \left(\frac{0.19475}{1.1616} \right)^2 \right] = 1880.9 \text{ kJ/kg}$$

From Table C.4, for $h_2 = 1880.9$ kJ/kg, $t_2 = 1700.6$ K. From the perfect gas equation of state,

$$\rho_2 = \frac{(5.669)10^5}{(287.04)(1700.6)} = 1.1614 \text{ kg/m}^3$$

The agreement between the assumed value of $\rho_2 = 1.1616$ kg/m^3 and the calculated value of $\rho_2 = 1.1614$ kg/m^3 is good enough so that no additional applications of the secant method are needed.

(6) *Calculation of* V_2. From equation 7.70,

$$V_2 = \frac{(0.19475)(1850)}{(1.1614)} = 310.22 \text{ m/s}$$

The following table presents the results for the two initial trials, and the results for the final trial obtained by the secant method.

Trial	ρ_2, kg/m^3	p_2, N/m$^2 \cdot 10^5$	h_2, kJ/kg	t_2, K	ρ_2, kg/m^3
1	1.0366	5.534	1868.6	1690.6	1.1404
2	1.1404	5.648	1879.1	1699.2	1.1580
3	1.1616	5.669	1880.9	1700.6	1.1614

A comparison of the results obtained for the *perfect gas* analysis with those obtained for the *imperfect gas* analysis indicates the following percent errors for the former.

Property	Perfect Gas	Imperfect Gas	Error, Percent
t_2, K	1859.6	1700.6	9.35
p_2, N/m$^2 \cdot 10^5$	5.534	5.669	−2.38
ρ_2, kg/m^3	1.0366	1.1614	−10.76
V_2, m/s	347.57	310.22	12.04

The errors tabulated above are not large, but they are certainly significant.

7-6 OBLIQUE SHOCK WAVES

The preceding sections are concerned with shock waves that are normal to the flow direction. In many practical cases, however, a supersonic flow is caused to change its flow direction and, as a consequence, a shock wave is formed that is inclined to the initial flow direction; such shock waves are termed *oblique shock waves*.

Figure 7.16 illustrates the case where a stationary oblique shock wave is produced by the frictionless surface of a solid wall changing the flow direction of a supersonic stream. Let V_1 and V_2, respectively, denote the stream velocities entering and leaving the oblique shock wave; ε is the angle between V_1 and the shock wave and is called the *wave angle*; β is the angle between V_2 and the shock wave, and $\delta = \varepsilon - \beta$ is termed the *flow deflection angle*. It will be assumed that the flow is a steady *two-dimensional* planar adiabatic flow, that no external work is involved, and that the effects of body forces are negligible.

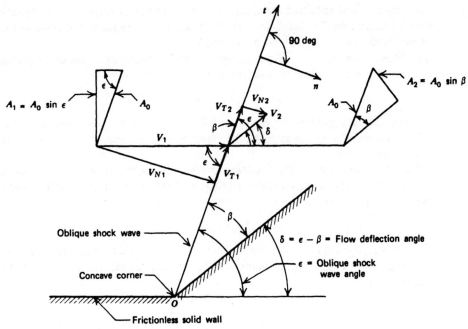

Figure 7.16 Velocity components for an oblique shock wave.

The components of the velocities V_1 and V_2 resolved in the tangential and normal directions to the oblique shock wave are denoted, respectively, by V_{T1}, V_{T2}, V_{N1}, and V_{N2}. From Fig. 7.16, it follows that

$$V_1^2 = V_{T1}^2 + V_{N1}^2 \qquad (7.71)$$

$$V_2^2 = V_{T2}^2 + V_{N2}^2 \qquad (7.72)$$

The velocity components tangential and normal to the oblique shock wave may be expressed in terms of the angles ε and β. Thus,

$$V_{T1} = V_1 \cos \varepsilon \qquad (7.73)$$

$$V_{N1} = V_1 \sin \varepsilon \qquad (7.74)$$

$$V_{T2} = V_2 \cos \beta \qquad (7.75)$$

$$V_{N2} = V_2 \sin \beta \qquad (7.76)$$

where $\beta = \varepsilon - \delta$.

(1) *Continuity equation*. Let A_0 denote a unit area parallel to the oblique shock wave; then the areas perpendicular to V_1 and V_2, denoted by A_1 and A_2, respectively, are given by

$$A_1 = A_0 \sin \varepsilon \qquad \text{and} \qquad A_2 = A_0 \sin \beta \qquad (7.77)$$

From Table 3.1, the continuity equation for the flow through the shock wave yields

$$\dot{m} = A_1 \rho_1 V_1 = A_2 \rho_2 V_2 = \text{constant} \qquad (7.78)$$

Substituting for A_1 and A_2 from equation 7.77 gives

$$A_0 \rho_1 V_1 \sin \varepsilon = A_0 \rho_2 V_2 \sin \beta \qquad (7.79)$$

From equations 7.74, 7.76, and 7.79, it follows that

$$\rho_1 V_{N1} = \rho_2 V_{N2} \qquad (7.80)$$

The requirement specified by equation 7.80 is identical physically to equation 7.2 derived in Section 7–3 for a normal shock wave; for the latter the velocities normal to the shock wave are denoted by V_1 and V_2.

(2) *Momentum equation.* Application of the momentum equation, equation 7.3, in the n direction, that is, *normal* to the oblique shock wave, gives

$$p_1 + \rho_1 V_{N1}{}^2 = p_2 + \rho_2 V_{N2}{}^2 \tag{7.81}$$

The requirement specified by equation 7.81 is the same as that specified by equation 7.5 for a normal shock wave; for the latter V_1 and V_2 denote the velocities normal to the shock wave.

Since there is no change in the static pressure in the direction *parallel* or *tangential* to the oblique shock wave, the application of the momentum equation parallel to the oblique shock wave gives

$$\rho_1 V_{N1} V_{T1} = \rho_2 V_{N2} V_{T2} \tag{7.82}$$

Since $\rho_1 V_{N1} = \rho_2 V_{N2}$ (see equation 7.80), it is concluded from equation 7.82 that the tangential velocity components of the oblique shock wave are equal to each other; that is,

$$V_{T1} = V_{T2} = V_T \tag{7.83}$$

From equations 7.73 and 7.75, since $V_{T1} = V_{T2}$, the velocity ratio V_2/V_1 is given by

$$\frac{V_2}{V_1} = \frac{V_{T2}}{\cos \beta} \cdot \frac{\cos \varepsilon}{V_{T1}} = \frac{\cos \varepsilon}{\cos \beta} \tag{7.84}$$

It follows from equations 7.80, 7.74, 7.76, and 7.84 that

$$\frac{\rho_2}{\rho_1} = \frac{V_{N1}}{V_{N2}} = \frac{V_1 \sin \varepsilon}{V_2 \sin \beta} = \frac{\cos \beta \sin \varepsilon}{\cos \varepsilon \sin \beta} = \frac{\tan \varepsilon}{\tan \beta} \tag{7.85}$$

(3) *Energy equation.* From Table 3.1, equation 3.164, the energy equation for the adiabatic flow of a gas (neglecting body forces, $g\,dz = 0$), is given by

$$h_1 + \frac{V_1{}^2}{2} = h_2 + \frac{V_2{}^2}{2} = H = \text{constant}. \tag{7.86}$$

Substituting for $V_1{}^2$ and $V_2{}^2$ from equations 7.71 and 7.72 into equation 7.86, and noting that $V_{T1} = V_{T2}$ yields

$$h_1 + \frac{V_{N1}{}^2}{2} = h_2 + \frac{V_{N2}{}^2}{2} = \text{constant} \tag{7.87}$$

Equation 7.87 states that the energy equation for an oblique shock wave is identical with that for a normal shock wave, but with V_{N1} and V_{N2} replacing V_1 and V_2 (see equation 7.6).

It is shown above that the continuity, momentum, and energy equations have identical forms for normal and oblique shock waves. It may be concluded, therefore, that all of the relationships derived previously for the normal shock wave are applicable to the normal components of the velocities V_1 and V_2 in front of and in back of an oblique shock wave, respectively, *provided that the normal velocity component V_{N1} is supersonic.*

Since $V_{N2} < V_{N1}$ (V_{N2} must be subsonic) and $V_{T1} = V_{T2}$, the velocity V_2 leaving the oblique shock wave is deflected from the direction of V_1; that is, *the fluid stream is deflected toward the oblique shock wave* (see Fig. 7.16).

(4) *Equation of state.* To complete the system of equations, the equations of state

for the flowing gas are required. Thus,

$$\rho = \rho(p,t) \quad \text{and} \quad h = h(p,t) \tag{7.88}$$

Equations 7.73, 7.74, 7.75, 7.76, 7.80, 7.81, 7.83, 7.87, and 7.88 (two equations) comprise a set of ten equations for the ten properties p_2, ρ_2, t_2, h_2, V_2, V_{N2}, V_{T2}, V_{N1}, V_{T1}, and ε, in terms of the known upstream conditions p_1, ρ_1, t_1, h_1, V_1, and δ. The aforementioned set of equations is considerably more involved than the set of equations developed in Section 7-3 for the normal shock wave because of the extra degree of freedom introduced into the physical process by the *flow deflection angle δ*. The oblique shock wave has two degrees of freedom: (1) the *upstream conditions*, and (2) the *flow deflection angle δ*; whereas there is only one degree of freedom in the normal shock wave, namely the *upstream conditions*. The preceding equations of the present section are quite general. When the equations of state for the flowing fluid are specified, the changes produced in the fluid properties caused by its flow through an *oblique shock wave* are determinable.

7-7 OBLIQUE SHOCK WAVES IN PERFECT GASES

In the present section, the general equations for *oblique shock waves*, derived in Section 7-6, are applied to the flow of a *perfect gas*.

7-7(a) Governing Equations

The equations of state for a perfect gas are given by equations 7.23 and 7.24, which are repeated below for convenience.

$$p = \rho R t \tag{7.23}$$

$$h = c_p t \tag{7.24}$$

The energy equation for the adiabatic flow of a perfect gas through an oblique shock wave is obtained from equation 7.87. Thus,

$$c_p t_1 + \frac{V_{N1}^2}{2} = c_p t_2 + \frac{V_{N2}^2}{2} \tag{7.89}$$

The continuity equation 7.80 and the momentum equation 7.81 retain their forms.

Because the thermodynamic properties of the gas after the oblique shock wave depend on the *upstream conditions* and the *flow deflection angle δ*, it is desirable to express the pertinent relationships in terms of those variables. It is beyond the scope of this book to present their derivations; they may be found in the standard works on supersonic aerodynamics.[3, 4]

7-7(b) Property Ratios Across an Oblique Shock Wave

The ratio of the static pressure in back of the oblique shock wave p_2 to the static pressure in front of it p_1 is given by

$$\frac{p_2}{p_1} = \frac{2\gamma}{\gamma + 1} M_1^2 \sin^2 \varepsilon - \frac{\gamma - 1}{\gamma + 1} \tag{7.90}$$

The density ratio ρ_2/ρ_1 is given by

$$\frac{\rho_2}{\rho_1} = \frac{\tan \varepsilon}{\tan \beta} = \frac{V_{N1}}{V_{N2}} = \frac{(\gamma + 1)M_1^2 \sin^2 \varepsilon}{2 + (\gamma - 1)M_1^2 \sin^2 \varepsilon} \tag{7.91}$$

The velocity ratio V_2/V_1 is given by

$$\frac{V_2}{V_1} = \frac{\sin \varepsilon}{\sin \beta} \left[\frac{2}{(\gamma + 1)M_1^2 \sin^2 \varepsilon} + \frac{\gamma - 1}{\gamma + 1} \right] \tag{7.92}$$

The Mach number after the oblique shock wave M_2 is obtained from

$$\frac{\tan \varepsilon}{\tan \beta} = \frac{2}{\gamma + 1} \left(\frac{1}{M_2^2 \sin^2 \beta} + \frac{\gamma - 1}{2} \right) \tag{7.93}$$

The oblique shock wave angle ε is calculated from

$$\frac{1}{\tan \delta} = \left(\frac{\gamma + 1}{2} \frac{M_1^2}{M_1^2 \sin^2 \varepsilon - 1} - 1 \right) \tan \varepsilon \tag{7.94}$$

The preceding equations may be employed for calculating the properties of the flowing gas in back of an oblique shock wave, for fixed initial conditions, provided one final condition is known. Ordinarily, either the flow deflection angle δ or the static pressure p_2 is given and, from that information, the changes because of the oblique shock wave may be calculated. It is apparent from equations 7.90 to 7.94 that the intensity of an oblique shock wave depends not only on the initial Mach number M_1, but also on the wave angle ε. If M_1 is fixed, then the shock wave intensity depends entirely on ε, and unless $M_1 \sin \varepsilon \geq 1$, there will be no shock wave. It should be noted that, when $\varepsilon = \pi/2$ (normal shock wave), the equations transform to those presented in Section 7–4 for the normal shock wave; the latter is, of course, merely a special case of the oblique shock wave. Tables for aiding in the computation of changes in the flow properties because of an oblique shock wave are presented in Table C.12 in the Appendix, and in References 5 and 7.

Figures 7.17 and 7.18 present the oblique shock wave angle ε and the downstream Mach number M_2, respectively, as functions of the initial Mach number M_1 for different values of the flow deflection angle δ, for $\gamma = 1.40$. It is seen from Figs. 7.17 and 7.18 that, for the same values of M_1 and δ, two different sets of values for ε and M_2 are possible; that is, one of two oblique shock waves having different strengths are possible. Which of the two occurs in practice depends on the other boundary conditions for the flow. Experiments show that the weaker of the two shock waves occurs most frequently. In special cases where the downstream pressure is extremely high, the stronger of the two shock waves may occur; for example, the shock wave on the spike of a supersonic inlet for an air-breathing jet engine when no flow is allowed to pass through the inlet. Figure 7.19 presents the static pressure ratio p_2/p_1 as a function of M_1 for different values of the flow deflection angle δ, for $\gamma = 1.4$.

The entropy change for a shock wave is given by equation 7.60, which is repeated here for convenience. Thus,

$$\Delta s = c_v \ln \left[\frac{p_2}{p_1} \left(\frac{\rho_1}{\rho_2} \right)^\gamma \right] \tag{7.60}$$

For an oblique shock wave, p_2/p_1 is given by equation 7.90, and ρ_1/ρ_2 is given by equation 7.91. Since both p_2/p_1 and ρ_2/ρ_1 are functions of the wave angle ε, the entropy change Δs is also a function of ε, and increases with ε. If $\sin \varepsilon = 1/M = \sin \alpha$, where α is the Mach angle [see Section 3–8(b)], then p_2/p_1 and ρ_2/ρ_1 are both unity, and $\Delta s = 0$. Furthermore, if $\varepsilon = \pi/2$ (normal shock wave), then Δs is a maximum.

The flow deflection angle δ is zero when $M_2 = M_1$ (acoustic or Mach wave) and also when $\varepsilon = \pi/2$ (normal shock wave). Between those limits the value of δ is positive, so that it has a maximum value, which is denoted by δ_m. The wave angle ε_m

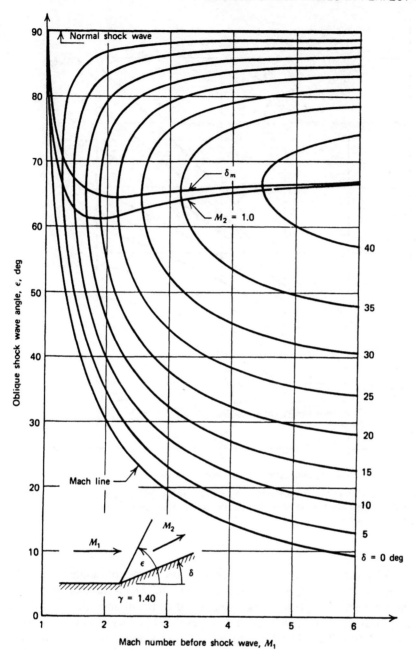

Figure 7.17 Shock wave angle ε as a function of the initial Mach number M_1 for different values of the flow deflection angle δ for $\gamma = 1.4$.

corresponding to δ_m, the maximum possible flow deflection angle for a given initial Mach number M_1, is determined from[3]

$$\sin^2 \varepsilon_m = \frac{1}{\gamma M_1^2}\left\{\frac{\gamma + 1}{4}M_1^2 - 1\right.$$

$$\left. + \left[(\gamma + 1)\left(1 + \frac{\gamma - 1}{2}M_1^2 + \frac{\gamma + 1}{16}M_1^4\right)\right]^{1/2}\right\} \quad (7.95)$$

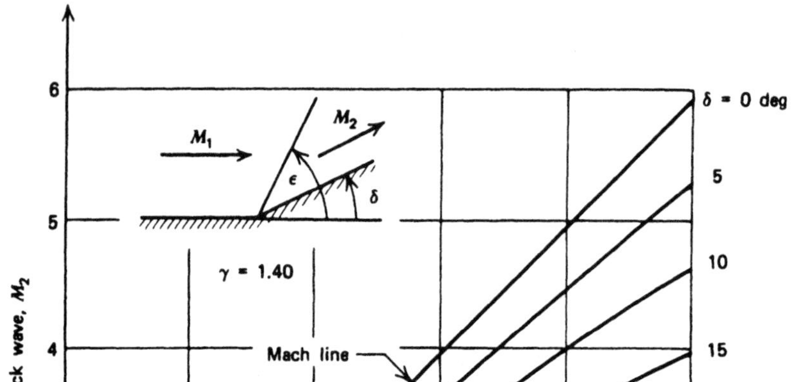

Figure 7.18 Mach number downstream of an oblique shock wave M_2 as a function of the initial Mach number M_1 for different values of the flow deflection angle δ for $\gamma = 1.4$.

Figure 7.20 presents the maximum possible flow deflection angle δ_m and the corresponding wave angle ε_m as functions of the initial Mach number M_1 for $\gamma = 1.4$. If the flow deflection angle δ is larger than δ_m, then the oblique shock wave becomes detached from the corner. Increasing δ causes the shock wave to move upstream, as illustrated in Fig. 7.21. The limiting value of δ_m where $M_1 = \infty$, for $\gamma = 1.4$, is $\delta_m = 45.37$ deg.

There is a particular value of ε, denoted by ε_s, for which the flow leaves the oblique shock wave with the sonic speed; that is, $M_2 = 1$. That particular value of ε_s is obtained from[3]

$$\sin^2 \varepsilon_s = \frac{1}{\gamma M_1^2} \left\{ \frac{\gamma + 1}{4} M_1^2 - \frac{3 - \gamma}{4} \right.$$
$$\left. + \left[(\gamma + 1) \left(\frac{9 + \gamma}{16} - \frac{3 - \gamma}{8} M_1^2 + \frac{\gamma + 1}{16} M_1^4 \right) \right]^{1/2} \right\} \quad (7.96)$$

Two FORTRAN computer programs are presented below for computing the properties of oblique shock waves: subroutine OSHOCKL and subroutine OSHOCKP. Subroutine OSHOCKL computes the limiting values ε_m and δ_m, and the corresponding

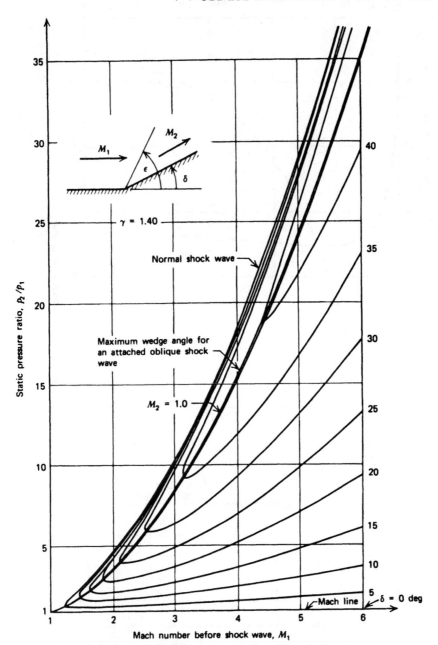

Figure 7.19 Static pressure ratio across an oblique shock wave p_2/p_1 as a function of the initial Mach number M_1 for different values of the flow deflection angle δ for $\gamma = 1.40$.

values of M_2 and p_2/p_1, as functions of M_1. It also computes δ_s and ε_s, which are the values of δ and ε when $M_2 = 1$, as well as p_2/p_1 corresponding to $M_2 = 1$. Subroutine OSHOCKP computes ε, M_2, and p_2/p_1 as functions of M_1, with the flow deflection angle δ as a parameter. It computes those properties for both strong and weak shock waves. Because of space limitations, only one of the tables generated by those programs is included in this book; that is Table C.12, which presents the wave angle ε as a function of M_1 for weak oblique shock waves, for $\gamma = 1.4$. Figures 7.17 to 7.19 are, however, based on those tables. For Mach numbers from 1.00 to 3.75, the last two columns in each row of Table C.12 present the maximum wave angle ε_m and the maximum flow deflection angle δ_m, respectively. For M from 3.80 to 4.40, the last column presents ε_m.

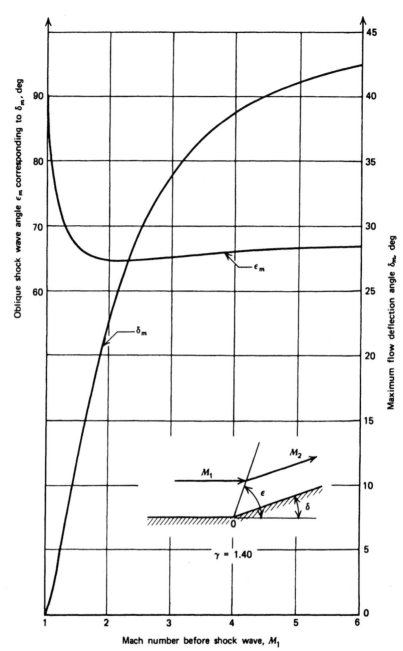

Figure 7.20 Maximum flow deflection angle δ_m and the corresponding wave angle ε_m for oblique shock waves as a function of the upstream Mach number M_1 for $\gamma = 1.40$.

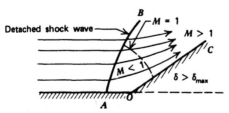

Figure 7.21 Illustration of the detachment of an oblique shock wave because of the flow deflection angle δ being larger than the maximum value δ_{max}.

```
      SUBROUTINE OSHOCKL

C     LIMITING PROPERTIES FOR AN OBLIQUE SHOCK WAVE IN A PERFECT GAS

      REAL M1,MM,MX,MS $ DIMENSION J(9),DM(9)
      NAMELIST /DATA/ G,J,DM
      DF(E)=ATAN((MM*SIN(E)**2-1.0)/(TAN(E)*(G2*MM+1.0-MM*SIN(E)**2)))
      AMF(E,D)=1.0/(SIN(E-D)*SQRT(G2*TAN(E)/TAN(E-D)-G5))
      PF(E)=G6*(MM*SIN(E)**2-G7) $ READ (5,DATA)
      G1=G+1.0 $ G2=G1/2.0 $ G3=G1/4.0 $ G4=G1/16.0 $ G5=(G-1.0)/2.0
      G6=G/G2 $ G7=G5/G $ G8=(3.0-G)/4.0 $ G9=(G+9.0)/16.0
      G10=(3.0-G)/8.0 $ RAD=57.2957795 $ WRITE (6,2000) G $ I=1 $ L=0
      M1=1.0 $ DX=0.0 $ EX=90.0 $ MX=1.0 $ PX=1.0 $ DS=0.0 $ ES=90.0
      PS=1.0 $ WRITE (6,2010) M1,DX,EX,MX,PX,DS,ES,PS

C     CALCULATE LIMITING PROPERTIES

   40 N=J(I) $ DO 50 K=1,N $ M1=M1+DM(I) $ MM=M1**2
      EX=ASIN(SQRT((G3*MM-1.0+SQRT(G1*(1.0+G5*MM+G4*MM**2)))/(G*MM)))
      DX=DF(EX) $ MX=AMF(EX,DX) $ PX=PF(EX) $ DX=DX*RAD $ EX=EX*RAD
      ES=ASIN(SQRT((G3*MM-G8+SQRT(G1*(G9-G10*MM+G4*MM**2)))/(G*MM)))
      DS=DF(ES) $ PS=PF(ES) $ DS=DS*RAD $ ES=ES*RAD
      WRITE (6,2010) M1,DX,EX,MX,PX,DS,ES,PS $ L=L+1
      IF(L.LT.50) GO TO 50 $ WRITE (6,2000) G $ L=0
   50 CONTINUE $ I=I+1 $ IF((I.EQ.10).OR.(J(I).EQ.0)) RETURN $ GO TO 40

 2000 FORMAT (1H1,3X,66HLIMITING VALUES OF PROPERTIES FOR AN OBLIQUE SHO
     1CK WAVE.   GAMMA = ,F5.2//4X,2HM1,5X,2HDX,6X,2HEX,8X,2HMX,8X,2HPX,
     28X,2HDS,8X,2HES,8X,2HPS/1H )
 2010 FORMAT (1H ,F5.2,7F10.5)
      END




      SUBROUTINE OSHOCKP

C     OBLIQUE SHOCK WAVE IN A PERFECT GAS

      REAL M1,MM $ DIMENSION J(9),DM(9),D(22),E(22),T(3)
      NAMELIST /DATA/ II,IJ,G,J,DM
      AF(EE)=MM*SIN(EE)**2-1.0 $ BF(EE)=G2*MM-AF(EE) $ READ (5,DATA)
      G1=G+1.0 $ G2=G1/2.0 $ G3=G1/4.0 $ G4=G1/16.0 $ G5=(G-1.0)/2.0
      G6=G/G2 $ G7=G5/G $ T(1)=6H ANGLE $ T(2)=6H    M2 $ T(3)=6H P2/P1
      NE=1 $ NF=21 $ DEND=40.0 $ DELD=2.0 $ RAD=57.2957795

      DO 100 ND=1,NE $ I=1 $ L=0 $ LL=0
      D(1)=DEND*FLOAT(ND-1) $ M1=1.0 $ DO 40 M=2,NF $ D(M)=D(M-1)+DELD
   40 CONTINUE $ WRITE(6,2000)T(II),G,(D(M),M=1,NF) $ WRITE (6,2010) M1
   50 N=J(I) $ DO 90 K=1,N $ M1=M1+DM(I) $ WRITE (6,2010) M1 $ L=L+1
      AL=ASIN(1.0/M1)*RAD $ IF(IJ.EQ.2)AL=90.0 $ MM=M1**2 $ NN=21 $ NA=0
      EX=ASIN(SQRT((G3*MM-1.0+SQRT(G1*(1.0+G5*MM+G4*MM**2)))/(G*MM)))
      DX=ATAN(AF(EX)/(BF(EX)*TAN(EX)))*RAD
      IF(D(1).GT.DX+1.0)GO TO 60 $ EE=(AL-(-1.0)**IJ*D(1))/RAD $ LL=LL+1
      DO 70 M=1,NN $ DD=D(M) $ IF (DD.LT.DX) GO TO 60
      DD=DX $ E(M+1)=DX $ NN=M $ IF (NN.LT.NF) NA=1
   60 IF (DD.GT.0.0) GO TO 65 $ E(M)=AL $ IK=2*II+IJ $ GO TO 68
   65 TND=TAN(DD/RAD) $ A=AF(EE) $ B=BF(EE) $ C=2.0*MM*SIN(EE)*COS(EE)
      H=1.0/(((TND*B)**2+A**2) $ FE=EE-ATAN(A/(B*TND))
      FEP=1.0-C*H*TND*(A+B) $ DE=-FE/FEP $ EE=EE+DE
      IF (ABS(DE).GT.0.000001) GO TO 60 $ E(M)=EE*RAD
   68 IF (II.EQ.2) E(M)=1.0/(SIN((E(M)-D(M))/RAD)*SQRT(G2*TAN(E(M)/RAD)/
     1TAN((E(M)-D(M))/RAD)-G5))
      IF ((IK.EQ.6).AND.(DD.EQ.0.0)) E(M)=SQRT((1.0+G5*MM)/(G*MM-G5))
      IF (II.EQ.3) E(M)=G6*(MM*SIN(E(M)/RAD)**2-G7)
   70 CONTINUE $ NP=NN+NA $ WRITE (6,2020) (E(M),M=1,NP)
   80 IF (L.LT.50) GO TO 90 $ WRITE (6,2000) T(II),G,(D(M),M=1,NF) $ L=0
   90 CONTINUE $ I=I+1 $ IF((I.EQ.10).OR.(J(I).EQ.0))GO TO 100 $GO TO 50
  100 CONTINUE $ RETURN

 2000 FORMAT (1H1,40X,A6,37H FOR AN OBLIQUE SHOCK WAVE.  GAMMA = ,F5.2//
     153X,27HFLOW TURNING ANGLE, DEGREES//4X,2HM1,1X,21F6.1/1H )
 2010 FORMAT (1H ,F5.2)
 2020 FORMAT (1H+,6X,21F6.2)
      END
```

The input to subroutine OSHOCKL comprises G, the value of the specific heat ratio γ, J(1), J(2), ..., J(9), which specify the number of entries in each section of the table, and DM(1), etc., which specify the Mach number increment in each section of the table [see Section 4–4(b)]. The same input data are required by subroutine OSHOCKP, with the two additional parameters II and IJ. II = 1 denotes that the wave angle ε is to be calculated, II = 2 specifies that the Mach number M_2 after the shock wave is to be computed, and II = 3 denotes that the pressure ratio p_2/p_1 is to be determined. IJ = 1 specifies the weak shock wave solution, and IJ = 2 denotes the strong shock wave solution. Each combination of II and IJ results in a complete table; six combinations are possible. Table C.12 corresponds to the following input data.

II = 1, IJ = 1, G = 1.4, J(1) = 100, J(2) = 80, J(3) = 28, J(4) = 20, J(5) = 10, J(6) = 10, J(7) = 0, DM(1) = 0.01, DM(2) = 0.02, DM(3) = 0.05, DM(4) = 0.25, DM(5) = 0.5, DM(6) = 1.0

The same data apply to subroutine OSHOCKL, except that II and IJ are not specified.

Example 7.8.   Air having an initial Mach number $M_1 = 3.0$, a free-stream static pressure $p_1 = 1.0133 \cdot 10^5$ N/m², and an initial static temperature $t_1 = 300$ K is deflected through an angle $\delta = 15$ deg by a frictionless surface (see Fig. 7.16). Assuming that a weak oblique shock wave occurs, calculate (a) the static pressure p_2, (b) the density ρ_2, (c) the static temperature t_2, (d) the change in stagnation pressure ΔP, (e) the velocity V_2, and (f) the Mach number M_2.

Solution

From Figs. 7.17 and 7.18, for $\delta = 15$ deg and $M_1 = 3.0$, $\varepsilon = 32.2$ deg and $M_2 = 2.25$.
(a)   From equation 7.90, for $M_1 = 3.0$ and $\varepsilon = 32.2$ deg,

$$\frac{p_2}{p_1} = \frac{2\gamma}{\gamma + 1}\left(M_1{}^2 \sin^2 \varepsilon - \frac{\gamma - 1}{2\gamma}\right) = \frac{2.8}{2.4}\left[(3)^2 \sin^2 (32.2) - \frac{0.4}{2.8}\right] = 2.787$$

$$p_2 = 2.787(1.0133)10^5 = 2.824 \cdot 10^5 \text{ N/m}^2 \ (40.96 \text{ lbf/in.}^2)$$

(b)   From equation 7.85,

$$\frac{\rho_2}{\rho_1} = \frac{\tan \varepsilon}{\tan \beta} = \frac{\tan \varepsilon}{\tan (\varepsilon - \delta)} = \frac{\tan (32.2)}{\tan (32.2 - 15)} = 2.034$$

From equation 7.23,

$$\rho_1 = \frac{p_1}{Rt_1} = \frac{(1.0133)10^5}{(287.04)(300)} = 1.1767 \text{ kg/m}^3$$

$$\rho_2 = 2.034 \, \rho_1 = 2.034(1.1767) = 2.393 \text{ kg/m}^3 \ (0.1494 \text{ lbm/ft}^3)$$

(c)   From equation 7.23,

$$\frac{t_2}{t_1} = \frac{p_2}{p_1}\frac{\rho_1}{\rho_2} = \frac{2.80}{2.034} = 1.377$$

$$t_2 = 1.377(300) = 413.1 \text{ K} \ (743.6 \text{ R})$$

(d)   From Table C.6 (isentropic flow), for $M_1 = 3.0$ and $M_2 = 2.25$, $p_1/P_1 = 0.027224$ and $p_2/P_2 = 0.086492$. Consequently,

$$P_1 = \frac{(1.0133)10^5}{0.027224} = 37.22 \cdot 10^5 \text{ N/m}^2$$

$$\frac{P_2}{P_1} = \frac{P_2}{p_2}\frac{p_2}{p_1}\frac{p_1}{P_1} = \frac{(2.80)(0.027224)}{(0.086492)} = 0.8813$$

$$P_2 = (0.8813)(37.22)10^5 = 32.80 \cdot 10^5 \text{ N/m}^2$$

$$\Delta P = P_2 - P_1 = (32.80 - 37.22)10^5 = -4.42 \cdot 10^5 \text{ N/m}^2 \text{ (64.11 lbf/in.}^2)$$

(e)   From equation 7.92,

$$\frac{V_2}{V_1} = \frac{\sin \varepsilon}{\sin \beta}\left[\frac{2}{(\gamma + 1)M_1^2 \sin^2 \varepsilon} + \frac{\gamma - 1}{\gamma + 1}\right]$$

$$\frac{V_2}{V_1} = \frac{\sin (32.2)}{\sin (32.2 - 15)}\left[\frac{2}{2.4(3)^2 \sin^2 (32.2)} + \frac{0.4}{2.4}\right] = 0.8880$$

$$V_1 = M_1 a_1 = M_1(\gamma R t_1)^{1/2} = 3.0[1.4(287.04)(300)]^{1/2} = 1042 \text{ m/s}$$

$$V_2 = (0.8880)(1042) = 925.3 \text{ m/s} \text{ (3036 ft/sec)}$$

(f) $a_2 = (\gamma R t_2)^{1/2} = [1.4(287.04)(413.1)]^{1/2} = 407.4 \text{ m/s}$

$$M_2 = \frac{V_2}{a_2} = \frac{925.3}{407.4} = 2.271$$

7-7(c)   The Rankine-Hugoniot Equation for an Oblique Shock Wave

In Section 7-4(c), the density ratio ρ_2/ρ_1 across a normal shock wave is related to the corresponding pressure ratio p_2/p_1 (see equation 7.55, the *Rankine-Hugoniot equation*). That equation is obtained by combining the equation of state for a *perfect gas* with the continuity, momentum, and energy equations. If V_1 and V_2 are replaced by the normal velocity components V_{N1} and V_{N2}, respectively, for an oblique shock wave, an identical relationship is obtained between ρ_2/ρ_1 and p_2/p_1, for an oblique shock wave. Hence, the Rankine-Hugoniot equation, equation 7.55, is valid for an *oblique shock wave*. Consequently, the pressure ratio, density ratio, and temperature ratio across an *oblique* shock wave are related in exactly the same manner as they are for a *normal* shock wave.

7-7(d)   Prandtl's Velocity Equation for an Oblique Shock Wave

For a normal shock wave [see equation 7.58, Section 7-4(d)], the Prandtl velocity equation is

$$V_1 V_2 = a^{*2} \tag{7.97}$$

A similar relationship may be derived for the product $V_{N1} V_{N2}$ for an oblique shock wave.[5] Thus,

$$V_{N1} V_{N2} = a^{*2} - \frac{\gamma - 1}{\gamma + 1} V_T^2 \tag{7.98}$$

Equation 7.98 may be regarded as the general form of the *Prandtl* equation derived in Section 7–4(d). For a normal shock wave, $V_T = 0$, and equation 7.98 reduces to equation 7.97.

7–7(e) Analysis of Oblique Shock Waves by Employing the Normal Shock Wave Relationships

The governing equations for an oblique shock wave, expressed in terms of the normal velocity components V_{N1} and V_{N2}, have the identical forms as the normal shock wave equations (see Section 7–6). Consequently, all of the results obtained for the normal shock wave are applicable to the normal components of the velocities V_1 and V_2 in front of and in back of an oblique shock wave, respectively. In particular, Table C.11 for the *normal shock wave* may be employed for relating the flow properties in front of and in back of an oblique shock wave.

To apply Table C.11 to the oblique shock wave, *define*

$$M_1' \equiv M_1 \sin \varepsilon \tag{7.99}$$

Then all of the property ratios such as p_2/p_1, etc., in Table C.11 are valid for the oblique shock wave. From Fig. 7.16, it is seen that

$$M_2 = \frac{M_2'}{\sin (\varepsilon - \delta)} \tag{7.100}$$

where M_2' is the Mach number corresponding to M_1' in the normal shock wave tables. Hence, M_2 for the oblique shock wave may be determined from the normal shock wave tables and equation 7.100.

Obviously, the oblique shock wave angle ε must be known before the normal shock wave tables may be employed. Figure 7.17 and Table C.12 present ε as a function of M_1 and δ for $\gamma = 1.4$. Consequently, by employing Table C.11 in conjunction with either Fig. 7.17 or Table C.12, the properties of oblique shock waves may be determined.

Example 7.9. Solve Example 7.8 employing the normal shock wave tables.

Solution

From Table C.12, for $\delta = 15$ deg and $M_1 = 3.0$, $\varepsilon = 32.26$ deg. From equation 7.99,

$$M_1' = M_1 \sin \varepsilon = 3.0 \sin (32.26) = 1.6013$$

From Table C.11, for $M_1' = 1.6013$,

$$M_2' = 0.6681, \quad \frac{p_2}{p_1} = 2.8249, \quad \frac{\rho_2}{\rho_1} = 2.0339, \quad \frac{t_2}{t_1} = 1.3889, \quad \frac{P_2}{P_1} = 0.8947$$

From Example 7.8, $\rho_1 = 1.1767$ kg/m^3, $P_1 = 37.22 \cdot 10^5$ N/m^2, and $V_1 = 1042$ m/s.
(a) $p_2 = (2.8249)(1.0133)10^5 = 2.862 \cdot 10^5$ N/m^2 (41.52 lbf/in.^{2})
(b) $\rho_2 = (2.0339)(1.1767) = 2.393$ kg/m^3 (0.1494 lbm/ft^3)
(c) $t_2 = (1.3889)(300) = 416.7$ K (750.1 R)
(d) $P_2 = (0.8947)(37.22)10^5 = 33.31 \cdot 10^5$ N/m^2
     $\Delta P = P_2 - P_1 = (33.31 - 37.22)10^5 = -3.91 \cdot 10^5$ N/m^2 (-56.71 lbf/in.^{2})
(f) From equation 7.100,

$$M_2 = \frac{M_2'}{\sin (\varepsilon - \delta)} = \frac{0.6681}{\sin (32.26 - 15)} = 2.252$$

(e) $a_2 = (\gamma R t_2)^{1/2} = [1.4(287.04)(416.7)]^{1/2} = 409.2$ m/s
$V_2 = M_2 a_2 = (2.252)(409.2) = 921.5$ m/s (3023 ft/sec)

7–8 OBLIQUE SHOCK WAVES IN IMPERFECT GASES

In Section 7–7, the properties of oblique shock waves in a *perfect gas* are discussed. The influence of *imperfect gas* effects on the flow properties for an oblique shock wave are considered in the present section.

Substituting the continuity equation, equation 7.80, into the momentum equation, equation 7.81, and the energy equation, equation 7.87, we obtain

$$p_2 = p_1 + \rho_1 V_{N1}{}^2 \left(1 - \frac{\rho_1}{\rho_2}\right) \tag{7.101}$$

and

$$h_2 = h_1 + \frac{V_{N1}{}^2}{2}\left[1 - \left(\frac{\rho_1}{\rho_2}\right)^2\right] \tag{7.102}$$

The continuity equation, equation 7.80, may be written in the form

$$V_{N2} = \frac{\rho_1 V_{N1}}{\rho_2} \tag{7.103}$$

The equations of state for the gas are given by equations 7.88, which are repeated here.

$$\rho = \rho(p,t) \quad \text{and} \quad h = h(p,t) \tag{7.104}$$

Equations 7.101 to 7.104 comprise a set of five equations in terms of the flow properties in front of the oblique shock wave, at point 1, and p_2, ρ_2, t_2, h_2, and V_{N2} in back of the oblique shock wave. The aforementioned set of equations is identical to the set developed in Section 7–5 for a normal shock wave in an imperfect gas, with V_1 and V_2 in those latter equations replaced by V_{N1} and V_{N2}, respectively. Consequently, the numerical procedure employed in Example 7.7 may be applied directly to the oblique shock wave.

However, because of the additional degree of freedom permitted by the *flow deflection angle* δ, the *wave angle* ε must also be determined iteratively. If a value for ε is assumed, equations 7.73, 7.74, 7.72, and 7.76, respectively, yield

$$V_{N1} = V_1 \sin \varepsilon \tag{7.105}$$

$$V_T = V_{T1} = V_{T2} = V_1 \cos \varepsilon \tag{7.106}$$

$$V_2 = (V_T{}^2 + V_{N2}{}^2)^{1/2} \tag{7.107}$$

$$\varepsilon = \delta + \sin^{-1}\left(\frac{V_{N2}}{V_2}\right) \tag{7.108}$$

For an assumed value of ε, equations 7.105 and 7.106 determine V_{N1} and V_T. With V_{N1} specified, the set of equations for p_2, ρ_2, h_2, t_2, and V_{N2} may be solved by iteration. Then equation 7.107 may be employed for determining V_2, and ε may be computed from equation 7.108. If the assumed and computed values of ε agree to within an acceptable tolerance, the problem is solved. If not, a new value of ε is assumed, and the above procedure is repeated until satisfactory convergence is attained.

Example 7.10 demonstrates that the numerical method presented in this section is easy to apply, and achieves rapid convergence to the desired value of ε.

Example 7.10. A symmetrical wedge having a 30 deg half-angle is placed on the centerline of the wind tunnel discussed in Example 4.6. The flow properties in the test section are: $H = 2252.414$ kJ/kg, $T = 2000$ K, $M = 6.0$, $p_1 = 1696.4$ N/m^2, $h_1 = 274.445$ kJ/kg, $t_1 = 273.23$ K, $V_1 = 1989.0$ m/s, $\rho_1 = 0.021630$ kg/m^3, and $a_1 = 331.42$ m/s. Calculate the shock wave angle ε and the flow properties behind the oblique shock wave, assuming that air is a thermally perfect gas, taking into account the variation of the specific heat of the air with the temperature.

Solution

The free-stream conditions and the flow deflection angle δ are specified. For a first approximation, determine the properties of the oblique shock wave for a perfect gas. From Table C.4, at $t_1 = 273.23$ K, $\gamma = 1.4005$; assume $\gamma = 1.40$. From Table C.12, the wave angle $\varepsilon = 40.79$ deg. From equation 7.99,

$$M_1' = M_1 \sin \varepsilon = 6.0 \sin 40.79 = 3.91973$$

From Table C.11, for $M_1' = 3.91973$,

$$M_2' = 0.43719, \qquad \frac{\rho_2}{\rho_1} = 4.5266, \qquad \frac{t_2}{t_1} = 3.9232, \qquad \frac{p_2}{p_1} = 17.760$$

$$V_{N1} = V_1 \sin \varepsilon = 1989.0 \sin 40.79 = 1299.4 \text{ m/s}$$

From equation 7.100,

$$M_2 = \frac{M_2'}{\sin (\varepsilon - \delta)} = \frac{0.43719}{\sin (40.79 - 30)} = 2.3353$$

$$p_2 = 17.760(1696.4) = 30,128 \text{ N/m}^2$$

$$t_2 = 3.9232(273.23) = 1071.9 \text{ K}$$

$$\rho_2 = 4.5266(0.021630) = 0.097910 \text{ kg/m}^3$$

$$a_2 = (\gamma R t_2)^{1/2} = [1.4(287.04)(1071.9)]^{1/2} = 656.32 \text{ m/s}$$

$$V_2 = M_2 a_2 = 2.3353(656.32) = 1532.7 \text{ m/s}$$

As pointed out earlier, an iterative solution is required for taking into account the *imperfect gas* effects. The initial conditions and the flow deflection angle are specified. A trial value for ε is chosen, then the problem becomes analogous to the iterative procedure employed in Example 7.7 for a normal shock wave in a real gas. When the iteration has converged, ε is calculated from equation 7.108 and compared with the assumed value of ε. The entire procedure is then repeated until the desired convergence is attained. The numerical procedure is summarized below.

Assume $\varepsilon = 40.79$ deg. From equations 7.105 and 7.106,

$$V_{N1} = V_1 \sin \varepsilon = 1989.0 \sin 40.79 = 1299.4 \text{ m/s}$$

$$V_T = V_{T1} = V_1 \cos \varepsilon = 1989.0 \cos 40.79 = 1505.9 \text{ m/s}$$

From equation 7.101,

$$p_2 = 1696.4 + (0.021630)(1299.4)^2 \left(1 - \frac{0.021630}{\rho_2} \right)$$

$$p_2 = 1696.4 + 36,521 \left(1 - \frac{0.021630}{\rho_2} \right) \text{ N/m}^2 \qquad \text{(a)}$$

From equation 7.102,

$$h_2 = 274.445 + \frac{(1299.4)^2}{2(1000)}\left[1 - \left(\frac{0.021630}{\rho_2}\right)^2\right]$$

$$h_2 = 274.445 + 844.220\left[1 - \left(\frac{0.021360}{\rho_2}\right)^2\right] \text{kJ/kg} \qquad (b)$$

Substituting $\rho_2 = 0.097910$ kg/m³ into equations (a) and (b) above, we obtain $p_2 = 30{,}149$ N/m² and $h_2 = 1077.463$ kJ/kg, respectively. From Table C.4, for $h_2 = 1077.463$ kJ/kg, $t_2 = 1026.4$ K. From the equation of state for a thermally perfect gas,

$$\rho_2 = \frac{p_2}{Rt_2} = \frac{30{,}149}{(287.04)(1026.4)} = 0.10233 \text{ kg/m}^3 \qquad (c)$$

Equation (c) gives the value of ρ_2 for the second trial. The second trial is performed in the same manner as the first trial and yields a second calculated value of $\rho_2 = 0.10321$ kg/m³. As is done in Example 7.7, the secant method [see Appendix A–4(b)] is applied to the two sets of assumed and calculated values of ρ_2 to yield a third assumed value of $\rho_2 = 0.10343$ kg/m³. Employing the latter value of ρ_2, the third calculated value of $\rho_2 = 0.10341$ kg/m³, which is close enough to the third assumed value of ρ_2. Hence, convergence has been attained.

The values of ρ_2, p_2, h_2 and t_2 for the three trials are tabulated below.

Trial	ρ_2, kg/m³	p_2, N/m²	h_2, kJ/kg	t_2, K	ρ_2, kg/m³
1	0.097910	30,149	1077.463	1026.4	0.10233
2	0.10233	30,498	1080.946	1029.5	0.10321
3	0.10343	30,580	1081.744	1030.2	0.10341

From equation 7.103,

$$V_{N2} = \frac{\rho_1 V_{N1}}{\rho_2} = \frac{(0.021630)(1299.4)}{0.10343} = 271.74 \text{ m/s}$$

From equation 7.107,

$$V_2 = (V_T{}^2 + V_{N2}{}^2)^{1/2} = [(1505.9)^2 + (271.74)^2]^{1/2} = 1530.2 \text{ m/s}$$

$$a_2 = (\gamma Rt_2)^{1/2} = [1.4(287.04)(1030.2)]^{1/2} = 643.42 \text{ m/s}$$

$$M_2 = \frac{V_2}{a_2} = \frac{1530.2}{643.42} = 2.3782$$

From equation 7.108,

$$\varepsilon = \delta + \sin^{-1}\left(\frac{V_{N2}}{V_2}\right) = 30 + \sin^{-1}\left(\frac{271.74}{1530.2}\right) = 40.23 \text{ deg}$$

The computed value of $\varepsilon = 40.23$ deg agrees reasonably well with the assumed value of $\varepsilon = 40.79$ deg. If greater accuracy is desired, the entire procedure may be repeated for $\varepsilon = 40.23$ deg.

The following table compares the results for the *imperfect gas* with the results for the *perfect gas*.

Property	Perfect Gas	Imperfect Gas	Error, Percent
ε, deg	40.79	40.23	1.39
p_2, N/m^2	30,128	30,580	-1.48
t_2, K	1071.9	1030.2	4.05
ρ_2, kg/m^3	0.097910	0.10343	-5.34
V_2, m/s	1532.7	1530.2	0.16
M_2	2.3353	2.3782	-1.80

The errors tabulated above are smaller than those obtained in Example 7.7 for a normal shock wave, because there is a smaller variation in the specific heat of the air because of the smaller temperature range involved.

7–9 THE HODOGRAPH SHOCK POLAR

In the analysis of steady two-dimensional planar isentropic flow, a transformation may be made that interchanges the role of the independent variables (x and y) and the dependent variables (u and v). From the transformation we obtain a set of partial differential equations for x and y as functions of u and v; the transformation is known as the *hodograph transformation*.[1-4, 8] The results of such an analysis are often presented in the uv plane, which is called the *hodograph plane*, instead of in the xy plane. Although the hodograph method is not employed extensively in this book, it is employed for presenting the solutions to two important flows: the oblique shock wave, and the Prandtl-Meyer expansion wave (see Section 8–7).

Since the tangential velocity components $V_{T1} = V_{T2} = V_T$, the relationship among V_1, V_2, V_{N1}, V_{N2}, ε and δ for an oblique shock wave may be illustrated graphically by combining the velocity triangles before and after the shock wave in the manner illustrated in Fig. 7.22. The result is a right triangle having its hypotenuse equal to V_1 and closed by the sides V_T and V_{N1}, where V_{N1} is perpendicular to V_T. Let u_2 and v_2 denote the Cartesian coordinates of the final velocity V_2; the combined velocity diagrams may then be constructed in the uv, or hodograph, plane, as illustrated in Fig. 7.23.

It is apparent from Fig. 7.23 that, if V_1 is held fixed while the shock wave angle ε is varied, a series of right triangles will be formed, each of which has V_1 as its hypotenuse. Simultaneously, the magnitude and direction of the velocity vector V_2 changes with the shock wave angle ε, and its end point describes a curve. The locus of the positions of the end point of the velocity vector V_2 as ε is varied is called the *shock polar*. Each different value of V_1 gives rise to a different shock polar.

It is easier to develop the shock polar by plotting the equation for the locus of the end point of V_2 than by employing the graphical construction outlined above. It is customary to express the equation for the shock polar in terms of u_2 and v_2, the Cartesian components of V_2. Let the u axis be coincident with the direction of V_1, as in Fig. 7.23; then

$$V_T = V_{T1} = V_{T2} = V_1 \cos \varepsilon \qquad (7.109)$$

$$V_{N1} = V_1 \sin \varepsilon \qquad (7.110)$$

$$V_{N2} = V_1 \sin \varepsilon - \frac{v_2}{\cos \varepsilon} \qquad (7.111)$$

Hence,

$$\tan \varepsilon = \frac{V_1 - u_2}{v_2} \qquad (7.112)$$

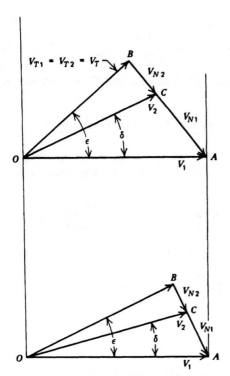

Figure 7.22 Effect of the wave angle ε on the velocity V_2 leaving an oblique shock wave for a fixed value of the velocity V_1 entering the shock wave.

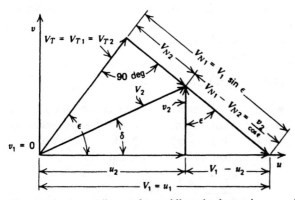

Figure 7.23 Vector diagram for an oblique shock wave in terms of the Cartesian velocity components u and v of the velocity vectors.

Substituting for V_T, V_{N1}, and V_{N2} from equations 7.109, 7.110, and 7.111, respectively, into equation 7.98, yields

$$V_1{}^2 \sin^2 \varepsilon - V_1 v_2 \tan \varepsilon = a^{*2} - \frac{\gamma - 1}{\gamma + 1} V_1{}^2 \cos^2 \varepsilon \qquad (7.113)$$

To eliminate ε from equation 7.113, the conventional trigonometric expressions for $\sin^2 \varepsilon$ and $\cos^2 \varepsilon$ are combined with equation 7.112. Thus,

$$\sin^2 \varepsilon = \frac{\tan^2 \varepsilon}{1 + \tan^2 \varepsilon} = \frac{(V_1 - u_2)^2}{v_2{}^2 + (V_1 - u_2)^2} \qquad (7.114)$$

and

$$\cos^2 \varepsilon = 1 - \sin^2 \varepsilon = \frac{v_2{}^2}{v_2{}^2 + (V_1 - u_2)^2} \qquad (7.115)$$

Substituting equations 7.112, 7.114, and 7.115 into equation 7.113 and rearranging gives

$$v_2{}^2 \left(a^{*2} + \frac{2}{\gamma + 1} V_1{}^2 - V_1 u_2 \right) = (V_1 - u_2)^2 (V_1 u_2 - a^{*2}) \qquad (7.116)$$

Equation 7.116 gives the relationship between u_2, v_2, and V_1 for an oblique shock wave. Equation 7.116 may be made dimensionless by dividing each of its terms by a^{*4}. The result is

$$\left(\frac{v_2}{a^*} \right)^2 \left[1 + \frac{2}{\gamma + 1} \left(\frac{V_1}{a^*} \right)^2 - \frac{V_1}{a^*} \frac{u_2}{a^*} \right] = \left(\frac{V_1}{a^*} - \frac{u_2}{a^*} \right)^2 \left(\frac{V_1}{a^*} \frac{u_2}{a^*} - 1 \right) \qquad (7.117)$$

Let $M_1^* = V_1/a^*$, $u_2^* = u_2/a^*$, and $v_2^* = v_2/a^*$; then equation 7.117 transforms to

$$v_2^{*2} \left(1 + \frac{2}{\gamma + 1} M_1^{*2} - M_1^* u_2^* \right) = (M_1^* - u_2^*)^2 (M_1^* u_2^* - 1) \qquad (7.118)$$

Solving equation 7.118 for v_2^{*2} yields

$$v_2^{*2} = (M_1^* - u_2^*)^2 \, \frac{M_1^* u_2^* - 1}{\dfrac{2}{\gamma + 1} M_1^{*2} - M_1^* u_2^* + 1} \qquad (7.119)$$

Equation 7.119 is the equation for the *shock polar*. By means of that equation, the dimensionless velocity component v_2^* may be plotted as a function of u_2^*, for a given value of M_1^*. Each value of M_1^*, of course, gives rise to a separate curve. Figure 7.24 illustrates the case where $M_1^* = 2.0$, for $\gamma = 1.4$. The curve is a *strophoid* and is symmetrical with respect to the u_2^* axis.

When $v_2^* = 0$, equation 7.119 reduces to

$$(M_1^* - u_2^*)^2 (M_1^* u_2^* - 1) = 0 \qquad (7.120)$$

Equation 7.120 shows that when $v_2^* = 0$, the shock polar crosses the u_2^* axis at the points D and E in Fig. 7.24. If v_2^* becomes infinite, then $M_1^* u_2^*$ approaches the value $1 + 2M_1^{*2}/(\gamma + 1)$. Hence, for all values of $u_2^* > M_1^*$, the dimensionless velocity component u_2^* approaches the asymptote defined by the equation

$$u_2^* = \frac{1}{M_1^*} + \frac{2}{\gamma + 1} M_1^* \qquad (7.121)$$

Refer to Fig. 7.24. The intersection of a ray drawn from the origin, such as OC, with the shock polar curve gives the values of v_2^* and u_2^* for the flow deflection angle δ, which is the angle made by the ray and the u_2^* axis. If $u_2^* = M_1^*$, then $v_2^* = 0$, which indicates that the oblique shock wave is very weak and there is no deflection of the flow; the initial dimensionless speed $M_1^* = V_1/a^*$ is then given by the ray OD (see Fig. 7.24). If $M_1^* u_2^* = 1$, then again $v_2^* = 0$, as shown earlier. In that case, $V_1 V_2 = a^{*2}$, and the condition corresponds to a normal shock wave. It is represented by point E in Fig. 7.24.

A ray drawn in the direction of OC intersects the shock polar in the three points A, B, and C. Since the intersection of a ray from the origin with the shock polar represents the values of u_2^* and v_2^*, the point C has no physical significance because

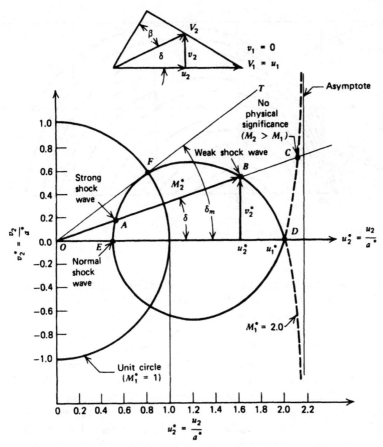

State A, Strong shock wave
  B, Weak shock wave (one that usually occurs)
  C, $M_2 > M_1$ (no physical significance)
  D, $u_2^* = M_1^*$ ($v_2^* = 0$), (no shock wave)
  E, $V_1 V_2 = a^{*2}$ ($v_2^* = 0$, normal shock wave)

Figure 7.24   Shock polar for $M_1^* = 2.0$ and $\gamma = 1.4$.

it corresponds to the impossible condition $M_2 > M_1$. Consequently, all portions of the shock polar to the right of D have no physical significance, and only that portion of the shock polar, the *closed loop*, to the left of point D is considered to be the shock polar.

The intersection points A and B made by the ray OC with the shock polar correspond to the two possible solutions of the shock polar equation (equation 7.119) that satisfy the continuity, momentum, and energy equations for the oblique shock wave. Point A corresponds to a strong oblique shock wave, and point B corresponds to a weak oblique shock wave. Experiments show that the weak shock wave, that corresponding to ray OB, is generally produced.

Figure 7.25 presents the complete family of hodograph shock polars for $\gamma = 1.4$.

For a given value of M_1^* there is a maximum flow deflection angle δ_m that corresponds to the angle made by the ray from the origin, which is tangent to the shock polar, as illustrated by the ray OT in Fig. 7.24. If the required flow deflection angle for a given value of M_1^* is larger than δ_m, then there cannot be an oblique shock

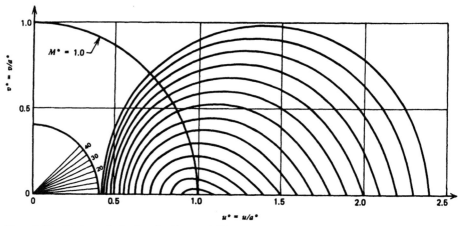

Figure 7.25 Hodograph shock polar for $\gamma = 1.40$.

wave attached to the body causing the flow deflection. Instead, a detached shock wave will form ahead of the body, as illustrated in Fig. 7.21. The shock wave detaches from the corner O and moves upstream to some position, such as point A, in Fig. 7.21, where equilibrium conditions are satisfied. Since there is no deflection of the flow at point A in Fig. 7.21, the shock wave at that point is a normal shock wave, and the flow downstream to it is subsonic. A subsonic region is formed in the vicinity of the corner, and the streamlines are no longer parallel to OC, but each has a different curvature. Consequently, the pressure is nonuniform in the region $BAOC$. Because of the variation in pressure, as one proceeds outward along the shock wave from point A in Fig. 7.21, a point is reached where the flow behind the shock wave is sonic. Beyond that point, the downstream flow is supersonic. To satisfy the above conditions, the shock wave must be curved, as illustrated in Fig. 7.21. The problem of establishing the exact position and configuration of the shock wave AB in Fig. 7.21 is extremely complicated.

Example 7.11. Air at a Mach number of 3.0 and a static temperature of 290 K is deflected through an angle $\delta = 20$ deg by a frictionless surface. Using the hodograph shock polar presented in Fig. 7.25, determine (a) the Mach number, and (b) the velocity behind the attached oblique shock wave. Assume $\gamma = 1.40$.

Solution

(a) From Table C.6 (isentropic flow), for $M_1 = 3.0$, $M_1^* = 1.964$. In Fig. 7.25, locate the shock polar passing through $M_1^* = u_1^* = 1.964$ and $v_1^* = 0.0$. Follow this curve upward until it intersects a straight line drawn from the origin at 20 deg. At that point, determine u_2^* and v_2^*. Thus,

$$u_2^* = 1.53 \quad \text{and} \quad v_2^* = 0.56$$

$$M_2^* = (u_2^{*2} + v_2^{*2})^{1/2} = [(1.53)^2 + (0.56)^2]^{1/2} = 1.6293$$

The above result may also be obtained directly by measuring the distance from the origin to point 2. From Table C.6, for $M_2^* = 1.6293$,

$$M_2 = 1.9920$$

As a check, from Fig. 7.18, for $M_1 = 3.0$, $M_2 = 1.99$.
(b) From Table C.6 (isentropic flow), for $M_1 = 3.0$ and $M^* = 1.0$, $t_1/T = 0.35714$

and $t^*/T = 0.83333$. Thus,

$$T = \frac{290}{0.35714} = 812.0 \text{ K}$$

$$t^* = (0.83333)(812.0) = 676.7 \text{ K}$$

$$a^* = (\gamma R t^*)^{1/2} = [1.4(287.04)(676.7)]^{1/2} = 521.5 \text{ m/s}$$

$$V_2 = M_2^* a^* = (1.6293)(521.5) = 849.6 \text{ m/s } (2787.5 \text{ ft/sec})$$

7–10 SUPERSONIC FLOW OVER WEDGES AND CONES

In the preceding sections the inherent features of an oblique shock wave that forms in a *two-dimensional planar* concave corner are discussed. In the present section, the discussion is extended to the supersonic flow over a stationary two-dimensional planar wedge in a free stream. In addition, the characteristic features for the conical shock wave attached to the tip of a cone in a supersonic flow are analyzed; the latter is a *two-dimensional axisymmetric flow*.

7–10(a) Flow over a Wedge

Consider the case where an infinitely long smooth symmetrical wedge is immersed in a uniform supersonic flow with the vertex of the wedge facing upstream, with the central plane of the wedge parallel to the approaching flow; the configuration is illustrated schematically in Fig. 7.26. The *wedge angle*, denoted by δ_w, is defined as one-half of the total included angle between the top and bottom surfaces of the wedge. Hence, δ_w is the same as the flow deflection angle δ, defined in Section 7–6. Because the wedge is assumed to be symmetrical about the plane 0–0, the flows over its top and bottom surfaces are symmetrical (see Fig. 7.26). Consequently, the flow pattern for the complete wedge is obtained by studying the flow over only one of its surfaces; the width of the wedge perpendicular to the plane of the paper is assumed to be infinite. If the wedge is unsymmetrical with respect to the direction of the incident flow, the flows over the top and bottom surfaces must be studied separately, because the flow deflection angle is different on the two surfaces.

It is evident that the supersonic flow over a symmetrical wedge with an attached oblique shock wave at its vertex is quite similar to the supersonic flow into a frictionless concave corner with an oblique shock wave emanating from that corner, discussed in the preceding sections. It will be recalled that for a given value of the incident flow Mach number M_1, there is a maximum value for the flow deflection angle δ_m that will keep the shock wave attached to the corner; the condition corresponds to that denoted by point F in the shock polar (Fig. 7.24). The corresponding value of the maximum wave angle ε_m is given by equation 7.95, or it may be read from Fig. 7.20. For the wedge, the conditions are completely analogous. If the wedge angle δ_w is smaller than or equal to the maximum flow deflection angle δ_m, the oblique shock wave will become attached to the vertex of the wedge, and the supersonic flow will be deflected through the flow deflection angle $\delta = \delta_w$. After passing through the oblique shock wave, the flow is deflected so that the streamlines are parallel to the surfaces of the wedge. Because of the compression of the gas by the oblique shock wave, the streamlines flowing parallel to the wedge are spaced closer together than they are in front of the oblique shock wave. The properties of the flow in back of the oblique shock wave and the wave angle ε (see Fig. 7.26a) depend on the Mach number M_1 and the wedge angle δ_w. For gases with $\gamma = 1.4$, it is pointed

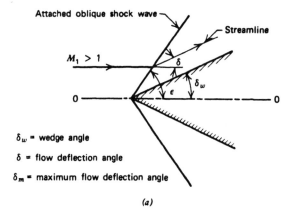

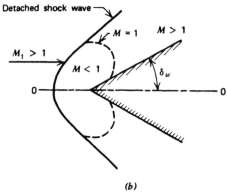

Figure 7.26 Flow over a symmetrical wedge. (a) Attached oblique shock wave, $\delta_w < \delta_m$. (b) Detached shock wave, $\delta_w > \delta_m$.

out in Section 7–7(b) that the limiting value of $\delta_m = 45.37$ deg and corresponds to the condition $M_1 = \infty$. Consequently, if the included angle of the symmetrical wedge is larger than 90.74 deg (i.e., $2\delta_m$), the oblique shock wave cannot be attached to the apex of the wedge for any free-stream Mach number M_1.

If the wedge angle δ_w is larger than δ_m for a given value of M_1, the oblique shock wave becomes detached, as illustrated in Fig. 7.26b. The distance that the *detached shock wave*, also called a *head* or *bow wave*, moves upstream from the vertex of the wedge depends on M_1 and δ_w. The portion of the detached shock wave immediately in front of the apex of the wedge is normal to M_1, and the flow characteristics in that region approximate those for a normal shock wave. On either side of that small normal shock wave, the shock wave becomes oblique and the strength of the shock wave decreases.

Because the strength of the shock wave increases as the wave angle ε increases (see equation 7.90), the *shock drag* for a body in a supersonic flow increases with the wave angle ε. Hence, if a body is to have a low drag in a supersonic stream, it should not produce a shock wave with a large wave angle ε; that is, the nose of the body should be sharp rather than blunt so that the flow is turned through only a small flow deflection angle.

In an actual case the wedge would be of finite length and form the nose of some object. Consequently, the flow after passing over the wedge is deflected again by the changes in the shape of the wetted surface downstream from the nose, as illustrated

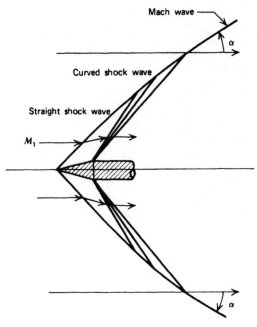

Figure 7.27   Flow over a finite wedge.

schematically in Fig. 7.27, where it is shown that continuous expansion waves emanate from the wedge-plate junction. Because of the flow deflection at that junction, the pressure and velocity conditions behind the shock wave are affected. For that reason experiments with finite wedges show that the shock wave attached to the apex becomes curved, that its inclination approaches the Mach angle α in a relatively short distance from the wedge, and that the flow at a small distance from the wedge is essentially undisturbed by the presence of the wedge. If the flow after passing through the shock wave is supersonic, additional shock waves may form, owing to the deflections of that flow. As a matter of fact, irregularities on the surface of the wedge may caused such shock waves. In most cases the latter are very weak, and each shock wave is inclined at an angle not greatly different from the Mach angle corresponding to the local flow Mach number.

For an unsymmetrical wedge, which is equivalent to a symmetrical wedge at an angle of attack α with respect to the free stream, the oblique shock wave is attached as long as neither of the two flow deflection angles δ_1 and δ_2, illustrated in Fig. 7.28, is larger than the maximum flow deflection angle δ_m corresponding to the Mach number M_1 for the incident flow.

The procedure for determining the flow properties for the flow over a wedge is similar to that employed in Sections 7–7 and 7–8 for determining the flow properties for the supersonic flow into a concave corner.

1. For the given value of M_1, determine the value of δ_m from Fig. 7.20, or by solving equations 7.94 and 7.95.
2. If $\delta_w < \delta_m$, the shock wave is attached to the wedge and the properties p_2, ρ_2, M_2, etc., may be calculated by the methods presented in Sections 7–7 and 7–8.

Example 7.12.   Standard sea-level air approaches a symmetrical wedge with $M_1 = 2.0$. The included angle of the wedge is 30 deg. Calculate (a) the shock wave angle ε, (b) the Mach number M_2 in back of the shock wave, (c) the static pressure of the air

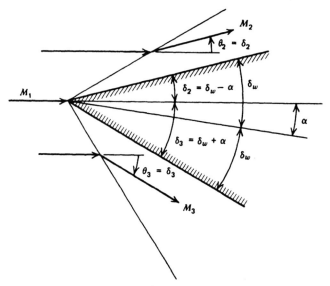

Figure 7.28   Flow over an unsymmetrical wedge.

flowing over the surface of the wedge, (d) the change in the stagnation pressure of the air after flowing through the shock wave, and (e) the static temperature of the gas flowing over the wedge. Assume $\gamma = 1.40$. The static pressure and temperature of the air are $1.0133 \cdot 10^5$ N/m^2 and 288.15 K, respectively.

Solution

From Fig. 7.20, for $M_1 = 2.0$, $\delta_m = 26$ deg. Since $\delta_w = (30/2) = 15$ deg, the shock wave is attached.

(a)   From Table C.12, for $M_1 = 2.0$ and $\delta = 15$ deg, $\varepsilon = 45.34$ deg. (Check by Fig. 7.17, $\varepsilon = 45.3$ deg).

(b)   From equation 7.99,

$$M_1' = M_1 \sin \varepsilon = 2.0 \sin (45.34) = 1.4226$$

From Table C.11, for $M_1' = 1.4226$,

$$M_2' = 0.73039 \qquad \frac{p_2}{p_1} = 2.1944 \qquad \frac{P_2}{P_1} = 0.95237 \qquad \frac{t_2}{t_1} = 1.26933$$

From equation 7.100,

$$M_2 = \frac{M_2'}{\sin (\varepsilon - \delta)} = \frac{(0.73039)}{\sin (45.34 - 15)} = 1.4459$$

(Check by Fig. 7.18, $M_2 = 1.44$).

(c)   From part (b),

$$p_2 = 2.1944(1.0133)10^5 = 2.224 \cdot 10^5 \text{ N/m}^2 \ (32.26 \text{ lbf/in.}^2)$$

(d)   From part (b),

$$\Delta P = P_2 - P_1 = P_1\left(\frac{P_2}{P_1} - 1\right) = P_1(0.95237 - 1) = (-0.04763)P_1$$

From Table C.6, for $M_1 = 2.0$, $p_1/P_1 = 0.1278$. Thus,

$$P_1 = \frac{(1.0133)10^5}{(0.1278)} = 7.929 \cdot 10^5 \text{ N/m}^2$$

$$\Delta P = (-0.04763)(7.929)10^5 = -0.3777 \cdot 10^5 \text{ N/m}^2 \ (-5.48 \text{ lbf/in.}^2)$$

(e) From part (b),

$$t_2 = 1.26933(288.15) = 365.8 \quad (658.4 \text{ R})$$

7–10(b) Flow over a Cone

The phenomena associated with the supersonic flow of a gas over a cone are of interest because the noses of supersonic missiles and the fuselages of supersonic aircraft are conical in shape. Furthermore, the cone may be utilized as a supersonic diffuser for decelerating a supersonic flow.

The theoretical analysis of the supersonic flow over a cone at zero angle of attack involves solving the equations for steady two-dimensional axisymmetric flow. Numerical solutions of the differential equations have been obtained for a limited number of cases, and the theoretical results are in good agreement with experiment.[9-12] The governing equations, the numerical solution procedure, and a numerical example are presented in Section 16–5.

The flow model is illustrated in Fig. 7.29. The basic assumptions are as follows:

1. The flow is axisymmetric with respect to the axis of the cone. Consequently, the solution on any plane containing the cone axis is the same as that on any other plane. Consider the yz plane.
2. The flow is steady and isentropic before and after the shock wave; the latter is attached to the vertex of the cone.
3. Since no characteristic length appears in the physical model, the spatial variables y and z may appear only in the combination y/z. For a coordinate system located

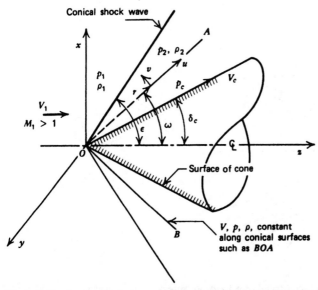

Figure 7.29 Supersonic flow over a symmetrical cone at zero angle of attack.

at the vertex of the cone, $y/z = \tan \omega$. The properties of the fluid may, therefore, vary only with ω, and they are constant along any ray from the vertex of the cone. Consequently, the shock wave surface itself has a symmetrical conical configuration.

The flow described by assumption 3 above is termed a *conical flow*. Because of the aforementioned assumptions, the analysis of the flow over a cone may be based on the study of a meridional section with the spherical coordinates r, ω, as illustrated in Fig. 7.29. Conservation of mass requires that the mass flow rate for a stream tube entering the shock wave remains constant as the flow expands over the cone. Because the cross-sectional flow area increases with radius in an axisymmetric flow, the streamlines must converge to satisfy continuity. The streamlines, therefore, curve toward the cone surface, as illustrated in Fig. 7.30a. On the other hand, the streamlines for the flow behind the shock wave attached to a wedge are parallel to the wedge surface, as shown in Fig. 7.30b.

It is shown in Section 7–10(a) that, in the case of the supersonic flow over a wedge, the pressure, velocity, and density are uniform behind the shock wave. The latter

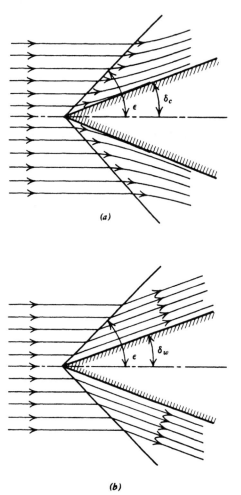

(a)

(b)

Figure 7.30 Comparison of the streamline patterns for supersonic flow over a symmetrical wedge and a symmetrical cone at zero angle of attack. (a) Cone. (b) Wedge.

condition does not occur in the case of a cone because of the aforementioned convergence of the streamlines toward the surface of the cone. As a consequence, there is an isentropic compression (*diffusion*) of the gas between the conical shock wave, where the conditions are denoted by the subscript 2, and the cone surface, where the conditions are denoted by the subscript *c*.

Figure 7.31 presents the *conical shock wave angle* ε as a function of the initial Mach number M_1 for different values of the cone angle δ_c, for $\gamma = 1.40$. The line

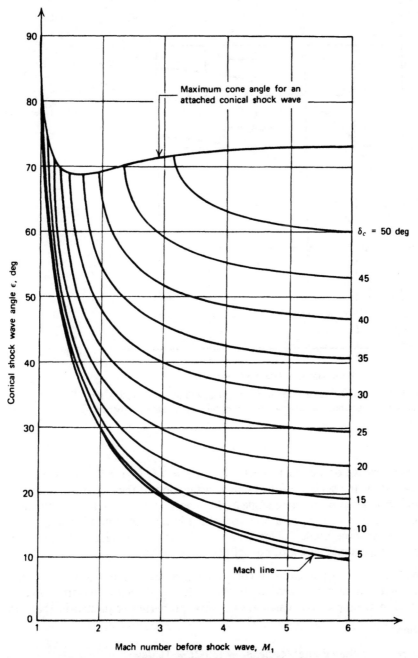

Figure 7.31   Shock wave angle ε for a conical shock wave as a function of the initial Mach number M_1 for different values of the cone angle δ_c for $\gamma = 1.40$.

across the top of the curves is the locus of the conditions for which the conical shock wave becomes detached from the apex of the cone, for different values of M_1 and δ_c. The conical shock wave angle ε at which the shock wave becomes detached, called the *detachment angle*, is larger for a cone than it is for a wedge, as is readily seen by comparing Figs. 7.17 and 7.31. Figure 7.32 presents the maximum values of the wedge angle δ_w and the cone angle δ_c for which it is possible for a shock wave to be attached

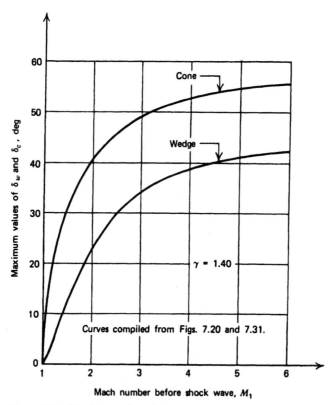

Figure 7.32 Maximum value of the wedge angle δ_w and the cone angle δ_c for an attached oblique shock wave as a function of the initial Mach number M_1 for $\gamma = 1.40$.

to either a wedge or a cone, respectively, for $\gamma = 1.4$. Figure 7.33 presents the Mach number M_c at the surface of the cone as a function of the initial Mach number M_1 for different values of the cone angle δ_c, for $\gamma = 1.40$. Figure 7.34 presents the surface static pressure ratio p_c/p_1 as a function of the initial Mach number M_1 for different values of the cone angle δ_c, for $\gamma = 1.40$. Kopal[11] and Sims[12] present extensive tables for supersonic flow around right circular cones.

Example 7.13. A right circular cone having a total included angle of 30 deg is placed in an air stream at zero angle of attack. For the same free-stream conditions specified in Example 7.12, calculate the flow properties requested in that example.

Solution

For a 30 deg included angle cone, $\delta_c = (30/2) = 15$ deg.
(a) From Fig. 7.31, for $M_1 = 2.0$ and $\delta_c = 15$ deg, $\varepsilon = 33.9$ deg.

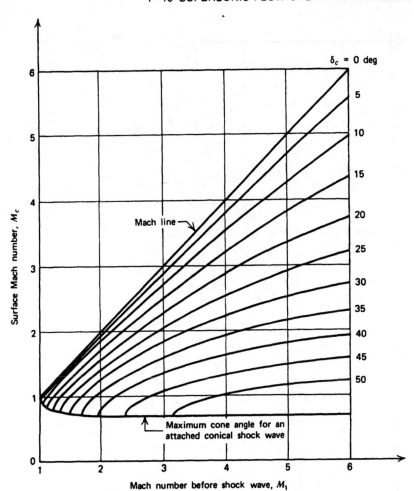

Figure 7.33 Surface Mach number M_c for a conical shock wave as a function of the initial Mach number M_1 for different values of the cone angle δ_c for $\gamma = 1.40$.

(b) From Fig. 7.33, $M_c = 1.70$.

(c) From Fig. 7.34, $p_c/p_1 = 1.57$. Hence, for $p_1 = 1.0133 \cdot 10^5$ N/m^2,

$$p_c = (1.57)(1.0133)10^5 = 1.591 \cdot 10^5 \text{ N/m}^2 \text{ (23.07 lbf/in.}^2)$$

(d) From Example 7.12, part (d), $P_1 = 7.929 \cdot 10^5$ N/m^2. From Table C.6, for $M_c = 1.70$, $p_c/P_c = 0.20259$. Thus,

$$P_c = \frac{(1.591)10^5}{(0.20259)} = 7.853 \cdot 10^5 \text{ N/m}^2$$

$$\Delta P = P_c - P_1 = (7.853 - 7.929)10^5 = -0.0760 \cdot 10^5 \text{ N/m}^2 \, (-1.10 \text{ lbf/in.}^2)$$

(e) From Table C.6, for $M_1 = 2.0$ and $M_c = 1.70$, $t_1/T = 0.55556$ and $t_c/T = 0.63371$. Since the flow is adiabatic, $T = $ constant. Thus,

$$t_c = t_1 \left(\frac{T}{t_1} \right) \left(\frac{t_c}{T} \right) = \frac{(288.15)(0.63371)}{(0.55556)} = 328.7 \text{ K (591.6 R)}$$

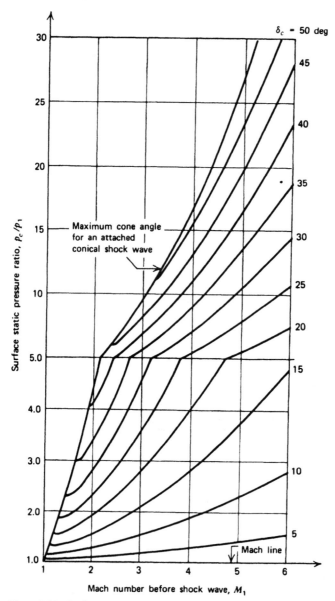

Figure 7.34 Surface static pressure ratio p_c/p_1 for a conical shock wave as a function of the initial Mach number M_1 for different values of the cone angle δ_c for $\gamma = 1.40$.

Comparing the above results with those obtained in Example 7.12, it is seen that, for the same free-stream conditions, the shock wave attached to a cone is considerably weaker than the shock wave attached to a wedge having the same included angle.

7-11 REFLECTION AND INTERACTION OF OBLIQUE SHOCK WAVES

It is shown in the preceding sections that when a supersonic flow encounters a concave corner, an oblique shock wave is propagated from that corner and instantaneously deflects the flow behind the shock wave toward the latter. The present section is concerned with the rules governing the reflection and interaction of oblique shock waves.

7–11(a) Reflection from a Wall

Figure 7.35a illustrates the flow of a gas between two frictionless surfaces AOB and DCE. The portion DC of the lower surface is parallel to the upper surface, but at point C there is a *concave* corner and the portion CE of the lower surface makes the angle δ_1 with the incident flow, which is parallel to surfaces AB and DC. The Mach number for the incident flow is M_1. Because of the concave corner an oblique shock wave emanates from the corner C and impinges upon the surface AB at point O. All of the streamlines after passing through the shock wave CO are turned toward the shock wave; that is, they are deflected toward the surface AB. The physical boundary condition, however, requires that the flow be parallel to the surfaces OB and CE. As a consequence, an oblique shock wave OF is reflected from point O that is of sufficient strength to turn the flow so that it becomes parallel to surface OB. For the flow, $M_1 > M_2 > M_3$.

When the reflected shock wave OF strikes surface CE, another shock wave reflection occurs. In principle such reflections can continue as long as the upstream Mach number and the required flow deflection angle permit an *attached oblique shock* wave to be formed (see Fig. 7.32). Such reflections are called *regular reflections*.

When an attached oblique shock wave cannot be formed at the wall, the shock wave becomes normal to the wall and "curves out" to become tangent to the incident shock wave, as illustrated in Fig. 7.35b. Such reflections are termed *Mach reflections*. The shock wave from the wall to the junction at point G is called a *Mach shock wave*. A *slip line*, which is a line across which some of the flow properties are discontinuous, is formed at the junction G. A subsonic flow region is formed immediately downstream from the Mach shock wave. The analysis of Mach reflection is not well developed.

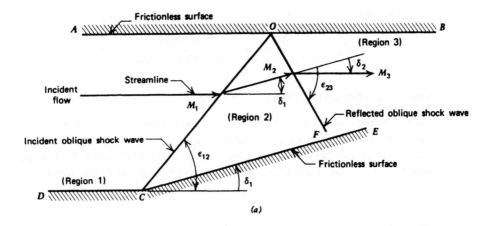

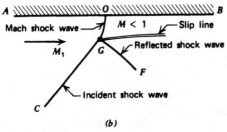

Figure 7.35 Reflection of an oblique shock wave from a wall. (a) Regular reflection. (b) Mach reflection.

Example 7.14.   Air, assumed to be a perfect gas, flows between two parallel frictionless walls. The initial flow Mach number is $M_1 = 4.0$. The lower wall abruptly turns into the flow with a turning angle $\delta_1 = 20$ deg, as illustrated in Fig. 7.35a. Determine (a) the wave angle ε_{12} of the initial oblique shock wave from the lower wall, and the Mach number M_2 after that shock wave, and (b) the wave angle ε_{23} of the oblique shock wave reflected from the upper wall, and the Mach number M_3 after that reflected shock wave.

Solution

(a)   From Table C.12, for $M_1 = 4.0$ and $\delta_1 = 20$ deg, $\varepsilon_{12} = 32.5$ deg. From equation 7.99,

$$M_1' = M_1 \sin \varepsilon_{12} = 4.0 \sin 32.5 = 2.149$$

From Table C.11, for $M_1' = 2.149$, $M_2' = 0.5541$. From equation 7.100,

$$M_2 = \frac{M_2'}{\sin (\varepsilon_{12} - \delta_1)} = \frac{0.5541}{\sin (32.5 - 20)} = 2.560$$

(b)   When the incident shock wave CO impinges on the upper wall AOB, the flow angle behind that shock wave (i.e., of M_2) is 20 deg in the direction toward that wall. To satisfy the solid wall boundary condition, a shock wave must be reflected having a deflection angle $\delta = -20$ deg relative to the direction of the flow in Region 2. Hence, $\delta_2 = -20$ deg. From Table C.12, for $M_2 = 2.56$ and $\delta_2 = -20$ deg, the wave angle ε_{23} is

$$\varepsilon_{23} = -42.11 \text{ deg } (\textit{relative to the flow direction in Region 2})$$

From equation 7.99,

$$M_2' = M_2 \sin \varepsilon_{23} = 2.560 \sin (42.11) = 1.717$$

From Table C.11, for $M_2' = 1.717$, $M_3' = 0.63621$. From equation 7.100,

$$M_3 = \frac{M_3'}{\sin (\varepsilon_{23} - \delta_2)} = \frac{0.63621}{\sin (42.11 - 20)} = 1.690$$

7–11(b)   Reflection from a Free Pressure Boundary

Figure 7.36 illustrates the intersection of an oblique shock wave with a free pressure boundary, such as a jet of gas; the static pressure at the boundary of the jet is p_0. The incident wave AO turns the flow toward the boundary through the flow deflection angle δ, and raises the static pressure of the gas flowing out of the shock wave, so

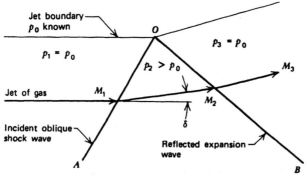

Figure 7.36   Reflection of an oblique shock wave from a free pressure boundary.

that $p_2 > p_o$. The physical boundary condition at point O, however, requires that the pressure remain equal to p_o. Because $p_2 > p_o$, a wave OB is reflected back from point O into the flowing jet of gas in a manner that returns the static pressure of the gas to the value p_o. Since $p_3 = p_o$, and the reflected wave reduces the static pressure of the gas from $p_2 > p_o$ to $p_3 = p_o$, that wave is an expansion wave. Expansion waves are discussed in Chapter 8. At this point, it is noted that expansion waves turn the flow *away* from the wave, so that the boundary of the jet turns outward, as indicated in Fig. 7.36.

7–11(c)   Neutralization of an Oblique Shock Wave

Consider the neutralization, or cancellation, of an oblique shock wave at a solid wall, as illustrated in Fig. 7.37. The incident shock wave AO turns the flow through the flow deflection angle δ toward the wave, and raises the static pressure of the flowing gas. At the intersection of the shock wave with the wall at the corner O, the physical boundary condition requires that the flow be parallel to the wall. But if the portion of the wall, denoted by OD, is turned through the angle δ with respect to the wall CO, then the aforementioned physical boundary condition is satisfied, and there is no reflected wave. Consequently, providing the surface OD with the appropriate turning angle δ achieves *wave cancellation*.

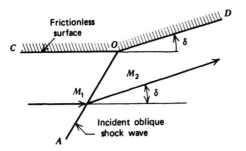

Figure 7.37   Neutralization of an oblique shock wave.

7–11(d)   Intersection of Oblique Shock Waves

Figure 7.38a illustrates a plane in which two oblique shock waves OA and OB intersect. Each of the shock waves OA and OB turns the initially parallel uniform flow inward, as indicated. In general, the oblique shock waves OA and OB may be of unequal strength, so that the fluid properties in Regions 2 and 3 are not identical. The physical boundary condition at point O is that the static pressures in the Regions 4 and 5 must be equal (i.e., mechanical equilibrium must be established between Regions 4 and 5). In general, the remaining flow properties in Regions 4 and 5 will be different. Thus, the streamline through the point O divides the Regions 4 and 5, and there are differences in the properties of the fluid on each side of this streamline. A streamline that divides the flow field into two regions having different fluid properties is called a *slip line*. The transmitted oblique shock waves OC and OD adjust themselves so that $p_4 = p_5$. In general, the conditions in Regions 4 and 5 may be determined by a trial and error procedure that matches p_4 and p_5. The wave pattern illustrated in Fig. 7.38a is termed *regular intersection*.

In some cases, there is no solution for the transmitted oblique shock waves that satisfies all of the flow conditions. In such a case, a more complex shock pattern

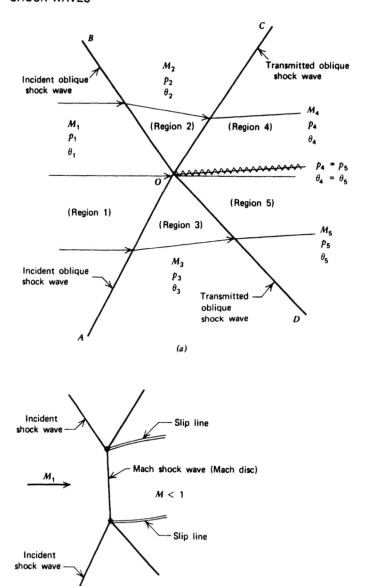

Figure 7.38 Intersection of two oblique shock waves. (a) Regular intersection.
(b) Mach intersection.

involving normal shock waves develops, similar to the Mach reflection discussed in
Section 7–11(a). The aforementioned type of intersection, termed *Mach intersection*,
is illustrated in Fig. 7.38b.

Example 7.15 Air, assumed to be a perfect gas, flows between two horizontal
parallel frictionless plates at a Mach number of 4.0 and a static pressure of $1.0 \cdot 10^5$
N/m^2. At a given cross section, denoted by $x - x$, the upper wall turns downward
10 deg into the flow, and the lower wall turns upward 20 deg into the flow, as illustrated in Fig. 7.39a. Determine the flow properties M and p in the Regions 4 and 5,
and the net flow deflection angle β.

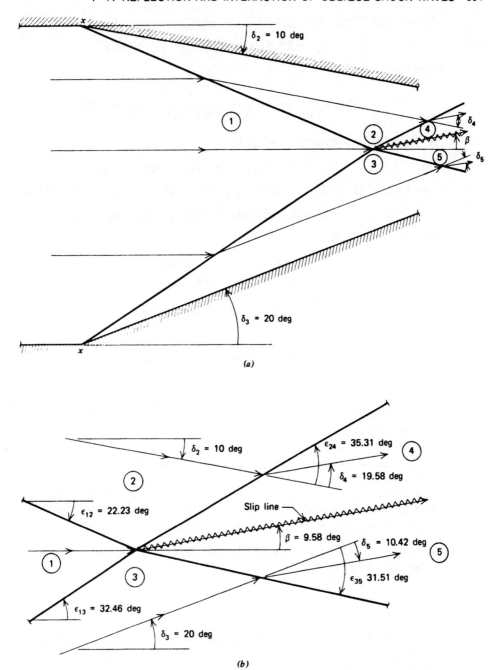

Figure 7.39 Intersection of two oblique shock waves (Example 7.15). (*a*) Flow model. (*b*) Shock waves, flow deflection angles, wave angles, and slip line.

Solution

From Table C.12, for $M_1 = 4.0$, and $\delta_2 = 10$ deg and $\delta_3 = 20$ deg, respectively, $\varepsilon_{12} = 22.23$ deg and $\varepsilon_{13} = 32.46$ deg. For the upper wall, equation 7.99 gives

$$M_1' = M_1 \sin \varepsilon_{12} = 4.0 \sin (22.23) = 1.5133$$

From Table C.11 for normal shock waves, for $M_1' = 1.5133$, $p_2/p_1 = 2.5051$ and

$M_2' = 0.6964$. Thus,

$$p_2 = 2.5051(1.0)10^5 = 2.5051 \cdot 10^5 \text{ N/m}^2$$

From equation 7.100,

$$M_2 = \frac{M_2'}{\sin(\varepsilon_{12} - \delta_2)} = \frac{0.6964}{\sin(22.23 - 10)} = 3.2874$$

For the lower wall, equation 7.99 gives

$$M_1' = M_1 \sin \varepsilon_{13} = 4.0 \sin(32.46) = 2.1468$$

From Table C.11, for $M_1' = 2.1468$, $p_3/p_1 = 5.2103$, and $M_3' = 0.5544$. Thus,

$$p_3 = 5.2103(1.0)10^5 = 5.2103 \cdot 10^5 \text{ N/m}^2$$

From equation 7.100,

$$M_3 = \frac{M_3'}{\sin(\varepsilon_{13} - \delta_3)} = \frac{0.5544}{\sin(32.46 - 20)} = 2.5695$$

The solution cannot be continued unless one condition in either Region 4 or 5 is specified. Consequently, an iterative computation is required. The most convenient parameter to vary is the net flow deflection angle β, which is assumed to be upward because the lower wall produces more initial turning than the upper wall. The following procedure is applicable.

1. Assume β, and calculate $\delta_4 = 10 + \beta$ and $\delta_5 = 20 - \beta$. Note that δ_4 and ε_{24} are measured relative to δ_2, and δ_5 and ε_{35} are measured relative to δ_3.
2. From Table C.12, find ε_{24} and ε_{35}.
3. Calculate M_2' and M_3' from equation 7.99.
4. From Table C.11, determine p_4/p_2, M_4', p_5/p_3, and M_5'.
5. Calculate p_4 and p_5.
6. If p_4 and p_5 agree to the desired tolerance, the solution is complete. Calculate M_4 and M_5 from equation 7.100. If p_4 and p_5 disagree, choose another value of β and repeat the procedure until convergence is obtained.

Steps 1 to 5 above are illustrated below for the first trial of the solution.

1. Assume $\beta = 5$ deg. Then,

$$\delta_4 = 10 + 5 = 15 \text{ deg}$$

$$\delta_5 = 20 - 5 = 15 \text{ deg}$$

2. From Table C.12, for $\delta_4 = 15$ deg and $M_2 = 3.2874$, $\varepsilon_{24} = 30.36$ deg. From Table C.12, for $\delta_5 = 15$ deg and $M_3 = 2.5695$, $\varepsilon_{35} = 36.16$ deg.
3. From equation 7.99,

$$M_2' = M_2 \sin \varepsilon_{24} = 3.2874 \sin(30.36) = 1.6616$$

$$M_3' = M_3 \sin \varepsilon_{35} = 2.5695 \sin(36.16) = 1.5161$$

4. From Table C.11, for $M_2' = 1.6616$, $p_4/p_2 = 2.8823$, and for $M_3' = 1.5161$, $p_5/p_3 = 2.5150$. Values of M_4' and M_5' are not needed until after the final trial.
5. From the results of step 4, we obtain

$$p_4 = 2.8823(2.5051)10^5 = 7.2204 \cdot 10^5 \text{ N/m}^2$$

$$p_5 = 2.5150(5.2103)10^5 = 13.104 \cdot 10^5 \text{ N/m}^2$$

6. $\Delta p = p_5 - p_4 = (13.104 - 7.2204)10^5 = 5.884 \cdot 10^5 \text{ N/m}^2$. The results of the first trial are presented in column (1) of the table at the end of the problem.

The value of Δp, from step 6, is obviously unacceptable. Hence, a second trial value of β is assumed. Since $p_5 > p_4$, the flow from Region 3 into Region 5 is turned through too large of a net flow deflection angle. Thus, β should be increased for the second trial. A value of $\beta = 10$ deg is assumed.

The results for the second trial are presented in column (2) of the accompanying table. The values of p_4 and p_5 corresponding to $\beta = 10$ deg are $10.3844 \cdot 10^5 \text{ N/m}^2$ and $9.8391 \cdot 10^5 \text{ N/m}^2$, respectively, which yield a value of $\Delta p = -0.5453 \cdot 10^5 \text{ N/m}^2$.

A third *trial* value for β is obtained by the secant method [see Appendix A-4(b)]. For the present application of the secant method, the assumed parameter is β and the calculated parameter is $\Delta p = p_5 - p_4$, which should be zero. The equation of the locus of Δp versus β, obtained from the results of the first two trials, is

$$\frac{\Delta p - 5.884 \cdot 10^5}{\beta - 5} = \frac{[(-0.5453) - 5.884]10^5}{(10 - 5)} = -1.2859 \cdot 10^5$$

$$\Delta p = (12.313 - 1.2859\beta)10^5$$

Setting $\Delta p = 0$ and solving for β gives the third trial value of $\beta = 9.5756$ deg.

For $\beta = 9.5756$ deg, repeating the solution yields $\Delta p = -0.0397 \cdot 10^5 \text{ N/m}^2$, which is sufficiently close to zero. The results obtained for the third trial are presented in column (3) of the accompanying table.

After the last trial, the Mach numbers in Regions 4 and 5 may be determined. From Table C.11, at $M_2' = 1.9001$, $M_4' = 0.5956$, and at $M_3' = 1.3429$, $M_5' = 0.7650$. From equation 7.100,

$$M_4 = \frac{M_4'}{\sin(\varepsilon_{24} - \delta_4)} = \frac{0.5956}{\sin(35.31 - 19.58)} = 2.1969$$

$$M_5 = \frac{M_5'}{\sin(\varepsilon_{35} - \delta_5)} = \frac{0.7650}{\sin(31.51 - 10.42)} = 2.1260$$

The flow angle in Region 5 is $\beta = 9.58$ deg, and the pressure in Regions 4 and 5 is $p_4 = p_5 = 0.5 \cdot (10.1341 + 10.0944)10^5 = 10.1143 \cdot 10^5 \text{ N/m}^2$. Figure 7.39b presents the corresponding wave pattern.

Trial	(1)	(2)	(3)
β, deg	5.0	10.0	9.5756
δ_4, deg	15.0	20.0	19.5756
ε_{24}, deg	30.36	35.79	35.31
M_2'	1.6616	1.9225	1.9001
p_4/p_2	2.8823	4.1453	4.0454
M_4'			0.5956
p_4, N/m$^2 \cdot 10^5$	7.2204	10.3844	10.1341
δ_5, deg	15.0	10.0	10.4244
ε_{35}, deg	36.16	31.10	31.51
M_3'	1.5161	1.3272	1.3429
p_5/p_3	2.5150	1.8884	1.9374
M_5'			0.7650
p_5, N/m$^2 \cdot 10^5$	13.104	9.8391	10.0944
Δp, N/m$^2 \cdot 10^5$	5.884	-0.5453	-0.0397

7–11(e)   Summary of the Rules for the Interaction of Oblique Shock Waves

1. An oblique shock wave turns the flow toward the wave.
2. Oblique shock waves from successive corners converge toward each other.
3. An oblique shock wave reflects from a solid boundary as an oblique shock wave.
4. An oblique shock wave reflects from a free pressure boundary as an expansion wave.
5. Oblique shock wave reflections may be cancelled by turning the wall away from the flow direction.

7–12   SUPERSONIC WIND TUNNELS

A wind tunnel is an apparatus for producing in a laboratory flow conditions that simulate those of actual flight. Wind tunnels may be classified loosely as subsonic, transonic, and supersonic, depending on the Mach number in the *test section*. The general features of supersonic wind tunnels are discussed below, with particular attention being devoted to the operation of the diffuser and to the associated starting problem.

Figure 7.40 illustrates schematically the general arrangement of the components for four different methods for supplying gas, usually air, to a wind tunnel.

1. High-pressure gas storage tanks for a *blow-down wind tunnel*.
2. A compression and diffusion system for a *continuous flow wind tunnel*.
3. A vacuum tank or a vacuum pump for an *atmospheric inlet wind tunnel*.
4. A shock tube for a *high-pressure, high-enthalpy wind tunnel*.

The essential features of any wind tunnel are a source of gas, a converging-diverging nozzle for providing a supersonic flow, a test section, and an exhaust system. The available time for testing a model in a wind tunnel, called the *testing time*, depends on the source of the gas and the type of exhaust gas system. In the *blow-down wind tunnel* (see Fig. 7.40a), the testing time depends on the capacity of the gas storage tanks, whereas in the *shock tube driven wind tunnel* (see Fig. 7.40d), it is determined by the size of the shock tube. If the exhaust system is a vacuum tank, the testing time is determined by the volume of the exhaust tank. For an *atmospheric inlet wind tunnel* operated by a vacuum pump (see Fig. 7.40c), the testing time is essentially infinite, but the size of the test section and its flow Mach number depend on the capacity of the vacuum pump. The only practical apparatus for obtaining long testing times in a large wind tunnel is the *continuous flow tunnel*, illustrated in Fig. 7.40b. The limiting factor for a continuous flow wind tunnel is the power available for driving the compressor. In the case of a supersonic flow tunnel having a large test section, the power required may be several hundred megawatts. The characteristics of the nozzle and the diffuser for a continuous flow wind tunnel are discussed below.

7–12(a)   Fixed Geometry Supersonic Wind Tunnel

Figure 7.41a illustrates schematically the nozzle, diffuser, and test section of an ideal continuous flow supersonic wind tunnel; that is, one in which the flow is isentropic throughout. The operating characteristics of the nozzle are the same as those of the De Laval nozzle discussed in Section 4–7. The test section is essentially a constant area isentropic flow region, and the diffuser is basically a reversed De Laval nozzle. Figure 7.41b illustrates diagrammatically the static pressure distribution between

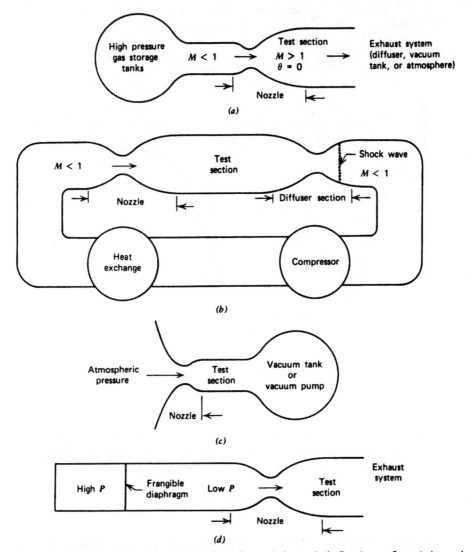

Figure 7.40 Several types of wind tunnels. (*a*) Blow-down wind tunnel. (*b*) Continuous flow wind tunnel. (*c*) Atmospheric inlet wind tunnel. (*d*) Shock tube driven wind tunnel.

Stations 1 and 4. The power required for maintaining steady flow in the tunnel is that needed for overcoming the boundary layer effects and any energy losses associated with the model being investigated.

Two problems arise, however, that complicate the operation of the above arrangement. Because of the energy losses associated with the frictional effects between the throat of the nozzle and the throat of the diffuser, there is a *stagnation pressure* drop between those two stations. If the Mach number is unity in the throats of the nozzle and the diffuser, then, since the mass flow rate $\dot{m}^*$ is constant, equation 4.137, Table 4.6, yields

$$\dot{m}_N^* = \frac{\psi^* P_N A_N}{\sqrt{\gamma R T_N}} = \dot{m}_D^* = \frac{\psi^* P_D A_D}{\sqrt{\gamma R T_D}} \tag{7.122}$$

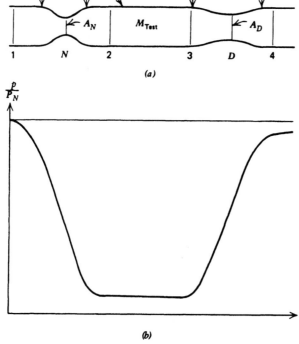

Figure 7.41 Ideal supersonic wind tunnel. (*a*) Geometry. (*b*) Pressure distribution.

where the subscripts N and D denote the conditions at the throats of the nozzle and the diffuser, respectively. For the adiabatic flow of a perfect gas, $T_N = T_D$, and γ, R, and ψ^* are constants. In that case, equation 7.122 yields

$$\frac{A_D}{A_N} = \frac{P_N^*}{P_D^*} > 1.0 \tag{7.123}$$

Equation 7.123 shows that the *diffuser throat* must be larger than the *nozzle throat*. If the diffuser throat is too small, a supersonic flow cannot be attained in the nozzle exit section, nor in the test section, because the diffuser throat would be the first cross section of the wind tunnel to *choke* during the *start-up* of the tunnel. On the other hand, if the diffuser throat is too large, supersonic flow will persist into the diffuser divergence and a shock wave will form therein.

The following more serious problem occurs during the start-up of the wind tunnel. The flow initiated by the compressor simultaneously lowers the pressure at the diffuser exit and raises the pressure at the diffuser inlet. The latter flow process is sufficiently slow and may be regarded as being a steady flow throughout the wind tunnel; that is, during start-up the flow in the wind tunnel is quasi-steady. The static pressures at the different stations of the wind tunnel may be referred to the nozzle inlet stagnation pressure P_N. Figure 7.42 illustrates schematically the steps occurring in the start-up of a wind tunnel nozzle. Assume that A_D is large enough so that the flow does not choke at the diffuser throat. As the pressure ratio p_4/P_N is decreased, the nozzle attains the sequence of states described in Section 4–7(c), until the normal shock wave positions itself at the nozzle exit plane, as illustrated by curve e in Fig. 7.42. When that condition occurs, the drop in the stagnation pressure attains its largest value, because the Mach number of the flow in front of the normal shock wave is

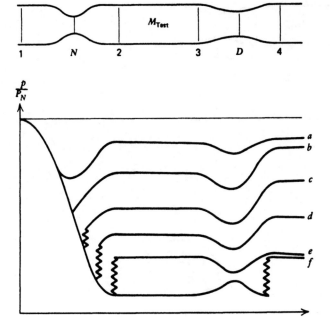

Figure 7.42   Pressure distributions for several stages during the start-up of a supersonic wind tunnel.

the test section Mach number M_{Test}; the latter is the largest Mach number for the flow in the wind tunnel circuit.

As the pressure ratio p_4/P_N is decreased further, the shock wave moves very rapidly through the test section and the diffuser inlet and positions itself in the diffuser divergence at an area that is slightly larger than that of the test section. When that occurs, the shock is said to have been "swallowed" by the diffuser. Curve f in Fig. 7.42 and curve a in Fig. 7.43 illustrate that particular operating condition. The pressure ratio p_4/P_N may now be increased, by adjusting the compressor pressure ratio, so that the shock wave is moved upstream toward the diffuser throat where the shock strength is a minimum, and the corresponding stagnation pressure drop is likewise a minimum; see curve c in Fig. 7.43. The location of the shock wave is determined by the flow conditions in the portion of the wind tunnel downstream from the diffuser throat.

The *inlet area ratio* of the diffuser A_3/A_D is smaller than the *nozzle area ratio* A_2/A_N, since $A_D > A_N$. Hence, if $M_D > 1.0$ and the normal shock wave positions itself at the diffuser throat, there is a stagnation pressure drop that must be compensated for by the compressor. Curve c in Fig. 7.43 illustrates diagrammatically the largest obtainable stagnation pressure recovery for a fixed geometry supersonic diffuser. In practice, the normal shock wave would be positioned slightly downstream from the diffuser throat because of the ever-present small fluctuations in the flow that may cause a shock wave that is located exactly at the throat to move slightly upstream into the diffuser inlet where its position would be unstable. The shock wave would continue moving through the diffuser inlet and locate itself once again in the nozzle divergence. Consequently, the entire procedure for swallowing the shock wave must be repeated.

Example 7.16.   Consider a fixed geometry supersonic diffuser operating with air as the working fluid. Neglecting all losses except the stagnation pressure drop for the flow associated with the normal shock wave, calculate the following as a function of

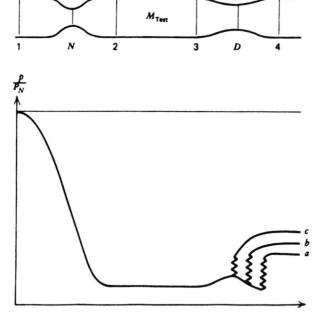

Figure 7.43 Pressure distributions in a fixed geometry supersonic diffuser as the pressure ratio p_4/P_N is increased.

the test section Mach number M_{Test}: (a) the ratio of the minimum diffuser throat area to the nozzle throat area, A_D/A_N, (b) the Mach number at the diffuser throat when the shock wave is positioned at that location, and (c) the corresponding stagnation pressure ratio P_D/P_N.

Solution

(a) The ratio A_D/A_N is obtained from equation 7.123, where the value of P_N/P_D is obtained from Table C.11 (normal shock wave) corresponding to M_{Test}. For example, for $M_{\text{Test}} = 3.0$, Table C.11 gives $P_D/P_N = 0.32834$, and

$$\frac{A_D}{A_N} = \frac{1}{0.32834} = 3.0456$$

which is the minimum area ratio that would permit the shock wave to be swallowed.
(b) When the shock wave has been swallowed, the flow in the diffuser inlet is supersonic, the Mach number at the diffuser entrance is the test section Mach number M_{Test}, and the Mach number at the diffuser throat is supersonic (i.e., $A_D > A^*$). Thus,

$$\frac{A_D}{A^*} = \frac{A_D}{A_N} = f(M_D)$$

where $A^* = A_N$ for isentropic flow in the nozzle and test section. From Table C.6, for $A/A^* = 3.0456$, $M_D = 2.635$.
(c) The stagnation pressure ratio P_D/P_N when the shock wave is located at the diffuser throat is that for a normal shock wave corresponding to M_D. Hence, for $M_{\text{Test}} = 3.0$, $M_D = 2.635$, and from Table C.11 for $M_D = 2.635$, $P_D/P_N = 0.44762$.

Figure 7.44 presents plots of A_D/A_N, M_D, and P_D/P_N, for $\gamma = 1.4$, as functions of the test section Mach number M_{Test} for a fixed geometry supersonic diffuser. Also shown is P_y/P_x, the stagnation pressure ratio for a normal shock wave at the test

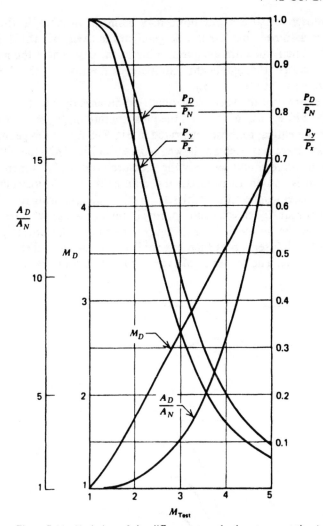

Figure 7.44 Variation of the diffuser-to-nozzle throat area ratio A_D/A_N, the diffuser throat Mach number M_D, the stagnation pressure ratio P_y/P_x, and the stagnation pressure ratio P_D/P_N with the test section Mach number M_{Test} for $\gamma = 1.40$.

section Mach number M_{Test}. The stagnation pressure ratio P_D/P_N for a fixed geometry supersonic diffuser is seen to decrease rapidly as the test section Mach number increases. In fact, the ratio P_D/P_N is not much larger than the ratio P_y/P_x for a normal shock wave located in the test section when $M = M_{Test}$. Because of that severe decrease in stagnation pressure, fixed geometry supersonic diffusers are, in general, not employed for large supersonic wind tunnels.

7–12(b) Variable Geometry Supersonic Wind Tunnel

The basic features of a fixed geometry continuous flow wind tunnel are discussed in Section 7–12(a), where it is shown that one of the major problems with a fixed geometry system is the large stagnation pressure drop associated with the shock wave in the diffuser. The pressure drop caused by the shock wave may be significantly decreased by employing a variable geometry system. In practice, either the nozzle throat area or the diffuser throat area, or both, may be made variable. In the following discussion, it is assumed that the nozzle geometry is fixed, but the diffuser throat area is variable. The results are also applicable to the case where the nozzle geometry is variable, because the significant parameter is the ratio of the diffuser throat area to the nozzle throat area A_D/A_N.

During start-up, the throat of the variable geometry diffuser is opened to an area larger than that specified by equation 7.123. As the pressure ratio p_4/P_N is decreased, the sequence of events is identical to those illustrated in Fig. 7.42 for a fixed geometry diffuser. Once the shock wave has been swallowed (curve f in Fig. 7.42), the diffuser throat area is decreased, while simultaneously the pressure ratio p_4/P_N is increased so that the shock wave is caused to move to the diffuser throat. Decreasing the diffuser throat area A_D increases the diffuser inlet area ratio A_3/A_D, causing the Mach number in the diffuser throat to decrease and approach unity as A_D approaches A_N. In the limit, the Mach number in the diffuser throat becomes unity. Accordingly, the normal shock wave at the diffuser throat then has vanishing strength, and the entire flow field becomes isentropic. The sequence of events is illustrated in Fig. 7.45.

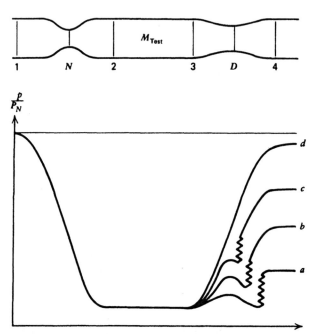

Figure 7.45 Pressure distributions in a variable geometry supersonic diffuser as the pressure ratio p_4/P_N is increased and the area ratio A_D/A_N is decreased.

The completely isentropic flow described above is, of course, physically unattainable in a practical wind tunnel because of viscous effects encountered by the flow. The flow in the diffuser is in the direction of an unfavorable pressure gradient (i.e., $dp/dx > 0$) with respect to the boundary layer, so that the energy loss because of viscous effects may become significant. In addition, if the Mach number at the diffuser throat is unity, any slight disturbance in the flow may cause the sonic point to move temporarily into the diffuser convergence, which is an unstable mode of operation. If that occurs, the diffuser would *unstart*, a normal shock wave would appear in the nozzle divergence, and the start-up procedure would have to be repeated. For the above reasons, the Mach number in the diffuser throat must be slightly larger than unity, with a corresponding stagnation pressure recovery somewhat less than unity.

The major conclusion from the foregoing discussion is that the *diffuser area ratio* A_D/A_N is the most significant parameter influencing the start-up phenomena in the wind tunnel. Figure 7.44 presents A_D/A_N, P_D/P_N, P_y/P_x, and M_D as functions of M_{Test}, for gases with $\gamma = 1.40$.

7–13 INLETS FOR AIR-BREATHING PROPULSION ENGINES[13, 14]

The efficient diffusion (flow compression) of the air entering an air-breathing propulsion engine is essential to its efficient functioning, irrespective of whether it propels a vehicle at either a subsonic or a supersonic speed. For example, the air entering the combustor of a ramjet engine must be decelerated for combustion reasons to a Mach number not exceeding approximately 0.2. For a turbojet engine to perform satisfactorily, the air supplied to the inlet face of the axial flow compressor should have a Mach number less than approximately 0.4 and a fairly uniform flow pattern. The air flowing through the intake duct of the engine must experience the smallest possible decrease in its stagnation pressure. Furthermore, the flow of the air over the external surfaces of the air intake should not produce a large external drag. For brevity, the intake-diffuser system for an air-breathing propulsion engine is called an *inlet*.

If the velocity of the free-stream air relative to the inlet is supersonic, the diffusion process is complicated by the formation of shock waves.

At first blush, flow diffusion appears to be merely the reverse of flow expansion. It is, however, considerably more difficult to achieve an efficient diffusion than an efficient expansion, because in diffusion the fluid flows in the direction of a *positive pressure gradient*; in a flow expansion the flow is in the direction of a *negative pressure gradient*. The difficulty in obtaining an efficient diffusion arises from the interaction of the main stream flow with the boundary layer.

A flow passage that converts the kinetic energy associated with a *subsonic gas flow* ($M < 1$) into a static pressure rise is termed a *subsonic diffuser*, and one for decelerating a *supersonic gas flow* ($M > 1$) to approximately the local acoustic speed ($M \approx 1$) is called a *supersonic diffuser*. If a vehicle is propelled at a supersonic speed by either a ramjet or a turbojet engine, the complete diffusion of the free-stream air is achieved in two steps: (1) a supersonic diffusion to $M \approx 1$, and (2) a subsonic diffusion from $M \approx 1$ to $M < 1$.

Inlets may be classified into two main groups: (1) subsonic inlets, and (2) supersonic inlets. They may be further subdivided into *external-compression inlets* in which the flow compression takes place in the free stream, and *internal-compression inlets* wherein the flow compression takes place inside of the inlet.

7–13(a) Performance Criteria for Inlets

There are several different criteria employed for expressing the performance of an inlet. Three of the more common ones are discussed below. The diffusion process is illustrated schematically on the *hs* plane in Fig. 7.46.

The *stagnation pressure ratio* η_P, defined as the ratio of the stagnation pressure at the diffuser exit P_2 to the free-stream stagnation pressure P_o, is one of the most important performance criteria for an inlet, because the stagnation pressure of the gases at the nozzle entrance determines the nozzle pressure ratio, and hence the

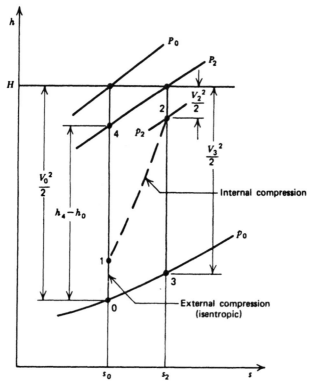

Figure 7.46 Diffusion process plotted on the *hs* plane.

nozzle thrust (see Section 4–10). *By definition,*

$$\eta_P \equiv \frac{P_2}{P_o} \tag{7.124}$$

The *kinetic energy efficiency* η_{KE} is defined as the ratio of the kinetic energy $V_3^2/2$, which would be obtained by expanding the diffused air isentropically from state 2 to the free-stream static pressure p_o, to the free-stream kinetic energy $V_o^2/2$. Thus,

$$\eta_{KE} \equiv \frac{V_3^2/2}{V_o^2/2} = \frac{H - h_3}{H - h_o} \tag{7.125}$$

The above definition of η_{KE} is analogous to the kinetic energy efficiency for nozzles defined in Section 4–8(b).

A third performance criterion is the *adiabatic diffuser efficiency* η_D, defined as

$$\eta_D \equiv \frac{h_4 - h_o}{V_o^2/2} = \frac{h_4 - h_o}{H - h_o} \tag{7.126}$$

where point 4 is a fictitious state corresponding to an isentropic compression from the free-stream static pressure p_o to the actual stagnation pressure P_2.

7–13(b) Subsonic Inlets

All air-breathing propulsion engines for propelling aircraft at subsonic flight speeds are equipped with subsonic inlets. The velocity of the *captured air* inducted into such an inlet is subsonic throughout its flow path. In general, the flow compression of that air involves two steps: (1) an external compression of the free-stream air, from Station 0 to Station 1 in Fig. 7.47, and (2) an internal compression of the air flowing in the duct, from Station 1 to Station 2 in Fig. 7.48. Subsonic inlets may be grouped into two basic types.

1. External-compression subsonic inlets.
2. Internal-compression subsonic inlets.

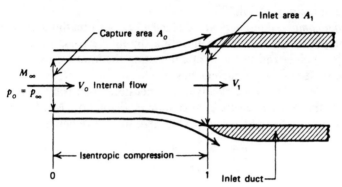

Figure 7.47   Subsonic external compression inlet.

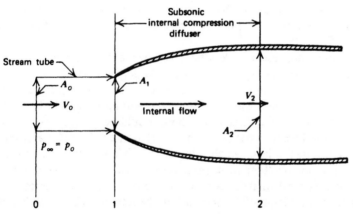

Figure 7.48   Subsonic internal compression inlet.

Figure 7.47 illustrates schematically a *subsonic external-compression inlet*; for simplicity it is assumed to be of circular cross section. Its general appearance is a constant-area duct with a contoured intake lip. Because the internal flow passage has a constant area, all of the diffusion of the entering air (*the internal flow*) occurs upstream from the inlet (Station 1). The flow stream tube entering the inlet has the capture area A_o far removed from A_1. At A_o, $p_\infty = p_o$, and the cross-sectional area of the stream tube increases to $A_1 > A_o$ at the entrance to the inlet. All of the transformation of the kinetic energy of the air into static pressure rise takes place external to the inlet where there are no solid surfaces. Hence, the *external diffusion is isentropic* ($ds = 0$).

Because the capture area $A_o < A_1$, some of the flowing air is deflected as it approaches the inlet and is accelerated as it flows over the *intake lip*; the flow is said to "*spill over*" the inlet. Because of the increased local Mach number resulting from the acceleration of the flow, the "spilled" flow produces an external drag. It is important that the local Mach number in the vicinity of the intake lip be less than unity to avoid forming shock waves with their attendant large external drag. Consequently, the type of diffuser illustrated in Fig. 7.47 is unsatisfactory for application in the high-subsonic-speed range.

There is no satisfactory analytical procedure for designing a subsonic external-compression diffuser; the current theory is quite similar to the airfoil theory employed in analyzing incompressible flow.

Subsonic internal-compression inlets are utilized in all air-breathing engines. The diffusion is accomplished in the duct connecting the intake cross section and the burner in a ramjet engine, and in the duct connecting the intake cross section and the compressor inlet face in a gas-turbine engine.

It is shown in Table 4.4, that for a steady one-dimensional isentropic flow, the flow area change dA/A, the static pressure change dp/p, and the flow Mach number M are related by equation 4.51, which is repeated here for convenience. Thus,

$$\frac{dA}{A} = \frac{1 - M^2}{\gamma M^2} \frac{dp}{p} \tag{4.51}$$

Although equation 4.51, strictly speaking, applies to an isentropic flow ($ds = 0$), it may be employed for obtaining qualitative information regarding an adiabatic flow ($ds > 0$). The equation shows that the area ratio A_2/A_1 for a duct that is to diffuse a subsonic flow must exceed unity; that is, the flow streamlines must diverge. Hence, a subsonic internal-compression inlet is a diverging duct, such as that illustrated schematically in Fig. 7.48.

Subsonic internal-compression inlets have received extensive study but, because of the lack of an adequate theory, their design is largely empirical.

It is apparent that, to prevent the energy losses in an internal-compression diffuser from becoming excessive, the positive pressure gradient dp/dx should be kept small; that is, the area increase in the direction of flow should be gradual. However, if the pressure gradient is to be kept small, then the length of the diffuser for accomplishing a specified reduction in the flow Mach number may become so large that the corresponding decrease in the stagnation pressure because of *wall* (*skin*) *friction* may be excessive. For a conical internal-compression diffuser, the maximum semiangle of the divergence for preventing flow separation phenomena is between 5 and 7 deg for a substantially incompressible flow ($M_o < 0.4$); it decreases approximately as $(1 - M_o^2)$ for higher subsonic Mach numbers.[14] Because of the restrictions on the available space and weight for the diffuser, it is rarely possible to utilize such small divergence angles. Consequently, the design of a subsonic internal-compression diffuser is usually a compromise between the stagnation pressure recovery and the length available for the diffuser.

7-13(c) Normal Shock Wave Supersonic Inlets

All air-breathing propulsion engines that propel vehicles at supersonic speeds, except the SCRAMJET* engine, require that the air flow leave the inlet with a subsonic velocity. The design of an efficient supersonic inlet is critical, since neither the external nor the internal compression of a supersonic air stream can be accomplished

---

* Acronym for Supersonic Combustion **Ram**jet, a type of engine that has received study and some technology development. It permits operating the combustion chamber with a supersonic velocity.

isentropically because of the formation of shock waves. As pointed out in Section 7–4(e), shock waves are inevitably accompanied by an increase in the entropy of the flowing air and, therefore, by energy dissipation. One of the important design problems is to produce a supersonic inlet having a shock wave pattern that will cause only a relatively small increase in the entropy of the diffused air. In addition, the supersonic diffusion should be accomplished with a minimum of external drag. In most cases those two requirements are conflicting, so that a practical design is a compromise between achieving the minimum stagnation pressure loss and the minimum external drag. The inlet also must be capable of operating stably over a range of *angles of attack* and at conditions other than the *design point* without serious adverse effects on either the stagnation pressure recovery or the external drag.

A normal shock wave supersonic inlet is illustrated schematically in Fig. 7.49. The *design point*, illustrated in Fig. 7.49a, occurs when the normal shock wave is attached to the inlet lip. The stagnation pressure ratio at the design point is that for a normal shock wave at the design Mach number; that pressure ratio is presented in Fig. 7.44 by the curve labeled P_y/P_x, for $\gamma = 1.40$. The latter figure shows that for a low supersonic flight Mach number (below $M_o = 1.8$), the normal shock wave inlet furnishes a subsonic flow to the subsonic internal-compression diffuser that has a reasonable stagnation pressure ratio. The subsonic flow leaving the normal shock wave is further compressed in flowing through the subsonic internal-compression diffuser.

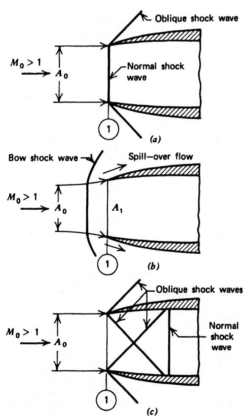

Figure 7.49 Operating characteristics of a normal shock wave supersonic inlet. (*a*) Design condition. (*b*) Effect of increasing the back pressure. (*c*) Effect of decreasing the back pressure.

If the engine requires a smaller mass rate of air flow than the design value, the excess air must be *spilled-over* the lip (leading edges) of the diffuser. To accomplish the spill-over of the air, the normal shock wave *detaches* from the diffuser lip, as illustrated in Fig. 7.49b, and a *bow shock wave* forms upstream to the inlet. The spillage of the air over the lip of the inlet increases the external drag.

We may assume that the portion of the bow shock wave immediately in front of the inlet is a normal shock wave. Hence, the flow properties in front of and behind the bow shock wave may be computed by means of the equations for a normal shock wave [see Section 7-4(a) and Table C.11].

If the back pressure in the outlet section of the diffuser is decreased, the normal shock wave is *swallowed*; that is, the shock wave locates itself inside the diffuser, as illustrated in Fig. 7.49c. Oblique shock waves form inside the inlet, and the stagnation pressure ratio of the inlet decreases.

The normal shock wave supersonic inlet gives reasonable performance at flight Mach numbers less than approximately 1.80. At values of M_o larger than 1.80, the stagnation pressure recovery of the normal shock wave inlet decreases rapidly.

7-13(d) The Converging-Diverging Supersonic Inlet

It would be natural to expect that reversing the flow of air through a converging-diverging nozzle, a device for obtaining large supersonic exhaust velocities with high efficiency (see Section 4-7), would achieve an efficient supersonic internal compression of the reversed supersonic air flow. Such an arrangement is illustrated schematically in Fig. 7.50. Unfortunately, it is practically impossible to achieve shockless diffusion with such a device because of the adverse interactions between the main

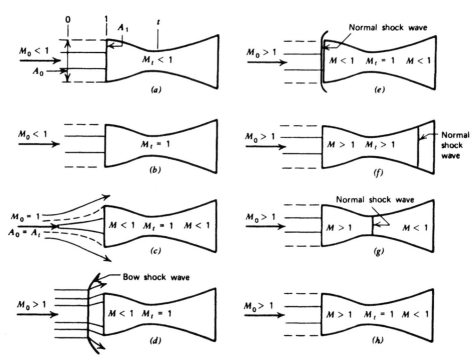

Figure 7.50 Effect of increasing the free-stream Mach number on the flow conditions for a converging-diverging diffuser.

stream flow and the boundary layer as a consequence of the positive pressure gradient (see Section 5–10). Furthermore, the presence of shock waves in the converging-diverging passage causes a large increase in the entropy of the air so that the main stream flow is no longer isentropic. Even if it were possible to design a converging-diverging flow passage to achieve a substantially isentropic diffusion at the *design point*, its poor off-design operating characteristics would make it impractical.[13] A fixed geometry converging-diverging passage can give a substantially isentropic diffusion, because of its fixed *contraction ratio*, at only one operating point, the design point. Moreover, the *starting* of such a diffuser, as explained in Section 7–12, presents a critical problem.

The starting problem for a supersonic inlet is quite similar to that for the supersonic wind tunnel discussed in Section 7–12. The sequence of conditions attained during start-up is illustrated schematically in Fig. 7.50, for a fixed geometry converging-diverging diffuser. At low subsonic flight speeds, the entire flow is subsonic, as illustrated in Fig. 7.50a. The *capture area*, denoted by A_o, depends on the mass flow rate requirements of the engine, and may be greater than, equal to, or less than the inlet area A_1. As M_o is increased, or the engine mass flow rate increases, the operating condition illustrated in Fig. 7.50b prevails, where $M_t = 1$ and $M_o < 1$. The streamline pattern ahead of the inlet is no longer influenced by the conditions downstream of the diffuser throat, but depends only on M_o and the inlet area ratio A_1/A_t. When the flight Mach number M_o reaches unity, the capture area A_o equals the throat area A_t, as illustrated in Fig. 7.50c. Considerable drag because of spillage of air over the lips of the inlet may occur if A_1/A_t is very large.

Further increases in the flight Mach number to supersonic values cause a bow shock wave to form upstream of the inlet, as illustrated in Fig. 7.50d. Both *spillage drag* and *shock drag* are present. When the flight Mach number reaches a sufficiently high value, a normal shock wave is located at the inlet entrance, as illustrated in Fig. 7.50e. A slight increase in M_o causes the normal shock wave to enter the inlet convergence, a position that is unstable, so that the shock wave procedes through the inlet throat and positions itself in the inlet divergence, as illustrated in Fig. 7.50f. At that point, the inlet is started. As in the start-up of the supersonic wind tunnel diffuser, discussed in Section 7–12, the inlet back pressure may be increased, thereby causing the shock wave to move toward the diffuser throat, where its strength decreases. The maximum stagnation pressure recovery is attained when the normal shock wave is located exactly at the throat; a location where the shock position is unstable. The design point is, therefore, chosen so that the normal shock wave will locate itself slightly downstream from the throat, as illustrated in Fig. 7.50g. Should the back pressure be increased sufficiently, the shock wave will be *disgorged* from the diffuser, and it will be necessary to repeat the start-up procedure.

As in the case of the fixed geometry wind tunnel diffuser discussed in Section 7–12(b), the Mach number at the diffuser throat, in Fig. 7.50f, is supersonic. Instead of increasing the back pressure to drive the shock wave toward the throat, the flight Mach number M_o may be decreased, thus decreasing the Mach number at the diffuser throat. When M_o is lowered to the value corresponding to isentropic choked flow in the diffuser inlet, which is that given in Table C.6 for the area ratio A_1/A_t, the Mach number in the throat of the inlet is $M_t = 1.0$. If the back pressure is adjusted properly, the inlet divergence is subsonic throughout, and the entire flow through the diffuser is isentropic, as illustrated in Fig. 7.50h. The last-mentioned method for starting the inlet is known as *overspeeding*.

The diffuser inlet area ratio A_t/A_1 corresponding to a normal shock wave attached to the inlet entrance plane for a given value of the free-stream Mach number M_o, as illustrated in Fig. 7.50e, may be obtained by first determining the subsonic Mach number immediately downstream of the normal shock wave, from Table C.11, and then determining, from Table C.6, the isentropic area ratio required for accelerating that subsonic flow to $M = 1$. Figure 7.51 presents that area ratio A_t/A_1 as a function of M_o, for $\gamma = 1.40$, for the operating condition where the normal shock wave is attached to the inlet lip. Operation at any point in the region below the curve corresponds to a situation where the shock wave is detached in front of the inlet. Operation at any point in the region above the curve corresponds to a condition where the shock wave is swallowed. For example, Fig. 7.51 shows that for an area ratio $A_t/A_1 = 0.70$, the shock wave becomes attached at a flight Mach number of $M_o = 3.2$.

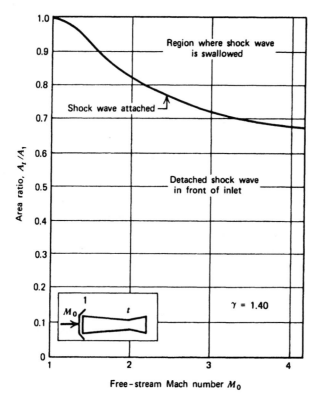

Figure 7.51   Diffuser inlet area ratio A_t/A_1 for attaching a normal shock wave to the diffuser entrance plane.

Figure 7.52 presents A_t/A_1 as a function of M_o, for $\gamma = 1.40$, for isentropic flow throughout the diffuser and a throat Mach number $M_t = 1.0$; the values are taken from Table C.6. A point located in the region below that curve corresponds to an inlet throat area that is too small for passing all of the captured flow at that particular flight Mach number: a normal shock wave would propagate from the diffuser throat through the inlet and position itself upstream from the inlet entrance, thus unstarting the inlet. For any operating point above the curve in Fig. 7.52, the flow in the inlet convergence would be supersonic up to and including the throat. A normal shock wave would position itself somewhere in the inlet divergence, the location depending on the back pressure presented by the remainder of the engine.

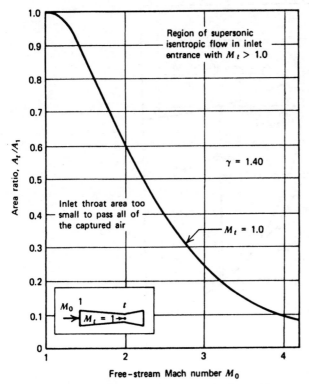

Figure 7.52 Diffuser inlet area ratio A_t/A_1 for isentropic supersonic flow in the inlet convergence and $M_t = 1.0$.

The curve presented in Fig. 7.52 applies only after the inlet is started. Consequently, an operating point on or above the curve presented in Fig. 7.51 must first be obtained before Fig. 7.52 is applicable. The two curves of Figs. 7.51 and 7.52 are presented together in Fig. 7.53 to illustrate the overall operating features of the converging-diverging supersonic inlet.

Refer to Fig. 7.53 and consider an inlet having a fixed area ratio $A_t/A_1 = 0.70$. During start-up, only the upper curve in Fig. 7.53 is physically meaningful. The horizontal line ab traces the operation of the inlet as M_o is increased. At point b the shock wave is swallowed, and the lower curve becomes physically meaningful. If the flight Mach number M_o is decreased, the operating point moves from b toward c. In the region between b and c, the shock wave is swallowed and is located in the inlet divergence. At point c, the shock wave is located at the inlet throat where $M_t = 1.0$, and the shock strength is vanishingly small; that point corresponds to Fig. 7.50h. If M_o is decreased further, the shock wave is disgorged. Thus, there is a hysteresis effect, in that the shock wave is swallowed at a flight Mach number corresponding to point b, but the shock wave is not disgorged until the flight Mach number corresponds to point c.

At point c in Fig. 7.53, $M_o = 1.80$. Consequently, a fixed geometry inlet having $A_t/A_1 = 0.7$, designed for shock free operation at $M_o = 1.8$, would have to be overspeeded to $M_o = 3.2$ to start the inlet. To do that, a propulsion system would be required that is capable of propelling the vehicle to $M_o = 3.2$ with an unstarted inlet, a totally unrealistic situation. Overspeeding for starting the subject type of inlet is, therefore, practical only for very low supersonic design Mach numbers. In fact, the

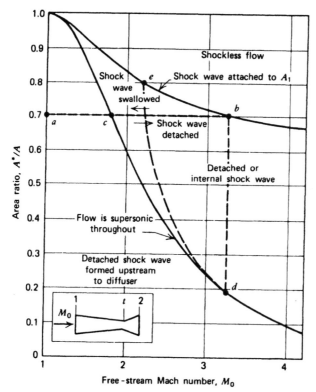

Figure 7.53 Area ratio for detached, attached, and swallowed normal shock waves as a function of the free-stream Mach number for the converging-diverging diffuser.

upper curve in Fig. 7.53 approaches the asymptotic limit of $A_t/A_1 \approx 0.60$. From the lower curve, at $A_t/A_1 = 0.60$, $M_o = 2.0$. Consequently, for a design flight Mach number of 2.0, the inlet would have to be overspeeded to a Mach number of infinity to start it, indicating that for design Mach numbers above 2.0, the subject type of inlet is impossible.

A procedure for eliminating the overspeeding problem is to employ a variable geometry inlet. Consider an inlet designed to operate at $M_o = 3.2$. If the area ratio A_t/A_1 of the variable geometry inlet is adjusted so that its value is equal to that at point b in Fig. 7.53, the inlet will start at $M_o = 3.2$. A strong shock wave will, however, position itself in the inlet divergence. If the area of the variable throat is then reduced, the operating point for the inlet will move vertically downward from point b toward point d. When $A_t/A_1 = 0.2$ approximately, the inlet operates at point d, and the entire flow through the inlet is isentropic. Alternately, if the area ratio A_t/A_1 is adjusted initially to a value larger than that corresponding to point b in Fig. 7.53, for example, point e, the inlet starts at a lower Mach number. As the flight speed is increased, the area of the variable throat is reduced simultaneously, so that the operating path of the inlet follows a path such as curve ed in Fig. 7.53. It is apparent that starting a *variable geometry inlet* is a straightforward procedure.

The major drawback to the practical application of a variable geometry converging-diverging passage as an axisymmetric inlet is its mechanical complexity. For a two-dimensional planar inlet, the mechanical design can be accomplished, and such inlets have been constructed. For applications requiring an axisymmetric inlet, which are quite common, the conical shock wave or spike inlet discussed in Section 7–13(e) has proven to be quite successful.

7-13(e) Conical Shock Wave or Spike Inlet

It was pointed out by K. Oswatitsch that the large decrease in stagnation pressure across a normal shock wave may be reduced by decelerating the free-stream air by means of one or more oblique shock waves, followed by a *weak* normal shock wave.[15] By employing that principle, an efficient external compression of the air is achieved before the air flows into a subsonic internal-compression diffuser.

Figure 7.54 illustrates schematically the essential features of a conical shock wave supersonic inlet employing the aforementioned external compression principle. A central body, or *conical spike*, is placed inside an efficient subsonic internal-compression diffuser, so that the system is *axisymmetric*. The conical nose protrudes through the inlet section of the subsonic diffuser into the free-stream air.

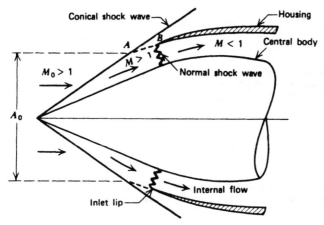

Figure 7.54 Schematic diagram of a conical shock wave (Oswatitsch type) supersonic inlet.

When a supersonic flow impinges on the cone, a conical shock wave is produced [see Section 7-10(b)], as illustrated in Fig. 7.54. The compressed air leaving the conical shock wave enters the subsonic internal-compression diffuser through the annular opening formed between the surface of the *central body* and the *housing* of the subsonic diffuser. Theoretically, a *weak normal shock wave* is formed at the lip of the subsonic diffuser, and the air enters that diffuser with a subsonic velocity.

It is evident from Fig. 7.54 that the conical central body deflects the supersonic airstream from its initial flow direction, so that the weak normal shock wave is perpendicular to the average direction of the streamlines at the inlet to the subsonic diffuser, and not to the direction of the free-stream air. The supersonic diffusion is accomplished externally before the air enters the subsonic diffuser. The annular flow area for the *internal flow* is smaller than A_o; the external compression achieved by the conical shock wave is, therefore, quite strong. The configuration illustrated in Fig. 7.54 is called an *external-compression inlet*, because all of the *supersonic* diffusion takes place external to the inlet housing.

The position of the normal shock wave with respect to the inlet has a profound influence on the performance of the inlet. There are three distinct operating conditions under which the spike inlet for an air-breathing engine can operate. Refer to Fig. 7.55. When the normal shock wave is positioned at the inlet lip, as illustrated in Fig. 7.55a, the operation is said to be *critical*. When the exit pressure in the subsonic internal-compression diffuser is too low for maintaining the normal shock

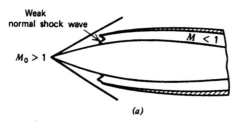

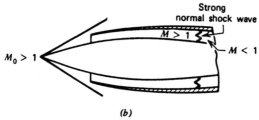

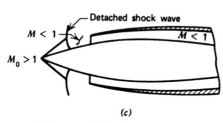

Figure 7.55 Three principal operating modes for supersonic external compression diffusers. (*a*) Critical operation. (*b*) Supercritical operation. (*c*) Subcritical operation.

wave at the lip of the inlet, the air flows into the subsonic diffuser with a supersonic speed, and a strong normal shock wave forms in the diverging portion of the subsonic diffuser, as illustrated in Fig. 7.55*b*. That type of operation is said to be *supercritical*. If the exit pressure from the subsonic diffuser exceeds the static pressure that can be accommodated by the engine downstream to the diffuser, the normal shock wave is expelled from the diffuser and moves upstream toward the vertex of the conical center body, as illustrated in Fig. 7.55*c*. Behind the normal shock wave the flow is subsonic and, because the normal shock wave is detached from the inlet lip, there is *spill-over* of the incoming air over the inlet housing, thereby increasing the external drag. The operation of the diffuser under the latter condition is said to be *subcritical*.

In general, for an air-breathing engine, the position of the normal shock wave with reference to the inlet and the amount of *spill-over* air depend on the flight Mach number, the fuel-air ratio, the combustion efficiency, and the area of the nozzle exit cross section.

In general, the energy losses encountered with a conical shock wave inlet may be grouped into two classes: (1) those pertinent to the external shock wave compression in the supersonic inlet, and (2) those caused by wall friction and separation phenomena in the subsonic internal-compression diffuser (see Section 5–10). It is desirable for the normal shock wave in front of the subsonic diffuser to be weak in order to reduce the tendency for the flow to separate as it flows in the direction of a positive pressure gradient in the subsonic diffuser.

Figure 7.56 illustrates schematically a *mixed compression inlet*, wherein a portion of the internal flow is supersonic. Because the flow entering the inlet is supersonic, a series of oblique shock waves forms in the annular passage between the inlet housing and centerbody, which are terminated by a normal shock wave. The normal shock wave positions itself in the diverging flow passage at a location determined by the back pressure created by the remainder of the engine. By choosing the design point so that the normal shock wave is downstream from the minimum cross section, the supersonic portion of the diffuser is not influenced by the flow fluctuations in the remainder of the flow system, provided those fluctuations are not large enough to force the normal shock wave into the converging portion of the duct. If the latter occurs, the shock wave will be disgorged, and the diffuser operation becomes subcritical (see Fig. 7.55c).

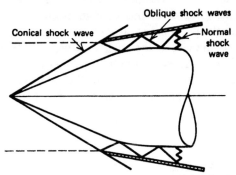

Figure 7.56 Mixed compression supersonic inlet.

The starting problem discussed in Section 7–13(d) for the converging-diverging supersonic inlet must also be considered for the conical shock wave or spike inlet. The problem is much less severe, however, since the Mach number at the entrance to the inlet housing is lower than the free-stream Mach number because of the formation of the conical shock wave. In addition, the central spike may be translated in the axial direction, thereby varying both the inlet entrance area and the minimum flow area. By the proper matching of the aforementioned flow areas as a function of spike position, the inlet may be started without difficulty. During off-design operation (i.e., flight Mach numbers other than the design value and angles of attack other than zero), the spike may be translated so that the conical shock wave is always positioned at the inlet lip.

The maximum stagnation pressure recovery, at zero angle of attack, has been emphasized as a performance criterion. Inlets for air-breathing propulsion engines must, however, give satisfactory performance over the anticipated range of angles of attack. Experience has demonstrated that the spike inlet has satisfactory performance characteristics with regard to variations in the angle of attack. Ordinarily, a decrease of approximately 20 per cent in stagnation pressure recovery results at an angle of attack of 5 deg.

Another factor to be considered is the *external drag* of the inlet. A spike inlet introduces a drag component known as the *additive drag* that arises from the pressure forces acting on the stream tube *AB* illustrated in Fig. 7.54; the latter drag is a *net force* acting in the direction opposite to the flight direction. By positioning the conical shock wave at the lip of the inlet, the additive drag may be eliminated. The longer the distance between the conical shock wave and the inlet lip is, the larger the additive drag is.

In general, high stagnation pressure recovery with a spike inlet is accompanied by a large external drag, and vice versa. The external drag is high because the large cowling angles employed may induce flow separation on the exterior housing. By careful aerodynamic design, however, the external drag may be kept at tolerable values.

Larger stagnation pressure recoveries are obtainable if several successive weak shock waves instead of one relatively strong conical shock wave are utilized for decelerating the air. To increase the number of oblique shock waves, several "breaks" are made on the projecting cone, as illustrated schematically in Fig. 7.57. In the extreme case the spike configuration is a curved surface giving an infinite number of infinitely weak shock waves, that is, Mach waves (see Section 8–6). In theory, such a surface achieves *shockless* deceleration of a supersonic flow to the *sonic* speed; that is, the compression is *isentropic*. A spike of this configuration is termed an *isentropic spike*. Unfortunately, an isentropic spike is satisfactory for only one flight Mach number and is very sensitive to changes in the angle of attack. Furthermore, at high flight Mach numbers, the flow is turned through large angles to enter the subsonic internal-compression diffuser, which may increase the additive drag to prohibitive values and may also affect the pattern of the flow into the subsonic diffuser.

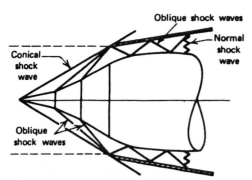

Figure 7.57 Multiple oblique shock wave supersonic inlet.

It appears that a spike inlet for air-breathing engines for propelling vehicles at Mach numbers up to $M_o = 2.0$ need have only one conical shock wave. For flight Mach numbers above $M_o = 2.0$, two or more oblique shock waves may be required for achieving the desired stagnation pressure recovery. Increasing the number of oblique shock waves makes the diffusion system more sensitive to changes in the flight Mach number and in the angle of attack. It appears that two oblique shock waves are sufficient for a design flight Mach number of approximately $M_o = 3.0$.

Figure 7.58 presents the stagnation pressure recovery ratio P_2/P_o as a function of the free-stream Mach number M_o for several types of supersonic diffusers.[15]

Example 7.17. A two-dimensional planar inlet for an air-breathing engine for supersonic aircraft may consist of a wedge forebody leading into a cowl. A cross section through such an inlet is illustrated in Fig. 7.59. The purpose of the forebody is to generate an oblique shock wave, thus slowing down, or *diffusing*, the external flow

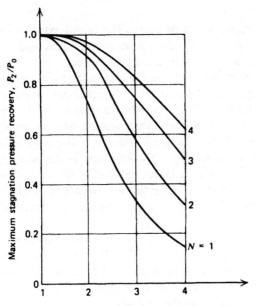

Figure 7.58 Maximum stagnation pressure recovery for multiple oblique shock wave diffusers as a function of free-stream Mach number M_0 (Based on Reference 15).

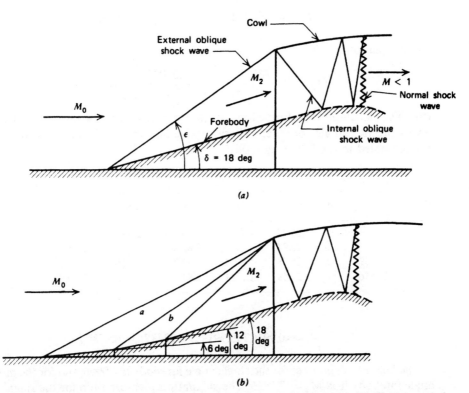

Figure 7.59 Forebodies for planar supersonic inlets. (a) Single angle forebody. (b) Multiple angle forebody.

to a lower Mach number and higher static pressure. The initial external oblique shock wave is followed by a series of internal oblique shock waves and a single normal shock wave for reducing the velocity to a subsonic value.

One performance criterion of a supersonic inlet is the stagnation pressure recovery obtainable with its shock wave system (see equation 7.124); the latter is expressible in terms of the stagnation pressure ratio P_2/P_o, which is a measure of the entropy increase for the air (see equation 3.88). The stagnation pressure ratio for an oblique shock wave depends on the flow deflection angle δ (the larger the value of δ, the smaller the value of P_2/P_o). A series of smaller flow deflection angles will produce a series of weak oblique shock waves, with a larger value of P_2/P_o than would be produced from a single wedge having the same total flow deflection angle in one step. The multiple angle forebody type of inlet is illustrated schematically in Fig. 7.59b.

To illustrate the higher stagnation pressure recovery of the multiple angle forebody inlet, consider an inlet designed to operate at $M_1 = 3.0$. The total flow deflection angle desired on the forebody is 18 deg. Calculate the final Mach number and the forebody stagnation pressure ratio for (a) a single flow deflection angle of 18 deg, and (b) for a series of three flow deflection angles of 6 deg each.

Solution

(a)   First, consider the single angle forebody. From Table C.12, for $M_o = 3.0$ and $\delta = 18$ deg, $\varepsilon = 35.47$ deg. From equation 7.99,

$$M_o' = M_o \sin \varepsilon = 3.0 \sin (18) = 1.7408$$

From Table C.11, for $M_o' = 1.7408$,

$$M_2' = 0.63032 \quad \text{and} \quad \frac{P_2}{P_o} = 0.83853$$

From equation 7.100,

$$M_2 = \frac{M_2'}{\sin (\varepsilon - \delta)} = \frac{0.63032}{\sin (35.47 - 18)} = 2.0996$$

(b)   The calculations performed in part (a) are repeated below three times for equal flow deflection angles of 6 deg. Those results, and the results obtained in part (a), are presented in the following table.

Part	M	δ	ε	M_o'	M_2'	M_2	P_2/P_o
(a)	3	18	35.47	1.7408	0.63032	2.0996	0.83852
(b)	3	6	23.94	1.2173	0.83162	2.6999	0.99102
	2.6999	6	26.58	1.2081	0.83718	2.3816	0.99199
	2.3816	6	29.59	1.1761	0.85744	2.1426	0.99487

$$\frac{P_2}{P_o} = (0.99102)(0.99199)(0.99487) = 0.97804$$

The final Mach number for the single angle forebody is 2.0996, and for the multiple angle forebody it is $M_2 = 2.1426$. The stagnation pressure ratio for the single angle forebody is 0.83852, which corresponds to a stagnation pressure loss of 16.15 percent.

For the multiple angle forebody, $P_2/P_o = 0.97804$, and the corresponding stagnation pressure loss is 2.20 percent. Consequently, the flow for the multiple angle forebody is almost isentropic, and a considerable performance increase is obtained compared to that for the single angle forebody.

7–14 SUMMARY

Under certain conditions a *normal shock wave* may be formed perpendicular to the direction of the flow of a body of compressible fluid, which is far removed from solid surfaces. The shock wave is an exceedingly rapid irreversible compression of the gas, and it is very thin, only a few mean free paths in thickness. In the relative coordinate system, where the observer moves at the same speed as the shock wave, the latter appears to be stationary with a supersonic flow ($M_1 > 1$) of gas entering it and a subsonic flow ($M_2 < 1$) leaving it. The reduction in the kinetic energy associated with the gas flowing through the normal shock wave goes into compressing the gas from p_1 to $p_2 > p_1$, raising the gas temperature from t_1 to $t_2 > t_1$, increasing its entropy from s_1 to $s_2 > s_1$, while the stagnation temperature $T_1 = T_2 = T =$ constant. Furthermore, the strength of the normal shock wave $\beta = (p_2 - p_1)/p_1$ increases as M_1 is increased.

If the flowing fluid is a perfect gas, some rather simple quantitative relationships may be derived relating the conditions p_1, ρ_1, t_1, s_1, and V_1 in front of the normal shock wave to p_2, ρ_2, t_2, s_2, and V_2 behind it. The aforementioned flow properties may be related to M_1 and M_2.

The entropy change for the normal shock wave, expressed by $\Delta s/c_v$, depends on the pressure ratio p_2/p_1 and the specific heat ratio for the gas γ. Hence, for a given gas, $\Delta s/c_v$ depends only on p_2/p_1. In terms of the shock strength $\beta = (p_2 - p_1)/p_1$, it is shown (see equation 7.63) that for a *weak shock wave*, the entropy increase is insignificant, indicating that a weak shock wave produces an efficient compression of the gas. In other words, if M_1 is not significantly larger than unity (a weak shock wave condition), then no significant increase in entropy (or decrease in stagnation pressure) is experienced by the gas in flowing through the normal shock wave.

For a normal shock wave in a perfect gas, the flow property ratios (M_2, p_2/p_1, ρ_2/ρ_1, t_2/t_1, P_2/P_1, and P_2/p_1) are functions of M_1 and γ. Consequently, for a given gas, they are functions solely of M_1. Table C.11 presents those variables as functions of M_1 for different values of γ.

A body that moves in a frictionless fluid at supersonic speed and causes shock waves to form produces a drag force that is known as either *shock* or *wave drag* [see Section 7–4(f)].

If a planar supersonic flow ($M_1 > 1$) has its direction changed by a frictionless plane that makes a *concave corner* with the initial flow direction, the final flow velocity V_2 makes the *flow deflection angle* δ with the initial flow velocity V_1. An oblique shock wave emanates from the corner and makes the wave angle ε with V_1 (see Fig. 7.16). Because the initial flow is turned through the angle δ, the oblique shock wave has an additional degree of freedom compared to the normal shock wave; that is, the flow properties behind the oblique shock wave depend not only on the upstream conditions but also on δ. Consequently, the governing equations for an oblique shock wave are more complicated than those for a normal shock wave. The continuity, momentum, and energy equations, when expressed in terms of the normal components of the gas velocities in front of and in back of the oblique shock wave, V_{N1} and V_{N2}, respectively, are identical to those for the normal shock wave. Consequently, the relations derived for the normal shock wave are applicable to the oblique shock wave, if V_1 and V_2 for the normal shock wave are replaced by V_{N1} and V_{N2}. From a general point of view, the normal shock wave may be regarded as a special case of the oblique shock wave where the wave angle $\varepsilon = \pi/2$.

The strength of an oblique shock wave depends not only on M_1 but also on the flow deflection angle δ, and there is a maximum value of δ, denoted by δ_m, for which the oblique shock wave is attached to the concave corner; the corresponding wave angle is $\varepsilon = \varepsilon_m$. If $\delta > \delta_m$ the oblique shock wave detaches from the concave corner, and further increase in δ causes the shock wave to move upstream and change its configuration (see Fig. 7.21).

Table C.11, for the normal shock wave, may be employed for the oblique shock wave, by replacing M_1 for the oblique shock wave by $M_1' = M_1 \sin \varepsilon$. The corresponding value for $M_2 = M_2'/\sin(\varepsilon - \delta)$.

The results obtained for the oblique shock wave may be applied directly to the steady two-dimensional planar supersonic flow over a wedge. If the wedge is unsymmetrical with respect to the incident flow, the flows over its top and bottom surfaces must be studied separately. If the semiangle of a symmetrical wedge δ_w is smaller than the maximum flow deflection angle δ_m, the oblique shock wave becomes attached to the vertex of the wedge, and the flow in back of the oblique shock wave becomes parallel to the surfaces of the wedge. If $\delta_w > \delta_m$, the shock wave is detached and a *bow shock wave* moves upstream from the vertex; the distance it moves upstream depends on M_1 and δ_w.

The flow over a cone is a two-dimensional axisymmetric flow. Solutions of the basic differential equations may be obtained by numerical techniques. Because of the continuity requirement for the flow behind a *conical shock wave* attached to the vertex of a cone, the streamlines are curved and converge toward the surface of the cone. Consequently, the pressure, velocity, and density behind the conical shock wave are not uniform as is the case for the attached oblique shock wave for a wedge. Because of the convergence of the streamlines, there is an isentropic compression of the gas between the conical shock wave and the surface of the cone. The detachment angle for a cone is larger than it is for a wedge (see Fig. 7.32).

Oblique shock waves may be reflected from solid surfaces and jet boundaries, and interact with each other. The rules for the reflection, interaction, and cancellation of oblique shock waves are summarized in Section 7-11(e).

REFERENCES

1. M. J. Zucrow, *Aircraft and Missile Propulsion*, Vol. 1, Chap. 3, Wiley, New York, 1958.
2. A. H. Shapiro, *The Dynamics and Thermodynamics of Compressible Fluid Flow*, Chap. 5, Ronald Press, New York, 1953.
3. A. Ferri, *Elements of Aerodynamics of Supersonic Flow*, Macmillan, New York, 1949.
4. R. Courant and K. O. Friedrichs, *Supersonic Flow and Shock Waves*, Interscience Publishers, New York, 1948.
5. Ames Research Staff, "Equations, Tables and Charts for Compressible Flow," NACA Report 1135, 1953.
6. M. Linzer and D. F. Hornig, "Structure of Shock Fronts in Argon and Nitrogen," *Physics of Fluids*, Vol. 6, No. 12, pp. 1661–1668, December 1963.
7. J. H. Keenan and J. Kaye, *Gas Tables*, Wiley, New York, 1945.
8. R. von Mises, *Mathematical Theory of Compressible Fluid Flow*, Academic Press, New York, 1958.
9. G. I. Taylor and J. W. Maccoll, "The Air Pressure on a Cone Moving at High Speeds," *Proceedings of the Royal Society of London*, Series A, Vol. 139, pp. 279–311, 1933.
10. J. W. Maccoll, "The Conical Shock Wave Formed by a Cone Moving at High Speed," *Proceedings of the Royal Society of London*, Series A, Vol. 159, pp. 459–472, 1937.

11. Z. Kopal, "Tables of Supersonic Flow around Cones," Report No. 1, Massachusetts Institute of Technology, Department of Electrical Engineering, Center of Analysis, 1947.
12. J. Sims, "Tables for Supersonic Flow around Right Circular Cones at Zero Angle of Attack," NASA SP–3004, 1964.
13. M. J. Zucrow, *Aircraft and Missile Propulsion*, Vol. 1, Chap. 5, Wiley, New York, 1958.
14. J. V. Foa, *Elements of Flight Propulsion*, Chap. 8, Wiley, New York, 1960.
15. K. Oswatitsch, "The Efficiency of Shock Diffusers," NACA TM 1140, 1947.

PROBLEMS

1. A stream of air flowing with $M = 4$, $p = 1.0 \cdot 10^5$ N/m^2, and $t = 300$ K passes through a stationary normal shock wave. Calculate the following properties behind the shock wave, assuming that air behaves as a perfect gas with $c_p = 1004.8$ J/kg-K: (a) p, (b) t, (c) P, (d) T, (e) V, and (f) M.

2. Work Problem 1, taking into account the variation of specific heat with temperature.

3. A shock wave created by an explosion is traveling into the surrounding atmosphere at a velocity of 1700 m/s. The static properties of the atmosphere are $p_o = 1.0135 \cdot 10^5$ N/m^2 and $t_o = 290$ K. Assume that the air behaves as a perfect gas with $c_p = 1004.8$ J/kg-K. Calculate the following properties relative to a stationary observer behind the shock wave: (a) p, (b) t, (c) M, (d) P, and (e) T.

4. Work Problem 3, taking into account the variation of c_p with t.

5. Air is flowing at 180 m/s through a constant-area pipe. The end of the pipe is suddenly closed, and a normal shock wave propagates back into the pipe, which is 30 m long. How long will it be before the effect of the closing of the pipe is felt at the pipe inlet? Assume that the air behaves as a perfect gas with a constant specific heat.

6. Calculate the time asked for in Problem 5, taking into account the effect of temperature on the specific heat of air.

7. A Pitot tube placed in a supersonic wind tunnel flowing air indicates a stagnation pressure $P = 6.550 \cdot 10^5$ N/m^2. The Mach number in the tunnel $M = 4.0$. Calculate the tunnel free-stream static pressure assuming that the air behaves as a perfect gas.

8. Air is flowing in a converging-diverging nozzle with an area ratio of 10. The inlet stagnation pressure $P = 7.0 \cdot 10^5$ N/m^2, and the pressure in the exhaust region is 1 atm. A normal shock wave is standing in the nozzle divergence. Determine the area ratio at which the normal shock wave occurs. Assume $\gamma = 1.40$.

9. The nozzle described in Problem 8 discharges into a receiver that is initially evacuated. As the receiver is filled by the nozzle exhaust gases, the effective back pressure presented to the nozzle increases. Eventually a normal shock wave appears at the nozzle exit plane and propagates upstream as the back pressure increases. Calculate the location (i.e., the area ratio) of the normal shock wave as a function of the pressure in the receiver.

10. Air, assumed to be a perfect gas, flows from the storage tanks of a blow-down wind tunnel where $t = 290$ K and $p = 70 \cdot 10^5$ N/m^2. A symmetrical wedge having a semiangle $\alpha = 15$ deg is placed in the test section where $M = 3.0$. Calculate the following properties on the surface of the wedge: (a) t, (b) p, (c) M, (d) the wave angle ε, and (e) the minimum Mach number for which the shock wave will remain attached to the wedge. Assume $\gamma = 1.40$.

11. The wedge described in Problem 10 is moved to a location where the wave angle ε is observed optically to be 30 deg. At that point, calculate (a) the free-stream values of M, p, and t, and (b) the same flow properties on the surface of the wedge.

12. Work Problems 10 and 11 employing the hodograph shock polar presented in Fig. 7.25.

13. The leading edge of a control fin on a reentry body may be approximated as a wedge having an included angle of 10 deg. Calculate the flow properties on the surface of the wedge, at zero angle of attack, when the free-stream velocity is 1500 m/s and the altitude is 30,000 m (a) assuming that air behaves as a perfect gas, and (b) taking into account the variation of specific heat with temperature.

14. A wedge having a total included angle of 30 deg is placed in a wind tunnel where $M = 3.0$. Determine the maximum angle of attack of the wedge for which the oblique shock waves will remain attached.

15. A right circular cone with a total included angle of 30 deg is placed in the wind tunnel described in Problems 10 and 11. Work Problems 10 and 11 for the aforementioned symmetrical cone.

16. A two-dimensional supersonic inlet comprises two 15 deg wedges placed on the top and bottom surfaces of a rectangular flow passage. An oblique shock wave propagates into the inlet from the leading edge of both wedges, as illustrated in Fig. 7.39a (where the values of δ are different). Calculate the flow properties M, p, and t behind the initial shock waves and the transmitted shock waves if $M_o = 3.0$. Assume that $\gamma = 1.40$, $p_o = 1.0135 \cdot 10^5$ N/m^2, and $t_o = 300$ K.

17. The inlet described in Problem 16 is oriented at a 5 deg angle of attack. Determine the flow properties M, p, and t, behind the initial and transmitted shock waves for the conditions specified in Problem 16.

18. A variable geometry converging-diverging inlet is to be designed for flight at $M_o = 2.0$. Calculate the area ratio for achieving the design condition of isentropic flow throughout, and also the area ratio required for swallowing the starting shock wave. Sketch the flow field just before the shock wave is swallowed, and for operation at the design condition, noting the values of the Mach number at all points of interest. Assume that air behaves as a perfect gas.

19. A two-dimensional planar wind tunnel nozzle is designed to give a parallel uniform flow at a Mach number $M = 3.0$ when air is the flowing fluid ($\gamma = 1.40$). The test gas is supplied from a blow-down air supply initially at a pressure of $70 \cdot 10^5$ N/m^2, and the nozzle exhausts to the atmosphere ($p_o = 1$ atm). During operation, the pressure in the supply reservoir decreases. At what supply pressure will oblique shock waves first appear in the exhaust jet? A test region extending one diameter downstream and 10 percent of the diameter in height is required, in which the flow is shock free. What is the minimum supply pressure for obtaining the desired test region?

20. Air flows in a rectangular passage with a Mach number $M = 3.0$. The bottom wall of the passage abruptly turns into the flow with the angle δ, thereby propagating an oblique shock wave into the flow. Determine the maximum value of δ for regular reflection of the initial shock wave, for (a) one, (b) two, (c) three, and (d) four reflections.

21. A stream of air flows in an insulated pipe having a diameter of 0.10 m at the mass flow rate of 1.0 kg/s. The stagnation temperature is 295 K, and the average friction coefficient $\bar{f} = 0.002$. (a) At the pipe inlet, the pressure is $0.140 \cdot 10^5$ N/m^2. Find the Mach number, velocity and stagnation pressure. (b) At the exit section, find the maximum length and the corresponding pressure for shockless flow. (c) For the length found in part (b), what must the pressure be at the exit section for a normal shock wave to form at the inlet?

22. An isentropic converging-diverging nozzle designed for $M = 2.2$ is supplied with air from a reservoir at $p = 8.0 \cdot 10^5$ N/m^2 and $t = 300$ R. The nozzle exit is connected to a perfectly insulated pipe of length L, diameter D, and average friction coefficient $\bar{f} = 0.0025$. The pipe discharges into a space where the pressure is maintained at $3.5 \cdot 10^5$ N/m^2. Calculate the mass flux G, and the ratio L/D for a normal shock to stand at (a) the nozzle throat, (b) the nozzle exit, and (c) the duct exit.

23. Find the speed of an explosion wave that causes the atmospheric air temperature to rise from 280 K to 320 K immediately after the passage of the wave.

24. Air flows through a frictionless converging-diverging nozzle. The area of the exit section A_e is four times the area of the throat section A_t, and the ratio of the stagnation pressure at the entrance to the static pressure in the exit section is 3. The semiangle of the divergence is 10 deg. Assume that air is a perfect gas, that $\gamma = 1.4$ and is constant, one-dimensional flow, and that the only loss is due to the formation of a normal shock wave in the diverging portion of the nozzle. Calculate (a) the Mach number M_x where the normal shock occurs, (b) the area A_x where the shock wave occurs, and (c) the increase in entropy.

25. A subsonic diffuser operating under isentropic conditions with air has an inlet area of $0.15\,m^2$. The inlet conditions are $V_1 = 240$ m/s, $t_1 = 300$ K, $p_1 = 0.70 \cdot 10^5$ N/m^2. The velocity leaving the diffuser is 120 m/s. Calculate (a) the mass flow rate, (b) the stagnation temperature at the exit, (c) the stagnation pressure at the exit, (d) the static pressure at the exit, (e) the entropy change across the diffuser, and (f) the exit area.

26. Air at 310 K and $1.030 \cdot 10^5$ N/m^2 enters a diffuser with a velocity of 245 m/s. The diffuser is to be designed to reduce the velocity of the air to 60 m/s. The mass flow rate through the diffuser is 13.6 kg/s. Find, assuming isentropic flow, (a) the inlet diameter, (b) the outlet diameter, and (c) the rise in static pressure.

27. Air enters an isentropic diffuser with a Mach number of 3.6 and is decelerated to a Mach number of 2. The diffuser passes a mass flow rate of 15 kg/s. The initial static pressure and temperature of the air are $1.030 \cdot 10^5$ N/m^2 and 310 K, respectively. Calculate, assuming $\gamma = 1.4$, (a) the area of diffuser inlet, (b) the area of the diffuser exit, (c) the stagnation temperature for the inlet air, (d) the stagnation pressure for the inlet air, (e) the stagnation temperature for the outlet air, (f) the stagnation pressure for the outlet air, and (g) the static pressure of the outlet air.

28. A normal shock wave diffuser is employed for diffusing a supersonic stream from the free-stream Mach number M_o to the subsonic Mach number M_b. Plot the stagnation pressure ratio for the shock wave P_h/P_o as a function of M_o.

29. A normal shock wave is formed ahead of the diffuser of a turbojet engine operating at $M = 1.3$ at 15,000 m. Calculate the stagnation pressure and Mach number just after the shock wave.

30. A ramjet having a two-dimensional wedge inlet operates at a Mach number of 2.40. The semiangle of the wedge is 15 deg. A normal shock wave occurs at the inlet to the subsonic diffuser. The isentropic efficiency of the subsonic diffuser is 0.90. Calculate the stagnation pressure ratio of the diffuser assuming $\gamma = 1.4$ and that the flow between the oblique shock wave and the normal shock wave is isentropic.

31. Calculate the improvement, in percent, in the stagnation pressure ratio obtained in Problem 30 if a second oblique shock wave is formed by a second wedge having an angle of 10 deg located downstream from the vertex of the first wedge (total half angle = 25 deg).

32. In Example 7.17 a multiple oblique shock wave inlet is described where the total turning angle of 18 deg is accomplished by three flow deflection angles of 6 deg each, and the corresponding stagnation pressure recovery $\eta_p = 0.97804$. Determine the stagnation pressure recovery of a multiple oblique shock wave inlet operating at the same conditions, for (a) two equal flow deflection angles of 9 deg each, and (b) four equal flow deflection angles of 4.5 deg each.

8
expansion waves

8–1 PRINCIPAL NOTATION FOR CHAPTER 8

The notation for Chapter 3 and the additional notation presented below apply to Chapter 8.

b = $(\gamma + 1)/(\gamma - 1)$.
u^* = u/a^*, dimensionless velocity component.
v^* = v/a^*, dimensionless velocity component.
V velocity magnitude.
V_N component of velocity normal to a wave.
V_T component of velocity tangential to a wave.

Greek Letters

α $= \sin^{-1}(1/M)$, Mach angle.

$\bar{\alpha}$ average Mach angle.

δ $= v_2 - v_1$, flow deflection angle.

$\bar{\delta}$ average flow deflection angle.

θ streamline angle.

v Prandtl-Meyer angle; $v = \delta$ when $M_1 = 1$.

Subscripts

1   conditions before an expansion wave.

2   conditions after an expansion wave.

8-2 INTRODUCTION

It is shown in Section 7-7 that a steady, uniform, two-dimensional supersonic flow into a concave corner (see Fig. 7.16) is turned through the flow deflection angle δ by an oblique shock wave, which behaves as a *flow discontinuity*. In passing through the oblique shock wave, the entropy of the fluid is increased. If the corner is *convex* (see Fig. 8.1), and an oblique shock wave were to emanate from it, the fluid would be required to expand in flowing through the discontinuity (shock wave) with an accompanying decrease in its entropy, a requirement that violates the second law of thermodynamics. Consequently, the wave emanating from the convex corner is an *expansion wave*.

This chapter is concerned with the theory pertinent to expansion waves in a steady two-dimensional planar supersonic flow.

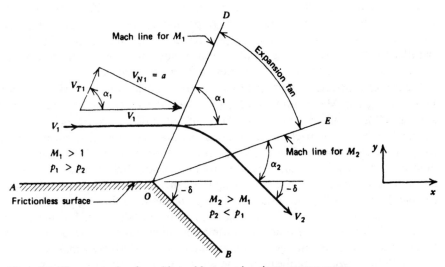

Figure 8.1 Flow expansion from M_1 to M_2 around a sharp convex corner.

8-3 SIMPLE WAVES[1-4]

Expansion waves and compression waves are phenomena that may appear in supersonic flows. It is shown in Chapter 7 that if the initial flow is steady and uniform, oblique shock waves may be analyzed by applying one-dimensional techniques.

Certain types of expansion waves in a steady uniform parallel two-dimensional flow may also be analyzed by one-dimensional techniques. Such expansion waves are called either *simple waves* or *Prandtl-Meyer waves*, after L. Prandtl and T. Meyer who originated the theory of such expansion waves in 1908.[1]

In this section, the general features and theory of simple waves are discussed, and their governing equations are derived. In Section 8–4, solutions to the latter equations are obtained for a perfect gas, and in Section 8–5 their solutions for an imperfect gas are presented.

8–3(a) General Features of Simple Waves

It is pointed out in Section 3–8(b) that if a small pressure disturbance is propagated from a *geometric point* into a supersonic flow field, its influence is confined to the region enclosed within the Mach cone (see Fig. 3.15); for the latter, the Mach angle α is given by

$$\alpha = \sin^{-1}\left(\frac{1}{M}\right)$$

The strength of the small disturbance considered in Section 3–8(b) is comparable to that of a sound wave. Moreover, because a point source does not occupy any finite space, it cannot produce a finite disturbance. Any changes in the properties of the fluid in back of the Mach cone must, therefore, be infinitesimally small. Consequently, a Mach line in a supersonic flow may be regarded as the limit approached by a shock wave as its strength $\beta = (p_2 - p_1)/p_1$ approaches zero.

The Mach angle α has physical significance only for either a sonic flow ($M = 1$) or a supersonic flow ($M > 1$), because $\sin \alpha$ cannot exceed unity. For a sonic flow, $\alpha = 90$ deg, and the Mach cone transforms into a plane perpendicular to the flow velocity. Figure 8.2 illustrates schematically a steady uniform two-dimensional flow in the xy plane. The streamlines are parallel to the x axis, and the Mach lines for a sonic flow are straight lines perpendicular to the streamlines; that is, they are parallel to the y axis. For a steady uniform supersonic flow, the Mach lines are straight lines inclined to the streamlines at the Mach angle α.

Figure 8.1 illustrates schematically a steady uniform two-dimensional supersonic flow, having the Mach number M_1, that wets the frictionless surface AO, where surface AO is parallel to the x axis. At point O the flow encounters a sharp convex corner formed by the frictionless surfaces AO and OB; note that surface OB *slopes away* from the incident flow, that has the uniform velocity V_1. Obviously, if the flow is eventually to become parallel to the surface OB, it must accelerate after encountering the disturbance caused by the convex corner O; that is, *the fluid must expand*. Consequently, the static pressure p_2 of the fluid flowing parallel to surface OB must be smaller than p_1, the static pressure for the fluid flowing parallel to surface AO. The steady uniform two-dimensional supersonic flow around a convex corner described in the foregoing is known as a *Prandtl-Meyer expansion*.

The Prandtl-Meyer analysis for the steady uniform supersonic flow of a compressible fluid around a sharp convex corner is based on the following assumptions and observations.

1. The flow is steady two-dimensional and isentropic throughout the flow field.
2. Before the compressible fluid begins turning around the corner, all of the streamlines are straight and parallel to the surface AO (Fig. 8.1).
3. All of the flow properties in the flow field upstream to where the turn commences have constant values.

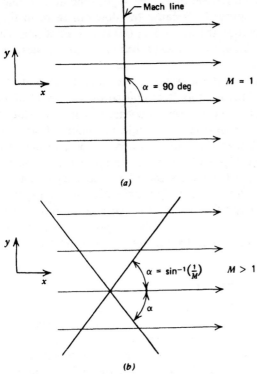

Figure 8.2 Mach lines in steady uniform sonic and supersonic flows. (*a*) Sonic flow. (*b*) Supersonic flow.

4. All of the flow properties in the flow field are constant after the turning of the fluid is completed, so that the streamlines for the turned flow are parallel to the surface *OB*.
5. All of the flow properties along every Mach line emanating from the corner *O* are constant; consequently, each Mach line is a straight line.

No justification for item 5 is presented in this chapter. In Section 12–6, however, where a rigorous mathematical basis is employed for developing the concept of simple waves, it is shown that such waves have the following intrinsic features:

1. A simple wave region is one in which the velocity magnitude V and the streamline angle θ are not independent [i.e., $V = V(\theta)$].
2. In a simple wave region, the Mach lines of one family are straight lines, and along each Mach line of that family, the flow properties are constant.
3. If a region of uniform flow adjoins a nonuniform region, then the nonuniform region must be a simple wave region.

According to item 1, the expansion region, called the *expansion fan* (see Fig. 8.1), associated with the supersonic flow around a convex corner is a simple wave region. Accordingly, each Mach line inside of the expansion fan must be a straight line along which the flow properties are uniform. To completely specify the expansion flow field, the relationship $V = V(\theta)$ must be determined.

Since the disturbance produced by the corner (see Fig. 8.1) is very weak, a Mach wave or Mach line *OD* is propagated radially outward from the corner into the incident supersonic flow. Upstream to the Mach line *OD*, which makes the Mach angle $\alpha = \sin^{-1} (1/M_1)$ with the direction of V_1, the supersonic flow is unaware of the disturbance because of the corner O. Because the flow upstream to the Mach line OD has constant flow properties, the Mach line OD is a straight line. The fluid velocity perpendicular to OD is equal to the local speed of sound a_1, because $V_{N1} = V_1 \sin \alpha_1 = V_1/M_1 = a_1$.

After crossing the Mach line OD, the flow turns and accelerates, and all of the changes in the static pressure of the fluid are transmitted by a series of Mach lines (radial lines) emanating outward from the convex corner O. After crossing the last of that series of Mach lines the expansion of the fluid is completed, and the fluid flows parallel to the surface OB with the velocity V_2. As a consequence of the expansion, the final Mach number $M_2 > M_1$. Let OE denote the final Mach line of the expansion fan; the Mach line OE makes the Mach angle $\alpha_2 = \sin^{-1} (1/M_2)$ with the final direction of the fluid stream. Consequently, in the *expansion fan* bounded by the Mach lines OD and OE, the fluid is accelerated from V_1 to V_2 (or its Mach number changes from M_1 to M_2), and its static pressure is decreased from p_1 to p_2. Moreover, the fluid stream is deflected through the flow deflection angle $-\delta$.

8–3(b) Governing Equation for a Prandtl-Meyer Flow

Figure 8.3 illustrates the turning of a steady uniform supersonic flow through an infinitesimal angle $-d\delta$; the angle δ is termed the *flow deflection angle*. Note that the coordinate system specified in the figure defines δ as being positive when it increases in the counterclockwise direction, which is conventional. Line OD is assumed to be a straight Mach line that turns the flow through the negative flow deflection angle $-d\delta$ while accelerating it from V to $V + dV$.

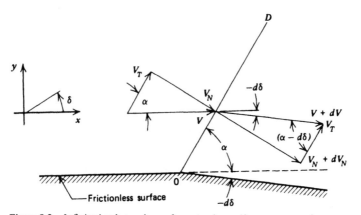

Figure 8.3 Infinitesimal turning of a steady uniform supersonic stream.

The flow deflection process is analyzed in a manner similar to that employed in analyzing the oblique shock wave in Section 7–6. The velocities V and $V + dV$ are resolved into their tangential and normal components, as illustrated in Fig. 8.3. As in the case of an oblique shock wave, there is no pressure gradient along the Mach wave, so that the tangential velocity component has the same value on both sides of the wave (see equation 7.83). Hence,

$$V_T = V \cos \alpha = (V + dV) \cos (\alpha - d\delta) \qquad (8.1)$$

Expanding $\cos(\alpha - d\delta)$ gives

$$\cos(\alpha - d\delta) = \cos\alpha\cos d\delta + \sin\alpha\sin d\delta = \cos\alpha + d\delta\sin\alpha \qquad (8.2)$$

where, for the infinitesimal flow deflection angle $d\delta$, $\cos d\delta \approx 1$ and $\sin d\delta \approx d\delta$. Substituting equation 8.2 into equation 8.1, and dropping higher-order terms, yields

$$\frac{dV}{V} = -\tan\alpha\, d\delta \qquad (8.3)$$

From Fig. 8.4, it is seen that

$$\tan\alpha = \frac{1}{\sqrt{M^2 - 1}} \qquad (8.4)$$

Combining equations 8.3 and 8.4 gives

$$d\delta = -\sqrt{M^2 - 1}\,\frac{dV}{V} \qquad (8.5)$$

Equation 8.5 is the *governing differential equation* for a *Prandtl-Meyer flow*. Its integration requires knowledge of the relationship between the velocity V and the Mach number M. For a perfect gas, the integration of equation 8.5 is presented in Section 8–4. A procedure for taking into account imperfect gas effects is presented in Section 8–5.

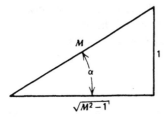

Figure 8.4   Mach angle geometry.

8–4 SIMPLE WAVES IN A PERFECT GAS

In this section, equation 8.5 is integrated for the flow of a perfect gas. Thus, from the definition of the Mach number M and the speed of sound a in a perfect gas (see Section 1–17),

$$V^2 = M^2 a^2 = M^2 \gamma R t \qquad (8.6)$$

For the steady isentropic flow considered here, the stagnation temperature T is constant. From equation 3.181, Table 3.3,

$$\frac{T}{t} = 1 + \frac{\gamma - 1}{2}M^2 \qquad (8.7)$$

Combining equations 8.6 and 8.7 gives

$$V^2 = \frac{\gamma R T M^2}{1 + \dfrac{\gamma - 1}{2}M^2} = \frac{a_o^2 M^2}{1 + \dfrac{\gamma - 1}{2}M^2} \qquad (8.8)$$

where $a_o = (\gamma R T)^{1/2}$ is the stagnation acoustic speed, and is a constant for the flow. Equation 8.8 is the required relationship between V and M. It is valid only for a perfect gas.

Differentiating equation 8.8 logarithmically gives

$$\frac{dV}{V} = \frac{1}{1 + \dfrac{\gamma - 1}{2} M^2} \frac{dM}{M} \qquad (8.9)$$

Eliminating dV/V from equation 8.5 by means of equation 8.9 yields

$$d\delta = -\frac{\sqrt{M^2 - 1}}{1 + \dfrac{\gamma - 1}{2} M^2} \frac{dM}{M} \qquad (8.10)$$

Equation 8.10 is the governing differential equation for the simple wave flow of a perfect gas, in terms of δ and M.

8-4(a) The Prandtl-Meyer Angle

The integration of equation 8.10 yields

$$\delta = -\sqrt{\frac{\gamma + 1}{\gamma - 1}} \tan^{-1} \sqrt{\frac{\gamma - 1}{\gamma + 1} (M^2 - 1)} + \tan^{-1} \sqrt{M^2 - 1} + \text{constant} \qquad (8.11)$$

The constant of integration may be determined if the initial Mach number M_1 and the flow deflection angle δ_1 are known. In particular, if the initial Mach number M_1 is unity and the initial flow deflection angle δ_1 is zero, as illustrated in Fig. 8.5,

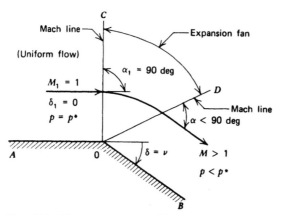

Figure 8.5 Flow expansion from $M = 1$ to $M = M$.

the constant in equation 8.11 is zero. For that special case, the symbol δ is replaced by the symbol v, and equation 8.11 becomes

$$v = -\sqrt{b} \tan^{-1} \sqrt{\frac{(M^2 - 1)}{b}} + \tan^{-1} \sqrt{M^2 - 1} \qquad (8.12)$$

where

$$b = \frac{\gamma + 1}{\gamma - 1}$$

The flow deflection angle v, given by equation 8.12, is known as the *Prandtl-Meyer angle*; it is the angle through which an initially sonic flow must be deflected to expand it to the supersonic Mach number M. For clockwise rotation of the flow, as assumed in the derivation, v, as obtained from equation 8.12 is always negative. Because equation 8.10 is derived for a steady two-dimensional planar flow, the results given by equations 8.11 and 8.12 are restricted to that type of flow. Figure 8.6 presents the Prandtl-Meyer angle v as a function of the Mach number M for $\gamma = 1.4$. Table C.13 presents v as a function of the Mach number M for several values of γ.

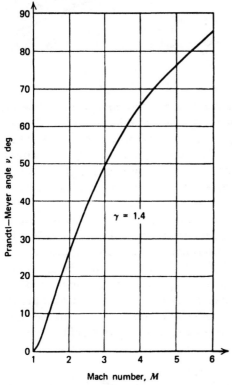

Figure 8.6 The Prandtl-Meyer angle v as a function of the Mach number M for $\gamma = 1.4$.

The flow deflection angle δ that expands a flowing fluid from an initial Mach number M_1 larger than unity to a final Mach number $M_2 > M_1$ is given by $\delta = v_2 - v_1$. The equivalent physical processes are illustrated schematically in Fig. 8.7. It is assumed that a sonic flow ($M_o = 1$, $v_o = 0$) is expanded to the actual initial Mach number M_1 and turned through the flow deflection angle v_1, in Region A. It is then imagined that a fictitious flow is expanded from $M_o = 1$ to M_2 and turned through the flow deflection angle v_2, as illustrated in Region B. In Region C, the expansion process in Region A is subtracted from that in Region B. The result is an expansion from M_1 to M_2 with a flow deflection angle $\delta = v_2 - v_1$. From equation 8.12, δ is given by

$$\delta = v_2 - v_1 = -\sqrt{b}\left\{\tan^{-1}\left[\frac{(M_2{}^2 - 1)}{b}\right]^{1/2} - \tan^{-1}\left[\frac{(M_1{}^2 - 1)}{b}\right]^{1/2}\right\}$$
$$+ \left[\tan^{-1}(M_2{}^2 - 1)^{1/2} - \tan^{-1}(M_1{}^2 - 1)^{1/2}\right] \quad (8.13)$$

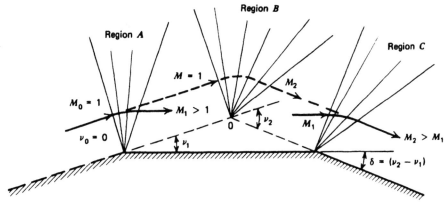

Figure 8.7 Expansion with a supersonic initial Mach number.

Once the Mach number for the flow has been determined for a given flow deflection angle δ from equation 8.13, the remaining flow variables may either be calculated by applying the isentropic flow equations, or obtained directly from Table C.6. In solving numerical problems, instead of employing equation 8.13, it is more straightforward to determine M_2 for a given M_1 and δ, by recalling that $\delta = (v_2 - v_1)$, and employing either Fig. 8.6 or Table C.13. The solution of equation 8.13 for M_2 generally requires a time-consuming iterative procedure.

The results developed above apply to the flow around a corner wherein the flow turns in a clockwise (or negative angular) direction, as illustrated in Fig. 8.1. In a real flow the fluid may turn around a convex corner in a counterclockwise (or positive angular) direction, as illustrated in Fig. 8.8. Such a flow may be analyzed in the same manner as is the case of clockwise turning, by assuming an infinitesimal turning angle ($+d\delta$) in equation 8.1. The remainder of the analysis is identical to that discussed above, with the change in the sign of $d\delta$ carrying through to the final result. Hence, for a Prandtl-Meyer flow with counterclockwise turning of the flow, equation 8.12 changes to

$$v = +\sqrt{b}\,\tan^{-1}\left[\frac{(M^2 - 1)}{b}\right]^{1/2} - \tan^{-1}(M^2 - 1)^{1/2} \qquad (8.14)$$

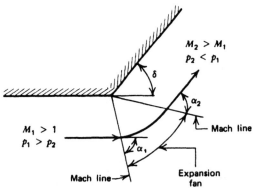

Figure 8.8 Counterclockwise Prandtl-Meyer expansion.

Equations 8.12 and 8.14 may, therefore, be combined to yield

$$\pm v = \sqrt{b} \tan^{-1} \left[\frac{(M^2 - 1)}{b} \right]^{1/2} - \tan^{-1} (M^2 - 1)^{1/2} \qquad (8.15)$$

where the $+v$ refers to counterclockwise turning, and the $-v$ refers to clockwise turning. With the aforementioned interpretation, graphical and tabular results for the Prandtl-Meyer angle v as a function of the Mach number M may be employed for turning in either direction; the $+$ or $-$ sign being chosen according to the direction of the turning for the flow.

The flow deflection angle δ is related to the initial and final Prandtl-Meyer angles v_1 and v_2. For either clockwise or counterclockwise turning,

$$\delta = v_2 - v_1 \qquad (8.16)$$

where v and δ are negative for clockwise turning and positive for counterclockwise turning.

A FORTRAN computer program, subroutine PMEYER, for determining the Prandtl-Meyer angle is presented below. The input comprises $G(1), G(2), \ldots, G(5)$, which are five values of the specific heat ratio γ, and the parameters $J(I)$ and $DM(I)$ discussed in Section 4-4(b). The results presented in Table C.13 are obtained with the following specification.

$G(1) = 1.1, G(2) = 1.2, G(3) = 1.3, G(4) = 1.4, G(5) = 1.67,$
$J(1) = 50, J(2) = 40, J(3) = 40, J(4) = 4, J(5) = 15, J(6) = 0,$
$DM(1) = 0.02, DM(2) = 0.05, DM(3) = 0.1, DM(4) = 0.5, DM(5) = 1.0$

```
      SUBROUTINE PMEYER

C     PRANDTL-MEYER ANGLE FOR A PERFECT GAS

      REAL M $ DIMENSION G(5),J(9),DM(9),PM(5),B(5) $ DATA INF/6H INFIN/
      NAMELIST /DATA/ G,J,DM $ READ (5,DATA)
      DO 20 KK=1,5 $ B(KK)=(G(KK)+1.0)/(G(KK)-1.0) $ PM(KK)=0.0
   20 CONTINUE $ M=1.0 $ WRITE (6,2000) (G(I),I=1,5)
      WRITE (6,2010) M,(PM(I),I=1,5) $ I=1 $ L=0 $ RAD=57.29518

C     CALCULATE THE PRANDTL-MEYER ANGLE

   30 N=J(I) $ DO 50 K=1,N $ M=M+DM(I) $ A=M**2-1.0 $ DO 40 KK=1,5
      PM(KK)=(SQRT(B(KK))*ATAN(SQRT(A/B(KK)))-ATAN(SQRT(A)))*RAD
   40 CONTINUE $ WRITE (6,2010) M,(PM(KK),KK=1,5) $ L=L+1
      IF (L.LT.50) GO TO 50 $ WRITE (6,2000) (G(KK),KK=1,5) $ L=0
   50 CONTINUE $ I=I+1 $ IF((I.EQ.10).OR.(J(I).EQ.0)) GO TO 60 $GO TO 30
   60 DO 70 KK=1,5 $ PM(KK)=90.0*(SQRT(B(KK))-1.0)
   70 CONTINUE $ WRITE (6,2020) INF,(PM(KK),KK=1,5) $ RETURN

 2000 FORMAT (1H1,21X,19HPRANDTL-MEYER ANGLE//4X,1HM,4X,5(4HG = ,F4.2,
     13X)/1H )
 2010 FORMAT (1H ,F5.2,5F11.4)
 2020 FORMAT (A6,5F11.4)
      END
```

Example 8.1. A uniform flow of air having a Mach number of unity expands around a convex (expansion) corner that turns the flow through an angle of 20 deg counterclockwise (i.e., in the positive angular direction). Calculate the final Mach number.

Solution

Since the initial flow is sonic, the initial Prandtl-Meyer angle v_1 is zero. From equation 8.16, $\delta = v_2 - v_1$. Thus,

$$v_2 = v_1 + \delta = 0 + 20 = 20 \text{ deg}$$

From Table C.13, for $v_2 = 20$ deg,

$$M_2 = 1.775$$

Example 8.2. It is desired to expand a uniform sonic stream of air to a Mach number of 2.0 by turning it around a sharp convex corner. What is the required turning angle?

Solution

From Table C.13, for $M_2 = 2.0$, $v_2 = 26.5$ deg. Since $v_1 = 0$ for sonic flow, equation 8.16 yields

$$\delta = v_2 - v_1 = 26.5 - 0 = 26.5 \text{ deg}$$

A turning angle of either 26.5 deg or -26.5 deg will expand an initially sonic flow to a Mach number of 2.0.

Example 8.3. Air at a Mach number of 1.4 expands around a sharp convex corner that turns it through an angle of 20 deg. What is the final Mach number?

Solution

From Table C.13, for $M_1 = 1.4$, $v_1 = 9$ deg. From equation 8.16, $\delta = v_2 - v_1$. Hence,

$$v_2 = v_1 + \delta = 9 + 20 = 29 \text{ deg}$$

From Table C.13, for $v_2 = 29$ deg,

$$M_2 = 2.096$$

Example 8.4. A perfect gas having $\gamma = 1.4$ approaches a sharp convex corner with a Mach number $M_1 = 2.0$. After expanding around the corner, the final direction of the flow is -10 deg (clockwise) from its initial direction. The initial static pressure and temperature of the gas are $p_1 = 1.0133 \cdot 10^5$ N/m^2 and $t_1 = 290$ K. Assuming that the expansion process is isentropic, determine (a) the final Mach number, (b) the final static pressure, and (c) the final static temperature.

Solution

(a) From Table C.13, for $M_1 = 2.0$, $v_1 = -26.38$ deg for clockwise turning. From equation 8.16, $\delta = v_2 - v_1$. Thus,

$$v_2 = v_1 + \delta = -26.38 - 10 = -36.38 \text{ deg}$$

From Table C.13, for $v_2 = 36.38$ deg,

$$M_2 = 2.383$$

(b) Since the flow is isentropic, the stagnation pressure P is constant. From Table C.6, for $M_1 = 2.0$ and $M_2 = 2.383$, $p_1/P = 0.12780$ and $p_2/P = 0.06992$. Thus,

$$p_2 = p_1 \left(\frac{P}{p_1}\right)\left(\frac{p_2}{P}\right) = \frac{(1.0133)10^5(0.06992)}{0.12780} = 0.5544 \cdot 10^5 \text{ N/m}^2 \text{ (8.040 lbf/in.}^2)$$

(c) Since the flow is adiabatic, the stagnation temperature T is constant. From

Table C.6, for $M_1 = 2.0$ and $M_2 = 2.383$, $t_1/T = 0.55556$ and $t_2/T = 0.46760$. Thus,

$$t_2 = t_1 \left(\frac{T}{t_1}\right)\left(\frac{t_2}{T}\right) = \frac{(290)(0.46760)}{0.55556} = 244.1 \text{ K (439.4 R)}$$

Example 8.5.   The flow described in Example 8.4 is expanded through a 10 deg (counterclockwise) turn (see Fig. 8.8). Determine the Mach number after the turn.

Solution

(a)   From Table C.13, for $M_1 = 2.0$, $v_1 = 26.38$ deg for counterclockwise turning. From equation 8.16, $\delta = v_2 - v_1$. Thus,

$$v_2 = v_1 + \delta = 26.38 + 10 = 36.38 \text{ deg}$$

From Table C.13, for $v_2 = 36.38$ deg,

$$M_2 = 2.383$$

Example 8.6.   Figure 8.9a illustrates diagrammatically a two-dimensional double-wedge profile that moves through the atmosphere, at sea level, at zero angle of attack with $M = 3.0$. Calculate the lift and drag per unit length ℓ of the wedge. The top slanted surfaces of the wedge are each 1.0 m wide.

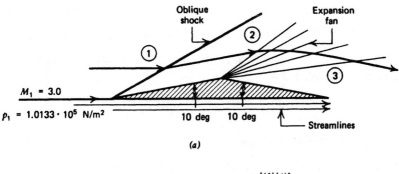

(a)

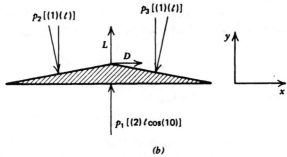

(b)

Figure 8.9   Sketch for Example 8.6. (a) Flow pattern. (b) Forces acting on the wedge.

Solution

An oblique shock wave forms at the leading edge, a Prandtl-Meyer expansion fan forms at the top corner, the underside flow remains parallel to the x axis, and the intersection of the two flows at the trailing edge may be neglected, since it does not affect the pressure on the wedge. The aforementioned phenomena are illustrated in Fig. 8.9a. A force diagram is presented in Fig. 8.9b, where L and D are the aerodynamic lift and drag forces, respectively, exerted by the fluid on the wedge.

From Table C.12, for $M_1 = 3.0$ and $\delta = 10$ deg, $\varepsilon = 27.38$ deg. From equation 7.99,

$$M'_1 = M_1 \sin \varepsilon = 3.0 \sin (27.38) = 1.380$$

From Table C.11, for $M_1 = 1.380$, $M'_2 = 0.74829$ and $p_2/p_1 = 2.055$, where $p_1 = 1.0133 \cdot 10^5$ N/m². Hence, $p_2 = (2.055)(1.0133)10^5 = 2.082 \cdot 10^5$ N/m². From equation 7.100,

$$M_2 = \frac{M'_2}{\sin (\varepsilon - \delta)} = \frac{0.74829}{\sin (27.38 - 10)} = 2.505$$

The flow deflection angle δ for the Prandtl-Meyer expansion is 20 deg. From Table C.13, for $M = 2.505$, $v_2 = 39.24$ deg. Thus,

$$v_3 = v_2 + \delta = 39.24 + 20 = 59.24 \text{ deg}$$

From Table C.13, for $v = 59.24$ deg, $M_3 = 3.545$. From Table C.6, for $M_2 = 2.505$ and $M_3 = 3.545$, $p_2/P_2 = 0.05808$ and $p_3/P_3 = 0.01230$. Since the expansion is isentropic, $P_2 = P_3$. Thus,

$$p_3 = p_2 \left(\frac{p_3}{P_3}\right)\left(\frac{P_2}{p_2}\right) = \frac{(2.082)10^5(0.01230)}{(0.05808)} = 0.4409 \cdot 10^5 \text{ N/m}^2$$

Refer to Fig. 8.9b. The lift force L for a wedge having a length ℓ measured in meters is

$$L = -(2.082)10^5[(1)\ell] \cos (10) - (0.4409)10^5[(1)\ell] \cos (10)$$

$$+ (1.0133)10^5[(2) \ell \cos (10)] = -48,876 \, \ell \text{ N}$$

Similarly, the drag force D is

$$D = (2.082)10^5[(1)\ell] \sin (10) - (0.4409)10^5[(1)\ell] \sin (10) = 28,497 \, \ell \text{ N}$$

8-4(b) Maximum Prandtl-Meyer Angle

The maximum value of the Prandtl-Meyer angle $\pm v$ for a given flow is obtained when the final Mach number is infinite. If $M = \infty$ is substituted in equation 8.15, the maximum Prandtl-Meyer angle, denoted by $\pm v_{max}$, is obtained. Thus,

$$\pm v_{max} = \lim_{M \to \infty} \pm v = \frac{\pi}{2}(\sqrt{b} - 1) \qquad (8.17a)$$

For gases with $\gamma = 1.4$, $\pm v_{max}$ is

$$\pm v_{max} = 90(\sqrt{6} - 1) = 130.45 \text{ deg} \qquad (8.17b)$$

The theory, therefore, indicates that a supersonic flow may be accelerated and turned through quite large angles around a sharp convex corner without the flow separation phenomena that would occur in a subsonic flow (see Section 5–10). It should be realized, however, that the Prandtl-Meyer analysis assumes that the fluid is perfect ($\mu = 0$). In the case of a real fluid, the flow deflection angles that are obtainable without flow separation from the surface are smaller than those predicted by the Prandtl-Meyer theory. Moreover, the maximum Prandtl-Meyer angle $\pm v_{max}$ corresponds to $p = 0$. At zero static pressure, the assumption that the fluid is a continuum is no longer valid. Figure 8.10 presents the calculated values of the maximum flow deflection angle $\delta_{max} = v_{max} - v_1$ as a function of the initial Mach number M_1, for $\gamma = 1.40$.

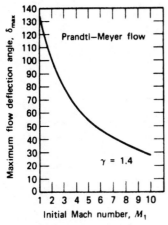

Figure 8.10 The maximum flow deflection angle δ_{max} for a two-dimensional flow as a function of the initial Mach number M for $\gamma = 1.4$.

8-4(c) Expanding Supersonic Flow over Successive Corners and Convex Surfaces

Consider the situation where a steady uniform supersonic flow is expanded over several consecutive sharp convex corners, as illustrated in Fig. 8.11a. Since the flow is initially parallel to the surface AB, a Mach line BL_1 emanates from the corner B, making the angle α_1 with the initial flow direction; $\alpha_1 = \sin^{-1}(1/M_1)$. The expansion of the fluid as it flows around corner B causes an expansion fan to be formed that is bounded by the Mach lines BL_1 and BL_2, the latter making the angle α_2 with the direction of the expanded flow; $\alpha_2 = \sin^{-1}(1/M_2)$. Between the corners B and C, the flow streamlines are parallel to the frictionless surface BC until the corner C is reached. At C there is the Mach line CL_1 making the angle α_2 with the gas flowing upstream to it; that is, CL_1 is parallel to BL_2. Beyond C the conditions repeat themselves.

The preceding results may be applied directly to the steady uniform supersonic flow of a compressible fluid over a convex curved surface such as that illustrated in Fig. 8.11b. Referring to Fig. 8.11a, if the several distances BC, CD, etc., are made infinitesimally small, as illustrated in Fig. 8.11b, the successive sharp corners disappear and a smooth frictionless convex surface BD is formed. At each point N on the convex surface BD, the flow deflection angle δ_N is known. Knowing δ_N, the value of M_N and the Mach angle α_N may be calculated. Accordingly, the expansion of a compressible fluid flowing over a convex curved surface may be analyzed in a manner similar to that employed for analyzing the steady uniform supersonic flow around a convex corner.

Example 8.7. Air flows over a succession of three convex corners having flow deflection angles of 5 deg, 10 deg, and 20 deg. The initial Mach number is 1.40. Determine the Mach number after each corner.

Solution

From Table C.13, for $M_1 = 1.40$, the initial Prandtl-Meyer angle v_1 is 9 deg. After the first corner, the Prandtl-Meyer angle v_2, is

$$v_2 = v_1 + \delta_1 = 9 + 5 = 14 \text{ deg}$$

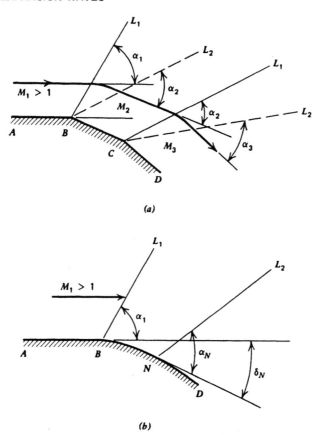

Figure 8.11 Expansive flow over successive corners and convex surfaces. (a) Expansive flow over successive corners. (b) Expansive flow over a convex surface.

From Table C.13, for $v = 14$ deg,

$$M_2 = 1.571$$

After the second corner, the Prandtl-Meyer angle v_3 is

$$v_3 = v_2 + \delta_2 = 14 + 10 = 24 \text{ deg}$$

From Table C.13, for $v = 24$ deg,

$$M_3 = 1.915$$

After the third corner, the Prandtl-Meyer angle v_4 is given by

$$v_4 = v_3 + \delta_3 = 24 + 20 = 44 \text{ deg}$$

From Table C.13, for $v = 44$ deg, we obtain

$$M_4 = 2.718$$

As a check, the total flow deflection angle is 35 deg. Thus,

$$v_4 = v_1 + \sum \delta = 9 + 35 = 44 \text{ deg}$$

From Table C.13, for $v = 44$ deg, the Mach number is

$$M_4 = 2.718$$

8-5 SIMPLE WAVES IN IMPERFECT GASES

The governing equation for a simple wave flow, equation 8.5, may be written in the following form. Thus,

$$d\delta = -\frac{\sqrt{M^2 - 1}}{V} dV \tag{8.5}$$

In Section 8-4, the above equation is integrated for the case of a *perfect gas*. In the present section, a method is presented for integrating equation 8.5 for an *imperfect gas*.

In Section 4-5, the principal sources of imperfect gas effects are discussed. It is pointed out that for an imperfect gas which behaves as a simple system, its equation of state may be written in the following form.

$$\rho = \rho(h,s), \qquad t = t(h,s), \qquad a = a(h,s), \qquad \text{etc.} \tag{8.18}$$

Simple wave flows, by definition, are isentropic, so that equation 8.18 may be written as

$$\rho = \rho(h), \qquad t = t(h), \qquad a = a(h), \qquad \text{etc.} \tag{8.19}$$

From equation 3.164, Table 3.1, the energy equation for an adiabatic flow with no body forces ($g\,dz = 0$), is

$$h + \frac{V^2}{2} = H = \text{constant} \tag{8.20}$$

Equation 8.20 shows that the enthalpy h is a unique function of the flow velocity V. Combining equations 8.19 and 8.20 yields

$$\rho = \rho(V), \qquad t = t(V), \qquad a = a(V), \qquad \text{etc.} \tag{8.21}$$

But $M = V/a$, and $a = a(V)$. Hence,

$$M = M(V) \tag{8.22}$$

Equation 8.5 may, therefore, be written in the form

$$d\delta = -\frac{\sqrt{M^2 - 1}}{V} dV = f(V)\,dV \tag{8.23}$$

where

$$f(V) \equiv \frac{\sqrt{M^2 - 1}}{V} \tag{8.24}$$

Equation 8.23 may be integrated either numerically or graphically for any fluid if its equation of state is known, for a given set of initial conditions. Thus,

$$\delta_2 - \delta_1 = \int_{V_1}^{V_2} f(V)\,dV \tag{8.25}$$

The results, of course, are valid only for the particular fluid and for the specific initial conditions. General solutions, such as those presented in Table C.13, are not possible for imperfect gases. Each particular problem must be solved as a special case.

One procedure for integrating equation 8.25 is presented below.

1. Determine the initial flow conditions V_1, h_1, and δ_1.
2. Select several closely spaced values of V, increasing from V_1, subject to the limitation that $V \leq V'_{max}$, where V'_{max} is given by equation 3.185, Table 3.4.

3. For each value of V, calculate h by means of equation 8.20. Then, determine the properties of interest from equation 8.19, in particular the speed of sound a, from which we obtain $M = V/a$.

4. Calculate $f(V)$ from equation 8.24 for each chosen value of V in step 2, and plot $f(V)$ as a function of V. The area under the curve of $f(V)$ versus V is the value of the integral in equation 8.25. The numerical integration may be performed by any one of the standard techniques, such as actually measuring the area under the curve, the trapezoidal rule, Simpson's rule, etc. (see Appendix A–5). The solution is complete when the desired value of δ_2 is obtained.

The above procedure is simple and straightforward, and is readily adapted to the digital computer.

Example 8.8. Consider the blow-down wind tunnel described in Examples 4.3 and 4.4. The flowing medium is air at a stagnation temperature of 2000 K and a stagnation pressure of $35 \cdot 10^5$ N/m². The wind tunnel is a two-dimensional planar flow passage with a sharp cornered throat, as illustrated in Fig. 8.12. Assume that the flow at the throat, where the Mach number M is unity, is a steady uniform flow. In expanding around the sharp corner, the flow is turned through an angle of 45 deg. Calculate the flow properties after the expansion fan, assuming that the air is a thermally perfect gas, but take into account the variation of the specific heat of the air with temperature.

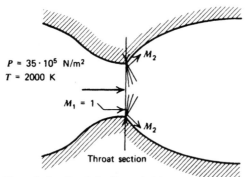

Figure 8.12 Sketch for Example 8.8.

Solution

For a thermally perfect gas, $p = \rho R t$ and $a^2 = \gamma R t$. Since $c_p = c_p(t)$, the caloric equation of state for air is represented by Table C.4. The stagnation enthalpy corresponding to a stagnation temperature of 2000 K is $H = 2252.414$ kJ/kg, and the corresponding relative stagnation pressure is $P_r = 1520.8$. The flow properties of air at the sonic condition are determined in Example 4.3. Thus, $V_t = 806.52$ m/s, $h_t = 1927.181$ kJ/kg, $t_t = 1738.3$ K, and $p_t = 19.073 \cdot 10^5$ N/m².

As a guide, the solution based on assuming that air is a perfect gas is obtained first. At the temperature $t_t = 1738.3$ K, Table C.4 gives $\gamma = 1.3037$. Hence, assume that $\gamma = 1.30$. For $M_1 = 1.0$, $v_1 = 0$ deg, and for $\delta = 45$ deg,

$$v_2 = v_1 + \delta = 0 + 45 = 45 \text{ deg}$$

From Table C.13, for $v_2 = 45$ deg, $M_2 = 2.5653$. From Table C.6, for $M_t = 1.0$ and

$M_2 = 2.5653,$

$$\frac{t_t}{T} = 0.86957 \qquad \frac{p_t}{P} = 0.54573 \qquad \frac{t_2}{T} = 0.50332 \qquad \frac{p_2}{P} = 0.051180$$

$$t_2 = t_t \left(\frac{T}{t_t}\right)\left(\frac{t_2}{T}\right) = \frac{(1738.3)(0.50332)}{(0.86957)} = 1006.2 \text{ K}$$

$$p_2 = p_t \left(\frac{P}{p_t}\right)\left(\frac{p_2}{P}\right) = \frac{(19.073)10^5(0.051180)}{(0.54573)} = 1.7887 \cdot 10^5 \text{ N/m}^2$$

$$a_2 = (\gamma R t_2)^{1/2} = [1.3(287.04)(1006.2)]^{1/2} = 612.75 \text{ m/s}$$

$$V_2 = M_2 a_2 = 2.5653(612.75) = 1571.9 \text{ m/s}$$

Hence, the air velocity increases from 806.52 m/s to 1571.9 m/s in flowing through the expansion fan, assuming that the air is a perfect gas.

For the imperfect gas, the procedure outlined in the preceding section is followed.

1. The initial flow conditions are $V_1 = 806.52$ m/s, $h_1 = 1927.181$ kJ/kg, $p_1 = 19.073 \cdot 10^5$ N/m², $t_1 = 1738.3$ K, and $\delta_1 = 45$ deg.
2. Choose a velocity increment of 100 m/s.
3. From equation 8.20, for $H = 2252.414$ kJ/kg,

$$h = 2252.414 - \frac{V^2}{2(1000)} \text{ kJ/kg} \tag{a}$$

where V is in meters per second. With h known, t may be found from Table C.4 by interpolation. For a thermally perfect gas,

$$a = (\gamma R t)^{1/2} \tag{b}$$

Then,

$$M = \frac{V}{a} \tag{c}$$

4. Determine $f(V)$ from equation 8.24, and integrate equation 8.25 by the trapezoidal rule [see Appendix A–5(a)].

The results are presented in the table at the end of the problem. The first row (Point 1) presents the initial data. For the second row,

$$V_2 = V_1 + 100 = 806.52 + 100 = 906.52 \text{ m/s}$$

From equation (a),

$$h_2 = 2252.414 - \frac{(906.52)^2}{2000} = 1841.525 \text{ kJ/kg}$$

From Table C.4, for $h_2 = 1841.525$ kJ/kg, $t_2 = 1668.6$ K and $\gamma = 1.3056$. From equation (b),

$$a_2 = [1.3056(287.04)(1668.6)]^{1/2} = 790.77 \text{ m/s}$$

From equation (c),

$$M_2 = \frac{906.52}{790.77} = 1.1464$$

The above results are presented in columns (2) to (5) in the aforementioned table.

The values presented in the remaining rows of the table are determined in the same manner for each velocity increment of 100 m/s.

The function $f(V)$ corresponding to each velocity listed in the table is determined from equation 8.24. Thus,

$$f(V) = \frac{(M^2 - 1)^{1/2}}{V} \tag{d}$$

For the initial point, $M_1 = 1$ and $f(V_1) = 0$. For the second row in the table,

$$f(V_2) = \frac{[(1.1464)^2 - 1]^{1/2}}{906.52} = 6.1837 \cdot 10^{-4} \text{ s/m}$$

The values of $f(V)$ are presented in column (6) of the table.

After $f(V)$ is computed, equation 8.25 is integrated numerically for δ, employing the trapezoidal rule [see Appendix A–5(a)]. Thus, for each increment in V, corresponding to the rows in the table,

$$\delta_{i+1} = \delta_i + \frac{1}{2}[f(V_{i+1}) + f(V_i)]100 \text{ rad} \tag{e}$$

For example, the value of δ_2 corresponding to row 2, where the velocity is $V_2 = 906.52$ m/s, is

$$\delta_2 = \delta_1 + \frac{1}{2}[f(V_2) + f(V_1)]100$$

$$\delta_2 = 0.0 + \frac{1}{2}[6.1837 \cdot 10^{-4} + 0.0]100 = 0.03092 \text{ rad} = 1.7716 \text{ deg}$$

The values of δ corresponding to the tabulated values of V are presented in columns (8) and (9) of the table, in radians and degrees, respectively.

Rows 8 and 9 in the table bracket the desired flow deflection angle of 45 deg. The solution at $\delta = 45$ deg, obtained by linear interpolation between those two rows, is presented in row 10.

From Table C.4, for $t_1 = 1738.3$ K and $t_2 = 959.95$ K, $p_{r1} = 828.84$ and $p_{r2} = 71.534$. Thus,

$$p_2 = p_1 \left(\frac{p_{r2}}{p_{r1}}\right) = \frac{(19.073 \cdot 10^5)(71.534)}{828.84} = 1.6461 \cdot 10^5 \text{ N/m}^2$$

The errors caused by assuming that air is a perfect gas are presented below for the different pertinent properties of the flowing gas.

Property	Perfect Gas	Imperfect Gas	Error, Percent
M	2.5653	2.6037	−1.47
V, m/s	1571.9	1581.6	−0.61
p, N/m$^2 \cdot 10^5$	1.7887	1.6461	8.66
t, K	1006.2	959.95	4.82

Although the errors caused by assuming that air is a perfect gas are not large in the present example, they would be significant in the design of a wind tunnel.

Results from the Numerical Procedure for Solving Example 8.8.

Point	(1) V, m/s	(2) h, kJ/kg	(3) t, K	(4) a, m/s	(5) M	(6) $f(V) \cdot 10^4$ s/m	(7) $f(V)\,dV$	(8) δ, rad	(9) δ, deg
1	806.52	1927.181	1738.3	806.52	1.0000	0.0000	——	0.00000	0.0000
2	906.52	1841.525	1668.6	790.77	1.1464	6.1837	0.03092	0.03092	1.7716
3	1006.5	1745.873	1590.3	772.74	1.3025	8.2916	0.07238	0.10330	5.9185
4	1106.5	1640.221	1503.3	752.16	1.4711	9.7508	0.09021	0.19351	11.0874
5	1206.5	1524.569	1407.3	728.75	1.6556	10.936	0.10343	0.29694	17.0136
6	1306.5	1398.917	1302.0	702.15	1.8607	12.010	0.11473	0.41167	23.5870
7	1406.5	1263.265	1186.9	671.87	2.0934	13.076	0.12543	0.53710	30.7736
8	1506.5	1117.613	1061.4	637.06	2.3648	14.225	0.13651	0.67361	38.5947
9	1606.5	961.961	924.78	596.83	2.6918	15.556	0.14891	0.82252	47.1266
10	1581.6	1001.701	959.95	607.44	2.6307	——	——	——	45.0000

8-6 SUPERSONIC FLOW OVER A CONCAVE SURFACE

The steady uniform supersonic flow of a compressible fluid into a concave corner, and over either a wedge or a cone, is characterized by abrupt changes in the direction of the incident flow and the formation of shock waves (see Chapter 7). Now consider the case of the steady uniform supersonic flow of a compressible fluid over a *concave surface* designed so that the changes in the direction of the incident flow occur gradually instead of abruptly, as illustrated schematically in Fig. 8.13a.

To obtain a clearer picture of the pertinent phenomenon, consider first the case where the flow is deflected through a series of consecutive small angles formed by successive concave corners. Figure 8.13b illustrates the general case for two successive concave corners. The change in the flow direction caused by the configuration illustrated in Fig. 8.13b causes the flow to be compressed in moving from the region ABL to region LCD; the streamlines become more closely spaced. At the concave corner B, a weak oblique shock wave BL is formed that, because of the assumed gradual deflection of the flow, is substantially the same as a Mach wave. The corresponding Mach line BL makes the angle $\alpha_1 = \sin^{-1}(1/M_1)$ with the initial flow direction.

In crossing the Mach line BL the flow Mach number is decreased somewhat from the initial value M_1 to the smaller value M_2. Between B and C the flow is parallel to the surface BC. At C, because of the small deflection of the flow, the Mach line CL is generated. Since $M_2 < M_1$, the angle α_2 is larger than α_1. Accordingly, the two Mach waves represented by BL and CL must intersect and, consequently, strengthen each other. In general, whenever such waves intersect, a new and stronger wave is created that will have the same wave angle as that for the oblique compression shock wave that would occur if the flow were deflected through the total flow deflection angle in one step.

The considerations presented above are readily applied to the case illustrated in Fig. 8.13a where the flow direction is changed gradually by a concave surface. When the flow experiences the first change in its direction, a Mach wave is formed and, with each successive infinitesimal change in direction, other Mach waves are generated. Because of the gradual change in the direction of the supersonic flow by the concave surface, an infinite number of Mach waves are generated; only a few are illustrated in Fig. 8.13a.

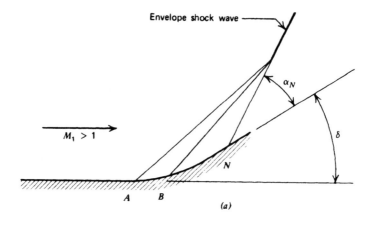

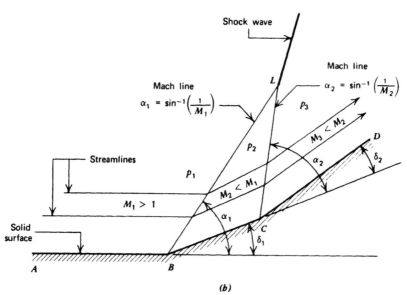

Figure 8.13 Compressive flow into successive corners and over a concave surface. (*a*) Compressive flow over a concave surface. (*b*) Compressive flow into successive corners.

Each consecutive Mach line must be steeper than the one preceding it. Accordingly, all of the Mach lines must intersect and consequently strengthen each other, thereby producing an envelope across which there is a finite change in the flow conditions. Across the envelope the phenomenon is no longer continuous; for that reason it is termed an *envelope shock wave*. The envelope shock wave need not necessarily intersect the surface causing the deflection of the flow.

The foregoing demonstrates that a concave surface produces a change in the flow direction of a supersonic stream and, no matter how gradual that change is, the Mach waves must intersect and strengthen each other in some region of the flow field. If the envelope shock wave does not intersect the deflecting surface, then there is an isentropic compression of the fluid in the region, external to the boundary layer, adjacent to the concave surface.

The preceding illustrates that there is a basic difference between the expansion and the compression of a supersonic flow. *In an expansion, the Mach waves do not intersect each other. In a compression, on the other hand, the Mach waves always intersect and strengthen each other.*

Example 8.9. A uniform stream of air having a Mach number of 2.0 flows into a sharp concave corner having a flow deflection angle of 10 deg. Calculate the Mach number and the static pressure ratio for the flow after the corner by (a) assuming that an oblique shock wave is propagated into the flow from the corner, and (b) approximating the oblique shock wave by a continuous isentropic compression. Calculate the percent error in the Mach number and the pressure ratio for the isentropic approximation. Assume $\gamma = 1.4$.

Solution

(a) From Fig. 7.17, for $M_1 = 2.0$ and $\delta = 10$ deg, $\varepsilon = 39.5$ deg. Because of the difficulty of obtaining accurate values for M_2 and p_2/p_1 from Figs. 7.18 and 7.19 at small flow deflection angles, the normal shock wave table (Table C.11) is employed for determining M_2 and p_2/p_1. From equation 7.99,

$$M_1' = M_1 \sin \varepsilon = 2.0 \sin (39.5) = 1.272$$

From Table C.11, for $M_1' = 1.272$, $p_2/p_1 = 1.721$ and $M_2' = 0.80057$. From equation 7.100,

$$M_2 = \frac{M_2'}{\sin (\varepsilon - \delta)} = \frac{0.80057}{\sin (39.5 - 10)} = 1.626$$

(b) From Table C.13, for $M_1 = 2.0$, $v_1 = 26.5$ deg. Thus,

$$v_2 = v_1 - \delta = 26.5 - 10 = 16.5 \text{ deg}$$

From Table C.13, for $v_2 = 16.5$ deg, $M_2 = 1.655$. Since the Prandtl-Meyer compression is isentropic, $P_2/P_1 = 1.0$. From Table C.6, for $M_1 = 2.0$ and $M_2 = 1.655$, $p_1/P_1 = 0.1278$ and $p_2/P_2 = 0.2168$. Hence,

$$\frac{p_2}{p_1} = \frac{p_2}{P_2} \frac{P_1}{p_1} = \frac{0.2168}{0.1278} = 1.696$$

The percent error in the Mach number M_2 is

$$\text{error} = 100 \frac{(1.655 - 1.626)}{1.626} = 1.78 \text{ percent}$$

The percent error in static pressure ratio p_2/p_1 is

$$\text{error} = 100 \frac{(1.696 - 1.721)}{1.721} = -1.45 \text{ percent}$$

8-7 HODOGRAPH CHARACTERISTICS FOR SIMPLE WAVE FLOW

In Section 7-9, the hodograph plane is employed as a means for presenting the essential features of oblique shock waves in a useful compact form. Figure 7.25 presents a complete *shock polar* for $\gamma = 1.4$.

Equation 8.11, which is repeated here for convenience, relates the flow deflection angle δ for a Prandtl-Meyer expansion and the Mach number M (see Fig. 8.1).

$$\pm \delta = \sqrt{b} \tan^{-1} [(M^2 - 1)/b]^{1/2} - \tan^{-1} (M^2 - 1)^{1/2} + \text{constant} \quad (8.26)$$

In equation 8.26, $+\delta$ denotes a counterclockwise turn of the expanded flow, and $-\delta$ a clockwise turn. Equation 8.26 shows that the Mach number and the flow deflection angle are uniquely related for a simple wave flow. Furthermore, since the flow is uniform, Mach lines are straight lines along which the flow properties are constant. The entire flow field may be determined once the relationship between the flow deflection angle and the Mach number has been established.

It is of advantage to relate the flow deflection angle δ to the dimensionless velocity $M^* = V/a^*$ instead of the Mach number M; note that when M approaches infinity, M^* attains the limiting value $[(\gamma + 1)/(\gamma - 1)]^{1/2} = \sqrt{b}$ (see equation 3.107). From equation 3.189, Table 3.4,

$$M^2 = \frac{\dfrac{2}{\gamma + 1} M^{*2}}{1 - \dfrac{\gamma - 1}{\gamma + 1} M^{*2}} \tag{8.27}$$

Substituting equation 8.27 into equation 8.26 yields

$$\pm \delta = \sqrt{b} \tan^{-1} \left(\frac{M^{*2} - 1}{b - M^{*2}} \right)^{1/2} - \tan^{-1} \left(\frac{M^{*2} - 1}{1 - M^{*2}/b} \right)^{1/2} + \text{constant} \tag{8.28}$$

If $M^* = 1$ when $\delta = 0$, the constant in equation 8.28 is equal to zero. For that special case, the flow deflection angle δ is equal to the Prandtl-Meyer angle v. Hence,

$$\pm v = \sqrt{b} \tan^{-1} \left(\frac{M^{*2} - 1}{b - M^{*2}} \right)^{1/2} - \tan^{-1} \left(\frac{M^{*2} - 1}{1 - M^{*2}/b} \right)^{1/2} \tag{8.29}$$

In equation 8.29, $M^* = (u^{*2} + v^{*2})^{1/2}$ and $v = \tan^{-1}(v^*/u^*)$, where $u^* = u/a^*$ and $v^* = v/a^*$ are the dimensionless velocity components.

Figure 8.14 presents the dimensionless velocity M^* as a function of the Prandtl-Meyer angle v, in the hodograph plane.

In the general case, where $M_1^* > 1$ and $\delta_1 \neq 0$, equation 8.28 is applicable, and the value of the constant is determined by the specific values of M_1^* and δ_1; those values locate the initial point in the hodograph plane. The remainder of the flow process is determined by solving equation 8.28 and plotting the results in the hodograph plane, as illustrated in Fig. 8.15. For a given set of initial conditions, either expanding or compressing flows may occur, and which type of flow does occur depends on the geometry of the solid boundary. The resulting curves plotted in the hodograph plane are called *hodograph characteristics*.

For the present it is sufficient to state that the *hodograph characteristics* for a uniform steady two-dimensional planar isentropic flow are epicycloids;[3] in this case the epicycloid is the curve generated by rolling a circle of radius $(b - 1)/2$ on the circumference of a circle of radius $M^* = 1$. The starting point for the epicycloid in Fig. 8.14 is $(M^* = 1, v = 0)$. The starting points for the epicycloids in Fig. 8.15 are $(M_1^* > 1, \delta_1 > 0)$. Two distinct hodograph characteristics pass through each point in the hodograph plane; for example, curves A and B in Fig. 8.15 correspond to the same initial point (M_1^*, δ_1). Four flows are possible from a given initial point.

1. An expanding counterclockwise flow corresponding to the portion of curve A between points i and a.
2. An expanding clockwise flow corresponding to the portion of curve B between points i and b.
3. A compressing clockwise flow corresponding to the portion of curve A between points i and c.
4. A compressing counterclockwise flow corresponding to the portion of curve B between points i and d.

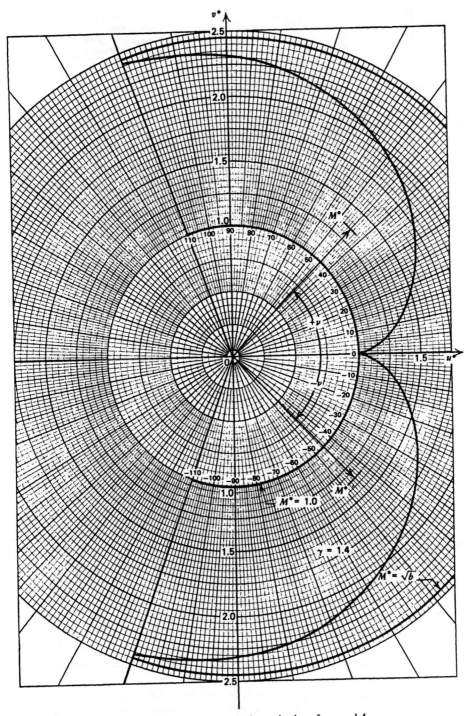

Figure 8.14 The Prandtl-Meyer angle γ in the hodograph plane for $\gamma = 1.4$

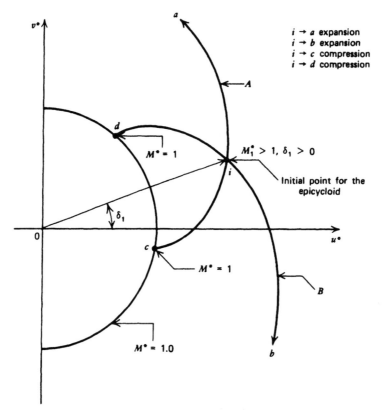

Figure 8.15 Arbitrary initial conditions in the hodograph plane.

A flowing gas may expand in either direction along a hodograph characteristic until either a specified flow deflection angle δ or a specified dimensionless velocity M^* is attained. The maximum attainable flow velocity for the gas after an expansion is the maximum isentropic speed V'_{max} (see equation 3.185, Table 3.4). Conversely, a compression of the gas may be produced in either direction along a hodograph characteristic until a specified flow deflection angle δ or a specified dimensionless velocity M^* is attained. For compressing flows, the limiting flow velocity for a Prandtl-Meyer compression is the sonic velocity, that is, $M^* = 1.0$.

The curves presented in Fig. 8.15, when rotated about the origin so that the $M^* = 1$ points (points c and d) coincide with the point ($M_1^* = 1, \delta = 0$), are identical to those for the Prandtl-Meyer angle v presented in Fig. 8.14. Consequently, the hodograph characteristics from any arbitrary initial point in the hodograph plane may be obtained by an appropriate rotation of the hodograph characteristics for the Prandtl-Meyer angle v until the desired initial conditions are located on the rotated Prandtl-Meyer characteristic.

The complete hodograph plane may be constructed in either of the two ways: first, by solving equation 8.28 for a range of initial conditions, or second, by rotating the curves for the Prandtl-Meyer angle through an appropriate range of angles. Figure 8.16 presents the complete hodograph plane for $\gamma = 1.4$. Distinct hodograph characteristics, for both clockwise and counterclockwise turning, are initiated from the sonic circle ($M_1^* = 1.0$) for initial flow deflection angles from -60 deg to 60 deg in increments of 10 deg. Flows involving angles that fall outside of the range of Fig. 8.16 may be transformed to the region presented in Fig. 8.16 by a rotation of the

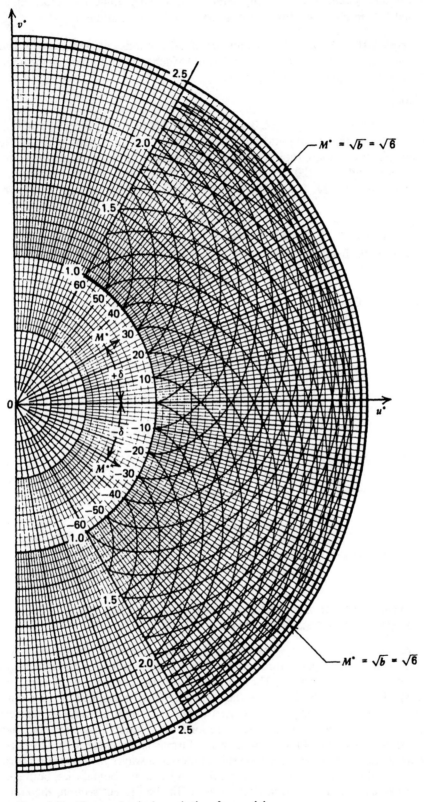

Figure 8.16 The complete hodograph plane for $\gamma = 1.4$.

447

coordinate system. The foregoing demonstrates that graphical solutions may be obtained for simple wave flows with arbitrary initial conditions.

Example 8.10.   Air at a Mach number of 2.0 expands counterclockwise around a 20 deg expansion corner. Determine the final Mach number employing the hodograph characteristics presented in Fig. 8.16.

Solution

From Table C.6, for $M_1 = 2.0$ and $\gamma = 1.4$, $M_1^* = 1.633$. With $M_1^* = 1.633$ and $\delta_1 = 0$, locate the initial point of the expansion in the hodograph plane of Fig. 8.16. Then follow the hodograph curve passing through the initial point through a flow deflection angle of $+20$ deg and locate the final point after the expansion, as illustrated in Fig. 8.17. The radius from the origin to that final point is M_2^*. Thus,

$$M_2^* = 1.92$$

From Table C.6, for $M_2^* = 1.92$, $M_2 = 2.82$.
   As a check, from Table C.13, for $M_1 = 2.0$, $\nu_1 = 26.5$ deg. Then,

$$\nu_2 = \nu_1 + \delta = 26.5 + 20 = 46.5 \text{ deg}$$

From Table C.13, for $\nu_2 = 46.5$ deg, $M_2 = 2.835$, which checks.

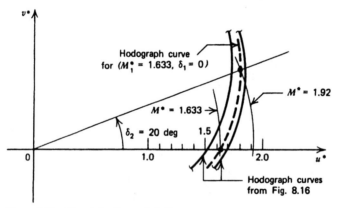

Figure 8.17   Sketch for Example 8.10.

8-8 APPROXIMATION OF CONTINUOUS EXPANSION WAVES BY FINITE AMPLITUDE DISCRETE WAVES

In the preceding sections, it is shown that the supersonic expansion of a steady uniform flow takes place through a series of infinitesimal expansion waves that constitute an expansion fan. The infinitesimal waves comprising the expansion fan are Mach lines, which diverge as they emanate from the source of the disturbance [i.e., from the convex (expansion) corner or surface]. In direct contrast, compression waves converge, eventually coalescing to form an oblique shock wave. In Chapter 7, it is shown that an oblique shock wave forms immediately at a sharp concave (compression) corner, and the discrete compression waves may be traced through a flow field; their interaction with each other and with the boundaries, both solid and free pressure, are discussed in Section 7–11. In the present section, the concept of discrete expansion waves is introduced, and the rules governing their interactions are developed.

Figure 8.18a illustrates a typical continuous expansion fan, and Fig. 8.18b illustrates a discrete expansion wave that approximates the continuous expansion fan. The discrete expansion wave is defined in such a manner that it produces the same values of flow deflection angle δ and Mach number M_2 downstream of it as are created by the continuous expansion fan. Hence, the flow properties before and after the continuous expansion fan are correctly determined by the discrete expansion wave. Obviously, the flow properties within the expansion fan are incorrect. For an infinitesimal flow deflection angle, however, the two approaches are identical because the discrete expansion wave and the continuous expansion fan both approach a Mach wave in the limit.

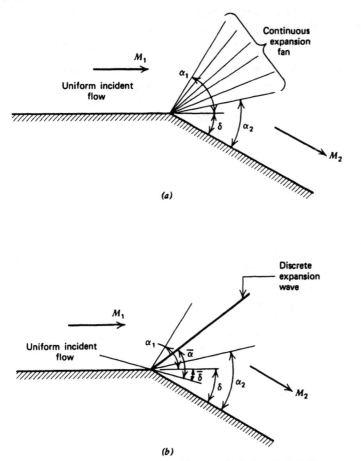

Figure 8.18 Approximation of a continuous expansion fan by discrete expansion waves. (a) Continuous expansion fan. (b) Discrete expansion waves.

The location of the discrete expansion wave is defined in terms of the average Mach angle $\bar{\alpha}$, which is measured relative to the average direction of the fluid flow around the corner, that is, $\bar{\delta} = \delta/2$. It is seen in Fig. 8.18b that the flow is deflected away from the discrete expansion wave as it flows through the wave. By contrast, as shown in Chapter 7, the flow through an oblique shock wave is always turned toward the compressive (shock) wave. There are, however, at least three definitions of $\bar{\alpha}$ in current use. They are:

$$\bar{\alpha} \equiv \frac{1}{2}(\alpha_1 + \alpha_2) \tag{8.30}$$

$$\bar{\alpha} \equiv \sin^{-1}\left(\frac{1}{\bar{M}}\right) = \sin^{-1}\left(\frac{2}{M_1 + M_2}\right) \tag{8.31}$$

$$\bar{\alpha} \equiv \sin^{-1}\left(\frac{\bar{a}}{\bar{V}}\right) \tag{8.32}$$

where the *bar* over a symbol designates its *mean* value. There is some basis for favoring the definition of $\bar{\alpha}$ given by equation 8.31, and it is the one recommended. For small flow deflection angles, however, the values of $\bar{\alpha}$ obtained from the above three definitions are not significantly different.

The concept of a discrete expansion wave serves as a means for obtaining an approximate solution for complex interactions of continuous expansion waves.

Example 8.11. Air at a Mach number of 1.4 turns about a 20 deg convex corner. Determine (a) the properties of the Prandtl-Meyer expansion fan, and (b) the locations of two discrete expansion waves of equal flow deflection angle that approximate the properties of the continuous expansion fan (see Fig. 8.19).

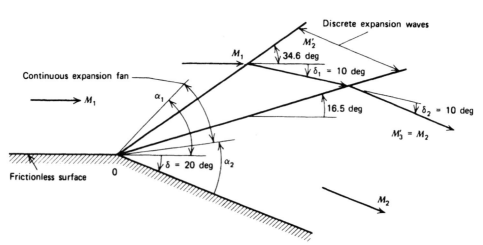

Figure 8.19   Sketch for Example 8.11.

Solution

(a)   From Table C.13, for $M_1 = 1.4$, $v_1 = 9$ deg. Thus,

$$v_2 = v_1 + \delta = 9 + 20 = 29 \text{ deg}$$

From Table C.13, for $v = 29$ deg,

$$M_2 = 2.096$$

(b)   For two discrete expansion waves of equal flow turning angle, $\delta_1 = \delta_2 = 10$ deg. Thus, for the first expansion wave, $\delta_1 = 10$ deg, and

$$v_2' = v_1 + \delta_1 = 9 + 10 = 19 \text{ deg}$$

From Table C.13, for $v_2 = 19$ deg,

$$M_2' = 1.741$$

The average flow deflection angle is $\bar{\delta} = \delta/2 = 5$ deg. From equation 8.31

$$\bar{\alpha}_1 = \sin^{-1}\left(\frac{2}{M_1 + M_2'}\right) = \sin^{-1}\left(\frac{2}{1.4 + 1.741}\right) = 39.6 \text{ deg}$$

Consequently, the first discrete expansion wave makes an angle of $(39.6 - 5) = 34.6$ deg with respect to the initial flow direction.

For the second discrete expansion wave, $\delta_2 = 10$ deg, and

$$v_3' = v_2' + \delta_2 = 19 + 10 = 29 \text{ deg}$$

From Table C.13, for $v_3' = 29$ deg,

$$M_3' = 2.096$$

The average flow deflection angle for the second turn is $\bar{\delta} = \delta_2/2 = 5$ deg after the first flow deflection angle $\delta_1 = 10$ deg, or 15 deg with respect to the direction of the initial flow. The average wave angle is

$$\bar{\alpha}_2 = \sin^{-1}\left(\frac{2}{M_2' + M_3'}\right) = \sin^{-1}\left(\frac{2}{1.741 + 2.096}\right) = 31.5 \text{ deg}$$

Hence, the second discrete expansion wave makes an angle of $(31.5 - 15) = 16.5$ deg with respect to the direction of the initial flow. The results are illustrated in Fig. 8.19.

8-8(a) Reflection of an Expansion Wave Intersecting a Wall

Figure 8.20 illustrates schematically the steady uniform supersonic flow of a compressible fluid parallel to a flat frictionless wall. A discrete expansion wave AO propagates into the flowing fluid, impinges on the flat wall at point O, and is reflected. Consider the streamline pq. As it crosses the expansion wave AO, the streamline is deflected away from the wall through the angle δ. Since the physical requirement is that the flow remain parallel to the wall, the strength of the reflected wave OB must be equal to that of AO so that it will deflect the streamline to become parallel to the wall. In general, the angle of reflection r is not equal to the angle of incidence i. Since the reflected wave must turn the flow back toward the wall, which means away from the reflected wave, the latter must be an expansion wave. The

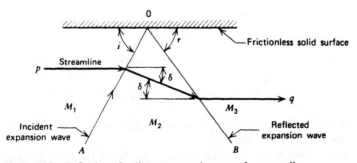

Figure 8.20 Reflection of a discrete expansion wave from a wall.

aforementioned process is illustrated in the hodograph plane in Fig. 8.21, where state 1 corresponds to the initial flow, state 2 corresponds to the flow after the incident expansion wave, and state 3 corresponds to the flow after the reflected expansion wave. It is evident that the final flow angle is parallel to the wall, and that the flow has been accelerated to a larger Mach number.

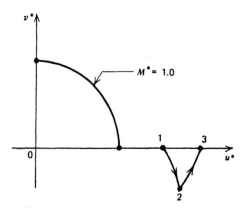

Figure 8.21 Wave reflection in the hodograph plane.

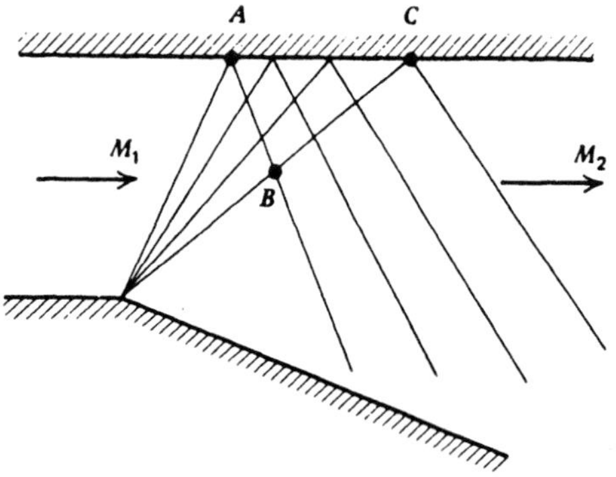

Figure 8.22 Reflection of a continuous expansion fan from a wall.

The process illustrated schematically in Fig. 8.21 is exact for a Mach wave and approximate for a wave of finite amplitude. Figure 8.22 illustrates diagrammatically the real process for a continuous expansion fan of finite amplitude. Each Mach wave is reflected from the surface AC. In the region ABC, however, waves of both families are present and they interact with each other. Such a region cannot be analyzed by the concepts of simple wave flow, but must be analyzed by two-dimensional techniques such as those presented in Chapters 12 and 16.

Example 8.12. Air at a Mach number of 2.0 flows between two parallel frictionless walls. At a given point, the lower wall has a sharp expansion corner of 20 deg. Approximate the resulting expansion fan by a discrete expansion wave, and calculate the properties of the discrete expansion wave that is reflected when the incident expansion wave impinges on the upper wall.

Solution

From Table C.13, for $M_1 = 2.0$, $v_1 = 26.5$ deg. Thus,

$$v_2 = v_1 + \delta = 26.5 + 20 = 46.5 \text{ deg}$$

From Table C.13, for $v = 46.5$ deg,

$$M_2 = 2.84$$

The average flow deflection angle is $\bar{\delta} = \delta/2 = 10$ deg, and the average Mach angle is

$$\bar{\alpha} = \sin^{-1}\left(\frac{2}{M_1 + M_2}\right) = \sin^{-1}\left(\frac{2}{2.0 + 2.84}\right) = 24.5 \text{ deg}$$

The discrete expansion wave, therefore, makes an angle of $(24.5 - 10) = 14.5$ deg with respect to the initial flow direction.

Since the incident wave is an expansion wave that turns the flow 20 deg away from the wall, the reflected wave must, therefore, be an expansion wave turning the flow back parallel to the wall. Thus,

$$v_3 = v_2 + \delta_2 = 46.5 + 20 = 66.5 \text{ deg}$$

From Table C.13, for $v = 66.5$ deg,

$$M_3 = 4.06$$

The average flow deflection angle for the reflected discrete expansion wave is $\bar{\delta} = \delta_2/2 = 10$ deg, and the average Mach angle is

$$\bar{\alpha} = \sin^{-1}\left(\frac{2}{M_2 + M_3}\right) = \sin^{-1}\left(\frac{2}{2.84 + 4.06}\right) = 16.9 \text{ deg}$$

Hence, the reflected wave makes an angle of $(16.9 + 10) = 26.9$ deg with respect to the wall.

8–8(b) Reflection of an Expansion Wave Intersecting a Free Pressure Boundary

Figure 8.23 illustrates schematically a discrete expansion wave that intersects the boundary of a free jet; the static pressure at the boundary of the jet is specified. The discrete expansion wave AO intersects the jet boundary, where the jet pressure p_1 is equal to the ambient pressure p_o. Because of the expansion wave AO, the pressure in Region 2 is smaller than p_o, and the flow is turned away from the expansion wave. When the expansion wave AO intersects the jet boundary at point O, the pressure in the jet is decreased below p_o. Since point O is on the jet boundary, a wave is reflected from it in a manner that will return the fluid pressure in Region 3 to the ambient pressure. Consequently, the reflected wave OB must be a *compression wave*, that is, an oblique shock wave. Since compression waves turn the flow toward the wave, the boundary of the free jet is turned inward, as illustrated in Fig. 8.23. In actual flows, the aforementioned process is spread out because of the diffuse nature of the actual expansion fan. Nevertheless, the essential features of the reflection process are illustrated schematically in Fig. 8.23.

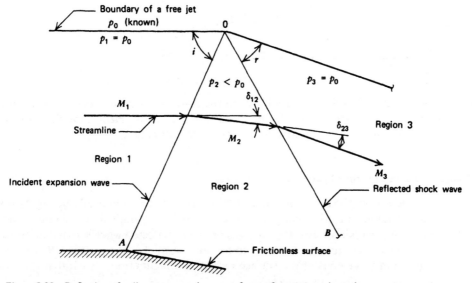

Figure 8.23 Reflection of a discrete expansion wave from a free pressure boundary.

Example 8.13. Air having $M = 2$ flows between two parallel walls at 1 atm pressure. At a given axial location, the upper wall ends and the upper surface of the flow field is exposed to the atmosphere. At the same axial location, the lower wall turns 20 deg away from the flow direction, thus propagating a continuous expansion fan into the flow. Assume that the expansion fan is a single discrete expansion wave. An oblique shock wave is reflected from the jet boundary, as illustrated in Fig. 8.23. Determine the total deflection angle for the flow from its initial direction after the oblique shock wave. Assume that $\gamma = 1.4$.

Solution

From Table C.13, for $M_1 = 2.0$, $\nu_1 = 26.4$ deg. For clockwise turning

$$\nu_2 = \nu_1 + \delta = -26.4 - 20 = -46.4 \text{ deg}$$

From Table C.13, for $\nu_2 = -46.4$ deg,

$$M_2 = 2.84$$

The average flow deflection angle $\bar{\delta} = \delta/2 = -10$ deg, and the average Mach angle is

$$\bar{\alpha} = \sin^{-1}\left(\frac{2}{M_1 + M_2}\right) = \sin^{-1}\left(\frac{2}{2.0 + 2.84}\right) = 24.5 \text{ deg}$$

The expansion wave, therefore, makes an angle of $(24.5 - 10) = 14.5$ deg with respect to the initial flow direction. From Table C.6, for $M_1 = 2.0$ and $M_2 = 2.84$, $p_1/P_1 = 0.1278$ and $p_2/P_2 = 0.0347$. Since the expansion wave is isentropic, $P_1 = P_2$. Hence,

$$\frac{p_2}{p_1} = \frac{p_2}{P_2}\frac{P_1}{p_1} = \frac{0.0347}{0.1278} = 0.2715$$

$$p_2 = (0.2715)p_1 = 0.2715 \text{ atm}$$

Because the incident wave is an expansion wave, it lowers the air pressure behind it to 0.2715 atm. The reflected wave at the free jet boundary must be an oblique shock wave of sufficient strength to raise the static pressure of the air behind the reflected wave to 1 atm. Thus,

$$\frac{p_3}{p_2} = \frac{p_{\text{atm}}}{p_2} = \frac{1.0}{0.2715} = 3.683$$

From Fig. 7.19, for $M_2 = 2.84$ and $p_3/p_2 = 3.683$,

$$\delta_{23} = 20 \text{ deg}$$

Note that δ_{23} is measured relative to the flow direction in Region 2 (see Fig. 8.23), which is -20 deg relative to the initial flow direction. Hence, the flow direction in Region 3 is -40 deg relative to the initial flow direction.

8-8(c) Neutralization of an Expansion Wave

Consider now the physical situation illustrated diagrammatically in Fig. 8.24 where the direction of a frictionless wall is changed by an angle δ at the point where the incident expansion wave AO impinges on it. No reflected wave is required for causing the streamlines crossing wave AO to become parallel to the surface OB. The incident expansion wave ends at the surface because its inclination is such that the reflected wave is *neutralized* or *canceled*.

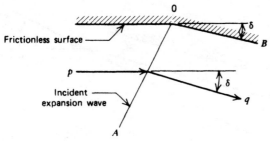

Figure 8.24 Neutralization of a discrete expansion wave.

8–8(d) Intersection of Expansion Waves

Figure 8.25 illustrates diagrammatically a flow passage that causes two expansion waves to intersect. In Fig. 8.25 the expansion wave AB emanates from the lower corner A and is intersected at B by the expansion wave DB emanating from the upper corner D.

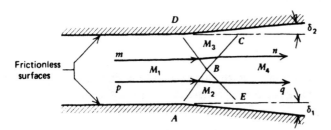

Figure 8.25 Intersection of two discrete expansion waves.

Consider the two parallel streamlines mn and pq, which are deflected by the expansion waves. Streamline pq is first deflected downward through the angle δ_1 by the expansion wave ABC, and is then deflected upward through the angle δ_2 by the expansion wave DBE. Similarly, streamline mn is first deflected upward through the angle δ_2 and then downward by δ_1. Since both of the streamlines are deflected through the same total angle when they reach the region corresponding to M_4, they are again parallel to each other; obviously, if $\delta_1 = \delta_2$, the final direction of the streamlines mn and pq is identical with their original direction upstream to the expansion waves.

The values of M_2, M_3, and M_4 may be determined directly from tables for given values of M_1, δ_1, and δ_2. The average direction of the expansion wave between any two regions is determined from the value of the average Mach angle $\bar{\alpha}$ (see equation 8.31). The average Mach angle $\bar{\alpha}$ is measured from the average directions of the velocities in the two regions.

Example 8.14. Air at a Mach number of 2.0 flows between two parallel frictionless walls. At a given location, both walls turn 15 deg away from the initial flow direction, as illustrated in Fig. 8.26. Treat the continuous expansion waves as discrete expansion waves, and determine the Mach number in Region 4. Draw the resulting wave pattern to scale.

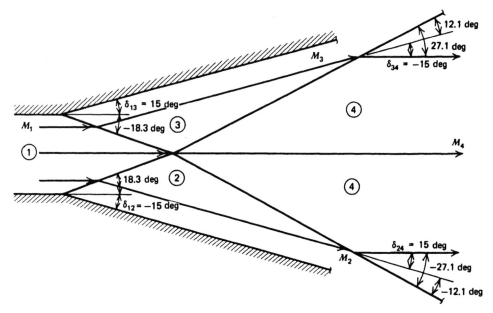

Figure 8.26 Sketch for Example 8.14.

Solution

Since both walls turn the initial flow through equal flow deflection angles, the two initial expansion waves have the same properties. Consider the wave from the upper wall. From Table C.13, for $M_1 = 2.0$, $v_1 = 26.4$ deg. Then,

$$v_3 = v_1 + \delta_{13} = 26.4 + 15 = 41.4 \text{ deg}$$

From Table C.13, for $v_3 = 41.4$ deg, $M_3 = 2.60$. The average flow deflection angle from Region 1 to Region 3 is $\bar{\delta}_{13} = \delta_{13}/2 = 7.5$ deg. The average Mach angle is

$$\bar{\alpha}_{13} = \sin^{-1}\left(\frac{2}{M_1 + M_3}\right) = \sin^{-1}\left(\frac{2}{2.0 + 2.60}\right) = 25.8 \text{ deg}$$

Hence, the initial expansion wave makes an angle of $(-25.8 + 7.5) = -18.3$ deg with respect to the initial flow direction. A mirror image wave is propagated from the lower wall at an angle of 18.3 deg with respect to the initial flow direction. Those waves are illustrated diagrammatically in Fig. 8.26.

At the point of intersection of the two initial expansion waves, they must be transmitted with sufficient strength to turn the flows in Regions 2 and 3 back to the initial flow direction. If the initial expansion waves are of different strengths, the two transmitted waves are also of different strengths, so that the flow direction behind the transmitted waves cannot be parallel to the initial flow direction. Example 7.15 presents a problem involving the interaction of two oblique shock waves of unequal strengths. The analysis for expansion waves of unequal strengths is quite similar.

In the present case, both transmitted expansion waves must turn the flow through a flow deflection angle of 15 deg. Thus, for the wave that is transmitted into the lower half-plane, $\delta_{24} = 15$ deg with respect to the flow direction in Region 2. From the results obtained above, $M_2 = M_3 = 2.60$, and $v_2 = v_3 = 41.4$ deg. Thus,

$$v_4 = v_2 + \delta_{24} = 41.4 + 15 = 56.4 \text{ deg}$$

From Table C.13, for $v_4 = 56.4$ deg, $M_4 = 3.37$. The average flow deflection angle from Region 2 to Region 4, relative to the flow direction in Region 2, is $\bar{\delta}_{24} = \delta_{24}/2 = 7.5$ deg. The average Mach angle is

$$\bar{\alpha}_{24} = \sin^{-1}\left(\frac{2}{M_2 + M_4}\right) = \sin^{-1}\left(\frac{2}{2.60 + 3.37}\right) = 19.6 \text{ deg}$$

The transmitted wave, therefore, makes an angle of $(-19.6 + 7.5) = -12.1$ deg with respect to the flow direction in Region 2, and an angle of $(-12.1 - 15) = -27.1$ deg with respect to the flow direction in Region 1. The results are presented in Fig. 8.26.

8-8(e) Summary of Observations Concerning the Interaction of Expansion Waves

The following observations apply to expansion waves.

1. An expansion wave turns the flow away from itself.
2. Expansion waves diverge from each other.
3. An expansion wave reflects from a solid boundary as an expansion wave.
4. An expansion wave reflects from a free pressure boundary (i.e., a free jet) as an oblique shock wave.
5. Expansion wave reflections from a wall may be cancelled by turning the wall into the flow direction.

For an isentropic compression wave that is governed by simple wave flow theory, all of the above observations are reversed.

8-9 SUMMARY

A steady uniform planar supersonic flow of a compressible fluid may be turned in a direction *away* from the initial flow direction by causing it to flow around a sharp *convex corner*. As the fluid flows around the corner, it accelerates from $M_1 > 1$ to $M_2 > M_1$ and consequently expands, so that after the new flow direction is established, $V_2 > V_1$ and $p_2 < p_1$. Because the disturbance produced by the corner is very weak, Mach lines are propagated outwardly from it. For a steady uniform initial flow, the Mach lines are straight lines. The turning of the flow is bounded by the two Mach lines corresponding to M_1 and M_2, and they make the Mach angles $\alpha_1 = \sin^{-1}(1/M_1)$ and $\alpha_2 = \sin^{-1}(1/M_2)$ with respect to the flow directions in Regions 1 and 2, respectively. Between the two bounding Mach lines are an infinite number of Mach lines emanating from the convex corner, and along each Mach line the flow properties are constant; they vary, of course, from Mach line to Mach line. The region between the bounding Mach lines is called the *expansion fan* (see Fig. 8.1).

The type of flow just described is known as a *Prandtl-Meyer expansion*. The angle δ through which the expanding flow is turned is called the *flow deflection angle*, and the corresponding expansion waves are called *simple waves*. If the flowing fluid is a perfect gas, quantitative relationships may be derived between the variables pertinent to the aforementioned type of expansion.

In the special case of the flow of a perfect gas having an initial Mach number $M_1 = 1$ (sonic flow), the initial Mach line emanating from the corner is perpendicular to V_1 and the corresponding flow deflection angle is termed the *Prandtl-Meyer angle*, which is denoted by v; that is, $\delta = v$. Table C.13 presents values of v as a function of M, for several values of γ. The flow deflection angle for an expansion from $M_1 > 1$ to $M_2 > M_1$ is given by $\delta = v_2 - v_1$. According to the Prandtl-Meyer theory, the maximum possible value of the Prandtl-Meyer angle ($v = v_{max}$) corresponds to that for which $M_2 = \infty$; for a perfect gas with $\gamma = 1.4$, that value is $\pm v_{max} = 130.45$ deg. In the actual expansion of a gas around a convex corner, however, the maximum possible flow deflection without the flow separating from the wetted surface would be much smaller.

As in the case of the oblique shock wave (see Section 7-9), the Prandtl-Meyer expansion may also be plotted on the hodograph plane (u^*v^* plane). The dimensionless velocity $M^* = \sqrt{u^{*2} + v^{*2}}$ is plotted for different initial values of M^* and δ in Fig. 8.16.

No matter how gradually the direction of a supersonic flow is altered by a continuous *concave* surface, even though only Mach waves are propagated from the surface, a shock wave is eventually formed at some distance from the concave deflecting surface. The shock wave arises from the Mach waves emanating from the surface intersecting and strengthening each other so that they coalesce, forming an *envelope shock*. In contradistinction, the Mach waves propagated from a *convex* corner diverge and cannot intersect each other. Consequently, an expansion diffuses instead of strengthens the initial disturbance.

The expansion of a compressible fluid flowing over a continuous convex surface may be analyzed as a series of Prandtl-Meyer expansions by subdividing the surface so that a series of convex corners is formed. The number of those corners is then increased indefinitely so that the flow deflection angles decrease in magnitude and eventually a continuous surface is obtained.

The governing differential equation for a simple wave flow for an imperfect gas may be integrated numerically if the gas behaves as a simple system. However, no general solution is possible. For an imperfect gas, a numerical procedure for analyzing the simple wave flow is described in Section 8-5 and is applied in Example 8.8.

A procedure for approximating continuous expansion waves by finite amplitude discrete expansion waves is presented in Section 8-8. General observations pertaining to the intersection of simple waves with a solid wall, a free pressure boundary, and other simple waves, and also their cancellation, are summarized in Section 8-8(e).

REFERENCES

1. T. Meyer, "Uber Zweidimensionale Bewegungsvorgange in einen Gas, das mit Uberschallgeschwindigkeit strömt," Ph. D. Dissertation, Göttingen, 1908. Included in *Foundation of High Speed Aerodynamics*, Dover Publications, New York, 1951.
2. M. J. Zucrow, *Aircraft and Missile Propulsion*, Vol. 1, Chap. 3, Wiley, New York, 1958.
3. A. H. Shapiro, *The Dynamics and Thermodynamics of Compressible Fluid Flow*, Chap. 15, Ronald Press, New York, 1954.
4. R. Courant and K. O. Friedrichs, *Supersonic Flow and Shock Waves*, Macmillan, New York, 1949.

PROBLEMS

1. Air flowing over a horizontal frictionless surface approaches a sharp convex corner with an initial Mach number $M_1 = 1.0$. The initial pressure and temperature are $1.0 \cdot 10^5$ N/m^2 and 300 K, respectively. Calculate the Mach number, pressure, and temperature after expanding around the corner for a flow deflection angle of (a) 15 deg clockwise, (b) 30 deg clockwise,

(c) 60 deg clockwise, (d) 15 deg counterclockwise, (e) 30 deg counterclockwise, and (f) 60 deg counterclockwise. In parts a, b, and c, the air flows above the surface, and in parts d, e, and f, the flow is below the surface. Assume $\gamma = 1.4$.

2. Air with an initial Mach number $M_1 = 1.0$ flows around a sharp convex corner. The initial pressure and temperature are $1.0 \cdot 10^5$ N/m^2 and 300 K, respectively. Calculate the required flow deflection angle, and the corresponding pressure and temperature, to obtain a final Mach number of (a) 2.0, (b) 3.0, (c) 4.0, and (d) 5.0. Assume $\gamma = 1.4$.

3. Consider a smooth frictionless surface inclined at an angle $\theta = 30$ deg with respect to the horizontal. Air flows from left to right above the surface with an initial Mach number $M_1 = 2.0$. Calculate the Mach number and flow angle for the flow after a sharp corner having the flow deflection angle (a) 15 deg clockwise, (b) 30 deg clockwise, (c) 60 deg clockwise, and (d) 15 deg counterclockwise. Assume $\gamma = 1.4$.

4. Air flows above a frictionless surface having a sharp corner. The flow angle and Mach number downstream from the corner are -60 deg and 4.0, respectively. Calculate the upstream Mach number and flow angle for a flow deflection angle of (a) 15 deg clockwise, (b) 30 deg clockwise, (c) 60 deg clockwise, and (d) 15 deg counterclockwise. Assume $\gamma = 1.4$.

5. A wind tunnel nozzle is designed to yield a parallel uniform flow of air with a Mach number $M = 3.0$. The stagnation pressure of the air supply reservoir $P = 70 \cdot 10^5$ N/m^2, and the nozzle exhausts into the atmosphere. Calculate the flow angle at the exit lip of the nozzle if the atmospheric pressure $p_o = 1.0$ atm. Assume $\gamma = 1.4$.

6. For the wind tunnel described in Problem 5, determine the stagnation pressure of the air supply for which the flow angle at the exit lip is zero. Describe what happens when the air supply pressure drops below that value.

7. Consider the symmetric double wedge profile illustrated in Fig. 8.9. The airfoil is located in an airstream with a free-stream Mach number $M = 3.0$, and a free-stream static pressure $p = 1.0133 \cdot 10^5$ N/m^2. Calculate the lift and drag per unit width of the wedge for an angle of attack of (a) 5 deg clockwise, (b) 10 deg clockwise, (c) 15 deg clockwise, (d) 5 deg counterclockwise, (e) 10 deg counterclockwise, and (f) 15 deg counterclockwise. Plot the results, as a function of angle of attack. Assume $\gamma = 1.4$.

8. Consider a diamond shaped body composed of two of the double wedge profiles illustrated in Fig. 8.9, placed on top of each other. Calculate the lift and drag per unit width for that body for the angles of attack specified in Problem 7. Plot the results as a function of angle of attack.

9. Air at a Mach number $M = 2.0$ flows above a frictionless horizontal surface that makes a clockwise 90 deg turn around a cylindrical body joined tangentially to the horizontal surface. The radius of the cylinder is 1.0 m. Determine the Mach number M and flow angle θ at a point 2 m from the center of the cylinder along a 45 deg line through the cylinder axis. Assume $\gamma = 1.4$.

10. Air at a Mach number $M_1 = 1.0$ flows above a frictionless horizontal surface. Plot a streamline of the flow for a flow deflection angle of 90 deg clockwise. Assume $\gamma = 1.4$.

11. Plot the streamline that is initially 1.0 m above the surface for the flow described in Problem 9.

12. Air with an initial Mach number $M_1 = 2.0$ flows over three sharp corners in succession, having clockwise turning angles of 5, 10, and 15, deg respectively. Calculate the Mach number and flow angle after each of the three corners. Plot the flow field illustrating the expansion fans.

13. The Prandtl-Meyer angle v is related to the Mach number M by equation 8.15. A fluid with $\gamma = 1.25$ and an initial Mach number $M_1 = 1.0$ flows over a sharp corner having a flow deflection angle $\delta = 45$ deg. Calculate the Mach number after the turn employing the secant method [see Appendix A–4(b)].

14. Air flows over the control surface of a reentry vehicle. The control surface is the diamond shaped body described in Problem 8. The flow properties on the first portion of the upper surface are $V = 1500$ m/s, $t = 2500$ K, and $p = 4$ atm. Calculate the flow properties after the air passes over the corner at the top of the profile, taking into account the variation of the specific heat of the air with temperature.

15. A perfect gas ($\gamma = 1.4$) flows into a concave corner with an initial Mach number $M_1 = 2.0$. Calculate the Mach number after the turn for flow deflection angles δ from zero up to the maximum angle for which the oblique shock wave remains attached to the corner (or for which a Prandtl-Meyer compression yields a sonic flow) for (a) an oblique shock wave, and (b) a Prandtl-Meyer compression. Plot M_2 versus δ for both the oblique shock wave and the Prandtl-Meyer compression.

16. Work Problem 1 employing the hodograph plane presented in Fig. 8.16.

17. Work Problem 2 employing the hodograph plane presented in Fig. 8.16.

18. Work Problem 3 employing the hodograph plane presented in Fig. 8.16.

19. Work Problem 4 employing the hodograph plane presented in Fig. 8.16.

20. Air flowing over a frictionless surface with an initial Mach number $M_1 = 2.0$ makes a 30 deg clockwise turn. Determine (a) the properties of the Prandtl-Meyer expansion fan, and (b) the locations and flow properties for three discrete expansion waves having equal turning angles that turn the flow through the same total flow deflection angle.

21. Air is flowing from left to right between two parallel frictionless walls with a Mach number $M = 3.0$. The lower wall has a sharp corner that turns the flow 15 deg clockwise. Determine the properties of the single discrete expansion wave that approximates the continuous expansion fan. Calculate the properties of the discrete expansion wave that is reflected from the upper wall when the incident discrete expansion wave impinges on that wall. Assume $\gamma = 1.4$.

22. The upper wall of the flow passage described in Problem 21 is terminated at the axial location where the sharp convex corner occurs in the lower wall. The static pressure in the ambient region at the end of the upper wall is equal to that in the initial parallel uniform flow. Determine the properties of the oblique shock wave that is reflected from the jet boundary when the discrete expansion wave hits that boundary.

23. Air at the sonic condition flows between two parallel frictionless surfaces, as illustrated in Fig. 8.26. The flow deflection angles are both 30 deg. Consider the initial expansion waves as discrete waves and determine the Mach number in Region 4.

24. In Problem 23, the flow deflection angle on the upper wall is 30 deg and that on the lower wall is 20 deg. Calculate the Mach number and flow angle in Regions 2, 3, and 4, treating the continuous expansion fans as discrete expansion waves.

9

mass addition, combustion waves, and generalized steady one-dimensional flow

9–1 PRINCIPAL NOTATION FOR CHAPTER 9

The notation presented in Section 3–1 applies to the flows discussed in this chapter. The following additional notation is also applicable.

δD generalized body force.

G $= \dot{m}/A$, mass flux.

H $= h + V^2/2 + gz$, stagnation enthalpy.

$d\dot{m}$ differential mass flow rate of mass addition stream.

$\dot{r}$ $= (1 + kV)cp^n$, linear burning rate for a solid propellant.

u internal energy.

Δu° energy of combustion.

WP wetted perimeter of burning solid propellant perforation.

y $= V_{ix}/V$, ratio of x component of velocity of mass addition stream to velocity of main stream.

Greek

Λ generalized flow function defined by equation 9.113.

ρ_P density of solid propellant.

ψ $= 1 + \dfrac{\gamma - 1}{2} M^2$

Subscripts

e denotes aft-end of a solid propellant grain.

i denotes mass addition stream.

o denotes head-end of a solid propellant grain.

Superscripts

$*$ denotes critical (i.e., sonic) condition.

9–2 INTRODUCTION

Chapters 4 to 8 are concerned with the analysis of simple flows, that is, flows in which a single *driving potential* is responsible for the changes in the flow properties. In addition to other simple flows that are not discussed in those chapters, there are complex flows in which *two or more* driving potentials act simultaneously. In this chapter, the following additional one-dimensional flows of practical significance are considered.

1. Simple mass addition.
2. Combustion waves.
3. Generalized steady one-dimensional flow.

Because of the complexity of the equations describing these flow processes, the integration of the pertinent differential equations is performed in most cases by applying numerical techniques. For each of the above flows, the general features of the relevant methods of numerical analysis are discussed.

9–3 FLOW WITH MASS ADDITION

All of the flows considered in the previous chapters have the common feature that the mass flow rate is constant at each cross section of the flow passage being considered. In this section, the equations are derived for determining the effects, for compressible fluids, of *mass addition* to the main stream flow. Only simple mass addition is considered; that is, a flow wherein the sole *driving potential* causing the changes in the flow properties is the rate of the mass addition $d\dot{m}$. Hence, $\delta Q = \delta W = dA = \delta F_i = \delta D = g\,dz = 0$. The special case where the mass addition is normal to the flow direction of the main stream is considered in detail, and the results are applied to the interior ballistics of a solid propellant rocket motor.

9-3(a)   Governing Equations for Flow with Simple Mass Addition

Figure 9.1 illustrates the physical model for simple mass addition. Mass is injected at the rate $d\dot{m}$ through the top of the control surface A. In practice, the mass addition process may take various forms, such as the gases produced by the burning of the internal surface (passage walls) of a hollow solid propellant grain, or the uniform flow of a gaseous coolant through a porous wall in the case of transpiration cooling. The significant feature of the mass addition process is that the mass of fluid added to the main stream at the rate $d\dot{m}$ has the flow properties p_i, ρ_i, h_i, V_i, etc., while those for the main stream flow, which has the mass flow rate $\dot{m}$, are p, ρ, h, V, etc. The two gas streams are assumed to mix completely, and the mixed stream, having the mass flow rate $\dot{m} + d\dot{m}$, leaves the control volume $\mathscr{V}$ with the uniform properties $p + dp$, $\rho + d\rho$, etc. For *simple mass addition*, it is assumed further that the two streams of gas have identical values of molecular weight, specific heat, and stagnation enthalpy (i.e., $R = R_i$, $c_p = c_{pi}$, and $H = H_i$), and they are perfect gases.

The equations governing simple mass addition are obtained by applying the governing equations for the flow of a fluid (see Table 3.1) to the control volume $\mathscr{V}$ illustrated in Fig. 9.1, where the flow area A remains constant.

(1)   *Continuity equation*

$$\dot{m} = \rho A V \tag{9.1}$$

Differentiating equation 9.1 yields

$$\frac{d\dot{m}}{\dot{m}} = \frac{d\rho}{\rho} + \frac{dV}{V} \tag{9.2}$$

where $d\dot{m}$ is the rate of mass addition to the main stream flow for which the mass flow rate is $\dot{m}$.

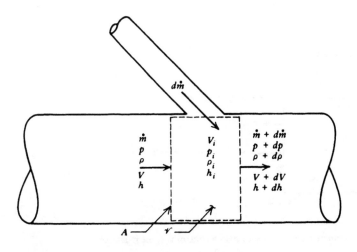

$$\delta Q = \delta W = dA = \delta F_i = \delta D = g\,dz = 0$$

Figure 9.1   Physical model for simple mass addition.

(2) Energy equation

Because the main stream and the injected gas stream have identical stagnation enthalpies, the stagnation enthalpy of the combined stream remains constant. Thus,

$$H = h + \frac{V^2}{2} = \text{constant} \tag{9.3}$$

For a perfect gas, $h = c_p t$, so that equation 9.3 becomes

$$c_p T = c_p t + \frac{V^2}{2} = \text{constant} \tag{9.4}$$

Differentiating equation 9.4 and introducing the definitions $c_p = \gamma R/(\gamma - 1)$, $a^2 = \gamma R t$, and $M = V/a$ into the result gives

$$\frac{dt}{t} + (\gamma - 1)M^2 \frac{dV}{V} = 0 \tag{9.5}$$

(3) Momentum equation

Figure 9.2 illustrates schematically the forces and momentum fluxes acting on the control volume $\mathcal{V}$ in the x direction, where V_{ix} is the x component of V_i. Applying the momentum equation gives

$$pA - (p + dp)A = (\dot{m} + d\dot{m})(V + dV) - \dot{m}V - d\dot{m}\,V_{ix} \tag{9.6}$$

By definition, the parameter y is

$$y \equiv \frac{V_{ix}}{V} \tag{9.7}$$

If $V_{ix} = 0$, then $y = 0$, and the mass is added so that it enters the main gas stream in a direction normal to the main stream velocity V. If $V_{ix} = V$, then $y = 1$, and the mass addition has the same velocity as the main gas stream. Finally, if $V_{ix} > V$, then $y > 1$, and the mass addition has a larger momentum and kinetic energy per unit mass than the main gas stream. Devices having $y > 1$ are utilized for pumping a low kinetic energy stream with a gas having a higher specific kinetic energy. Substituting equations 9.1 and 9.7 into equation 9.6, dividing by A, and neglecting the

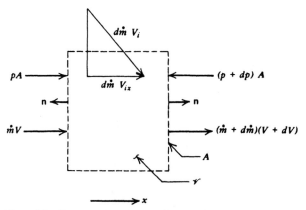

Figure 9.2 Forces and momentum fluxes acting on a control volume in simple mass addition.

higher-order term $d\dot{m}\, dV$, gives

$$dp + \rho V\, dV + \rho V^2(1 - y)\frac{d\dot{m}}{\dot{m}} = 0 \tag{9.8}$$

Dividing equation 9.8 by p, introducing $p/\rho = a^2/\gamma$, and simplifying, yields

$$\frac{dp}{p} + \gamma M^2\frac{dV}{V} + \gamma M^2(1 - y)\frac{d\dot{m}}{\dot{m}} = 0 \tag{9.9}$$

Logarithmic differentiation of the equation of state for a perfect gas, $p = \rho Rt$, gives

$$\frac{dp}{p} = \frac{d\rho}{\rho} + \frac{dt}{t} \tag{9.10}$$

Differentiating $M = V/a = V/\sqrt{\gamma Rt}$ yields

$$\frac{dM}{M} = \frac{dV}{V} - \frac{1}{2}\frac{dt}{t} \tag{9.11}$$

Differentiating the definition of the stagnation pressure P (see equation 3.182) gives

$$\frac{dP}{P} = \frac{dp}{p} + \frac{\gamma M^2}{1 + \dfrac{\gamma - 1}{2}M^2}\frac{dM}{M} \tag{9.12}$$

Differentiating the definition of the impulse function $\mathscr{F}$ (see equation 3.139) gives

$$\frac{d\mathscr{F}}{\mathscr{F}} - \frac{dp}{p} - \frac{2\gamma M^2}{1 + \gamma M^2}\frac{dM}{M} = 0 \tag{9.13}$$

From equation 3.178, Table 3.2, for $dT = 0$,

$$\frac{ds}{c_p} = -\frac{\gamma - 1}{\gamma}\frac{dP}{P} \tag{9.14}$$

9–3(b) Effect of Mass Addition on the Flow Properties

Equations 9.2, 9.5, and 9.9 to 9.14 comprise a system of eight equations involving the nine property ratios $d\dot{m}/\dot{m}$, dM/M, dV/V, dt/t, dp/p, $d\rho/\rho$, dP/P, $d\mathscr{F}/\mathscr{F}$, and ds/c_p. Those eight equations may be solved simultaneously, employing the method presented in Section 4–4(c) for isentropic flow, to yield eight relationships between the independent variable $d\dot{m}/\dot{m}$ and the eight dependent variables dM/M, dp/p, etc. Those equations are presented in Table 9.1.

It is seen from Table 9.1, equations 9.15 to 9.22, that the influence coefficients of $d\dot{m}/\dot{m}$ include both M and y as variables. Consequently, whether the change in a dependent variable dM/M, dp/p, etc., is either positive or negative because of the simple mass addition depends not only on whether the flow is subsonic or supersonic, but also on the magnitude of the parameter y, defined by equation 9.7. For example, the term $[(1 + \gamma M^2) - y\gamma M^2]$ in equations 9.15, 9.18, and 9.19 of Table 9.1, is positive for values of y between 0 and $(1 + \gamma M^2)/\gamma M^2$, and negative for values of y larger than $(1 + \gamma M^2)/\gamma M^2$. For most cases of engineering interest, the value of y is such that the terms involving y in the numerators of equations 9.15 to 9.21 are always positive. For such cases the effect of the mass addition $d\dot{m}$ on the change in the dependent variables depends only on whether the main stream flow is either subsonic or supersonic.

Table 9.1 Influence Coefficients for Simple Mass Addition for a Perfect Gas

$$\frac{dM}{M} = \frac{\left(1 + \frac{\gamma - 1}{2} M^2\right)}{1 - M^2} \left[(1 + \gamma M^2) - y\gamma M^2\right] \frac{d\dot{m}}{\dot{m}} \tag{9.15}$$

$$\frac{dp}{p} = -\frac{\gamma M^2}{1 - M^2} \left[2\left(1 + \frac{\gamma - 1}{2} M^2\right)(1 - y) + y\right] \frac{d\dot{m}}{\dot{m}} \tag{9.16}$$

$$\frac{d\rho}{\rho} = -\frac{1}{1 - M^2} \left[(\gamma + 1)M^2 - y\gamma M^2\right] \frac{d\dot{m}}{\dot{m}} \tag{9.17}$$

$$\frac{dt}{t} = -\frac{(\gamma - 1)M^2}{1 - M^2} \left[(1 + \gamma M^2) - y\gamma M^2\right] \frac{d\dot{m}}{\dot{m}} \tag{9.18}$$

$$\frac{dV}{V} = \frac{1}{1 - M^2} \left[(1 + \gamma M^2) - y\gamma M^2\right] \frac{d\dot{m}}{\dot{m}} \tag{9.19}$$

$$\frac{dP}{P} = -\gamma M^2(1 - y) \frac{d\dot{m}}{\dot{m}} \tag{9.20}$$

$$\frac{d\mathscr{F}}{\mathscr{F}} = y \frac{\gamma M^2}{1 + \gamma M^2} \frac{d\dot{m}}{\dot{m}} \tag{9.21}$$

$$\frac{ds}{c_p} = -\frac{\gamma - 1}{\gamma} \frac{dP}{P} = (\gamma - 1)M^2(1 - y) \frac{d\dot{m}}{\dot{m}} \tag{9.22}$$

Equation 9.20 shows that the direction of the change in the stagnation pressure dP/P depends only on whether y is less than or larger than 1.0. Equation 9.21 shows that mass addition $d\dot{m}$ always produces a force acting on the passage walls in the direction opposite to the flow direction for the main stream, except in the special case when $y = 0$ (mass addition normal to the main stream); in the latter case no force is exerted on those walls. Equation 9.22 shows that the direction of the entropy change for the flowing gas is opposite to that for the corresponding change in its stagnation pressure.

If $y < 1.0$, all of the factors involving y in the influence coefficients in equations 9.15 to 9.22 are positive. For that special case, the effect of the mass addition $d\dot{m}$ on the dependent variables is summarized in Table 9.2. Similar results may be obtained for any value of y.

Table 9.2 shows that mass addition causes the Mach number to increase for a subsonic flow and to decrease for a supersonic flow. It is apparent, therefore, that the main stream flow chokes at a Mach number of unity, and that no further mass addition to it is possible without altering the upstream conditions. Consequently, a continuous transition from subsonic to supersonic flow is impossible if the sole driving potential is mass addition.

9–3(c) Mass Addition Normal to the Main Stream

There are many applications of engineering interest in which the mass addition $d\dot{m}$ is normal to the flow direction for the main stream (i.e., $y = 0$), as in the two examples cited in Section 9–3(a). Accordingly, the equations governing mass addition normal to the main stream are obtained from equations 9.15 to 9.22 by setting $y = 0$. The resulting equations are all exact differential equations that may be integrated between

Table 9.2 Effect of the Mass Addition $d\dot{m}$ on the Flow Properties for a Perfect Gas (for $y < 1.0$)

Property Ratio	$M < 1$	$M > 1$
$\dfrac{dM}{M}$	+	−
$\dfrac{dp}{p}$	−	+
$\dfrac{d\rho}{\rho}$	−	+
$\dfrac{dt}{t}$	−	+
$\dfrac{dV}{V}$	+	−
$\dfrac{dP}{P}$	−	−
$\dfrac{d\mathscr{F}}{\mathscr{F}}$	$\geqq 0$	$\geqq 0$
$\dfrac{ds}{c_p}$	+	+

a general point in the flow and some reference state to yield a set of algebraic equations for the property ratios of the main stream flow. To illustrate, integrate equation 9.15, Table 9.1, employing the *critical state*, denoted by the superscript *, as the reference state. Thus,

$$\int_{\dot{m}}^{\dot{m}^*} \frac{d\dot{m}}{\dot{m}} = \int_{M}^{1} \frac{1 - M^2}{M(1 + \gamma M^2)\left(1 + \dfrac{\gamma - 1}{2} M^2\right)} \, dM$$

$$\frac{\dot{m}}{\dot{m}^*} = \frac{G}{G^*} = \frac{M\left[2(\gamma + 1)\left(1 + \dfrac{\gamma - 1}{2} M^2\right)\right]^{1/2}}{1 + \gamma M^2} \tag{9.23}$$

where $G = \dot{m}/A$ denotes the *mass flux*.

The remaining property ratios may be obtained in like manner by integrating equations 9.16 to 9.22. It is easier, however, to obtain the ratios t/t^*, p/p^*, etc., directly from the original governing equations. For example, from the definition of the stagnation temperature (see equation 3.181, Table 3.2) and, by assumption, recalling that $T = $ constant, we obtain

$$\frac{t}{t^*} = \frac{\gamma + 1}{2\left(1 + \dfrac{\gamma - 1}{2} M^2\right)} \tag{9.24}$$

From the definition $V = Ma$,

$$\frac{V}{V^*} = M^* = M \left[\frac{\gamma + 1}{2\left(1 + \frac{\gamma - 1}{2} M^2\right)} \right]^{1/2} \tag{9.25}$$

From $\dot{m} = \rho A V = pAV/Rt$,

$$\frac{p}{p^*} = \frac{\gamma + 1}{1 + \gamma M^2} \tag{9.26}$$

From $p = \rho Rt$,

$$\frac{\rho}{\rho^*} = \frac{2\left(1 + \frac{\gamma - 1}{2} M^2\right)}{1 + \gamma M^2} \tag{9.27}$$

From the definition of the stagnation pressure (see equation 3.182, Table 3.2)

$$\frac{P}{P^*} = \frac{\gamma + 1}{1 + \gamma M^2} \left[\left(\frac{2}{\gamma + 1}\right)\left(1 + \frac{\gamma - 1}{2} M^2\right)\right]^{\gamma/(\gamma - 1)} \tag{9.28}$$

The change in entropy may be obtained from equations 9.14 and 9.28. Thus,

$$s - s^* = -R \ln \frac{P}{P^*} \tag{9.29}$$

For mass addition normal to the stream, $d\mathcal{F} = 0$ and $\mathcal{F} = $ constant.

Table C.14 presents values of the property ratios M^*, t/t^*, p/p^*, ρ/ρ^*, P/P^*, and G/G^* as functions of the main stream flow Mach number M. Table 9.3 summarizes the form of Table C.14, and also presents the limiting values for the aforementioned property ratios.

Table 9.3 Limiting Values of the Property Ratios for Simple Mass Addition Normal to the Main Stream for a Perfect Gas

M	M^*	$\dfrac{t}{t^*}$	$\dfrac{p}{p^*}$	$\dfrac{\rho}{\rho^*}$	$\dfrac{P}{P^*}$	$\dfrac{\dot{G}}{G^*}$
0	0	$\left(\dfrac{\gamma + 1}{2}\right)$	$(\gamma + 1)$	2	$(\gamma + 1)\left(\dfrac{2}{\gamma + 1}\right)^{\gamma/(\gamma - 1)}$	0
1	1	1	1	1	1	1
∞	$\left(\dfrac{\gamma + 1}{\gamma - 1}\right)^{1/2}$	0	0	$\left(\dfrac{\gamma - 1}{\gamma}\right)$	∞	$\dfrac{(\gamma^2 - 1)^{1/2}}{\gamma}$

A FORTRAN computer program, subroutine MASS, for calculating the property ratios for simple mass addition for a perfect gas with $y = 0$ is presented below. The input to this program is the same as that discussed in Section 4–4(b). The results presented in Table C.14 for $\gamma = 1.4$ are obtained by the following specifications.

G = 1.4, J(1) = 50, J(2) = 50, J(3) = 10, J(4) = 20, J(5) = 12, J(6) = 5, J(7) = 0,
DM(1) = 0.01, DM(2) = 0.02, DM(3) = 0.05, DM(4) = 0.1, DM(5) = 0.5, DM(6) = 1.0

```
       SUBROUTINE MASS

C      MASS ADDITION PROPERTY RATIOS FOR A PERFECT GAS

       REAL M,MS $ DIMENSION J(9),DM(9) $ DATA INF/6H INFIN/
       NAMELIST /DATA/ G,J,DM $ READ (5,DATA)
       G1=(G-1.0)/2.0 $ G2=2.0/(G+1.0) $ G3=G+1.0 $ G4=G/(G-1.0)
       M=0.0 $ MS=0.0 $ T=1.0/G2 $ P=G3 $ R=2.0 $ PO=G3*G2**G4 $ GG=0.0
       WRITE (6,2000) G $ WRITE (6,2020) M,MS,T,P,R,PO,GG $ I=1 $ L=0

C      CALCULATE PROPERTY RATIOS

   40  N=J(I) $ DO 50 K=1,N $ M=M+DM(I) $ C=1.0+G1*M**2 $ MS=M/SQRT(G2*C)
       T=1.0/(G2*C) $ P=G3/(1.0+G*M**2) $ R=P/T $ PO=P/T**G4 $ GG=R*MS
       WRITE (6,2020) M,MS,T,P,R,PO,GG $ L=L+1 $ IF (L.LT.50) GO TO 50
       WRITE (6,2000) G $ L=0
   50  CONTINUE $ I=I+1 $ IF((I.EQ.10).OR.(J(I).EQ.0)) GO TO 60 $GO TO 40
   60  MS=SQRT((G+1.0)/(G-1.0)) $ T=0.0 $ P=0.0 $ R=(G-1.0)/G
       GG=SQRT(G**2-1.0)/G
       WRITE (6,2030) INF,MS,T,P,R,INF,GG $ RETURN

 2000  FORMAT (1H1,14X,40HMASS ADDITION NORMAL TO STREAM,  GAMMA =,F5.2//
      14X,1HM,6X,2HM*,7X,4HT/T*,9X,4HP/P*,9X,4HR/R*,9X6HPO/PO*7X4HG/G*/)
 2020  FORMAT (1H ,F5.2,F10.5,5E13.5)
 2030  FORMAT (A6,F10.5,3E13.5,3X,A6,E17.5)
       END
```

9–3(d) Application to Solid Propellant Rocket Motors

Figure 9.3 illustrates schematically the general features of a rocket motor equipped
with an internal-burning solid propellant grain. The cross-sectional area of the
perforation in the grain, denoted by A, is the area through which the combustion
gas flows toward the propulsive nozzle of the rocket motor; the perforation is parallel
to the longitudinal axis of the rocket motor. The cross-sectional area A is not a
constant while the grain is burning, even if its value is the same at all cross sections
along the grain, because the burning surface recedes in the direction normal to itself.

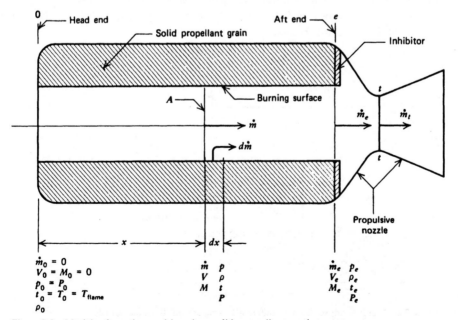

Figure 9.3 Model of an internal-burning solid propellant rocket motor.

The *linear burning rate* $\dot{r}$, the rate at which the burning surface recedes, is generally very small (0.001 to 0.5 m/s, approximately) compared to the velocity of the combustion gas flowing toward the propulsive nozzle; the latter velocity is ordinarily several hundred meters per second. No serious error, therefore, is introduced by neglecting the unsteady effects caused by the variation of A with time in determining the interior ballistics of a solid propellant rocket motor, and the gas flow is treated as quasi-steady.

The end of the propellant grain located at the nozzle end of the rocket motor is called the *aft end*, and the end of the grain located at the closed end of the casing enclosing the grain is called the *head end*. The head-end and aft-end surfaces of the grain are assumed to be inhibited from burning, so that the only burning surface is that comprising the perforation in the grain.

It is an empirically demonstrated fact that the surface of the grain burns normal to itself at a rate that depends on the propellant composition, the propellant temperature, the static pressure p acting on the thin combustion zone at the propellant surface, and the velocity V of the main *core flow*. The *linear burning rate* $\dot{r}$ is correlated by an empirical equation of the following form:

$$\dot{r} = (1 + kV)cp^n \qquad (9.30)$$

where k is an *erosive burning* coefficient, c is a coefficient that reflects the effect of the propellant temperature, and n is the pressure exponent. The parameters k, c, and n are determined experimentally for each propellant formulation.

Figure 9.4 illustrates schematically a differential element of the burning surface of a solid propellant. The burning occurs along the length dx of the grain in an extremely thin combustion zone adjacent to the surface, releasing combustion gases having no axial velocity. Hence, it may be assumed that $y = 0$. The combustion gas mixes immediately with the main gas stream. The stagnation enthalpy of the flowing fluid is determined by the chemical energy released by the combustion process. If the propellant composition is uniform, then all of the combustion gases will have the same level of chemical energy, and the stagnation enthalpy will be constant throughout the flow passage. From Fig. 9.4, the rate of mass addition $d\dot{m}$ is obtained as the product of the linear burning rate $\dot{r}$, the area of the burning surface dA, and the density ρ_P of the solid propellant. Hence,

$$d\dot{m} = \dot{r}\rho_P \, dA \qquad (9.31)$$

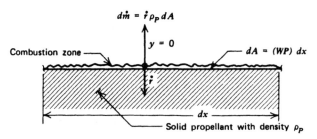

Figure 9.4 Mass addition from a differential element of burning surface.

The burning surface area dA is given by

$$dA = (WP)\,dx \qquad (9.32)$$

where (WP) is the wetted perimeter of the flow passage. The shape of the perforation in the grain is determined by the *surface area* versus *burned distance* relationship required for satisfying the mission requirements for the vehicle propelled by the rocket motor. Figure 9.5 illustrates several simple perforation shapes.[1] The mass addition rate is obtained by combining equations 9.30, 9.31, and 9.32. Thus,

$$d\dot{m} = (1 + kV)cp^n\rho_P(WP)\,dx \qquad (9.33)$$

Equation 9.33 shows that the mass addition rate $d\dot{m}$ is coupled to the properties of the flow field through the velocity V and the pressure p; these, in turn, depend on the history of the mass addition in the flow passage. Because of the aforementioned coupling, it is not possible, in general, to obtain a closed form solution for the flow properties inside a solid propellant grain, even if the flow area is constant. An iterative procedure involving numerical integration must be employed.

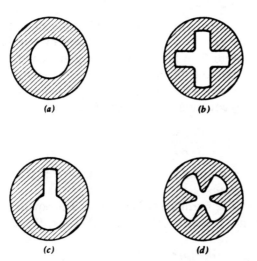

Figure 9.5 Simple perforation shapes for internal-burning solid propellant grains. (*a*) Cylindrical. (*b*) Cruciform. (*c*) Keyhole. (*d*) Star.

The most direct numerical procedure is to assume the pressure at the head end of the grain, $p_o = P_o$ (see Fig. 9.3). The stagnation temperature of the combustion gas is equal to the combustion flame temperature, $T = T_o = T_{\text{flame}} = \text{constant}$, and is a calculable function of the propellant composition and combustion pressure (see Section 14–4). From the thermal equation of state for a perfect gas,

$$\rho_o = \frac{P_o}{RT} \qquad (9.34)$$

At the head end of the grain (see Fig. 9.3), $M_o = 0$. From Table 9.3, the reference critical conditions may be determined. Thus,

$$P^* = \left(\frac{1}{\gamma + 1}\right)\left(\frac{\gamma + 1}{2}\right)^{\gamma/(\gamma - 1)} P_o \qquad (9.35)$$

$$t^* = \left(\frac{2}{\gamma + 1}\right) T \tag{9.36}$$

$$V^* = a^* = (\gamma R t^*)^{1/2} \tag{9.37}$$

$$\rho^* = \frac{1}{2}\rho_o \tag{9.38}$$

To express G^*, the *critical mass flux*, in terms of ρ_o and a_o, we write

$$G^* = \rho^* V^* = \frac{\rho_o}{2}\left(\frac{2\gamma R T}{\gamma + 1}\right)^{1/2} = \frac{\rho_o a_o}{[2(\gamma + 1)]^{1/2}} \tag{9.39}$$

From equation 9.33, the mass flux $G = \dot{m}/A$ is given by

$$G = \frac{1}{A}\int_o^{\dot{m}} d\dot{m} = \frac{(WP)\rho_P c}{A}\int_o^x (1 + kV)p^n \, dx \tag{9.40}$$

Equation 9.40 must be integrated numerically; equations 9.25 and 9.26 may be employed for determining V and p as functions of M, and equation 9.23 is employed for determining M from the ratio G/G^*.

When equation 9.40 has been integrated from the head end to the aft end of the propellant grain, the total mass flow rate $\dot{m}_e$ for the gas produced by burning the grain is known. The stagnation pressure P_e at the aft end of the grain may be determined from equation 9.28. For steady flow, the mass flow rate of combustion gas $\dot{m}_e$ must equal the mass flow rate $\dot{m}_t$ crossing the nozzle throat t. Assuming isentropic flow in the nozzle, $\dot{m}_t = \dot{m}^*$ is given by equation 4.137, Table 4.6, where $P = P_e$. Thus,

$$\dot{m}_t = \frac{\Gamma P_e A_t}{\sqrt{\gamma R T}} \tag{9.41}$$

If $\dot{m}_e$ and $\dot{m}_t$ do not agree within an acceptable tolerance, a new value is assumed for P_o and the entire procedure is repeated.

For the special case where the erosive burning coefficient $k = 0$, in equation 9.33, and the propellant burning rate is not pressure dependent, that is, $n = 0$ and $\dot{r} = c$, equation 9.40 becomes

$$G = \frac{(WP)\rho_P \dot{r}}{A} x \tag{9.42}$$

In the above special case, G is linearly proportional to x, and the distribution of the flow properties along the perforation in the grain may be determined directly from Table C.14.

As mentioned above, in the general case, the flow area along the grain perforation is not constant. Consequently, both area change and mass addition are present simultaneously, and the assumption that the only driving potential is simple mass addition is invalid. The generalized flow equations derived in Section 9–5 must then be solved numerically. The iterative approach discussed above must still be employed, however, for determining the head-end pressure and the total mass flow rate of the combustion gas.

Example 9.1.   The solid propellant rocket motor illustrated in Fig. 9.3 has a solid propellant grain with a constant-area cylindrical perforation having a diameter $D = 0.025$ m, a constant linear burning rate $\dot{r} = 0.025$ m/s, and a propellant density $\rho_P = 2500$ kg/m^3. The combustion gases have $\gamma = 1.2$, $R = 320$ J/kg-K, and a flame temperature $T = 3000$ K. The grain length $L = 0.30$ m and the throat area of the exhaust nozzle $A_t = 0.00030 \text{ m}^2$. Calculate (a) the propellant consumption rate $\dot{m}_e$, (b) the nozzle inlet Mach number M_e and the stagnation pressure P_e, (c) the head-end stagnation pressure P_o, (d) the head-end density ρ_o, (e) the maximum mass flux G^*, and (f) the static pressure drop between the grain head-end and aft-end. Denote the head-end properties by the subscript o, and the grain exit and nozzle entrance properties by the subscript e.

Solution

(a) From equation 9.42, which applies when $\dot{r}$ is constant,

$$\dot{m}_e = G_e A_e = \dot{r}\rho_P(WP)x_e = \dot{r}\rho_P\pi DL$$

$$\dot{m}_e = (0.025)(2500)\pi(0.025)(0.30) = 1.4726 \text{ kg/s (3.2487 lbm/sec)}$$

(b) $\dfrac{A_e}{A^*} = \dfrac{\pi D^2}{4A_t} = \dfrac{\pi(0.025)^2}{4(0.00030)} = 1.6363$

From Table C.6, for $A_e/A^* = 1.6363$, $M_e = 0.39382$. From equation 9.41 (where $\Gamma = 0.71043$ from Table C.15),

$$P_e = \frac{\dot{m}_e\sqrt{\gamma RT}}{\Gamma A_t} = \frac{(1.4726)[1.2(320)(3000)]^{1/2}}{(0.71043)(0.0003)} = 74.160 \cdot 10^5 \text{ N/m}^2 \text{ (1075.6 lbf/in.}^2\text{)}$$

(c) From Table C.14, for $M_e = 0.39382$, $P_e/P^* = 1.1483$. Thus,

$$P^* = \frac{(74.160)10^5}{1.1483} = 64.582 \cdot 10^5 \text{ N/m}^2$$

From Table C.14, for $M_o = 0.0$, $P_o/P^* = 1.2418$. Hence,

$$P_0 = (1.2418)(64.582)10^5 = 80.198 \cdot 10^5 \text{ N/m}^2 \text{ (1163.2 lbf/in.}^2\text{)}$$

(d) $\rho_o = \dfrac{P_o}{RT} = \dfrac{(80.198)10^5}{(320)(3000)} = 8.3540 \text{ kg/m}^3 \text{ (0.52152 lbm/ft}^3\text{)}$

(e) From Table C.14, for $M_e = 0.39382$, $G_e/G^* = 0.70171$.

$$G_e = \frac{\dot{m}_e}{A_e} = \frac{4\dot{m}_e}{\pi D^2} = \frac{4(1.4726)}{\pi(0.025)^2} = 3000.0 \text{ kg/s-m}^2$$

$$G^* = \frac{(3000.0)}{(0.70171)} = 4275.3 \text{ kg/s-m}^2 \text{ (6.0795 lbm/sec-ft}^2\text{)}$$

(f) From Table C.6, for $M_e = 0.39382$, $p_e/P_e = 0.91178$. Thus,

$$p_e = (0.91178)(74.160)10^5 = 67.618 \cdot 10^5 \text{ N/m}^2$$

$$\Delta p = p_e - P_o = (67.618 - 80.198)10^5 = -12.58 \cdot 10^5 \text{ N/m}^2 \text{ (182.5 lbf/in.}^2\text{)}$$

9-4 COMBUSTION WAVES

The phenomenon of chemical reaction behind (or within) a moving wave propagating into a premixed combustible gas mixture is a classical topic of combustion science. This section presents the general features of such wave-driven chemical reactions, called *combustion waves*.

Figure 9.6 illustrates schematically a long tube containing a premixed combustible gas mixture. If the combustible gas mixture is ignited at the left end of the tube, a planar combustion wave propagates to the right into the unburned *reactants*,

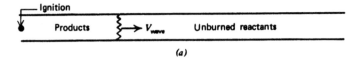

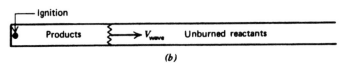

Figure 9.6 Combustion waves in a premixed gaseous system in a long tube. (a) Open end tube. (b) Closed end tube.

leaving burned *products* behind the wave. Two different types of combustion waves are observed experimentally: *deflagrations*, which propagate at subsonic speeds with respect to the reactants, and *detonations*, which propagate at supersonic speeds with respect to the reactants.

In a stationary reference system, the combustion wave is an unsteady flow process. In the relative coordinate system, however, where the observer moves with the same speed as the combustion wave, the combustion wave appears to be stationary, as illustrated schematically in Fig. 9.7; that is, it is a *steady combustion wave*.

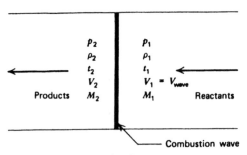

Figure 9.7 Steady combustion wave.

Table 9.4 summarizes the general characteristics of detonation and deflagration waves.

Whether a detonation wave or a deflagration wave occurs in a given situation depends on several variables, but the principal factors are the composition of the gaseous reactants, and whether the tube is open or closed at the ignition end.

The items of prime interest in studying combustion waves are the wave propagation speed and the flow properties of the products of combustion. For a detonation wave, the wave propagation speed is termed the *detonation velocity*, and for a deflagration wave it is termed the *flame speed*.

The analysis of a combustion-driven wave is quite similar to that employed for the normal shock wave in Section 7–3, the difference being that shock waves are adiabatic while combustion waves are diabatic because of the thermal energy released by the chemical reaction. The combustion wave may be regarded as a generalization

Table 9.4 General Characteristics of
Detonation and Deflagration Waves

	Detonation	Deflagration
M_1	>1	<1
M_2	$\leqq 1$	<1
$\dfrac{V_2}{V_1}$	<1	>1
$\dfrac{p_2}{p_1}$	>1	<1
$\dfrac{\rho_2}{\rho_1}$	>1	<1
$\dfrac{t_2}{t_1}$	>1	<1

of the normal shock wave so that it includes energy release, which is equivalent to a simple heat addition. Consequently, the analysis of combustion waves presented herein is basically an extension of the analysis presented in Section 7–3 for the normal shock wave.

9-4(a) Governing Equations for Combustion Waves

The governing equations for the normal shock wave derived in Section 7–3 are directly applicable to combustion waves, if the specific internal energy u is defined so that it includes the chemical energy of formation for the reactants and products. Hence, equations 7.2, 7.5, and 7.8, which are repeated below and renumbered, are applicable.

$$G = \rho_1 V_1 = \rho_2 V_2 \tag{9.43}$$

$$p_1 + \rho_1 V_1{}^2 = p_2 + \rho_2 V_2{}^2 \tag{9.44}$$

$$u_1' + p_1 v_1 + \frac{V_1{}^2}{2} = u_2' + p_2 v_2 + \frac{V_2{}^2}{2} \tag{9.45}$$

where

$$u' \equiv u + u^\circ \tag{9.46}$$

In equation 9.46, u is the sensible internal energy (i.e., $\int c_v \, dt$), and u° is the afore-mentioned *energy of formation*, which is the energy required to form the gaseous species from the naturally occurring elements (see Section 14–3). Equation 9.45 becomes

$$u_1 + p_1 v_1 + \frac{V_1{}^2}{2} + \Delta u^\circ = u_2 + p_2 v_2 + \frac{V_2{}^2}{2} \tag{9.47}$$

where

$$\Delta u^\circ = u_1^\circ - u_2^\circ$$

is the energy released by the chemical reaction, and is called the *energy of combustion*. For a simple system, the caloric equation of state is given by

$$u = u(p, \rho) \tag{9.48}$$

Combining equations 9.43, 9.44, and 9.47, we obtain the equation for the *Hugoniot curve*, called an *H-curve*, for a diabatic flow (see equation 7.11). Thus,

$$u_2 - (u_1 + \Delta u^\circ) = \frac{1}{2}(p_1 + p_2)(v_1 - v_2) \qquad (9.49)$$

Equation 9.49, in conjunction with equation 9.48, specifies the locus of the allowable final states for the products behind a combustion wave for a given initial state and energy of combustion. Figure 9.8 illustrates the aforementioned locus, plotted in the *pv* plane; it is known as the Hugoniot curve. Equations 9.43 and 9.44 may be combined to yield the equation for a *Rayleigh line*, called an *R-line* [see Section 6-3(b)]. Thus,

$$G^2 = -\frac{(p_2 - p_1)}{(v_2 - v_1)} = -\left(\frac{dp}{dv}\right)_{R\text{-}line} \qquad (9.50)$$

The combustion products must, therefore, satisfy simultaneously the equations for both the *H-curve* and the *R-line*.

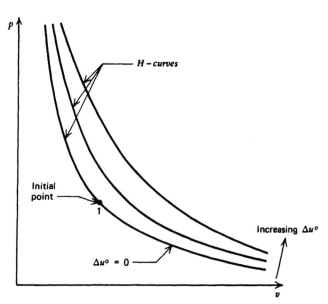

Figure 9.8   Sketch of several Hugoniot curves for diabatic flow.

9–4(b) The Hugoniot Curve for Diabatic Flow

Figure 9.8 illustrates schematically three different Hugoniot curves, corresponding to three different values for Δu°, plotted in the *pv* plane. From the configurations of those curves, it appears that two types of combustion waves are possible: (1) those for which the pressure p and the density ρ increase, called a *detonation wave*, and (2) those for which p and ρ decrease, called a *deflagration wave*. The lowest *H-curve*, for $\Delta u^\circ = 0$, is the adiabatic *H-curve* for a normal shock wave derived in Section 7–3. It is shown there that for a normal shock wave the states on the portion of the *H-curve* below the initial point, point 1, violate the second law of thermodynamics for an isolated system, and are, therefore, physically unattainable. No such restriction exists, however, in the case of a diabatic flow.

It is of interest to determine the direction of the absolute velocity of the burned gases relative to a stationary observer. From Fig. 9.7,

$$V_{\text{Products}} = V_1 - V_2 = V_1 \left(1 - \frac{V_2}{V_1} \right) \tag{9.51}$$

Substituting for V_2/V_1 from equation 9.43 into equation 9.51, we obtain

$$V_{\text{Products}} = V_1 \left(1 - \frac{\rho_1}{\rho_2} \right) \tag{9.52}$$

For a detonation wave, $\rho_2/\rho_1 > 1$, and

$$0 \leqq V_{\text{Products}} \leqq V_1 \tag{9.53}$$

For a deflagration wave, $\rho_2/\rho_1 < 1$, and

$$V_{\text{Products}} < 0 \tag{9.54}$$

The products of combustion, therefore, flow behind a detonation wave at a lesser speed than the detonation velocity. For a deflagration wave, the products of combustion flow in the direction opposite to that for the wave.

Figure 9.9 illustrates schematically an *H-curve* for diabatic flow, subdivided into five separate regions. Points J and K are the points of tangency of the *H-curve*

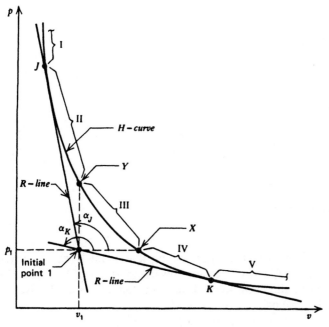

Figure 9.9 Regions of special interest on a Hugoniot curve for diabatic flow.

with the *R-lines*, through point 1, having the maximum and minimum slopes, respectively. At point Y, the specific volume is the same as it is at point 1, and at point X, the static pressure is the same as it is at point 1. For an adiabatic *H-curve*, points X and Y coincide with point 1.

Denote the angle between the v-axis and an R-line by α, as illustrated in Fig. 9.10. From equation 9.50,

$$\tan \alpha = \left(\frac{dp}{dv}\right)_{R\text{-line}} = -G^2 = -\rho^2 V^2 \tag{9.55}$$

Hence,

$$V = \frac{\sqrt{-\tan \alpha}}{\rho} \tag{9.56}$$

In the regions of the H-curve labeled I, II, IV, and V, $\tan \alpha$ is negative, and in those regions equation 9.56 yields physically realizable values for V. In Region III, however, $\tan \alpha$ is positive, so that V is imaginary. Consequently, in Region III, the mathematical solutions have no physical meaning, and the physically meaningful portions of the H-curve are Regions I, II, IV, and V.

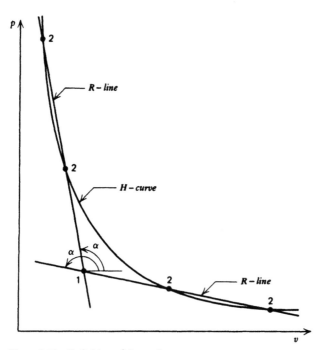

Figure 9.10 Definition of the angle α.

Now consider point Y in Fig. 9.9. Approaching that point from the physically meaningful Regions I and II, α is decreasing toward the limit of 90 deg and $\tan \alpha$ is increasing in a negative direction toward the limit $\tan \alpha = -\infty$. Consequently, at point Y, equation 9.56 yields

$$V_Y = \frac{\sqrt{-(-\infty)}}{\rho_Y} = \infty \tag{9.57}$$

Thus, at point Y the velocity of the gaseous combustion products behind the combustion wave is infinite. At point X, $\alpha = 180$ deg and $\tan \alpha = -0$. From equation 9.56, we obtain

$$V_X = \frac{\sqrt{-(-0)}}{\rho_X} = 0 \tag{9.58}$$

Hence, at point X the velocity of the gaseous combustion products behind the combustion wave is zero.

Point J and K are called the *Chapman-Jouguet points*. At those points, the *H-curve* is tangent to an *isentrope* (i.e., a line of constant entropy). That fact may be demonstrated by differentiating the equation of the *H-curve*, equation 9.49, introducing equation 1.64, and solving for $(ds/dp)_H$, where the subscript H denotes that the differentiation is along the *H-curve*. Thus,

$$du_H = -\frac{1}{2}(p_1 + p)\, dv_H + \frac{1}{2}(v_1 - v)\, dp_H \tag{9.59}$$

Equation 1.64 is repeated and renumbered below for convenience.

$$t\, ds = du + p\, dv \tag{9.60}$$

Combining equations 9.59, 9.60, and 9.50, we obtain (see equation 7.16)

$$2t\left(\frac{ds}{dp}\right)_H = (v_1 - v)\left[1 + \frac{G^2}{(dp/dv)_H}\right] \tag{9.61}$$

At points J and K, the *H-curve* is tangent to the *R-line*. Hence,

$$\left(\frac{dp}{dv}\right)_H = \left(\frac{dp}{dv}\right)_R = -G^2 \tag{9.62}$$

The term in brackets in equation 9.61 is, therefore, zero at points J and K, and at those points

$$\left(\frac{ds}{dp}\right)_H = 0 \tag{9.63}$$

Consequently, at points J and K, $ds = 0$, indicating that s = constant at those two points on the *H-curve*. From equation 9.50, at points J and K,

$$\left(\frac{dp}{dv}\right)_R = -\rho^2\left(\frac{dp}{d\rho}\right)_s = -\rho^2 a^2 = -G^2 = -\rho^2 V^2 \tag{9.64}$$

Therefore,

$$V = a \qquad (at\ points\ J\ and\ K) \tag{9.65}$$

Accordingly, at the Chapman-Jouguet points J and K, the velocity of the gaseous combustion products is exactly sonic.

In Region I, as point 2 moves away from point J, the pressure p and the density ρ increase toward ∞. From equation 9.43,

$$V_2 = \frac{G}{\rho_2} \tag{9.66}$$

Hence, as ρ_2 approaches ∞, V_2 decreases toward zero. In Region V, as point 2 moves away from point K, p and ρ decrease toward zero. From equation 9.66, as ρ_2 approaches zero, V_2 increases toward ∞.

Figure 9.11 summarizes the types of combustion waves corresponding to the regions on the *H-curve* discussed above. Region III is not shown because no real processes occur in that region. The remaining four regions are identified as follows.

Region I: strong detonation, $M_1 > 1$ and $M_2 < 1$.
Region II: weak detonation, $M_1 > 1$ and $M_2 > 1$.

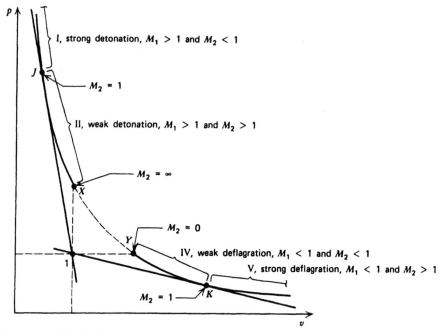

Figure 9.11 Combustion regions on the Hugoniot curve.

Region IV: weak deflagration, $M_1 < 1$ and $M_2 < 1$.
Region V: strong deflagration, $M_1 < 1$ and $M_2 > 1$.

Region V is also a region wherein no physical processes are attainable, as may be demonstrated by the following argument. The release of combustion energy is the equivalent of a heat addition process. Consequently, states in Region V require that the subsonic flow entering the combustion wave be accelerated to supersonic speeds in flowing through the combustion zone. The analysis of the Rayleigh line, however, demonstrates that a subsonic flow cannot be accelerated to a supersonic speed by heat addition alone [see Section 6–3(d)]. It is, therefore, concluded that since supersonic velocities behind the combustion wave are not possible, the states in Region V are not physically realizable.

Weak detonations corresponding to Region II cannot be ruled out absolutely, but they appear much less probable than strong detonations. If a detonation wave is considered to be an adiabatic shock wave followed by heat addition caused by combustion, the path of the process might appear to be in accordance with the one illustrated in Fig. 9.12. Point A represents the fluid state immediately after passing through the shock wave. From point A, the fluid must move along the *R-line* to point B or C. Since the flow at point A is subsonic, arguments similar to that presented above for eliminating Region V show that point B, which corresponds to a subsonic flow, would represent a final state on the *R-line* process, whereas point C could not because it corresponds to a supersonic flow. Consequently, Region II corresponds to a physically unrealizable process.

Summarizing, the regions of the Hugoniot curve that correspond to physically attainable processes for a diabatic flow are those indicated in Fig. 9.13. Strong detonations occur in Region I or at the Chapman-Jouguet point J, and weak deflagrations occur in Region IV or at the Chapman-Jouguet point K.

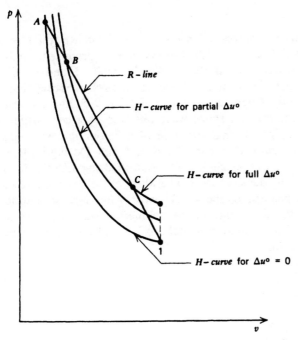

Figure 9.12 Schematic illustrations of Hugoniot curves for no reaction ($\Delta u' = 0$), partial reaction (partial $\Delta u°$), and complete reaction (full $\Delta u°$).

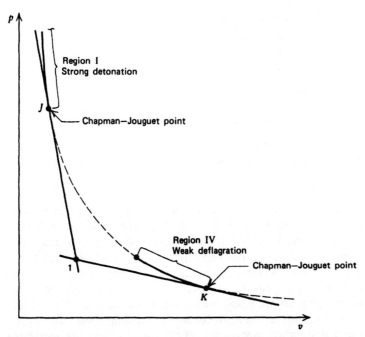

Figure 9.13 Physically attainable regions of the Hugoniot curve for diabatic flow.

9-4(c) Detonation Waves

In Section 9-4(b), detonation waves are shown to occur in Region I or at the Chapman-Jouguet point J (see Fig. 9.13). Detonations in premixed combustible gas mixtures were first observed by Bertholet, Vieille, Mallard, and Le Chatelier in the latter part of the nineteenth century. Chapman, in 1899, proposed that steady detonation waves occur at the Chapman-Jouguet point. In 1905, Jouguet postulated that a detonation wave comprises a shock wave followed by a combustion wave, which is indeed the case. The internal structure of a detonation wave has been described analytically by Zeldovich,[2] von Neumann,[3] and Döring.[4] Figure 9.14 illustrates schematically the Zeldovich-von Neumann-Döring (ZND) detonation wave structure. The pressure, density, and temperature of the gas mixture increase abruptly because of the passage of the shock wave. After a short ignition lag, deflagration occurs and the final state of the detonation wave is achieved. The general features of the ZND detonation model are in good qualitative agreement with experimental observations.

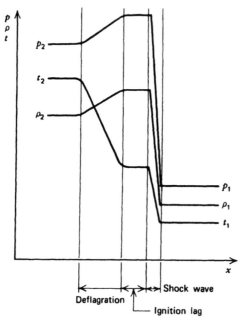

Figure 9.14 Structure of a *ZND* detonation wave.

Whether a detonation wave or a deflagration wave is produced in a given combustion situation depends on whether the end of the tube wherein ignition occurs is open or closed (see Fig. 9.6). The released combustion energy generates compression waves that propagate in both directions. In a closed-end tube, however, those waves reflect from the closed end, overtake the combustion wave and strengthen it until it becomes a fully developed detonation wave. For an open-end tube (see Fig. 9.6) the left-running compression waves leave the tube, so that there is no reinforcement of the combustion wave. Accordingly, a deflagration wave is produced. The formation of a detonation wave also depends on the fuel/oxidizer ratio, also called the mixture ratio. There is a definite upper and a definite lower limit to what constitutes a detonable mixture ratio; those limiting mixture ratios are called the *detonation limits*. If the mixture ratio is outside of the detonation limits, no detonation wave will either form or propagate, even when all of the other factors are favorable to the formation of a detonation wave. Belles[5] postulated a model for predicting the detonation limits that is in reasonable agreement with the experimental observations.

The detonation velocity and the flow properties of the gaseous mixture behind the detonation wave may be determined by plotting the *H-curve* and the tangent *R-line*, and determining the Chapman-Jouguet point (see point *J*, Fig. 9.13). An approximate algebraic technique may be developed as follows. At the Chapman-Jouguet point *J*, $M_2 = 1$. Hence, substituting $V_2 = a_2$ into equation 9.43 yields

$$V_1 = \frac{\rho_2 a_2}{\rho_1} \tag{9.67}$$

Combining equations 9.43 and 9.44, solving for ρ_2/ρ_1, noting that $p_1 \ll p_2$ for a detonation wave, and introducing $V_2 = a_2$, we obtain

$$\frac{\rho_2}{\rho_1} = 1 + \frac{p_2}{\rho_2 V_2^2}\left(1 - \frac{p_1}{p_2}\right) \approx 1 + \frac{p_2}{\rho_2 a_2^2} \tag{9.68}$$

Combining equations 9.67 and 9.68 yields

$$V_1 = a_2\left(1 + \frac{p_2}{\rho_2 a_2^2}\right) \tag{9.69}$$

Solving equation 9.49 for u_2 and noting that $p_1 \ll p_2$ gives

$$u_2 \approx u_1 + \Delta u^\circ + \frac{1}{2}\frac{p_2}{\rho_2}\left(\frac{\rho_2}{\rho_1} - 1\right) \tag{9.70}$$

Substituting equation 9.68 into equation 9.70, we obtain

$$u_2 = u_1 + \Delta u^\circ + \frac{p_2^2}{2\rho_2^2 a_2^2} \tag{9.71}$$

In the case of an imperfect gas, equation 9.71 must be solved numerically. The values of p_2 and ρ_2 are assumed and those for u_2 and a_2 are calculated from the equation of state. The latter values are then substituted into equation 9.71 until it is satisfied. When state 2 is determined, V_1 is calculated from equation 9.69. Table 9.5 compares the calculated and measured detonation velocities for several gaseous mixtures.

For a perfect gas, $p = \rho Rt$, $u = c_v t$, and $a^2 = \gamma Rt$. In that case, equation 9.71 may be rewritten as

$$c_{v2}t_2 = c_{v1}t_1 + \Delta u^\circ + \frac{R_2 t_2}{2\gamma_2} \tag{9.72}$$

Hence,

$$t_2 = \frac{2\gamma_2}{\gamma_2 + 1}\left(\frac{c_{v1}}{c_{v2}}t_1 + \frac{\Delta u^\circ}{c_{v2}}\right) \tag{9.73}$$

Equation 9.73 may be solved directly for t_2 for given values of the energy of combustion Δu°.

Table 9.5 A Comparison of Calculated and Measured
Detonation Velocities

Gaseous Mixture	p_2, atm	t_2, K	V_1, m/s Calculated	V_1, m/s Measured
$(2H_2 + O_2)$	18.05	3583	2806	2819
$(2H_2 + O_2) + 3O_2$	15.3	2970	1925	1922
$(2H_2 + O_2) + 3N_2$	15.63	3003	2033	2055
$(2H_2 + O_2) + 4H_2$	15.97	2976	3627	3527

$p_1 = 1$ atm and $t_1 = 291$ K

(Taken from Reference 6)

Example 9.2. Estimate the detonation velocity V_1 for the premixed gaseous mixture comprising $2H_2 + O_2 + 3N_2$. The initial temperature and pressure of the reactants are $t_1 = 298.15$ K and $p_1 = 1$ atm, respectively.

Solution

The detonation velocity may be estimated by neglecting the dissociation of the gaseous products formed behind the detonation wave and assuming that the product gases behave as would a mixture of perfect gases. In view of the aforementioned assumptions, equations 9.67 to 9.73 are applicable.

To apply equations 9.67 to 9.73, the molecular weights and specific heats of the reactant gases in front of and the gaseous combustion products in back of the detonation wave must be known. In addition, the energy of combustion Δu° for the reactants must be known. The properties of a mixture of perfect gases are discussed in Section 1–15(h).

The energy of combustion is the chemical energy released by the chemical reaction, and it is converted into thermal energy. The procedure for taking into account the chemical energy of a substance is discussed in Section 14–3.

Neglecting dissociation, the combustible gas mixture, the reactants, may be assumed to satisfy the following chemical reaction

$$2H_2 + O_2 + 3N_2 = 2H_2O + 3N_2 \tag{a}$$

where the quantities 2, 1, 3, 2, and 3 are taken as moles. For an adiabatic reaction (i.e., one in which there is no heat transfer with the surroundings), it is shown in Section 14–4 that the *absolute enthalpy of the reactants* H_R must equal the *absolute enthalpy of the products* H_p, where the *absolute enthalpy* H is equal to the sum of the sensible enthalpy change $(H^\circ - H^\circ_{ref})$ above the reference temperature, and the energy of formation $\Delta H^\circ_{f,ref}$, which is the chemical energy required for forming the chemical species from its stable elements at the reference temperature. Table D.1, Volume 2, presents values of $(H^\circ - H^\circ_{298})$ and $\Delta H^\circ_{f,298}$ for the species in equation (a). Thus,

$$H = \sum_{i=1}^{n} N_i[(H^\circ - H^\circ_{298}) + \Delta H^\circ_{f,298}] = \sum_{i=1}^{n} N_i H_i = N \sum_{i=1}^{n} X_i H_i \tag{b}$$

Selected portions of Table D.1, Volume 2, for the gaseous species of interest in the present problem are presented in the following table.

Species	$\Delta H^\circ_{f,298},$ kJ/kmol	$T,$ K	$c_p,$ kJ/kmol-K	$H^\circ - H^\circ_{298},$ kJ/kmol
H_2	0	298	28.859	0
O_2	0	298	29.354	0
N_2	0	298	29.122	0
		3000	37.066	92,753
		3100	37.124	96,462
		3200	37.178	100,178
H_2O	−241,827	298	33.561	0
		3000	55.711	126,353
		3100	56.006	131,939
		3200	56.280	137,554

The energy of combustion Δu° is defined as the amount of energy that must be removed from the products of combustion by heat transfer to change the temperature of the products to the temperature of the reactants. Thus,

$$H_R(t_R) = H_P(t_R) + \Delta u^\circ \qquad \text{(c)}$$

For the *reactants* in equation (a) above, the total number of moles $N_R = (2 + 1 + 3) = 6$. Thus,

$$X_{H_2} = \frac{2}{6} = 0.33333, \qquad X_{O_2} = \frac{1}{6} = 0.16667, \qquad X_{N_2} = \frac{3}{6} = 0.5$$

For the *products of combustion*, $N_P = (2 + 3) = 5$, and

$$X_{H_2O} = \frac{2}{5} = 0.40, \qquad X_{N_2} = \frac{3}{5} = 0.60$$

From equation 1.141a, the molecular weight $\bar{m}$ of a gas mixture is

$$\bar{m} = \sum_{i=1}^{n} X_i \bar{m}_i$$

Consequently,

$$\bar{m}_1 = (0.33333)(2.016) + (0.16667)(32.0) + (0.5)(28.0) = 20.005$$

$$\bar{m}_2 = (0.4)(18.016) + (0.6)(28.0) = 24.006$$

From equation (b) and selected portions of Table D.1, Volume 2,

$$H_R(298) = 2(0 + 0) + 1(0 + 0) + 3(0 + 0) = 0$$

$$H_P(298) = 2(0 - 241,827/1000) + 3(0 + 0) = -483.654 \text{ kJ}$$

$H_R(298)$ is zero for this special case where the reactants are naturally occurring elements and the initial temperature t_R is the reference temperature $t_{ref} = 298.15$ K. In general, $H_R(t_R)$ is nonzero. Substituting for H_R and H_P into equation (c), we obtain

$$\Delta u^\circ = 0 - (-483.654) = 483.654 \text{ kJ}$$

On a molar basis,

$$\Delta \bar{u}^\circ = \frac{\Delta u^\circ}{N_P} = \frac{483.654}{5} = 96.731 \text{ kJ/mol}$$

The specific heat $\bar{c}_p$ on a molar basis is given by equation 1.144d.

$$\bar{c}_p = \sum_{i=1}^n X_i \bar{c}_{pi} \tag{d}$$

For the reactants, at $t = 298.15$ K,

$$\bar{c}_{p1} = (0.33333)(28.859) + (0.16667)(29.354) + (0.5)(29.122) = 29.073 \text{ J/mol-K}$$

For a perfect gas, $\bar{c}_v = \bar{c}_p - \bar{R}$. Thus,

$$\bar{c}_{v1} = 29.073 - 8.314 = 20.759 \text{ J/mol-K}$$

For the products, t_2 is unknown. Consequently, an iterative procedure is employed wherein t_2 is assumed, $\bar{c}_{p2}$, $\bar{c}_{v2}$, and γ_2 are calculated, and t_2 is determined from equation 9.73. The value of t_2 is iterated to convergence. For the first iteration, assume $t_2 = 3000$ K. From equation (d) and selected portions of Table D.1, Volume 2, we obtain

$$\bar{c}_{p2} = (0.4)(55.711) + (0.6)(37.066) = 44.524 \text{ J/mol-K}$$

$$\bar{c}_{v2} = 44.524 - 8.314 = 36.210 \text{ J/mol-K}$$

$$\gamma_2 = \frac{\bar{c}_{p2}}{\bar{c}_{v2}} = \frac{44.524}{36.210} = 1.230$$

Equation 9.73 may be rewritten in terms of molar properties. Thus,

$$t_2 = \frac{2\gamma_2}{\gamma_2 + 1}\left[\frac{(\bar{c}_{v1}/\bar{m}_1)}{(\bar{c}_{v2}/\bar{m}_2)}t_1 + \frac{\Delta\bar{u}^\circ}{\bar{c}_{v2}}\right] \tag{e}$$

Substituting the appropriate values into equation (e), we obtain

$$t_2 = \frac{2(1.230)}{2.230}\left[\frac{(20.759/20.005)}{(36.210/24.006)}(298.15) + \frac{96,731}{36.210}\right] = 3173.3 \text{ K}$$

For the second iteration, assume $t_2 = 3200$ K.

$$\bar{c}_{p2} = (0.4)(56.280) + (0.6)(37.178) = 44.819 \text{ J/mol-K}$$

$$\bar{c}_{v2} = 44.819 - 8.314 = 36.505 \text{ J/mol-K}$$

$$\gamma_2 = \frac{44.819}{36.505} = 1.228$$

$$t_2 = \frac{2(1.228)}{2.228}\left[\frac{(20.759/20.005)}{(36.505/24.006)}(298.15) + \frac{96,731}{36.505}\right] = 3145.3 \text{ K}$$

The value of t_2 for the third iteration is obtained from the secant method [see Appendix A–4(d)]. Thus,

$$\frac{t_2 - 3145.3}{t_2 - 3200} = \frac{3173.3 - 3145.3}{3000 - 3200} = -0.14000 \tag{f}$$

Solving equation (f) gives $t_2 = 3152.0$ K. Then,

$$\bar{c}_{p2} = (0.4)(56.149) + (0.6)(37.152) = 44.751 \text{ J/mol-K}$$

$$\bar{c}_{v2} = 44.751 - 8.314 = 36.437 \text{ J/mol-K}$$

$$\gamma_2 = \frac{44.751}{36.437} = 1.228$$

$$t_2 = \frac{2(1.228)}{2.228}\left[\frac{(20.759/20.005)}{(36.437/24.006)}(298.15) + \frac{96,731}{36.437}\right] = 3151.2 \text{ K}$$

Thus, $t_2 = 3151.2$ K.

From equation 9.68, for a perfect gas,

$$\frac{\rho_2}{\rho_1} = 1 + \frac{p_2}{\rho_2 a_2{}^2} = 1 + \frac{p_2\rho_2}{\rho_2\gamma_2 p_2} = 1 + \frac{1}{\gamma_2} = \frac{\gamma_2 + 1}{\gamma_2} = \frac{1.228 + 1}{1.228} = 1.814$$

For a perfect gas, $a_2 = \sqrt{\gamma_2 R_2 t_2}$.

$$R_2 = \frac{\bar{R}}{\bar{m}_2} = \frac{8314.3}{24.006} = 346.34 \text{ J/kg-K}$$

$$a_2 = [1.228(346.34)(3151.2)]^{1/2} = 1157.7 \text{ m/s}$$

From equation 9.67, we obtain

$$V_1 = \frac{\rho_2 a_2}{\rho_1} = (1.814)(1157.7) = 2100.0 \text{ m/s}$$

From the perfect gas law,

$$\frac{p_2}{p_1} = \frac{\rho_2 R_2 t_2}{\rho_1 R_1 t_1}$$

$$R_1 = \frac{8314.3}{20.005} = 415.60 \text{ J/kg-K}$$

$$\frac{p_2}{p_1} = (1.814)\frac{(346.34)(3151.2)}{(415.60)(298.15)} = 15.977$$

$$p_2 = 15.977(1.0) = 15.977 \text{ atm}$$

The results obtained above are approximate because dissociation of the combustion products is neglected. Because of the presence of the diluent N_2 in the present example, the temperature of the products of combustion is reduced below that which would be obtained for the mixture $2H_2 + O_2$. Consequently, dissociation is less important and the above results should be reasonable approximations. From Table 9.5, the measured detonation velocity is 2055 m/s. The calculated velocity of 2100.0 m/s is, therefore, only 2.2 percent too large, which is good agreement despite the approximations involved.

9-4(d) Deflagration Waves

It is shown in Section 9-4(b) that deflagrations correspond to states on the weak deflagration branch of the Hugoniot curve (see Fig. 9.13). The flame speed of a deflagration wave is controlled by a combination of heat conduction and mass diffusion, whereas the velocity of a detonation wave is controlled by the abrupt pressure, density, and temperature increases associated with the normal shock wave. Therefore, the deflagration flame speed depends on several factors, such as whether the process is laminar or turbulent, and also on the details of the reaction mechanism.

Several theories have been developed for predicting the speeds of deflagration waves. Basically, three types of theories have been proposed: (1) thermal theories, which are based on rate controlled heat transfer, (2) diffusion theories, which are based on rate controlled mass transfer, and (3) comprehensive theories, which are a combination of rate controlled heat and mass transfer. The thermal and diffusion theories are oversimplifications of the actual process. Both theories predict an increase in flame speed with an increase in flame temperature, which is in agreement with experimental observations. The analyses developed by Zeldovich and Frank-Kamenetskii,[7] Zeldovich and Semenov,[8] Semenov,[9] and Zeldovich[10] are representative of the comprehensive theories, and there is reasonable agreement between those theories and experiment.

Finally, it is noted that there are definite lower and upper limits to the mixture ratio that will support a steady deflagration wave. Those limits are called the *flammability limits*. For a given gaseous mixture the flammability limits are usually wider than the corresponding *detonation limits*. Flammability limits depend on the initial pressure and temperature of the combustible mixture. The comprehensive theory developed by Spalding[11] is representative of the analytical approach for predicting flammability limits.

9–5 GENERALIZED STEADY ONE-DIMENSIONAL FLOW[12, 13]

A *simple flow* is a steady one-dimensional flow in which a *single driving potential* is responsible for the changes in the flow properties. Isentropic flow with simple area change is considered in Chapter 4, simple frictional flow is discussed in Chapter 5, simple diabatic flow is discussed in Chapter 6, and simple mass addition is analyzed in Section 9–3. When the working fluid is a perfect gas, algebraic equations may be derived for those simple flows. Those equations relate all of the flow properties to the flow Mach number. When imperfect gas effects must be taken into account, straightforward numerical techniques may be applied for solving a specific problem. Shock waves, isothermal flow in a constant-area duct with friction, isothermal flow in a variable-area duct, and detonation and deflagration waves are, strictly speaking, not simple flows. Nevertheless, such flows may be analyzed in a straightforward manner, yielding results analogous to those obtained for simple flows.

In the general case for a steady one-dimensional flow, however, several driving potentials may be present simultaneously, and no straightforward closed form solutions for the governing equations are obtainable, even for a perfect gas. Such flows are called *generalized steady one-dimensional flows* and are discussed in the present section.

9–5(a) Governing Equations for Generalized Steady One-Dimensional Flow

Figure 9.15 illustrates schematically the physical model for a generalized steady one-dimensional flow. The independent driving potentials considered are:

1. Area change dA.
2. Wall friction δF_f.
3. Heat transfer δQ.
4. Work δW.
5. Mass addition $d\dot{m}$.
6. Body forces caused by gravity $\rho g A \, dz$.
7. Other body forces, drag of entrained particles, etc., denoted by δD.
8. Chemical reactions through their effect on the fluid equations of state.

The continuity and momentum equations for generalized steady one-dimensional flow are obtainable from Table 3.1. Thus,

(1) *Continuity equation* (equation 3.158)

$$\dot{m} = \rho A V \qquad (9.74)$$

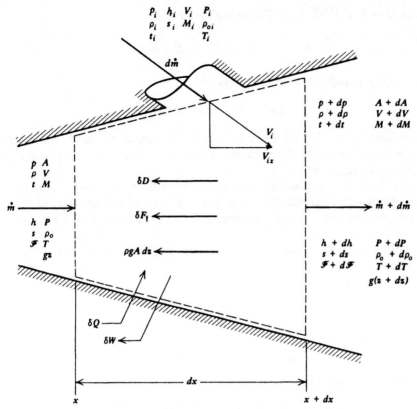

Figure 9.15 Physical model for generalized steady one-dimensional flow.

$$\frac{d\dot{m}}{\dot{m}} = \frac{d\rho}{\rho} + \frac{dA}{A} + \frac{dV}{V} \tag{9.75}$$

(2) *Momentum equation* (equation 3.160)

$$dp + \rho V \, dV + \rho g \, dz + \frac{\rho V^2}{2}\left(\frac{4\mathfrak{f} \, dx}{\mathscr{D}}\right) + \frac{\delta D}{A} + \rho V^2(1 - y)\frac{d\dot{m}}{\dot{m}} = 0 \tag{9.76}$$

where the mass addition term is obtained from equation 9.8.

(3) *Energy equation.* The energy equation is obtained by applying equation 2.88 (see Table 2.1) to the differential control volume illustrated in Fig. 9.15. Thus,

$$\delta W - \delta Q + (\dot{m} + d\dot{m})\left[h + dh + \frac{V^2}{2} + d\left(\frac{V^2}{2}\right) + g(z + dz)\right]$$
$$- \dot{m}\left(h + \frac{V^2}{2} + gz\right) - d\dot{m}\left(h_i + \frac{V_i^2}{2} + gz_i\right) = 0 \tag{9.77}$$

Combining terms, neglecting products of differentials, and dividing by $\dot{m}$, we obtain

$$\delta W - \delta Q + dh + d\left(\frac{V^2}{2}\right) + g \, dz$$
$$+ \left[\left(h + \frac{V^2}{2} + gz\right) - \left(h_i + \frac{V_i^2}{2} + gz_i\right)\right]\frac{d\dot{m}}{\dot{m}} = 0 \tag{9.78}$$

The term in brackets arises because of the difference in the stagnation enthalpy H of the main stream and the stagnation enthalpy H_i of the mass addition stream, where $H = h + V^2/2 + gz$. Define the parameter

$$dH_i \equiv (H - H_i)\frac{d\dot{m}}{\dot{m}} \tag{9.79}$$

Substituting equation 9.79 into equation 9.78 gives

$$\delta W - \delta Q + dH + dH_i = 0 \tag{9.80}$$

Equations 9.75, 9.76, and 9.80 are valid for the flow of any fluid. Those equations must be supplemented by the equation of state of the fluid, which may be in algebraic, graphical, or tabular form.

For a simple thermodynamic system, the seven fluid properties p, ρ, t, h, u, s, and a are completely determined by specifying any two of the seven properties. Thus, the equation of state for the fluid provides five relationships between those seven fluid properties.

The following additional flow properties may also be of interest in certain situations.

(4) *Mach number*

$$M = \frac{V}{a} \tag{9.81}$$

(5) *Impulse function*

$$\mathscr{F} = pA + \dot{m}V = A(p + \rho V^2) \tag{9.82}$$

(6) *Stagnation enthalpy*

$$H = (h)_{V=0} \quad \text{(adiabatic process)} \tag{9.83}$$

(7) *Stagnation temperature*

$$T = (t)_{V=0} \quad \text{(isentropic process)} \tag{9.84}$$

(8) *Stagnation pressure*

$$P = (p)_{V=0} \quad \text{(isentropic process)} \tag{9.85}$$

(9) *Stagnation density*

$$\rho_o = (\rho)_{V=0} \quad \text{(isentropic process)} \tag{9.86}$$

Equations 9.75, 9.76, and 9.80 to 9.86, in conjunction with the five relationships provided by the equation of state, comprise a system of 14 relationships between the 14 fluid properties p, ρ, t, h, u, s, a, V, M, H, P, T, ρ_o, and $\mathscr{F}$, and the seven independent driving potentials dA, δQ, δW, $(4\mathfrak{f}\, dx/\mathscr{D})$, δD, $\rho gA\, dz$, and $d\dot{m}$. To obtain a solution to that system of equations, the seven independent driving potentials must be calculable. The area change dA is known from the geometry of the flow passage, and so are the length x and elevation z of the passage. The five remaining driving potentials δQ, δW, $\delta F_\mathfrak{f}$, δD, and $d\dot{m}$ depend on the fluid properties. Models must, therefore, be developed relating those driving potentials to the fluid properties. The model for $\delta F_\mathfrak{f}$, derived in Section 3–5(d), is included in equation 9.76, where the friction coefficient $\mathfrak{f}$ depends on the fluid properties. For a burning solid propellant surface, a model for $d\dot{m}$ is developed in Section 9–3(d). No specific models are developed in this book for relating δQ, δW, and δD to the fluid properties.

The system of equations derived in this section for the generalized steady one-dimensional flow of an arbitrary compressible fluid is quite involved, and their solutions for specific initial and boundary conditions are obtainable only by employing numerical integration techniques. The aforementioned equations are applied to the generalized steady one-dimensional flow of a perfect gas in Section 9–5(b).

9–5(b) Generalized Steady One-Dimensional Flow of a Perfect Gas

Equations 9.75 to 9.86 may be simplified in the case of the flow of a perfect gas, for which $p = \rho Rt$, $h = c_p t$, $a^2 = \gamma Rt$, and the effects of gravity are negligible. Equation 9.80 may then be rewritten in the form

$$\delta Q - \delta W - dH_i = dH = c_p \, dT \tag{9.87}$$

Dividing both sides of equation 9.87 by $c_p t$, and introducing the definition of the stagnation temperature T, equation 3.181, we obtain

$$\frac{\delta Q - \delta W - dH_i}{c_p t} = \left(1 + \frac{\gamma - 1}{2} M^2\right) \frac{dT}{T} \tag{9.88}$$

Equation 9.88 shows that the effects of heat transfer, work, and the difference between H and H_i are reflected directly in the stagnation temperature change dT.

The appropriate form of the momentum equation is obtained from equation 3.170, Table 3.2. Thus,

$$\frac{dp}{p} + \gamma M^2 \frac{dM}{M} + \frac{\gamma M^2}{2} \frac{dt}{t} + \frac{\gamma M^2}{2}\left[\left(\frac{4\mathfrak{f}\,dx}{\mathscr{D}}\right) + \frac{2\,\delta D}{\gamma M^2 pA}\right]$$

$$+ \gamma M^2 (1 - y) \frac{d\dot{m}}{\dot{m}} = 0 \quad (9.89)$$

where the mass addition term is obtained from equation 9.9.

For a perfect gas, $p = \rho Rt$ and $M = V/(\gamma Rt)^{1/2}$. Logarithmic differentiation yields

$$\frac{dp}{p} = \frac{d\rho}{\rho} + \frac{dt}{t} \tag{9.90}$$

$$\frac{dM}{M} = \frac{dV}{V} - \frac{1}{2}\frac{dt}{t} \tag{9.91}$$

From equations 3.74, 3.78, 3.139, and 3.176, respectively, for a perfect gas,

$$\frac{dT}{T} = \frac{dt}{t} + \frac{(\gamma - 1)M^2}{1 + \dfrac{\gamma - 1}{2} M^2} \frac{dM}{M} \tag{9.92}$$

$$\frac{dP}{P} = \frac{dp}{p} + \frac{\gamma M^2}{1 + \dfrac{\gamma - 1}{2} M^2} \frac{dM}{M} \tag{9.93}$$

$$\frac{d\mathscr{F}}{\mathscr{F}} = \frac{dp}{p} + \frac{dA}{A} + \frac{2\gamma M^2}{1 + \gamma M^2} \frac{dM}{M} \tag{9.94}$$

$$\frac{ds}{c_p} = \frac{dt}{t} - \frac{\gamma - 1}{\gamma}\frac{dp}{p} \tag{9.95}$$

9–5(c) Influence Coefficients

Equations 9.75 and 9.89 to 9.95 comprise a set of eight equations relating the changes in the eight flow properties, or dependent variables, dp/p, $d\rho/\rho$, dt/t, dV/V, dM/M,

dP/P, $d\mathscr{F}/\mathscr{F}$, and ds/c_p, to the four independent variables or *driving potentials* dA/A, dT/T, $(4\mathfrak{f}\,dx/\mathscr{D} + 2\,\delta D/\gamma M^2 pA)$, and $d\dot{m}/\dot{m}$. Those eight equations are linear in the derivatives of the flow properties and the driving potentials. Consequently, the eight equations may be solved simultaneously, as is done in Section 4–4(c) for isentropic flow, to obtain equations relating the change in each flow property (dependent variable) to the driving potentials. In matrix form, those equations are

$$
\begin{bmatrix}
0 & 1 & 0 & 1 & 0 & 0 & 0 & 0 \\
1 & 0 & \dfrac{\gamma M^2}{2} & 0 & \gamma M^2 & 0 & 0 & 0 \\
1 & -1 & -1 & 0 & 0 & 0 & 0 & 0 \\
0 & 0 & \dfrac{1}{2} & -1 & 1 & 0 & 0 & 0 \\
0 & 0 & 1 & 0 & \dfrac{(\gamma-1)M^2}{\psi} & 0 & 0 & 0 \\
1 & 0 & 0 & 0 & \dfrac{\gamma M^2}{\psi} & -1 & 0 & 0 \\
-1 & 0 & 0 & 0 & -\dfrac{2\gamma M^2}{1+\gamma M^2} & 0 & 1 & 0 \\
\dfrac{\gamma-1}{\gamma} & 0 & -1 & 0 & 0 & 0 & 0 & 1
\end{bmatrix}
\begin{bmatrix}
\dfrac{dp}{p} \\[2pt]
\dfrac{d\rho}{\rho} \\[2pt]
\dfrac{dt}{t} \\[2pt]
\dfrac{dV}{V} \\[2pt]
\dfrac{dM}{M} \\[2pt]
\dfrac{dP}{P} \\[2pt]
\dfrac{d\mathscr{F}}{\mathscr{F}} \\[2pt]
\dfrac{ds}{c_p}
\end{bmatrix}
=
\begin{bmatrix}
\dfrac{d\dot{m}}{\dot{m}} - \dfrac{dA}{A} \\[2pt]
K + L \\[2pt]
0 \\[2pt]
0 \\[2pt]
\dfrac{dT}{T} \\[2pt]
0 \\[2pt]
\dfrac{dA}{A} \\[2pt]
0
\end{bmatrix}
\tag{9.96}
$$

where

$$
\psi \equiv 1 + \frac{\gamma-1}{2}M^2 \tag{9.97}
$$

$$
K \equiv -\frac{\gamma M^2}{2}\left[\left(\frac{4\mathfrak{f}\,dx}{\mathscr{D}}\right) + \frac{2\,\delta D}{\gamma M^2 pA}\right] \tag{9.98}
$$

$$
L \equiv -\gamma M^2(1-y)\frac{d\dot{m}}{\dot{m}} \tag{9.99}
$$

Equation 9.96 may be written as

$$
Ax = b \tag{9.100}
$$

where A is the 8×8 coefficient matrix, x is the column matrix containing the property changes, and b is the column matrix containing the driving potentials.

The changes in the eight dependent variables may be obtained from equation 9.96 by applying Cramer's rule (see Appendix A–3). Thus,

$$
x_j = \frac{\det(A^j)}{\det(A)} \qquad (j = 1, \ldots 8) \tag{9.101}
$$

where $\det(A^j)$ is $\det(A)$ with the jth column of A replaced by b. The value of $\det(A)$ is

$$
\det(A) = \frac{1 - M^2}{1 + \dfrac{\gamma-1}{2}M^2} \tag{9.102}
$$

Solving equation 9.96 for dM/M, we obtain

$$
\frac{(1 - M^2)}{1 + \dfrac{\gamma - 1}{2} M^2} \frac{dM}{M} =
\begin{vmatrix}
0 & 1 & 0 & 1 & \left(\dfrac{d\dot{m}}{\dot{m}} - \dfrac{dA}{A}\right) & 0 & 0 & 0 \\
1 & 0 & \dfrac{\gamma M^2}{2} & 0 & (K + L) & 0 & 0 & 0 \\
1 & -1 & -1 & 0 & 0 & 0 & 0 & 0 \\
0 & 0 & \dfrac{1}{2} & -1 & 0 & 0 & 0 & 0 \\
0 & 0 & 1 & 0 & \dfrac{dT}{T} & 0 & 0 & 0 \\
1 & 0 & 0 & 0 & 0 & -1 & 0 & 0 \\
-1 & 0 & 0 & 0 & \dfrac{dA}{A} & 0 & 1 & 0 \\
\dfrac{\gamma - 1}{\gamma} & 0 & -1 & 0 & 0 & 0 & 0 & 1
\end{vmatrix}
\tag{9.103}
$$

Solving equation 9.103 for dM/M yields

$$
\begin{aligned}
\frac{dM}{M} = \frac{\left(1 + \dfrac{\gamma - 1}{2} M^2\right)}{1 - M^2} & \left\{ -\frac{dA}{A} + \frac{\gamma M^2}{2}\left[\left(\frac{4\mathfrak{f}\, dx}{\mathscr{D}}\right) + \frac{2\, \delta D}{\gamma M^2 pA}\right] \right. \\
& \left. + \frac{(1 + \gamma M^2)}{2}\frac{dT}{T} + \left[(1 + \gamma M^2) - \gamma\gamma M^2\right]\frac{d\dot{m}}{\dot{m}} \right\}
\end{aligned}
\tag{9.104}
$$

In a similar manner, the changes in the seven remaining flow properties may be determined. The results are summarized in Table 9.6. Any one of the changes in the flow properties listed in the first column of Table 9.6 may be related to the independent driving potentials listed across the top of the table by multiplying the driving potential by the tabulated expression located at the intersection of the row containing the desired flow property change and the column containing the desired driving potential. For a generalized steady one-dimensional flow, all of the driving potentials may contribute to the change in each flow property. The entries in Table 9.6 are called *influence coefficients*, and they are the partial derivatives of each flow property with respect to each driving potential. They determine the influence of each driving potential on the change in each flow property.

In general, the differential equations summarized in Table 9.6 must be integrated numerically for specific initial and boundary conditions. Actually, the integration of only equation 9.104 requires employing numerical methods, because the values of the remaining dependent variables may be obtained from much simpler forms of the governing equations. Thus, from equation 9.87,

$$
T_2 = T_1 + \frac{Q - W - \Delta H_i}{c_p}
\tag{9.105}
$$

The ratio p_2/p_1 may be obtained from equation 3.167, which is repeated below.

$$
\dot{m} = ApM\left[\frac{\gamma}{RT}\left(1 + \frac{\gamma - 1}{2} M^2\right)\right]^{1/2}
\tag{3.167}
$$

Applying equation 3.167 to any two cross sections A_1 and A_2 of a flow passage,

Table 9.6 Influence Coefficients for the Generalized Steady One-Dimensional Flow of a Perfect Gas

Change in Flow Property	Driving Potentials			
	$\dfrac{dA}{A}$	$\left[\left(\dfrac{4f\,dx}{\mathscr{D}}\right)+\dfrac{2\,\delta D}{\gamma M^2 pA}\right]$	$\dfrac{dT}{T}$	$\dfrac{d\dot{m}}{\dot{m}}$
$\dfrac{dM}{M}$	$-\dfrac{\psi}{1-M^2}$	$\dfrac{\gamma M^2\psi}{2(1-M^2)}$	$\dfrac{(1+\gamma M^2)\psi}{2(1-M^2)}$	$\dfrac{\psi[(1+\gamma M^2)-\gamma\gamma M^2]}{1-M^2}$
$\dfrac{dp}{p}$	$\dfrac{\gamma M^2}{1-M^2}$	$-\dfrac{\gamma M^2[1+(\gamma-1)M^2]}{2(1-M^2)}$	$-\dfrac{\gamma M^2\psi}{1-M^2}$	$-\dfrac{\gamma M^2[2\psi(1-y)+y]}{1-M^2}$
$\dfrac{d\rho}{\rho}$	$\dfrac{M^2}{1-M^2}$	$-\dfrac{\gamma M^2}{2(1-M^2)}$	$-\dfrac{\psi}{1-M^2}$	$-\dfrac{[(\gamma+1)M^2-\gamma\gamma M^2]}{1-M^2}$
$\dfrac{dt}{t}$	$\dfrac{(\gamma-1)M^2}{1-M^2}$	$-\dfrac{\gamma(\gamma-1)M^4}{2(1-M^2)}$	$\dfrac{(1-\gamma M^2)\psi}{1-M^2}$	$\dfrac{(\gamma-1)M^2[(1+\gamma M^2)-\gamma\gamma M^2]}{1-M^2}$
$\dfrac{dV}{V}$	$-\dfrac{1}{1-M^2}$	$\dfrac{\gamma M^2}{2(1-M^2)}$	$\dfrac{\psi}{1-M^2}$	$\dfrac{[(1+\gamma M^2)-\gamma\gamma M^2]}{1-M^2}$
$\dfrac{dP}{P}$	0	$-\dfrac{\gamma M^2}{2}$	$-\dfrac{\gamma M^2}{2}$	$-\gamma M^2(1-y)$
$\dfrac{d\mathscr{F}}{\mathscr{F}}$	$\dfrac{1}{1+\gamma M^2}$	$-\dfrac{\gamma M^2}{2(1+\gamma M^2)}$	0	$\dfrac{\gamma\gamma M^2}{1+\gamma M^2}$
$\dfrac{ds}{c_p}$	0	$\dfrac{(\gamma-1)M^2}{2}$	ψ	$(\gamma-1)M^2(1-y)$

we obtain

$$\frac{p_2}{p_1}=\frac{\dot{m}_2}{\dot{m}_1}\frac{A_1}{A_2}\frac{M_1}{M_2}\left[\frac{T_2\left(1+\dfrac{\gamma-1}{2}M_1{}^2\right)}{T_1\left(1+\dfrac{\gamma-1}{2}M_2{}^2\right)}\right]^{1/2} \tag{9.106}$$

Similarly, from the definition of the stagnation temperature (equation 3.181, Table 3.3)

$$\frac{t_2}{t_1}=\frac{T_2\left(1+\dfrac{\gamma-1}{2}M_1{}^2\right)}{T_1\left(1+\dfrac{\gamma-1}{2}M_2{}^2\right)} \tag{9.107}$$

From the definition $M\equiv V/(\gamma Rt)^{1/2}$,

$$\frac{V_2}{V_1}=\frac{M_2}{M_1}\left(\frac{t_2}{t_1}\right)^{1/2} \tag{9.108}$$

From the thermal equation of state for a perfect gas, $p=\rho Rt$, we obtain

$$\frac{\rho_2}{\rho_1}=\frac{p_2}{p_1}\frac{t_1}{t_2} \tag{9.109}$$

Applying the definition of the stagnation pressure (equation 3.182, Table 3.3) at cross sections A_1 and A_2, we obtain

$$\frac{P_2}{P_1} = \frac{p_2}{p_1} \left(\frac{1 + \dfrac{\gamma - 1}{2} M_2^{2}}{1 + \dfrac{\gamma - 1}{2} M_1^{2}} \right)^{\gamma/(\gamma - 1)} \tag{9.110}$$

Likewise, from the definition of the impulse function,

$$\frac{\mathcal{F}_2}{\mathcal{F}_1} = \frac{p_2 A_2 (1 + \gamma M_2^{2})}{p_1 A_1 (1 + \gamma M_1^{2})} \tag{9.111}$$

The solution to a generalized steady one-dimensional flow problem usually proceeds as follows.

1. The initial and boundary conditions are established, and models are developed for the driving potentials.
2. Equation 9.104 is integrated for the first step along the flow passage by any standard numerical integration algorithm, for example, the Runge-Kutta method presented in Appendix A–6(d).
3. Equations 9.105 to 9.111 are applied for determining the remaining flow properties.
4. Steps 2 and 3 are repeated for subsequent steps along the flow passage until the flow properties have been determined in the region of interest.

Example 9.3.   Air at a Mach number $M = 0.9$ enters a subsonic diffuser hăving a semiangle $\alpha = 7$ deg and an inlet radius $y = 0.25$ m. Set up a numerical algorithm employing the fourth-order Runge-Kutta method [see Appendix A–6(d)], for the determination of M versus x, taking into account the presence of wall friction and area change. Assume $\gamma = 1.40$ and $\bar{f} = 0.01$. Perform the calculations for the first step $\Delta x = 0.01$ m to illustrate the numerical procedure.

Solution

For flow with friction and area change, $\delta D = dT = d\dot{m} = 0$. Consequently, equation 9.104 becomes

$$\frac{dM}{dx} = f(M,x) = \frac{M \left(1 + \dfrac{\gamma - 1}{2} M^2 \right)}{1 - M^2} \left(-\frac{1}{A} \frac{dA}{dx} + \frac{\gamma M^2}{2} \frac{4\bar{f}}{D} \right) \tag{a}$$

For a conical flow passage,

$$y(x) = \frac{D}{2} = a + bx = y_i + (\tan \alpha)x \tag{b}$$

For a flow passage having a circular cross section,

$$\frac{dA}{A} = \frac{d(\pi y^2)}{\pi y^2} = 2 \frac{dy}{y} \tag{c}$$

Substituting equation (b) into equation (c), we obtain

$$\frac{dA}{A} = 2 \frac{d[y_i + (\tan \alpha)x]}{y} = \frac{2 \tan \alpha}{y} dx \tag{d}$$

Substituting equations (b) and (d) into equation (a) yields

$$\frac{dM}{dx} = f(M,x) = \frac{M \left(1 + \dfrac{\gamma - 1}{2} M^2 \right)}{(1 - M^2) y(x)} (-2 \tan \alpha + \gamma M^2 \bar{f}) \tag{e}$$

For the present problem, $\gamma = 1.40$, $\alpha = 7$ deg, $y_i = 0.25$ m, and $\mathfrak{f} = 0.01$. Substituting those values into equation (e), we obtain

$$\frac{dM}{dx} = \frac{M(1 + 0.2M^2)}{(1 - M^2)y(x)}(-0.24556912 + 0.014M^2) \tag{f}$$

where

$$y(x) = (0.25 + 0.12278456x) \text{ m} \tag{g}$$

The fourth-order Runge-Kutta algorithm, presented in Appendix A–6(d), is repeated below for convenience, where y has been replaced by M.

$$M_{n+1} = M_n + \frac{h}{6}(m_1 + 2m_2 + 2m_3 + m_4) \tag{h}$$

$$m_1 = f(x_n, M_n) \tag{i}$$

$$m_2 = f\left(x_n + \frac{h}{2}, M_n + \frac{h}{2}m_1\right) \tag{j}$$

$$m_3 = f\left(x_n + \frac{h}{2}, M_n + \frac{h}{2}m_2\right) \tag{k}$$

$$m_4 = f(x_n + h, M_n + hm_3) \tag{l}$$

For the first step, $x_1 = 0.0$, $y_i = 0.25$, $h = 0.01$ m, and $M_1 = 0.90$. Substituting those values into equation (i) gives

$$m_1 = \frac{(0.9)[1 + (0.2)(0.9)^2]}{[1 - (0.9)^2](0.25)}\{-0.24556912 + 0.014(0.9)^2\} = -5.15698555$$

For the calculation of m_2,

$$x_1 + \frac{h}{2} = 0.0 + \frac{(0.01)}{2} = 0.005 \text{ m}$$

$$y\left(x_1 + \frac{h}{2}\right) = 0.25 + (0.12278456)(0.005) = 0.25061392 \text{ m}$$

$$M_1 + \frac{h}{2}m_1 = 0.9 + \frac{(0.01)}{2}(-5.15698555) = 0.87421507$$

Substituting the above values into equation (j) yields

$$m_2 = \frac{(0.87421507)[1 + (0.2)(0.87421507)^2]}{[1 - (0.87421507)^2](0.25061392)}[-0.24556912 + (0.014)(0.87421507)^2]$$

$$m_2 = -4.00649688$$

Repeating the above steps for m_3 gives

$$M_1 + \frac{h}{2}m_2 = 0.9 + \frac{(0.01)}{2}(-4.00649688) = 0.87996752$$

$$m_3 = \frac{(0.87996752)[1 + 0.2(0.87996752)^2]}{[1 - (0.87996752)^2](0.25061392)}$$

$$[-0.24556912 + (0.014)(0.87996752)^2]$$

$$m_3 = -4.21803678$$

Finally, repeating the above steps for m_4, we obtain

$$x_1 + h = 0.0 + 0.01 = 0.01 \text{ m}$$

$$y(x_1 + h) = 0.25 + (0.12278456)(0.01) = 0.25122785 \text{ m}$$

$$M_1 + hm_3 = 0.9 + (0.01)(-4.21803678) = 0.85781963$$

$$m_4 = \frac{(0.85781963)[1 + 0.2(0.85781963)^2]}{[1 - (0.85781963)^2](0.25122785)}$$

$$[-0.24556912 + (0.014)(0.85781963)^2]$$

$$m_4 = -3.48878692$$

Substituting the values of m_1, m_2, m_3, and m_4 into equation (h) yields the value of M_2 at $x_2 = 0.01$ m.

$$M_2 = 0.9 + \frac{(0.01)}{6}[-5.15698555 - 2(4.00649688) - 2(4.21803678) - 3.48878692]$$

$$M_2 = 0.9 + \frac{(0.01)(-25.09483979)}{6} = 0.85817527$$

The above procedure, although somewhat tedious for hand calculation, is readily programmed for computer calculation.

9–5(d)   General Features of Generalized Steady One-Dimensional Flow of a Perfect Gas

In Section 9–5(c), it is shown that the Mach number change dM/M for a generalized steady one-dimensional flow of a perfect gas is determined by equation 9.104. To obtain a solution to the latter equation, it must be integrated numerically for specific initial and boundary conditions. However, the general features of such flows may be determined directly from equation 9.104.

Rewrite equation 9.104 in the following form:

$$\frac{dM}{M} = \frac{\Lambda}{1 - M^2} \tag{9.112}$$

where, *by definition*,

$$\Lambda \equiv \left(1 + \frac{\gamma - 1}{2} M^2\right)\left\{-\frac{dA}{A} + \frac{\gamma M^2}{2}\left[\left(\frac{4\mathfrak{f}\,dx}{\mathscr{D}}\right) + \frac{2\,\delta D}{\gamma M^2 pA}\right]\right.$$

$$\left. + \frac{(1 + \gamma M^2)}{2}\frac{dT}{T} + [(1 + \gamma M^2) - y\gamma M^2]\frac{d\dot{m}}{\dot{m}}\right\} \tag{9.113}$$

From equation 9.112, it is seen that the effect of the driving potentials on the direction of the change of the flow Mach number M depends not only on whether the initial flow is subsonic ($M < 1$) or supersonic ($M > 1$), but also on the sign of Λ. For a simple flow, the sign of Λ depends uniquely on the sign of the particular single driving potential. But, for a generalized flow, the sign of Λ is determined by the contributions of all of the different driving potentials to the magnitude of Λ. Table 9.7 summarizes all of the possibilities. The general behavior and the limiting conditions for a generalized steady one-dimensional flow may be determined from Table 9.7.

Consider the case where Λ in equation 9.112 is always *negative*. If the initial flow is subsonic ($M < 1$), M continually decreases. Similarly, if the initial flow is supersonic

Table 9.7 Relationship between Λ and dM

Λ	dM		
	$M < 1$	$M = 1$	$M > 1$
$-$	$-$	∞	$+$
0	0	$\dfrac{0}{0}$	0
$+$	$+$	∞	$-$

($M > 1$), M increases continuously. Hence, there are no limitations (other than $M = 0$ or $M = \infty$) on the Mach number when Λ is negative. Figure 9.16a illustrates schematically the characteristics of a generalized steady one-dimensional flow with Λ negative. Examples are isentropic flow in a diverging passage and simple diabatic flow with cooling.

When Λ is zero, it follows from equation 9.112 that M remains constant. In a simple flow, all of the flow properties would remain constant but, in a generalized flow, there is no such restriction. For example, in a flow with friction, the passage walls may be designed to diverge at such a rate that the effects of dA/A and ($4\mathfrak{f}\, dx/\mathscr{D}$) exactly cancel. From equation 9.113, that condition gives

$$\frac{dA}{A} = \frac{\gamma M^2}{2}\left(\frac{4\mathfrak{f}\, dx}{\mathscr{D}}\right) \tag{9.114}$$

Similarly, the Mach number in a combustor may be kept constant by diverging the walls at a rate that cancels the effects because of heat addition. In that case,

$$\frac{dA}{A} = \frac{(1 + \gamma M^2)}{2}\frac{dT}{T} \tag{9.115}$$

When Λ is positive, a flow that is initially subsonic increases its Mach number as it proceeds downstream, but the Mach number of a flow that is initially supersonic decreases. Hence, both subsonic and supersonic flows proceed toward the limiting state where $M = 1$ and, at that state, the flow is choked. In the studies of the simple flows, it is shown that when choking occurs ($M = 1$), a readjustment in the steady-state flow pattern must occur if further changes are made in the pertinent driving potentials. A flow that is subsonic upstream from the choking location experiences a decrease in mass flow rate. Those that are initially supersonic develop shock waves and, if the additional driving potential is strong enough, the entire flow may become subsonic. Similar effects occur in a generalized flow, which chokes in exactly the same manner as does a simple flow. Figure 9.16b illustrates schematically the Mach number M as a function of the length x of the flow passage, for flow with $\Lambda > 0$. Examples are isentropic flow in a converging passage, simple frictional flow, simple diabatic flow with heating, and simple mass addition.

So far, the three possibilities $\Lambda < 0$, $\Lambda = 0$, and $\Lambda > 0$ have been examined. There are flows in which Λ changes sign during the flow process. Figure 9.16c illustrates schematically the case where Λ changes from negative to positive. The initial part of the flow is identical to that illustrated in Fig. 9.16a. After Λ changes sign, the flow proceeds toward the location where $M = 1$ and choking occurs. Examples

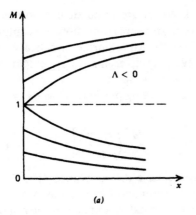

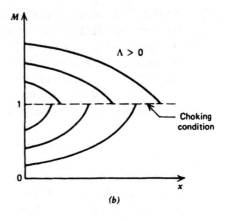

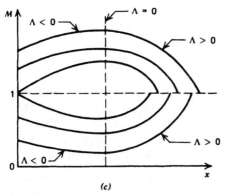

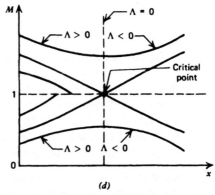

Figure 9.16 Relationship between M and x for different values of Λ. (a) Λ negative. (b) Λ positive (c) Λ changes from negative to positive. (d) Λ changes from positive to negative.

are isentropic flow in a diverging-converging passage, and cooling followed by a converging passage or a frictional flow.

Finally, Fig. 9.16d illustrates schematically the situation where Λ changes from positive to negative. Both subsonic and supersonic flows proceed toward a Mach number of unity when Λ is positive and, if Λ remains positive, both types of flow will eventually choke. If Λ passes through zero while the flow Mach number is not unity, then the Mach numbers for both subsonic and supersonic flows proceed away from the sonic speed. In the special case where $\Lambda = 0$ at the same location where $M = 1$, the critical point illustrated on Fig. 9.16d, then there are two possibilities: (1) the flow may proceed as a subsonic flow, or (2) it may proceed as a supersonic flow. Which of those two possibilities actually occurs depends on the downstream physical boundary conditions at the exit cross section of the flow passage [see Section 4–3(e)]. Hence, at the special location where $\Lambda = 0$ and $M = 1$ simultaneously, a continuous transition may proceed from subsonic to supersonic flow, or from supersonic to subsonic flow. It is shown in Chapter 7 that an *abrupt transition* from supersonic to subsonic flow may occur in a normal shock wave. It is shown in

Section 7–3, however, that expansion shock waves are impossible. Consequently, the only allowable transitions between subsonic and supersonic flow are continuous transitions where $\Lambda = 0$ and $M = 1$ simultaneously. An important practical example wherein such a transition occurs is the converging-diverging nozzle employed for accelerating an isentropic subsonic flow to supersonic speeds.

When only friction is present in a converging-diverging passage, $\delta D = dT = d\dot{m} = 0$, and equation 9.113 reduces to

$$\Lambda = \left(1 + \frac{\gamma - 1}{2} M^2\right)\left[-\frac{dA}{A} + \frac{\gamma M^2}{2}\left(\frac{4f\, dx}{\mathscr{D}}\right)\right] \tag{9.116}$$

In the converging portion of the aforementioned nozzle, $dA < 0$. Since $(4f\, dx/\mathscr{D}) > 0$, Λ is positive throughout the entire converging region, and M remains subsonic. At the nozzle throat, $dA_t = 0$, but $(4f\, dx/\mathscr{D}) > 0$, so that Λ remains positive and M must still be less than unity. The transition to supersonic flow occurs where $\Lambda = 0$ and, in that case, equation 9.116 reduces to

$$\frac{dA}{A} = \frac{\gamma M^2}{2}\left(\frac{4f\, dx}{\mathscr{D}}\right) > 0 \tag{9.117}$$

Equation 9.117 shows that at the critical location, $dA > 0$. Accordingly, the critical location is in the diverging portion of the nozzle, downstream from the nozzle throat.

Similar results may be demonstrated for heat addition and mass addition in a converging-diverging passage. Heat removal causes the critical location to be in the converging portion of the nozzle. When more than two driving potentials are acting simultaneously, the analysis becomes more complicated, but the general features of the flow may be determined from equation 9.112 in a similar manner. Reference 14 presents the conditions necessary for a continuous transition through the sonic speed for other combinations of dA, dT, $(4f\, dx/\mathscr{D})$, and $d\dot{m}/\dot{m}$.

Example 9.4.   Air flows in an insulated circular duct with wall friction. Determine the contour of the duct wall required to maintain a constant Mach number within the duct. Calculate the area ratio of the duct required to decrease the static pressure to 80 percent of its initial value.

Solution

From equations 9.112 and 9.113, for $dM = \delta D = dT = d\dot{m} = 0$, we obtain

$$\frac{dM}{M} = \frac{\left(1 + \dfrac{\gamma - 1}{2} M^2\right)}{1 - M^2}\left[-\frac{dA}{A} + \frac{\gamma M^2}{2}\left(\frac{4f\, dx}{\mathscr{D}}\right)\right] = 0 \tag{a}$$

Rearranging equation (a) yields

$$\frac{dA}{A} = \frac{\gamma M^2}{2}\left(\frac{4f\, dx}{\mathscr{D}}\right) \tag{b}$$

For a circular duct, $\mathscr{D} = D$, and

$$A = \frac{\pi D^2}{4} \tag{c}$$

$$\frac{dA}{A} = 2\frac{dD}{D} \tag{d}$$

Substituting equation (d) into equation (b) gives

$$dD = \gamma M^2 \mathfrak{f} \, dx$$

Integrating equation (e), we obtain

$$D_2 - D_1 = \gamma M^2 \mathfrak{f}(x_2 - x_1) \tag{f}$$

Consequently, the duct is a diverging conical passage.

From equation 9.106, for $\dot{m}_1 = \dot{m}_2$, $M_1 = M_2$, and $T_1 = T_2$, we obtain

$$\frac{p_2}{p_1} = \frac{A_1}{A_2} \tag{g}$$

From equation (g), it is apparent that the product pA remains constant in the subject type of flow. For $p_2 = 0.80p_1$,

$$\frac{A_2}{A_1} = \frac{p_1}{p_2} = \frac{1}{0.80} = 1.25$$

Example 9.5. A ramjet combustor is to be designed so that the Mach number remains constant during the combustion process. For a combustor having a circular cross section, determine the contour of the combustor wall, neglecting the effects of the fuel on the mass flow rate and specific heats of the air flowing through the combustor. Determine the pressure ratio of a combustor with $M = 0.5$ for a stagnation temperature increase of 100 percent.

Solution

From equations 9.112 and 9.113, for $dM = (4\mathfrak{f} \, dx/\mathscr{D}) = \delta D = d\dot{m} = 0$, we obtain

$$\frac{dM}{M} = \frac{\left(1 + \dfrac{\gamma - 1}{2} M^2\right)}{1 - M^2}\left[-\frac{dA}{A} + \frac{(1 + \gamma M^2)}{2}\frac{dT}{T}\right] = 0 \tag{a}$$

Rearranging equation (a) gives

$$\frac{dA}{A} = \frac{(1 + \gamma M^2)}{2}\frac{dT}{T} \tag{b}$$

Integrating equation (b) yields

$$\ln\left(\frac{A_2}{A_1}\right) = \frac{(1 + \gamma M^2)}{2}\ln\left(\frac{T_2}{T_1}\right) \tag{c}$$

Equation (c) may be rewritten as

$$\frac{A_2}{A_1} = \left(\frac{T_2}{T_1}\right)^{(1 + \gamma M^2)/2} \tag{d}$$

Solving for D_2/D_1 gives

$$\frac{D_2}{D_1} = \left(\frac{A_2}{A_1}\right)^{1/2} = \left(\frac{T_2}{T_1}\right)^{(1 + \gamma M^2)/4} \tag{e}$$

Accordingly, the duct is a diverging passage, where the value of D depends on the local value of T.

From equation 9.106, for $\dot{m}_1 = \dot{m}_2$ and $M_1 = M_2$, we obtain

$$\frac{p_2}{p_1} = \frac{A_1}{A_2}\left(\frac{T_2}{T_1}\right)^{1/2} \tag{f}$$

Substituting for A_1/A_2 from equation (d) into equation (f) gives

$$\frac{p_2}{p_1} = \left(\frac{T_1}{T_2}\right)^{(1+\gamma M^2)/2}\left(\frac{T_2}{T_1}\right)^{1/2} = \left(\frac{T_2}{T_1}\right)^{-\gamma M^2/2} \tag{g}$$

From equation (g), for $\gamma = 1.4$, $M = 0.5$, and $T_2/T_1 = 2.0$, we obtain

$$\frac{p_2}{p_1} = (2.0)^{-(1.4)(0.5)^2/2} = (2.0)^{-0.1750} = 0.8858$$

9–5(e) Numerical Integration for Generalized Steady One-Dimensional Flow

In Section 9–5(a), the equations governing the generalized steady one-dimensional flow of a fluid are derived. Those results are applied to the flow of a perfect gas in Sections 9–5(b) and 9–5(c). The governing equations comprise a set of first-order nonlinear nonhomogeneous ordinary differential equations, which are too complex to be integrated in closed form. They may, however, be integrated by any of the standard numerical methods for integrating such equations.[15, 16] The modified Euler method and the Runge-Kutta method, presented in Appendix A–6, may be employed. There are, however, more sophisticated predictor-corrector algorithms that generally yield a better compromise between computing time and accuracy.[15, 16]

From Table 9.7, it is seen that there is a singularity where the flow velocity becomes sonic, because of the term $1/(1 - M^2)$ in equation 9.112. Although the singularity does not appear explicitly in the equations derived in Section 9–5(a), it is there nevertheless. In the case of either a perfect gas or an imperfect gas, special techniques are required for calculations passing through the singular point. Several methods have been employed for either dealing with or for eliminating the singularity. Some of them are discussed briefly in the following paragraphs.

One of the earliest methods is based on linearizing the differential equations in the vicinity of the sonic point and integrating the resulting linear equations in closed form to obtain an approximate solution that is valid in the neighborhood of the singularity. Standard numerical integration methods are employed both upstream to and downstream from the linearized flow region.

A second method employs any standard numerical integration method until the Mach number approaches unity. At some predetermined initial value of M (e.g., $M = 0.95$), the derivatives of the properties are held constant, and the flow properties are extrapolated across the singularity to a predetermined final value of M (e.g., $M = 1.05$).

A third method for determining dM/dx at the singularity is based on applying L'Hospital's rule* to equation 9.112 at the sonic condition, and then employing that value of dM/dx for extrapolating across the singular point. Equation 9.112 may be rewritten as

$$\frac{dM}{dx} = \frac{\Lambda'(x)}{1 - M^2} \tag{9.118}$$

---

* According to L'Hospital's rule, if the ratio $f(x)/F(x)$ is an indeterminate of the form $0/0$ for $x = a$, but if $f'(x)/F'(x)$ is not indeterminate at $x = a$, then

$$\lim_{x \to a} \frac{f(x)}{F(x)} = \frac{f'(a)}{F'(a)}$$

where

$$\Lambda'(x) \equiv M\left(1 + \frac{\gamma - 1}{2}M^2\right)\left\{-\frac{d(\ln A)}{dx} + \frac{\gamma M^2}{2}\left(\frac{4\mathfrak{f}}{\mathscr{D}} + \frac{2\,\delta D/dx}{\gamma M^2 pA}\right)\right.$$

$$\left. + \frac{(1 + \gamma M^2)}{2}\frac{d(\ln T)}{dx} + \left[(1 + \gamma M^2) - \gamma\gamma M^2\right]\frac{d(\ln \dot{m})}{dx}\right\} \quad (9.119)$$

where A, $\mathfrak{f}$, $\mathscr{D}$, etc., are assumed to be known functions of x. Applying L'Hospital's rule to equation 9.118 to determine dM/dx at the singularity or critical state, denoted by an asterisk superscript, we obtain

$$\left(\frac{dM}{dx}\right)^* = \lim_{M \to 1}\left(\frac{dM}{dx}\right) = \frac{(d\Lambda'/dx)^*}{-2(dM/dx)^*} \quad (9.120)$$

which may be rewritten as

$$\left[\left(\frac{dM}{dx}\right)^*\right]^2 + \frac{1}{2}\left(\frac{d\Lambda'}{dx}\right)^* = 0 \quad (9.121)$$

Since M appears explicitly in $\Lambda'(x)$, the derivative $(d\Lambda'/dx)^*$ contains terms involving $(dM/dx)^*$. Consequently, equation 9.121 is a linear quadratic equation for $(dM/dx)^*$, which yields two values for $(dM/dx)^*$. Which of those two values applies in a particular situation depends on the downstream physical boundary conditions [see Section 4–3(e)] and, depending on the value of $(dM/dx)^*$, the flow may proceed along either the subsonic or the supersonic branch of the curve $M = f(x)$ downstream from the location where $M = 1$. The method just described may be applied only when the driving potentials depend explicitly on x alone; that is, $A(x)$, $T(x)$, $\mathfrak{f}(x)$, $\delta D(x)$, and $\dot{m}(x)$. In general, the driving potentials depend explicitly on the flow properties, and, in that case, the method is inapplicable.

Reference 14 presents some results obtained by the method described in the preceding paragraph. Figure 9.17 presents the Mach number M for adiabatic flow

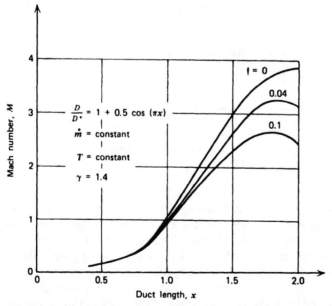

Figure 9.17 Adiabatic flow with friction in a converging-diverging duct (taken from Reference 14).

in a converging-diverging duct with friction, where the duct diameter is specified by $D/D^* = 1.0 + 0.5 \cos(\pi x)$, for values of $\mathfrak{f} = 0$, 0.04, and 0.1. The figure shows that friction reduces the Mach number below its corresponding isentropic value ($\mathfrak{f} = 0$) at every cross section of the duct, that the throat Mach number (at $x = 1.0$) is less than unity, and that the cross section where $M = 1$ is located downstream from the throat. Near the duct exit, the area change term dA becomes smaller than the friction term ($4\mathfrak{f}\, dx/\mathcal{D}$), and the Mach number decreases even though the area continues to increase. Figure 9.18 presents the Mach number distributions in a diverging parabolic duct with $A_2/A_1 = 1.5$ and zero slope at the exit, for different linear stagnation temperature distributions along the duct where $T_2/T_1 = 1.0$, 1.1, and 1.2. Also shown is the Mach number distribution for the case of $T_2/T_1 = 1.1$ and $\mathfrak{f} = 0.02$. Figure 9.18 shows that the effect of heat addition is to reduce the flow Mach number below its corresponding isentropic value ($T_2/T_1 = 1$), the amount of the reduction increasing with an increase in the amount of the heat addition. The combined action of friction and heat transfer, the curve for $T_2/T_1 = 1.1$ and $\mathfrak{f} = 0.02$, reduces the Mach number more than does heat transfer alone.

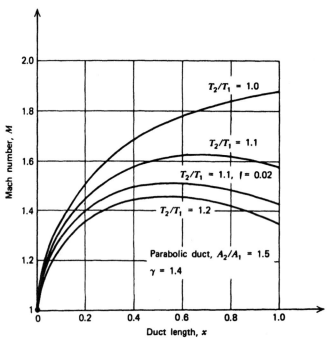

Figure 9.18 Supersonic flow in a divergent duct with a linear increase in stagnation temperature (taken from Reference 14).

The fourth method to be discussed eliminates the singularity at the sonic location. Furthermore, it is applicable to both perfect and imperfect gases, and the driving potentials may assume any form whatsoever. The method is based on choosing a different set of driving potentials. In Section 9–5(c), the driving potentials included the area change dA/A. In the present method, the static pressure distribution $p(x)$ replaces the driving potential $A(x)$. The area A thus becomes a dependent variable, which must be obtained from the solution to the governing equations. The resulting set of differential equations does not have a singularity at the sonic location. Consequently, any standard numerical integration algorithm may be employed for integrating the equations through the transonic region. Because the area distribution is usually known and the pressure distribution is unknown, an iterative approach must be employed wherein a trial value is assumed for the pressure distribution. The governing equations are then solved to obtain the corresponding area distribution, and the latter is compared to the known area distribution. If the known and calculated area distributions do not agree, the assumed pressure distribution is modified and the governing equations are solved for the modified pressure distribution. When the calculated area distribution agrees with the known area distribution to within a desired tolerance, the solution is complete. This procedure is time-consuming because the differential equations must be solved several times at each solution point. Consequently, the subject method should be employed only in the transonic region; for example, from $M = 0.9$ to $M = 1.1$. Outside of that region, equation 9.112 may be solved quite easily.

A set of influence coefficients, similar to those presented in Table 9.6, may be derived for the method of analysis that employs a specified pressure distribution by replacing the first column in the coefficient matrix in equation 9.96 by the coefficients of dA/A, removing dA/A from the column matrix containing the nonhomogeneous terms, and adding dp/p as required to that column matrix. The singularity in the equations presented in Table 9.6 arises from the determinant of the coefficient matrix, det (A), given by equation 9.102. For the *specified pressure distribution method*,

$$\det (A) = -\frac{\gamma M^2}{1 + \dfrac{\gamma - 1}{2} M^2} \tag{9.122}$$

and no singularity appears in the resulting equations.

Instead of developing the influence coefficients for the specified pressure distribution method, which are valid only for the flow of a perfect gas, a method of integration is presented below that is applicable to both perfect gases and imperfect gases. First, assume the pressure distribution.

$$p = p(x) \qquad (assumed\ to\ be\ known) \tag{9.123}$$

Then solve equation 9.76 for dV. Thus,

$$dV = -\frac{dp}{\rho V} - \frac{V}{2}\left(\frac{4f\,dx}{\mathscr{D}}\right) - V(1 - y)\frac{d\dot{m}}{\dot{m}} - \frac{g}{V}dz - \frac{\delta D}{\dot{m}} \tag{9.124}$$

Solving equation 9.80 for dh, we obtain

$$dh = \delta Q - \delta W - dH_i - V\,dV \tag{9.125}$$

From the equation of state for the fluid, in algebraic, graphical, or tabular form,

$$\rho = \rho(p,h) \tag{9.126}$$

From the continuity equation,

$$A = \frac{\dot{m}}{\rho V} \tag{9.127}$$

Any other thermodynamic property of interest (i.e., t, u, s, and a) may be obtained from the equation of state for the fluid. Equations 9.124 and 9.125 comprise a set of coupled first-order nonlinear nonhomogeneous ordinary differential equations that may be solved by any standard numerical integration algorithm. There are no singularities present, and the numerical integration proceeds smoothly. An iterative procedure is employed for modifying the assumed pressure distribution so that the area determined by equation 9.127 approaches the known area distribution. Equations 9.124 to 9.127 apply to the flow of any fluid. In a given flow, detailed models must be developed for relating $\mathfrak{f}$, $d\dot{m}$, δQ, δW, and dH_i to the flow properties. The foregoing method of analysis has been employed successfully in a variety of complex nozzle problems; for example, flow with chemical reaction, and flow wherein suspended particles are carried by the gas (see Sections 14-6 and 14-7, respectively).

The final method to be discussed is the determination of the solution to a *steady one-dimensional flow* problem as the asymptotic solution at large time of the *unsteady one-dimensional flow* problem having the same boundary conditions as the steady flow. The addition of a second independent variable, the time, greatly complicates the governing equations and the numerical solution. However, the singularity at the sonic point is not present in the unsteady flow equations, which is a definite advantage over the steady flow equations. The equations governing unsteady one-dimensional and quasi-one-dimensional flow with friction, heat transfer, and mass addition are derived in Section 19-3.

9-5(f) Simple Flows

A generalized steady one-dimensional flow, being the most general steady one-dimensional flow process possible, includes all of the simple flows as special cases. For example, assuming that all of the driving potentials in Table 9.6 are zero except for dA/A, we obtain the differential equations for isentropic flow with simple area change. Similarly, the second column in Table 9.6 corresponds to simple frictional flow (i.e., the Fanno line), the third column corresponds to simple diabatic flow (i.e., the Rayleigh line), and the fourth column corresponds to simple mass addition. The listings in those four columns correspond identically with the influence coefficients presented in Tables 4.4, 5.3, 6.2, and 9.1 for the aforementioned simple flows.

For all of the simple flows, the differential equations summarized in Table 9.6 are exact differential equations, which may be integrated to yield algebraic equations for all of the property ratios. Table 9.8 summarizes the integrated results. Column 1 (simple area change) is equivalent to Table 4.2, where P, T, and ρ_0 in Table 4.2 are replaced in terms of p^*, t^*, and ρ^* by employing equations 4.139, 4.140, and 4.141, respectively. The second column (simple frictional flow) in Table 9.7 corresponds to Table 5.5, the third column (simple diabatic flow) corresponds to Table 6.4, and the fourth column (simple mass addition, $y = 0$) corresponds to equations 9.24 to 9.29.

Tables 9.6 and 9.8 summarize the major relationships pertinent to the steady one-dimensional flow of a perfect gas.

9-6 SUMMARY

This chapter is concerned with three principal topics; (1) steady one-dimensional flow with simple mass addition, (2) one-dimensional combustion waves, and (3) generalized steady one-dimensional flow.

Simple mass addition is defined as a steady one-dimensional flow wherein the sole driving potential for producing changes in the flow properties is the rate of mass addition $d\dot{m}$. The governing equations for simple mass addition are obtained from the continuity, momentum, and energy equations, assuming that the fluids in the main and in the injected streams are the same perfect gas. The influence coefficients of $d\dot{m}/\dot{m}$ in the governing differential equations include both the Mach number M and the *injection parameter* $y = V_{ix}/V$ as parameters. Consequently, a

Table 9.8 Equations for Simple One-Dimensional Flow of a Perfect Gas

	Simple Area Change	Simple Frictional Flow	Simple Diabatic Flow	Simple Mass Addition, y = 0
$\dfrac{A}{A^*}$	$\dfrac{1}{M}\left[\left(\dfrac{2}{\gamma+1}\right)\psi\right]^{(\gamma+1)/2(\gamma-1)}$	1	1	1
$\dfrac{T}{T^*}$	1	1	$\dfrac{2(\gamma+1)M^2\psi}{(1+\gamma M^2)^2}$	1
$\dfrac{4\bar{f}L^*}{\mathscr{D}}$	0	$\dfrac{1-M^2}{\gamma M^2}+\left(\dfrac{\gamma+1}{2\gamma}\right)\ln\left\{M^2\left[\left(\dfrac{2}{\gamma+1}\right)\psi\right]^{-1}\right\}$	0	0
$\dfrac{\dot{m}}{\dot{m}^*}$	1	1	1	$\dfrac{M\sqrt{2(\gamma+1)}\,\psi}{1+\gamma M^2}$
$\dfrac{p}{p^*}$	$\left[\left(\dfrac{2}{\gamma+1}\right)\psi\right]^{-\gamma/(\gamma-1)}$	$\dfrac{1}{M}\left[\left(\dfrac{2}{\gamma+1}\right)\psi\right]^{-1/2}$	$\dfrac{\gamma+1}{1+\gamma M^2}$	$\dfrac{\gamma+1}{1+\gamma M^2}$
$\dfrac{\rho}{\rho^*}$	$\left[\left(\dfrac{2}{\gamma+1}\right)\psi\right]^{-1/(\gamma-1)}$	$\dfrac{1}{M}\left[\left(\dfrac{2}{\gamma+1}\right)\psi\right]^{1/2}$	$\dfrac{1+\gamma M^2}{(\gamma+1)M^2}$	$\dfrac{2\psi}{1+\gamma M^2}$
$\dfrac{t}{t^*}$	$\left[\left(\dfrac{2}{\gamma+1}\right)\psi\right]^{-1}$	$\left[\left(\dfrac{2}{\gamma+1}\right)\psi\right]^{-1}$	$\dfrac{(\gamma+1)^2 M^2}{(1+\gamma M^2)^2}$	$\left[\left(\dfrac{2}{\gamma+1}\right)\psi\right]^{-1}$
$\dfrac{V}{V^*}$	$M\left[\left(\dfrac{2}{\gamma+1}\right)\psi\right]^{-1/2}$	$M\left[\left(\dfrac{2}{\gamma+1}\right)\psi\right]^{-1/2}$	$\dfrac{(\gamma+1)M^2}{1+\gamma M^2}$	$M\left[\left(\dfrac{2}{\gamma+1}\right)\psi\right]^{-1/2}$
$\dfrac{P}{P^*}$	1	$\dfrac{1}{M}\left[\left(\dfrac{2}{\gamma+1}\right)\psi\right]^{(\gamma+1)/2(\gamma-1)}$	$\dfrac{\gamma+1}{1+\gamma M^2}\left[\left(\dfrac{2}{\gamma+1}\right)\psi\right]^{\gamma/(\gamma-1)}$	$\dfrac{\gamma+1}{1+\gamma M^2}\left[\left(\dfrac{2}{\gamma+1}\right)\psi\right]^{\gamma/(\gamma-1)}$
$\dfrac{\mathscr{F}}{\mathscr{F}^*}$	$\dfrac{1+\gamma M^2}{M\sqrt{2(\gamma+1)}\,\psi}$	$\dfrac{1+\gamma M^2}{M\sqrt{2(\gamma+1)}\,\psi}$	1	1
$\dfrac{s-s^*}{c_p}$	0	$\ln\left\{M^{(\gamma-1)/\gamma}\left[\left(\dfrac{2}{\gamma+1}\right)\psi\right]^{-(\gamma+1)/2\gamma}\right\}$	$\ln\left\{M^2\left(\dfrac{\gamma+1}{1+\gamma M^2}\right)^{(\gamma+1)/\gamma}\right\}$	$\ln\left\{\left[\left(\dfrac{2}{\gamma+1}\right)\psi\right]^{-1}\left(\dfrac{1+\gamma M^2}{\gamma+1}\right)^{(\gamma-1)/\gamma}\right\}$

change in dM/M, dp/p, etc., is either positive or negative, depending not only on whether the main stream flow is subsonic or supersonic, but also on the magnitude of y. For $y < 1$, the case of interest here, if $M < 1$, mass addition increases the main stream Mach number M, and if $M > 1$, it decreases M. For either subsonic or supersonic flow, mass addition causes choking of the main stream flow when $M = 1$. At choking, no further mass addition is possible unless the upstream conditions are altered.

Many useful applications of simple mass addition are characterized by $y = 0$; that is, the mass addition $d\dot{m}$ is *normal* to the direction of the main fluid stream. For $y = 0$, the governing differential equations are all exact differential equations and may be integrated between a reference state, such as the *critical state* where $M = 1$, and a general point in the flow field. In Section 9–3(d) the developments for mass addition with $y = 0$ are applied to a solid propellant rocket motor equipped with an internal-burning solid propellant grain.

Combustion waves, planar waves propagated in premixed combustible gas mixtures (*reactants*) contained in long tubes, are of two types; *deflagration waves*, which propagate with a *flame speed* that is subsonic with respect to the burned reactants, and *detonation waves*, which propagate with a *detonation velocity* that is supersonic with respect to the reactants. In a stationary reference system a combustion wave is an unsteady flow process but, in a relative coordinate system, it is a steady combustion wave. Whether a deflagration wave or a detonation wave occurs under a given set of conditions depends primarily on the composition of the gaseous reactants and whether their ignition occurs at the open or closed end of the tube. The locus of the allowable final states for the products for a combustion wave for a given initial state and *energy of combustion* $\Delta u°$, plotted in the pv plane, is called the *Hugoniot curve* or *H-curve*. Each different value of $\Delta u°$ gives rise to a separate *H-curve*. The combustion products must simultaneously satisfy the equations for both the *H-curve* and the *Rayleigh line* (*R-line*). Not all of the regions on the *H-curve* are attainable physically. Those attainable are strong detonations including the Chapman-Jouget point J, and weak deflagrations including the Chapman-Jouget point K (see Fig. 9.13). At the points J and K, the *H-curve* is tangent to an *isentrope*.

In the case of perfect gases, the *detonation velocity* for gaseous fuel-oxidizer mixtures may be calculated with reasonable accuracy (see Table 9.5). For an imperfect gas, numerical methods are necessary (see Example 9.2).

There are definite lower and upper limits, called the *flammability limits*, to the gaseous mixture ratio that will support a steady deflagration wave. The flammability limits are usually wider than the corresponding *detonation limits*. Several theories for predicting the flame speed of deflagration waves are presented in the references.

A generalized steady one-dimensional flow of a compressible fluid is one in which all of the following seven independent *driving potentials* act simultaneously; (1) area change dA, (2) wall friction δF_f, (3) heat addition δQ, (4) work δW, (5) mass addition $d\dot{m}$, (6) gravitational body force $\rho g A\, dz$, and (7) other body forces, drag of entrained particles, submerged bodies, etc., denoted by δD. A system of 14 equations is obtained relating the 14 fluid properties p, ρ, t, h, u, s, a, V, M, H, P, T, ρ_o, and $\mathscr{F}$, and the above seven independent driving potentials. To obtain a solution for the aforementioned system of equations, the seven independent driving potentials must be calculable. The system of equations derived for the generalized steady one-dimensional flow of an arbitrary compressible fluid is quite complicated. Their solutions for specific initial and boundary conditions are obtainable only by applying numerical methods.

A great simplification is introduced by assuming that the flowing fluid is a perfect gas. Table 9.6 presents the influence coefficients for the generalized steady one-dimensional flow of a perfect gas, and Table 9.8 presents the equations for simple one-dimensional flow of a perfect gas. The general features of the generalized steady one-dimensional flow of a perfect gas are discussed in detail in Section 9–5(d). It is shown that there is a singularity when the flow has a Mach number of unity because of the term $1/(\Gamma - M^2)$ in the expression for dM/M (see equation 9.112). Procedures are presented in Section 9–5(e) for dealing with that singularity.

REFERENCES

1. M. J. Zucrow, *Aircraft and Missile Propulsion*, Vol. II, pp. 468–524, Wiley, New York, 1964.
2. Y. B. Zeldovich, "On the Theory of the Propagation of Detonation in Gaseous Systems" (in Russian), *Journal of Experimental and Theoretical Physics (USSR)*, Vol. 10, p. 542, 1940. Translated as NACA TM 1261, November 1950.
3. J. von Neumann, "Theory of Detonation Waves," OSRD Report 549, 1942.
4. W. Döring, "Über den Detonationsvorgang in Gasen," *Annalen der Physik*, Vol. 43, pp. 421–436, 1943.
5. F. E. Belles, "Detonability and Chemical Kinetics: Prediction of Limits of Detonability of Hydrogen," *Seventh Symposium (International) on Combustion*, pp. 745–751, Butterworths, London, 1959.
6. B. Lewis and G. von Elbe, *Combustion, Flames and Explosions of Gases*, Second Edition, pp. 524–528, Academic Press, New York, 1961.
7. Y. B. Zeldovich and D. A. Frank-Kamenetskii (in Russian), *Doklady Akademy Nauk USSR*, Vol. 19, p. 693, 1938.
8. Y. B. Zeldovich and N. Semenov, *Journal of Experimental and Theoretical Physics (USSR)* (in Russian), Vol. 10, p. 1116, 1940.
9. N. N. Semenov, "Thermal Theory of Combustion and Explosion. III—Theory of Normal Flame Propagation" (in Russian), *Progress in Physical Sciences (USSR)*, Vol. 24, p. 433, 1940. Translated as NACA TM 1026, September 1942.
10. Y. B. Zeldovich, "Theory of Flame Propagation" (in Russian), *Journal of Physical Chemistry (USSR)*, Vol. 22, p. 27, 1948. Translated as NACA TM 1282, June 1951.
11. D. B. Spalding, "A Theory of Inflammability Limits and Flame Quenching," *Proceedings of the Royal Society of London*, Vol. 240A, pp. 83–100, 1957.
12. A. H. Shapiro and W. R. Hawthorne, "The Mechanics and Thermodynamics of Steady One-Dimensional Gas Flow," *Journal of Applied Mechanics*, Vol. 14, No. 4, pp. A-317 to A-336, December 1947.
13. A. H. Shapiro, *The Dynamics and Thermodynamics of Compressible Fluid Flow*, Vol. 1, Chap. 8, Ronald Press, New York, 1953.
14. E. W. Beans, "Computer Solution to Generalized One-Dimensional Flow," *Journal of Space-craft and Rockets*, Vol. 7, No. 12, pp. 1460–1464, December 1970.
15. B. Carnahan, H. A. Luther, and J. O. Wilkes, *Applied Numerical Methods*, Wiley, New York, 1969.
16. S. D. Conte and C. de Boor, *Elementary Numerical Analysis*, McGraw-Hill, New York, 1972.

PROBLEMS

1. Show that for simple mass addition normal to the main stream in a constant-area solid propellant grain

$$P_o = p + \rho V^2$$

where P_o is the pressure at the head end of the grain.

2. The solid propellant grain described in Example 9.1 develops several longitudinal cracks along the grain that increase the burning surface area by 50 percent. Calculate the quantities requested in Example 9.1 for the cracked grain.

3. A piece of the igniter hardware lodges in the throat section of the solid propellant rocket

motor described in Example 9.1, effectively decreasing the throat area by 50 percent. Calculate the quantities requested in Example 9.1 for the rocket motor with the blocked throat.

4. Example 9.1 presents an analysis of a solid propellant rocket motor with a constant linear burning rate. The linear burning rate of a solid propellant is actually a function of the local static pressure (see equation 9.30). The resulting analysis, however, becomes extremely complicated. A compromise procedure is based on the assumption that the linear burning rate is a function of the head-end static pressure. Thus,

$$\dot{r} = cP_o{}^n$$

Work Example 9.1 based on this linear burning rate model, where $c = 0.0000100$, $n = 0.5$, P_o is in newtons per square meter, and $\dot{r}$ is in meters per second.

5. Work Example 9.1 employing the linear burning rate model specified by equation 9.30, with $k = 0$, $c = 0.0000100$, $n = 0.5$, p in newtons per square meter, and $\dot{r}$ in meters per second.

6. Estimate the detonation velocity V_1 for the premixed gaseous mixture comprising $2H_2 + O_2$, neglecting the effects of dissociation. The initial temperature and pressure of the reactants are $t_1 = 298.15$ K and $p_1 = 1$ atm, respectively.

7. Estimate the detonation velocity V_1 for the premixed gaseous mixture comprising $2H_2 + O_2 + 4H_2$, neglecting the effects of dissociation. The initial temperature and pressure of the reactants are $t_1 = 298.15$ K and $p_1 = 1$ atm, respectively.

8. Work Problem 6 accounting for the effects of dissociation (see Section 14–4).

9. Work Problem 7 accounting for the effects of dissociation (see Section 14–4).

10. Continue Example 9.3 up to the location where $M = 0.10$. Calculate the property ratios p_2/p_1, t_2/t_1, and V_2/V_1 at that point, and compare the results for an isentropic diffusion to (a) the same area, and (b) the same Mach number.

11. A solid propellant grain consists of a cylindrical passage in which the Mach number increases to the value M_1. To reduce erosive burning effects associated with high velocities, the remainder of the grain is to be contoured to maintain a constant Mach number. Determine the required contour. Consider the same propellant properties specified in Example 9.1. For $M_1 = 0.5$, calculate the area ratio of the grain flow passage required to double the propellant mass flow rate, and the static pressure ratio across the variable area portion of the grain.

12. Isothermal flow in a variable area passage, treated in Section 6–7, is a generalized one-dimensional flow for which an exact solution exists. Consequently, such a flow may be employed as a test case for the generalized one-dimensional flow analysis presented in Section 9–5. Consider a test case in which air enters a conical convergence having a semiamgle of 5 deg. Heat is added through the passage wall to maintain the static temperature constant. The initial properties are $M_1 = 0.20$, $t_1 = 300$ K, and $p_1 = 10 \cdot 10^5$ N/m^2. Employ the analysis presented in Section 6–7 to determine the required heat transfer along the passage, in joules per kilogram, and the corresponding flow properties. With that heat transfer specified, employ the generalized flow analysis presented in Section 9–5, and numerically integrate for the flow properties along the passage. Compare the numerically determined results with the exact values obtained from the isothermal flow exact analysis.

13. Isothermal flow in a constant-area passage with friction, discussed in Section 5–7, is a generalized one-dimensional flow for which an exact solution exists. Consequently, such a flow may be employed as a test case for the generalized one-dimensional flow analysis presented in Section 9–5. As a test case, consider air flowing in a constant-area passage having a friction coefficient $f = 0.005$. Heat transfer through the passage wall maintains the static temperature constant. The flow properties at the passage inlet are $M_1 = 0.10$, $t_1 = 300$ K, and $p_1 = 5 \cdot 10^5$ N/m^2. From the exact analysis presented in Section 5–7, determine the heat transfer along the passage, in joules per kilogram, and the corresponding flow properties. With that heat transfer specified, numerically integrate for the flow properties along the passage, employing the generalized one-dimensional flow analysis presented in Section 9–5. Compare the numerically determined results with the exact values obtained from the isothermal flow exact analysis.

10

general features of the steady multidimensional adiabatic flow of an inviscid compressible fluid

10-1 PRINCIPAL NOTATION FOR CHAPTER 10

a	speed of sound.
$\mathbf{B}$	vector body force per unit mass.
$d\mathbf{A}$	vector differential area.
$\mathbf{F}$	vector force.
g	acceleration because of gravity.
h	specific enthalpy.
H	stagnation enthalpy.
$\mathbf{i, j, k}$	unit vectors in x, y, and z directions, respectively.
p	static pressure.
Q	heat.
$\mathbf{r}$	radius vector.
s	specific entropy.
t	time.
T	static temperature.
u, v, w	velocity components in x, y, and z directions, respectively.
V	$= (u^2 + v^2 + w^2)^{1/2}$, magnitude of velocity.
$\mathbf{V}$	velocity vector.
W	work.

Greek Letters

Γ	$= \oint \mathbf{V} \cdot d\mathbf{l}$, fluid circulation.
δ	0 for planar flow, and 1 for axisymmetric flow.
ζ	$= \nabla \times \mathbf{V} = 2\omega$, vorticity.
ρ	static density.
ϕ	velocity potential function.
ψ	stream function.

ω $= \frac{1}{2}\nabla \times \mathbf{V}$, fluid rotation.

Ω   angular velocity.

Subscripts

r, θ, z velocity component in that direction.

x, y, z, t partial derivative with respect to the subscript.

Other notation and vector identities

δ inexact differential.

∇ vector del operator $= \mathbf{i}\frac{\partial}{\partial x} + \mathbf{j}\frac{\partial}{\partial y} + \mathbf{k}\frac{\partial}{\partial z}$.

∇f gradient of $f = \mathbf{i}\frac{\partial f}{\partial x} + \mathbf{j}\frac{\partial f}{\partial y} + \mathbf{k}\frac{\partial f}{\partial z}$.

$\nabla \cdot \mathbf{V}$ divergence of $\mathbf{V} = \frac{\partial V_x}{\partial x} + \frac{\partial V_y}{\partial y} + \frac{\partial V_z}{\partial z}$.

$\nabla \times \mathbf{V}$ curl of $\mathbf{V} \equiv \mathbf{i}\left(\frac{\partial V_z}{\partial y} - \frac{\partial V_y}{\partial z}\right) + \mathbf{j}\left(\frac{\partial V_x}{\partial z} - \frac{\partial V_z}{\partial x}\right) + \mathbf{k}\left(\frac{\partial V_y}{\partial x} - \frac{\partial V_x}{\partial y}\right)$.

$\frac{D(\ )}{Dt}$ $= \frac{\partial(\ )}{\partial t} + (\mathbf{V} \cdot \nabla)(\ )$, substantial or material derivative, denoting differentiation following the motion of a fluid particle.

$\frac{\partial}{\partial(\ )}$ partial derivative with respect to ( ).

$(\mathbf{V} \cdot \nabla)\mathbf{V}$ $= \nabla\left(\frac{V^2}{2}\right) - \mathbf{V} \times (\nabla \times \mathbf{V})$.

$\nabla \cdot (\rho\mathbf{V})$ $= (\mathbf{V} \cdot \nabla)\rho + \rho\nabla \cdot \mathbf{V}$.

x, y, z Cartesian coordinates.

r, θ, z cylindrical coordinates.

10–2 INTRODUCTION

Chapter 2 is concerned with deriving the governing equations for the flow of a fluid, based on the concept of a rigid nonaccelerating control volume. The equations are derived in both integral and differential form, and are assembled in Tables 2.1 and 2.2, respectively, for ease of reference.

Chapters 3 to 9 are concerned with obtaining solutions to those equations for the special case of steady uniform flow at each cross section of a flow passage, commonly termed *one-dimensional flow*, under conditions where different *driving potentials* cause the changes in the flow properties. In particular, Chapter 3 presents the general features of the steady one-dimensional flow of a compressible fluid.

The present chapter is concerned with the general features of the *steady multidimensional adiabatic inviscid flow of a compressible fluid*.[1-7] The differential forms of the governing (vector) equations are employed for deriving the corresponding equations in the Cartesian and cylindrical coordinate systems. Several general concepts are discussed (rotation, circulation, vorticity, velocity potential, and stream function), and some important theorems are presented (Kelvin's theorem, Crocco's theorem, and the Helmholtz vorticity theorems).

10–3 GOVERNING DIFFERENTIAL EQUATIONS FOR THE STEADY MULTIDIMENSIONAL ADIABATIC INVISCID FLOW OF A COMPRESSIBLE FLUID

From Table 2.2 we obtain the following differential equations governing the flow of a compressible fluid through a rigid nonaccelerating control volume.

Continuity equation

$$\rho_t + \nabla \cdot (\rho \mathbf{V}) = 0 \tag{10.1}$$

Momentum equation

$$\rho \frac{D\mathbf{V}}{Dt} + \nabla p - \rho \mathbf{B} - d\mathbf{F}_{shear} = 0 \tag{10.2}$$

Energy equation

$$\delta \dot{W}_{shaft} + \delta \dot{W}_{shear} - \delta \dot{Q} + \rho \frac{D}{Dt}\left(h + \frac{V^2}{2} + gz\right) - p_t = 0 \tag{10.3}$$

Entropy equation *

$$\rho \frac{Ds}{Dt} \geqq \delta \dot{Q}/T \tag{10.4}$$

Consider the steady ($\partial/\partial t = 0$) adiabatic ($\delta \dot{Q} = 0$) inviscid ($d\mathbf{F}_{shear} = \delta \dot{W}_{shear} = 0$) flow of a compressible fluid in the absence of body forces ($\mathbf{B} = gz = 0$) and external work ($\delta W = 0$). It is shown in Chapter 14, that if the flowing fluid is either in chemical equilibrium (i.e., the chemical reaction rates are infinite) or its chemical composition is frozen (i.e., the chemical reaction rates are zero), then all of the flow processes are reversible.

For the steady adiabatic inviscid flow of a compressible fluid in the absence of body forces and external work, the following mathematical conditions apply.

$$\mathbf{V}_t = \rho_t = p_t = \mathbf{B} = g\,dz = d\mathbf{F}_{shear} = \delta \dot{Q} = \delta \dot{W}_{shear} = \delta \dot{W}_{shaft} = 0 \tag{10.5}$$

In addition, for steady flow (see Section 10–1),

$$\frac{D(\ )}{Dt} = (\mathbf{V} \cdot \nabla)(\ ) \tag{10.6}$$

Substituting the conditions listed in equation 10.5 into equations 10.1 to 10.4 transforms that set of equations into the following set. Thus,

$$\nabla \cdot (\rho \mathbf{V}) = 0 \tag{10.7}$$

$$\rho \frac{D\mathbf{V}}{Dt} + \nabla p = 0 \tag{10.8}$$

$$\rho \frac{D}{Dt}\left(h + \frac{V^2}{2}\right) = 0 \tag{10.9}$$

$$\rho \frac{Ds}{Dt} = 0 \tag{10.10}$$

For describing the flow field of a compressible fluid, the equations of state for the latter fluid are necessary in addition to the preceding set of four equations. The general forms of these equations of state for a compressible fluid are as follows.

$$T = T(p,\rho) \tag{10.11}$$

$$h = h(p,\rho) \tag{10.12}$$

---

* In this section, t denotes the time and T denotes the static temperature.

10-3(a) Cartesian Coordinate System

Equations 10.7 to 10.10 are the vector equations governing the steady adiabatic inviscid flow of a compressible fluid in the absence of body forces and external work. A more detailed study of the flow field under the aforementioned restrictions requires that the flow variables be related to a suitable coordinate system. The one most commonly employed is the Cartesian coordinate system, and it is considered first.

Figure 10.1a illustrates a velocity vector **V** in the Cartesian coordinate system. Let the point $P(x,y,z)$ denote an arbitrarily selected point in the flow field. The vector velocity for the fluid at the point $P(x,y,z)$ is $\mathbf{V}(x,y,z)$, which has the velocity components u, v, and w parallel to the x, y, and z axes, respectively. Let the *unit vectors* parallel to those respective axes be denoted by **i**, **j**, and **k**.

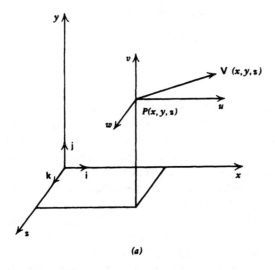

(a)

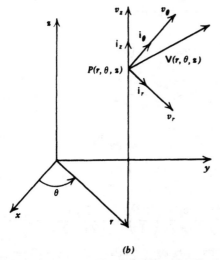

(b)

Figure 10.1 The velocity vector in the Cartesian and cylindrical coordinate systems. (a) Cartesian coordinates. (b) Cylindrical coordinates.

Equations 10.7 to 10.10, when expressed in Cartesian coordinates, become

$$(\rho u)_x + (\rho v)_y + (\rho w)_z = 0 \tag{10.13}$$

$$uu_x + vu_y + wu_z + \frac{1}{\rho} p_x = 0 \tag{10.14}$$

$$uv_x + vv_y + wv_z + \frac{1}{\rho} p_y = 0 \tag{10.15}$$

$$uw_x + vw_y + ww_z + \frac{1}{\rho} p_z = 0 \tag{10.16}$$

$$uh_x + vh_y + wh_z + u\left(\frac{V^2}{2}\right)_x + v\left(\frac{V^2}{2}\right)_y + w\left(\frac{V^2}{2}\right)_z = 0 \tag{10.17}$$

$$us_x + vs_y + ws_z = 0 \tag{10.18}$$

where the notation $\partial(\rho u)/\partial x = (\rho u)_x$, etc., is employed.

10-3(b) Cylindrical Coordinate System

The flow inside a circular duct, termed an *internal* flow, and the flow around a body of revolution, termed an *external* flow, possess axial symmetry because of the circular cross sections of the solid boundaries. Figure 10.1*b* illustrates the cylindrical coordinate system employed for studying such types of flows. The coordinates are r, θ, and z, where r is the radius to the projection of the arbitrary point $P(x,y,z)$ on the xy plane, z is its distance above the xy plane, and θ is the angle in the xy plane between the radius r and the x axis. The unit vectors parallel to the *cylindrical coordinates* r, θ, and z are denoted by $\mathbf{i}_r$, $\mathbf{i}_\theta$, and $\mathbf{i}_z$, respectively. The velocity vector for the fluid at the point $P(r,\theta,z)$ is $\mathbf{V}(r,\theta,z)$. The components of $\mathbf{V}$ with respect to the cylindrical coordinates are accordingly v_r, v_θ, and v_z, respectively.

Equations 10.7 to 10.10 may be expressed in the cylindrical coordinate system by applying the vector operators to that coordinate system. In the following equations, partial differentiation is written out explicitly to avoid confusion, and subscripts are employed for denoting the velocity components. The vector equations 10.7 to 10.10, when expressed in cylindrical coordinates, are transformed to

$$\frac{1}{r}\frac{\partial}{\partial r}(r\rho v_r) + \frac{1}{r}\frac{\partial}{\partial \theta}(\rho v_\theta) + \frac{\partial}{\partial z}(\rho v_z) = 0 \tag{10.19}$$

$$v_r\frac{\partial v_r}{\partial r} + \frac{v_\theta}{r}\frac{\partial v_r}{\partial \theta} - \frac{v_\theta^2}{r} + v_z\frac{\partial v_r}{\partial z} + \frac{1}{\rho}\frac{\partial p}{\partial r} = 0 \tag{10.20}$$

$$v_r\frac{\partial v_\theta}{\partial r} + \frac{v_\theta}{r}\frac{\partial v_\theta}{\partial \theta} + \frac{v_r v_\theta}{r} + v_z\frac{\partial v_\theta}{\partial z} + \frac{1}{\rho r}\frac{\partial p}{\partial \theta} = 0 \tag{10.21}$$

$$v_r\frac{\partial v_z}{\partial r} + \frac{v_\theta}{r}\frac{\partial v_z}{\partial \theta} + v_z\frac{\partial v_z}{\partial z} + \frac{1}{\rho}\frac{\partial p}{\partial z} = 0 \tag{10.22}$$

$$v_r\frac{\partial h}{\partial r} + \frac{v_\theta}{r}\frac{\partial h}{\partial \theta} + v_z\frac{\partial h}{\partial z} + v_r\frac{\partial}{\partial r}\left(\frac{V^2}{2}\right) + \frac{v_\theta}{r}\frac{\partial}{\partial \theta}\left(\frac{V^2}{2}\right) + v_z\frac{\partial}{\partial z}\left(\frac{V^2}{2}\right) = 0 \tag{10.23}$$

$$v_r\frac{\partial s}{\partial r} + \frac{v_\theta}{r}\frac{\partial s}{\partial \theta} + v_z\frac{\partial s}{\partial z} = 0 \tag{10.24}$$

10–3(c) Two-Dimensional Planar and Axisymmetric Flows

For a two-dimensional *planar* flow, the velocity component $w = 0$ and all of the partial derivatives $\partial(\ )/\partial z = 0$. Similarly, for a two-dimensional *axisymmetric* flow, the velocity component $v_\theta = 0$ and $\partial(\ )/\partial\theta = 0$. The corresponding coordinate systems are illustrated in Fig. 10.2. Substituting the conditions $w = \partial(\ )/\partial z = 0$ into equations 10.13 to 10.18, we obtain a set of equations for a two-dimensional planar flow expressed in Cartesian coordinates. In an analogous manner, substituting $v_\theta = \partial(\ )/\partial\theta = 0$ into equations 10.19 to 10.24 gives the corresponding equations for a two-dimensional axisymmetric flow, expressed in cylindrical coordinates.

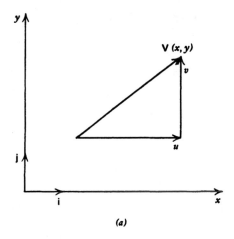

(a)

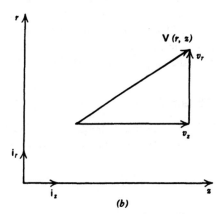

(b)

Figure 10.2 Two-dimensional coordinate systems.
(a) Planar. (b) Axisymmetric.

The latter set of equations may be transformed to the notation of the Cartesian coordinate system by means of the following transformation equations. Thus,

$$v_r = v, \qquad v_z = u, \qquad r = y, \qquad \text{and} \qquad z = x \tag{10.25}$$

If the set of equations for a two-dimensional planar flow (equations 10.13 to 10.18) are compared with the transformed set of equations for a two-dimensional axisymmetric flow expressed in Cartesian coordinate notation, it is seen that those two sets of equations are identical except for the additional term $\rho v/y$ in the continuity equation for the axisymmetric flow. Those two sets of equations are exactly identical if the term $\delta \rho v/y$ is added to the continuity equation, where $\delta = 0$ for the two-dimensional planar flow and $\delta = 1$ for the two-dimensional axisymmetric flow.

Consequently, the following single set of equations, in Cartesian coordinate notation, is applicable to both two-dimensional planar flow and two-dimensional axisymmetric flow.

Continuity equation

$$(\rho u)_x + (\rho v)_y + \delta \frac{\rho v}{y} = 0 \tag{10.26}$$

where $\delta = 0$ for planar flow, and $\delta = 1$ for axisymmetric flow.

Momentum equations

$$uu_x + vu_y + \frac{1}{\rho} p_x = 0 \tag{10.27}$$

$$uv_x + vv_y + \frac{1}{\rho} p_y = 0 \tag{10.28}$$

Energy equation

$$uh_x + vh_y + u \left(\frac{V^2}{2} \right)_x + v \left(\frac{V^2}{2} \right)_y = 0 \tag{10.29}$$

Entropy equation

$$us_x + vs_y = 0 \tag{10.30}$$

The flows governed by the set of equations 10.26 to 10.30 are termed two-dimensional flows because only two coordinates, x and y, are needed for specifying the flow field. For clarity, the two types of two-dimensional flow discussed above will be referred to hereafter, as has been done in the foregoing, as *two-dimensional planar flow* and *two-dimensional axisymmetric flow*; the former refers to a flow in the xy plane (Fig. 10.2a), and the latter to a flow in the rz plane (Fig. 10.2b).

10–3(d) Streamline, Pathline, Fluid Line, and Stream Tube

By definition, a *streamline* is a line that at all of its points is tangent to the velocity vector for the flowing fluid, as illustrated in Fig. 10.3. A *pathline* is *defined* as the

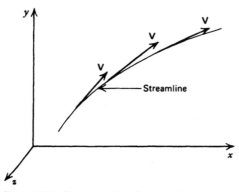

Figure 10.3 Characteristics of a streamline.

locus of the spatial positions traversed by a fluid particle as it moves in the flow field. For a steady flow the streamlines and pathlines are identical.

Figure 10.4 illustrates a streamline passing through the point P in the Cartesian coordinate system. At the point P, a differential element of the streamline, denoted by $d\mathbf{l}$, is given by

$$d\mathbf{l} = \mathbf{i}\, dx + \mathbf{j}\, dy + \mathbf{k}\, dz$$

The vector $d\mathbf{l}$ is a vector tangent to the velocity vector $\mathbf{V}$, at the point P. The components of $\mathbf{V}$ in the directions parallel to the x, y, and z coordinate axes are u, v, and w, respectively.

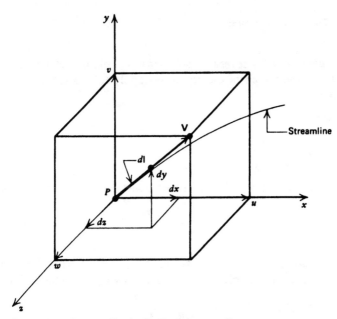

Figure 10.4 A streamline in the Cartesian coordinate system.

If a tetrahedron is drawn having the velocity vector $\mathbf{V}$ as its diagonal and its faces or sides parallel to the Cartesian coordinate planes, as illustrated in Fig. 10.4, then it is seen that

$$\frac{dx}{u} = \frac{dy}{v} = \frac{dz}{w} \tag{10.31a}$$

Rearranging equation 10.31a yields

$$\frac{dy}{dx} = \frac{v}{u}, \qquad \frac{dy}{dz} = \frac{v}{w}, \qquad \frac{dx}{dz} = \frac{u}{w} \tag{10.31b}$$

Equations 10.31a and 10.31b are the equations for a streamline in Cartesian coordinates.

By definition, a *fluid line* is a line passing through a set of identifiable fluid particles.

Figure 10.5 illustrates a closed curve C in the fluid, which is a fluid line. All of the streamlines passing through the curve C form a *stream tube*. Since all of the generators of the stream tube are streamlines, no mass may cross the surface of a stream tube. A stream tube of infinitesimal cross section is called a *stream filament*.

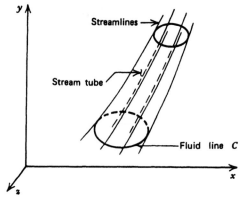

Figure 10.5 Streamlines forming a stream tube.

Example 10.1. The equations for several two-dimensional planar velocity fields are presented below. Determine which of the equations apply to possible fields for an incompressible fluid.

(a) $\quad \mathbf{V} = -\mathbf{i}\Omega y + \mathbf{j}\Omega x$

(b) $\quad \mathbf{V} = -\mathbf{i}\dfrac{\Gamma y}{x^2 + y^2} + \mathbf{j}\dfrac{\Gamma x}{x^2 + y^2}$

where Γ denotes the circulation, which is assumed constant [see Section 10–4(a)].

(c) $\quad \mathbf{V} = \mathbf{i}xe^x + \mathbf{j}ye^y$

Solution

For an incompressible flow, $\rho = $ constant throughout the flow field. Hence, for a two-dimensional planar incompressible flow, the continuity equation (equation 10.7) reduces to

$$\nabla \cdot \mathbf{V} = \frac{\partial u}{\partial x} + \frac{\partial v}{\partial y} = u_x + v_y = 0$$

where u and v denote the x and y components of $\mathbf{V}$, respectively.
(a) For this case, $u = -\Omega y$ and $v = \Omega x$. Hence,

$$\nabla \cdot \mathbf{V} = u_x + v_y = 0 - 0 = 0$$

Consequently, equation (a) represents a possible incompressible flow.
(b) In this case,

$$u = -\frac{\Gamma y}{(x^2 + y^2)} \quad \text{and} \quad v = \frac{\Gamma x}{(x^2 + y^2)}$$

where Γ is a constant. Hence,

$$\nabla \cdot \mathbf{V} = u_x + v_y = -\frac{2xy\Gamma}{(x^2 + y^2)^2} + \frac{2xy\Gamma}{(x^2 + y^2)^2} = 0$$

Equation (b), therefore, represents a possible incompressible flow.
(c) For equation (c), $u = xe^x$ and $v = ye^y$. Hence,

$$\nabla \cdot \mathbf{V} = u_x + v_y = xe^x + e^x + ye^y + e^y \neq 0$$

Consequently, equation (c) does not represent a possible incompressible flow. If equation (c) represents a velocity field, it would be that for a compressible fluid.

Summarizing, equation (a) represents the velocity field for a *forced vortex* [see Section 10-4(e)], and equation (b) represents the velocity field for a *free vortex* [see Section 10-4(d)].

10-4 CIRCULATION, ROTATION, AND VORTICITY

Many of the general features of a multidimensional flow may be studied conveniently in terms of the parameters known as *circulation, rotation,* and *vorticity*. In this section, those parameters are defined, and some examples illustrating their use are presented.

10-4(a) Circulation

Figure 10.6 illustrates a simple closed spatial curve C in a flow field. Each element of arc on curve C may be considered to be an infinitesimal vector $d\mathbf{l}$ having the magnitude $d\ell$, and the direction of the tangent to the curve C. By *definition*, the *circulation* Γ around the closed curve C is given by

$$\Gamma \equiv \oint_C \mathbf{V} \cdot d\mathbf{l} = \oint_C V \, d\ell \, \cos \alpha \tag{10.32}$$

where α is the angle between $\mathbf{V}$ and $d\mathbf{l}$. By convention, the *positive* direction for the integration of the scalar product $\mathbf{V} \cdot d\mathbf{l}$ is *counterclockwise*.

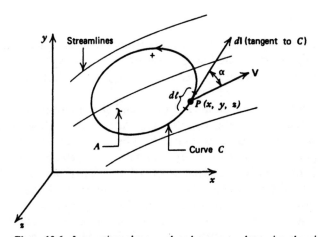

Figure 10.6 Integration along a closed curve to determine the circulation.

To illustrate the significance of the circulation Γ, consider a two-dimensional flow where both the closed curve C and the velocity vector $\mathbf{V}$ lie in the xy plane. For $\mathbf{V}$ and $d\mathbf{l}$, we may write

$$\mathbf{V} = \mathbf{i}u + \mathbf{j}v \qquad \text{and} \qquad d\mathbf{l} = \mathbf{i} \, dx + \mathbf{j} \, dy$$

Forming the scalar product $\mathbf{V} \cdot d\mathbf{l}$, we obtain for the circulation Γ,

$$\Gamma = \oint_C (u \, dx + v \, dy) \tag{10.33}$$

The circulation Γ may be expressed in terms of an integral over the surface A bounded by the closed curve C by applying *Stokes' theorem*. The latter states that if there is a surface A having the closed curve C for its perimeter, then

$$\oint_C \mathbf{V} \cdot d\mathbf{l} = \int_A (\nabla \times \mathbf{V}) \cdot d\mathbf{A} \qquad (Stokes'\ theorem) \qquad (10.34)$$

Applying Stokes' theorem to equation 10.32 yields

$$\Gamma = \oint_C \mathbf{V} \cdot d\mathbf{l} = \int_A (\nabla \times \mathbf{V}) \cdot d\mathbf{A} \qquad (10.35)$$

where $\nabla \times \mathbf{V} = \operatorname{curl} \mathbf{V} = \mathbf{i}(w_y - v_z) + \mathbf{j}(u_z - w_x) + \mathbf{k}(v_x - u_y)$. Equation 10.35 transforms the line integral around the closed curve C into a surface integral over the surface A enclosed by the curve C.

For a *two-dimensional flow* in the xy plane, $\mathbf{V} = \mathbf{i}u + \mathbf{j}v$, and

$$\nabla \times \mathbf{V} = \operatorname{curl} \mathbf{V} = \mathbf{k}(v_x - u_y) \qquad (10.36)$$

Let dA denote a differential element of the area A enclosed by the curve C. Since dA lies entirely within the xy plane, $d\mathbf{A} = \mathbf{k}\, dx\, dy$. Substituting for $\mathbf{V}$ and $d\mathbf{A}$ into equation 10.35, we obtain

$$\Gamma = \int_A (v_x - u_y)\, dx\, dy \qquad (10.37)$$

Taking the limit of Γ as A approaches 0, equation 10.37 yields

$$\lim_{A \to 0} \frac{\Gamma}{A} = \frac{d\Gamma}{dA} = (v_x - u_y) \qquad (10.38)$$

10-4(b) Rotation

The *fluid rotation* at a point is *defined* as the instantaneous average angular velocity of two mutually perpendicular *fluid lines* [see Section 10–3(d)] passing through that point. That definition of fluid rotation is a generalization of the definition of solid body rotation.

For the latter, if P denotes a point on a solid body and Q is an arbitrary neighboring point, then when the body rotates with the angular velocity ω about an axis through P, the curl of the velocity vector at point Q will be identical to that at point P, because all of the particles of a solid body maintain fixed relationships with respect to each other. Consequently, the specification of the angular velocity of one line on the solid body completely specifies the angular velocities of all lines.

This is not the case, however, for a fluid or a deformable body. In a flowing fluid all of the fluid lines are capable of becoming displaced relative to one another, and such displacements may occur. Consequently, in the case of a fluid, we are concerned with the *mean* or *average* angular velocity of a fluid element about a point, and that mean angular velocity is called the *rotation of the fluid* or, briefly, the *rotation*; it is denoted by the vector $\boldsymbol{\omega}$.

Figure 10.7 illustrates schematically the motion of a fluid particle in the xy plane. At the time t, the centroid of the fluid particle is located at the point O and its Cartesian velocity components, with respect to the x and y axes, are denoted by u and v, respectively. The lines OA and OB, in Figure 10.7, are two mutually perpendicular *fluid lines* that emanate from the fluid particle at point O. As a consequence of the motion of the fluid particle and the fluid lines, at the later time $t + dt$, the points O, A, and B have moved upward and to the right, as illustrated in Fig. 10.7, to new locations. Because of the *velocity gradients* in the fluid, the fluid particles at the points A and B move at different speeds than does the fluid particle at point O. Consequently, the fluid lines OA and OB make the angles $d\alpha$ and $-d\beta$ with respect

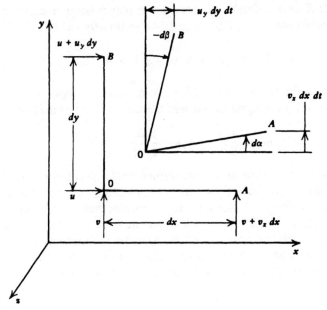

Figure 10.7 Schematic illustration of the rotation of a fluid.

to the x and y axes, respectively. The angle $d\beta$ must be defined as being *negative* because a positive value for $u_y \, dy$ results in a clockwise rotation of OB; that is, in the negative angular direction.

By *definition*, the component of fluid rotation ω_z in a plane perpendicular to the z axis is given by

$$\omega_z \equiv \frac{1}{2}\left(\frac{d\alpha}{dt} + \frac{d\beta}{dt}\right) \qquad (10.39)$$

where the subscript Z attached to ω does not signify partial differentiation, but that the rotation ω is measured in a plane perpendicular to the z axis. For an infinitesimal time interval dt, the angles $d\alpha$ and $d\beta$ will be so small that $\tan(d\alpha) \cong d\alpha$ and $\tan(d\beta) \cong d\beta$. From Figure 10.7, it is seen that

$$d\alpha = \frac{v_x \, dx \, dt}{dx} = v_x \, dt \qquad (10.40)$$

$$d\beta = \frac{-u_y \, dy \, dt}{dy} = -u_y \, dt \qquad (10.41)$$

Substituting equations 10.40 and 10.41 into equation 10.39, we obtain

$$\omega_z = \frac{1}{2}(v_x - u_y) \qquad (10.42)$$

In a similar manner the components of the *rotation vector* **ω** may be obtained in the yz and xz planes. The results are

$$\omega_X = \frac{1}{2}(w_y - v_z) \qquad (10.43)$$

$$\omega_Y = \frac{1}{2}(u_z - w_x) \qquad (10.44)$$

where the subscripts X and Y denote the rotation in planes perpendicular to those directions. Adding equations 10.42, 10.43, and 10.44 vectorially yields

$$\boldsymbol{\omega} = \mathbf{i}\omega_X + \mathbf{j}\omega_Y + \mathbf{k}\omega_Z = \frac{1}{2}\nabla \times \mathbf{V} \tag{10.45}$$

By comparing equations 10.38 and 10.42 for two-dimensional flow, it is apparent that the mean fluid rotation is equal to one half the circulation per unit area. Thus,

$$\omega_Z = \frac{1}{2}(v_x - u_y) = \frac{1}{2}\frac{d\Gamma}{dA} \tag{10.46}$$

Flows in which the fluid rotation $\boldsymbol{\omega} = 0$ are termed *irrotational flows*. Figure 10.8a illustrates the rotation of two fluid lines with time for the case where the average rotation of two initially perpendicular fluid lines remains zero so that the flow is *irrotational*. The situation is one where each of the fluid lines is distorted so that there is a change in the relative positions of the particles without any net rotation being produced. Figure 10.8b, on the other hand, illustrates the case where the two fluid lines experience a net fluid rotation with time; such a flow is said to be *rotational*.

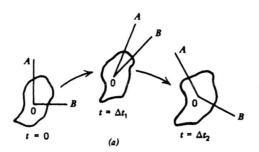

(a)

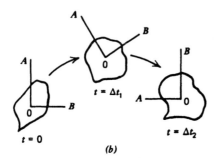

(b)

Figure 10.8   Schematic illustration of irrotational and rotational motions in a fluid. (a) Irrotational motion. (b) Rotational motion.

10–4(c)   Vorticity and the Vortex Tube

The vector $\nabla \times \mathbf{V}$ defines, for each value of t and at each point in a flow field, a vector that is equal to $2\boldsymbol{\omega}$. *By definition*, $\nabla \times \mathbf{V}$ is the *vorticity vector* or, briefly, the *vorticity*, and is denoted by ζ. Hence,

$$\zeta = \nabla \times \mathbf{V} = 2\boldsymbol{\omega} \tag{10.47}$$

In Cartesian coordinates,

$$\zeta = \mathbf{i}\zeta_X + \mathbf{j}\zeta_Y + \mathbf{k}\zeta_Z = \mathbf{i}(w_y - v_z) + \mathbf{j}(u_z - w_x) + \mathbf{k}(v_x - u_y) \qquad (10.48)$$

where the subscripts X, Y, and Z denote the component of ζ in the x, y, and z directions, respectively. From equations 10.46 and 10.48, it is seen that the vorticity for a two-dimensional flow is equal to the circulation per unit area. Thus,

$$\zeta_z = \frac{d\Gamma}{dA} = 2\omega_z \qquad (10.49)$$

By definition, a *vortex line* is a line in the fluid that at all of its points is tangent to the vorticity vector. Figure 10.9 illustrates a vortex line. Along a vortex line the components ω_X, ω_Y, and ω_Z of the fluid rotation vector $\boldsymbol{\omega}$, defined by equation 10.45, are related to the corresponding components dx, dy, and dz of the tangent vector to the vortex line by the relationships

$$\frac{dx}{\omega_X} = \frac{dy}{\omega_Y} = \frac{dz}{\omega_Z} \qquad (10.50)$$

Equation 10.50 is the equation for a vortex line in Cartesian coordinates.

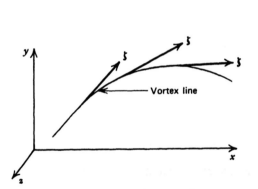

Figure 10.9   Characteristics of a vortex line.

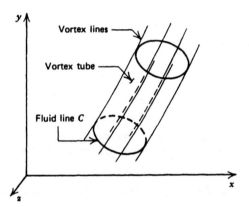

Figure 10.10   Vortex lines forming a vortex tube.

Figure 10.10 illustrates a closed curve C in the fluid, which is a fluid line. All of the vortex lines passing through curve C form a *vortex tube*. A vortex tube of infinitesimal cross section is called a *vortex filament*.

The foregoing shows that the three parameters, *circulation*, *rotation*, and *vorticity*, are uniquely related. Consequently, knowing any one of the three determines the other two. In Sections 10–6, 10–7, and 10–8, some general theorems are developed relating the circulation, rotation, and vorticity to the other flow properties.

10-4(d) The Free Vortex

A particularly simple example of irrotational motion is afforded by the free vortex; that is, a circular flow where the magnitude of the fluid velocity is inversely proportional to the distance from the center of the vortex. Figure 10.11 illustrates the characteristics of a free vortex, which is a reasonable approximation to the bathtub drain vortex, the flow in a spinning internal-burning tubular solid propellant rocket motor, and the tornado vortex.

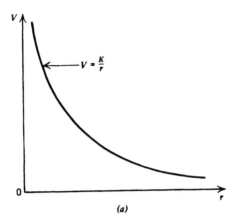

(a)

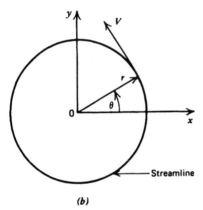

(b)

Figure 10.11 Flow characteristics of a free vortex. (a) Velocity distribution. (b) Streamline pattern.

The magnitude of the fluid velocity in a free vortex is given by

$$V = \frac{K}{r} \tag{10.51}$$

Figure 10.11a illustrates the velocity distribution in a free vortex. Equation 10.51 is *singular* at the origin, where it predicts an infinite velocity, which is, of course, unattainable physically. From Table 3.4, it is apparent that even the maximum isentropic speed V'_{max} for a gas, which is unattainable physically, has a finite value. In the free vortex of a real fluid, the central region of the vortex generally contains some low-speed fluid, which frequently behaves as a *forced vortex* [see Section 10–4(e)]. In the regions at a distance from the center, however, the free vortex model is a good approximation to a free vortex in a real fluid.

Figure 10.11b illustrates a streamline in the flow field of a free vortex. The circulation around the streamline is

$$\Gamma = \oint_C \mathbf{V} \cdot d\mathbf{l} = \int_0^{2\pi} \frac{K}{r}(r\, d\theta) = 2\pi K = \text{constant} \tag{10.52}$$

From equation 10.46, the magnitude of the rotation of the fluid ω_z in the xy plane at the *origin* is

$$\omega_z = \frac{1}{2}\frac{d\Gamma}{dA} = \lim_{r \to 0}\frac{\Gamma}{2A} = \lim_{r \to 0}\frac{2\pi K}{2\pi r^2} = \infty \tag{10.53}$$

Equation 10.53 indicates that ω_z approaches infinity as the radius of the curve approaches zero. As pointed out above, that result is physically meaningless.

Figure 10.12 illustrates a closed curve located entirely within the region where the fluid velocity V is finite and smaller than the maximum isentropic speed. The flow model is, therefore, realistic. Consider two streamlines of a free vortex corresponding to the velocities V_1 and V_2; the radii to V_1 and V_2 are denoted by R_1 and R_2, respectively, and they make the angles θ_1 and θ_2, respectively, with the x axis. By extending R_1 to the streamline corresponding to V_2, the closed curve $abcd$ is formed. The direction for positive integration around that closed curve is indicated in Fig. 10.12.

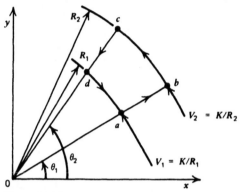

Figure 10.12 A closed curve ($abcd$) that does not enclose the origin of a free vortex.

The circulation Γ around the closed curve $abcd$ is (see equation 10.32)

$$\Gamma = \oint_C \mathbf{V} \cdot d\mathbf{l} = V_2[R_2(\theta_2 - \theta_1)] + V_1[R_1(\theta_1 - \theta_2)] \tag{10.54}$$

Introducing equation 10.51 into equation 10.54 yields

$$\Gamma = 0 \tag{10.55}$$

Since R_1, R_2, θ_1, and θ_2 are arbitrary, equation 10.55 is a general result that applies to any closed curve not enclosing the origin. Hence, the rotation for this case is

$$\omega_z = \frac{1}{2}\frac{d\Gamma}{dA} = 0 \tag{10.56}$$

Equation 10.56 shows that the flow external to the center of a free vortex is an irrotational flow.

10-4(e) The Forced Vortex

A simple example of a rotational flow is the *forced vortex*, which is a circular flow where the magnitude of the fluid velocity at a point is directly proportional to the distance of the point from the center of the vortex. Figure 10.13 illustrates the characteristics of a forced vortex. A practical example of a forced vortex is the flow field that occurs when an end-burning solid propellant rocket motor is rotated about its longitudinal axis. As discussed in Section 10-4(d), the flow in the vicinity of the core of a free vortex quite frequently behaves as a forced vortex.

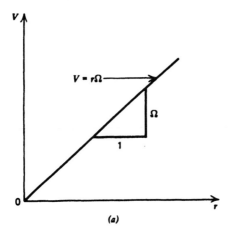

(a)

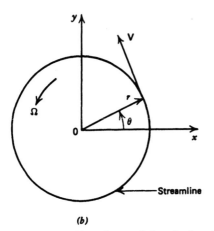

(b)

Figure 10.13 Flow characteristics of a forced vortex.
(a) Velocity distribution. (b) Streamline pattern.

The magnitude of the *tangential velocity* of the fluid, denoted by V, is given by

$$V = r\Omega \tag{10.57}$$

where Ω is the angular velocity of the forced vortex. Figure 10.13a illustrates the velocity distribution in a forced vortex.

Figure 10.13b illustrates a closed circular streamline at the radius r in a forced vortex; the center of the circular streamline, denoted by O, is coincident with the origin of the coordinate axes. The angle between the radius r and the x axis is θ, and V is the vector velocity along the streamline. The circulation is given by

$$\Gamma = \oint_C \mathbf{V} \cdot d\mathbf{l} = \int_0^{2\pi} (r\Omega)(r\,d\theta) \tag{10.58}$$

Hence,

$$\Gamma = 2\pi r^2 \Omega \tag{10.59}$$

Accordingly, from equation 10.46, the fluid rotation is

$$\omega_z = \frac{1}{2}\frac{d\Gamma}{dA} = \lim_{r \to 0} \frac{\Gamma}{2A} = \lim_{r \to 0} \frac{2\pi r^2 \Omega}{2\pi r^2} = \Omega = \text{constant} \tag{10.60}$$

Hence, the flow is *rotational*. In fact, the rotation is constant throughout the fluid; that is, it rotates as a solid body. Consequently, the forced vortex is sometimes called the *solid body vortex*.

Example 10.2.   Consider the equations for the possible velocity fields presented in Example 10.1. Which of them apply to possible irrotational flow fields?

(a) $\mathbf{V} = -\mathbf{i}\Omega y + \mathbf{j}\Omega x$

(b) $\mathbf{V} = \dfrac{-\mathbf{i}\Gamma y}{(x^2 + y^2)} + \dfrac{\mathbf{j}\Gamma x}{(x^2 + y^2)}$

(c) $\mathbf{V} = \mathbf{i}x e^x + \mathbf{j}y e^y$

Solution

For an irrotational flow field, $\omega = 0$. For two-dimensional flow, this requires that

$$\omega_z = u_y - v_x = 0$$

(a) $\omega_z = -\Omega - \Omega = -2\Omega$

Hence, (a) is not a possible irrotational flow field.

(b) $\omega_z = \dfrac{\Gamma[-(x^2 + y^2) + 2y^2]}{(x^2 + y^2)^2} - \dfrac{\Gamma[(x^2 + y^2) - 2x^2]}{(x^2 + y^2)^2} = \dfrac{0}{(x^2 + y^2)^2}$

Thus, (b) is a possible irrotational flow everywhere except at the origin, which is a singular point in the flow field.

(c) $\omega_z = 0 + 0 = 0$

Equation (c) is a possible irrotational flow. However, as shown in Example 10.1, for the flow to exist, the fluid would have to be compressible to satisfy the continuity equation.

10-5 EULER'S MOMENTUM EQUATION FOR THE STEADY ADIABATIC INVISCID FLOW OF A COMPRESSIBLE FLUID

For the type of flow under consideration the mathematical restrictions are those pertaining to equations 10.7 to 10.10, which are repeated here for convenience. Thus,

$$\mathbf{V}_t = \rho_t = p_t = \mathbf{B} = g\,dz = d\mathbf{F}_{shear} = \delta Q = \delta W_{shear} = \delta W_{shaft} = 0 \quad (10.61)$$

The momentum equation subject to the above restrictions is equation 10.8. Thus, noting that for steady flow, $D\mathbf{V}/Dt = (\mathbf{V} \cdot \nabla)\mathbf{V}$ (see equation 10.6), equation 10.8 becomes

$$(\mathbf{V} \cdot \nabla)\mathbf{V} + \frac{1}{\rho}\nabla p = 0 \quad (10.62)$$

Equation 10.62 is *Euler's (momentum) equation* for steady flow.

Introduce the following vector identity (see Vector Identities in Section 10–1).

$$(\mathbf{V} \cdot \nabla)\mathbf{V} = \nabla\left(\frac{V^2}{2}\right) - \mathbf{V} \times (\nabla \times \mathbf{V}) \quad (10.63)$$

Substituting equation 10.63 in equation 10.62 gives

$$\mathbf{V} \times (\nabla \times \mathbf{V}) = \frac{1}{\rho}\nabla p + \nabla\left(\frac{V^2}{2}\right) \quad (10.64)$$

Equation 10.64 is the sum of several vectors in space, as illustrated schematically in Fig. 10.14. A scalar equation may be obtained by summing the components of equation 10.64 in the direction of the arbitrary vector $d\mathbf{r} = \mathbf{i}\, dx + \mathbf{j}\, dy + \mathbf{k}\, dz$. Thus,

$$d\mathbf{r} \cdot (\text{equation } 10.64) = 0 \tag{10.65}$$

Performing the scalar product indicated in equation 10.65 and expanding the result in Cartesian coordinates yields

$$[(w\, dy - v\, dz)(w_y - v_z) + (u\, dz - w\, dx)(u_z - w_x) + (v\, dx - u\, dy)(v_x - u_y)$$
$$= \frac{1}{\rho}(p_x\, dx + p_y\, dy + p_z\, dz) + \left[\left(\frac{V^2}{2}\right)_x dx + \left(\frac{V^2}{2}\right)_y dy + \left(\frac{V^2}{2}\right)_z dz\right] \tag{10.66}$$

where, as before $w_y = \partial w/\partial y$, etc. Referring to equations 10.42, 10.43, and 10.44, it is seen that the left side of equation 10.66 contains the terms $(w_y - v_z) = 2\omega_x$, etc. Also, the two terms on the right side are equal to $(1/\rho)\, dp$ and $d(V^2/2)$, respectively. Substituting the latter equivalent expressions into equation 10.66, we obtain

$$2[(w\, dy - v\, dz)\omega_X + (u\, dz - w\, dx)\omega_Y + (v\, dx - u\, dy)\omega_Z]$$
$$= \frac{1}{\rho}\, dp + d\left(\frac{V^2}{2}\right) \tag{10.67}$$

Equation 10.67 is valid for any direction $d\mathbf{r}$ within the flow field.

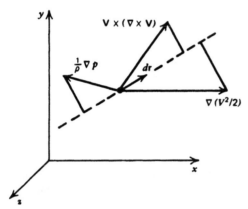

Figure 10.14 Vector representation of Euler's equation for the steady adiabatic inviscid flow of a compressible fluid.

10–5(a) Steady Flow along a Streamline

Substituting the equation for a streamline, that is, equation 10.31b, into equation 10.67, gives

$$\frac{1}{\rho}\, dp + d\left(\frac{V^2}{2}\right) = 0 \tag{10.68}$$

Equation 10.68 is the Euler equation for a streamline. It is valid for a steady inviscid flow in the absence of body forces. Referring to equation 3.9, it is seen that equation 10.68 is the Bernoulli equation for steady one-dimensional flow in the absence of a body force. Hence, in the steady multidimensional inviscid flow of a compressible fluid, the Bernoulli equation is valid along a streamline.

Integrating equation 10.68 yields

$$\int \frac{dp}{\rho} + \frac{V^2}{2} = \text{constant} \qquad (\textit{along a streamline}) \qquad (10.69)$$

where, as before, the constant is the Bernoulli constant; the latter will, in general, vary from streamline to streamline, depending on the characteristics of the flow field under consideration. If all of the streamlines originate in a region of uniform flow, the Bernoulli constant will have the same value throughout the flow field.

10–5(b) Steady Irrotational Flow

For an irrotational flow, ω is zero, which requires that

$$\omega_X = \omega_Y = \omega_Z = 0 \qquad (10.70)$$

Substituting equation 10.70 into 10.67, we obtain, for steady irrotational flow,

$$\frac{1}{\rho} dp + d\left(\frac{V^2}{2}\right) = 0 \qquad (10.71)$$

which is again the Bernoulli equation. There are no restrictions, however, on the direction of the application of equation 10.71. Hence, in a steady irrotational flow, the Bernoulli equation applies between any two points in the flow. When integrated, equation 10.71 yields

$$\int \frac{dp}{\rho} + \frac{V^2}{2} = \text{constant} \qquad (\textit{throughout the flow field}) \qquad (10.72)$$

Equation 10.72 is the same mathematical result that is obtained in Section 10–5(a) for the flow along streamlines. As pointed out in Section 10–5(a), if the flow originates in a region where the flow is uniform, which is an irrotational flow, then the Bernoulli constant has the same value throughout the flow field.

Summarizing, a steady inviscid flow, in the absence of body forces, which is initially irrotational behaves as a flow that is irrotational throughout; that the foregoing is indeed the case is demonstrated in Section 10–6.

10–6 KELVIN'S THEOREM

It is of interest to determine what happens to the circulation on a closed fluid line, such as that illustrated in Fig. 10.15, as it moves through the flow field. As the curve

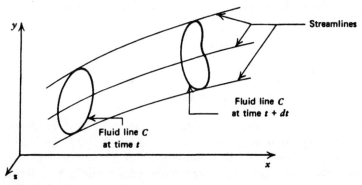

Figure 10.15   Flow model for Kelvin's theorem.

C moves with the motion of the fluid, it remains a closed curve because of the continuity of the flow. Hence, at a later time $t + dt$, curve C will have moved to a different location and its shape will have probably changed, but it will still be a closed fluid line. A question to be resolved, therefore, is the change in the circulation around the curve C as a function of time. It should be recalled that the same fluid particles comprise curve C at all times.

The time derivative of the property of a particle located on the closed fluid line C is, by definition, a substantial derivative. Hence, the interest here is in the substantial derivative of the circulation Γ around curve C, which is denoted by $D\Gamma/Dt$. Hence,

$$\frac{D\Gamma}{Dt} = \frac{D}{Dt} \oint_C \mathbf{V} \cdot d\mathbf{l} \tag{10.73}$$

where $\mathbf{V}$ is the vector velocity at a point on curve C at any instant. Differentiating the scalar product forming the right side of equation 10.73 yields

$$\frac{D\Gamma}{Dt} = \oint_C \left[\frac{D\mathbf{V}}{Dt} \cdot d\mathbf{l} + \mathbf{V} \cdot \frac{D(d\mathbf{l})}{Dt} \right] \tag{10.74}$$

The first term on the right side of equation 10.74 arises from the change in the velocity around the fluid line (curve C), and the second term arises from the change in its location. From equation 10.8,

$$\oint_C \frac{D\mathbf{V}}{Dt} \cdot d\mathbf{l} = -\oint_C \frac{1}{\rho} \nabla p \cdot d\mathbf{l} = -\oint_C \frac{dp}{\rho} \tag{10.75}$$

If it is assumed that either $\rho = \rho(p)$ or $\rho = $ constant, then the term dp/ρ is an exact differential, and the value of the line integral is independent of the path [see Section 1–7(a)]; in particular, its value is zero for a closed path.

The second term in equation 10.74 may be transformed by interchanging the order of the substantial derivative operator and the differential operator. Thus,

$$\oint_C \mathbf{V} \cdot \frac{D(d\mathbf{l})}{Dt} = \oint_C \mathbf{V} \cdot d\left(\frac{D\mathbf{l}}{Dt} \right) \tag{10.76}$$

The vector $\mathbf{l}$ is the position vector of the fluid particles; its time derivative $(D\mathbf{l}/Dt)$ is, therefore, the fluid velocity $\mathbf{V}$. Hence, equation 10.76 becomes

$$\oint_C \mathbf{V} \cdot \frac{D(d\mathbf{l})}{Dt} = \oint_C \mathbf{V} \cdot d\mathbf{V} = \oint_C d\left(\frac{V^2}{2} \right) = 0 \tag{10.77}$$

Again, for a closed curve, the line integral of the exact differential is zero.

Substituting equations 10.75 and 10.77 in equation 10.74, we obtain

$$\frac{D\Gamma}{Dt} = 0 \tag{10.78}$$

Integrating the last equation yields

$$\Gamma = \oint_C \mathbf{V} \cdot d\mathbf{l} = \text{constant} \tag{10.79}$$

Equation 10.79 is *Kelvin's theorem*. According to equation 10.79, if the closed curve around which the integration is conducted is displaced through the fluid by the particles comprising a closed fluid line, the circulation around that closed curve is invariant with time. Consequently, if the circulation is initially zero throughout the flow field, it will remain zero for all time. Since a flow for which $\Gamma = 0$ is an *irrotational flow*, equation 10.79 indicates that if a flow is initially irrotational, the irrotational motion will persist indefinitely. On the other hand, a flow that is initially rotational will remain rotational with undiminished intensity.

The theorem may be readily extended so that it applies to flows in the presence of a conservative force field; that is, where the external force is derivable from a *potential*.

One major consequence of Kelvin's theorem lies in its application to flows that are initially uniform, and thus are irrotational. For such initial conditions, the entire flow must be irrotational, a characteristic that may be employed for simplifying the analysis of such flows.

In summary, the conclusions derived from Kelvin's theorem are restricted to inviscid flows for which the density is either a function of the pressure only or is constant, and to the condition that the net external force acting on the fluid is derivable from a potential.

10-7 CROCCO'S THEOREM

The general relationship known as *Crocco's theorem* provides a connection between the rotation or vorticity of a flowing fluid and the gradients in its entropy and stagnation enthalpy. Its derivation is presented below for the steady inviscid flow of a compressible fluid.

Equation 10.64, which is a form of Euler's equation, is repeated and renumbered here for convenience.

$$\mathbf{V} \times (\nabla \times \mathbf{V}) = \frac{1}{\rho}\nabla p + \nabla\left(\frac{V^2}{2}\right) \tag{10.80}$$

As pointed out in Section 10-5, equation 10.64 applies to the steady inviscid flow of a compressible fluid in the absence of body forces.

For a pure substance, the fluid properties are related by equation 1.65, which is repeated and renumbered here for convenience*

$$T\,ds = dh - \frac{1}{\rho}\,dp \tag{10.81}$$

Equation 10.81 is valid for a homogeneous fluid experiencing either a reversible or an irreversible process, and is applicable in each of the three directions of a Cartesian coordinate system. Hence,

$$Ts_x = h_x - \frac{1}{\rho}p_x \tag{10.82}$$

$$Ts_y = h_y - \frac{1}{\rho}p_y \tag{10.83}$$

$$Ts_z = h_z - \frac{1}{\rho}p_z \tag{10.84}$$

where, as before, $s_x = \partial s/\partial x$, etc. Vectorial addition of the last three equations yields

$$T\nabla s = \nabla h - \frac{1}{\rho}\nabla p \tag{10.85}$$

---

* In this section, t denotes the time and T denotes the static temperature.

The procedure employed for deriving equation 10.85 is applicable to any total differential equation that is valid throughout a field of flow (i.e., independent of the path).

By *definition*, the stagnation enthalpy of a flowing fluid is

$$H \equiv h + \frac{V^2}{2} \qquad (10.86)$$

Differentiating the latter expression and placing the result in gradient form gives

$$\nabla H = \nabla h + \nabla \left(\frac{V^2}{2}\right) \qquad (10.87)$$

Substituting equations 10.85 and 10.87 into equation 10.80 and employing the definitions of the rotation ω and the vorticity ζ (from equation 10.47), we obtain the following form of *Crocco's theorem*. Thus,

$$\mathbf{V} \times (\nabla \times \mathbf{V}) = \mathbf{V} \times \zeta = \mathbf{V} \times (2\omega) = \nabla H - T\,\nabla s \qquad (10.88)$$

where ∇H and ∇s are the gradients for the stagnation enthalpy and the entropy of the fluid, respectively.

For the steady adiabatic inviscid flow of a compressible fluid in the absence of external work and body forces, the values of H and s remain constant along each streamline in the fluid. Thus, if $\nabla H \neq 0$ and $\nabla s \neq 0$, then ∇H and ∇s must be gradients normal to the streamlines. If those gradients exist, they must be present in the initial flow, and the flow is said to be *rotational*. Moreover, by virtue of Kelvin's theorem, the flow will remain rotational with undiminished intensity throughout the flow field.

Equation 10.88 applies to a point in the flow field. Crocco's theorem may be restricted to the flow along a streamline by forming the scalar product of the velocity vector $\mathbf{V}$ and equation 10.88. Thus,

$$\mathbf{V} \cdot [\mathbf{V} \times (\nabla \times \mathbf{V})] = (\mathbf{V} \cdot \nabla)H - T(\mathbf{V} \cdot \nabla)s \qquad (10.89)$$

By *definition*, $\mathbf{V} \times (\nabla \times \mathbf{V})$ is a vector perpendicular to $\mathbf{V}$, so that the scalar product on the left side of equation 10.89 is equal to zero. Recalling that $D(\ )/Dt = (\mathbf{V} \cdot \nabla)(\ )$ for a steady flow, equation 10.89 may be written as

$$\frac{DH}{Dt} - T\frac{Ds}{Dt} = 0 \qquad (10.90)$$

Hence, for the steady flow along a streamline,

$$\frac{DH}{Dt} = T\frac{Ds}{Dt} \qquad (10.91)$$

Equation 10.91 shows that if there are no dissipative effects in the flow along a streamline, so that $Ds/Dt = 0$, then

$$\frac{DH}{Dt} = 0 \qquad (10.92)$$

Integrating equation 10.92 yields

$$H = h + \frac{V^2}{2} = \text{constant} \qquad (\textit{along a streamline}) \qquad (10.93)$$

Equation 10.93 is a generalization of equation 3.38 for the steady one-dimensional adiabatic inviscid flow of a compressible fluid in the absence of external work and body forces.

If all of the streamlines in a flow field originate in a region of uniform flow, then the stagnation enthalpy H will have the same value at every point in the flow field.

Several special cases of Crocco's theorem for *steady flow* are discussed in the following subsections.

10-7(a) Homenergetic Flow

By definition, a steady *homenergetic* flow is one wherein the stagnation enthalpy has the same constant value along all of the streamlines; that is, $\nabla H = 0$. In that case, each particle in the flow field has the same value of stagnation enthalpy, and equation 10.88 reduces to

$$\mathbf{V} \times 2\boldsymbol{\omega} = -T\nabla s \tag{10.94}$$

Hence, a homenergetic flow is rotational if there is an entropy gradient normal to the streamlines.

An example of such a flow is the flow field behind the curved shock wave that forms in front of a blunt-nosed body moving through the atmosphere with a supersonic speed. It should be noted that the flow upstream from the shock wave is uniform, and thus irrotational but, in flowing through the curved shock wave, the flow becomes rotational because the shock wave produces an entropy gradient normal to the streamlines.

10-7(b) Isentropic and Homentropic Flow

By definition, an *isentropic* flow is one wherein the entropy is constant along each streamline, but its value may vary from streamline to streamline. A *homentropic* flow is one wherein the entropy has the same constant value on every streamline; that is, $\nabla s = 0$. Hence, in a homentropic flow all of the fluid particles have the same value of entropy at all points in the flow field.

For a homentropic flow, equation 10.88 reduces to

$$\mathbf{V} \times 2\boldsymbol{\omega} = \nabla H \tag{10.95}$$

Equation 10.95 shows that a flow is rotational if there is a gradient in the stagnation enthalpy. In most, if not all, practical compressible flows the creation of a stagnation enthalpy gradient is accompanied by the formation of an entropy gradient. Consequently, the flow described by equation 10.95 is of little practical interest.

10-7(c) Homentropic Homenergetic Flow

For flows in which both H and s are uniform throughout the flow field, equation 10.88 reduces to

$$\mathbf{V} \times (2\boldsymbol{\omega}) = 0 \tag{10.96}$$

The vector product $\mathbf{V} \times 2\boldsymbol{\omega}$ is given by

$$\mathbf{V} \times 2\boldsymbol{\omega} = 2V\omega \sin \alpha \tag{10.97}$$

where α is the angle between $\mathbf{V}$ and $\boldsymbol{\omega}$. Equation 10.96 is satisfied if any one of the following three situations occurs.

1. $\mathbf{V} = 0$; there is no flow.
2. $\boldsymbol{\omega} = 0$; the flow is irrotational.
3. $\mathbf{V}$ is parallel to $\boldsymbol{\omega}$, so that $\sin \alpha$ is 0.

Case 1 is, of course, trivial. In employing equation 10.96, care must be exercised in determining which of the Cases 2 or 3 is applicable.

Figure 10.16 illustrates the vectors $\mathbf{V}$, $\nabla \times \mathbf{V}$, and $\mathbf{V} \times (\nabla \times \mathbf{V})$ for the case where ω is not zero. The vector product $\mathbf{V} \times (2\omega)$ is a vector in the direction perpendicular to the plane containing the two vectors $\mathbf{V}$ and 2ω. Thus, as illustrated in Fig. 10.16, $\mathbf{V} \times (2\omega)$ is a vector perpendicular to $\mathbf{V}$. If the flow is two-dimensional (i.e., planar or axisymmetric), $\mathbf{V}$ and $\nabla \times \mathbf{V}$ both lie in the same plane, the xy plane, but as indicated above, $\mathbf{V} \times (2\omega)$ must point in the z direction. The latter requires that there be a z component of velocity, which violates the assumption that the flow is two-dimensional. It is inferred, therefore, that ω must be zero. Hence, a two-dimensional homentropic *homenergetic* flow is irrotational.

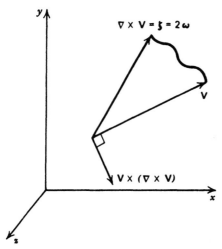

Figure 10.16 The vectors $\mathbf{V}$, $\nabla \times \mathbf{V}$, and $\mathbf{V} \times (\nabla \times \mathbf{V})$ in the Cartesian coordinate system.

A three-dimensional flow, however, may have z components of velocity. Accordingly, such a flow may or may not be irrotational, depending on whether or not the initial flow is irrotational. Consequently, if the initial flow is irrotational, then the entire flow will remain irrotational. However, if the initial flow is rotational, then according to Kelvin's theorem (Section 10-6), the rotation will remain constant. The vorticity ζ, or the rotation ω, must then be parallel to $\mathbf{V}$. Such a flow is called a *Beltrami flow*.

10-8 THE HELMHOLTZ VORTICITY THEOREMS

In Section 10-4(c), the vorticity of a fluid is defined as

$$\zeta = \nabla \times \mathbf{V} = 2\omega \qquad (10.98)$$

In Section 10-4(c), it is shown that the vorticity is equal to the circulation per unit area. In the present section, some general results concerning vorticity are discussed.

A *vortex line*, illustrated in Fig. 10.9, is defined as a line that is everywhere tangent to the vorticity vector. The analogy of that definition to that for a streamline is apparent. Vortex lines are field lines in the field $\nabla \times \mathbf{V}$, just as streamlines are field lines in the field $\mathbf{V}$. A *vortex tube*, illustrated in Fig. 10.17, is defined as a tube consisting of all of the vortex lines passing through a closed fluid curve C that is not parallel to the vortex lines; the analogy to a stream tube is apparent.

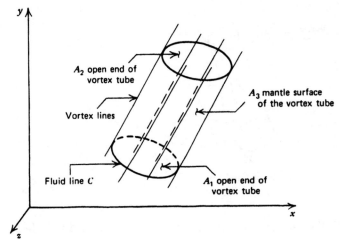

Figure 10.17   A vortex tube.

Equation 10.35, which is repeated here, relates the circulation and the vorticity.

$$\Gamma = \oint_C \mathbf{V} \cdot d\mathbf{l} = \int_A \zeta \cdot d\mathbf{A} \tag{10.35}$$

If the curve C is shrunk to a point, the line integral becomes zero. In that case, the surface A becomes a closed surface enclosing the volume $\mathscr{V}$ of the vortex tube, and equation 10.35 gives

$$\int_A \zeta \cdot d\mathbf{A} = 0 \tag{10.99}$$

For the vortex tube illustrated in Fig. 10.17, equation 10.99 yields

$$\int_{A_1 + A_2 + A_3} \zeta \cdot d\mathbf{A} = 0 \tag{10.100}$$

On the mantle surface A_3, the vorticity ζ is parallel to that surface by definition, so that $\zeta \cdot d\mathbf{A}$ is zero. Equation 10.100 reduces, therefore, to

$$\int_{A_1} \zeta \cdot d\mathbf{A}_1 = \int_{A_2} \zeta \cdot d\mathbf{A}_2 \tag{10.101}$$

Thus, in general,

$$\int_A \zeta \cdot d\mathbf{A} = \text{constant} \tag{10.102}$$

where A is a surface enclosed by the vortex tube. Again, it is apparent that there is an analogy to a stream tube, where $\int_A \rho \mathbf{V} \cdot d\mathbf{A}$ is constant. We may, therefore, conclude that the vorticity of a fluid flow has properties that are analogous to the mass rate of flow per unit area $G = \rho \mathbf{V}$.

The above considerations lead to the two vorticity theorems credited to Helmholtz.

1. The vorticity of a vortex tube is constant.
2. Vortex lines are material lines in flows that preserve circulation.

The first theorem follows directly from equation 10.102.

The second theorem may be proved by considering the flow model illustrated in Fig. 10.18. At the time t, the vortex tube has a closed fluid line C inscribed on its mantle surface. Since ζ is parallel to the surface of the vortex tube, Γ is zero for curve C. At time $t + dt$, the closed curve C' comprises the same fluid particles that comprised curve C, since it is a fluid line by definition. For a flow in which the circulation is

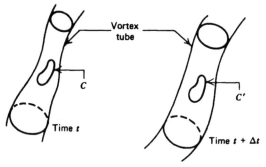

Figure 10.18 A vortex tube at the times t and $t + dt$.

preserved, Kelvin's theorem states that $D\Gamma/Dt = 0$ on any closed fluid line. Hence,

$$\oint_C \mathbf{V} \cdot d\mathbf{l} = 0 \qquad (at\ time\ t + dt) \tag{10.103}$$

Hence, ζ is tangent to C' at time $t + dt$, and C' is the element of a vortex tube. The last statement is true for any closed curve on the mantle surface of a vortex tube. The generators of the vortex tube, which are vortex lines, are, therefore, also material lines.

A conclusion based on the Helmholtz vortex theorems is that a vortex tube cannot end within a fluid, but it must either be a closed curve (e.g., a smoke ring) or reach a solid boundary.

10-9 THERMODYNAMICS OF THE STEADY ADIABATIC INVISCID FLOW OF A COMPRESSIBLE FLUID

The energy equation for the steady multidimensional flow of a compressible fluid in the absence of heat transfer, shear forces, work, and body forces, is given by equation 10.9, which may be rewritten in the form

$$\frac{D}{Dt}\left(h + \frac{V^2}{2}\right) = 0 \tag{10.104}$$

In terms of the stagnation enthalpy $H = h + V^2/2$, equation 10.104 may be rewritten as

$$\frac{DH}{Dt} = 0 \tag{10.105}$$

Hence,

$$H = h + \frac{V^2}{2} = \text{constant} \qquad (along\ a\ streamline) \tag{10.106}$$

Equation 10.106 is a generalization of equation 3.38, which applies to steady one-dimensional flow [see Section 3–6(b)].

If a flow originates in a region of uniform flow where H is constant throughout, then (since H remains constant along a streamline), the stagnation enthalpy H is constant throughout the entire flow field. As stated in Section 10–7(a), that flow is said to be *homenergetic*. Homenergetic flows occur frequently in practice. Examples are the external flow over bodies with or without shock waves, and the many internal flows where the flow originates in a region of uniform flow.

10-9(a) Equations of State

In addition to the basic conservation laws, it is necessary to know the equations of state for the fluid. For an inviscid fluid, those equations are the thermal and the caloric equations of state. For a pure substance, those equations, in functional form, are expressed by

$$t = t(p,\rho) \tag{10.107}$$

$$h = h(p,\rho) \tag{10.108}$$

Equations 10.107 and 10.108 may either take the form of suitable tables and graphs or, in simple cases, they may be represented by algebraic equations. For a perfect gas those equations have the following particularly simple forms (see Section 1–15).

$$p = \rho R t \tag{10.109}$$

$$h = c_p t \tag{10.110}$$

For nondissipative flows, such as those discussed in this chapter, the entropy remains constant along a streamline. Since $s = s(p,\rho)$, then

$$\rho = \rho(p) \quad \text{(along a streamline)} \tag{10.111}$$

For a perfect gas, equation 10.111 is given by

$$p\rho^{-\gamma} = \text{constant} \quad \text{(along a streamline)} \tag{10.112}$$

For a flow that originates in a uniform flow region, equations 10.111 and 10.112 become

$$\rho = \rho(p) \quad \text{(throughout the flow field)} \tag{10.113}$$

and

$$p\rho^{-\gamma} = \text{constant} \quad \text{(throughout the flow field)} \tag{10.114}$$

The speed of sound a is an important parameter in the study of the flow of compressible fluids. For a pure substance, a is a thermodynamic property given by the functional relationship

$$a = a(p,\rho) \tag{10.115}$$

For an isentropic flow, $\rho = \rho(p)$ along streamlines, and therefore

$$a = a(p) \quad \text{(along streamlines)} \tag{10.116}$$

If the flow is homentropic [$\nabla s = 0$, see Section 10–7(b)], then

$$a = a(p) \quad \text{(throughout the flow field)} \tag{10.117}$$

For an isentropic flow, p and V are related through the Bernoulli equation (equation 10.69), which is repeated here. Thus,

$$\int \frac{dp}{\rho} + \frac{V^2}{2} = \text{constant} \quad \text{(along a streamline)} \tag{10.69}$$

Since $\rho = \rho(p)$ along a streamline, equation 10.69 yields

$$V = V(p) \quad \text{(along a streamline)} \tag{10.118}$$

and equation 10.116 becomes

$$a = a(V) \quad \text{(along a streamline)} \tag{10.119}$$

For a homentropic flow, p and V are related through the Bernoulli equation (equation 10.72), which is repeated below.

$$\int \frac{dp}{\rho} + \frac{V^2}{2} = \text{constant} \qquad (\textit{throughout the flow field}) \qquad (10.72)$$

Since $\rho = \rho(p)$ throughout the flow field for homentropic flows, equation 10.72 yields

$$V = V(p) \qquad (\textit{throughout the flow field}) \qquad (10.120)$$

and equation 10.117 becomes

$$a = a(V) \qquad (\textit{throughout the flow field}) \qquad (10.121)$$

An example of equation 10.121 is given by equation 3.84 for the perfect gas.

$$a_o^2 = a^2 + \frac{\gamma - 1}{2} V^2 = \text{constant} \qquad (\textit{throughout the flow field}) \qquad (3.84)$$

10–9(b) The Speed of Sound Equation

The speed of sound a is defined by equation 1.180, which is repeated and renumbered here for convenience. Thus,

$$a^2 = \left(\frac{\partial p}{\partial \rho} \right)_s \qquad (10.122)$$

For an isentropic flow (i.e., a flow with constant values of entropy along each of its streamlines), the partial derivative expressed in equation 10.122 may be written as a total derivative. Hence,

$$dp = a^2 \, d\rho \qquad (\textit{along a streamline}) \qquad (10.123)$$

If the time derivatives following a particle are determined, we obtain the substantial (or material) derivatives given below.

$$\frac{Dp}{Dt} - a^2 \frac{D\rho}{Dt} = 0 \qquad (10.124)$$

In some applications, equation 10.124 may be employed in place of the energy equation (equation 10.9) when the flow is an isentropic flow, thus eliminating enthalpy derivatives from the governing equations.

In other applications, equation 10.124 is employed for eliminating the density derivatives from the governing equations. Density derivatives appear only in the continuity equation (equation 10.1), which may be written in the form

$$\frac{D\rho}{Dt} + \rho \nabla \cdot V = 0 \qquad (10.125)$$

Substituting equation 10.124 into equation 10.125 yields the following alternate form for the continuity equation. Thus,

$$\frac{Dp}{Dt} + \rho a^2 \, \nabla \cdot \mathbf{V} = 0 \qquad (10.126)$$

10–10 THE VELOCITY POTENTIAL FUNCTION

The governing equations for a steady irrotational flow [see Section 10–5(b)] may be combined and reduced to a single partial differential equation in terms of the *velocity potential function* ϕ; the latter function and the aforementioned partial differential equation are discussed in this section.

10–10(a) Definition of the Velocity Potential Function

According to Kelvin's theorem, for an irrotational flow,

$$\Gamma \equiv \oint_c \mathbf{V} \cdot d\mathbf{l} = 0 \qquad (10.127)$$

From the properties of exact differentials discussed in Section 1–7(a), the product $\mathbf{V} \cdot d\mathbf{l}$ is the exact differential of some function. Denote this function by ϕ, which is termed the *velocity potential function* for reasons that will become apparent later. Thus,

$$d\phi \equiv \mathbf{V} \cdot d\mathbf{l} \qquad (10.128)$$

Since $\mathbf{V} = \mathbf{V}(x,y,z)$ in the Cartesian coordinate system, $\phi = \phi(x,y,z)$. Expanding equation 10.128 gives

$$d\phi = \phi_x\, dx + \phi_y\, dy + \phi_z\, dz = u\, dx + v\, dy + w\, dz \qquad (10.129)$$

where $\phi_x = \partial\phi/\partial x$, etc. Since dx, dy, and dz are arbitrary, their coefficients in equation 10.129 must be equal. Hence,

$$u = \phi_x, \qquad v = \phi_y, \qquad \text{and } w = \phi_z \qquad (10.130)$$

Equation 10.130 shows that $\mathbf{V}$ is equal to the *gradient* of ϕ. Thus,

$$\mathbf{V} = \nabla\phi = \mathbf{i}u + \mathbf{j}v + \mathbf{k}w \qquad (10.131)$$

Thus, knowledge of the single variable ϕ completely specifies the velocity field for an irrotational flow. Conversely, for the velocity potential ϕ to exist, the flow must be irrotational.

For any scalar,

$$\nabla \times \nabla (\text{scalar}) = 0 \qquad (10.132)$$

Hence,

$$\nabla \times \nabla\phi = \nabla \times \mathbf{V} = \boldsymbol{\zeta} = 0 \qquad (10.133)$$

Equation 10.133 shows that the vorticity is zero, as it must be for an irrotational flow.

10–10(b) The Equations of Motion in Terms of ϕ

The governing equations may be expressed in terms of ϕ instead of $\mathbf{V}$ by employing equation 10.130. The continuity equation for steady flow, with density derivatives eliminated (equation 10.126), becomes

$$(\mathbf{V} \cdot \nabla)p + \rho a^2\, \nabla \cdot \mathbf{V} = 0 \qquad (10.134)$$

Euler's momentum equation for steady irrotational flow ($\nabla \times \mathbf{V} = 0$) may be obtained from equation 10.64. Thus,

$$\nabla p + \rho\, \nabla\left(\frac{V^2}{2}\right) = 0 \qquad (10.135)$$

Forming the scalar product of $\mathbf{V}$ and equation 10.135 gives

$$(\mathbf{V} \cdot \nabla)p + \rho(\mathbf{V} \cdot \nabla)\left(\frac{V^2}{2}\right) = 0 \qquad (10.136)$$

Combining equations 10.134 and 10.136 to eliminate ∇p yields

$$(\mathbf{V} \cdot \nabla)\left(\frac{V^2}{2}\right) - a^2 \nabla \cdot \mathbf{V} = 0 \tag{10.137}$$

Equation 10.137 is an important result that is generally called the *gas dynamic equation*.

Expanding equation 10.137 in Cartesian coordinates yields

$$(u^2 - a^2)u_x + (v^2 - a^2)v_y + (w^2 - a^2)w_z + uv(u_y + v_x)$$
$$+ uw(w_x + u_z) + vw(w_y + v_z) = 0 \tag{10.138}$$

In terms of ϕ, as defined by equation 10.130, equation 10.138 becomes

$$(\phi_x{}^2 - a^2)\phi_{xx} + (\phi_y{}^2 - a^2)\phi_{yy} + (\phi_z{}^2 - a^2)\phi_{zz} + 2\phi_x\phi_y\phi_{xy}$$
$$+ 2\phi_x\phi_z\phi_{xz} + 2\phi_y\phi_z\phi_{yz} = 0 \tag{10.139}$$

Equation 10.139 is the governing partial differential equation for the velocity potential ϕ for steady multidimensional irrotational flow.

Equation 10.139 is a nonlinear partial differential equation. It is a *quasi-linear* differential equation because it is linear in the highest-order derivatives of ϕ, in this case the second derivatives. The coefficients of the derivatives are functions of the velocity components u, v, and w, and also of the acoustic speed a. Excepting some special flows having simple conditions of symmetry, it is extremely difficult to obtain exact solutions to equation 10.139.

For steady two-dimensional planar or axisymmetric flow, equations 10.138 and 10.139 become

$$(u^2 - a^2)u_x + (v^2 - a^2)v_y + 2uvu_y - \delta\frac{a^2 v}{y} = 0 \tag{10.140}$$

and

$$(\phi_x{}^2 - a^2)\phi_{xx} + (\phi_y{}^2 - a^2)\phi_{yy} + 2\phi_x\phi_y\phi_{xy} - \delta\frac{a^2\phi_y}{y} = 0 \tag{10.141}$$

where $\delta = 0$ for planar flow and $\delta = 1$ for axisymmetric flow. For a two-dimensional irrotational flow, $\omega = 0$, which requires that

$$u_y - v_x = 0 \tag{10.142}$$

Equation 10.142 is employed in deriving equations 10.140 and 10.141. Solutions to equations 10.140 and 10.142 for several problems of general interest are presented in Chapters 11, 12, and 16.

In the case of a steady three-dimensional incompressible flow, $a^2 = \infty$. Dividing equation 10.139 by a^2 and setting $a^2 = \infty$ gives

$$\phi_{xx} + \phi_{yy} + \phi_{zz} = \nabla^2\phi = 0 \tag{10.143}$$

Equation 10.143 is *Laplace's equation*. It applies only to a steady irrotational *incompressible* flow.

10–10(c) General Features of the Potential Equation

Consider equation 10.141, which is the potential equation for a steady two-dimensional irrotational flow. If it is possible to find a function $\phi(x,y)$ for the flow field, all of the flow properties are determinable; the function $\phi(x,y)$ must, of course, satisfy the *boundary conditions*. At any point in the flow field, if $\phi(x,y)$ is known, then the velocity **V** of the fluid at that point may be calculated. Furthermore, by employing the energy equation, the equations of state, and the relationship between the fluid properties for an isentropic process, we may compute p, ρ, h, t, and a.

In general, the *boundary conditions* for the flow of a fluid are the known values of **V**, p, ρ, h, t, and a in a specified region of the flow field. If that region is one where the flow is undisturbed by the presence of a body immersed in the fluid, then the boundary conditions are those of the *undisturbed free stream*; they are usually denoted by $\mathbf{V}_\infty$, p_∞, ρ_∞, h_∞, t_∞, and a_∞. If the flow is a uniform flow parallel to the x axis at infinity, that boundary condition is expressed as $(\phi_x)_\infty = V_\infty$ and $(\phi_y)_\infty = 0$.

In the vicinity of a body of known shape immersed in the fluid, the boundary condition stipulates that the flow must be tangent to the surface of the body. In other words, the gradient of ϕ, given by $\nabla\phi$, must be parallel to the surface of the body along the curve that defines the cross section of the body in the plane under consideration. In the case when the flow is two-dimensional, the geometry of the body must also be two-dimensional.

Equation 10.141 for a two-dimensional flow is a nonhomogeneous second-order partial differential equation, linear in terms of the second derivatives, and belongs to the general type[8]

$$Au_{xx} + Bu_{xy} + Cu_{yy} + Du_x + Eu_y + Fu = 0 \qquad (10.144)$$

where u is any function of x and y; $u = u(x,y)$. Equation 10.144 may be classified into three distinct types, depending on whether the criterion $(B^2 - 4AC)$ is positive, negative, or zero. If $(B^2 - 4AC)$ is positive, the equation is *hyperbolic*; if it is negative, the equation is *elliptic*; and if it is zero, the equation is *parabolic*.

It is convenient to divide equation 10.141 by $-a^2$ and rewrite it in the form

$$\left(1 - \frac{u^2}{a^2}\right)\phi_{xx} - \frac{2uv}{a^2}\,\phi_{xy} + \left(1 - \frac{v^2}{a^2}\right)\phi_{yy} + \delta\frac{v}{y} = 0 \qquad (10.145)$$

For equation 10.145, the criterion $(B^2 - 4AC)$ is given by

$$(B^2 - 4AC) = \frac{4u^2v^2}{a^2} - 4\left(1 - \frac{u^2}{a^2}\right)\left(1 - \frac{v^2}{a^2}\right) = \frac{V^2}{a^2} - 1 = M^2 - 1 \quad (10.146)$$

For *subsonic* flow ($M < 1$), $(B^2 - 4AC) < 0$, indicating that equation 10.141 is *elliptic*. For *sonic* flow ($M = 1$), $(B^2 - 4AC) = 0$, indicating that the equation is *parabolic*. For *supersonic* flow ($M > 1$), $(B^2 - 4AC) > 1$, indicating that the potential equation is *hyperbolic*.

The differences in the type of the potential equation for subsonic, sonic, and supersonic flow are related to the physical differences in the flow fields corresponding to those three types of flow. If the flow field comprises more than one type of flow, the flow equations likewise are of more than one type, and the analysis of the flow field becomes complicated. That condition is encountered in *transonic* flow, where changes occur in the type of the equations governing the flow field.

Example 10.3. Several possible velocity potential functions are given below. Which ones are valid if the flow is incompressible?

(a) $\phi = -\Gamma \tan^{-1}(x/y)$
(b) $\phi = x + y$
(c) $\phi = \ln xyz$

Solution

If a velocity potential exists, then $\mathbf{V} = \nabla\phi$, $\nabla \cdot \mathbf{V} = \nabla^2\phi = 0$, and $\omega = \nabla \times \mathbf{V} = \nabla \times \nabla\phi = 0$ is automatically satisfied. Hence, a function is a possible velocity potential function if $\nabla^2\phi = 0$.

(a) $\quad \nabla^2\phi = \phi_{xx} + \phi_{yy} = \dfrac{\partial}{\partial x}\left[\dfrac{-\Gamma y}{(x^2 + y^2)}\right] + \dfrac{\partial}{\partial y}\left[\dfrac{\Gamma x}{(x^2 + y^2)}\right]$

$\quad \nabla^2\phi = \dfrac{2xy}{(x^2 + y^2)^2} + \dfrac{-2xy}{(x^2 + y^2)} = 0$

Equation (a) defines a possible velocity potential function. In fact, it yields the velocity field specified in part (b) of Examples 10.1 and 10.2, which is a free vortex flow field.

(b) $\quad \nabla^2\phi = \phi_{xx} + \phi_{yy} = \dfrac{\partial}{\partial x}(1) + \dfrac{\partial}{\partial y}(1) = 0$

Equation (b) defines a possible velocity potential function.

(c) $\quad \nabla^2\phi = \phi_{xx} + \phi_{yy} + \phi_{zz} = \dfrac{\partial}{\partial x}\left(\dfrac{yz}{xyz}\right) + \dfrac{\partial}{\partial y}\left(\dfrac{xz}{xyz}\right) + \dfrac{\partial}{\partial z}\left(\dfrac{xy}{xyz}\right)$

$\quad \nabla^2\phi = -\left(\dfrac{1}{x^2} + \dfrac{1}{y^2} + \dfrac{1}{z^2}\right) \neq 0$

Equation (c) does not define a possible velocity potential function.

Example 10.4. Consider the free vortex discussed in Section 10–4(d). The velocity field is given by

$$\mathbf{V} = \frac{-\mathbf{i}\Gamma y}{(x^2 + y^2)} + \frac{\mathbf{j}\Gamma x}{(x^2 + y^2)}$$

where the circulation Γ is a constant. Derive the velocity potential function for the above flow field.

Solution

For an irrotational flow field, $u = \phi_x$ and $v = \phi_y$. Thus,

$$\phi_x = u = \frac{-\Gamma y}{(x^2 + y^2)}$$

$$\phi_1 = \int \frac{-\Gamma y}{(x^2 + y^2)}\, dx + c_1(y) = -\Gamma \tan^{-1}\left(\frac{x}{y}\right) + c_1(y)$$

In a similar manner,

$$\phi_y = v = \frac{\Gamma x}{(x^2 + y^2)}$$

$$\phi_2 = \int \frac{\Gamma x}{(x^2 + y^2)}\, dy + c_2(x) = \Gamma \tan^{-1}\left(\frac{y}{x}\right) + c_2(x)$$

Comparing ϕ_1 and ϕ_2, it is seen that ϕ_2 does not contain a function of y alone, hence $c_1(y)$ in the expression for ϕ_1 may be chosen equal to zero. Likewise, $c_2(x)$ in the expression for ϕ_2 may be chosen equal to zero. From the properties of inverse trigonometric functions,

$$\tan^{-1} a + \tan^{-1} b = \tan^{-1}\frac{a + b}{1 - ab}$$

Subtracting ϕ_1 from ϕ_2 yields

$$\phi_2 - \phi_1 = \Gamma \tan^{-1}\left(\frac{y}{x}\right) + \Gamma \tan^{-1}\left(\frac{x}{y}\right) = \Gamma \tan^{-1}\left(\frac{\frac{y}{x} + \frac{x}{y}}{1 - 1}\right)$$

$$= \Gamma \tan^{-1}(\infty) = (2n + 1)\frac{\pi}{2}$$

where $n = 0, 1, 2, \ldots$ Thus, ϕ_1 and ϕ_2 differ only by a constant, so both are suitable velocity potential functions for the free vortex.

Lines of constant ϕ are given by

$$y = \left[\tan\left(\frac{\phi}{\Gamma}\right)\right]x$$

which are straight lines directed radially outward from the origin. In Section 10–11, it is shown that lines of constant ϕ are perpendicular to streamlines. Hence, the streamlines for the free vortex are concentric circles.

10–11 THE STREAM FUNCTION

In Section 10–10, the irrotationality condition is shown to be sufficient for defining the velocity potential function ϕ. For a steady two-dimensional flow, another point function, termed the *stream function*, may be defined; it is defined in the present section.

10–11(a) Definition of the Stream Function

It is pointed out in Section 10–10 that for a velocity potential function to exist, the flow field must be irrotational. Equation 10.142, renumbered below, is the irrotationality condition for a steady two-dimensional flow. Thus,

$$\frac{\partial}{\partial x}(-v) + \frac{\partial}{\partial y}(u) = 0 \tag{10.147}$$

Equation 10.147 suggests that a function ϕ, defined by $u = \phi_x$ and $v = \phi_y$, satisfies the irrotationality condition. For a steady two-dimensional planar flow ($\delta = 0$), equation 10.26, the continuity equation, becomes

$$\frac{\partial}{\partial x}(\rho u) + \frac{\partial}{\partial y}(\rho v) = 0 \tag{10.148}$$

Equation 10.148 suggests the introduction of a function ψ, termed the *stream function*, which is defined so that it satisfies equation 10.148 identically. Thus,

$$\psi_y \equiv \rho u \quad \text{and} \quad \psi_x \equiv -\rho v \tag{10.149}$$

Substituting equation 10.149 into equation 10.148 and interchanging the order of differentiation verifies that equation 10.149 satisfies equation 10.148.

The only necessary condition for the existence of the stream function for a two-dimensional planar flow is that the flow be steady.

For a steady two-dimensional axisymmetric flow [see Section 10–3(c)], the continuity equation may be written in the form

$$(y\rho u)_x + (y\rho v)_y = 0 \tag{10.150}$$

Hence, the stream function for a steady two-dimensional axisymmetric flow is defined by

$$\psi_x \equiv -y\rho v \quad \text{and} \quad \psi_y \equiv y\rho u \quad (10.151)$$

10–11(b) The Equations of Motion in Terms of ψ

For a steady two-dimensional irrotational flow, the irrotationality condition, equation 10.142, Euler's equation, equation 10.135, and the speed of sound equation, equation 10.122, may be combined, and the result expressed in terms of ψ. Thus,

$$(\psi_y{}^2 - y^{2\delta}\rho^2 a^2)\psi_{xx} + (\psi_x{}^2 - y^{2\delta}\rho^2 a^2)\psi_{yy} - 2\psi_x\psi_y\psi_{xy} - \delta y a^2 \psi_y = 0 \quad (10.152)$$

where, as before, $\delta = 0$ for a two-dimensional planar flow, and $\delta = 1$ for a two-dimensional axisymmetric flow. Equation 10.152 for determining ψ is complicated by the appearance of the density ρ in the coefficients of the partial derivatives ψ_{xx} and ψ_{yy}, because for a compressible flow ρ is a function of V. Another complication is the appearance of y^2 in the coefficients of the partial derivatives ψ_{xx} and ψ_{yy} in the case of an axisymmetric flow. For those reasons, the solution to equation 10.152 is not often·pursued. Moreover, the information usually desired may be obtained with less difficulty by solving equation 10.141.

10–11(c) Physical Interpretation of the Stream Function

The physical significance of the stream function may be illustrated by referring to the two-dimensional planar flow field illustrated in Fig. 10.19. The mass rate of flow between points A and B on two different streamlines is given by

$$d\dot{m} = (\rho u\, dy - \rho v\, dx)D \quad (10.153)$$

where D is the width of the flow passage, and the negative sign is introduced before the differential dx because the integration from A to O is in the negative x direction. For a two-dimensional flow, $\psi = \psi(x,y)$, and

$$d\psi = \psi_x\, dx + \psi_y\, dy \quad (10.154)$$

Substituting equation 10.149 into equation 10.154 gives

$$d\psi = (\rho u\, dy - \rho v\, dx) \quad (10.155)$$

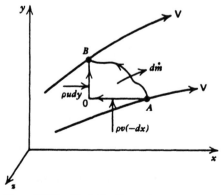

Figure 10.19 Model for illustrating the physical significance of the stream function.

Comparing equations 10.153 and 10.155, we obtain

$$d\dot{m} = D \, d\psi \qquad (10.156)$$

Integration of equation 10.156 between points A and B yields

$$\dot{m}_{A-B} = (\psi_B - \psi_A)D \qquad (10.157)$$

Hence, the mass flow rate between the two points A and B is proportional to the difference in the values of ψ between those points.

For a two-dimensional axisymmetric flow,

$$d\dot{m} = 2\pi(y\rho u \, dy - y\rho v \, dx) = 2\pi \, d\psi \qquad (10.158)$$

and

$$\dot{m}_{A-B} = 2\pi(\psi_B - \psi_A) \qquad (10.159)$$

From equation 10.31(b), the equation of a streamline in a two-dimensional flow is

$$u \, dy - v \, dx = 0 \qquad (10.160)$$

Combining equations 10.160 and 10.155 gives

$$d\psi = \rho(u \, dy - v \, dx) = 0 \quad \text{(along a streamline)} \qquad (10.161)$$

Integrating equation 10.161, we obtain

$$\psi = \text{constant} \quad \text{(along a streamline)} \qquad (10.162)$$

Hence, ψ has a constant value along a streamline, and the difference in the values of ψ between two streamlines is proportional to the mass rate of flow of fluid between the two streamlines.

Finally, from equation 10.161 it follows that

$$\left(\frac{dy}{dx}\right)_{\psi = \text{constant}} = \frac{v}{u} \qquad (10.163)$$

10–11(d) Relationship between ϕ and ψ

In a two-dimensional flow, the lines of constant ϕ may be obtained from equation 10.129. Thus,

$$d\phi = u \, dx + v \, dy = 0 \qquad (10.164)$$

and

$$\left(\frac{dy}{dx}\right)_{\phi = \text{constant}} = -\frac{u}{v} \qquad (10.165)$$

Comparing equations 10.163 and 10.165, it is apparent that the lines of constant ϕ are *orthogonal* to the lines of constant ψ.

Example 10.5.   Consider the forced vortex discussed in Section 10–4(e). The velocity field is given by

$$\mathbf{V} = -\mathbf{i}\Omega y + \mathbf{j}\Omega x$$

Derive the stream function for this flow field.

Solution

For incompressible flow, $u = -\psi_y$ and $v = \psi_x$. Thus,

$$\psi_y = -u = \Omega y$$

$$\psi = \int \Omega y \, dy + c_1(x) = \frac{\Omega y^2}{2} + c_1(x) \qquad \text{(a)}$$

In a similar manner,

$$\psi_x = v = \Omega x$$

$$\psi = \int \Omega x \, dx + c_2(y) = \frac{\Omega x^2}{2} + c_2(y) \tag{b}$$

Comparing equations (a) and (b), it is apparent that we may choose $c_1(x) = \Omega x^2/2$ and $c_2(x) = \Omega y^2/2$. Thus,

$$\psi(x,y) = \frac{\Omega(x^2 + y^2)}{2} + c$$

where c is a pure constant.

Note that the lines of constant ψ, which are streamlines, are concentric circles with centers at the origin.

10–12 SUMMARY

In Chapter 10, the differential forms of the governing equations for fluid flow derived in Chapter 2 are restricted to the conditions $\mathbf{V}_t = \rho_t = p_t = \mathbf{B} = g \, dz = d\mathbf{F}_{shear} = \delta Q = \delta W_{shear} = \delta W_{shaft} = 0$. The resulting equations are the vector equations governing the steady adiabatic inviscid flow of a compressible fluid; for convenience of reference they are assembled in Table 10.1. From those equations the corresponding equations for steady multidimensional adiabatic inviscid flow of a compressible fluid are derived in Cartesian and cylindrical coordinates; they are presented in Table 10.2. It is shown (see Table 10.2) that one set of equations specifies the flow fields for both *two-dimensional planar flow* ($\delta = 0$) and *two-dimensional axisymmetric flow* ($\delta = 1$).

Table 10.1   Vector Equations Governing the Steady Adiabatic Inviscid Flow of a Compressible Fluid*

Continuity equation

$$\nabla \cdot (\rho \mathbf{V}) = 0 \tag{10.166}$$

Momentum or dynamic equation

$$\rho \frac{D\mathbf{V}}{Dt} + \nabla p = 0 \tag{10.167}$$

Energy equation

$$\rho \frac{D}{Dt}\left(h + \frac{V^2}{2} \right) = 0 \tag{10.168a}$$

Speed of sound equation

$$\frac{Dp}{Dt} - a^2 \frac{D\rho}{Dt} = 0 \tag{10.168b}$$

Entropy equation

$$\rho \frac{Ds}{Dt} = 0 \tag{10.169}$$

Equations of state

$$t = t(p,\rho) \quad \text{and} \quad h = h(p,\rho) \tag{10.170}$$

---

* For the integral and differential forms of the *general equations*, see Tables 2.1 and 2.2.

Table 10.2 Governing Equations for the Steady Multidimensional Adiabatic Inviscid Flow of a Compressible Fluid

Cartesian coordinates $(x, y, \text{and } z)$

Continuity equation

$$(\rho u)_x + (\rho v)_y + (\rho w)_z = 0 \qquad (10.171)$$

Momentum or dynamic equations

$$uu_x + vu_y + wu_z + \frac{1}{\rho} p_x = 0 \qquad (10.172a)$$

$$uv_x + vv_y + wv_z + \frac{1}{\rho} p_y = 0 \qquad (10.172b)$$

$$uw_x + vw_y + ww_z + \frac{1}{\rho} p_z = 0 \qquad (10.172c)$$

Energy equation

$$uh_x + vh_y + wh_z + u\left(\frac{V^2}{2}\right)_x + v\left(\frac{V^2}{2}\right)_y + w\left(\frac{V^2}{2}\right)_z = 0 \qquad (10.173a)$$

Speed of sound equation

$$up_x + vp_y + wp_z - a^2\left(u\rho_x + v\rho_y + w\rho_z\right) = 0 \qquad (10.173b)$$

Entropy equation

$$us_x + vs_y + ws_z = 0 \qquad (10.174)$$

Cylindrical coordinates $(r, \theta, \text{and } z)$

Continuity equation

$$\frac{1}{r}\frac{\partial}{\partial r}(r\rho v_r) + \frac{1}{r}\frac{\partial}{\partial \theta}(\rho v_\theta) + \frac{\partial}{\partial z}(\rho v_z) = 0 \qquad (10.175)$$

Momentum equations

$$v_r \frac{\partial v_r}{\partial r} + \frac{v_\theta}{r}\frac{\partial v_r}{\partial \theta} - \frac{v_\theta^2}{r} + v_z \frac{\partial v_r}{\partial z} + \frac{1}{\rho}\frac{\partial p}{\partial r} = 0 \qquad (10.176a)$$

$$v_r \frac{\partial v_\theta}{\partial r} + \frac{v_\theta}{r}\frac{\partial v_\theta}{\partial \theta} + \frac{v_r v_\theta}{r} + v_z \frac{\partial v_\theta}{\partial z} + \frac{1}{\rho r}\frac{\partial p}{\partial \theta} = 0 \qquad (10.176b)$$

$$v_r \frac{\partial v_z}{\partial r} + \frac{v_\theta}{r}\frac{\partial v_z}{\partial \theta} + v_z \frac{\partial v_z}{\partial z} + \frac{1}{\rho}\frac{\partial p}{\partial z} = 0 \qquad (10.176c)$$

Energy equation

$$v_r \frac{\partial h}{\partial r} + \frac{v_\theta}{r}\frac{\partial h}{\partial \theta} + v_z \frac{\partial h}{\partial z} + v_r \frac{\partial}{\partial r}\left(\frac{V^2}{2}\right) + \frac{v_\theta}{r}\frac{\partial}{\partial \theta}\left(\frac{V^2}{2}\right) + v_z \frac{\partial}{\partial z}\left(\frac{V^2}{2}\right) = 0 \qquad (10.177a)$$

Speed of sound equation

$$v_r \frac{\partial p}{\partial r} + \frac{v_\theta}{r}\frac{\partial p}{\partial \theta} + v_z \frac{\partial p}{\partial z} - a^2\left(v_r \frac{\partial \rho}{\partial r} + \frac{v_\theta}{r}\frac{\partial \rho}{\partial \theta} + v_z \frac{\partial \rho}{\partial z}\right) = 0 \qquad (10.177b)$$

Entropy equation

$$v_r \frac{\partial s}{\partial r} + \frac{v_\theta}{r}\frac{\partial s}{\partial \theta} + v_z \frac{\partial s}{\partial z} = 0 \qquad (10.178)$$

Table 10.2 (*Continued*)

Two-dimensional planar and axisymmetric flows

Continuity equation

$$(\rho u)_x + (\rho v)_y + \delta \frac{\rho v}{y} = 0 \tag{10.179}$$

Planar flow, $\delta = 0$. Axisymmetric flow, $\delta = 1$.

Momentum or dynamic equations

$$uu_x + vu_y + \frac{1}{\rho} p_x = 0 \tag{10.180a}$$

$$uv_x + vv_y + \frac{1}{\rho} p_y = 0 \tag{10.180b}$$

Energy equation

$$uh_x + vh_y + u\left(\frac{V^2}{2}\right)_x + v\left(\frac{V^2}{2}\right)_y = 0 \tag{10.181a}$$

Speed of sound equation

$$up_x + vp_y - a^2(u\rho_x + v\rho_y) = 0 \tag{10.181b}$$

Entropy equation

$$us_x + vs_y = 0 \tag{10.182}$$

Transformation equations (from r and z to y and x)

$$v = v_r, \qquad u = v_z, \qquad y = r, \qquad x = z \tag{10.183}$$

Geometric and flow field parameters

Equations for the streamline

$$\frac{dy}{dx} = \frac{v}{u}; \qquad \frac{dz}{dy} = \frac{w}{v}; \qquad \frac{dx}{dz} = \frac{u}{w} \tag{10.184}$$

Circulation

$$\Gamma = \oint_C \mathbf{V} \cdot d\mathbf{l} = \oint_C V \, d\ell \cos \alpha = \int_A (\nabla \times \mathbf{V}) \cdot d\mathbf{A} \tag{10.185}$$

Rotation

$$\boldsymbol{\omega} = \frac{1}{2} \nabla \times \mathbf{V} \tag{10.186}$$

Vorticity

$$\boldsymbol{\zeta} = \nabla \times \mathbf{V} = 2\boldsymbol{\omega} = \mathbf{i}(w_y - v_z) + \mathbf{j}(u_z - w_x) + \mathbf{k}(v_x - u_y) \tag{10.187}$$

$$\zeta_z = \frac{d\Gamma}{dA} = 2\omega_z \tag{10.188}$$

The parameters known as the *circulation* Γ, *rotation* $\boldsymbol{\omega}$, and *vorticity* $\boldsymbol{\zeta}$ are defined and related to each other (see Table 10.2). Thus, the vorticity is equal to the circulation per unit area, and the rotation is equal to one half the vorticity. A flow for which $\boldsymbol{\omega} \neq 0$ is termed a *rotational flow*, and one for which $\boldsymbol{\omega} = 0$ is called an *irrotational flow*. A simple example of an irrotational flow is the *free vortex*, which has a singular point at the origin [see Section 10-4(d)]. An example of a rotational flow is the *forced vortex* [see Section 10-4(e)].

In the case of a steady multidimensional inviscid flow, the Bernoulli equation is valid along a streamline. The magnitude of the Bernoulli constant will, in general, vary from streamline to streamline, depending on the characteristics of the flow field. If the flow originates in a region of uniform flow, however, then the Bernoulli constant has the same value throughout the flow field; the latter statement is also true for a steady irrotational flow.

Kelvin's theorem (see Section 10–6) shows that if a flow is initially irrotational, it will remain irrotational with time, and if it is initially rotational, it will remain rotational with unchanging intensity. The latter two conclusions are restricted to inviscid flows for which the density is either a function of the pressure or is constant, and for which the net external force acting on the fluid is derivable from a *potential*.

For a steady homenergetic flow the stagnation enthalpy H is uniform throughout the flow field (i.e., $\nabla H = 0$). Crocco's theorem (see Section 10–7) shows that if there is an entropy gradient ∇s in such a flow field, then the flow is *rotational* ($\omega \neq 0$). For a homentropic steady flow the entropy is uniform throughout the flow field ($\nabla s = 0$). If there is a gradient in stagnation enthalpy ∇H in such a flow field, then the flow is *rotational* ($\omega \neq 0$).

For an *irrotational* flow the knowledge of a single variable, the velocity potential ϕ, specifies the velocity field. Only if a flow is irrotational ($\omega = 0$) can there be a velocity potential. A single governing partial differential equation is derived for steady multidimensional irrotational flow, in terms of the velocity potential ϕ (see Table 10.3).

For a steady two-dimensional flow, a point function, termed the *stream function* ψ, may be defined. The only necessary condition for ψ to exist for a two-dimensional

Table 10.3 Governing Equations for the Steady Multidimensional Irrotational Flow of a Compressible Fluid

---

Irrotationality condition

$$\omega = i\omega_X + j\omega_Y + k\omega_Z = 0$$

Two-dimensional flow

$$\omega_z = u_y - v_x = 0 \tag{10.189}$$

Gas dynamic equation

$$(\mathbf{V} \cdot \nabla)\left(\frac{V^2}{2}\right) - a^2 \nabla \cdot \mathbf{V} = 0 \tag{10.190}$$

Cartesian coordinates in terms of u, v, and w

$$(u^2 - a^2)u_x + (v^2 - a^2)v_y + (w^2 - a^2)w_z + uv(u_y + v_x)$$
$$+ uw(w_x + u_z) + vw(w_y + v_z) = 0 \tag{10.191}$$

Cartesian coordinates in terms of the velocity potential ϕ

$$(\phi_x{}^2 - a^2)\phi_{xx} + (\phi_y{}^2 - a^2)\phi_{yy} + (\phi_z{}^2 - a^2)\phi_{zz}$$
$$+ 2\phi_x\phi_y\phi_{xy} + 2\phi_x\phi_z\phi_{xz} + 2\phi_y\phi_z\phi_{yz} = 0 \tag{10.192}$$

Two-dimensional planar ($\delta = 0$) and axisymmetric ($\delta = 1$) flow in terms of u and v

$$(u^2 - a^2)u_x + (v^2 - a^2)v_y + 2uvu_y - \delta\frac{a^2 v}{y} = 0 \tag{10.193}$$

Two-dimensional planar ($\delta = 0$) and axisymmetric ($\delta = 1$) flow in terms of the velocity potential ϕ

$$(\phi_x{}^2 - a^2)\phi_{xx} + (\phi_y{}^2 - a^2)\phi_{yy} + 2\phi_x\phi_y\phi_{xy} - \delta\frac{a^2 \phi_y}{y} = 0 \tag{10.194}$$

---

flow is that the flow be steady. The stream function ψ has a constant value along a streamline and the difference in the values of ψ between two streamlines is proportional to the mass rate of flow of fluid between them. It is shown in Section 10-10(d) that the lines of constant ϕ for a steady two-dimensional flow are orthogonal to the lines of constant ψ.

REFERENCES

1. A. H. Shapiro, *The Dynamics and Thermodynamics of Compressible Fluid Flow*, Vol. 1, Ronald Press, New York, 1953.
2. R. von Mises, *Mathematical Theory of Compressible Fluid Flow*, Academic Press, New York, 1958.
3. R. Courant and K. O. Friedrichs, *Supersonic Flow and Shock Waves*, Interscience Publishers, New York, 1958.
4. K. Oswatitsch, *Gas Dynamics*, Academic Press, New York, 1956.
5. J. A. Owczarek, *Fundamentals of Gas Dynamics*, International Textbook Co., Scranton, Pa., 1964.
6. H. S. Tsien, "The Equations of Gas Dynamics," *Fundamentals of Gas Dynamics*, Section A, High Speed Aerodynamics and Jet Propulsion, Vol. III, Princeton University Press, Princeton, N.J., 1958.
7. L. I. Sedov, *Two-Dimensional Problems in Hydrodynamics and Aerodynamics* (translated from the Russian by C. K. Chu, H. Cohen, and B. Seckler), Chaps. VIII and IX, Interscience Publishers, New York, 1964.
8. W. Kaplan, *Advanced Calculus*, Addison-Wesley, Reading, Mass., 1952.

PROBLEMS

1. Show that the unit vectors in the cylindrical coordinate system illustrated in Fig. 10.1*b* are related to the unit vectors in the Cartesian coordinate system illustrated in Fig. 10.1*a* by the following equations.

$$\mathbf{i}_r = \mathbf{i} \cos \theta + \mathbf{j} \sin \theta$$

$$\mathbf{i}_\theta = -\mathbf{i} \sin \theta + \mathbf{j} \cos \theta$$

$$\mathbf{i}_z = \mathbf{k}$$

2. Figure 10.20 illustrates the unit vectors and the velocity vector in the spherical coordinate system. Show that those unit vectors are related to the unit vectors in the Cartesian coordinate system illustrated in Fig. 10.1*a* by the following equations.

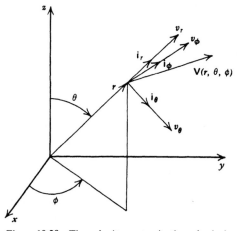

Figure 10.20 The velocity vector in the spherical coordinate system.

$$\mathbf{i}_r = \mathbf{i} \sin \theta \cos \phi + \mathbf{j} \sin \theta \sin \phi + \mathbf{k} \cos \theta$$

$$\mathbf{i}_\theta = \mathbf{i} \cos \theta \cos \phi + \mathbf{j} \cos \theta \sin \phi - \mathbf{k} \sin \theta$$

$$\mathbf{i}_\phi = -\mathbf{i} \sin \phi + \mathbf{j} \cos \phi$$

3. Show that in the cylindrical coordinate system

$$\frac{\partial}{\partial \theta}(\mathbf{i}_r) = \mathbf{i}_\theta \qquad \frac{\partial}{\partial \theta}(\mathbf{i}_\theta) = -\mathbf{i}_r$$

4. Show that in the spherical coordinate system,

$$\frac{\partial}{\partial \theta}(\mathbf{i}_r) = \mathbf{i}_\theta \qquad \frac{\partial}{\partial \theta}(\mathbf{i}_\theta) = -\mathbf{i}_r$$

$$\frac{\partial}{\partial \phi}(\mathbf{i}_r) = \mathbf{i}_\phi \sin \theta \qquad \frac{\partial}{\partial \phi}(\mathbf{i}_\theta) = \mathbf{i}_\phi \cos \theta$$

$$\frac{\partial}{\partial \phi}(\mathbf{i}_\phi) = -\mathbf{i}_r \sin \theta - \mathbf{i}_\theta \cos \theta$$

5. Write out the expanded form of the vector operator ∇ in the Cartesian, cylindrical, and spherical coordinate systems.

6. Expand ∇p and $\nabla \cdot \mathbf{V}$ in the Cartesian, cylindrical, and spherical coordinate systems.

7. Expand the operator $(\mathbf{V} \cdot \nabla)$ in the Cartesian, cylindrical, and spherical coordinate systems.

8. Expand $\nabla \times \mathbf{V}$ in the Cartesian, cylindrical, and spherical coordinate systems.

9. Expand Dp/Dt and $D\mathbf{V}/Dt$ in the Cartesian, cylindrical, and spherical coordinate systems.

10. Write out the expanded forms of the continuity, momentum, energy, and entropy equations in the Cartesian coordinate system.

11. Write out the expanded forms of the continuity, momentum, energy, and entropy equations in the cylindrical coordinate system.

12. Write out the expanded forms of the continuity, momentum, energy, and entropy equations in the spherical coordinate system.

13. Verify that the transformation specified by equation 10.25 transforms equations 10.19 to 10.24 into equations 10.26 to 10.30

14. Verify the following vector identity by expanding the terms into components in the Cartesian, cylindrical, and spherical coordinate systems.

$$(\mathbf{V} \cdot \nabla)\mathbf{V} = \nabla\left(\frac{V^2}{2}\right) - \mathbf{V} \times (\nabla \times \mathbf{V})$$

15. Which of the following velocity fields are possible flows for an incompressible fluid?
    (a) $\mathbf{V} = \mathbf{i}(x + y) + \mathbf{j}(x - y + z) + \mathbf{k}(x + y + 3)$
    (b) $\mathbf{V} = \mathbf{i}xy + \mathbf{j}yz + \mathbf{k}(yz + z^2)$
    (c) $\mathbf{V} = \mathbf{i}x(y + z) + \mathbf{j}y(x + z) + \mathbf{k}[-(x + y)z - z^2]$
    (d) $\mathbf{V} = \mathbf{i}xyzt + \mathbf{j}(-xyzt^2) + \mathbf{k}[(z^2/2)(xt^2 - yt)]$

16. Which of the velocity fields specified in Problem 15 are irrotational?

17. With the two velocity components given, determine the third velocity component so that the continuity equation is satisfied. Which of the resulting velocity fields are irrotational? Assume that the fluid density is constant.
    (a) $u = x^2 + y^2 + z^2, v = -xy - yx - xz, w = ?$
    (b) $u = \ln(y^2 + z^2), v = \sin(x^2 + y^2), w = ?$

18. List the necessary requirements for the existence of a velocity potential function.

19. Examine the following functions to determine if they could represent the velocity potential for an incompressible inviscid flow.
    (a) $f = x + y + z$
    (b) $f = x + xy + xyz$
    (c) $f = x^2 + y^2 + z^2$

    (d) $f = x^2z - y^2 - z^2$

    (e) $f = \sin(x + y + z)$

    (f) $f = \ln x$

20. List the necessary requirements for the existence of a stream function.

21. Derive an expression for the stream function for the free vortex discussed in Section 10–4(d).

22. Derive equation 10.152.

23. The stream function $\psi = U_\infty \sin\theta\,(R^2/r - r)$ represents an incompressible inviscid flow about an infinite cylinder of radius R, where U_∞ is the velocity of the free stream, and θ is measured clockwise from the stagnation point on the front of the cylinder. Determine expressions for v_r and v_θ

24. An infinitely long cylinder of radius R_1 is rotating coaxially with the angular velocity ω_1 within a hollow infinitely long cylindrical tube of radius R_2, which is rotating with the angular velocity ω_2. The velocity distribution of the incompressible fluid between the two cylinders is

$$V(r) = \frac{(\omega_2 R_2{}^2 - \omega_1 R_1{}^2)r - R_1{}^2 R_2{}^2(\omega_2 - \omega_1)/r}{R_2{}^2 - R_1{}^2}$$

Determine an expression for the vorticity at a point in the fluid. Under what conditions, if any, would the vorticity be zero?

11
introduction to flow with small perturbations

11-1 PRINCIPAL NOTATION FOR CHAPTER 11

The notation of Chapter 10 (Section 10-1) is applicable here. The following additional notation is employed in the present chapter.

$\tilde{a}$ dimensional local speed of sound.
a_∞ free-stream speed of sound.
h amplitude of a wavy wall.
C_p pressure coefficient.
ℓ wave length of a wavy wall.
M_∞ free-stream Mach number.
p perturbation pressure.
$\tilde{p}$ dimensional pressure.
u x direction perturbation velocity.
$\tilde{u}$ x direction dimensional velocity.
U_∞ x direction free-stream velocity.
v y direction perturbation velocity.

$\tilde{v}$ y direction dimensional velocity.
$\mathbf{V}_\infty$ free stream velocity vector.
w z direction perturbation velocity.
$\tilde{w}$ z direction dimensional velocity.

Greek letters

α $= \sin^{-1}(1/M)$, Mach angle.
δ 0 for planar flow, 1 for axisymmetric flow.
ρ perturbation density.
$\tilde{\rho}$ dimensional density.
ϕ perturbation velocity potential.
Φ velocity potential.

Subscripts

∞ denotes free-stream value of a property.

11–2 INTRODUCTION

The equations governing the multidimensional adiabatic inviscid flow of a compressible fluid in the absence of work and body forces are discussed in Chapter 10. Those equations comprise a set of coupled nonlinear partial differential equations, and their general solutions are unknown. In fact, closed form solutions are available for only a few very simple types of flow. In general, numerical techniques are employed for obtaining their solutions for specific initial and boundary conditions. For small disturbances in a stationary gas, and for flows where the velocity deviates only slightly from a parallel uniform flow, *small perturbation techniques* may be employed for linearizing the governing equations, thereby simplifying the mathematical procedures for solving the resulting equations.

A major objective of this book is the development of the mathematical techniques for solving nonlinear flows. Consequently, relatively minor attention is devoted to the consideration of flows with small perturbations. However, in view of the usefulness of the method of small perturbations, the linearized equations governing irrotational flows obtained by applying the latter method are developed in the present chapter, primarily for the purpose of illustrating the approximations involved. Several books present studies of linearized flow over airfoils, so that subject is not discussed in this book.

For illustrative purposes, small perturbation techniques are applied to the propagation of a planar acoustic wave, and to the classic problem of the flow over a wavy wall. The solution to the latter problem illustrates several general features of subsonic and supersonic flows.

Two additional problems of considerable practical importance wherein the method of small perturbations yields useful results are discussed in Chapter 15. Those two problems are the three-dimensional acoustic wave motion in a rigid cylindrical cavity, and the steady two-dimensional irrotational transonic flow in the throat of a converging-diverging nozzle.

11–3 PROPAGATION OF A PLANAR ACOUSTIC WAVE

An important illustration of the application of the technique of *small perturbations* (or small disturbances) to a body of an inviscid compressible fluid, such as air that is at rest, is provided by the planar propagation of an acoustic wave in the latter gas. Assume that the compressible fluid extends indefinitely in all directions so that no solid boundaries are involved. The properties of the fluid when it is at rest, or in the *undisturbed state*, are denoted by ρ_∞, p_∞, a_∞, and $\mathbf{V}_\infty = 0$.

In the present section, the governing equation is derived for the propagation of a planar wave in a uniform stationary mass of gas under the conditions prescribed above. The resulting equation is known as the *one-dimensional wave equation*. It is applicable to a small pressure disturbance of arbitrary shape, and its general solution is termed the *D'Alembert solution*.[1–3]

11–3(a) The One-Dimensional Wave Equation

Consider a compressible fluid at rest, having the uniform rest properties ρ_∞, p_∞, a_∞, and $\mathbf{V}_\infty = 0$. Assume that a weak pressure disturbance, such as an *acoustic wave*, is propagated in the fluid, with a propagation speed that is fast enough to warrant assuming adiabatic wave propagation; this assumption is in good agreement with the physics for the situation. Because the fluid is assumed to be inviscid, its motion because of the small pressure disturbance may be assumed to be isentropic. Furthermore, for a gas in an acoustic field, body forces are negligible.

For the isentropic process under consideration, the continuity, momentum, and speed of sound equations are the following.

(a) *Continuity equation* (equation 10.1)

$$\tilde{\rho}_t + \nabla \cdot (\tilde{\rho}\mathbf{V}) = 0 \tag{11.1}$$

(b) *Momentum equation* (equation 10.8)

$$\tilde{\rho}\frac{D\mathbf{V}}{Dt} + \nabla\tilde{p} = 0 \tag{11.2}$$

(c) *Speed of sound equation* (equation 10.124)

$$\frac{D\tilde{p}}{Dt} - \tilde{a}^2 \frac{D\tilde{\rho}}{Dt} = 0 \tag{11.3}$$

Combining equations 11.1 and 11.3, we obtain

$$\tilde{\rho}\tilde{a}^2 \nabla \cdot \mathbf{V} + \frac{D\tilde{p}}{Dt} = 0 \tag{11.4}$$

In the above equations the *tilde* over a letter denotes that it is a *dimensional* flow property ($\tilde{\rho}$, $\tilde{p}$, $\tilde{u}$, etc.). A letter without the tilde is employed for denoting a *perturbation* or *disturbance* in the pertinent flow property.

The aforementioned planar disturbance of the fluid causes motion of its particles parallel to the x axis, so that the velocity vector $\mathbf{V}$ at an arbitrary point in the fluid is given by the vector equation

$$\mathbf{V} = \mathbf{i}\tilde{u} + \mathbf{j}(0) + \mathbf{k}(0) = \mathbf{i}\tilde{u} \tag{11.5}$$

Accordingly, $\nabla \cdot \mathbf{V}$ is given by

$$\nabla \cdot \mathbf{V} = \left(\mathbf{i}\frac{\partial}{\partial x} + \mathbf{j}\frac{\partial}{\partial y} + \mathbf{k}\frac{\partial}{\partial z}\right) \cdot (\mathbf{i}\tilde{u}) = \tilde{u}_x \tag{11.6}$$

The substantial derivative of a general property N is given by (see *Vector Identities*, Section 2–1)

$$\frac{DN}{dt} = \frac{\partial N}{\partial t} + (\mathbf{V} \cdot \nabla)N = N_t + (\mathbf{i}\tilde{u}) \cdot \left(\mathbf{i}\frac{\partial N}{\partial x} + \mathbf{j}\frac{\partial N}{\partial y} + \mathbf{k}\frac{\partial N}{\partial z}\right) \tag{11.7}$$

or

$$\frac{DN}{Dt} = N_t + \tilde{u}N_x \tag{11.8}$$

From equation 11.8, it follows that

$$\frac{DV}{Dt} = \mathbf{i}(\tilde{u}_t + \tilde{u}\tilde{u}_x) \tag{11.9}$$

$$\frac{D\tilde{p}}{Dt} = \tilde{p}_t + \tilde{u}\tilde{p}_x \tag{11.10}$$

$$\frac{D\tilde{\rho}}{Dt} = \tilde{\rho}_t + \tilde{u}\tilde{\rho}_x \tag{11.11}$$

Substituting equations 11.9, 11.10, and 11.11 into equations 11.2 and 11.4 yields

$$\tilde{\rho}\tilde{u}_t + \tilde{\rho}\tilde{u}\tilde{u}_x + \tilde{p}_x = 0 \tag{11.12}$$

$$\tilde{\rho}\tilde{a}^2\tilde{u}_x + \tilde{p}_t + \tilde{u}\tilde{p}_x = 0 \tag{11.13}$$

Equations 11.12 and 11.13 apply to unsteady one-dimensional planar flow fields.

As stated above, the perturbation quantities are denoted by the same letters as those employed for the dimensional properties, except without the *tilde*. Then by definition, the perturbation quantities are

$$\tilde{u} = u, \qquad \tilde{a} = a_\infty + a, \qquad \tilde{p} = p_\infty + p, \qquad \tilde{\rho} = \rho_\infty + \rho \tag{11.14}$$

where u is very small, and a, p, and ρ are much smaller than the corresponding *undisturbed values* a_∞, p_∞, and ρ_∞, respectively; also, in the undisturbed state, $u_\infty = 0$. Substituting the above perturbation quantities into equations 11.12 and 11.13, and neglecting all products of perturbations and their derivatives yields

$$\rho_\infty u_t + p_x = 0 \tag{11.15}$$

$$\rho_\infty a_\infty^2 u_x + p_t = 0 \tag{11.16}$$

Assume that the second derivatives are continuous, and differentiate equation 11.15 with respect to x and equation 11.16 with respect to t. Combining the results and eliminating u_{xt} from the set of two equations yields the single equation

$$p_{tt} = a_\infty^2 p_{xx} \tag{11.17}$$

Equation 11.17 is the classical *one-dimensional wave equation* that governs the one-dimensional propagation of weak pressure disturbances.

The one-dimensional wave equation has wide application in engineering and physics. It is the governing partial differential equation for such phenomena as the vibrating membrane, the longitudinal vibrations of a long bar, and the transverse vibrations of a beam.[3]

11–3(b) The D'Alembert Solution of the Wave Equation

The general features associated with the propagation of a weak pressure disturbance are now investigated. Consider a general solution of equation 11.17 having the following form, known as the *D'Alembert* (1747) solution.[1-3]

$$p(x,t) = f(x - a_\infty t) + g(x + a_\infty t) \tag{11.18}$$

That equation 11.18 is a solution of equation 11.17 may be verified by direct substitution. It is seen that equation 11.18 yields two solutions, or waves; the *f-waves* obtained from the solution $p(x,t) = f(x - a_\infty t)$, and the *g-waves* obtained from the solution $p(x,t) = g(x + a_\infty t)$. Either or both waves may occur in a given flow situation. Equation 11.17 is a linear differential equation; consequently, the solution for $p(x,t)$ may be constructed by superposing the *f-waves* and the *g-waves*. The general characteristics of such waves are described below.

First, consider a pressure disturbance caused only by the *f-wave*. Thus,

$$p(x,t) = f(x - a_\infty t) \tag{11.19}$$

Figure 11.1 illustrates an arbitrary pressure disturbance at the initial time t_o. Consider the arbitrary point x_o. At the time t_o, the amplitude of the pressure disturbance at x_o is given by

$$p(x_o,t_o) = f(x_o - a_\infty t_o) \tag{11.20}$$

At some later time $t_1 > t_o$, that same amplitude will occur at some downstream point x_1. Thus,

$$p(x_o,t_o) = p(x_1,t_1) \tag{11.21}$$

For equation 11.21 to be true, equation 11.19 requires that

$$f(x_o - a_\infty t_o) = f(x_1 - a_\infty t_1) \tag{11.22}$$

The requirement expressed by equation 11.22 is satisfied only if

$$x_o - a_\infty t_o = x_1 - a_\infty t_1 \tag{11.23}$$

Solving the last equation for x_1 yields

$$x_1 = x_o + a_\infty(t_1 - t_o) \tag{11.24}$$

Hence, the point on the initial pressure disturbance (see Fig. 11.1) has propagated to the right an amount equal to the distance $a_\infty(t_1 - t_o)$ during the time interval $(t_1 - t_o)$. Accordingly, the *velocity of propagation* of the wave is given by

$$V_{propagation} = \frac{a_\infty(t_1 - t_o)}{(t_1 - t_o)} = a_\infty \tag{11.25}$$

where a_∞ is the *speed of sound in the undisturbed fluid.*

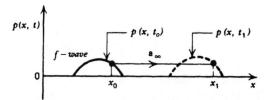

Figure 11.1   Propagation of a planar pressure disturbance.

The selection of the point x_o on the initial pressure disturbance is arbitrary. Consequently, every point of curve $p(x,t_o)$, for the initial pressure disturbance, will move to the right by the same amount. Consequently, the entire pressure disturbance moves to the right at the speed a_∞ with no change in the wave shape.

The function $f(x - a_\infty t)$ represents, therefore, a *right-running wave* that propagates an initial disturbance to the right at the speed a_∞, and that preserves the wave form. Similar reasoning shows that the function $g(x + a_\infty t)$ represents a *left-running wave* that propagates an initial disturbance to the left at the speed a_∞, also preserving the wave form.

A general pressure disturbance, such as that represented by equation 11.18, is, therefore, the sum of the left- and right-running waves. Such waves are called *traveling waves*. If the propagation path is of finite length, reflected waves occur at the boundaries, and *standing wave forms* may be generated by the superposition of the two traveling waves. The rules for wave reflection given in Section 7–11 for compression waves, and in Section 8–8 for expansion waves, must be obeyed at the boundaries. Accordingly, waves of the same type (either compression or expansion) reflect from a solid boundary (i.e., a closed-end tube), and waves of the opposite type reflect from a fixed pressure boundary (i.e., an open-end tube).

The foregoing concepts of wave propagation form the basis of the science of acoustics. In general, acoustic waves may propagate in all directions and the boundary conditions are more complex (for example, energy absorption, dispersion, and phase shift may occur).* The basic features of the wave propagation process are, however, illustrated by the D'Alembert solution.

11–4 FLOW WITH SMALL PERTURBATIONS

In the present section, the linearized equations for steady two-dimensional irrotational flow are derived. A single equation is obtained that is valid for both subsonic and supersonic flows, and a separate equation is derived for the case of transonic flow. The derivations of those equations follow, in general, the technique presented in Reference 4. The resulting equations are valid for both planar and axisymmetric flows.

11–4(a) Linearization of the Governing Equations

Consider an undisturbed infinite compressible medium which constitutes a uniform flow field. Let $\mathbf{V}_\infty$ denote the vector velocity at a point in such a flow field. Assume that the direction of the velocity of the uniform flow field, termed the *free-stream velocity*, is parallel to the x axis of a Cartesian coordinate system, as illustrated in Fig. 11.2a. In that case,

$$\mathbf{V}_\infty = \mathbf{i}U_\infty + \mathbf{j}(0) + \mathbf{k}(0) = \mathbf{i}U_\infty \qquad (11.26)$$

where U_∞ is the *free-stream velocity* for the undisturbed uniform flow. The fluid properties p, ρ, and a, in the undisturbed flow field, are likewise uniform and their values are denoted by p_∞, ρ_∞, and a_∞, respectively.

Assume that a body of good aerodynamic shape, such as a thin airfoil, is inserted into the uniform flow field, as illustrated in Fig. 11.2b. The obstruction caused by the body causes the streamlines of the flow in the vicinity of the body to deviate a small amount from the undisturbed direction of the uniform flow field; that is, from the direction parallel to U_∞, which is parallel to the x axis.

At this point, it is desirable to introduce the notation discussed in Section 11–3(a) for the rectangular components of the instantaneous velocity vector $\mathbf{V}$ for a fluid

---

* The three-dimensional wave equation is derived in Section 15–3, and that equation is solved in Section 15–4 for the acoustic wave motion in a rigid cylindrical cavity.

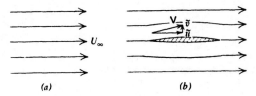

Figure 11.2 Uniform and perturbed flows. (*a*) Uniform flow. (*b*) Perturbed flow.

particle. Accordingly, **V** is defined by

$$\mathbf{V} = \mathbf{i}\tilde{u} + \mathbf{j}\tilde{v} + \mathbf{k}\tilde{w} \tag{11.27}$$

where the *tilde* over a letter denotes that it is a *dimensional velocity component*, and the symbols u, v, and w denote the *perturbation* velocity components.

For the steady irrotational flow of a compressible fluid, the *gas dynamic equation* (see equation 10.190, Table 10.3) applies. It is repeated and renumbered here for convenience. Thus,

$$(\mathbf{V} \cdot \nabla)\left(\frac{V^2}{2}\right) - a^2 \nabla \cdot \mathbf{V} = 0 \tag{11.28}$$

The corresponding equation in Cartesian coordinates (see equation 10.191) is repeated below, expressed in the notation for the Cartesian components of **V** (see equation 11.27 above). Thus,

$$a^2(\tilde{u}_x + \tilde{v}_y + \tilde{w}_z) = \tilde{u}^2\tilde{u}_x + \tilde{v}^2\tilde{v}_y + \tilde{w}^2\tilde{w}_z + \tilde{u}\tilde{v}(\tilde{u}_y + \tilde{v}_x) \\ + \tilde{u}\tilde{w}(\tilde{u}_z + \tilde{w}_x) + \tilde{v}\tilde{w}(\tilde{v}_z + \tilde{w}_y) \tag{11.29}$$

From equation 3.186, the energy equation for a perfect gas, it follows that

$$a_o^2 = a^2 + \frac{\gamma - 1}{2}V^2 = a^2 + \frac{\gamma - 1}{2}(\tilde{u}^2 + \tilde{v}^2 + \tilde{w}^2) = a_\infty^2 + \frac{\gamma - 1}{2}U_\infty^2 \tag{11.30}$$

For the perturbed flow caused by the presence of the thin airfoil, let u, v, and w denote the *perturbation velocities* parallel to the x, y, z coordinate axes, respectively. The perturbation velocities are assumed to be very small compared to the undisturbed uniform velocity U_∞, so that the ratios u/U_∞, v/U_∞, and w/U_∞ are considerably smaller than unity. Hence, at a point near the body, the velocity vector **V** is given by

$$\mathbf{V} = \mathbf{i}(U_\infty + u) + \mathbf{j}v + \mathbf{k}w \tag{11.31}$$

where

$$\tilde{u} = U_\infty + u, \quad \tilde{v} = v, \quad \tilde{w} = w \tag{11.32}$$

Substituting equation 11.32 into equation 11.29 transforms the gas dynamic equation into the following *perturbation equation*. Thus,

$$a^2(u_x + v_y + w_z) = (U_\infty + u)^2 u_x + v^2 v_y + w^2 w_z + (U_\infty + u)v(u_y + v_x) \\ + vw(v_z + w_y) + w(U_\infty + u)(u_z + w_x) \tag{11.33}$$

Substituting equation 11.32 into equation 11.30 yields

$$a^2 = a_\infty^2 - \frac{\gamma - 1}{2}(2uU_\infty + u^2 + v^2 + w^2) \tag{11.34}$$

Substituting equation 11.34 for a^2 into equation 11.33, dividing by a_∞^2, and rearranging, yields

$$
(1 - M_\infty^2)u_x + v_y + w_z
$$

$$
= M_\infty^2 \left[(\gamma + 1)\frac{u}{U_\infty} + \frac{\gamma + 1}{2}\left(\frac{u}{U_\infty}\right)^2 + \frac{\gamma - 1}{2}\frac{v^2 + w^2}{U_\infty^2} \right]u_x
$$

$$
+ M_\infty^2 \left[(\gamma - 1)\frac{u}{U_\infty} + \frac{\gamma + 1}{2}\left(\frac{v}{U_\infty}\right)^2 + \frac{\gamma - 1}{2}\frac{u^2 + w^2}{U_\infty^2} \right]v_y
$$

$$
+ M_\infty^2 \left[(\gamma - 1)\frac{u}{U_\infty} + \frac{\gamma + 1}{2}\left(\frac{w}{U_\infty}\right)^2 + \frac{\gamma - 1}{2}\frac{u^2 + v^2}{U_\infty^2} \right]w_z
$$

$$
+ M_\infty^2 \left[\frac{v}{U_\infty}\left(1 + \frac{u}{U_\infty}\right)(u_y + v_x) + \frac{w}{U_\infty}\left(1 + \frac{u}{U_\infty}\right)(u_z + w_x) + \frac{vw}{U_\infty^2}(w_y + v_z) \right]
$$

$$
(11.35)
$$

where $M_\infty = U_\infty/a_\infty$. Equation 11.35 is the complete *gas dynamic equation* expressed in terms of the perturbation velocities. Its left side is linear, but the right-hand side is nonlinear.

Assume that the perturbation velocities u, v, and w are small compared with U_∞, so that all of the products of the perturbation velocities appearing in equation 11.35 may be set equal to zero. Equation 11.35 then reduces to

$$
(1 - M_\infty^2)u_x + v_y + w_z = M_\infty^2(\gamma + 1)\frac{u}{U_\infty}u_x
$$

$$
+ M_\infty^2(\gamma - 1)\frac{u}{U_\infty}(v_y + w_z) + M_\infty^2\frac{v}{U_\infty}(u_y + v_x) + M_\infty^2\frac{w}{U_\infty}(u_z + w_x) \quad (11.36)
$$

Although equation 11.36 is considerably simpler than equation 11.35, it is still highly nonlinear. Consequently, additional simplifications are needed before attempting its solution.

Assume further that all of the products of the perturbation velocities and the derivatives of the perturbation velocities are negligible. Consequently, all of the terms on the right-hand side of equation 11.36 vanish, and the following linear equation is obtained.

$$
(1 - M_\infty^2)u_x + v_y + w_z = 0 \quad (11.37)
$$

Equation 11.37 is valid for both subsonic and supersonic flows.

For transonic flow, since M_∞ is close to unity, the coefficient $(1 - M_\infty^2)$ on the left-hand side of equation 11.36 becomes very small. Consequently, the first term on the right-hand side of that equation may be of the same order of magnitude as the first term on the left-hand side, and should, therefore, be retained. All terms, after the first, on the right-hand side of equation 11.36 may, therefore, be neglected in the case of a transonic flow. Hence, for a transonic flow,

$$
(1 - M_\infty^2)u_x + v_y + w_z = M_\infty^2(\gamma + 1)\left(\frac{u}{U_\infty}\right)u_x \quad (11.38)
$$

Equations 11.37 and 11.38 are frequently termed the *small perturbation equations*. As stated earlier, equation 11.37 is valid for both subsonic and supersonic flows, and equation 11.38 is valid for transonic flows; providing, of course, that the perturbation velocities are small compared to the undisturbed free-stream velocity U_∞.

In the case of *hypersonic flows*, the free-stream Mach number M_∞ is very large, and other methods are required for linearizing the gas dynamic equation.

In the case of a *two-dimensional planar* flow, the perturbation velocity component

$w = 0$, and its partial derivative w_z may be deleted from equations 11.37 and 11.38.

For *two-dimensional axisymmetric* flow, the equation corresponding to equation 11.29 is equation 10.193, with $\delta = 1$. The latter equation is similar to equation 11.29, except that the velocity component $\tilde{w}$ and its derivatives are zero, and the additional term $a^2\tilde{v}/y$ is present. Repeating the steps employed in deriving equations 11.37 and 11.38, we obtain the following linearized equation for a *two-dimensional axisymmetric* flow.

$$(1 - M_\infty^2)u_x + v_y + \frac{v}{y} = 0 \tag{11.39}$$

Equation 11.39 is valid for both subsonic and supersonic two-dimensional axisymmetric isentropic flows.

For a two-dimensional axisymmetric *transonic* flow, the following equation is applicable.

$$(1 - M_\infty^2)u_x + v_y + \frac{v}{y} = M_\infty^2(\gamma + 1)\left(\frac{u}{U_\infty}\right)u_x \tag{11.40}$$

Except for the term v/y in equations 11.39 and 11.40, those equations are identical to equations 11.37 and 11.38 for two-dimensional planar flow, for which $w_z = 0$. Consequently, a single set of equations may be employed for both planar and axisymmetric two-dimensional flows. Thus,

$$(1 - M_\infty^2)u_x + v_y + \delta\frac{v}{y} = 0 \tag{11.41}$$

and

$$(1 - M_\infty^2)u_x + v_y + \delta\frac{v}{y} = M_\infty^2(\gamma + 1)\left(\frac{u}{U_\infty}\right)u_x \tag{11.42}$$

where, as before, $\delta = 0$ for planar flows, and $\delta = 1$ for axisymmetric flows. Equations 11.41 and 11.42 are the *perturbation equations* for *two-dimensional flows*.

11–4(b) Perturbation Velocity Potential

The flows under consideration are assumed to be irrotational. Consequently, as shown in Section 10–10, there must be a velocity potential Φ such that $\mathbf{V} = \nabla\Phi$. Thus,

$$\tilde{u} = \Phi_x, \qquad \tilde{v} = \Phi_y, \qquad \tilde{w} = \Phi_z \tag{11.43}$$

For the case at hand

$$\tilde{u} = U_\infty + u, \qquad \tilde{v} = v, \qquad \tilde{w} = w \tag{11.44}$$

Thus, the velocity potential Φ may be defined by

$$\Phi = U_\infty x + \phi \tag{11.45}$$

where ϕ is the *perturbation velocity potential*. Thus,

$$\tilde{u} = \Phi_x = U_\infty + \phi_x = U_\infty + u \tag{11.46}$$

$$\tilde{v} = \Phi_y = \phi_y = v \tag{11.47}$$

$$\tilde{w} = \Phi_z = \phi_z = w \tag{11.48}$$

In terms of the perturbation velocity potential ϕ, equations 11.41 and 11.42 become, respectively,

$$(1 - M_\infty^2)\phi_{xx} + \phi_{yy} + \delta\frac{\phi_y}{y} = 0 \tag{11.49}$$

and

$$(1 - M_\infty^2)\phi_{xx} + \phi_{yy} + \delta \frac{\phi_y}{y} = \frac{M_\infty^2(\gamma + 1)}{U_\infty} \phi_x\phi_{xx} \qquad (11.50)$$

For the flow to be irrotational, $\nabla \times \mathbf{V} = 2\boldsymbol{\omega} = 0$. That requirement is met identically by the velocity potential defined by equation 11.45. Equation 11.49, which is linear, may be applied to subsonic and supersonic flows outside of the transonic flow regime. Equation 11.50, which still retains a nonlinear term on the right-hand side, is applicable to the transonic flow regime.

11–4(c) Classification of the Perturbation Equations

In Section 10–10(c), the potential equation was classified as being either of the elliptic, parabolic, or hyperbolic type, depending on whether the flow is subsonic, sonic, or supersonic, respectively. The perturbation equations, equations 11.49 and 11.50, being linearized forms of the potential equation may be classified by evaluating the parameter $B^2 - 4AC$, where A, B, and C are the coefficients of ϕ_{xx}, ϕ_{xy}, and ϕ_{yy}, respectively, in those equations. Thus, $A = (1 - M_\infty^2)$, $B = 0$, and $C = 1$, and

$$B^2 - 4AC = 4(M_\infty^2 - 1) \qquad (11.51)$$

Hence, the perturbation equations may be classified as follows:

$$\begin{array}{ll} M_\infty < 1 & \text{elliptic} \\ M_\infty = 1 & \text{parabolic} \\ M_\infty > 1 & \text{hyperbolic} \end{array} \qquad (11.52)$$

The most significant difference between the solutions to elliptic and hyperbolic equations is that the solution to an elliptic equation depends continuously on all of the boundary conditions. The solution to a hyperbolic equation, on the other hand, does not depend on all of the boundary conditions continuously, but in a solution there are regions of influence and domains of dependence. These general features of elliptic and hyperbolic equations are present in the solutions to the perturbation equations.

11–4(d) Linearization of the Boundary Condition

For an inviscid flow, the boundary condition to be applied at the surface of a solid boundary is that the direction of the flow velocity vector must be tangent to the solid surface. In other words, the velocity vector is everywhere at right angles to the normal to the solid boundary. In general, the equation of a surface in three-dimensional space is given by

$$f(x,y,z) = 0 \qquad (11.53)$$

Figure 11.3 illustrates the boundary condition at a solid wall; since $f(x,y,z) = 0$ on the solid surface, ∇f is normal to the surface, where ∇f is the *gradient of f*. The required *tangency condition* is given by

$$\mathbf{V} \cdot \nabla f = 0 \qquad (11.54)$$

In Cartesian coordinates the corresponding equation is

$$\tilde{u}f_x + \tilde{v}f_y + \tilde{w}f_z = 0 \qquad (11.55)$$

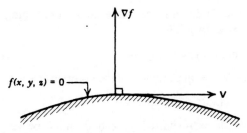

Figure 11.3 Solid wall boundary condition.

In terms of the perturbation velocities u, v, and w, the above boundary condition becomes

$$(U_\infty + u)f_x + vf_y + wf_z = 0 \qquad (11.56)$$

For flows with small perturbations, the boundary condition specified by equation 11.56 may be simplified. To illustrate, consider a two-dimensional planar flow about a thin body that causes a small perturbation in the flow. In that case the perturbation velocity component $w = 0$. Hence, equation 11.56 becomes

$$(U_\infty + u)f_x + vf_y = 0 \qquad (11.57)$$

Rearranging equation 11.57 yields

$$\frac{v}{U_\infty + u} = -\frac{f_x}{f_y} \qquad (11.58)$$

By employing the approximation that $u \ll U_\infty$, u may be neglected relative to U_∞, and equation 11.58 may be rewritten in the form

$$\frac{v}{U_\infty} = -\frac{f_x}{f_y} \qquad (11.59)$$

For the assumed two-dimensional surface, $f(x,y) = 0$. Hence, from equation 11.53 we obtain the following boundary condition on the surface.

$$df = 0 = f_x \, dx + f_y \, dy \qquad (11.60)$$

Rearranging equation 11.60 yields

$$\left(\frac{dy}{dx}\right)_{\text{Surface}} = -\frac{f_x}{f_y} \qquad (11.61)$$

Combining equations 11.59 and 11.61, we obtain

$$\frac{v}{U_\infty} = \left(\frac{dy}{dx}\right)_{\text{Surface}} \qquad (11.62)$$

In equation 11.62 the term v/U_∞ is the slope of the streamline, and $(dy/dx)_{\text{Surface}}$ represents the slope of the surface of the body. Hence, at a point $P(x,y)$ on the surface of the body, the perturbation velocity v is equal to the product of U_∞ and the slope of the surface dy/dx at that point.

As pointed out earlier, the entire analysis is predicated on the condition that the body is so thin that the perturbation velocities are very small compared to U_∞. Consequently, at the point $P(x,y)$, the ordinate y cannot be significantly different from zero.

Further simplification of equation 11.62 may be achieved by expanding $v(x,y)$ at a fixed value of x into a Taylor series in y. Thus,

$$v(x,y) = v(x,0) + v_y(x,0)y + \cdots \tag{11.63}$$

If the surface of the body is specified by $y = \eta(x)$, then v on the body surface is given by

$$v[x,\eta(x)] = v(x,0) + v_y(x,0)\eta + \cdots \tag{11.64}$$

For the flow to experience only small perturbations, the body must be very thin. Consequently, because η is small, all of the terms involving η may be neglected. Hence, equation 11.64 becomes

$$v[x,\eta(x)] = v(x,0) \tag{11.65}$$

Consequently, equation 11.62 becomes

$$\frac{v(x,0)}{U_\infty} = \left(\frac{dy}{dx}\right)_{\text{Surface}} \tag{11.66}$$

In terms of the perturbation velocity potential ϕ, defined by equation 11.47, equation 11.66 becomes

$$\phi_y(x,0) = U_\infty \left(\frac{dy}{dx}\right)_{\text{Surface}} \tag{11.67}$$

Equation 11.67 presents the linearized boundary condition at a solid wall for a two-dimensional planar flow. It is not applicable to a two-dimensional axisymmetric flow (see Reference 4).

11–4(e) Linearization of the Pressure Coefficient

In Section 3–13(b) the pressure coefficient C_p for the flow of a perfect gas is given by equation 3.157, which is repeated and renumbered here for convenience.

$$C_p = \frac{2}{\gamma M_\infty^2} \left\{ \left[1 + \frac{\gamma - 1}{2} M_\infty^2 \left(1 - \frac{V^2}{V_\infty^2} \right) \right]^{\gamma/(\gamma-1)} - 1 \right\} \tag{11.68}$$

In terms of the perturbation velocities u, v, and w, equation 11.68 becomes

$$C_p = \frac{2}{\gamma M_\infty^2} \left\{ \left[1 + \frac{\gamma - 1}{2} M_\infty^2 \left(1 - \frac{(U_\infty + u)^2 + v^2 + w^2}{U_\infty^2} \right) \right]^{\gamma/(\gamma-1)} - 1 \right\} \tag{11.69}$$

Rearranging equation 11.69 yields

$$C_p = \frac{2}{\gamma M_\infty^2} \left\{ \left[1 - \frac{\gamma - 1}{2} M_\infty^2 \left(\frac{2u}{U_\infty} + \frac{u^2 + v^2 + w^2}{U_\infty^2} \right) \right]^{\gamma/(\gamma-1)} - 1 \right\} \tag{11.70}$$

In equation 11.70, expand the expression enclosed by the square brackets by the binomial theorem and neglect the cubes and higher powers of the perturbation velocities. The result is

$$C_p = -\left[\frac{2u}{U_\infty} + (1 - M_\infty^2) \left(\frac{u}{U_\infty} \right)^2 + \frac{v^2 + w^2}{U_\infty^2} \right] \tag{11.71}$$

For a flow with small perturbations,

$$\frac{u}{U_\infty} \ll 1 \quad \text{and} \quad \frac{v}{U_\infty} \ll 1 \tag{11.72}$$

For consistency with the derivations of equations 11.49 and 11.50, a first-order approximation is desired, so that only the first term of equation 11.71 needs to be retained. Consequently, the *linearized pressure coefficient* for a two-dimensional planar flow is

$$C_p = -\frac{2u}{U_\infty} \tag{11.73}$$

11–5 TWO-DIMENSIONAL FLOW PAST A WAVY WALL

A classical example of linearized flow is the two-dimensional planar flow over a wavy wall, studied by Ackeret.[5] The wall geometry is illustrated in Fig. 11.4, where the boundary surface over which the fluid flows is defined by the equation

$$y = h \cos\left(\frac{2\pi x}{\ell}\right) \tag{11.74}$$

where h is the *amplitude* of the waves in the wall, and ℓ is their *wavelength*. In this section, solutions are presented for both subsonic and supersonic flow over such a wavy wall.

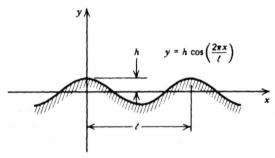

Figure 11.4 Geometry of a wavy wall.

The equation for the perturbation velocity potential ϕ, for both subsonic and supersonic flow, is equation 11.49. For planar flow, since $\delta = 0$, equation 11.49 becomes

$$(1 - M_\infty^2)\phi_{xx} + \phi_{yy} = 0 \tag{11.75}$$

It is evident from equation 11.52, that equation 11.75 is of the elliptic type for subsonic flow, and of the hyperbolic type for supersonic flow.

Equation 11.75 is subject to the following boundary conditions. As y approaches infinity

$$u(x,\infty) = \phi_x(x,\infty) \text{ and } v(x,\infty) = \phi_y(x,\infty) \text{ both remain finite} \tag{11.76}$$

The linearized boundary condition applicable to the surface of the wavy wall is (see equation 11.67)

$$v(x,0) = \phi_y(x,0) = U_\infty \left(\frac{dy}{dx}\right)_{\text{Surface}} \tag{11.77}$$

Hence,

$$\phi_y(x,0) = -U_\infty \left(\frac{2\pi h}{\ell}\right) \sin\left(\frac{2\pi x}{\ell}\right) \tag{11.78}$$

In addition, the flow is periodic with respect to x.

The analysis presented below assumes that the fluid is inviscid. In the presence of a solid boundary, a fluid adheres to the boundary, and a boundary layer is developed wherein viscous effects are not negligible (see Section 5–10). In many situations, especially in the presence of a favorable pressure gradient (i.e., one where the pressure decreases in the flow direction), the boundary layer is very thin. In such cases, an inviscid analysis may predict the flow field outside of the boundary layer with sufficient accuracy. For the wavy wall flow considered here, however, the pressure gradient is zero, and the boundary layer would grow in thickness along the wall. Eventually the boundary layer thickness would become much larger than the amplitude of the wall waviness, and an inviscid analysis of the flow near the wavy wall would be meaningless. With the foregoing reservations in mind, the solution for the inviscid flow over a wavy wall presented in this section must be interpreted as the limiting flow field for very thin boundary layers, and the general trends that are observed are qualitative rather than quantitative. Despite these qualifications, the general features observed for the subsonic and supersonic flows over a wavy wall are of great utility in interpreting more complex flow fields.

11–5(a) Subsonic Flow

The linearized potential equation governing subsonic flow is equation 11.75, where the coefficient $(1 - M_\infty{}^2)$ is positive. The x coordinate may be redefined so that the coefficient $(1 - M_\infty{}^2)$ in equation 11.75 is absorbed. For example, let

$$x \equiv \bar{x}\sqrt{1 - M_\infty{}^2} \tag{11.79}$$

Then equation 11.75 becomes

$$\phi_{\bar{x}\bar{x}} + \phi_{yy} = 0 \tag{11.80}$$

Equation 11.80 is the *two-dimensional Laplace equation*. It is well known that the solutions to equation 11.80 depend continuously on the boundary conditions, and such a dependence should, therefore, be expected for the two-dimensional subsonic flow past a wavy wall.

One of the most powerful methods for solving a linear partial differential equation is to assume that the solution is a product of several independent functions, each function depending on only one of the independent variables.[6] Separation of variables is then employed to solve for the assumed functions. The justification for assuming a product form for the solution is based solely on finding a solution that satisfies the boundary conditions of the problem. If such a solution is found, then the method is applicable. If not, the solution must be sought by other methods.

A product solution is now investigated for equation 11.75. Thus, assume a solution of the form

$$\phi(x,y) = X(x)Y(y) \tag{11.81}$$

Substituting equation 11.81 into equation 11.75 gives

$$(1 - M_\infty{}^2)X''Y + XY'' = 0 \tag{11.82}$$

where X'' and Y'' denote the second derivatives of the functions X and Y with respect to x and y, respectively. Dividing equation 11.82 through by $(1 - M_\infty{}^2)XY$ yields

$$\frac{X''}{X} = -\frac{Y''}{(1 - M_\infty{}^2)Y} \tag{11.83}$$

In equation 11.83, the term on the left-hand side is a function only of x, while the term on the right-hand side is a function solely of y. Moreover, the equation must hold for arbitrary values of x and y. The equality expressed by equation 11.83

can hold only if each side of equation 11.83 is equal to the same constant. Let $-\beta^2$ denote that constant. Introducing the constant $-\beta^2$ we obtain

$$\frac{X''}{X} = -\frac{Y''}{(1 - M_\infty^2)Y} = -\beta^2 = \text{constant} \qquad (11.84)$$

The negative sign preceding β^2 was chosen to satisfy the periodicity of the wall with respect to the x axis.

Since the functions X and Y are independent of each other, equation 11.84 yields the following two equations:

$$X'' + \beta^2 X = 0 \qquad (11.85)$$

$$Y'' - \beta^2(1 - M_\infty^2)Y = 0 \qquad (11.86)$$

The general solutions of equations 11.85 and 11.86 are

$$X = A \sin \beta x + B \cos \beta x \qquad (11.87)$$

$$Y = C \exp\left[-\beta\sqrt{(1 - M_\infty^2)}y\right] + D \exp\left[\beta\sqrt{(1 - M_\infty^2)}y\right] \qquad (11.88)$$

The boundary condition at $y = \infty$ (see equation 11.76) requires that the velocity components u and v remain finite as $y \to \infty$. Consequently, in equation 11.88, the coefficient $D = 0$. Applying the boundary condition on the solid surface, that is, equation 11.78, with $y = 0$ yields

$$\phi_y(x,0) = -U_\infty \left(\frac{2\pi h}{\ell}\right) \sin\left(\frac{2\pi x}{\ell}\right) = X(x)Y'(0)$$

$$= (A \sin \beta x + B \cos \beta x)(-C\beta\sqrt{1 - M_\infty^2}) \qquad (11.89)$$

Since no cosine function appears on the left-hand side of equation 11.89, the coefficient $B = 0$. Equation 11.89, therefore, reduces to

$$U_\infty \left(\frac{2\pi h}{\ell}\right) \sin\left(\frac{2\pi x}{\ell}\right) = AC\beta\sqrt{1 - M_\infty^2} \sin \beta x \qquad (11.90)$$

The coefficients of the sine functions in equation 11.90 are both constant, signifying that both sine functions depend solely on x, and that the arguments of the sine functions and also their coefficients must be equal. Hence,

$$\beta = \frac{2\pi}{\ell} \qquad \text{and} \qquad AC = \frac{U_\infty h}{\sqrt{1 - M_\infty^2}} \qquad (11.91)$$

The final form of the solution for $\phi(x,y)$ is, accordingly,

$$\phi(x,y) = \frac{U_\infty h}{\sqrt{1 - M_\infty^2}} \sin\left(\frac{2\pi x}{\ell}\right) \exp\left(-\frac{2\pi}{\ell}\sqrt{1 - M_\infty^2}y\right) \qquad (11.92)$$

Hence, the perturbation velocity components u and v are given by

$$u(x,y) = \phi_x(x,y) = \frac{U_\infty}{\sqrt{1 - M_\infty^2}}\left(\frac{2\pi h}{\ell}\right) \cos\left(\frac{2\pi x}{\ell}\right) \exp\left(-\frac{2\pi}{\ell}\sqrt{1 - M_\infty^2}y\right) \qquad (11.93)$$

$$v(x,y) = \phi_y(x,y) = -U_\infty \left(\frac{2\pi h}{\ell}\right) \sin\left(\frac{2\pi x}{\ell}\right) \exp\left(-\frac{2\pi}{\ell}\sqrt{1 - M_\infty^2}y\right) \qquad (11.94)$$

The equation of a streamline in a two-dimensional flow is obtained from equation 10.184 which, for the flow with small perturbations under consideration here, becomes

$$\frac{dy}{dx} = \frac{\tilde{v}}{\tilde{u}} = \frac{v}{U_\infty + u} \qquad (11.95)$$

By substituting equations 11.93 and 11.94 into equation 11.95, we obtain an equation that cannot be integrated. Equation 10.95 may, however, be further simplified in view of the assumption of small perturbations. Thus, rewrite equation 11.95 in the following form.

$$\frac{dy}{dx} = \frac{v}{U_\infty + u \pm M_\infty{}^2 u} = \frac{v}{U_\infty + (1 - M_\infty{}^2)u \mp M_\infty{}^2 u} \tag{11.96}$$

For low subsonic flows, the term $M_\infty{}^2 u$ may be neglected in comparison with the term $(1 - M_\infty{}^2)u$. Introducing equations 11.93 and 11.94 into equation 11.96 and making the aforementioned assumption yields

$$dy \left[1 + \left(\frac{2\pi h}{\ell} \right) \sqrt{1 - M_\infty{}^2} \cos \left(\frac{2\pi x}{\ell} \right) \exp \left(-\frac{2\pi}{\ell} \sqrt{1 - M_\infty{}^2} y \right) \right]$$
$$= \left(\frac{2\pi h}{\ell} \right) \sin \left(\frac{2\pi x}{\ell} \right) \exp \left(-\frac{2\pi}{\ell} \sqrt{1 - M_\infty{}^2} y \right) dx \tag{11.97}$$

Rearranging equation 11.97 gives the following equation for dy. Thus,

$$dy = h \, d \left[\cos \left(\frac{2\pi x}{\ell} \right) \exp \left(-\frac{2\pi}{\ell} \sqrt{1 - M_\infty{}^2} y \right) \right] \tag{11.98}$$

Integrating equation 11.98, we obtain

$$y = h \cos \left(\frac{2\pi x}{\ell} \right) \exp \left(-\frac{2\pi}{\ell} \sqrt{1 - M_\infty{}^2} y \right) + \text{constant} \tag{11.99}$$

When y approaches zero, the exponential term in equation 11.99 approaches unity and, in that case, equation 11.99 becomes

$$y = h \cos \left(\frac{2\pi x}{\ell} \right) + \text{constant} \tag{11.100}$$

Comparing equation 11.74, which defines the wall contour, with equation 11.100 shows that the constant of integration in equation 11.99 must be zero. Thus, equation 11.99, which is the equation for the streamlines, becomes

$$y = \left[h \exp \left(-\frac{2\pi}{\ell} \sqrt{1 - M_\infty{}^2} y \right) \right] \cos \left(\frac{2\pi x}{\ell} \right) \tag{11.101}$$

Equation 11.101 shows that the streamlines above the surface of the wavy wall (i.e., $y > 0$) oscillate in phase with the wall oscillation. The amplitude of the waviness of the streamlines, however, decreases exponentially with increasing distance from the wall. Figure 11.5 presents selected streamlines for the subsonic flow over a wavy wall. At an infinite distance from the wall (i.e., $y \to \infty$), the flow is uniform and parallel to the x axis. The effect of the wall disturbance is propagated in all directions, illustrating that in a subsonic flow, the flow field depends continuously on all of the boundary conditions.

Combining equations 11.73 and 11.93, the following equation is obtained for the linearized pressure coefficient C_p. Thus,

$$C_p = -\frac{2u}{U_\infty} = -\frac{2}{\sqrt{1 - M_\infty{}^2}} \left(\frac{2\pi h}{\ell} \right) \cos \left(\frac{2\pi x}{\ell} \right) \exp \left(-\frac{2\pi}{\ell} \sqrt{1 - M_\infty{}^2} y \right) \tag{11.102}$$

The pressure coefficient at the wall surface, $y = 0$, is given by

$$(C_p)_{\text{Surface}} = -\frac{2}{\sqrt{1 - M_\infty{}^2}} \left(\frac{2\pi h}{\ell} \right) \cos \left(\frac{2\pi x}{\ell} \right) \tag{11.103}$$

Equation 11.103 indicates that the wall pressure distribution is harmonic, and that the peak pressures occur at the bottom of the troughs in the wall, as illustrated in Fig. 11.6. The pressure variation on the wall is symmetrical about the pressure peaks, and it is 180 deg out of phase with the variations in the contour of the wall, with the result that there is no drag force on the wall.

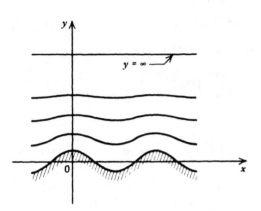

Figure 11.5   Streamline pattern for subsonic flow over a wavy wall.

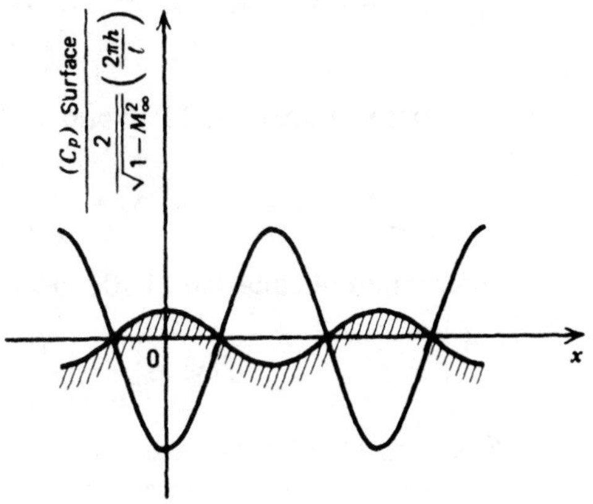

Figure 11.6   Wall pressure distribution for subsonic flow over a wavy wall.

The results developed in this section demonstrate several important general features of a subsonic flow. The behavior of the flow is quite different, however, if the flow is supersonic.

11–5(b) Supersonic Flow

For a supersonic flow $M_\infty > 1$, so that the coefficient $(1 - M_\infty^2)$ in equation 11.75 is negative. There is no change, of course, in equation 11.74 for the geometry of the surface of the wall. Neither are there any changes in the equations for the boundary conditions (equations 11.76, 11.77, and 11.78).

For supersonic flow, the linearized potential equation, equation 11.75, transforms to the following equation

$$\phi_{xx} = \frac{1}{M_\infty^2 - 1} \phi_{yy} \qquad (11.104)$$

where the coefficient $1/(M_\infty^2 - 1)$ is positive. Equation 11.104 is of the same form as the *wave equation* (equation 11.17) that governs an acoustic field.

Hence, it is to be expected that the D'Alembert solution of the wave equation (Section 10–3) should be applicable to equation 10.104. In the present case, the D'Alembert solution takes the following form.

$$\phi(x,y) = f(x - \sqrt{M_\infty^2 - 1}\,y) + g(x + \sqrt{M_\infty^2 - 1}\,y) \qquad (11.105)$$

It may be verified by direct substitution that equation 11.105 is a solution of equation 11.104. The form of the functions f and g must be such that they satisfy the boundary conditions.

First, consider only the function f, so the function g is set equal to zero. The boundary condition at the wall, that the flow must be tangent to the wall, is given by equation 11.78, which is repeated here for convenience.

$$\phi_y(x,0) = -U_\infty \left(\frac{2\pi h}{\ell}\right) \sin\left(\frac{2\pi x}{\ell}\right) \qquad (11.78)$$

In terms of the function f in equation 11.105, the aforementioned boundary condition is given by

$$\phi_y(x,y) = f' \frac{\partial(x - \sqrt{M_\infty^2 - 1}y)}{\partial y} = -\sqrt{M_\infty^2 - 1}f' \tag{11.106}$$

where the notation f' denotes the total derivative of the function f with respect to its argument $(x - \sqrt{M_\infty^2 - 1}y)$. Thus, at $y = 0$,

$$\phi_y(x,0) = -\sqrt{M_\infty^2 - 1}f'(x) \tag{11.107}$$

Equating equations 11.78 and 11.107 and solving for $f'(x)$ yields

$$f'(x) = \frac{U_\infty}{\sqrt{M_\infty^2 - 1}} \left(\frac{2\pi h}{\ell}\right) \sin\left(\frac{2\pi x}{\ell}\right) \tag{11.108}$$

Integration of equation 11.108 yields

$$f(x) = -\frac{U_\infty h}{\sqrt{M_\infty^2 - 1}} \cos\left(\frac{2\pi x}{\ell}\right) + \text{constant} \tag{11.109}$$

Equation 11.109 yields the functional form of the function f. The function f required in equation 11.105 is obtained by replacing the argument x in equation 11.109 by the argument $(x - \sqrt{M_\infty^2 - 1}y)$. Thus, equation 11.105 becomes

$$\phi(x,y) = -\frac{U_\infty h}{\sqrt{M_\infty^2 - 1}} \cos\left[\left(\frac{2\pi}{\ell}\right)\left(x - \sqrt{M_\infty^2 - 1}y\right)\right] + \text{constant} \tag{11.110}$$

That equation 11.110 is a solution may be verified by substituting it into equation 11.104. Note that equation 11.110 satisfies the boundary conditions at $y = \infty$ (equation 11.76).

The perturbation velocity components u and v are given by

$$u(x,y) = \phi_x(x,y) = \frac{U_\infty}{\sqrt{M_\infty^2 - 1}} \left(\frac{2\pi h}{\ell}\right) \sin\left[\left(\frac{2\pi}{\ell}\right)(x - \sqrt{M_\infty^2 - 1}y)\right] \tag{11.111}$$

$$v(x,y) = \phi_y(x,y) = -U_\infty \left(\frac{2\pi h}{\ell}\right) \sin\left[\left(\frac{2\pi}{\ell}\right)(x - \sqrt{M_\infty^2 - 1}y)\right] \tag{11.112}$$

The equation of a streamline in a two-dimensional planar flow is given by equation 11.95, which is repeated below.

$$\frac{dy}{dx} = \frac{\tilde{v}}{\tilde{u}} = \frac{v}{U_\infty + u} \tag{11.95}$$

Upon substituting equations 11.111 and 11.112 into equation 11.95, the slope of the streamlines is given by

$$\frac{dy}{dx} = \frac{-\left(\frac{2\pi h}{\ell}\right) \cos\left[\left(\frac{2\pi}{\ell}\right)(x - \sqrt{M_\infty^2 - 1}y)\right]}{1 + \frac{1}{\sqrt{M_\infty^2 - 1}} \cos\left[\left(\frac{2\pi}{\ell}\right)(x - \sqrt{M_\infty^2 - 1}y)\right]} \tag{11.113}$$

From equation 11.113, it is seen that all of the streamlines have the same slope along the lines on which the parameter $(x - \sqrt{M_\infty^2 - 1}y)$ is a constant. Referring to equation 11.105, it is seen that the function f is constant along such lines and from equations 11.111 and 11.112, the velocity components u and v are constant on those lines. Hence, all of the flow properties are constant along those lines.

Consequently, the flow pattern along the wavy wall is propagated unchanged in form out into the flow field along the lines specified by

$$x - \sqrt{M_\infty^2 - 1}\, y = \text{constant} \tag{11.114}$$

Differentiating equation 11.114 and solving for the slope of those lines along which $f = \text{constant}$ yields

$$\left(\frac{dy}{dx}\right)_{f = \text{constant}} = \frac{1}{\sqrt{M_\infty^2 - 1}} = \tan\alpha \tag{11.115}$$

where α is the Mach angle, illustrated in Fig. 11.7.

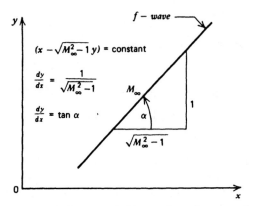

Figure 11.7   Relationship between lines of constant f and the Mach angle α.

Hence, *the lines of constant f are left-running Mach lines.* Accordingly, in the case of the supersonic flow over a wavy wall, the disturbances caused by the waviness in the wall are propagated into the compressible fluid along Mach lines. For a flow from left to right, the disturbances generated by the wavy wall are propagated downstream along left-running Mach lines. If the solution had been obtained for the *g-waves* instead of the *f-waves*, the aforementioned lines would be right-running Mach lines. It is apparent, however, that right-running Mach lines that originate at infinity, where there is no disturbance, cannot carry perturbations in the velocities of a flow that moves from left to right above a wavy wall. Hence, the function g is *zero* for such a flow. Only the family of *f-waves* is a solution to the flow from left to right over the wavy wall. For a flow in the lower halfplane, beneath the wavy wall, only the *g family*, or right-running Mach lines, would be present. For a flow from right to left, the aforementioned results are, of course, reversed.

Figure 11.8 illustrates selected streamlines for the supersonic flow over a wavy wall. It is seen that the perturbations in the surface are propagated undiminished to infinity along left-running Mach lines and that there is no attenuation in the amplitude of the disturbance caused by the wavy wall. Reference to equation 11.101 for the subsonic flow case indicates that the latter equation contains an exponential attenuation term. The supersonic flow is not symmetrical about a vertical line, as is the subsonic flow.

Introducing equation 11.111 into equation 11.73, the linearized pressure coefficient for supersonic flow is obtained. Thus,

$$C_p = -\frac{2u}{U_\infty} = -\frac{2}{\sqrt{M_\infty^2 - 1}}\left(\frac{2\pi h}{\ell}\right)\sin\left[\left(\frac{2\pi}{\ell}\right)(x - \sqrt{M_\infty^2 - 1}\, y)\right] \tag{11.116}$$

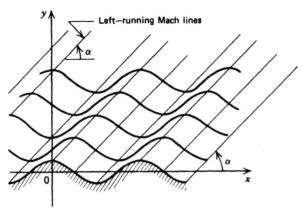

Figure 11.8 Streamline pattern for supersonic flow over a wavy wall.

The pressure coefficient at the wall surface, $y = 0$, is given by

$$(C_p)_{\text{Surface}} = -\frac{2}{\sqrt{M_\infty^2 - 1}}\left(\frac{2\pi h}{\ell}\right)\sin\left(\frac{2\pi x}{\ell}\right) \qquad (11.117)$$

Equation 11.117 shows that the pressure distribution at the wall is harmonic, with the peak pressures occuring at one quarter of a wavelength after the peak in the wall amplitude, as illustrated in Fig. 11.9, because the sine function in equation 11.117 lags the cosine function in equation 11.74 by one quarter of a wavelength. Hence, the pressure is equal to p_∞ at the bottom of the troughs in the wall, increases from left to right to a maximum value midway up the front side of the wall, and then decreases to p_∞ at the peaks in the wall. The pressure continues to decrease to a minimum value midway down the back side of the wall, and then increases to p_∞ at the troughs in the wall. The pressure, therefore, is always larger than p_∞ on the back surfaces of the wavy wall and less than p_∞ on the forward surfaces of the wavy wall. As a consequence, there is a drag on the wall from left to right. It will be recalled that according to the analysis, there is no drag for the subsonic flow over a wavy wall. Equation 11.116 also shows that for the supersonic flow over a wavy wall, the wall pressure distribution is propagated undiminished out to infinity along the

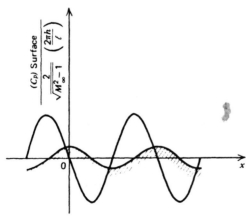

Figure 11.9 Wall pressure distribution for supersonic flow over a wavy wall.

left-running Mach lines, and that the flow upstream of a Mach line is unaware of the downstream disturbances.

11-5(c) Comparison of Subsonic and Supersonic Flow over a Wavy Wall

Several distinctive features of subsonic and supersonic flows are made apparent by the study of the linearized flow over a wavy wall. The streamline pattern is seen to have the same harmonic variation as the wall surface. In subsonic flow, the streamlines are symmetrical about vertical lines, indicating that the wall disturbances are propagated uniformly in all directions. In supersonic flow, the streamline pattern is unsymmetrical, and the wall disturbances are propagated directly along the *downstream family* of Mach lines. The intensity of the disturbance due to the waviness of the wall, as evidenced by both the amplitude of the streamline oscillation and the amplitude of the pressure coefficient variation, diminishes in subsonic flow as one moves away from the wall, and the effect of the disturbance vanishes completely at infinity. By contrast, the disturbance in the case of a supersonic flow is propagated with undiminished intensity out to infinity. The variation in the wall pressure coefficient is 180 deg out of phase with the wall oscillation in subsonic flow, so that no wall drag is developed. In supersonic flow, the variation in the pressure coefficient lags the wall oscillation by 90 deg, resulting in a wall drag force in the flow direction. The amplitude of the pressure coefficient variation increases in subsonic flow as the Mach number increases, whereas in supersonic flow, it decreases as the Mach number increases. The results discussed in the foregoing demonstrate vividly some of the differences in the physical effects produced by subsonic and supersonic compressible flows. They also give an insight into the type of initial and boundary conditions that may be specified for each type of flow.

11-5(d) Transonic Flow

As pointed out in Section 10-10(c), transonic flow is complicated by the fact that the entire flow field cannot be described by a single type of partial differential equation. The transonic flow over a wavy wall has been investigated by Kaplan.[7] The mathematical details are too involved to warrant inclusion here. It will only be mentioned that the solution technique is based on expanding the perturbation velocity potential into a power series. This technique, the power series expansion, is illustrated in Section 15-5, where it is employed for obtaining the solution for the transonic flow in the throat region of a converging-diverging nozzle.

In general, the transonic flow over bodies and inside flow passages is a branch of fluid mechanics that is not well developed either analytically or experimentally.[8]

11-6 SUMMARY

For small disturbances in a stationary gas, and for flows where the velocity deviates only slightly from a parallel uniform flow, the method of *small perturbations* may be employed for linearizing the governing equations. Simpler mathematical procedures may then be employed for obtaining solutions to the resulting linearized equations. The method of small perturbations is applicable to many situations, for example, acoustic fields, flow over thin airfoils, and flow in the transonic region of converging-diverging nozzles.

The propagation of an acoustic wave is governed by the wave equation, which is applicable to a small disturbance of arbitrary shape. The general solution of the one-dimensional wave equation for an initially undisturbed uniform gas is known as the *D'Alembert solution*, which shows that acoustic waves propagate at the speed of sound of the undisturbed fluid a_∞, and that the wave form is preserved during the process. Two families of waves are propagated in a one-dimensional acoustic field: right-running waves and left-running waves. Such waves are called *traveling waves*. When the propagation path is of finite length, the traveling waves are reflected at the boundaries, and *standing waves* may be formed by the superposition of the two traveling waves. Waves of the same type (either compression or expansion) reflect from a solid boundary, and waves of the opposite type reflect from a fixed pressure boundary. The D'Alembert solution of the wave equation illustrates the basic features of the propagation of acoustic waves.

The perturbation equations for steady two-dimensional flow are derived. A single equation is obtained that is valid for both subsonic flow and supersonic flow, and a separate equation is obtained for transonic flow. The small perturbation equation governing both subsonic flow and supersonic flow is a linear partial differential equation, whereas the small perturbation equation governing transonic flow retains one nonlinear term. A perturbation velocity potential is defined, the boundary condition for a flow with small perturbations is derived, and the pressure coefficient for a flow with small perturbations is obtained. The perturbation equation may be classified as elliptic for subsonic flow, parabolic for sonic flow, and hyperbolic for supersonic flow.

The solution to the classical problem of flow over a two-dimensional wavy wall is presented. That solution illustrates the general features of both subsonic flow and supersonic flow. In subsonic flow, the intensity of the disturbances because of the presence of the wavy wall diminishes as we move away from the wall and vanishes completely at infinity. The disturbances are propagated uniformly in all directions. In supersonic flow, however, the disturbances are propagated with undiminished intensity out to infinity along the Mach lines of the supersonic flow. The flow over a wavy wall demonstrates the physical differences produced by subsonic and supersonic flows, and gives insight into the type of initial and boundary conditions that may be specified for each type of flow.

Two additional problems of practical interest wherein the method of small perturbations is applicable are three-dimensional acoustic wave motion and the steady two-dimensional irrotational transonic flow in the throat of a converging-diverging nozzle. Those problems are discussed in Chapter 15.

REFERENCES

1. R. von Mises, *Mathematical Theory of Compressible Fluid Flow*, pp. 34–38, Academic Press, New York, 1958.
2. C. R. Wylie, *Advanced Engineering Mathematics*, Chap. 7, McGraw-Hill, New York, 1951.
3. L. A. Pipes and L. R. Harvill, *Applied Mathematics for Engineers and Physicists*, Chap. 12, McGraw-Hill, New York, 1970.
4. H. W. Liepmann and A. Roshko, *Elements of Gas Dynamics*, Chap. 4, Wiley, New York, 1957.
5. J. Ackeret, "Über Luftkräfte bei sehr grossen Geschwindigkeiten insbesondere bei ebenen Strömungen," *Helvetica Physica Acta*, Vol. 1, pp. 301–322, 1928.
6. R. V. Churchill, *Fourier Series and Boundary Value Problems*, Second Edition, McGraw-Hill, New York, 1969.
7. C. Kaplan, "On a Solution of the Nonlinear Differential Equations for Transonic Flow Past a Wave-Shaped Wall," NACA TN 2383, 1951.

8. R. D. Flack and H. D. Thompson, "Comparison of Pressure and LDV Velocity Measurements with Predictions in Transonic Flow," *Journal of the American Institute of Aeronautics and Astronautics*, Vol. 13, No. 1, pp. 53–59, January 1975.

PROBLEMS

1. Equation 11.17 is the classical one-dimensional wave equation for the pressure perturbation in a uniform region at rest. Show that the corresponding velocity perturbation also satisfies the one-dimensional wave equation.

2. In Section 11–3(b), it is shown that the *f-wave* in the D'Alembert solution of the wave equation represents a right-running wave that propagates an initial disturbance to the right at the speed a_∞, and preserves the wave form. Show that the *g-wave* represents a left-running wave that propagates an initial disturbance to the left at the speed a_∞, and preserves the wave form.

3. A symmetrical triangular shaped pressure disturbance having a peak pressure of 1000 N/m² above the pressure in the undisturbed fluid and a width of 0.01 m is propagating to the right in a closed cylindrical tube 1 m long containing air at a temperature of 300 K. Sketch the pressure distribution in the tube for selected instants of time, giving careful consideration to the wave reflection process at the closed ends of the tube.

4. The propagation of a planar acoustic wave in a uniform region is governed by the one-dimensional wave equation that is derived in Section 11–3(a). Derive the corresponding three-dimensional wave equation for an acoustic disturbance in a three-dimensional region.

5. The longitudinal acoustic wave motion in a long cylindrical tube of length L is governed by the one-dimensional wave equation. The general solution to that equation is the D'Alembert solution discussed in Section 11–3(b). The specific forms of the functions $f(x - a_x t)$ and $g(x + a_\infty t)$ may be determined by solving the wave equation by the separation of variables procedure presented in Section 11–5(a) for subsonic flow over a wavy wall. Thus, assume that

$$p(x,t) = X(x)T(t)$$

Substitute the above expression into equation 11.17, and show that the functions $X(x)$ and $T(t)$ satisfy the following ordinary differential equations:

$$X(x) = Ae^{i\lambda x} + Be^{-i\lambda x}$$

$$T(t) = Ce^{i\lambda a \cdot t} + De^{-i\lambda a \cdot t}$$

where λ is the separation constant. Initial and boundary conditions are required to simplify the results further.

For a tube closed at both ends, $u(0) = u(L) = 0$, and $u_t(0) = u_t(L) = 0$. Combine those boundary conditions with equations 11.15 and 11.16, and show that

$$X(x) = A\left[\exp\left(i\frac{n\pi x}{L}\right) + \exp\left(-i\frac{n\pi x}{L}\right)\right] = 2A\cos\left(\frac{n\pi x}{L}\right)$$

where $\lambda = n\pi/L$ $(n = 0,1,2, \ldots)$.

Standing harmonic wave patterns are obtained when $C = D$. In that case, show that

$$T(t) = 2C\cos\omega t$$

where $\omega = an\pi/L$ is the frequency in radians per second, and $f = \omega/2\pi$ is the frequency in cycles per second. n is called the wave number, where $n = 1$ is the first harmonic or fundamental mode, $n = 2$ is the second harmonic or first overtone, etc. The remaining constants of integration are determined by the initial condition

$$p(x,0) = f(x)$$

where $f(x)$ is the acoustic pressure distribution at $t = 0$.

6. The linearized pressure coefficient for the two-dimensional planar flow of a compressible fluid is $C_p = -2u/U_\infty$, where u is the perturbation velocity component in the x direction and U_∞ is the undisturbed free-stream velocity. Show that the corresponding linearized pressure coefficient for an incompressible fluid is $C_p = -2u_i/U_\infty$, where u_i is the perturbation velocity component for the incompressible fluid.

7. The lift and drag coefficients C_L and C_D, respectively, are defined by

$$C_L = \frac{L/A}{\frac{1}{2}\rho_\infty V_\infty^2} \quad \text{and} \quad C_D = \frac{D/A}{\frac{1}{2}\rho_\infty V_\infty^2}$$

where L and D are the lift and the drag forces, respectively, A is some characteristic area associated with the body, and $\rho_\infty V_\infty^2/2$ is the incompressible dynamic pressure. The lift and drag forces are obtained by integrating the pressure forces over the surface. For a two-dimensional planar body of width w, we obtain

$$L = \int_{A_{lower}} (p - p_\infty)w\, dx - \int_{A_{upper}} (p - p_\infty)w\, dx$$

$$D = \int_{A_{left}} (p - p_\infty)w\, dy - \int_{A_{right}} (p - p_\infty)w\, dy$$

where A_{lower} and A_{upper} refer to the lower and upper surfaces, respectively, and A_{left} and A_{right} refer to the left-facing and right-facing surfaces, respectively. The pressure coefficient C_p is defined as

$$C_p = \frac{p - p_\infty}{\frac{1}{2}\rho_\infty V_\infty^2}$$

Determine expressions for the lift and drag coefficients in terms of the pressure coefficient.

8. A thin wavy wall is employed as an airfoil. Neglecting leading edge effects, trailing edge effects, and end effects, determine an expression for the lift coefficient and drag coefficient per wave length for subsonic flow (see Problem 7).

9. For the airfoil described in Problem 8, determine an expression for the lift coefficient and drag coefficient per wave length for supersonic flow.

10. Consider a weakly disturbed two-dimensional planar subsonic flow between a wavy wall and a straight wall separated by the distance H. Determine the velocity potential for this flow and the pressure distribution along the wavy wall.

11. Consider a weakly disturbed two-dimensional planar subsonic flow between two wavy walls separated by the distance $2H$, where the peaks and troughs of the two profiles occur at the same axial locations. Determine the velocity potential for this flow and the pressure distribution along each of the two walls and the centerline.

12. *Similarity laws* are rules relating the pressure field for one flow with that for a related flow. The question of the possibility of relating the pressure field for subsonic flow over a thin airfoil to the pressure field for incompressible flow over a similar airfoil naturally arises, because of the large number of solutions available for incompressible flow for a wide variety of boundary conditions. One such set of similarity rules relating subsonic flow and incompressible flow over thin airfoils is called the *Goethert rules*.

    Show that the potential equation for two-dimensional planar incompressible flow is given by

$$\bar{\phi}_{xx} + \bar{\phi}_{yy} = 0$$

where $\bar{\phi}$ is the perturbation velocity potential for incompressible flow. The above equation is Laplace's equation.

    Define the following transformation equations.

$$\bar{x} = c_1 x, \quad \bar{y} = c_2 y, \quad \bar{\phi} = c_3 \phi, \quad \bar{U}_\infty = c_4 U_\infty$$

where c_1 to c_4 are constants to be determined so that the flows are similar, and the bar over a quantity denotes the incompressible flow value of that quantity. Show that the perturbation equation for two-dimensional planar subsonic flow (equation 11.49 with $\delta = 0$) transforms to Laplace's equation if

$$\frac{c_1}{c_2} = \frac{1}{\sqrt{1 - M_\infty^2}}$$

Show that to satisfy the linearized boundary condition (equation 11.66) for the incompressible flow, we require

$$\frac{c_2 c_4}{c_3} = -\frac{1}{\sqrt{1 - M_\infty^2}}$$

The pressure coefficients for linearized subsonic flow and incompressible flow are given by $C_p = -2u/U_\infty$ and $\bar{C}_p = -2\bar{u}/\bar{U}_\infty$, respectively. Show that

$$C_p = \frac{\bar{C}_p}{1 - M_\infty^2}$$

Show that the lift coefficient is given by

$$C_L = \bar{C}_L(1 - M_\infty^2)$$

The above transformation equations require a linear stretching of the dimensions of the airfoil. Such a transformation is called an *affine transformation*, and the corresponding similarity rules are known as the Goethert rules.

12

introduction to the method of characteristics with application to steady two-dimensional irrotational supersonic flow

12-1 PRINCIPAL NOTATION FOR CHAPTER 12

The notation of Chapter 10 (see Section 10-1) is applicable here. The following additional notation is employed in the present chapter.

C denotes characteristic curve $+$ or $-$.
Q,R,S,T coefficients in finite difference equations.

Greek Letters

α $= \sin^{-1}(1/M)$, Mach angle.
Γ initial-value line.
δ equals 0 for planar flow, and 1 for axisymmetric flow.
θ $= \tan^{-1}(v/u)$, flow or streamline angle.
λ slope of a characteristic curve.
σ arbitrary parameters in the method of characteristics.

Subscripts

x,y partial derivative with respect to that variable.
$1,2,4$ points in finite difference network.
$+$ denotes a positive, or left-running, characteristic.
$-$ denotes a negative, or right-running, characteristic.

12-2 INTRODUCTION

Chapter 10 discusses the partial differential equations that govern the steady flow of an adiabatic inviscid compressible fluid, and Chapter 11 presents the application of linearization of those equations for analyzing weakly disturbed flows. Linearization techniques yield useful results for uniform flows subjected to small perturbations. There are, however, important practical flow fields that cannot be approximated by a weakly disturbed uniform flow field. Such flow fields must, therefore, be analyzed by methods that obtain solutions to the pertinent nonlinear partial differential equations.

Only a few exact solutions are known to the nonlinear partial differential equations of fluid flow; one exact solution is that for a spherical source flow, which is essentially a one-dimensional flow in spherical coordinates (see Section 4-9). In general, the solutions to these nonlinear partial differential equations are based on applying numerical methods of analysis. Furthermore, the numerical method of analysis that is best suited for analyzing a specific problem depends on the type of partial differential equation governing the flow field.

As pointed out in Section 10-10(c), for a subsonic flow the governing equations are elliptic, and their solutions throughout the flow field depend on all of the boundary conditions; that is, those upstream and downstream from the region of interest. Numerical methods for solving subsonic flow problems are not well developed. Frequently, relaxation techniques are employed, and that numerical technique, at times, encounters serious problems of stability and convergence. Some success has been obtained with the application of the method of integral relations to subsonic flows.[1-3] Several investigators have determined the solution for a steady subsonic flow field by obtaining the asymptotic solution at large times for an equivalent unsteady flow problem having the same boundary conditions (see Reference 4-7); the latter time-dependent method changes the mathematical type of the governing partial differential equations from elliptic to hyperbolic [see Section 10-10(c)]. For hyperbolic equations, the flow properties at each point in the flow field depend on those properties in a finite region of the upstream flow field, but are independent of the downstream conditions. Thus, *marching-type* numerical methods may be applied for obtaining the solutions to such flow fields. The most accurate marching type method applicable to quasi-linear partial differential equations is the *method of characteristics*, which is the subject of the present chapter.

As shown in Section 10–10(c), the governing equations for a steady supersonic flow field are hyperbolic, and the method of characteristics is applicable to such a flow field.

A flow field in which there are regions of both subsonic flow and supersonic flow is called a *mixed flow field*. In the subsonic flow region the governing equations are elliptic, in the supersonic flow region they are hyperbolic, and they are parabolic where the flow is sonic. It is very difficult indeed to obtain solutions for mixed flow problems. The method of integral relations and also the time-dependent method may be applied to some types of mixed flow.

In the present chapter, the general theory of the *method of characteristics*, which is the most accurate numerical technique for solving hyperbolic partial differential equations, is discussed. The method of characteristics is applied to steady two-dimensional, planar and axisymmetric, *irrotational* supersonic flow. A complete numerical algorithm is developed for determining the flow properties at an interior point in a supersonic flow field. In Chapter 16, numerical algorithms for determining the flow properties at boundary points in a supersonic flow field and examples that illustrate the implementation of the numerical methods are presented. In that chapter several applications of the method of characteristics to practical engineering problems are presented.

Simple waves are a special type of flow that occurs frequently in practical compressible flow problems. The Prandtl-Meyer expansion fan discussed in Section 8–3 is a practical example of a simple wave in a steady two-dimensional irrotational supersonic flow field. The expansion waves described in Section 13–4(b) are practical examples of a simple wave in an unsteady one-dimensional planar homentropic flow field. The properties of simple waves are reviewed in Section 12–6, and proofs of those properties are presented.

12–3 GENERAL THEORY OF THE METHOD OF CHARACTERISTICS

In many engineering problems the governing differential equations are a set of *quasi-linear nonhomogeneous partial differential equations of the first order* for functions of two independent variables. A quasi-linear partial differential equation of the first order is defined as one that may be *nonlinear in the dependent variables*, but is *linear in their first partial derivatives*. When the pertinent governing equations are hyperbolic, the method of characteristics may be employed for obtaining their solution.

The concept of characteristics may be introduced from several points of view.

1. From a physical point of view, a *characteristic curve*, for brevity termed a *characteristic*, is defined as the path of propagation of a physical disturbance. In Section 3–8(b) it is shown that in a supersonic flow field, disturbances are propagated along the Mach lines for the flow. It is shown in Section 12–4 that the *Mach lines are the characteristics* for a supersonic flow.

2. From a purely heuristic point of view, a characteristic is defined as a curve along which the governing partial differential equations can be manipulated into total differential equations.

3. From a more rigorous mathematical point of view, a characteristic is a curve across which the derivatives of a physical property may be discontinuous, while the property itself remains continuous. Thus, regions of flow having continuous properties and derivatives within each region, but a discontinuity in derivatives at their interface, may be joined together along a characteristic. For example, the Prandtl-Meyer expansion fan is joined to the uniform upstream and downstream flow regions along Mach lines [see Section 8–3(a)].

4. From the most rigorous mathematical point of view, a characteristic is defined as a curve along which the governing partial differential equations reduce to an *interior operator*, that is, a total differential equation. That *interior operator* is known as the *compatibility equation*. Accordingly, along a characteristic, the dependent variables may not be specified arbitrarily, because they must satisfy the compatibility equation.

These concepts may be employed for developing numerical procedures for solving hyperbolic partial differential equations. For simplicity, the theory will be developed first for a single partial differential equation, and then for a system of two partial differential equations. The theory is extended to systems of any number of partial differential equations in Appendix B. The method of characteristics for three independent variables is introduced in Chapter 20.

12–3(a) A Single Partial Differential Equation

The general features of the method of characteristics are developed in this section by each of the three approaches outlined in Items 2, 3, and 4 of Section 12–3. Consider a linear combination of the first-order partial derivatives of the function $f(x,y)$ (that is, a first-order partial differential equation). Thus,

$$af_x + bf_y + c = 0 \qquad (12.1)$$

In equation 12.1, the coefficients a and b, and the nonhomogeneous term c, may be functions of x, y, and f [$a = a(x,y,f)$, etc.]. The characteristic and compatibility equations corresponding to equation 12.1 are derived below.

Consider *first* the heuristic approach presented as Item 2 in Section 12–3. Rearrange equation 12.1 to read as follows.

$$a\left(f_x + \frac{b}{a}f_y\right) + c = 0 \qquad (12.2)$$

If the dependent variable $f(x,y)$ is restricted to being a *continuous function* (no restrictions are placed on the derivatives of f), then the total differential df is given by

$$df = f_x \, dx + f_y \, dy \qquad (12.3)$$

Equation 12.3 may be rearranged to give the derivative df/dx. Thus,

$$\frac{df}{dx} = \left(f_x + \frac{dy}{dx}f_y\right) \qquad (12.4)$$

Let the *characteristic* for equation 12.1 be *defined* as the curve in the xy plane that has the slope, at each of its points, given by

$$\frac{dy}{dx} = \frac{b}{a} = \lambda(x,y,f) \qquad \text{(characteristic equation)} \qquad (12.5)$$

where the slope λ of the characteristic may depend on x, y, and f, because, in general the coefficients a and b may be functions of the latter variables. It remains to demonstrate that the curve specified by equation 12.5 has the properties attributed to a characteristic.

Accepting for the moment that equation 12.5 does indeed define a characteristic for the partial differential equation 12.1, then $dy/dx = b/a$ along that characteristic, and equation 12.4 may be written as

$$\frac{df}{dx} = \left(f_x + \frac{b}{a} f_y \right) \qquad (along\ the\ characteristic) \qquad (12.6)$$

Substituting the left-hand side of equation 12.6 for the term in parentheses in equation 12.2 gives

$$a\,df + c\,dx = 0 \qquad (compatibility\ equation) \qquad (12.7)$$

Equation 12.7 is a total differential equation, valid along, and only along, the characteristic specified by equation 12.5. Consequently, equation 12.7 is the *compatibility equation* corresponding to equation 12.1. The existence of equation 12.7, valid along the characteristic specified by equation 12.5, justifies the choice of equation 12.5 as the defining equation for the characteristic.

A characteristic is also defined (in Item 3, Section 12–3) as a curve along which the derivatives of the dependent variables may be discontinuous. To demonstrate that the curve defined by equation 12.5 does have that property, consider equations 12.1 and 12.3 as two equations for determining the partial derivatives f_x and f_y. Thus,

$$af_x + bf_y = -c$$
$$dx\,f_x + dy\,f_y = df \qquad\qquad (12.8)$$

Equations 12.8 may be solved simultaneously for f_x and f_y by applying Cramer's rule (see Appendix A, Section A–3). Thus, for f_y,

$$f_y = \frac{\begin{vmatrix} a & -c \\ dx & df \end{vmatrix}}{\begin{vmatrix} a & b \\ dx & dy \end{vmatrix}} = \frac{a\,df + c\,dx}{a\,dy - b\,dx} \qquad (12.9)$$

By virtue of equation 12.5, the denominator of equation 12.9 is zero along the curve specified by equation 12.5, and from equation 12.7, the numerator must also be zero along that curve. Accordingly, along the characteristic specified by equation 12.5,

$$f_y = \frac{0}{0} \qquad (12.10)$$

Hence, the partial derivative f_y is indeterminate and may be discontinuous along the characteristic. Note that equation 12.10 states only that the derivative *may* be discontinuous, not that it *must* be discontinuous. The latter statement is important because the method of characteristics is also applicable to flows that have continuous derivatives. Solving for f_x gives

$$f_x = \frac{\begin{vmatrix} -c & b \\ df & dy \end{vmatrix}}{\begin{vmatrix} a & b \\ dx & dy \end{vmatrix}} = \frac{-(b\,df + c\,dy)}{a\,dy - b\,dx} \qquad (12.11)$$

Here too, the denominator of equation 12.11 is equal to zero along the characteristic, and it can be shown that the numerator is likewise zero, by substituting for b from equation 12.5 in terms of a, dx, and dy. Consequently, f_x is also indeterminate, and may, therefore, be discontinuous along the characteristic.

From a more rigorous mathematical point of view, a characteristic is defined (in Item 4, Section 12–3) as a curve along which the governing partial differential equation reduces to an *interior operator*, that is, a total differential relationship along the characteristic. The governing partial differential equation, equation 12.1, may be viewed as the projection of the vector ∇f in the direction of the arbitrary vector $\mathbf{W}$, where $\mathbf{W}$ is defined by

$$\mathbf{W} = \mathbf{i}a + \mathbf{j}b \qquad (12.12)$$

Equation 12.1 may, therefore, be written in the form

$$af_x + bf_y = \mathbf{W} \cdot \nabla f = d_{\mathbf{W}}f = -c \qquad (12.13)$$

The notation $d_{\mathbf{W}}f$ denotes the derivative of f in the direction of the vector $\mathbf{W}$.

Accordingly, the following question may be posed. Is there a curve that is everywhere tangent to the vector $\mathbf{W}$, so that equation 12.13 is an *interior operator* along that curve? Such a curve, if it exists, is called a *characteristic*, and the interior operator specified by equation 12.13 is termed the *compatibility equation*. If a characteristic tangent to $\mathbf{W}$ exists throughout the region of the xy plane that is of interest, then at every point of the characteristic, the normal, $\mathbf{N} = \mathbf{i}n_x + \mathbf{j}n_y$, to the characteristic must be perpendicular everywhere to the vector $\mathbf{W}$. Hence, at a point on the characteristic,

$$\mathbf{N} \cdot \mathbf{W} = n_x a + n_y b = 0 \qquad (12.14)$$

Figure 12.1 illustrates the vectors $\mathbf{W}$ and $\mathbf{N}$, and the characteristic curve C. From equation 12.14, the tangent to the normal vector $\mathbf{N}$ is in the direction

$$\frac{n_y}{n_x} = -\frac{a}{b} \qquad (12.15)$$

The slope of the characteristic C is everywhere normal to $\mathbf{N}$; that is, the slope of curve C is given by the negative reciprocal of the slope of $\mathbf{N}$. Thus,

$$\frac{dy}{dx} = \frac{b}{a} \qquad (along\ the\ characteristic\ C) \qquad (12.16)$$

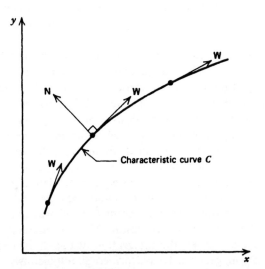

Figure 12.1 Relationship between the vectors $\mathbf{W}$ and $\mathbf{N}$, and the characteristic curve C.

Equation 12.16 is the same as equation 12.5. The integral of equation 12.16 defines the desired characteristic along which the interior operator, defined by equation 12.13, is applicable.

It is desirable to express equation 12.13 in a form that shows explicitly that $d_{\mathbf{w}}f$ is an interior operator (i.e., a total differential relationship). For a continuous function $f(x,y)$,

$$\frac{df}{dx} = f_x + \frac{dy}{dx} f_y \tag{12.17}$$

Along the characteristic C specified by equation 12.16, equation 12.17 may be written as

$$\frac{df}{dx} = f_x + \frac{b}{a} f_y \tag{12.18}$$

Equation 12.13 may be rewritten in the form

$$d_{\mathbf{w}}f = a\left(f_x + \frac{b}{a} f_y\right) = -c \tag{12.19}$$

Substituting the left-hand side of equation 12.18 for the term in parentheses in equation 12.19 yields

$$d_{\mathbf{w}}f = a\frac{df}{dx} = -c \tag{12.20}$$

Equation 12.20 may be rewritten to give

$$a\,df + c\,dx = 0 \qquad \text{(compatibility equation)} \tag{12.21}$$

Equation 12.21 is, therefore, the desired interior operator that applies along the characteristic C, and is identical to equation 12.7 determined previously.

In the preceding discussions, the concept of a characteristic is developed from the three different points of view presented at the beginning of Section 12–3.

Partial differential equations for which characteristics exist are known as *hyperbolic* equations. The function $f(x,y)$ may be determined along a characteristic by solving the compatibility equation, equation 12.7, which is a total differential equation relating df to dx along the characteristic. Hence, the original partial differential equation, equation 12.1, may be replaced by the equivalent system of a characteristic along which a compatibility equation is valid. Such a replacement is the basis of the method of characteristics.

In the general case where the coefficients a and b depend on the unknown function f, so that λ depends on f, equation 12.5 cannot be integrated until f is known. However, f is determined by integrating equation 12.7 along the characteristic obtained by integrating equation 12.5. In general, equations 12.5 and 12.7 must be solved simultaneously. If a, b, and c are complicated expressions, which is ordinarily the case, it is highly probable that the integration will have to be performed numerically.

To solve equation 12.1, initial data must be specified. Figure 12.2 illustrates an *initial-value line* $\Gamma_o(x,y)$ in the xy plane along which $f(x,y) = f_o(x,y)$ is specified. Starting at any point $P(x,y)$ on Γ_o, equation 12.5 is integrated to determine the characteristic C in the xy plane that passes through the *initial-value point* P. The value of the function f at any selected point, such as point Q on characteristic C, is then determined by integrating the compatibility equation 12.7 along characteristic C from point P to point Q. As stated earlier, the integration of the characteristic and compatibility equations must, in general, be performed simultaneously, and usually by numerical techniques.

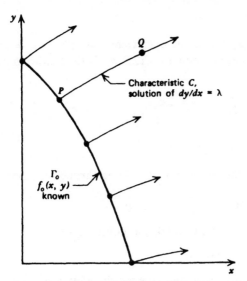

Figure 12.2 Characteristics for a first-order partial differential equation.

By selecting different points, such as point P along the initial-value curve Γ_o, the entire xy plane may be covered with characteristics, along each of which the values of the dependent variable $f(x,y)$ may be determined. If the solution is performed numerically, then values are obtained only at discrete points. By increasing the number of points selected on the initial-value line Γ_o, more characteristics are extended into the xy plane. Consequently, if the spacing of those discrete points is made sufficiently close, a practically continuous distribution is obtained.

It is important to keep in mind the assumptions and restrictions that must be observed in applying the method of characteristics. It is applicable only to quasi-linear hyperbolic partial differential equations. Although developed here for first-order equations, it can be applied in some cases to higher-order equations. Otherwise, there are no restrictions on the equations. They may be both nonlinear and non-homogenous. They are, however, subject to the important *restriction that the dependent variables be continuous in the region of interest.* It is mandatory that the latter condition be satisfied. It should be noted, however, that there are no restrictions on the continuity of the derivatives of the dependent variable.

In the case of a single partial differential equation, the method of characteristics is always applicable. For systems comprising more than one partial differential equation, further limitations on the applicability of the method are encountered. In Section 12-3(b), the method of characteristics is extended to a system of two equations. In Appendix B, the method is extended to a system of any number of equations.

Example 12.1.   Consider the first-order partial differential equation

$$u_x + 2xu_y - 3x^2 = 0 \qquad\qquad (a)$$

The initial condition for $u(x,y)$ is given by

$$u(0,y) = 5y + 10 \qquad\qquad (b)$$

Employing the method of characteristics, determine (a) the equation for the characteristic passing through the point (2,4), (b) the compatibility equation valid along that characteristic, and (c) the value of $u(2,4)$.

Solution

(a)   The functional relation is $u(x,y)$, and the coefficients are $a = 1$, $b = 2x$, and $c = -3x^2$. Substituting those values into equation 12.5 yields the following characteristic equation.

$$\frac{dy}{dx} = \frac{2x}{1} = 2x \tag{c}$$

Integrating equation (c) yields

$$y = x^2 + C_1 \tag{d}$$

The value of the constant C_1 for the characteristic passing through the point $(2,4)$ is, from equation (d),

$$4 = (2)^2 + C_1, \qquad C_1 = 0$$

Hence, the equation of the characteristic through the point $(2,4)$ is

$$y = x^2 \tag{e}$$

(b)   The compatibility equation corresponding to the partial differential equation (a) is obtained from equation 12.7. Thus,

$$(1)\, du + (-3x^2)\, dx = 0 \tag{f}$$

Integrating equation (f) gives

$$u = x^3 + C_2 \tag{g}$$

The value of the constant C_2 for the compatibility equation that is valid along the characteristic specified by equation (e) is determined from the initial condition, equation (b). The initial condition requires that the value of y corresponding to $x = 0$ be determined on the characteristic specified by equation (e). Hence, from equation (e),

$$y = (0)^2 = 0 \tag{h}$$

From equation (h) it follows that the characteristic passing through the point $(2,4)$ crosses the initial-value line at the point $(0,0)$. Substituting that point into equation (b) yields

$$u(0,0) = 5(0) + 10 = 10 \tag{i}$$

Substituting the initial value $u(0,0) = 10$ into equation (g) gives

$$10 - (0)^3 = C_2, \qquad C_2 = 10 \tag{j}$$

Consequently, from equation (g), the compatibility equation that is valid along the characteristic passing through the point $(2,4)$ is

$$u = x^3 + 10 \tag{k}$$

(c)   The value of $u(2,4)$ is found from equation (k). Thus,

$$u(2,4) = (2)^3 + 10 = 18$$

12–3(b) System of Two Partial Differential Equations

Consider a system comprised of two partial differential equations, denoted by L_1 and L_2, involving the two dependent variables $u(x,y)$ and $v(x,y)$. Thus,

$$\begin{aligned}
L_1 &= a_{11}u_x + b_{11}u_y + a_{12}v_x + b_{12}v_y + c_1 = 0 \\
L_2 &= a_{21}u_x + b_{21}u_y + a_{22}v_x + b_{22}v_y + c_2 = 0
\end{aligned} \tag{12.22}$$

In equation 12.22 the a's, b's, and c's may be functions of x, y, u, and v. It is desired to find an equivalent system of characteristic and compatibility equations that may replace equations 12.22.

Since equations 12.22 are coupled in the dependent variables $u(x,y)$ and $v(x,y)$, both equations must be considered simultaneously. To do that, form the *differential operator L*, which is a *linear sum* of L_1 and L_2. Thus,

$$L = \sigma_1 L_1 + \sigma_2 L_2 = 0 \tag{12.23}$$

In equation 12.23 the parameters σ_1 and σ_2 are arbitrary functions, or weighting factors, which are to be determined later. Substituting equations 12.22 into equation 12.23 and putting the result into a form that is analagous to equation 12.2 yields the following expression. Thus,

$$(a_{11}\sigma_1 + a_{21}\sigma_2)\left[u_x + \frac{(b_{11}\sigma_1 + b_{21}\sigma_2)}{(a_{11}\sigma_1 + a_{21}\sigma_2)} u_y \right]$$

$$+ (a_{12}\sigma_1 + a_{22}\sigma_2)\left[v_x + \frac{(b_{12}\sigma_1 + b_{22}\sigma_2)}{(a_{12}\sigma_1 + a_{22}\sigma_2)} v_y \right] + (c_1\sigma_1 + c_2\sigma_2) = 0 \tag{12.24}$$

Since the dependent variables must be continuous,

$$\frac{du}{dx} = u_x + \lambda u_y \quad \text{and} \quad \frac{dv}{dx} = v_x + \lambda v_y \tag{12.25}$$

where $\lambda = dy/dx$. The *characteristics* are defined as the curves having the slopes

$$\lambda = \frac{b_{11}\sigma_1 + b_{21}\sigma_2}{a_{11}\sigma_1 + a_{21}\sigma_2} \quad \text{and} \quad \lambda = \frac{b_{12}\sigma_1 + b_{22}\sigma_2}{a_{12}\sigma_1 + a_{22}\sigma_2} \tag{12.26}$$

Along the characteristics specified by equation 12.26, equation 12.24 becomes

$$(a_{11}\sigma_1 + a_{21}\sigma_2)\, du + (a_{12}\sigma_1 + a_{22}\sigma_2)\, dv + (c_1\sigma_1 + c_2\sigma_2)\, dx = 0 \tag{12.27}$$

Equation 12.27 is the *compatibility equation* for equations 12.22.

Equations 12.26 may be rearranged with the arbitrary parameters σ_1 and σ_2 considered as the unknown variables. Thus,

$$\sigma_1(a_{11}\lambda - b_{11}) + \sigma_2(a_{21}\lambda - b_{21}) = 0$$
$$\sigma_1(a_{12}\lambda - b_{12}) + \sigma_2(a_{22}\lambda - b_{22}) = 0 \tag{12.28}$$

For equations 12.28 to have a solution for σ_1 and σ_2 other than the trivial one $\sigma_1 = \sigma_2 = 0$, the determinant of the coefficients of σ_1 and σ_2 must vanish. Thus,

$$\begin{vmatrix} (a_{11}\lambda - b_{11}) & (a_{21}\lambda - b_{21}) \\ (a_{12}\lambda - b_{12}) & (a_{22}\lambda - b_{22}) \end{vmatrix} = 0 \tag{12.29}$$

Expanding the determinant in equation 12.29 gives the following equation for λ.

$$a\lambda^2 + b\lambda + c = 0 \tag{12.30}$$

In equation 12.30, $a = (a_{11}a_{22} - a_{12}a_{21})$, $b = (-a_{22}b_{11} - a_{11}b_{22} + a_{12}b_{21} + a_{21}b_{12})$, and $c = (b_{11}b_{22} - b_{12}b_{21})$. Since a, b, and c are functions of the coefficients of the original system of equations (equations 12.22), they are, of course, also functions of x, y, u, and v.

The *type* of a partial differential equation may be classified according to the magnitude of the term $(b^2 - 4ac)$. If $(b^2 - 4ac) < 0$, no real solutions exist for λ, and the characteristics are imaginary. A set of differential equations that results in imaginary characteristics is termed *elliptic*. If $(b^2 - 4ac) = 0$, one real characteristic passes through each point, and the set of differential equations is *parabolic*. If $(b^2 - 4ac) > 0$, two real characteristics pass through each point, and the set of differential equations is *hyperbolic*. The discussions that follow are restricted to hyperbolic systems of differential equations, for brevity termed *hyperbolic systems* [see also Section 10–10(c)].

For a hyperbolic system, equation 12.30 has two distinct roots, or solutions: λ_+ and λ_-. Hence, there are *two characteristics* that satisfy the following two ordinary differential equations.

$$\frac{dy}{dx} = \lambda_+ \qquad \text{and} \qquad \frac{dy}{dx} = \lambda_- \qquad (12.31)$$

Since the roots λ_+ and λ_- are functions of x, y, u, and v, the location of the characteristics in a specific problem depends on the particular values of u and v. Assuming that a solution for u and v exists (whether known or to be determined), the equations $dy/dx = \lambda_+(u,v,x,y)$ and $dy/dx = \lambda_-(u,v,x,y)$, where u and v are functions of x and y, are two ordinary differential equations of the first order that define two families of characteristics in the xy plane.

Return now to equation 12.27, which is the compatibility equation for the system. The arbitrary parameters σ_1 and σ_2 may be eliminated in terms of λ_+ and λ_- by solving either of equations 12.28 for one of the σ's in terms of the other. That result may then be substituted into equation 12.27, and the remaining σ may be cancelled out, since it appears in every term, thus eliminating both σ_1 and σ_2. Accordingly, two equations are obtained; one for λ_+ and one for λ_-. They are the two compatibility equations relating du, dv, dx, and dy along the characteristics given by equations 12.31. We may, therefore, replace the original system of partial differential equations (i.e., equations 12.22) by equations 12.27 and 12.31.

The integration of equations 12.31 determines two characteristics in the xy plane passing through each point thereof; one corresponding to λ_+ and the other corresponding to λ_-. As in the case of a single equation [see Section 12–3(a)], initial data must either be specified or calculable along some initial-value curve Γ_o lying in the xy plane. The solution is then propagated from those initial points along the characteristics into the xy plane. Along each C_+ and C_- characteristic, corresponding to λ_+ and λ_-, respectively, one relationship between du, dv, dx, and dy is provided by the compatibility equation (equation 12.27). Because that equation involves both u and v, it alone cannot be solved for both u and v along a single characteristic. At the intersection of a C_+ and a C_- characteristic, however, two equations relating u and v are available, one compatibility equation that is valid on the C_+ and another that is valid on the C_- characteristic. At the point of intersection there are, therefore, sufficient relationships available for solving for both u and v.

Figure 12.3 illustrates schematically these concepts. The initial-value line is denoted by Γ_o, and u and v are known along Γ_o. At each point on Γ_o, a C_+ and a C_- characteristic may be extended out into the xy plane by integrating equation 12.27. The C_+ and C_- characteristics emanating from a particular point, such as A, do not intersect. However, a C_+ and C_- characteristic may also be extended out from point B. The C_- characteristic emanating from point A intersects the C_+ characteristic emanating from point B at the point D. The two compatibility equations valid along AD and BD may therefore be solved simultaneously for u and v at the point D.

The foregoing procedure may be applied at any two points lying on Γ_o, up to and including the end points. A new initial-value line, denoted by Γ_1, is thereby produced, and along Γ_1 the solution is known. The entire process applied along Γ_o may be repeated along Γ_1, yielding a second solution line Γ_2, etc. It is possible to continue the

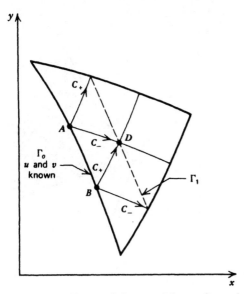

Figure 12.3 Characteristic network for two first-order
partial differential equations.

process until the entire region of interest has been covered, or the full extent of the
influence of the initial-value line Γ_o has been reached. To continue beyond that point
requires the specification of boundary data in addition to the initial data. In mathematics, an initial-value problem such as that described above is known as a *Cauchy
problem*.

In general, because of the nonlinear nature of the original partial differential
equations and the coupling between the characteristic and compatibility equations,
the solution must be obtained numerically at a selected number of discrete points. By
varying the point spacing along Γ_o, any desired number of solution points may be
obtained, so that a nearly continuous solution for the dependent variables u and v
may be approached. The following two considerations, which are pointed out in
Section 12–3(a), must be kept in mind; (1) the variables u and v must be continuous in
order that equation 12.25 be valid, a requirement that limits the applicability of the
method of characteristics to regions where all of the dependent variables are continuous, but excludes regions where there are discontinuities, such as shock waves;
(2) no restrictions are placed on the derivatives of the dependent variables.

12–3(c) Domain of Dependence and Range of Influence

The considerations presented in Section 12–3(b) lead to the concepts of a *domain of
dependence* and a *range of influence*.

Figure 12.4 illustrates the domain of dependence of a point P, plotted in the xy
plane. It is the region in the xy plane bounded by the outermost characteristics from
the initial-value line Γ_o that pass through the point P. Thus, it is the region wherein
the solution of the initial-value problem may be obtained.

Figure 12.5 illustrates the range of influence of a point Q located on the initial-value
line Γ_o. It is the region in the xy plane containing all of the points that are influenced
by the initial data at point Q. The range of influence is composed of all of the points
having a domain of dependence containing the point Q. It is, therefore, the region
between the two outermost characteristics passing through point Q.

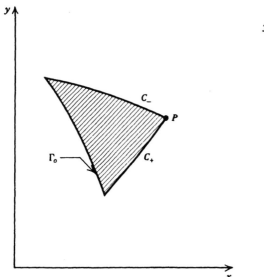

Figure 12.4 Domain of dependence.

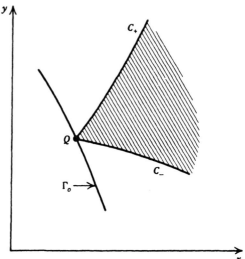

Figure 12.5 Range of influence.

For a solution to be possible, the initial-value line cannot be a characteristic at any place unless the initial data are given along two intersecting characteristic curves. The latter type of problem is known as the *Goursat problem*. Several types of domains having different types of initial-value lines may be analyzed. The present discussion is concerned, however, only with such domains in which no part of the initial-value line is a characteristic.

By applying the method of characteristics as developed in the foregoing, it is possible to solve many complicated systems of partial differential equations, provided the system is quasi-linear and hyperbolic. Equations of the latter type are encountered frequently in fluid flow problems.

12–4 APPLICATION TO STEADY TWO-DIMENSIONAL, PLANAR AND AXISYMMETRIC, IRROTATIONAL SUPERSONIC FLOW

The general theory of the method of characteristics is now applied to develop a numerical procedure for analyzing the flow field for a steady two-dimensional, planar and axisymmetric, irrotational supersonic flow. It is shown that the characteristics, as stated at the beginning of Section 12–3, are *Mach lines*, and that there are sufficient compatibility equations for replacing the original system of governing partial differential equations.

A complete numerical algorithm, called a *unit process*, is developed for an interior point of the flow. A numerical example is presented illustrating the application of the unit process for an *interior point* of a supersonic flow field.

12–4(a) Governing Equations

The equations governing the steady two-dimensional, planar and axisymmetric, irrotational flow of a compressible fluid are the *gas dynamic equation* (equation 10.193), the *irrotationality condition* (equation 10.189), and the *speed of sound relationship* (equation 10.121). Those equations are summarized in Table 12.1.

Table 12.1   Governing Equations for Steady Two-Dimensional Irrotational Flow

$$(u^2 - a^2)u_x + (v^2 - a^2)v_y + 2uvu_y - \frac{\delta a^2 v}{y} = 0 \qquad (10.193)$$

$$u_y - v_x = 0 \qquad (10.189)$$

$$a = a(V) = a(u,v) \qquad (\textit{throughout the flow field}) \qquad (10.121)$$

In equation 10.193, $\delta = 0$ for planar flow, and $\delta = 1$ for axisymmetric flow.

Equation 10.121 is obtained from the general thermodynamic properties of a homenergetic flow, and it places no restrictions on the specific thermal and caloric equations of state for the fluid [see Section 10-9(a)]. All that is required is that the functional relationship $a = a(V)$ be known. It may be algebraic, as in equation 3.84 for a perfect gas, or in tabular form, as in the case of either a frozen gas mixture with variable specific heats, or an equilibrium gas mixture having a variable composition and variable specific heats (see Sections 14-4 and 14-5).

12-4(b) Characteristic Equation

The procedure presented in Section 12-3(b) is now applied to the system of two partial differential equations comprising equations 10.193 and 10.189. Multiplying those two equations by the unknown parameters σ_1 and σ_2, respectively, and adding the result gives the following equation, which is analogous to equation 12.23. Thus,

$$\sigma_1 \,(\text{equation } 10.193) + \sigma_2 \,(\text{equation } 10.189) = 0$$

Substituting equations 10.193 and 10.189 into the above equation yields

$$u_x[\sigma_1(u^2 - a^2)] + u_y[\sigma_1(2uv) + \sigma_2] + v_x[-\sigma_2]$$
$$+ v_y[\sigma_1(v^2 - a^2)] + [-\sigma_1 \delta a^2 v/y] = 0 \qquad (12.32)$$

Factoring out the coefficients of the partial derivatives of u and v in the x direction, and rearranging, the following expression is obtained.

$$\sigma_1(u^2 - a^2)\left[u_x + \frac{\sigma_1(2uv) + \sigma_2}{\sigma_1(u^2 - a^2)}u_y\right]$$
$$+ (-\sigma_2)\left[v_x + \frac{\sigma_1(v^2 - a^2)}{-\sigma_2}v_y\right] - \frac{\sigma_1 \delta a^2 v}{y} = 0 \qquad (12.33)$$

From equation 12.25, if $u(x,y)$ and $v(x,y)$ are assumed to be continuous functions then

$$\frac{du}{dx} = u_x + \lambda u_y \qquad \text{and} \qquad \frac{dv}{dx} = v_x + \lambda v_y \qquad (12.34)$$

where, as before, $\lambda = dy/dx$. In accordance with equation 12.26, the characteristics corresponding to equation 12.33 have the slopes λ that are equal to the coefficients of u_y and v_y in equation 12.33. Thus

$$\lambda = \frac{\sigma_1(2uv) + \sigma_2}{\sigma_1(u^2 - a^2)} \qquad \text{and} \qquad \lambda = \frac{\sigma_1(v^2 - a^2)}{-\sigma_2} \qquad (12.35)$$

Substituting equations 12.34 and 12.35 into equation 12.33 transforms the latter to read as follows.

$$\sigma_1(u^2 - a^2)\,du - \sigma_2\,dv - \left(\frac{\sigma_1 \delta a^2 v}{y}\right)dx = 0 \qquad (12.36)$$

Equation 12.36 is analogous to equation 12.27, and is a total differential equation. Accordingly, it is the *compatibility equation* corresponding to the system of equations 10.193 and 10.189. Equation 12.36 is valid when λ is given by equations 12.35.

It remains to derive expressions for λ and to eliminate the unknown parameters σ_1 and σ_2 from equation 12.36. Equations 12.35 may be rewritten in the following forms. Thus,

$$\sigma_1[(u^2 - a^2)\lambda - 2uv] + \sigma_2(-1) = 0$$
$$\sigma_1(v^2 - a^2) + \sigma_2(\lambda) = 0 \tag{12.37}$$

Equations 12.37 are analogous to equations 12.28. Accordingly, the solution for σ_1 and σ_2, other than the trivial one $\sigma_1 = \sigma_2 = 0$, requires that the determinant of the coefficients of σ_1 and σ_2 vanish; that is,

$$\begin{vmatrix} [(u^2 - a^2)\lambda - 2uv] & -1 \\ (v^2 - a^2) & \lambda \end{vmatrix} = 0 \tag{12.38}$$

Expanding the determinant and setting it equal to zero yields

$$(u^2 - a^2)\lambda^2 - 2uv\lambda + (v^2 - a^2) = 0 \tag{12.39}$$

Equation 12.39 is analogous to equation 12.30.

Solving equation 12.39 for the values of λ along which equation 12.36 is valid yields

$$\lambda_\pm = \left(\frac{dy}{dx}\right)_\pm = \frac{uv \pm a^2\sqrt{M^2 - 1}}{u^2 - a^2} \tag{12.40}$$

Equation 12.40 yields two results for λ, denoted by the subscripts $+$ and $-$, corresponding to the positive and negative signs, respectively, preceding the square root sign. Those two ordinary differential equations specify two curves in the xy plane, which are the characteristics, and they are real when the Mach number $M > 1$. Consequently, the method of characteristics is applicable to a steady two-dimensional, planar and axisymmetric, *supersonic* flow, but *not* to a *subsonic* flow.

From the geometries presented in Figs. 12.6 and 12.7, an alternate form of equation 12.40 may be obtained by expressing u and v in terms of the velocity magnitude V and the flow or streamline angle θ, and by expressing M in terms of the Mach angle α. From Fig. 12.6,

$$u = V\cos\theta, \qquad v = V\sin\theta, \qquad \theta = \tan^{-1}\left(\frac{v}{u}\right) \tag{12.41}$$

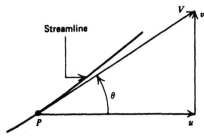

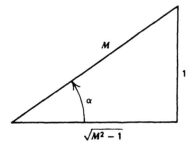

Figure 12.6 Relationship between the velocity components u and v, and the velocity magnitude V and the flow angle θ.

Figure 12.7 Relationship between the Mach angle α and the Mach number M.

From Fig. 12.7,

$$\alpha = \sin^{-1}\left(\frac{1}{M}\right), \qquad M = \text{cosec } \alpha, \qquad \sqrt{M^2 - 1} = \cot \alpha \qquad (12.42)$$

Substituting equations 12.41 and 12.42 into equation 12.40 gives

$$\lambda_{\pm} = \frac{V^2 \sin \theta \cos \theta \pm a^2 \cot \alpha}{V^2 \cos^2 \theta - a^2} = \frac{\sin \theta \cos \theta \pm \cot \alpha (1/M^2)}{\cos^2 \theta - (1/M^2)} \qquad (12.43)$$

$$\lambda_{\pm} = \frac{\sin \theta \cos \theta \pm \cot \alpha \sin^2 \alpha}{\cos^2 \theta - \sin^2 \alpha} = \frac{\sin \theta \cos \theta \pm \sin \alpha \cos \alpha}{\cos^2 \theta - \sin^2 \alpha} \qquad (12.44)$$

Applying the standard trigonometric identities, equation 12.44 becomes

$$\left(\frac{dy}{dx}\right)_{\pm} = \lambda_{\pm} = \tan(\theta \pm \alpha) \qquad (12.45)$$

where θ is the flow angle and α is the Mach angle.

Figure 12.8 illustrates the geometry of the C_+ and C_- characteristics in the xy plane. It shows that the C_+ and C_- characteristics are symmetrical about the streamline, making the angles $+\alpha$ and $-\alpha$, respectively, with respect to the streamline angle θ; that is, the angle made by the tangent to the streamline at point P with the x axis. The property of a characteristic, it will be recalled, is that it is a curve along which information at a point in a flow field is propagated into the downstream region of the flow field. It is evident that for the case illustrated in Fig. 12.8, the

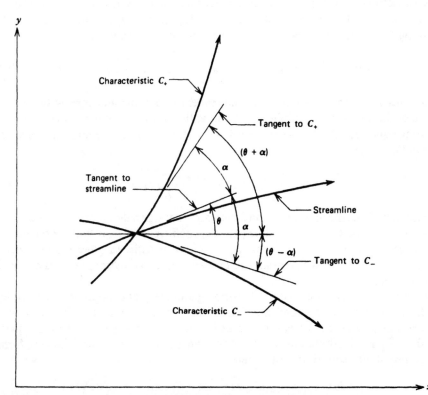

Figure 12.8   Characteristics as the Mach lines at a point in a flow field.

information at point P influences the region located between the C_+ and C_- characteristics. The latter are identical to the *Mach lines* [discussed in Section 3–8(b)]. Hence, the characteristics have both a physical and a mathematical significance; they are the carriers of information throughout a supersonic flow field.

12–4(c) Compatibility Equation

Equation 12.36 is the compatibility equation for the supersonic flow being studied. It governs the relationship between u and v along a characteristic (Mach line). To apply equation 12.36, the parameters σ_1 and σ_2 must be eliminated, which is accomplished by solving equation 12.37 for σ_2 in terms of σ_1. Thus,

$$\sigma_2 = \sigma_1[(u^2 - a^2)\lambda - 2uv]$$

and (12.46)

$$\sigma_2 = -\sigma_1 \frac{v^2 - a^2}{\lambda}$$

Equations 12.46 are not independent, as may be seen by substituting equation 12.39 into either one of the equations 12.46 to obtain the other. Thus, either of the equations 12.46 may be substituted into equation 12.36 to eliminate σ_1 and σ_2.

Substituting the first of equations 12.46 into equation 12.36 and dividing by σ_1 yields the following result (recall that σ_1 cannot be zero if equation 12.38 is valid).

$$(u^2 - a^2)\,du_\pm + [2uv - (u^2 - a^2)\lambda_\pm]\,dv_\pm - \left(\frac{\delta a^2 v}{y}\right) dx_\pm = 0 \quad (12.47)$$

The subscript $\pm$ indicates that the differentials du, dv, and dx, are to be evaluated along the C_+ or C_- characteristic, respectively; the corresponding slope is $\lambda_\pm$. Substituting the second of equations 12.46 into equation 12.36 yields

$$(u^2 - a^2)\,du_\pm + (v^2 - a^2)\lambda_\pm^{-1}\,dv_\pm - \left(\frac{\delta a^2 v}{y}\right) dx_\pm = 0 \quad (12.48)$$

Equations 12.47 and 12.48 are not independent equations, but merely two different forms of the single compatibility equation valid on each of the characteristics. Substituting equation 12.40 into equation 12.47 yields yet another equivalent form of the *compatibility equation*. Thus

$$du_\pm + \lambda_\mp\,dv_\pm - \delta\left(\frac{a^2 v}{y}\right) dx_\pm = 0 \quad (12.49)$$

In equation 12.49, the inverted $\mp$ on λ indicates that when du, dv, and dx are calculated along a C_+ characteristic, the value of λ, the coefficient of dv, is that for the local C_- characteristic, and vice versa along the C_- characteristic. In other words, when the *upper subscripts* of the differentials are chosen, the *upper subscript* of λ is applicable, and vice versa.

Equation 12.47 is employed for developing a numerical procedure for determining the flow properties of a supersonic flow field. For completing the discussion of the compatibility equation, however, two of its other forms are mentioned. If equation 12.40 for $\lambda_\pm$ is substituted into either equation 12.47 or 12.48, the following form of the compatibility equation is obtained.

$$\left(\frac{dv}{du}\right)_\pm = \frac{uv \pm a^2\sqrt{M^2 - 1}}{a^2 - v^2} - \delta\frac{a^2 v}{a^2 - v^2}\frac{1}{y}\left(\frac{dy}{du}\right)_\pm \quad (12.50)$$

Substituting the definitions of V, θ, and α, given by equations 12.41 and 12.42, into equation 12.50 converts the compatibility equation to the following form.

$$\frac{1}{V}\left(\frac{dV}{d\theta}\right)_{\pm} = \pm\tan\alpha + \delta\frac{\sin\alpha\tan\alpha\sin\theta}{\sin(\theta\pm\alpha)}\frac{1}{y}\left(\frac{dy}{d\theta}\right)_{\pm} \qquad (12.51)$$

The order of the signs, $\pm$, in equations 12.50 and 12.51 is important. For example, in equation 12.51, if the positive, or upper, subscript of the differentials $(dV/d\theta)$ is chosen, the upper signs must be chosen for the remaining terms [i.e., $+\tan\alpha$, $\sin(\theta+\alpha)$, and $(dy/d\theta)_{+}$], and vice versa.

Any'one of equations 12.47 to 12.51 may be employed as the compatibility equation that is valid along the characteristics for a steady two-dimensional, planar and axisymmetric, irrotational supersonic flow. In the finite difference algorithms presented in Section 12-5, equation 12.47 is employed. Table 12.2 presents the aforementioned characteristic and compatibility equations.

Table 12.2 Characteristic and Compatibility Equations for Steady Two-Dimensional Irrotational Supersonic Flow

---

Characteristic equation

$$\left(\frac{dy}{dx}\right)_{\pm} = \lambda_{\pm} = \tan(\theta\pm\alpha) \qquad (Mach\ lines) \qquad (12.45)$$

Compatibility equation

$$(u^2 - a^2)\,du_{\pm} + [2uv - (u^2 - a^2)\lambda_{\pm}]\,dv_{\pm} - (\delta a^2 v/y)\,dx_{\pm} = 0 \qquad (along\ Mach\ lines) \qquad (12.47)$$

12-5 NUMERICAL IMPLEMENTATION OF THE METHOD OF CHARACTERISTICS

In Section 12-4, the characteristic and compatibility equations are derived for a steady two-dimensional irrotational supersonic flow. In the present section, a numerical procedure is developed for solving those equations. The general features of the numerical integration algorithm, the modified Euler *predictor-corrector method*, are discussed, and the finite difference equations are determined.

12-5(a) Numerical Integration Procedure

Equation 12.45 defines one C_{+} and one C_{-} characteristic (i.e., two Mach lines) in the xy plane, while the compatibility equation, equation 12.47, provides a differential relationship between the velocity components u and v that is valid along each characteristic. Equations 12.45 and 12.47 are nonlinear total differential equations, and their solutions are obtained by applying finite difference techniques.

Note that equation 12.45 defines two characteristics passing through each point in the flow field. Equation 12.47 specifies one relationship between the velocity components u and v on each of the two characteristics. To obtain two independent relationships between u and v at a point in the flow field, a network must be devised wherein two characteristics intersect at a common point. At the point of intersection, there is one relationship between u and v on each of the intersecting characteristics, so that two relationships between u and v are obtained at that point. Figure 12.9 illustrates a network wherein the C_{-} characteristic through point 1 and the C_{+} characteristic through point 2 intersect at point 4, thus permitting the simultaneous application of the compatibility equation (equation 12.47) at point 4, the point of

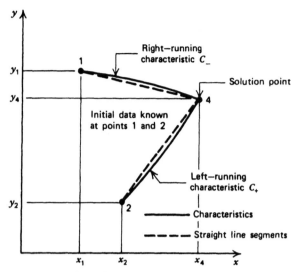

Figure 12.9 Finite difference network for applying the method of characteristics to a steady two-dimensional irrotational supersonic flow.

intersection, which is an *interior point*. The procedure must be modified to obtain the flow properties at a *wall point*, an *axis point*, and a *free pressure boundary point* (see Section 16–3)

In general, a finite difference method is employed for integrating equations 12.45 and 12.47. Consequently, in constructing the finite difference *grid*, or *network*, the portion of a characteristic connecting two points of the grid, such as points 1 and 4 and points 2 and 4 in Fig. 12.9, is replaced by dashed straight lines, the slope of the dashed line being determined in an average sense between the endpoints of the pertinent segment of the characteristic.

The numerical method employed in the present section for integrating equations 12.45 and 12.47 is the *modified Euler predictor-corrector method*[8,9] [see Appendix A–6(e)], which is a second-order method for integrating total differential equations. The corrector algorithm presented in Appendix A–6(e) is based on the *average coefficient method*, wherein the numerical values of the coefficients of the differential equations are determined as the average of the values of the coefficients at the initial points and the solution point. Hoffman[10] showed that more accurate results are obtained by employing a corrector algorithm based on the *average property method*, wherein the numerical values of the coefficients of the differential equations are determined based on the average values of the properties at the initial points and the solution point. The basic features of the modified Euler predictor-corrector algorithm based on the average property method are summarized below.

Consider the ordinary differential equation

$$\frac{dy}{dx} = f(x,y) \tag{12.52}$$

which may be rewritten in the form

$$dy = f(x,y)\,dx \tag{12.53}$$

The problem is to integrate equation 12.53 numerically from a known starting point i, designated as (x_i, y_i), where $y_i = y(x_i)$. A predicted value of the solution at $x_{i+1} = x_i + \Delta x$, denoted by $y^0(x_i + \Delta x) = y^0_{i+1}$, is obtained from the Euler *predictor algo-*

rithm [see Appendix A-6(c)], termed the *predictor* for brevity. Thus,

$$y_{i+1}^0 = y_i + f(x_i,y_i)\,\Delta x \qquad (predictor\ algorithm) \qquad (12.54)$$

where Δx is the step size of the finite difference algorithm. It is shown in Appendix A-6(c), that the accumulated error for successive applications of the Euler predictor algorithm varies linearly with the step size Δx.

The accuracy of the solution obtained by the Euler predictor may be improved by employing y_i and y_{i+1}^0 for estimating the value of $y_{i+1/2} = y(x_i + \Delta x/2)$, and then replacing $f(x_i,y_i)$ in equation 12.54 by the value of $f(x,y)$ determined at the midpoint of the interval. Thus,

$$y_{i+1}^1 = y_i + f\left(x_i + \frac{\Delta x}{2}, \frac{y_i + y_{i+1}^0}{2}\right)\Delta x \qquad (corrector\ algorithm) \quad (12.55)$$

where $y_{i+1}^1 = y^1(x_i + \Delta x)$ is the corrected value of the solution at x_{i+1}. Equation 12.55 is termed the *corrector algorithm*, or for brevity the *corrector*. In Appendix A-6(e) it is shown that the accumulated error for the corrector is *second order* in step size, and the solution converges rapidly with a reduction in the step size.*

The predictor-corrector method may be further refined by replacing the predictor value y_{i+1}^0 in equation 12.55 by the corrector value y_{i+1}^1, and repeating equation 12.55, thus obtaining y_{i+1}^2; where the superscript 2 denotes the *second application* of the corrector algorithm. Continued application of this procedure leads to the following algorithm.

$$y_{i+1}^n = y_i + f\left(x_i + \frac{\Delta x}{2}, \frac{y_i + y_{i+1}^{n-1}}{2}\right)\Delta x \qquad \left(\begin{array}{c} corrector\ algorithm \\ with\ interation \end{array}\right) \quad (12.56)$$

where y_{i+1}^n is the value of y after n applications of the corrector. Equation 12.56 is known as the *modified Euler predictor-corrector method with interation.*

Considerations of numerical stability generally place an upper limit on the allowable step size Δx. For isentropic flows, the permissible maximum step size is so large that it is rarely a consideration in the numerical computations. There is no assurance, however, that the iteration indicated in equation 12.56 is worthwhile. In many cases, a single application of the corrector may yield a sufficiently accurate result. Some accuracy studies illustrating the foregoing remarks are presented in Section 17-6.

12-5(b) Direct and Inverse Marching Methods

In the preceding section, the general features of the modified Euler integration method are discussed. Two different procedures for applying that method for the integration of the characteristic and compatibility equations are possible: the *direct marching method* and the *inverse marching method*. Figure 12.10 illustrates schematically the direct marching method, and Fig. 12.11 illustrates the inverse marching method.

In the direct marching method, continuous families of left- and right-running characteristics are followed throughout the flow field. By applying the unit processes directly from any two solution points previously determined, such as points 1 and 2 in Fig. 12.10, the next point in the network (i.e., point 4) may be determined.

In the inverse marching method, the solution is obtained on successive lines, termed *solution lines*, which are not characteristics. Typically, those lines are perpendicular to the mean flow direction, such as lines AB and CD in Fig. 12.11. The points at which the fluid properties are to be determined on a solution line, such as line AB in

---

* Accuracy criteria for numerical methods of integration are discussed in Section 17-6.

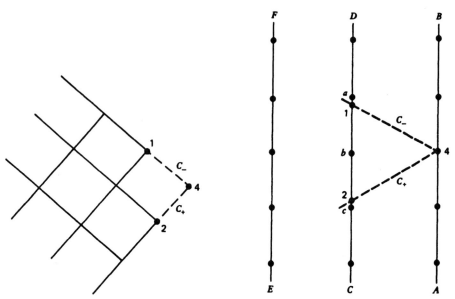

Figure 12.10 Direct marching method.   **Figure 12.11** Inverse marching method.

Fig. 12.11, are prespecified in some prescribed manner (e.g., equally spaced across the flow field). The unit processes are then obtained by projecting rearward-running characteristics from a point on the solution line, such as point 4, to intersect the previous solution line, denoted by *CD*. The fluid properties at points 1 and 2 are found by interpolation along the known solution line *CD*. To insure numerical stability, the interpolating functions must include points located outside of the intersections of the rearward-running characteristics and the last known solution line as, for example, the points *a* and *c* on the solution line *CD*. In general, the location of the points of intersection, points 1 and 2, are determined by iteration.

The direct marching method has the obvious advantage that the location of points 1 and 2, and the flow properties at those points, are known without iteration and interpolation. The inverse marching method has a slight advantage in the computational logic, however, because the number and the location of the solution points may be specified beforehand. In general, the direct marching method is the more accurate of the two methods.

12–5(c) Finite Difference Equations

Figure 12.9 illustrates schematically the finite difference grid for the unit process for determining the flow properties at an *interior point*. As noted earlier in Section 12–5, in constructing the finite difference grids for applying the numerical integration algorithm, the portion of a characteristic curve connecting two points in the network is replaced by a straight dashed line connecting those points.

Table 12.3 presents the finite difference equations corresponding to the characteristic and compatibility equations presented in Table 12.2, where the subscripts + and − denote left- and right-running characteristics, respectively. According to the Euler method, the finite difference equations are obtained from the differential equations by replacing the differentials dx, dy, du, and dv by the differences Δx, Δy, Δu, and Δv. The coefficients λ, Q, R, and S are then determined at the initial points for the predictor and in an average manner (either average property or average coefficient) for the corrector.

Table 12.3   Finite Difference Equations for Steady Two-Dimensional Irrotational Supersonic Flow

$$\Delta y_\pm = \lambda_\pm \, \Delta x_\pm \tag{12.57}$$

$$Q_\pm \, \Delta u_\pm + R_\pm \, \Delta v_\pm - S_\pm \, \Delta x_\pm = 0 \tag{12.58}$$

$$\lambda_\pm = \tan(\theta \pm \alpha) \tag{12.59}$$

$$Q = u^2 - a^2 \tag{12.60}$$

$$R = 2uv - (u^2 - a^2)\lambda \tag{12.61}$$

$$S = \delta \frac{a^2 v}{y} \tag{12.62}$$

$+$ or $-$ denotes C_+ or C_- characteristic, respectively

12-5(d) Unit Process for an Interior Point

A point located in the interior of a supersonic flow field, such as point 4 in Fig. 12.10, is termed an *interior point*. The finite difference grid for determining the location of point 4 and the fluid properties at that point is illustrated schematically in Fig. 12.9. The interior point (point 4) is located at the intersection of the C_- and C_+ characteristics emanating into the flow field from the initial-value points 1 and 2, respectively. The locations of points 1 and 2, as well as the flow properties at those points, are assumed to be known; points 1 and 2 are termed *initial-value points*. The problem is to develop a method for establishing the location and flow properties at the interior point (point 4) from the known data at the initial-value points (points 1 and 2).

The problem is solved by integrating the two characteristic equations, equations 12.45, to obtain the characteristics 14 and 24, integrating the two compatibility equations, equations 12.47, which are valid along the characteristics 14 and 24, and solving the four resulting difference equations simultaneously for x_4, y_4, u_4, and v_4.

The application of the numerical method of analysis discussed in Section 12-5(a) for determining the location and flow properties at an interior point of a supersonic flow field is discussed here in general terms. The purpose is to describe the analytic method without complicating its description with the details of the pertinent numerical processes. Example 12.2 illustrates the calculation procedure in its entirety.

The location of the solution point, point 4, is determined by writing equation 12.57 of Table 12.3 in finite difference form in terms of points 1, 2, and 4. Thus,

$$(y_4 - y_2) = \lambda_+(x_4 - x_2) \tag{12.63}$$

$$(y_4 - y_1) = \lambda_-(x_4 - x_1) \tag{12.64}$$

Equations 12.63 and 12.64 may be rewritten as

$$y_4 - \lambda_+ x_4 = y_2 - \lambda_+ x_2 \tag{12.65}$$

$$y_4 - \lambda_- x_4 = y_1 - \lambda_- x_1 \tag{12.66}$$

Equations 12.65 and 12.66 may be solved simultaneously for x_4 and y_4. The slopes λ_+ and λ_- in equations 12.65 and 12.66 are given by equation 12.59. Thus,

$$\lambda_+ = \tan(\theta_+ + \alpha_+) \quad \text{and} \quad \lambda_- = \tan(\theta_- - \alpha_-) \tag{12.67}$$

where

$$\theta_\pm = \tan^{-1}\left(\frac{v_\pm}{u_\pm}\right) \tag{12.68}$$

$$V_{\pm} = \sqrt{u_{\pm}^2 + v_{\pm}^2} \tag{12.69}$$

$$a_{\pm} = a(V_{\pm}) \qquad (\textit{from equation 10.121}) \tag{12.70}$$

$$M_{\pm} = \frac{V_{\pm}}{a_{\pm}} \tag{12.71}$$

$$\alpha_{\pm} = \sin^{-1}\left(\frac{1}{M_{\pm}}\right) \tag{12.72}$$

Consequently, both $\theta_{\pm}$ and $\alpha_{\pm}$, and hence $\lambda_{\pm}$, may be determined for specified values of $u_{\pm}$ and $v_{\pm}$.

The compatibility equation, equation 12.58 of Table 12.3, when expressed in finite difference form in terms of points 1, 2, and 4, yields

$$Q_+(u_4 - u_2) + R_+(v_4 - v_2) - S_+(x_4 - x_2) = 0 \tag{12.73}$$

$$Q_-(u_4 - u_1) + R_-(v_4 - v_1) - S_-(x_4 - x_1) = 0 \tag{12.74}$$

Equations 12.73 and 12.74 may be rewritten as

$$Q_+u_4 + R_+v_4 = T_+ \tag{12.75}$$

$$Q_-u_4 + R_-v_4 = T_- \tag{12.76}$$

where

$$T_+ = S_+(x_4 - x_2) + Q_+u_2 + R_+v_2 \tag{12.77}$$

$$T_- = S_-(x_4 - x_1) + Q_-u_1 + R_-v_1 \tag{12.78}$$

Equations 12.75 and 12.76 may be solved simultaneously for u_4 and v_4. The coefficients appearing in equations 12.75 and 12.76 are given by

$$Q_+ = (u_+^2 - a_+^2) \tag{12.79}$$

$$R_+ = (2u_+v_+ - Q_+\lambda_+) \tag{12.80}$$

$$S_+ = \delta\frac{a_+^2v_+}{y_+} \tag{12.81}$$

$$Q_- = (u_-^2 - a_-^2) \tag{12.82}$$

$$R_- = (2u_-v_- - Q_-\lambda_-) \tag{12.83}$$

$$S_- = \delta\frac{a_-^2v_-}{y_-} \tag{12.84}$$

Consequently, $Q_{\pm}$, $R_{\pm}$, and $T_{\pm}$ may be determined for specified values of $u_{\pm}$, $v_{\pm}$, and $y_{\pm}$.

Table 12.4 presents the computational equations for steady two-dimensional irrotational supersonic flow.

Table 12.4 Computational Equations for Steady Two-Dimensional Irrotational Supersonic Flow

$y_4 - \lambda_+x_4 = y_2 - \lambda_+x_2$	(12.85)
$y_4 - \lambda_-x_4 = y_1 - \lambda_-x_1$	(12.86)
$Q_+u_4 + R_+v_4 = T_+$	(12.87)
$Q_-u_4 + R_-v_4 = T_-$	(12.88)

For the *Euler predictor algorithm*, the values of $u_\pm$, $v_\pm$, and $y_\pm$ are given by

$$u_+ = u_2 \qquad v_+ = v_2 \qquad y_+ = y_2 \qquad (12.89)$$

$$u_- = u_1 \qquad v_- = v_1 \qquad y_- = y_1 \qquad (12.90)$$

Substituting those values into the computational equations and solving those equations simultaneously yields the predicted values x_4^0, y_4^0, u_4^0, and v_4^0.

For the *Euler corrector algorithm*, the values of $u_\pm$, $v_\pm$, and $y_\pm$ are given by

$$u_+ = \frac{u_2 + u_4}{2} \qquad v_+ = \frac{v_2 + v_4}{2} \qquad y_+ = \frac{y_2 + y_4}{2} \qquad (12.91)$$

$$u_- = \frac{u_1 + u_4}{2} \qquad v_- = \frac{v_1 + v_4}{2} \qquad y_- = \frac{y_1 + y_4}{2} \qquad (12.92)$$

The average values of $\theta_\pm$, $V_\pm$, $a_\pm$, $M_\pm$, and $\alpha_\pm$ are determined by substituting the above values of $u_\pm$ and $v_\pm$ into equations 12.68 to 12.72. Substituting those values into the computational equations and solving those equations simultaneously yields the corrected values x_4^1, y_4^1, u_4^1, and v_4^1.

The *Euler corrector algorithm with iteration* (equation 12.56) is applied next to improve the accuracy of the solution values. To accomplish that objective, the results of the previous corrector (i.e., x_4^1, y_4^1, u_4^1, and v_4^1) are substituted into the computational equations so that new values are obtained for the average properties $u_\pm$, $v_\pm$, etc. The corrector algorithm is then repeated employing those average values of the properties, thereby obtaining a second set of corrector values, x_4^2, y_4^2, u_4^2, and v_4^2. The iteration algorithm may be applied as many times as is required for obtaining the desired degree of convergence. The determinations for convergence after n applications of the corrector are made as follows.

$$\left| P^n - P^{n-1} \right| \leq \text{(specified tolerance)} \qquad (12.93)$$

where P represents x_4, y_4, u_4, and v_4. Typical values of the tolerance are 0.0001 m for x_4 and y_4 and 0.1 m/s for u_4 and v_4.

For an axisymmetric flow, the coefficient S, given by equation 12.62, is indeterminate on the x axis, where both v and y are zero, as illustrated in Fig. 12.12. In that case, the ratio v_2/y_2 is approximated by the ratio v_1/y_1 for the predictor. For the corrector, S_+ is based on the average values for y_+ and v_+, which are nonzero.

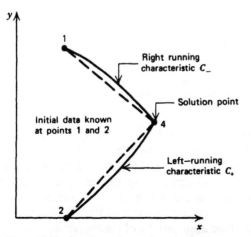

Figure 12.12   Interior point on the axis of symmetry.

Figure 12.13 presents a block diagram illustrating the *modified Euler predictor-corrector algorithm* for the unit process for an interior point. That block diagram is essentially the flow diagram for a computer implementation of the algorithm. A FORTRAN computer program for implementing the numerical integration procedure is presented in Section 16–3(b).

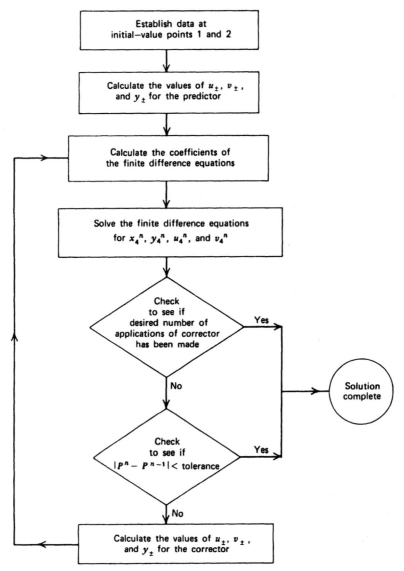

Figure 12.13 Block diagram illustrating the interior point unit process.

Example 12.2. A perfect gas with $\gamma = 1.2$ and $R = 320.0$ J/kg-K is flowing in an axisymmetric propulsive nozzle having a stagnation pressure $P = 70.0 \cdot 10^5$ N/m² and a stagnation temperature $T = 3330$ K. The location and properties at two neighboring points in the flow field (points 1 and 2 in Fig. 12.9) are given in Table 12.5. Employing the unit process for an interior point, calculate the location of and the flow properties at the downstream intersection (point 4 of Fig. 12.9) of the characteristics through points 1 and 2 for three applications of the corrector.

Table 12.5 Initial-Value Data for an
Interior Point (Example 12.2)

	Point 1	Point 2
x, m	0.2500	0.2750
y, m	0.5000	0.4000
u, m/s	3000.0	3030.0
v, m/s	525.0	430.0

Solution

The computations for the predictor and the first application of the corrector are presented below, and the results are presented in Table 12.6.

Table 12.6 Values of the Solution, for Successive Trials, for an
Interior Point (Example 12.2)

	(0)	(1)	(2)	(3)
λ_+	0.34620	0.34727	0.34693	0.34691
λ_-	−0.022579	−0.032664	−0.033507	−0.033522
$Q_+ \cdot 10^{-6}$, m^2/s^2	8.8388	8.8802	8.8742	8.8743
$R_+ \cdot 10^{-6}$, m^2/s^2	−0.45419	−0.41676	−0.42396	−0.42409
$S_+ \cdot 10^{-6}$, m^2/s^3	367.80	332.02	332.26	332.31
$T_+ \cdot 10^{-9}$, m^3/s^3	26.685	26.814	26.793	26.793
$Q_- \cdot 10^{-6}$, m^2/s^2	8.6488	8.7846	8.7787	8.7787
$R_- \cdot 10^{-6}$, m^2/s^2	3.3453	3.2278	3.2227	3.2226
$S_- \cdot 10^{-6}$, m^2/s^3	368.72	335.48	335.77	335.81
$T_- \cdot 10^{-9}$, m^3/s^3	27.811	28.144	28.124	28.124
x_4, m	0.54463	0.53605	0.53566	0.53565
y_4, m	0.49335	0.49066	0.49043	0.49042
u_4, m/s	3042.2	3040.4	3040.5	3040.5
v_4, m/s	448.5	444.7	444.6	444.6

Column (0)—predictor values.   Column (2)—first iteration of corrector.
Column (1)—corrector values.   Column (3)—second iteration of corrector.

(a) *The speed of sound equation.* For a perfect gas, the speed of sound is given by equation 3.186, which is repeated below.

$$a^2 = a_o^{\,2} - \frac{\gamma - 1}{2} V^2 = \gamma RT - \frac{\gamma - 1}{2} V^2 \qquad (3.186)$$

For the present problem

$$a = \left[1.2(320.0)(3330.0) - \frac{1.2 - 1}{2} V^2 \right]^{1/2} = (1{,}278{,}720 - 0.1 \, V^2)^{1/2} \text{ m/s} \qquad (a)$$

(b) *Calculate the coefficients for the predictor.* For the predictor, set the values of the flow properties $u_\pm$, $v_\pm$, and $y_\pm$ equal to their values at points 2 and 1, respectively. Thus,

$$u_+ = 3030.0 \text{ m/s} \qquad v_+ = 430.0 \text{ m/s} \qquad y_+ = 0.4000 \text{ m}$$

$$u_- = 3000.0 \text{ m/s} \qquad v_- = 525.0 \text{ m/s} \qquad y_- = 0.5000 \text{ m}$$

From equations 12.68 to 12.72:

$$V_+ = (u_+^2 + v_+^2)^{1/2} = [(3030.0)^2 + (430.0)^2]^{1/2} = 3060.4 \text{ m/s}$$

$$\theta_+ = \tan^{-1}\left(\frac{v_+}{u_+}\right) = \tan^{-1}\left(\frac{430.0}{3030.0}\right) = 8.077 \text{ deg}$$

$$a_+ = [1,278,720 - 0.1(3060.4)^2]^{1/2} = 584.93 \text{ m/s}$$

$$\alpha_+ = \sin^{-1}\left(\frac{a_+}{V_+}\right) = \sin^{-1}\left(\frac{584.93}{3060.4}\right) = 11.019 \text{ deg}$$

$$V_- = (u_-^2 + v_-^2)^{1/2} = [(3000.0)^2 + (525.0)^2]^{1/2} = 3045.6 \text{ m/s}$$

$$\theta_- = \tan^{-1}\left(\frac{v_-}{u_-}\right) = \tan^{-1}\left(\frac{525.0}{3000.0}\right) = 9.927 \text{ deg}$$

$$a_- = [1,278,720 - 0.1(3045.6)^2]^{1/2} = 592.59 \text{ m/s}$$

$$\alpha_- = \sin^{-1}\left(\frac{a_-}{V_-}\right) = \sin^{-1}\left(\frac{592.59}{3045.6}\right) = 11.220 \text{ deg}$$

Substituting the above values into equations 12.67 and 12.79 to 12.84, we obtain:

$$\lambda_+ = \tan(\theta_+ + \alpha_+) = \tan(8.077 + 11.019) = 0.34620$$

$$Q_+ = u_+^2 - a_+^2 = (3030.0)^2 - (584.93)^2 = 8.8388 \cdot 10^6 \text{ m}^2/\text{s}^2$$

$$R_+ = 2u_+v_+ - Q_+\lambda_+ = 2.0(3030.0)(430.0) - (8.8388 \cdot 10^6)(0.34620)$$

$$R_+ = -0.45419 \cdot 10^6 \text{ m}^2/\text{s}^2$$

$$S_+ = a_+^2 v_+/y_+ = (584.93)^2(430.0)/0.4000 = 367.80 \cdot 10^6 \text{ m}^2/\text{s}^3$$

$$\lambda_- = \tan(\theta_- - \alpha_-) = \tan(9.927 - 11.220) = -0.022579$$

$$Q_- = u_-^2 - a_-^2 = (3000.0)^2 - (592.59)^2 = 8.6488 \cdot 10^6 \text{ m}^2/\text{s}^2$$

$$R_- = 2u_-v_- - Q_-\lambda_- = 2(3000.0)(525.0) - (8.6488 \cdot 10^6)(-0.022579)$$

$$R_- = 3.3453 \cdot 10^6 \text{ m}^2/\text{s}^2$$

$$S_- = a_-^2 v_-/y_- = (592.59)^2(525.0)/0.5000 = 368.72 \cdot 10^6 \text{ m}^2/\text{s}^3$$

(c) *Determination of x_4, y_4, u_4, and v_4 for the predictor.* Two equations for the determination of x_4 and y_4 are obtained by substituting the values of the coefficients calculated in part (b) into equations 12.85 and 12.86. Thus,

$$y_4 - 0.34620x_4 = 0.4000 - (0.34620)(0.2750) \tag{b}$$

$$y_4 + 0.022579x_4 = 0.5000 + (0.022579)(0.2500) \tag{c}$$

Solving equations (b) and (c) simultaneously, we obtain

$$x_4 = 0.54463 \text{ m} \quad \text{and} \quad y_4 = 0.49335 \text{ m}$$

From equations 12.77 and 12.78, the values of T_+ and T_- for the predictor are obtained. Thus,

$$T_+ = (367.80 \cdot 10^6)(0.54463 - 0.2750) + (8.8388 \cdot 10^6)(3030.0)$$
$$- (0.45419 \cdot 10^6)(430.0) = 26.685 \cdot 10^9 \text{ m}^3/\text{s}^3$$

$$T_- = (368.72 \cdot 10^6)(0.54463 - 0.2500) + (8.6488 \cdot 10^6)(3000.0)$$
$$+ (3.3453 \cdot 10^6)(525.0) = 27.811 \cdot 10^9 \text{ m}^3/\text{s}^3$$

Substituting the values of $Q_\pm$, $R_\pm$, and $T_\pm$ into equations 12.87 and 12.88 yields the following two equations for the determination of u_4 and v_4.

$$(8.8388 \cdot 10^6)u_4 - (0.45419 \cdot 10^6)v_4 = 26.685 \cdot 10^9 \qquad \text{(d)}$$

$$(8.6488 \cdot 10^6)u_4 + (3.3453 \cdot 10^6)v_4 = 27.811 \cdot 10^9 \qquad \text{(e)}$$

Solving equations (d) and (e) simultaneously yields

$$u_4 = 3042.2 \text{ m/s} \qquad \text{and} \qquad v_4 = 448.5 \text{ m/s}$$

The application of the Euler predictor algorithm is complete. The column in Table 12.6 labeled (0) presents the predictor values for the terms $\lambda_\pm$, x_4, y_4, $Q_\pm$, etc.; that is, the terms that are calculated by the predictor portion of the numerical algorithm. (d) *Calculation of the coefficients for the corrector.* One application of the corrector algorithm is presented in detail below. Before the corrector algorithm may be applied, the average values of the flow properties must be calculated (see equations 12.91 and 12.92), employing the initial and predictor values of u, v, and y. Thus:

$$u_+ = \frac{u_2 + u_4}{2} = \frac{3030.0 + 3042.2}{2} = 3036.1 \text{ m/s}$$

$$v_+ = \frac{v_2 + v_4}{2} = \frac{430.0 + 448.5}{2} = 439.3 \text{ m/s}$$

$$y_+ = \frac{y_2 + y_4}{2} = \frac{0.4000 + 0.49335}{2} = 0.44668 \text{ m}$$

$$u_- = \frac{u_1 + u_4}{2} = \frac{3000.0 + 3042.2}{2} = 3021.1 \text{ m/s}$$

$$v_- = \frac{v_1 + v_4}{2} = \frac{525.0 + 448.5}{2} = 486.8 \text{ m/s}$$

$$y_- = \frac{y_1 + y_4}{2} = \frac{0.5000 + 0.49335}{2} = 0.49668 \text{ m}$$

From equations 12.68 to 12.72:

$$V_+ = [(3036.1)^2 + (439.3)^2]^{1/2} = 3067.7 \text{ m/s}$$

$$\theta_+ = \tan^{-1}\left(\frac{439.3}{3036.1}\right) = 8.233 \text{ deg}$$

$$a_+ = [1{,}278{,}720 - 0.1(3067.7)^2]^{1/2} = 581.07 \text{ m/s}$$

$$\alpha_+ = \sin^{-1}\left(\frac{581.07}{3067.7}\right) = 10.919 \text{ deg}$$

$$V_- = [(3021.1)^2 + (486.8)^2]^{1/2} = 3060.1 \text{ m/s}$$

$$\theta_- = \tan^{-1}\left(\frac{486.8}{3021.1}\right) = 9.154 \text{ deg}$$

$$a_- = [1{,}278{,}720 - 0.1(3060.1)^2]^{1/2} = 585.06 \text{ m/s}$$

$$\alpha_- = \sin^{-1}\left(\frac{585.06}{3060.1}\right) = 11.022 \text{ deg}$$

Substituting the above values into equations 12.67 and 12.79 to 12.84 yields:

$$\lambda_+ = \tan(8.233 + 10.919) = 0.34727$$

$$Q_+ = (3036.1)^2 - (581.07)^2 = 8.8802 \cdot 10^6 \; m^2/s^2$$

$$R_+ = 2.0(3036.1)(439.3) - (8.8802 \cdot 10^6)(0.34727) = -0.41676 \cdot 10^6 \; m^2/s^2$$

$$S_+ = (581.07)^2(439.3)/0.44668 = 332.02 \cdot 10^6 \; m^2/s^3$$

$$\lambda_- = \tan(9.154 - 11.022) = -0.032664$$

$$Q_- = (3021.1)^2 - (585.06)^2 = 8.7846 \cdot 10^6 \; m^2/s^2$$

$$R_- = 2.0(3021.1)(486.8) - (8.7846 \cdot 10^6)(-0.032664) = 3.2278 \cdot 10^6 \; m^2/s^2$$

$$S_- = (585.06)^2(486.8)/0.49668 = 335.48 \cdot 10^6 \; m^2/s^3$$

(e) *Determination of* x_4, y_4, u_4, *and* v_4 *for the corrector.* From equations 12.85 and 12.86, respectively,

$$y_4 - 0.34727x_4 = 0.4000 - (0.34727)(0.2750) \tag{f}$$

$$y_4 + 0.032664x_4 = 0.5000 + (0.032664)(0.2500) \tag{g}$$

Solving equations (f) and (g) simultaneously for x_4 and y_4, we obtain

$$x_4 = 0.53605 \; m \qquad \text{and} \qquad y_4 = 0.49066 \; m$$

From equations 12.77 and 12.78, respectively, we obtain

$$T_+ = (332.02 \cdot 10^6)(0.53605 - 0.2750) + (8.8802 \cdot 10^6)(3030.0)$$
$$- (0.41676 \cdot 10^6)(430.0) = 26.814 \cdot 10^9 \; m^3/s^3$$

$$T_- = (335.48 \cdot 10^6)(0.53605 - 0.2500) + (8.7846 \cdot 10^6)(3000.0)$$
$$+ (3.2278 \cdot 10^6)(525.0) = 28.144 \cdot 10^9 \; m^3/s^3$$

Substituting for $Q_\pm$, $R_\pm$, and $T_\pm$ into equations 12.87 and 12.88, we obtain

$$(8.8802 \cdot 10^6)u_4 - (0.41676 \cdot 10^6)v_4 = 26.814 \cdot 10^9 \tag{h}$$

$$(8.7846 \cdot 10^6)u_4 + (3.2278 \cdot 10^6)v_4 = 28.144 \cdot 10^9 \tag{i}$$

Solving equations (h) and (i) simultaneously for u_4 and v_4 yields

$$u_4 = 3040.4 \; m/s \qquad \text{and} \qquad v_4 = 444.7 \; m/s$$

The application of the Euler corrector algorithm is complete. The column labeled (1) in Table 12.6 presents the results of applying the corrector algorithm.

(f) *Iteration of the corrector.* Two iterations of the corrector algorithm are presented in columns (2) and (3) in Table 12.6. Those results are obtained by twice repeatings steps (d) and (e), each with a new set of corrector values for x_4, y_4, u_4, and v_4. It is evident from columns (1), (2), and (3) of Table 12.6 that successive applications of the corrector have only a small effect on the final results.

The thermodynamic properties of the flowing fluid, such as p, ρ, and t, may be calculated from the flow velocity V and the stagnation properties of the fluid.

12-6 SIMPLE WAVES[11]

A special type of flow known as a *simple wave flow* occurs frequently in practical compressible flow problems. The Prandtl-Meyer expansion fan discussed in Section 8–3 is a practical example of a simple wave flow in a steady two-dimensional irrotational supersonic flow field. Simple waves also occur frequently in unsteady one-dimensional planar homentropic flow fields [see Section 13–4(b)]. Simple waves may occur when the governing partial differential equations are homogeneous and the coefficients depend only on the dependent variables. In this section the properties of simple waves are reviewed, and proofs of those properties are presented.

Simple waves in steady two-dimensional irrotational supersonic flow have the following properties.

1. A simple wave is defined as one in which V and θ (or u and v) are not independent [i.e., $V = V(\theta)$ or $u = u(v)$].
2. In a simple wave, the characteristics (Mach lines) of one family are straight lines having constant properties.
3. If a region of uniform flow adjoins a nonuniform region, then the nonuniform region must be a simple wave region.

In a steady flow, a simple wave region is possible only in a two-dimensional planar irrotational supersonic flow. For the latter type of flow the governing equations are equations 10.193 and 10.189; they are repeated here for convenience. Thus,

$$(u^2 - a^2)u_x + (v^2 - a^2)v_y + 2uvv_x = 0 \tag{10.193}$$

$$u_y - v_x = 0 \tag{10.189}$$

According to property 1 above,

$$u = u(v) \qquad \text{and} \qquad v = v(u) \tag{12.94}$$

Differentiating the relationship $u = u(v)$ with respect to x yields,

$$u_x = u_v v_x = v_x \left(\frac{du}{dv}\right) \tag{12.95}$$

where $u_v = du/dv$, since u is a function of v alone. Differentiating the relationship $v = v(u)$ with respect to y yields

$$v_y = v_u u_y = u_y \left(\frac{dv}{du}\right) = v_x \left(\frac{dv}{du}\right) \tag{12.96}$$

where u_y is replaced by v_x by employing the irrotationality condition, equation 10.189.

Substituting equations 12.95 and 12.96 into equation 10.193 and noting that, in general, $v_x \neq 0$ so that it may be divided out of the result, we obtain

$$(u^2 - a^2)\left(\frac{du}{dv}\right) + (v^2 - a^2)\left(\frac{dv}{du}\right) + 2uv = 0 \tag{12.97}$$

Solving equation 12.97 for (dv/du) yields

$$\left(\frac{dv}{du}\right)_\pm = \frac{uv \pm a^2\sqrt{M^2 - 1}}{a^2 - v^2} \tag{12.98}$$

Equation 12.98 is identical to equation 12.50, which is the compatibility equation for a steady two-dimensional planar ($\delta = 0$) irrotational supersonic flow. It is valid only along the characteristics for a supersonic flow. Hence, the relationship $u = u(v)$ is possible only along characteristics. More specifically, since $u = u(v)$ uniquely, the $+$ and $-$ signs in equation 12.98 both cannot hold simultaneously. The relationship $u = u(v)$ can, therefore, hold only along either a right-running characteristic or a left-running characteristic. On the other characteristic, u and v are independent of each other.

Now consider property 2 listed above. For a continuous flow,

$$dv = v_x \, dx + v_y \, dy \tag{12.99}$$

Along a line of constant v, which is also a line of constant u, since $u = u(v)$,

$$dv = v_x \, dx + v_y \, dy = 0 \tag{12.100}$$

and

$$\left(\frac{dy}{dx}\right)_v = -\frac{v_x}{v_y} \tag{12.101}$$

Substituting equation 12.96 into equation 12.101 yields

$$\left(\frac{dy}{dx}\right)_v = -\left(\frac{du}{dv}\right) \tag{12.102}$$

Solving equation 12.97 for $-(du/dv)$, we obtain

$$\left(\frac{dy}{dx}\right)_v = \frac{uv \pm a^2\sqrt{M^2 - 1}}{u^2 - a^2} \tag{12.103}$$

Equation 12.103 is identical to equation 12.40, which is the equation for the characteristics. Hence, u and v and, therefore, all of the flow properties, can be constant only on characteristics. If M and θ are constant, then $\lambda_{\pm} = \tan(\theta \pm \alpha)$ is constant, and the characteristics (Mach lines) are straight lines. Once again, the $+$ and $-$ signs in equation 12.103 both cannot hold simultaneously. Consequently, only one family of characteristics are straight lines along which the flow properties remain constant.

The proof of statement 3 listed above requires employing the hodograph plane discussed in Section 8–7. In the hodograph plane, the solution to equation 12.98 yields a fixed set of curves independent of the geometry of the flow field (see Fig. 8.16). Those curves are called the hodograph characteristics. They are denoted by Γ_+ and Γ_- corresponding to the $+$ and $-$ sign, respectively, in equation 12.98. The slopes of those hodograph characteristics are given by equation 12.98. The slopes of the physical characteristics C_+ and C_- (i.e., the Mach lines) are given by equation 12.40.

$$\left(\frac{dy}{dx}\right)_{\pm} = \frac{uv \pm a^2\sqrt{M^2 - 1}}{(u^2 - a^2)} \tag{12.40}$$

If $(dy/dx)_+$ from equation 12.40 is multiplied by $(dv/du)_-$ from equation 12.98, or $(dy/dx)_-$ is multiplied by $(dv/du)_+$, the product in both cases is -1. Thus,

$$\left(\frac{dy}{dx}\right)_{\pm}\left(\frac{dv}{du}\right)_{\mp} = -1 \tag{12.104}$$

Equation 12.104 shows that the slopes of the physical characteristics C_+ and C_- are the negative reciprocals of the slopes of the hodograph characteristics of the opposite family, Γ_- and Γ_+, respectively. In other words, the physical and the hodograph characteristics of opposite families are orthogonal, a fact that may often be employed advantageously in obtaining a graphical solution to a steady planar irrotational supersonic flow problem.

The flow in a simple wave region is governed by equation 12.98 which, upon substituting for u and v in terms of V and θ (see Fig. 12.6) and for M in terms of α (see Fig. 12.7) becomes (see equation 8.3),

$$\frac{dV}{d\theta} = \pm V \tan \alpha \tag{12.105}$$

In Section 8–3, equation 12.105 is integrated for a perfect gas to give (see equation 8.11, where θ is replaced by δ)

$$\pm \theta = \sqrt{b} \tan^{-1} [(M^2 - 1)/b]^{1/2} - \tan^{-1} (M^2 - 1)^{1/2} + \text{constant} \quad (12.106)$$

Equation 12.106 is the equation that defines the hodograph characteristics in Fig. 8.16. Hence, the entire simple wave region in the physical plane is mapped into a single hodograph characteristic in the hodograph plane. The initial condition (i.e., M_1 and δ_1) and whether the turning is clockwise or counterclockwise determines the particular hodograph characteristic corresponding to a given physical plane. In Section 8–4, it is shown that clockwise turning, in which the left-running C_+ characteristics are straight lines, maps onto the Γ_- hodograph characteristic, and vice versa. Consequently, for a simple wave, all of the characteristics of one family in the physical plane lie on a single characteristic of the opposite family in the hodograph plane. That feature leads to a simple proof of statement 3 made at the beginning of this section.

In Fig. 12.14 the line AB represents a line of uniform, or constant, properties in the xy plane, which is the *physical plane*. The line AB maps into a point in the hodograph plane, such as point C in Fig. 12.15. Since line AB is a C_+ characteristic, the

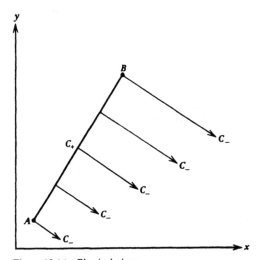

Figure 12.14 Physical plane.

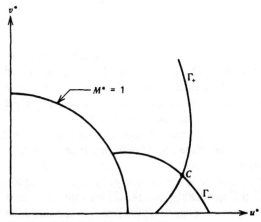

Figure 12.15 Hodograph plane.

data along line AB are propagated into the xy plane along the C_- characteristics crossing line AB. There are two possibilities. First, the surrounding physical field is uniform, in which case the entire physical plane maps into the single point C in the hodograph plane. Second, the solution for all of the physical characteristics C_- that cross C_+ proceeds along the single hodograph characteristic Γ_- that emanates from the point C in the hodograph plane. Each different point on Γ_- defines a different C_+ characteristic along which the flow properties are constant, since an entire C_+ characteristic maps into a single point on Γ_-. Hence, all of the C_- characteristics crossing line AB either enter a region of uniform flow or they define a series of straight C_+ characteristics along each of which the flow has constant properties; that is, a simple wave region. The behavior just described continues until it is interrupted by some incompatible boundary condition in the simple wave region that generates physical characteristics of the opposite family.

The above discussion substantiates the statement regarding the third property of simple wave flows; that if a region (or line) of uniform flow adjoins a nonuniform region, then the latter region must be a simple wave region.

Figure 12.16 illustrates a Prandtl-Meyer flow, which is the most familiar example of a simple wave flow. Figure 12.17 illustrates the corresponding hodograph plane for the simple wave expansion fan. The simple wave expansion fan region is adjoined by Regions 1 and 2, in each of which the flow properties have constant values.

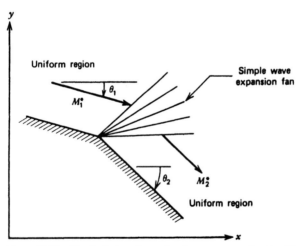

Figure 12.16 Physical plane for a simple wave expansion fan (Prandtl-Meyer flow).

Figure 12.18 illustrates schematically the destruction of a simple wave by the presence of solid boundaries. Because of the presence of the upper wall, reflected waves emanate from the boundary AC, and Region ABC contains simultaneously waves of both families. In such a situation, the only recourse for analyzing the flow is to apply the numerical analysis developed in Sections 12-4 and 12-5.

In summary, it is shown that a region of uniform flow can only be adjoined to a simple wave region, and that the simple wave region will continue until destroyed by incompatible boundary conditions.

From a purely mathematical point of view, the unique properties of simple waves are due to the special nature of the governing partial differential equations. The necessary conditions for the occurrence of simple waves are that the coefficients of

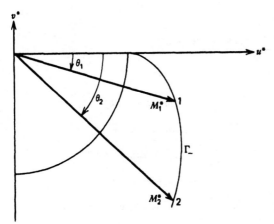

Figure 12.17 Hodograph plane for a simple wave expansion fan (Prandtl-Meyer flow).

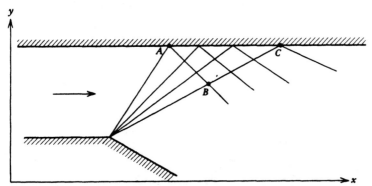

Figure 12.18 Destruction of a simple wave region by reflected waves from a solid boundary.

the governing partial differential equations depend only on the dependent variables (u and v in this case), and that the governing equations are homogeneous. Such equations are said to be *reducible*. Both of those conditions are satisfied by equations 10.189 and 10.193. Any flow for which the governing equations satisfy the aforementioned conditions can have simple wave regions. In particular, the governing equations for unsteady one-dimensional planar homentropic flow satisfy the aforementioned conditions, and simple waves do appear frequently in such flows [see Section 13-4(b)].

Example 12.3. Consider the two-dimensional planar flow of a parallel uniform stream of air initially at $M = 1.0$ over the cylindrical afterbody illustrated in Fig. 12.19a. Employing simple wave flow concepts and the generalized hodograph diagram of Fig. 8.16, construct graphically the expansion waves through a turning angle of -60 deg at 5 deg intervals.

Solution

For the given initial conditions, $M_1^* = 1.0$ and $\theta_1 = 0$ deg. Those conditions determine the particular Γ_- characteristic that describes this simple wave expansion; the Γ_- characteristic is illustrated in the hodograph plane in Fig. 12.19b. Rays from

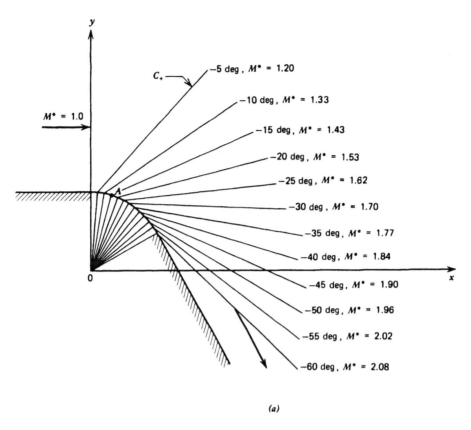

(a)

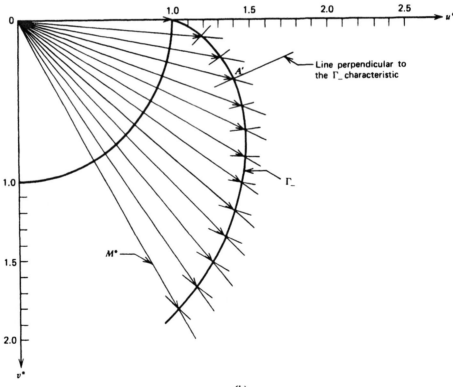

(b)

Figure 12.19 Simple wave flow over a cylinder. (*a*) Physical plane. (*b*) Hodograph plane.

the origin of both the physical plane and the hodograph plane are drawn at 5 deg intervals. The intersection of each such ray in the hodograph plane with the Γ_- *solution characteristic* determines the image point in the hodograph plane of the C_+ Mach line originating at the corresponding point on the cylinder in the physical plane, since the flow angles at the physical points are equal to the flow angles in the hodograph plane. For example, point A in the physical plane is located at a point on the cylinder where the turning angle is -15 deg. Point A' in the hodograph plane is the point on the Γ_- solution characteristic where $\delta = -15$ deg. From equation 12.104, the slope of each C_+ characteristic corresponding to a particular point on the Γ_- characteristic is perpendicular to the Γ_- characteristic at that point. Segments of these perpendiculars to the Γ_- solution characteristic are drawn on Fig. 12.19b. Lines parallel to these perpendiculars are drawn through the corresponding points on the cylinder in the physical plane illustrated in Fig. 12.19a, and emanate into the flow field as straight Mach lines.

The length of a ray from the origin O to a point on the Γ_- *solution characteristic* in Fig. 12.19b gives the value of M^* for the corresponding C_+ characteristic. Those values of M^* are indicated for the corresponding angles in Fig. 12.19a. By the aforementioned procedure, the flow field presented in Fig. 12.19a is constructed from the hodograph plane of Fig. 12.19b.

The above results are also obtainable in a simple straightforward analytical manner by employing Table C.13. Comparison of the graphical and tabular results shows agreement in the values of M^* within ± 0.01 at all points. Graphical constructions are quite useful for developing an insight into the details of a particular flow field. Analytical techniques employing a computer, however, are considerably faster and more accurate. Consequently, in practice, little, if any, use is made of the graphical techniques that were formerly employed for analyzing supersonic flow fields.

12-7 SUMMARY

The *method of characteristics* is a procedure for replacing a set of quasi-linear nonhomogeneous partial differential equations by a set of *compatibility equations* (interior operators) that is valid on surfaces called *characteristic surfaces*. For a problem in two independent dimensions (e.g., two space dimensions, or one space dimension and time), the characteristic surfaces are characteristic lines, called *characteristics*, and the compatibility equations are total differential relationships that are valid along the characteristics. Consequently, finite difference methods for solving the characteristic and compatibility equations are considerably simpler than methods for solving the original set of partial differential equations. The method of characteristics is applicable only to hyperbolic partial differential equations.

In the present chapter, the general theory of the method of characteristics for problems in two independent variables is presented. Characteristics are shown to have several properties. From a physical point of view, they are the path of propagation of physical disturbances. From a purely heuristic point of view, they are curves along which the governing partial differential equations can be manipulated into total differential equations. They are also curves across which the derivatives of a physical property may be discontinuous, while the property itself remains continuous. From a rigorous mathematical point of view, a characteristic is defined as a curve along which the governing partial differential equations reduce to an interior operator, that is, a total differential equation. The general theory is developed first for a single differential equation, for which characteristics always exist, and then for a system of two equations. In the latter case, characteristics exist only if the governing partial differential equations are hyperbolic. The theory is extended to any number of partial differential equations, for two independent variables, in Appendix B. The method of characteristics for three independent variables is introduced in Chapter 20 in Volume 2.

The general theory of the method of characteristics is applied to the case of steady two-dimensional, planar and axisymmetric, irrotational supersonic flow. The characteristics for this case are the Mach lines of the flow. A complete numerical algorithm, called a *unit process*, is developed for an interior point of a flow field. A numerical example is presented illustrating the application of the unit process for an interior point.

Numerical algorithms are developed in Chapter 16 for the boundary points of a flow (e.g., a solid wall, an axis of symmetry, and a free pressure boundary), and numerical examples illustrating those unit processes are presented. FORTRAN computer programs for implementing all of the unit processes are presented, and several applications to engineering problems are discussed.

REFERENCES

1. O. M. Belotserkovskii, "Flow with a Detached Shock Wave about a Symmetric Profile," *Journal of Applied Mathematics and Mechanics*, Vol. 22, pp. 279–296, 1958.
2. O. M. Belotserkovskii and P. I. Chuskin, "The Numerical Solution of Problems in Gas Dynamics," *Basic Developments in Fluid Mechanics*, Vol. 1, edited by M. Holt, pp. 1–123, Academic Press, New York, 1965.
3. S. G. Liddle and R. D. Archer, "Transonic Flow in Nozzles Using the Method of Integral Relations," *Journal of Spacecraft and Rockets*, Vol. 8, No. 7, pp. 722–728, July 1971.
4. G. Moretti and M. Abbett, "A Time-Dependent Computational Method for Blunt Body Flows," *AIAA Journal*, Vol. 4, No. 12, pp. 2136–2141, December 1966.
5. J. D. Anderson, "A Time-Dependent Analysis for Vibrational and Chemical Nonequilibrium Nozzle Flows," *AIAA Journal*, Vol. 8, No. 3, pp. 545–550, March 1970.
6. R. Serra, "Determination of Internal Gas Flows by a Transient Numerical Technique," *AIAA Journal*, Vol. 10, No. 5, pp. 603–611, May 1972.
7. M. C. Cline, "The Computation of Steady, Two-Dimensional Transonic Nozzle Flow by a Time Dependent Method," *AIAA Journal*, Vol. 12, No. 2, pp. 500–501, February 1974.
8. Bruce Carnahan, H. A. Luther, and James O. Wilkes, *Applied Numerical Methods*, Chap. 6, John Wiley, New York, 1969.
9. S. D. Conte and C. de Boor, *Elementary Numerical Analysis*, Second Edition, Chap. 6, McGraw-Hill, New York, 1972.
10. J. D. Hoffman, "Accuracy Studies of the Numerical Method of Characteristics for Axisymmetric, Steady Supersonic Flows," *Journal of Computational Physics*, Vol. 11, No. 2, pp. 210–239, February 1973.
11. R. Courant and K. O. Friedrichs, *Supersonic Flow and Shock Waves*, pp. 59–62, Interscience Publishers, New York, 1948.

PROBLEMS

1. Given the first-order partial differential equation

$$u_x + u_y = 1$$

and the initial condition $u(0,y) = y$ for all y, determine (a) the characteristic equation, (b) the compatibility equation, and (c) the value of $u(2,4)$.

2. Consider the first-order partial differential equation

$$uu_x + u_y - u = 0$$

with the initial condition $u(x,0) = x + 10$. Determine (a) the characteristic equation, (b) the compatibility equation, and (c) the value of $u(5,10)$.

3. The velocity of a particle is described by the following first-order partial differential equation.

$$uyu_x + uxu_y + xy = 0$$

The particle velocity distribution at $y = 0$ is $u(x,0) = x^3 - 54$. Determine (a) the character-istic equation, (b) the compatibility equation, and (c) the y coordinate and the velocity of the particle that was initially located at $(4,0)$ when that particle reaches $x = 10$.

4. The linearized equations governing the unsteady one-dimensional motion of small distur-bances (acoustic waves) in a constant-area passage are given by equations 11.15 and 11.16, which are repeated below,

$$\rho_\infty u_t + p_x = 0$$

$$\rho_\infty a_\infty^2 u_x + p_t = 0$$

where u and p are the perturbation velocity and pressure, respectively, and the subscript ∞ denotes the uniform undisturbed free stream. Employ the general theory of the method of characteristics to determine (a) the characteristic equations for such a flow field, and (b) the corresponding compatibility equations. (c) If the fluid is air with $p_\infty =$ one atm and $t_\infty = 300$ K, sketch on the xt plane (x abscissa and t ordinate) the range of influence of a disturbance created at $x = 10$ m and $t = 100$ s.

5. *Long wave motions* denote motions in an incompressible fluid having a free surface, where the vertical displacements of the fluid particles are small compared with their horizontal displacements. Such motions occur in tidal waves, storm surges, and tsunamis, where the fluid depth remains essentially constant and the fluid velocities are small, so that the resulting wave motions are linear. When long wave motions occur in estuaries, rivers, and canals, the fluid depth may change appreciably and the fluid velocities may become significant. The resulting waves are called bores, and the wave motions exhibit nonlinear behavior.

One-dimensional long wave motions occur in channels where the side boundaries are parallel. When the side boundaries are not parallel, the one-dimensional approximation may still yield useful results. The latter case is analogous to the one-dimensional flow approxima-tion employed in Chapter 4 for the flow of a fluid in a variable-area passage. The equations governing long wave motions are obtained by applying the continuity and momentum equations to the differential control volume illustrated in Fig. 12.20, where h denotes the surface height, w denotes the width of the channel, and u denotes the fluid velocity. The

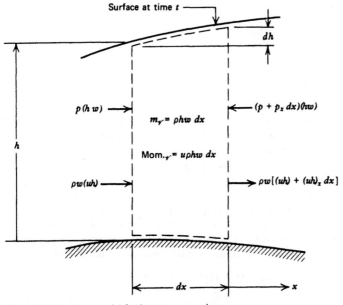

Figure 12.20 Flow model for long wave motions.

pressure forces and the mass flow rates at the control surface are indicated, as are the mass and momentum stored within the control volume. Terms of order higher than the first have been neglected. The pressure change across the control volume is caused by the change in the hydrostatic pressure, that is, $dp/dh = -\rho g$, where g is the local acceleration caused by gravity. (a) For the flow model illustrated in Fig. 12.20, show that the partial differential equations governing long wave motions are

$$h_t + uh_x + hu_x = 0 \qquad\qquad (a)$$

$$u_t + uu_x + gh_x = 0 \qquad\qquad (b)$$

Equations (a) and (b) comprise a system of two quasi-linear partial differential equations for the determination of h and u. Employing the general theory of the method of characteristics, determine (b) the characteristic equations, and (c) the corresponding compatibility equations.

In the study of tidal motions, u is small and h is large and effectively constant. For that case, (d) show that the linearized forms of equations (a) and (b) are

$$h_t + h_\infty u_x = 0 \qquad\qquad (c)$$

$$u_t + gh_x = 0 \qquad\qquad (d)$$

where h_∞ is the undisturbed fluid depth. Employing the general theory of the method of characteristics, determine (e) the characteristic equations, and (f) the corresponding compatibility equations. Note the similarity of linearized long wave motions to one-dimensional acoustic wave motions considered in Problem 12.4.

6. Section 12–5(d) presents a numerical procedure for implementing the unit process for an interior point. That procedure is based on the *average property method* discussed in Section 12–5(a). Develop an equivalent numerical procedure for implementing the unit process for an interior point based on the *average coefficient method* discussed in Section 12–5(a).

7. Example 12.2 illustrates the unit process for an interior point based on the *average property method* developed in Section 12–5(d). Work Example 12.2 employing the unit process based on the *average coefficient method* developed in Problem 12.7.

8. Consider the nozzle described in Example 12.2. The location and flow properties at two neighboring points in the flow field are given in the following table.

	Point 1	Point 2
x, m	0.5000	0.5100
y, m	0.6000	0.5000
u, m/s	1300.0	1270.0
v, m/s	350.0	315.0

Employing the unit process for an interior point described in Section 12–5(d), calculate the location of and the flow properties at the downstream intersection of the characteristics for three applications of the corrector.

9. The unit process for a point in the interior of a supersonic flow field is described in Section 12–5(d) and illustrated in Fig. 12.9. For a point on a solid boundary, illustrated in Fig. 12.21, point 1 of the interior point unit process is outside of the flow field. Consequently, the characteristic equation for right-running Mach line 14 and the corresponding compatibility equation are not applicable at a solid boundary. At a solid boundary, however, two boundary conditions are available to replace the lost characteristic and compatibility equations. The location and slope of the solid boundary must be specified. Thus,

$$y_{wall}(x) = f(x) \qquad (specified) \qquad\qquad (a)$$

$$\tan \theta_w(x) = \frac{df(x)}{dx} \qquad (specified) \qquad\qquad (b)$$

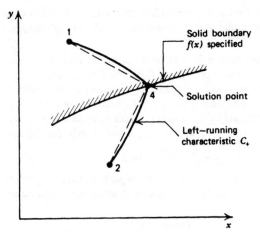

Figure 12.21 Unit process for a wall point.

Develop a numerical procedure for implementing the unit process for a wall point, analogous to the procedure presented in Section 12–5(d) for an interior point.

10. Consider the nozzle described in Example 12.2. The location and flow properties at a point near the wall are $x = 1.000$ m, $y = 1.400$ m, $u = 1600$ m/s, and $v = 400$ m/s. A conical wall having an angle of 15 deg passes through the point $x = 1.000$ m and $y = 1.500$ m. Employing the unit process for a wall point developed in Problem 12.9, calculate the location and flow properties for the point located at the intersection of the left-running characteristic through the given data point and the nozzle wall.

11. The unit process for a point on the axis of symmetry of a supersonic flow field is illustrated in Fig. 12.22. Point 2 is a mirror image of point 1, and the unit process for an interior point may be applied to determine the location of and the flow properties at point 4. The unit process for an axis of symmetry point may be simplified, however, by noting that on an axis of symmetry, $v_4 = y_4 = 0$. Consequently, the characteristic equation for right-running Mach line 14 and the corresponding compatibility equation suffice to determine x_4 and u_4. Develop a numerical procedure for implementing the unit process for an axis of symmetry point, analogous to the procedure presented in Section 12–5(d) for an interior point.

12. Consider the nozzle described in Example 12.2. At a point near the axis of symmetry, $x = 1.000$ m, $y = 0.030$ m, $u = 1500$ m/s, and $v = 25$ m/s. Employing the unit process for an

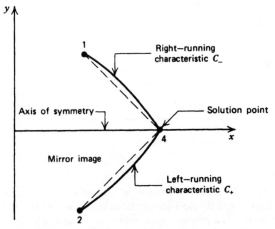

Figure 12.22 Unit process for an axis of symmetry point.

axis of symmetry point developed in Problem 12.11, calculate the location of and the flow properties at the intersection of the right-running characteristic through the given point and the axis of symmetry.

13. Figure 12.23 illustrates the unit process for a point on the boundary of a free jet, called a free pressure boundary point. Point 1 is outside of the flow field. Consequently, the characteristic equation for right-running Mach line 14 and the corresponding compatibility equation are not applicable at a free pressure boundary point. At the boundary of a free jet, however, the static pressure p_4 of the jet is equal to the ambient pressure p_o. Since the jet flow is isentropic, the velocity V_4 is a unique function of p_4, and thus p_o. Consequently, the equation

$$V_4 = \sqrt{u_4{}^2 + v_4{}^2} = f(p_o) \qquad (a)$$

replaces the compatibility equation. The location of the jet boundary may be determined in either of two ways. First, line 34 is a streamline and must satisfy the relationship

$$\frac{dy}{dx} = \frac{v}{u} \quad (along\ line\ 34) \qquad (b)$$

Equation (b) may be employed in place of the characteristic equation along line 14. Second, the mass flow rate across line 23 must equal the mass flow rate along line 24.

$$\dot{m}_{23} = \dot{m}_{34} \qquad (c)$$

Equation (c) may be employed in place of the characteristic equation along line 14. Develop a numerical procedure for implementing the unit process for a free pressure boundary point, analogous to the procedure presented in Section 12–5(d) for an interior point.

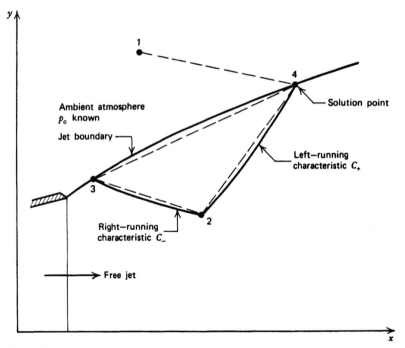

Figure 12.23   Unit process for a free pressure boundary point.

14. The gas described in Example 12.2 is flowing as a free jet in a region where the ambient pressure is 14,500 N/m². The locations of and the flow properties at two points in the jet are given in the following table, where point 3 is a previous point on the jet boundary.

	Point 2	Point 3
x, m	2.0000	1.9500
y, m	1.4500	1.4450
u, m/s	2650.0	2640.0
v, m/s	1125.0	1120.0

Determine the location of and the flow properties at the intersection of the left-running characteristic through the given point and the free jet boundary, employing the unit process developed in Problem 12.13.

13

the method of characteristics applied to unsteady one-dimensional homentropic flow

13-1 PRINCIPAL NOTATION FOR CHAPTER 13

a speed of sound.
$C_{\pm}$ denotes physical characteristics.
h static specific enthalpy.
M Mach number.
p absolute static pressure.
s specific entropy.
t time.
T absolute static temperature.
u velocity of fluid.

Greek Letters

γ specific heat ratio.
$\Gamma_{\pm}$ denotes state characteristics.
δ denotes planar, cylindrical, or spherical one-dimensional flow (see Table 13.1).
$\lambda_{\pm}$ slope of physical characteristics.
ρ static density.

Subscripts

$+$ denotes C_{+} and Γ_{+} characteristics.
$-$ denotes C_{-} and Γ_{-} characteristics.

Superscripts

$*$ denotes nondimensional property.

13-2 INTRODUCTION

There are several practical fluid mechanical systems in which unsteady flow phenomena occur, that is, the flow properties of the fluid vary rapidly with time. For such flows the time t is one of the variables determining the properties of the fluid. Consequently, the flow field for an unsteady flow, being time dependent, cannot be described by the equations discussed in Chapters 3 to 12 because their derivations assume that the flow is either steady or quasi-steady. Accordingly, this chapter is concerned with the analysis of unsteady one-dimensional homentropic flow.

The nineteenth-century literature pertaining to the unsteady flow of gases deals primarily with developing a basic understanding of wave phenomena rather than with the solutions to specific problems. Since World War II, however, there has been an increased interest in both the understanding and applications of wave phenomena. Much has been published about the unsteady flow of gases, and several excellent books are available.[1-5] Reference 1 is the classic on that subject.

The analytical methods pertinent to the unsteady propagation of very small pressure disturbances in a gas, the field of acoustics, is well developed. Sections 11-3 and 15-3 present the governing equations for an acoustic field, and Section 15-4 discusses the acoustics of a closed rigid hollow cylinder.

Unsteady flows in which there are large amplitude changes in the flow properties of the flowing gas occur during the start-up, shutdown, and transient operation of turbomachines (such as axial and radial flow compressors and turbines), steam and gas turbine power-plants, and continuous flow propulsion engines (such as turbojets, turbofans, ramjets, and liquid and solid propellant rocket engines). In the case of reciprocating or piston type machinery (air compressors, pumps, Otto and Diesel cycle internal combustion engines), the basic operating cycle for the piston machine is based on unsteady flow processes. Unsteady flow phenomena are also encountered in the propagation of explosive waves and detonations, and in the operation of a shock tube.

In this chapter the governing equations are derived for the unsteady one-dimensional homentropic flow of a compressible fluid, in the absence of work and body forces, and the corresponding characteristic and compatibility equations are also derived. The general features of unsteady one-dimensional continuous flow are discussed, and a numerical algorithm is presented for the determination of an *interior point* in an unsteady one-dimensional homentropic flow.

In Chapter 19, the unsteady generalized quasi-one-dimensional flow of a compressible flow is considered. The general features of unsteady one-dimensional flow with discontinuities are discussed, and numerical algorithms are developed for the boundary points of such a flow. Numerical examples are presented for illustrating the application of those unit processes to specific unsteady one-dimensional and quasi-one-dimensional flow problems.

13-3 MATHEMATICAL ANALYSIS OF UNSTEADY ONE-DIMENSIONAL HOMENTROPIC FLOW

A truly *one-dimensional flow* is one wherein the flow properties depend only on one space coordinate x and the time t.* Consequently, the pathlines must be straight lines. Accordingly, planar flow in a constant-area duct, flow with cylindrical symmetry, and flow with spherical symmetry are *one-dimensional flows*. Flow in a duct having a slowly varying cross section where the duct height is small in comparison with the radius of curvature of the axis of the duct is called a *quasi-one-dimensional flow*.

An *unsteady homentropic flow* is a flow for which the entropy remains constant throughout the flow field for all time. An *unsteady generalized quasi-one-dimensional flow* is one that includes the effects of area change, friction, heat transfer, and mass addition.

The governing equations for unsteady one-dimensional homentropic flow are derived in the present section, and the corresponding characteristic and compatibility equations are determined. The governing equations for unsteady generalized quasi-one-dimensional flow are derived in Section 19-3(c).

13-3(a) Governing Equations

The governing equations for unsteady one-dimensional homentropic flow are obtained from Table 2.2 by assuming that $\mathbf{B} = d\mathbf{F}_{shear} = \delta \dot{W} = \delta \dot{Q} = g\, dz = 0$. Thus:

(1) *Continuity equation*

$$\rho_t + \nabla \cdot (\rho \mathbf{V}) = 0 \qquad (13.1)$$

(2) *Momentum equation*

$$\rho \frac{D\mathbf{V}}{Dt} + \nabla p = 0 \qquad (13.2)$$

---

* In Chapters 13 and 19 (Chapter 19 appears in Volume 2), the time is denoted by t, and the absolute static temperature is denoted by T.

(3)   *Energy equation*

$$\rho \frac{D}{Dt}\left(h + \frac{V^2}{2}\right) - p_t = 0 \tag{13.3}$$

(4)   *Entropy equation*

$$\frac{Ds}{Dt} = 0 \tag{13.4}$$

Figure 13.1 illustrates the Cartesian, cylindrical, and spherical coordinate systems for a one-dimensional flow. In all three cases, x denotes the spatial coordinate and u the flow velocity. Expressing equation 13.1 in these three coordinate systems, we obtain the following forms for the continuity equation.

$$\rho_t + u\rho_x + \rho u_x = 0 \qquad (Cartesian) \tag{13.5a}$$

$$\rho_t + u\rho_x + \rho u_x + \frac{\rho u}{x} = 0 \qquad (cylindrical) \tag{13.5b}$$

$$\rho_t + u\rho_x + \rho u_x + \frac{2\rho u}{x} = 0 \qquad (spherical) \tag{13.5c}$$

The above three equations may be written as the following single continuity equation

$$\rho_t + u\rho_x + \rho u_x + \frac{\delta \rho u}{x} = 0 \tag{13.6}$$

where $\delta = 0$ for a planar flow, $\delta = 1$ for a cylindrical flow, and $\delta = 2$ for a spherical flow.

Equation 13.2, the vector momentum equation, when expressed in any of these three coordinate systems, yields the following common momentum equation.

$$\rho u_t + \rho u u_x + p_x = 0 \tag{13.7}$$

Since the flow is assumed to be homentropic, the speed of sound definition (see equation 1.180) may be employed in place of the energy equation as a governing equation. Hence

$$a^2 = \left(\frac{\partial p}{\partial \rho}\right)_s = \frac{dp}{d\rho} \qquad (for\ homentropic\ flow) \tag{13.8}$$

where

$$a = a(p) \qquad (for\ homentropic\ flow) \tag{13.9}$$

Equation 13.8 may be written as the *substantial derivative*.

$$\frac{Dp}{Dt} - a^2 \frac{D\rho}{Dt} = 0 \tag{13.10}$$

Equation 13.10 yields the following common speed of sound equation for all three coordinate systems.

$$p_t + u p_x - a^2(\rho_t + u\rho_x) = 0 \tag{13.11}$$

The entropy equation, equation 13.4, is identically satisfied for a homentropic flow.

The continuity equation, equation 13.6, may be simplified further for homentropic flow by substituting for $\rho_t + u\rho_x$ from equation 13.11. The result is

$$\rho a^2 u_x + p_t + u p_x + \delta \frac{\rho u a^2}{x} = 0 \tag{13.12}$$

Equations 13.7 and 13.12 comprise a set of two equations for determining the two flow properties u and p. Those equations are summarized in Table 13.1.

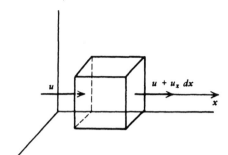

(a)

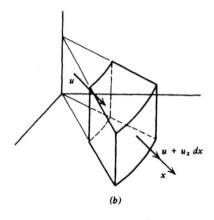

(b)

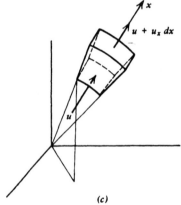

(c)

Figure 13.1 Coordinate systems for unsteady one-dimensional flow. (a) Cartesian coordinates. (b) Cylindrical coordinates. (c) Spherical coordinates.

Table 13.1 Governing Equations for Unsteady One-Dimensional Homentropic Flow

$$\rho u_t + \rho u u_x + p_x = 0 \tag{13.7}$$

$$\rho a^2 u_x + p_t + u p_x + \delta \frac{\rho u a^2}{x} = 0 \tag{13.12}$$

$\delta = 0$ for planar one-dimensional flow
$\delta = 1$ for cylindrical one-dimensional flow
$\delta = 2$ for spherical one-dimensional flow

For a homentropic flow, the density ρ is a unique function of the pressure p throughout the flow (see equation 10.113). Thus

$$\rho = \rho(p) \qquad (\textit{for homentropic flow}) \tag{13.13}$$

13-3(b) Characteristic and Compatibility Equations

The characteristic and compatibility equations corresponding to equations 13.7 and 13.12 are determined by multiplying those equations by the unknown parameters σ_1 and σ_2, respectively, and summing. Thus,

$$\sigma_1(13.7) + \sigma_2(13.12) = 0$$

Performing the above summation yields

$$u_x(\rho u \sigma_1 + \rho a^2 \sigma_2) + u_t(\rho \sigma_1) + p_x(\sigma_1 + u\sigma_2)$$

$$+ p_t(\sigma_2) + \left(\sigma_2 \delta \frac{\rho u a^2}{x}\right) = 0 \tag{13.14}$$

Factoring out the coefficients of u_x and p_x, we obtain

$$(\rho u \sigma_1 + \rho a^2 \sigma_2)\left(u_x + \frac{\rho \sigma_1}{\rho u \sigma_1 + \rho a^2 \sigma_2} u_t\right)$$

$$+ (\sigma_1 + u\sigma_2)\left(p_x + \frac{\sigma_2}{\sigma_1 + u\sigma_2} p_t\right) + \left(\sigma_2 \delta \frac{\rho u a^2}{x}\right) = 0 \tag{13.15}$$

If $u(x,t)$ and $p(x,t)$ are assumed to be continuous functions, then

$$\frac{du}{dx} = u_x + \lambda u_t \qquad \text{and} \qquad \frac{dp}{dx} = p_x + \lambda p_t \tag{13.16}$$

where $\lambda = dt/dx$. Accordingly, the slopes of the characteristics, $dt/dx = \lambda$, are the coefficients of u_t and p_t. Thus,

$$\lambda = \frac{\sigma_1}{u\sigma_1 + a^2\sigma_2} = \frac{\sigma_2}{\sigma_1 + u\sigma_2} \tag{13.17}$$

Substituting equations 13.16 into equation 13.15 yields

$$(\rho u \sigma_1 + \rho a^2 \sigma_2)\, du + (\sigma_1 + u\sigma_2)\, dp + \sigma_2 \delta \frac{\rho u a^2}{x}\, dx = 0 \tag{13.18}$$

Equations 13.17 are the *characteristic equations* for unsteady one-dimensional homentropic flow, and equation 13.18 is the corresponding *compatibility equation*. It is necessary, however, to eliminate the unknown parameters σ_1 and σ_2 from equations 13.17 and 13.18.

Solving equations 13.17 for σ_1 and σ_2, we obtain

$$\sigma_1(u\lambda - 1) + \sigma_2(a^2\lambda) = 0 \tag{13.19}$$

$$\sigma_1(\lambda) + \sigma_2(u\lambda - 1) = 0 \tag{13.20}$$

For equations 13.19 and 13.20 to have a solution other than the trivial one $\sigma_1 = \sigma_2 = 0$, the determinant of the coefficient matrix for those equations must vanish. Thus,

$$\begin{vmatrix} (u\lambda - 1) & a^2\lambda \\ \lambda & (u\lambda - 1) \end{vmatrix} = 0 \tag{13.21}$$

Expanding the above determinant, we obtain

$$(u\lambda - 1)^2 - a^2\lambda^2 = 0 \tag{13.22}$$

Solving equation 13.22 for λ yields

$$u\lambda - 1 = \pm a\lambda \tag{13.23}$$

$$\left(\frac{dt}{dx}\right)_{\pm} = \lambda_{\pm} = \frac{1}{u \pm a} \tag{13.24}$$

Figure 13.2 illustrates schematically the characteristics in the xt plane; that is, the *physical plane*. Two characteristics exist, the C_+ and C_- characteristics, corresponding to the $+$ and $-$ signs, respectively, in equation 13.24. Four possible cases are illustrated in Fig. 13.2, depending on whether the fluid velocity u is positive or negative,

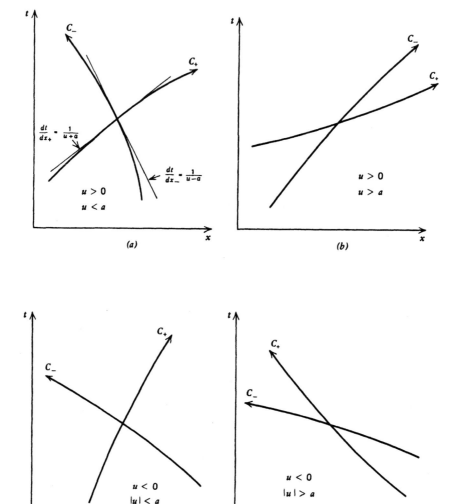

Figure 13.2 Characteristics for unsteady one-dimensional homentropic flow. (*a*) Subsonic flow from left to right. (*b*) Supersonic flow from left to right. (*c*) Subsonic flow from right to left. (*d*) Supersonic flow from right to left.

and whether u is either less than or greater than the local speed of sound a; that is, whether the flow is subsonic or supersonic. In each of the four cases, λ is real and the characteristics exist. Hence, the equations governing the unsteady one-dimensional homentropic flow of a compressible fluid are hyperbolic for both subsonic and supersonic flows.

One of the properties of a characteristic curve is that it is the path along which information is propagated through a flow field (see Section 12-3). Along the C_+ and C_- characteristics, the speed of propagation is

$$\left(\frac{dx}{dt}\right)_\pm = \frac{1}{\lambda_\pm} = u \pm a \tag{13.25}$$

Hence, the propagation speed along the $C_\pm$ characteristics is equal to the speed of sound relative to the moving fluid. In that regard, the $C_\pm$ characteristics in unsteady flow are analogous to the Mach lines in a steady supersonic flow, and are known as the *Mach lines of the unsteady flow*. Consequently, in an unsteady one-dimensional homentropic flow, two characteristic curves pass through each point of the flow field; the right- and left-running characteristics (Mach lines), illustrated in Fig. 13.2.

The compatibility equations that are valid along the characteristic curves are obtained from equation 13.18, by solving equation 13.19 or 13.20 for σ_1 in terms of σ_2 and substituting the result into equation 13.18. From equation 13.20,

$$\sigma_1 = -\sigma_2 \frac{u\lambda - 1}{\lambda} \tag{13.26}$$

Substituting equation 13.26 into equation 13.18 and dividing by σ_2 yields the following result.

$$\rho\left[-\frac{u(u\lambda - 1)}{\lambda} + a^2\right] du + \left[-\frac{(u\lambda - 1)}{\lambda} + u\right] dp + \delta \frac{\rho u a^2}{x} dx = 0 \tag{13.27}$$

Recalling that $(u\lambda - 1)^2 = a^2\lambda^2$ and $\lambda_\pm = 1/(u \pm a)$ along the characteristics, equation 13.27 may be simplified to yield

$$dp_\pm \pm \rho a\, du_\pm = -\delta \frac{\rho u a^2}{x} dt_\pm \tag{13.28}$$

where the upper subscript $+$ attached to the differentials dp_+, $du_+^{'}$, and dt_+ and the upper sign $+$ in the coefficient $\pm \rho a$ correspond to the C_+ characteristic, and the lower subscript $-$ and the lower sign $-$ correspond to the C_- characteristic.

Summarizing, for unsteady one-dimensional homentropic flow, two distinct characteristics exist; the right- and left-running Mach lines, and there exists one compatibility equation valid along each of the two characteristics. Consequently, the derived system of characteristic and compatibility equations (equations 13.24 and 13.28) may replace the original system of partial differential equations (equations 13.7 and 13.12). Table 13.2 presents these characteristic and compatibility equations.

To apply the characteristic and compatibility equations presented in Table 13.2, a characteristic grid must be devised wherein the two characteristic curves intersect at a common point, so that the two compatibility equations may be solved simultaneously at the point of intersection for the flow properties u and p. Figure 13.3 illustrates schematically the intersection of the two characteristic curves at a common solution point.

Table 13.2   Characteristic and Compatibility Equations for Unsteady
One-Dimensional Homentropic Flow

Characteristic equation

$$\left(\frac{dt}{dx}\right)_{\pm} = \lambda_{\pm} = \frac{1}{u \pm a} \qquad (Mach\ lines) \tag{13.24}$$

Compatibility equation

$$dp_{\pm} \pm \rho a\, du_{\pm} = -\delta \frac{\rho u a^2}{x} dt_{\pm} \qquad (along\ Mach\ lines) \tag{13.28}$$

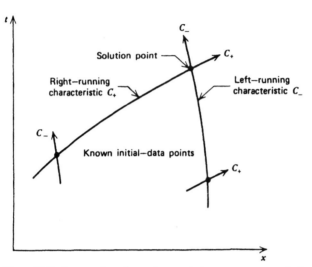

Figure 13.3   Intersection of two characteristics at a common solution point.

13-4 GENERAL FEATURES OF UNSTEADY
ONE-DIMENSIONAL HOMENTROPIC FLOW

The equations governing unsteady one-dimensional homentropic flow are derived in
Section 13–3. In the present section, the general features of unsteady one-dimensional
flow are described by considering the special case of unsteady one-dimensional *planar*
homentropic flow; that is, a flow in which $\delta = 0$. Consideration of the latter special
case is important to the understanding of the general case; that is, an unsteady
generalized quasi-one-dimensional flow including area change, friction, heat transfer,
and mass addition. The general case is discussed in Chapter 19 in Volume 2.

Unsteady one-dimensional planar homentropic flow occurs frequently in engineering. For example, simple waves, discussed in Chapter 8 for steady two-
dimensional planar irrotational supersonic flow, may also occur in unsteady
one-dimensional planar homentropic flow. The shock tube, discussed in Section
19–5(c), and the Ludwieg tube, discussed in Section 19–7(c), are examples of a flow
field containing regions of planar homentropic flow and regions of simple wave flow.

If the flowing fluid is a perfect gas, the solution for the flow properties in a planar
homentropic flow is obtained in a relatively simple form; its application to a variety
of elementary wave processes is presented in the present section.[1-5]

13-4(a) The u^*a^*, or State, Plane for a Perfect Gas

For a perfect gas, the compatibility equation for planar homentropic flow, equation 13.28, may be integrated in closed form. The results are much simpler if the solution is expressed in terms of the velocity u and the speed of sound a instead of the velocity u and the pressure p.

For a perfect gas, $p = \rho RT$ and $a^2 = \gamma RT = \gamma p/\rho$, where T denotes the static temperature of the gas, and if it experiences an isentropic process [see Section 1–15(e), equations 1.126 and 1.127]

$$Tp^{-(\gamma-1)/\gamma} = \text{constant} \qquad \text{and} \qquad p\rho^{-\gamma} = \text{constant} \tag{13.29}$$

Substituting for $T = a^2/\gamma R$ in equation 13.29 gives

$$a^2 p^{-(\gamma-1)/\gamma} = \text{constant} \tag{13.30}$$

Logarithmic differentiation of equation 13.30 yields

$$-\frac{\gamma-1}{\gamma}\frac{dp}{p} + 2\frac{da}{a} = 0 \tag{13.31}$$

Solving equation 13.31 for dp, we obtain

$$dp = \frac{2\gamma}{\gamma-1}\left(\frac{p}{a}\right)da \tag{13.32}$$

Substituting for dp in equation 13.28 from equation 13.32 yields

$$\frac{2\gamma}{\gamma-1}\left(\frac{p}{a}\right)da_\pm \pm \rho a\, du_\pm = 0 \tag{13.33}$$

Substituting for $a^2 = \gamma p/\rho$ into equation 13.33 and rearranging, we obtain

$$da_\pm \pm \frac{\gamma-1}{2}du_\pm = 0 \tag{13.34}$$

Integrating equation 13.34 yields

$$a_\pm \pm \frac{\gamma-1}{2}u_\pm = \text{constant} \tag{13.35}$$

Equation 13.35 is made dimensionless by dividing through by a reference velocity, for example, the speed of sound a_o in some initial undisturbed portion of the flow field. Dividing equation 13.35 by a_o gives

$$a_\pm^* \pm \frac{\gamma-1}{2}u_\pm^* = \text{constant} \tag{13.36}$$

where $a^* = a/a_o$ and $u^* = u/a_o$. Equation 13.36 represents a straight line on the u^*a^* plane, called the *state plane*, having the slope $\pm(\gamma-1)/2$. Note that a^* denotes the normalized speed of sound; it does not denote the sonic velocity where $M = 1$.

The flow properties T, p, and ρ corresponding to the value of a^* determined from equation 13.36 may be calculated from equation 13.29. Thus,

$$p_\pm^* = (T_\pm^*)^{\gamma/(\gamma-1)} = (a_\pm^*)^{2\gamma/(\gamma-1)} \tag{13.37}$$

$$\rho_\pm^* = (a_\pm^*)^{2/(\gamma-1)} \tag{13.38}$$

where $p^* = p/p_o$, $T^* = T/T_o$, and $\rho^* = \rho/\rho_o$.

Equation 13.36 determines two families of straight lines on the state plane, termed the *state plane characteristics*; the latter are denoted by Γ_+ and Γ_-, where the subscripts + and − correspond to the C_+ and C_- *physical characteristics*, respectively. For $\gamma = 1.40$, the slopes of the state plane characteristics are ± 0.20. By employing a grid spacing where one unit along the a^* axis corresponds to five units along the u^* axis, the slopes of the state plane characteristics are ± 1.0. Note that the slope of the Γ_+ characteristic is $-(\gamma - 1)/2$, and that of the Γ_- characteristic is $+(\gamma - 1)/2$.

Two state plane characteristics pass through each point in the state plane, as illustrated in Fig. 13.4. Four possible types of flow may, therefore, issue from a given initial point, such as point i in Fig. 13.4.

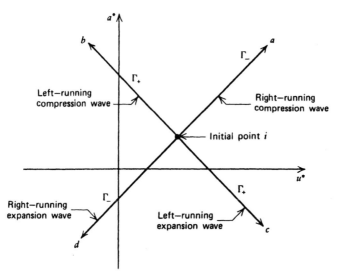

Figure 13.4   Characteristics in the state plane.

1. A right-running compression wave, corresponding to the Γ_- characteristic from point i to a.
2. A left-running compression wave, corresponding to the Γ_+ characteristic from point i to b.
3. A left-running expansion wave (also called a *rarefaction* wave), corresponding to the Γ_+ characteristic from point i to c.
4. A right-running expansion wave, corresponding to the Γ_- characteristic from point i to d.

In a particular flow field, the direction taken from the initial point i in the state plane depends on the direction in which the wave travels (right- or left-running) and the type of wave (compression or expansion).

The complete state plane may be constructed by drawing lines having the slope ± 1 through all points in the u^*a^* state plane. Figure 13.5 presents selected uniformly spaced state plane characteristics that cover the region $-1.5 \leq u^* \leq 1.5$ and $0.6 \leq a^* \leq 1.4$ of the state plane.

The location of the right- and left-running characteristics in the physical plane (i.e., the xt plane), called the *physical characteristics*, is determined by integrating equation 13.24. Along right-running, or C_+, characteristics, the quantity $a_+^* + (\gamma - 1)u_+^*/2$ is a constant, and along left-running, or C_-, characteristics, the quantity

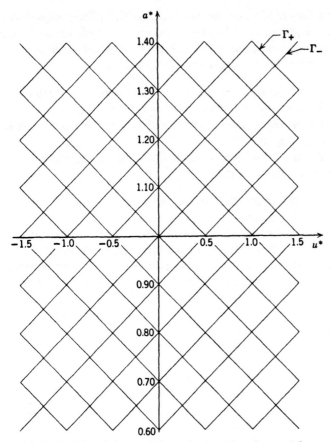

Figure 13.5   Selected characteristics in the state plane for $\gamma = 1.4$.

$a^*_\pm - (\gamma - 1)u^*_\pm/2$ is a constant. Those constants are frequently called *Riemann invariants*, after the mathematician B. Riemann.

Hence, the solution to an unsteady one-dimensional planar homentropic flow reduces to determining the location of the characteristics in the physical plane. Along the characteristics the quantities $a^*_\pm \pm (\gamma - 1)u^*_\pm/2$ are constants.

13-4(b) Uniform Flow Regions, Simple Wave Regions, and Nonsimple Wave Regions

A *uniform flow region* is one in which all of the thermodynamic properties of the fluid are uniform and its velocity is constant. A stationary region where $u = 0$ is a special case of a uniform flow region. A uniform flow region is said to be *doubly degenerate* in the state plane, because a region in the physical plane maps onto a single point in the state plane; that is, the u^*a^* plane. Regions of uniform flow occur quite frequently in unsteady flows.

A *simple wave region* is one wherein there are waves of only one family; that is, either right- or left-running waves. A simple wave region is said to be *singly degenerate* in the state plane, because a region in the physical plane maps onto a line in the state plane. The general features of simple waves are reviewed below.

The properties of simple waves are discussed for steady two-dimensional planar homentropic (i.e., irrotational) flow in Section 12–6. It is shown there that simple waves may occur in a flow field when the governing partial differential equations are *reducible*;[1] that is, the equations are homogeneous and the coefficients of the partial derivatives depend only on the dependent variables. It is pointed out in Section 12–6, that a *simple wave flow* has the following special properties.

1. A simple wave region is defined as one wherein the dependent variables (u and p in the present case) are not independent [$u = u(p)$, $a = a(p)$, etc.].
2. In a simple wave region, the characteristics (Mach lines) of one family are straight lines having constant properties.
3. If a region of uniform flow adjoins a nonuniform region, then the nonuniform region must be a simple wave region.

It is shown in the present section that simple waves may occur in *unsteady one-dimensional planar homentropic flow*.

The governing equations for unsteady one-dimensional planar *homentropic* flow, equations 13.7 and 13.12 with $\delta = 0$, are homogeneous, and the coefficients of the partial derivatives depend only on the flow properties u and p. Furthermore, since the entropy s is uniform throughout the flow field, all of the thermodynamic properties are unique functions of a *single* thermodynamic property, for example, the pressure p. Consequently, equations 13.7 and 13.12 are reducible, and simple waves may occur in the flow field of an unsteady one-dimensional planar homentropic flow.

From Property 3 above, a simple wave region adjoins a uniform flow region. Regions of uniform flow occur, as stated earlier, quite frequently in unsteady flows; consequently, simple wave regions also occur frequently.

The propagation of a pressure disturbance into a stationary region, discussed in Section 13–4(c), results in the formation of a simple wave region. The reflection of pressure disturbances at the boundaries of simple wave regions results in waves of both families, which interact and create a *nonsimple wave region*. Nonsimple wave regions in the physical plane map into regions in the state plane. Examples of non-simple wave regions are discussed in Sections 13–4(d) and 13–4(e). A more complex unsteady flow field, one involving uniform flow regions, simple wave regions, non-simple wave regions, and discontinuities such as shock waves and contact surfaces, is produced in the shock tube, discussed in Section 19–5(c).

13–4(c) Simple Compression and Expansion Waves

Figure 13.6*a* illustrates schematically a piston at rest in a long tube wherein there is a stationary gas ($u = 0$). At the time $t = 0$, the piston accelerates smoothly to the right. The initial motion of the piston produces a small pressure disturbance that propagates to the right into the undisturbed fluid with the speed of sound a, as illustrated schematically in the xt plane in Fig. 13.6*b*. Hence, the velocity of the front of the wave is $(dx/dt)_+ = a$, because a right-running wave is a C_+ characteristic.

The wave created by the initial piston motion imparts a pressure increment $dp > 0$ and a velocity increment $du > 0$ to the initially undisturbed gas, and is, therefore, a *compression wave*. For an isentropic process, if $dp > 0$ then $da > 0$. The initial infinitesimal compression wave is illustrated on the state plane in Fig. 13.6*c*. As the piston continues to accelerate, successive compression waves are generated at the face of the piston and move into the disturbed gas. The latter has a velocity u and speed of sound a larger than the corresponding values for the undisturbed gas. Consequently, the speeds of propagation, $(dx/dt)_+ = u + a$, of the consecutive waves increase, and each wave travels faster than the preceding wave.

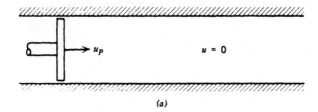

(a)

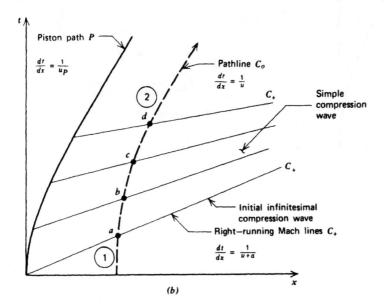

(b)

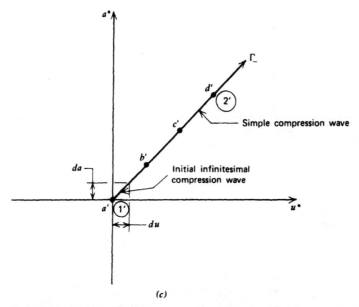

(c)

Figure 13.6 Right-running waves created by the motion of a piston in a gas. (a) Physical arrangement. (b) Compression wave in the physical plane. (c) Compression wave on the state plane.

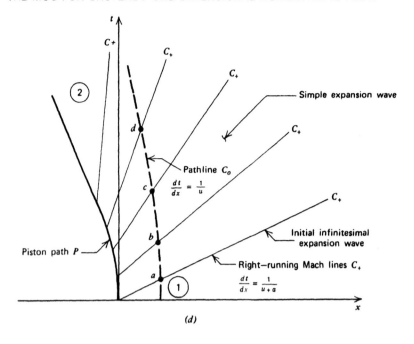

(d)

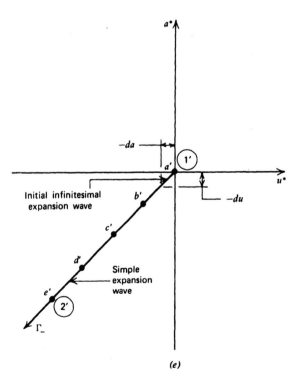

(e)

Figure 13.6 (*Continued*) (*d*) Expansion wave in the physical plane. (*e*) Expansion wave on the state plane.

The slopes of the C_+ characteristics, or right-running Mach lines, given by $(dt/dx)_+ = 1/(u + a)$, become more and more horizontal, and the waves converge, as illustrated in Fig. 13.6b. A typical *pathline*, or particle path, is indicated in Fig. 13.6b.

The process illustrated in Fig. 13.6b comprises a continuous *compression wave*. It is a *simple* compression wave, since all of the disturbances propagate in the direction of increasing values of x, that is, from the left into a region of uniform flow. Since the fluid particles enter the wave from the right, the compression wave illustrated in Fig. 13.6b is called a *right-facing* compression wave. Figure 13.6c illustrates the simple compression wave on the state plane; that is, the u^*a^* plane. The uniform regions 1 and 2 in the physical plane map onto points 1' and 2', respectively, in the state plane. Selected C_+ characteristics on the pathline, denoted by a, b, etc., are indicated in the physical plane in Fig. 13.6b, and their corresponding images on the state plane are denoted by a', b', etc., in Fig. 13.6c.

Figure 13.6d illustrates schematically the situation where the piston is accelerated smoothly in the direction of increasing values of $-x$, that is, to the left away from the gas enclosed by the tube. The initial motion of the piston produces a small pressure disturbance that propagates to the right with the speed of sound a. The velocity of the front of the wave, which is a C_+ characteristic, is $(dx/dt)_+ = a$. The wave created by the initial piston motion imparts a pressure increment $dp < 0$, a velocity increment $du < 0$, and a speed of sound increment $da < 0$ to the undisturbed gas in the tube. Hence, the wave is an *expansion wave*. The initial infinitesimal expansion wave is illustrated in the state plane in Fig. 13.6e. As the piston accelerates away from the gas, successive expansion waves are propagated into the disturbed fluid, which has an increasing negative velocity $-u$ and a decreasing speed of sound a. Accordingly, the speed of propagation, $(dx/dt)_+ = u + a$, of each consecutive expansion wave is smaller than that of the preceding wave. As a result, the slope, $(dt/dx)_+ = 1/(u + a)$, of each successive right-running Mach line C_+ becomes steeper, and the waves diverge, as illustrated in Fig. 13.6d. A typical pathline is illustrated in Fig. 13.6d.

The wave pattern illustrated in Fig. 13.6d is an *expansion wave*. Because all of the disturbances originate and propagate from the left into a region of undisturbed fluid, the wave pattern illustrated in Fig. 13.6d is a *simple expansion wave*. It is termed a *right-facing* expansion wave because the fluid particles enter the wave from the right side. Figure 13.6e illustrates schematically the simple expansion wave on the state plane.

Figures 13.7a and 13.7b illustrate schematically a *left-facing* compression wave and a *left-facing* expansion wave, respectively. The corresponding state planes are illustrated in Figs. 13.7c and 13.7d.

From Figs. 13.6 and 13.7, the following rules for simple compression and expansion waves are obtained.

Rule 1. *Simple compression waves* accelerate the fluid in the direction of the propagation of the wave. The pressure, velocity, and speed of sound increase, thus increasing the wave propagation speed, and the waves converge. Pathlines bend *toward* the waves.

Rule 2. *Simple expansion waves* accelerate the fluid in the direction opposite to the propagation of the wave. The pressure and speed of sound decrease while the velocity increases in the opposite direction, thereby decreasing the wave propagation speed. The waves diverge, and the pathlines bend *away* from the waves.

The simple compression and expansion waves discussed above are analogous to the simple compression and expansion waves for steady two-dimensional planar homentropic flow, discussed in Sections 8–4(c) and 8–6.

As pointed out in Section 13–4(a), the properties of a simple wave region may also be described in terms of the state plane characteristics $\Gamma_{\pm}$. For a simple right-facing C_+ wave, the flow properties are constant along the C_+ physical characteristics. Each C_+ characteristic is thus a straight line, which maps onto a point in the state plane, as illustrated in Fig. 13.4. The flow properties change in a right-facing C_+ wave only across the C_+ waves; that is, along the C_- physical characteristics that cross them. All states in a simple right-facing C_+ wave, therefore, map onto a single Γ_- characteristic in the state plane. Similarly, all states in a simple left-facing C_- wave map onto a single Γ_+ characteristic in the state plane.

For a specified piston motion $u_P = f(t)$, the piston path P may be determined, and the corresponding value of $u = u_P$ is, therefore, specified at each point along P. Knowing $u^* = u/a_o$ along the piston path P, the corresponding values of a^* may be

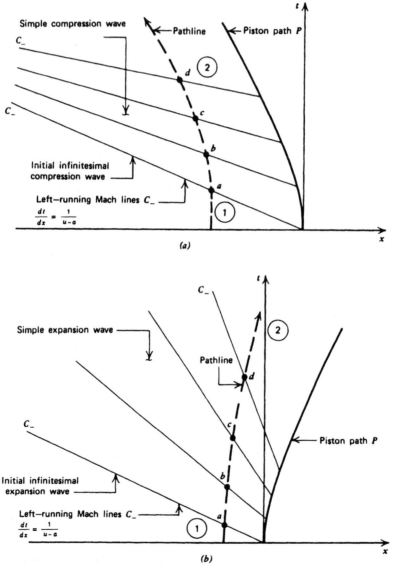

Figure 13.7 Left-running waves created by the motion of a piston. (a) Compression waves. (b) Expansion waves.

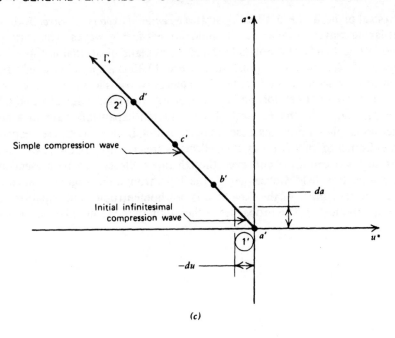

(c)

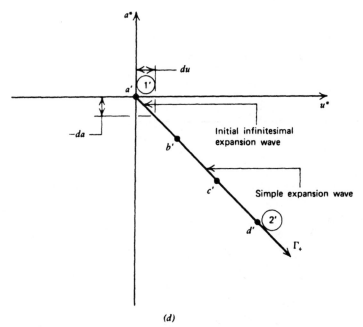

(d)

Figure 13.7 *(Continued)* *(c)* Compression wave on the state plane. *(d)* Expansion wave on the state plane.

calculated from equation 13.36, and the slope of the corresponding Mach line emanating from each point on the piston path may be determined from equation 13.24. Hence, the simple wave region emanating from the surface of a piston moving in a continuous manner may be determined completely.

When the piston is acclerated instantaneously to a finite velocity, a *centered wave* results. Figure 13.8*a* illustrates a centered compression wave, which is a *shock wave* in the physical plane, and Fig. 13.8*b* illustrates a *centered expansion wave*. Both waves are right-facing waves. Similar results hold for left-facing waves. The state plane illustrated in Fig. 13.6*e* corresponds to the physical plane illustrated in Fig. 13.8*b*.

The *centered expansion wave* illustrated in Fig. 13.8*b* is a special case of a simple expansion wave. The origin of a centered expansion wave is a singular point, where multiple values are obtained for the flow properties at a single value of *x* and *t*. Such a wave is analogous to the Prandtl-Meyer centered expansion wave in a steady two-dimensional planar flow, discussed in Section 8–4. Because a centered expansion wave issues from a geometric point, such effects as area change, friction, heat transfer, and mass addition cannot occur, even though those effects may be present in the remainder of the flow field. Consequently, the flow through the origin is homentropic even when the remainder of the flow field is not homentropic, and equations 13.24 and 13.28 are applicable. The properties at the origin of the centered expansion wave

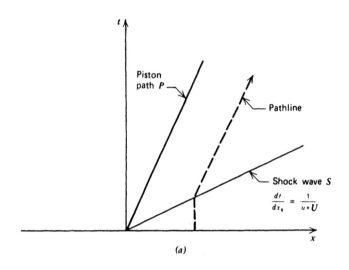

(a)

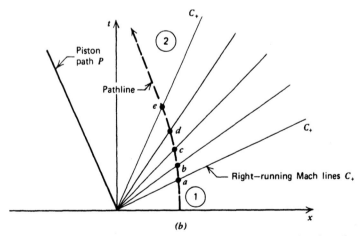

(b)

Figure 13.8 Centered waves created by the instantaneous acceleration of a piston. (a) Centered compression, or shock, wave. (b) Centered expansion wave.

are, therefore, those corresponding to the case where continuous expansion waves emanate from a point.

The direction of the Mach lines emanating from the origin of a centered expansion wave may be determined from the characteristic equation for the Mach lines, equation 13.24. If the flow field is homentropic, then a simple expansion wave develops from the Mach lines emanating from the origin. Even if the flow field is not homentropic, the solution described above for the flow properties at the origin of the centered expansion wave is still valid. A nonsimple expansion wave develops, however, from the Mach lines emanating from the singular point.

Example 13.1.   Air, assumed to be a perfect gas, fills a 3.0 m long tube at a pressure $p_o = 2.00 \cdot 10^5$ N/m^2 and a static temperature $T_o = 300$ K, as illustrated in Fig. 13.9a. The right-hand end of the tube is suddenly opened to the surrounding atmosphere, which is at a pressure of $1.00 \cdot 10^5$ N/m^2. A centered expansion wave propagates into the tube. Calculate (a) u, a, and M along the Mach lines corresponding to $p = 2.00$, 1.75, 1.50, 1.25, and $1.00 \cdot 10^5$ N/m^2, and (b) the time required for the first expansion wave to reach the left end of the tube. Plot the corresponding Mach lines on the xt plane. For air, $\gamma = 1.40$ and $R = 287.04$ J/kg-K.

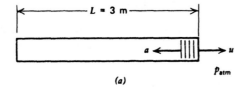

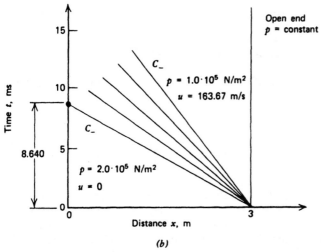

Figure 13.9   Illustrations for Example 13.1. (a) Physical arrangement. (b) Physical plane.

Solution

(a)   Choose the reference properties p_o, T_o, ρ_o, and a_o to be those of the undisturbed air in the tube before the expansion wave is initiated. Thus,

$$a_o = (\gamma R T_o)^{1/2} = [1.4(287.04)(300)]^{1/2} = 347.21 \text{ m/s}$$

A left-running expansion wave propagates into the tube from the right end, corresponding to line *i-c* in Fig. 13.4, which is a Γ_+ characteristic. From equation 13.36,

$$a^* + 0.2u^* = 1.0 \tag{a}$$

From equation 13.37,

$$a^* = (p^*)^{(\gamma - 1)/2\gamma} = (p^*)^{0.14286} \tag{b}$$

For $p = 1.00 \cdot 10^5$ N/m^2, $p^* = 1.00/2.00 = 0.5$. From equation (b),

$$a^* = (0.5)^{0.14286} = 0.90572$$

and

$$a = a^*a_o = (0.90572)(347.21) = 314.48 \text{ m/s}$$

From equation (a),

$$u^* = 5.0(1.0 - a^*) = 5.0(1.0 - 0.90572) = 0.47140$$

and

$$u = u^*a_o = (0.47140)(347.21) = 163.67 \text{ m/s}$$

From the definition of the Mach number,

$$M = \frac{u}{a} = \frac{u^*}{a^*} = \frac{0.47140}{0.90572} = 0.52047$$

When plotting results on the xt plane, it is convenient to nondimensionalize the distance x by the tube length L, and the time t by the ratio (L/a_o). Thus, $x^* = x/L$ and $t^* = (ta_o/L)$, and

$$\lambda_{\pm}^* = \frac{dt^*}{dx_{\pm}^*} = \frac{d(ta_o/L)}{d(x/L)} = a_o \frac{dt}{dx} = a_o\lambda_{\pm} = \frac{a_o}{u \pm a} = \frac{1}{u^* \pm a^*} \tag{c}$$

where $\lambda_{\pm}^*$ denotes the corresponding slope in the x^*t^* plane. For the present problem,

$$\lambda_{-}^* = \frac{1}{u^* - a^*} = \frac{1}{0.47140 - 0.90572} = -2.3025$$

The results for the remaining values of p specified in part (a) are presented in the following table.

p, N/m$^2 \cdot 10^5$	p^*	a^*	a, m/s	u^*	u, m/s	M	λ^*
2.00	1.00000	1.00000	347.21	0.00000	0.0	0.00000	−1.0000
1.75	0.8750	0.98110	340.65	0.09450	32.81	0.09632	−1.1279
1.50	0.7500	0.95973	333.23	0.20135	69.91	0.20980	−1.3186
1.25	0.6250	0.93506	324.66	0.32470	112.74	0.34725	−1.6384
1.00	0.50000	0.90572	314.48	0.47140	163.67	0.52047	−2.3025

Figure 13.9b illustrates the five waves considered above on the x^*t^* plane. The actual expansion wave is a continuous wave having an infinite number of Mach lines, of which the above are five particular ones.

(b) From Fig. 13.9b, the equation of the first expansion wave to reach the left end is

$$\frac{dt}{dx_-} = \frac{t - 0}{0 - x} = \lambda_- = \frac{\lambda_-^*}{a_o}$$

and the time t corresponding to $x = L$ is

$$t = -\frac{L\lambda_-^*}{a_o} = -\frac{(3.0)(-1.0)}{347.21} = 8.640 \text{ ms}$$

13-4(d) Reflection of Compression and Expansion Waves

When either a continuous compression or expansion wave impinges upon a boundary of a flow region, a reflected wave is generated. Waves may be reflected from the closed ends and open ends of a duct, shock waves, and contact surfaces. The present section discusses the reflections from the closed and open ends of a duct. The reflections and interactions with shock waves and with contact surfaces are discussed in Sections 19-5(a) and 19-5(b), respectively.

Figure 13.10a illustrates schematically the flow model for analyzing the intersection of a right-running compression wave of infinitesimal size with a closed end. The

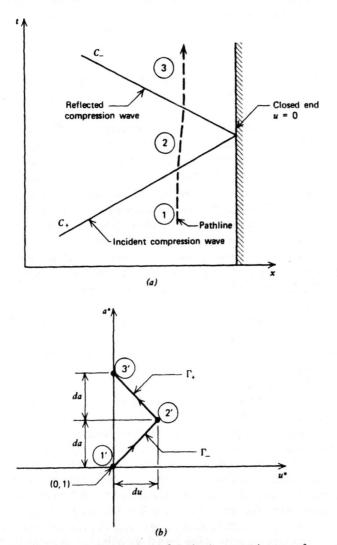

Figure 13.10 Reflection of an infinitesimal compression wave from a closed end. (a) Physical plane. (b) State plane.

incident compression wave imparts a pressure increment $dp > 0$ and a velocity increment $du > 0$ to the fluid particles. At the solid wall, however, the fluid velocity u must remain equal to zero. To satisfy that condition, a reflected wave is required that imparts a velocity increment $du < 0$ to the fluid particle adjacent to the solid wall; that is, the velocity increment acts in the same direction as the reflected wave. It follows from Rule 1, Section 13-4(c), that the reflected wave must be a compression wave. Figure 13.10*b* illustrates schematically the reflected compression wave in the state plane; the incident compression wave changes the state from point 1' to 2' along a Γ_- characteristic, and the reflected compression wave must follow the Γ_+ characteristic from point 2' to 3' so that the fluid velocity is zero at the wall. The uniform Regions 1, 2, and 3 in the physical plane, map onto points 1', 2', and 3', respectively, in the state plane.

If the incident compression wave comprises several infinitesimal waves thereby forming a *simple compression wave*, the incident wave imparts the finite changes Δu and Δp to the fluid. Figure 13.11*a* illustrates schematically, in the physical plane, a simple compression wave impinging upon the closed end of a tube. A simple compression wave is reflected from the closed end and a nonsimple wave interaction region is formed adjacent to the wall where waves of both families are present. Figure 13.11*b* illustrates the aforementioned waves in the corresponding state plane, where the uniform regions 1, 2, and 3 map onto points 1', 2', and 3', respectively. The incident simple compression wave region maps onto line 1'-2' in the state plane, and the shaded nonsimple wave region in the physical plane maps onto the shaded region in the state plane.

In Fig. 13.11, several selected physical and state characteristics are illustrated. The points *a*, *b*, etc., in the physical plane map onto points *a'*, *b'*, etc., respectively, in the state plane. Since the state plane characteristics Γ_+ and Γ_- are fixed, they are independent of the physical plane C_+ and C_- characteristics. Points *a*, *b*, *c*, and *d* in the physical plane are determined solely by the incident simple compression wave. Point *e* is located at the intersection of the C_+ characteristic through point *b* with the wall, where $u = 0$. Point *e'* is, therefore, located at the intersection of the Γ_+ characteristic through point *b'* with the *a** axis, where $u^* = 0$. Similarly, point *f* is located at the intersection of the C_+ and C_- characteristics through points *c* and *e*, respectively, and point *f'* is located at the intersection of the Γ_+ and Γ_- characteristics through points *c'* and *e'*, respectively. The locations of the remaining points illustrated in Fig. 13.11 are determined in a similar manner.

Because the simple compression wave interacts continuously with the wall, the entropy of the fluid remains constant in the nonsimple wave interaction region. Consequently, the flow remains homentropic. The compression wave emerging from the nonsimple wave interaction region must be a *simple* compression wave because it is adjacent to a uniform flow region, Region 2.

Figure 13.12*a* illustrates schematically, in the physical plane, the intersection of an infinitesimal expansion wave with the closed end of a duct. The incident expansion wave imparts a pressure increment $dp < 0$ and a velocity increment $du < 0$ to the fluid particles. Since the boundary condition at the solid wall is that the flow velocity must remain equal to zero, a reflected wave that imparts a velocity increment $du > 0$ to the fluid particle adjacent to the solid wall is generated. Since the velocity increment acts in the direction opposite to that of the propagation of the reflected wave, the reflected wave must be an expansion wave [see Rule 2 in Section 13-4(c)]. Figure 13.12*b* illustrates the incident and reflected expansion waves in the state plane. The incident expansion wave changes the state of the fluid from point 1' to 2' along the Γ_- characteristic, and the reflected wave must follow the Γ_+ characteristic from point 2'

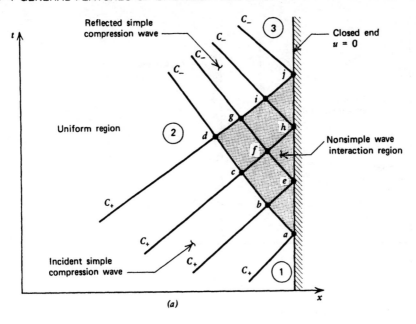

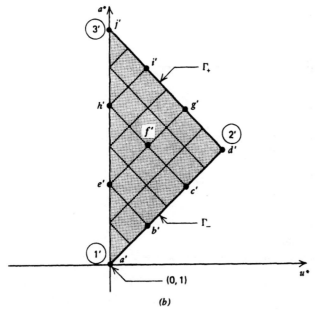

Figure 13.11 Reflection of a simple compression wave from a closed end. (a) Physical plane. (b) State plane.

to 3′ so that the flow velocity will be zero. The uniform Regions 1, 2, and 3 in the physical plane map onto points 1′, 2′, and 3′, respectively, in the state plane.

Figure 13.13 illustrates a simple expansion wave, which comprises several infinitesimal expansion waves, impinging on the closed end of a duct. As in the case of the simple compression wave (see Fig. 13.11a), a nonsimple wave interaction region is formed adjacent to the wall. The reflected wave in this case is a simple expansion wave. The foregoing demonstrates that the reflected waves from the closed end of a duct are of the same type as the incident waves.

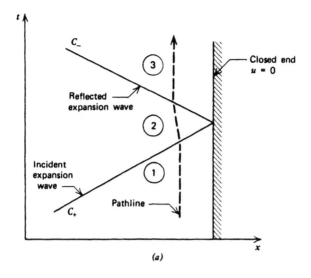

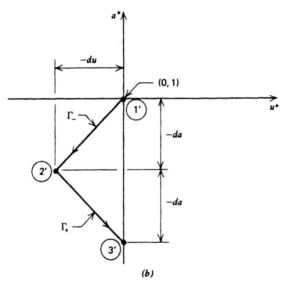

Figure 13.12   Reflection of an infinitesimal expansion wave from a closed end.
(a) Physical plane. (b) State plane.

Waves may also reflect from an open end of a duct. Four different cases are possible: the flow at the open end may be either an inflow or outflow, and it may be either subsonic or supersonic. In the case of an outflow, the interaction at the open end should take into account the unsteady flow effects that are present in the exhaust region; those effects may be two- or three-dimesional. For a low subsonic outflow velocity, a reasonable approximation to the actual phenomena may be obtained by neglecting the flow field external to the open end of the duct and assuming that the pressure at the end of the duct is equal to the ambient pressure. For high subsonic outflow from the duct, the aforementioned approximation may be unrealistic, and a more realistic approximation must be developed. In the case of supersonic outflow, the outflow velocity is larger than the propagation velocity of the reflected waves, which travel at the speed of sound relative to the fluid, so that such waves would be

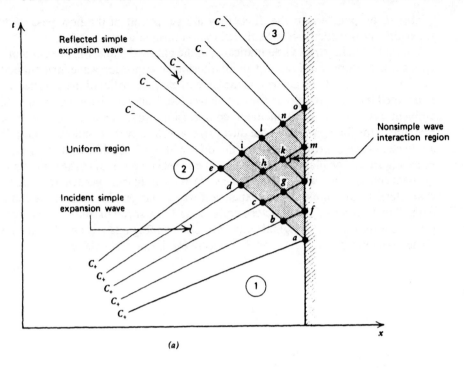

(a)

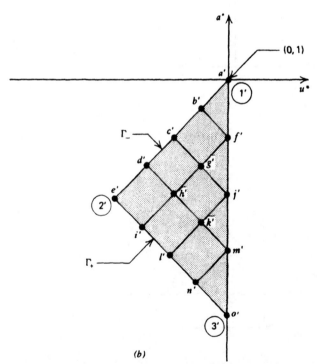

(b)

Figure 13.13 Reflection of a simple expansion wave from a closed end. (a) Physical plane. (b) State plane.

unable to propagate into the duct; they are swept out of the flow passage by the supersonic flow velocity. Hence, there are no reflected waves.

Figure 13.14a illustrates schematically, in the physical plane, the open end of a duct with a low subsonic outflow, and an infinitesimal compresion wave incident upon the open end of the duct. To determine the general characteristics of the reflected wave, it is assumed that the pressure in the exit plane of the duct remains equal to the constant ambient pressure. For an incident compression wave, $dp > 0$ and $du > 0$. The pressure of the fluid particle at the duct exit plane, however, remains constant. Hence, a reflected wave is propagated into the duct with $dp < 0$; that is, it is an *expansion wave*. Figure 13.14b illustrates the incident and reflected waves in the state plane. The incident compression wave changes the state of the fluid from point 1' to 2' along a Γ_- characteristic, and the reflected expansion wave changes the state of the fluid from point 2' to 3' along a Γ_+ characteristic. At state 3' the speed of sound a, and hence the pressure p, returns to its original value. The uniform Regions 1, 2, and 3 in the physical plane map onto points 1', 2', and 3', respectively, in the state plane.

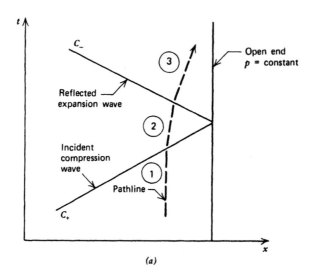

(a)

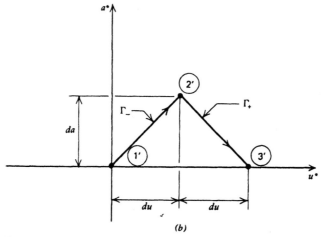

(b)

Figure 13.14 Reflection of an infinitesimal compression wave from an open end. (a) Physical plane. (b) State plane.

When a simple compression wave of finite amplitude impinges on an open end, a nonsimple wave interaction region is formed, and a simple expansion wave is reflected, as illustrated in Fig. 13.15. Consequently, Region 1'-2'-3' in Fig. 13.14b becomes a region of finite size.

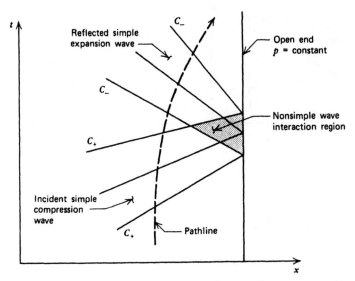

Figure 13.15 Reflection of a simple compression wave from an open end.

Figure 13.16a illustrates schematically the case where an infinitesimal expansion wave is reflected from an open end, and the duct exit plane pressure p is constant. The incident wave imparts $dp < 0$ and $du < 0$ to the fluid particles. At the tube exit plane a wave is reflected with $dp > 0$ to maintain the pressure constant; that is, the reflected wave is a *compression wave*. The incident and reflected waves are illustrated in the state plane in Fig. 13.16b. The state of the fluid is changed from point 1' to 2' along the Γ_- characteristic, which is an expansion wave. The reflected compression wave changes the fluid state from point 2' to point 3' where the pressure is equal to the initial pressure at the duct exit. The uniform Regions 1, 2, and 3, in Fig. 13.16a, map onto the points 1', 2', and 3', respectively, in the state plane.

Figure 13.17 illustrates the reflection of a simple expansion wave from an open end, where the static pressure p is constant. A nonsimple wave interaction region is formed at the open end, and a simple compression wave is reflected from the open end. In this case, Region 1'-2'-3' in Fig. 13.16b becomes a region of finite size.

By similar reasoning, analogous results may be obtained for subsonic inflow. For supersonic inflow, waves from within the duct cannot propagate to the inlet of the duct, so wave reflection is meaningless.

Summarizing, the following rules of wave reflection apply at the end of a duct.

Rule 1. When a wave is incident on a *closed end*, a wave of the same type is reflected. Thus, incident compression waves reflect as compression waves and incident expansion waves reflect as expansion waves.

Rule 2. When a wave is incident on an *open end* where the pressure p in the exit plane is constant and the flow is subsonic, a wave of the opposite type is reflected. If the flow is supersonic, no waves are reflected.

These results hold at the end of the duct for both outflow and inflow.

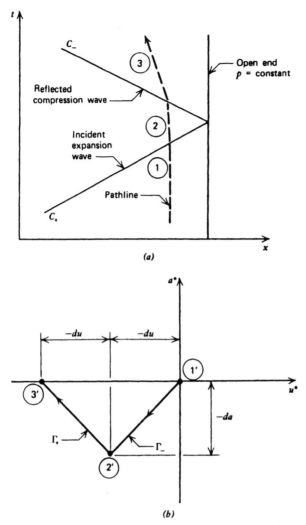

Figure 13.16 Reflection of an infinitesimal expansion wave from an open end. (a) Physical plane. (b) State plane.

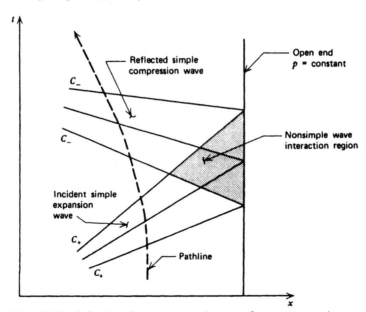

Figure 13.17 Reflection of a simple expansion wave from an open end.

13-4(e) Intersection of Continuous Waves

When continuous waves intersect, the result of their interaction depends on the type of waves involved (i.e., compression or expansion), and the direction of propagation of the incident waves (i.e., whether both travel in the same or opposite directions). Expansion waves traveling in the same direction diverge (see Fig. 13.6d), and no interaction can occur. Compression waves of the same family, on the other hand, converge, and may ultimately coalesce to form a shock wave. The coalescence of compression waves to form a shock wave is discussed in Section 19-5(a).

Figure 13.18a illustrates, in the physical plane, the interaction of two simple compression waves traveling in opposite directions. Figure 13.18b illustrates the interaction of the two compression waves in the state plane. In the interaction region, waves of both families are present, so that the region where the interaction occurs is a nonsimple wave region. The points in the state plane corresponding to points in the physical plane are determined by following the appropriate C_+ and C_- characteristics, in the physical plane, and the corresponding Γ_+ and Γ_- characteristics in the state plane. The flow in the interaction region is homentropic; consequently, the transmitted compression waves are simple waves since they are adjacent to the uniform regions 2 and 3.

Figure 13.19a illustrates, in the physical plane, the interaction of two simple expansion waves traveling in opposite directions, and Fig. 13.19b illustrates their interaction in the state plane.

Figure 13.20a illustrates, in the physical plane, the intersection of a right-running simple compression wave and a left-running simple expansion wave and, in Fig. 13.20b, they are illustrated in the state plane. The corresponding points in the physical and state planes are determined by following the appropriate C_+ and Γ_+ characteristics. The incident simple compression wave emerges from the nonsimple wave interaction region as a transmitted simple compression wave, and the incident simple expansion wave emerges as a transmitted simple expansion wave. Corresponding points in the physical and state planes are labeled a, b, etc., and a', b', etc.

13-5 NUMERICAL IMPLEMENTATION OF THE METHOD OF CHARACTERISTICS

In Section 13-4, the characteristic and compatibility equations are derived for an unsteady one-dimensional homentropic flow. When the flowing fluid is a perfect gas with constant specific heats, the compatibility equation may be integrated in closed form to obtain a simple algebraic equation relating the velocity u and the speed of sound a (see equation 13.35). For a simple wave flow, the Mach lines of one family are straight lines along which the flow properties are constant. Consequently, for the simple wave flow of a perfect gas with constant specific heats, both the characteristic equation and the compatibility equation may be solved without resorting to numerical integration. For the nonsimple wave flow of a perfect gas, however, the characteristics must be determined by numerical integration of the characteristic equation, which is straightforward because the integrated form of the compatibility equation is still applicable.

When the flowing fluid is other than a perfect gas with constant specific heats, however, the compatibility equation must also be integrated numerically. In simple wave regions, the characteristics are straight lines even for the flow of such a gas, and may be determined without numerical integration once the solution to the compatibility equation is determined numerically.

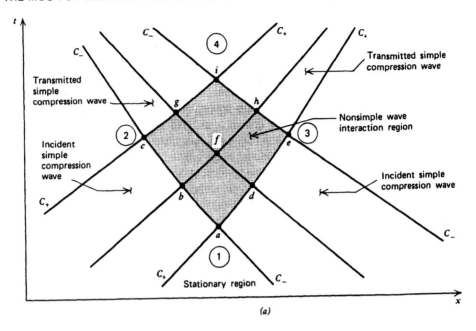

(a)

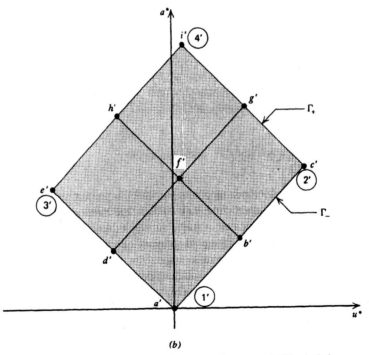

(b)

Figure 13.18 Intersection of two simple compression waves. (a) Physical plane.
(b) State plane.

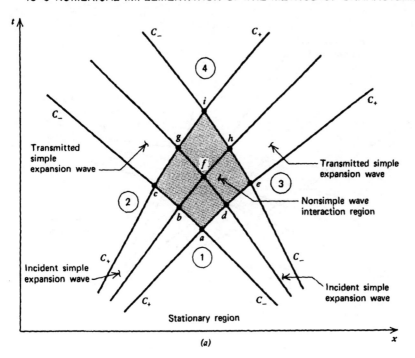

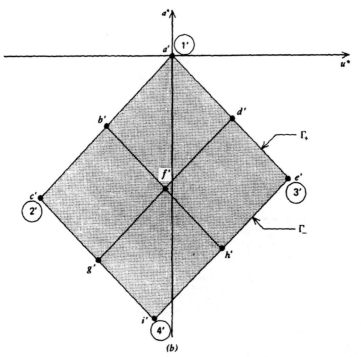

Figure 13.19   Intersection of two simple expansion waves. (*a*) Physical plane. (*b*) State plane.

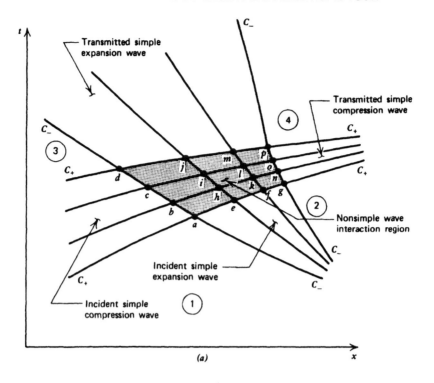

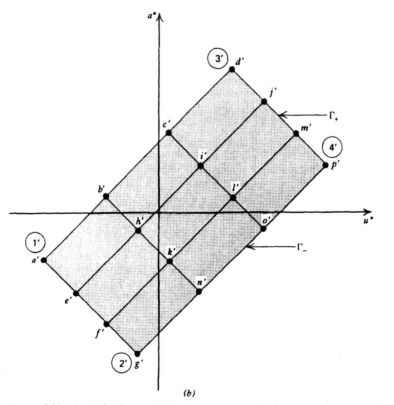

Figure 13.20 A nonsimple wave interaction between a simple compression wave and a simple expansion wave. (*a*) Physical plane. (*b*) State plane.

In the general case where the flowing fluid is other than a perfect gas and the flow is a nonsimple wave flow, both the characteristic equation and the compatibility equation must be integrated numerically. A numerical procedure for integrating those equations is developed below. A complete numerical algorithm is developed for the unit process for determining the flow properties at an interior point of the flow field, and a numerical example illustrating that unit process is presented.

13-5(a) Numerical Integration Procedure

The numerical integration procedure employed for integrating the characteristic and compatibility equations for unsteady one-dimensional homentropic flow is the *modified Euler predictor-corrector method*, which is discussed in detail in Section 12–5(a).

Two different procedures for applying that method are discussed in Section 12–5(b); they are the *direct marching method* and the *inverse marching method*. The direct marching method is employed in Section 12–5(d) for developing the unit process for an interior point for steady two-dimensional irrotational supersonic flow. In the study of unsteady flows, however, it is often advantageous to determine the solution at prespecified points located on lines of constant time, termed *t-lines*.[6] Consequently, that is the approach taken in the following discussion.

Figure 13.21 presents the finite difference grid for an interior point based on the inverse marching method. The location of the solution point, point 4, is prespecified;

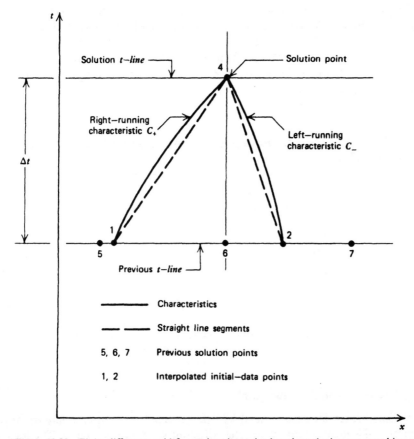

Figure 13.21 Finite difference grid for an interior point based on the inverse marching method.

for example, as equally spaced points along the solution lines. The two characteristics intersecting at point 4 are extended rearward to intersect the initial-value line at points 1 and 2. The values of the flow properties at points 1 and 2 are determined by interpolation employing the previous solution points, points 5, 6, and 7. The interpolation reduces the accuracy of the numerical results.

Further, to insure that the solution is numerically stable, the Courant-Friedrichs-Lewy (CFL) stability criterion[7] must be satisfied. The criterion requires that the initial-data points (points 1 and 2 in Fig. 13.21) fall between the previous solution points (points 5, 6, and 7) that are employed in the interpolation for determining the flow properties at points 1 and 2. Consequently, the interpolation is based on flow properties outside of the domain of dependence of point 4, which further reduces the accuracy of the numerical integration. That is the price that must be paid, however, for employing the inverse marching method.

13-5(b) Finite Difference Equations

Figure 13.21 illustrates schematically the finite difference grid for the unit process for an interior point. The portions of the characteristic curves connecting two points are approximated by straight dashed line segments connecting those points. The location and flow properties for the solution points on the previous *t-line*, points 5, 6, and 7, are given, and the location of point 4 is specified. The location of points 1 and 2 are determined by iteration.

According to the Euler method, the finite difference equations corresponding to the differential equations are determined by replacing the differentials dt, dx, du, and dp by the differences Δt, Δx, Δu, and Δp. The resulting finite difference equations are presented in Table 13.3, where the subscripts $+$ and $-$ denote right- and left-running characteristics, respectively. The coefficients λ, Q, and S are determined at the initial points for the predictor, and in an average manner (either average property or average coefficient) for the corrector.

Table 13.3 Finite Difference Equations for Unsteady One-Dimensional Homentropic Flow

$$\Delta t_{\pm} = \lambda_{\pm}\, \Delta x_{\pm} \tag{13.39}$$

$$\Delta p_{\pm} \pm Q_{\pm}\, \Delta u_{\pm} = S_{\pm}\, \Delta t_{\pm} \tag{13.40}$$

$$\lambda_{\pm} = \frac{1}{u_{\pm} \pm a_{\pm}} \tag{13.41}$$

$$Q = \rho a \tag{13.42}$$

$$S = -\delta \frac{\rho u a^2}{x} \tag{13.43}$$

$+$ or $-$ denotes C_+ or C_- characteristic curve, respectively

13-5(c) Unit Process for an Interior Point

Figure 13.21 illustrates schematically, in the xt plane, the finite difference grid for determining the location and properties of point 4, an interior point. The locations of the initial-data points, points 1 and 2, are determined by writing equation 13.39, of Table 13.3, in finite difference form in terms of points 1, 2, and 4. Thus,

$$\Delta t = \lambda_+ (x_4 - x_1) \tag{13.44a}$$

$$\Delta t = \lambda_- (x_4 - x_2) \tag{13.44b}$$

where $\Delta t = \Delta t_+ = \Delta t_-$ for an inverse marching method employing the *t-lines* as solution lines. Since Δt and x_4 are known, equations 13.44 may be solved for x_1 and x_2. From Table 13.3, the slopes λ_+ and λ_- are given by equation 13.41. Thus,

$$\lambda_+ = \frac{1}{u_+ + a_+} \quad \text{and} \quad \lambda_- = \frac{1}{u_- - a_-} \tag{13.45}$$

where

$$a_\pm = a(p_\pm) \quad (\textit{from equation 13.9}) \tag{13.46}$$

Consequently, $\lambda_\pm$ may be determined for specified values of $u_\pm$ and $p_\pm$.

The values of u and p at points 1 and 2 may be determined by linear interpolation between the previous solution points. For example,

$$u_1 = u_6 + \left(\frac{u_5 - u_6}{x_5 - x_6}\right)(x_1 - x_6) = m_u x_1 + b_u \tag{13.47}$$

where

$$m_u = \frac{u_5 - u_6}{x_5 - x_6} \tag{13.48}$$

$$b_u = u_6 - m_u x_6 \tag{13.49}$$

Similar linear interpolation formulas may be written for u_2, p_1, and p_2.

The compatibility equation (equation 13.40 of Table 13.3), which is valid along the characteristics, may be expressed in finite difference form in terms of points 1, 2, and 4, as follows.

$$p_4 + Q_+ u_4 = T_+ \tag{13.50}$$

$$p_4 - Q_- u_4 = T_- \tag{13.51}$$

where

$$T_+ = S_+ \Delta t + p_1 + Q_+ u_1 \tag{13.52}$$

$$T_- = S_- \Delta t + p_2 - Q_- u_2 \tag{13.53}$$

Equations 13.50 and 13.51 may be solved simultaneously for u_4 and p_4. The coefficients appearing in those equations are given by

$$Q_+ = \rho_+ a_+ \tag{13.54}$$

$$S_+ = -\delta \frac{\rho_+ u_+ a_+^2}{x_+} \tag{13.55}$$

$$Q_- = \rho_- a_- \tag{13.56}$$

$$S_- = -\delta \frac{\rho_- u_- a_-^2}{x_-} \tag{13.57}$$

where

$$\rho_\pm = \rho(p_\pm) \quad (\textit{from equation 13.13}) \tag{13.58}$$

Table 13.4 presents the computational equations for unsteady one-dimensional homentropic flow.

Table 13.4 Computational Equations for Unsteady One-Dimensional Homentropic Flow

$$\Delta t = \lambda_+ (x_4 - x_1) \tag{13.59}$$

$$\Delta t = \lambda_- (x_4 - x_2) \tag{13.60}$$

$$p_4 + Q_+ u_4 = T_+ \tag{13.61}$$

$$p_4 - Q_- u_4 = T_- \tag{13.62}$$

For the *Euler predictor algorithm*, the values of u_+ and p_+ are those of the initial-data points, points 1 and 2. The locations and flow properties at those points, however, are not known for an indirect marching method. Consequently, an iterative procedure is required to determine the locations and flow properties for the initial-data points. For the first iteration, assume

$$u_+ = u_5, \qquad p_+ = p_5, \qquad u_- = u_7, \qquad p_- = p_7 \tag{13.63}$$

Substituting those values into the computational equations, the predicted values $x_1{}^0$, $u_1{}^0$, $p_1{}^0$, $x_2{}^0$, $p_2{}^0$, and $u_2{}^0$ may be determined. For the second and subsequent iterations, assume

$$u_+ = u_1{}^n, \qquad p_+ = p_1{}^n, \qquad u_- = u_2{}^n, \qquad p_- = p_2{}^n \tag{13.64}$$

where n denotes the previous iteration. Substituting those values into the computational equations yields the corrected values $x_1{}^{n+1}$, $u_1{}^{n+1}$, $p_1{}^{n+1}$, $x_2{}^{n+1}$, $u_2{}^{n+1}$, and $p_2{}^{n+1}$. The iteration procedure may be repeated a specified number of times, or repeated until a specified convergence tolerance is achieved.

Once the locations and flow properties at the initial-data points, points 1 and 2, are determined, the flow properties at the solution point, point 4, may be determined. For the *Euler predictor algorithm*, the values of u_+ and p_+ are given by equation 13.64. Substituting those values into the computational equations yields the predicted values $u_4{}^0$ and $p_4{}^0$.

For the *Euler corrector algorithm*, the values of u_+ and p_+ are given by

$$u_+ = \frac{u_1 + u_4}{2} \qquad p_+ = \frac{p_1 + p_4}{2} \tag{13.65}$$

$$u_- = \frac{u_2 + u_4}{2} \qquad p_- = \frac{p_2 + p_4}{2} \tag{13.66}$$

As discussed for the Euler predictor algorithm, an iteration is required to determine the locations and the flow properties for the initial-data points. During that iteration, the values of u_4 and p_4 remain fixed at the values determined by the Euler predictor, while the values of x_1, u_1, p_1, x_2, u_2, and p_2 vary. Once those values are determined, the values of u_+ and p_+ given by equations 13.65 and 13.66 are substituted into the computational equations, yielding the corrected values $u_4{}^1$ and $p_4{}^1$.

The *Euler corrector algorithm with iteration* (equation 12.56) may then be applied to improve the accuracy of the solution values. That is accomplished by substituting the corrected values $u_4{}^1$ and $p_4{}^1$ into the computational equations and repeating the Euler corrector algorithm to obtain a second set of corrector values, $u_4{}^2$ and $p_4{}^2$. The iteration algorithm may be terminated after a specified number of applications, or it may be applied as many times as required for obtaining a desired degree of convergence.

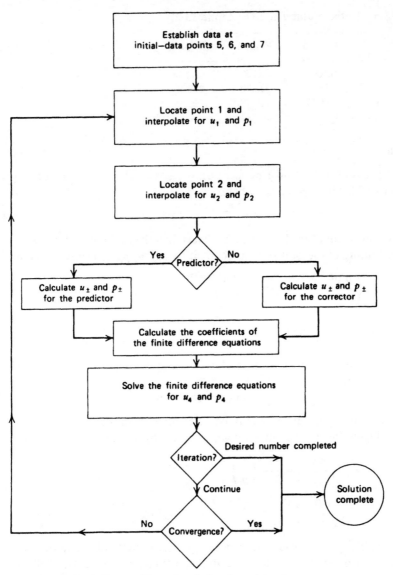

Figure 13.22 Block diagram illustrating the interior point unit process.

Figure 13.22 presents a block diagram illustrating the modified Euler predictor-corrector algorithm for the unit process for an interior point. Figure 13.22 may serve as the flow chart for the computer implementation of the algorithm.

Example 13.2. A small high-pressure solid propellant rocket motor experiences unsteady flow effects during the blow-down of the combustion chamber after propellant burn-out. The combustion gases may be treated as a perfect gas with $\gamma = 1.20$ and $R = 320.0$ J/kg-K. The pressure and temperature at the beginning of the blow-down are $p_i = 690 \cdot 10^5$ N/m^2 and $T_i = 3330.0$ K, respectively. At a given instant of time during the blow-down, the flow properties at three adjacent locations in the combustion chamber (points 5, 6, and 7 in Fig. 13.21) are those presented in Table 13.5. Calculate the flow properties at point 4 (see Fig. 13.21) at a time 0.020 ms later, assuming that the flow is planar and homentropic.

Table 13.5 Initial-Value Data (Example 13.2)

	Point 5	Point 6	Point 7
x, m	0.1000	0.1250	0.1500
u, m/s	188.30	235.30	282.50
p, N/m$^2 \cdot 10^5$	111.77	110.78	109.54

Solution

(a) *Equations of state.* For a perfect gas, the speed of sound equation is given by equation 13.30, which is repeated below.

$$a^2 p^{-(\gamma - 1)/\gamma} = \text{constant} \tag{13.30}$$

The constant in equation 13.30 may be determined from the initial values of a and p at the beginning of the blow-down, denoted by a_i and p_i, respectively. Thus,

$$a = a_i \left(\frac{p}{p_i}\right)^{(\gamma-1)/2\gamma} = \sqrt{\gamma R T_i} \left(\frac{p}{p_i}\right)^{(\gamma-1)/2\gamma}$$

$$a = [(1.20)(320.0)(3330.0)]^{1/2} \left(\frac{p}{p_i}\right)^{0.0833333} = 1130.81 \left(\frac{p}{p_i}\right)^{0.0833333} \text{ m/s} \quad \text{(a)}$$

For a perfect gas, the density ρ is related to the pressure p by equation 13.29, which is repeated below.

$$p\rho^{-\gamma} = \text{constant} \tag{13.29}$$

The constant in equation 13.29 may be determined from the initial values of p and T at the beginning of the blow-down, denoted by p_i and T_i, respectively. Thus,

$$\rho = \rho_i \left(\frac{p}{p_i}\right)^{1/\gamma} = \frac{p_i}{RT_i} \left(\frac{p}{p_i}\right)^{1/\gamma}$$

$$\rho = \frac{(690.0)10^5}{(320.0)(3330.0)} \left(\frac{p}{p_i}\right)^{0.833333} = 64.7523 \left(\frac{p}{p_i}\right)^{0.833333} \text{ kg/m}^3 \quad \text{(b)}$$

(b) *Interpolating polynomials.* Interpolating polynomials are required along lines 5–6 and 6–7. For u along line 5–6, we obtain, from equations 13.47, 13.48, and 13.49,

$$u = m_u x + b_u = \left(\frac{u_5 - u_6}{x_5 - x_6}\right) x + (u_6 - m_u x_6)$$

$$u = \frac{(188.30 - 235.30)}{(0.1000 - 0.1250)} + [235.30 - (1880.00)(0.1250)]$$

$$u = 1880.00x + 0.300 \text{ m/s} \quad \text{(c)}$$

where u is in m/s and x is in m. Similarly,

$$p = (-39.6000x + 115.730) \cdot 10^5 \text{ N/m}^2 \quad \text{(d)}$$

Along line 6–7,

$$u = 1888.00x - 0.700 \text{ m/s} \quad \text{(e)}$$

$$p = (-49.6000x + 116.980) \cdot 10^5 \text{ N/m}^2 \quad \text{(f)}$$

(c) *Locate point 1 and calculate the coefficients for the predictor.* For the first pass through the predictor, set the values of the average flow properties along the C_+

characteristic 14 equal to those at point 5. Thus,

$$u_+ = 188.30 \text{ m/s} \qquad p_+ = 111.77 \cdot 10^5 \text{ N/m}^2$$

From equation (a), we obtain

$$a_+ = 1130.81 \left(\frac{111.77}{690.00}\right)^{0.0833333} = 971.65 \text{ m/s}$$

Substituting the values for u_+ and a_+ into equation 13.45, we obtain

$$\lambda_+ = \frac{1}{u_+ + a_+} = \frac{1}{188.30 + 971.65} = 0.86210 \text{ ms/m} \tag{g}$$

From equation 13.59,

$$x_1 = x_4 - \frac{\Delta t}{\lambda_+} = 0.1250 - \frac{0.020}{0.86210} = 0.10180 \text{ m} \tag{h}$$

Since x_1 lies between x_5 and x_6, u_1 and p_1 may be determined from equations (c) and (d), respectively. Thus,

$$u_1 = 191.69 \text{ m/s} \qquad p_1 = 111.70 \cdot 10^5 \text{ N/m}^2$$

The above procedure yields tentative values for the location and flow properties at point 1. Those values may be improved by repeating the above steps iteratively, each time employing the most recent values of u_1 and p_1. The results of two passes, including the first pass, are summarized in the following table.

Pass	u_+, m/s	p_+, N/m$^2 \cdot 10^5$	a_+, m/s	x_1, m
1	188.30	111.77	971.65	0.10180
2	191.69	111.70	971.60	0.10173

From equation (b), we obtain

$$\rho_+ = 64.7523 \left(\frac{111.70}{690.00}\right)^{0.833333} = 14.199 \text{ kg/m}^3$$

Substituting the values of ρ_+, a_+, u_1, and p_1 into equations 13.54 and 13.52 gives

$$Q_+ = \rho_+ a_+ = (14.199)(971.60) = 13{,}796 \text{ N-s/m}^3 \tag{i}$$

$$T_+ = p_1 + Q_+ u_1 = 111.70 \cdot 10^5 + (13{,}796)(191.69) = 138.14 \cdot 10^5 \text{ N/m}^2 \tag{j}$$

(d) *Locate point 2 and calculate the coefficients for the predictor.* For the first pass through the predictor, set the values of the average flow properties along the C_- characteristic 24 equal to those at point 7. Then,

$$a_- = 1130.81 \left(\frac{109.54}{690.00}\right)^{0.0833333} = 970.02 \text{ m/s}$$

Substituting into equation 13.45 gives

$$\lambda_- = \frac{1}{u_- - a_-} = \frac{1}{282.50 - 970.02} = -1.4545 \text{ ms/m} \tag{k}$$

From equation 13.60,

$$x_2 = x_4 - \frac{\Delta t}{\lambda_-} = 0.1250 - \frac{0.020}{-1.4545} = 0.13875 \text{ m} \tag{1}$$

Since x_2 lies between x_6 and x_7, x_2 may be substituted into equations (e) and (f) to yield

$$u_2 = 261.26 \text{ m/s} \qquad p_2 = 110.10 \cdot 10^5 \text{ N/m}^2$$

Repeating the above procedure, we obtain the following results.

Pass	u_-, m/s	p_-, N/m² · 10⁵	a_-, m/s	x_2, m
1	282.50	109.54	970.02	0.13875
2	261.26	110.10	970.43	0.13918
3	262.08	110.08	970.42	0.13917

From equation (b), $\rho_- = 14.027 \text{ kg/m}^3$. From equations 13.56 and 13.53, we obtain

$$Q_- = \rho_- a_- = (14.027)(970.42) = 13{,}612 \text{ N-s/m}^3 \tag{m}$$

$$T_- = p_2 - Q_- u_2 = 110.08 \cdot 10^5 - (13{,}612)(262.08) = 74.403 \cdot 10^5 \text{ N/m}^2 \tag{n}$$

(e) *Determine u_4 and p_4 for the predictor.* The values of the coefficients determined in steps (c) and (d) may be substituted into equations 13.61 and 13.62 to yield two equations for the determination of u_4 and p_4. Thus,

$$p_4 + 13{,}796 u_4 = 138.14 \cdot 10^5 \tag{o}$$

$$p_4 - 13{,}612 u_4 = 74.403 \cdot 10^5 \tag{p}$$

Solving equations (o) and (p) simultaneously, we obtain

$$u_4 = 232.56 \text{ m/s} \qquad p_4 = 106.06 \cdot 10^5 \text{ N/m}^2$$

The application of the Euler predictor algorithm is complete. The results obtained for the predictor are presented in column (0) in Table 13.6.

Table 13.6   Values of the Solution, for Successive Trials, for the Interior Point Unit Process (Example 13.2)

	0	1	2	3
λ_+, ms/m	0.85963	0.84656	0.84649	0.84649
x_1, m	0.10173	0.10137	0.10137	0.10137
λ_-, ms/m	-1.4118	-1.3863	-1.3864	-1.3864
x_2, m	0.13917	0.13943	0.13943	0.13943
Q_+, N-s/m³	13,796	13,477	13,478	13,478
$T_+ \cdot 10^{-5}$, N/m²	138.14	137.44	137.44	137.44
Q_-, N-s/m³	13,612	13,383	13,384	13,384
$T_- \cdot 10^{-5}$, N/m²	74.403	74.927	74.924	74.924
u_4, m/s	232.56	232.73	232.73	232.73
p_4, N/m² · 10⁵	106.06	106.07	106.07	106.07

(f) *Locate point 1 and calculate the coefficients for the corrector.* For the application of the Euler corrector algorithm, average values of the flow properties along the C_+ and C_- characteristics are employed. Point 1 must be located iteratively. For the first iteration for the location of point 1,

$$u_+ = \frac{u_1 + u_4}{2} = \frac{191.69 + 232.56}{2} = 212.13 \text{ m/s} \tag{p}$$

$$p_+ = \frac{p_1 + p_4}{2} = \frac{(111.70 + 106.06) \cdot 10^5}{2} = 108.88 \cdot 10^5 \text{ N/m}^2 \tag{q}$$

where u_1 and p_1 for the first trial are the final values obtained for the predictor. Substituting into equations (a), (g), and (h), we obtain the values of a_+, λ_+, and x_1. Improved values for u_1 and p_1 may be determined from the interpolating polynomials, and the procedure may be repeated iteratively to obtain improved values for x_1. The following results are obtained.

Pass	u_1, m/s	p_1, N/m$^2 \cdot 10^5$	u_+, m/s	p_+, N/m$^2 \cdot 10^5$	a_+, m/s	x_1, m
1	191.69	111.70	212.13	108.88	969.53	0.10137
2	190.87	111.72	211.72	108.89	969.54	0.10137

From equation (b), $\rho_+ = 13.901 \text{ kg/m}^3$ and, from equations (i) and (j), we obtain

$$Q_+ = 13,477 \text{ N-s/m}^3 \qquad T_+ = 137.44 \cdot 10^5 \text{ N/m}^2$$

(g) *Locate point 2 and calculate the coefficients for the corrector.* Employing average properties along line 24, in a manner analogous to that presented in part (f) for line 14, we obtain the following results.

Pass	u_2, m/s	p_2, N/m$^2 \cdot 10^5$	u_-, m/s	p_-, N/m$^2 \cdot 10^5$	a_-, m/s	x_2, m
1	262.08	110.08	247.32	108.07	968.93	0.13943
2	262.55	110.06	247.56	108.06	968.92	0.13943

From equation (b), $\rho_- = 13.812 \text{ kg/m}^3$ and, from equations (m) and (n), we obtain

$$Q_- = 13,383 \text{ N-s/m}^3 \qquad T_- = 74.927 \cdot 10^5 \text{ N/m}^2$$

(h) *Determine u_4 and p_4 for the corrector.* Substituting the values of the coefficients into equations 13.61 and 13.62, we obtain

$$p_4 + 13,477u_4 = 137.44 \cdot 10^5$$

$$p_4 - 13,383u_4 = 74.927 \cdot 10^5$$

Solving the above equations simultaneously yields

$$u_4 = 232.73 \text{ m/s} \qquad p_4 = 106.07 \cdot 10^5 \text{ N/m}^2$$

The application of the modified Euler corrector algorithm is complete. The results are presented in column (1) of Table 13.6.

(i) *Iteration of the corrector*. The results obtained by two additional applications of the corrector are presented in columns (2) and (3) of Table 13.6. From those results, it is clear that iteration of the corrector has an almost negligible effect on the final results.

13–6 SUMMARY

In this chapter the governing equations are derived for unsteady one-dimensional homentropic flow in the absence of work and body forces. The method of characteristics is applied to those equations, and the corresponding characteristic and compatibility equations are obtained. Those equations are valid for both subsonic flow and supersonic flow.

The general features of unsteady one-dimensional flow are demonstrated by considering the special case of unsteady one-dimensional *planar* homentropic flow. For a perfect gas with constant specific heats, the compatibility equation may be integrated in closed form. On the state plane (i.e., the ua plane), the solution to the compatibility equation, called the *state plane characteristics*, yields two families of straight lines having the slopes $\pm(\gamma - 1)/2$. The problem solution then reduces to determining the location of the characteristics in the physical plane along which the state plane characteristics are applicable.

In unsteady one-dimensional flow, uniform flow regions, simple wave regions, and nonsimple wave regions occur. Those regions are patched together along the characteristics of the flow. Regions of uniform flow and simple wave regions occur frequently in unsteady one-dimensional flow.

Simple compression and expansion waves occur when a disturbance moves into an undisturbed fluid. Simple compression waves accelerate the fluid in the direction of the propagation of the wave, and increase the pressure, velocity, and sound speed of the fluid, thus increasing the wave propagation speed. Compression waves converge, and the pathlines bend toward the waves. Simple expansion waves accelerate the fluid in the direction opposite to the propagation of the wave, and decrease the pressure and sound speed of the fluid. Consequently, the wave propagation speed decreases, the waves diverge, and the pathlines bend away from the waves. When a wave is incident on a closed end, a wave of the same type is reflected. When a wave is incident on an open end where the pressure is constant and the flow is subsonic, a wave of the opposite type is reflected. If the flow is supersonic, no waves are reflected. When simple compression and expansion waves intersect, nonsimple wave regions are formed.

The general theory of the method of characteristics is applied to the case of unsteady one-dimensional homentropic flow. A complete numerical algorithm, called a unit process, is developed for an interior point of a flow field. A numerical example is presented illustrating the application of the unit process for an interior point.

Generalized unsteady one-dimensional and quasi-one-dimensional flow, including the effects of friction, heat transfer, and mass addition, is discussed in Chapter 19 (Volume 2). Numerical algorithms are developed for the interior and boundary points of the flow, and numerical examples illustrating those unit processes are presented. FORTRAN computer programs for implementing the numerical algorithms are presented, and several applications to engineering problems are discussed.

REFERENCES

1. R. Courant and K. O. Friedrichs, *Supersonic Flow and Shock Waves*, Chap. III, Interscience Publishers, New York, 1958.
2. G. Rudinger, *Wave Diagrams for Nonsteady Flow in Ducts*, Van Nostrand, Princeton, N.J., 1955.

3. K. P. Stanyukovich, *Unsteady Motion of Continuous Media*, Pergamon Press, New York, 1960.

4. A. H. Shapiro, *The Dynamics and Thermodynamics of Compressible Fluid Flow*, Vol. II, Chaps. 23, 24, and 25, Ronald Press, New York, 1953.

5. J. A. Owczarek, *Fundamentals of Gas Dynamics*, Chaps. 7 and 8, International Textbook Co., Scranton, Pa., 1964.

6. D. R. Hartree, "Some Practical Methods of Using Characteristics in the Calculation of Nonsteady Compressible Flow," U.S. Atomic Energy Commission, Report No. AECU-2713, 1953.

7. R. Courant, K. O. Friedrichs, and H. Lewy, "Uberdie Partiellen Differenzialgleichungen der Mathematisehen Physik," *Mathematische Annalen*, Vol. 100, pp. 32–74, 1928.

PROBLEMS

1. Figure 13.13 illustrates schematically the physical plane and the state plane for the reflection of a right-running simple expansion wave from a closed end. Example 13.1 presents the construction of a left-running centered expansion wave propagating into a 3.0 m long tube filled with still air. Figure 13.5 illustrates selected characteristics in the state plane for $\gamma = 1.40$. Combine the above concepts and data and construct the state plane for the five discrete expansion waves considered in Example 13.1. Determine the flow properties u, p, and M at the points where the reflected waves emerge from the nonsimple wave interaction region (points analogous to points e, h, k, m, and n in Fig. 13.13a).

2. A numerical procedure for implementing the unit process for an interior point is presented in Section 13–5(c). That procedure is based on the *average property method* discussed in Section 12–5(a). Develop an equivalent unit process for an interior point based on the *average coefficient method* described in Section 12–5(a).

3. The unit process for an interior point, based on the *average property method*, is illustrated by a numerical example in Example 13.2. Work Example 13.2, employing the unit process based on the average coefficient method developed in Problem 13.2. The initial data are as follows.

	Point 1	Point 2
x, m	0.10137	0.13943
u, m/s	190.87	262.55
p, N/m$^2 \cdot 10^5$	111.72	110.06

4. Consider the rocket motor described in Example 13.2. The location and flow properties at three neighboring points in the flow field at a given instant of time are presented in the following table.

	Point 5	Point 6	Point 7
x, m	0.2000	0.2250	0.2500
u, m/s	315.0	365.0	420.0
p, N/m$^2 \cdot 10^5$	90.50	89.00	87.00

Employing the unit process for an interior point described in Section 13–5(c), determine the location and flow properties at point 4 (see Fig. 13.21) at a time 0.015 ms later.

5. The unit process for a point in the interior of an unsteady one-dimensional flow field is described in Section 13–5(c) and illustrated in Fig. 13.21. For a point on a solid boundary, illustrated in Fig. 13.23, point 5 is outside of the flow field. Figure 13.23a illustrates a stationary solid boundary, and Fig. 13.23b illustrates a moving solid boundary. Consequently, the characteristic equation for right-running characteristic 14 and the corresponding compatibility equation are not applicable at a solid boundary. However, the location and velocity

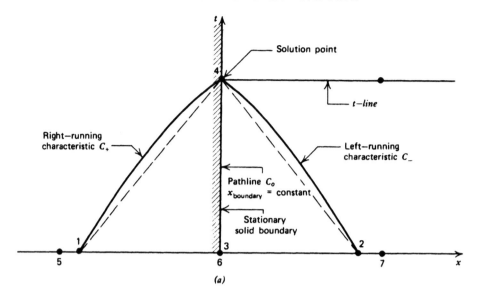

(a)

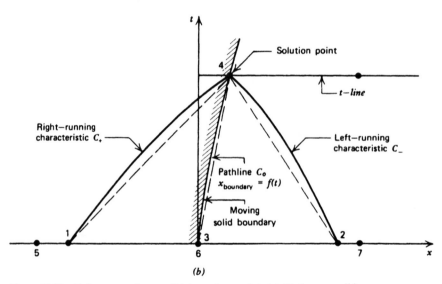

(b)

Figure 13.23 Unit process for a solid boundary point. (a) Stationary solid boundary. (b) Moving solid boundary.

of the solid boundary are known, thus supplying two conditions for replacing the lost characteristic and compatibility equations.

$$x_{boundary} = f(t) \qquad (specified) \qquad (a)$$

$$u_{boundary} = \frac{dx_{boundary}}{dt} \qquad (specified) \qquad (b)$$

Develop a numerical procedure for implementing the unit process for a solid boundary point, analogous to the procedure presented in Section 13–5(c) for an interior point.

6. Consider the rocket motor described in Example 13.2. The location and flow properties at the head end and at a point adjacent to the head end are given in the following table.

	Point 6	Point 7
x, m	0.000	0.025
u, m/s	0.0	50.0
p, N/m$^2 \cdot 10^5$	114.0	113.0

Employing the unit process for a solid boundary point developed in Problem 13.5, calculate the flow properties at the head end at a time 0.020 ms later.

7. The unit process for a point at the open end of a flow passage with outflow is illustrated in Fig. 13.24. For supersonic outflow, illustrated in Fig. 13.24b, both characteristics fall within the flow field, and the unit process for an interior point may be employed to determine the flow properties at point 4. Note that at point 4, $M_4 > 1$ and $p_4 > p_o$. For subsonic outflow,

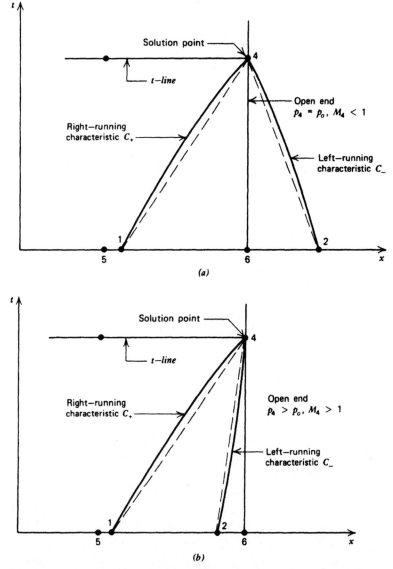

Figure 13.24   Unit process for outflow from an open end. (a) Subsonic outflow. (b) Supersonic outflow.

illustrated in Fig. 13.24a, $M_4 < 1$ and $p_4 = p_o$, and left-running characteristic 24 falls outside of the flow field. Consequently, the characteristic equation for left-running Mach line 24 and the corresponding compatibility equation are not applicable at an open end with subsonic outflow. However, the condition $p_4 = p_o$ replaces the lost compatibility equation. Develop a numerical procedure for implementing the unit process for an open end with subsonic outflow, analogous to the procedure presented in Section 13-5(c) for an interior point.

8. Under what conditions, if any, would the unit process for subsonic outflow at an open end developed in Problem 13.7 be applicable for subsonic inflow?

9. Describe briefly the physical model for subsonic inflow at an open end, and list the governing equations that are applicable.

10. Describe briefly the unit process for supersonic inflow at an open end.

11. Air is flowing out of a constant-area tube into the atmosphere. The flow properties at a given instant of time at the tube exit ($x = 1.000$ m) and an adjacent point inside the tube are given in the following table.

	Point 5	Point 6
x, m	0.950	1.000
u, m/s	160.0	165.0
p, N/m$^2 \cdot 10^5$	1.400	1.013

Employ the unit process for subsonic outflow at an open end developed in Problem 13.7, and determine the flow properties at the exit of the tube at a time 0.030 ms later.

A
numerical analysis

A–1 INTRODUCTION

The material in this Appendix is a brief introduction to the theory and implementation of the numerical methods employed in this book. Because the emphasis here is on the application of the methods, no rigorous proofs are presented. Numerical examples are included for illustrating the application of each of the numerical algorithms. The material included herein presents only a brief introduction to the methods of numerical analysis. A list of references is presented so that the interested reader may pursue the subject more deeply.

A–2 APPROXIMATION AND INTERPOLATION

Approximation is the process of approximating a general function by a simpler function that may then be employed in mathematical operations such as differentiation and integration. Approximating functions may also be employed for *interpolation*; that is, the determination of the value of a function at a point where its value is unknown from its known values at a discrete set of neighboring points. Approximating functions may be polynomials, trigonometric functions, exponential functions, etc. The polynomial is the most common approximating function, and the application of a simple interpolating polynomial is illustrated in the present section, for both univariate and bivariate interpolations. For a detailed discussion of the properties of the interpolating polynomial, refer to References 1–7.

In a *univariate* interpolating polynomial, the dependent variable depends on only one independent variable, for example, $y = f(x)$,

$$y = a + bx + cx^2 + dx^3 + \cdots \tag{A.1}$$

where the coefficients a, b, etc., are chosen so that the approximating function $y(x)$ represents the actual function as closely as possible. When the number of known data points $n + 1$ equals the number of coefficients in equation A.1 (i.e., the largest power of x is x^n), a unique interpolating polynomial results. In that case the coefficients are determined by substituting the sets of x, y pairs into equation A.1 to determine $n + 1$ linear algebraic equations for the coefficients. Those equations may be solved by the methods discussed in Section A–3.

When more than $n + 1$ sets of x, y pairs are employed, the interpolating polynomial must approximate those points in an average sense. For example, if a second-order interpolating polynomial is desired, equation A.1 becomes

$$y = a + bx + cx^2 \tag{A.2}$$

If m sets of x, y pairs are available, they may be substituted into equation A.2 to obtain the following m residual equations.

$$w_1 = a + bx_1 + cx_1{}^2 - y_1 \tag{A.3a}$$

$$w_2 = a + bx_2 + cx_2{}^2 - y_2 \tag{A.3b}$$

$$\cdots\cdots\cdots\cdots\cdots\cdots\cdots\cdots\cdots$$

$$w_m = a + bx_m + cx_m{}^2 - y_m \tag{A.3m}$$

The values of a, b, and c are chosen to make the values of the w_m as small as possible.

The *method of least squares* says that the best interpolating polynomial is obtained when the sum of the squares of the residuals is a minimum. That is,

$$\sum_{i=1}^{m} w_m{}^2 = w_1{}^2 + w_2{}^2 + \cdots + w_m{}^2 = f(a,b,c)$$

must be a minimum. The condition that $f(a,b,c)$ is a minimum is satisfied if partial derivatives with respect to a, b, and c are each zero. Consequently,

$$\frac{\partial f}{\partial a} = 2w_1 \frac{\partial w_1}{\partial a} + 2w_2 \frac{\partial w_2}{\partial a} + \cdots + 2w_m \frac{\partial w_m}{\partial a}$$

$$= 2w_1 + 2w_2 + \cdots + 2w_m = 0 \tag{A.4a}$$

where $\partial w_1/\partial a = \partial w_2/\partial a = \cdots = 1$. In a similar manner, we obtain

$$\frac{\partial f}{\partial b} = 2w_1 x_1 + 2w_2 x_2 + \cdots + 2w_m x_m = 0 \tag{A.4b}$$

$$\frac{\partial f}{\partial c} = 2w_1 x_1{}^2 + 2w_2 x_2{}^2 + \cdots + 2w_m x_m{}^2 = 0 \tag{A.4c}$$

Substituting equations A.3 into equations A.4 and collecting the coefficients of a, b, and c, we obtain

$$am + b \sum_{i=1}^{m} x_i + c \sum_{i=1}^{m} x_i^2 = \sum_{i=1}^{m} y_i \tag{A.5a}$$

$$a \sum_{i=1}^{m} x_i + b \sum_{i=1}^{m} x_i^2 + c \sum_{i=1}^{m} x_i^3 = \sum_{i=1}^{m} x_i y_i \tag{A.5b}$$

$$a \sum_{i=1}^{m} x_i^2 + b \sum_{i=1}^{m} x_i^3 + c \sum_{i=1}^{m} x_i^4 = \sum_{i=1}^{m} x_i^2 y_i \tag{A.5c}$$

Equations A.5 are three linear algebraic equations for a, b, and c that may be solved simultaneously by the methods discussed in Section A–3.

The method of least squares may be applied to higher-order interpolating polynomials and to other forms of approximating functions. In all cases, the number of linear equations for the unknown coefficients equals the number of the coefficients that are to be determined.

In a *bivariate* interpolating polynomial, the dependent variable depends on two independent variables, for example, $z = f(x,y)$. The procedure presented above for univariate interpolation may be applied in the same manner for determining bivariate interpolating polynomials. For example, for a second-order bivariate interpolating polynomial,

$$z = a + bx + cx^2 + dy + ey^2 + fxy \tag{A.6}$$

The known data values are substituted into equation A.6 to obtain the residual equations. Those residual equations are then squared and summed, and the resulting equation is differentiated with respect to a, b, c, etc., to determine six linear non-homogeneous algebraic equations for the six coefficients in equation A.6.

The method of least squares may be applied to any type of approximating function with any number of independent variables in the same manner as illustrated above.

Example A.1. In Table 1.6, Section 1–16(a), values of the coefficients for a fourth-order interpolating polynomial (see equation 1.158) are presented for the constant pressure specific heat c_p versus temperature t for several gases. Determine the values of the coefficients for a second-order interpolating polynomial for c_p versus t for N_2, for the temperature range 1000 K $\leq t \leq$ 5000 K. As data points, employ the c_p values from Table D.1k (see Volume 2) at $t = 1000$ K and every 500 K thereafter up to and including 5000 K.

Solution

The following nine data points in Table A.1 are obtained from Table D.1k.

Table A.1 Data Points for the Interpolating Polynomial for c_p versus t for N_2

t, K	c_p, J/mol-K	$\dfrac{c_p}{R}$	t, K	c_p, J/mol-K	$\dfrac{c_p}{R}$
1000	32.698	3.93272	3500	37.329	4.48971
1500	34.800	4.18554	4000	37.569	4.51858
2000	36.018	4.33203	4500	37.799	4.54624
2500	36.687	4.41250	5000	37.950	4.56440
3000	37.066	4.45808			

The second-order interpolating polynomial is given by

$$c_p = (a + bt + ct^2)R \qquad (a)$$

Equations A.5 are solved to determine the values a, b, and c. The coefficients of a, b, and c and the nonhomogeneous terms in equation A.5 are computed as follows.

$$\sum_{i=1}^{9} t_i = 1000 + 1500 + \cdots + 5000 = 27 \cdot 10^3$$

$$\sum_{i=1}^{9} t_i^2 = (1000)^2 + (1500)^2 + \cdots + (5000)^2 = 96 \cdot 10^6$$

$$\sum_{i=1}^{9} t_i^3 = (1000)^3 + (1500)^3 + \cdots + (5000)^3 = 378 \cdot 10^9$$

$$\sum_{i=1}^{9} t_i^4 = (1000)^4 + (1500)^4 + \cdots + (5000)^4 = 1583.25 \cdot 10^{12}$$

$$\sum_{i=1}^{9} \left(\frac{c_p}{R}\right)_i = 3.93272 + 4.18554 + \cdots + 4.56440 = 39.43980$$

$$\sum_{i=1}^{9} t_i \left(\frac{c_p}{R}\right)_i = (1000)(3.93272) + (1500)(4.18554) + \cdots + (5000)(4.56440)$$
$$= 120.34897 \cdot 10^3$$

$$\sum_{i=1}^{9} t_i^2 \left(\frac{c_p}{R}\right)_i = (1000)^2(3.93272) + (1500)^2(4.18554) + \cdots + (5000)^2(4.56440)$$
$$= 431.84674 \cdot 10^6$$

Substituting the above values into equation A.5, we obtain

$$9a + 27 \cdot 10^3 b + 96 \cdot 10^6 c = 39.43980$$

$$27 \cdot 10^3 a + 96 \cdot 10^6 b + 378 \cdot 10^9 c = 120.34897 \cdot 10^3$$

$$96 \cdot 10^6 a + 378 \cdot 10^9 b + 1583.25 \cdot 10^{12} c = 431.84674 \cdot 10^6$$

The above system of three linear algebraic equations is solved for a, b, and c by Cramer's rule in Example A.2. The results are

$$c_p = (3.587000 + 0.453813 \cdot 10^{-3}t - 0.053085 \cdot 10^{-6}t^2)R \qquad (b)$$

The errors in the calculated values from equation (b) are presented in Table A.2, and the corresponding curve fit is illustrated in Fig. A.1.

Table A.2 Comparison of Initial-Data Points and the Interpolating Polynomial

t, K	$\dfrac{c_p}{R}$, Table D.1k	$\dfrac{c_p}{R}$, equation (b)	Error, Percent
1000	3.93272	3.98773	1.40
2000	4.33203	4.28229	−1.15
3000	4.45808	4.47067	0.28
4000	4.51858	4.55289	0.76
5000	4.56440	4.52894	−0.78

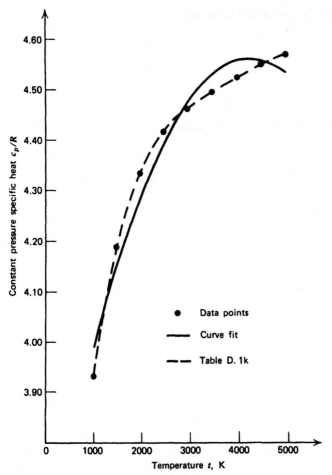

Figure A.1   Comparison of the initial-data points and the curve fit for c_p/R versus t for N_2.

A–3 THE SOLUTION OF SYSTEMS OF LINEAR ALGEBRAIC EQUATIONS

Systems of linear algebraic equations occur frequently in numerical analysis. For example, such systems occur in the least squares fitting of interpolating polynomials (see Appendix A–2) and in the solution of linear difference equations. Many powerful matrix methods exist for solving such systems of equations. The fundamental method employed for *direct solutions* is the *Gauss elimination method. Iterative methods* start with an initial approximation, and improve that approximation by a suitable algorithm. Matrix methods are discussed in References 1–7.

Cramer's rule presents a straightforward method for obtaining the solution of systems of linear algebraic equations. When the number of equations reaches more than three or four, Cramer's rule becomes impractical because of the large amount of computation involved, unless the majority of the coefficients in the equations are zero. However, for systems of equations in which the coefficients are specified as functions or parameters instead of as numerical values, Cramer's rule must be applied. Such a situation arises when solving systems of difference equations in functional form to obtain explicit equations for the unknown variables, and when solving systems of differential equations for influence coefficients [see Section 9–5(c)].

Consider the following system of linear equations.

$$a_{11}x_1 + a_{12}x_2 + \cdots + a_{1n}x_n = b_1 \qquad \text{(A.7a)}$$

$$a_{21}x_1 + a_{22}x_2 + \cdots + a_{2n}x_n = b_2 \qquad \text{(A.7b)}$$

$$\cdots\cdots\cdots\cdots\cdots\cdots\cdots\cdots\cdots\cdots\cdots\cdots$$

$$a_{n1}x_1 + a_{n2}x_2 + \cdots + a_{nn}x_n = b_n \qquad \text{(A.7n)}$$

By definition, the matrices A, x, and b are defined as

$$A \equiv \begin{bmatrix} a_{11} & a_{12} & \cdots & a_{1n} \\ a_{21} & a_{22} & & a_{2n} \\ \vdots & \vdots & & \vdots \\ a_{n1} & a_{n2} & \cdots & a_{nn} \end{bmatrix}, \quad x \equiv \begin{bmatrix} x_1 \\ x_2 \\ \vdots \\ x_n \end{bmatrix}, \quad b \equiv \begin{bmatrix} b_1 \\ b_2 \\ \vdots \\ b_n \end{bmatrix}$$

Then the system of equations A.7a–A.7n, when written in matrix form, reduces to

$$Ax = b \qquad \text{(A.8)}$$

where $A = (a_{ij})$ is the $n \times n$ matrix comprising the coefficients in equations A.7, x is the n-vector x_1, x_2, etc., and b is the n-vector b_1, b_2, etc.

Denote the value of the determinant of the matrix A by det (A). If det $(A) \neq 0$, then Cramer's rule states that

$$x_j = \frac{\det (A^j)}{\det (A)} \qquad (j = 1, \ldots, n) \qquad \text{(A.9)}$$

where A^j is the matrix obtained by replacing the jth column of A by **b**.

The determination of the value of det (A) is accomplished by expanding the determinant by minors. A minor M_{ij} of the $n \times n$ matrix A is the determinant of the $(n - 1) \times (n - 1)$ matrix obtained from A by deleting the ith row and the jth column. Expanding the determinant of A by minors along the ith row, we obtain

$$\det (A) = a_{i1}(-1)^{i+1}M_{i1} + a_{i2}(-1)^{i+2}M_{i2} + \cdots + a_{in}(-1)^{i+n}M_{in} \quad \text{(A.10)}$$

Alternately, expanding along the jth column yields

$$\det (A) = a_{1j}(-1)^{1+j}M_{1j} + a_{2j}(-1)^{2+j}M_{2j} + \cdots + a_{nj}(-1)^{n+j}M_{nj} \quad \text{(A.11)}$$

By employing either equation A.10 or A.11, the determinant of order n may be expressed as the sum of n determinants of order $n - 1$. By applying the procedure recursively, det (A) will eventually be expressed as the sum of determinants of order 1, which is simply the sum of the corresponding elements. The aforementioned procedure is practical for determinants of low order, or for determinants in which most of the elements are zero so that most of the summands drop out.

Example A.2.   Three linear algebraic equations are obtained in Example A.1 for the coefficients a, b, and c of the interpolating polynomial for the specific heat c_p of N_2 as a function of temperature t. Determine the values of a, b, and c by applying Cramer's rule.

Solution

In matrix form, those equations may be written as

$$\begin{vmatrix} 9 & 27 \cdot 10^3 & 96 \cdot 10^6 \\ 27 & 96 \cdot 10^3 & 378 \cdot 10^6 \\ 96 & 378 \cdot 10^3 & 1583.25 \cdot 10^6 \end{vmatrix} \begin{vmatrix} a \\ b \\ c \end{vmatrix} = \begin{vmatrix} 39.43980 \\ 120.34897 \\ 431.84674 \end{vmatrix}$$

To apply Cramer's rule, det (A) must be computed. Therefore,

$$\det(A) = 9 \begin{vmatrix} 96 \cdot 10^3 & 378 \cdot 10^6 \\ 378 \cdot 10^3 & 1583.25 \cdot 10^6 \end{vmatrix} - 27 \begin{vmatrix} 27 \cdot 10^3 & 96 \cdot 10^6 \\ 378 \cdot 10^3 & 1583.25 \cdot 10^6 \end{vmatrix}$$

$$+ 96 \begin{vmatrix} 27 \cdot 10^3 & 96 \cdot 10^6 \\ 96 \cdot 10^3 & 378 \cdot 10^6 \end{vmatrix}$$

$$\det(A) = 9(151{,}992 - 142{,}884)10^9 - 27(42{,}747.75 - 36{,}288)10^9$$

$$+ 96(10{,}206 - 9216)10^9 = 2598.75 \cdot 10^9$$

The coefficient a is determined by replacing the first column of A with the non-homogeneous terms, and calculating det (A^j)/det (A). Therefore,

$$a = \frac{\det(A^j)}{\det(A)} = \frac{\begin{vmatrix} 39.43980 & 27 \cdot 10^3 & 96 \cdot 10^6 \\ 120.34897 & 96 \cdot 10^3 & 378 \cdot 10^6 \\ 431.84674 & 378 \cdot 10^3 & 1583.25 \cdot 10^6 \end{vmatrix}}{2598.75 \cdot 10^9}$$

Solving for det (A^j) in the manner illustrated above for det (A), we obtain

$$a = \frac{9321.715 \cdot 10^9}{2598.75 \cdot 10^9} = 3.587000$$

In a similar manner, replacing the second column of A with the nonhomogeneous terms and solving for b gives

$$b = \frac{\begin{vmatrix} 9 & 39.43980 & 96 \cdot 10^6 \\ 27 & 120.34897 & 378 \cdot 10^6 \\ 96 & 431.84674 & 1583.25 \cdot 10^6 \end{vmatrix}}{2598.75 \cdot 10^9} = \frac{1179.346 \cdot 10^6}{2598.75 \cdot 10^9} = 0.453813 \cdot 10^{-3}$$

Accordingly, for c, we obtain

$$c = \frac{\begin{vmatrix} 9 & 27 \cdot 10^3 & 39.43980 \\ 27 & 96 \cdot 10^3 & 120.34897 \\ 96 & 378 \cdot 10^3 & 431.84674 \end{vmatrix}}{2598.75 \cdot 10^9} = \frac{-137.9542 \cdot 10^3}{2598.75 \cdot 10^9} = -0.053085 \cdot 10^{-6}$$

A–4 THE SOLUTION OF NONLINEAR ALGEBRAIC EQUATIONS

A problem that arises frequently in numerical analysis is the determination of the roots of a nonlinear function $f(x) = 0$. The nonlinear function $f(x)$ may be either an algebraic expression, such as a polynomial or transcendental function, or it may be a complete numerical algorithm comprising many steps. In a few cases, such as a factorable polynomial, it is possible to determine the exact roots of $f(x)$. In general, however, the solution must be determined iteratively by some numerical technique, and an approximate solution is obtained. In this section, two methods are presented for the solution of nonlinear equations: the *Newton-Raphson method* and the *secant method*.

A–4(a) The Newton Raphson Method

Consider a nonlinear algebraic equation $f(x) = 0$, for which the first derivative $f'(x)$ is calculable. Assume that an approximate value of the root, denoted by x_i, is known, where the exact root is $x_i + h$. Applying Taylor's formula [see Appendix

A–6(a)], we may write

$$f(x_i + h) = 0 = f(x_i) + hf'(x_i) + \frac{h^2}{2} f''(x_i + \zeta h)(0 \leq \zeta \leq 1) \quad \text{(A.12)}$$

If h is relatively small, the term containing h^2 may be neglected in comparison with the term containing h, so that equation A.12 reduces to

$$f(x_i) + hf'(x_i) = 0 \quad \text{(A.13)}$$

Solving equation A.13 for h yields

$$h = -\frac{f(x_i)}{f'(x_i)} \quad \text{(A.14)}$$

From equation A.14 we obtain the improved approximation

$$x_{i+1} = x_i + h = x_i - \frac{f(x_i)}{f'(x_i)} \quad \text{(A.15)}$$

The root determined by equation A.15 is not exact because of the approximation that the second-order term in equation A.12 equals zero. Equation A.15 may be applied repetitively to yield successively closer approximations to the exact value of the root. The algorithm expressed by equation A.15 is known as the *Newton-Raphson* method.

A–4(b)   The Secant Method

Consider a nonlinear algorithm of the form $f(x) = C$, where the algorithm comprises any number of steps, including the solution of differential equations, the use of tabular data, the use of graphical data, etc. Let x_{i-1} and x_i denote two values of x in the neighborhood of the exact root. The approximate roots x_i and x_{i-1} may either bracket the exact root, or they may lie on either side of the exact root. Substituting those two values into the nonlinear algorithm $f(x)$ yields the two values $f(x_{i-1})$ and $f(x_i)$. Plotting those two pairs of x, $f(x)$ values in the $f(x)$ versus x plane furnishes a functional relationship between x and $f(x)$, as illustrated in Fig. A.2. Assume that the functional relationship between $f(x)$ and x is linear, and that the straight line between the two x, $f(x)$ pairs intersects the line $f(x) = C$, thereby determining

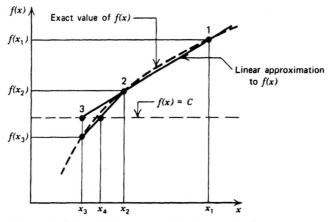

Figure A.2   Schematic illustration of the secant method.

an improved approximation to the exact root, x_{i+1}. Therefore,

$$\frac{f(x_{i+1}) - f(x_i)}{x_{i+1} - x_i} = \frac{f(x_i) - f(x_{i-1})}{x_i - x_{i-1}} = m_i \qquad (A.16)$$

where m_i is the slope of the straight line relating x and $f(x)$. Let $f(x_{i+1}) = C$, the desired value of $f(x)$, and solve equation A.16 for the corresponding value of x_{i+1}. The result is

$$x_{i+1} = x_i + \frac{C - f(x_i)}{m_i} \qquad (A.17)$$

The improved root determined by equation A.17 is not exact unless the relationship between x and $f(x)$ is linear, as assumed in equation A.16. Consequently, equation A.17 must be applied repetitively to obtain improved approximations to the exact root. The algorithm specified by equation A.17 is known as the *secant method*.

In the special case where $f(x) = 0$ and $x_{i-1} \rightarrow x_i$, the slope m_i in equation A.16 approaches $f'(x_i)$. Substituting those values into equation A.17, we obtain

$$x_{i+1} = x_i - \frac{f(x_i)}{f'(x_i)} \qquad (A.18)$$

which is the Newton-Raphson algorithm (see equation A.15).

Both the Newton-Raphson method and the secant method are applicable to systems of nonlinear equations.

Example A.3.   The relationship between the Mach number M and the area ratio $\varepsilon = A/A^*$ for steady one-dimensional isentropic flow is given by equation 3.169 (see Table 3.2), repeated below.

$$\varepsilon = \frac{1}{M}\left[\frac{2}{\gamma + 1}\left(1 + \frac{\gamma - 1}{2}M^2\right)\right]^{(\gamma + 1)/2(\gamma - 1)} \qquad (3.169)$$

The determination of the value of ε corresponding to a given value of M is straightforward. The reverse calculation, that is, the determination of the value of M corresponding to a given value for ε, requires an iterative solution of equation 3.169. To illustrate, calculate the value of M, in the supersonic flow regime, corresponding to $\varepsilon = 10.0$, for $\gamma = 1.40$, by (a) the Newton-Raphson method, and (b) the secant method.

Solution

Equation 3.169 may be written as

$$\varepsilon = \frac{1}{M}\left(\frac{2}{\gamma + 1}\right)^{(\gamma + 1)/2(\gamma - 1)}\left(1 + \frac{\gamma - 1}{2}M^2\right)^{(\gamma + 1)/2(\gamma - 1)} = \frac{a}{M}(1 + bM^2)^c \quad \text{(a)}$$

where

$$c = \frac{\gamma + 1}{2(\gamma - 1)}, \qquad a = \left(\frac{2}{\gamma + 1}\right)^c, \qquad b = \frac{\gamma - 1}{2} \qquad \text{(b)}$$

(a)   For the Newton-Raphson method, define $f(M)$ as follows.

$$f(M) = M\varepsilon - a(1 + bM^2)^c = 0 \qquad \text{(c)}$$

Differentiating equation (c) yields

$$f'(M) = \varepsilon - 2abcM(1 + bM^2)^{c-1} \qquad \text{(d)}$$

Substituting equations (c) and (d) into equation A.15, we obtain

$$M_{i+1} = M_i - \frac{f(M)}{f'(M)} \tag{e}$$

For $\gamma = 1.40$,

$$c = \frac{2.4}{2(0.4)} = 3.0, \qquad a = \left(\frac{2}{2.4}\right)^3 = 0.57870370, \qquad b = \frac{0.4}{2} = 0.2$$

Substituting the values for ε, a, b, and c into equations (c) and (d) gives

$$f(M) = 10M - 0.57870370(1 + 0.2M^2)^{3.0} \tag{f}$$

$$f'(M) = 10.0 - 0.69444444M(1 + 0.2M^2)^{2.0} \tag{g}$$

To obtain the value of M corresponding to the supersonic branch of $f(M) = 0$, the first trial for $f(M)$ must be supersonic. Assume $M_1 = 5.00$. Then,

$$f(M_1) = 10(5.0) - (0.57870370)[1 + 0.2(5.0)^2]^{3.0} = -75.0$$

$$f'(M_1) = 10.0 - (0.69444444)(5.0)[1.0 + 0.2(5.0)^2]^{2.0} = -115.0$$

Substituting the above values into equation (e) gives

$$M_2 = 5.0 - \frac{-75.0}{-115.0} = 4.34782609$$

Repeating the above steps successively, we obtain

$$f(M_2) = -19.75356661, \qquad f'(M_2) = -59.00745120, \qquad M_3 = 4.01306215$$

$$f(M_3) = -3.38866969, \qquad f'(M_3) = -39.65127766, \qquad M_4 = 3.92760035$$

$$f(M_4) = -0.17872610, \qquad f'(M_4) = -35.51906555, \qquad M_5 = 3.92256852$$

$$f(M_5) = -0.00058922, \qquad f'(M_5) = -35.28509122, \qquad M_6 = 3.92255182$$

$$f(M_6) = -0.00000003, \qquad f'(M_6) = -35.28431645, \qquad M_7 = 3.92255182$$

Consequently, for $\varepsilon = 10,\cdot M = 3.92255182$.

(b) For the secant method, define $f(M)$ as follows.

$$f(M) = \frac{a}{M}(1 + bM^2)^c = \varepsilon \tag{h}$$

Substituting equation (h) into equation A.17, we obtain

$$M_{i+1} = M_i + \frac{\varepsilon - f(M_i)}{m_i} \tag{i}$$

where, from equation A.16,

$$m_i = \frac{f(M_i) - f(M_{i-1})}{M_i - M_{i-1}} \tag{j}$$

Substituting the values of ε, a, b, and c into equation (h) gives

$$f(M) = \frac{(0.57870370)}{M}(1 + 0.2M^2)^{3.0} = 10.0 \tag{k}$$

To obtain the value of M corresponding to the supersonic branch of the solution, choose $M_1 = 5.0$ and $M_2 = 5.1$. Then,

$$f(M_1) = \frac{(0.57870370)}{5.0}[1 + 0.2(5.0)^2]^{3.0} = 25.0$$

$$f(M_2) = \frac{(0.57870370)}{5.1}[1 + 0.2(5.1)^2]^{3.0} = 27.06957090$$

Substituting the above values into equation (j), we obtain

$$m_2 = \frac{27.06957090 - 25.0}{5.1 - 5.0} = 20.69570900$$

Substituting for m_2 into equation (i) yields

$$M_3 = 5.1 + \frac{10.0 - 27.06957090}{20.69570900} = 4.27521207$$

Repeating the above steps, we obtain

$f(M_3) = 13.65822106,$	$m_3 = 16.26036150,$	$M_4 = 4.05023423$
$f(M_4) = 11.20921291,$	$m_4 = 10.88555277,$	$M_5 = 3.93915003$
$f(M_5) = 10.15030606,$	$m_5 = 9.53247050,$	$M_6 = 3.92338223$
$f(M_6) = 10.00747220,$	$m_6 = 9.05858117,$	$M_7 = 3.92255736$
$f(M_7) = 10.00004974,$	$m_7 = 8.99832822,$	$M_8 = 3.92255183$
$f(M_8) = 10.00000003,$	$m_8 = 8.99077591,$	$M_9 = 3.92255183$

Both methods, of course, yield the same result. The rates of convergence are comparable. The secant method requires less computation, even though two more iterations are required, because each iteration requires a single calculation of $f(M)$, whereas the Newton-Raphson method requires the calculation of both $f(M)$ and $f'(M)$ for each iteration.

A–5 INTEGRATION

The need for *numerical integration*, or *numerical quadrature*, arises when the integral

$$I = \int_a^b f(x)\, dx \tag{A.19}$$

cannot be evaluated exactly, either because the expression for $f(x)$ is too complicated, or the values of $f(x)$ are known only at a discrete number of points. The latter case is the one considered here. Two straightforward methods are presented below for calculating the value of I; they are the *trapezoidal rule* and *Simpson's rule*.

Consider the $n + 1$ discrete points illustrated in Fig. A.3a, where the solid curve represents the exact value of $f(x)$. The value of the integral in equation A.19 is the area under the solid curve. An approximating function, discussed in Appendix A–2, may be fitted to the set of discrete data points. The approximating function may then be substituted into equation A.19 and integrated to furnish an approximate value for I.

A–5(a) The Trapezoidal Rule

The simplest approximating function to $f(x)$ is a series of straight lines joining the $n + 1$ discrete data points, illustrated by the curve in Fig. A.3b. The area of each

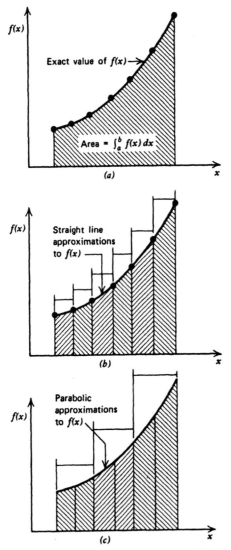

Figure A.3   Determination of the value of the integral
$I = \int_a^b f(x)\,dx$. (a) Exact value of the integral. (b) The
trapezoidal rule. (c) Simpson's rule.

trapezoid is simply the width of the base multiplied by the average of the heights of the two sides. The total area under the approximating function, which is the approximate value of I, is

$$I = \sum_{i=2}^{n+1} (x_i - x_{i-1}) \frac{[f(x_i) + f(x_{i-1})]}{2} \tag{A.20}$$

When the interval $x_i - x_{i-1} = h$ is constant, equation A.20 becomes

$$I = \frac{h}{2} \sum_{i=1}^{n+1} C_i f(x_i) \tag{A.21}$$

where $C_i = 1, 2, 2, \ldots, 2, 2, 1$. The truncation error of the trapezoidal rule is second order in h.

A-5(b) Simpson's Rule

When the $n + 1$ discrete data points are equally spaced and odd in number, a more accurate quadrature algorithm known as Simpson's rule is applicable. The data points are grouped into sets of three x, $f(x)$ pairs, and an approximating function comprising a parabola is applied to each such set of three points, as illustrated in Fig. A.3c. Thus, if there are $n + 1$ discrete data points, $n/2$ parabolas are obtained. Each parabola is integrated over its region of applicability, determining, thereby, the corresponding incremental area. By summing the total area under all $n/2$ parabolas, the approximate value of I is obtained. The derivation of the algorithm is presented in the references. The result is

$$I = \frac{h}{3} \sum_{i=1}^{n+1} C_i f(x_i) \tag{A.22}$$

where $C_i = 1, 4, 2, \ldots, 2, 4, 1$. The truncation error is fourth order in h.

Example A. 4. Compute the value of the integral

$$I = \int_2^5 \ln x \, dx \tag{a}$$

(a) by exact integration, (b) by the trapezoidal rule with seven points, and (c) by Simpson's rule with seven points.

Solution

(a) $\quad I = \int_2^5 \ln x \, dx = x(\ln x - 1)\Big|_2^5$

$\quad I = 5[\ln (5) - 1] - 2[\ln (2) - 1] = 3.66089520$

(b) Divide the interval of integration into six equal increments each of width $h = 0.5$. Determine the values of $f(x) = \ln x$ at each point of the subdivision. The values are tabulated in the following table.

x	$\ln x$	x	$\ln x$
2.0	0.69314718	4.0	1.38629436
2.5	0.91629073	4.5	1.50407740
3.0	1.09861229	5.0	1.60943791
3.5	1.25276297		

Applying the trapezoidal rule, equation A.21, we obtain

$$I = \frac{0.5}{2} [0.69314718 + 2(0.91629073 + 1.09861229 + 1.25276297$$

$$+ 1.38629436 + 1.50407740) + 1.60943791] = 3.65466515$$

(c) Applying Simpson's rule, equation A.22, we obtain

$$I = \frac{0.5}{3} [0.69314718 + 4(0.91629073 + 1.25276297 + 1.50407740)$$

$$+ 2(1.09861229 + 1.38629436) + 1.60943791] = 3.66082046$$

Even with the coarse spacing employed, the result obtained by the trapezoidal rule is in error by only -0.17 percent, while Simpson's rule is in error by only -0.0020 percent.

A-6 THE NUMERICAL SOLUTION OF ORDINARY DIFFERENTIAL EQUATIONS

Problems in engineering and science are described by the basic laws of nature, which are often expressed as ordinary differential equations. Techniques have been developed for determining the closed form solutions of some types of ordinary differential equations. Most differential equations cannot, however, be solved in closed form. Consequently, numerical methods are employed for obtaining their solutions. Several good numerical methods are available for solving ordinary differential equations. Some of them are described in this section.

A-6(a) General Considerations

The differential equation considered is the first-order differential equation of the form

$$\frac{dy}{dx} = y' = f(x,y) \tag{A.23}$$

with the specified initial value

$$y(x_o) = y_o \tag{A.24}$$

The function $f(x,y)$ may be linear or nonlinear. Differential equations of higher order than first may always be written as a system of first-order differential equations, which may be solved by applying the integration algorithm to each of the equations in parallel. Consequently, the methods presented here for integrating equation A.23 may be applied to higher-order equations and to systems of such equations.

The most fundamental approach for obtaining a numerical solution to an ordinary differential equation involves expressing the unknown function in a Taylor series, and employing the differential equation for determining the required derivatives of the series. The Taylor series expansion about the point (x_o, y_o) for a function $y(x)$ is

$$y(x_o + h) = y(x_o) + hy'[x_o, y(x_o)] + \frac{h^2}{2!} y''[x_o, y(x_o)] + \cdots \tag{A.25}$$

That approach, known as Taylor's method, is discussed in Section A-6(b).

The *error* of a numerical method is the difference between the numerical result and the exact solution. The Taylor series in equation A.25 may be written as Taylor's formula with a remainder. Therefore,

$$y(x_o + h) = y(x_o) + hy'[x_o, y(x_o)] + \cdots + \frac{h^n}{n!} y^n[x_o, y(x_o)] + R_{n+1}$$

where

$$R_{n+1} = \frac{h^{n+1}}{(n+1)!} y^{n+1}[x_o + \zeta h, y(x_o + \zeta h)] \quad 0 \leqq \zeta \leqq 1$$

When terms through order n are retained in the Taylor series approximation, the method is said to have nth-order accuracy. In that case, the *local truncation error* R_{n+1} is $(n + 1)$th order.

When a numerical integration algorithm is applied repetitively m times to cover a range of the independent variable, the total accumulated error E is equal to the sum of the local truncation error R_{n+1} at each step. Hence,

$$E = \text{total accumulated error} = \sum_{i=1}^{m} (R_{n+1})_i \propto mh^{n+1}$$

The number of steps m required for integrating the independent variable over a specified range of values is inversely proportional to the step size h; that is, $m \propto 1/h$. Consequently,

$$E \propto mh^{n+1} \propto \left(\frac{1}{h}\right) h^{n+1} = h^n$$

The total accumulated error E is, therefore, proportional to h^n for a method having nth-order accuracy.

For the same step size h, the higher-order methods are, in general, more accurate than those of lower order.

Numerical methods for solving differential equations may be divided into two groups: (1) *explicit methods* in which the solution at the unknown point is expressed in terms of known points only, and (2) *implicit methods* in which the solution at the unknown point is expressed in terms of known points and the unknown point itself. Those two methods correspond to a Taylor series forward expansion about the known point, and a Taylor series backward expansion about the unknown point, respectively. Taylor's method, Euler's method, the Runge-Kutta methods, and the modified Euler predictor-corrector method, described in Sections A–6(b) to A–6(e), respectively, are all explicit methods. The implicit Euler method presented in Section A–6(f) is an example of an implicit method.

Numerical integration methods are also classified by the number of known data points that are required for calculating the solution at the next point. *Single step* methods require data at a single known point to obtain the solution at the unknown point. All of the methods discussed in this section are single step methods.

Multistep methods make use of the solution at two or more known points. Interpolating polynomials are fitted to the known points and the unknown point, the interpolating polynomial is integrated in closed form, and the resulting equation is solved for the value of the dependent variable at the unknown point. Multistep methods are not self-starting. Consequently, a single step method must be employed to determine the initial starting values. Moreover, multistep methods are generally less accurate than single step methods of the same order. The multistep methods, however, require less computing time than the single step methods of the same order. References 1–7 present detailed discussions of the more common multistep methods.

An important property of a numerical integration method that must receive careful attention is its *stability*. The term stability as employed in the literature is somewhat ambiguous. In general, a numerical integration method is said to be unstable if the errors introduced into the calculations from any source (initial-data errors, truncation errors, or round-off errors) grow at an exponential rate as the calculation proceeds. Certain equations with very specific initial conditions cannot be solved in a stable manner by any numerical integration method. Such equations are said to be *inherently unstable*. Another type of instability, *partial instability*, depends on the equation being solved, the specific initial conditions, and the particular numerical method being employed. The occurrence of partial instability depends on the step size h. For many algorithms, a stability analysis may be conducted for determining a stability criterion for establishing the maximum step size h for achieving a stable solution [see Sections A–6(c) and A–6(e)]. In other cases, the stability limits have to be determined by numerical experimentation. Numerical instability is an ever-present problem, and it must be given careful consideration in solving a differential equation numerically.

A–6(b)   Taylor's Method

The most straightforward approach for approximating the solution $y(x)$ of an ordinary differential equation is to express the solution $y(x)$ in the form of a Taylor series expansion (see equation A.25) about the initial point x_o. Therefore,

$$y(x_o + h) = y(x_o) + hy'[x_o,y(x_o)] + \frac{h^2}{2!}y''[x_o,y(x_o)] + \cdots \qquad \text{(A.25)}$$

Substituting the initial condition, equation A.24, into equation A.23, we obtain

$$y'[x_o,y(x_o)] = f[x_o,y(x_o)] \qquad \text{(A.26)}$$

The higher-order derivatives in equation A.25 may be obtained by applying the chain rule of partial differentiation to $y' = f(x,y)$. Thus,

$$y'' = f'(x,y) = \frac{df}{dx} = \frac{\partial f}{\partial x} + \frac{\partial f}{\partial y}\frac{dy}{dx} = f_x + f_y f \qquad \text{(A.27a)}$$

$$y''' = f''(x,y) = f_{xx} + 2f_{xy}f + f_{yy}f^2 + f_x f_y + f_y^2 f \qquad \text{(A.27b)}$$

In the above manner, the derivatives of y of any order may be expressed in terms of $f(x,y)$ and its partial derivatives. Unless $f(x,y)$ is a very simple function, however, the expression for the higher-order derivatives become more and more complicated. Consequently, only a limited number of the terms in equation A.25 may be employed, and equation A.25 becomes an approximation to the exact solution.

Taylor's method of order n is obtained from equation A.25 by truncating the series after the term containing h^n. Thus,

$$y(x_o + h) = y(x_o) + hy'[x_o,y(x_o)] + \cdots + \frac{h^n}{n!}y^n[x_o,y(x_o)] \qquad \text{(A.28)}$$

where the higher-order derivatives are determined from equation A.27. The order of the accumulated error in Taylor's method of order n is n. Consequently, for $n = 1$, the accumulated error decreases linearly with the step size h, for $n = 2$ the accumulated error decreases with the square of the step size h, etc.

Taylor's method is impractical for orders higher than the second or third because of the complications introduced by the determination of the higher-order derivatives. Taylor's method is of considerable theoretical significance, however, because the practical higher-order methods attempt to achieve the same accuracy as Taylor's method without calculating the higher-order derivatives.

A–6(c)   Euler's Method

For $n = 1$, Taylor's method, equation A.28, becomes

$$y(x_o + h) = y(x_o) + hy'[x_o,y(x_o)] \qquad \text{(A.29)}$$

which is known as *Euler's method*. Figure A.4 illustrates the geometrical interpretation of Euler's method. At the point n, equation A.29 yields the increment $\Delta y_n = hy'[x_n,y(x_n)]$ to y_n. In other words, the slope of the solution curve at point n is extended forward to x_{n+1} to determine y_{n+1}. The procedure may be applied repetitively to determine approximations y_n to $y(x_n)$. The accumulated error of Euler's method decreases linearly with the step size h; that is, it is a first-order method.

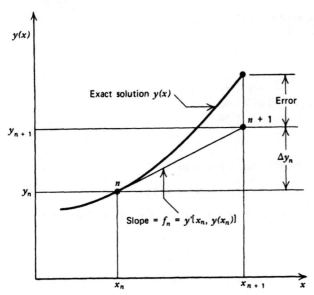

Figure A.4   Euler's method.

Carnahan, et al.[2] present a stability analysis for the Euler method, where it is shown that a stable solution results if the step size h satisfies the inequality

$$h \frac{\partial f}{\partial y} < 0$$

If $\partial f / \partial y$ is negative, the solution is clearly stable. If $\partial f / \partial y$ is positive, the solution will tend toward instability. By keeping the step size sufficiently small so that $h \, \partial f / \partial y$ is close to zero, the error might be kept under control, at least for a small range of x.

Euler's method is the simplest and most straightforward numerical integration algorithm for ordinary differential equations. The accuracy of Euler's method, however, is poor unless an extremely small step size h is employed. Consequently, Euler's method is not employed very often in practice.

Example A.5.   Consider the differential equation

$$\frac{dy}{dx} = y'(x,y) = f(x,y) = x + y \tag{a}$$

with the initial condition

$$y(0) = 1 \tag{b}$$

Calculate the solution at $x = 0.5$ by (a) the closed form exact solution, and (b) Euler's method with $h = 0.1$.

Solution

(a)   The closed form solution of equation (a) with the initial condition specified by equation (b) is

$$y(x) = 2e^x - x - 1 \tag{c}$$

Equation (c) may be confirmed as the solution of equation (a) by direct substitution. The solution to equation (c) is presented in Table A.3.

Table A.3 The Exact Solution for
$y' = x + y$ with $y(0) = 1$

x	y	x	y
0	1.0	0.3	1.39971762
0.1	1.11034184	0.4	1.58364940
0.2	1.24280552	0.5	1.79744254

(b) Euler's method is given by equation A.29.

$$y_{n+1} = y_n + hy'_n \qquad (d)$$

where $y_{n+1} = y(x_o + h)$, $y_n = y(x_o)$, and $y'_n = y'[x_n, y(x_n)]$. Substituting equation (a) into equation (d), we obtain

$$y_{n+1} = y_n + h(x_n + y_n) \qquad (e)$$

The results obtained by applying equation (e) repetitively from $x = 0$ to $x = 0.5$ with $h = 0.1$ are presented in Table A.4, along with the error that is seen to be increasing monotonically.

Table A.4 Numerical Solution Obtained by the Euler Method

n	x_n	y_n	f_n	hf_n	Error
0	0.0	1.0	1.0	0.1	
1	0.1	1.1	1.2	0.12	−0.01034184
2	0.2	1.22	1.42	0.142	−0.02280552
3	0.3	1.362	1.662	0.1662	−0.03771762
4	0.4	1.5282	1.9282	0.19282	−0.05544940
5	0.5	1.72102			−0.07642254

A–6(d) Runge-Kutta Methods

Euler's method is impractical for most applications because it requires a very small step size h to obtain reasonable accuracy, and so are Taylor's higher-order methods because of the complications introduced by the higher-order derivatives. The Runge-Kutta methods attempt to obtain higher-order accuracy without calculating the higher-order derivatives by determining the function $f(x, y)$ at selected points within each step. Runge-Kutta methods may be derived for any order.

The simplest of the Runge-Kutta methods is the second-order method. The recursion formula for that method is

$$y_{n+1} = y_n + \frac{h}{2}(m_1 + m_2) \qquad (A.30)$$

where

$$m_1 = f(x_n, y_n) \qquad (A.31)$$

$$m_2 = f(x_n + h, y_n + hm_1) \qquad (A.32)$$

The geometrical interpretation of equation A.30 is illustrated in Fig. A.5. By means of Euler's method, the value y_a is obtained by projecting the slope $m_1 = f(x_n, y_n)$

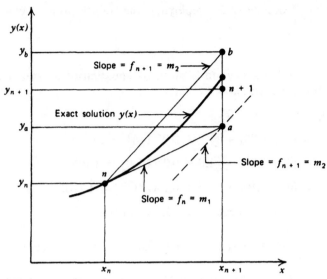

Figure A.5   The second-order Runge-Kutta method.

forward from point n to x_{n+1}. The slope at point a, $f(x_{n+1},y_a) = f(x_{n+1},y_n + hm_1) = m_2 = f_{n+1}$, is then projected forward from point n to obtain the value y_b. The average of the values of y_a and y_b is then taken as the solution for y_{n+1}, leading to the algorithm presented in equation A.30. Variations of the aforementioned method may be obtained by weighting the values of y_a and y_b by different amounts before computing their average value. Equation A.30 is, however, the most common form of the Runge-Kutta method of second order.

The most popular Runge-Kutta method is the fourth-order algorithm presented in equation A.33.

$$y_{n+1} = y_n + \frac{h}{6}(m_1 + 2m_2 + 2m_3 + m_4) \tag{A.33}$$

where

$$m_1 = f(x_n, y_n) \tag{A.34a}$$

$$m_2 = f\left(x_n + \frac{h}{2}, y_n + \frac{h}{2}m_1\right) \tag{A.34b}$$

$$m_3 = f\left(x_n + \frac{h}{2}, y_n + \frac{h}{2}m_2\right) \tag{A.34c}$$

$$m_4 = f(x_n + h, y_n + hm_3) \tag{A.34d}$$

Equation A.33 may be interpreted as a weighted average of y' calculated at x_n, two values of y' calculated at $x_n + h/2$, and y' calculated at $x_{n+1} = x_n + h$. Variations of the expressions presented in equation A.34 are obtained for different values of the weights assigned to the various terms. The algorithm presented above is one of the most common of the Runge-Kutta fourth-order methods.

The Runge-Kutta methods are always stable for a sufficiently small step size h. They are the most popular numerical methods for solving initial-value problems. In some applications, however, such as the *stiff* equations that occur in nonequilibrium chemically reacting flows (see Section 14–6), the allowable step size h may be so small as to make the method impractical.

Example A.6. Work Example A.5 employing the fourth-order Runge-Kutta method.

Solution

The fourth-order Runge-Kutta method is given by equations A.33 and A.34. For the first step, $x_o = 0$, $y_o = 1$, and $h = 0.1$. Thus,

$$m_1 = 0 + 1 = 1.0$$

$$m_2 = (0 + 0.05) + [1.0 + 0.05(1.0)] = 1.10$$

$$m_3 = (0 + 0.05) + [1.0 + 0.05(1.10)] = 1.105$$

$$m_4 = (0 + 0.10) + [1.0 + 0.10(1.105)] = 1.2105$$

Substituting the above values into equation A.33, we obtain

$$y_1 = 1.0 + \left(\frac{0.1}{6}\right)[1.0 + 2(1.10) + 2(1.105) + 1.2105] = 1.110342$$

The above results, and the results obtained at succeeding steps, are presented in Table A.5. The Runge-Kutta method is accurate to four more significant figures than the Euler method. The comparison clearly illustrates the advantage of a fourth-order method over a first-order method.

Table A.5 Numerical Solution Obtained by the Fourth-Order Runge-Kutta Method

n	x_n	y_n	m_1	m_2	m_3	m_4	Error
0	0	1.0	1.0	1.1	1.105	1.2105	
1	0.1	1.11034167	1.21034167	1.32085875	1.32638461	1.44298013	−0.00000017
2	0.2	1.24280515	1.44280515	1.56494540	1.57105242	1.69991039	−0.00000037
3	0.3	1.39971700	1.69971700	1.83470285	1.84145214	1.98386221	−0.00000062
4	0.4	1.58364849	1.98364849	2.13283091	2.14029003	2.29767749	−0.00000091
5	0.5	1.79744128					−0.00000126

A–6(e) The Modified Euler Predictor-Corrector Method

A *predictor-corrector* method is one that employs an explicit algorithm of a specified order to *predict* a temporary value of the solution, and then employs an algorithm of higher order to *correct* the predicted solution to obtain the final value of the solution. Consequently, two or more numerical integration algorithms are employed in succession.

The *modified Euler predictor-corrector* algorithm is the simplest of the predictor-corrector methods. For the predictor step, Euler's method is employed (see equation A.29). Therefore,

$$y_{n+1} = y_n + hf(x_n,y_n) \tag{A.35}$$

Euler's method is Taylor's method of first order. To approximate Taylor's method of second order, the second derivative $y''[x_n,y(x_n)] = f'[x_n,y(x_n)]$ must be estimated.

The derivative function $f(x,y)$ may be expanded in a forward Taylor series (see equation A.25) about the initial point x_n, y_n to obtain

$$f(x_n + h) = f(x_n) + hf'[x_n,y(x_n)] + \frac{h^2}{2!} f''[x_n,y(x_n)] + \cdots \tag{A.36}$$

Neglecting terms of order h^2 and higher, equation A.36 may be solved for $f'[x_n, y(x_n)] = y''[x_n, y(x_n)]$. Thus,

$$y''[x_n, y(x_n)] = f'[x_n, y(x_n)] = \frac{f(x_n + h) - f(x_n)}{h} \tag{A.37}$$

Equation A.37 is, of course, only an approximation to $y''[x_n, y(x_n)]$. Substituting the latter value of $y''[x_n, y(x_n)]$ into Taylor's formula, equation A.28, and neglecting terms of order h^3 and higher, we obtain

$$y(x_n + h) = y(x_n) + \frac{h}{2} \{ f(x_n, y_n) + f[x_n + h, y(x_n + h)] \} \tag{A.38}$$

Equation A.38 may be employed as a *corrector* algorithm, where $y(x_n + h)$ is obtained from the predictor algorithm, equation A.35.

The modified Euler predictor-corrector algorithm is the same as the Runge-Kutta method of second order, equation A.30. The corrector algorithm may be applied only once, or it may be applied repetitively for determining successively corrected values of $y(x_n + h)$. When the corrector is applied repetitively, the algorithm is called the *modified Euler predictor-corrector method with iteration*. Both the corrector and the corrector with iteration are second-order methods.

The modified Euler predictor-corrector method is numerically stable if the step size satisfies the condition

$$h \frac{\partial f}{\partial y} < 2 \tag{A.39}$$

Consequently, for $\partial f / \partial y$ negative, the method is always stable. For $\partial f / \partial y$ positive, h must be small enough so that the error is kept under control.

Example A.7. Work Example A.5 employing the modified Euler predictor-corrector method.

Solution

The modified Euler predictor-corrector method comprises equations A.35 and A.36. Therefore,

$$y_{n+1}^p = y_n + h f_n \tag{a}$$

$$y_{n+1}^c = y_n + \frac{h}{2}(f_n + f_{n+1}) \tag{b}$$

where the superscripts p and c denote the predictor and corrector steps, respectively, and

$$f_n = x_n + y_n \tag{c}$$

$$f_{n+1} = x_{n+1} + y_{n+1}^p \tag{d}$$

For the first step

$$f_0 = x_0 + y_0 = 0.0 + 1.0 = 1.0$$

$$y_1^p = y_0 + h f_0 = 1.0 + (0.1)(1.0) = 1.1$$

$$f_1 = x_1 + y_1^p = 0.1 + 1.1 = 1.2$$

$$y_1^c = y_0 + \frac{h}{2}(f_0 + f_1) = 1.0 + \frac{(0.1)}{2}(1.0 + 1.2) = 1.11$$

Table A.6 presents the results obtained by applying the above procedure from $x = 0.0$ to $x = 0.5$. The modified Euler predictor-corrector method yields nearly two more significant figures than the Euler method, but approximately three less significant figures than the fourth-order Runge-Kutta method.

Table A.6   Numerical Solution Obtained by the Modified Euler Predictor-Corrector Method

n	x_n	y_n	f_n	y_{n+1}^p	f_{n+1}	Error
0	0	1.0	1.0	1.1	1.2	
1	0.1	1.11	1.21	1.231	1.431	-0.00034184
2	0.2	1.24205	1.44205	1.386255	1.686255	-0.00075552
3	0.3	1.39846525	1.69846525	1.56831178	1.96831178	-0.00125237
4	0.4	1.58180410	1.98180410	1.77998451	2.27998451	-0.00184530
5	0.5	1.79489353				-0.00254901

Example A.8.   Solve the differential equation presented in Example A.5 by the modified Euler predictor-corrector method with one iteration at each step.

Solution

The computational equations are the same as equations (a) to (d) in Example A.7, with the exception that the corrector, equation (b), is applied twice. The predictor and corrector results for the first step are identical to those presented in Example A.7. Iteration of those results yields

$$f_1^c = x_1 + y_1^c = 0.1 + 1.11 = 1.21$$

$$y_1^c = y_0 + \frac{h}{2}(f_0 + f_1^c) = 1.0 + \frac{(0.1)}{2}(1.0 + 1.21) = 1.1105$$

The results obtained by the application of the modified Euler predictor-corrector method with one iteration are presented in Table A.7. For this particular example, the magnitude of the error is approximately halved by the application of the corrector. No general conclusions regarding the application of iteration are possible. In some cases, iteration is worthwhile, in others it has an insignificant effect, and in some cases it actually leads to a less accurate solution. In a practical application, numerical experimentation is required for determining the value of iteration.

Table A.7   Numerical Results Obtained by the Modified Euler Predictor-Corrector Method with Iteration

n	x_n	y_n	f_n	y_{n+1}^p	f_{n+1}^p	y_{n+1}^c	f_{n+1}^c	Error
0	0	1.0	1.0	1.1	1.2	1.11	1.21	
1	0.1	1.1105	1.2105	1.23155	1.43155	1.2426025	1.4426025	0.00015816
2	0.2	1.24315513	1.44315513	1.38747064	1.68747064	1.39968641	1.69968641	0.00034961
3	0.3	1.40029720	1.70029720	1.57032692	1.97032692	1.58382841	1.98382841	0.00057958
4	0.4	1.58450348	1.98450348	1.78295383	2.28295383	1.79787635	2.29787635	0.00085408
5	0.5	1.79862247						0.00117993

A-6(f) The Implicit Euler Method

The Euler method presented in Section A–6(c) is an *explicit* method obtained from the Taylor series expansion for $y(x)$ about the known point x_n. The *implicit Euler method* is similar to the explicit Euler method, except that the Taylor series expansion for $y(x)$ is a backward expansion about the forward point $x_n + h$. Therefore,

$$y(x_n) = y(x_n + h) + (-h)y'[x_n + h, y(x_n + h)]$$
$$+ \frac{(-h)^2}{2!} y''[x_n + h, y(x_n + h)] + \cdots \quad (A.40)$$

Truncating the series after the first-order term and solving for $y(x_n + h)$, we obtain

$$y(x_n + h) = y(x_n) + hy'[x_n + h, y(x_n + h)] \quad (A.41)$$

Equation A.41 is the algorithm for the first-order *implicit Euler method*.

The implicit feature of the method is apparent in equation A.41, where the value of $y'[x_n + h, y(x_n + h)]$ depends on the unknown value $y(x_n + h)$. When the derivative function $y'(x, y) = f(x, y)$ is linear in y, the value of $y(x_n + h)$ on the right-hand side of equation A.41 may be transferred to the left-hand side of the equation and combined with the $y(x_n + h)$ appearing there, thereby furnishing an equation for the direct determination of $y(x_n + h)$ (see Example A.9). When $f(x, y)$ is nonlinear in y, other approaches are required.

One approach for solving equation A.41 when $f(x, y)$ is nonlinear in y is an iterative procedure wherein a trial value of $y(x_n + h)$ is assumed, and equation A.41 is solved to obtain an improved value for $y(x_n + h)$. The improved value, or some value determined by a weighted combination of the trial and improved values, is then substituted into equation A.41 to obtain a second improved value for $y(x_n + h)$. The procedure is repeated to convergence.

A second approach for solving equation A.41 when $f(x, y)$ is nonlinear in y is to expand the derivative function $f(x, y)$ in a forward Taylor series about the known point x_n. Therefore,

$$f(x_n + h) = f(x_n) + h \frac{df}{dx}(x_n) + \cdots \quad (A.42)$$

where

$$\frac{df}{dx}(x_n) = \left(\frac{\partial f}{\partial x} + \frac{\partial f}{\partial y} \frac{dy}{dx} \right)_{x_n} = f_x(x_n) + f_y(x_n) \frac{y(x_n + h) - y(x_n)}{h} \quad (A.43)$$

Substituting equation A.43 into equation A.42 yields

$$f(x_n + h) = f(x_n) + hf_x(x_n) + [y(x_n + h) - y(x_n)]f_y(x_n) \quad (A.44)$$

Substituting equation A.44 into equation A.41, we obtain

$$y(x_n + h) = y(x_n) + h\{f(x_n) + hf_x(x_n) + [y(x_n + h) - y(x_n)]f_y(x_n)\} \quad (A.45)$$

The right-hand side of equation A.45 is linear in $y(x_n + h)$. Consequently, rearranging equation A.45 yields an equation for the direct determination of $y(x_n + h)$. The determination of the partial derivatives f_x and f_y may be complicated if $f(x, y)$ is a complicated expression.

Example A.9. Work Example A.5 employing the implicit Euler method.

Solution

The algorithm for the implicit Euler method is equation A.41. Therefore,

$$y_{n+1} = y_n + hy'_{n+1} \tag{a}$$

For the present problem,

$$y'_{n+1} = x_{n+1} + y_{n+1} \tag{b}$$

Since $y'(x,y)$ is linear in y, equation (b) may be substituted into equation (a), which may be solved directly for y_{n+1} to obtain

$$y_{n+1} = y_n + h(x_{n+1} + y_{n+1})$$

$$y_{n+1} = \frac{y_n + hx_{n+1}}{1 - h} \tag{c}$$

For $h = 0.1$, equation (c) becomes

$$y_{n+1} = \frac{y_n + (0.1)x_{n+1}}{0.9} \tag{d}$$

For the first step,

$$y_1 = \frac{y_0 + (0.1)x_1}{0.9} = \frac{1.0 + (0.1)(0.1)}{0.9} = 1.12222222$$

Repeating the above steps until $x = 0.5$, we obtain Table A.8. The magnitude of the error is approximately equal to that of the Euler method (see Example A.5), although the sign is reversed. The opposite sign of the error between an explicit and an implicit method is an inherent feature of those methods when they are applied to a differential equation having a solution that is monotonic in character. The solution of the present differential equation is a monotonically increasing function [see equation (c) in Example A.5]. For such a function, explicit methods generally underpredict the solution and implicit methods generally overpredict the solution. For monotonically decreasing functions, those trends are reversed.

Table A.8   Numerical Solution Obtained by the Implicit Euler Method

n	x_n	y_n	Error
0	0	1.0	
1	0.1	1.12222222	0.01188038
2	0.2	1.26913580	0.02633028
3	0.3	1.44348422	0.04376660
4	0.4	1.64831580	0.06466640
5	0.5	1.88701756	0.08957502

A–7 SUMMARY

A brief introduction to numerical analysis is presented in this Appendix. Table A.9 summarizes the numerical algorithms discussed.

Table A.9 Summary of Numerical Algorithms

Interpolating polynomial.

$$y = a + bx + cx^2 + \cdots \tag{A.1}$$

Cramer's rule. Given $Ax = b$,

$$x_j = \frac{\det(A^j)}{\det(A)} \qquad (j = 1, \ldots, n) \tag{A.9}$$

The Newton-Raphson method. Given $f(x) = 0$,

$$x_{i+1} = x_i - \frac{f(x_i)}{f'(x_i)} \tag{A.15}$$

The secant method. Given $f(x) = C$,

$$x_{i+1} = x_i + \frac{C - f(x_i)}{m_i} \tag{A.16}$$

$$m_i = \frac{f(x_i) - f(x_{i-1})}{x_i - x_{i-1}} \tag{A.17}$$

The trapezoidal rule. Given $I = \int_a^b f(x)\,dx$,

$$I = \frac{h}{2} \sum_{i=1}^{n+1} C_i f(x_i) \qquad (C_i = 1, 2, \ldots, 2, 1) \tag{A.21}$$

Simpson's rule. Given $I = \int_a^b f(x)\,dx$,

$$I = \frac{h}{3} \sum_{i=1}^{n+1} C_i f(x_i) \qquad (C_i = 1, 4, 2, \ldots, 2, 4, 1) \tag{A.22}$$

Taylor's method. Given $dy/dx = y'(x,y) = f(x,y)$, where $y(0) = y_o$,

$$y_{n+1} = y_n + hy'_n + \frac{h^2}{2!} y''_n + \cdots + \frac{h^n}{n!} y_n{}^n \tag{A.28}$$

Euler's method.

$$y_{n+1} = y_n + hy'_n \tag{A.29}$$

Fourth-order Runge-Kutta method.

$$y_{n+1} = y_n + \frac{h}{6}(m_1 + 2m_2 + 2m_3 + m_4) \tag{A.33}$$

$$m_1 = f(x_n, y_n) \qquad m_2 = f\left(x_n + \frac{h}{2}, y_n + \frac{h}{2} m_1\right) \tag{A.34}$$

$$m_3 = f\left(x_n + \frac{h}{2}, y_n + \frac{h}{2} m_2\right) \qquad m_4 = f(x_n + h, y_n + hm_3) \tag{A.34}$$

The modified Euler predictor-corrector method.

$$y_{n+1}^p = y_n + hf_n \tag{A.35}$$

$$y_{n+1}^c = y_n + \frac{h}{2}(f_n + f_{n+1}) \tag{A.38}$$

$$y_{n+1}^c = y_n + \frac{h}{2}(f_n + f_{n+1}^c) \qquad \text{(with iteration)}$$

The first-order implicit Euler method.

$$y_{n+1} = y_n + hf_{n+1} \tag{A.41}$$

$$y_{n+1} = y_n + h\{f(x_n) + hf_x(x_n) + [y_{n+1} - y_n]f_y(x_n)\} \tag{A.45}$$

REFERENCES

1. S. D. Conte and C. de Boor, *Elementary Numerical Analysis*, Second Edition, McGraw-Hill, New York, 1972.
2. B. Carnahan, H. A. Luther, and J. O. Wilkes, *Applied Numerical Methods*, Wiley, New York, 1969.
3. R. W. Hamming, *Numerical Methods for Scientists and Engineers*, McGraw-Hill, New York, 1962.
4. J. B. Scarborough, *Numerical Mathematical Analysis*, The Johns Hopkins Press, Baltimore, Md., 1958.
5. F. B. Hildebrand, *Introduction to Numerical Analysis*, McGraw-Hill, New York, 1956.
6. P. K. Henrici, *Elements of Numerical Analysis*, Wiley, New York, 1964.
7. D. McCracken and W. S. Dorn, *Numerical Methods and Fortran Programming*, Wiley, New York, 1964.
8. W. E. Milne, *Numerical Solution of Differential Equations*, Wiley, New York, 1953.
9. P. K. Henrici, *Discrete Variable Methods for Ordinary Differential Equations*, Wiley, New York, 1962.
10. R. D. Richtmyer and K. W. Morton, *Difference Methods for Initial-Value Problems*, Second Edition, Interscience, New York, 1967.
11. L. Fox, Editor, *Numerical Solution of Ordinary and Partial Differential Equations*, Addison-Wesley, Reading, Mass., 1962.
12. G. E. Forsythe and W. R. Wasow, *Finite-Difference Methods for Partial Differential Equations*, Wiley, New York, 1960.
13. D. Greenspan, *Discrete Numerical Methods in Physics and Engineering*, Academic Press, New York, 1974.

B
the method of characteristics in two independent variables

In many engineering problems the governing equations are a system of quasi-linear nonhomogeneous partial differential equations of the first order for functions of two independent variables. A quasi-linear partial differential equation of the first order is defined as one that is nonlinear in the dependent variables, but linear in the first partial derivatives of the dependent variables. Such a system of n equations may be written as*

$$L_i = a_{ij}\frac{\partial u^j}{\partial x} + b_{ij}\frac{\partial u^j}{\partial y} + c_i = 0 \qquad (i, j = 1, \ldots, n) \tag{B.1}$$

where the superscript j identifies a particular dependent variable, and the coefficients a_{ij}, b_{ij}, and c_i depend on x, y, $u^1, \ldots, u^n$. When expanded, this system of equations is

$$
\begin{aligned}
L_1 &= a_{11}u_x^1 + a_{12}u_x^2 + \cdots + a_{1n}u_x^n + b_{11}u_y^1 + b_{12}u_y^2 + \cdots + b_{1n}u_y^n + c_1 = 0 \\
L_2 &= a_{21}u_x^1 + a_{22}u_x^2 + \cdots + a_{2n}u_x^n + b_{21}u_y^1 + b_{22}u_y^2 + \cdots + b_{2n}u_y^n + c_2 = 0 \\
&\;\;\vdots \qquad\qquad\qquad\qquad\qquad\qquad\qquad\qquad\qquad\qquad\qquad\qquad \\
L_n &= a_{n1}u_x^1 + a_{n2}u_x^2 + \cdots + a_{nn}u_x^n + b_{n1}u_y^1 + b_{n2}u_y^2 + \cdots + b_{nn}u_y^n + c_n = 0
\end{aligned} \tag{B.2}
$$

When such systems are hyperbolic, the desired solution may be obtained by the method of characteristics. Chapter 12 presents in detail the development of the theory for a single partial differential equation [Section 12–3(a)], and for two partial differential equations [Section 12–3(b)]. Those developments are not repeated in the subject Appendix.

By analogy to the case of two equations discussed in Section 12–3(b), the method of characteristics may be extended to a system of n differential equations. The governing equations for such a system are equations B.1. Equations B.1 may be combined linearly to form the following sum.

$$L = \sigma_i L_i = \sigma_1 L_1 + \sigma_2 L_2 + \cdots + \sigma_n L_n = 0 \tag{B.3}$$

---

* In accordance with accepted convention, when an index is repeated, summation is carried out with respect to that index.

Putting equation B.3 into the form of equation 12.27 results in the *compatibility equation* pertinent to the system.

$$(a_{ij}\sigma_i)\,du^j + c_i\sigma_i\,dx = 0 \qquad (i, j = 1, \ldots, n) \tag{B.4}$$

Equation B.4 applies along the *characteristic curves* specified by

$$a_{ij}\lambda = b_{ij}\sigma_i \qquad (i, j = 1, \ldots, n) \tag{B.5}$$

Solving equation B.5 for σ_i yields

$$\sigma_i(a_{ij}\lambda - b_{ij}) = 0 \qquad (i, j = 1, \ldots, n) \tag{B.6}$$

For equation B.6 to have a solution other than the trivial solution $\sigma_i = 0$, the determinant of the coefficients of σ_i must vanish. Therefore,

$$|(a_{ij}\lambda - b_{ij})| = 0 \qquad (i, j = 1, \ldots, n) \tag{B.7}$$

The expanded determinant results in an algebraic equation of the nth degree for $\lambda = dy/dx$, giving n roots λ_j $(j = 1, \ldots, n)$, which determine n characteristic directions. If all n roots are distinct and real, the system is totally hyperbolic. In that case there are n families of characteristic curves defined by the ordinary differential equations

$$\frac{dy}{dx} = \lambda_j \qquad (j = 1, \ldots, n) \tag{B.8}$$

Once the λ_j are determined from equation B.7, the σ_i may be determined from equation B.5. Those results may be substituted into the general compatibility equation, equation B.4, to determine the compatibility equations for the system. Equations B.4 and B.8 may replace the original system of partial differential equations, equations B.1.

The initial-value problem may be now formulated for the above system of hyperbolic partial differential equations. Assume a curve Γ_o is given in the xy plane, and that continuous values of u^j are prescribed along Γ_o, as illustrated in Fig. B.1. The problem is to determine, in the neighborhood of Γ_o, a solution u^j of the system of governing equations that has the prescribed initial values along Γ_o. By replacing the original system of governing equations by the characteristic system, the problem is reduced to determining the solution of a set of total differential equations along the characteristic curves.

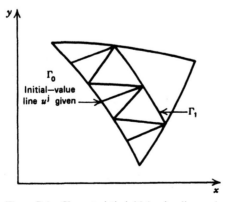

Figure B.1 Characteristic initial-value line and solution network.

In general, those equations are nonlinear, and each is coupled with the others. For that reason, an iterative solution is required. The compatibility equations, each of which is valid along one of the characteristic curves, may be written in finite difference form, as can the equations of the characteristic curves. By moving along the characteristic curves, the initial values of u^j along Γ_o may be extended into the domain enclosed by the outermost characteristic curves passing through the initial-data curve Γ_o. By continuing in small steps along Γ_o, a new curve Γ_1 may be obtained with all of the values of u^j determined along that curve, as illustrated in Fig. B.1.

The foregoing considerations result in the concepts of *domain of dependence* and *range of influence*. Figure B.2 illustrates the domain of dependence of a point P, which is the region in the xy plane bounded by the outermost characteristic curves passing through the point P on the initial-value line Γ_o, and is the region wherein the solution of the initial-value problem may be determined. Figure B.3 illustrates the range of influence of the point Q on the initial-value line Γ_o, which is the region in the xy plane containing all of the points that are influenced by the initial data at point Q. The range of influence comprises all of the points having a domain of dependence containing the point Q. Therefore, it is the region between the two outermost characteristic curves passing through point Q.

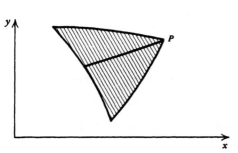

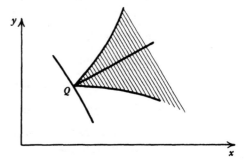

Figure B.2 Domain of dependence of point P.   **Figure B.3** Range of influence of point Q.

For a solution to be possible, the initial-value line cannot be characteristic at any place unless initial data are given along two intersecting characteristic curves. Several types of domains having different types of initial-value lines may be analyzed. The present discussion is concerned, however, only with domains in which the initial-value line is nowhere characteristic.

By applying the method of characteristics as presented in the foregoing discussion, it is possible to solve many complicated systems of partial differential equations, provided that the system is quasi-linear and hyperbolic. The latter conditions are frequently encountered in fluid flow problems.

C
tables

Tables C.1 to C.15 are presented in this Appendix. An index to the tables is presented below.

In the computer generated tables presented in this Appendix, the following convention is employed for specifying the power of 10 associated with a table entry.

$$.XXXXX \pm N = .XXXXX \cdot 10^{\pm N}$$

For example,

$$.12924 + 1 = 0.12924 \cdot 10^{+1} = 1.2924$$
$$.77234 - 1 = 0.77234 \cdot 10^{-1} = 0.077234$$

Table C.1 Dimensional Formulas and Units for Physical Quantities of Interest in Gas Dynamics

| Quantity | Dimensional Formula | | System of Units | | | |
	(M,L,T)	(F,L,T)	EE	EA	EG	SI
Angular distance			rad	rad	rad	rad
Angular velocity	$1/T$	$1/T$	rad/sec	rad/sec	rad/sec	rad/s
Angular acceleration	$1/T^2$	$1/T^2$	rad/sec^2	rad/sec^2	rad/sec^2	rad/s^2
Area	L^2	L^2	ft^2	ft^2	ft^2	m^2
Density	M/L^3	FT^2/L^4	lbm/ft^3	lbm/ft^3	slug/ft^3	kg/m^3
Dynamic viscosity	M/LT	FT/L^2	lbf-sec/ft^2	pdl-sec/ft^2	lbf-sec/ft^2	N-s/m^2
Energy	ML^2/T^2	FL	ft-lbf	ft-pdl	ft-lbf	J
Enthalpy	L^2/T^2	FL/M	ft-lbf/lbm	ft-pdl/lbm	ft-lbf/slug	J/kg
Force	ML/T^2	F	lbf	pdl	lbf	N
Heat	ML^2/T^2	FL	ft-lbf	ft-pdl	ft-lbf	J
Impulse	ML/T	FT	lbf-sec	pdl-sec	lbf-sec	N-s
Kinematic viscosity	L^2/T	L^2/T	ft^2/sec	ft^2/sec	ft^2/sec	m^2/s
Linear distance	L	L	ft	ft	ft	m
Linear velocity	L/T	L/T	ft/sec	ft/sec	ft/sec	m/s
Linear acceleration	L/T^2	L/T^2	ft/sec^2	ft/sec^2	ft/sec^2	m/s^2
Mass	M	FT^2/L	lbm	lbm	slug	kg
Moment	ML^2/T^2	FL	ft-lbf	ft-pdl	ft-lbf	m-N
Momentum	ML/T	FT	lbm-ft/sec	lbm-ft/sec	slug-ft/sec	kg-m/s
Pressure	M/LT^2	F/L^2	lbf/ft^2	pdl/ft^2	lbf/ft^2	N/m^2
Power	ML^2/T^3	FL/T	ft-lbf/sec	ft-pdl/sec	ft-lbf/sec	J/s
Time	T	T	sec	sec	sec	s
Torque	ML^2/T^2	FL	ft-lbf	ft-pdl	ft-lbf	m-N
Weight	ML/T^2	F	lbf	pdl	lbf	N
Work	ML^2/T^2	FL	ft-lbf	ft-pdl	ft-lbf	J

Table C.2 Universal Physical Constants

c = velocity of light in vacuo = $2.997295 \cdot 10^8$ m/s (186,282.4 mi/sec)

g_o = standard acceleration of gravity = 9.80665 m/s^2 (32.174 ft/sec^2)

h = Planck's constant = $6.626196 \cdot 10^{-34}$ J-s

k = Boltzmann's constant = $1.38054 \cdot 10^{-23}$ J/K

N_A = Avogadro's number = $6.02252 \cdot 10^{23}$ particles/mol

$\bar{R}$ = universal gas constant = 8.31434 J/mol-K (1545.3 ft-lbf/lbmole-R)

Table C.3   Thermodynamic Constants for Common Gases

Gas	Chemical Symbol	Molecular Weight	Specific Heat Ratio γ at 298.15 K (77 F)	EE units		SI units	
				Gas Constant $R = \bar{R}/\bar{m}$ (ft-lbf/ lbm-R)	Specific Heat c_p at 77 F (Btu/ lbm-R)	Gas Constant $R = \bar{R}/\bar{m}$ (J/kg-K)	Specific Heat c_p at 298.15 K (J/kg-K)
Air		28.964	1.400	53.352	0.2398	287.06	1004.0
Argon	A	39.944	1.658	38.687	0.1253	208.15	524.61
Carbon monoxide	CO	28.010	1.398	55.170	0.249	296.83	1042.5
Carbon dioxide	CO_2	44.010	1.288	35.112	0.202	188.92	845.73
Helium	He	4.000	1.659	386.33	1.250	2078.2	5233.5
Hydrogen	H_2	2.016	1.405	766.52	3.419	4124.2	14315.0
Methane	CH_4	16.043	1.304	96.322	0.531	518.25	2223.2
Nitrogen	N_2	28.013	1.400	55.164	0.248	296.80	1038.3
Oxygen	O_2	32.000	1.395	48.291	0.219	259.82	916.90
Water	H_2O	18.016	1.329	85.774	0.445	461.50	1863.1

Table C.4   Thermodynamic Properties of Air in SI Units

t	h	u	ϕ	p_r	v_r	c_p	c_v	γ	a
K	kJ/kg	kJ/kg	kJ/kg-K			kJ/kg-K	kJ/kg-K		m/s
50	48.554	34.202	4.8839	.18039-2	.79556+7	1.0317	.7447	1.3854	141.01
55	53.709	37.922	4.9822	.25404-2	.62142+7	1.0303	.7432	1.3862	147.93
60	58.857	41.635	5.0718	.34709-2	.49617+7	1.0289	.7418	1.3869	154.55
65	63.997	45.341	5.1541	.46235-2	.40352+7	1.0275	.7405	1.3876	160.90
70	69.132	49.040	5.2302	.60272-2	.33336+7	1.0262	.7392	1.3883	167.01
75	74.259	52.732	5.3009	.77121-2	.27913+7	1.0249	.7379	1.3890	172.92
80	79.381	56.419	5.3670	.97095-2	.23649+7	1.0237	.7367	1.3896	178.63
85	84.496	60.099	5.4291	.12052-1	.20244+7	1.0225	.7355	1.3903	184.17
90	89.606	63.773	5.4875	.14772-1	.17488+7	1.0213	.7343	1.3909	189.55
95	94.710	67.442	5.5427	.17903-1	.15230+7	1.0202	.7332	1.3915	194.79
100	99.808	71.106	5.5950	.21482-1	.13361+7	1.0192	.7321	1.3920	199.89
105	104.902	74.764	5.6447	.25543-1	.11799+7	1.0181	.7311	1.3926	204.87
110	109.990	78.417	5.6920	.30123-1	.10481+7	1.0171	.7301	1.3931	209.73
115	115.073	82.065	5.7372	.35260-1	.93613+6	1.0162	.7292	1.3936	214.48
120	120.152	85.708	5.7804	.40991-1	.84026+6	1.0153	.7282	1.3941	219.13
125	125.226	89.347	5.8218	.47356-1	.75763+6	1.0144	.7274	1.3946	223.69
130	130.296	92.982	5.8616	.54394-1	.68599+6	1.0136	.7265	1.3951	228.16
135	135.361	96.613	5.8999	.62145-1	.62352+6	1.0127	.7257	1.3955	232.54
140	140.423	100.239	5.9367	.70650-1	.56878+6	1.0120	.7250	1.3959	236.84
145	145.481	103.862	5.9722	.79951-1	.52056+6	1.0113	.7242	1.3963	241.07
150	150.536	107.482	6.0064	.90090-1	.47790+6	1.0106	.7235	1.3967	245.22
155	155.587	111.098	6.0396	.10111+0	.44000+6	1.0099	.7229	1.3971	249.31
160	160.635	114.710	6.0716	.11306+0	.40621+6	1.0093	.7222	1.3974	253.33
165	165.680	118.320	6.1027	.12597+0	.37595+6	1.0087	.7217	1.3977	257.29
170	170.722	121.927	6.1328	.13990+0	.34878+6	1.0081	.7211	1.3980	261.18
175	175.761	125.531	6.1620	.15489+0	.32429+6	1.0076	.7206	1.3983	265.02
180	180.798	129.133	6.1904	.17099+0	.30215+6	1.0071	.7201	1.3986	268.81
185	185.832	132.732	6.2180	.18824+0	.28209+6	1.0067	.7196	1.3989	272.54
190	190.864	136.329	6.2448	.20669+0	.26385+6	1.0062	.7192	1.3991	276.22
195	195.895	139.924	6.2709	.22639+0	.24723+6	1.0058	.7188	1.3993	279.86
200	200.923	143.517	6.2964	.24739+0	.23204+6	1.0055	.7185	1.3995	283.44
205	205.949	147.109	6.3212	.26974+0	.21814+6	1.0052	.7181	1.3997	286.98
210	210.975	150.699	6.3454	.29349+0	.20538+6	1.0049	.7178	1.3998	290.48
215	215.998	154.287	6.3691	.31869+0	.19364+6	1.0046	.7176	1.4000	293.93
220	221.021	157.875	6.3922	.34539+0	.18283+6	1.0044	.7173	1.4001	297.34
225	226.042	161.461	6.4147	.37364+0	.17284+6	1.0042	.7171	1.4002	300.71
230	231.062	165.046	6.4368	.40350+0	.16361+6	1.0040	.7170	1.4003	304.05
235	236.082	168.630	6.4584	.43502+0	.15505+6	1.0039	.7168	1.4004	307.34
240	241.101	172.214	6.4795	.46826+0	.14711+6	1.0037	.7167	1.4005	310.60
245	246.119	175.798	6.5002	.50327+0	.13973+6	1.0037	.7166	1.4005	313.83
250	251.137	179.381	6.5205	.54011+0	.13286+6	1.0036	.7166	1.4006	317.02
255	256.155	182.963	6.5404	.57883+0	.12645+6	1.0036	.7165	1.4006	320.17

Table C.4   Thermodynamic Properties of Air in SI Units (*Continued*)

t	h	u	φ	p_r	v_r	c_p	c_v	γ	a
K	kJ/kg	kJ/kg	kJ/kg-K			kJ/kg-K	kJ/kg-K		m/s
260	261.173	186.546	6.5599	.61949+0	.12046+6	1.0036	.7165	1.4006	323.30
265	266.191	190.129	6.5790	.66216+0	.11487+6	1.0036	.7166	1.4006	326.39
270	271.209	193.712	6.5977	.70688+0	.10963+6	1.0036	.7166	1.4005	329.45
275	276.228	197.295	6.6161	.75372+0	.10472+6	1.0037	.7167	1.4005	332.48
280	281.246	200.879	6.6342	.80275+0	.10012+6	1.0038	.7168	1.4004	335.48
285	286.266	204.463	6.6520	.85401+0	.95787+5	1.0040	.7169	1.4004	338.46
290	291.286	208.048	6.6695	.90758+0	.91714+5	1.0041	.7171	1.4003	341.40
295	296.307	211.634	6.6866	.96352+0	.87879+5	1.0043	.7173	1.4002	344.32
300	301.329	215.221	6.7035	.10219+1	.84264+5	1.0045	.7175	1.4000	347.21
305	306.353	218.809	6.7201	.10827+1	.80853+5	1.0048	.7177	1.3999	350.08
310	311.377	222.398	6.7365	.11462+1	.77631+5	1.0050	.7180	1.3998	352.92
315	316.403	225.989	6.7525	.12122+1	.74585+5	1.0053	.7183	1.3996	355.73
320	321.430	229.581	6.7684	.12810+1	.71702+5	1.0056	.7186	1.3994	358.52
325	326.459	233.175	6.7840	.13525+1	.68971+5	1.0059	.7189	1.3993	361.29
330	331.489	236.770	6.7995	.14269+1	.66383+5	1.0063	.7193	1.3991	364.03
335	336.522	240.367	6.8145	.15041+1	.63928+5	1.0067	.7196	1.3989	366.75
340	341.556	243.966	6.8294	.15843+1	.61596+5	1.0071	.7200	1.3986	369.45
345	346.592	247.568	6.8441	.16676+1	.59380+5	1.0075	.7204	1.3984	372.12
350	351.631	251.171	6.8586	.17540+1	.57274+5	1.0079	.7209	1.3982	374.78
355	356.671	254.776	6.8729	.18436+1	.55268+5	1.0084	.7213	1.3979	377.41
360	361.714	258.384	6.8870	.19365+1	.53359+5	1.0089	.7218	1.3976	380.02
365	366.760	261.995	6.9009	.20327+1	.51539+5	1.0094	.7223	1.3974	382.62
370	371.808	265.608	6.9146	.21324+1	.49804+5	1.0099	.7229	1.3971	385.19
375	376.859	269.223	6.9282	.22355+1	.48148+5	1.0104	.7234	1.3968	387.74
380	381.912	272.842	6.9416	.23423+1	.46566+5	1.0110	.7240	1.3965	390.27
385	386.969	276.463	6.9548	.24527+1	.45055+5	1.0116	.7246	1.3961	392.79
390	392.028	280.087	6.9679	.25668+1	.43611+5	1.0122	.7252	1.3958	395.28
395	397.091	283.715	6.9808	.26848+1	.42229+5	1.0128	.7258	1.3955	397.76
400	402.156	287.345	6.9935	.28067+1	.40906+5	1.0134	.7264	1.3951	400.22
405	407.225	290.979	7.0061	.29326+1	.39640+5	1.0141	.7271	1.3948	402.66
410	412.297	294.616	7.0186	.30625+1	.38426+5	1.0148	.7277	1.3944	405.09
415	417.373	298.256	7.0309	.31967+1	.37263+5	1.0155	.7284	1.3940	407.49
420	422.452	301.900	7.0430	.33351+1	.36147+5	1.0162	.7292	1.3936	409.89
425	427.535	305.548	7.0551	.34778+1	.35075+5	1.0169	.7299	1.3933	412.26
430	432.621	309.199	7.0670	.36250+1	.34047+5	1.0177	.7306	1.3929	414.62
435	437.711	312.854	7.0787	.37768+1	.33059+5	1.0184	.7314	1.3924	416.96
440	442.805	316.513	7.0904	.39331+1	.32110+5	1.0192	.7322	1.3920	419.29
445	447.903	320.176	7.1019	.40942+1	.31197+5	1.0200	.7330	1.3916	421.60
450	453.005	323.843	7.1133	.42601+1	.30319+5	1.0208	.7338	1.3912	423.90
455	458.111	327.513	7.1246	.44309+1	.29474+5	1.0216	.7346	1.3907	426.18
460	463.221	331.188	7.1357	.46067+1	.28661+5	1.0224	.7354	1.3903	428.44
465	468.335	334.868	7.1468	.47877+1	.27877+5	1.0233	.7363	1.3898	430.70
470	473.454	338.551	7.1577	.49738+1	.27122+5	1.0242	.7371	1.3894	432.93
475	478.577	342.239	7.1686	.51653+1	.26395+5	1.0250	.7380	1.3889	435.16
480	483.704	345.931	7.1793	.53622+1	.25693+5	1.0259	.7389	1.3885	437.37
485	488.836	349.628	7.1900	.55647+1	.25017+5	1.0268	.7398	1.3880	439.57
490	493.973	353.329	7.2005	.57727+1	.24363+5	1.0278	.7407	1.3875	441.75
495	499.114	357.035	7.2109	.59865+1	.23733+5	1.0287	.7417	1.3870	443.92
500	504.260	360.746	7.2213	.62062+1	.23124+5	1.0296	.7426	1.3865	446.08
505	509.410	364.461	7.2315	.64318+1	.22536+5	1.0306	.7436	1.3860	448.22
510	514.566	368.182	7.2417	.66636+1	.21968+5	1.0316	.7445	1.3855	450.35
515	519.726	371.907	7.2518	.69015+1	.21419+5	1.0325	.7455	1.3850	452.47
520	524.891	375.637	7.2617	.71457+1	.20887+5	1.0335	.7465	1.3845	454.58
525	530.061	379.372	7.2716	.73963+1	.20374+5	1.0345	.7475	1.3840	456.67
530	535.236	383.112	7.2814	.76535+1	.19877+5	1.0355	.7485	1.3835	458.76
535	540.417	386.857	7.2912	.79173+1	.19395+5	1.0366	.7495	1.3829	460.83
540	545.602	390.607	7.3008	.81880+1	.18930+5	1.0376	.7506	1.3824	462.89
545	550.792	394.362	7.3104	.84655+1	.18479+5	1.0386	.7516	1.3819	464.94
550	555.988	398.123	7.3199	.87501+1	.18042+5	1.0397	.7527	1.3814	466.98
555	561.189	401.889	7.3293	.90418+1	.17618+5	1.0407	.7537	1.3808	469.00
560	566.396	405.660	7.3386	.93408+1	.17208+5	1.0418	.7548	1.3803	471.02
565	571.607	409.437	7.3479	.96473+1	.16810+5	1.0429	.7559	1.3797	473.02
570	576.825	413.219	7.3571	.99613+1	.16424+5	1.0440	.7569	1.3792	475.02
575	582.047	417.006	7.3662	.10283+2	.16050+5	1.0451	.7580	1.3786	477.00
580	587.275	420.799	7.3753	.10612+2	.15687+5	1.0462	.7591	1.3781	478.98
585	592.509	424.598	7.3843	.10950+2	.15334+5	1.0473	.7603	1.3775	480.94
590	597.748	428.402	7.3932	.11295+2	.14992+5	1.0484	.7614	1.3770	482.90
595	602.993	432.211	7.4020	.11649+2	.14660+5	1.0495	.7625	1.3764	484.84
600	608.243	436.027	7.4108	.12011+2	.14338+5	1.0507	.7636	1.3759	486.77
605	613.499	439.848	7.4195	.12382+2	.14024+5	1.0518	.7648	1.3753	488.70
610	618.761	443.674	7.4282	.12761+2	.13720+5	1.0529	.7659	1.3748	490.61
615	624.029	447.507	7.4368	.13150+2	.13424+5	1.0541	.7671	1.3742	492.52
620	629.302	451.345	7.4453	.13547+2	.13137+5	1.0552	.7682	1.3736	494.42
625	634.581	455.189	7.4538	.13953+2	.12857+5	1.0564	.7694	1.3731	496.30
630	639.866	459.039	7.4622	.14368+2	.12585+5	1.0576	.7705	1.3725	498.18
635	645.157	462.894	7.4706	.14793+2	.12321+5	1.0587	.7717	1.3719	500.05
640	650.453	466.756	7.4789	.15228+2	.12063+5	1.0599	.7729	1.3714	501.92
645	655.756	470.623	7.4872	.15672+2	.11813+5	1.0611	.7741	1.3708	503.77
650	661.064	474.496	7.4954	.16126+2	.11569+5	1.0623	.7752	1.3702	505.61
655	666.378	478.375	7.5035	.16590+2	.11332+5	1.0635	.7764	1.3697	507.45
660	671.699	482.260	7.5116	.17065+2	.11101+5	1.0646	.7776	1.3691	509.28
665	677.025	486.151	7.5196	.17549+2	.10876+5	1.0658	.7788	1.3685	511.10

Table C.4 Thermodynamic Properties of Air in SI Units (*Continued*)

t	h	u	φ	p_r	v_r	c_p	c_v	γ	a
K	kJ/kg	kJ/kg	kJ/kg-K			kJ/kg-K	kJ/kg-K		m/s
670	682.357	490.048	7.5276	.18045+2	.10657+5	1.0670	.7800	1.3680	512.91
675	687.695	493.951	7.5356	.18551+2	.10444+5	1.0682	.7812	1.3674	514.71
680	693.039	497.860	7.5435	.19067+2	.10236+5	1.0694	.7824	1.3669	516.51
685	698.389	501.775	7.5513	.19595+2	.10034+5	1.0706	.7836	1.3663	518.30
690	703.746	505.697	7.5591	.20134+2	.98363+4	1.0718	.7848	1.3657	520.08
695	709.108	509.624	7.5668	.20685+2	.96439+4	1.0730	.7860	1.3652	521.85
700	714.476	513.557	7.5745	.21247+2	.94563+4	1.0743	.7872	1.3646	523.62
705	719.850	517.496	7.5822	.21821+2	.92733+4	1.0755	.7884	1.3640	525.38
710	725.231	521.441	7.5898	.22407+2	.90949+4	1.0767	.7897	1.3635	527.13
715	730.617	525.392	7.5973	.23005+2	.89209+4	1.0779	.7909	1.3629	528.87
720	736.010	529.350	7.6049	.23615+2	.87511+4	1.0791	.7921	1.3624	530.61
725	741.408	533.313	7.6123	.24238+2	.85854+4	1.0803	.7933	1.3618	532.34
730	746.813	537.283	7.6198	.24874+2	.84237+4	1.0815	.7945	1.3613	534.07
735	752.223	541.258	7.6271	.25522+2	.82660+4	1.0827	.7957	1.3607	535.78
740	757.640	545.240	7.6345	.26184+2	.81119+4	1.0840	.7969	1.3602	537.49
745	763.063	549.227	7.6418	.26858+2	.79616+4	1.0852	.7981	1.3596	539.20
750	768.492	553.221	7.6491	.27547+2	.78147+4	1.0864	.7994	1.3591	540.90
755	773.927	557.221	7.6563	.28249+2	.76714+4	1.0876	.8006	1.3585	542.59
760	779.368	561.227	7.6635	.28965+2	.75313+4	1.0888	.8018	1.3580	544.27
765	784.815	565.239	7.6706	.29694+2	.73945+4	1.0900	.8030	1.3575	545.95
770	790.268	569.257	7.6777	.30439+2	.72609+4	1.0912	.8042	1.3569	547.63
775	795.727	573.280	7.6848	.31197+2	.71303+4	1.0924	.8054	1.3564	549.29
780	801.192	577.310	7.6918	.31971+2	.70027+4	1.0936	.8066	1.3559	550.95
785	806.663	581.346	7.6988	.32759+2	.68780+4	1.0948	.8078	1.3553	552.61
790	812.140	585.388	7.7057	.33563+2	.67561+4	1.0960	.8090	1.3548	554.26
795	817.623	589.436	7.7127	.34382+2	.66369+4	1.0972	.8102	1.3543	555.90
800	823.112	593.490	7.7195	.35216+2	.65204+4	1.0984	.8114	1.3538	557.54
805	828.607	597.550	7.7264	.36066+2	.64065+4	1.0996	.8125	1.3532	559.18
810	834.108	601.615	7.7332	.36932+2	.62951+4	1.1008	.8137	1.3527	560.80
815	839.615	605.687	7.7400	.37815+2	.61861+4	1.1019	.8149	1.3522	562.42
820	845.127	609.764	7.7467	.38714+2	.60796+4	1.1031	.8161	1.3517	564.04
825	850.646	613.848	7.7534	.39629+2	.59753+4	1.1043	.8173	1.3512	565.65
830	856.170	617.937	7.7601	.40562+2	.58733+4	1.1054	.8184	1.3507	567.26
835	861.700	622.032	7.7668	.41512+2	.57735+4	1.1066	.8196	1.3502	568.86
840	867.236	626.133	7.7734	.42479+2	.56759+4	1.1078	.8207	1.3497	570.46
845	872.778	630.239	7.7799	.43463+2	.55803+4	1.1089	.8219	1.3492	572.05
850	878.325	634.352	7.7865	.44466+2	.54867+4	1.1101	.8230	1.3487	573.64
855	883.878	638.470	7.7930	.45487+2	.53952+4	1.1112	.8242	1.3483	575.22
860	889.437	642.593	7.7995	.46526+2	.53055+4	1.1123	.8253	1.3478	576.79
865	895.002	646.723	7.8059	.47583+2	.52178+4	1.1135	.8264	1.3473	578.37
870	900.572	650.858	7.8124	.48660+2	.51318+4	1.1146	.8276	1.3468	579.93
875	906.148	654.998	7.8188	.49755+2	.50477+4	1.1157	.8287	1.3464	581.50
880	911.729	659.145	7.8251	.50870+2	.49653+4	1.1168	.8298	1.3459	583.06
885	917.316	663.296	7.8314	.52005+2	.48846+4	1.1179	.8309	1.3454	584.61
890	922.908	667.454	7.8377	.53159+2	.48055+4	1.1190	.8320	1.3450	586.16
895	928.506	671.616	7.8440	.54333+2	.47280+4	1.1201	.8331	1.3445	587.71
900	934.109	675.784	7.8503	.55528+2	.46522+4	1.1212	.8342	1.3441	589.25
905	939.718	679.958	7.8565	.56743+2	.45778+4	1.1223	.8352	1.3437	590.78
910	945.332	684.137	7.8627	.57980+2	.45049+4	1.1233	.8363	1.3432	592.32
915	950.951	688.321	7.8688	.59237+2	.44335+4	1.1244	.8374	1.3428	593.85
920	956.576	692.510	7.8749	.60516+2	.43636+4	1.1254	.8384	1.3424	595.37
925	962.205	696.705	7.8811	.61816+2	.42950+4	1.1265	.8394	1.3419	596.89
930	967.840	700.904	7.8871	.63139+2	.42278+4	1.1275	.8405	1.3415	598.41
935	973.480	705.109	7.8932	.64483+2	.41619+4	1.1285	.8415	1.3411	599.92
940	979.125	709.319	7.8992	.65850+2	.40973+4	1.1295	.8425	1.3407	601.43
945	984.776	713.535	7.9052	.67240+2	.40339+4	1.1306	.8435	1.3403	602.94
950	990.431	717.755	7.9112	.68653+2	.39718+4	1.1315	.8445	1.3399	604.44
955	996.091	721.980	7.9171	.70089+2	.39109+4	1.1325	.8455	1.3395	605.94
960	1001.756	726.210	7.9230	.71549+2	.38512+4	1.1335	.8465	1.3391	607.44
965	1007.426	730.445	7.9289	.73033+2	.37926+4	1.1345	.8475	1.3387	608.93
970	1013.101	734.684	7.9348	.74540+2	.37351+4	1.1354	.8484	1.3383	610.42
975	1018.781	738.929	7.9406	.76073+2	.36787+4	1.1364	.8494	1.3379	611.90
980	1024.465	743.178	7.9464	.77630+2	.36235+4	1.1373	.8503	1.3376	613.38
985	1030.154	747.432	7.9522	.79211+2	.35692+4	1.1382	.8512	1.3372	614.86
990	1035.847	751.690	7.9580	.80819+2	.35160+4	1.1392	.8521	1.3368	616.34
995	1041.545	755.953	7.9637	.82452+2	.34638+4	1.1401	.8530	1.3365	617.81
1000	1047.248	760.220	7.9694	.84110+2	.34125+4	1.1410	.8539	1.3361	619.28
1010	1058.666	768.768	7.9808	.87504+2	.33130+4	1.1427	.8556	1.3355	622.21
1020	1070.101	777.333	7.9921	.91007+2	.32170+4	1.1444	.8573	1.3348	625.13
1030	1081.553	785.915	8.0032	.94620+2	.31245+4	1.1460	.8590	1.3341	628.03
1040	1093.022	794.513	8.0143	.98344+2	.30354+4	1.1477	.8607	1.3335	630.92
1050	1104.507	803.128	8.0253	.10218+3	.29494+4	1.1493	.8623	1.3329	633.80
1060	1116.008	811.759	8.0362	.10614+3	.28665+4	1.1509	.8639	1.3322	636.66
1070	1127.526	820.406	8.0470	.11021+3	.27866+4	1.1526	.8655	1.3316	639.50
1080	1139.059	829.069	8.0578	.11441+3	.27094+4	1.1542	.8671	1.3310	642.34
1090	1150.609	837.749	8.0684	.11873+3	.26350+4	1.1557	.8687	1.3304	645.16
1100	1162.174	846.444	8.0790	.12318+3	.25631+4	1.1573	.8703	1.3298	647.97
1110	1173.755	855.154	8.0894	.12777+3	.24936+4	1.1589	.8718	1.3292	650.76
1120	1185.351	863.880	8.0998	.13248+3	.24266+4	1.1604	.8734	1.3286	653.54
1130	1196.963	872.622	8.1102	.13733+3	.23617+4	1.1619	.8749	1.3281	656.31
1140	1208.590	881.378	8.1204	.14232+3	.22991+4	1.1634	.8764	1.3275	659.07
1150	1220.232	890.150	8.1306	.14745+3	.22386+4	1.1649	.8779	1.3269	661.82
1160	1231.889	898.937	8.1407	.15273+3	.21800+4	1.1664	.8794	1.3264	664.55
1170	1243.560	907.738	8.1507	.15815+3	.21234+4	1.1679	.8809	1.3258	667.27
1180	1255.247	916.554	8.1606	.16373+3	.20686+4	1.1694	.8823	1.3253	669.98

Table C.4 Thermodynamic Properties of Air in SI Units (*Continued*)

t	h	u	φ	p_r	v_r	c_p	c_v	γ	a
K	kJ/kg	kJ/kg	kJ/kg-K			kJ/kg-K	kJ/kg-K		m/s
1190	1266.948	925.585	8.1705	.16946+3	.20156+4	1.1708	.8838	1.3248	672.68
1200	1278.663	934.230	8.1803	.17535+3	.19643+4	1.1722	.8852	1.3242	675.36
1210	1290.392	943.089	8.1900	.18140+3	.19146+4	1.1737	.8866	1.3237	678.04
1220	1302.136	951.962	8.1997	.18761+3	.18665+4	1.1751	.8680	1.3232	680.70
1230	1313.894	960.849	8.2093	.19399+3	.18199+4	1.1765	.8894	1.3227	683.36
1240	1325.665	969.751	8.2188	.20054+3	.17748+4	1.1778	.8908	1.3222	686.00
1250	1337.450	978.666	8.2283	.20727+3	.17310+4	1.1792	.8922	1.3217	688.63
1260	1349.249	987.594	8.2377	.21417+3	.16887+4	1.1806	.8935	1.3212	691.25
1270	1361.061	996.536	8.2470	.22125+3	.16476+4	1.1819	.8949	1.3207	693.86
1280	1372.887	1005.491	8.2563	.22851+3	.16078+4	1.1832	.8962	1.3203	696.46
1290	1384.726	1014.460	8.2655	.23597+3	.15691+4	1.1845	.8975	1.3198	699.05
1300	1396.578	1023.442	8.2747	.24361+3	.15317+4	1.1858	.8988	1.3193	701.64
1310	1408.443	1032.436	8.2838	.25145+3	.14953+4	1.1871	.9001	1.3189	704.21
1320	1420.320	1041.444	8.2928	.25949+3	.14601+4	1.1884	.9014	1.3184	706.77
1330	1432.211	1050.464	8.3018	.26773+3	.14258+4	1.1897	.9027	1.3180	709.32
1340	1444.114	1059.497	8.3107	.27618+3	.13926+4	1.1909	.9039	1.3175	711.86
1350	1456.030	1068.543	8.3196	.28484+3	.13604+4	1.1922	.9052	1.3171	714.39
1360	1467.958	1077.600	8.3284	.29371+3	.13291+4	1.1934	.9064	1.3167	716.92
1370	1479.898	1086.670	8.3371	.30280+3	.12986+4	1.1946	.9076	1.3162	719.43
1380	1491.851	1095.753	8.3458	.31211+3	.12691+4	1.1959	.9088	1.3158	721.94
1390	1503.815	1104.847	8.3544	.32165+3	.12404+4	1.1971	.9100	1.3154	724.43
1400	1515.792	1113.953	8.3630	.33141+3	.12125+4	1.1982	.9112	1.3150	726.92
1410	1527.780	1123.071	8.3716	.34141+3	.11854+4	1.1994	.9124	1.3146	729.40
1420	1539.780	1132.201	8.3800	.35165+3	.11590+4	1.2006	.9135	1.3142	731.87
1430	1551.792	1141.342	8.3885	.36213+3	.11334+4	1.2017	.9147	1.3138	734.33
1440	1563.815	1150.495	8.3968	.37286+3	.11085+4	1.2029	.9158	1.3134	736.79
1450	1575.849	1159.659	8.4052	.38383+3	.10843+4	1.2040	.9170	1.3130	739.23
1460	1587.895	1168.834	8.4135	.39507+3	.10607+4	1.2051	.9181	1.3126	741.67
1470	1599.951	1178.021	8.4217	.40656+3	.10378+4	1.2062	.9192	1.3123	744.10
1480	1612.019	1187.218	8.4299	.41831+3	.10155+4	1.2073	.9203	1.3119	746.52
1490	1624.098	1196.427	8.4380	.43034+3	.99380+3	1.2084	.9214	1.3115	748.93
1500	1636.188	1205.646	8.4461	.44263+3	.97268+3	1.2095	.9225	1.3112	751.34
1510	1648.288	1214.876	8.4541	.45521+3	.95212+3	1.2106	.9235	1.3108	753.73
1520	1660.399	1224.117	8.4621	.46807+3	.93210+3	1.2116	.9246	1.3104	756.12
1530	1672.520	1233.368	8.4701	.48121+3	.91260+3	1.2127	.9256	1.3101	758.50
1540	1684.652	1242.630	8.4780	.49464+3	.89362+3	1.2137	.9267	1.3097	760.88
1550	1696.795	1251.902	8.4858	.50837+3	.87513+3	1.2147	.9277	1.3094	763.24
1560	1708.947	1261.184	8.4936	.52241+3	.85712+3	1.2158	.9287	1.3091	765.60
1570	1721.110	1270.476	8.5014	.53674+3	.83957+3	1.2168	.9297	1.3087	767.95
1580	1733.282	1279.779	8.5091	.55139+3	.82247+3	1.2178	.9307	1.3084	770.30
1590	1745.465	1289.091	8.5168	.56635+3	.80580+3	1.2187	.9317	1.3081	772.64
1600	1757.657	1298.413	8.5245	.58164+3	.78956+3	1.2197	.9327	1.3077	774.97
1610	1769.859	1307.745	8.5321	.59726+3	.77373+3	1.2207	.9337	1.3074	777.29
1620	1782.071	1317.086	8.5396	.61320+3	.75829+3	1.2217	.9346	1.3071	779.60
1630	1794.293	1326.437	8.5472	.62948+3	.74324+3	1.2226	.9356	1.3068	781.91
1640	1806.523	1335.798	8.5546	.64610+3	.72856+3	1.2235	.9365	1.3065	784.22
1650	1818.763	1345.168	8.5621	.66307+3	.71425+3	1.2245	.9375	1.3062	786.51
1660	1831.013	1354.547	8.5695	.68039+3	.70028+3	1.2254	.9384	1.3059	788.80
1670	1843.271	1363.935	8.5768	.69807+3	.68666+3	1.2263	.9393	1.3056	791.08
1680	1855.539	1373.333	8.5842	.71611+3	.67337+3	1.2272	.9402	1.3053	793.36
1690	1867.816	1382.739	8.5915	.73452+3	.66040+3	1.2281	.9411	1.3050	795.63
1700	1880.102	1392.155	8.5987	.75330+3	.64774+3	1.2290	.9420	1.3047	797.89
1710	1892.396	1401.579	8.6059	.77247+3	.63539+3	1.2299	.9429	1.3044	800.15
1720	1904.699	1411.012	8.6131	.79202+3	.62333+3	1.2308	.9437	1.3041	802.40
1730	1917.011	1420.453	8.6202	.81196+3	.61155+3	1.2316	.9446	1.3039	804.64
1740	1929.332	1429.903	8.6273	.83230+3	.60006+3	1.2325	.9454	1.3036	806.88
1750	1941.660	1439.362	8.6344	.85304+3	.58883+3	1.2333	.9463	1.3033	809.11
1760	1953.998	1448.829	8.6414	.87419+3	.57787+3	1.2342	.9471	1.3031	811.33
1770	1966.343	1458.305	8.6484	.89576+3	.56716+3	1.2350	.9480	1.3028	813.55
1780	1978.697	1467.788	8.6554	.91774+3	.55670+3	1.2358	.9488	1.3025	815.76
1790	1991.059	1477.280	8.6623	.94016+3	.54648+3	1.2366	.9496	1.3023	817.97
1800	2003.430	1486.780	8.6692	.96300+3	.53650+3	1.2374	.9504	1.3020	820.17
1810	2015.808	1496.288	8.6761	.98629+3	.52674+3	1.2382	.9512	1.3018	822.37
1820	2028.194	1505.804	8.6829	.10100+4	.51721+3	1.2390	.9520	1.3015	824.56
1830	2040.588	1515.327	8.6897	.10342+4	.50789+3	1.2398	.9528	1.3013	826.74
1840	2052.990	1524.859	8.6964	.10588+4	.49878+3	1.2406	.9535	1.3010	828.92
1850	2065.399	1534.398	8.7032	.10839+4	.48988+3	1.2413	.9543	1.3008	831.09
1860	2077.816	1543.945	8.7098	.11095+4	.48117+3	1.2421	.9551	1.3005	833.26
1870	2090.241	1553.499	8.7165	.11356+4	.47266+3	1.2428	.9558	1.3003	835.42
1880	2102.673	1563.061	8.7231	.11621+4	.46434+3	1.2436	.9565	1.3001	837.57
1890	2115.112	1572.630	8.7297	.11891+4	.45620+3	1.2443	.9573	1.2998	839.72
1900	2127.559	1582.206	8.7363	.12167+4	.44823+3	1.2450	.9580	1.2996	841.87
1910	2140.013	1591.790	8.7428	.12447+4	.44045+3	1.2458	.9587	1.2994	844.01
1920	2152.474	1601.381	8.7494	.12732+4	.43283+3	1.2465	.9595	1.2992	846.14
1930	2164.943	1610.979	8.7558	.13023+4	.42537+3	1.2472	.9602	1.2989	848.27
1940	2177.418	1620.584	8.7623	.13319+4	.41808+3	1.2479	.9609	1.2987	850.39
1950	2189.900	1630.196	8.7687	.13620+4	.41094+3	1.2486	.9616	1.2985	852.51
1960	2202.390	1639.815	8.7751	.13926+4	.40396+3	1.2493	.9622	1.2983	854.63
1970	2214.886	1649.441	8.7814	.14238+4	.39712+3	1.2500	.9629	1.2981	856.73
1980	2227.389	1659.074	8.7878	.14556+4	.39043+3	1.2506	.9636	1.2979	858.84
1990	2239.898	1668.713	8.7941	.14879+4	.38388+3	1.2513	.9643	1.2977	860.93
2000	2252.414	1678.359	8.8003	.15208+4	.37747+3	1.2520	.9649	1.2975	863.03
2050	2315.093	1726.686	8.8313	.16940+4	.34735+3	1.2552	.9681	1.2965	873.42
2100	2377.928	1775.170	8.8616	.18825+4	.32020+3	1.2582	.9712	1.2955	883.68
2150	2440.913	1823.803	8.8912	.20873+4	.29566+3	1.2611	.9741	1.2947	893.84

Table C.4 Thermodynamic Properties of Air in SI Units (*Continued*)

t	h	u	φ	p_r	v_r	c_p	c_v	γ	a
K	kJ/kg	kJ/kg	kJ/kg-K			kJ/kg-K	kJ/kg-K		m/s
2200	2504.040	1872.579	8.9203	.23094+4	.27343+3	1.2639	.9769	1.2938	903.88
2250	2567.302	1921.490	8.9487	.25499+4	.25327+3	1.2666	.9795	1.2930	913.81
2300	2630.694	1970.531	8.9766	.28098+4	.23495+3	1.2691	.9821	1.2923	923.64
2350	2694.210	2019.695	9.0039	.30904+4	.21826+3	1.2715	.9845	1.2916	933.37
2400	2757.844	2068.977	9.0307	.33928+4	.20304+3	1.2738	.9868	1.2909	942.99
2450	2821.590	2118.373	9.0570	.37182+4	.18913+3	1.2760	.9890	1.2902	952.53
2500	2885.445	2167.876	9.0828	.40679+4	.17640+3	1.2782	.9911	1.2896	961.96
2550	2949.404	2217.483	9.1081	.44432+4	.16473+3	1.2802	.9932	1.2890	971.31
2600	3013.462	2267.190	9.1330	.48455+4	.15401+3	1.2821	.9951	1.2884	980.57
2650	3077.616	2316.993	9.1574	.52762+4	.14416+3	1.2840	.9970	1.2879	989.75
2700	3141.862	2366.887	9.1814	.57367+4	.13509+3	1.2858	.9988	1.2874	998.84
2750	3206.196	2416.870	9.2050	.62285+4	.12673+3	1.2876	1.0005	1.2869	1007.85
2800	3270.616	2466.939	9.2282	.67532+4	.11901+3	1.2892	1.0022	1.2864	1016.78
2850	3335.119	2517.090	9.2511	.73124+4	.11187+3	1.2909	1.0038	1.2859	1025.64
2900	3399.702	2567.321	9.2735	.79077+4	.10526+3	1.2924	1.0054	1.2855	1034.41
2950	3464.362	2617.631	9.2956	.85408+4	.99140+2	1.2940	1.0069	1.2850	1043.12
3000	3529.098	2668.015	9.3174	.92135+4	.93459+2	1.2955	1.0084	1.2846	1051.75

Table C.4 Thermodynamic Properties of Air in EE Units

t	h	u	φ	p_r	v_r	c_p	c_v	γ	a
R	Btu/ lbm	Btu/ lbm	Btu/ lbm-R			Btu/ lbm-R	Btu/ lbm-R		ft/sec
100	23.338	16.482	1.1925	.26337-2	.14066+5	.2460	.1775	1.3863	487.79
110	25.796	18.255	1.2159	.37069-2	.10994+5	.2457	.1771	1.3871	511.75
120	28.251	20.024	1.2372	.50620-2	.87823+4	.2453	.1768	1.3879	534.65
130	30.703	21.790	1.2569	.67395-2	.71461+4	.2450	.1764	1.3886	556.64
140	33.151	23.553	1.2750	.87812-2	.59065+4	.2446	.1761	1.3893	577.80
150	35.595	25.312	1.2919	.11231-1	.49481+4	.2443	.1758	1.3901	598.23
160	38.037	27.068	1.3076	.14133-1	.41942+4	.2440	.1755	1.3907	618.01
170	40.475	28.821	1.3224	.17534-1	.35919+4	.2437	.1752	1.3914	637.18
180	42.911	30.571	1.3363	.21482-1	.31042+4	.2434	.1749	1.3920	655.80
190	45.344	32.318	1.3495	.26026-1	.27046+4	.2432	.1746	1.3927	673.92
200	47.774	34.063	1.3620	.31216-1	.23736+4	.2429	.1743	1.3932	691.57
210	50.202	35.805	1.3738	.37103-1	.20968+4	.2426	.1741	1.3938	708.79
220	52.627	37.545	1.3851	.43740-1	.18634+4	.2424	.1738	1.3944	725.62
230	55.050	39.282	1.3959	.51180-1	.16649+4	.2422	.1736	1.3949	742.06
240	57.471	41.017	1.4062	.59479-1	.14948+4	.2420	.1734	1.3954	758.16
250	59.889	42.750	1.4160	.68692-1	.13483+4	.2417	.1732	1.3958	773.92
260	62.306	44.481	1.4255	.78877-1	.12212+4	.2416	.1730	1.3963	789.37
270	64.720	46.210	1.4346	.90090-1	.11103+4	.2414	.1728	1.3967	804.53
280	67.133	47.937	1.4434	.10239+0	.10131+4	.2412	.1726	1.3971	819.41
290	69.544	49.663	1.4519	.11584+0	.92744+3	.2410	.1725	1.3975	834.03
300	71.954	51.387	1.4600	.13050+0	.85165+3	.2409	.1723	1.3978	848.40
310	74.362	53.109	1.4679	.14643+0	.78430+3	.2407	.1722	1.3982	862.52
320	76.768	54.830	1.4756	.16370+0	.72421+3	.2406	.1720	1.3985	876.42
330	79.174	56.550	1.4830	.18236+0	.67041+3	.2405	.1719	1.3988	890.10
340	81.578	58.269	1.4901	.20248+0	.62207+3	.2404	.1718	1.3990	903.57
350	83.981	59.986	1.4971	.22414+0	.57850+3	.2403	.1717	1.3993	916.84
360	86.383	61.703	1.5039	.24739+0	.53910+3	.2402	.1716	1.3995	929.92
370	88.784	63.418	1.5105	.27231+0	.50337+3	.2401	.1715	1.3997	942.82
380	91.184	65.133	1.5169	.29896+0	.47089+3	.2400	.1714	1.3999	955.54
390	93.584	66.847	1.5231	.32742+0	.44128+3	.2399	.1714	1.4000	968.08
400	95.983	68.561	1.5292	.35775+0	.41422+3	.2399	.1713	1.4002	980.46
410	98.381	70.273	1.5351	.39003+0	.38944+3	.2398	.1713	1.4003	992.68
420	100.779	71.986	1.5409	.42433+0	.36669+3	.2398	.1712	1.4004	1004.75
430	103.177	73.698	1.5465	.46072+0	.34576+3	.2397	.1712	1.4005	1016.67
440	105.574	75.410	1.5520	.49929+0	.32647+3	.2397	.1712	1.4005	1028.44
450	107.971	77.121	1.5574	.54011+0	.30866+3	.2397	.1712	1.4006	1040.08
460	110.368	78.833	1.5627	.58325+0	.29218+3	.2397	.1711	1.4006	1051.58
470	112.765	80.544	1.5678	.62880+0	.27691+3	.2397	.1711	1.4006	1062.95
480	115.162	82.256	1.5729	.67683+0	.26273+3	.2397	.1712	1.4006	1074.19
490	117.560	83.967	1.5778	.72743+0	.24955+3	.2397	.1712	1.4005	1085.30
500	119.957	85.679	1.5827	.78068+0	.23727+3	.2398	.1712	1.4005	1096.30
510	122.355	87.391	1.5874	.83667+0	.22582+3	.2398	.1712	1.4004	1107.18
520	124.753	89.104	1.5921	.89547+0	.21513+3	.2398	.1713	1.4003	1117.94
530	127.151	90.817	1.5966	.95718+0	.20513+3	.2399	.1713	1.4002	1128.59
540	129.550	92.530	1.6011	.10219+1	.19577+3	.2399	.1714	1.4000	1139.14
550	131.950	94.244	1.6055	.10897+1	.18699+3	.2400	.1714	1.3999	1149.58
560	134.350	95.959	1.6098	.11606+1	.17875+3	.2401	.1715	1.3997	1159.92
570	136.751	97.674	1.6141	.12348+1	.17101+3	.2401	.1716	1.3996	1170.15
580	139.153	99.390	1.6183	.13124+1	.16372+3	.2402	.1717	1.3994	1180.29
590	141.555	101.107	1.6224	.13935+1	.15686+3	.2403	.1718	1.3992	1190.33
600	143.959	102.825	1.6264	.14780+1	.15039+3	.2404	.1719	1.3989	1200.28
610	146.364	104.544	1.6304	.15663+1	.14428+3	.2405	.1720	1.3987	1210.14
620	148.769	106.265	1.6343	.16582+1	.13852+3	.2406	.1721	1.3984	1219.90
630	151.176	107.986	1.6382	.17540+1	.13306+3	.2407	.1722	1.3982	1229.58

Table C.4   Thermodynamic Properties of Air in EE Units (*Continued*)

t	h	u	ϕ	p_r	v_r	c_p	c_v	γ	a
	Btu/ lbm	Btu/ lbm	Btu/ lbm-R			Btu/ lbm-R	Btu/ lbm-R		ft/sec
640	153.584	109.708	1.6419	.18538+1	.12790+3	.2409	.1723	1.3979	1239.18
650	155.993	111.432	1.6457	.19576+1	.12301+3	.2410	.1724	1.3976	1248.69
660	158.404	113.157	1.6494	.20656+1	.11837+3	.2411	.1726	1.3973	1258.12
670	160.816	114.883	1.6530	.21778+1	.11398+3	.2413	.1727	1.3969	1267.46
680	163.229	116.611	1.6566	.22944+1	.10980+3	.2414	.1729	1.3966	1276.73
690	165.644	118.340	1.6601	.24155+1	.10583+3	.2416	.1730	1.3963	1285.93
700	168.060	120.071	1.6636	.25411+1	.10205+3	.2417	.1732	1.3959	1295.04
710	170.478	121.804	1.6670	.26715+1	.98459+2	.2419	.1733	1.3955	1304.08
720	172.898	123.538	1.6704	.28067+1	.95037+2	.2421	.1735	1.3951	1313.05
730	175.320	125.274	1.6737	.29468+1	.91775+2	.2422	.1737	1.3947	1321.95
740	177.743	127.012	1.6770	.30920+1	.88664+2	.2424	.1739	1.3943	1330.78
750	180.168	128.751	1.6803	.32423+1	.85695+2	.2426	.1740	1.3939	1339.54
760	182.595	130.492	1.6835	.33980+1	.82860+2	.2428	.1742	1.3935	1348.23
770	185.024	132.236	1.6867	.35591+1	.80151+2	.2430	.1744	1.3930	1356.86
780	187.455	133.981	1.6898	.37257+1	.77560+2	.2432	.1746	1.3926	1365.42
790	189.887	135.728	1.6929	.38980+1	.75083+2	.2434	.1748	1.3921	1373.92
800	192.322	137.478	1.6960	.40761+1	.72711+2	.2436	.1750	1.3917	1382.36
810	194.759	139.229	1.6990	.42601+1	.70439+2	.2438	.1753	1.3912	1390.73
820	197.199	140.983	1.7020	.44502+1	.68263+2	.2440	.1755	1.3907	1399.04
830	199.640	142.739	1.7049	.46465+1	.66176+2	.2443	.1757	1.3902	1407.30
840	202.084	144.497	1.7079	.48492+1	.64175+2	.2445	.1759	1.3897	1415.49
850	204.530	146.257	1.7108	.50583+1	.62254+2	.2447	.1762	1.3892	1423.63
860	206.978	148.020	1.7136	.52740+1	.60410+2	.2449	.1764	1.3887	1431.71
870	209.429	149.785	1.7165	.54966+1	.58638+2	.2452	.1766	1.3881	1439.74
880	211.882	151.552	1.7193	.57260+1	.56935+2	.2454	.1769	1.3876	1447.71
890	214.337	153.322	1.7220	.59625+1	.55298+2	.2457	.1771	1.3871	1455.63
900	216.795	155.095	1.7248	.62062+1	.53724+2	.2459	.1774	1.3865	1463.50
910	219.256	156.870	1.7275	.64573+1	.52209+2	.2462	.1776	1.3860	1471.32
920	221.719	158.647	1.7302	.67159+1	.50750+2	.2464	.1779	1.3854	1479.08
930	224.184	160.427	1.7329	.69822+1	.49345+2	.2467	.1781	1.3848	1486.79
940	226.653	162.210	1.7355	.72563+1	.47992+2	.2470	.1784	1.3843	1494.46
950	229.124	163.996	1.7381	.75388+1	.46687+2	.2472	.1787	1.3837	1502.07
960	231.597	165.784	1.7407	.78286+1	.45429+2	.2475	.1789	1.3831	1509.64
970	234.074	167.574	1.7433	.81272+1	.44216+2	.2478	.1792	1.3825	1517.17
980	236.553	169.368	1.7458	.84343+1	.43045+2	.2480	.1795	1.3819	1524.64
990	239.035	171.164	1.7483	.87501+1	.41915+2	.2483	.1798	1.3814	1532.07
1000	241.519	172.963	1.7508	.90747+1	.40824+2	.2486	.1801	1.3808	1539.46
1010	244.007	174.765	1.7533	.94083+1	.39771+2	.2489	.1803	1.3802	1546.80
1020	246.497	176.570	1.7558	.97511+1	.38752+2	.2492	.1806	1.3796	1554.10
1030	248.990	178.378	1.7582	.10103+2	.37768+2	.2495	.1809	1.3789	1561.36
1040	251.486	180.188	1.7606	.10465+2	.36817+2	.2498	.1812	1.3783	1568.57
1050	253.985	182.002	1.7630	.10837+2	.35896+2	.2501	.1815	1.3777	1575.74
1060	256.487	183.818	1.7654	.11218+2	.35006+2	.2503	.1818	1.3771	1582.88
1070	258.992	185.638	1.7677	.11610+2	.34144+2	.2506	.1821	1.3765	1589.97
1080	261.500	187.460	1.7701	.12011+2	.33310+2	.2509	.1824	1.3759	1597.02
1090	264.011	189.285	1.7724	.12424+2	.32503+2	.2512	.1827	1.3753	1604.04
1100	266.525	191.114	1.7747	.12847+2	.31721+2	.2516	.1830	1.3746	1611.01
1110	269.042	192.945	1.7769	.13281+2	.30963+2	.2519	.1833	1.3740	1617.95
1120	271.562	194.780	1.7792	.13726+2	.30229+2	.2522	.1836	1.3734	1624.85
1130	274.086	196.617	1.7814	.14183+2	.29517+2	.2525	.1839	1.3728	1631.72
1140	276.612	198.458	1.7837	.14651+2	.28827+2	.2528	.1842	1.3721	1638.55
1150	279.141	200.302	1.7859	.15130+2	.28158+2	.2531	.1845	1.3715	1645.34
1160	281.674	202.149	1.7881	.15622+2	.27509+2	.2534	.1848	1.3709	1652.10
1170	284.209	203.999	1.7902	.16126+2	.26879+2	.2537	.1852	1.3702	1658.83
1180	286.748	205.852	1.7924	.16642+2	.26267+2	.2540	.1855	1.3696	1665.52
1190	289.290	207.709	1.7946	.17171+2	.25674+2	.2544	.1858	1.3690	1672.18
1200	291.835	209.568	1.7967	.17713+2	.25098+2	.2547	.1861	1.3684	1678.80
1210	294.383	211.431	1.7988	.18268+2	.24538+2	.2550	.1864	1.3677	1685.40
1220	296.935	213.297	1.8009	.18836+2	.23995+2	.2553	.1867	1.3671	1691.96
1230	299.489	215.166	1.8030	.19418+2	.23467+2	.2556	.1871	1.3665	1698.49
1240	302.047	217.038	1.8051	.20014+2	.22953+2	.2559	.1874	1.3659	1704.99
1250	304.608	218.913	1.8071	.20623+2	.22454+2	.2563	.1877	1.3652	1711.46
1260	307.173	220.792	1.8092	.21247+2	.21969+2	.2566	.1880	1.3646	1717.90
1270	309.740	222.674	1.8112	.21886+2	.21498+2	.2569	.1883	1.3640	1724.31
1280	312.311	224.559	1.8132	.22539+2	.21039+2	.2572	.1887	1.3634	1730.69
1290	314.885	226.447	1.8152	.23207+2	.20593+2	.2575	.1890	1.3627	1737.05
1300	317.462	228.339	1.8172	.23891+2	.20159+2	.2579	.1893	1.3621	1743.37
1310	320.042	230.234	1.8192	.24590+2	.19736+2	.2582	.1896	1.3615	1749.67
1320	322.625	232.132	1.8211	.25305+2	.19325+2	.2585	.1900	1.3609	1755.94
1330	325.212	234.033	1.8231	.26036+2	.18925+2	.2588	.1903	1.3603	1762.18
1340	327.802	235.937	1.8250	.26783+2	.18535+2	.2592	.1906	1.3597	1768.40
1350	330.395	237.845	1.8270	.27547+2	.18156+2	.2595	.1909	1.3591	1774.59
1360	332.992	239.756	1.8289	.28328+2	.17786+2	.2598	.1912	1.3585	1780.76
1370	335.591	241.670	1.8308	.29126+2	.17426+2	.2601	.1916	1.3579	1786.90
1380	338.194	243.587	1.8327	.29941+2	.17075+2	.2604	.1919	1.3573	1793.01
1390	340.800	245.507	1.8346	.30774+2	.16733+2	.2608	.1922	1.3567	1799.10
1400	343.409	247.431	1.8364	.31625+2	.16400+2	.2611	.1925	1.3561	1805.17
1410	346.022	249.358	1.8383	.32495+2	.16075+2	.2614	.1928	1.3555	1811.21
1420	348.637	251.288	1.8401	.33383+2	.15759+2	.2617	.1932	1.3549	1817.23
1430	351.256	253.221	1.8420	.34290+2	.15450+2	.2620	.1935	1.3543	1823.23
1440	353.878	255.157	1.8438	.35216+2	.15149+2	.2623	.1938	1.3538	1829.20
1450	356.503	257.097	1.8456	.36162+2	.14855+2	.2627	.1941	1.3532	1835.15
1460	359.131	259.040	1.8474	.37127+2	.14568+2	.2630	.1944	1.3526	1841.08

Table C.4 Thermodynamic Properties of Air in EE Units (*Continued*)

t	h	u	φ	p_r	v_r	c_p	c_v	γ	a
R	Btu/ lbm	Btu/ lbm	Btu/ lbm-R			Btu/ lbm-R	Btu/ lbm-R		ft/sec
1470	361.763	260.985	1.8492	.38113+2	.14289+2	.2633	.1947	1.3521	1846.99
1480	364.397	262.934	1.8510	.39119+2	.14016+2	.2636	.1950	1.3515	1852.88
1490	367.034	264.886	1.8528	.40145+2	.13750+2	.2639	.1954	1.3509	1858.74
1500	369.675	266.841	1.8545	.41193+2	.13490+2	.2642	.1957	1.3504	1864.59
1510	372.319	268.800	1.8563	.42262+2	.13237+2	.2645	.1960	1.3498	1870.41
1520	374.966	270.761	1.8580	.43353+2	.12989+2	.2648	.1963	1.3493	1876.22
1530	377.615	272.725	1.8598	.44466+2	.12747+2	.2651	.1966	1.3487	1882.00
1540	380.268	274.692	1.8615	.45601+2	.12511+2	.2654	.1969	1.3482	1887.76
1550	382.924	276.663	1.8632	.46759+2	.12281+2	.2657	.1972	1.3477	1893.51
1560	385.583	278.636	1.8649	.47940+2	.12055+2	.2660	.1975	1.3471	1899.24
1570	388.245	280.612	1.8666	.49144+2	.11835+2	.2663	.1978	1.3466	1904.95
1580	390.910	282.592	1.8683	.50372+2	.11620+2	.2666	.1981	1.3461	1910.64
1590	393.578	284.574	1.8700	.51624+2	.11410+2	.2669	.1984	1.3456	1916.31
1600	396.248	286.559	1.8717	.52901+2	.11205+2	.2672	.1987	1.3451	1921.96
1610	398.922	288.547	1.8734	.54202+2	.11004+2	.2675	.1989	1.3446	1927.60
1620	401.598	290.538	1.8750	.55528+2	.10808+2	.2678	.1992	1.3441	1933.22
1630	404.278	292.532	1.8767	.56880+2	.10616+2	.2681	.1995	1.3436	1938.82
1640	406.960	294.528	1.8783	.58257+2	.10429+2	.2684	.1998	1.3431	1944.41
1650	409.645	296.528	1.8799	.59661+2	.10246+2	.2686	.2001	1.3426	1949.98
1660	412.333	298.530	1.8816	.61091+2	.10067+2	.2689	.2004	1.3422	1955.54
1670	415.023	300.535	1.8832	.62548+2	.98913+1	.2692	.2006	1.3417	1961.07
1680	417.717	302.543	1.8848	.64033+2	.97198+1	.2695	.2009	1.3412	1966.60
1690	420.413	304.553	1.8864	.65545+2	.95521+1	.2697	.2012	1.3408	1972.10
1700	423.111	306.566	1.8880	.67085+2	.93881+1	.2700	.2014	1.3403	1977.60
1710	425.813	308.582	1.8896	.68653+2	.92276+1	.2703	.2017	1.3399	1983.07
1720	428.517	310.601	1.8911	.70250+2	.90705+1	.2705	.2020	1.3394	1988.54
1730	431.223	312.622	1.8927	.71877+2	.89168+1	.2708	.2022	1.3390	1993.99
1740	433.932	314.645	1.8943	.73533+2	.87664+1	.2710	.2025	1.3386	1999.42
1750	436.644	316.671	1.8958	.75218+2	.86192+1	.2713	.2027	1.3381	2004.84
1760	439.358	318.700	1.8974	.76935+2	.84751+1	.2715	.2030	1.3377	2010.25
1770	442.075	320.731	1.8989	.78681+2	.83340+1	.2718	.2032	1.3373	2015.64
1780	444.794	322.765	1.9004	.80459+2	.81959+1	.2720	.2035	1.3369	2021.02
1790	447.516	324.801	1.9020	.82269+2	.80606+1	.2723	.2037	1.3365	2026.39
1800	450.240	326.839	1.9035	.84110+2	.79282+1	.2725	.2040	1.3361	2031.74
1810	452.966	328.880	1.9050	.85982+2	.77987+1	.2727	.2042	1.3358	2037.10
1820	455.695	330.923	1.9065	.87888+2	.76717+1	.2730	.2044	1.3354	2042.43
1830	458.425	332.968	1.9080	.89828+2	.75473+1	.2732	.2046	1.3350	2047.75
1840	461.158	335.016	1.9095	.91801+2	.74255+1	.2734	.2049	1.3346	2053.06
1850	463.894	337.065	1.9110	.93807+2	.73061+1	.2736	.2051	1.3343	2058.35
1860	466.631	339.117	1.9124	.95849+2	.71892+1	.2739	.2053	1.3339	2063.63
1870	469.371	341.171	1.9139	.97924+2	.70746+1	.2741	.2055	1.3336	2068.89
1880	472.113	343.228	1.9154	.10000+3	.69623+1	.2743	.2057	1.3332	2074.14
1890	474.857	345.286	1.9168	.10218+3	.68523+1	.2745	.2060	1.3329	2079.38
1900	477.603	347.347	1.9183	.10437+3	.67445+1	.2747	.2062	1.3325	2084.60
1910	480.351	349.410	1.9197	.10659+3	.66388+1	.2749	.2064	1.3322	2089.80
1920	483.102	351.475	1.9212	.10884+3	.65352+1	.2752	.2066	1.3318	2095.00
1930	485.854	353.542	1.9226	.11114+3	.64336+1	.2754	.2068	1.3315	2100.18
1940	488.609	355.611	1.9240	.11347+3	.63340+1	.2756	.2070	1.3311	2105.34
1950	491.366	357.682	1.9254	.11584+3	.62364+1	.2758	.2072	1.3308	2110.49
1960	494.125	359.756	1.9268	.11825+3	.61407+1	.2760	.2074	1.3305	2115.63
1970	496.886	361.831	1.9282	.12070+3	.60468+1	.2762	.2077	1.3301	2120.76
1980	499.649	363.909	1.9296	.12318+3	.59547+1	.2764	.2079	1.3298	2125.87
1990	502.415	365.938	1.9310	.12571+3	.58644+1	.2766	.2081	1.3295	2130.97
2000	505.182	368.070	1.9324	.12828+3	.57758+1	.2768	.2083	1.3292	2136.05
2020	510.723	372.240	1.9352	.13355+3	.56036+1	.2772	.2087	1.3285	2146.19
2040	516.272	376.418	1.9379	.13898+3	.54379+1	.2776	.2091	1.3279	2156.27
2060	521.828	380.603	1.9406	.14458+3	.52783+1	.2780	.2095	1.3273	2166.30
2080	527.393	384.797	1.9433	.15037+3	.51246+1	.2784	.2099	1.3266	2176.29
2100	532.966	388.999	1.9460	.15633+3	.49765+1	.2788	.2103	1.3260	2186.22
2120	538.547	393.208	1.9486	.16248+3	.48338+1	.2792	.2107	1.3254	2196.11
2140	544.135	397.425	1.9512	.16882+3	.46962+1	.2796	.2111	1.3248	2205.95
2160	549.731	401.650	1.9538	.17535+3	.45635+1	.2800	.2114	1.3242	2215.75
2180	555.334	405.883	1.9564	.18208+3	.44355+1	.2804	.2118	1.3237	2225.50
2200	560.945	410.122	1.9590	.18901+3	.43120+1	.2807	.2122	1.3231	2235.21
2220	566.564	414.370	1.9615	.19616+3	.41928+1	.2811	.2125	1.3225	2244.87
2240	572.189	418.624	1.9641	.20351+3	.40777+1	.2815	.2129	1.3220	2254.49
2260	577.822	422.886	1.9666	.21108+3	.39666+1	.2818	.2133	1.3214	2264.06
2280	583.463	427.155	1.9690	.21887+3	.38593+1	.2822	.2136	1.3209	2273.59
2300	589.110	431.431	1.9715	.22688+3	.37556+1	.2825	.2140	1.3204	2283.09
2320	594.764	435.715	1.9740	.23513+3	.36554+1	.2829	.2143	1.3199	2292.54
2340	600.426	440.005	1.9764	.24361+3	.35585+1	.2832	.2147	1.3193	2301.95
2360	606.094	444.302	1.9788	.25234+3	.34648+1	.2836	.2150	1.3188	2311.32
2380	611.769	448.606	1.9812	.26131+3	.33742+1	.2839	.2154	1.3183	2320.65
2400	617.450	452.916	1.9836	.27053+3	.32866+1	.2843	.2157	1.3178	2329.94
2420	623.139	457.234	1.9859	.28000+3	.32019+1	.2846	.2160	1.3173	2339.19
2440	628.834	461.558	1.9883	.28974+3	.31198+1	.2849	.2164	1.3169	2348.41
2460	634.535	465.838	1.9906	.29975+3	.30404+1	.2852	.2167	1.3164	2357.59
2480	640.243	470.225	1.9929	.31002+3	.29635+1	.2856	.2170	1.3159	2366.73
2500	645.958	474.568	1.9952	.32058+3	.28891+1	.2859	.2173	1.3155	2375.84
2520	651.679	478.918	1.9975	.33141+3	.28170+1	.2862	.2176	1.3150	2384.91
2540	657.406	483.274	1.9998	.34254+3	.27471+1	.2865	.2180	1.3145	2393.94
2560	663.139	487.636	2.0020	.35396+3	.26794+1	.2868	.2183	1.3141	2402.94

Table C.4 Thermodynamic Properties of Air in EE Units (*Continued*)

t	h	u	φ	p_r	v_r	c_p	c_v	γ	a
R	Btu/ lbm	Btu/ lbm	Btu/ lbm-R			Btu/ lbm-R	Btu/ lbm-R		ft/sec
2580	668.878	492.004	2.0042	.36568+3	.26138+1	.2871	.2186	1.3137	2411.91
2600	674.624	496.379	2.0065	.37770+3	.25502+1	.2874	.2189	1.3132	2420.84
2620	680.375	500.759	2.0087	.39004+3	.24885+1	.2877	.2192	1.3128	2429.74
2640	686.133	505.145	2.0108	.40270+3	.24287+1	.2880	.2195	1.3124	2438.60
2660	691.896	509.537	2.0130	.41568+3	.23707+1	.2883	.2198	1.3120	2447.44
2680	697.665	513.935	2.0152	.42899+3	.23144+1	.2886	.2200	1.3116	2456.24
2700	703.440	518.339	2.0173	.44263+3	.22598+1	.2889	.2203	1.3112	2465.00
2720	709.221	522.749	2.0195	.45662+3	.22068+1	.2892	.2206	1.3108	2473.74
2740	715.007	527.164	2.0216	.47096+3	.21553+1	.2894	.2209	1.3104	2482.44
2760	720.799	531.584	2.0237	.48565+3	.21054+1	.2897	.2212	1.3100	2491.12
2780	726.596	536.010	2.0258	.50071+3	.20569+1	.2900	.2214	1.3096	2499.76
2800	732.399	540.442	2.0279	.51613+3	.20098+1	.2903	.2217	1.3092	2508.37
2820	738.207	544.879	2.0299	.53193+3	.19640+1	.2905	.2220	1.3088	2516.96
2840	744.020	549.321	2.0320	.54811+3	.19196+1	.2908	.2223	1.3085	2525.51
2860	749.839	553.769	2.0340	.56468+3	.18764+1	.2911	.2225	1.3081	2534.03
2880	755.663	558.222	2.0361	.58164+3	.18344+1	.2913	.2228	1.3077	2542.53
2900	761.492	562.680	2.0381	.59901+3	.17936+1	.2916	.2230	1.3074	2551.00
2920	767.326	567.143	2.0401	.61679+3	.17539+1	.2918	.2233	1.3070	2559.43
2940	773.166	571.611	2.0421	.63498+3	.17153+1	.2921	.2235	1.3067	2567.85
2960	779.010	576.085	2.0440	.65360+3	.16778+1	.2923	.2238	1.3063	2576.23
2980	784.859	580.563	2.0460	.67265+3	.16413+1	.2926	.2240	1.3060	2584.58
3000	790.713	585.046	2.0480	.69214+3	.16058+1	.2928	.2243	1.3057	2592.91
3020	796.572	589.534	2.0499	.71207+3	.15712+1	.2931	.2245	1.3054	2601.22
3040	802.436	594.026	2.0519	.73246+3	.15376+1	.2933	.2248	1.3050	2609.49
3060	808.305	598.524	2.0538	.75330+3	.15049+1	.2935	.2250	1.3047	2617.74
3080	814.178	603.026	2.0557	.77462+3	.14730+1	.2938	.2252	1.3044	2625.96
3100	820.056	607.533	2.0576	.79642+3	.14420+1	.2940	.2255	1.3041	2634.16
3120	825.938	612.044	2.0595	.81870+3	.14118+1	.2942	.2257	1.3038	2642.33
3140	831.825	616.560	2.0614	.84147+3	.13824+1	.2945	.2259	1.3035	2650.48
3160	837.717	621.080	2.0632	.86474+3	.13538+1	.2947	.2261	1.3032	2658.60
3180	843.613	625.605	2.0651	.88852+3	.13259+1	.2949	.2264	1.3029	2666.70
3200	849.513	630.134	2.0669	.91282+3	.12987+1	.2951	.2266	1.3026	2674.77
3220	855.418	634.668	2.0688	.93764+3	.12722+1	.2953	.2268	1.3023	2682.82
3240	861.327	639.206	2.0706	.96300+3	.12464+1	.2956	.2270	1.3020	2690.85
3260	867.240	643.748	2.0724	.98890+3	.12213+1	.2958	.2272	1.3017	2698.85
3280	873.157	648.294	2.0742	.10154+4	.11968+1	.2960	.2274	1.3015	2706.83
3300	879.079	652.845	2.0760	.10424+4	.11729+1	.2962	.2276	1.3012	2714.78
3320	885.005	657.399	2.0778	.10699+4	.11496+1	.2964	.2278	1.3009	2722.71
3340	890.934	661.958	2.0796	.10981+4	.11268+1	.2966	.2280	1.3006	2730.62
3360	896.868	666.520	2.0814	.11268+4	.11047+1	.2968	.2282	1.3004	2738.51
3380	902.806	671.087	2.0831	.11562+4	.10830+1	.2970	.2284	1.3001	2746.37
3400	908.748	675.658	2.0849	.11861+4	.10619+1	.2972	.2286	1.2999	2754.21
3420	914.693	680.232	2.0866	.12167+4	.10414+1	.2974	.2288	1.2996	2762.03
3440	920.643	684.810	2.0884	.12478+4	.10213+1	.2976	.2290	1.2994	2769.83
3460	926.596	689.392	2.0901	.12797+4	.10017+1	.2978	.2292	1.2991	2777.60
3480	932.553	693.978	2.0918	.13121+4	.98257+0	.2979	.2294	1.2989	2785.36
3500	938.513	698.568	2.0935	.13452+4	.96390+0	.2981	.2296	1.2986	2793.09
3520	944.478	703.161	2.0952	.13790+4	.94568+0	.2983	.2298	1.2984	2800.80
3540	950.446	707.758	2.0969	.14134+4	.92788+0	.2985	.2299	1.2981	2808.49
3560	956.418	712.359	2.0986	.14485+4	.91051+0	.2987	.2301	1.2979	2816.16
3580	962.393	716.963	2.1003	.14843+4	.89354+0	.2989	.2303	1.2977	2823.81
3600	968.372	721.571	2.1019	.15208+4	.87697+0	.2990	.2305	1.2975	2831.44
3620	974.354	726.182	2.1036	.15580+4	.86078+0	.2992	.2306	1.2972	2839.05
3640	980.340	730.796	2.1052	.15959+4	.84496+0	.2994	.2308	1.2970	2846.64
3660	986.329	735.414	2.1069	.16346+4	.82951+0	.2995	.2310	1.2968	2854.21
3680	992.321	740.036	2.1085	.16740+4	.81441+0	.2997	.2312	1.2966	2861.76
3700	998.317	744.660	2.1101	.17141+4	.79966+0	.2999	.2313	1.2964	2869.29
3720	1004.316	749.288	2.1118	.17551+4	.78524+0	.3000	.2315	1.2962	2876.80
3740	1010.319	753.920	2.1134	.17967+4	.77115+0	.3002	.2316	1.2959	2884.29
3760	1016.324	758.554	2.1150	.18392+4	.75737+0	.3004	.2318	1.2957	2891.76
3780	1022.333	763.192	2.1166	.18825+4	.74390+0	.3005	.2320	1.2955	2899.22
3800	1028.345	767.833	2.1182	.19265+4	.73073+0	.3007	.2321	1.2953	2906.65
3820	1034.360	772.477	2.1197	.19714+4	.71785+0	.3008	.2323	1.2951	2914.07
3840	1040.379	777.124	2.1213	.20171+4	.70526+0	.3010	.2324	1.2949	2921.47
3860	1046.400	781.774	2.1229	.20637+4	.69294+0	.3011	.2326	1.2948	2928.84
3880	1052.425	786.428	2.1244	.21111+4	.68090+0	.3013	.2327	1.2946	2936.21
3900	1058.452	791.084	2.1260	.21593+4	.66911+0	.3014	.2329	1.2944	2943.55
3920	1064.482	795.743	2.1275	.22084+4	.65758+0	.3016	.2330	1.2942	2950.87
3940	1070.516	800.405	2.1291	.22585+4	.64630+0	.3017	.2332	1.2940	2958.18
3960	1076.552	805.071	2.1306	.23094+4	.63526+0	.3019	.2333	1.2938	2965.47
3980	1082.591	809.739	2.1321	.23612+4	.62446+0	.3020	.2335	1.2936	2972.74
4000	1088.633	814.409	2.1336	.24139+4	.61389+0	.3022	.2336	1.2935	2980.00
4100	1118.884	837.805	2.1411	.26918+4	.56428+0	.3029	.2343	1.2926	3016.01
4200	1149.203	861.268	2.1484	.29945+4	.51961+0	.3035	.2350	1.2918	3051.61
4300	1179.585	884.795	2.1555	.33236+4	.47930+0	.3041	.2356	1.2910	3086.81
4400	1210.028	908.382	2.1625	.36809+4	.44285+0	.3047	.2362	1.2903	3121.61
4500	1240.528	932.027	2.1694	.40679+4	.40982+0	.3053	.2367	1.2896	3156.04
4600	1271.083	955.726	2.1761	.44866+4	.37983+0	.3058	.2373	1.2889	3190.10
4700	1301.691	979.479	2.1827	.49387+4	.35256+0	.3063	.2378	1.2883	3223.81
4800	1332.350	1003.281	2.1891	.54263+4	.32771+0	.3068	.2383	1.2877	3257.17
4900	1363.056	1027.132	2.1955	.59513+4	.30502+0	.3073	.2387	1.2872	3290.20
5000	1393.809	1051.030	2.2017	.65158+4	.28428+0	.3078	.2392	1.2866	3322.90

Table C.5   Properties of the U.S. Standard Atmosphere*

Z, m	t, K	p, N/m^2	ρ, kg/m^3	g, m/s^2	$\mu \cdot 10^5$, kg/m-s^2	$k \cdot 10^6$, kcal/m-s-K	a, m/s
-1000	294.65	$1.1393+5$	$1.3470+0$	9.8097	1.8206	6.1748	344.11
500	291.50	1.0748	1.2849	9.8082	1.8050	6.1140	342.21
0	288.15	1.0133	1.2250	9.8066	1.7894	6.0530	340.29
500	284.90	$9.5461+4$	1.1673	9.8051	1.7737	5.9919	338.37
1000	281.65	8.9876	1.1117	9.8036	1.7579	5.9305	336.44
1500	278.40	8.4560	1.0581	9.8020	1.7420	5.8690	334.49
2000	275.15	7.9501	1.0066	9.8005	1.7260	5.8073	332.53
2500	271.91	7.4692	$9.5695-1$	9.7989	1.7099	5.7454	330.56
3000	268.66	7.0121	9.0925	9.7974	1.6938	5.6833	328.58
3500	265.41	6.5780	8.6340	9.7959	1.6775	5.6210	326.59
4000	262.17	6.1660	8.1935	9.7943	1.6612	5.5586	324.59
4500	258.92	5.7753	7.7704	9.7928	1.6448	5.4959	322.57
5000	255.68	5.4048	7.3643	9.7912	1.6282	5.4331	320.55
6000	249.19	4.7218	6.6011	9.7882	1.5949	5.3068	316.45
7000	242.70	4.1105	5.9002	9.7851	1.5612	5.1798	312.31
8000	236.22	3.5652	5.2579	9.7820	1.5271	5.0520	308.11
9000	229.73	3.0801	4.6706	9.7789	1.4926	4.9235	303.85
10,000	223.25	2.6500	4.1351	9.7759	1.4577	4.7942	299.53
11,000	216.77	2.2700	3.6480	9.7728	1.4223	4.6642	295.14
12,000	216.65	1.9399	3.1194	9.7697	1.4216	4.6617	295.07
13,000	216.65	1.6580	2.6660	9.7667	1.4216	4.6617	295.07
14,000	216.65	1.4170	2.2786	9.7636	1.4216	4.6617	295.07
15,000	216.65	1.2112	1.9475	9.7605	1.4216	4.6617	295.07
16,000	216.65	1.0353	1.6647	9.7575	1.4216	4.6617	295.07
17,000	216.65	$8.8497+3$	1.4230	9.7544	1.4216	4.6617	295.07
18,000	216.65	7.5652	1.2165	9.7513	1.4216	4.6617	295.07
19,000	216.65	6.4675	1.0400	9.7483	1.4216	4.6617	295.07
20,000	216.65	5.5293	$8.8910-2$	9.7452	1.4216	4.6617	295.07
21,000	217.58	4.7289	7.5715	9.7422	1.4267	4.6804	295.70
22,000	218.57	4.0475	6.4510	9.7391	1.4322	4.7004	296.38
23,000	219.57	3.4669	5.5006	9.7361	1.4376	4.7204	297.05
24,000	220.56	2.9717	4.6938	9.7330	1.4430	4.7403	297.72
25,000	221.55	2.5492	4.0084	9.7300	1.4484	4.7602	298.39
26,000	222.54	2.1884	3.4257	9.7269	1.4538	4.7800	299.06
27,000	223.54	1.8800	2.9298	9.7239	1.4592	4.7999	299.72
28,000	224.53	1.6162	2.5076	9.7208	1.4646	4.8197	300.39
29,000	225.52	1.3904	2.1478	9.7178	1.4699	4.8395	301.05
30,000	226.51	1.1970	1.8410	9.7147	1.4753	4.8593	301.71
31,000	227.50	1.0313	1.5792	9.7117	1.4806	4.8790	302.37
32,000	228.49	$8.8906+2$	1.3555	9.7086	1.4859	4.8988	303.03
33,000	230.97	7.6731	1.1573	9.7056	1.4992	4.9481	304.67
34,000	233.74	6.6341	$9.8874-3$	9.7026	1.5140	5.0031	306.49
35,000	236.51	5.7459	8.4634	9.6995	1.5287	5.0579	308.30
36,000	239.28	4.9852	7.2579	9.6965	1.5433	5.1125	310.10
37,000	242.05	4.3325	6.2355	9.6935	1.5578	5.1670	311.89
38,000	244.82	3.7714	5.3666	9.6904	1.5723	5.2213	313.67
39,000	247.58	3.2882	4.6267	9.6874	1.5866	5.2577	315.43
40,000	250.35	2.8714	3.9957	9.6844	1.6009	5.3295	317.19
42,000	255.88	2.1997	2.9948	9.6783	1.6293	5.4370	320.67
44,000	261.40	1.6950	2.2589	9.6723	1.6573	5.5438	324.12
46,000	266.93	1.3134	1.7141	9.6662	1.6851	5.5601	327.52
48,000	270.65	1.0230	1.3167	9.6602	1.7037	5.7214	329.80

Table C.5 Properties of the U.S. Standard Atmosphere* (*Continued*)

Z, m	t, K	p, N/m²	ρ, kg/m³	g, m/s²	μ · 10⁵, kg/m-s²	k · 10⁶, kcal/m-s-K	a, m/s
50,000	270.65	7.9779 + 1	1.0269	9.6542	1.7037	5.7214	329.80
55,000	265.59	4.2752	5.6075 − 4	9.6391	1.6784	5.6245	326.70
60,000	255.77	2.2461	3.0592	9.6241	1.6287	5.4349	320.61
65,000	239.28	1.1446	1.6665	9.6091	1.5433	5.1125	310.10
70,000	219.70	5.5205 + 0	8.7535 − 5	9.5941	1.4383	4.7230	297.14
75,000	200.15	2.4904	4.335	9.579	1.329	4.327	283.61
80,000	180.65	1.0366	1.999	9.564	1.216	3.925	269.44
85,000	180.65	4.1250 − 1	7.955 − 6	9.550	1.216	3.925	269.44
90,000	180.65	1.6438	3.170	9.535	1.216	3.925	269.44
95,000	195.51	6.8012 − 2	1.211	9.520			
100,000	210.02	3.0075	4.974 − 7	9.505			

* Data taken from "U.S. Standard Atmosphere, 1962," Superintendent of Documents, U.S. Government Printing Office, Washington, D.C., 1962.

Table C.6 Isentropic Flow ($\gamma = 1.40$)

M	M*	$\dfrac{t}{T}$	$\dfrac{p}{P}$	$\dfrac{\rho}{\rho_o}$	$\dfrac{\mathscr{F}}{\mathscr{F}^*}$	$\left(\dfrac{p}{P}\right)\left(\dfrac{A}{A^*}\right)$	$\dfrac{A}{A^*}$
.00	.00000	.10000+1	.10000+1	.10000+1	INFIN	INFIN	INFIN
.01	.01095	.99998+0	.99993+0	.99995+0	.45649+2	.57870+2	.57874+2
.02	.02191	.99992+0	.99972+0	.99980+0	.22834+2	.28934+2	.28942+2
.03	.03286	.99982+0	.99937+0	.99955+0	.15232+2	.19288+2	.19301+2
.04	.04381	.99968+0	.99888+0	.99920+0	.11435+2	.14465+2	.14481+2
.05	.05476	.99950+0	.99825+0	.99875+0	.91584+1	.11571+2	.11591+2
.06	.06570	.99928+0	.99748+0	.99820+0	.76428+1	.96416+1	.96659+1
.07	.07664	.99902+0	.99658+0	.99755+0	.65620+1	.82631+1	.82915+1
.08	.08758	.99872+0	.99553+0	.99681+0	.57529+1	.72292+1	.72616+1
.09	.09851	.99838+0	.99435+0	.99596+0	.51249+1	.64248+1	.64613+1
.10	.10944	.99800+0	.99303+0	.99502+0	.46236+1	.57813+1	.58218+1
.11	.12035	.99759+0	.99158+0	.99398+0	.42146+1	.52546+1	.52992+1
.12	.13126	.99713+0	.98998+0	.99284+0	.38747+1	.48156+1	.48643+1
.13	.14217	.99663+0	.98826+0	.99160+0	.35881+1	.44441+1	.44969+1
.14	.15306	.99610+0	.98640+0	.99027+0	.33432+1	.41255+1	.41824+1
.15	.16395	.99552+0	.98441+0	.98884+0	.31317+1	.38494+1	.39103+1
.16	.17482	.99491+0	.98228+0	.98731+0	.29474+1	.36077+1	.36727+1
.17	.18569	.99425+0	.98003+0	.98569+0	.27855+1	.33943+1	.34635+1
.18	.19654	.99356+0	.97765+0	.98398+0	.26422+1	.32047+1	.32779+1
.19	.20739	.99283+0	.97514+0	.98218+0	.25146+1	.30349+1	.31123+1
.20	.21822	.99206+0	.97250+0	.98028+0	.24004+1	.28820+1	.29635+1
.21	.22904	.99126+0	.96973+0	.97829+0	.22976+1	.27437+1	.28293+1
.22	.23984	.99041+0	.96685+0	.97620+0	.22046+1	.26178+1	.27076+1
.23	.25063	.98953+0	.96383+0	.97403+0	.21203+1	.25029+1	.25968+1
.24	.26141	.98861+0	.96070+0	.97177+0	.20434+1	.23975+1	.24956+1
.25	.27217	.98765+0	.95745+0	.96942+0	.19732+1	.23005+1	.24027+1
.26	.28291	.98666+0	.95408+0	.96698+0	.19088+1	.22109+1	.23173+1
.27	.29364	.98563+0	.95060+0	.96446+0	.18496+1	.21279+1	.22385+1
.28	.30435	.98456+0	.94700+0	.96185+0	.17950+1	.20508+1	.21656+1
.29	.31504	.98346+0	.94329+0	.95916+0	.17446+1	.19790+1	.20979+1
.30	.32572	.98232+0	.93947+0	.95638+0	.16979+1	.19119+1	.20351+1
.31	.33637	.98114+0	.93554+0	.95352+0	.16546+1	.18491+1	.19765+1
.32	.34701	.97993+0	.93150+0	.95058+0	.16144+1	.17902+1	.19219+1
.33	.35762	.97868+0	.92736+0	.94756+0	.15769+1	.17349+1	.18707+1
.34	.36822	.97740+0	.92312+0	.94446+0	.15420+1	.16827+1	.18229+1
.35	.37879	.97609+0	.91877+0	.94128+0	.15094+1	.16335+1	.17780+1
.36	.38935	.97473+0	.91433+0	.93803+0	.14789+1	.15871+1	.17358+1
.37	.39988	.97335+0	.90979+0	.93470+0	.14503+1	.15431+1	.16961+1
.38	.41039	.97193+0	.90516+0	.93130+0	.14236+1	.15014+1	.16587+1
.39	.42087	.97048+0	.90043+0	.92782+0	.13985+1	.14618+1	.16234+1
.40	.43133	.96899+0	.89561+0	.92427+0	.13749+1	.14242+1	.15901+1
.41	.44177	.96747+0	.89071+0	.92066+0	.13527+1	.13883+1	.15587+1
.42	.45218	.96592+0	.88572+0	.91697+0	.13318+1	.13542+1	.15289+1
.43	.46257	.96434+0	.88065+0	.91322+0	.13122+1	.13216+1	.15007+1

Table C.6   Isentropic Flow ($\gamma = 1.40$) (*Continued*)

M	M*	$\dfrac{t}{T}$	$\dfrac{p}{P}$	$\dfrac{\rho}{\rho_o}$	$\dfrac{\mathscr{F}}{\mathscr{F}^*}$	$\left(\dfrac{p}{P}\right)\left(\dfrac{A}{A^*}\right)$	$\dfrac{A}{A^*}$
.44	.47293	.96272+0	.87550+0	.90940+0	.12937+1	.12905+1	.14740+1
.45	.48326	.96108+0	.87027+0	.90551+0	.12763+1	.12607+1	.14487+1
.46	.49357	.95940+0	.86496+0	.90157+0	.12598+1	.12322+1	.14246+1
.47	.50385	.95769+0	.85958+0	.89756+0	.12443+1	.12050+1	.14018+1
.48	.51410	.95595+0	.85413+0	.89349+0	.12296+1	.11788+1	.13801+1
.49	.52433	.95418+0	.84861+0	.88936+0	.12158+1	.11537+1	.13595+1
.50	.53452	.95238+0	.84302+0	.88517+0	.12027+1	.11295+1	.13398+1
.51	.54469	.95055+0	.83737+0	.88093+0	.11903+1	.11063+1	.13212+1
.52	.55483	.94869+0	.83165+0	.87663+0	.11786+1	.10840+1	.13034+1
.53	.56493	.94681+0	.82588+0	.87228+0	.11675+1	.10625+1	.12865+1
.54	.57501	.94489+0	.82005+0	.86788+0	.11571+1	.10417+1	.12703+1
.55	.58506	.94295+0	.81417+0	.86342+0	.11471+1	.10217+1	.12549+1
.56	.59507	.94098+0	.80823+0	.85892+0	.11378+1	.10024+1	.12403+1
.57	.60505	.93898+0	.80224+0	.85437+0	.11289+1	.98381+0	.12263+1
.58	.61501	.93696+0	.79621+0	.84978+0	.11205+1	.96580+0	.12130+1
.59	.62492	.93491+0	.79013+0	.84514+0	.11126+1	.94840+0	.12003+1
.60	.63481	.93284+0	.78400+0	.84045+0	.11050+1	.93155+0	.11882+1
.61	.64466	.93073+0	.77784+0	.83573+0	.10979+1	.91525+0	.11767+1
.62	.65448	.92861+0	.77164+0	.83096+0	.10912+1	.89946+0	.11656+1
.63	.66427	.92646+0	.76540+0	.82616+0	.10848+1	.88416+0	.11552+1
.64	.67402	.92428+0	.75913+0	.82132+0	.10788+1	.86932+0	.11451+1
.65	.68374	.92208+0	.75283+0	.81644+0	.10731+1	.85493+0	.11356+1
.66	.69342	.91986+0	.74650+0	.81153+0	.10678+1	.84096+0	.11265+1
.67	.70307	.91762+0	.74014+0	.80659+0	.10627+1	.82739+0	.11179+1
.68	.71268	.91535+0	.73376+0	.80162+0	.10579+1	.81422+0	.11097+1
.69	.72225	.91306+0	.72735+0	.79661+0	.10534+1	.80141+0	.11018+1
.70	.73179	.91075+0	.72093+0	.79158+0	.10492+1	.78896+0	.10944+1
.71	.74129	.90841+0	.71448+0	.78652+0	.10451+1	.77685+0	.10873+1
.72	.75076	.90606+0	.70803+0	.78143+0	.10414+1	.76507+0	.10806+1
.73	.76019	.90369+0	.70155+0	.77632+0	.10378+1	.75360+0	.10742+1
.74	.76958	.90129+0	.69507+0	.77119+0	.10345+1	.74243+0	.10681+1
.75	.77894	.89888+0	.68857+0	.76604+0	.10314+1	.73155+0	.10624+1
.76	.78825	.89644+0	.68207+0	.76086+0	.10284+1	.72095+0	.10570+1
.77	.79753	.89399+0	.67556+0	.75567+0	.10257+1	.71061+0	.10519+1
.78	.80677	.89152+0	.66905+0	.75046+0	.10231+1	.70053+0	.10471+1
.79	.81597	.88903+0	.66254+0	.74523+0	.10208+1	.69070+0	.10425+1
.80	.82514	.88652+0	.65602+0	.73999+0	.10185+1	.68110+0	.10382+1
.81	.83426	.88400+0	.64951+0	.73474+0	.10165+1	.67173+0	.10342+1
.82	.84335	.88146+0	.64300+0	.72947+0	.10145+1	.66259+0	.10305+1
.83	.85239	.87890+0	.63650+0	.72419+0	.10128+1	.65366+0	.10270+1
.84	.86140	.87633+0	.63000+0	.71891+0	.10112+1	.64493+0	.10237+1
.85	.87037	.87374+0	.62351+0	.71361+0	.10097+1	.63640+0	.10207+1
.86	.87929	.87114+0	.61703+0	.70831+0	.10083+1	.62806+0	.10179+1
.87	.88818	.86852+0	.61057+0	.70300+0	.10070+1	.61991+0	.10153+1
.88	.89703	.86589+0	.60412+0	.69768+0	.10059+1	.61193+0	.10129+1
.89	.90583	.86324+0	.59768+0	.69236+0	.10049+1	.60413+0	.10108+1
.90	.91460	.86059+0	.59126+0	.68704+0	.10040+1	.59650+0	.10089+1
.91	.92332	.85791+0	.58486+0	.68172+0	.10032+1	.58903+0	.10071+1
.92	.93201	.85523+0	.57848+0	.67640+0	.10025+1	.58171+0	.10056+1
.93	.94065	.85253+0	.57211+0	.67108+0	.10019+1	.57455+0	.10043+1
.94	.94925	.84982+0	.56578+0	.66576+0	.10014+1	.56753+0	.10031+1
.95	.95781	.84710+0	.55946+0	.66044+0	.10009+1	.56066+0	.10021+1
.96	.96633	.84437+0	.55317+0	.65513+0	.10006+1	.55392+0	.10014+1
.97	.97481	.84162+0	.54691+0	.64982+0	.10003+1	.54732+0	.10008+1
.98	.98325	.83887+0	.54067+0	.64452+0	.10001+1	.54085+0	.10003+1
.99	.99165	.83611+0	.53446+0	.63923+0	.10000+1	.53451+0	.10001+1
1.00	1.00000	.83333+0	.52828+0	.63394+0	.10000+1	.52828+0	.10000+1
1.01	1.00831	.83055+0	.52213+0	.62866+0	.10000+1	.52218+0	.10001+1
1.02	1.01658	.82776+0	.51602+0	.62339+0	.10001+1	.51619+0	.10003+1
1.03	1.02481	.82496+0	.50994+0	.61813+0	.10003+1	.51031+0	.10007+1
1.04	1.03300	.82215+0	.50389+0	.61289+0	.10005+1	.50454+0	.10013+1
1.05	1.04114	.81934+0	.49787+0	.60765+0	.10008+1	.49888+0	.10020+1
1.06	1.04925	.81651+0	.49189+0	.60243+0	.10012+1	.49332+0	.10029+1
1.07	1.05731	.81368+0	.48595+0	.59722+0	.10016+1	.48787+0	.10039+1
1.08	1.06533	.81085+0	.48005+0	.59203+0	.10020+1	.48250+0	.10051+1
1.09	1.07331	.80800+0	.47418+0	.58686+0	.10025+1	.47724+0	.10064+1
1.10	1.08124	.80515+0	.46835+0	.58170+0	.10031+1	.47207+0	.10079+1
1.11	1.08913	.80230+0	.46257+0	.57655+0	.10036+1	.46698+0	.10095+1
1.12	1.09699	.79944+0	.45682+0	.57143+0	.10043+1	.46199+0	.10113+1
1.13	1.10479	.79657+0	.45111+0	.56632+0	.10050+1	.45708+0	.10132+1
1.14	1.11256	.79370+0	.44545+0	.56123+0	.10057+1	.45225+0	.10153+1

Table C.6 Isentropic Flow ($\gamma = 1.40$) (*Continued*)

M	M*	$\dfrac{t}{T}$	$\dfrac{p}{P}$	$\dfrac{\rho}{\rho_o}$	$\dfrac{\mathscr{F}}{\mathscr{F}*}$	$\left(\dfrac{p}{P}\right)\left(\dfrac{A}{A*}\right)$	$\dfrac{A}{A*}$
1.15	1.12029	.79083+0	.439A3+0	.55616+0	.10065+1	.44751+0	.10175+1
1.16	1.12797	.78795+0	.43425+0	.55112+0	.10073+1	.44284+0	.10198+1
1.17	1.13561	.78506+0	.42872+0	.54609+0	.10081+1	.43825+0	.10222+1
1.18	1.14321	.78218+0	.42322+0	.54108+0	.10090+1	.43374+0	.10248+1
1.19	1.15077	.77929+0	.41778+0	.53610+0	.10099+1	.42930+0	.10276+1
1.20	1.15828	.77640+0	.41238+0	.53114+0	.10108+1	.42493+0	.10304+1
1.21	1.16575	.77350+0	.40702+0	.52620+0	.10118+1	.42063+0	.10334+1
1.22	1.17319	.77061+0	.40171+0	.52129+0	.10128+1	.41640+0	.10366+1
1.23	1.18057	.76771+0	.39645+0	.51640+0	.10138+1	.41224+0	.10398+1
1.24	1.18792	.76481+0	.39123+0	.51154+0	.10149+1	.40814+0	.10432+1
1.25	1.19523	.76190+0	.38606+0	.50670+0	.10159+1	.40411+0	.10468+1
1.26	1.20249	.75900+0	.38093+0	.50189+0	.10170+1	.40014+0	.10504+1
1.27	1.20972	.75610+0	.375A6+0	.49710+0	.10182+1	.39622+0	.10542+1
1.28	1.21690	.75319+0	.37083+0	.49234+0	.10193+1	.39237+0	.10581+1
1.29	1.22404	.75029+0	.365A5+0	.48761+0	.10205+1	.38858+0	.10621+1
1.30	1.23114	.74738+0	.36091+0	.48290+0	.10217+1	.38484+0	.10663+1
1.31	1.23819	.74448+0	.35603+0	.47822+0	.10229+1	.38116+0	.10706+1
1.32	1.24521	.74158+0	.35119+0	.47357+0	.10241+1	.37754+0	.10750+1
1.33	1.25218	.73867+0	.34640+0	.46895+0	.10254+1	.37396+0	.10796+1
1.34	1.25912	.73577+0	.34166+0	.46436+0	.10267+1	.37044+0	.10842+1
1.35	1.26601	.73287+0	.33697+0	.45980+0	.10279+1	.36697+0	.10890+1
1.36	1.27286	.72997+0	.33233+0	.45526+0	.10292+1	.36355+0	.10940+1
1.37	1.27968	.72707+0	.32773+0	.45076+0	.10306+1	.36018+0	.10990+1
1.38	1.28645	.72418+0	.32319+0	.44628+0	.10319+1	.35686+0	.11042+1
1.39	1.29318	.72128+0	.31869+0	.44184+0	.10332+1	.35359+0	.11095+1
1.40	1.29987	.71839+0	.31424+0	.43742+0	.10346+1	.35036+0	.11149+1
1.41	1.30652	.71550+0	.30984+0	.43304+0	.10360+1	.34717+0	.11205+1
1.42	1.31313	.71262+0	.30549+0	.42869+0	.10373+1	.34403+0	.11262+1
1.43	1.31970	.70973+0	.30118+0	.42436+0	.10387+1	.34093+0	.11320+1
1.44	1.32623	.70685+0	.29693+0	.42007+0	.10401+1	.33788+0	.11379+1
1.45	1.33272	.70398+0	.29272+0	.41581+0	.10415+1	.33486+0	.11440+1
1.46	1.33917	.70110+0	.28856+0	.41158+0	.10430+1	.33189+0	.11501+1
1.47	1.34558	.69824+0	.28445+0	.40739+0	.10444+1	.32896+0	.11565+1
1.48	1.35195	.69537+0	.28039+0	.40322+0	.10458+1	.32606+0	.11629+1
1.49	1.35828	.69251+0	.27637+0	.39909+0	.10473+1	.32321+0	.11695+1
1.50	1.36458	.68966+0	.27240+0	.39498+0	.10487+1	.32039+0	.11762+1
1.51	1.37083	.68680+0	.26848+0	.39091+0	.10502+1	.31761+0	.11830+1
1.52	1.37705	.68396+0	.26461+0	.38688+0	.10516+1	.31487+0	.11899+1
1.53	1.38322	.68112+0	.26078+0	.38287+0	.10531+1	.31216+0	.11970+1
1.54	1.38936	.67828+0	.25700+0	.37890+0	.10546+1	.30949+0	.12042+1
1.55	1.39546	.67545+0	.25326+0	.37495+0	.10560+1	.30685+0	.12116+1
1.56	1.40152	.67262+0	.24957+0	.37105+0	.10575+1	.30424+0	.12190+1
1.57	1.40755	.66980+0	.24593+0	.36717+0	.10590+1	.30167+0	.12266+1
1.58	1.41353	.66699+0	.24233+0	.36332+0	.10605+1	.29913+0	.12344+1
1.59	1.41948	.66418+0	.23878+0	.35951+0	.10620+1	.29662+0	.12422+1
1.60	1.42539	.66138+0	.23527+0	.35573+0	.10635+1	.29414+0	.12502+1
1.61	1.43127	.65858+0	.23181+0	.35198+0	.10650+1	.29170+0	.12584+1
1.62	1.43710	.65579+0	.22839+0	.34827+0	.10665+1	.28928+0	.12666+1
1.63	1.44290	.65301+0	.22501+0	.34458+0	.10680+1	.28690+0	.12750+1
1.64	1.44866	.65023+0	.22168+0	.34093+0	.10695+1	.28454+0	.12836+1
1.65	1.45439	.64746+0	.21839+0	.33731+0	.10710+1	.28221+0	.12922+1
1.66	1.46008	.64470+0	.21515+0	.33372+0	.10725+1	.27991+0	.13010+1
1.67	1.46573	.64194+0	.21195+0	.33017+0	.10740+1	.27764+0	.13100+1
1.68	1.47135	.63919+0	.20879+0	.32664+0	.10755+1	.27540+0	.13190+1
1.69	1.47693	.63645+0	.20567+0	.32315+0	.10770+1	.27318+0	.13283+1
1.70	1.48247	.63371+0	.20259+0	.31969+0	.10785+1	.27099+0	.13376+1
1.71	1.48798	.63099+0	.19956+0	.31626+0	.10800+1	.26883+0	.13471+1
1.72	1.49345	.62827+0	.19656+0	.31287+0	.10815+1	.26669+0	.13567+1
1.73	1.49889	.62556+0	.19361+0	.30950+0	.10830+1	.26457+0	.13665+1
1.74	1.50429	.62285+0	.19070+0	.30617+0	.10845+1	.26248+0	.13764+1
1.75	1.50966	.62016+0	.18782+0	.30287+0	.10860+1	.26042+0	.13865+1
1.76	1.51499	.61747+0	.18499+0	.29959+0	.10875+1	.25837+0	.13967+1
1.77	1.52029	.61479+0	.18219+0	.29635+0	.10890+1	.25636+0	.14070+1
1.78	1.52555	.61211+0	.17944+0	.29315+0	.10905+1	.25436+0	.14175+1
1.79	1.53078	.60945+0	.17672+0	.28997+0	.10920+1	.25239+0	.14282+1
1.80	1.53598	.60680+0	.17404+0	.28682+0	.10935+1	.25044+0	.14390+1
1.81	1.54114	.60415+0	.17140+0	.28370+0	.10950+1	.24851+0	.14499+1
1.82	1.54626	.60151+0	.16879+0	.28061+0	.10965+1	.24661+0	.14610+1
1.83	1.55136	.59888+0	.16622+0	.27756+0	.10980+1	.24472+0	.14723+1
1.84	1.55642	.59626+0	.16369+0	.27453+0	.10995+1	.24286+0	.14836+1
1.85	1.56145	.59365+0	.16119+0	.27153+0	.11009+1	.24102+0	.14952+1

Table C.6   Isentropic Flow ($\gamma = 1.40$) (*Continued*)

M	M^*	$\dfrac{t}{T}$	$\dfrac{p}{P}$	$\dfrac{\rho}{\rho_o}$	$\dfrac{\mathscr{F}}{\mathscr{F}^*}$	$\left(\dfrac{p}{P}\right)\left(\dfrac{A}{A^*}\right)$	$\dfrac{A}{A^*}$
1.86	1.56644	.59104+0	.15873+0	.26857+0	.11024+1	.23920+0	.15069+1
1.87	1.57140	.58845+0	.15631+0	.26563+0	.11039+1	.23739+0	.15187+1
1.88	1.57633	.58586+0	.15392+0	.26272+0	.11054+1	.23561+0	.15308+1
1.89	1.58123	.58329+0	.15156+0	.25984+0	.11068+1	.23385+0	.15429+1
1.90	1.58609	.58072+0	.14924+0	.25699+0	.11083+1	.23211+0	.15553+1
1.91	1.59092	.57816+0	.14695+0	.25417+0	.11097+1	.23038+0	.15677+1
1.92	1.59572	.57561+0	.14470+0	.25138+0	.11112+1	.22868+0	.15804+1
1.93	1.60049	.57307+0	.14247+0	.24861+0	.11126+1	.22699+0	.15932+1
1.94	1.60523	.57054+0	.14028+0	.24588+0	.11141+1	.22532+0	.16062+1
1.95	1.60993	.56802+0	.13813+0	.24317+0	.11155+1	.22367+0	.16193+1
1.96	1.61460	.56551+0	.13600+0	.24049+0	.11170+1	.22203+0	.16326+1
1.97	1.61925	.56301+0	.13390+0	.23784+0	.11184+1	.22042+0	.16461+1
1.98	1.62386	.56051+0	.13184+0	.23521+0	.11198+1	.21882+0	.16597+1
1.99	1.62844	.55803+0	.12981+0	.23262+0	.11213+1	.21724+0	.16735+1
2.00	1.63299	.55556+0	.12780+0	.23005+0	.11227+1	.21567+0	.16875+1
2.02	1.64201	.55064+0	.12389+0	.22499+0	.11255+1	.21259+0	.17160+1
2.04	1.65090	.54576+0	.12009+0	.22004+0	.11283+1	.20957+0	.17451+1
2.06	1.65967	.54091+0	.11640+0	.21519+0	.11311+1	.20661+0	.17750+1
2.08	1.66833	.53611+0	.11282+0	.21045+0	.11339+1	.20371+0	.18056+1
2.10	1.67687	.53135+0	.10935+0	.20580+0	.11366+1	.20088+0	.18369+1
2.12	1.68530	.52663+0	.10599+0	.20126+0	.11393+1	.19809+0	.18690+1
2.14	1.69362	.52194+0	.10273+0	.19681+0	.11420+1	.19537+0	.19018+1
2.16	1.70183	.51730+0	.99562-1	.19247+0	.11447+1	.19270+0	.19354+1
2.18	1.70992	.51269+0	.96495-1	.18821+0	.11474+1	.19008+0	.19698+1
2.20	1.71791	.50813+0	.93522-1	.18405+0	.11500+1	.18751+0	.20050+1
2.22	1.72579	.50361+0	.90640-1	.17998+0	.11526+1	.18499+0	.20409+1
2.24	1.73357	.49912+0	.87846-1	.17600+0	.11552+1	.18252+0	.20777+1
2.26	1.74125	.49468+0	.85139-1	.17211+0	.11578+1	.18010+0	.21153+1
2.28	1.74882	.49027+0	.82515-1	.16830+0	.11603+1	.17772+0	.21538+1
2.30	1.75629	.48591+0	.79973-1	.16458+0	.11628+1	.17539+0	.21931+1
2.32	1.76366	.48158+0	.77509-1	.16095+0	.11653+1	.17310+0	.22333+1
2.34	1.77093	.47730+0	.75122-1	.15739+0	.11678+1	.17086+0	.22744+1
2.36	1.77811	.47305+0	.72810-1	.15391+0	.11703+1	.16866+0	.23164+1
2.38	1.78519	.46885+0	.70570-1	.15052+0	.11727+1	.16649+0	.23593+1
2.40	1.79218	.46468+0	.68399-1	.14720+0	.11751+1	.16437+0	.24031+1
2.42	1.79907	.46056+0	.66297-1	.14395+0	.11775+1	.16229+0	.24479+1
2.44	1.80587	.45647+0	.64261-1	.14078+0	.11798+1	.16024+0	.24936+1
2.46	1.81258	.45242+0	.62288-1	.13768+0	.11821+1	.15823+0	.25403+1
2.48	1.81921	.44841+0	.60378-1	.13465+0	.11844+1	.15626+0	.25880+1
2.50	1.82574	.44444+0	.58528-1	.13169+0	.11867+1	.15432+0	.26367+1
2.52	1.83219	.44051+0	.56736-1	.12879+0	.11890+1	.15242+0	.26865+1
2.54	1.83855	.43662+0	.55000-1	.12597+0	.11912+1	.15055+0	.27372+1
2.56	1.84483	.43277+0	.53319-1	.12321+0	.11934+1	.14871+0	.27891+1
2.58	1.85103	.42895+0	.51692-1	.12051+0	.11956+1	.14691+0	.28420+1
2.60	1.85714	.42517+0	.50115-1	.11787+0	.11978+1	.14513+0	.28960+1
2.62	1.86318	.42143+0	.48589-1	.11530+0	.11999+1	.14339+0	.29511+1
2.64	1.86913	.41772+0	.47110-1	.11278+0	.12021+1	.14168+0	.30073+1
2.66	1.87501	.41406+0	.45679-1	.11032+0	.12042+1	.13999+0	.30647+1
2.68	1.88081	.41043+0	.44292-1	.10792+0	.12062+1	.13834+0	.31233+1
2.70	1.88653	.40683+0	.42950-1	.10557+0	.12083+1	.13671+0	.31830+1
2.72	1.89218	.40328+0	.41650-1	.10328+0	.12103+1	.13511+0	.32440+1
2.74	1.89775	.39976+0	.40391-1	.10104+0	.12123+1	.13354+0	.33061+1
2.76	1.90325	.39627+0	.39172-1	.98851-1	.12143+1	.13199+0	.33695+1
2.78	1.90868	.39282+0	.37992-1	.96714-1	.12163+1	.13047+0	.34342+1
2.80	1.91404	.38941+0	.36848-1	.94626-1	.12182+1	.12897+0	.35001+1
2.82	1.91933	.38603+0	.35741-1	.92587-1	.12202+1	.12750+0	.35674+1
2.84	1.92455	.38268+0	.34669-1	.90594-1	.12221+1	.12605+0	.36359+1
2.86	1.92970	.37937+0	.33631-1	.88648-1	.12240+1	.12463+0	.37058+1
2.88	1.93479	.37610+0	.32625-1	.86747-1	.12258+1	.12323+0	.37771+1
2.90	1.93981	.37286+0	.31651-1	.84889-1	.12277+1	.12185+0	.38498+1
2.92	1.94477	.36965+0	.30708-1	.83075-1	.12295+1	.12049+0	.39238+1
2.94	1.94966	.36647+0	.29795-1	.81302-1	.12313+1	.11916+0	.39993+1
2.96	1.95449	.36333+0	.28910-1	.79571-1	.12331+1	.11785+0	.40763+1
2.98	1.95925	.36022+0	.28054-1	.77879-1	.12348+1	.11655+0	.41547+1
3.00	1.96396	.35714+0	.27224-1	.76226-1	.12366+1	.11528+0	.42346+1
3.10	1.98661	.34223+0	.23449-1	.68517-1	.12450+1	.10921+0	.46573+1
3.20	2.00786	.32808+0	.20228-1	.61654-1	.12530+1	.10359+0	.51210+1
3.30	2.02781	.31466+0	.17477-1	.55541-1	.12605+1	.98371-1	.56286+1
3.40	2.04656	.30193+0	.15125-1	.50093-1	.12676+1	.93526-1	.61837+1
3.50	2.06419	.28986+0	.13111-1	.45233-1	.12743+1	.89018-1	.67896+1
3.60	2.08077	.27840+0	.11385-1	.40894-1	.12807+1	.84818-1	.74501+1
3.70	2.09639	.26752+0	.99029-2	.37017-1	.12867+1	.80897-1	.81691+1

Table C.6 Isentropic Flow ($\gamma = 1.40$) (*Continued*)

M	M*	$\dfrac{t}{T}$	$\dfrac{p}{P}$	$\dfrac{\rho}{\rho_0}$	$\dfrac{\mathscr{F}}{\mathscr{F}^*}$	$\left(\dfrac{p}{P}\right)\left(\dfrac{A}{A^*}\right)$	$\dfrac{A}{A^*}$
3.80	2.11111	.25720+0	.86290-2	.33549-1	.12924+1	.77234-1	.89506+1
3.90	2.12499	.24740+0	.75320-2	.30445-1	.12978+1	.73806-1	.97990+1
4.00	2.13809	.23810+0	.65861-2	.27662-1	.13029+1	.70595-1	.10719+2
4.10	2.15046	.22925+0	.57690-2	.25164-1	.13077+1	.67582-1	.11715+2
4.20	2.16215	.22085+0	.50621-2	.22921-1	.13123+1	.64752-1	.12792+2
4.30	2.17321	.21286+0	.44494-2	.20903-1	.13167+1	.62091-1	.13955+2
4.40	2.18368	.20525+0	.39176-2	.19087-1	.13208+1	.59587-1	.15210+2
4.50	2.19360	.19802+0	.34553-2	.17449-1	.13247+1	.57227-1	.16562+2
4.60	2.20300	.19113+0	.30526-2	.15971-1	.13285+1	.55000-1	.18018+2
4.70	2.21192	.18457+0	.27012-2	.14635-1	.13320+1	.52898-1	.19583+2
4.80	2.22038	.17832+0	.23943-2	.13427-1	.13354+1	.50911-1	.21264+2
4.90	2.22842	.17235+0	.21256-2	.12333-1	.13386+1	.49031-1	.23067+2
5.00	2.23607	.16667+0	.18900-2	.11340-1	.13416+1	.47251-1	.25000+2
5.10	2.24334	.16124+0	.16832-2	.10439-1	.13446+1	.45564-1	.27070+2
5.20	2.25026	.15605+0	.15013-2	.96204-2	.13473+1	.43963-1	.29283+2
5.30	2.25685	.15110+0	.13411-2	.88753-2	.13500+1	.42444-1	.31649+2
5.40	2.26313	.14637+0	.11997-2	.81965-2	.13525+1	.41000-1	.34175+2
5.50	2.26913	.14184+0	.10748-2	.75775-2	.13549+1	.39628-1	.36869+2
5.60	2.27484	.13751+0	.96430-3	.70124-2	.13572+1	.38321-1	.39740+2
5.70	2.28030	.13337+0	.86635-3	.64959-2	.13594+1	.37077-1	.42797+2
5.80	2.28552	.12940+0	.77941-3	.60233-2	.13615+1	.35892-1	.46050+2
5.90	2.29051	.12560+0	.70214-3	.55904-2	.13635+1	.34761-1	.49507+2
6.00	2.29528	.12195+0	.63336-3	.51936-2	.13655+1	.33682-1	.53180+2
6.50	2.31626	.10582+0	.38547-3	.36427-2	.13740+1	.28962-1	.75134+2
7.00	2.33333	.92593-1	.24156-3	.26088-2	.13810+1	.25156-1	.10414+3
7.50	2.34738	.81633-1	.15543-3	.19040-2	.13867+1	.22046-1	.14184+3
8.00	2.35907	.72464-1	.10243-3	.14135-2	.13915+1	.19473-1	.19011+3
8.50	2.36889	.64725-1	.68984-4	.10658-2	.13955+1	.17321-1	.25109+3
9.00	2.37722	.58140-1	.47386-4	.81504-3	.13989+1	.15504-1	.32719+3
9.50	2.38433	.52493-1	.33141-4	.63134-3	.14019+1	.13957-1	.42113+3
10.00	2.39046	.47619-1	.23563-4	.49483-3	.14044+1	.12628-1	.53594+3
11.00	2.40040	.39683-1	.12448-4	.31369-3	.14085+1	.10480-1	.84191+3
12.00	2.40804	.33557-1	.69222-5	.20628-3	.14117+1	.88342-2	.12762+4
13.00	2.41404	.28736-1	.40223-5	.13998-3	.14141+1	.75461-2	.18761+4
14.00	2.41883	.24876-1	.24278-5	.97597-4	.14161+1	.65195-2	.26854+4
15.00	2.42272	.21739-1	.15146-5	.69680-4	.14177+1	.56883-2	.37552+4
16.00	2.42591	.19157-1	.97309-6	.50795-4	.14191+1	.50061-2	.51446+4
17.00	2.42857	.17007-1	.64147-6	.37719-4	.14202+1	.44393-2	.69205+4
18.00	2.43081	.15198-1	.43272-6	.28473-4	.14211+1	.39634-2	.91593+4
19.00	2.43270	.13661-1	.29800-6	.21813-4	.14219+1	.35600-2	.11946+5
20.00	2.43432	.12346-1	.20908-6	.16935-4	.14226+1	.32150-2	.15377+5
INFIN	2.44949	.0	.0	.0	.14289+1	.0	INFIN

Table C.6 Isentropic Flow ($\gamma = 1.30$)

M	M*	$\dfrac{t}{T}$	$\dfrac{p}{P}$	$\dfrac{\rho}{\rho_0}$	$\dfrac{\mathscr{F}}{\mathscr{F}^*}$	$\left(\dfrac{p}{P}\right)\left(\dfrac{A}{A^*}\right)$	$\dfrac{A}{A^*}$
.00	.00000	.10000+1	.10000+1	.10000+1	INFIN	INFIN	INFIN
.01	.01072	.99999+0	.99994+0	.99995+0	.46631+2	.58522+2	.58526+2
.02	.02145	.99994+0	.99974+0	.99980+0	.23324+2	.29261+2	.29268+2
.03	.03217	.99987+0	.99942+0	.99955+0	.15559+2	.19506+2	.19518+2
.04	.04289	.99976+0	.99896+0	.99920+0	.11679+2	.14629+2	.14644+2
.05	.05361	.99963+0	.99838+0	.99875+0	.93536+1	.11702+2	.11721+2
.06	.06433	.99946+0	.99766+0	.99820+0	.78051+1	.97512+1	.97740+1
.07	.07504	.99927+0	.99682+0	.99755+0	.67007+1	.83573+1	.83840+1
.08	.08575	.99904+0	.99585+0	.99681+0	.58738+1	.73118+1	.73423+1
.09	.09646	.99879+0	.99475+0	.99596+0	.52320+1	.64986+1	.65329+1
.10	.10716	.99850+0	.99353+0	.99502+0	.47196+1	.58479+1	.58860+1
.11	.11785	.99819+0	.99217+0	.99397+0	.43014+1	.53154+1	.53574+1
.12	.12855	.99784+0	.99069+0	.99283+0	.39539+1	.48716+1	.49174+1
.13	.13925	.99747+0	.98909+0	.99160+0	.36607+1	.44961+1	.45457+1
.14	.14991	.99707+0	.98736+0	.99026+0	.34102+1	.41741+1	.42275+1
.15	.16059	.99664+0	.98551+0	.98883+0	.31939+1	.38950+1	.39522+1
.16	.17125	.99617+0	.98353+0	.98731+0	.30053+1	.36507+1	.37118+1
.17	.18191	.99568+0	.98143+0	.98568+0	.28396+1	.34351+1	.35001+1
.18	.19256	.99516+0	.97921+0	.98397+0	.26929+1	.32434+1	.33123+1
.19	.20320	.99461+0	.97687+0	.98216+0	.25622+1	.30718+1	.31446+1

Table C.6   Isentropic Flow ($\gamma = 1.30$) (*Continued*)

M	M*	$\dfrac{t}{T}$	$\dfrac{p}{P}$	$\dfrac{\rho}{\rho_o}$	$\dfrac{\mathscr{F}}{\mathscr{F}^*}$	$\left(\dfrac{p}{P}\right)\left(\dfrac{A}{A^*}\right)$	$\dfrac{A}{A^*}$
.20	.21384	.99404+0	.97441+0	.98026+0	.24452+1	.29174+1	.29940+1
.21	.22446	.99343+0	.97183+0	.97826+0	.23398+1	.27776+1	.28581+1
.22	.23507	.99279+0	.96914+0	.97618+0	.22445+1	.26505+1	.27349+1
.23	.24567	.99213+0	.96633+0	.97400+0	.21580+1	.25344+1	.26227+1
.24	.25627	.99143+0	.96341+0	.97173+0	.20792+1	.24280+1	.25202+1
.25	.26685	.99071+0	.96037+0	.96937+0	.20072+1	.23300+1	.24262+1
.26	.27742	.98996+0	.95722+0	.96693+0	.19411+1	.22396+1	.23396+1
.27	.28797	.98918+0	.95397+0	.96440+0	.18803+1	.21558+1	.22598+1
.28	.29852	.98838+0	.95060+0	.96178+0	.18242+1	.20779+1	.21859+1
.29	.30905	.98754+0	.94713+0	.95907+0	.17724+1	.20054+1	.21174+1
.30	.31956	.98668+0	.94355+0	.95629+0	.17244+1	.19377+1	.20537+1
.31	.33007	.98579+0	.93986+0	.95341+0	.16799+1	.18744+1	.19943+1
.32	.34056	.98487+0	.93608+0	.95046+0	.16385+1	.18150+1	.19389+1
.33	.35103	.98393+0	.93220+0	.94742+0	.15999+1	.17591+1	.18871+1
.34	.36149	.98296+0	.92821+0	.94431+0	.15639+1	.17065+1	.18385+1
.35	.37193	.98196+0	.92413+0	.94111+0	.15303+1	.16569+1	.17930+1
.36	.38236	.98093+0	.91995+0	.93784+0	.14989+1	.16101+1	.17502+1
.37	.39277	.97988+0	.91568+0	.93449+0	.14694+1	.15657+1	.17099+1
.38	.40316	.97880+0	.91132+0	.93106+0	.14418+1	.15237+1	.16719+1
.39	.41354	.97769+0	.90687+0	.92756+0	.14158+1	.14838+1	.16361+1
.40	.42390	.97656+0	.90233+0	.92399+0	.13915+1	.14458+1	.16023+1
.41	.43424	.97541+0	.89771+0	.92034+0	.13686+1	.14097+1	.15704+1
.42	.44456	.97422+0	.89300+0	.91663+0	.13470+1	.13753+1	.15401+1
.43	.45486	.97301+0	.88821+0	.91284+0	.13267+1	.13425+1	.15115+1
.44	.46514	.97178+0	.88334+0	.90899+0	.13075+1	.13112+1	.14843+1
.45	.47541	.97052+0	.87839+0	.90507+0	.12894+1	.12812+1	.14586+1
.46	.48565	.96924+0	.87336+0	.90109+0	.12724+1	.12525+1	.14341+1
.47	.49587	.96793+0	.86827+0	.89704+0	.12563+1	.12250+1	.14109+1
.48	.50607	.96659+0	.86310+0	.89292+0	.12410+1	.11987+1	.13888+1
.49	.51625	.96524+0	.85785+0	.88875+0	.12266+1	.11734+1	.13678+1
.50	.52641	.96386+0	.85255+0	.88452+0	.12130+1	.11491+1	.13479+1
.52	.54666	.96102+0	.84174+0	.87588+0	.11880+1	.11033+1	.13107+1
.54	.56682	.95809+0	.83068+0	.86701+0	.11655+1	.10608+1	.12770+1
.56	.58689	.95507+0	.81939+0	.85794+0	.11454+1	.10213+1	.12464+1
.58	.60686	.95196+0	.80790+0	.84866+0	.11273+1	.98448+0	.12186+1
.60	.62673	.94877+0	.79620+0	.83920+0	.11112+1	.95007+0	.11932+1
.62	.64650	.94546+0	.78433+0	.82956+0	.10966+1	.91783+0	.11702+1
.64	.66616	.94212+0	.77230+0	.81975+0	.10836+1	.88756+0	.11492+1
.66	.68572	.93867+0	.76012+0	.80979+0	.10720+1	.85909+0	.11302+1
.68	.70517	.93514+0	.74782+0	.79969+0	.10616+1	.83225+0	.11129+1
.70	.72451	.93153+0	.73540+0	.78945+0	.10524+1	.80691+0	.10972+1
.72	.74374	.92785+0	.72289+0	.77910+0	.10441+1	.78295+0	.10831+1
.74	.76285	.92409+0	.71029+0	.76864+0	.10369+1	.76024+0	.10703+1
.76	.78184	.92027+0	.69764+0	.75808+0	.10304+1	.73870+0	.10589+1
.78	.80072	.91637+0	.68493+0	.74743+0	.10248+1	.71823+0	.10486+1
.80	.81947	.91241+0	.67218+0	.73671+0	.10199+1	.69876+0	.10395+1
.82	.83810	.90838+0	.65942+0	.72593+0	.10156+1	.68021+0	.10315+1
.84	.85661	.90429+0	.64665+0	.71509+0	.10120+1	.66252+0	.10245+1
.86	.87499	.90014+0	.63388+0	.70420+0	.10089+1	.64563+0	.10185+1
.88	.89324	.89593+0	.62113+0	.69329+0	.10064+1	.62948+0	.10134+1
.90	.91136	.89166+0	.60842+0	.68234+0	.10043+1	.61402+0	.10092+1
.92	.92936	.88734+0	.59575+0	.67139+0	.10027+1	.59922+0	.10058+1
.94	.94722	.88297+0	.58313+0	.66042+0	.10015+1	.58502+0	.10032+1
.96	.96495	.87855+0	.57058+0	.64946+0	.10006+1	.57140+0	.10014+1
.98	.98254	.87408+0	.55811+0	.63851+0	.10002+1	.55831+0	.10004+1
1.00	1.00000	.86957+0	.54573+0	.62759+0	.10000+1	.54573+0	.10000+1
1.02	1.01732	.86501+0	.53344+0	.61669+0	.10001+1	.53362+0	.10003+1
1.04	1.03451	.86041+0	.52126+0	.60583+0	.10006+1	.52197+0	.10014+1
1.06	1.05156	.85577+0	.50919+0	.59501+0	.10013+1	.51074+0	.10030+1
1.08	1.06847	.85109+0	.49724+0	.58424+0	.10022+1	.49991+0	.10054+1
1.10	1.08524	.84638+0	.48542+0	.57353+0	.10033+1	.48946+0	.10083+1
1.12	1.10187	.84164+0	.47374+0	.56288+0	.10047+1	.47937+0	.10119+1
1.14	1.11836	.83686+0	.46220+0	.55231+0	.10063+1	.46962+0	.10160+1
1.16	1.13471	.83206+0	.45081+0	.54181+0	.10080+1	.46020+0	.10208+1

Table C.6 Isentropic Flow ($\gamma = 1.30$) (*Continued*)

M	M^*	$\dfrac{t}{T}$	$\dfrac{p}{P}$	$\dfrac{\rho}{\rho_o}$	$\dfrac{\mathscr{F}}{\mathscr{F}^*}$	$\left(\dfrac{p}{P}\right)\left(\dfrac{A}{A^*}\right)$	$\dfrac{A}{A^*}$
1.18	1.15091	.82723+0	.43958+0	.53139+0	.10099+1	.45108+0	.10262+1
1.20	1.16698	.82237+0	.42850+0	.52106+0	.10119+1	.44226+0	.10321+1
1.22	1.18290	.81749+0	.41759+0	.51082+0	.10141+1	.43372+0	.10386+1
1.24	1.19869	.81259+0	.40685+0	.50068+0	.10165+1	.42544+0	.10457+1
1.26	1.21432	.80766+0	.39628+0	.49065+0	.10189+1	.41742+0	.10533+1
1.28	1.22982	.80272+0	.38588+0	.48071+0	.10215+1	.40964+0	.10616+1
1.30	1.24517	.79777+0	.37566+0	.47089+0	.10241+1	.40209+0	.10703+1
1.32	1.26039	.79280+0	.36562+0	.46118+0	.10269+1	.39476+0	.10797+1
1.34	1.27545	.78781+0	.35576+0	.45159+0	.10297+1	.38764+0	.10896+1
1.36	1.29038	.78282+0	.34609+0	.44211+0	.10327+1	.38073+0	.11001+1
1.38	1.30516	.77781+0	.33660+0	.43276+0	.10357+1	.37401+0	.11111+1
1.40	1.31981	.77280+0	.32730+0	.42353+0	.10387+1	.36748+0	.11227+1
1.42	1.33430	.76778+0	.31819+0	.41443+0	.10419+1	.36112+0	.11349+1
1.44	1.34866	.76275+0	.30927+0	.40546+0	.10451+1	.35494+0	.11477+1
1.46	1.36288	.75773+0	.30053+0	.39662+0	.10483+1	.34892+0	.11610+1
1.48	1.37696	.75269+0	.29198+0	.38791+0	.10516+1	.34306+0	.11750+1
1.50	1.39089	.74766+0	.28361+0	.37933+0	.10549+1	.33735+0	.11895+1
1.55	1.42512	.73509+0	.26352+0	.35849+0	.10634+1	.32372+0	.12284+1
1.60	1.45848	.72254+0	.24457+0	.33849+0	.10721+1	.31091+0	.12712+1
1.65	1.49099	.71004+0	.22675+0	.31935+0	.10808+1	.29887+0	.13180+1
1.70	1.52265	.69759+0	.21003+0	.30107+0	.10897+1	.28753+0	.13690+1
1.75	1.55347	.68522+0	.19436+0	.28365+0	.10986+1	.27682+0	.14243+1
1.80	1.58348	.67295+0	.17972+0	.26706+0	.11075+1	.26671+0	.14841+1
1.85	1.61268	.66077+0	.16605+0	.25129+0	.11164+1	.25715+0	.15486+1
1.90	1.64108	.64872+0	.15331+0	.23633+0	.11252+1	.24808+0	.16182+1
1.95	1.66871	.63679+0	.14147+0	.22215+0	.11340+1	.23949+0	.16929+1
2.00	1.69558	.62500+0	.13046+0	.20874+0	.11427+1	.23133+0	.17732+1
2.10	1.74710	.60187+0	.11079+0	.18408+0	.11597+1	.21620+0	.19514+1
2.20	1.79577	.57937+0	.93934-1	.16213+0	.11763+1	.20248+0	.21556+1
2.30	1.84173	.55757+0	.79548-1	.14267+0	.11923+1	.19000+0	.23885+1
2.40	1.88511	.53648+0	.67308-1	.12546+0	.12078+1	.17860+0	.26535+1
2.50	1.92605	.51613+0	.56923-1	.11029+0	.12226+1	.16818+0	.29545+1
2.60	1.96468	.49652+0	.48129-1	.96932-1	.12368+1	.15861+0	.32954+1
2.70	2.00113	.47767+0	.40696-1	.85197-1	.12504+1	.14980+0	.36811+1
2.80	2.03553	.45956+0	.34420-1	.74898-1	.12634+1	.14169+0	.41165+1
2.90	2.06799	.44218+0	.29156-1	.65869-1	.12758+1	.13419+0	.46073+1
3.00	2.09863	.42553+0	.24663-1	.57957-1	.12876+1	.12725+0	.51598+1
3.10	2.12756	.40958+0	.20900-1	.51028-1	.12988+1	.12082+0	.57807+1
3.20	2.15489	.39432+0	.17729-1	.44961-1	.13095+1	.11484+0	.64776+1
3.30	2.18070	.37972+0	.15055-1	.39648-1	.13196+1	.10928+0	.72586+1
3.40	2.20510	.36576+0	.12800-1	.34995-1	.13293+1	.10410+0	.81328+1
3.50	2.22817	.35242+0	.10896-1	.30918-1	.13385+1	.99263-1	.91098+1
3.60	2.25000	.33967+0	.92883-2	.27345-1	.13472+1	.94744-1	.10200+2
3.70	2.27066	.32749+0	.79288-2	.24211-1	.13555+1	.90516-1	.11416+2
3.80	2.29022	.31586+0	.67783-2	.21460-1	.13634+1	.86554-1	.12769+2
3.90	2.30875	.30474+0	.58034-2	.19044-1	.13709+1	.82837-1	.14274+2
4.00	2.32632	.29412+0	.49765-2	.16920-1	.13781+1	.79346-1	.15944+2
4.50	2.40162	.24768+0	.23633-2	.95417-2	.14090+1	.64723-1	.27387+2
5.00	2.46021	.21053+0	.11686-2	.55508-2	.14333+1	.53704-1	.45957+2
5.50	2.50643	.18059+0	.60114-3	.33288-2	.14527+1	.45217-1	.75220+2
6.00	2.54337	.15625+0	.32104-3	.20546-2	.14683+1	.38555-1	.12010+3
6.50	2.57329	.13629+0	.17754-3	.13027-2	.14809+1	.33238-1	.18722+3
7.00	2.59779	.11976+0	.10140-3	.84667-3	.14914+1	.28932-1	.28534+3
7.50	2.61807	.10596+0	.59651-4	.56296-3	.15000+1	.25400-1	.42581+3
8.00	2.63503	.94340-1	.36059-4	.38222-3	.15073+1	.22469-1	.62312+3
8.50	2.64934	.84477-1	.22346-4	.26452-3	.15134+1	.20011-1	.89551+3
9.00	2.66151	.76046-1	.14168-4	.18631-3	.15186+1	.17932-1	.12656+4
9.50	2.67195	.68788-1	.91737-5	.13336-3	.15231+1	.16157-1	.17612+4
10.00	2.68095	.62500-1	.60555-5	.96887-4	.15270+1	.14631-1	.24161+4
11.00	2.69561	.52219-1	.27793-5	.53224-4	.15333+1	.12158-1	.43743+4
12.00	2.70692	.44248-1	.13558-5	.30641-4	.15382+1	.10259-1	.75665+4
13.00	2.71583	.37951-1	.69708-6	.18368-4	.15420+1	.87698-2	.12581+5
14.00	2.72295	.32895-1	.37516-6	.11405-4	.15451+1	.75816-2	.20209+5
15.00	2.72874	.28777-1	.21015-6	.73027-5	.15476+1	.66184-2	.31494+5
INFIN	2.76887	.0	.0	.0	.15650+1	.0	INFIN

Table C.6 Isentropic Flow ($\gamma = 1.20$)

M	M^*	$\dfrac{t}{T}$	$\dfrac{p}{P}$	$\dfrac{\rho}{\rho_o}$	$\dfrac{\mathscr{F}}{\mathscr{F}^*}$	$\left(\dfrac{p}{P}\right)\left(\dfrac{A}{A^*}\right)$	$\dfrac{A}{A^*}$
.00	.00000	.10000+1	.10000+1	.10000+1	INFIN	INFIN	INFIN
.01	.01049	.99999+0	.99994+0	.99995+0	.47679+2	.59202+2	.59206+2
.02	.02098	.99996+0	.99976+0	.99980+0	.23848+2	.29601+2	.29608+2
.03	.03146	.99991+0	.99946+0	.99955+0	.15907+2	.19733+2	.19744+2
.04	.04195	.99984+C	.99904+0	.99920+0	.11940+2	.14799+2	.14814+2
.05	.05243	.99975+0	.99850+0	.99875+0	.95620+1	.11839+2	.11857+2
.06	.06292	.99964+0	.99784+0	.99820+0	.79784+1	.98653+1	.98866+1
.07	.07340	.99951+0	.99707+0	.99755+0	.68488+1	.84554+1	.84803+1
.08	.08388	.99936+0	.99617+0	.99681+0	.60030+1	.73979+1	.74264+1
.09	.09435	.99919+0	.99515+0	.99596+0	.53463+1	.65754+1	.66074+1
.10	.10483	.99900+0	.99402+0	.99501+0	.48221+1	.59173+1	.59529+1
.11	.11530	.99879+0	.99277+0	.99397+0	.43942+1	.53788+1	.54180+1
.12	.12577	.99856+0	.99140+0	.99283+0	.40385+1	.49300+1	.49727+1
.13	.13623	.99831+0	.98992+0	.99159+0	.37384+1	.45502+1	.45965+1
.14	.14669	.99804+0	.98832+0	.99026+0	.34819+1	.42246+1	.42745+1
.15	.15714	.99776+0	.98661+0	.98883+0	.32640+1	.39424+1	.39959+1
.16	.16760	.99745+0	.98478+0	.98730+0	.30672+1	.36954+1	.37526+1
.17	.17804	.99712+0	.98283+0	.98567+0	.28974+1	.34775+1	.35382+1
.18	.18848	.99677+0	.98078+0	.98396+0	.27470+1	.32837+1	.33481+1
.19	.19891	.99640+0	.97861+0	.98214+0	.26131+1	.31103+1	.31783+1
.20	.20934	.99602+0	.97633+0	.98024+0	.24931+1	.29542+1	.30258+1
.21	.21977	.99561+0	.97394+0	.97824+0	.23850+1	.28130+1	.28882+1
.22	.23018	.99518+0	.97145+0	.97615+0	.22873+1	.26845+1	.27634+1
.23	.24059	.99474+0	.96884+0	.97396+0	.21985+1	.25672+1	.26498+1
.24	.25099	.99427+0	.96613+0	.97169+0	.21176+1	.24597+1	.25459+1
.25	.26139	.99379+0	.96331+0	.96933+0	.20436+1	.23607+1	.24507+1
.26	.27177	.99329+0	.96038+0	.96687+0	.19757+1	.22694+1	.23630+1
.27	.28215	.99276+0	.95735+0	.96433+0	.19132+1	.21847+1	.22821+1
.28	.29252	.99222+0	.95422+0	.96171+0	.18555+1	.21061+1	.22072+1
.29	.30288	.99166+0	.95099+0	.95899+0	.18022+1	.20329+1	.21377+1
.30	.31324	.99108+0	.94766+0	.95619+0	.17529+1	.19646+1	.20731+1
.31	.32358	.99048+0	.94423+0	.95330+0	.17070+1	.19006+1	.20129+1
.32	.33391	.98986+0	.94070+0	.95034+0	.16644+1	.18407+1	.19567+1
.33	.34424	.98923+0	.93708+0	.94728+0	.16246+1	.17843+1	.19041+1
.34	.35455	.98857+0	.93336+0	.94415+0	.15875+1	.17313+1	.18549+1
.35	.36486	.98790+0	.92955+0	.94094+0	.15528+1	.16812+1	.18087+1
.36	.37515	.98721+0	.92565+0	.93765+0	.15204+1	.16340+1	.17652+1
.37	.38543	.98649+0	.92166+0	.93427+0	.14900+1	.15892+1	.17243+1
.38	.39570	.98577+0	.91758+0	.93083+0	.14614+1	.15468+1	.16858+1
.39	.40596	.98502+0	.91341+0	.92730+0	.14346+1	.15066+1	.16494+1
.40	.41621	.98425+0	.90915+0	.92370+0	.14094+1	.14684+1	.16151+1
.41	.42644	.98347+0	.90482+0	.92003+0	.13857+1	.14320+1	.15826+1
.42	.43667	.98267+0	.90040+0	.91628+0	.13634+1	.13973+1	.15519+1
.43	.44688	.98185+0	.89590+0	.91246+0	.13423+1	.13642+1	.15228+1
.44	.45707	.98101+0	.89132+0	.90858+0	.13225+1	.13327+1	.14952+1
.45	.46726	.98015+0	.88667+0	.90462+0	.13037+1	.13025+1	.14690+1
.46	.47743	.97928+0	.88194+0	.90060+0	.12860+1	.12736+1	.14441+1
.47	.48758	.97839+0	.87713+0	.89651+0	.12693+1	.12459+1	.14205+1
.48	.49773	.97748+0	.87226+0	.89235+0	.12534+1	.12194+1	.13980+1
.49	.50786	.97655+0	.86731+0	.88814+0	.12385+1	.11940+1	.13766+1
.50	.51797	.97561+0	.86230+0	.88385+0	.12243+1	.11695+1	.13563+1
.52	.53815	.97367+0	.85207+0	.87511+0	.11982+1	.11234+1	.13185+1
.54	.55828	.97167+0	.84159+0	.86613+0	.11748+1	.10807+1	.12841+1
.56	.57835	.96959+0	.83088+0	.85694+0	.11537+1	.10410+1	.12529+1
.58	.59835	.96745+0	.81994+0	.84753+0	.11348+1	.10040+1	.12245+1
.60	.61826	.96525+0	.80880+0	.83792+0	.11179+1	.96941+0	.11986+1
.62	.63811	.96298+0	.79746+0	.82812+0	.11026+1	.93704+0	.11750+1
.64	.65790	.96065+0	.78595+0	.81814+0	.10889+1	.90666+0	.11536+1
.66	.67761	.95826+0	.77428+0	.80800+0	.10767+1	.87809+0	.11341+1
.68	.69725	.95580+0	.76245+0	.79771+0	.10657+1	.85117+0	.11164+1
.70	.71681	.95329+0	.75049+0	.78727+0	.10559+1	.82576+0	.11003+1
.72	.73630	.95071+0	.73842+0	.77670+0	.10472+1	.80174+0	.10858+1
.74	.75570	.94808+0	.72624+0	.76601+0	.10395+1	.77899+0	.10726+1
.76	.77503	.94539+0	.71397+0	.75520+0	.10327+1	.75741+0	.10609+1
.78	.79427	.94265+0	.70162+0	.74430+0	.10266+1	.73692+0	.10503+1
.80	.81342	.93985+0	.68921+0	.73332+0	.10214+1	.71743+0	.10409+1
.82	.83249	.93700+0	.67675+0	.72225+0	.10169+1	.69887+0	.10327+1
.84	.85147	.93409+0	.66425+0	.71112+0	.10130+1	.68117+0	.10255+1
.86	.87036	.93113+0	.65174+0	.69994+0	.10097+1	.66427+0	.10192+1
.88	.88917	.92813+0	.63921+0	.68871+0	.10069+1	.64813+0	.10140+1
.90	.90787	.92507+0	.62668+0	.67744+0	.10047+1	.63268+0	.10096+1
.92	.92649	.92196+0	.61417+0	.66615+0	.10029+1	.61789+0	.10061+1
.94	.94501	.91881+0	.60168+0	.65484+0	.10016+1	.60371+0	.10034+1
.96	.96344	.91562+0	.58923+0	.64353+0	.10007+1	.59010+0	.10015+1

Table C.6 Isentropic Flow ($\gamma = 1.20$) *(Continued)*

M	M^*	$\dfrac{t}{T}$	$\dfrac{p}{P}$	$\dfrac{\rho}{\rho_o}$	$\dfrac{\mathscr{F}}{\mathscr{F}^*}$	$\left(\dfrac{p}{P}\right)\left(\dfrac{A}{A^*}\right)$	$\dfrac{A}{A^*}$
.98	.98177	.91236+0	.57682+0	.63222+0	.10002+1	.57703+0	.10004+1
1.00	1.00000	.90909+0	.56447+0	.62092+0	.10000+1	.56447+0	.10000+1
1.02	1.01813	.90576+0	.55219+0	.60964+0	.10002+1	.55239+0	.10004+1
1.04	1.03616	.90240+0	.53999+0	.59839+0	.10006+1	.54076+0	.10014+1
1.06	1.05409	.89899+0	.52787+0	.58718+0	.10014+1	.52956+0	.10032+1
1.08	1.07192	.89554+0	.51585+0	.57602+0	.10024+1	.51875+0	.10056+1
1.10	1.08965	.89206+0	.50393+0	.56490+0	.10037+1	.50833+0	.10087+1
1.12	1.10727	.88854+0	.49211+0	.55385+0	.10052+1	.49827+0	.10125+1
1.14	1.12479	.88499+0	.48042+0	.54286+0	.10069+1	.48854+0	.10169+1
1.16	1.14220	.88140+0	.46685+0	.53194+0	.10089+1	.47915+0	.10220+1
1.18	1.15950	.87778+0	.45741+0	.52110+0	.10110+1	.47006+0	.10276+1
1.20	1.17670	.87413+0	.44611+0	.51035+0	.10133+1	.46126+0	.10340+1
1.22	1.19379	.87044+0	.43495+0	.49969+0	.10157+1	.45274+0	.10409+1
1.24	1.21077	.86673+0	.42394+0	.48913+0	.10183+1	.44449+0	.10485+1
1.26	1.22764	.86299+0	.41308+0	.47867+0	.10211+1	.43649+0	.10567+1
1.28	1.24440	.85922+0	.40238+0	.46831+0	.10240+1	.42873+0	.10655+1
1.30	1.26105	.85543+0	.39184+0	.45807+0	.10270+1	.42120+0	.10749+1
1.32	1.27759	.85161+0	.38147+0	.44794+0	.10302+1	.41389+0	.10850+1
1.34	1.29402	.84777+0	.37126+0	.43793+0	.10334+1	.40679+0	.10957+1
1.36	1.31034	.84391+0	.36123+0	.42804+0	.10367+1	.39990+0	.11071+1
1.38	1.32654	.84003+0	.35136+0	.41828+0	.10402+1	.39319+0	.11191+1
1.40	1.34264	.83612+0	.34167+0	.40864+0	.10437+1	.38668+0	.11317+1
1.42	1.35862	.83220+0	.33216+0	.39914+0	.10473+1	.38033+0	.11450+1
1.44	1.37449	.82825+0	.32283+0	.38978+0	.10510+1	.37416+0	.11590+1
1.46	1.39024	.82429+0	.31368+0	.38055+0	.10548+1	.36815+0	.11736+1
1.48	1.40588	.82032+0	.30471+0	.37146+0	.10586+1	.36230+0	.11890+1
1.50	1.42141	.81633+0	.29593+0	.36251+0	.10625+1	.35660+0	.12050+1
1.55	1.45973	.80629+0	.27475+0	.34076+0	.10724+1	.34297+0	.12483+1
1.60	1.49734	.79618+0	.25472+0	.31993+0	.10826+1	.33016+0	.12962+1
1.65	1.53424	.78601+0	.23581+0	.30001+0	.10930+1	.31810+0	.13490+1
1.70	1.57043	.77580+0	.21801+0	.28102+0	.11036+1	.30674+0	.14070+1
1.75	1.60591	.76555+0	.20130+0	.26295+0	.11143+1	.29600+0	.14704+1
1.80	1.64068	.75529+0	.18564+0	.24579+0	.11251+1	.28584+0	.15398+1
1.85	1.67476	.74502+0	.17100+0	.22953+0	.11359+1	.27622+0	.16153+1
1.90	1.70813	.73475+0	.15734+0	.21415+0	.11468+1	.26709+0	.16975+1
1.95	1.74081	.72451+0	.14463+0	.19962+0	.11576+1	.25842+0	.17868+1
2.00	1.77281	.71429+0	.13281+0	.18593+0	.11684+1	.25018+0	.18837+1
2.10	1.83478	.69396+0	.11169+0	.16095+0	.11899+1	.23485+0	.21027+1
2.20	1.89410	.67385+0	.93626-1	.13894+0	.12110+1	.22090+0	.23594+1
2.30	1.95083	.65402+0	.78263-1	.11966+0	.12317+1	.20817+0	.26598+1
2.40	2.00507	.63452+0	.65262-1	.10285+0	.12519+1	.19649+0	.30108+1
2.50	2.05688	.61538+0	.54310-1	.88254-1	.12715+1	.18577+0	.34205+1
2.60	2.10636	.59666+0	.45119-1	.75619-1	.12906+1	.17589+0	.38983+1
2.70	2.15359	.57837+0	.37431-1	.64718-1	.13090+1	.16676+0	.44550+1
2.80	2.19865	.56054+0	.31019-1	.55338-1	.13267+1	.15830+0	.51033+1
2.90	2.24165	.54318+0	.25685-1	.47286-1	.13439+1	.15046+0	.58578+1
3.00	2.28266	.52632+0	.21256-1	.40386-1	.13604+1	.14317+0	.67354+1
3.10	2.32177	.50994+0	.17585-1	.34484-1	.13762+1	.13638+0	.77554+1
3.20	2.35907	.49407+0	.14546-1	.29441-1	.13915+1	.13004+0	.89402+1
3.30	2.39464	.47870+0	.12033-1	.25137-1	.14061+1	.12412+0	.10315+2
3.40	2.42857	.46382+0	.99565-2	.21466-1	.14202+1	.11859+0	.11910+2
3.50	2.46093	.44944+0	.82418-2	.18338-1	.14336+1	.11340+0	.13759+2
3.60	2.49180	.43554+0	.68260-2	.15673-1	.14466+1	.10853+0	.15899+2
3.70	2.52125	.42212+0	.56573-2	.13402-1	.14589+1	.10396+0	.18376+2
3.80	2.54935	.40917+0	.46924-2	.11468-1	.14708+1	.99657-1	.21238+2
3.90	2.57617	.39667+0	.38955-2	.98205-2	.14822+1	.95607-1	.24543+2
4.00	2.60177	.38462+0	.32371-2	.84165-2	.14931+1	.91790-1	.28355+2
4.50	2.71360	.33058+0	.13051-2	.39480-2	.15411+1	.75642-1	.57959+2
5.00	2.80306	.28571+0	.54399-3	.19040-2	.15799+1	.63290-1	.11634+3
5.50	2.87525	.24845+0	.23518-3	.94661-3	.16115+1	.53653-1	.22813+3
6.00	2.93406	.21739+0	.10555-3	.48552-3	.16374+1	.46006-1	.43587+3
6.50	2.98240	.19139+0	.49145-4	.25678-3	.16589+1	.39846-1	.81078+3
7.00	3.02251	.16949+0	.23708-4	.13987-3	.16767+1	.34819-1	.14687+4
7.50	3.05608	.15094+0	.11827-4	.78356-4	.16916+1	.30668-1	.25930+4
8.00	3.08440	.13514+0	.60899-5	.45065-4	.17043+1	.27204-1	.44671+4
8.50	3.10847	.12158+0	.32299-5	.26566-4	.17151+1	.24286-1	.75192+4
9.00	3.12909	.10989+0	.17610-5	.16025-4	.17243+1	.21806-1	.12383+5
9.50	3.14686	.99751-1	.98513-6	.98759-5	.17323+1	.19682-1	.19979+5
10.00	3.16228	.90909-1	.56447-6	.62092-5	.17393+1	.17850-1	.31623+5
11.00	3.18752	.76336-1	.19787-6	.25921-5	.17506+1	.14870-1	.75152+5
12.00	3.20713	.64935-1	.74968-7	.11545-5	.17595+1	.12572-1	.16770+6
13.00	3.22265	.55866-1	.30401-7	.54417-6	.17665+1	.10764-1	.35407+6
14.00	3.23512	.48544-1	.13086-7	.26957-6	.17721+1	.93171-2	.71200+6
15.00	3.24529	.42553-1	.59374-8	.13953-6	.17767+1	.81417-2	.13713+7
INFIN	3.31662	.0	.0	.0	.18091+1	.0	INFIN

Table C.6 Isentropic Flow ($\gamma = 1.10$)

M	M^*	$\dfrac{t}{T}$	$\dfrac{p}{P}$	$\dfrac{\rho}{\rho_o}$	$\dfrac{\mathscr{F}}{\mathscr{F}^*}$	$\left(\dfrac{p}{P}\right)\left(\dfrac{A}{A^*}\right)$	$\dfrac{A}{A^*}$
.00	.00000	.10000+1	.10000+1	.10000+1	INFIN	INFIN	INFIN
.01	.01025	.10000+1	.99995+0	.99995+0	.48800+2	.59912+2	.59915+2
.02	.02049	.99996+0	.99978+0	.99980+0	.24408+2	.29956+2	.29962+2
.03	.03074	.99996+0	.99951+0	.99955+0	.16281+2	.19970+2	.19980+2
.04	.04099	.99992+0	.99912+0	.99920+0	.12220+2	.14977+2	.14991+2
.05	.05123	.99988+0	.99863+0	.99875+0	.97852+1	.11982+2	.11998+2
.06	.06148	.99982+0	.99802+0	.99820+0	.81640+1	.99844+1	.10004+2
.07	.07172	.99976+0	.99731+0	.99755+0	.70074+1	.85578+1	.85809+1
.08	.08196	.99968+0	.99649+0	.99681+0	.61413+1	.74878+1	.75142+1
.09	.09220	.99960+0	.99556+0	.99596+0	.54689+1	.66555+1	.66852+1
.10	.10244	.99950+0	.99452+0	.99501+0	.49319+1	.59897+1	.60227+1
.11	.11268	.99940+0	.99337+0	.99397+0	.44936+1	.54449+1	.54812+1
.12	.12292	.99928+0	.99211+0	.99283+0	.41292+1	.49909+1	.50305+1
.13	.13315	.99916+0	.99075+0	.99159+0	.38216+1	.46067+1	.46497+1
.14	.14339	.99902+0	.98928+0	.99025+0	.35588+1	.42773+1	.43237+1
.15	.15362	.99888+0	.98771+0	.98882+0	.33316+1	.39919+1	.40416+1
.16	.16385	.99872+0	.98603+0	.98729+0	.31336+1	.37421+1	.37951+1
.17	.17407	.99856+0	.98424+0	.98566+0	.29594+1	.35217+1	.35781+1
.18	.18430	.99838+0	.98235+0	.98394+0	.28052+1	.33257+1	.33855+1
.19	.19452	.99820+0	.98036+0	.98213+0	.26677+1	.31504+1	.32135+1
.20	.20473	.99800+0	.97826+0	.98022+0	.25446+1	.29926+1	.30591+1
.21	.21495	.99780+0	.97606+0	.97822+0	.24336+1	.28498+1	.29197+1
.22	.22516	.99759+0	.97376+0	.97612+0	.23332+1	.27200+1	.27933+1
.23	.23537	.99736+0	.97136+0	.97393+0	.22420+1	.26014+1	.26781+1
.24	.24557	.99713+0	.96886+0	.97165+0	.21588+1	.24927+1	.25729+1
.25	.25577	.99688+0	.96626+0	.96928+0	.20827+1	.23927+1	.24763+1
.26	.26597	.99663+0	.96356+0	.96682+0	.20129+1	.23004+1	.23874+1
.27	.27616	.99637+0	.96077+0	.96427+0	.19486+1	.22149+1	.23054+1
.28	.28635	.99610+0	.95788+0	.96163+0	.18893+1	.21355+1	.22294+1
.29	.29654	.99581+0	.95489+0	.95891+0	.18344+1	.20616+1	.21590+1
.30	.30672	.99552+0	.95181+0	.95609+0	.17835+1	.19926+1	.20935+1
.31	.31690	.99522+0	.94864+0	.95320+0	.17363+1	.19280+1	.20324+1
.32	.32707	.99491+0	.94537+0	.95021+0	.16923+1	.18675+1	.19754+1
.33	.33723	.99458+0	.94202+0	.94715+0	.16513+1	.18106+1	.19220+1
.34	.34739	.99425+0	.93857+0	.94400+0	.16130+1	.17570+1	.18720+1
.35	.35755	.99391+0	.93504+0	.94076+0	.15772+1	.17065+1	.18251+1
.36	.36770	.99356+0	.93142+0	.93745+0	.15437+1	.16589+1	.17810+1
.37	.37785	.99320+0	.92771+0	.93406+0	.15122+1	.16137+1	.17395+1
.38	.38799	.99283+0	.92392+0	.93059+0	.14827+1	.15710+1	.17003+1
.39	.39812	.99245+0	.92004+0	.92704+0	.14550+1	.15304+1	.16634+1
.40	.40825	.99206+0	.91608+0	.92341+0	.14289+1	.14918+1	.16285+1
.41	.41837	.99167+0	.91204+0	.91971+0	.14043+1	.14552+1	.15955+1
.42	.42849	.99126+0	.90792+0	.91593+0	.13811+1	.14202+1	.15643+1
.43	.43860	.99084+0	.90373+0	.91208+0	.13593+1	.13869+1	.15346+1
.44	.44870	.99041+0	.89945+0	.90816+0	.13387+1	.13551+1	.15066+1
.45	.45880	.98998+0	.89510+0	.90417+0	.13192+1	.13247+1	.14799+1
.46	.46889	.98953+0	.89068+0	.90010+0	.13008+1	.12956+1	.14546+1
.47	.47897	.98908+0	.88619+0	.89597+0	.12834+1	.12677+1	.14306+1
.48	.48904	.98861+0	.88162+0	.89177+0	.12669+1	.12410+1	.14077+1
.49	.49911	.98814+0	.87698+0	.88751+0	.12513+1	.12154+1	.13859+1
.50	.50918	.98765+0	.87228+0	.88318+0	.12366+1	.11908+1	.13652+1
.52	.52928	.98666+0	.86267+0	.87433+0	.12093+1	.11444+1	.13266+1
.54	.54935	.98563+0	.85281+0	.86524+0	.11848+1	.11015+1	.12916+1
.56	.56938	.98456+0	.84270+0	.85592+0	.11628+1	.10616+1	.12597+1
.58	.58939	.98346+0	.83237+0	.84637+0	.11430+1	.10244+1	.12307+1
.60	.60936	.98232+0	.82182+0	.83661+0	.11252+1	.98966+0	.12042+1
.62	.62929	.98114+0	.81106+0	.82665+0	.11092+1	.95716+0	.11801+1
.64	.64919	.97993+0	.80011+0	.81650+0	.10948+1	.92668+0	.11582+1
.66	.66905	.97868+0	.78898+0	.80617+0	.10819+1	.89803+0	.11382+1
.68	.68887	.97740+0	.77769+0	.79567+0	.10703+1	.87104+0	.11200+1
.70	.70866	.97609+0	.76625+0	.78502+0	.10599+1	.84559+0	.11035+1
.72	.72840	.97473+0	.75466+0	.77422+0	.10506+1	.82153+0	.10886+1
.74	.74810	.97335+0	.74295+0	.76329+0	.10424+1	.79876+0	.10751+1
.76	.76776	.97193+0	.73112+0	.75223+0	.10351+1	.77717+0	.10630+1
.78	.78738	.97048+0	.71919+0	.74107+0	.10287+1	.75668+0	.10521+1
.80	.80695	.96899+0	.70717+0	.72980+0	.10231+1	.73720+0	.10425+1
.82	.82647	.96747+0	.69507+0	.71844+0	.10182+1	.71865+0	.10339+1
.84	.84595	.96592+0	.68291+0	.70700+0	.10140+1	.70098+0	.10265+1
.86	.86538	.96434+0	.67070+0	.69550+0	.10105+1	.68411+0	.10200+1
.88	.88477	.96272+0	.65844+0	.68394+0	.10075+1	.66801+0	.10145+1
.90	.90410	.96108+0	.64616+0	.67232+0	.10051+1	.65260+0	.10100+1
.92	.92338	.95940+0	.63385+0	.66068+0	.10032+1	.63786+0	.10063+1
.94	.94262	.95769+0	.62154+0	.64900+0	.10017+1	.62373+0	.10035+1
.96	.96180	.95595+0	.60924+0	.63731+0	.10006+1	.61018+0	.10015+1

Table C.6 Isentropic Flow ($\gamma = 1.10$) (*Continued*)

M	M*	$\dfrac{t}{T}$	$\dfrac{p}{P}$	$\dfrac{\rho}{\rho_o}$	$\dfrac{\mathscr{F}}{\mathscr{F}*}$	$\left(\dfrac{p}{P}\right)\left(\dfrac{A}{A*}\right)$	$\dfrac{A}{A*}$
.98	.98093	.95418+0	.59695+0	.62561+0	.10002+1	.59717+0	.10004+1
1.00	1.00000	.95238+0	.58468+0	.61391+0	.10000+1	.58468+0	.10000+1
1.02	1.01902	.95055+0	.57245+0	.60223+0	.10002+1	.57266+0	.10004+1
1.04	1.03799	.94869+0	.56026+0	.59056+0	.10007+1	.56110+0	.10015+1
1.06	1.05689	.94681+0	.54813+0	.57892+0	.10015+1	.54997+0	.10034+1
1.08	1.07575	.94489+0	.53606+0	.56732+0	.10027+1	.53924+0	.10059+1
1.10	1.09454	.94295+0	.52406+0	.55577+0	.10041+1	.52889+0	.10092+1
1.12	1.11328	.94098+0	.51214+0	.54427+0	.10058+1	.51890+0	.10132+1
1.14	1.13195	.93898+0	.50032+0	.53283+0	.10077+1	.50926+0	.10179+1
1.16	1.15057	.93696+0	.48858+0	.52145+0	.10099+1	.49994+0	.10232+1
1.18	1.16913	.93491+0	.47695+0	.51016+0	.10122+1	.49093+0	.10293+1
1.20	1.18762	.93284+0	.46543+0	.49894+0	.10146+1	.48221+0	.10360+1
1.22	1.20606	.93073+0	.45403+0	.48782+0	.10176+1	.47377+0	.10435+1
1.24	1.22443	.92861+0	.44275+0	.47679+0	.10206+1	.46559+0	.10516+1
1.26	1.24273	.92646+0	.43160+0	.46586+0	.10237+1	.45767+0	.10604+1
1.28	1.26098	.92428+0	.42058+0	.45504+0	.10270+1	.44999+0	.10699+1
1.30	1.27916	.92208+0	.40971+0	.44433+0	.10305+1	.44254+0	.10801+1
1.32	1.29727	.91966+0	.39698+0	.43374+0	.10341+1	.43531+0	.10911+1
1.34	1.31532	.91762+0	.38839+0	.42326+0	.10378+1	.42829+0	.11027+1
1.36	1.33330	.91535+0	.37796+0	.41292+0	.10417+1	.42147+0	.11151+1
1.38	1.35121	.91306+0	.36769+0	.40270+0	.10456+1	.41484+0	.11282+1
1.40	1.36906	.91075+0	.35758+0	.39262+0	.10497+1	.40840+0	.11421+1
1.42	1.38684	.90841+0	.34763+0	.38268+0	.10540+1	.40213+0	.11568+1
1.44	1.40454	.90606+0	.33785+0	.37288+0	.10583+1	.39603+0	.11722+1
1.46	1.42219	.90369+0	.32824+0	.36322+0	.10627+1	.39009+0	.11884+1
1.48	1.43976	.90129+0	.31880+0	.35371+0	.10672+1	.38431+0	.12055+1
1.50	1.45726	.89886+0	.30953+0	.34435+0	.10717+1	.37868+0	.12234+1
1.55	1.50070	.89276+0	.28712+0	.32161+0	.10835+1	.36521+0	.12720+1
1.60	1.54369	.68652+0	.26583+0	.29985+0	.10957+1	.35256+0	.13263+1
1.65	1.58623	.68018+0	.24565+0	.27909+0	.11083+1	.34066+0	.13868+1
1.70	1.62830	.87374+0	.22658+0	.25932+0	.11212+1	.32942+0	.14539+1
1.75	1.66992	.86721+0	.20862+0	.24057+0	.11344+1	.31881+0	.15282+1
1.80	1.71106	.86059+0	.19175+0	.22281+0	.11477+1	.30877+0	.16103+1
1.85	1.75172	.85388+0	.17594+0	.20605+0	.11613+1	.29925+0	.17009+1
1.90	1.79191	.84710+0	.16117+0	.19026+0	.11750+1	.29022+0	.18007+1
1.95	1.83161	.84025+0	.14739+0	.17542+0	.11888+1	.28163+0	.19107+1
2.00	1.87083	.83333+0	.13459+0	.16151+0	.12027+1	.27346+0	.20318+1
2.10	1.94780	.81934+0	.11171+0	.13634+0	.12306+1	.25824+0	.23117+1
2.20	2.02282	.80515+0	.92185-1	.11449+0	.12586+1	.24436+0	.26507+1
2.30	2.09586	.79083+0	.75665-1	.95678-1	.12865+1	.23165+0	.30615+1
2.40	2.16699	.77640+0	.61791-1	.79567-1	.13142+1	.21996+0	.35597+1
2.50	2.23607	.76190+0	.50224-1	.65918-1	.13416+1	.20918+0	.41650+1
2.60	2.30324	.74738+0	.40643-1	.54380-1	.13687+1	.19921+0	.49015+1
2.70	2.36849	.73287+0	.32756-1	.44696-1	.13954+1	.18996+0	.57992+1
2.80	2.43183	.71839+0	.26301-1	.36611-1	.14215+1	.18136+0	.68955+1
2.90	2.49329	.70398+0	.21045-1	.29894-1	.14472+1	.17334+0	.82366+1
3.00	2.55289	.68966+0	.16786-1	.24340-1	.14723+1	.16585+0	.98800+1
3.10	2.61067	.67545+0	.13351-1	.19766-1	.14969+1	.15884+0	.11897+2
3.20	2.66667	.66138+0	.10591-1	.16013-1	.15208+1	.15226+0	.14377+2
3.30	2.72091	.64746+0	.83817-2	.12946-1	.15442+1	.14608+0	.17429+2
3.40	2.77345	.63371+0	.66195-2	.10446-1	.15670+1	.14027+0	.21191+2
3.50	2.82431	.62016+0	.52180-2	.84140-2	.15892+1	.13480+0	.25834+2
3.60	2.87355	.60680+0	.41065-2	.67675-2	.16108+1	.12964+0	.31569+2
3.70	2.92120	.59365+0	.32271-2	.54361-2	.16318+1	.12476+0	.38660+2
3.80	2.96730	.58072+0	.25330-2	.43618-2	.16522+1	.12015+0	.47433+2
3.90	3.01190	.56802+0	.19861-2	.34965-2	.16720+1	.11578+0	.58294+2
4.00	3.05505	.55556+0	.15560-2	.28008-2	.16912+1	.11164+0	.71749+2
4.50	3.25042	.49689+0	.45594-3	.91757-3	.17790+1	.93849-1	.20584+3
5.00	3.41565	.44444+0	.13366-3	.30073-3	.18542+1	.79882-1	.59767+3
5.50	3.55553	.39801+0	.39704-4	.99756-4	.19184+1	.68722-1	.17309+4
6.00	3.67423	.35714+0	.12058-4	.33761-4	.19732+1	.59674-1	.49490+4
6.50	3.77532	.32129+0	.37653-5	.11719-4	.20201+1	.52245-1	.13875+5
7.00	3.86174	.28986+0	.12134-5	.41861-5	.20603+1	.46079-1	.37976+5
7.50	3.93598	.26230+0	.40429-6	.15414-5	.20950+1	.40912-1	.10119+6
8.00	4.00000	.23810+0	.13940-6	.58547-6	.21250+1	.36542-1	.26214+6
8.50	4.05551	.21680+0	.49739-7	.22942-6	.21510+1	.32819-1	.65982+6
9.00	4.10385	.19802+0	.18357-7	.92701-7	.21738+1	.29623-1	.16137+7
9.50	4.14614	.18141+0	.70011-8	.38593-7	.21937+1	.26861-1	.38366+7
10.00	4.18330	.16667+0	.27564-8	.16538-7	.22112+1	.24459-1	.88736+7
11.00	4.24515	.14184+0	.46765-9	.32969-8	.22404+1	.20513-1	.43864+8
12.00	4.29407	.12195+0	.88725-10	.72755-9	.22635+1	.17435-1	.19651+9
13.00	4.33333	.10582+0	.18632-10	.17607-9	.22821+1	.14992-1	.80464+9
14.00	4.36527	.92593-1	.42888-11	.46319-10	.22972+1	.13022-1	.30362+10
15.00	4.39155	.81633-1	.10728-11	.13141-10	.23096+1	.11412-1	.10638+11
INFIN	4.58258	.0	.0	.0	.24004+1	.0	INFIN

Table C.7 Adiabatic Flow with Friction in a Constant-Area Duct (Fanno Line, $\gamma = 1.40$)

M	$M^* = \dfrac{\rho^*}{\rho}$	$\dfrac{t}{t^*}$	$\dfrac{p}{p^*}$	$\dfrac{P}{P^*}$	$\dfrac{\mathscr{F}}{\mathscr{F}^*}$	$\dfrac{4fL^*}{\mathscr{D}}$
.00	.00000	.12000+1	INFIN	INFIN	INFIN	INFIN
.01	.01095	.12000+1	.10954+3	.57874+2	.45649+2	.71344+4
.02	.02191	.11999+1	.54770+2	.28942+2	.22834+2	.17784+4
.03	.03286	.11998+1	.36512+2	.19301+2	.15232+2	.78708+3
.04	.04381	.11996+1	.27382+2	.14481+2	.11435+2	.44035+3
.05	.05476	.11994+1	.21903+2	.11591+2	.91584+1	.28002+3
.06	.06570	.11991+1	.18251+2	.96659+1	.76428+1	.19303+3
.07	.07664	.11988+1	.15642+2	.82915+1	.65620+1	.14066+3
.08	.08758	.11985+1	.13684+2	.72616+1	.57529+1	.10672+3
.09	.09851	.11981+1	.12162+2	.64613+1	.51249+1	.83496+2
.10	.10944	.11976+1	.10944+2	.58218+1	.46236+1	.66922+2
.11	.12035	.11971+1	.99466+1	.52992+1	.42146+1	.54688+2
.12	.13126	.11966+1	.91156+1	.48643+1	.36747+1	.45408+2
.13	.14217	.11960+1	.84123+1	.44969+1	.35881+1	.38207+2
.14	.15306	.11953+1	.78093+1	.41824+1	.33432+1	.32511+2
.15	.16395	.11946+1	.72866+1	.39103+1	.31317+1	.27932+2
.16	.17482	.11939+1	.68291+1	.36727+1	.29474+1	.24198+2
.17	.18569	.11931+1	.64253+1	.34635+1	.27855+1	.21115+2
.18	.19654	.11923+1	.60662+1	.32779+1	.26422+1	.18543+2
.19	.20739	.11914+1	.57448+1	.31123+1	.25146+1	.16375+2
.20	.21822	.11905+1	.54554+1	.29635+1	.24004+1	.14533+2
.21	.22904	.11895+1	.51936+1	.28293+1	.22976+1	.12956+2
.22	.23984	.11885+1	.49554+1	.27076+1	.22046+1	.11596+2
.23	.25063	.11874+1	.47378+1	.25968+1	.21203+1	.10416+2
.24	.26141	.11863+1	.45383+1	.24956+1	.20434+1	.93865+1
.25	.27217	.11852+1	.43546+1	.24027+1	.19732+1	.84834+1
.26	.28291	.11840+1	.41851+1	.23173+1	.19088+1	.76876+1
.27	.29364	.11828+1	.40279+1	.22385+1	.18496+1	.69832+1
.28	.30435	.11815+1	.38820+1	.21656+1	.17950+1	.63572+1
.29	.31504	.11801+1	.37460+1	.20979+1	.17446+1	.57989+1
.30	.32572	.11788+1	.36191+1	.20351+1	.16979+1	.52993+1
.31	.33637	.11774+1	.35002+1	.19765+1	.16546+1	.48507+1
.32	.34701	.11759+1	.33887+1	.19219+1	.16144+1	.44467+1
.33	.35762	.11744+1	.32840+1	.18707+1	.15769+1	.40821+1
.34	.36822	.11729+1	.31853+1	.18229+1	.15420+1	.37520+1
.35	.37879	.11713+1	.30922+1	.17780+1	.15094+1	.34525+1
.36	.38935	.11697+1	.30042+1	.17358+1	.14789+1	.31801+1
.37	.39988	.11680+1	.29209+1	.16961+1	.14503+1	.29320+1
.38	.41039	.11663+1	.28420+1	.16587+1	.14236+1	.27054+1
.39	.42087	.11646+1	.27671+1	.16234+1	.13985+1	.24983+1
.40	.43133	.11628+1	.26958+1	.15901+1	.13749+1	.23085+1
.41	.44177	.11610+1	.26280+1	.15587+1	.13527+1	.21344+1
.42	.45218	.11591+1	.25634+1	.15289+1	.13318+1	.19744+1
.43	.46257	.11572+1	.25017+1	.15007+1	.13122+1	.18272+1
.44	.47293	.11553+1	.24428+1	.14740+1	.12937+1	.16915+1
.45	.48326	.11533+1	.23865+1	.14487+1	.12763+1	.15664+1
.46	.49357	.11513+1	.23326+1	.14246+1	.12598+1	.14509+1
.47	.50385	.11492+1	.22809+1	.14018+1	.12443+1	.13441+1
.48	.51410	.11471+1	.22313+1	.13801+1	.12296+1	.12453+1
.49	.52433	.11450+1	.21838+1	.13595+1	.12158+1	.11539+1
.50	.53452	.11429+1	.21381+1	.13398+1	.12027+1	.10691+1
.52	.55483	.11384+1	.20519+1	.13034+1	.11786+1	.91742+0
.54	.57501	.11339+1	.19719+1	.12703+1	.11571+1	.78663+0
.56	.59507	.11292+1	.18975+1	.12403+1	.11378+1	.67357+0
.58	.61501	.11244+1	.18282+1	.12130+1	.11205+1	.57568+0
.60	.63481	.11194+1	.17634+1	.11882+1	.11050+1	.49082+0
.62	.65448	.11143+1	.17026+1	.11656+1	.10912+1	.41720+0
.64	.67402	.11091+1	.16456+1	.11451+1	.10788+1	.35330+0
.66	.69342	.11038+1	.15919+1	.11265+1	.10678+1	.29785+0
.68	.71268	.10984+1	.15413+1	.11097+1	.10579+1	.24978+0
.70	.73179	.10929+1	.14935+1	.10944+1	.10492+1	.20814+0
.72	.75076	.10873+1	.14482+1	.10806+1	.10414+1	.17215+0
.74	.76958	.10815+1	.14054+1	.10681+1	.10345+1	.14112+0
.76	.78825	.10757+1	.13647+1	.10570+1	.10284+1	.11447+0

Table C.7 Adiabatic Flow with Friction in a Constant-Area Duct (Fanno Line, $\gamma = 1.40$) (*Continued*)

M	$M^* = \dfrac{\rho^*}{\rho}$	$\dfrac{t}{t^*}$	$\dfrac{p}{p^*}$	$\dfrac{P}{P^*}$	$\dfrac{\mathscr{F}}{\mathscr{F}^*}$	$\dfrac{4fL^*}{\mathscr{D}}$
.78	.80677	.10698+1	.13261+1	.10471+1	.10231+1	.91672-1
.80	.82514	.10638+1	.12893+1	.10382+1	.10185+1	.72290-1
.82	.84335	.10578+1	.12542+1	.10305+1	.10145+1	.55932-1
.84	.86140	.10516+1	.12208+1	.10237+1	.10112+1	.42256-1
.86	.87929	.10454+1	.11889+1	.10179+1	.10083+1	.30965-1
.88	.89703	.10391+1	.11583+1	.10129+1	.10059+1	.21795-1
.90	.91460	.10327+1	.11291+1	.10089+1	.10040+1	.14512-1
.92	.93201	.10263+1	.11011+1	.10056+1	.10025+1	.89133-2
.94	.94925	.10198+1	.10743+1	.10031+1	.10014+1	.48154-2
.96	.96633	.10132+1	.10485+1	.10014+1	.10006+1	.20571-2
.98	.98325	.10066+1	.10238+1	.10003+1	.10001+1	.49470-3
1.00	1.00000	.10000+1	.10000+1	.10000+1	.10000+1	.0
1.02	1.01658	.99331+0	.97711+0	.10003+1	.10001+1	.45869-3
1.04	1.03300	.98658+0	.95507+0	.10013+1	.10005+1	.17685-2
1.06	1.04925	.97982+0	.93383+0	.10029+1	.10012+1	.38379-2
1.08	1.06533	.97302+0	.91335+0	.10051+1	.10020+1	.65846-2
1.10	1.08124	.96618+0	.89359+0	.10079+1	.10031+1	.99350-2
1.12	1.09699	.95932+0	.87451+0	.10113+1	.10043+1	.13823-1
1.14	1.11256	.95244+0	.85608+0	.10153+1	.10057+1	.18188-1
1.16	1.12797	.94554+0	.83826+0	.10198+1	.10073+1	.22977-1
1.18	1.14321	.93861+0	.82103+0	.10248+1	.10090+1	.28142-1
1.20	1.15828	.93168+0	.80436+0	.10304+1	.10108+1	.33638-1
1.22	1.17319	.92473+0	.78822+0	.10366+1	.10128+1	.39426-1
1.24	1.18792	.91777+0	.77258+0	.10432+1	.10149+1	.45471-1
1.26	1.20249	.91080+0	.75743+0	.10504+1	.10170+1	.51739-1
1.28	1.21690	.90383+0	.74274+0	.10581+1	.10193+1	.58201-1
1.30	1.23114	.89686+0	.72848+0	.10663+1	.10217+1	.64832-1
1.32	1.24521	.88989+0	.71465+0	.10750+1	.10241+1	.71607-1
1.34	1.25912	.88292+0	.70122+0	.10842+1	.10267+1	.78504-1
1.36	1.27286	.87596+0	.68818+0	.10940+1	.10292+1	.85503-1
1.38	1.28645	.86901+0	.67551+0	.11042+1	.10319+1	.92586-1
1.40	1.29987	.86207+0	.66320+0	.11149+1	.10346+1	.99738-1
1.42	1.31313	.85514+0	.65122+0	.11262+1	.10373+1	.10694+0
1.44	1.32623	.84822+0	.63958+0	.11379+1	.10401+1	.11419+0
1.46	1.33917	.84133+0	.62825+0	.11501+1	.10430+1	.12146+0
1.48	1.35195	.83445+0	.61722+0	.11629+1	.10458+1	.12875+0
1.50	1.36458	.82759+0	.60648+0	.11762+1	.10487+1	.13605+0
1.55	1.39546	.81054+0	.58084+0	.12116+1	.10560+1	.15427+0
1.60	1.42539	.79365+0	.55679+0	.12502+1	.10635+1	.17236+0
1.65	1.45439	.77695+0	.53421+0	.12922+1	.10710+1	.19023+0
1.70	1.48247	.76046+0	.51297+0	.13376+1	.10785+1	.20780+0
1.75	1.50966	.74419+0	.49295+0	.13865+1	.10860+1	.22504+0
1.80	1.53598	.72816+0	.47407+0	.14390+1	.10935+1	.24189+0
1.85	1.56145	.71238+0	.45623+0	.14952+1	.11009+1	.25832+0
1.90	1.58609	.69686+0	.43936+0	.15553+1	.11083+1	.27433+0
1.95	1.60993	.68162+0	.42339+0	.16193+1	.11155+1	.28989+0
2.00	1.63299	.66667+0	.40825+0	.16875+1	.11227+1	.30500+0
2.10	1.67687	.63762+0	.38024+0	.18369+1	.11366+1	.33385+0
2.20	1.71791	.60976+0	.35494+0	.20050+1	.11500+1	.36091+0
2.30	1.75629	.58309+0	.33200+0	.21931+1	.11628+1	.38623+0
2.40	1.79218	.55762+0	.31114+0	.24031+1	.11751+1	.40989+0
2.50	1.82574	.53333+0	.29212+0	.26367+1	.11867+1	.43198+0
2.60	1.85714	.51020+0	.27473+0	.28960+1	.11978+1	.45259+0
2.70	1.88653	.48820+0	.25878+0	.31830+1	.12083+1	.47182+0
2.80	1.91404	.46729+0	.24414+0	.35001+1	.12182+1	.48976+0
2.90	1.93981	.44743+0	.23066+0	.38498+1	.12277+1	.50652+0
3.00	1.96396	.42857+0	.21822+0	.42346+1	.12366+1	.52216+0
3.10	1.98661	.41068+0	.20672+0	.46573+1	.12450+1	.53678+0
3.20	2.00786	.39370+0	.19608+0	.51210+1	.12530+1	.55044+0
3.30	2.02781	.37760+0	.18621+0	.56286+1	.12605+1	.56323+0
3.40	2.04656	.36232+0	.17704+0	.61837+1	.12676+1	.57521+0
3.50	2.06419	.34783+0	.16851+0	.67896+1	.12743+1	.58643+0
3.60	2.08077	.33408+0	.16055+0	.74501+1	.12807+1	.59695+0
3.70	2.09639	.32103+0	.15313+0	.81691+1	.12867+1	.60684+0

Table C.7   Adiabatic Flow with Friction in a Constant-Area Duct (Fanno Line, $\gamma = 1.40$) (*Continued*)

M	$M^* = \dfrac{\rho^*}{\rho}$	$\dfrac{t}{t^*}$	$\dfrac{p}{p^*}$	$\dfrac{P}{P^*}$	$\dfrac{\mathscr{F}}{\mathscr{F}^*}$	$\dfrac{4fL^*}{\mathscr{D}}$
3.80	2.11111	.30864+0	.14620+0	.89506+1	.12924+1	.61612+0
3.90	2.12499	.29688+0	.13971+0	.97990+1	.12978+1	.62485+0
4.00	2.13809	.28571+0	.13363+0	.10719+2	.13029+1	.63306+0
4.50	2.19360	.23762+0	.10833+0	.16562+2	.13247+1	.66763+0
5.00	2.23607	.20000+0	.89443-1	.25000+2	.13416+1	.69380+0
5.50	2.26913	.17021+0	.75012-1	.36869+2	.13549+1	.71400+0
6.00	2.29528	.14634+0	.63758-1	.53180+2	.13655+1	.72988+0
6.50	2.31626	.12698+0	.54823-1	.75134+2	.13740+1	.74254+0
7.00	2.33333	.11111+0	.47619-1	.10414+3	.13810+1	.75280+0
7.50	2.34738	.97959-1	.41731-1	.14184+3	.13867+1	.76121+0
8.00	2.35907	.86957-1	.36860-1	.19011+3	.13915+1	.76819+0
8.50	2.36889	.77670-1	.32787-1	.25109+3	.13955+1	.77404+0
9.00	2.37722	.69767-1	.29348-1	.32719+3	.13989+1	.77899+0
9.50	2.38433	.62992-1	.26419-1	.42113+3	.14019+1	.78320+0
10.00	2.39046	.57143-1	.23905-1	.53594+3	.14044+1	.78683+0
11.00	2.40040	.47619-1	.19838-1	.84191+3	.14085+1	.79270+0
12.00	2.40804	.40268-1	.16723-1	.12762+4	.14117+1	.79721+0
13.00	2.41404	.34483-1	.14284-1	.18761+4	.14141+1	.80074+0
14.00	2.41883	.29851-1	.12341-1	.26854+4	.14161+1	.80356+0
15.00	2.42272	.26087-1	.10768-1	.37552+4	.14177+1	.80584+0
INFIN	2.44949	.0	.0	INFIN	.14289+1	.82151+0

Table C.7   Adiabatic Flow with Friction in a Constant-Area Duct (Fanno Line, $\gamma = 1.30$)

M	$M^* = \dfrac{\rho^*}{\rho}$	$\dfrac{t}{t^*}$	$\dfrac{p}{p^*}$	$\dfrac{P}{P^*}$	$\dfrac{\mathscr{F}}{\mathscr{F}^*}$	$\dfrac{4fL^*}{\mathscr{D}}$
.00	.00000	.11500+1	INFIN	INFIN	INFIN	INFIN
.02	.02145	.11499+1	.53617+2	.29268+2	.23324+2	.19155+4
.04	.04289	.11497+1	.26806+2	.14644+2	.11679+2	.47443+3
.06	.06433	.11494+1	.17868+2	.97740+1	.78051+1	.20805+3
.08	.08575	.11489+1	.13396+2	.73423+1	.58738+1	.11508+3
.10	.10716	.11483+1	.10716+2	.58860+1	.47196+1	.72202+2
.12	.12855	.11475+1	.89269+1	.49174+1	.39539+1	.49020+2
.14	.14991	.11466+1	.76486+1	.42275+1	.34102+1	.35120+2
.16	.17125	.11456+1	.66895+1	.37118+1	.30053+1	.26157+2
.18	.19256	.11444+1	.59432+1	.33123+1	.26929+1	.20058+2
.20	.21384	.11431+1	.53459+1	.29940+1	.24452+1	.15732+2
.22	.23507	.11417+1	.48569+1	.27349+1	.22445+1	.12562+2
.24	.25627	.11401+1	.44491+1	.25202+1	.20792+1	.10177+2
.26	.27742	.11385+1	.41038+1	.23396+1	.19411+1	.83413+1
.28	.29852	.11366+1	.38076+1	.21859+1	.18242+1	.69035+1
.30	.31956	.11347+1	.35507+1	.20537+1	.17244+1	.57594+1
.32	.34056	.11326+1	.33257+1	.19389+1	.16385+1	.48370+1
.34	.36149	.11304+1	.31271+1	.18385+1	.15639+1	.40848+1
.36	.38236	.11281+1	.29503+1	.17502+1	.14989+1	.34653+1
.38	.40316	.11256+1	.27920+1	.16719+1	.14418+1	.29507+1
.40	.42390	.11230+1	.26493+1	.16023+1	.13915+1	.25200+1
.42	.44456	.11204+1	.25202+1	.15401+1	.13470+1	.21572+1
.44	.46514	.11175+1	.24026+1	.14843+1	.13075+1	.18499+1
.46	.48565	.11146+1	.22951+1	.14341+1	.12724+1	.15882+1
.48	.50607	.11116+1	.21965+1	.13888+1	.12410+1	.13645+1
.50	.52641	.11084+1	.21056+1	.13479+1	.12130+1	.11724+1
.55	.57687	.11001+1	.19070+1	.12614+1	.11552+1	.80035+0
.60	.62673	.10911+1	.17409+1	.11932+1	.11112+1	.54086+0
.65	.67596	.10815+1	.15999+1	.11395+1	.10777+1	.35856+0
.70	.72451	.10713+1	.14786+1	.10972+1	.10524+1	.23048+0
.75	.77236	.10605+1	.13731+1	.10644+1	.10335+1	.14129+0

Table C.7 Adiabatic Flow with Friction in a Constant-Area Duct (Fanno Line, $\gamma = 1.30$) (*Continued*)

M	$M^* = \dfrac{\rho^*}{\rho}$	$\dfrac{t}{t^*}$	$\dfrac{p}{p^*}$	$\dfrac{P}{P^*}$	$\dfrac{\mathscr{F}}{\mathscr{F}^*}$	$\dfrac{4fL^*}{\mathscr{D}}$
.80	.81947	.10493+1	.12804+1	.10395+1	.10199+1	.80445-1
.85	.86581	.10376+1	.11984+1	.10214+1	.10104+1	.40528-1
.90	.91136	.10254+1	.11251+1	.10092+1	.10043+1	.16229-1
.95	.95610	.10129+1	.10594+1	.10022+1	.10010+1	.36750-2
1.00	1.00000	.10000+1	.10000+1	.10000+1	.10000+1	.0
1.05	1.04305	.98681+0	.94608+0	.10021+1	.10009+1	.30567-2
1.10	1.08524	.97334+0	.89689+0	.10083+1	.10033+1	.11218-1
1.15	1.12655	.95963+0	.85183+0	.10184+1	.10071+1	.23239-1
1.20	1.16698	.94572+0	.81040+0	.10321+1	.10119+1	.38160-1
1.25	1.20652	.93165+0	.77217+0	.10495+1	.10177+1	.55236-1
1.30	1.24517	.91743+0	.73679+0	.10703+1	.10241+1	.73884-1
1.35	1.28293	.90311+0	.70394+0	.10948+1	.10312+1	.93648-1
1.40	1.31981	.88872+0	.67337+0	.11227+1	.10387+1	.11417+0
1.45	1.35579	.87428+0	.64485+0	.11543+1	.10467+1	.13516+0
1.50	1.39089	.85981+0	.61817+0	.11895+1	.10549+1	.15640+0
1.55	1.42512	.84536+0	.59318+0	.12284+1	.10634+1	.17771+0
1.60	1.45848	.83092+0	.56972+0	.12712+1	.10721+1	.19895+0
1.65	1.49099	.81654+0	.54765+0	.13180+1	.10808+1	.22001+0
1.70	1.52265	.80223+0	.52687+0	.13690+1	.10897+1	.24081+0
1.75	1.55347	.78801+0	.50726+0	.14243+1	.10986+1	.26128+0
1.80	1.58346	.77389+0	.48873+0	.14841+1	.11075+1	.28137+0
1.85	1.61268	.75989+0	.47120+0	.15486+1	.11164+1	.30103+0
1.90	1.64108	.74603+0	.45459+0	.16182+1	.11252+1	.32025+0
1.95	1.66871	.73231+0	.43885+0	.16929+1	.11340+1	.33901+0
2.00	1.69558	.71875+0	.42390+0	.17732+1	.11427+1	.35728+0
2.20	1.79577	.66628+0	.37103+0	.21556+1	.11763+1	.42547+0
2.40	1.88511	.61695+0	.32728+0	.26535+1	.12078+1	.48599+0
2.60	1.96468	.57100+0	.29063+0	.32954+1	.12368+1	.53938+0
2.80	2.03553	.52849+0	.25963+0	.41165+1	.12634+1	.58638+0
3.00	2.09863	.48936+0	.23318+0	.51598+1	.12876+1	.62774+0
3.20	2.15489	.45347+0	.21044+0	.64776+1	.13095+1	.66419+0
3.40	2.20510	.42063+0	.19075+0	.81328+1	.13293+1	.69637+0
3.60	2.25000	.39063+0	.17361+0	.10200+2	.13472+1	.72485+0
3.80	2.29022	.36323+0	.15860+0	.12769+2	.13634+1	.75011+0
4.00	2.32632	.33824+0	.14539+0	.15944+2	.13761+1	.77259+0
5.00	2.46021	.24211+0	.98408-1	.45957+2	.14333+1	.85428+0
6.00	2.54337	.17969+0	.70649-1	.12010+3	.14683+1	.90370+0
7.00	2.59779	.13772+0	.53016-1	.28534+3	.14914+1	.93548+0
8.00	2.63503	.10849+0	.41172-1	.62312+3	.15073+1	.95699+0
9.00	2.66151	.87452-1	.32858-1	.12656+4	.15186+1	.97216+0
10.00	2.68095	.71875-1	.26810-1	.24161+4	.15270+1	.98323+0
11.00	2.69561	.60052-1	.22278-1	.43743+4	.15333+1	.99154+0
12.00	2.70692	.50885-1	.18798-1	.75665+4	.15382+1	.99793+0
13.00	2.71583	.43643-1	.16070-1	.12581+5	.15420+1	.10030+1
14.00	2.72295	.37829-1	.13893-1	.20209+5	.15451+1	.10070+1
15.00	2.72874	.33094-1	.12128-1	.31494+5	.15476+1	.10102+1
INFIN	2.76887	.0	.0	INFIN	.15650+1	.10326+1

Table C.7 Adiabatic Flow with Friction in a Constant Area Duct (Fanno Line, $\gamma = 1.20$)

M	$M^* = \dfrac{\rho^*}{\rho}$	$\dfrac{t}{t^*}$	$\dfrac{p}{p^*}$	$\dfrac{P}{P^*}$	$\dfrac{\mathscr{F}}{\mathscr{F}^*}$	$\dfrac{4fL^*}{\mathscr{D}}$
.00	.00000	.11000+1	INFIN	INFIN	INFIN	INFIN
.02	.02098	.11000+1	.52439+2	.29608+2	.23848+2	.20754+4
.04	.04195	.10998+1	.26218+2	.14814+2	.11940+2	.51419+3
.06	.06242	.10996+1	.17477+2	.98866+1	.79784+1	.22558+3
.08	.08388	.10993+1	.13106+2	.74264+1	.60030+1	.12483+3
.10	.10483	.10989+1	.10483+2	.59529+1	.48221+1	.78365+2
.12	.12577	.10984+1	.87338+1	.49727+1	.40385+1	.53236+2
.14	.14669	.10978+1	.74842+1	.42745+1	.34619+1	.38165+2
.16	.16760	.10972+1	.65467+1	.37526+1	.30672+1	.28444+2
.18	.18848	.10964+1	.58173+1	.33481+1	.27470+1	.21827+2
.20	.20934	.10956+1	.52336+1	.30258+1	.24931+1	.17133+2
.22	.23018	.10947+1	.47558+1	.27634+1	.22873+1	.13691+2
.24	.25099	.10937+1	.43575+1	.25459+1	.21176+1	.11100+2
.26	.27177	.10926+1	.40203+1	.23630+1	.19757+1	.91056+1
.28	.29252	.10914+1	.37311+1	.22072+1	.18555+1	.75424+1
.30	.31324	.10902+1	.34804+1	.20731+1	.17529+1	.62978+1
.32	.33391	.10889+1	.32609+1	.19567+1	.16644+1	.52938+1
.34	.35455	.10874+1	.30671+1	.18549+1	.15875+1	.44744+1
.36	.37515	.10859+1	.28947+1	.17652+1	.15204+1	.37992+1
.38	.39570	.10843+1	.27403+1	.16858+1	.14614+1	.32380+1
.40	.41621	.10827+1	.26013+1	.16151+1	.14094+1	.27680+1
.42	.43667	.10809+1	.24754+1	.15519+1	.13634+1	.23717+1
.44	.45707	.10791+1	.23609+1	.14952+1	.13225+1	.20357+1
.46	.47743	.10772+1	.22563+1	.14441+1	.12860+1	.17495+1
.48	.49773	.10752+1	.21603+1	.13980+1	.12534+1	.15044+1
.50	.51797	.10732+1	.20719+1	.13563+1	.12243+1	.12940+1
.55	.56831	.10677+1	.18787+1	.12681+1	.11640+1	.88550+0
.60	.61826	.10618+1	.17174+1	.11986+1	.11179+1	.59992+0
.65	.66777	.10554+1	.15805+1	.11436+1	.10826+1	.39872+0
.70	.71681	.10486+1	.14629+1	.11003+1	.10559+1	.25696+0
.75	.76537	.10414+1	.13607+1	.10666+1	.10360+1	.15793+0
.80	.81342	.10338+1	.12710+1	.10409+1	.10214+1	.90155-1
.85	.86093	.10259+1	.11916+1	.10222+1	.10112+1	.45540-1
.90	.90787	.10176+1	.11208+1	.10096+1	.10047+1	.18284-1
.95	.95424	.10089+1	.10573+1	.10023+1	.10011+1	.41512-2
1.00	1.00000	.10000+1	.10000+1	.10000+1	.10000+1	.0
1.05	1.04514	.99077+0	.94797+0	.10022+1	.10010+1	.34711-2
1.10	1.08965	.98127+0	.90054+0	.10087+1	.10037+1	.12772-1
1.15	1.13350	.97152+0	.85709+0	.10194+1	.10079+1	.26528-1
1.20	1.17670	.96154+0	.81715+0	.10340+1	.10133+1	.43674-1
1.25	1.21922	.95135+0	.78030+0	.10525+1	.10197+1	.63381-1
1.30	1.26105	.94098+0	.74618+0	.10749+1	.10270+1	.84996-1
1.35	1.30219	.93043+0	.71451+0	.11013+1	.10351+1	.10801+0
1.40	1.34264	.91973+0	.68502+0	.11317+1	.10437+1	.13200+0
1.45	1.38238	.90890+0	.65749+0	.11662+1	.10529+1	.15666+0
1.50	1.42141	.89796+0	.63174+0	.12050+1	.10625+1	.18173+0
1.55	1.45973	.88692+0	.60759+0	.12483+1	.10724+1	.20699+0
1.60	1.49734	.87580+0	.58490+0	.12962+1	.10826+1	.23229+0
1.65	1.53424	.86461+0	.56354+0	.13490+1	.10930+1	.25749+0
1.70	1.57043	.85337+0	.54340+0	.14070+1	.11036+1	.28249+0
1.75	1.60591	.84211+0	.52438+0	.14704+1	.11143+1	.30721+0
1.80	1.64068	.83082+0	.50638+0	.15398+1	.11251+1	.33158+0
1.85	1.67476	.81952+0	.48934+0	.16153+1	.11359+1	.35554+0
1.90	1.70813	.80823+0	.47317+0	.16975+1	.11468+1	.37907+0
1.95	1.74081	.79696+0	.45781+0	.17868+1	.11576+1	.40213+0
2.00	1.77281	.78571+0	.44320+0	.18837+1	.11684+1	.42470+0
2.20	1.89410	.74124+0	.39134+0	.23594+1	.12110+1	.50987+0
2.40	2.00507	.69797+0	.34810+0	.30108+1	.12519+1	.58675+0
2.60	2.10636	.65632+0	.31159+0	.38983+1	.12906+1	.65570+0
2.80	2.19865	.61659+0	.28044+0	.51033+1	.13267+1	.71734+0
3.00	2.28266	.57895+0	.25363+0	.67354+1	.13604+1	.77238+0
3.20	2.35907	.54348+0	.23038+0	.89402+1	.13915+1	.82154+0
3.40	2.42857	.51020+0	.21008+0	.11910+2	.14202+1	.86548+0
3.60	2.49180	.47909+0	.19227+0	.15899+2	.14466+1	.90481+0
3.80	2.54935	.45008+0	.17655+0	.21238+2	.14708+1	.94008+0
4.00	2.60177	.42308+0	.16261+0	.28355+2	.14931+1	.97177+0

Table C.7 Adiabatic Flow with Friction in a Constant-Area
Duct (Fanno Line, $\gamma = 1.20$) (*Continued*)

M	$M^* = \dfrac{\rho^*}{\rho}$	$\dfrac{t}{t^*}$	$\dfrac{p}{p^*}$	$\dfrac{P}{P^*}$	$\dfrac{\mathscr{F}}{\mathscr{F}^*}$	$\dfrac{4fL^*}{\mathscr{D}}$
5.00	2.80306	.31429+0	.11212+0	.11634+3	.15799+1	.10896+1
6.00	2.93406	.23913+0	.81502-1	.43587+3	.16374+1	.11632+1
7.00	3.02251	.18644+0	.61684-1	.14687+4	.16767+1	.12115+1
8.00	3.08440	.14865+0	.48194-1	.44671+4	.17043+1	.12447+1
9.00	3.12909	.12088+0	.38631-1	.12383+5	.17243+1	.12683+1
10.00	3.16228	.10000+0	.31623-1	.31623+5	.17393+1	.12857+1
11.00	3.18752	.83969-1	.26343-1	.75152+5	.17506+1	.12988+1
12.00	3.20713	.71429-1	.22272-1	.16770+6	.17595+1	.13090+1
13.00	3.22265	.61453-1	.19069-1	.35407+6	.17665+1	.13170+1
14.00	3.23512	.53398-1	.16506-1	.71200+6	.17721+1	.13234+1
15.00	3.24529	.46809-1	.14424-1	.13713+7	.17767+1	.13286+1
INFIN	3.31662	.0	.0	INFIN	.18091+1	.13647+1

Table C.7 Adiabatic Flow with Friction in a Constant-Area
Duct (Fanno Line, $\gamma = 1.10$)

M	$M^* = \dfrac{\rho^*}{\rho}$	$\dfrac{t}{t^*}$	$\dfrac{p}{p^*}$	$\dfrac{P}{P^*}$	$\dfrac{\mathscr{F}}{\mathscr{F}^*}$	$\dfrac{4fL^*}{\mathscr{D}}$
.00	.00000	.10500+1	INFIN	INFIN	INFIN	INFIN
.02	.02049	.10500+1	.51234+2	.29962+2	.24408+2	.22644+4
.04	.04099	.10499+1	.25616+2	.14991+2	.12220+2	.56117+3
.06	.06148	.10498+1	.17077+2	.10004+2	.81640+1	.24629+3
.08	.08196	.10497+1	.12807+2	.75142+1	.61413+1	.13636+3
.10	.10244	.10495+1	.10244+2	.60227+1	.49319+1	.85650+2
.12	.12292	.10492+1	.85361+1	.50305+1	.41292+1	.58220+2
.14	.14339	.10490+1	.73157+1	.43237+1	.35588+1	.41765+2
.16	.16385	.10487+1	.64002+1	.37951+1	.31336+1	.31149+2
.18	.18430	.10483+1	.56881+1	.33855+1	.28052+1	.23921+2
.20	.20473	.10479+1	.51184+1	.30591+1	.25446+1	.18790+2
.22	.22516	.10475+1	.46521+1	.27933+1	.23332+1	.15027+2
.24	.24557	.10470+1	.42634+1	.25729+1	.21588+1	.12193+2
.26	.26597	.10465+1	.39345+1	.23874+1	.20129+1	.10011+2
.28	.28635	.10459+1	.36525+1	.22294+1	.18893+1	.82991+1
.30	.30672	.10453+1	.34080+1	.20935+1	.17835+1	.69357+1
.32	.32707	.10447+1	.31940+1	.19754+1	.16923+1	.58352+1
.34	.34739	.10440+1	.30051+1	.18720+1	.16130+1	.49365+1
.36	.36770	.10432+1	.28372+1	.17810+1	.15437+1	.41955+1
.38	.38799	.10425+1	.26869+1	.17003+1	.14827+1	.35791+1
.40	.40825	.10417+1	.25516+1	.16285+1	.14289+1	.30624+1
.42	.42849	.10408+1	.24291+1	.15643+1	.13811+1	.26265+1
.44	.44870	.10399+1	.23177+1	.15066+1	.13387+1	.22567+1
.46	.46889	.10390+1	.22159+1	.14546+1	.13008+1	.19412+1
.48	.48904	.10380+1	.21226+1	.14077+1	.12669+1	.16710+1
.50	.50918	.10370+1	.20367+1	.13652+1	.12366+1	.14387+1
.55	.55937	.10344+1	.18492+1	.12753+1	.11735+1	.98709+0
.60	.60936	.10314+1	.16927+1	.12042+1	.11252+1	.67049+0
.65	.65913	.10283+1	.15601+1	.11480+1	.10881+1	.44682+0
.70	.70866	.10249+1	.14462+1	.11035+1	.10599+1	.28874+0
.75	.75794	.10213+1	.13474+1	.10689+1	.10387+1	.17796+0
.80	.80695	.10174+1	.12609+1	.10425+1	.10231+1	.10187+0
.85	.85567	.10134+1	.11843+1	.10231+1	.10122+1	.51601-1
.90	.90410	.10091+1	.11162+1	.10100+1	.10051+1	.20776-1
.95	.95221	.10047+1	.10551+1	.10024+1	.10012+1	.47306-2
1.00	1.00000	.10000+1	.10000+1	.10000+1	.10000+1	.0
1.05	1.04745	.99514+0	.95007+0	.10023+1	.10011+1	.39784-2
1.10	1.09454	.99010+0	.90458+0	.10092+1	.10041+1	.14682-1
1.15	1.14127	.98488+0	.86296+0	.10205+1	.10087+1	.30583-1
1.20	1.18762	.97948+0	.82474+0	.10360+1	.10148+1	.50497-1
1.25	1.23359	.97391+0	.78950+0	.10559+1	.10221+1	.73497-1
1.30	1.27916	.96819+0	.75690+0	.10801+1	.10305+1	.98851-1

Table C.7 Adiabatic Flow with Friction in a Constant-Area Duct (Fanno Line, $\gamma = 1.10$) (*Continued*)

M	$M^* = \dfrac{\rho^*}{\rho}$	$\dfrac{t}{t^*}$	$\dfrac{p}{p^*}$	$\dfrac{P}{P^*}$	$\dfrac{\mathscr{F}}{\mathscr{F}^*}$	$\dfrac{4fL^*}{\mathscr{D}}$
1.35	1.32431	.96231+0	.72665+0	.11088+1	.10397+1	.12598+0
1.40	1.36906	.95628+0	.69850+0	.11421+1	.10497+1	.15442+0
1.45	1.41337	.95012+0	.67223+0	.11802+1	.10605+1	.18380+0
1.50	1.45726	.94382+0	.64767+0	.12234+1	.10717+1	.21383+0
1.55	1.50070	.93740+0	.62464+0	.12720+1	.10835+1	.24426+0
1.60	1.54369	.93085+0	.60300+0	.13263+1	.10957+1	.27490+0
1.65	1.58623	.92419+0	.58264+0	.13868+1	.11083+1	.30560+0
1.70	1.62830	.91743+0	.56343+0	.14539+1	.11212+1	.33623+0
1.75	1.66992	.91057+0	.54528+0	.15282+1	.11344+1	.36669+0
1.80	1.71106	.90361+0	.52810+0	.16103+1	.11477+1	.39689+0
1.85	1.75172	.89657+0	.51182+0	.17009+1	.11613+1	.42676+0
1.90	1.79191	.88945+0	.49637+0	.18007+1	.11750+1	.45627+0
1.95	1.83161	.88226+0	.48169+0	.19107+1	.11888+1	.48536+0
2.00	1.87083	.87500+0	.46771+0	.20318+1	.12027+1	.51400+0
2.20	2.02282	.84541+0	.41794+0	.26507+1	.12586+1	.62368+0
2.40	2.16695	.81522+0	.37621+0	.35597+1	.13142+1	.72507+0
2.60	2.30324	.78475+0	.34072+0	.49015+1	.13687+1	.81818+0
2.80	2.43183	.75431+0	.31018+0	.68955+1	.14215+1	.90337+0
3.00	2.55289	.72414+0	.28365+0	.98800+1	.14723+1	.98117+0
3.20	2.66667	.69444+0	.26042+0	.14377+2	.15208+1	.10522+1
3.40	2.77345	.66540+0	.23992+0	.21191+2	.15670+1	.11170+1
3.60	2.87355	.63714+0	.22172+0	.31569+2	.16108+1	.11762+1
3.80	2.96730	.60976+0	.20549+0	.47433+2	.16522+1	.12303+1
4.00	3.05505	.58333+0	.19094+0	.71749+2	.16912+1	.12798+1
5.00	3.41565	.46667+0	.13663+0	.59767+3	.18542+1	.14723+1
6.00	3.67423	.37500+0	.10206+0	.49490+4	.19732+1	.16005+1
7.00	3.86174	.30435+0	.78811-1	.37976+5	.20603+1	.16889+1
8.00	4.00000	.25000+0	.62500-1	.26214+6	.21250+1	.17517+1
9.00	4.10385	.20792+0	.50665-1	.16137+7	.21738+1	.17976+1
10.00	4.18330	.17500+0	.41833-1	.88736+7	.22112+1	.18321+1
11.00	4.24515	.14894+0	.35084-1	.43864+8	.22404+1	.18585+1
12.00	4.29407	.12805+0	.29820-1	.19651+9	.22635+1	.18792+1
13.00	4.33333	.11111+0	.25641-1	.80464+9	.22821+1	.18957+1
14.00	4.36527	.97222-1	.22272-1	.30362+10	.22972+1	.19089+1
15.00	4.39155	.85714-1	.19518-1	.10638+11	.23096+1	.19198+1
INFIN	4.58256	.0	.0	INFIN	.24004+1	.19970+1

Table C.8 Isothermal Flow with Friction in a Constant-Area Duct ($\gamma = 1.40$)

M	$\dfrac{V}{V_L} = \dfrac{p_L}{p} = \dfrac{\rho_L}{\rho}$	$\dfrac{T}{T_L}$	$\dfrac{P}{P_L}$	$\dfrac{\mathscr{F}}{\mathscr{F}_L}$	$\dfrac{4fL_{max}}{\mathscr{D}}$
.00	.0	.87500+0	INFIN	INFIN	INFIN
.02	.23664-1	.87507+0	.26488+2	.21141+2	.17772+4
.04	.47329-1	.87528+0	.13255+2	.10588+2	.43933+3
.06	.70993-1	.87563+0	.88493+1	.70784+1	.19212+3
.08	.94657-1	.87612+0	.66500+1	.53295+1	.10589+3
.10	.11832+0	.87675+0	.53334+1	.42849+1	.66160+2
.12	.14199+0	.87752+0	.44581+1	.35925+1	.44699+2
.14	.16565+0	.87843+0	.38352+1	.31012+1	.31847+2
.16	.18931+0	.87948+0	.33698+1	.27318+1	.23573+2
.18	.21298+0	.88067+0	.30096+1	.24551+1	.17953+2
.20	.23664+0	.88200+0	.27230+1	.22342+1	.13975+2
.22	.26031+0	.88347+0	.24899+1	.20510+1	.11066+2
.24	.28397+0	.88508+0	.22970+1	.19027+1	.88830+1
.26	.30764+0	.88683+0	.21350+1	.17791+1	.72087+1
.28	.33130+0	.88872+0	.19974+1	.16749+1	.59013+1
.30	.35496+0	.89075+0	.18791+1	.15861+1	.48650+1

Table C.8 Isothermal Flow with Friction in a Constant-Area Duct ($\gamma = 1.40$) (*Continued*)

M	$\dfrac{V}{V_L} = \dfrac{p_L}{p} = \dfrac{\rho_L}{\rho}$	$\dfrac{T}{T_L}$	$\dfrac{P}{P_L}$	$\dfrac{\mathscr{F}}{\mathscr{F}_L}$	$\dfrac{4fL_{max}}{\mathscr{D}}$
.32	.37863+0	.89292+0	.17768+1	.15099+1	.40331+1
.34	.40229+0	.89523+0	.16874+1	.14440+1	.33578+1
.36	.42596+0	.89768+0	.16090+1	.13868+1	.28046+1
.38	.44962+0	.90027+0	.15398+1	.13369+1	.23479+1
.40	.47329+0	.90300+0	.14784+1	.12931+1	.19682+1
.42	.49695+0	.90587+0	.14237+1	.12546+1	.16507+1
.44	.52062+0	.90888+0	.13749+1	.12207+1	.13840+1
.46	.54428+0	.91203+0	.13311+1	.11908+1	.11591+1
.48	.56794+0	.91532+0	.12918+1	.11643+1	.96873+0
.50	.59161+0	.91875+0	.12565+1	.11410+1	.80732+0
.55	.65077+0	.92794+0	.11827+1	.10937+1	.50207+0
.60	.70993+0	.93800+0	.11259+1	.10593+1	.29895+0
.65	.76909+0	.94894+0	.10823+1	.10347+1	.16552+0
.70	.82825+0	.96075+0	.10495+1	.10178+1	.80848-1
.75	.88741+0	.97344+0	.10255+1	.10071+1	.30949-1
.80	.94657+0	.98700+0	.10092+1	.10015+1	.62566-2
.8452	.10000+1	.10000+1	.10000+1	.10000+1	.0
.85	.10057+1	.10014+1	.99931+0	.10000+1	.65124-4
.90	.10649+1	.10167+1	.99528+0	.10020+1	.75854-2
.95	.11241+1	.10329+1	.99649+0	.10068+1	.25338-1
1.00	.11832+1	.10500+1	.10025+1	.10142+1	.50758-1
1.05	.12424+1	.10679+1	.10131+1	.10236+1	.81931-1
1.10	.13015+1	.10867+1	.10280+1	.10349+1	.11741+0
1.15	.13607+1	.11064+1	.10471+1	.10478+1	.15610+0
1.20	.14199+1	.11270+1	.10703+1	.10621+1	.19715+0
1.25	.14790+1	.11484+1	.10975+1	.10776+1	.23990+0
1.30	.15382+1	.11707+1	.11288+1	.10941+1	.28386+0
1.35	.15973+1	.11939+1	.11642+1	.11117+1	.32861+0
1.40	.16565+1	.12180+1	.12039+1	.11301+1	.37385+0
1.45	.17157+1	.12429+1	.12478+1	.11493+1	.41933+0
1.50	.17748+1	.12687+1	.12962+1	.11691+1	.46486+0
1.55	.18340+1	.12954+1	.13492+1	.11896+1	.51029+0
1.60	.18931+1	.13230+1	.14069+1	.12107+1	.55550+0
1.65	.19523+1	.13514+1	.14697+1	.12323+1	.60039+0
1.70	.20115+1	.13807+1	.15378+1	.12543+1	.64489+0
1.75	.20706+1	.14109+1	.16113+1	.12768+1	.68894+0
1.80	.21298+1	.14420+1	.16906+1	.12997+1	.73250+0
1.85	.21889+1	.14739+1	.17760+1	.13229+1	.77555+0
1.90	.22481+1	.15067+1	.18678+1	.13465+1	.81804+0
1.95	.23073+1	.15404+1	.19663+1	.13703+1	.85998+0
2.00	.23664+1	.15750+1	.20720+1	.13945+1	.90134+0
2.20	.26031+1	.17220+1	.25741+1	.14936+1	.10610+1
2.40	.28397+1	.18830+1	.32263+1	.15959+1	.12114+1
2.60	.30764+1	.20580+1	.40646+1	.17007+1	.13532+1
2.80	.33130+1	.22470+1	.51332+1	.18074+1	.14868+1
3.00	.35496+1	.24500+1	.64848+1	.19157+1	.16131+1
3.20	.37863+1	.26670+1	.81821+1	.20252+1	.17325+1
3.40	.40229+1	.28980+1	.10299+2	.21358+1	.18458+1
3.60	.42596+1	.31430+1	.12922+2	.22472+1	.19535+1
3.80	.44962+1	.34020+1	.16152+2	.23593+1	.20559+1
4.00	.47329+1	.36750+1	.20104+2	.24721+1	.21537+1
5.00	.59161+1	.52500+1	.56043+2	.30426+1	.25839+1
6.00	.70993+1	.71750+1	.13937+3	.36201+1	.29398+1
7.00	.82825+1	.94500+1	.31322+3	.42016+1	.32429+1
8.00	.94657+1	.12075+2	.64633+3	.47857+1	.35065+1
9.00	.10649+2	.15050+2	.12419+4	.53714+1	.37397+1
10.00	.11832+2	.18375+2	.22477+4	.59583+1	.39488+1
11.00	.13015+2	.22050+2	.38679+4	.65461+1	.41382+1
12.00	.14199+2	.26075+2	.63759+4	.71345+1	.43112+1
13.00	.15382+2	.30450+2	.10129+5	.77234+1	.44700+1
14.00	.16565+2	.35175+2	.15562+5	.83127+1	.46182+1
15.00	.17748+2	.40250+2	.23309+5	.89023+1	.47557+1
INFIN	INFIN	INFIN	INFIN	INFIN	INFIN

Table C.9 Frictionless Flow with Heat Transfer in a Constant-Area Duct (Rayleigh Line, $\gamma = 1.40$)

M	$M^* = \dfrac{\rho^*}{\rho}$	$\dfrac{t}{t^*}$	$\dfrac{p}{p^*}$	$\dfrac{P}{P^*}$	$\dfrac{T}{T^*}$
.00	.00000	.0	.24000+1	.12679+1	.0
.01	.00024	.57584−3	.23997+1	.12678+1	.47988−3
.02	.00096	.23014−2	.23987+1	.12675+1	.19180−2
.03	.00216	.51710−2	.23970+1	.12671+1	.43099−2
.04	.00383	.91749−2	.23946+1	.12665+1	.76482−2
.05	.00598	.14300−1	.23916+1	.12657+1	.11922−1
.06	.00860	.20529−1	.23880+1	.12647+1	.17119−1
.07	.01168	.27841−1	.23836+1	.12636+1	.23223−1
.08	.01522	.36212−1	.23787+1	.12623+1	.30215−1
.09	.01922	.45616−1	.23731+1	.12608+1	.38075−1
.10	.02367	.56020−1	.23669+1	.12591+1	.46777−1
.11	.02856	.67393−1	.23600+1	.12573+1	.56297−1
.12	.03388	.79698−1	.23526+1	.12554+1	.66606−1
.13	.03962	.92896−1	.23445+1	.12533+1	.77675−1
.14	.04578	.10695+0	.23359+1	.12510+1	.89471−1
.15	.05235	.12181+0	.23267+1	.12486+1	.10196+0
.16	.05931	.13743+0	.23170+1	.12461+1	.11511+0
.17	.06666	.15377+0	.23067+1	.12434+1	.12888+0
.18	.07439	.17078+0	.22959+1	.12406+1	.14324+0
.19	.08247	.18841+0	.22845+1	.12377+1	.15814+0
.20	.09091	.20661+0	.22727+1	.12346+1	.17355+0
.21	.09969	.22533+0	.22604+1	.12314+1	.18943+0
.22	.10879	.24452+0	.22477+1	.12281+1	.20574+0
.23	.11821	.26413+0	.22345+1	.12247+1	.22244+0
.24	.12792	.28411+0	.22209+1	.12213+1	.23948+0
.25	.13793	.30440+0	.22069+1	.12177+1	.25684+0
.26	.14821	.32496+0	.21925+1	.12140+1	.27446+0
.27	.15876	.34573+0	.21777+1	.12102+1	.29231+0
.28	.16955	.36667+0	.21626+1	.12064+1	.31035+0
.29	.18058	.38774+0	.21472+1	.12025+1	.32855+0
.30	.19183	.40887+0	.21314+1	.11985+1	.34686+0
.31	.20329	.43004+0	.21154+1	.11945+1	.36525+0
.32	.21495	.45119+0	.20991+1	.11904+1	.38369+0
.33	.22678	.47228+0	.20825+1	.11863+1	.40214+0
.34	.23879	.49327+0	.20657+1	.11822+1	.42056+0
.35	.25096	.51413+0	.20487+1	.11779+1	.43894+0
.36	.26327	.53482+0	.20314+1	.11737+1	.45723+0
.37	.27572	.55529+0	.20140+1	.11695+1	.47541+0
.38	.28828	.57553+0	.19964+1	.11652+1	.49346+0
.39	.30095	.59549+0	.19787+1	.11609+1	.51134+0
.40	.31373	.61515+0	.19608+1	.11566+1	.52903+0
.41	.32658	.63448+0	.19428+1	.11523+1	.54651+0
.42	.33951	.65346+0	.19247+1	.11480+1	.56376+0
.43	.35251	.67205+0	.19065+1	.11437+1	.58076+0
.44	.36556	.69025+0	.18882+1	.11394+1	.59748+0
.45	.37865	.70804+0	.18699+1	.11351+1	.61393+0
.46	.39178	.72538+0	.18515+1	.11308+1	.63007+0
.47	.40493	.74228+0	.18331+1	.11266+1	.64589+0
.48	.41810	.75871+0	.18147+1	.11224+1	.66139+0
.49	.43127	.77466+0	.17962+1	.11182+1	.67655+0
.50	.44444	.79012+0	.17778+1	.11141+1	.69136+0
.52	.47075	.81955+0	.17409+1	.11059+1	.71990+0
.54	.49696	.84695+0	.17043+1	.10979+1	.74695+0
.56	.52302	.87227+0	.16678+1	.10901+1	.77249+0
.58	.54887	.89552+0	.16316+1	.10826+1	.79648+0
.60	.57447	.91670+0	.15957+1	.10753+1	.81892+0
.62	.59978	.93584+0	.15603+1	.10682+1	.83983+0
.64	.62477	.95298+0	.15253+1	.10615+1	.85920+0
.66	.64941	.96816+0	.14908+1	.10550+1	.87708+0
.68	.67366	.98144+0	.14569+1	.10489+1	.89350+0
.70	.69751	.99290+0	.14235+1	.10431+1	.90850+0
.72	.72093	.10026+1	.13907+1	.10376+1	.92212+0
.74	.74392	.10106+1	.13585+1	.10325+1	.93442+0
.76	.76645	.10171+1	.13270+1	.10278+1	.94546+0

Table C.9 Frictionless Flow with Heat Transfer in a Constant-Area Duct (Rayleigh Line, $\gamma = 1.40$) (*Continued*)

M	$M^* = \dfrac{\rho^*}{\rho}$	$\dfrac{t}{t^*}$	$\dfrac{p}{p^*}$	$\dfrac{P}{P^*}$	$\dfrac{T}{T^*}$
.78	.78853	.10220+1	.12961+1	.10234+1	.95528+0
.80	.81013	.10255+1	.12658+1	.10193+1	.96395+0
.82	.83125	.10276+1	.12362+1	.10157+1	.97152+0
.84	.85190	.10285+1	.12073+1	.10124+1	.97807+0
.86	.87207	.10283+1	.11791+1	.10095+1	.98363+0
.88	.89175	.10269+1	.11515+1	.10070+1	.98828+0
.90	.91097	.10245+1	.11246+1	.10049+1	.99207+0
.92	.92970	.10212+1	.10984+1	.10031+1	.99506+0
.94	.94797	.10170+1	.10728+1	.10017+1	.99729+0
.96	.96577	.10121+1	.10479+1	.10008+1	.99883+0
.98	.98311	.10064+1	.10236+1	.10002+1	.99971+0
1.00	1.00000	.10000+1	.10000+1	.10000+1	.10000+1
1.02	1.01645	.99304+0	.97698+0	.10002+1	.99973+0
1.04	1.03246	.98554+0	.95456+0	.10008+1	.99895+0
1.06	1.04804	.97755+0	.93275+0	.10017+1	.99769+0
1.08	1.06320	.96913+0	.91152+0	.10031+1	.99601+0
1.10	1.07795	.96031+0	.89087+0	.10049+1	.99392+0
1.12	1.09230	.95115+0	.87078+0	.10070+1	.99148+0
1.14	1.10626	.94169+0	.85123+0	.10095+1	.98871+0
1.16	1.11984	.93196+0	.83222+0	.10124+1	.98564+0
1.18	1.13305	.92200+0	.81374+0	.10157+1	.98230+0
1.20	1.14589	.91185+0	.79576+0	.10194+1	.97872+0
1.22	1.15838	.90153+0	.77827+0	.10235+1	.97492+0
1.24	1.17052	.89108+0	.76127+0	.10279+1	.97092+0
1.26	1.18233	.88052+0	.74473+0	.10328+1	.96675+0
1.28	1.19382	.86988+0	.72865+0	.10380+1	.96243+0
1.30	1.20499	.85917+0	.71301+0	.10437+1	.95798+0
1.32	1.21585	.84843+0	.69780+0	.10497+1	.95341+0
1.34	1.22642	.83766+0	.68301+0	.10561+1	.94873+0
1.36	1.23669	.82689+0	.66863+0	.10629+1	.94398+0
1.38	1.24669	.81613+0	.65464+0	.10701+1	.93914+0
1.40	1.25641	.80539+0	.64103+0	.10777+1	.93425+0
1.42	1.26587	.79469+0	.62779+0	.10856+1	.92931+0
1.44	1.27507	.78405+0	.61491+0	.10940+1	.92434+0
1.46	1.28402	.77346+0	.60237+0	.11028+1	.91933+0
1.48	1.29273	.76294+0	.59018+0	.11120+1	.91431+0
1.50	1.30120	.75250+0	.57831+0	.11215+1	.90928+0
1.55	1.32142	.72680+0	.55002+0	.11473+1	.89669+0
1.60	1.34031	.70174+0	.52356+0	.11756+1	.88419+0
1.65	1.35800	.67738+0	.49880+0	.12066+1	.87184+0
1.70	1.37455	.65377+0	.47562+0	.12402+1	.85971+0
1.75	1.39007	.63095+0	.45390+0	.12767+1	.84784+0
1.80	1.40462	.60894+0	.43353+0	.13159+1	.83628+0
1.85	1.41829	.58774+0	.41440+0	.13581+1	.82504+0
1.90	1.43112	.56734+0	.39643+0	.14033+1	.81414+0
1.95	1.44319	.54774+0	.37954+0	.14516+1	.80358+0
2.00	1.45455	.52893+0	.36364+0	.15031+1	.79339+0
2.10	1.47533	.49356+0	.33454+0	.16162+1	.77406+0
2.20	1.49383	.46106+0	.30864+0	.17434+1	.75613+0
2.30	1.51035	.43122+0	.28551+0	.18860+1	.73954+0
2.40	1.52515	.40384+0	.26478+0	.20451+1	.72421+0
2.50	1.53846	.37870+0	.24615+0	.22218+1	.71006+0
2.60	1.55046	.35561+0	.22936+0	.24177+1	.69700+0
2.70	1.56131	.33439+0	.21417+0	.26343+1	.68494+0
2.80	1.57114	.31486+0	.20040+0	.28731+1	.67380+0
2.90	1.58008	.29687+0	.18788+0	.31359+1	.66350+0
3.00	1.58824	.28028+0	.17647+0	.34245+1	.65398+0
3.10	1.59568	.26495+0	.16604+0	.37408+1	.64516+0
3.20	1.60250	.25078+0	.15649+0	.40871+1	.63699+0
3.30	1.60877	.23766+0	.14773+0	.44655+1	.62940+0
3.40	1.61453	.22549+0	.13966+0	.48783+1	.62236+0
3.50	1.61983	.21419+0	.13223+0	.53280+1	.61580+0
3.60	1.62474	.20369+0	.12537+0	.58173+1	.60970+0
3.70	1.62928	.19390+0	.11901+0	.63488+1	.60401+0

Table C.9   Frictionless Flow with Heat Transfer in a
Constant-Area Duct (Rayleigh Line, $\gamma = 1.40$) (*Continued*)

M	$M^* = \dfrac{\rho^*}{\rho}$	$\dfrac{t}{t^*}$	$\dfrac{p}{p^*}$	$\dfrac{P}{P^*}$	$\dfrac{T}{T^*}$
3.80	1.63348	.18478+0	.11312+0	.69256+1	.59870+0
3.90	1.63739	.17627+0	.10765+0	.75505+1	.59373+0
4.00	1.64103	.16831+0	.10256+0	.82268+1	.58909+0
4.50	1.65588	.13540+0	.81772-1	.12502+2	.56982+0
5.00	1.66667	.11111+0	.66667-1	.18634+2	.55556+0
5.50	1.67474	.92719-1	.55363-1	.27211+2	.54473+0
6.00	1.68093	.78487-1	.46693-1	.38946+2	.53633+0
6.50	1.68579	.67263-1	.39900-1	.54683+2	.52970+0
7.00	1.68966	.58264-1	.34483-1	.75414+2	.52438+0
7.50	1.69279	.50943-1	.30094-1	.10229+3	.52004+0
8.00	1.69536	.44910-1	.26490-1	.13662+3	.51647+0
8.50	1.69750	.39883-1	.23495-1	.17992+3	.51349+0
9.00	1.69930	.35650-1	.20979-1	.23388+3	.51098+0
9.50	1.70082	.32053-1	.18846-1	.30041+3	.50885+0
10.00	1.70213	.28972-1	.17021-1	.38161+3	.50702+0
11.00	1.70423	.24003-1	.14085-1	.59774+3	.50407+0
12.00	1.70582	.20207-1	.11846-1	.90405+3	.50181+0
13.00	1.70707	.17243-1	.10101-1	.13267+4	.50005+0
14.00	1.70806	.14885-1	.87146-2	.18963+4	.49865+0
15.00	1.70886	.12979-1	.75949-2	.26488+4	.49752+0
INFIN	1.71429	.0	.0	INFIN	.48980+0

Table C.9   Frictionless Flow with Heat Transfer in a
Constant-Area Duct (Rayleigh Line, $\gamma = 1.30$)

M	$M^* = \dfrac{\rho^*}{\rho}$	$\dfrac{t}{t^*}$	$\dfrac{p}{p^*}$	$\dfrac{P}{P^*}$	$\dfrac{T}{T^*}$
.00	.00000	.0	.23000+1	.12552+1	.0
.02	.00092	.21138-2	.22988+1	.12548+1	.18382-2
.04	.00367	.84289-2	.22952+1	.12539+1	.73312-2
.06	.00824	.18867-1	.22893+1	.12523+1	.16415-1
.08	.01460	.33300-1	.22810+1	.12500+1	.28984-1
.10	.02270	.51551-1	.22705+1	.12471+1	.44894-1
.12	.03251	.73402-1	.22577+1	.12437+1	.63966-1
.14	.04396	.96596-1	.22429+1	.12397+1	.85987-1
.16	.05698	.12684+0	.22259+1	.12351+1	.11072+0
.18	.07151	.15782+0	.22070+1	.12300+1	.13790+0
.20	.08745	.19120+0	.21863+1	.12245+1	.16726+0
.22	.10473	.22662+0	.21639+1	.12185+1	.19849+0
.24	.12325	.26373+0	.21398+1	.12121+1	.23131+0
.26	.14292	.30216+0	.21142+1	.12053+1	.26541+0
.28	.16364	.34156+0	.20873+1	.11983+1	.30050+0
.30	.18532	.38159+0	.20591+1	.11909+1	.33629+0
.32	.20785	.42189+0	.20298+1	.11834+1	.37250+0
.34	.23114	.46217+0	.19995+1	.11756+1	.40886+0
.36	.25510	.50213+0	.19684+1	.11677+1	.44512+0
.38	.27963	.54150+0	.19365+1	.11596+1	.48106+0
.40	.30464	.58002+0	.19040+1	.11515+1	.51647+0

Table C.9 Frictionless Flow with Heat Transfer in a
Constant-Area Duct (Rayleigh Line, $\gamma = 1.30$) (*Continued*)

M	$M^* = \dfrac{\rho^*}{\rho}$	$\dfrac{t}{t^*}$	$\dfrac{p}{p^*}$	$\dfrac{P}{P^*}$	$\dfrac{T}{T^*}$
.42	.33004	.61748+0	.18710+1	.11434+1	.55115+0
.44	.35575	.65369+0	.18375+1	.11352+1	.58494+0
.46	.38169	.68849+0	.18038+1	.11271+1	.61769+0
.48	.40778	.72173+0	.17699+1	.11191+1	.64928+0
.50	.43396	.75329+0	.17358+1	.11111+1	.67960+0
.55	.49937	.82437+0	.16508+1	.10919+1	.74937+0
.60	.56403	.88370+0	.15668+1	.10739+1	.80993+0
.65	.62724	.93119+0	.14846+1	.10574+1	.86105+0
.70	.68845	.96728+0	.14050+1	.10426+1	.90294+0
.75	.74729	.99279+0	.13285+1	.10299+1	.93614+0
.80	.80349	.10088+1	.12555+1	.10193+1	.96139+0
.85	.85690	.10163+1	.11860+1	.10109+1	.97952+0
.90	.90745	.10166+1	.11203+1	.10049+1	.99143+0
.95	.95514	.10108+1	.10583+1	.10012+1	.99799+0
1.00	1.00000	.10000+1	.10000+1	.10000+1	.10000+1
1.05	1.04212	.98506+0	.94524+0	.10012+1	.99823+0
1.10	1.08162	.96686+0	.89390+0	.10049+1	.99334+0
1.15	1.11860	.94613+0	.84582+0	.10112+1	.98593+0
1.20	1.15320	.92353+0	.80084+0	.10199+1	.97653+0
1.25	1.18557	.89956+0	.75876+0	.10312+1	.96556+0
1.30	1.21583	.87470+0	.71942+0	.10451+1	.95342+0
1.35	1.24412	.84929+0	.68264+0	.10616+1	.94041+0
1.40	1.27057	.82365+0	.64825+0	.10809+1	.92679+0
1.45	1.29532	.79803+0	.61609+0	.11028+1	.91279+0
1.50	1.31847	.77261+0	.58599+0	.11276+1	.89858+0
1.55	1.34014	.74755+0	.55781+0	.11552+1	.88430+0
1.60	1.36044	.72297+0	.53142+0	.11858+1	.87008+0
1.65	1.37947	.69896+0	.50669+0	.12195+1	.85600+0
1.70	1.39731	.67560+0	.48350+0	.12563+1	.84215+0
1.75	1.41405	.65291+0	.46173+0	.12964+1	.82856+0
1.80	1.42978	.63095+0	.44129+0	.13400+1	.81529+0
1.85	1.44456	.60971+0	.42208+0	.13872+1	.80237+0
1.90	1.45846	.58922+0	.40400+0	.14381+1	.78982+0
1.95	1.47154	.56948+0	.38699+0	.14929+1	.77765+0
2.00	1.48387	.55047+0	.37097+0	.15518+1	.76587+0
2.20	1.52660	.48151+0	.31541+0	.18325+1	.72269+0
2.40	1.56079	.42293+0	.27097+0	.21970+1	.68551+0
2.60	1.58848	.37326+0	.23496+0	.26644+1	.65370+0
2.80	1.61115	.33110+0	.20550+0	.32583+1	.62649+0
3.00	1.62992	.29518+0	.18110+0	.40074+1	.60320+0
3.20	1.64561	.26446+0	.16070+0	.49467+1	.58319+0
3.40	1.65885	.23804+0	.14350+0	.61181+1	.56592+0
3.60	1.67010	.21522+0	.12887+0	.75714+1	.55096+0
3.80	1.67975	.19540+0	.11633+0	.93656+1	.53794+0
4.00	1.68607	.17810+0	.10550+0	.11570+2	.52656+0
5.00	1.71642	.11784+0	.68657-1	.32063+2	.48675+0
6.00	1.73222	.83349-1	.48117-1	.81794+2	.46386+0
7.00	1.74189	.61922-1	.35549-1	.19133+3	.44961+0
8.00	1.74822	.47754-1	.27316-1	.41341+3	.44017+0
9.00	1.75259	.37921-1	.21637-1	.83339+3	.43361+0
10.00	1.75573	.30826-1	.17573-1	.15823+4	.42888+0
11.00	1.75805	.25543-1	.14529-1	.28529+4	.42535+0
12.00	1.75985	.21507-1	.12221-1	.49191+4	.42266+0
13.00	1.76121	.18354-1	.10421-1	.81586+4	.42055+0
14.00	1.76231	.15846-1	.89914-2	.13079+5	.41888+0
15.00	1.76320	.13817-1	.78365-2	.20350+5	.41752+0
INFIN	1.76923	.0	.0	INFIN	.40826+0

Table C.9   Frictionless Flow with Heat Transfer in a
Constant-Area Duct (Rayleigh Line, $\gamma = 1.20$)

M	$M^* = \dfrac{\rho^*}{\rho}$	$\dfrac{t}{t^*}$	$\dfrac{p}{p^*}$	$\dfrac{P}{P^*}$	$\dfrac{T}{T^*}$
.00	.00000	.0	.22000+1	.12418+1	.0
.02	.00088	.19341−2	.21989+1	.12415+1	.17584−2
.04	.00351	.77143−2	.21958+1	.12407+1	.70142−2
.06	.00789	.17274−1	.21905+1	.12392+1	.15710−1
.08	.01397	.30506−1	.21832+1	.12371+1	.27750−1
.10	.02174	.47259−1	.21739+1	.12345+1	.43006−1
.12	.03114	.67348−1	.21626+1	.12313+1	.61314−1
.14	.04213	.90554−1	.21494+1	.12276+1	.82483−1
.16	.05464	.11663+0	.21344+1	.12235+1	.10630+0
.18	.06861	.14530+0	.21177+1	.12188+1	.13252+0
.20	.08397	.17627+0	.20992+1	.12137+1	.16089+0
.22	.10064	.20924+0	.20792+1	.12082+1	.19114+0
.24	.11853	.24390+0	.20578+1	.12023+1	.22301+0
.26	.13756	.27993+0	.20349+1	.11960+1	.25620+0
.28	.15765	.31700+0	.20108+1	.11895+1	.29044+0
.30	.17870	.35482+0	.19856+1	.11827+1	.32547+0
.32	.20063	.39308+0	.19592+1	.11757+1	.36100+0
.34	.22334	.43149+0	.19320+1	.11684+1	.39680+0
.36	.24675	.46978+0	.19039+1	.11610+1	.43261+0
.38	.27076	.50770+0	.18751+1	.11535+1	.46821+0
.40	.29530	.54502+0	.18456+1	.11459+1	.50340+0
.42	.32028	.58152+0	.18157+1	.11383+1	.53798+0
.44	.34562	.61703+0	.17853+1	.11306+1	.57179+0
.46	.37125	.65136+0	.17545+1	.11229+1	.60468+0
.48	.39709	.68438+0	.17235+1	.11153+1	.63650+0
.50	.42308	.71598+0	.16923+1	.11078+1	.66716+0
.55	.48826	.78810+0	.16141+1	.10895+1	.73812+0
.60	.55307	.84969+0	.15363+1	.10722+1	.80026+0
.65	.61679	.90042+0	.14599+1	.10563+1	.85315+0
.70	.67884	.94046+0	.13854+1	.10420+1	.89686+0
.75	.73881	.97037+0	.13134+1	.10296+1	.93178+0
.80	.79638	.99097+0	.12443+1	.10191+1	.95854+0
.85	.85137	.10032+1	.11784+1	.10109+1	.97791+0
.90	.90365	.10081+1	.11156+1	.10049+1	.99072+0
.95	.95319	.10067+1	.10562+1	.10012+1	.99781+0
1.00	1.00000	.10000+1	.10000+1	.10000+1	.10000+1
1.05	1.04412	.98884+0	.94705+0	.10012+1	.99805+0
1.10	1.08564	.97407+0	.89723+0	.10050+1	.99267+0
1.15	1.12466	.95642+0	.85041+0	.10114+1	.98446+0
1.20	1.16129	.93652+0	.80645+0	.10204+1	.97399+0
1.25	1.19565	.91493+0	.76522+0	.10321+1	.96172+0
1.30	1.22787	.89211+0	.72655+0	.10466+1	.94807+0
1.35	1.25808	.86846+0	.69030+0	.10640+1	.93339+0
1.40	1.28640	.84429+0	.65632+0	.10843+1	.91798+0
1.45	1.31294	.81989+0	.62447+0	.11077+1	.90207+0
1.50	1.33784	.79547+0	.59459+0	.11342+1	.88587+0
1.55	1.36119	.77121+0	.56657+0	.11640+1	.86954+0
1.60	1.38310	.74726+0	.54028+0	.11973+1	.85323+0
1.65	1.40368	.72372+0	.51558+0	.12342+1	.83704+0
1.70	1.42301	.70068+0	.49239+0	.12749+1	.82106+0
1.75	1.44118	.67820+0	.47059+0	.13196+1	.80536+0
1.80	1.45827	.65634+0	.45008+0	.13686+1	.78999+0
1.85	1.47435	.63512+0	.43078+0	.14220+1	.77499+0
1.90	1.48950	.61457+0	.41260+0	.14802+1	.76039+0
1.95	1.50377	.59470+0	.39547+0	.15435+1	.74621+0
2.00	1.51724	.57551+0	.37931+0	.16122+1	.73246+0
2.20	1.56404	.50542+0	.32315+0	.19483+1	.68186+0
2.40	1.60162	.44534+0	.27806+0	.24050+1	.63806+0
2.60	1.63213	.39406+0	.24144+0	.30206+1	.60041+0
2.80	1.65719	.35029+0	.21138+0	.38465+1	.56811+0
3.00	1.67797	.31284+0	.18644+0	.49512+1	.54036+0
3.20	1.69536	.28069+0	.16556+0	.64250+1	.51647+0
3.40	1.71006	.25297+0	.14793+0	.83867+1	.49582+0
3.60	1.72257	.22895+0	.13291+0	.10991+2	.47789+0

Table C.9 Frictionless Flow with Heat Transfer in a Constant-Area Duct (Rayleigh Line, $\gamma = 1.20$) (*Continued*)

M	$M^* = \dfrac{\rho^*}{\rho}$	$\dfrac{t}{t^*}$	$\dfrac{p}{p^*}$	$\dfrac{P}{P^*}$	$\dfrac{T}{T^*}$
3.80	1.73330	.20806+0	.12003+0	.14440+2	.46226+0
4.00	1.74257	.16979+0	.10891+0	.18991+2	.44858+0
5.00	1.77419	.12591+0	.70968-1	.73640+2	.40062+0
6.00	1.79186	.89187-1	.49774-1	.26619+3	.37297+0
7.00	1.80268	.66319-1	.36789-1	.87595+3	.35571+0
8.00	1.80977	.51176-1	.28278-1	.26211+4	.34427+0
9.00	1.81466	.40654-1	.22403-1	.71813+4	.33632+0
10.00	1.81818	.33058-1	.18182-1	.18182+5	.33058+0
11.00	1.82079	.27399-1	.15048-1	.42929+5	.32630+0
12.00	1.82276	.23073-1	.12658-1	.95311+5	.32303+0
13.00	1.82434	.19694-1	.10795-1	.20044+6	.32047+0
14.00	1.82557	.17004-1	.93141-2	.40178+6	.31843+0
15.00	1.82657	.14828-1	.81181-2	.77180+6	.31678+0
INFIN	1.83333	.0	.0	INFIN	.30556+0

Table C.9 Frictionless Flow with Heat Transfer in a Constant-Area Duct (Rayleigh Line, $\gamma = 1.10$)

M	$M^* = \dfrac{\rho^*}{\rho}$	$\dfrac{t}{t^*}$	$\dfrac{p}{p^*}$	$\dfrac{P}{P^*}$	$\dfrac{T}{T^*}$
.00	.00000	.0	.21000+1	.12278+1	.0
.02	.00084	.17624-2	.20991+1	.12276+1	.16786-2
.04	.00335	.70312-2	.20963+1	.12267+1	.66969-2
.06	.00753	.15751-1	.20917+1	.12254+1	.15004-1
.08	.01335	.27831-1	.20853+1	.12235+1	.26514-1
.10	.02077	.43146-1	.20772+1	.12212+1	.41112-1
.12	.02977	.61539-1	.20673+1	.12183+1	.58651-1
.14	.04029	.82826-1	.20557+1	.12149+1	.78959-1
.16	.05229	.10680+0	.20425+1	.12111+1	.10184+0
.18	.06570	.13322+0	.20277+1	.12069+1	.12708+0
.20	.08046	.16184+0	.20115+1	.12022+1	.15445+0
.22	.09650	.19241+0	.19938+1	.11972+1	.18369+0
.24	.11375	.22465+0	.19749+1	.11918+1	.21457+0
.26	.13213	.25828+0	.19547+1	.11861+1	.24681+0
.28	.15157	.29302+0	.19333+1	.11801+1	.28016+0
.30	.17197	.32861+0	.19108+1	.11738+1	.31437+0
.32	.19327	.36478+0	.18874+1	.11673+1	.34919+0
.34	.21537	.40126+0	.18631+1	.11606+1	.38436+0
.36	.23820	.43781+0	.18380+1	.11538+1	.41966+0
.38	.26166	.47420+0	.18122+1	.11468+1	.45488+0
.40	.28571	.51020+0	.17857+1	.11397+1	.48980+0
.42	.31024	.54563+0	.17587+1	.11326+1	.52423+0
.44	.33518	.58030+0	.17313+1	.11254+1	.55801+0
.46	.36046	.61404+0	.17035+1	.11182+1	.59099+0
.48	.38601	.64672+0	.16754+1	.11111+1	.62302+0
.50	.41176	.67820+0	.16471+1	.11040+1	.65398+0
.55	.47665	.75105+0	.15757+1	.10867+1	.72610+0
.60	.54155	.81465+0	.15043+1	.10702+1	.78982+0
.65	.60573	.86844+0	.14337+1	.10550+1	.84455+0
.70	.66862	.91234+0	.13645+1	.10412+1	.89018+0
.75	.72973	.94668+0	.12973+1	.10291+1	.92695+0
.80	.78873	.97203+0	.12324+1	.10189+1	.95537+0
.85	.84538	.98916+0	.11701+1	.10108+1	.97609+0
.90	.89952	.99894+0	.11105+1	.10049+1	.98990+0
.95	.95107	.10023+1	.10538+1	.10012+1	.99761+0
1.00	1.00000	.10000+1	.10000+1	.10000+1	.10000+1
1.05	1.04632	.99301+0	.94905+0	.10013+1	.99785+0
1.10	1.09009	.98206+0	.90090+0	.10051+1	.99188+0
1.15	1.13138	.96788+0	.85548+0	.10116+1	.98274+0

Table C.9   Frictionless Flow with Heat Transfer in a
Constant-Area Duct (Rayleigh Line, $\gamma = 1.10$) (*Continued*)

M	$M^* = \dfrac{\rho^*}{\rho}$	$\dfrac{t}{t^*}$	$\dfrac{p}{p^*}$	$\dfrac{P}{P^*}$	$\dfrac{T}{T^*}$
1.20	1.17028	.95108+0	.81269+0	.10209+1	.97101+0
1.25	1.20690	.93222+0	.77241+0	.10331+1	.95719+0
1.30	1.24134	.91179+0	.73452+0	.10482+1	.94175+0
1.35	1.27373	.89020+0	.69889+0	.10665+1	.92507+0
1.40	1.30418	.86780+0	.66540+0	.10880+1	.90747+0
1.45	1.33281	.84488+0	.63391+0	.11129+1	.88924+0
1.50	1.35971	.82170+0	.60432+0	.11415+1	.87061+0
1.55	1.38501	.79844+0	.57649+0	.11739+1	.85177+0
1.60	1.40881	.77529+0	.55031+0	.12104+1	.83288+0
1.65	1.43119	.75236+0	.52569+0	.12512+1	.81407+0
1.70	1.45226	.72978+0	.50251+0	.12967+1	.79546+0
1.75	1.47210	.70762+0	.48069+0	.13472+1	.77712+0
1.80	1.49080	.68595+0	.46012+0	.14030+1	.75912+0
1.85	1.50842	.66482+0	.44074+0	.14646+1	.74151+0
1.90	1.52505	.64426+0	.42245+0	.15326+1	.72433+0
1.95	1.54074	.62429+0	.40519+0	.16073+1	.70760+0
2.00	1.55556	.60494+0	.38889+0	.16894+1	.69136+0
2.20	1.60721	.53370+0	.33207+0	.21061+1	.63130+0
2.40	1.64885	.47200+0	.28626+0	.27086+1	.57899+0
2.60	1.68279	.41890+0	.24893+0	.35811+1	.53360+0
2.80	1.71072	.37329+0	.21820+0	.48508+1	.49487+0
3.00	1.73394	.33406+0	.19266+0	.67106+1	.46132+0
3.20	1.75342	.30024+0	.17123+0	.94531+1	.43235+0
3.40	1.76990	.27098+0	.15311+0	.13523+2	.40725+0
3.60	1.78395	.24556+0	.13756+0	.19599+2	.38542+0
3.80	1.79602	.22339+0	.12436+0	.28710+2	.36635+0
4.00	1.80645	.20395+0	.11290+0	.42425+2	.34964+0
5.00	1.84211	.13573+0	.73684-1	.32233+3	.29086+0
6.00	1.86207	.96314-1	.51724-1	.25081+4	.25684+0
7.00	1.87432	.71695-1	.38251-1	.18432+5	.23557+0
8.00	1.88235	.55363-1	.29412-1	.12336+6	.22145+0
9.00	1.88790	.44002-1	.23307-1	.74236+6	.21163+0
10.00	1.89189	.35793-1	.18919-1	.40131+7	.20453+0
11.00	1.89485	.29673-1	.15660-1	.19579+8	.19924+0
12.00	1.89711	.24993-1	.13174-1	.86816+8	.19519+0
13.00	1.89888	.21336-1	.11236-1	.35260+9	.19202+0
14.00	1.90026	.18424-1	.96953-2	.13217+10	.18950+0
15.00	1.90141	.16066-1	.84507-2	.46058+10	.18746+0
INFIN	1.90909	.0	.0	INFIN	.17355+0

Table C.10   Isothermal Frictionless Flow in
a Variable-Area Duct ($\gamma = 1.40$)

M	$\dfrac{p}{p^*}$	$\dfrac{P}{P^*}$	$\dfrac{A}{A^*}$	$\dfrac{\mathscr{F}}{\mathscr{F}^*}$	$\dfrac{T}{T^*}$
.00	.20138+1	.10638+1	INFIN	INFIN	.83333+0
.02	.20132+1	.10638+1	.24836+2	.20845+2	.83340+0
.04	.20115+1	.10638+1	.12429+2	.10440+2	.83360+0
.06	.20087+1	.10638+1	.82973+1	.69794+1	.83393+0
.08	.20048+1	.10638+1	.62352+1	.52550+1	.83440+0
.10	.19997+1	.10638+1	.50007+1	.42250+1	.83500+0
.12	.19936+1	.10638+1	.41801+1	.35422+1	.83573+0
.14	.19863+1	.10638+1	.35960+1	.30579+1	.83660+0
.16	.19780+1	.10638+1	.31598+1	.26975+1	.83760+0
.18	.19686+1	.10638+1	.28221+1	.24198+1	.83873+0
.20	.19581+1	.10637+1	.25534+1	.22000+1	.84000+0
.22	.19467+1	.10637+1	.23350+1	.20223+1	.84140+0
.24	.19342+1	.10636+1	.21542+1	.18761+1	.84293+0
.26	.19207+1	.10635+1	.20025+1	.17542+1	.84460+0

Table C.10 Isothermal Frictionless Flow in
a Variable-Area Duct ($\gamma = 1.40$) (*Continued*)

M	$\dfrac{p}{p^*}$	$\dfrac{P}{P^*}$	$\dfrac{A}{A^*}$	$\dfrac{\mathscr{F}}{\mathscr{F}^*}$	$\dfrac{T}{T^*}$
.28	.19062+1	.10634+1	.18736+1	.16514+1	.84640+0
.30	.18906+1	.10632+1	.17629+1	.15639+1	.84833+0
.32	.18745+1	.10631+1	.16671+1	.14888+1	.85040+0
.34	.18572+1	.10628+1	.15836+1	.14238+1	.85260+0
.36	.18391+1	.10626+1	.15104+1	.13674+1	.85493+0
.38	.18202+1	.10623+1	.14458+1	.13182+1	.85740+0
.40	.18004+1	.10620+1	.13886+1	.12750+1	.86000+0
.42	.17798+1	.10616+1	.13377+1	.12371+1	.86273+0
.44	.17585+1	.10611+1	.12924+1	.12036+1	.86560+0
.46	.17365+1	.10606+1	.12519+1	.11741+1	.86860+0
.48	.17138+1	.10600+1	.12156+1	.11481+1	.87173+0
.50	.16905+1	.10593+1	.11831+1	.11250+1	.87500+0
.55	.16295+1	.10573+1	.11158+1	.10784+1	.88375+0
.60	.15652+1	.10547+1	.10648+1	.10444+1	.89333+0
.65	.14982+1	.10513+1	.10269+1	.10202+1	.90375+0
.70	.14290+1	.10472+1	.99967+0	.10036+1	.91500+0
.75	.13583+1	.10421+1	.98160+0	.99306+0	.92708+0
.80	.12866+1	.10361+1	.97156+0	.98750+0	.94000+0
.8452	.12214+1	.10297+1	.96874+0	.98601+0	.95238+0
.85	.12144+1	.10289+1	.96877+0	.98603+0	.95375+0
.90	.11422+1	.10206+1	.97274+0	.98796+0	.96833+0
.95	.10706+1	.10110+1	.98319+0	.99276+0	.98375+0
1.00	.10000+1	.10000+1	.10000+1	.10000+1	.10000+1
1.05	.93076+0	.98761+0	.10232+1	.10093+1	.10171+1
1.10	.86329+0	.97376+0	.10530+1	.10205+1	.10350+1
1.15	.79792+0	.95838+0	.10898+1	.10332+1	.10537+1
1.20	.73492+0	.94147+0	.11339+1	.10472+1	.10733+1
1.40	.51069+0	.85853+0	.13987+1	.11143+1	.11600+1
1.60	.33554+0	.75344+0	.18626+1	.11937+1	.12600+1
1.80	.20846+0	.63276+0	.26650+1	.12815+1	.13733+1
2.00	.12246+0	.50618+0	.40831+1	.13750+1	.15000+1
2.20	.68017-1	.38421+0	.66828+1	.14727+1	.16400+1
2.40	.35722-1	.27590+0	.11664+2	.15736+1	.17933+1
2.60	.17739-1	.18699+0	.21682+2	.16769+1	.19600+1
2.80	.83291-2	.11941+0	.42879+2	.17821+1	.21400+1
3.00	.36979-2	.71758-1	.90142+2	.18889+1	.23333+1
3.20	.15523-2	.40542-1	.20131+3	.19969+1	.25400+1
4.20	.87365-5	.91175-3	.27253+5	.25492+1	.37733+1
5.20	.12125-7	.42665-5	.15860+8	.31135+1	.53400+1
6.20	.41496-11	.42374-8	.38869+11	.36839+1	.72400+1
7.20	.35021-15	.91641-12	.39659+15	.42579+1	.94733+1
8.20	.72884-20	.44139-16	.16732+20	.48341+1	.12040+2
9.20	.37405-25	.48212-21	.29059+25	.54120+1	.14940+2
10.20	.47338-31	.12112-26	.20711+31	.59908+1	.16173+2
11.20	.14773-37	.70776-33	.60437+37	.65705+1	.21740+2
12.20	.11369-44	.97039-40	.72095+44	.71508+1	.25640+2
13.20	.21576-52	.31439-47	.35111+52	.77316+1	.29873+2
14.20	.10097-60	.24206-55	.69743+60	.83127+1	.34440+2
INFIN	.0	.0	INFIN	INFIN	INFIN

Table C.11   Normal Shock Waves ($\gamma = 1.40$)

M_1	M_2	$\dfrac{V_1}{V_2} = \dfrac{\rho_2}{\rho_1}$	$\dfrac{t_2}{t_1}$	$\dfrac{p_2}{p_1}$	$\dfrac{P_2}{P_1} = \dfrac{A_1^*}{A_2^*}$	$\dfrac{P_2}{p_1}$
1.00	1.00000	.10000+1	.10000+1	.10000+1	.10000+1	.18929+1
1.01	.99013	.10167+1	.10066+1	.10234+1	.10000+1	.19152+1
1.02	.98052	.10334+1	.10132+1	.10471+1	.99999+0	.19379+1
1.03	.97115	.10502+1	.10198+1	.10710+1	.99997+0	.19610+1
1.04	.96203	.10671+1	.10263+1	.10952+1	.99992+0	.19844+1
1.05	.95313	.10840+1	.10328+1	.11196+1	.99985+0	.20083+1
1.06	.94445	.11009+1	.10393+1	.11442+1	.99975+0	.20325+1
1.07	.93598	.11179+1	.10458+1	.11690+1	.99961+0	.20570+1
1.08	.92771	.11349+1	.10522+1	.11941+1	.99943+0	.20819+1
1.09	.91965	.11520+1	.10586+1	.12194+1	.99920+0	.21072+1
1.10	.91177	.11691+1	.10649+1	.12450+1	.99893+0	.21328+1
1.11	.90408	.11862+1	.10713+1	.12708+1	.99860+0	.21588+1
1.12	.89656	.12034+1	.10776+1	.12968+1	.99821+0	.21851+1
1.13	.88922	.12206+1	.10840+1	.13230+1	.99777+0	.22118+1
1.14	.88204	.12378+1	.10903+1	.13495+1	.99726+0	.22388+1
1.15	.87502	.12550+1	.10966+1	.13762+1	.99669+0	.22661+1
1.16	.86816	.12723+1	.11029+1	.14032+1	.99605+0	.22937+1
1.17	.86145	.12896+1	.11092+1	.14304+1	.99535+0	.23217+1
1.18	.85488	.13069+1	.11154+1	.14578+1	.99457+0	.23500+1
1.19	.84846	.13243+1	.11217+1	.14854+1	.99372+0	.23786+1
1.20	.84217	.13416+1	.11280+1	.15133+1	.99280+0	.24075+1
1.21	.83601	.13590+1	.11343+1	.15414+1	.99180+0	.24367+1
1.22	.82999	.13764+1	.11405+1	.15698+1	.99073+0	.24663+1
1.23	.82408	.13938+1	.11468+1	.15984+1	.98958+0	.24961+1
1.24	.81830	.14112+1	.11531+1	.16272+1	.98836+0	.25263+1
1.25	.81264	.14286+1	.11594+1	.16562+1	.98706+0	.25568+1
1.26	.80709	.14460+1	.11657+1	.16855+1	.98568+0	.25875+1
1.27	.80164	.14634+1	.11720+1	.17150+1	.98422+0	.26186+1
1.28	.79631	.14808+1	.11783+1	.17448+1	.98268+0	.26500+1
1.29	.79108	.14983+1	.11846+1	.17748+1	.98107+0	.26816+1
1.30	.78596	.15157+1	.11909+1	.18050+1	.97937+0	.27136+1
1.31	.78093	.15331+1	.11972+1	.18354+1	.97760+0	.27459+1
1.32	.77600	.15505+1	.12035+1	.18661+1	.97575+0	.27784+1
1.33	.77116	.15680+1	.12099+1	.18970+1	.97382+0	.28112+1
1.34	.76641	.15854+1	.12162+1	.19282+1	.97182+0	.28444+1
1.35	.76175	.16028+1	.12226+1	.19596+1	.96974+0	.28778+1
1.36	.75718	.16202+1	.12290+1	.19912+1	.96758+0	.29115+1
1.37	.75269	.16376+1	.12354+1	.20230+1	.96534+0	.29455+1
1.38	.74829	.16549+1	.12418+1	.20551+1	.96304+0	.29798+1
1.39	.74396	.16723+1	.12482+1	.20874+1	.96065+0	.30144+1
1.40	.73971	.16897+1	.12547+1	.21200+1	.95819+0	.30492+1
1.41	.73554	.17070+1	.12612+1	.21528+1	.95566+0	.30844+1
1.42	.73144	.17243+1	.12676+1	.21858+1	.95306+0	.31198+1
1.43	.72741	.17416+1	.12741+1	.22190+1	.95039+0	.31555+1
1.44	.72345	.17589+1	.12807+1	.22525+1	.94765+0	.31915+1
1.45	.71956	.17761+1	.12872+1	.22862+1	.94484+0	.32278+1
1.46	.71574	.17934+1	.12938+1	.23202+1	.94196+0	.32643+1
1.47	.71198	.18106+1	.13003+1	.23544+1	.93901+0	.33011+1
1.48	.70829	.18278+1	.13069+1	.23888+1	.93600+0	.33382+1
1.49	.70466	.18449+1	.13136+1	.24234+1	.93293+0	.33756+1
1.50	.70109	.18621+1	.13202+1	.24583+1	.92979+0	.34133+1
1.51	.69758	.18792+1	.13269+1	.24934+1	.92659+0	.34512+1
1.52	.69413	.18963+1	.13336+1	.25288+1	.92332+0	.34894+1
1.53	.69073	.19133+1	.13403+1	.25644+1	.92000+0	.35279+1
1.54	.68739	.19303+1	.13470+1	.26002+1	.91662+0	.35667+1
1.55	.68410	.19473+1	.13538+1	.26362+1	.91319+0	.36057+1
1.56	.68087	.19643+1	.13606+1	.26725+1	.90970+0	.36450+1
1.57	.67768	.19812+1	.13674+1	.27090+1	.90615+0	.36846+1
1.58	.67455	.19981+1	.13742+1	.27458+1	.90255+0	.37244+1
1.59	.67147	.20149+1	.13811+1	.27828+1	.89890+0	.37646+1
1.60	.66844	.20317+1	.13880+1	.28200+1	.89520+0	.38050+1
1.61	.66545	.20485+1	.13949+1	.28574+1	.89145+0	.38456+1
1.62	.66251	.20653+1	.14018+1	.28951+1	.88765+0	.38866+1
1.63	.65962	.20820+1	.14088+1	.29330+1	.88381+0	.39278+1
1.64	.65677	.20986+1	.14158+1	.29712+1	.87992+0	.39693+1
1.65	.65396	.21152+1	.14228+1	.30096+1	.87599+0	.40110+1

Table C.11 Normal Shock Waves ($\gamma = 1.40$) (*Continued*)

M_1	M_2	$\dfrac{V_1}{V_2} = \dfrac{\rho_2}{\rho_1}$	$\dfrac{t_2}{t_1}$	$\dfrac{p_2}{p_1}$	$\dfrac{P_2}{P_1} = \dfrac{A_1^*}{A_2^*}$	$\dfrac{P_2}{p_1}$
1.66	.65119	.21316+1	.14299+1	.30482+1	.87201+0	.40531+1
1.67	.64847	.21484+1	.14369+1	.30870+1	.86800+0	.40953+1
1.68	.64579	.21649+1	.14440+1	.31261+1	.86394+0	.41379+1
1.69	.64315	.21813+1	.14512+1	.31654+1	.85985+0	.41807+1
1.70	.64054	.21977+1	.14583+1	.32050+1	.85572+0	.42238+1
1.71	.63798	.22141+1	.14655+1	.32448+1	.85156+0	.42672+1
1.72	.63545	.22304+1	.14727+1	.32848+1	.84736+0	.43108+1
1.73	.63296	.22467+1	.14800+1	.33250+1	.84312+0	.43547+1
1.74	.63051	.22629+1	.14873+1	.33655+1	.83886+0	.43989+1
1.75	.62809	.22791+1	.14946+1	.34062+1	.83457+0	.44433+1
1.76	.62570	.22952+1	.15019+1	.34472+1	.83024+0	.44880+1
1.77	.62335	.23113+1	.15093+1	.34884+1	.82589+0	.45330+1
1.78	.62104	.23273+1	.15167+1	.35298+1	.82151+0	.45782+1
1.79	.61875	.23433+1	.15241+1	.35714+1	.81711+0	.46237+1
1.80	.61650	.23592+1	.15316+1	.36133+1	.81268+0	.46695+1
1.81	.61428	.23751+1	.15391+1	.36554+1	.80823+0	.47155+1
1.82	.61209	.23909+1	.15466+1	.36978+1	.80376+0	.47618+1
1.83	.60993	.24067+1	.15541+1	.37404+1	.79927+0	.48084+1
1.84	.60780	.24224+1	.15617+1	.37832+1	.79476+0	.48552+1
1.85	.60570	.24381+1	.15693+1	.38262+1	.79023+0	.49023+1
1.86	.60363	.24537+1	.15770+1	.38695+1	.78569+0	.49497+1
1.87	.60158	.24693+1	.15847+1	.39130+1	.78112+0	.49973+1
1.88	.59957	.24848+1	.15924+1	.39568+1	.77655+0	.50452+1
1.89	.59758	.25003+1	.16001+1	.40008+1	.77196+0	.50934+1
1.90	.59562	.25157+1	.16079+1	.40450+1	.76736+0	.51418+1
1.91	.59368	.25310+1	.16157+1	.40894+1	.76274+0	.51905+1
1.92	.59177	.25463+1	.16236+1	.41341+1	.75812+0	.52394+1
1.93	.58988	.25616+1	.16314+1	.41790+1	.75349+0	.52886+1
1.94	.58802	.25767+1	.16394+1	.42242+1	.74884+0	.53381+1
1.95	.58618	.25919+1	.16473+1	.42696+1	.74420+0	.53878+1
1.96	.58437	.26069+1	.16553+1	.43152+1	.73954+0	.54378+1
1.97	.58258	.26220+1	.16633+1	.43610+1	.73488+0	.54881+1
1.98	.58082	.26369+1	.16713+1	.44071+1	.73021+0	.55386+1
1.99	.57907	.26518+1	.16794+1	.44534+1	.72555+0	.55894+1
2.00	.57735	.26667+1	.16875+1	.45000+1	.72087+0	.56404+1
2.02	.57397	.26962+1	.17038+1	.45938+1	.71153+0	.57433+1
2.04	.57068	.27255+1	.17203+1	.46885+1	.70218+0	.58473+1
2.06	.56747	.27545+1	.17369+1	.47842+1	.69284+0	.59523+1
2.08	.56433	.27833+1	.17536+1	.48808+1	.68351+0	.60583+1
2.10	.56128	.28119+1	.17705+1	.49783+1	.67420+0	.61654+1
2.12	.55829	.28402+1	.17875+1	.50768+1	.66492+0	.62735+1
2.14	.55538	.28683+1	.18046+1	.51762+1	.65567+0	.63827+1
2.16	.55254	.28962+1	.18219+1	.52765+1	.64645+0	.64929+1
2.18	.54977	.29238+1	.18393+1	.53778+1	.63727+0	.66042+1
2.20	.54706	.29512+1	.18569+1	.54800+1	.62814+0	.67165+1
2.22	.54441	.29784+1	.18746+1	.55831+1	.61905+0	.68298+1
2.24	.54182	.30053+1	.18924+1	.56872+1	.61002+0	.69442+1
2.26	.53930	.30319+1	.19104+1	.57922+1	.60105+0	.70597+1
2.28	.53683	.30584+1	.19285+1	.58981+1	.59214+0	.71762+1
2.30	.53441	.30845+1	.19468+1	.60050+1	.58329+0	.72937+1
2.32	.53205	.31105+1	.19652+1	.61128+1	.57452+0	.74122+1
2.34	.52974	.31362+1	.19838+1	.62215+1	.56581+0	.75319+1
2.36	.52749	.31617+1	.20025+1	.63312+1	.55718+0	.76525+1
2.38	.52528	.31869+1	.20213+1	.64418+1	.54862+0	.77742+1
2.40	.52312	.32119+1	.20403+1	.65533+1	.54014+0	.78969+1
2.42	.52100	.32367+1	.20595+1	.66658+1	.53175+0	.80207+1
2.44	.51894	.32612+1	.20788+1	.67792+1	.52344+0	.81455+1
2.46	.51691	.32855+1	.20982+1	.68935+1	.51521+0	.82713+1
2.48	.51493	.33095+1	.21178+1	.70088+1	.50707+0	.83982+1
2.50	.51299	.33333+1	.21375+1	.71250+1	.49901+0	.85261+1
2.52	.51109	.33569+1	.21574+1	.72421+1	.49105+0	.86551+1
2.54	.50923	.33803+1	.21774+1	.73602+1	.48318+0	.87851+1
2.56	.50741	.34034+1	.21976+1	.74792+1	.47540+0	.89161+1
2.58	.50562	.34263+1	.22179+1	.75991+1	.46772+0	.90482+1
2.60	.50387	.34490+1	.22383+1	.77200+1	.46012+0	.91813+1
2.62	.50216	.34714+1	.22590+1	.78418+1	.45263+0	.93155+1

Table C.11 Normal Shock Waves ($\gamma = 1.40$) (*Continued*)

M_1	M_2	$\dfrac{V_1}{V_2} = \dfrac{\rho_2}{\rho_1}$	$\dfrac{t_2}{t_1}$	$\dfrac{p_2}{p_1}$	$\dfrac{P_2}{P_1} = \dfrac{A_1^*}{A_2^*}$	$\dfrac{P_2}{p_1}$
2.64	.50048	.34937+1	.22797+1	.79645+1	.44522+0	.94506+1
2.66	.49883	.35157+1	.23006+1	.80882+1	.43792+0	.95869+1
2.68	.49722	.35374+1	.23217+1	.82128+1	.43070+0	.97241+1
2.70	.49563	.35590+1	.23429+1	.83383+1	.42359+0	.98624+1
2.72	.49408	.35803+1	.23642+1	.84648+1	.41657+0	.10002+2
2.74	.49256	.36015+1	.23858+1	.85922+1	.40965+0	.10142+2
2.76	.49107	.36224+1	.24074+1	.87205+1	.40283+0	.10283+2
2.78	.48960	.36431+1	.24292+1	.88498+1	.39610+0	.10426+2
2.80	.48817	.36636+1	.24512+1	.89800+1	.38946+0	.10569+2
2.82	.48676	.36838+1	.24733+1	.91111+1	.38293+0	.10714+2
2.84	.48538	.37039+1	.24955+1	.92432+1	.37649+0	.10859+2
2.86	.48402	.37238+1	.25179+1	.93762+1	.37014+0	.11006+2
2.88	.48269	.37434+1	.25405+1	.95101+1	.36389+0	.11154+2
2.90	.48138	.37629+1	.25632+1	.96450+1	.35773+0	.11302+2
2.92	.48010	.37821+1	.25861+1	.97808+1	.35167+0	.11452+2
2.94	.47884	.38012+1	.26091+1	.99175+1	.34570+0	.11603+2
2.96	.47760	.38200+1	.26322+1	.10055+2	.33982+0	.11754+2
2.98	.47638	.38387+1	.26555+1	.10194+2	.33404+0	.11907+2
3.00	.47519	.38571+1	.26790+1	.10333+2	.32834+0	.12061+2
3.02	.47402	.38754+1	.27026+1	.10474+2	.32274+0	.12216+2
3.04	.47287	.38935+1	.27264+1	.10615+2	.31723+0	.12372+2
3.06	.47174	.39114+1	.27503+1	.10758+2	.31180+0	.12529+2
3.08	.47063	.39291+1	.27744+1	.10901+2	.30646+0	.12687+2
3.10	.46953	.39466+1	.27986+1	.11045+2	.30121+0	.12846+2
3.12	.46846	.39639+1	.28230+1	.11190+2	.29605+0	.13006+2
3.14	.46741	.39811+1	.28475+1	.11336+2	.29097+0	.13167+2
3.16	.46637	.39981+1	.28722+1	.11483+2	.28597+0	.13329+2
3.18	.46535	.40149+1	.28970+1	.11631+2	.28106+0	.13492+2
3.20	.46435	.40315+1	.29220+1	.11780+2	.27623+0	.13656+2
3.22	.46336	.40479+1	.29471+1	.11930+2	.27148+0	.13821+2
3.24	.46240	.40642+1	.29724+1	.12081+2	.26681+0	.13987+2
3.26	.46144	.40803+1	.29979+1	.12232+2	.26222+0	.14155+2
3.28	.46051	.40963+1	.30234+1	.12385+2	.25771+0	.14323+2
3.30	.45959	.41120+1	.30492+1	.12538+2	.25328+0	.14492+2
3.32	.45868	.41276+1	.30751+1	.12693+2	.24892+0	.14662+2
3.34	.45779	.41431+1	.31011+1	.12848+2	.24463+0	.14834+2
3.36	.45691	.41583+1	.31273+1	.13005+2	.24043+0	.15006+2
3.38	.45605	.41734+1	.31537+1	.13162+2	.23629+0	.15180+2
3.40	.45520	.41884+1	.31802+1	.13320+2	.23223+0	.15354+2
3.42	.45436	.42032+1	.32069+1	.13479+2	.22823+0	.15530+2
3.44	.45354	.42179+1	.32337+1	.13639+2	.22431+0	.15706+2
3.46	.45273	.42323+1	.32607+1	.13800+2	.22045+0	.15884+2
3.48	.45194	.42467+1	.32878+1	.13962+2	.21667+0	.16062+2
3.50	.45115	.42609+1	.33151+1	.14125+2	.21295+0	.16242+2
3.60	.44741	.43296+1	.34537+1	.14953+2	.19531+0	.17156+2
3.70	.44395	.43949+1	.35962+1	.15805+2	.17919+0	.18095+2
3.80	.44073	.44568+1	.37426+1	.16680+2	.16447+0	.19060+2
3.90	.43774	.45156+1	.38928+1	.17578+2	.15103+0	.20051+2
4.00	.43496	.45714+1	.40469+1	.18500+2	.13876+0	.21068+2
4.10	.43236	.46245+1	.42048+1	.19445+2	.12756+0	.22111+2
4.20	.42994	.46749+1	.43666+1	.20413+2	.11733+0	.23179+2
4.30	.42767	.47229+1	.45322+1	.21405+2	.10800+0	.24273+2
4.40	.42554	.47685+1	.47017+1	.22420+2	.99481−1	.25393+2
4.50	.42355	.48119+1	.48751+1	.23458+2	.91698−1	.26539+2
4.60	.42168	.48532+1	.50523+1	.24520+2	.84586−1	.27710+2
4.70	.41992	.48926+1	.52334+1	.25605+2	.78086−1	.28907+2
4.80	.41826	.49301+1	.54184+1	.26713+2	.72140−1	.30130+2
4.90	.41670	.49659+1	.56073+1	.27845+2	.66699−1	.31379+2
5.00	.41523	.50000+1	.58000+1	.29000+2	.61716−1	.32653+2
5.10	.41384	.50326+1	.59966+1	.30178+2	.57151−1	.33954+2
5.20	.41252	.50637+1	.61971+1	.31380+2	.52966−1	.35280+2
5.30	.41127	.50934+1	.64014+1	.32605+2	.49126−1	.36632+2
5.40	.41009	.51218+1	.66097+1	.33853+2	.45600−1	.38009+2
5.50	.40897	.51489+1	.68218+1	.35125+2	.42361−1	.39412+2
5.60	.40791	.51749+1	.70378+1	.36420+2	.39383−1	.40841+2
5.70	.40690	.51998+1	.72577+1	.37738+2	.36643−1	.42296+2
5.80	.40594	.52236+1	.74814+1	.39080+2	.34120−1	.43777+2
5.90	.40503	.52464+1	.77091+1	.40445+2	.31795−1	.45283+2

Table C.11 Normal Shock Waves ($\gamma = 1.40$) (*Continued*)

M_1	M_2	$\dfrac{V_1}{V_2} = \dfrac{\rho_2}{\rho_1}$	$\dfrac{t_2}{t_1}$	$\dfrac{p_2}{p_1}$	$\dfrac{P_2}{P_1} = \dfrac{A_1^*}{A_2^*}$	$\dfrac{P_2}{p_1}$
6.00	.40416	.52683+1	.79406+1	.41833+2	.29651-1	.46815+2
6.50	.40038	.53651+1	.91564+1	.49125+2	.21148-1	.54862+2
7.00	.39736	.54444+1	.10469+2	.57000+2	.15351-1	.63553+2
7.50	.39491	.55102+1	.11879+2	.65458+2	.11329-1	.72887+2
8.00	.39289	.55652+1	.13387+2	.74500+2	.84878-2	.82865+2
8.50	.39121	.56117+1	.14991+2	.84125+2	.64492-2	.93488+2
9.00	.38980	.56512+1	.16693+2	.94333+2	.49639-2	.10475+3
9.50	.38860	.56850+1	.18492+2	.10512+3	.38664-2	.11666+3
10.00	.38758	.57143+1	.20387+2	.11650+3	.30448-2	.12922+3
11.00	.38592	.57619+1	.24471+2	.14100+3	.19451-2	.15626+3
12.00	.38466	.57987+1	.28943+2	.16783+3	.12866-2	.18587+3
13.00	.38368	.58276+1	.33805+2	.19700+3	.87709-3	.21806+3
14.00	.38289	.58507+1	.39055+2	.22850+3	.61380-3	.25282+3
15.00	.38226	.58696+1	.44694+2	.26233+3	.43953-3	.29016+3
16.00	.38174	.58851+1	.50722+2	.29850+3	.32119-3	.33008+3
17.00	.38131	.58980+1	.57138+2	.33700+3	.23899-3	.37257+3
18.00	.38095	.59088+1	.63944+2	.37783+3	.18072-3	.41763+3
19.00	.38065	.59180+1	.71139+2	.42100+3	.13865-3	.46527+3
20.00	.38039	.59259+1	.78722+2	.46650+3	.10777-3	.51548+3
21.00	.38016	.59327+1	.86694+2	.51433+3	.64778-4	.56827+3
22.00	.37997	.59387+1	.95055+2	.56450+3	.67414-4	.62364+3
23.00	.37980	.59438+1	.10381+3	.61700+3	.54140-4	.68158+3
24.00	.37965	.59484+1	.11294+3	.67183+3	.43877-4	.74209+3
25.00	.37952	.59524+1	.12247+3	.72900+3	.35859-4	.80518+3
INFIN	.37796	.60000+1	INFIN	INFIN	.0	INFIN

Table C.11 Normal Shock Waves ($\gamma = 1.30$)

M_1	M_2	$\dfrac{V_1}{V_2} = \dfrac{\rho_2}{\rho_1}$	$\dfrac{t_2}{t_1}$	$\dfrac{p_2}{p_1}$	$\dfrac{P_2}{P_1} = \dfrac{A_1^*}{A_2^*}$	$\dfrac{P_2}{p_1}$
1.00	1.00000	.10000+1	.10000+1	.10000+1	.10000+1	.18324+1
1.02	.98049	.10349+1	.10104+1	.10457+1	.99999+0	.18746+1
1.04	.96192	.10702+1	.10206+1	.10922+1	.99992+0	.19183+1
1.06	.94422	.11058+1	.10307+1	.11397+1	.99975+0	.19634+1
1.08	.92733	.11416+1	.10407+1	.11881+1	.99942+0	.20099+1
1.10	.91120	.11777+1	.10506+1	.12374+1	.99891+0	.20578+1
1.12	.89578	.12141+1	.10605+1	.12876+1	.99817+0	.21070+1
1.14	.88102	.12507+1	.10703+1	.13387+1	.99719+0	.21575+1
1.16	.86688	.12876+1	.10801+1	.13907+1	.99595+0	.22092+1
1.18	.85333	.13246+1	.10898+1	.14436+1	.99441+0	.22622+1
1.20	.84033	.13618+1	.10995+1	.14974+1	.99258+0	.23164+1
1.22	.82785	.13993+1	.11092+1	.15521+1	.99043+0	.23718+1
1.24	.81585	.14368+1	.11189+1	.16077+1	.98796+0	.24283+1
1.26	.80432	.14746+1	.11286+1	.16642+1	.98517+0	.24861+1
1.28	.79322	.15125+1	.11383+1	.17217+1	.98204+0	.25449+1
1.30	.78253	.15505+1	.11480+1	.17800+1	.97858+0	.26050+1
1.32	.77224	.15886+1	.11578+1	.18392+1	.97478+0	.26661+1
1.34	.76232	.16268+1	.11676+1	.18994+1	.97064+0	.27283+1
1.36	.75274	.16651+1	.11774+1	.19604+1	.96618+0	.27917+1
1.38	.74351	.17035+1	.11872+1	.20224+1	.96139+0	.28561+1
1.40	.73459	.17419+1	.11971+1	.20852+1	.95627+0	.29217+1
1.42	.72597	.17804+1	.12070+1	.21490+1	.95084+0	.29883+1
1.44	.71764	.18189+1	.12170+1	.22136+1	.94510+0	.30559+1
1.46	.70958	.18574+1	.12271+1	.22792+1	.93905+0	.31247+1
1.48	.70179	.18960+1	.12372+1	.23457+1	.93272+0	.31945+1
1.50	.69425	.19346+1	.12473+1	.24130+1	.92610+0	.32654+1
1.55	.67642	.20310+1	.12730+1	.25854+1	.90838+0	.34471+1
1.60	.65992	.21272+1	.12991+1	.27635+1	.88911+0	.36353+1
1.65	.64463	.22230+1	.13257+1	.29472+1	.86847+0	.38300+1
1.70	.63041	.23185+1	.13529+1	.31365+1	.84664+0	.40311+1
1.75	.61718	.24133+1	.13805+1	.33315+1	.82380+0	.42385+1
1.80	.60484	.25074+1	.14087+1	.35322+1	.80014+0	.44522+1

Table C.11 Normal Shock Waves ($\gamma = 1.30$) (*Continued*)

M_1	M_2	$\dfrac{V_1}{V_2} = \dfrac{\rho_2}{\rho_1}$	$\dfrac{t_2}{t_1}$	$\dfrac{p_2}{p_1}$	$\dfrac{P_2}{P_1} = \dfrac{A_1^*}{A_2^*}$	$\dfrac{P_2}{p_1}$
1.85	.59330	.26007+1	.14375+1	.37385+1	.77582+0	.46723+1
1.90	.58251	.26932+1	.14668+1	.39504+1	.75102+0	.48986+1
1.95	.57238	.27846+1	.14968+1	.41680+1	.72589+0	.51312+1
2.00	.56288	.28750+1	.15274+1	.43913+1	.70057+0	.53700+1
2.05	.55394	.29643+1	.15586+1	.46202+1	.67521+0	.56150+1
2.10	.54553	.30524+1	.15905+1	.48548+1	.64992+0	.58662+1
2.15	.53760	.31392+1	.16230+1	.50950+1	.62481+0	.61236+1
2.20	.53011	.32248+1	.16562+1	.53409+1	.59998+0	.63873+1
2.25	.52304	.33091+1	.16900+1	.55924+1	.57552+0	.66570+1
2.30	.51635	.33920+1	.17245+1	.58496+1	.55150+0	.69330+1
2.35	.51001	.34735+1	.17597+1	.61124+1	.52799+0	.72151+1
2.40	.50400	.35536+1	.17956+1	.63809+1	.50504+0	.75033+1
2.45	.49831	.36324+1	.18321+1	.66550+1	.48269+0	.77978+1
2.50	.49290	.37097+1	.18694+1	.69348+1	.46098+0	.80983+1
2.55	.48775	.37855+1	.19073+1	.72202+1	.43994+0	.84050+1
2.60	.48286	.38600+1	.19459+1	.75113+1	.41958+0	.87178+1
2.65	.47820	.39330+1	.19853+1	.78080+1	.39993+0	.90368+1
2.70	.47377	.40045+1	.20253+1	.81104+1	.38099+0	.93619+1
2.75	.46954	.40747+1	.20661+1	.84185+1	.36276+0	.96931+1
2.80	.46550	.41434+1	.21075+1	.87322+1	.34525+0	.10030+2
2.85	.46164	.42107+1	.21497+1	.90515+1	.32844+0	.10374+2
2.90	.45796	.42766+1	.21925+1	.93765+1	.31233+0	.10723+2
2.95	.45444	.43411+1	.22361+1	.97072+1	.29691+0	.11079+2
3.00	.45107	.44043+1	.22804+1	.10043+2	.28216+0	.11441+2
3.10	.44475	.45265+1	.23711+1	.10733+2	.25463+0	.12183+2
3.20	.43895	.46435+1	.24648+1	.11445+2	.22958+0	.12949+2
3.30	.43360	.47555+1	.25613+1	.12180+2	.20686+0	.13740+2
3.40	.42867	.48625+1	.26607+1	.12937+2	.18631+0	.14555+2
3.50	.42411	.49648+1	.27630+1	.13717+2	.16775+0	.15395+2
3.60	.41989	.50625+1	.28681+1	.14520+2	.15102+0	.16259+2
3.70	.41597	.51559+1	.29763+1	.15345+2	.13596+0	.17147+2
3.80	.41233	.52451+1	.30873+1	.16193+2	.12242+0	.18060+2
3.90	.40893	.53303+1	.32012+1	.17063+2	.11025+0	.18997+2
4.00	.40577	.54118+1	.33181+1	.17957+2	.99326-1	.19959+2
4.50	.39275	.57678+1	.39462+1	.22761+2	.59395-1	.25133+2
5.00	.38319	.60526+1	.46476+1	.28130+2	.36127-1	.30915+2
5.50	.37597	.62822+1	.54225+1	.34065+2	.22427-1	.37308+2
6.00	.37039	.64687+1	.62710+1	.40565+2	.14225-1	.44309+2
6.50	.36600	.66218+1	.71930+1	.47630+2	.92176-2	.51919+2
7.00	.36248	.67485+1	.81886+1	.55261+2	.60978-2	.60138+2
7.50	.35962	.68543+1	.92579+1	.63457+2	.41139-2	.68966+2
8.00	.35726	.69434+1	.10401+2	.72217+2	.28271-2	.78403+2
8.50	.35529	.70190+1	.11618+2	.81543+2	.19765-2	.88449+2
9.00	.35364	.70837+1	.12908+2	.91435+2	.14041-2	.99103+2
9.50	.35223	.71393+1	.14272+2	.10189+3	.10125-2	.11037+3
10.00	.35103	.71875+1	.15710+2	.11291+3	.74021-3	.12224+3
INFIN	.33968	.76667+1	INFIN	INFIN	.0	INFIN

Table C.11 Normal Shock Waves ($\gamma = 1.20$)

M_1	M_2	$\dfrac{V_1}{V_2} = \dfrac{\rho_2}{\rho_1}$	$\dfrac{t_2}{t_1}$	$\dfrac{p_2}{p_1}$	$\dfrac{P_2}{P_1} = \dfrac{A_1^*}{A_2^*}$	$\dfrac{P_2}{p_1}$
1.00	1.00000	.10000+1	.10000+1	.10000+1	.10000+1	.17716+1
1.02	.98046	.10366+1	.10072+1	.10441+1	.99999+0	.18109+1
1.04	.96181	.10736+1	.10143+1	.10890+1	.99992+0	.18517+1
1.06	.94397	.11111+1	.10213+1	.11348+1	.99974+0	.18939+1
1.08	.92691	.11490+1	.10283+1	.11815+1	.99941+0	.19374+1
1.10	.91057	.11873+1	.10352+1	.12291+1	.99888+0	.19822+1
1.12	.89491	.12260+1	.10420+1	.12775+1	.99813+0	.20282+1
1.14	.87989	.12651+1	.10488+1	.13268+1	.99712+0	.20755+1
1.16	.86546	.13046+1	.10555+1	.13770+1	.99584+0	.21240+1
1.18	.85160	.13444+1	.10622+1	.14281+1	.99425+0	.21736+1
1.20	.83827	.13846+1	.10689+1	.14800+1	.99235+0	.22244+1

Table C.11 Normal Shock Waves ($\gamma = 1.20$) (*Continued*)

M_1	M_2	$\dfrac{V_1}{V_2} = \dfrac{\rho_2}{\rho_1}$	$\dfrac{t_2}{t_1}$	$\dfrac{p_2}{p_1}$	$\dfrac{P_2}{P_1} = \dfrac{A_1^*}{A_2^*}$	$\dfrac{P_2}{p_1}$
1.22	.82545	.14251+1	.10756+1	.15328+1	.99012+0	.22764+1
1.24	.81310	.14660+1	.10822+1	.15865+1	.98755+0	.23294+1
1.26	.80121	.15071+1	.10889+1	.16410+1	.98463+0	.23836+1
1.28	.78974	.15485+1	.10955+1	.16964+1	.98136+0	.24389+1
1.30	.77867	.15902+1	.11022+1	.17527+1	.97773+0	.24952+1
1.32	.76799	.16322+1	.11088+1	.18099+1	.97373+0	.25526+1
1.34	.75768	.16745+1	.11155+1	.18679+1	.96938+0	.26110+1
1.36	.74771	.17170+1	.11222+1	.19268+1	.96467+0	.26705+1
1.38	.73807	.17597+1	.11289+1	.19866+1	.95960+0	.27311+1
1.40	.72875	.18027+1	.11357+1	.20473+1	.95417+0	.27926+1
1.42	.71974	.18458+1	.11425+1	.21088+1	.94840+0	.28552+1
1.44	.71100	.18892+1	.11493+1	.21712+1	.94229+0	.29188+1
1.46	.70255	.19328+1	.11561+1	.22345+1	.93585+0	.29834+1
1.48	.69435	.19765+1	.11630+1	.22986+1	.92908+0	.30490+1
1.50	.68641	.20204+1	.11699+1	.23636+1	.92200+0	.31156+1
1.55	.66757	.21308+1	.11873+1	.25300+1	.90298+0	.32865+1
1.60	.65009	.22420+1	.12051+1	.27018+1	.88222+0	.34635+1
1.65	.63381	.23539+1	.12231+1	.28791+1	.85991+0	.36466+1
1.70	.61864	.24663+1	.12415+1	.30618+1	.83624+0	.38357+1
1.75	.60447	.25789+1	.12602+1	.32500+1	.81141+0	.40308+1
1.80	.59121	.26918+1	.12793+1	.34436+1	.78562+0	.42319+1
1.85	.57877	.28048+1	.12987+1	.36427+1	.75907+0	.44390+1
1.90	.56710	.29177+1	.13186+1	.38473+1	.73196+0	.46519+1
1.95	.55612	.30304+1	.13388+1	.40573+1	.70446+0	.48708+1
2.00	.54578	.31429+1	.13595+1	.42727+1	.67674+0	.50955+1
2.05	.53603	.32549+1	.13806+1	.44936+1	.64897+0	.53261+1
2.10	.52682	.33664+1	.14021+1	.47200+1	.62129+0	.55626+1
2.15	.51812	.34773+1	.14240+1	.49518+1	.59383+0	.58049+1
2.20	.50989	.35876+1	.14464+1	.51891+1	.56672+0	.60530+1
2.25	.50209	.36971+1	.14692+1	.54318+1	.54005+0	.63070+1
2.30	.49469	.38058+1	.14925+1	.56800+1	.51393+0	.65667+1
2.35	.48767	.39135+1	.15162+1	.59336+1	.48842+0	.68323+1
2.40	.48100	.40203+1	.15404+1	.61927+1	.46360+0	.71036+1
2.45	.47465	.41261+1	.15650+1	.64573+1	.43952+0	.73808+1
2.50	.46861	.42308+1	.15901+1	.67273+1	.41622+0	.76637+1
2.55	.46285	.43343+1	.16156+1	.70027+1	.39373+0	.79525+1
2.60	.45737	.44368+1	.16417+1	.72836+1	.37209+0	.82470+1
2.65	.45213	.45380+1	.16681+1	.75700+1	.35131+0	.85473+1
2.70	.44714	.46379+1	.16951+1	.78618+1	.33139+0	.88533+1
2.75	.44236	.47367+1	.17225+1	.81591+1	.31234+0	.91651+1
2.80	.43779	.48341+1	.17505+1	.84618+1	.29415+0	.94827+1
2.85	.43342	.49302+1	.17788+1	.87700+1	.27681+0	.98061+1
2.90	.42924	.50250+1	.18077+1	.90836+1	.26032+0	.10135+2
2.95	.42523	.51184+1	.18370+1	.94027+1	.24465+0	.10470+2
3.00	.42139	.52105+1	.18669+1	.97273+1	.22979+0	.10811+2
3.10	.41417	.53906+1	.19279+1	.10393+2	.20239+0	.11509+2
3.20	.40751	.55652+1	.19909+1	.11060+2	.17791+0	.12231+2
3.30	.40136	.57343+1	.20559+1	.11789+2	.15613+0	.12975+2
3.40	.39566	.58980+1	.21228+1	.12520+2	.13683+0	.13743+2
3.50	.39038	.60562+1	.21916+1	.13273+2	.11978+0	.14534+2
3.60	.38547	.62091+1	.22624+1	.14047+2	.10476+0	.15347+2
3.70	.38090	.63567+1	.23351+1	.14844+2	.91556-1	.16184+2
3.80	.37665	.64992+1	.24098+1	.15662+2	.79973-1	.17043+2
3.90	.37267	.66367+1	.24865+1	.16502+2	.69829-1	.17926+2
4.00	.36895	.67692+1	.25651+1	.17364+2	.60958-1	.18831+2
4.50	.35355	.73642+1	.29877+1	.22000+2	.30934-1	.23702+2
5.00	.34214	.78571+1	.34595+1	.27182+2	.15856-1	.29148+2
5.50	.33345	.82671+1	.39807+1	.32909+2	.82705-2	.35166+2
6.00	.32669	.86087+1	.45514+1	.39182+2	.44076-2	.41759+2
6.50	.32134	.88947+1	.51716+1	.46000+2	.24044-2	.48925+2
7.00	.31703	.91356+1	.58413+1	.53364+2	.13434-2	.56664+2
7.50	.31352	.93396+1	.65605+1	.61273+2	.76849-3	.64976+2
8.00	.31061	.95135+1	.73293+1	.69727+2	.44981-3	.73862+2
8.50	.30818	.96626+1	.81476+1	.78727+2	.26912-3	.83322+2
9.00	.30613	.97912+1	.90155+1	.88273+2	.16439-3	.93354+2
9.50	.30439	.99027+1	.99330+1	.98364+2	.10241-3	.10396+3
10.00	.30289	.10000+2	.10900+2	.10900+3	.64993-4	.11514+3
INFIN	.28868	.11000+2	INFIN	INFIN	.0	INFIN

Table C.11   Normal Shock Waves ($\gamma = 1.10$)

M_1	M_2	$\dfrac{V_1}{V_2} = \dfrac{\rho_2}{\rho_1}$	$\dfrac{t_2}{t_1}$	$\dfrac{p_2}{p_1}$	$\dfrac{P_2}{P_1} = \dfrac{A_1^*}{A_2^*}$	$\dfrac{P_2}{p_1}$
1.00	1.00000	.10000+1	.10000+1	.10000+1	.10000+1	.17103+1
1.02	.98043	.10384+1	.10038+1	.10423+1	.99999+0	.17469+1
1.04	.96168	.10774+1	.10075+1	.10855+1	.99992+0	.17847+1
1.06	.94370	.11170+1	.10112+1	.11295+1	.99974+0	.18239+1
1.08	.92645	.11572+1	.10148+1	.11743+1	.99940+0	.18643+1
1.10	.90987	.11980+1	.10183+1	.12200+1	.99886+0	.19060+1
1.12	.89394	.12394+1	.10219+1	.12665+1	.99809+0	.19488+1
1.14	.87862	.12813+1	.10254+1	.13139+1	.99706+0	.19929+1
1.16	.86387	.13236+1	.10289+1	.13621+1	.99573+0	.20380+1
1.18	.84966	.13669+1	.10324+1	.14111+1	.99409+0	.20843+1
1.20	.83596	.14104+1	.10358+1	.14610+1	.99213+0	.21316+1
1.22	.82275	.14546+1	.10392+1	.15117+1	.98981+0	.21800+1
1.24	.80999	.14992+1	.10427+1	.15632+1	.98713+0	.22295+1
1.26	.79768	.15444+1	.10461+1	.16156+1	.98408+0	.22801+1
1.28	.78578	.15901+1	.10495+1	.16688+1	.98065+0	.23316+1
1.30	.77428	.16362+1	.10529+1	.17229+1	.97683+0	.23842+1
1.32	.76315	.16829+1	.10564+1	.17778+1	.97262+0	.24378+1
1.34	.75238	.17301+1	.10598+1	.18335+1	.96802+0	.24924+1
1.36	.74195	.17777+1	.10632+1	.18901+1	.96303+0	.25479+1
1.38	.73185	.18258+1	.10667+1	.19475+1	.95765+0	.26045+1
1.40	.72206	.18743+1	.10701+1	.20057+1	.95188+0	.26620+1
1.42	.71257	.19233+1	.10736+1	.20648+1	.94573+0	.27205+1
1.44	.70336	.19727+1	.10770+1	.21247+1	.93920+0	.27799+1
1.46	.69442	.20226+1	.10805+1	.21855+1	.93230+0	.28403+1
1.48	.68575	.20729+1	.10840+1	.22471+1	.92503+0	.29016+1
1.50	.67732	.21236+1	.10876+1	.23095+1	.91741+0	.29639+1
1.55	.65728	.22521+1	.10964+1	.24693+1	.89688+0	.31237+1
1.60	.63860	.23830+1	.11055+1	.26343+1	.87437+0	.32892+1
1.65	.62114	.25161+1	.11146+1	.28045+1	.85006+0	.34605+1
1.70	.60479	.26514+1	.11239+1	.29800+1	.82418+0	.36374+1
1.75	.58946	.27886+1	.11334+1	.31607+1	.79694+0	.38200+1
1.80	.57505	.29277+1	.11431+1	.33467+1	.76857+0	.40082+1
1.85	.56148	.30685+1	.11530+1	.35379+1	.73930+0	.42020+1
1.90	.54870	.32109+1	.11630+1	.37343+1	.70935+0	.44014+1
1.95	.53663	.33548+1	.11732+1	.39360+1	.67893+0	.46062+1
2.00	.52523	.35000+1	.11837+1	.41429+1	.64826+0	.48166+1
2.05	.51443	.36464+1	.11943+1	.43550+1	.61752+0	.50325+1
2.10	.50420	.37939+1	.12052+1	.45724+1	.58690+0	.52539+1
2.15	.49450	.39424+1	.12163+1	.47950+1	.55657+0	.54808+1
2.20	.48528	.40918+1	.12275+1	.50229+1	.52667+0	.57131+1
2.25	.47652	.42419+1	.12391+1	.52560+1	.49733+0	.59509+1
2.30	.46818	.43926+1	.12508+1	.54943+1	.46868+0	.61942+1
2.35	.46023	.45439+1	.12628+1	.57379+1	.44081+0	.64428+1
2.40	.45266	.46957+1	.12749+1	.59867+1	.41381+0	.66970+1
2.45	.44543	.48477+1	.12874+1	.62407+1	.38775+0	.69565+1
2.50	.43853	.50000+1	.13000+1	.65000+1	.36269+0	.72215+1
2.55	.43193	.51524+1	.13129+1	.67645+1	.33866+0	.74924+1
2.60	.42562	.53049+1	.13260+1	.70343+1	.31570+0	.77678+1
2.65	.41958	.54574+1	.13393+1	.73093+1	.29363+0	.80490+1
2.70	.41379	.56097+1	.13529+1	.75895+1	.27304+0	.83357+1
2.75	.40825	.57619+1	.13667+1	.78750+1	.25335+0	.86277+1
2.80	.40293	.59138+1	.13808+1	.81657+1	.23474+0	.89252+1
2.85	.39782	.60653+1	.13951+1	.84617+1	.21719+0	.92281+1
2.90	.39292	.62165+1	.14096+1	.87629+1	.20069+0	.95363+1
2.95	.38821	.63671+1	.14244+1	.90693+1	.18520+0	.98500+1
3.00	.38368	.65172+1	.14394+1	.93810+1	.17070+0	.10169+2
3.10	.37512	.68156+1	.14702+1	.10020+2	.14450+0	.10823+2
3.20	.36719	.71111+1	.15019+1	.10680+2	.12179+0	.11499+2
3.30	.35983	.74034+1	.15346+1	.11361+2	.10223+0	.12197+2
3.40	.35297	.76920+1	.15682+1	.12063+2	.85495−1	.12916+2
3.50	.34657	.79767+1	.16029+1	.12786+2	.71258−1	.13656+2
3.60	.34060	.82573+1	.16385+1	.13530+2	.59208−1	.14418+2
3.70	.33501	.85334+1	.16751+1	.14294+2	.49059−1	.15202+2
3.80	.32978	.88049+1	.17127+1	.15080+2	.40546−1	.16007+2
3.90	.32487	.90716+1	.17513+1	.15887+2	.33433−1	.16834+2

Table C.11 Normal Shock Waves ($\gamma = 1.10$) (*Continued*)

M_1	M_2	$\dfrac{V_1}{V_2} = \dfrac{\rho_2}{\rho_1}$	$\dfrac{t_2}{t_1}$	$\dfrac{p_2}{p_1}$	$\dfrac{P_2}{P_1} = \dfrac{A_1^*}{A_2^*}$	$\dfrac{P_2}{p_1}$
4.00	.32026	.93333+1	.17908+1	.16714+2	.27512-1	.17682+2
4.50	.30092	.10565+2	.20034+1	.21167+2	.10142-1	.22245+2
5.00	.28630	.11667+2	.22408+1	.26143+2	.36550-2	.27346+2
5.50	.27499	.12642+2	.25030+1	.31643+2	.13096-2	.32984+2
6.00	.26608	.13500+2	.27901+1	.37667+2	.47217-3	.39160+2
6.50	.25893	.14253+2	.31021+1	.44214+2	.17272-3	.45872+2
7.00	.25311	.14913+2	.34390+1	.51286+2	.64456-4	.53122+2
7.50	.24833	.15492+2	.38006+1	.58881+2	.24625-4	.60909+2
8.00	.24434	.16000+2	.41875+1	.67000+2	.96510-5	.69233+2
8.50	.24098	.16447+2	.45991+1	.75643+2	.38844-5	.78094+2
9.00	.23814	.16842+2	.50357+1	.84810+2	.16061-5	.87493+2
9.50	.23570	.17190+2	.54972+1	.94500+2	.68210-6	.97428+2
10.00	.23360	.17500+2	.59637+1	.10471+3	.29741-6	.10790+3
INFIN	.21320	.21000+2	INFIN	INFIN	.0	INFIN

Table C.12 Wave Angle ε for Oblique Shock Waves ($\gamma = 1.40$)

M_1									Flow turning angle δ, degrees												
	.0	2.0	4.0	6.0	8.0	10.0	12.0	14.0	16.0	18.0	20.0	22.0	24.0	26.0	28.0	30.0	32.0	34.0	36.0	38.0	40.0
1.00																					
1.01	81.93	85.36	.05																		
1.02	78.64	83.49	.14																		
1.03	76.14	82.08	.26																		
1.04	74.06	80.93	.40																		
1.05	72.25	79.94	.56																		
1.06	70.63	79.06	.73																		
1.07	69.16	78.27	.91																		
1.08	67.81	77.56	1.10																		
1.09	66.55	76.90	1.30																		
1.10	65.38	76.30	1.52																		
1.11	64.28	75.73	1.73																		
1.12	63.23	75.21	1.96																		
1.13	62.25	70.93	74.72	2.19																	
1.14	61.31	68.71	74.26	2.43																	
1.15	60.41	67.00	73.82	2.67																	
1.16	59.55	65.56	73.41	2.92																	
1.17	58.73	64.28	73.02	3.17																	
1.18	57.94	63.11	72.66	3.42																	
1.19	57.18	62.05	72.31	3.68																	
1.20	56.44	61.05	71.96	3.94																	
1.21	55.74	60.12	68.09	71.66	4.21																
1.22	55.05	59.24	66.03	71.36	4.47																
1.23	54.39	58.40	64.47	71.07	4.74																
1.24	53.75	57.60	63.15	70.80	5.01																
1.25	53.13	56.84	61.99	70.54	5.29																
1.26	52.53	56.12	60.93	70.29	5.56																
1.27	51.94	55.42	59.97	70.05	5.83																
1.28	51.38	54.75	59.06	67.38	69.82	6.11															
1.29	50.82	54.10	58.22	65.05	69.60	6.39															
1.30	50.28	53.47	57.42	63.46	69.40	6.66															
1.31	49.76	52.87	56.67	62.16	69.19	6.94															
1.32	49.25	52.28	55.95	61.03	67.00	7.22															
1.33	48.75	51.72	55.26	60.02	68.62	7.49															
1.34	48.27	51.17	54.60	59.09	68.64	7.77															
1.35	47.79	50.63	53.97	58.23	66.91	68.47	8.05														
1.36	47.33	50.11	53.36	57.43	64.29	68.31	8.35														
1.37	46.88	49.61	52.77	56.68	62.70	68.15	8.60														
1.38	46.44	49.12	52.20	55.96	61.43	68.00	8.86														
1.39	46.01	48.64	51.65	55.28	60.34	67.65	9.15														
1.40	45.58	48.17	51.12	54.63	59.37	67.72	9.43														
1.41	45.17	47.72	50.60	54.01	58.48	67.56	9.70														
1.42	44.77	47.27	50.10	53.42	57.67	67.45	9.97														
1.43	44.37	46.84	49.61	52.84	56.91	63.95	67.33	10.25													
1.44	43.98	46.42	49.14	52.29	56.19	62.31	67.21	10.52													
1.45	43.60	46.00	48.68	51.76	55.52	61.05	67.10	10.79													
1.46	43.23	45.60	48.23	51.24	54.87	59.98	66.99	11.05													
1.47	42.86	45.20	47.79	50.74	54.26	59.04	66.88	11.32													
1.48	42.51	44.82	47.37	50.25	53.68	58.19	66.78	11.59													
1.49	42.16	44.44	46.95	49.78	53.11	57.41	66.66	11.85													
1.50	41.81	44.06	46.54	49.33	52.57	56.68	64.36	66.59	12.11												

M1	.0	2.0	4.0	6.0	8.0	10.0	12.0	14.0	16.0	18.0	20.0	22.0	24.0	26.0	28.0	30.0	32.0	34.0	36.0	38.0	40.0
1.51	41.47	43.70	46.15	48.86	52.05	56.00	62.42	66.50	12.37												
1.52	41.14	43.34	45.76	48.45	51.55	55.35	61.10	66.41	12.63												
1.53	40.81	42.99	45.38	48.03	51.06	54.74	60.02	66.33	12.89												
1.54	40.49	42.65	45.01	47.62	50.59	54.16	59.08	66.25	13.15												
1.55	40.16	42.32	44.64	47.21	50.13	53.60	58.24	66.17	13.40												
1.56	39.87	41.99	44.29	46.82	49.69	53.06	57.47	66.10	13.66												
1.57	39.56	41.66	43.94	46.44	49.26	52.55	56.77	66.03	13.91												
1.58	39.27	41.34	43.59	46.07	48.84	52.05	56.10	63.38	65.96	14.16											
1.59	38.97	41.03	43.26	45.70	48.43	51.58	55.48	61.73	65.89	14.41											
1.60	38.68	40.72	42.93	45.34	48.03	51.12	54.89	60.54	65.83	14.65											
1.61	38.40	40.42	42.61	44.99	47.64	50.67	54.33	59.55	65.77	14.90											
1.62	38.12	40.13	42.29	44.65	47.26	50.23	53.79	58.69	65.71	15.14											
1.63	37.84	39.84	41.98	44.32	46.89	49.81	53.28	57.91	65.65	15.38											
1.64	37.57	39.55	41.60	43.99	46.53	49.41	52.79	57.20	65.60	15.62											
1.65	37.31	39.27	41.36	43.67	46.18	49.01	52.31	56.54	65.55	15.86											
1.66	37.04	38.99	41.08	43.35	45.84	48.62	51.85	55.93	63.58	65.50	16.09										
1.67	36.78	38.72	40.79	43.04	45.50	48.24	51.41	55.34	61.80	65.45	16.32										
1.68	36.53	38.45	40.51	42.74	45.17	47.88	50.98	54.79	60.60	65.40	16.55										
1.69	36.28	38.19	40.23	42.44	44.85	47.52	50.57	54.27	59.63	65.36	16.78										
1.70	36.03	37.93	39.96	42.14	44.53	47.17	50.17	53.77	58.79	65.32	17.01										
1.71	35.79	37.67	39.69	41.86	44.22	46.82	49.78	53.29	58.05	65.28	17.24										
1.72	35.55	37.42	39.42	41.57	43.91	46.49	49.40	52.83	57.36	65.24	17.46										
1.73	35.31	37.17	39.16	41.30	43.61	46.16	49.03	52.39	56.73	65.20	17.68										
1.74	35.08	36.93	38.90	41.02	43.32	45.84	48.67	51.96	56.14	65.17	17.90										
1.75	34.85	36.69	38.65	40.76	43.03	45.55	48.32	51.55	55.59	62.94	65.13	18.12									
1.76	34.62	36.45	38.40	40.49	42.75	45.22	47.98	51.15	55.06	61.42	65.10	18.34									
1.77	34.40	36.22	38.16	40.23	42.48	44.92	47.64	50.76	54.57	60.34	65.07	18.55									
1.78	34.18	35.99	37.91	39.98	42.20	44.63	47.32	50.38	54.09	59.46	65.04	18.76									
1.79	33.96	35.76	37.68	39.73	41.94	44.34	47.00	50.02	53.63	58.69	65.01	18.97									
1.80	33.75	35.54	37.44	39.48	41.67	44.06	46.69	49.66	53.20	57.99	64.99	19.18									
1.81	33.54	35.32	37.21	39.24	41.42	43.78	46.36	49.31	52.78	57.36	64.96	19.39									
1.82	33.33	35.10	36.99	39.00	41.16	43.51	46.08	48.98	52.37	56.76	64.94	19.59									
1.83	33.12	34.89	36.76	38.76	40.91	43.24	45.79	48.65	51.98	56.23	64.91	19.80									
1.84	32.92	34.68	36.54	38.53	40.67	42.98	45.50	48.33	51.60	55.71	64.89	20.00									
1.85	32.72	34.47	36.32	38.30	40.42	42.72	45.22	48.01	51.23	55.23	62.10	64.87	20.20								
1.86	32.52	34.26	36.11	38.08	40.19	42.46	44.95	47.71	50.88	54.76	60.91	64.85	20.40								
1.87	32.33	34.06	35.90	37.86	39.95	42.21	44.68	47.41	50.53	54.32	59.99	64.83	20.59								
1.88	32.13	33.86	35.69	37.64	39.72	41.97	44.41	47.12	50.19	53.90	59.21	64.82	20.79								
1.89	31.94	33.66	35.48	37.42	39.49	41.73	44.15	46.83	49.86	53.49	58.52	64.80	20.98								
1.90	31.76	33.47	35.28	37.21	39.27	41.49	43.90	46.55	49.54	53.10	57.90	64.78	21.17								
1.91	31.57	33.28	35.08	37.00	39.05	41.26	43.65	46.28	49.23	52.72	57.33	64.77	21.36								
1.92	31.39	33.08	34.88	36.79	38.84	41.03	43.40	46.01	48.93	52.35	56.80	64.75	21.54								
1.93	31.21	32.90	34.69	36.59	38.62	40.80	43.16	45.74	48.63	52.00	56.30	64.74	21.73								
1.94	31.03	32.71	34.49	36.39	38.41	40.58	42.92	45.48	48.34	51.65	55.83	64.73	21.91								
1.95	30.85	32.53	34.30	36.19	38.20	40.36	42.69	45.23	48.06	51.32	55.38	62.86	64.72	22.09							
1.96	30.68	32.35	34.12	36.00	38.00	40.14	42.46	44.98	47.78	51.00	54.96	61.49	64.71	22.27							
1.97	30.51	32.17	33.93	35.80	37.80	39.93	42.23	44.74	47.51	50.68	54.55	60.53	64.70	22.45							
1.98	30.35	31.99	33.75	35.61	37.60	39.72	42.01	44.50	47.25	50.37	54.16	59.74	64.69	22.63							
1.99	30.17	31.82	33.57	35.43	37.40	39.52	41.79	44.26	46.99	50.08	53.78	59.06	64.68	22.80							
2.00	30.00	31.65	33.39	35.24	37.21	39.31	41.58	44.03	46.73	49.79	53.42	58.46	64.67	22.97							

Table C.12 Wave Angle ε for Oblique Shock Waves ($\gamma = 1.40$) (*Continued*)

M_1

Flow turning angle δ, degrees

M_1	.0	2.0	4.0	6.0	8.0	10.0	12.0	14.0	16.0	18.0	20.0	22.0	24.0	26.0	28.0	30.0	32.0	34.0	36.0	38.0	40.0
2.02	29.67	31.31	33.04	34.88	36.83	38.92	41.15	43.58	46.24	49.22	52.74	57.39	64.65	23.31							
2.04	29.35	30.98	32.70	34.52	36.46	38.53	40.75	43.14	45.76	48.69	52.09	56.46	64.64	23.65							
2.06	29.04	30.66	32.37	34.18	36.10	38.15	40.35	42.72	45.30	48.17	51.49	55.63	64.63	23.98							
2.08	28.74	30.34	32.04	33.84	35.75	37.79	39.97	42.31	44.86	47.68	50.91	54.87	61.28	64.63	24.30						
2.10	28.44	30.03	31.72	33.51	35.41	37.43	39.59	41.91	44.43	47.21	50.36	54.17	59.77	64.62	24.61						
2.12	28.14	29.73	31.41	33.19	35.08	37.09	39.23	41.53	44.02	46.75	49.84	53.52	58.62	64.62	24.92						
2.14	27.86	29.44	31.11	32.88	34.76	36.75	38.87	41.15	43.62	46.32	49.35	52.91	57.65	64.62	25.23						
2.16	27.58	29.15	30.81	32.57	34.44	36.42	38.53	40.79	43.23	45.89	48.87	52.31	56.61	64.62	25.52						
2.18	27.30	28.87	30.52	32.27	34.13	36.10	38.20	40.44	42.85	45.49	48.41	51.79	56.05	64.62	25.82						
2.20	27.04	28.59	30.24	31.98	33.83	35.79	37.87	40.09	42.49	45.09	47.98	51.28	55.36	62.70	26.10						
2.22	26.77	28.32	29.96	31.69	33.53	35.48	37.55	39.76	42.14	44.71	47.55	50.79	54.72	60.85	26.38						
2.24	26.51	28.06	29.69	31.42	33.24	35.16	37.24	39.44	41.79	44.34	47.15	50.32	54.12	59.63	26.66						
2.26	26.26	27.80	29.42	31.14	32.96	34.89	36.94	39.12	41.46	43.98	46.75	49.87	53.56	58.65	26.93						
2.28	26.01	27.54	29.16	30.87	32.69	34.60	36.64	38.81	41.13	43.64	46.37	49.44	53.03	57.82	64.64	27.19					
2.30	25.77	27.29	28.91	30.61	32.42	34.33	36.35	38.51	40.82	43.30	46.01	49.03	52.54	57.08	64.63	27.45					
2.32	25.53	27.05	28.66	30.35	32.15	34.05	36.07	38.22	40.51	42.97	45.65	48.63	52.06	56.41	64.62	27.71					
2.34	25.30	26.81	28.41	30.10	31.89	33.79	35.80	37.93	40.21	42.65	45.31	48.25	51.61	55.79	64.61	27.96					
2.36	25.07	26.58	28.17	29.86	31.64	33.53	35.53	37.65	39.91	42.34	44.97	47.88	51.18	55.22	61.97	28.20					
2.38	24.85	26.35	27.93	29.61	31.39	33.27	35.26	37.38	39.63	42.04	44.65	47.52	50.77	54.69	60.94	28.45					
2.40	24.62	26.12	27.70	29.38	31.15	33.02	35.01	37.11	39.35	41.75	44.34	47.17	50.37	54.18	59.66	64.71	28.68				
2.42	24.41	25.90	27.48	29.14	30.91	32.78	34.76	36.85	39.08	41.46	44.03	46.84	49.99	53.71	58.83	64.72	28.91				
2.44	24.19	25.68	27.25	28.92	30.68	32.54	34.51	36.60	38.82	41.18	43.73	46.52	49.62	53.26	58.11	64.74	29.14				
2.46	23.99	25.47	27.03	28.69	30.45	32.31	34.27	36.35	38.56	40.91	43.45	46.20	49.27	52.83	57.47	64.75	29.36				
2.48	23.78	25.26	26.82	28.47	30.23	32.08	34.03	36.10	38.30	40.65	43.16	45.90	48.93	52.43	56.88	64.77	29.58				
2.50	23.58	25.05	26.61	28.26	30.01	31.85	33.80	35.87	38.06	40.39	42.89	45.60	48.60	52.04	56.33	64.78	29.80				
2.52	23.38	24.85	26.40	28.05	29.79	31.63	33.58	35.63	37.82	40.14	42.62	45.31	48.28	51.66	55.83	64.27	30.01				
2.54	23.18	24.65	26.20	27.84	29.58	31.41	33.35	35.41	37.58	39.89	42.36	45.04	47.97	51.30	55.36	62.05	30.22				
2.56	22.99	24.45	26.00	27.64	29.37	31.20	33.14	35.18	37.35	39.65	42.11	44.76	47.67	50.96	54.91	60.94	30.42				
2.58	22.81	24.26	25.80	27.44	29.17	30.99	32.92	34.96	37.12	39.41	41.86	44.50	47.38	50.63	54.49	60.08	64.85	30.62			
2.60	22.62	24.07	25.61	27.24	28.97	30.79	32.71	34.75	36.90	39.19	41.62	44.24	47.10	50.31	54.09	59.35	64.87	30.81			
2.62	22.44	23.89	25.42	27.05	28.77	30.59	32.51	34.54	36.68	38.96	41.39	43.99	46.83	50.00	53.71	58.72	64.88	31.01			
2.64	22.26	23.70	25.24	26.86	28.58	30.39	32.31	34.33	36.47	38.74	41.16	43.75	46.56	49.70	53.34	58.11	64.90	31.19			
2.66	22.08	23.52	25.05	26.67	28.39	30.20	32.11	34.13	36.26	38.53	40.93	43.51	46.31	49.41	52.99	57.62	64.92	31.38			
2.68	21.91	23.35	24.87	26.49	28.20	30.01	31.92	33.93	36.06	38.32	40.71	43.28	46.05	49.12	52.66	57.14	64.94	31.56			
2.70	21.74	23.17	24.70	26.31	28.02	29.82	31.73	33.73	35.86	38.11	40.50	43.05	45.81	48.85	52.33	56.69	64.96	31.74			
2.72	21.57	23.00	24.52	26.13	27.84	29.64	31.54	33.55	35.67	37.91	40.29	42.83	45.57	48.59	52.02	56.26	64.97	31.92			
2.74	21.41	22.83	24.35	25.96	27.66	29.46	31.36	33.36	35.48	37.71	40.08	42.61	45.34	48.33	51.72	55.87	63.25	32.09			
2.76	21.24	22.67	24.18	25.79	27.49	29.28	31.18	33.18	35.29	37.52	39.88	42.40	45.11	48.08	51.44	55.49	62.00	32.26			
2.78	21.08	22.50	24.02	25.62	27.32	29.11	31.00	33.00	35.10	37.33	39.68	42.19	44.89	47.84	51.16	55.13	61.14	32.42			
2.80	20.92	22.34	23.85	25.45	27.15	28.94	30.83	32.82	34.92	37.14	39.49	41.99	44.68	47.60	50.89	54.79	60.43	32.59			
2.82	20.77	22.19	23.69	25.29	26.98	28.77	30.66	32.65	34.75	36.96	39.30	41.79	44.47	47.38	50.62	54.46	59.83	32.75			
2.84	20.62	22.03	23.54	25.13	26.82	28.61	30.49	32.48	34.57	36.78	39.12	41.60	44.26	47.15	50.37	54.14	59.29	32.91			
2.86	20.47	21.88	23.38	24.97	26.66	28.44	30.33	32.31	34.40	36.60	38.94	41.40	44.06	46.93	50.13	53.85	58.80	65.09	33.06		
2.88	20.32	21.73	23.23	24.82	26.50	28.29	30.17	32.15	34.23	36.43	38.76	41.23	43.86	46.72	49.89	53.55	58.35	65.11	33.21		
2.90	20.17	21.58	23.08	24.67	26.35	28.13	30.01	31.99	34.07	36.26	38.58	41.04	43.67	46.51	49.65	53.27	57.93	65.15	33.36		
2.92	20.03	21.43	22.93	24.52	26.20	27.98	29.85	31.83	33.91	36.10	38.41	40.87	43.47	46.31	49.43	53.00	57.54	65.16	33.51		
2.94	19.89	21.29	22.78	24.37	26.05	27.82	29.70	31.67	33.75	35.94	38.25	40.69	43.30	46.12	49.21	52.74	57.17	65.18	33.65		
2.96	19.75	21.15	22.64	24.22	25.90	27.67	29.55	31.52	33.59	35.78	38.08	40.52	43.12	45.92	49.00	52.49	56.83	65.20	33.80		
2.98	19.61	21.00	22.49	24.08	25.75	27.53	29.40	31.37	33.44	35.62	37.92	40.36	42.95	45.73	48.79	52.25	56.50	65.22	33.94		
3.00	19.47	20.87	22.35	23.94	25.61	27.38	29.25	31.22	33.29	35.47	37.76	40.19	42.78	45.55	48.59	52.01	56.18	63.67	65.24	34.07	

P1	.0	2.0	4.0	6.0	8.0	10.0	12.0	14.0	16.0	18.0	20.0	22.0	24.0	26.0	28.0	30.0	32.0	34.0	36.0	38.0	40.0
3.05	19.14	20.53	22.01	23.59	25.26	27.03	28.90	30.86	32.92	35.10	37.38	39.80	42.36	45.11	48.10	51.45	55.46	61.50	65.29	34.41	36.14
3.10	18.82	20.20	21.68	23.26	24.93	26.69	28.55	30.51	32.57	34.74	37.02	39.42	41.97	44.69	47.65	50.93	54.80	60.21	65.34	34.73	36.59
3.15	18.51	19.89	21.37	22.94	24.60	26.37	28.23	30.18	32.24	34.40	36.67	39.06	41.59	44.30	47.22	50.45	54.20	59.20	65.38	35.03	36.87
3.20	18.21	19.59	21.06	22.63	24.29	26.05	27.91	29.86	31.92	34.08	36.34	38.72	41.24	43.92	46.81	49.99	53.65	58.35	65.43	35.33	37.09
3.25	17.92	19.29	20.76	22.33	23.99	25.75	27.60	29.56	31.61	33.76	36.02	38.39	40.90	43.56	46.43	49.57	53.14	57.62	65.47	35.61	37.31
3.30	17.64	19.01	20.48	22.04	23.70	25.46	27.31	29.26	31.31	33.46	35.71	38.08	40.57	43.22	46.06	49.16	52.67	56.96	65.52	35.88	37.51
3.35	17.37	18.73	20.20	21.76	23.42	25.17	27.03	28.98	31.02	33.17	35.42	37.78	40.26	42.90	45.72	48.78	52.22	56.37	65.88	65.56	37.71
3.40	17.10	18.47	19.93	21.49	23.15	24.90	26.75	28.70	30.75	32.89	35.13	37.49	39.97	42.59	45.39	48.42	51.81	55.84	61.91	65.60	37.91
3.45	16.85	18.21	19.67	21.23	22.88	24.63	26.49	28.44	30.48	32.62	34.86	37.21	39.68	42.29	45.07	48.08	51.42	55.34	60.90	65.69	65.92
3.50	16.60	17.96	19.42	20.97	22.63	24.38	26.24	28.18	30.22	32.36	34.60	36.95	39.41	42.01	44.77	47.76	51.05	54.89	60.09	65.73	65.96
3.55	16.36	17.71	19.17	20.73	22.38	24.14	25.99	27.94	29.98	32.12	34.35	36.69	39.15	41.74	44.49	47.45	50.70	54.46	59.40	65.77	65.99
3.60	16.13	17.48	18.93	20.49	22.14	23.90	25.75	27.70	29.74	31.88	34.11	36.45	38.90	41.48	44.21	47.15	50.38	54.07	58.79	65.81	66.03
3.65	15.90	17.25	18.70	20.26	21.91	23.67	25.52	27.47	29.51	31.65	33.88	36.21	38.66	41.23	43.95	46.87	50.06	53.69	58.25	65.85	66.06
3.70	15.68	17.03	18.48	20.03	21.69	23.44	25.30	27.25	29.29	31.44	33.65	35.99	38.43	40.99	43.70	46.61	49.77	53.34	57.76	65.88	66.09
3.75	15.47	16.81	18.26	19.81	21.47	23.23	25.08	27.03	29.07	31.21	33.44	35.77	38.20	40.76	43.46	46.35	49.49	53.01	57.31	64.19	66.12
3.80	15.26	16.60	18.05	19.60	21.26	23.02	24.87	26.82	28.86	31.00	33.23	35.56	37.99	40.54	43.23	46.10	49.22	52.70	56.89	62.94	66.15
3.85	15.05	16.39	17.84	19.40	21.05	22.81	24.67	26.62	28.66	30.80	33.03	35.35	37.78	40.33	43.01	45.87	48.96	52.41	56.51	62.09	66.18
3.90	14.86	16.20	17.64	19.20	20.85	22.61	24.47	26.42	28.47	30.61	32.83	35.16	37.58	40.13	42.80	45.65	48.72	52.13	56.15	61.41	66.21
3.95	14.66	16.00	17.45	19.00	20.66	22.42	24.28	26.23	28.28	30.42	32.65	34.97	37.39	39.93	42.60	45.43	48.48	51.86	55.81	60.83	66.24
4.00	14.48	15.81	17.26	18.81	20.47	22.23	24.09	26.05	28.10	30.24	32.46	34.79	37.21	39.74	42.40	45.22	48.26	51.61	55.50	60.32	66.27
4.05	14.29	15.63	17.07	18.63	20.29	22.05	23.91	25.87	27.92	30.06	32.29	34.61	37.03	39.56	42.21	45.03	48.04	51.36	55.20	59.86	66.30
4.10	14.12	15.45	16.89	18.45	20.11	21.88	23.74	25.70	27.75	29.89	32.12	34.44	36.86	39.38	42.03	44.83	47.84	51.13	54.92	59.45	65.77
4.15	13.94	15.27	16.72	18.27	19.94	21.70	23.57	25.53	27.58	29.72	31.95	34.27	36.69	39.21	41.86	44.65	47.64	50.91	54.65	59.07	64.34
4.20	13.77	15.10	16.55	18.10	19.77	21.54	23.41	25.37	27.42	29.56	31.79	34.11	36.53	39.05	41.69	44.47	47.45	50.70	54.35	58.72	63.58
4.25	13.61	14.94	16.38	17.94	19.60	21.37	23.24	25.21	27.27	29.41	31.64	33.96	36.37	38.89	41.52	44.30	47.27	50.50	54.15	58.40	63.00
4.30	13.45	14.77	16.22	17.78	19.44	21.22	23.09	25.06	27.11	29.26	31.49	33.81	36.22	38.74	41.37	44.14	47.09	50.30	53.92	58.10	62.51
4.35	13.29	14.62	16.06	17.62	19.29	21.06	22.94	24.91	26.97	29.11	31.35	33.67	36.08	38.59	41.22	43.98	46.92	50.12	53.71	57.81	62.09
4.40	13.14	14.46	15.90	17.46	19.13	20.91	22.79	24.76	26.82	28.97	31.20	33.53	35.94	38.45	41.07	43.83	46.76	49.94	53.50	57.54	61.71
4.45	12.99	14.31	15.75	17.31	18.99	20.77	22.65	24.62	26.68	28.83	31.07	33.39	35.80	38.31	40.93	43.68	46.60	49.77	53.30	57.29	61.37
4.50	12.84	14.16	15.61	17.17	18.84	20.62	22.50	24.48	26.55	28.70	30.94	33.26	35.67	38.17	40.79	43.54	46.45	49.60	53.10	57.05	61.06
4.55	12.70	14.02	15.46	17.02	18.70	20.48	22.37	24.35	26.42	28.57	30.81	33.13	35.54	38.04	40.66	43.40	46.31	49.44	52.92	56.83	60.78
4.60	12.56	13.88	15.32	16.88	18.56	20.35	22.23	24.22	26.29	28.44	30.68	33.00	35.41	37.91	40.53	43.27	46.17	49.29	52.75	56.61	60.51
4.65	12.42	13.74	15.18	16.75	18.43	20.22	22.11	24.09	26.16	28.32	30.56	32.88	35.29	37.80	40.40	43.14	46.03	49.14	52.58	56.40	60.26
4.70	12.28	13.60	15.05	16.61	18.30	20.09	21.98	23.97	26.04	28.20	30.44	32.77	35.17	37.68	40.28	43.01	45.90	49.00	52.41	56.21	60.03
4.75	12.15	13.47	14.92	16.48	18.17	19.96	21.86	23.85	25.92	28.08	30.33	32.65	35.06	37.56	40.17	42.89	45.77	48.86	52.26	56.02	59.90
4.80	12.02	13.34	14.79	16.36	18.04	19.84	21.74	23.73	25.81	27.97	30.22	32.54	34.95	37.45	40.05	42.78	45.65	48.73	52.11	55.84	59.78
4.85	11.90	13.22	14.66	16.23	17.92	19.72	21.62	23.61	25.70	27.86	30.11	32.44	34.84	37.34	39.94	42.66	45.53	48.60	51.96	55.67	59.67
4.90	11.78	13.09	14.54	16.11	17.80	19.60	21.50	23.50	25.59	27.76	30.00	32.33	34.74	37.24	39.84	42.55	45.42	48.47	51.82	55.51	59.56
4.95	11.66	12.97	14.42	15.99	17.68	19.49	21.39	23.39	25.48	27.65	29.90	32.23	34.64	37.14	39.73	42.45	45.30	48.35	51.69	55.35	59.45
5.00	11.54	12.85	14.30	15.88	17.57	19.38	21.28	23.29	25.38	27.55	29.80	32.13	34.54	37.04	39.63	42.34	45.20	48.24	51.56	55.21	59.35
5.25	10.98	12.29	13.75	15.33	17.03	18.85	20.78	22.79	24.90	27.08	29.34	31.68	34.09	36.59	39.17	41.87	44.70	47.71	50.97	54.65	59.21
5.50	10.48	11.79	13.24	14.84	16.55	18.39	20.33	22.35	24.47	26.66	28.93	31.28	33.69	36.19	38.77	41.46	44.24	47.26	50.47	54.06	58.40
5.75	10.01	11.33	12.79	14.39	16.12	17.97	19.92	21.96	24.09	26.30	28.57	30.92	33.34	35.84	38.42	41.11	43.91	46.87	50.04	53.56	57.74
6.00	9.59	10.91	12.37	13.98	15.73	17.59	19.55	21.61	23.75	25.97	28.26	30.61	33.04	35.54	38.12	40.79	43.58	46.52	49.67	53.14	57.19
6.25	9.21	10.52	11.99	13.61	15.37	17.24	19.21	21.29	23.45	25.67	27.97	30.33	32.76	35.26	37.84	40.52	43.30	46.22	49.35	52.77	56.72
6.50	8.85	10.16	11.64	13.27	15.04	16.93	18.92	21.01	23.17	25.41	27.71	30.08	32.52	35.02	37.60	40.27	43.05	45.96	49.06	52.44	56.32
6.75	8.52	9.83	11.32	12.96	14.74	16.64	18.64	20.75	22.93	25.17	27.48	29.86	32.30	34.80	37.38	40.05	42.82	45.72	48.81	52.16	55.98
7.00	8.21	9.53	11.02	12.67	14.46	16.38	18.40	20.51	22.70	24.96	27.28	29.66	32.10	34.61	37.19	39.85	42.62	45.51	48.58	51.91	55.67
7.25	7.93	9.24	10.74	12.40	14.21	16.14	18.18	20.30	22.50	24.76	27.09	29.48	31.92	34.43	37.01	39.68	42.44	45.33	48.38	51.69	55.41
7.50	7.66	8.98	10.48	12.16	13.98	15.92	17.97	20.10	22.31	24.58	26.92	29.31	31.76	34.27	36.86	39.52	42.26	45.16	48.20	51.49	55.17

Table C.12 Wave Angle ε for Oblique Shock Waves ($\gamma = 1.40$) (*Continued*)

Flow turning angle δ, degrees

M_1	.0	2.0	4.0	6.0	8.0	10.0	12.0	14.0	16.0	18.0	20.0	22.0	24.0	26.0	28.0	30.0	32.0	34.0	36.0	38.0	40.0
7.75	7.41	8.73	10.24	11.93	13.76	15.72	17.78	19.92	22.14	24.42	26.76	29.16	31.61	34.13	36.71	39.37	42.13	45.00	48.04	51.31	54.96
8.00	7.18	8.50	10.02	11.71	13.56	15.53	17.60	19.76	21.98	24.27	26.62	29.02	31.48	34.00	36.58	39.24	41.99	44.86	47.89	51.15	54.77
8.25	6.96	8.28	9.81	11.51	13.37	15.35	17.44	19.60	21.84	24.14	26.49	28.90	31.36	33.88	36.46	39.12	41.87	44.74	47.76	51.00	54.60
8.50	6.76	8.08	9.61	11.33	13.20	15.19	17.29	19.46	21.71	24.01	26.37	28.78	31.24	33.77	36.35	39.01	41.76	44.62	47.64	50.86	54.44
8.75	6.56	7.89	9.43	11.15	13.04	15.04	17.15	19.33	21.58	23.89	26.26	28.67	31.14	33.67	36.25	38.91	41.66	44.52	47.52	50.74	54.30
9.00	6.38	7.71	9.25	10.99	12.88	14.90	17.02	19.21	21.47	23.79	26.16	28.58	31.05	33.57	36.16	38.82	41.57	44.42	47.42	50.63	54.17
9.25	6.21	7.54	9.09	10.84	12.74	14.77	16.90	19.10	21.36	23.69	26.06	28.48	30.96	33.49	36.08	38.73	41.48	44.33	47.33	50.53	54.06
9.50	6.04	7.38	8.94	10.69	12.61	14.65	16.78	18.99	21.27	23.60	25.97	28.40	30.88	33.41	36.00	38.66	41.40	44.25	47.24	50.44	53.95
9.75	5.89	7.22	8.79	10.56	12.48	14.53	16.68	18.90	21.18	23.51	25.89	28.32	30.80	33.33	35.93	38.58	41.33	44.17	47.16	50.35	53.85
10.00	5.74	7.08	8.65	10.43	12.37	14.43	16.58	18.81	21.09	23.43	25.82	28.25	30.73	33.27	35.86	38.52	41.26	44.10	47.09	50.27	53.76
11.00	5.22	6.56	8.17	9.99	11.96	14.06	16.24	18.50	20.81	23.16	25.56	28.01	30.50	33.04	35.63	38.29	41.03	43.87	46.84	50.01	53.46
12.00	4.78	6.14	7.77	9.63	11.64	13.77	15.98	18.26	20.58	22.96	25.37	27.82	30.32	32.86	35.46	38.12	40.86	43.69	46.66	49.80	53.23
13.00	4.41	5.78	7.44	9.33	11.38	13.54	15.77	18.07	20.41	22.79	25.22	27.68	30.18	32.73	35.33	37.99	40.72	43.56	46.51	49.65	53.06
14.00	4.10	5.48	7.17	9.09	11.16	13.35	15.60	17.91	20.27	22.66	25.09	27.56	30.07	32.62	35.22	37.88	40.62	43.45	46.40	49.53	52.92
15.00	3.82	5.21	6.93	8.88	10.99	13.19	15.46	17.79	20.15	22.56	24.99	27.47	29.98	32.53	35.13	37.80	40.53	43.36	46.31	49.43	52.81
16.00	3.58	4.98	6.73	8.71	10.84	13.06	15.35	17.68	20.06	22.47	24.91	27.39	29.90	32.46	35.06	37.73	40.46	43.29	46.23	49.35	52.72
17.00	3.37	4.78	6.55	8.56	10.71	12.95	15.25	17.60	19.98	22.40	24.84	27.32	29.84	32.40	35.00	37.67	40.40	43.23	46.17	49.28	52.64
18.00	3.18	4.61	6.40	8.43	10.60	12.86	15.17	17.52	19.91	22.33	24.79	27.27	29.79	32.35	34.95	37.62	40.35	43.18	46.12	49.22	52.58
19.00	3.02	4.45	6.27	8.32	10.51	12.78	15.10	17.46	19.86	22.28	24.74	27.22	29.74	32.31	34.91	37.58	40.31	43.13	46.07	49.18	52.53
20.00	2.87	4.31	6.15	8.22	10.43	12.71	15.04	17.41	19.81	22.24	24.70	27.18	29.71	32.27	34.88	37.54	40.28	43.10	46.04	49.14	52.48
21.00	2.73	4.19	6.04	8.14	10.36	12.65	14.99	17.36	19.77	22.20	24.66	27.15	29.67	32.24	34.85	37.51	40.24	43.07	46.00	49.10	52.44
22.00	2.61	4.07	5.95	8.06	10.29	12.60	14.94	17.32	19.73	22.17	24.63	27.12	29.65	32.21	34.82	37.48	40.22	43.04	45.98	49.07	52.41
23.00	2.49	3.97	5.87	8.00	10.24	12.55	14.90	17.29	19.70	22.14	24.60	27.09	29.62	32.19	34.80	37.46	40.20	43.02	45.95	49.05	52.38
24.00	2.39	3.88	5.79	7.94	10.19	12.51	14.87	17.26	19.67	22.11	24.58	27.07	29.60	32.17	34.78	37.44	40.17	42.99	45.93	49.02	52.36
25.00	2.29	3.79	5.73	7.88	10.15	12.47	14.84	17.23	19.65	22.09	24.56	27.05	29.58	32.15	34.76	37.42	40.16	42.98	45.91	49.00	52.33

Table C.13 Prandtl-Meyer Angle v, Degrees

M	$\gamma = 1.10$	$\gamma = 1.20$	$\gamma = 1.30$	$\gamma = 1.40$	$\gamma = 1.67$
1.00	.0000	.0000	.0000	.0000	.0000
1.02	.1441	.1374	.1313	.1257	.1127
1.04	.4034	.3843	.3669	.3510	.3142
1.06	.7339	.6984	.6661	.6367	.5689
1.08	1.1191	1.0638	1.0136	.9680	.8632
1.10	1.5492	1.4710	1.4004	1.3362	1.1891
1.12	2.0175	1.9136	1.8199	1.7350	1.5410
1.14	2.5189	2.3867	2.2676	2.1599	1.9146
1.16	3.0496	2.8863	2.7397	2.6073	2.3066
1.18	3.6062	3.4094	3.2331	3.0742	2.7143
1.20	4.1862	3.9535	3.7454	3.5582	3.1355
1.22	4.7873	4.5161	4.2743	4.0572	3.5683
1.24	5.4075	5.0956	4.8180	4.5693	4.0111
1.26	6.0452	5.6902	5.3749	5.0931	4.4623
1.28	6.6988	6.2984	5.9437	5.6271	4.9209
1.30	7.3670	6.9190	6.5229	6.1702	5.3856
1.32	8.0487	7.5507	7.1116	6.7213	5.8556
1.34	8.7428	8.1926	7.7086	7.2793	6.3300
1.36	9.4482	8.8437	8.3131	7.8434	6.8079
1.38	10.1643	9.5032	8.9242	8.4129	7.2888
1.40	10.8900	10.1702	9.5413	8.9869	7.7719
1.42	11.6248	10.8440	10.1635	9.5649	8.2568
1.44	12.3679	11.5240	10.7904	10.1463	8.7429
1.46	13.1188	12.2097	11.4212	10.7304	9.2297
1.48	13.8767	12.9003	12.0555	11.3168	9.7169
1.50	14.6413	13.5954	12.6927	11.9051	10.2041
1.52	15.4121	14.2946	13.3325	12.4948	10.6909
1.54	16.1884	14.9973	13.9744	13.0855	11.1769
1.56	16.9700	15.7033	14.6179	13.6768	11.6620
1.58	17.7565	16.4119	15.2628	14.2685	12.1459
1.60	18.5474	17.1230	15.9088	14.8602	12.6282
1.62	19.3424	17.8362	16.5554	15.4516	13.1089
1.64	20.1412	18.5512	17.2024	16.0425	13.5877
1.66	20.9434	19.2676	17.8496	16.6327	14.0644
1.68	21.7489	19.9852	18.4967	17.2218	14.5388
1.70	22.5572	20.7037	19.1434	17.8097	15.0109
1.72	23.3682	21.4229	19.7895	18.3962	15.4804
1.74	24.1815	22.1426	20.4348	18.9812	15.9474
1.76	24.9970	22.8625	21.0792	19.5644	16.4115
1.78	25.8144	23.5825	21.7224	20.1456	16.8729
1.80	26.6336	24.3022	22.3643	20.7248	17.3312
1.82	27.4543	25.0217	23.0047	21.3019	17.7866
1.84	28.2763	25.7406	23.6435	21.8766	18.2389
1.86	29.0995	26.4586	24.2805	22.4489	18.6880
1.88	29.9237	27.1762	24.9157	23.0187	19.1339
1.90	30.7487	27.8927	25.5488	23.5859	19.5766
1.92	31.5744	28.6080	26.1798	24.1503	20.0159
1.94	32.4006	29.3221	26.8086	24.7120	20.4520
1.96	33.2272	30.0348	27.4350	25.2708	20.8846
1.98	34.0540	30.7461	28.0590	25.8266	21.3138
2.00	34.8810	31.4558	28.6806	26.3795	21.7396
2.05	36.9482	33.2223	30.2229	27.7481	22.7891
2.10	39.0136	34.9769	31.7480	29.0968	23.8168
2.15	41.0756	36.7183	33.2549	30.4250	24.8228
2.20	43.1330	38.4453	34.7430	31.7322	25.8070
2.25	45.1843	40.1569	36.2115	33.0181	26.7697
2.30	47.2284	41.8525	37.6601	34.2824	27.7109
2.35	49.2643	43.5312	39.0883	35.5251	28.6310
2.40	51.2911	45.1926	40.4958	36.7461	29.5303
2.45	53.3079	46.8360	41.8825	37.9455	30.4090
2.50	55.3140	48.4610	43.2481	39.1232	31.2676
2.55	57.3086	50.0674	44.5927	40.2794	32.1065
2.60	59.2912	51.6547	45.9163	41.4143	32.9260
2.65	61.2612	53.2229	47.2188	42.5281	33.7267
2.70	63.2181	54.7717	48.5003	43.6210	34.5088
2.75	65.1614	56.3010	49.7610	44.6934	35.2730
2.80	67.0908	57.8107	51.0010	45.7454	36.0195

Table C.13 Prandtl-Meyer Angle *v*, Degrees (*Continued*)

M	$\gamma = 1.10$	$\gamma = 1.20$	$\gamma = 1.30$	$\gamma = 1.40$	$\gamma = 1.67$
2.85	69.0059	59.3008	52.2205	46.7775	36.7489
2.90	70.9063	60.7712	53.4196	47.7898	37.4615
2.95	72.7918	62.2221	54.5987	48.7828	38.1579
3.00	74.6622	63.6534	55.7578	49.7568	38.8384
3.05	76.5172	65.0652	56.8974	50.7122	39.5034
3.10	78.3566	66.4576	58.0176	51.6492	40.1535
3.15	80.1802	67.8308	59.1188	52.5683	40.7889
3.20	81.9880	69.1848	60.2011	53.4698	41.4101
3.25	83.7798	70.5198	61.2650	54.3540	42.0175
3.30	85.5555	71.8361	62.3106	55.2214	42.6114
3.35	87.3150	73.1336	63.3384	56.0722	43.1923
3.40	89.0582	74.4128	64.3486	56.9069	43.7604
3.45	90.7852	75.6736	65.3415	57.7258	44.3162
3.50	92.4959	76.9165	66.3174	58.5291	44.8600
3.55	94.1903	78.1414	67.2766	59.3174	45.3922
3.60	95.8684	79.3488	68.2195	60.0908	45.9129
3.65	97.5301	80.5387	69.1463	60.8498	46.4227
3.70	99.1756	81.7115	70.0574	61.5946	46.9217
3.75	100.8049	82.8673	70.9530	62.3256	47.4103
3.80	102.4179	84.0065	71.8334	63.0431	47.8887
3.85	104.0149	85.1291	72.6990	63.7474	48.3573
3.90	105.5957	86.2355	73.5500	64.4388	48.8163
3.95	107.1606	87.3259	74.3867	65.1176	49.2659
4.00	108.7095	88.4006	75.2094	65.7841	49.7065
4.10	111.7599	90.5035	76.8140	67.0813	50.5614
4.20	114.7478	92.5463	78.3659	68.3325	51.3829
4.30	117.6739	94.5307	79.8672	69.5399	52.1728
4.40	120.5391	96.4588	81.3200	70.7054	52.9326
4.50	123.3445	98.3322	82.7262	71.8310	53.6640
4.60	126.0911	100.1527	84.0878	72.9184	54.3685
4.70	128.7798	101.9223	85.4065	73.9693	55.0473
4.80	131.4118	103.6424	86.6841	74.9855	55.7018
4.90	133.9881	105.3149	87.9222	75.9684	56.3332
5.00	136.5100	106.9414	89.1225	76.9194	56.9426
5.10	138.9785	108.5233	90.2864	77.8401	57.5311
5.20	141.3948	110.0623	91.4154	78.7316	58.0998
5.30	143.7600	111.5598	92.5110	79.5953	58.6494
5.40	146.0752	113.0171	93.5744	80.4324	59.1810
5.50	148.3417	114.4358	94.6068	81.2439	59.6954
5.60	150.5604	115.8171	95.6096	82.0310	60.1934
5.70	152.7326	117.1622	96.5839	82.7947	60.6756
5.80	154.8592	118.4725	97.5307	83.5359	61.1429
5.90	156.9415	119.7491	98.4511	84.2556	61.5958
6.00	158.9805	120.9932	99.3462	84.9546	62.0350
6.10	160.9771	122.2057	100.2169	85.6338	62.4611
6.20	162.9326	123.3879	101.0641	86.2939	62.8747
6.30	164.8478	124.5407	101.8886	86.9357	63.2762
6.40	166.7237	125.6651	102.6914	87.5599	63.6662
6.50	168.5614	126.7620	103.4732	88.1672	64.0452
6.60	170.3618	127.8324	104.2347	88.7583	64.4136
6.70	172.1258	128.8771	104.9768	89.3337	64.7717
6.80	173.8544	129.8969	105.7001	89.8940	65.1202
6.90	175.5484	130.8927	106.4053	90.4399	65.4592
7.00	177.2086	131.8653	107.0929	90.9718	65.7892
7.10	178.8360	132.8153	107.7637	91.4902	66.1106
7.20	180.4314	133.7435	108.4182	91.9957	66.4236
7.30	181.9956	134.6505	109.0570	92.4886	66.7285
7.40	183.5292	135.5372	109.6805	92.9695	67.0258
7.50	185.0332	136.4040	110.2894	93.4387	67.3156
7.60	186.5083	137.2516	110.8841	93.8967	67.5982
7.70	187.9551	138.0806	111.4650	94.3438	67.8739
7.80	189.3745	138.8915	112.0326	94.7804	68.1429
7.90	190.7670	139.6850	112.5874	95.2070	68.4054
8.00	192.1333	140.4615	113.1298	95.6237	68.6618
8.50	198.5946	144.1066	115.6685	97.5712	69.8575
9.00	204.4919	147.3964	117.9496	99.3171	70.9263
9.50	209.8901	150.3783	120.0095	100.8904	71.8870

Table C.13 Prandtl-Meyer Angle v, Degrees (*Continued*)

M	$\gamma = 1.10$	$\gamma = 1.20$	$\gamma = 1.30$	$\gamma = 1.40$	$\gamma = 1.67$
10.00	214.8453	153.0920	121.8780	102.3152	72.7551
11.00	223.6162	157.6437	125.1365	104.7946	74.2618
12.00	231.1242	161.8625	127.8803	106.8775	75.5240
13.00	237.6127	165.3025	130.2207	108.6510	76.5962
14.00	243.2689	168.2782	132.2396	110.1787	77.5182
15.00	248.2386	170.8761	133.9982	111.5079	78.3193
16.00	252.6363	173.1630	135.5434	112.6747	79.0217
17.00	256.5531	175.1909	136.9116	113.7070	79.6425
18.00	260.0622	177.0010	138.1312	114.6266	80.1951
19.00	263.2228	178.6263	139.2251	115.4510	80.6902
20.00	266.0837	180.0934	140.2116	116.1941	81.1361
21.00	268.6848	181.4243	141.1058	116.8673	81.5400
22.00	271.0596	182.6368	141.9199	117.4801	81.9074
23.00	273.2359	183.7461	142.6642	118.0402	82.2431
24.00	275.2374	184.7646	143.3473	118.5540	82.5510
25.00	277.0840	185.7031	143.9764	119.0271	82.8344
INFIN	322.4318	208.4962	159.1987	130.4541	89.6639

Table C.14 Mass Addition Normal to the Main Stream ($\gamma = 1.40$)

M	M*	$\dfrac{t}{t^*}$	$\dfrac{p}{p^*}$	$\dfrac{\rho}{\rho^*}$	$\dfrac{P}{P^*}$	$\dfrac{G}{G^*}$
.00	.00000	.12000+1	.24000+1	.20000+1	.12679+1	.0
.01	.01095	.12000+1	.23997+1	.19998+1	.12678+1	.21906-1
.02	.02191	.11999+1	.23987+1	.19990+1	.12675+1	.43795-1
.03	.03286	.11998+1	.23970+1	.19978+1	.12671+1	.65650-1
.04	.04381	.11996+1	.23946+1	.19962+1	.12665+1	.87454-1
.05	.05476	.11994+1	.23916+1	.19940+1	.12657+1	.10919+0
.06	.06570	.11991+1	.23880+1	.19914+1	.12647+1	.13084+0
.07	.07664	.11988+1	.23836+1	.19883+1	.12636+1	.15239+0
.08	.08758	.11985+1	.23787+1	.19848+1	.12623+1	.17383+0
.09	.09851	.11981+1	.23731+1	.19808+1	.12608+1	.19513+0
.10	.10944	.11976+1	.23669+1	.19763+1	.12591+1	.21628+0
.11	.12035	.11971+1	.23600+1	.19714+1	.12573+1	.23727+0
.12	.13126	.11966+1	.23526+1	.19661+1	.12554+1	.25808+0
.13	.14217	.11960+1	.23445+1	.19604+1	.12533+1	.27870+0
.14	.15306	.11953+1	.23359+1	.19542+1	.12510+1	.29912+0
.15	.16395	.11946+1	.23267+1	.19476+1	.12486+1	.31931+0
.16	.17482	.11939+1	.23170+1	.19407+1	.12461+1	.33928+0
.17	.18569	.11931+1	.23067+1	.19333+1	.12434+1	.35900+0
.18	.19654	.11923+1	.22959+1	.19256+1	.12406+1	.37847+0
.19	.20739	.11914+1	.22845+1	.19175+1	.12377+1	.39767+0
.20	.21822	.11905+1	.22727+1	.19091+1	.12346+1	.41660+0
.21	.22904	.11895+1	.22604+1	.19003+1	.12314+1	.43524+0
.22	.23984	.11885+1	.22477+1	.18912+1	.12281+1	.45359+0
.23	.25063	.11874+1	.22345+1	.18818+1	.12247+1	.47163+0
.24	.26141	.11863+1	.22209+1	.18721+1	.12213+1	.48937+0
.25	.27217	.11852+1	.22069+1	.18621+1	.12177+1	.50679+0
.26	.28291	.11840+1	.21925+1	.18518+1	.12140+1	.52389+0
.27	.29364	.11828+1	.21777+1	.18412+1	.12102+1	.54066+0
.28	.30435	.11815+1	.21626+1	.18304+1	.12064+1	.55709+0
.29	.31504	.11801+1	.21472+1	.18194+1	.12025+1	.57319+0
.30	.32572	.11788+1	.21314+1	.18082+1	.11985+1	.58895+0
.31	.33637	.11774+1	.21154+1	.17967+1	.11945+1	.60436+0
.32	.34701	.11759+1	.20991+1	.17851+1	.11904+1	.61943+0
.33	.35762	.11744+1	.20825+1	.17732+1	.11863+1	.63414+0
.34	.36822	.11729+1	.20657+1	.17612+1	.11822+1	.64851+0
.35	.37879	.11713+1	.20487+1	.17490+1	.11779+1	.66253+0
.36	.38935	.11697+1	.20314+1	.17367+1	.11737+1	.67619+0

Table C.14 Mass Addition Normal to the Main Stream ($\gamma = 1.40$) (*Continued*)

M	M*	$\dfrac{t}{t^*}$	$\dfrac{p}{p^*}$	$\dfrac{\rho}{\rho^*}$	$\dfrac{P}{P^*}$	$\dfrac{G}{G^*}$
.37	.39988	.11680+1	.20140+1	.17243+1	.11695+1	.68950+0
.38	.41039	.11663+1	.19964+1	.17117+1	.11652+1	.70246+0
.39	.42087	.11646+1	.19787+1	.16990+1	.11609+1	.71508+0
.40	.43133	.11628+1	.19608+1	.16863+1	.11566+1	.72734+0
.41	.44177	.11610+1	.19428+1	.16734+1	.11523+1	.73926+0
.42	.45218	.11591+1	.19247+1	.16605+1	.11480+1	.75084+0
.43	.46257	.11572+1	.19065+1	.16475+1	.11437+1	.76207+0
.44	.47293	.11553+1	.18882+1	.16344+1	.11394+1	.77297+0
.45	.48326	.11533+1	.18699+1	.16213+1	.11351+1	.78353+0
.46	.49357	.11513+1	.18515+1	.16082+1	.11308+1	.79377+0
.47	.50385	.11492+1	.18331+1	.15951+1	.11266+1	.80367+0
.48	.51410	.11471+1	.18147+1	.15819+1	.11224+1	.81326+0
.49	.52433	.11450+1	.17962+1	.15687+1	.11182+1	.82253+0
.50	.53452	.11429+1	.17778+1	.15556+1	.11141+1	.83148+0
.52	.55483	.11384+1	.17409+1	.15292+1	.11059+1	.84847+0
.54	.57501	.11339+1	.17043+1	.15030+1	.10979+1	.86426+0
.56	.59507	.11292+1	.16678+1	.14770+1	.10901+1	.87891+0
.58	.61501	.11244+1	.16316+1	.14511+1	.10826+1	.89246+0
.60	.63481	.11194+1	.15957+1	.14255+1	.10753+1	.90494+0
.62	.65448	.11143+1	.15603+1	.14002+1	.10682+1	.91642+0
.64	.67402	.11091+1	.15253+1	.13752+1	.10615+1	.92693+0
.66	.69342	.11038+1	.14908+1	.13506+1	.10550+1	.93653+0
.68	.71268	.10984+1	.14569+1	.13263+1	.10489+1	.94525+0
.70	.73179	.10929+1	.14235+1	.13025+1	.10431+1	.95315+0
.72	.75076	.10873+1	.13907+1	.12791+1	.10376+1	.96027+0
.74	.76958	.10815+1	.13585+1	.12561+1	.10325+1	.96666+0
.76	.78825	.10757+1	.13270+1	.12335+1	.10278+1	.97235+0
.78	.80677	.10696+1	.12961+1	.12115+1	.10234+1	.97738+0
.80	.82514	.10636+1	.12658+1	.11899+1	.10193+1	.98181+0
.82	.84335	.10576+1	.12362+1	.11687+1	.10157+1	.98566+0
.84	.86140	.10516+1	.12073+1	.11481+1	.10124+1	.98897+0
.86	.87929	.10454+1	.11791+1	.11279+1	.10095+1	.99178+0
.88	.89703	.10391+1	.11515+1	.11082+1	.10070+1	.99412+0
.90	.91460	.10327+1	.11246+1	.10890+1	.10049+1	.99603+0
.92	.93201	.10263+1	.10984+1	.10703+1	.10031+1	.99753+0
.94	.94925	.10198+1	.10728+1	.10520+1	.10017+1	.99865+0
.96	.96633	.10132+1	.10479+1	.10342+1	.10008+1	.99941+0
.98	.98325	.10066+1	.10236+1	.10169+1	.10002+1	.99986+0
1.00	1.00000	.10000+1	.10000+1	.10000+1	.10000+1	.10000+1
1.02	1.01658	.99331+0	.97698+0	.98355+0	.10002+1	.99986+0
1.04	1.03300	.98658+0	.95456+0	.96754+0	.10008+1	.99947+0
1.06	1.04925	.97982+0	.93275+0	.95196+0	.10017+1	.99885+0
1.08	1.06533	.97302+0	.91152+0	.93680+0	.10031+1	.99800+0
1.10	1.08124	.96618+0	.89087+0	.92205+0	.10049+1	.99696+0
1.12	1.09699	.95932+0	.87078+0	.90770+0	.10070+1	.99573+0
1.14	1.11256	.95244+0	.85123+0	.89374+0	.10095+1	.99434+0
1.16	1.12797	.94554+0	.83222+0	.88016+0	.10124+1	.99279+0
1.18	1.14321	.93861+0	.81374+0	.86695+0	.10157+1	.99111+0
1.20	1.15828	.93168+0	.79576+0	.85411+0	.10194+1	.98930+0
1.22	1.17319	.92473+0	.77827+0	.84162+0	.10235+1	.98738+0
1.24	1.18792	.91777+0	.76127+0	.82948+0	.10279+1	.98535+0
1.26	1.20249	.91080+0	.74473+0	.81767+0	.10328+1	.98324+0
1.28	1.21690	.90383+0	.72865+0	.80618+0	.10380+1	.98104+0
1.30	1.23114	.89686+0	.71301+0	.79501+0	.10437+1	.97876+0
1.32	1.24521	.88989+0	.69780+0	.78415+0	.10497+1	.97643+0
1.34	1.25912	.88292+0	.68301+0	.77358+0	.10561+1	.97403+0
1.36	1.27286	.87596+0	.66863+0	.76331+0	.10629+1	.97158+0
1.38	1.28645	.86901+0	.65464+0	.75331+0	.10701+1	.96909+0
1.40	1.29987	.86207+0	.64103+0	.74359+0	.10777+1	.96657+0
1.42	1.31313	.85514+0	.62779+0	.73413+0	.10856+1	.96401+0
1.44	1.32623	.84822+0	.61491+0	.72493+0	.10940+1	.96142+0
1.46	1.33917	.84133+0	.60237+0	.71598+0	.11028+1	.95882+0
1.48	1.35195	.83445+0	.59016+0	.70727+0	.11120+1	.95620+0
1.50	1.36458	.82759+0	.57831+0	.69880+0	.11215+1	.95356+0
1.55	1.39546	.81054+0	.55002+0	.67858+0	.11473+1	.94694+0
1.60	1.42539	.79365+0	.52356+0	.65969+0	.11756+1	.94031+0
1.65	1.45439	.77695+0	.49880+0	.64200+0	.12066+1	.93372+0

Table C.14 Mass Addition Normal to the Main Stream ($\gamma = 1.40$) (*Continued*)

M	M*	$\dfrac{t}{t^*}$	$\dfrac{p}{p^*}$	$\dfrac{\rho}{\rho^*}$	$\dfrac{P}{P^*}$	$\dfrac{G}{G^*}$
1.70	1.48247	.76046+0	.47562+0	.62545+0	.12402+1	.92721+0
1.75	1.50966	.74419+0	.45390+0	.60993+0	.12767+1	.92078+0
1.80	1.53598	.72816+0	.43353+0	.59538+0	.13159+1	.91448+0
1.85	1.56145	.71238+0	.41440+0	.58171+0	.13581+1	.90832+0
1.90	1.58609	.69686+0	.39643+0	.56888+0	.14033+1	.90229+0
1.95	1.60993	.68162+0	.37954+0	.55681+0	.14516+1	.89643+0
2.00	1.63299	.66667+0	.36364+0	.54545+0	.15031+1	.89072+0
2.10	1.67687	.63762+0	.33454+0	.52467+0	.16162+1	.87981+0
2.20	1.71791	.60976+0	.30864+0	.50617+0	.17434+1	.86956+0
2.30	1.75629	.58309+0	.28551+0	.48965+0	.18860+1	.85997+0
2.40	1.79218	.55762+0	.26478+0	.47485+0	.20451+1	.85101+0
2.50	1.82574	.53333+0	.24615+0	.46154+0	.22218+1	.84265+0
2.60	1.85714	.51020+0	.22936+0	.44954+0	.24177+1	.83486+0
2.70	1.88653	.48820+0	.21417+0	.43869+0	.26343+1	.82761+0
2.80	1.91404	.46729+0	.20040+0	.42886+0	.28731+1	.82085+0
2.90	1.93981	.44743+0	.18788+0	.41992+0	.31359+1	.81456+0
3.00	1.96396	.42857+0	.17647+0	.41176+0	.34245+1	.80869+0
3.10	1.98661	.41068+0	.16604+0	.40432+0	.37408+1	.80322+0
3.20	2.00786	.39370+0	.15649+0	.39750+0	.40671+1	.79812+0
3.30	2.02781	.37760+0	.14773+0	.39123+0	.44655+1	.79335+0
3.40	2.04656	.36232+0	.13966+0	.38547+0	.48783+1	.78890+0
3.50	2.06419	.34783+0	.13223+0	.38017+0	.53280+1	.78473+0
3.60	2.08077	.33408+0	.12537+0	.37526+0	.58173+1	.78083+0
3.70	2.09639	.32103+0	.11901+0	.37072+0	.63488+1	.77718+0
3.80	2.11111	.30864+0	.11312+0	.36652+0	.69256+1	.77376+0
3.90	2.12499	.29688+0	.10765+0	.36261+0	.75505+1	.77054+0
4.00	2.13809	.28571+0	.10256+0	.35897+0	.82268+1	.76752+0
4.50	2.19360	.23762+0	.81772-1	.34412+0	.12502+2	.75487+0
5.00	2.23607	.20000+0	.66667-1	.33333+0	.18634+2	.74536+0
5.50	2.26913	.17021+0	.55363-1	.32526+0	.27211+2	.73806+0
6.00	2.29528	.14634+0	.46693-1	.31907+0	.38946+2	.73234+0
6.50	2.31626	.12698+0	.39900-1	.31421+0	.54683+2	.72780+0
7.00	2.33333	.11111+0	.34483-1	.31034+0	.75414+2	.72414+0
7.50	2.34738	.97959-1	.30094-1	.30721+0	.10229+3	.72114+0
8.00	2.35907	.86957-1	.26490-1	.30464+0	.13662+3	.71866+0
8.50	2.36869	.77670-1	.23495-1	.30250+0	.17992+3	.71656+0
9.00	2.37722	.69767-1	.20979-1	.30070+0	.23388+3	.71483+0
9.50	2.38433	.62992-1	.18846-1	.29918+0	.30041+3	.71333+0
10.00	2.39046	.57143-1	.17021-1	.29787+0	.38161+3	.71205+0
11.00	2.40040	.47619-1	.14085-1	.29577+0	.59774+3	.70998+0
12.00	2.40804	.40268-1	.11846-1	.29418+0	.90405+3	.70839+0
13.00	2.41404	.34483-1	.10101-1	.29293+0	.13267+4	.70714+0
14.00	2.41883	.29851-1	.87146-2	.29194+0	.18963+4	.70615+0
15.00	2.42272	.26087-1	.75949-2	.29114+0	.26486+4	.70535+0
INFIN	2.44949	.0	.0	.28571+0	INFIN	.69985+0

Table C.14 Mass Addition Normal to the Main Stream ($\gamma = 1.30$)

M	M*	$\dfrac{t}{t^*}$	$\dfrac{p}{p^*}$	$\dfrac{\rho}{\rho^*}$	$\dfrac{P}{P^*}$	$\dfrac{G}{G^*}$
.00	.00000	.11500+1	.23000+1	.20000+1	.12552+1	.0
.02	.02145	.11499+1	.22988+1	.19991+1	.12548+1	.42874-1
.04	.04289	.11497+1	.22952+1	.19963+1	.12539+1	.85623-1
.06	.06433	.11494+1	.22893+1	.19918+1	.12523+1	.12812+0
.08	.08575	.11489+1	.22810+1	.19854+1	.12500+1	.17025+0
.10	.10716	.11483+1	.22705+1	.19773+1	.12471+1	.21188+0
.12	.12855	.11475+1	.22577+1	.19675+1	.12437+1	.25291+0
.14	.14991	.11466+1	.22429+1	.19560+1	.12397+1	.29324+0
.16	.17125	.11456+1	.22259+1	.19430+1	.12351+1	.33275+0
.18	.19256	.11444+1	.22070+1	.19285+1	.12300+1	.37135+0
.20	.21384	.11431+1	.21863+1	.19125+1	.12245+1	.40897+0

M	M^*	$\dfrac{t}{t^*}$	$\dfrac{p}{p^*}$	$\dfrac{\rho}{\rho^*}$	$\dfrac{P}{P^*}$	$\dfrac{G}{G^*}$
.22	.23507	.11417+1	.21639+1	.18953+1	.12185+1	.44552+0
.24	.25627	.11401+1	.21398+1	.18767+1	.12121+1	.48095+0
.26	.27742	.11385+1	.21142+1	.18571+1	.12053+1	.51518+0
.28	.29852	.11366+1	.20873+1	.18364+1	.11983+1	.54818+0
.30	.31956	.11347+1	.20591+1	.18147+1	.11909+1	.57991+0
.32	.34056	.11326+1	.20298+1	.17921+1	.11834+1	.61033+0
.34	.36149	.11304+1	.19995+1	.17689+1	.11756+1	.63942+0
.36	.38236	.11281+1	.19684+1	.17449+1	.11677+1	.66718+0
.38	.40316	.11256+1	.19365+1	.17204+1	.11596+1	.69359+0
.40	.42390	.11230+1	.19040+1	.16954+1	.11515+1	.71866+0
.42	.44456	.11204+1	.18710+1	.16700+1	.11434+1	.74239+0
.44	.46514	.11175+1	.18375+1	.16443+1	.11352+1	.76481+0
.46	.48565	.11146+1	.18038+1	.16183+1	.11271+1	.78593+0
.48	.50607	.11116+1	.17699+1	.15922+1	.11191+1	.80578+0
.50	.52641	.11084+1	.17358+1	.15660+1	.11111+1	.82438+0
.55	.57687	.11001+1	.16508+1	.15006+1	.10919+1	.86566+0
.60	.62673	.10911+1	.15668+1	.14360+1	.10739+1	.89996+0
.65	.67596	.10815+1	.14846+1	.13728+1	.10574+1	.92793+0
.70	.72451	.10713+1	.14050+1	.13115+1	.10426+1	.95023+0
.75	.77236	.10605+1	.13285+1	.12527+1	.10299+1	.96754+0
.80	.81947	.10493+1	.12555+1	.11965+1	.10193+1	.98050+0
.85	.86581	.10376+1	.11860+1	.11431+1	.10109+1	.98971+0
.90	.91136	.10254+1	.11203+1	.10925+1	.10049+1	.99571+0
.95	.95610	.10129+1	.10583+1	.10449+1	.10012+1	.99899+0
1.00	1.00000	.10000+1	.10000+1	.10000+1	.10000+1	.10000+1
1.05	1.04305	.98681+0	.94524+0	.95788+0	.10012+1	.99911+0
1.10	1.08524	.97334+0	.89390+0	.91838+0	.10049+1	.99666+0
1.15	1.12655	.95963+0	.84582+0	.88140+0	.10112+1	.99294+0
1.20	1.16698	.94572+0	.80084+0	.84680+0	.10199+1	.98819+0
1.25	1.20652	.93165+0	.75876+0	.81443+0	.10312+1	.98263+0
1.30	1.24517	.91743+0	.71942+0	.78417+0	.10451+1	.97643+0
1.35	1.28293	.90311+0	.68264+0	.75588+0	.10616+1	.96975+0
1.40	1.31981	.88872+0	.64825+0	.72943+0	.10809+1	.96270+0
1.45	1.35579	.87428+0	.61609+0	.70468+0	.11028+1	.95540+0
1.50	1.39089	.85981+0	.58599+0	.68153+0	.11276+1	.94793+0
1.55	1.42512	.84536+0	.55781+0	.65986+0	.11552+1	.94037+0
1.60	1.45848	.83092+0	.53142+0	.63956+0	.11858+1	.93278+0
1.65	1.49099	.81654+0	.50669+0	.62053+0	.12195+1	.92520+0
1.70	1.52265	.80223+0	.48350+0	.60269+0	.12563+1	.91768+0
1.75	1.55347	.78801+0	.46173+0	.58595+0	.12964+1	.91025+0
1.80	1.58348	.77389+0	.44129+0	.57022+0	.13400+1	.90293+0
1.85	1.61268	.75989+0	.42208+0	.55544+0	.13872+1	.89575+0
1.90	1.64108	.74603+0	.40400+0	.54154+0	.14381+1	.88872+0
1.95	1.66871	.73231+0	.38699+0	.52846+0	.14929+1	.88184+0
2.00	1.69558	.71875+0	.37097+0	.51613+0	.15518+1	.87514+0
2.20	1.79577	.66628+0	.31541+0	.47340+0	.18325+1	.85011+0
2.40	1.88511	.61695+0	.27097+0	.43921+0	.21970+1	.82796+0
2.60	1.96468	.57100+0	.23498+0	.41152+0	.26644+1	.80851+0
2.80	2.03553	.52849+0	.20550+0	.38885+0	.32583+1	.79151+0
3.00	2.09863	.48936+0	.18110+0	.37008+0	.40074+1	.77666+0
3.20	2.15469	.45347+0	.16070+0	.35439+0	.49467+1	.76367+0
3.40	2.20510	.42063+0	.14350+0	.34115+0	.61181+1	.75228+0
3.60	2.25000	.39065+0	.12887+0	.32990+0	.75714+1	.74227+0
3.80	2.29022	.36323+0	.11633+0	.32025+0	.93656+1	.73344+0
4.00	2.32632	.33824+0	.10550+0	.31193+0	.11570+2	.72564+0
5.00	2.46021	.24211+0	.68657-1	.28358+0	.32063+2	.69767+0
6.00	2.54337	.17969+0	.48117-1	.26770+0	.81794+2	.68107+0
7.00	2.59779	.13772+0	.35549-1	.25811+0	.19133+3	.67053+0
8.00	2.63503	.10849+0	.27316-1	.25178+0	.41341+3	.66345+0
9.00	2.66151	.87452-1	.21637-1	.24741+0	.83339+3	.65849+0
10.00	2.68095	.71875-1	.17557-1	.24427+0	.15823+4	.65489+0
11.00	2.69561	.60052-1	.14529-1	.24195+0	.28529+4	.65219+0
12.00	2.70692	.50685-1	.12221-1	.24017+0	.49191+4	.65012+0
13.00	2.71583	.43643-1	.10421-1	.23879+0	.81586+4	.64850+0
14.00	2.72295	.37829-1	.89914-2	.23769+0	.13079+5	.64721+0
15.00	2.72874	.33094-1	.78365-2	.23680+0	.20350+5	.64616+0
INFIN	2.76887	.0	.0	.23077+0	INFIN	.63897+0

Table C.14 Mass Addition Normal to the Main Stream ($\gamma = 1.20$)

M	M^*	$\dfrac{t}{t^*}$	$\dfrac{p}{p^*}$	$\dfrac{\rho}{\rho^*}$	$\dfrac{P}{P^*}$	$\dfrac{G}{G^*}$
.00	.00000	.11000+1	.22000+1	.20000+1	.12418+1	.0
.02	.02098	.11000+1	.21989+1	.19991+1	.12415+1	.41933−1
.04	.04195	.10998+1	.21958+1	.19965+1	.12407+1	.83751−1
.06	.06292	.10996+1	.21905+1	.19921+1	.12392+1	.12534+0
.08	.08388	.10993+1	.21832+1	.19860+1	.12371+1	.16658+0
.10	.10485	.10989+1	.21739+1	.19783+1	.12345+1	.20738+0
.12	.12577	.10984+1	.21626+1	.19689+1	.12313+1	.24762+0
.14	.14669	.10978+1	.21494+1	.19579+1	.12276+1	.28720+0
.16	.16760	.10972+1	.21344+1	.19454+1	.12235+1	.32603+0
.18	.18848	.10964+1	.21177+1	.19314+1	.12188+1	.36403+0
.20	.20934	.10956+1	.20992+1	.19160+1	.12137+1	.40111+0
.22	.23018	.10947+1	.20792+1	.18994+1	.12082+1	.43720+0
.24	.25099	.10937+1	.20578+1	.18815+1	.12023+1	.47224+0
.26	.27177	.10926+1	.20349+1	.18624+1	.11960+1	.50616+0
.28	.29252	.10914+1	.20108+1	.18424+1	.11895+1	.53893+0
.30	.31324	.10902+1	.19856+1	.18213+1	.11827+1	.57050+0
.32	.33391	.10889+1	.19592+1	.17994+1	.11757+1	.60084+0
.34	.35455	.10874+1	.19320+1	.17767+1	.11684+1	.62992+0
.36	.37515	.10859+1	.19039+1	.17533+1	.11610+1	.65773+0
.38	.39570	.10843+1	.18751+1	.17292+1	.11535+1	.68426+0
.40	.41621	.10827+1	.18456+1	.17047+1	.11459+1	.70951+0
.42	.43667	.10809+1	.18157+1	.16797+1	.11383+1	.73347+0
.44	.45707	.10791+1	.17853+1	.16544+1	.11306+1	.75617+0
.46	.47743	.10772+1	.17545+1	.16287+1	.11229+1	.77761+0
.48	.49773	.10752+1	.17235+1	.16029+1	.11153+1	.79781+0
.50	.51797	.10732+1	.16923+1	.15769+1	.11078+1	.81680+0
.55	.56831	.10677+1	.16141+1	.15117+1	.10895+1	.85914+0
.60	.61826	.10618+1	.15363+1	.14469+1	.10722+1	.89457+0
.65	.66777	.10554+1	.14599+1	.13832+1	.10563+1	.92366+0
.70	.71681	.10486+1	.13854+1	.13212+1	.10420+1	.94703+0
.75	.76537	.10414+1	.13134+1	.12612+1	.10296+1	.96529+0
.80	.81342	.10338+1	.12443+1	.12036+1	.10191+1	.97905+0
.85	.86093	.10259+1	.11784+1	.11486+1	.10109+1	.98889+0
.90	.90787	.10176+1	.11156+1	.10963+1	.10049+1	.99535+0
.95	.95424	.10089+1	.10562+1	.10468+1	.10012+1	.99890+0
1.00	1.00000	.10000+1	.10000+1	.10000+1	.10000+1	.10000+1
1.05	1.04514	.99077+0	.94705+0	.95588+0	.10012+1	.99903+0
1.10	1.08965	.98127+0	.89723+0	.91436+0	.10050+1	.99633+0
1.15	1.13350	.97152+0	.85041+0	.87534+0	.10114+1	.99220+0
1.20	1.17670	.96154+0	.80645+0	.83871+0	.10204+1	.98691+0
1.25	1.21922	.95135+0	.76522+0	.80435+0	.10321+1	.98067+0
1.30	1.26105	.94098+0	.72655+0	.77213+0	.10466+1	.97369+0
1.35	1.30219	.93043+0	.69030+0	.74192+0	.10640+1	.96612+0
1.40	1.34264	.91973+0	.65632+0	.71360+0	.10843+1	.95811+0
1.45	1.38238	.90890+0	.62447+0	.68706+0	.11077+1	.94977+0
1.50	1.42141	.89796+0	.59459+0	.66216+0	.11342+1	.94120+0
1.55	1.45973	.88692+0	.56657+0	.63881+0	.11640+1	.93249+0
1.60	1.49734	.87580+0	.54028+0	.61690+0	.11973+1	.92371+0
1.65	1.53424	.86461+0	.51558+0	.59632+0	.12342+1	.91490+0
1.70	1.57043	.85337+0	.49239+0	.57699+0	.12749+1	.90613+0
1.75	1.60591	.84211+0	.47059+0	.55882+0	.13196+1	.89742+0
1.80	1.64068	.83082+0	.45008+0	.54173+0	.13686+1	.88882+0
1.85	1.67476	.81952+0	.43078+0	.52565+0	.14220+1	.88034+0
1.90	1.70813	.80823+0	.41260+0	.51050+0	.14802+1	.87200+0
1.95	1.74081	.79696+0	.39547+0	.49623+0	.15435+1	.86383+0
2.00	1.77281	.78571+0	.37931+0	.48276+0	.16122+1	.85584+0
2.20	1.89410	.74124+0	.32315+0	.43596+0	.19483+1	.82575+0
2.40	2.00507	.69797+0	.27806+0	.39838+0	.24050+1	.79878+0
2.60	2.10636	.65632+0	.24144+0	.36787+0	.30206+1	.77486+0
2.80	2.19865	.61659+0	.21138+0	.34281+0	.38465+1	.75373+0
3.00	2.28266	.57895+0	.18644+0	.32203+0	.49512+1	.73509+0
3.20	2.35907	.54348+0	.16556+0	.30464+0	.64250+1	.71866+0
3.40	2.42857	.51020+0	.14793+0	.28994+0	.83867+1	.70414+0
3.60	2.49180	.47909+0	.13291+0	.27743+0	.10991+2	.69130+0
3.80	2.54935	.45008+0	.12003+0	.26670+0	.14440+2	.67990+0
4.00	2.60177	.42308+0	.10891+0	.25743+0	.18991+2	.66976+0
5.00	2.80306	.31429+0	.70968−1	.22581+0	.73640+2	.63295+0

Table C.14 Mass Addition Normal to the Main Stream ($\gamma = 1.20$) (*Continued*)

M	M^*	$\dfrac{t}{t^*}$	$\dfrac{p}{p^*}$	$\dfrac{\rho}{\rho^*}$	$\dfrac{P}{P^*}$	$\dfrac{G}{G^*}$
6.00	2.93406	.23913+0	.49774-1	.20814+0	.26619+3	.61071+0
7.00	3.02251	.18644+0	.36789-1	.19732+0	.87595+3	.59642+0
8.00	3.08440	.14865+0	.28278-1	.19023+0	.26211+4	.58675+0
9.00	3.12909	.12088+0	.22403-1	.18534+0	.71813+4	.57993+0
10.00	3.16228	.10000+0	.18182-1	.18182+0	.18182+5	.57496+0
11.00	3.18752	.83969-1	.15048-1	.17921+0	.42929+5	.57123+0
12.00	3.20713	.71429-1	.12658-1	.17722+0	.95311+5	.56835+0
13.00	3.22265	.61453-1	.10795-1	.17566+0	.20044+6	.56610+0
14.00	3.23512	.53398-1	.93141-2	.17443+0	.40178+6	.56430+0
15.00	3.24529	.46809-1	.81181-2	.17343+0	.77180+6	.56284+0
INFIN	3.31662	.0	.0	.16667+0	INFIN	.55277+0

Table C.14 Mass Addition Normal to the Main Stream ($\gamma = 1.10$)

M	M^*	$\dfrac{t}{t^*}$	$\dfrac{p}{p^*}$	$\dfrac{\rho}{\rho^*}$	$\dfrac{P}{P^*}$	$\dfrac{G}{G^*}$
.00	.00000	.10500+1	.21000+1	.20000+1	.12278+1	.0
.02	.02049	.10500+1	.20991+1	.19992+1	.12276+1	.40970-1
.04	.04099	.10499+1	.20963+1	.19966+1	.12267+1	.81835-1
.06	.06148	.10498+1	.20917+1	.19925+1	.12254+1	.12249+0
.08	.08196	.10497+1	.20853+1	.19867+1	.12235+1	.16283+0
.10	.10244	.10495+1	.20772+1	.19792+1	.12212+1	.20276+0
.12	.12292	.10492+1	.20673+1	.19702+1	.12183+1	.24218+0
.14	.14339	.10490+1	.20557+1	.19597+1	.12149+1	.28100+0
.16	.16385	.10487+1	.20425+1	.19477+1	.12111+1	.31913+0
.18	.18430	.10483+1	.20277+1	.19343+1	.12069+1	.35648+0
.20	.20473	.10479+1	.20115+1	.19195+1	.12022+1	.39300+0
.22	.22516	.10475+1	.19938+1	.19035+1	.11972+1	.42859+0
.24	.24557	.10470+1	.19749+1	.18862+1	.11918+1	.46321+0
.26	.26597	.10465+1	.19547+1	.18679+1	.11861+1	.49680+0
.28	.28635	.10459+1	.19333+1	.18484+1	.11801+1	.52931+0
.30	.30672	.10453+1	.19108+1	.18280+1	.11738+1	.56069+0
.32	.32707	.10447+1	.18874+1	.18067+1	.11673+1	.59092+0
.34	.34739	.10440+1	.18631+1	.17846+1	.11606+1	.61997+0
.36	.36770	.10432+1	.18380+1	.17618+1	.11538+1	.64781+0
.38	.38799	.10425+1	.18122+1	.17383+1	.11468+1	.67445+0
.40	.40825	.10417+1	.17857+1	.17143+1	.11397+1	.69985+0
.42	.42849	.10408+1	.17587+1	.16898+1	.11326+1	.72404+0
.44	.44870	.10399+1	.17313+1	.16648+1	.11254+1	.74700+0
.46	.46889	.10390+1	.17035+1	.16395+1	.11182+1	.76876+0
.48	.48904	.10380+1	.16754+1	.16140+1	.11111+1	.76931+0
.50	.50918	.10370+1	.16471+1	.15882+1	.11040+1	.80869+0
.55	.55937	.10344+1	.15757+1	.15234+1	.10867+1	.85212+0
.60	.60936	.10314+1	.15043+1	.14585+1	.10702+1	.88872+0
.65	.65913	.10283+1	.14337+1	.13943+1	.10550+1	.91900+0
.70	.70866	.10249+1	.13645+1	.13314+1	.10412+1	.94350+0
.75	.75794	.10213+1	.12973+1	.12703+1	.10291+1	.96278+0
.80	.80695	.10174+1	.12324+1	.12113+1	.10189+1	.97743+0
.85	.85567	.10134+1	.11701+1	.11546+1	.10108+1	.98797+0
.90	.90410	.10091+1	.11105+1	.11005+1	.10049+1	.99494+0
.95	.95221	.10047+1	.10538+1	.10489+1	.10012+1	.99880+0
1.00	1.00000	.10000+1	.10000+1	.10000+1	.10000+1	.10000+1
1.05	1.04745	.99514+0	.94905+0	.95368+0	.10013+1	.99893+0
1.10	1.09454	.99010+0	.90090+0	.90991+0	.10051+1	.99593+0
1.15	1.14127	.98488+0	.85548+0	.86862+0	.10116+1	.99133+0
1.20	1.18762	.97946+0	.81269+0	.82972+0	.10209+1	.98540+0
1.25	1.23359	.97391+0	.77241+0	.79310+0	.10331+1	.97836+0
1.30	1.27916	.96819+0	.73452+0	.75866+0	.10482+1	.97044+0
1.35	1.32431	.96231+0	.69889+0	.72627+0	.10665+1	.96181+0
1.40	1.36906	.95628+0	.66540+0	.69582+0	.10880+1	.95261+0
1.45	1.41337	.95012+0	.63391+0	.66719+0	.11129+1	.94300+0

Table C.14 Mass Addition Normal to the Main Stream ($\gamma = 1.10$) (*Continued*)

M	M*	$\dfrac{t}{t^*}$	$\dfrac{p}{p^*}$	$\dfrac{\rho}{\rho^*}$	$\dfrac{P}{P^*}$	$\dfrac{G}{G^*}$
1.50	1.45726	.94382+0	.60432+0	.64029+0	.11415+1	.93306+0
1.55	1.50070	.93740+0	.57649+0	.61499+0	.11739+1	.92291+0
1.60	1.54369	.93085+0	.55031+0	.59119+0	.12104+1	.91262+0
1.65	1.58623	.92419+0	.52569+0	.56881+0	.12512+1	.90226+0
1.70	1.62830	.91743+0	.50251+0	.54774+0	.12967+1	.89189+0
1.75	1.66992	.91057+0	.48069+0	.52790+0	.13472+1	.88154+0
1.80	1.71106	.90361+0	.46012+0	.50920+0	.14030+1	.87127+0
1.85	1.75172	.89657+0	.44074+0	.49158+0	.14646+1	.86111+0
1.90	1.79191	.88945+0	.42245+0	.47495+0	.15326+1	.85107+0
1.95	1.83161	.88226+0	.40519+0	.45926+0	.16073+1	.84119+0
2.00	1.87083	.87500+0	.38889+0	.44444+0	.16894+1	.83148+0
2.20	2.02282	.84541+0	.33207+0	.39279+0	.21061+1	.79454+0
2.40	2.16695	.81522+0	.28626+0	.35115+0	.27086+1	.76091+0
2.60	2.30324	.78475+0	.24893+0	.31721+0	.35811+1	.73062+0
2.80	2.43183	.75431+0	.21820+0	.28928+0	.48508+1	.70347+0
3.00	2.55289	.72414+0	.19266+0	.26606+0	.67106+1	.67921+0
3.20	2.66667	.69444+0	.17123+0	.24658+0	.94531+1	.65753+0
3.40	2.77345	.66540+0	.15311+0	.23010+0	.13523+2	.63816+0
3.60	2.87355	.63714+0	.13765+0	.21605+0	.19599+2	.62082+0
3.80	2.96730	.60976+0	.12438+0	.20398+0	.28710+2	.60527+0
4.00	3.05505	.58333+0	.11290+0	.19355+0	.42425+2	.59130+0
5.00	3.41565	.46667+0	.73684-1	.15789+0	.32233+3	.53931+0
6.00	3.67423	.37500+0	.51724-1	.13793+0	.25081+4	.50679+0
7.00	3.86174	.30435+0	.38251-1	.12568+0	.18432+5	.48536+0
8.00	4.00000	.25000+0	.29412-1	.11765+0	.12336+6	.47059+0
9.00	4.10385	.20792+0	.23307-1	.11210+0	.74236+6	.46003+0
10.00	4.18330	.17500+0	.18919-1	.10811+0	.40131+7	.45225+0
11.00	4.24515	.14894+0	.15660-1	.10515+0	.19579+8	.44636+0
12.00	4.29407	.12805+0	.13174-1	.10289+0	.86816+8	.44180+0
13.00	4.33333	.11111+0	.11236-1	.10112+0	.35260+9	.43820+0
14.00	4.36527	.97222-1	.96953-2	.99723-1	.13217+10	.43532+0
15.00	4.39155	.85714-1	.84507-2	.98592-1	.46058+10	.43297+0
INFIN	4.58258	.0	.0	.90909-1	INFIN	.41660+0

Table C.15 Useful Functions of the Specific Heat Ratio γ

γ	$\dfrac{1}{\gamma-1}$	$\dfrac{\gamma}{\gamma-1}$	$\dfrac{\gamma-1}{\gamma+1}$	$\dfrac{\gamma-1}{\gamma}$	$\dfrac{2}{\gamma+1}$	$\left(\dfrac{2}{\gamma+1}\right)^{1/(\gamma-1)}$	$\left(\dfrac{2}{\gamma+1}\right)^{\gamma/(\gamma-1)}$	$(\gamma+1)\left(\dfrac{2}{\gamma+1}\right)^{(\gamma+1)/2(\gamma-1)}$	$\left(\dfrac{2}{\gamma+1}\right)^{(\gamma+1)/2(\gamma-1)}$	$\gamma\left(\dfrac{2}{\gamma+1}\right)^{(\gamma+1)/2(\gamma-1)}$
1.05	20.0000	21.0000	.02439	.04762	.97561	.61027	.59539	1.22054	.60278	.63292
1.06	16.6667	17.6667	.02913	.05660	.97087	.61101	.59321	1.22201	.60204	.63817
1.07	14.2857	15.2857	.03382	.06542	.96618	.61174	.59105	1.22348	.60131	.64340
1.08	12.5000	13.5000	.03846	.07407	.96154	.61247	.58891	1.22494	.60057	.64862
1.09	11.1111	12.1111	.04306	.08257	.95694	.61319	.58679	1.22638	.59984	.65383
1.10	10.0000	11.0000	.04762	.09091	.95238	.61391	.58468	1.22783	.59912	.65903
1.11	9.0909	10.0909	.05213	.09910	.94787	.61463	.58259	1.22926	.59839	.66422
1.12	8.3333	9.3333	.05660	.10714	.94340	.61534	.58051	1.23069	.59767	.66940
1.13	7.6923	8.6923	.06103	.11504	.93897	.61605	.57845	1.23211	.59696	.67456
1.14	7.1429	8.1429	.06542	.12281	.93458	.61676	.57641	1.23352	.59624	.67972
1.15	6.6667	7.6667	.06977	.13043	.93023	.61746	.57438	1.23492	.59553	.68486
1.16	6.2500	7.2500	.07407	.13793	.92593	.61816	.57237	1.23632	.59483	.69000
1.17	5.8824	6.8824	.07834	.14530	.92166	.61886	.57037	1.23771	.59412	.69512
1.18	5.5556	6.5556	.08257	.15254	.91743	.61955	.56839	1.23910	.59342	.70023
1.19	5.2632	6.2632	.08676	.15966	.91324	.62024	.56643	1.24047	.59272	.70534
1.20	5.0000	6.0000	.09091	.16667	.90909	.62092	.56447	1.24184	.59203	.71043
1.21	4.7619	5.7619	.09502	.17355	.90498	.62160	.56254	1.24321	.59133	.71551
1.22	4.5455	5.5455	.09910	.18033	.90090	.62228	.56061	1.24456	.59064	.72058
1.23	4.3478	5.3478	.10314	.18699	.89686	.62296	.55870	1.24591	.58996	.72565
1.24	4.1667	5.1667	.10714	.19355	.89286	.62363	.55681	1.24725	.58927	.73070
1.25	4.0000	5.0000	.11111	.20000	.88889	.62430	.55493	1.24859	.58859	.73574
1.26	3.8462	4.8462	.11504	.20635	.88496	.62496	.55306	1.24992	.58791	.74077
1.27	3.7037	4.7037	.11894	.21260	.88106	.62562	.55121	1.25124	.58724	.74579
1.28	3.5714	4.5714	.12281	.21875	.87719	.62628	.54937	1.25256	.58656	.75080
1.29	3.4483	4.4483	.12664	.22481	.87336	.62693	.54754	1.25387	.58589	.75580
1.30	3.3333	4.3333	.13043	.23077	.86957	.62759	.54573	1.25517	.58523	.76080
1.31	3.2258	4.2258	.13420	.23664	.86580	.62824	.54393	1.25647	.58456	.76578
1.32	3.1250	4.1250	.13793	.24242	.86207	.62888	.54214	1.25776	.58390	.77075
1.33	3.0303	4.0303	.14163	.24812	.85837	.62952	.54036	1.25905	.58324	.77571
1.34	2.9412	3.9412	.14530	.25373	.85470	.63016	.53860	1.26033	.58259	.78067
1.35	2.8571	3.8571	.14894	.25926	.85106	.63080	.53685	1.26160	.58193	.78561
1.36	2.7778	3.7778	.15254	.26471	.84746	.63143	.53511	1.26287	.58128	.79054
1.37	2.7027	3.7027	.15612	.27007	.84388	.63206	.53339	1.26413	.58063	.79547
1.38	2.6316	3.6316	.15966	.27536	.84034	.63269	.53167	1.26538	.57999	.80038
1.39	2.5641	3.5641	.16318	.28058	.83682	.63332	.52997	1.26663	.57934	.80529
1.40	2.5000	3.5000	.16667	.28571	.83333	.63394	.52828	1.26788	.57870	.81019

1.41	2.4390	3.4390	.17012	.29078	.82988	.63456	.52660	1.26911	.57807	.81507
1.42	2.3810	3.3810	.17355	.29577	.82645	.63517	.52494	1.27035	.57743	.81995
1.43	2.3256	3.3256	.17695	.30070	.82305	.63579	.52328	1.27157	.57680	.82482
1.44	2.2727	3.2727	.18033	.30556	.81967	.63640	.52164	1.27279	.57617	.82968
1.45	2.2222	3.2222	.18367	.31034	.81633	.63700	.52000	1.27401	.57554	.83453
1.46	2.1739	3.1739	.18699	.31507	.81301	.63761	.51838	1.27522	.57491	.83937
1.47	2.1277	3.1277	.19028	.31973	.80972	.63821	.51677	1.27642	.57429	.84420
1.48	2.0833	3.0833	.19355	.32432	.80645	.63881	.51517	1.27762	.57367	.84903
1.49	2.0408	3.0408	.19679	.32886	.80321	.63941	.51358	1.27881	.57305	.85384
1.50	2.0000	3.0000	.20000	.33333	.80000	.64000	.51200	1.28000	.57243	.85865
1.51	1.9608	2.9608	.20319	.33775	.79681	.64059	.51043	1.28118	.57182	.86345
1.52	1.9231	2.9231	.20635	.34211	.79365	.64118	.50887	1.28236	.57121	.86824
1.53	1.8868	2.8868	.20949	.34641	.79051	.64177	.50732	1.28353	.57060	.87302
1.54	1.8519	2.8519	.21260	.35065	.78740	.64235	.50579	1.28470	.56999	.87779
1.55	1.8182	2.8182	.21569	.35484	.78431	.64293	.50426	1.28586	.56939	.88255
1.56	1.7857	2.7857	.21875	.35897	.78125	.64351	.50274	1.28702	.56879	.88731
1.57	1.7544	2.7544	.22179	.36306	.77821	.64408	.50123	1.28817	.56819	.89205
1.58	1.7241	2.7241	.22481	.36709	.77519	.64466	.49973	1.28931	.56759	.89679
1.59	1.6949	2.6949	.22780	.37107	.77220	.64523	.49824	1.29045	.56699	.90152
1.60	1.6667	2.6667	.23077	.37500	.76923	.64579	.49677	1.29159	.56640	.90624
1.61	1.6393	2.6393	.23372	.37888	.76628	.64636	.49530	1.29272	.56581	.91095
1.62	1.6129	2.6129	.23664	.38272	.76336	.64692	.49383	1.29385	.56522	.91566
1.63	1.5873	2.5873	.23954	.38650	.76046	.64748	.49238	1.29497	.56463	.92035
1.64	1.5625	2.5625	.24242	.39024	.75758	.64804	.49094	1.29608	.56405	.92504
1.65	1.5385	2.5385	.24528	.39394	.75472	.64860	.48951	1.29720	.56347	.92972
1.66	1.5152	2.5152	.24812	.39759	.75188	.64915	.48808	1.29830	.56289	.93439
1.67	1.4925	2.4925	.25094	.40120	.74906	.64970	.48667	1.29940	.56231	.93905
1.68	1.4706	2.4706	.25373	.40476	.74627	.65025	.48526	1.30050	.56173	.94371
1.69	1.4493	2.4493	.25651	.40828	.74349	.65080	.48386	1.30160	.56116	.94836
1.70	1.4286	2.4286	.25926	.41176	.74074	.65134	.48248	1.30268	.56059	.95300

index

conversion factors*

	To convert from	to	multiply by
Length	foot (ft)	meter (m)	0.3048[†]
	meter (m)	foot (ft)	3.28084
	inch (in.)	meter (m)	0.0254[†]
	meter (m)	inch (in.)	39.3701
	mile (mi)	meter (m)	$1.609344 \cdot 10^{3}$[†]
	meter (m)	mile (mi)	$0.621371 \cdot 10^{-3}$
Mass	pound (lbm)	kilogram (kg)	0.45359237[†]
	kilogram (kg)	pound (lbm)	2.20462
Force	pound (lbf)	newton (N)	4.4482216152605[†]
	newton (N)	pound (lbf)	0.224809
Temperature	rankine (R)	kelvin (K)	5/9
	kelvin (K)	rankine (R)	1.8
Area	ft²	m²	0.09290304[†]
	m²	ft²	10.7639
	in.²	m²	$6.4516 \cdot 10^{-4}$[†]
	m²	in.²	1550.00
Volume	ft³	m³	0.028316846592[†]
	m³	ft³	35.3147
	in.³	m³	$1.6387064 \cdot 10^{-5}$[†]
	m³	in.³	$6.10237 \cdot 10^{4}$
Velocity	ft/sec	m/s	0.3048*
	m/s	ft/sec	3.28084
	mi/hr	m/s	0.44704*
	m/s	mi/hr	2.23694
Density	lbm/ft³	kg/m³	16.018463
	kg/m³	lbm/ft³	0.0624280
Specific volume	ft³/lbm	m³/kg	0.0624280
	m³/kg	ft³/lbm	16.018463
Pressure	lbf/in.²	N/m²	6894.7572
	N/m²	lbf/in.²	$1.45038 \cdot 10^{-4}$
	lbf/ft²	N/m²	47.880258
	N/m²	lbf/ft²	0.0208854
	atmosphere (atm)	N/m²	$1.01325 \cdot 10^{5}$
	atmosphere (atm)	lbf/in.²	14.696
	bar	N/m²	$1.0 \cdot 10^{5}$[†]
	N/m²	pascal (Pa)	1.0[†]

QM LIBRARY
(MILE END)

conversion factors* (cont.)

	To convert from	to	multiply by
Torque	ft-lbf	m-N	1.35582
	m-N	ft-lbf	0.737562
Energy	Btu	ft-lbf	778.16[†]
	joule (J)	N-m	1.0[†]
	calorie (cal)	joule (J)	4.184[†]
	Btu	joule (J)	1055.04
	joule (J)	Btu	$9.47831 \cdot 10^{-4}$
	ft-lbf	joule (J)	1.3558179
	joule (J)	ft-lbf	0.737562
	Btu/lbm	J/kg	2325.965
	J/kg	Btu/lbm	$4.29929 \cdot 10^{-4}$
	ft-lbf/lbm	J/kg	2.98907
	J/kg	ft-lbf/lbm	0.334552
Power	horsepower (hp)	ft-lbf /sec	550.0[†]
	watt (W)	J/s	1.0[†]
	horsepower (hp)	watt (W)	745.700
	watt (W)	horsepower (hp)	$1.34102 \cdot 10^{-3}$
Specific heats, gas constants	Btu/lbm-R	J/kg-K	4186.8
	J/kg-K	Btu/lbm-R	$2.3885 \cdot 10^{-4}$
	ft-lbf/lbm-R	J/kg-K	5.38038
	J/kg-K	ft-lbf/lbm-R	0.185860
Viscosity	lbf-sec/ft²	N-s/m²	47.8803
	N-s/m²	lbf-sec/ft²	0.0208854
	lbm/ft-sec	kg/m-s	1.48816
	kg/m-s	lbm/ft-sec	0.671971
	Stoke	m²/s	0.0001[†]
	Poise	N-s/m²	0.1[†]
Mass flux	lbm/sec-ft²	kg/s-m²	4.88099
	kg/s-m²	lbm/sec-ft²	0.204876
Specific impulse	lbf-sec/lbm	N-s/kg	9.806650
	N-s/kg	lbf-sec/lbm	0.101972

* The conversion factors presented in Table 2 were taken from E. A. Mechtly, "The International System of Units," NASA SP-7012, Second Revision, 1973.
† A dagger (†) following a number denotes that it is an exact definition.

QM LIBRARY
(MILE END)

WITHDRAWN
FROM STOCK
QMUL LIBRARY

WITHDRAWN
FROM STOCK
QMUL LIBRARY

Lightning Source UK Ltd.
Milton Keynes UK
UKOW020556240812

197938UK00002B/3/A

9 780471 984405